WATER CONSERVANCY AND CIVIL CONSTRUCTION
VOLUME 1

Water Conservancy and Civil Construction gathers the most cutting-edge research on:

- Water Conservancy Projects
- Civil Engineering
- Construction Technology and Process

The book is aimed at academics and engineers in water and civil engineering.

PROCEEDINGS OF THE 4TH INTERNATIONAL CONFERENCE ON HYDRAULIC, CIVIL AND CONSTRUCTION ENGINEERING (HCCE 2022), HARBIN, CHINA, 16–18 DECEMBER 2022

Water Conservancy and Civil Construction
Volume 1

Edited by

Saheed Adeyinka Oke
Central University of Technology Free State, South Africa

Fauziah Ahmad
Universiti Sains Malaysia

CRC Press
Taylor & Francis Group
Boca Raton London New York Leiden

CRC Press is an imprint of the
Taylor & Francis Group, an **informa** business
A BALKEMA BOOK

First published 2024
by CRC Press/Balkema
4 Park Square, Milton Park, Abingdon, Oxon, OX14 4RN

and by CRC Press/Balkema
2385 NW Executive Center Drive, Suite 320, Boca Raton FL 33431

CRC Press/Balkema is an imprint of the Taylor & Francis Group, an informa business

British Library Cataloguing-in-Publication Data
A catalogue record for this book is available from the British Library

Library of Congress Cataloging-in-Publication Data
A catalog record has been requested for this book

SET
ISBN: 978-1-032-51440-6 (hbk)
ISBN: 978-1-032-51464-2 (pbk)

Volume 1
ISBN: 978-1-032-58614-4 (hbk)
ISBN: 978-1-032-58615-1 (pbk)
ISBN: 978-1-003-45081-8 (ebk)
DOI: 10.1201/9781003450818

Volume 2
ISBN: 978-1-032-58618-2 (hbk)
ISBN: 978-1-032-58619-9 (pbk)
ISBN: 978-1-003-45083-2 (ebk)
DOI: 10.1201/9781003450832

Typeset in Times New Roman
by MPS Limited, Chennai, India

 ISBN: 978-1-032-58618-2

Table of Contents

Structural seismic design and safety assessment

Preface

The 2022 4th International Conference on Hydraulic, Civil and Construction Engineering (HCCE 2022) was held via virtual form in Harbin, China from December 16th–18th, 2022. Previous conferences of the past three years in this series were held in Guangzhou virtually or physically and it was agreed to hold the conference once a year. The purpose of this series of annual conferences is to establish and develop constant international collaboration.

In recent decades, interest in hydraulic and civil construction engineering problems has been flourishing all over the globe because of both the theoretical interest and practical requirements. Considering the trend, the fourth conference was organized in order to provide forums for developing research cooperation and to promote activities in the field of hydraulic and civil construction engineering.

Because solutions to hydraulic and civil construction engineering problems are needed in various applied fields, we entertained about 200 participants at the fourth conference and arranged various speeches and presentations which ranged from structural seismic resistance to smart city in the real world. Many researchers all over the world have contributed to the emerging technology of hydraulic and civil construction engineering. Assoc. Prof. Rohayu Che Omar from Universiti Tenaga Nasional, Malaysia addressed a keynote speech on Disaster Resilience Index and Indicators System for Managing Risks in Hazardous Terrain (DRIMS). She introduced to us that the methods for landslide mapping and landslide hazard assessment have experienced significant improvements during the last decade, but the requirements of the users have become more challenging, leading to improving landslide protection techniques to stabilize a landslide too costly in financial or environmental terms. This new context can only be managed with a better knowledge of the landslide mechanisms and their behavior. At the same time, different technical solutions, especially in the domain of risk mitigation, must be searched to guarantee the appropriate level of safety for the infrastructure and population.

HCCE 2022 has been endorsed by many international and national hydraulic civil engineering organizations and publishers. The papers collected and undergone peer review in the proceedings of HCCE 2022 are classified as follows: Engineering Structure, Intelligent Building, Smart City, Structural Seismic Resistance, Monitoring and Testing, Engineering Facility, etc.

Last but not the least is our gratitude. As editors we would like to express our sincere thanks to all the plenary and invited speakers, the members of the Program Committee and the Technical Committee for the success of the conference, which has given rise to this present volume of selected papers. We would also like to thank the CRC Press Balkema – Taylor & Francis Group for their effective work to make this volume published.

The Committee of HCCE 2022

Committee Members

Conference Chairman
Prof. Fadi HAGE CHEHADE, ISBA TP, *Grande Ecole d'Ingénieurs de Spécialisation en Génie Civil, France*

Program Committee
Prof. Tetsuya Hiraishi, *Kyoto University, Japan*
A. Prof. Hazem Samih Mohame, *Southwest Petroleum University, Egypt*
A. Prof. Mohammad Arif Kamal, *Aligarh Muslim University, India*
A. Prof. Aeslina Abdul Kadir, *Universiti Tun Hussein Onn Malaysia, Malaysia*
Asst. Prof. Hamza Soualhi, *University of Laghouat, Algeria*
Senior Lecturer Mohammadreza Vafaei, *Universiti Teknologi Malaysia, Malaysia*
Senior Lecturer Au Yong Cheong Peng, *University of Malaya, Malaysia*
Senior Lecturer Nor Hasanah Binti Abdul Shukor Lim, *Universiti Teknologi Malaysia UTM, Malaysia*
Senior Lecturer Libriati Zardasti, *Universiti Teknologi Malaysia, Malaysia*
Ph. D. Dayang Zulaika Binti Abang Hasbollah, *Universiti Teknologi Malaysia, Malaysia*

Technical Committee
Prof. Dr. Mohammad Bin Ismail, *Universiti Teknologi Malaysia, Malaysia*
Prof. Ir. Dr. Hj. Ramli Nazir, *Universiti Teknologi Malaysia, Malaysia*
Prof. Dr. Muhd Zaimi Bin Abd Majid, *Universiti Teknologi Malaysia, Malaysia*
Prof. Lu, Jane Wei-Zhen, *City University of Hong Kong, Hong Kong, China*
Prof. Mingqiao Zhu, *Hunan University of Science and Technology, China*
Prof. QingXin Ren, *Shenyang Jianzhu University, China*
Prof. Bing Li, *Shenyang Jianzhu University, China*
Prof. Jianhui Yang, *Henan Polytechnic University, China*
Prof. Changfeng Yuan, *Qingdao University of Technology, China*
A. Prof. Bon-Gang HWANG, *National University of Singapore, Singapore*
A. Prof. Zhu Yuan, *Southeast University, China*
A. Prof. Chaofeng Zeng, *Hunan University of Science and Technology, China*
A. Prof. Weijun Cen, *Hohai University, China*
Asst. Professor Dr. Shah Kwok Wei, *National University of Singapore, Singapore*
Dr. Shaoyun Pu, *Southeast University, China*
Dr. Zhongzheng Lyu, *Dalian University of Technology, China*
Dr. Mohd Rosli Mohd Hasan, *Universiti Sains Malaysia, Malaysia*
Dr. Kim Hung Mo, *University of Malaya, Malaysia*
Dr. Yuen Choon Wah, *University of Malaya, Malaysia*
Dr. Huzaifa Bin Hashim, *University of Malaya, Malaysia*
Dr. Suhana Koting, *University of Malaya, Malaysia*
Dr. Sharifah Akmam Syed Zakaria, *Universiti Sains Malaysia, Malaysia*
Dr. Xian Zhang, *Southeast University, China*
Dr. Zhiming Chao, *University of Warwick, UK*
Dr. Jun Xie, *Central South University, China*
Dr. Derek Ma, *University of Warwick, England*
Dr. Ning Xu, *Shanghai Ershiye Construction CO., LTD., China*
Dr. Hongchao Shi, *Chengdu Technological University, China*
Dr. Li He, *Wuhan University of Science and Technology, China*

Mechanical equipment and hydraulic engineering management

Water Conservancy and Civil Construction – Oke & Ahmad (Eds)
© 2024 The Author(s), ISBN: 978-1-032-58614-4

Viscoelastic mechanics and fatigue performance of recycled asphalt mixture with steel slag

Peng Guo
Wuhan Municipal Road & Bridge Co. Ltd., Wuhan, China

Feiyi Liu
Wuhan Hanyang Municipal Construction Group Co. Ltd., Wuhan, China

Ruonan Pang & Sifa Fang
Wuhan Municipal Road & Bridge Co. Ltd., Wuhan, China

Fan Shen*
Wuhan Institute of Technology Materials Science and Engineering, Wuhan, China

ABSTRACT: In order to reduce the diseases of recycled asphalt mixture pavement mixed with steel slag and extend its service life, a steel slag recycled asphalt mixture with different recycled asphalt pavement (RAP) dosages was prepared in this paper. The viscoelastic mechanical properties and fatigue properties of a steel slag recycled asphalt mixture were investigated using static mechanical creep and three-point bending experiments. The results showed that the addition of steel slag could effectively improve the viscoelastic mechanical properties of a recycled asphalt mixture. The viscoelastic mechanical properties of the recycled asphalt mixture with steel slag first decreased and then increased with the increase of RAP content, and when the RAP content was 30%, the elastic modulus E1, the parallel element η2/E2, and the delay element η1 reached the lowest values, which were 125.9 MPa, 236.3 s, and 108403 MPa • s, respectively. Steel slag, RAP, and effective asphalt film thickness jointly affected the fatigue performance of recycled asphalt mixture with steel slag. When the content of RAP was 20%, the fatigue sensitivity of steel slag recycled asphalt mixture was the lowest, and the k value was 3.280. The fatigue life of steel slag recycled asphalt mixture was the highest, and the n value was 6.188. When the ambient temperature was higher than 15°C, the fatigue sensitivity and fatigue life of recycled asphalt mixture with steel slag were slightly improved with the increase in temperature.

1 INTRODUCTION

China is the world's largest producer and consumer of sandstone, with an annual output of nearly 20 billion tons. However, after years of exploitation, China's natural aggregate resources are becoming increasingly scarce. Natural sandstone in some areas has been nearly exhausted, which has led to a rapid increase in the price of aggregate in recent years. China's severe resource situation has promoted the recycling of waste resources in the road industry (Han & Xiao 2013; Jin 2018; Zhong & Zuo 2021). Asphalt pavement recycling technology uses recycled asphalt pavement (RAP) to replace some aggregates in preparing an asphalt mixture, thereby realizing the recycling of RAP. However, in order to avoid the impact of

*Corresponding Author: shenf@wit.edu.cn

DOI: 10.1201/9781003450818-1

aged asphalt in RAP on the performance of asphalt mixtures, the content of RAP in recycled asphalt mixtures generally does not exceed 30% (Ding *et al.* 2021; Xiao *et al.* 2016; Yang *et al.* 2022). In order to improve the replacement rate of natural aggregate in recycled asphalt mixtures, it is an effective solution to replace natural aggregate with steel slag, which has a good particle gradation shape, excellent mechanical properties, and good adhesion with asphalt. The preparation of recycled asphalt mixture by using RAP and steel slag instead of natural aggregate can realize high-quality conversion and utilization of waste resources, which is conducive to fundamentally solve the problem of stone shortage in China (Song 2021; Xie 2013; Yan & Zhou 2020; Zhou & Zhang 2019). As one of the main materials for pavement construction, the mechanical behavior of the asphalt mixture is directly related to the occurrence of asphalt pavement diseases (Cui *et al.* 2016; Liu 2014; Xiang & Zhang 2007). Because of the mixing of new and old asphalt and the combination of various aggregates, the influence factors on the mechanical behavior of recycled asphalt mixture mixed with steel slag are more complicated than those of ordinary hot mix asphalt mixture (Liu *et al.* 2021). Therefore, in order to reduce the occurrence of pavement diseases caused by recycled asphalt mixture mixed with steel slag and prolong its service life, it is necessary to conduct in-depth research on the dynamic response and fatigue performance of asphalt mixture (Sun 2021; Yuan & Sun 2019).

In this paper, the recycled asphalt mixture mixed with steel slag was prepared. By changing the content of RAP and the experimental temperature, the experimental methods of creep test and three-point bending test under stress control mode were used to study the mechanical properties and fatigue properties of recycled asphalt mixture mixed with steel slag under different contents of RAP, as well as the fatigue properties of asphalt mixture at different temperatures.

2 RAW MATERIALS AND TEST PLAN

2.1 *Raw materials*

In this paper, I-D modified asphalt was used as asphalt, hot disintegration steel slag (with low activity of hot disintegration steel slag) and natural limestone were used as aggregate, limestone mineral powder was used as mineral powder, polyester fiber was used as fiber, and RAP adopted pavement milling materials from an old road section in Jiangxia District, Wuhan City, Hubei Province, and the asphalt stone ratio of the mixture was 5.9%. Among these, the technical indicators of steel slag met the technical requirements of JT/T 1086-2016 *Steel Slag Used in Asphalt Mixture* (Research Institute of Highway Ministry of Transport 2017); the technical indicators of limestone aggregate and limestone mineral powder met the technical requirements of JTG F40-2004 Technical Specifications for Construction of Highway Asphalt Pavements (Research Institute of Highway Ministry of Transport 2009); the technical indicators of polyester fiber met the technical requirements of JTT 533-2020 *Fiber for Asphalt Pavements* (Research institute of highway ministry of transport 2020). According to JTG E20-2011 *Standard Test Methods of Bitumen and Bituminous Mixtures for Highway Engineering* (Research Institute of Highway Ministry of Transport 2011), the effective asphalt content in RAP was 4.9%, measured by the combustion furnace method (Liu & Zhang 2019). The technical indexes of I-D modified asphalt, aged asphalt in RAP, and steel slag are shown in Tables 1, 2, and 3.

2.2 *Test plans*

2.2.1 *Study on viscoelastic mechanical properties*

Asphalt mixtures with steel slag (all aggregates were steel slag), ordinary asphalt mixtures (all aggregates were natural aggregate), recycled asphalt mixtures with RAP content of 30%,

Table 1. I-D Modified asphalt technical indexes.

Test Project		Test Result	Technical Requirement
25 °C Elastic Restitution / %		87	≥ 80
Softening Point TR&B / °C		80.5	≥ 75
Solubility / %		99.84	≥ 99
Flash Point / °C		269	≥ 230
135 °C Kinematic Viscosity / Pa • s		2.6	≤ 3
Prolongation (5°C, 5cm/min) / cm		32	≥ 20
Penetration Degree (25°C, 100g, 5s) / 0.1mm		53	40–60
Density (15°C) / (g/cm3)		1.011	Measured
Residue after TFOT (or RTFOT)	Quality Change / %	–0.779	±0.8
	25 °C Penetration Ratio / %	72	≥ 65
	5 °C Prolongation / cm	20	≥ 15
	Penetration Ratio / %	74	≥ 65

Table 2. Technical indexes of aged asphalt in RAP.

	Test Project	Test Result
Penetration Degree / 0.1 mm	Usage Time / a	15
	Softening Point / °C	76
	15°C	11
	25°C	20
	30°C	31
	Penetration Ratio / %	74
	15°C Extensibility / cm	38
	135°C Viscosity / Pa • s	1349

Table 3. Technical indexes of steel slag.

Test Project	Specification Requirement	Experimental Result
Needle and Plate Particle Content /%	(Aggregate ≥ 9.5 mm) ≤ 12	8.3
Adhesion	≥ 5	5
Immersion Expansion Rate /%	≤ 1.8	1.5
Abrasion Value /%/	≤ 33	21.5
Crush Value /%	≤ 22	15.4
Water Absorption /%	≤ 3	1.5
Apparent Relative Density /g/cm^3	–	3.189

and recycled asphalt mixtures with steel slag and RAP content of 10%, 20%, 30%, and 40%, respectively (aggregates were steel slag and RAP), were prepared. The bending creep tests of the above asphalt mixtures were carried out, and the viscoelastic mechanical properties of the asphalt mixtures were analyzed by the Burgers model.

2.2.2 *Study on fatigue performance of mixture with different RAP content*

The recycled asphalt mixture with steel slag was prepared when the RAP content in the asphalt mixture was 10%, 20%, 30%, and 40%, respectively, and the three-point bending test under stress control mode was carried out to analyze the fatigue performance of asphalt mixtures with different RAP contents.

2.2.3 *Study on fatigue performance of mixtures at different temperatures*
The recycled asphalt mixture mixed with steel slag was prepared when the RAP content in the asphalt mixture was 30%, and the three-point bending tests at -10 °C, 15 °C, 25 °C, and 60 °C were carried out to analyze the fatigue performance of asphalt mixtures at different temperatures.

3 RESULTS ANALYSIS

3.1 *Viscoelastic mechanical properties*

The strain response curve of the Burgers model obtained from the creep test is shown in Figure 1.

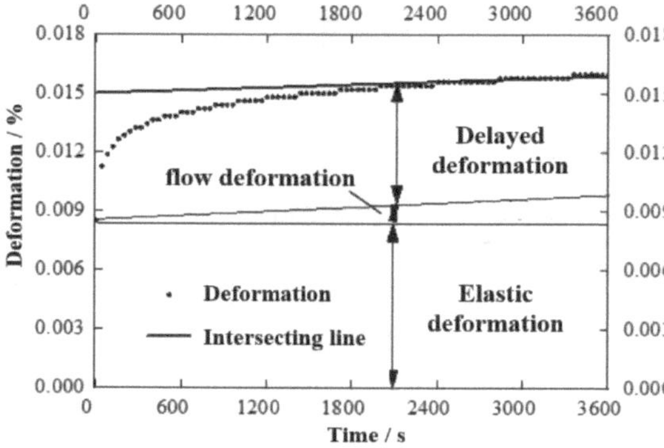

Figure 1. Burgers model strain response curve.

The graphical method was used to fit the strain response curve of the Burgers model. In the creep experiment, the input of stress is constant (Zhang 2006). According to the strain curve of the Burgess model in Figure 1, the graphical derivation and calculation formula are as follows:

When t is not 0, the input is a constant stress.

$$\sigma = \begin{cases} 0 & t < 0 \\ \sigma & t > 0 \end{cases} \tag{1}$$

In the formula, σ is the stress and t is the time.

The creep strain of specimen response is shown in Formula 2.

$$\varepsilon = \sigma_0 \left[\frac{1}{E_1} + \frac{t}{\eta_1} + \frac{1}{E_2} \left(1 - e^{-\frac{t}{\tau_r}} \right) \right] \tag{2}$$

where ε is the strain, σ_0 is the instantaneous stress, E_1 is the elastic modulus of the spring in the Burgers model, E_2 is the elastic modulus of the delay element spring in the Burgers model, η_1 is the viscosity of series dashpot in the Burgers model, and $\tau_r = \eta_1 / E_2$. When t=0, the strain is instantaneous elastic, which can be read out from the figure.

Taking the experimental group with 10% RAP as an example, the graphic method was used for calculation.

As shown in Figure 1, the instantaneous elastic strain ε_0 can be read from the curve in Figure 1. Therefore, the elastic modulus of the spring is as follows:

$$E_1 = \sigma_0 / \varepsilon_0 \tag{3}$$

E_1 = 2.499/0.0116 = 215.4 MPa.

Subtracting the instantaneous elastic strain from the Burgers model, the residual strain can be obtained from Formula 4.

$$\delta(t) = \varepsilon(t) - \varepsilon_0 = \sigma_0 \left[\frac{1}{\eta_1} + \frac{1}{E_2}\left(1 - e^{-\frac{t}{\tau_r}}\right) \right] \tag{4}$$

When the time is infinite, the relationship between strain and time is shown in Formula 5.

$$\delta(t)_{t\to\infty} = \sigma_0 \left(\frac{1}{E_2} + \frac{t}{\eta_1} \right) \tag{5}$$

Therefore, the slope $\tan\theta$ of the last section of the strain-time curve obtained the viscosity η_1 of the series dashpot can be obtained from the following formula.

$$\eta_1 = \frac{\sigma_0}{\tan\theta} \tag{6}$$

where η_1 = 2.499 / 0.000007 = 362953MPa, and to take $t = 0$ into the formula.

$$\delta(t)_{t\to\infty} = \frac{\sigma_0}{E_2} \tag{7}$$

It can be obtained that E_2 = 2.499/0.0241 = 103.8Mpa.

Next, in order to obtain the viscosity of the dashpot in the delay element, formula 4 is deformed and formula 8 is obtained.

$$\log\left\{ -\delta(t) + \frac{\sigma_0}{E_2} + \frac{\sigma t}{\eta_1} \right\} = \log \frac{\sigma_0}{E_2} - \frac{t}{\tau_r} \log e \tag{8}$$

The left part of Equation 8 is plotted for t, as shown in Figure 2 (a). The tangent at the end of the curve is intercepted to obtain the straight line in the logarithmic coordinate as shown in Figure 2 (b). From the straight line, the slope is $\tan\gamma$, and then Equation 9 is obtained.

$$\tan\gamma = \frac{\log e}{\tau_r} \tag{9}$$

Tan γ obtained from Figure 2 is −0.000004, τ_r = −loge/-0.00002379 = 365.1, and η_2 = −0.4343/−0.000004 = 123567 MPa · s.

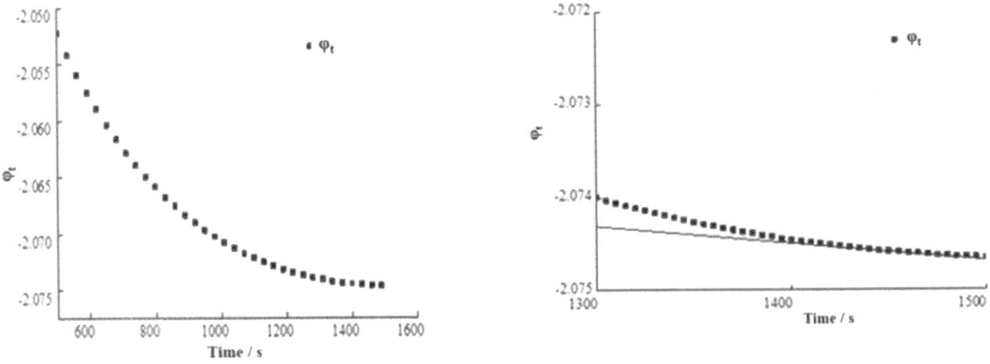

(a) Curve

(b) Tangent

Figure 2. Curve of $\tan\gamma = \log e/\tau_r$.

The parameter fitting results of recycled asphalt mixture with steel slag, common asphalt mixture, recycled asphalt mixture, and steel slag recycled asphalt mixture with different RAP contents are shown in Table 2 and Table 3.

Table 4. Parameters of the burgers model of recycled asphalt mixture mixed with steel slag.

Content of RAP/%	E_1/MPa	E_2/MPa	η_1/MPa · s	η_2/MPa · s	$\eta_2 \cdot E_2^{-1}$/s
10%	215.4	103.8	362953	123507	1189.9
20%	165.2	94.7	312375	34576	365.1
30%	125.9	154.0	108403	25030	236.3
40%	173.9	105.7	225880	39928	259.8

Table 5. Parameters of the burgers model for different types of asphalt mixture.

Mixture Type	E_1/MPa	E_2/MPa	η_1/MPa·s	η_2/MPa·s	$\eta_2 \cdot E_2^{-1}$/s
Recycled Asphalt Mixture with Steel Slag (30%RAP)	125.9	154.0	108403	25030	236.3
Asphalt Mixture with Steel Slag	199.3	86.5	398500	51959	600.7
Asphalt Mixture	180.2	105.8	201300	27028	255.4
Recycled Asphalt Mixture	123.1	109.7	287600	24583	224.1

The results showed that in the elastic deformation stage, the elastic modulus E1 reflected the ability of pavement to resist deformation under a high-speed load. In the stage of the deformation that could be recovered to a certain extent, the parallel elements in Burger's model jointly determined the strain development speed of the mixture. The speed was usually expressed by the numerical value of η_2/E_2, the larger η_2/E_2, the slower the strain development. In the unrecoverable stage of deformation, delay element η1 of the recycled asphalt mixture mixed with steel slag represented the ability of the mixture to resist unrecoverable deformation, which was usually reflected in the rutting resistance.

The results showed that when the content of RAP was in the range of 10–40%, the elastic modulus E1, η_2/E_2, and delay element η1 showed a change law of decreasing first and then increasing with the addition of RAP. When the content of RAP was 30%, it reached the lowest value, which was 125.9 MPa, 236.3 s, and 108403 MPa • s, respectively. According to the viscoelastic law of recycled asphalt mixture with steel slag obtained from the four parameters of the Burgers model, it could be known that when the RAP content was low, the proportion of aged asphalt wrapped on its surface in the mixture was relatively low, and the effect was not obvious. From the viewpoint of "Blackstone theory" (Ding, *et al.* 2015), RAP could be regarded as an internally damaged "aggregate". When the content of RAP was less than 30%, the viscoelastic property of the mixture decreased with the increase in the RAP content. When the RAP content was high, the proportion of aged asphalt increased, and its effect could not be ignored. At this point, the aged asphalt recovered the performance of some asphalt due to new asphalt blending. At this time, RAP was an aggregate with a certain asphalt film thickness. Under the adhesion between asphalt and aggregate, the viscoelastic energy of the mixture increased slightly.

The research results on the viscoelastic mechanical properties of different types of asphalt mixtures showed that the addition of steel slag could improve the viscoelastic mechanical properties of the asphalt mixture and recycled asphalt mixture. Among the four different types of asphalt mixtures, E1 and η_2/E_2 of asphalt mixture with steel slag and recycled asphalt mixture with steel slag were 199.3 MPa, 600.7 s, and 125.9 MPa, 236.3 s, respectively, higher than that asphalt mixture and recycled asphalt mixture. It was because the steel slag had high strength and its surface porous characteristics made it have a strong adhesion with asphalt, which was conducive to the improvement of the viscoelastic properties of the mixture.

3.2 *Fatigue performance of mixtures with different RAP contents*

In this paper, the *S-N* fatigue equation reflecting the fatigue performance of the mixture was obtained by linear regression after taking the natural logarithm of the stress level σ and the fatigue life N_f, respectively:

$$LgN_f = -kLg\sigma + n \tag{10}$$

where σ is the stress, N_f is the fatigue life, k is the reciprocal of the fatigue curve life exponential, and n is a constant.

The linear regression results of different RAP dosages are shown in Figure 3.

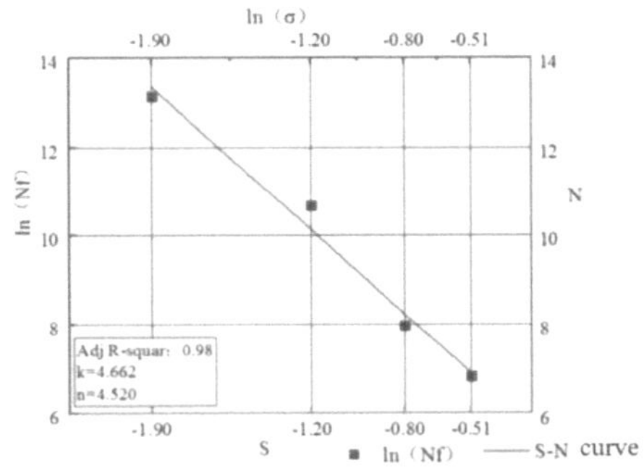

(a) 10% RAP

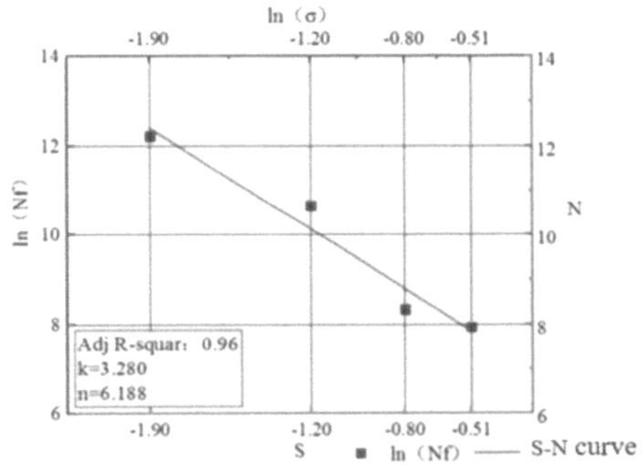

(b) 20% RAP

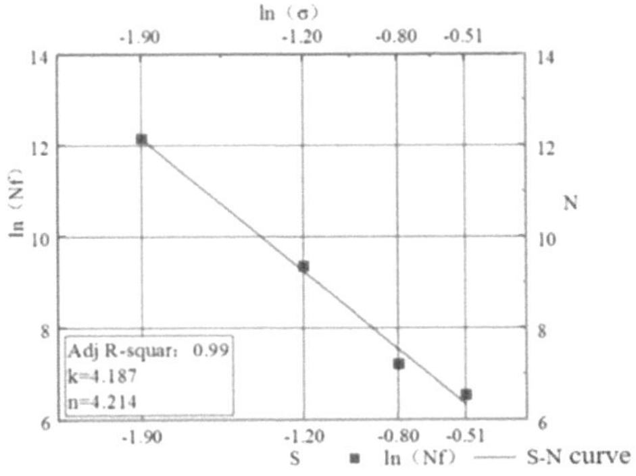

(c) 30% RAP

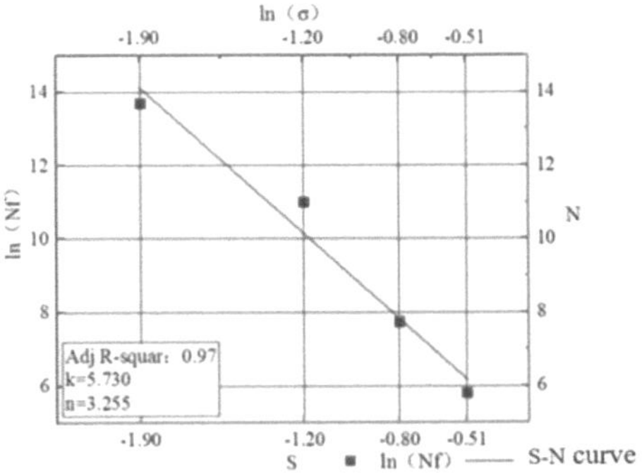

(d) 40% RAP

Figure 3. Regression diagram of the double logarithmic curve of asphalt mixture with different RAP dosage.

The research results showed that the fitting degree of the linear regression results of an asphalt mixture with different RAP contents was not less than 0.96. The k value obtained from the double logarithm fitting curve showed a change law of first decreasing and then increasing with the RAP content, and the n value showed a change law of first increasing and then decreasing with the RAP content. When the RAP content was 20%, the minimum k value was 3.280, the maximum n value was 6.188, and the fatigue life of the asphalt mixture was the highest. It was because the aggregate in the recycled asphalt mixture mixed with steel slag was composed of steel slag and RAP. The surface roughness of steel slag was high, which could improve the fatigue performance of the mixture. The internal damage of RAP would reduce the fatigue performance of the mixture. However, when the steel slag content was high, more fine steel slag aggregated with particle sizes less than 4.75 mm would be brought in. The oil absorption rate of fine steel slag aggregates was high, and the film thickness of RAP would be greatly reduced, resulting in uneven film thickness distribution in the mixture and reduced fatigue life. At the same time, as the asphalt in the mixture was composed of both aged asphalt and new asphalt, the increase in RAP would reduce the amount of new asphalt, the decrease in penetration of aged asphalt would increase the fatigue life of the mixture, and the decrease in new asphalt would reduce the fatigue life of the mixture.

3.3 *Fatigue properties of mixtures at different temperatures*

The three-point bending test results at different temperatures are shown in Figure 4.

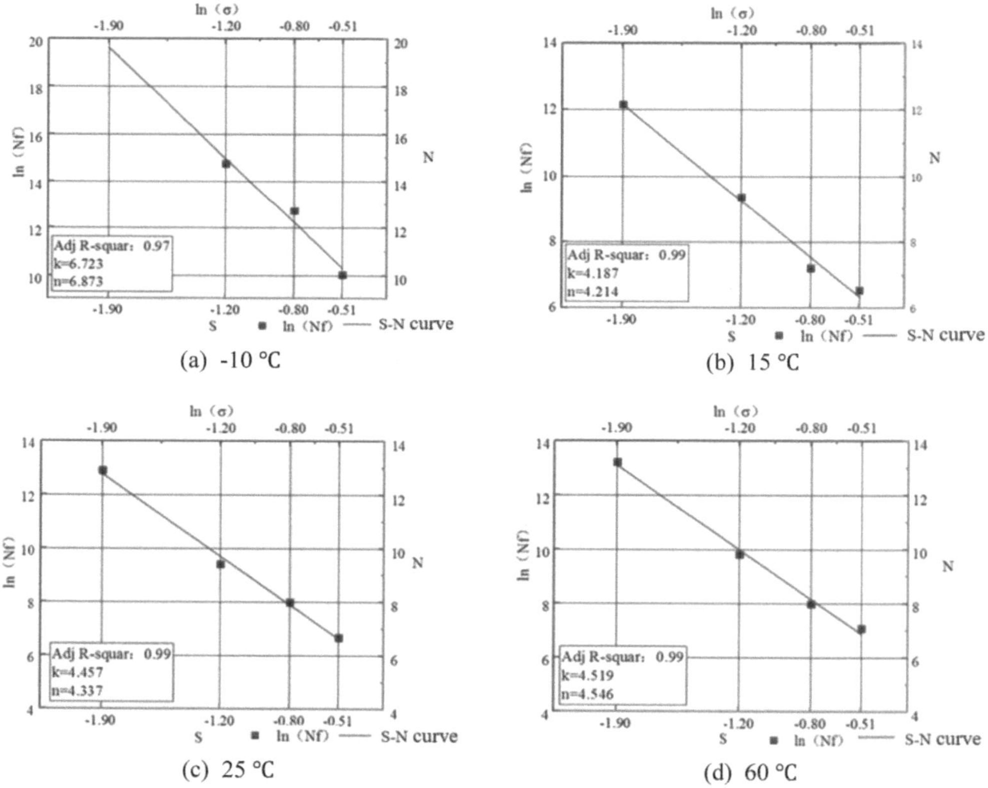

Figure 4. Regression diagram of the double logarithmic curve of recycled asphalt mixture mixed with steel slag at different temperatures.

The results showed that the fitting degree of the double logarithm fitting curve was not less than 0.96, and both the k and the n values showed a change law of decreasing first and then increasing with the temperature. When the temperature was −10 °C, the maximum *n* value was 6.873, indicating that the fatigue life of the asphalt mixture was the longest, and the maximum *k* value was 6.723, indicating that the asphalt mixture was the most sensitive to the change in stress level. It was because, at low temperatures, the material stiffness increased, the strain produced by bearing stress was small, and the fatigue life increased. However, with the increase in material stiffness, the material gradually tended to become brittle, and the material's flexibility decreases. Therefore, at the high-stress ratio, fatigue under higher stress decreased rapidly. As the temperature rose, asphalt mixtures became more flexible, the fatigue sensitivity of materials increased slightly, and the fatigue life also increased slightly.

4 CONCLUSIONS

The conclusions are as follows:

1) Adding steel slag could effectively improve the viscoelastic properties of a recycled asphalt mixture. After adding steel slag, the *E1* and η*2/E2* of the recycled asphalt mixture could reach 125.9 MPa and 236.3 s, respectively.
2) The viscoelastic performance of a recycled asphalt mixture with steel slag decreased first and then increased with the increased RAP content. When the RAP content was low, the proportion of aged asphalt wrapped on the surface of the mixture was relatively low, and the effect was not obvious. The viscoelastic performance of the mixture decreased slightly. When the RAP content was high, some aged asphalt recovered its performance under the blending action of new asphalt, and the viscoelastic performance of the mixture increased slightly.
3) The fatigue performance of recycled asphalt mixture with steel slag was affected by steel slag, RAP, and effective asphalt film thickness. When the RAP content was 20%, the fatigue sensitivity of the recycled asphalt mixture with steel slag was the lowest; the *k* value was 3.280; and the fatigue life was the highest; the *n* value was 6.188.
4) When the ambient temperature reached the lowest −10°C, the recycled asphalt mixture with steel slag tended to be brittle material. The asphalt mixture had the longest fatigue life, with an *n* value of 6.873, and the highest fatigue sensitivity, with a *k* value of 6.723. When the ambient temperature was higher than 15°C, the *k* value and *n* value of the recycled asphalt mixture with steel slag increased slightly, the fatigue sensitivity increased slightly, and the fatigue life also increased slightly.

ACKNOWLEDGMENT

The authors would like to thank The National Nature Science Foundation of China (52178248), The Internal Scientific Research Fund of Wuhan Institute of Technology, and The Science and Technology Project of Wuhan Municipal Construction Group Co. Ltd. (wszky201814) for providing funding for this experiment.

REFERENCES

Cui Xinzhuang, Huang Dan, Liu Lei, et al. A Review of Mechanics of Asphalt Pavement Disease[J]. *Journal of Shandong University (Engineering Science)*, 2016, 46 (05): 68–87.

Ding LongTing, Wang Xuancang, Zhang Mengyuan, Chen Zhao, Meng Jiaqi, Shao Xiansheng. Morphology and Properties Changes of Virgin and Aged Asphalt After Fusion[J]. *Construction and Building Materials*, 2021, 291.

Ding Qingjun, Zhao Mingyu, Shen Fan, et al. Characteristic Analysis of RAP Material Particle Composition Based on Gray System Theory [J]. *Journal of Building Materials*, 2015, 18 (04): 619–625.

Feipeng Xiao, Xiangdao Hou, Serji Amirkhanian, Kwang W. Kim. Superpave Evaluation of Higher RAP Contents Using WMA Technologies[J]. *Construction and Building Materials*, 2016, 112.

Han Jixian, Xiao Xuyu. Current Situation and Development Trend of Aggregate in China[J]. *China Concrete.* 2013 (09): 36–42.

Jin Xing. Application Status of Building Aggregate and Utilization of Slag[J]. *China Petroleum and Chemical Standard and Quality*, 2018, 38 (09): 96–97.

Liu Hui. *Viscoelastic Properties of Asphalt Mixture and its Effect on Pavement Response*[D]. Dalian University of Technology, 2014.

Liu Mingjin, Ke Wang, Li Chuangmin. Optimization Design of Mixture Ratio of AC-13 Asphalt Mixture Mixed with Steel Slag[J]. *Journal of Changsha University of Science & Technology(Natural Science)*, 2021, 18 (01): 24–32.

Liu Yanqiang, Zhang Jie. The Experimental Study on the Determination of the Asphalt Content Correction Coefficient by the Combustion Furnace Method [J]. *Shanxi Science & Technology of Communications*, 2019 (01): 28–30.

Research institute of highway ministry of transport. JT/T1086-2016, *Steel Slag Used in Asphalt Mixture [M]*. Beijing: China communication press, 2017.

Research Institute of Highway Ministry of Transport. JTG E20-2011 *Standard Test Methods of Bitumen and Bituminous Mixtures for Highway Engineering [M]*. Beijing: China communication press, 2011.

Research institute of highway ministry of transport. JTG F40-2004 *Technical Specifications for Construction of Highway Asphalt Pavements [M]*. Beijing: China Communication Press, 2009.

Research Institute of Highway Ministry of Transport. JTT 533-2020 *Fiber for Asphalt Pavements [M]*. Beijing: China communication press, 2020.

Song Juntao. Study on Performance of Hot Recycled Steel Slag Asphalt Mixture[J]. *Journal of China & Foreign Highway*, 2021, 41 (04): 287–291.

Sun Zhide, *Influence of Steel Slag Aggregates and Reclaimed Asphalt Pavement on Hot-mix Asphalt Pavement Performance [D]*. Xiangtan University, 2021.

Xiang Yuan, Zhang Lanjun. Current Status and Trend of Expressway Pavement Material Development[J]. *Technology of Highway and Transport*, 2007 (02): 75–79.

Xie Jun. *Research on the Preparation, Performance and Application of Basic Oxygen Furnace Slag Based Asphalt Concrete[D]*. Wuhan University Of Technology, 2013.

Yan Zhou, Zhou Yan. Analysis on the Properties of Modified Steel Slag Asphalt Mixture[J]. *Journal of Physics: Conference Series*, 2020, 1649 (1).

Yang Chao, Wu Shaopeng, Cui Peide, Amirkhanian Serji, Zhao Zenggang, Wang Fusong, Zhang Lei, Wei Minghua, Zhou Xinxing, Xie Jun. Performance Characterization and Enhancement Mechanism of Recycled Asphalt Mixtures Involving High RAP Content and Steel Slag[J]. *Journal of Cleaner Production*, 2022, 336.

Yuan Yin, Sun Yanna. Current Status of Research on Road Performance of Hot Mix Recycled Asphalt Mixture [J]. *Shanghai Highway*, 2019 (02): 94–97 + 106 + 7.

Zhang Xiaoning. *The principle and Application of Viscoelastic Mechanics of Asphalt and Asphalt mixture [M]*. Beijing: People's Communications Press, 2006: 100–137.

Zhong Mingran, Zuo Fuqin. Status Quo Analysis of Supply and Demand of Sand and Gravel Aggregate in China and Its Management Countermeasures[J]. *Journal of Guilin University of Aerospace Technology*, 2021, 26 (04): 504–509.

Zhou Yan, Zhang Hao. Study on Preparation and Performance of Steel Slag Asphalt Mixture Based on Steel Slag Aggregate[J]. *IOP Conference Series: Materials Science and Engineering*, 2019, 631 (2).

Water Conservancy and Civil Construction – Oke & Ahmad (Eds)
 ISBN: 978-1-032-58614-4

Local optimization of the cabin air duct of the Ro-Ro ship based on CFD technology

Jiahao Yang* & Lei Li*
Jiangsu University of Science and Technology, Jiangsu, P.R. China

ABSTRACT: Based on CFD technology, the typical air duct ventilation process of the passenger Ro-Ro vehicle cabin is simulated and analyzed, the local energy loss caused by the air duct structure is analyzed from the perspective of pressure drop, and the air duct structure is optimized for the local pressure drop caused by the rectangular air duct corner and rib plate structure in the typical air duct.

1 GENERAL INSTRUCTIONS

The design and manufacture of large high-end RoRo ships should fully consider the environmental comfort in the cabin, the air environment in the cabin will directly affect the work efficiency and mood of the personnel on board, and the pollutants in the cabin are more likely to pose a threat to the safety of passengers and transport personnel. Therefore, how to reduce the concentration of pollutants in the cabin of the Ro-Ro vehicle, improve the comfort of the cabin environment of the RO vehicle cabin, improve the crew's efficiency and ensure the crew's physical and mental health have become a research hotspot. However, the vehicle cabin has a large internal space, the cabin structure is constrained by the strength of the hull and the theme structure of the ship, the optimization space is small, and the air duct is one of the most important links of the cabin ventilation. The optimized design of the air duct has become a feasible solution to optimize the ventilation system.

The vehicle compartment relies on the fan to supply air to the air duct. The fresh air is sent to the cabin through the air duct, so the air duct is the key link of cabin ventilation, but the ship construction method is relatively rough. Due to the consideration of structural strength factors, the actual main structure of the air duct and the internal rib plate structure greatly affect the ventilation performance of the ventilation system, resulting in energy loss. Therefore, the structural optimization of the vehicle cabin air duct is of great significance in improving the ventilation performance of the cabin.

Li Banghua et al. conducted a ventilation simulation study on the air duct of the cargo hold of a car transport ship (Li 2020) and analyzed and optimized the duct structure from the angles of deck opening, air duct inlet inclination, and air duct outlet opening, but did not consider the influence of air duct angle and internal rib plate structure on the ventilation performance of the air duct. Based on CFD technology, this paper simulates and analyzes the real air duct model of the vehicle cabin of the RoRo ship, studies the influence of the tunnel angle and internal rib plate structure on the ventilation performance of the air duct, and studies the energy loss law in the process of air duct ventilation from the perspective of local pressure drop, and further optimizes the structure of the vehicle cabin air duct to improve the ventilation performance of the air duct (Duan 2018).

*Corresponding Authors: 15052689253@163.com and 2625736488@qq.com

DOI: 10.1201/9781003450818-2

2 NUMERICAL SIMULATION OF AIR DUCT VENTILATION

2.1 *Mesh model establishment*

There are many cabin air ducts on RoRo ships, and this paper selects typical vehicle air ducts for study (Shen 2017). The duct geometry is as follows:

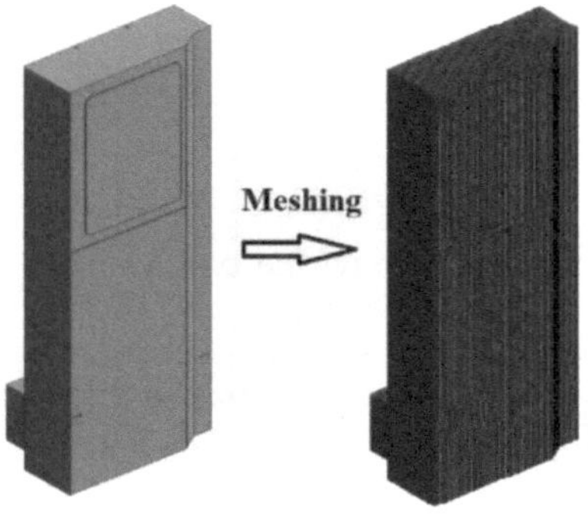

Figure 1. Typical air duct model.

The six-sided grid generation method is the main method to improve the simulation calculation accuracy and reduce the amount of calculation when meshing the three typical air ducts. The maximum size of the grid is 0.02 m, and the number of top air supply duct grids is 1584937.

2.2 *Input criteria determination and calculation model selection*

Since the wind duct structure contains many ribbed plate structures and the existence of air duct rotation angles, and the RNG k-ε equation considers turbulent vortex, low Reynolds number viscosity, and has good performance for transient flow and streamline bending, the RNG k-ε model is selected for calculation (Chen 2017). The SIMPLEC algorithm is selected to obtain better convergence, which has a fast convergence speed and is more suitable for solving steady-state problems.

The boundary conditions select the speed inlet and pressure outlet. According to the exhaust air volume of the fan and the inlet area, the inlet speed is 20 m/s, and the outlet pressure is atmospheric.

2.3 *Analysis of simulation results*

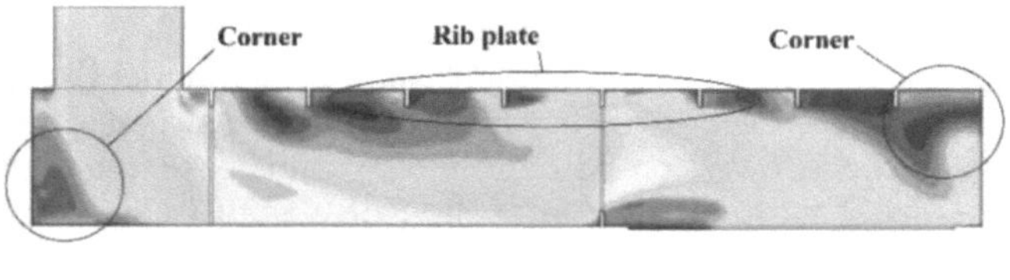

Figure 2. Top air duct simulation results.

It can be seen from the simulation results that at the corner of the air duct and the rib plate structure, the wind speed is low, the pressure is high, and the streamline is disordered.

3 THEORETICAL ANALYSIS OF AIR DUCT VENTILATION

The pressure reduction caused by energy loss when the fluid flows in the tube is caused by the collision and exchange of momentum between the fluid particles when overcoming internal friction and overcoming turbulence when the fluid flows, which is manifested in the pressure difference between the front and rear of the fluid flow, that is, the pressure drop. It includes frictional pressure drop Δpf, gravitational pressure drop Δpel, accelerated pressure drop Δpa and local resistance pressure drop Δpc, that is, flow pressure drop Δp = Δpf + Δpel + Δpa + Δpc.

3.1 *Frictional pressure drop analysis*

The frictional pressure drop is the pressure loss caused by the friction between the fluid flowing along the channel and the wall, and in this case, the pressure loss is caused by the friction between the air and the air duct wall. The frictional pressure drop calculation formula is:

$$\Delta pf = fL\rho v2/(2de) \tag{1}$$

In this research object, the factors such as air duct material and fluid density do not change, and the factors affecting the frictional pressure drop of the air duct are the length and equivalent diameter of the air duct. According to the formula, the larger the air duct length, the greater the frictional pressure drop. The larger the equivalent diameter of the air duct, the greater the friction pressure drop.

3.2 *Gravity pressure drop analysis*

Gravitational pressure drop, also known as lifting pressure drop, is a change in static pressure caused by different potential energy at different channel levels. The calculation formula is:

$$\Delta pel = g(z2 - z1) \tag{2}$$

The influencing factors of gravity pressure drop are only related to the fluid height difference and density because this paper's research object is the air duct. The air density is small, and the air duct height difference is not large. The influence of gravity pressure drop on the overall pressure drop can be ignored.

3.3 *Accelerated pressure drop analysis*

Accelerated pressure drop is a pressure drop caused by changes in the density or velocity of the fluid. The density can be approximately unchanged in the local area where the flow section changes. Only the velocity changes. The pressure drop generated at this point is called the local and accelerating pressure drop Δpa. The calculation formula is:

$$\Delta pa = \rho vdv = \rho(v22 - v21)/2 \tag{3}$$

In the vehicle cabin air duct, there are a large number of rib plate structures. The cross-sectional area of the air duct changes in many places, and the fluid flow rate also changes. The accelerated pressure drop accounts for a large part of the overall pressure drop, so we consider the acceleration pressure drop at the section change. The greater the section changes, the greater the flow velocity changes, and the greater the acceleration pressure drop.

3.4 *Shape resistance pressure drop analysis*

Shape resistance pressure drop is caused by the change of fluid movement direction or the runner's shape in a local area of the system. In this paper, it is mainly manifested as the shape resistance pressure drop brought by the corner of the air duct (Du 2020). Its calculation formula is:

$$\Delta pc = Kc\rho v2/2 \tag{4}$$

In this paper, the energy loss caused by the structure of the corner part of the air duct is mainly manifested.

4 DUCT STRUCTURE OPTIMIZATION

Through the simulation results and the analysis of air duct ventilation theory, it is found that the rib plate structure and air duct angle inside the air duct have a great influence on the ventilation performance of the air duct, causing the change of air duct section in the rib plate structure, resulting in energy loss. The flow separation phenomenon may occur locally (Liang 2021; Zhao 2018), and the rectangular corner structure of the corner part has too large a resistance pressure drop (He 2022). This paper focuses on optimizing the internal rib plate structure and rectangular corner of the air duct.

However, considering that in the process of ship construction, the strength of the hull structure is first guaranteed, and the internal rib plate and rectangular corner structure of the air duct cannot be directly deleted, nor can the size of the main body of the air duct be greatly changed, so this paper proposes an additive optimization idea. That is, during the construction process, the iron sheet is welded into the structure of the air duct rib plate. The corner of the air duct and the airflow contact surface is changed from the uneven rib plate structure to a flat iron sheet to reduce the accelerated and shape resistance pressure drop (Duan 2012).

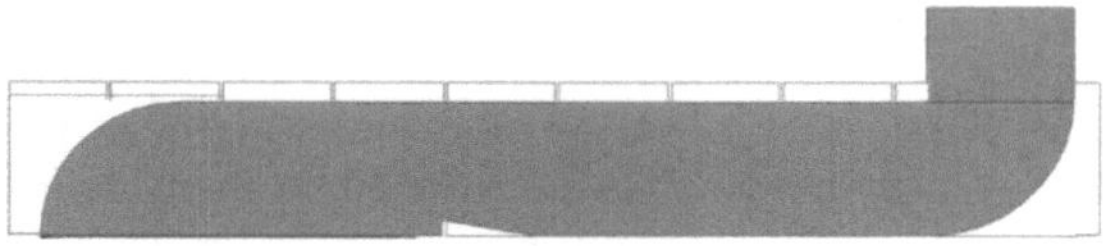

Figure 3. Comparison before and after model optimization.

5 COMPARATIVE ANALYSIS OF SIMULATION RESULTS BEFORE AND AFTER OPTIMIZATION

Aiming at the flow loss caused by the wind duct corner and rib plate structure, the three optimized air duct models are obtained after local optimization of the three typical air ducts. The optimized models are simulated and analyzed, respectively. In the simulation process, the optimized model selects the same meshing method, mesh size, calculation model, algorithm, and boundary conditions as the original model to ensure the consistency of the simulation parameters of the model before and after optimization to ensure the reliability of the comparison of results before and after optimization. The following is a comparison of simulation results before and after optimization:

From the simulation results, it can be observed that compared with the air duct before optimization, the wind speed streamline line of the optimized air duct for the corner and rib plate structure is smoother, there is no obvious streamline disorder, its speed distribution is more uniform, and the extremely low wind speed area is greatly reduced. It can be seen that after the optimization of the air duct, its ventilation performance has been greatly improved, and the flow loss has been effectively reduced.

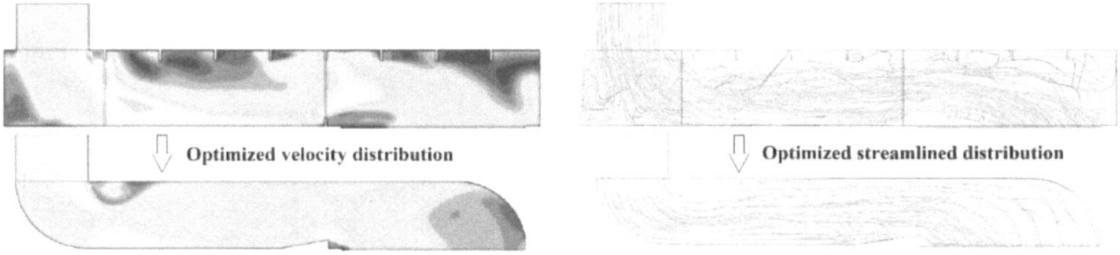

Figure 4. Comparison of simulation results before and after optimization.

6 CONCLUSION

In the existing cabin duct structure of the RoRo vehicle, the rectangular duct corner and many ribbed plate structures inside the duct will form a large shape resistance pressure drop and acceleration pressure drop during the ventilation process. The local wind speed is low, the pressure is large, and the flow is disordered, which greatly impacts the ventilation performance of the air duct.

Optimizing rectangular air duct corner to arc air duct angle and shielding optimization of ribbed plate structure can reduce pressure drop, reduce energy loss during ventilation, improve the uniformity of wind speed distribution inside the air duct, and avoid flow disorder.

REFERENCES

Chen Fuqiang. (2020). Pressure Drop Analysis and Aerodynamic Design of Compressor L-inlet duct. *Aerospace Science and Technology* 107: 1–14

Chen Lianfei. (2017). Numerical Simulation and Optimization Design of Air Duct Inlet Section. *Journal of Shenyang Institute of Technology (Natural Science Edition)*, 13(2): 124–127

Duan Zhipeng. (2012). Pressure Drop for Fully Developed Turbulent Flow in Circular and Noncircular Ducts. *Journal of Fluids Engineering*, 134:1–10

Duan Zhipeng. (2018). Numerical Simulation of Pressure Drop for Three-Dimensional Rectangular Microchannels. *Engineering Computations*, 35(6): 2234–2254

Du Xuzhi. (2020). The Effect of Bend Angle on Pressure Drop and Flow Behavior in a Corrugated Duct. *Acta Mechv*, 231: 3755–3777

He Zhuoyu. (2020). Numerical Value and Experimental Analysis of Local Resistance Loss of Marine Structure Wind Duct. *China Harbor Construction*, 40(1): 16–20.

Liang Qiufeng. (2021). Review of the Influence of Appendage and Tail Manipulation on the Wake Field of the Submarine. *Journal of Ordnance Equipment Engineering*, 42(10): 1–7.

Li Banghua, Zhang Zhanfei, *et al.* (2020). Numerical Simulation and Optimization Design of Air Duct Ventilation in the Cargo Compartment of An Automobile Carrier. *Ship and Ocean Engineering*, 49(3): 10–13.

Shen Xiaoxing. (2017). Design of Air Duct of Cargo Hold Structure of 6700 Vehicle Ro-ro Ship. *Ship Design Communications*, 2: 59–64.

Zhao Rui, Wang Chao, *et al.* (2018) Research Progress on Airflow Field Characteristics of Ship Surface. *Ship Mechanics*, 22(11): 1431–1444.

Water Conservancy and Civil Construction – Oke & Ahmad (Eds)
© 2024 The Author(s), ISBN: 978-1-032-58614-4

Collaborative optimization design of a storage and transportation launch box structure

Jiazheng Ding & Cungui Yu*
School of Mechanical Engineering, Nanjing University of Science and Technology, Nanjing, China

ABSTRACT: In this paper, a storage and transportation launch box is taken as the research object, and the multidisciplinary collaborative optimization method is used to optimize the design. This method can comprehensively consider the performance indicators under each working condition and realize the lightweight design of the structure. Firstly, the sensitivity analysis method is used to determine the main design variables of each working condition, and then the approximate model is constructed by the response surface method. Finally, the optimal design and calculation are based on the collaborative optimization method, and the lightweight result of the storage and transportation launch box structure is obtained. The weight of the optimized storage and transportation launch box is reduced by 11.1%, and the optimization effect is remarkable.

1 GENERAL INTRODUCTIONS

The rocket weapon system's storage and transportation launch box are important (Chen 2012). Box-type launch technology can centrally realize the functions of rocket storage, transportation, and launch. The lightweight research of weapons is an important direction of modern weapons research, which can effectively improve equipment mobility, realize long-distance delivery, and meet the characteristics of modern warfare, such as strong suddenness and fast combat operations (Shan 2022). Therefore, it is of great practical significance to carry out the lightweight design for the structural optimization design of the storage and transportation launch box and to meet the stiffness and strength requirements.

Domestic scholars have researched the optimization design of rocket weapon launch boxes. For example, some literature based on structural optimization theory and finite element analysis carried out topology optimization, shape optimization, and size optimization of rocket launcher launch boxes. The weight reduction design is realized on the premise of ensuring the rigidity and strength of the structure (Yu 2017; Yang 2018). One literature combines the radial basis neural network method, the non-inferior sorting genetic algorithm, and the fuzzy set theory to carry out the multi-objective optimization design of the orientation tubes of a launch box, which realized the lightweight of the launch box and improved the safety of airdrops (Hu 2014). One piece of literature established a random finite element model of the response surface of the artillery launch box and carried out a lightweight design based on reliability analysis (Cai 2017). The above literature research provides many references for the structural optimization design of the storage and transportation launch box. Still, some things could be improved, such as the coupling effect of the mutual influence between different working conditions, parameters, and indicators are not considered.

In this paper, a storage and transportation launch box is taken as the research object, and the multi-disciplinary collaborative optimization design method is adopted to comprehensively consider the influence of different working conditions and different design variables on

*Corresponding Author: yu_cungui@sina.com

DOI: 10.1201/9781003450818-3

the stiffness and strength indicators to research the weight reduction of the storage and transportation launch box structure.

2 OPTIMAL DESIGN MODEL ESTABLISHMENT

2.1 *Optimize the design model construction process*

The optimal design model for the storage and transportation launch box is constructed according to the following steps.

(1) We simplify the three-dimensional model of the storage and transportation launch box and build a finite element model.
(2) We carry out the pre-processing of the finite element model, including dividing the mesh, applying loads and constraints, etc., and performing finite element stiffness and strength analysis on the three typical working conditions of the storage and transportation launch box in the stacking condition, the marching condition and the launching condition.
(3) We use the optimal Latin hypercube method to perform multiple simulations for each working condition, obtain input-output data, and analyze the sensitivity of the design parameters.
(4) We construct an approximate model through the response surface method and verify the accuracy. When the accuracy meets the requirements, we proceed to the next step. Otherwise, we rebuild the approximate model.
(5) We build the optimal design model of the storage and transportation launch box based on the collaborative optimization method.
(6) We solve the optimization model by combining the multi-island genetic algorithm and the sequential quadratic programming method and analyze the results.

The construction process of the optimal design model of the storage and transportation launch box is shown in Figure 1.

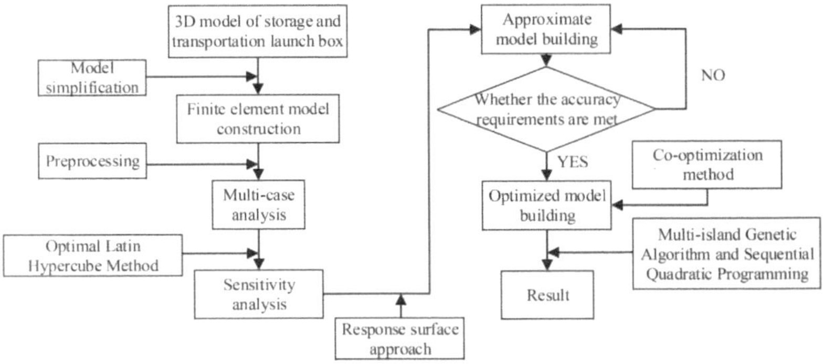

Figure 1. The construction process of the optimal storage and transportation launch box design model.

2.2 *Multidisciplinary collaborative optimization design method*

Collaborative optimization (CO) is a two-level hierarchical optimization strategy for solving complex engineering design problems (Rabeau 2007). It was originally proposed by Professor Kroo of Stanford University and others based on the All-at-once Method (AAO). The prominent feature of this kind of problem is that multiple disciplines are cross-coupling with each other (Wang 2017), and traditional algorithms, such as genetic algorithms, have poor solution effects.

The principle of the CO method is to decompose multidisciplinary problems into system level and multiple subsystem levels, and each subsystem is optimized in parallel without interfering with each other. System-level optimization coordinates the differences in the optimization results of various subsystems. After multiple iterations between system-level and subsystem-level optimization, a consistent optimization result is finally found (Hou 2017; Xia 2021). The framework of the collaborative optimization method is shown in Figure 2, where N is the number of subsystem-level optimizations.

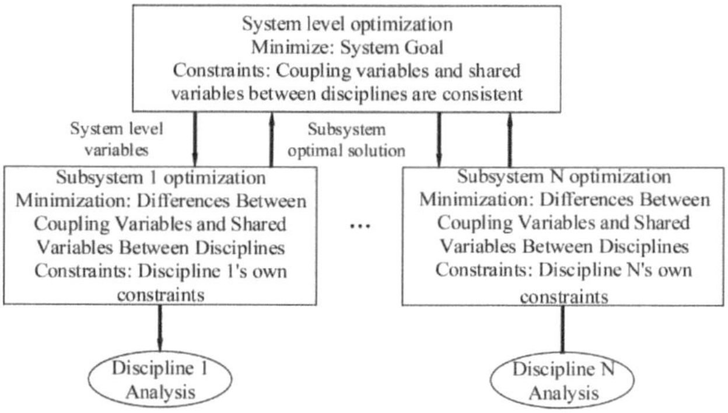

Figure 2. Co-optimization method framework.

2.3 *Construction of multi-working conditions optimization model*

There are stacking, marching, and launching working conditions in the actual use of the storage and transportation launch box. The loads and boundary conditions under different working conditions are different, resulting in different degrees of influence of the same design variable on the rigidity performance of different working conditions. There is a certain coupling relationship. Therefore, it is very suitable to use the CO method to solve the structural optimization problem of the storage and transportation launch box.

In this paper, the CO method is used to carry out the lightweight design of the launch box for storage and transportation. Different working conditions are regarded as subsystems, and parallel optimization is carried out with the constraints of each working condition as constraints. The system-level optimization objective is the weight of the launch box. The consistency constraint in the system-level optimization is iteratively solved for the constraint function. Finally, a structural weight minimization scheme that can meet the performance indicators of each working condition is obtained. The mathematical model is as follows.

The system-level optimization model is:

$$\begin{aligned}&\min \mathrm{M}\\&s.t.\ J_i(z)=\sum_{j=1}^{n_i}\lambda_j(z_{ij}-p_{ij})^2\leq s,\ i=1,2,3\\&\underline{x_{ij}}\leq z_{ij}\leq\overline{x_{ij}}\end{aligned} \tag{1}$$

The subsystem-level optimization model is:

$$\begin{aligned}&\min R_i(x)=\sum_{j=1}^{n_i}\lambda_j(x_{ij}-q_{ij})^2\\&s.t.\ H_{ik}(x)\leq 0\\&\underline{x_{ij}}\leq x_{ij}\leq\overline{x_{ij}}\end{aligned} \tag{2}$$

In Formulas (1) and (2), M is the optimization objective function, which represents the weight of the launch box; $J_i(z)$ and $R_i(x)$ are the consistency constraint functions, which represent the difference between the system-level design variables and the subsystem-level design variables; λ_j is the weight coefficient of the design variables; $p_{ij} = x^*_{ij}$, which represents the optimization result of the jth variable of the ith subsystem, and participates in the system-level optimization with a constant; $q_{ij} = z^*_{ij}$, which represents the system-level optimization result; $\underline{x_{ij}}$ and $\overline{x_{ij}}$ are the upper and lower limits of the design variables; n_i is the number of design variables in the subsystem. Since there will be conflicts between various working conditions, the optimal solution of one working condition may cause the remaining working conditions to fail to meet the constraints, so a relaxation factor s is introduced, indicating the consistency constraint relaxation range. When convergence is good, the value of s can be gradually reduced until convergence difficulties occur.

2.4 *Establishment of finite element model*

The geometric model of the storage and transportation launch box is shown in Figure 3. The box structure includes four splints, long-angle steels, vertical side plates, oblique stiffeners, and rear vertical plates. The box structure supports orientation tubes, and positioning rings are on it to limit the front and rear displacement. The locking mechanism fixes the rocket projectile. Within the orientation tube, the mass and moment of inertia are assigned realistically in the finite element model.

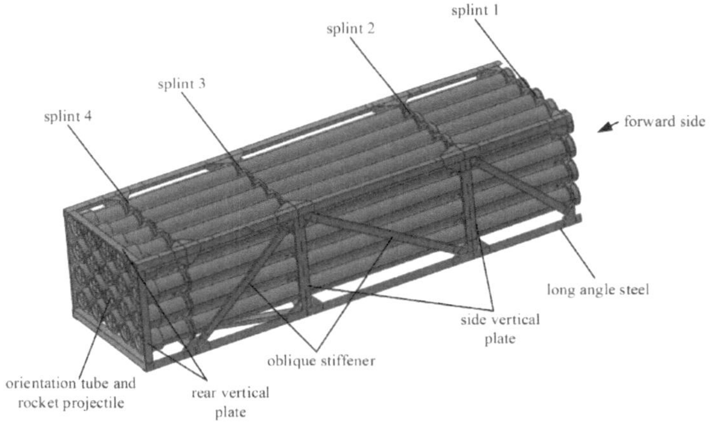

Figure 3. Geometric model of storage and transportation launch box.

The typical working conditions of the storage and transportation launch box include stacking, marching, and launching working conditions. Finite element models under different working conditions are established, respectively, and constraints and loads are added.

(1) Stacking working conditions. Rubber spacers support the angle steels at the bottom of the launch box, and two launch boxes are stacked. The bottom launch box is under the worst stress, so the finite element analysis of the bottom box is carried out;
(2) Marching working conditions. During the march, the launch box is fixed on the vehicle through the bottom positioning locking block. When the vehicle is bumpy, the working condition is the worst. At this time, the launch box is subjected to vertical acceleration in addition to gravity.

(3) Launching working conditions. During the launch of the rocket projectiles, the launch box is subjected to the impact force of the gas flow that changes with time in the direction of the air-facing surface.

Through finite element simulation calculation, the maximum stress value and maximum deformation of the storage and transportation launch box under various working conditions are obtained as its stiffness and strength performance indicators, which lays the foundation for subsequent sensitivity analysis and approximate model construction.

2.5 *Parameter sensitivity analysis*

The design of experiments (DOE) method is one of the most important statistical methods in the optimization process. It can effectively obtain the relationship between input parameters and output responses and is widely used in parameter sensitivity analysis and approximate model construction. Through the DOE analysis, the sensitivity of each design parameter to the output response is obtained, and the parameters that have a greater impact on the response function are screened out, thereby reducing the calculation in the optimization process and effectively improving the optimization efficiency.

We select the launch box structural parameters as the main design variables, including the thickness of the plywood 1 to 4, the thickness of the long angle steel, the thickness of the side vertical plate, the thickness of the rear vertical plate, and the thickness of the oblique stiffener, which represented by $x_1, x_2, \ldots, x_8$. The stiffness and strength of the structure are calculated by the simulation, which is expressed by the maximum stress value, which is represented by σ_m, and the maximum deformation amount, which is represented by s_m. This paper uses the Latin hypercube method to simulate and sample each working condition. 500 sample data for each working condition are obtained. The parameter sensitivity analysis is carried out. The results are shown in Table 1:

Table 1. Results of sensitivity analysis.

	Working condition 1		Working condition 2		Working condition 3	
Design variable	σ_m	s_m	σ_m	s_m	σ_m	s_m
x_1	2.19	0.19	5.62	3.01	–72.27	–59.84
x_2	5.08	0.02	–65.42	–25.56	–2.10	–12.49
x_3	–3.12	1.06	0.34	–41.74	–0.18	–0.96
x_4	–3.40	3.54	–0.16	7.51	–3.53	–1.04
x_5	–61.92	–38.59	23.91	–15.33	16.68	–23.10
x_6	3.99	0.03	0.01	5.19	2.37	1.58
x_7	4.79	–53.10	0.01	2.25	0.32	–0.38
x_8	–15.51	3.47	4.53	–8.92	2.55	0.61

unit: %

It can be seen from Table 1 that when the influence of different design variables on the stiffness and strength of the structure under each working condition is lower than 5%, the sensitivity of this variable is poor. After careful consideration, the three parameters with the highest degree of influence are screened for each working condition as their optimal design variables. Among them, the optimal design variables of working condition 1 are x_{15}, x_{17}, x_{18}. The optimal design variables of working condition 2 are x_{22}, x_{23}, x_{25}. The optimal design variables of working condition 3 are x_{31}, x_{32}, x_{35}. The design variables in the system-level optimization are $x_1, x_2, x_3, x_5, x_7, x_8$.

2.6 *Approximate model and error analysis*

In the process of co-optimization calculation, there is a nested relationship between system-level optimization and subsystem-level optimization, which leads to a geometric increase in the number of optimization iterations. Therefore, obtaining sufficient finite element simulation data through the DOE method and constructing an approximate model based on this method can effectively shorten the optimization design cycle. After the approximate model is constructed, the accuracy of the model is judged by the error analysis method. Common error evaluation indicators are root mean square error RMSE and coefficient of determination R^2. The mathematical formulas are:

$$RMSE = \left[\frac{\sum_{l=1}^{n_s}(y^l - \hat{y}^l)^2}{n_s}\right]^{1/2} \tag{3}$$

$$R^2 = 1 - \frac{\sum_{l=1}^{n_s}(y^l - \hat{y}^l)^2}{\sum_{l=1}^{n_s}(y^l - \bar{y})^2} \tag{4}$$

The root means square error RMSE represents the degree of difference between the predicted value and the true response. The smaller the value of RMSE, the higher the approximation of the model. The value of the coefficient of determination R2 is between 0 and 1, and the closer it is to 1, the approximate model the higher the accuracy. Due to the complex structure of the storage and transportation launch box, this paper uses the polynomial response surface method to construct an approximate model, which can fit complex response relationships and has good robustness. The corresponding approximate models were constructed for the three working conditions, and the error analysis was performed on them. The results are shown in Table 2. The constructed approximate models have high enough accuracy.

Table 2. Approximate model error analysis.

Working condition	Performance indicator	RMSE	R^2
1	σ_m	0.048	0.984
	s_m	0.003	0.999
2	σ_m	0.025	0.989
	s_m	0.098	0.893
3	σ_m	0.007	0.995
	s_m	0.013	0.998

3 OPTIMIZATION RESULTS AND ANALYSIS

According to the multi-working conditions optimization model shown in Formulas (1) and (2), the approximate model under the three operating conditions of the storage and transportation launch box is used as the optimization object to solve. In the system-level optimization, the multi-island genetic algorithm (MIGA) is used for the optimization solution; in the subsystem-level optimization, the sequential quadratic programming algorithm (SQP) is used for the optimization solution.

The final optimization results are shown in Figure 4. After 1001 iterations of convergence, the optimal solution for the weight M of the launch box is 210.263 kg, which is 11.1% lower than the initial value. The consistency constraints of the three subsystems all tend to be 0, which proves that the system level and the subsystem level reach the consistency requirements, and the incompatibility of the subsystem layer can be eliminated through the coordination of the system layer.

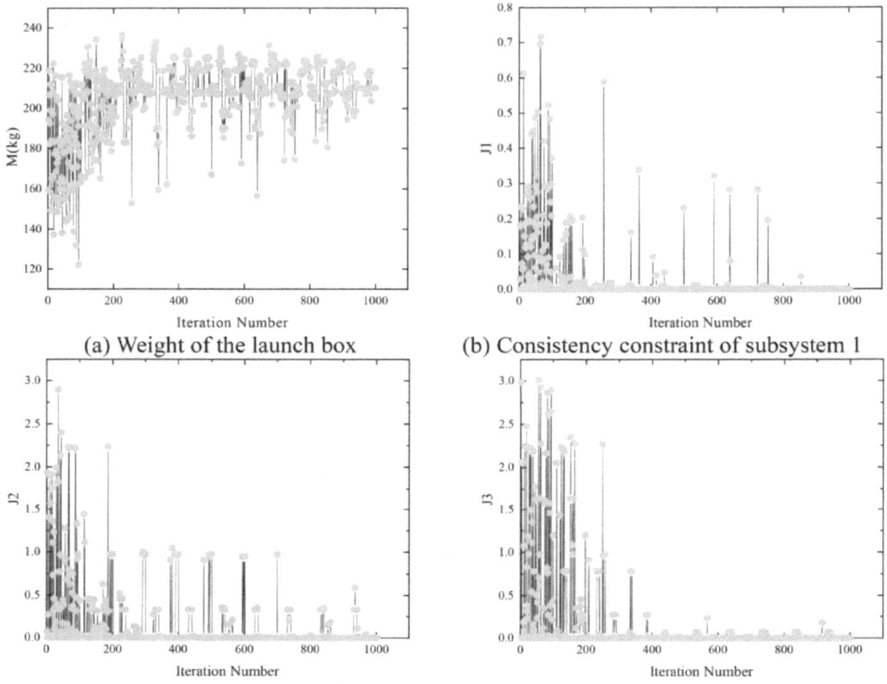

(a) Weight of the launch box

(b) Consistency constraint of subsystem 1

(c) Consistency constraint of subsystem 2

(d) Consistency constraint of subsystem 3

Figure 4. Optimizing results.

The optimal design variables calculated by the multi-working conditions collaborative optimization method are shown in Table 3. It can be seen that the maximum stress value and the maximum deformation amount have increased. Still, they are all within the allowable range, and the constraints can be satisfied in the three typical working conditions.

Table 3. Optimization results of each design variable.

Design variable	Initial value	Optimization result	Constraint	Before optimization	After optimization
x_1/mm	15	12.58	σ_{m1}/Mpa	70.01	89.42
x_2/mm	15	12.99	s_{m1}/mm	1.28	1.62
x_3/mm	15	10.41	σ_{m2}/Mpa	56.58	57.72
x_5/mm	5	4.08	s_{m2}/mm	0.59	0.66
x_7/mm	5	4.54	σ_{m3}/Mpa	122.86	122.215
x_8/mm	5	4.10	s_{m3}/mm	2.75	2.92

4 CONCLUSION

In this paper, lightweight research is carried out on the structure of a storage and transportation launch box. Based on the multi-disciplinary collaborative optimization method, an optimization model of multiple working conditions is established, and the optimization solution is carried out. The following conclusions are obtained:

(1) Through the sensitivity analysis of the finite element model of the storage and transportation launch box, the influence of different parameters on the stiffness and strength of the structure is obtained, and the key design variables of each working condition are screened out.
(2) The approximate model of the storage and transportation launch box is established by the response surface method, which greatly shortens the optimization design cycle and improves the optimization efficiency.
(3) Based on the multi-disciplinary collaborative optimization method, a structural optimization model of the storage and transportation launch box that comprehensively considers various working conditions can be established. The optimal solution that satisfies the constraints of each working condition is obtained by establishing consistency constraints. After optimization, the structural weight of the storage and transportation launch box is reduced by 11.1%, and the optimization effect is remarkable.

REFERENCES

Chunlai Shan, Pengke Liu, Bin Gu, *et al.* (2022). Application of Multilevel Optimization Algorithm in Artillery Integrated Design. *J. Acta Armamentarii*, 43(01), 11–19.

Cui-dong Yang, Zhang-yu Yan, Lei Han, *et al.* (2018). Study of Structural Optimization Method for Launching Canister of Rocket Launcher. *J. Journal of North University of China (Natural Science Edition)*, 39(01), 61–68.

De-yong Cai, Fu-jun Liu, Ke-wen Tian, *et al.* (2017). Reliability Analysis and Lightweight Design of the Rocket Launch Canister. *J. Journal of Machine Design*. 34(05), 81–85.

Jianguo Hu, Jianlin Zhong, Dawei Ma, *et al.* (2014). Multi-objective Optimization for the Orientator of Airdropping Launch Canister. *J. Ordnance Material Science and Engineering*, 37(03), 31–35.

Qirui Yu, Jun Li, Xin Zhao, *et al.* (2017). Optimization Design and Analysis on Rocket Launcher Canister Storage and Transport B on HyperWorks. *J. Ordnance Industry Automation*, 36(10), 80–83.

Rabeau, S., Dépincé, P., & Bennis, F. (2007). Collaborative Optimization of Complex Systems: A Multidisciplinary Approach. *International Journal on Interactive Design and Manufacturing (IJIDeM)*, 1 (4), 209–218.

Tianxiang Xia, Yueliang Lu, Bing Ke. (2021). Collaborative Optimization of Ram Air Turbine System Structures. *J. Journal of Nanjing University of Aeronautics & Astronautics*, 53(04), 583–590. DOI: 10.16356/j.1005-2615.2021.04.012.

Wang W, Gao F, Cheng Y, *et al.* (2017). Multidisciplinary Design Optimization for the Front Structure of an Electric Car Body-in-white based on an Improved Collaborative Optimization Method. *J. International Journal of Automotive Technology*, 18(6), 1007–1015.

Wenbin Hou, Chunlai Shan, Ye Yu, *et al.* (2017). Selection Method of Sharing Modules for Modular Product Family. *J. Journal of Hunan University (Natural Sciences)*, 44(02): 66–74. DOI: 10.16339/j.cnki.hdxbzkb.2017.02.010.

Yu Chen, Feng-yun Sun. (2012). Structure and Design of Transport Launcher. *J. Packing Engineering*, 33 (15), 132–135.

Water Conservancy and Civil Construction – Oke & Ahmad (Eds)
 ISBN: 978-1-032-58614-4

Water content and stability of cliff landslide under different rainfall intensity

Hui Li
Hebei Geo-Environment Monitoring, Shijiazhuang, Hebei, China
Hebei Key Laboratory of Geological Resources and Environment Monitoring and Protection, Shijiazhuang, Hebei, China

Jie Zhao
Hebei Geo-Environment Monitoring, Shijiazhuang, Hebei, China

Yi Pan
Hebei Geological Workers' University, Shijiazhuang, Hebei, China

Chenxi Li*, Xing Zhai & Yulong Li
Hebei Geo-Environment Monitoring, Shijiazhuang, Hebei, China
Hebei Key Laboratory of Geological Resources and Environment Monitoring and Protection, Shijiazhuang, Hebei, China

ABSTRACT: Rainfall is the primary influencing factor of cliff landslide water content and stability, which is the focus of geological disaster prevention and control. Taking the Nangaojiayu landslide in Jingxing County, the eastern foot of Taihang Mountain, as the research object, the development characteristics of the steep cliff landslide were discussed, and the relationship between rainfall and geological disaster was analyzed by combining the regional geological environment data and survey results. Based on this, the professional monitoring data were systematically combed, the relationship between water content, time, and rainfall was analyzed, and the stability of rainfall on landslides was studied. The results showed that: (1) Under the action of tectonic movement and external force, the spatial difference of a collapse, residual, and colluvial cliff landslide with a high content of large stone in the middle and rear parts and a high content of small cohesive soil in the front part was formed. (2) The change curve of water content in the shallow surface of the cliff landslide under the action of rainfall was in a normal distribution, which reflected the characteristics of rapid infiltration of groundwater in the middle and rear of the landslide and slow dissipation in the front, which was not conducive to the stability of the shallow layer of the landslide. Based on this, the mode of change in the relationship between water content and time was summarized. (3) The function equations of water content, time, and rainfall were established by using the Extremes function curve and Boltzmann function curve, which were basically applicable to the calculation and stability analysis of the water content of a cliff landslide. (4) When the rainfall intensity was less than 150 mm/d, the shallow sliding of the landslide was induced, and the impact on the overall stability was small, which provided a reference for the early warning and prediction of the landslide.

*Corresponding Author: hkylcx@126.com

DOI: 10.1201/9781003450818-4

1 INTRODUCTION

The steep cliff landform is common in the middle of the eastern foot of the Taihang Mountains, with more typical landforms such as Cangyan Mountain and Zhangshiyan Mountain in the Shijiazhuang area. Many slopes are made up of colluvial, residual, and slope deposits, which collapse and migrate during the formation of a steep cliff. The water content and stability of such slopes change under the influence of rainfall, which is particularly important for the prevention and control of geological disasters.

At present, domestic and foreign scholars have carried out much research on the water content and stability of landslides. Yuliza *et al.* (Habil *et al.* 2016; Yan *et al.* 2019) used professional monitoring, an electrical method, and frequency domain reflection analysis to measure the migration of groundwater and the change of water content on the slope. In terms of landslide water content and stability, Wang *et al.* (Feng *et al.* 2011; Gao *et al.* 2020; Lu & Wang 2019) used the triaxial shear test, ring shear test, direct shear test, and other shear tests to analyze soil water content parameters and study the relationship between water content and shear strength; Sinnappoo *et al.* (Elwood *et al.* 2021; Ivan & Sinnappoo 2022) discussed the influence of unsaturated soil water content change on matrix suction on landslide shear strength; Wei *et al.* (Guo *et al.* 2022; Wei *et al.* 2022) analyzed the impact of water content change on landslide stability by building a mechanical model, a numerical model, and a physical model. In terms of landslide water content and rainfall, Hakro *et al.* (Hakro & Harahap 2015) analyzed the increase of landslide water content and pore pressure under different rainfall intensities and their impact on landslide stability by building a landslide physical model; Abraham M. T. *et al.* (Abraham *et al.* 2020; Li *et al.* 2021) studied the relationship between rainfall, water content, and other multiple indicators by monitoring multiple indicators of landslides and explored the threshold value of landslide early warning indicators.

However, many studies were carried out on the basis of analyzing large amounts of monitoring data or physical models. In the absence of monitoring data, establishing a scientific and reasonable relationship between water content, time, and rainfall remains a challenge given the steep cliff landslides with large spatial differences in the fabric of sliding bodies. As a result, this paper analyzed the response relationship between rainfall and geological disasters by combining data from regional geological environment conditions, studied the shallow surface water content curve of landslides under the action of rainfall, and summarized the change model of the relationship between water content and time. The function equations of moisture content, time, and rainfall are fitted, and the influence of rainfall on the shallow slip was discussed, which will provide a reference for the prevention and control of geological disasters.

2 DEVELOPMENT CHARACTERISTICS OF CLIFF LANDSLIDE

2.1 *Tectonic movement of cliff landslide*

The research area is located in the famous fault zone at the eastern foot of the Taihang Mountains, bordering Jingxing Sag, and Zanhuang uplift. The Taihang fault uplift started in the Yanshan movement period and further rose in the Himalayan movement period, forming the current fault zone dominated by normal faults and subordinate to reverse faults. The cliffs are composed of limestone and quartz sandstone, and a large number of materials were produced under the collapse and denudation of Taihang Mountain, providing a material basis for the formation of the slope.

Under the conditions of great tectonic movement, the normal fault at the eastern foot of the Taihang Mountains descends from west to east, forming the western mountains and eastern plains, as well as lateral erosion and weathering by external forces, forming the slope of colluvial, residual, and slope deposits on the micro landform (Chen & Shen 1993; He *et al.*

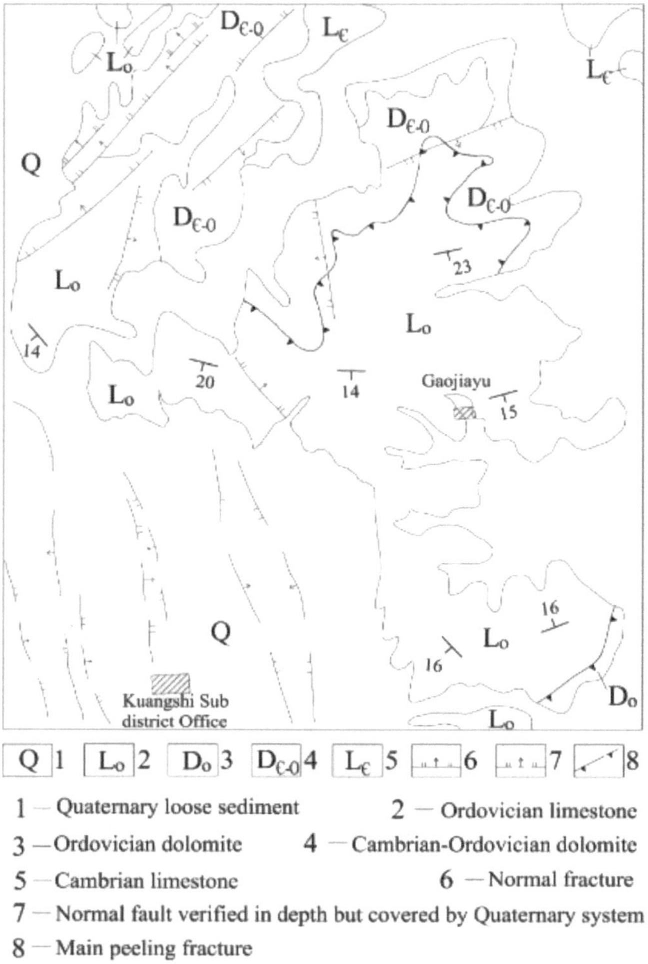

Figure 1. Geological structure of the study area.

2019) (Figure 1). It can be seen that the tectonic movement plays a decisive role in the formation of the cliff landslide, especially the secondary fault caused by the large fault. These microstructures control the formation of the cliff.

2.2 *Slope characteristics of cliff landslide*

Tectonic movement and external dynamic action are the material sources of the slope. The body of the slope contains a large number of rubble stones, whose lithology is consistent with that of the parent rock of the escarpment, which is all Great Wall sandstone. According to the survey, the middle and rear slopes of the colluvial, eluvial, and deluvial bodies with such spatial differences contain a large number of block stones with obvious angles and different particle sizes, which have good permeability, a high infiltration rate of surface water, and high shear strength (internal friction angle). The content of cohesive soil at the foot of the slope is high, and the permeability is poor. The groundwater infiltrated in the middle and rear parts is easy to store at the front edge of the slope, which is not conducive to the stability of the slope (Photo 1).

Photo 1. The material fabric of the front edge of the sliding body.

3 ANALYSIS OF WATER CONTENT CHANGE AND STABILITY OF LANDSLIDE

The study took the Nangaojiayu landslide as an example. Affected by artificial slope cutting, in 1996, under the continuous influence of heavy rainfall in flood season, cracks occurred in the middle and rear edges of the slope, and a tension crack appeared in the middle of the slope. The crack was 90 m long, generally 0.3 to 0.5 m wide, and nearly 1 m deep. It was distributed in an arc, and the slope was deformed. At this time, the landslide is stable. During the August 2000 rainstorm, the shallow deformation body of the landslide was cracked again, and then the cracks were buried. A drainage ditch was built on the slope, and a simple retaining wall was built in front of the slope. The landslide was stable again. To analyze the impact of rainfall on the landslide, professional monitoring equipment for water content, GNSS, deep displacement, etc. was installed in the middle and front of the landslide to analyze the change and stability of the landslide's water content from multiple dimensions, providing a reference for the prevention and control of similar landslides.

3.1 *Analysis of water content change*

3.1.1 *Analysis of water content monitoring data*

Figure 2 shows the change of water content in the shallow layer of the sliding mass under different rainfall intensities. The water content has increased twice. First, after the heavy rain in the first ten days of May, after six days of surface water infiltration, the water content of the soil mass with a buried depth of 0.3 m has increased from 16.8% to 22.9% and recovered to 17.6% after 32 days; the water content of the soil mass at the burial depth of 0.6 m increased from 10.0% to 21.9% and recovered to 10.6% after 35 days; and the water content of the soil mass at the burial depth of 1 m increased from 19.4% to 36.1% and returned to 19.0% after 39 days. Second, after the continuous rainfall in August, the soil water content at the depth of 0.3 m increased from 17.7% to 60.6% after 10 days and recovered to 17.4% after 104 days; the water content of the soil mass at the burial depth of 0.6 m increased from 10.0% to 21.9% after 11 days and recovered to 16.8% after 113 days; and the water content of the soil mass at a buried depth of 1 m increased from 18.1% to 61.3% after 18 days and recovered to 18.2% after 137 days.

This shows that the water content of a rainfall-type cliff landslide has the following characteristics: (1) The rainfall intensity has a significant impact on the water content. With

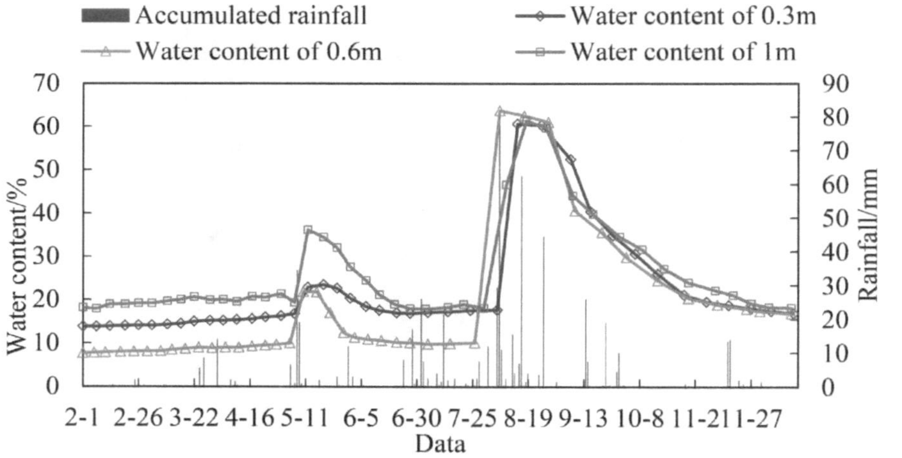

Figure 2. Relationship curve between soil water content and monthly rainfall.

the increase in rainfall intensity, the water content of the shallow surface layer of the slope increases, and the dissipation time also increases. The water content curve presents a positively skewed distribution. That is, the water content rises quickly and dissipates slowly, and the groundwater is stored on the slope. The main reason is the special fabric of the sliding mass. The content of gravel in the middle and rear parts is high, the particle size is large, and the precipitation is easy to penetrate; however, the front structure and weathering are mainly affected by cohesive soil with a small particle size, which is not conducive to the dissipation of water content in the slope. (2) When the rainfall on that day exceeds 35 mm, the water content of the shallow layer of the sliding mass changes significantly, indicating that the sensitivity of the water content of the landslide to rainfall is about 35 mm. (3) The water content in the shallow layer dissipates faster than that in the deep layer. In addition to the influence of surface evaporation, it is revealed that there is a local connection surface in the shallow part of the sliding mass; that is, there is shallow sliding in the landslide.

According to the curve of landslide water content and rainfall, as well as the curve of landslide water content and time, combined with the existing results, it can be inferred that under the same rainfall intensity and similar terrain, there are three modes of water content change with time, namely sudden rise and slow drop mode (Figure 3a), slow rise and sudden drop mode (Figure 3c), and the same rise and fall mode (Figure 3b) between them. In the sudden rise and slow fall mode, the water content changes in a positively skewed distribution, which means that the middle and rear parts of the landslide have a large catchment area and good permeability, while the front part has more cohesive soil and poor drainage, and the groundwater in the sliding body cannot be discharged in a short time, resulting in an increase in the gravity of the slope and a decrease in the shear strength, which is not conducive to the stability of the landslide and is more common in rubble landslides (Dong, Huang, Luo *et al.*, 2017). The slow rise and sudden drop mode is just the opposite of the mode. The water content changes in a negatively skewed distribution. The surface water in the middle and rear of the landslide is not easy to discharge, and it slowly infiltrates downward for a long time, but there is a good drainage space in the front, which is conducive to the stability of the landslide; in the same rise and fall mode, the water content changes in a normal distribution, the rising and falling trends of water content are consistent, and the stability is between a and c.

According to the curve of water content and time change, we can judge the change of water content with time from the terrain, material fabric, and other aspects during the investigation and exploration work, providing a reference for qualitative analysis of landslide stability and an early warning forecast.

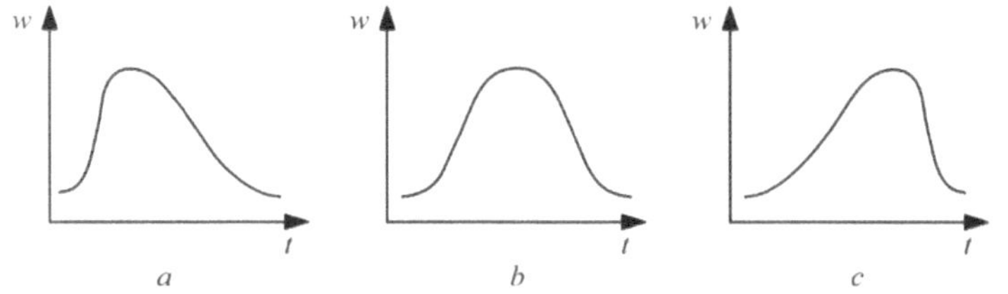

Figure 3. Relationship curve between water content and time.

3.1.2 *Functional equation of water content*

Due to the change in water content of the cliff landslide in the study area and the limitation of funds for professional monitoring equipment, it is not possible to promote the installation on a large scale. Therefore, according to the sudden rise and slow fall mode of the shallow layer water content of the landslide, combined with the existing water content monitoring data, the corresponding water content function equation is established to provide a theoretical reference for landslide prevention.

According to the curve of water content and time, the functional equation should be a peak function, so the Horton physical model (Liu *et al.* 2012) (Formula 1) of the soil infiltration model is taken as the basis to speculate on the form of the functional equation.

$$I(t) = I_f + (I_i - I_f)e^{-ct} \tag{1}$$

where I(t), I_f, and I_i are the infiltration rate, initial infiltration rate, and stable infiltration rate (mm/min), respectively; t is the infiltration time (min); and c is the parameter.

(1) The Equation of Moisture Content and Time Function

Based on the Horton model, the extremes function in the peak function is used to fit the data, and the related function among moisture content, rainfall, and time is discussed (Formula 2).

$$\begin{aligned} w &= w_0 + w_A e^{(-e^{-z}-z+1)} \\ z &= (t - 48.96)/9 \end{aligned} \tag{2}$$

where w is the moisture content (%), w_0 is the initial soil moisture content (%), t is the time (d), and w_A is the function of the relationship between moisture content and daily maximum rainfall.

(2) The Functional Equation of Moisture Content and Rainfall.

Through the statistical analysis of the change in water content of landslides during heavy rainfall from 2020 to 2022, it is found that the maximum value of water content is about 64%. At this time, with the increase in rainfall intensity, the increase in water content is very small (Figure 4). Therefore, in order to construct the relational function between water content and rainfall, the Boltzmann function curve is used to fit the two function equations (Formula 3).

$$w'_A = w_{max} + \frac{w_0 - w_{max}}{1 + e^{\frac{h-38.86}{2.17}}} \tag{3}$$

where w_{max} is the maximum water content of the slope in the study area, and h is the daily maximum rainfall (mm).

Through fitting, the functional equation can basically reflect the relationship between water content and rainfall, the correlation coefficient R^2 is 0.997, so the degree of the fitting is high.

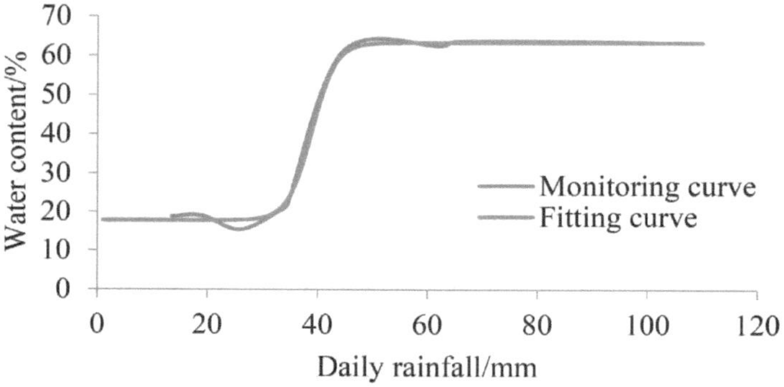

Figure 4. Relationship curve between water content and rainfall.

(3) The Function Equation of Water Content, Time, and Rainfall.

Combining Formulas 2 and 3, the function equations of water content, rainfall, and time in the study area are established (Formula 4).

$$w = w_0 + \left(w_{\max} - w_0 + \frac{w_0 - w_{\max}}{1 + e^{\frac{h-38.86}{2.17}}}\right) e^{\left(-e^{\frac{48.96-t}{16.02}} - \frac{t-48.96}{16.02} + 1\right)} \tag{4}$$

This paper chooses the change data of the soil water content at the sliding mass buried depth of 1 m from July 14 to December 15, 2020, to verify Formula 4. The initial soil water content indicates the average value of the water content in the first half of the year as 19.7%, the maximum daily rainfall is 82 mm, and the time is 163 days. According to the monitoring data, the maximum water content is 63.7%. The changing trend is essentially consistent after fitting (Figure 5). The establishment of an empirical formula for water content is helpful to the qualitative analysis of the stability of steep cliff landslides and provides a theoretical reference for the optimization of geological disasters, meteorological risk, early warning systems, and early warning and forecasting work.

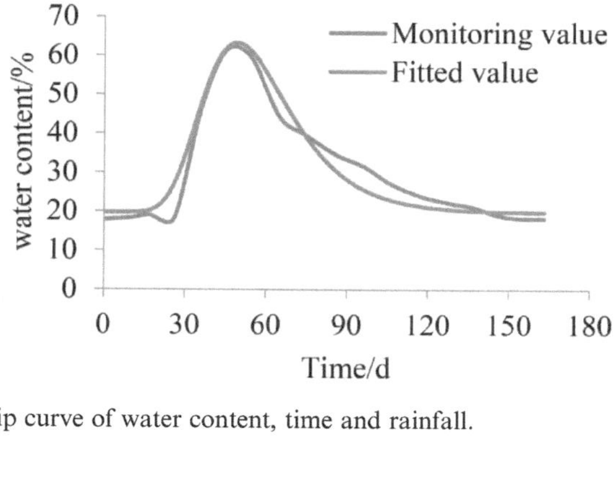

Figure 5. Relationship curve of water content, time and rainfall.

3.2 *Analysis of the landslide stability*

Figure 6 shows the relationship between daily rainfall and deep displacement from landslides. Under the influence of rainfall, the change in the deep displacement of the landslide is not obvious, indicating that the rainfall in a short period of time has little effect on the overall stability of the landslide. However, according to the dissipation time of water content, the increase in rainfall time will cause groundwater to continue to seep into the depth of the landslide, and the front edge of the landslide, dominated by clayey soil and overlying bedrock, will slow down the discharge of groundwater in the slope, which is not conducive to the stability of the landslide.

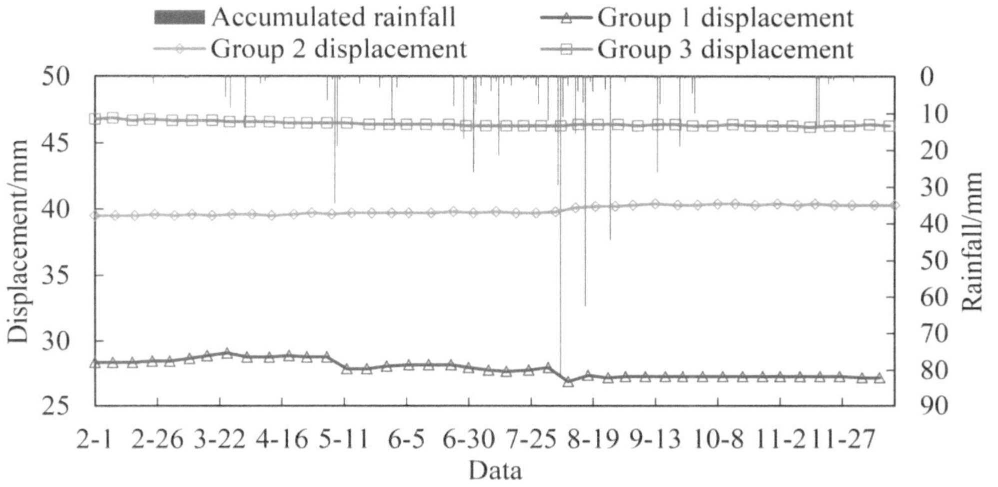

Figure 6. Relationship curve between landslide displacement and monthly rainfall.

3.3 *Stability analysis of the landslide under different rainfall conditions*

To further analyze the influence of rainfall on cliff landslides, based on the landslide engineering geological model, FLAC3d is used to build a numerical model to analyze the displacement and deformation of steep slopes under different rainfall intensities.

It can be seen from the figure that the surface displacement changes greatly in different parts of the landslide under different rainfall intensities. According to the inferred empirical formula of water content, in the natural state, the water content of the upper part of the slope is about 19.3%, and the change of the surface displacement of the landslide is small; when the rainfall intensity is 50 mm, the water content is about 63%, the displacement of the front edge of the landslide becomes larger, and the displacement of the trailing edge of the landslide is unchanged, and the influence of rainfall on the deformation of the front edge of the landslide begins to increase; when the rainfall intensity is greater than 50 mm, the water content in the upper part of the sliding mass is not increasing, but the duration is increasing, the surface water continues to infiltrate, and the displacement of the front edge is further increased at this time (Figure 7). Therefore, combined with the landslide monitoring data, it can be seen that when the daily rainfall is less than 150 mm, the overall stability of the landslide is good, and shallow sliding occurs in the middle and front of the landslide, which is also a key prevention part.

At the same time, according to the empirical formula for water content, when the daily rainfall exceeds 35 mm, the water content in the upper part of the sliding mass starts to increase. At this time, the meteorological risk warning for geological disasters should be done well according to the rainfall intensity and duration, combined with regional experience.

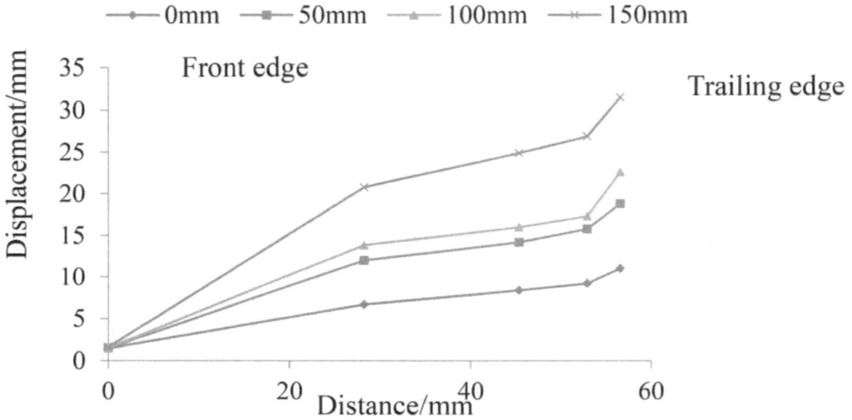

Figure 7. Displacement curve of landslide front under different rainfall intensity.

4 CONCLUSIONS

(1) The main influencing factor of the cliff landslide was tectonic movement, and the microstructure played a significant role in controlling it. When the internal force was combined with the external force, a landslide with different materials and fabric spaces was formed. The fabric characteristics of this type of sliding mass were a high content of block stones in the middle and rear, large particle size, high content of cohesive soil in the front, and small particle size of block stones, which had an obvious influence on the change in water content of the landslide.
(2) The monitoring data on water content were analyzed systematically. When the rainfall intensity exceeded 35 mm, the water content in the shallow layer of the landslide started to rise suddenly. Affected by the landslide fabric, it dissipated slowly in the later period, forming a positive skewness distribution curve, which was not conducive to the stability of the landslide. Based on this, the influence of landslide material (fabric), landforms, rainfall, etc. on water content was summarized. The water content time relationship curve was divided into the positively skewed distribution of sudden rise and slow fall mode, the negatively skewed distribution of slow rise and sudden drop mode, and the normal distribution of the same rise and fall mode. Different modes had different effects on landslide stability.
(3) According to the relationship curve of water content, time, and rainfall based on the extreme function curve and Boltzmann function curve, the empirical formula of the water content of a cliff landslide was established, which was basically in line with reality and provides a theoretical basis for geological disaster prevention.
(4) FLAC3d was used to analyze the surface displacement change of the landslide under different rainfall intensities. When the rainfall intensity was less than 150 mm, the landslide was mainly shallow sliding.

REFERENCES

Abraham M.T., Pradhan B., Satyam N., et al. 2020 IoT-based Geotechnical Monitoring of Unstable Slopes for Landslide Early Warning in the Darjeeling Himalayas. *Sensors*, vol 20: p2611.
Chen Xiuyu, Shen Lixin, 1993 Characteristics and Evolutionary Model of Extending Structures and Gliding Nappe Structures in Nanpozhuang-Jingxing Area. *Journal of Hebei College of Geology*, vol 16: pp 231–242.

Dong Hui, Huang Run-qiu, Luo Xiao, et al. 2017 Spatial Distribution and Variability of Infiltration Characteristics for Shallow Slope of Gravel Soil. *Chinese Journal of Geotechnical Engineering*, vol 39: pp 1501–1509.

Elwood D., Hendry M.T., Sattler K., et al. 2021 Quantifying the Contribution of Matric Suction on Changes in Stability and Displacement Rate of a Translational Landslide in Glaciolacustrine Clay. *Landslides*, vol 18: pp 1675–1689.

Feng Zh K., Nian T.K., Yu P. Ch, et al. 2011 Shear Test on Mixed Slide-zone Soils of Landslide under Different Water Content. *Advanced Materials Research*, vol 1359: pp 1208–1213.

Gao Kun, Lan Siqing, Lei Nengzhong, et al. 2020 Experimental Study on Correlation between Soil Moisture Content and Landslide Risk. *Journal of Experimental Mechanics*, vol 35: pp 300–308.

Guo Changbao, Liu Dingtao, Zhang Yongshuang, et al. 2022 Centrifuge Model Test of Reactivation Mechanism of Jiangdingya Ancient Landslide in Gansu Province. *Journal of Engineering Geology*, vol 30: pp 164–176.

Habil H., Munir M.M., Yuliza E., et al. 2016 Study of Soil Moisture Sensor for Landslide Early Warning System Experiment in Laboratory Scale, *Journal of Physics: Conference Series*, vol 739: p 012034.

Hakro M R, Harahap I H. 2015 Laboratory Experiments on Rainfall-induced Flowslide From Pore Pressure and Moisture Content Measurements. *Natural Hazards and Earth System Sciences Discussions*, vol 3: pp 1575–1613.

He Ping, Wang Qinchun, Wang Xiang. 2019 Study on Geological Characteristics and Geology Practical Value about Middle Taihang Mountains, *Journal of Hebei GEO University*, vol 42: pp 12–19.

Ivan G, Sinnappoo R. 2022 Effect of Water Content on Apparent Cohesion of Soils from Landslide Sites. *Geotechnics*, vol 2: pp 385–394.

Li Gao, Tan Jianmin, Wang Shimei, et al. 2021 Multi-index Monitoring and Comprehensive Early Warning of Landslides in Response to Rainfall: An Example of the Luo'ao Landslide in Southern Jiangxi Province. *Earth Science Frontiers*, vol 28: pp 283–294.

Liu Muxing, Nie Yan, Yu Jing. 2012 The Infiltration Process of Clay Soil Under Different Initial soil water contents. *Acta Ecologica Sinica*, vol 32: pp 871–878.

Lu C., Wang Y. Ch. 2019 Effects of Water Content and Shearing Rate on Residual Shear Stress. *Arabian Journal for Science and Engineering*, vol 44: pp 8915–8929.

Wei Zhanxi, Wu Yuanzhao, Xie Dongwu, et al. 2022 Research on Landslide Stability under Different Water Content Conditions based on the Dynamic Residual Strength. *Hydrogeology & Engineering Geology*, vol 49: pp 126–136.

Yan Ya-jing, Yan Yong-shuai, Zhao Gui-zhang, et al. 2019 Study on Moisture Migration in Natural Slope using High-density Electrical Resistivity Tomography Method. *Rock and Soil Mechanics*, vol 40: pp 2807–2814.

Water Conservancy and Civil Construction – Oke & Ahmad (Eds)
© 2024 The Author(s), ISBN: 978-1-032-58614-4

Effect of water deficit on potato yield and quality

Dandan Su
College of Agronomy and Agricultural Engineering, Liaocheng University, Liaocheng, China
College of Water Conservancy and Hydropower Engineering, Gansu Agricultural University, Lanzhou, Gansu, China

Shouchao Yu
College of Agronomy and Agricultural Engineering, Liaocheng University, Liaocheng, China

Hengjia Zhang*
College of Agronomy and Agricultural Engineering, Liaocheng University, Liaocheng, China
College of Water Conservancy and Hydropower Engineering, Gansu Agricultural University, Lanzhou, Gansu, China

Zeyi Wang & Dan Wen
College of Water Conservancy and Hydropower Engineering, Gansu Agricultural University, Lanzhou, Gansu, China

ABSTRACT: A potato with a shallow root system is extremely sensitive to soil water deficit, which has long been a major constraint in potato cultivation. Droughts have become more common in recent years because of global warming. As a result, potatoes, the world's fourth-largest food, became subject to severe water shortages, affecting both quality and quantity. This paper reviewed the impact of different levels of water deficit (mild, moderate, and severe) on potato yield and quality at various growth periods. We wish the current study could provide some theoretical support for the robust development of the potato industry.

1 INTRODUCTION

The potato (*Solanum tuberosum L.*) is the world's fourth-most important food crop; as of 2020, global potato production had reached 359.1 million tons in a harvested area of 16.5 million ha, demonstrating the potato's importance in ensuring food security. Asia, with a total production of 178.6 million t, became the world's largest potato-producing region, while China became the world's top potato producer with a total production of 0.782 billion t (AOSTAT 2020). In China, national potato production in 2010 was 0.766 billion t with a harvested area of 4.89 million ha, and by 2020, potato production will increase by 2.1% year-on-year (0.782 billion t) and harvest area will decrease by 13.7% year-on-year (4.22 million ha) (AOSTAT 2020). In recent years, China has developed irrigated agriculture and cultivated superior varieties. Its main potato production areas are in Gansu, Ningxia, the Southwest, Inner Mongolia, and the Northeast, mostly in arid and semi-arid areas. Although China is aggressively developing irrigated agriculture in areas where irrigation conditions are lacking, irrigation facilities for irrigated agriculture are insufficient, resulting in not only delayed irrigation or water shortages in planting areas but also short-term water deficits

*Corresponding Author: 596088683@qq.com

 DOI: 10.1201/9781003450818-5

during crop growth, which frequently cause yield reductions greater than the sum of other abiotic stress effects. In order to provide theoretical support for the wide application of water deficit technology in potato production, it is necessary to briefly review recent research on the effects of water deficit on potato yield and quality.

2 OVERVIEW OF POTATO WATER DEFICIT RESEARCH

At various growth stages of development, water deficit irrigation has been extensively studied when crops are exposed to a certain level of water shortage, with a view to maximizing water conservation conditions without significantly sacrificing crop yield and quality (Neupane & Guo 2019). The shallow root system of potatoes, mainly concentrated in the upper 30 cm of the soil layer (Lesczynski & Tanner 1976), leads to difficulties in adapting to soil moisture changes, making it very sensitive to soil moisture deficits, and water deficits become a major limiting factor for potato cultivation. Potato growth from seedling to maturity requires adequate soil moisture supply to meet growth requirements, and water deficit has different effects on potato yield and quality depending on duration and severity at different stages of growth and development. In recent years, droughts have occurred frequently due to the warming and drying of the climate, making the growing water shortage a serious challenge to food production, especially in arid and semi-arid regions where it is a key factor affecting potato yield and quality.

The potato growth period is relatively short, and the overall water consumption pattern is roughly parabolic, meaning that the intensity of water consumption is low in the early period, gradually increases in the middle, and then decreases in the later period (Tian *et al.* 2011). According to the relationship of potato yield formation, its whole growth period can be divided into four stages (seedling, tuber formation, tuber expansion, and starch accumulation). Although the definition of "water deficit" in crops varies widely in the scientific literature, we defined four levels of water deficit in potatoes to facilitate the analysis and summary of published research results in this paper, i.e., mild water deficit—soil moisture maintained at 60% to 70% of the field capacity (FC) or 65% to 80% for evapotranspiration (ET); moderate water deficit—soil moisture maintained at 50% to 60% FC or 50% to 65% ET; severe water deficit—soil moisture less than 50% FC or ET; and full irrigation—soil moisture generally greater than 70% FC or 80% ET.

3 EFFECT OF WATER DEFICIT ON POTATO YIELD

In Table 1, the effects of the same degree of water deficit at different growth periods of potatoes (seedling, tuber formation, tuber expansion, and starch accumulation), different degrees of water deficit (mild, moderate, and severe) at the same growth period, and different degrees of water deficit at different growth periods on potato yield are reviewed.

Table 1. Effect of different levels of water deficit on potato yield at different growth periods.

Growth Periods	Degree of Water Deficit	Yield Loss	References
Seedling Period	Mild Deficit	up to 24.63% 2%–5%	(Li *et al.* 2021; Wu *et al.* 2015; Xue *et al.* 2018;)
	Moderate Deficit	0.56% 12.08%	(Li *et al.* 2021; Xue *et al.* 2018)
Tuber Formation Period	Mild Deficit	4.89%–13.28%	(Du *et al.* 2017; Li *et al.* 2017, 2021; Xue *et al.* 2018)

(continued)

Table 1. Continued

Growth Periods	Degree of Water Deficit	Yield Loss	References
	Moderate Deficit	14.68%–22.93%	(Du *et al.* 2017; Li *et al.* 2017, 2021; Xue *et al.* 2018)
Tuber Expansion Period	Mild Deficit	10.90%–31.05%	(Du *et al.* 2017; Li *et al.* 2017, 2021; Xue *et al.* 2018)
	Moderate Deficit	16.91%–45.20%	(Du *et al.* 2017; Li *et al.* 2017, 2021; Xue *et al.* 2018)
Starch Accumulation Period	Mild Deficit	8.91%–28.16%	(Du *et al.* 2017; Li *et al.* 2017, 2021; Xue *et al.* 2018)
	Moderate Deficit	18.32% 27.92%	(Li *et al.* 2021; Xue *et al.* 2018)
Full Fertility Period	Mild Deficit	4.67%–8.71%	(Badr *et al.* 2010; Cao *et al.* 2019; Hassanpanah 2010)
	Moderate Deficit	22.09%–30.93%	(Badr *et al.* 2010; Cao *et al.* 2019; Hassanpanah 2010)
	Severe Deficit	43.69%–65.26%	(Badr *et al.* 2010; Cao *et al.* 2019; Zin *et al.* 2019)

3.1 *Seedling period*

Compared to full irrigation during the full fertility period, due to a significant increase in large potato weight and reasonable control of the number of potatoes per plant, a mild water deficit during the seedling period increased yield by 24.63% (Wu *et al.* 2015), while the study concluded that a mild water deficit during the seedling period only reduced potato yield by 2%–5%, while a moderate deficit during the seedling period reduced potato tuber yield by 0.56% and 12.08% (Li *et al.* 2021; Xue *et al.* 2018). However, water consumption during the seedling period only accounted for 10%–15% of the full fertility period (Tian *et al.* 2011), and the water-saving potential is not large in water-stressed potato growing areas. By controlling the number of potatoes and increasing the weight of tubers, mild water deficits can also increase yield and improve water utilization during the seedling period.

3.2 *Tuber formation period*

There are a large number of research results (Du *et al.* 2017; Jia *et al.* 2018; Li *et al.* 2017; 2021; Xue *et al.* 2018) in the potato tuber formation period. A mild water deficit at the tuber formation period only reduced 4.89%–13.28% yield, while a moderate water deficit led to a potato yield reduction of 14.68%–22.93%. However, during the potato tuber formation period, water consumption accounted for 23%–28% of the full fertility period, and the water-saving potential is large (Tian *et al.* 2011). Therefore, for potatoes with high water demand crops, a mild water deficit during the tuber formation period can reduce irrigation water to a greater extent without reducing yield significantly, and it can provide an irrigation water-saving strategy for potato cultivation in arid and semi-arid areas.

3.3 *Tuber expansion period*

This period is the largest soil water requirement for potatoes, accounting for about 45%–50% of the full potato fertility period (Tian *et al.* 2011). Although the potential for water savings during the tuber expansion period is greatest, a mild water deficit in this period can significantly reduce potato yield by 10.90%–31.05% (Du *et al.* 2017; Li *et al.* 2017; Li *et al.* 2021; Liu *et al.* 2018; Xue *et al.* 2018), and a moderate water deficit can significantly reduce yield by 16.91%–45.20% (Du *et al.* 2017; Li *et al.* 2017; Li *et al.* 2021; Xue *et al.* 2018). To achieve a consistent yield of potatoes, the tuber expansion period should be fully irrigated rather than water deficit.

3.4 *Starch accumulation period*

Many researchers have shown that during the starch accumulation period, a mild water deficit reduced potato yields by only 4.67%–8.71% (Du *et al.* 2017; Li *et al.* 2017; Li *et al.* 2021; Xue *et al.* 2018), while a moderate deficit treatment during this period significantly reduced potato yields by 18.32% (Xue *et al.* 2018) and 27.92% (Li *et al.* 2021). In addition, during the starch accumulation period, the water requirement of potatoes only accounts for about 10% of the full fertility period's water requirement (Tian *et al.* 2011). The water storage potential is small, and the water saving potential is not large, while the study shows that too much water in this period would cause an increase in the rate of bad potatoes (Feng *et al.* 2015), and a mild water deficit during the starch accumulation period does not significantly reduce the yield of potatoes. So, during the starch accumulation period, a mild water deficit could also be a water conservation model to improve the water utilization of potatoes.

3.5 *Full fertility period*

During the potato full fertility period, a mild water deficit at full fertility reduced yield by only 4.67%–8.71% (Badr *et al.* 2010; Cao *et al.* 2019; Hassanpanah 2010), which is an optimal irrigation pattern compared to a moderate water deficit at full fertility that significantly reduced yield by 22.09%–30.93% (Badr *et al.* 2010; Cao *et al.* 2019; Hassanpanah 2010; Zin *et al.* 2019), and a severe water deficit that significantly reduced yield by 43.69%–65.26% (Badr *et al.* 2010; Cao *et al.* 2019; Zin *et al.* 2019). For most potato growers, the use of mild water deficit irrigation during the potato's full fertility period is easy to operate and does not significantly reduce potato yields, making it the most efficient irrigation strategy with high water-saving potential.

4 EFFECT OF WATER DEFICIT ON POTATO QUALITY

The quality of potatoes is determined by their appearance quality, nutritional quality, and processing quality. Appearance quality and processing quality, along with its merits and demerits, directly determine its market value. There is a close relationship between nutritional quality and consumers' concerns about their nutritional balance. In the three aspects of the economic value of potatoes, only the water deficit does not significantly impact their quality, so as to increase farmers' incomes.

4.1 *Appearance quality*

Appearance quality mainly includes potato shape, a high rate of large and medium potatoes, shallow bud eyes, and the commercial potato rate. A lack of water during potato tuber formation and tuber expansion results in an exponential increase in cropping rate, with cropping rate increasing with the degree of deficit, whereas adequate irrigation reduces commercial potato rate (Zhang 2019). A lack of water during different growth periods can affect potato quality by increasing the proportion of potato bunches, whereas an abundance of soil moisture can reduce potato yield and quality (Kang *et al.* 2010).

4.2 *Nutritional quality*

Nutritional quality is the physicochemical composition of potato tubers, which are mainly composed of dry matter, starch, reducing sugars, protein, and ash. Wu concluded that in the range of field water holding capacity of 55% to 85%, there was a significant or highly significant positive correlation among potato tuber yield, tuber starch content, and soil water content (Wu 2007). The increased sugar content of potato tubers was caused by a water deficit (Eldredge *et al.* 1996). Furthermore, the sugar and protein contents of potato tubers decreased with a greater water deficit (Elhani *et al.* 2019).

4.3 *Processing quality*

The processing quality of potatoes is mainly determined by their dry matter, starch content, and reduced sugar. It has become increasingly important to the potato industry in recent years. The most serious impact on processing quality is known as physiological disorders of the end sugar of the tuber, i.e., high reducing sugar (glucose and fructose) content at one end of the tuber and the formation of black ends or a burnt appearance of the high-sugar tissue during frying (Thompson *et al.* 2008). Besides, the water deficit had the greatest effect on the sugar end of the potato tuber during the expansion period, while the water deficit during the tuber formation period had a slight effect on the sugar end (Shock Feibert & Saunders 1998).

5 CONCLUSION

In recent years, as global water scarcity has increased, arid and semi-arid regions have faced severe challenges in food production. According to the above literature, potatoes are easily affected by a water deficit at any time during their growth period. Among them, a mild water deficit will not significantly reduce the potato yield during the period of potato tuber formation. This is a more scientific way to conserve water and improve potato quality to a certain extent.

ACKNOWLEDGMENTS

This work was financially supported by the National Natural Science Foundation of China (No. 52269008), the Industrial Support Plan Project of Gansu Provincial Department of Education (No. 2022CYZC-51), and the Key Research and Planning Projects of Gansu Province (No. 18YF1NA073).

REFERENCES

Badr M.A., Abou Hussein S.D., El-Tohamy W.A., Gruda N.: Efficiency of Subsurface Drip Irrigation for Potato Production Under Different Dry Stress Conditions, *Gesunde Pflanzen* 62 (2), 63–70 (2010).

Cao Z.P., Liu Y.H., Zhang X.J., Shen B.Y., Qin S.H., et al.: Effect of Deficit Irrigation on Growth Yield and Water Use of Potato, *J. Journal of Agricultural Engineering* 35 (4), 114–123 (2019).

Du J, Zhang H.J., Ba Y.C., Yang X.T., et al.: Optimization of Water Production Function and Irrigation System for Potato Under Oasis Film Drip Irrigation to Regulate Deficit, *J. Agricultural Research in Arid Regions*. 35 (1), 158–164 + 177 (2017).

Eldredge E.P., Holmes Z.A., Mosle A.R., Shock C.C. and Stieber T.D.: Effects of Transitory Water Stress on Potato Tuber Stem-end Reducing Sugar and Fry Color. *American Potato Journal* 73 (11), 517–530 (1996).

Elhani S., Haddadi M., Csákvári E., Zantar S., Hamim A., et al.: Effects of Partial Root-zone Drying and Deficit Irrigation on Yield, Irrigation Water-use Efficiency and Some Potato (Solanum tuberosumL.) Quality Traits Under Glasshouse Conditions. *Agricultural Water Management* 224, 10574 (2019).

FAOSTAT. 2020, http://www.fao.org/faostat/en/#data/QC, last accessed 2022/8/25.

Feng D, Liu X.Y., Kang Y, Jiang S, Wan S, et al.: Effects of Different Soil Water Regulation of Subsurface Drip Irrigation on Potato Yield and Irrigation Water use Efficiency, *J. Water Conservation Irrigation* (8), 42–44 (2015).

Hassanpanah D.: Evaluation of Potato Cultivars for Resistance Against Water Deficit Stress Under In Vivo Conditions. *Potato Res* 53 (4), 383–392 (2010).
Jia L.G., Wu L, Chen Y, Yu J, Fan M.S.: Regulation of Potato Tuber Development by Water Deficit During Tuber Formation, *J. Journal of Inner Mongolia Agricultural University* 39 (2),13–19 (2018).
Kang Y.H., Zhao H.C., Gong X.C., Tian Z.M, Qiao H.M., et al.: Effects of Drought Stress on Potato Yield and Quality at Different Fertility Stages, *J. Anhui Agricultural Science* 8 (30), 16820–16822 (2010).
Lesczynski D and Tanner C.: Seasonal-Variation of Root Distribution of Irrigated, Field-Grown Russet Burbank Potato. *Am Potato J* 53 (2), 69–78 (1976).
Li F.Q., Deng H.L., Wang Y.C., Li X, Chen X.T., et al.: Potato Growth, Photosynthesis, Yield, and Quality Response to Regulated Deficit Drip Irrigation Under Film Mulching in a Cold and Arid Environment. *Sci Rep* 11 (1), 15888 (2021).
Li J, Zhang H.J., Zhou H.: Water Consumption Characteristics and Growth Dynamics of Potatoes Under Drip Irrigation with Soil Water Regulation Deficit Treatment, *J. Agricultural Research in Arid Regions* 35 (3), 80–87 (2017).
Liu J, Jia S.H., Liang Z.E.: Effect of Oasis Sub-membrane Drip Irrigation to Adjust Deficit on Potato Growth and Quality, *J. People's Yellow River* 40 (8), 152–156 (2018).
Neupane J and Guo W., Agronomic Basis and Strategies for Precision Water Management: A Review. *Agronomy-Basel* 9 (2), 87 (2019).
Shock C.C., Feibert E.B.G., Saunders L.D.: Potato Yield and Quality Response to Deficit Irrigation. *HortSci* 33 (4), 655–659 (1998).
Thompson A.L., Love S.L., Sowokinos J.R., Thornton M.K., ShockC.C.: Review of the Sugar End Disorder in Potato (*Solanum tuberosumL.*). *Potato Res* 85 (5), 375–386 (2008).
Tian Y, Huang Z.G., Yu X.Q.: Experimental Study on Water Requirement of Potato, *J. Modern Agricultural Science and Technology* (8), 91–92 + 94 (2011).
Wu L, Shi X.H., Yang H.Y., Qin Y.L., Jia L.G., et al.: Effect of Seedling Water Deficit on Potato Yield Formation, *J. China Potato* 29 (2), 80–84 (2015).
Wu X.W.: *Effect of Different Soil Moisture on Yield and Quality of No-till Straw-covered Potatoes at Early Growth Stage*, D. Master's thesis, Guangxi University (2007).
Xue D.X., Zhang H.J., Ba Y.C., Wang Y.C., Wang S.J.: Effect of Deficit-regulated Irrigation on Growth, Yield and Water use of Potato Under Film Drip Irrigation in a Desert Oasis, *J. Agricultural Research in Arid Regions* 236 (4), 109–116 + *132* (2018).
Zhang W.H.: *Effects of Different Fertility Water Regulation Deficits on Growth Characteristics, Yield and Quality of Potato Under Oasis Film Drip Irrigation*, D. Gansu Agricultural University (2019).
Zin E., Mattar M., Al-Ghobari H., Alazba A.: Water-Saving Irrigation Strategies in Potato Fields: Effects on Physiological Characteristics and Water Use in Arid Region. *Agronomy* 9 (4), 172 (2019).

Water Conservancy and Civil Construction – Oke & Ahmad (Eds)
© 2024 The Author(s), ISBN: 978-1-032-58614-4

The effects of water and nitrogen regulation on potato growth, development, and yield

Lintao Liu
College of Agronomy and Agricultural Engineering, Liaocheng University, Liaocheng, China
College of Water Conservancy and Hydropower Engineering, Gansu Agricultural University, Lanzhou, Gansu, China

Shouchao Yu
College of Agronomy and Agricultural Engineering, Liaocheng University, Liaocheng, China

Hengjia Zhang*
College of Agronomy and Agricultural Engineering, Liaocheng University, Liaocheng, China
College of Water Conservancy and Hydropower Engineering, Gansu Agricultural University, Lanzhou, Gansu, China

ABSTRACT: Water and nitrogen, as the main production factors of potatoes, are crucial for improving potato yield and quality. To understand the effects of water and nitrogen regulation on potato growth and yield at this stage and to provide a theoretical basis for subsequent in-depth research on water and nitrogen conservation and sustainable development, this paper systematically discussed the effects of water and nitrogen regulation on potato nitrogen fertilizer utilization, growth and development, photosynthesis, nutrient uptake and utilization, dry matter accumulation and distribution, yield and quality, and proposed future directions for water and nitrogen regulation in potato, intending to provide a theoretical reference for the efficient use of water and nitrogen in potato. Theoretical references were provided for the efficient use, quality, and yield improvement of potatoes.

1 INTRODUCTION

As one of the four major global food crops, potatoes play an important role in food security in developing countries. As of 2010, China's total potato production accounted for 22% of the global total, but the average yield per acre ranked only 93rd in the world. At present, the in-season utilization rate of fertilizers in China is 30–35% for nitrogen fertilizers, which is low, results in huge losses through various means, and also leads to serious consequences such as a decline in soil fertility, lower crop quality, and environmental pollution. The impact of different N levels on dry matter accumulation and N uptake in the potato mulch. In addition, regional water shortages are further limiting the development of potato production in China, so the search for the best water and nitrogen control solutions for potatoes has become a hot issue today.

Water and nitrogen play an irreplaceable role in the growth and yield formation of potatoes as major yield factors. Water and nitrogen regulation aims to use synergistic effects to promote the organic combination of water and nitrogen to improve crop yield and water and nitrogen use efficiency. Proper water and nitrogen regulation not only has an important impact on potato development, tuber formation, yield, and quality but also improves the soil ecology. Different water and nitrogen ratios have different effects on potato growth and tuber yield in

*Corresponding Author: 596088683@qq.com

DOI: 10.1201/9781003450818-6

different regions, and exploring water and nitrogen levels is the basis for screening N-efficient potato varieties, determining reasonable N fertilizer rates, and improving N fertilizer utilization efficiency. This paper, therefore, summarized the effects of water and nitrogen ratios on potato nitrogen utilization, growth, photosynthesis, nutrient uptake and utilization, dry matter accumulation and distribution, and yield, intending to find the best water and nitrogen ratios, improve water and nitrogen utilization, fully exploit the potential of potato production, and further improve the quality and efficiency of the potato industry.

2 EFFECTS OF WATER AND NITROGEN REGULATION ON N FERTILIZER USE EFFICIENCY IN POTATOES

Nitrogen is a component of many important organic compounds in potatoes (He 2015), and the level of N fertilizer utilization fully reflects the plant's N uptake and utilization status as well as the crop's ability to use N fertilizer to produce yield. Studies have shown that under the same water conditions, potato N fertilizer utilization and N fertilizer bias productivity gradually decreased with increasing N application (Chen *et al.* 2012). Under the same N application conditions, N fertilizer utilization decreased with increasing irrigation water, while N fertilizer bias productivity increased with increasing irrigation water (Chen *et al.* 2012; Zhang 2018). Dai (1998) found that the nitrogen utilization rate was 53% for potatoes irrigated deeply with water at bud emergence, while the nitrogen utilization rate was 43% for urea surface spreading followed by watering in season. Tang *et al.* (2021) found that the maximum water use efficiency of potatoes under different water and nitrogen coupling schemes was 85.9, 90.2, and 92.2 kg/(mm·ha) for dry, normal, and wet years, respectively, and the irrigation volumes to obtain the highest water use efficiency were 172, 107, and 87 mm, respectively.

3 EFFECTS OF WATER AND NITROGEN REGULATION ON N FERTILIZER USE EFFICIENCY ON POTATO GROWTH AND DEVELOPMENT AND PHOTOSYNTHESIS

3.1 *Effects on potato growth*

Potato growth and development are closely related to nitrogen content, and good nitrogen nutrition can slow down leaf senescence, leading to healthy plant growth and nutrient distribution. Studies have shown that the number of small potatoes and tubers in potatoes tended to decrease and then increase with increasing, while in contrast, the weight of large potatoes and total tubers tended to increase and then decrease with increasing N application (He 2015). With increasing N application, the number of potatoes per plant and the fresh weight of stems and leaves increase. In addition, studies have shown that a reasonable increase in nitrogen fertilization could help to increase the number of potatoes produced per plant, but that excessive nitrogen fertilization could lead to a decrease in the average weight of potatoes per plant, thus reducing yield (Tian *et al.* 2015). Potato emergence was significantly inhibited when N was applied at a rate of 180 kg/ha (Zhang *et al.* 2013). Therefore, N application rates should take the fertility and requirements of potatoes into account to avoid surplus stem and leaf growth and insufficient tuber growth.

3.2 *Effects on the photosynthetic properties of potatoes*

More than 90% of the dry matter in the harvested organs of the potato comes from photosynthetic products, and the number of photosynthetic products accumulated depends on the size of the leaf area, the photosynthetic capacity of the leaves, and the time the leaves spend working (Xu 2017). Existing studies showed that increased nitrogen fertilization in potatoes under normal and excess water conditions increased net photosynthetic rate, stomatal conductance, and chlorophyll content, thereby increasing the photosynthetic capacity of potatoes. Under water stress, however, root uptake of nitrogen was reduced due to lower root water uptake and leaf transpiration rates (Zhang *et al.* 2020). The effect of drought on photosynthesis was reduced by moderate N application under water deficit stress, while excessive N application inhibited photosynthesis and reduced yield. The reason for this might be that drought stress caused a decrease in photosynthetic efficiency and cytotoxicity (Mu & Chen 2021). Nitrogen application not only changed the osmotic pressure of the cells but also increased the synthesis of enzymes in the photosynthetic system, thereby enhancing photosynthesis (Wang *et al.* 2016). The expression of genes related to N metabolic pathways and hormone synthesis in the root system can improve plant adaptation to drought (Chen *et al.* 2019); and the expression of genes related to N metabolic pathways and hormone synthesis in the root system can improve plant adaptation to drought. It was found that the maximum net photosynthetic rate during potato tuber formation increased gradually with increasing levels of nitrogen (Zheng *et al.* 2010). However, excessive application of nitrogen fertilizer caused plant growth, mutual shading of the stems and leaves, reduced photosynthetic efficiency of the leaves, yellowing, and loss of light on the bottom leaves of the plant, delayed potato set, and reduced yield (Yang 2012).

3.3 *Effects of chlorophyll on potatoes*

The effect of nitrogen on photosynthesis is directly related to the chlorophyll content of the plant, which increases with nitrogen application, leading to an increase in the photosynthetic activity of the leaf pulp cells and the intensity of light absorption by the leaves, ultimately leading to an increase in the net photosynthetic rate. Existing studies showed that the chlorophyll content of potatoes increased with increasing nitrogen application under the same irrigation conditions, while the chlorophyll content of potatoes increased with increasing irrigation water under the same amount of follow-up nitrogen fertilizer (Gao 2017), indicating that irrigation and nitrogen application were both beneficial to the chlorophyll content of potatoes. This suggests that irrigation and nitrogen application are both beneficial to the chlorophyll content of potatoes (Gao 2017; Xie *et al.* 2003). Water promotes increased nitrogen uptake by the crop, and nitrogen is an important component of chlorophyll, so changes in its content in the crop will inevitably affect the chlorophyll content of the crop leaves (Gao 2017).

3.4 *Effects on the leaf area index of potatoes*

Nitrogen deficiency can lead to potato stem and leaf elongation and reduced dry matter content of the tubers, resulting in reduced leaf area and reduced yields in the population. Studies have shown that the leaf area index increased with increasing N application under the same irrigation conditions (Wei *et al.* 2017). Other studies showed that the leaf area index increased with increasing N application throughout the potato reproductive period and reached a maximum of 300 kg/ha (Yu *et al.* 2020). Under the same N fertilizer application conditions, the leaf area index of potatoes increased and then decreased as irrigation water increased. Under different water and nitrogen treatments, the leaf area index of potatoes varied between 0.1 and 2.7 $m^2 \cdot m^{-2}$ during the reproductive period, and the leaf area index increased by 2.05 for the high N treatment and 1.01 for the low N treatment at 45 days after planting compared to the no N treatment (Dai 1998).

4 EFFECTS OF WATER AND NITROGEN REGULATION ON NUTRIENT UPTAKE AND UTILIZATION IN POTATOES

Nitrogen uptake of potato plants under different water and nitrogen treatments shows an "S"-shaped increase and reaches a maximum at the tuber formation stage (He 2015). However, nitrogen accumulation throughout the plant does not increase with the amount of nitrogen applied. The rapid expansion of the tuber is therefore also a period of transfer of nitrogen accumulation from the stem and leaf (source) to the tuber (reservoir), which is also a critical period for the nitrogen nutrient requirements of potatoes. Studies have shown that the maximum N uptake was 0.08–0.23 g/plant for roots, 0.22–0.73 g/plant for stems, 0.57–1.46 g/plant for leaves, and 1.12–2.82 g/plant for tubers (Dai 1998). Finally, increased nitrogen fertilizer applications increased the nitrogen content of all potato organs. At different nitrogen levels, potato leaves had the highest nitrogen content, followed by the highest nitrogen content in stems, and the lowest nitrogen content in tubers.

5 EFFECTS OF WATER AND NITROGEN REGULATION ON DRY MATTER ACCUMULATION AND DISTRIBUTION IN POTATOES

Dry matter is the basis of yield, and a reasonable proportion of nitrogen is the key to high potato yields. An important reproductive period for potato dry matter accumulation is the tuber expansion period, when the growth of not only nutritional organs but also reproductive organs is required. Phased application of nitrogen fertilizer is beneficial for potato dry matter accumulation. Using nitrogen fertilizer as a basal fertilizer prevents excessive stem and leaf growth in the early stages of potato production and contributes to tuber formation. Fertilizer at the end of tuber formation can supplement the nutrient gap between the nutritional and growth organs and provide sufficient nutrients for the growth and development of the potato (Li 2014). The formation of the final potato yield is closely related to the distribution of dry matter in the organs and coordinated growth (Jiao *et al.* 2013). The direction of dry matter distribution determines the level of yield. Studies have shown that the dry matter mass of potato plants increased with increasing water irrigation (Dai 1998).

6 EFFECTS OF WATER AND NITROGEN REGULATION ON DRY MATTER ACCUMULATION AND DISTRIBUTION IN POTATOES

It was found that under drip irrigation, potato yield and individual potato quality showed a parabolic trend with increasing N application rates of 120 to 270 kg/ha, with yield and commercial potato rate reaching a maximum at 180 kg/ha (Zhou *et al.* 2004), while Jiao *et al.* (Liu *et al.* 2018) found that the best increase in yield was achieved with 120 kg/ha of pure N. Studies have shown that both the amount of irrigation water (Dai 1998) and the amount of nitrogen applied increased the yield of potato plants, mainly by increasing the weight of individual tubers (Chen *et al.* 2012). Excessive nitrogen fertilization caused potato plant growth and affected the transfer of photosynthetic products to the lower tubers, which in turn affected yield (Zhou *et al.* 2004). In addition, it has also been shown that under the same water conditions, the rate of growth of large and medium potatoes increased with both the amount of irrigation water and the amount of nitrogen applied (Liu *et al.* 2018).

7 CONCLUSIONS

Agricultural production is a complex process that is susceptible to external environmental influences. Future research on potato water and nitrogen control should focus on the following aspects, with clear water and fertilizer conservation objectives: (1) Systematic research on the impact of crop water and nitrogen control on potato yield and quality, taking external conditions into account such as quality, climate, region, and tillage practices, and selecting the best water and nitrogen control strategy that integrates yield, quality, and soil environment, is needed. (2) In areas where water resources are scarce, research should focus on water and nitrogen-saving strategies for potatoes and cultivate water-saving varieties. (3) The impact of water and nitrogen regulation on potatoes should be studied in terms of potato physiology to provide theoretical references for the efficient use of water and nitrogen in potatoes and improve quality and yield.

ACKNOWLEDGMENTS

This work was financially supported by the Industrial Support Plan Project of Gansu Provincial Department of Education (No. 2022CYZC-51), the National Natural Science Foundation of China (No. 52269008), and the Key Research and Planning Projects of Gansu Province (No. 18YF1NA073).

REFERENCES

Chen R, Meng M, Liang H, Zhang J and Wang Y 2012 Study of Potato Yield and N Fertilizer Utilization Characteristics Under Different Water and N Conditions. *Chinese Journal of Agronomy*. 28 (3), 6

Chen Y, Li C, Yi J, Yang Y and Gong M 2019. J. Sci. Transcriptome Response to Drought, Rehydration and Re-Dehydration in Potato. *Int. J. Mol.* 21 (1), 159

Dai Q 1998. A Brief Discussion on Nitrogen Fertilizer Utilization and Sustainable Development of Agriculture. *Inner Mongolia Agricultural Science and Technology*.

Gao F 2017. *Effect of Water and Nitrogen Coupling on Soil Water and Nitrogen Transport and Yield Quality of Potato Under Drip Irrigation*. Inner Mongolia Agricultural University

He W 2015. *Effect of Different Nitrogen Levels on Dry Matter Accumulation and Nitrogen Uptake in Mulched Potato*. Gansu Agricultural University

Jiao F, He H, Wei X and Zhao R 2013. J. Effect of Different Nitrogen Levels on Tuber Growth, Yield and Vitamin C Content of Potato. *Journal of Heilongjiang Bayi Agricultural Reclamation University*. 25 (04): 1–3

Li M 2014. *Effect of Nitrogen Fertilizer Management on Potato Tuber Growth and Development*. Northeast Agricultural University

Liu Z, Gao X, Zhang Y, Jiao J, Gu D and Gao J 2018. Effect of Under-film Drip Irrigation on Growth, Yield and Quality of Potatoes in Greenhouses. *Journal of Heilongjiang Bayi Agricunltural Reclamation University*. 30 (2), 5

Mu X and Chen Y 2021. The Physiological Response of Photosynthesis to Nitrogen Deficiency. *Plant Physiol.* 158, 76–82

Tang J, Xiao D, Wang J, Wang R, Bai H, Guo F and Liu J 2021. J. Optimization of Potato Water and Nitrogen Management Under Different Production Target Conditions. *Journal of Agricultural Engineering*. 37 (20): 108–116

Tian X, Wei Q, Liang X, Qi C and Shi Y 2015. Effect of Nitrogen on Yield and Quality of Potato Variety Dongnong 311. *Genomics and Applied Biology*. (9), 6

Wang J, Bingrui L, Li X, Zhu X, Zhu C and Jia H 2016. Evaluation of N Fertilizers Effects on Grape Based on the Expression of N Metabolic Genes. *Horticultural Plant Journal*. (5), 11

Wei Q, Cao M, Shi Y and Chen B 2017. Effect of Nitrogen Levels on Photosynthetic Characteristics and Yield of Potato Throughout the Reproductive Period. *Genomics and Applied Biology*, 36 (1), 7

Xie H, Shen R, Xu C and Qin Q 2003. Experimental Study on the Relationship between Water, Nitrogen Effect and Chlorophyll. *China Rural Water Conservancy and Hydropower*, (8), 4

Xu C 2017. *Mechanism of Drought Physiological Characteristics and Yield Effects of Nitrogen and Phosphorus Fertilization with Bacterial Fertilizer on Mulched Potato*. Inner Mongolia Agricultural University

Yang Y 2012. Effect of Nitrogen Fertilization on Potato Growth and Development. *Jilin Vegetable*. 000 (001)

Yu G, Zhang G, Wu L, Guo Z and Jie R 2020. Effect of Different Nitrogen Application on the Growth of A New Potato Variety 'NingYang 16'. *Chinese Melon and Vegetable*. 33 (2), 5

Zhang J 2018. *Study on the Water and Nitrogen Coupling Effect and Rational Utilization Mechanism of Potato at the Northern Foot of Yinshan Mountain*. China Agricultural University.

Zhang X, Chen F, Yuan A and Ma H 2013. Effect of Nitrogen, Phosphorus and Potassium Fertilization Levels on Potato Growth and Yield in the Northwest Arid Zone. *China Potato*. (4), 222–225

Zhang Z, Tariq A, Zeng F, Graciano C and Zhang B 2020. Nitrogen Application Mitigates Drought-induced Metabolic Changes in Alhagi sparsifolia Seedlings by Regulating Nutrient and Biomass Allocation Patterns. *Plant Physiol Bioch.*

Zheng S. L., Li G.P., Yuan J.C., Sun L., Liu L., Zhou H., Effect of Nitrogen Application Level on Photosynthetic Characteristics of Potato Tubers During Tuber Formation. *Northwest Journal of Agriculture*, (3), 6 (2010).

Zhou N, Zhang X, Qin Y, and Xu Q 2004. Effect of Different Drip Irrigation and N Application on Yield and Quality of Potato, *Soil Fertilizer*, (6), 3

Water Conservancy and Civil Construction – Oke & Ahmad (Eds)
© 2024 The Author(s), ISBN: 978-1-032-58614-4

Influence of joint location and temperature on the stability of tunnel surrounding rock

Chenchen Jiang*, Weiyu Tang*, Yimeng Xu* & Jiling Chen*
Power China Road Bridge Group Co., Ltd, Beijing, China

Xiangyang Chen*
Chongqing University, Chongqing, China

ABSTRACT: To investigate the effect of joints and high geothermal on the stability of tunnel surrounding rock, this paper took a certain tunnel of high geothermal in a highway in Honghe State, Jian (a) Yuan, as the engineering background. The stability of the tunnel surrounding rock under different joint locations and temperature conditions were discussed by using the combined methods of field testing and numerical simulation. The results revealed that the displacement of the tunnel section tended to an accelerating decrease trend as the distance between the joints and the tunnel vault increased, while it tended to an accelerating increase trend as the surrounding rock temperature increased. In all the simulation conditions, tunnel vault displacement was the largest, and the displacement of the left and right arch waists of the tunnel was the smallest. There were three distinct areas of tunnel displacement under the coupling effect of joint position and temperature. Simulation results can provide an important reference for the construction of high-ground temperature tunnels with joints.

1 INTRODUCTION

With the continuous development of China's economy, the mileage of highway construction in China has increased rapidly, and the advantages of tunnels crossing mountains are becoming more and more obvious. In recent years, tunnel construction has developed towards longer lengths and more complex geological conditions, and a large number of super-long tunnels and high geothermal tunnels have emerged. The high temperature affects the material and mechanical properties of the tunnel lining structure, leading to a reduction in the safety factor of the tunnel lining structure, and at the same time affects the tunnel construction efficiency, which has certain peculiarities. Therefore, it is of great practical significance to study the high geothermal tunnel.

Scholars have recently studied the stability of tunnels surrounding rocks with high geothermal tunnels. Li et al. (2016) explored the influence of temperature on tunnels surrounding rock structures and proposed relevant measures. Zheng et al. (2018) made a comparative analysis of the settlement and stress variation rule of the tunnel confining support structure under different temperature conditions and found that the lining stress was positively correlated with the surrounding rock temperature. Wang et al. (2016; 2019) studied the change rule of surrounding rock in the high-temperature tunnel, and the simulation results showed that the safety factor of the surrounding rock was negatively correlated with

*Corresponding Authors: 773568379@qq.com, 809623338@qq.com, 757354373@qq.com, 154700528@qq.com and zhangpgut@163.com

DOI: 10.1201/9781003450818-7

the temperature. Sun et al. (2018) analyzed the stress characteristics of tunnel secondary lining under different conditions of ground temperature and surrounding rock grade and concluded that the lining safety factor decreased with the increase in temperature and surrounding rock grade. Wu et al. (2019) studied the influence of temperature on tunnel lining structure by using FLAC3D software. Wu et al. (2020) used ANSYS software to simulate and study the deformation law of tunnel surrounding rock under high-temperature conditions, obtaining the change law of tunnel surrounding rock deformation with time. Hua et al. (2021) conducted field tests on the influence of high-temperature hot water on the safety of tunnel lining structures, and the results showed that the permeability coupling effect of high-temperature hot water reduced the stability of the tunnel's surrounding rock. It can be seen from the above studies that the influence of temperature on the tunnel lining structure is significant at present, but the influence of joint location and temperature on the tunnel lining structure has not been reported. Therefore, it is of great theoretical significance and engineering value to explore the influence of joint location and temperature on the stability of the tunnel's surrounding rock.

Based on a high-geothermal tunnel of the Jianyuan Expressway in Honghe Prefecture, Yunnan Province, this paper studied the deformation characteristics of the tunnel lining structure under different joint positions and temperature conditions by combining numerical simulation and field experiments to provide a reference for the design and construction of similar tunnel projects.

2 PROJECT PROFILE

A geothermal tunnel is located in Kele Village, Jiasha Township, Gejiu City, Honghe Prefecture. The total length of the tunnel is 3366 m, and the maximum buried depth is 643 m. The two-way double tunnel construction mode is adopted. The tunnel passes through the middle Triassic Gejiu Formation limestone and Yanshanian intrusive granites. The maximum temperature of the limestone section is 65.1 °C, and the maximum temperature of the granite section is 88.8 °C (Hu et al. 2020). The tunnel site is surrounded by the bottom of the sand waterfall springs, Nige, and Laohutan hot springs. The hot spring water temperature is 51.8–86.6 °C (Hu et al. 2021). The tunnel is currently the highest geothermal highway tunnel in China. Combined with the field temperature test results of the tunnel, the temperature curves at different distances from the tunnel exit section to the tunnel face are shown in Figure 1.

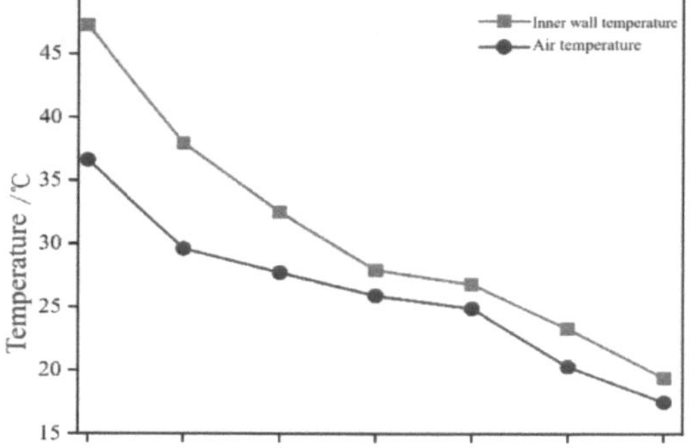

Figure 1. The temperature curves at different distances from the tunnel exit section to the tunnel face.

3 FIELD TEST

3.1 *Field test plan*

The LZ5K45 + 800–LZ5K45 + 840 section of the left tunnel of a high-temperature tunnel was selected for the field test. The total length of this section was 40 m, the tunnel surrounding rock was grade IV, the buried depth was 150 m, the surrounding rock temperature was 40 °C, and the distance between the joint and the tunnel crown was 0.5 m. The test section mileage was LZ5K45 + 810 and LZ5K45 + 825, respectively. The initial support measuring points were arranged at the arch crown A, arch waist B, and arch waist C of the section. The measuring point layout scheme is shown in Figure 2. The monitoring period was 30 days: once a day in the first 15 days and once every two days in the last 15 days.

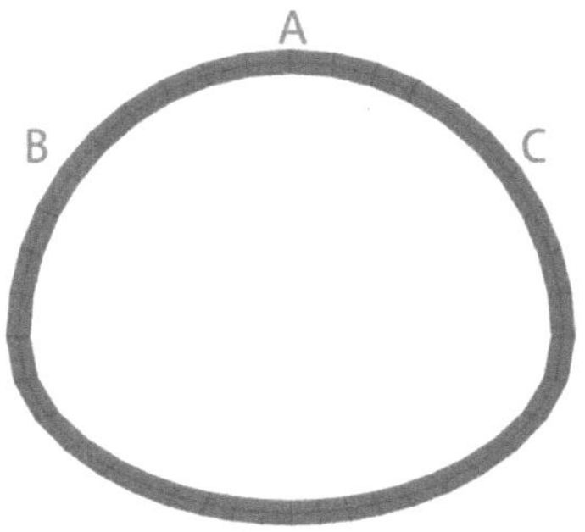

Figure 2. Measuring point layout scheme.

3.2 *Analysis of measured results*

Through long-term monitoring of the initial support of the tunnel, the arch crown settlement time history curves of the above two sections were obtained. Because of the field test, this paper only selected the displacement data at the location of arch crown A for analysis, and the results are shown in Figure 3.

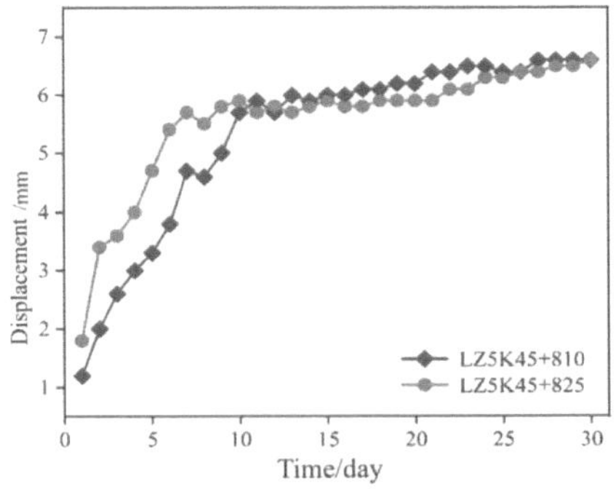

Figure 3. Displacement changes curve of tunnel initial support.

Figure 3 shows that the arch crown displacement changes of LZ5K45 + 810 and LZ5K45 + 825 sections are generally similar. Due to the complex rock strata at the tunnel site, different lithology, joints, and other factors, the displacement of the two sections had some differences, but the overall trend of the two section measuring points was consistent, which could provide good on-site verification for numerical simulation.

4 NUMERICAL SIMULATION ANALYSIS

In this paper, with the help of FLAC3D finite difference software and temperature stress coupling numerical simulation technology, the material parameters were inversely calculated from the tunnel field measured data, and the deformation law of the initial support of the high-temperature tunnel was simulated.

4.1 *Principle of heat-force coupling*

FLAC3D thermal analysis was based on the thermal balance equation derived from the principle of energy conservation, and the equation expression is (Liu et al. 2009):

$$-q_{i,j} + q_v = \rho C_v \frac{\alpha T}{\alpha t} \tag{1}$$

where $q_{i,j}$ is the heat flow, W/m^2; q_v is the intensity of body heat source, W/m^3; ρ is the density, kg/m^3; C_v is the heat in a fixed volume, J/(kg·°C); and T is the temperature, °C; t is the time, s.

The thermal stress coupling calculation in FLAC3D is a one-way coupling method. The simulation calculation process is to first calculate the temperature field of the model and then superimpose the temperature stress and the model self-weight stress (Tang et al. 2019). The temperature in the model leads to material deformation, which leads to the unilateral change of stress.

4.2 *Computational model*

The left tunnel, LZ5K45 + 800–LZ5K45 + 840, was selected for this study; the surrounding rock was grade IV; the buried depth was 150 m; and the initial support is shotcrete. According to Saint Venant's principle (Yuan et al. 2012), the influence range of tunnel excavation is about 3–5 times the tunnel diameter direction. This model selected three times the tunnel diameter. In conjunction with the tunnel diameter, 1:1 scale modeling was used. The geometric dimension of the tunnel model was 50 m × 50 m × 40 m, the model grid was divided into 30720 units and 30752 nodes, and the three-dimensional calculation model of the tunnel was shown in Figure. 4. Among them, the M-C constitutive model was adopted for the surrounding rock and the elastic constitutive model was adopted for the lining.

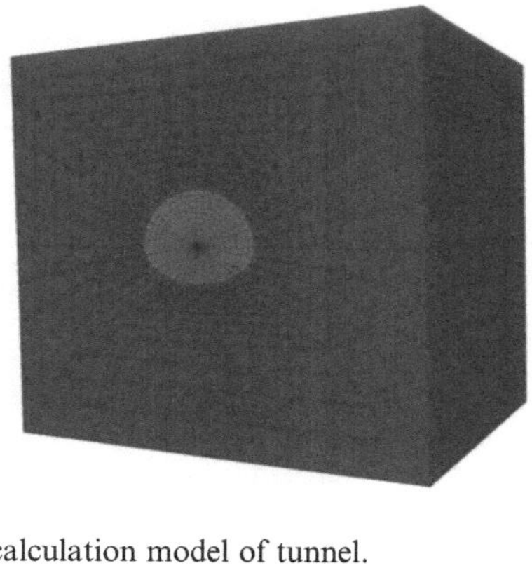

Figure 4. Three-dimensional calculation model of tunnel.

The first 10 m along the axial direction of the tunnel was selected as the analysis section, and the vault and left and right arch waist analysis points were set in this section. The deformation evolution law of the initial support of the tunnel was studied by extracting the deformation data from the analysis points.

4.3 *Material parameters and boundary conditions*

According to the field-measured data of the tunnel, the relevant parameters of the surrounding rock materials were obtained through inversion calculation. For more information, see Table 1.

Table 1. Material thermodynamic parameters.

Material	Elastic Modulus E/GPa	Poisson's Ratio ν	Density $\rho/(kg \cdot m^{-3})$	Thermal Conductivity $k/(W/m \cdot k)$	Specific Heat Capacity $c/(J/kg \cdot k)$	Expansion Coefficient/°C^{-1}
Surrounding Rock	10	0.25	2210	2.3	960	8E-5
Primary Lining	23	0.23	2120	1.9	915	1E-5

The boundary conditions of the numerical model were as follows: the surface was a free surface with X-direction displacement constrained on both sides, Y-direction displacement constrained on the longitudinal ends, and Z-direction displacement constrained on the lower bottom. The thermodynamic boundary was set at the temperature value corresponding to the working condition and was constant.

4.4 *Working condition of calculation*

In this paper, the tunnel simulation conditions under five different joint positions and five different temperature conditions were established. The distance between the joint and the tunnel crown was respectively 0.5, 1.0, 1.5, 2.0, and 2.5 m, and the tunnel temperature was respectively 20, 40, 60, 80, and 100 °C. The interface method in FLAC3D was used to establish the joints. The tunnel joint positions were shown in Figure 5. In order to reduce the influence of other factors on the numerical simulation results, the tunnel model size remained unchanged during the simulation.

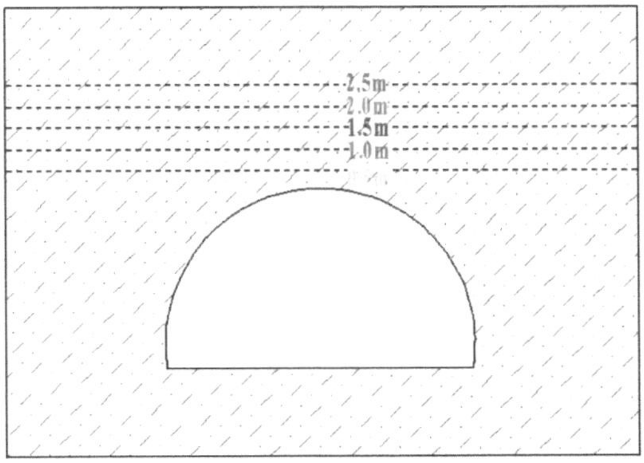

Figure 5. Tunnel joint location.

5 ANALYSIS OF THE INFLUENCE OF TUNNEL SURROUNDINGS ON ROCK STABILITY

5.1 *Effects of joint position on the stability of tunnel surrounding rock*

In order to study the influence of different joint positions on the stability of surrounding rock, the model set the tunnel surrounding rock temperature at 40 °C and the distance between the joint and the tunnel crown at 0.5, 1.0, 1.5, 2.0, and 2.5 m, respectively, for simulation calculations. LZ5K45 + 810 was selected as the monitoring section, and the variation rule of tunnel vault displacement with excavation time at different joint positions was obtained through simulation calculation, as shown in Figure 6. The displacement of measuring points A, B, and C of section 15 days from the tunnel excavation was calculated through simulation, as shown in Figure 7.

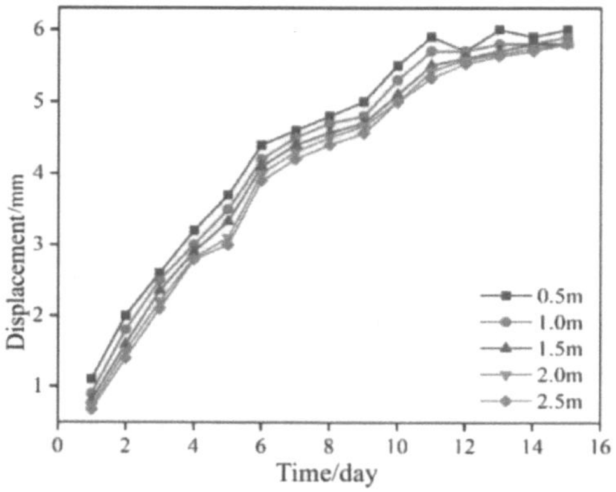

Figure 6. Tunnel vault displacement at different joint positions.

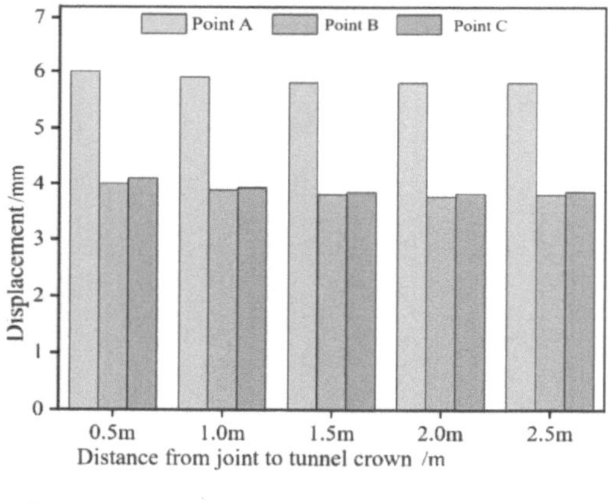

Figure 7. Displacement of tunnel arch crown and left and right arch waists at different joint positions.

1) Figure 6 shows that when the distance between the joint and the tunnel vault is 0.5 m, the vault displacement is 1.12 mm on the first day of tunnel excavation and 6.00 mm on the 15th day of tunnel excavation; when the distance between the joint and the tunnel vault is 2.51 m, the vault displacement is 0.78 mm on the first day and 5.8 mm on the 15th day after the tunnel excavation. For the same excavation time of the tunnel, the displacement of the tunnel vault decreases slowly with the increase in the distance between the joint and the tunnel vault. When the distance between the joint and the tunnel vault increases from 0.5 m to 2.5 m, the displacement of the tunnel vault on the 15th day decreases by 9.63%, 4.68%, 3.27%, and 2.15%, respectively. With the passage of time, the deformation of the tunnel's surrounding rock progressed through three stages: rapid deformation, slow deformation, and stable deformation.

2) Figure 7 shows that the displacement of tunnel arch crowns at different joint locations is greater than that of the left and right arch waists, which conforms to the general tunnel settlement law. The displacements of the tunnel vault's left arch waist and right arch waist decrease with the increase in the distance between the joint and the tunnel vault. This is because the plastic zone is easy to form by the close joint of the tunnel, while the fracture surface cannot be formed by the remote joint. The influence of the joint on tunnel stability can be ignored (Zhen et al. 2021). This result is consistent with the research results of Yang et al. (2020).

5.2 *Effects of temperature on stability of tunnel surrounding rock*

In order to study the influence of different temperatures on the stability of surrounding rock, the model set the distance between the joint and the tunnel crown as 0.5 m and the surrounding rock temperature as 20, 40, 60, 80, and 100 °C, respectively, for simulation calculation. LZ5K45 + 810 was selected as the monitoring section, and the variation rule of tunnel vault displacement with excavation time at different joint positions obtained through simulation calculation is shown in Figure 8. The displacement of measuring points A, B, and C at sections A, B, and C 15 days after tunnel excavation calculated by simulation is shown in Figure 9.

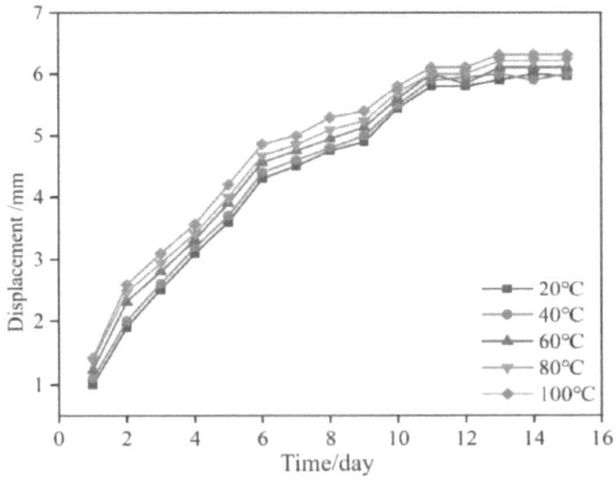

Figure 8. Variation of crown settlement of tunnels at different temperatures.

1) Figure 8 shows that when the surrounding rock temperature of the tunnel is 20 °C, the vault displacement on the first day of tunnel excavation is 1.05 mm, and the vault displacement on the 15th day is 5.96 mm; when the distance between the joint and the tunnel vault is 100 °C, the vault displacement is 1.42 mm on the first day of tunnel excavation and 6.30 mm on the 15th day. For the same tunnel excavation time, the displacement of the tunnel vault increases slowly with the increase in temperature of the tunnel's surrounding rock. When the distance between the joint and the tunnel vault increases from 20 °C to 100 °C, the displacement of the tunnel vault on the 15th day increases by 0.05 mm, 0.09 mm, 0.14 mm, and 0.23 mm, respectively.
2) Figure 9 shows that the displacement of the tunnel vault changes most significantly, which is significantly higher than that of other parts, and the left and right waist parts change the least. The displacement of the tunnel vault and left and right arch waists increases with the increase in temperature. This is because the high temperature causes thermal stress in the rock, which exceeds the ultimate strength of the internal mineral bonding points (Song et al. 2018), leading to the reduction of the rock's mechanical

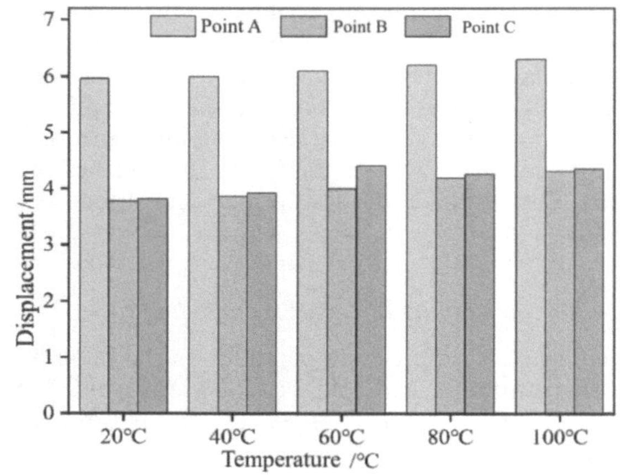

Figure 9. Variation of settlement of arch crown and left and right arch waists of tunnels at different temperatures.

properties. The surrounding rock of the tunnel face is more prone to deformation after tunnel excavation. It can be seen that the high temperature has a great impact on the stability of the tunnel surrounding the rock, which should be paid enough attention to.

5.3 *Effects of joint location and temperature coupling on the stability of tunnel surrounding rock*

In order to study the influence of different joint locations and temperature coupling on the stability of surrounding rock, the model set the distance between the joint and the tunnel crown as 0.5, 1.0, 1.5, 2.0, and 2.5 m, and the surrounding rock stability as 20, 40, 60, 80, and 100°C, respectively, for simulation calculation (Xi et al. 2020). LZ5K45 + 810 was selected as the monitoring section, and the simulated joint location and arch crown displacement under temperature coupling are shown in Figure 10.

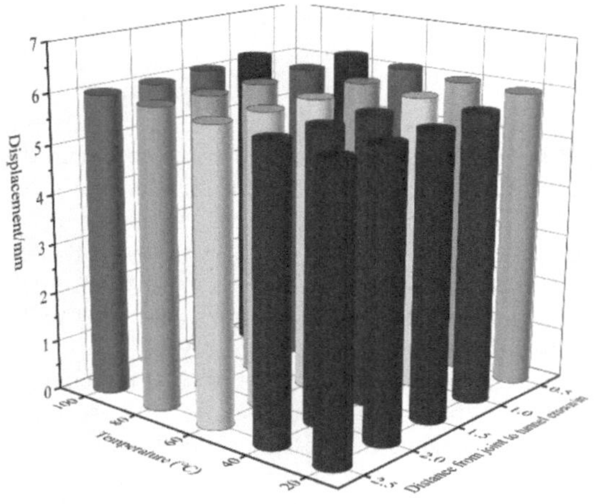

Figure 10. Displacement of arch crown under joint location and temperature coupling.

Figure 10 shows that the displacement of the tunnel vault decreases with the increase in the distance between the joint and the tunnel vault and increases with the increase in temperature. When the temperature was 100 °C and the distance between the joint and the tunnel vault was 0.5 m, the tunnel vault displacement was the largest. When the temperature was 20 °C and the distance between the joint and the tunnel vault was 2.5 m, the tunnel vault displacement was the smallest. According to the difference in the influence of two factors on the tunnel vault displacement, it was divided into three areas: (1) Area I (temperature was 100 °C, and the distance between the joint and the tunnel vault was 0.5 m): the tunnel vault displacement in this area was the largest, belonging to the rapid deformation area; (2) Zone II (the temperature was 60–80 °C and the distance between the joint and the tunnel crown was 1.0–1.5 m): the displacement of the tunnel crown in this area changed greatly, belonging to the slow deformation zone; (3) Zone III (the temperature was 20–40 °C and the distance between the joint and the tunnel crown was 2.0–2.5 m): the displacement of the tunnel crown in this area changed slowly, which belonged to a stable deformation zone. The above zoning can be used for safety prevention and control during tunnel construction and has important engineering value.

6 CONCLUSIONS

1) When the distance between the joint and the tunnel vault increased, the displacement of the tunnel vault and the left and right arch waists decreased rapidly; when the temperature of the surrounding rock rose, the displacement of tunnel measuring points increased rapidly.
2) Under different joint positions and different temperature conditions, the displacement changed at different positions of the tunnel section was different. The displacement of the tunnel vault was the largest, and the displacement of the left and right arch waists was the smallest.
3) The displacement variation under the joint position and temperature coupling could be divided into three regions: the displacement in the first region changed rapidly, the displacement in the second region changed slowly, and the displacement in the third region changed steadily.
4) The method of combining numerical simulation with field testing was adopted to ensure that the displacement of the numerical simulator coincides with that of the field monitoring. It is suggested that the tunnel support strength should be improved when the tunnel temperature exceeds 60 °C. According to the above analysis results, the tunnel support measures can be dynamically adjusted in combination with the specific joint position and temperature of the tunnel to ensure construction safety.

REFERENCES

Hua Yang, Liu Jinsong, Su Wei, *et al.* Analysis on the Safety Impact of High-temperature Hot Water Environment on Tunnel Lining Structure Based on FIB Calculation Method [J/OL]. *Railway standard design*: 1–8.

Hu Zheng, Ruan Niaofu. *Special Report on the Cause Analysis of High Ground Temperature in the Nige Tunnel and Feigu Tunnel of Jianshui Yuanyang Expressway Project in Honghe Prefecture [R]*. Guiyang: Powerchina Guiyang Survey, Design and Research Institute Co., Ltd., 2020.

Hu Zheng, Tian Maozhong, Guo Weixiang, *et al.* Study on the Geological Origin of Tunnel High Ground Temperature [J/OL]. *Journal of Southwest Jiaotong University*: 1–8.

Liu Wengang, Wang Ju, Zhou Hongwei, *et al.* Study on Thermal Mechanical Coupling Simulation of Granite in High-level Radioactive Waste Repository [J]. *Journal of Rock Mechanics and Engineering*, 2009, 28 (S1): 2875–288.

Li Guoliang, Cheng Lei, Wang Fei. Research on Key Technologies of High Temperature Tunnel Construction [J]. *Railway Standard Design*, 2016, 60 (06): 55–59.

Song Zhanping, Zhang Qiang, Zhao Keming, *et al.* Research on Optimization of Advance Construction of Double Heading Tunnel Based on On-site Monitoring and Numerical Analysis [J]. *Journal of Xi'an University of Architecture and Technology (Natural Science Edition)*, 2018, 50 (05): 654–661.

Sun Qiqing, Zheng Zongxi, Tan Yongjie. Analysis of the Mechanical Characteristics of the Secondary Lining of the High Temperature Tunnel [J]. *Journal of Railway Engineering*, 2018, 35 (04): 70–80.

Tang Xinghua, Wang Nianyan, Tong Jianjun, *et al.* Research on Initial Support Stress Field and Safety of High Rock Temperature Tunnel [J]. *Journal of Southwest Jiaotong University*, 2019, 54 (01): 32–38.

Wang Mingnian, Tang Xinghua, Wu Qiujun, *et al.* Evolution law of Surrounding Rock Supporting Structure Temperature Field of High Rock Temperature Tunnel [J]. *Journal of Railways*, 2016, 38 (11): 126–131.

Wang Nianyan, Wang Qiling, Hu Yunpeng, *et al.* Research on Mechanical Properties of Initial Tunnel Support Under High Temperature Environment [J]. *Journal of Railways*, 2019, 41 (11): 116–122.

Wu Biao, Peng Xuejun, Yuan Chao, *et al.* Discussion on Lining Structure Design and Construction Technology of High Temperature Tunnel [J]. *Journal of Hunan University of Science and Technology (Natural Science Edition)*, 2019, 34 (02): 18–24.

Wu Zhaofeng, Hu Huirong. Research on Deformation and Mechanical Behavior of Tunnel Lining Structure Under Fire and High Temperature [J]. *Modern Tunnel Technology*, 2020, 57 (06): 101–106.

Xi Baoping, Wu Yangchun, Zhao Yangsheng, *et al.* Strength Comparison of Granite Under Different Cooling Modes and Experimental Study on Characterization of Thermal Failure Ability [J]. *Journal of Rock Mechanics and Engineering*, 2020, 39 (02): 286–300.

Yuan Song, Wang Zhengzheng, Zhou Jiamei. Study on Boundary Value Range of Tunnel Seismic Dynamic Calculation [J]. *Journal of Civil Engineering*, 2012, 45 (11): 166–172.

Yuan Tie, Li Chang, Shen Zhijun, *et al.* Analysis of the Influence of Joints on the Stability of Mudstone Tunnels [J]. *Highway*, 2020, 65 (02): 288–294.

Zheng Yingren, Wang Yongfu, Wang Cheng, *et al.* Stability Analysis and Failure Law Discussion of Jointed Rock Tunnel–Lecture 1 of Tunnel Stability Analysis [J]. *Journal of Underground Space and Engineering*, 2011, 7 (04): 649–656.

Zheng Wen, Liu Naifei, Liu Xiaoping. Stress Characteristics of Supporting Structure of High Temperature Tunnel [J]. *Coal Field Geology and Exploration*, 2018, 46 (06): 138–143.

Water Conservancy and Civil Construction – Oke & Ahmad (Eds)
 ISBN: 978-1-032-58614-4

Effects of water and nitrogen coupling on yield, water, and fertilizer utilization rate and quality of potato

Tao Chen
College of Agronomy and Agricultural Engineering, Liaocheng University, Liaocheng, China
College of Water Conservancy and Hydropower Engineering, Gansu Agricultural University, Lanzhou, Gansu, China

Shouchao Yu
College of Agronomy and Agricultural Engineering, Liaocheng University, Liaocheng, China

Hengjia Zhang*
College of Agronomy and Agricultural Engineering, Liaocheng University, Liaocheng, China
College of Water Conservancy and Hydropower Engineering, Gansu Agricultural University, Lanzhou, Gansu, China

ABSTRACT: Water and nitrogen are closely related and mutually restricted factors in the process of crop growth and development, which are also two key factors limiting agricultural development and agricultural production efficiency. With climate change, the supply of water resources is becoming increasingly tight, and unreasonable fertilization has caused the expansion of farmland ecosystem pollution. Theoretical and technical research on reducing crop irrigation water and improving fertilizer use efficiency to achieve high crop quality and yields, as well as sustainable development, has received a lot of attention. As the fourth largest grain and feed crop, the potato ranks first in the world in terms of production area and total output value in China. It is of great significance to ensure the high quality and high yield of potatoes for food security and economic development. Potatoes are sensitive to soil moisture and nitrogen fertilizer. An excessive or small amount of water and nitrogen application will lead to a decrease in yield and nutritional quality. Reasonable nitrogen and water applications can promote nutrient absorption by potato stems, leaves, and stems, as well as the distribution of photosynthetic products, ultimately increasing yield and improving quality. At the same time, it can improve potato water and nitrogen use efficiency and reduce soil nitrogen loss and soil pollution.

1 INTRODUCTION

Water is the main factor limiting crop growth and yield, and it is also the main limiting factor in agricultural production. A shortage of water resources will also limit the sustainable development of agriculture (Cao *et al.* 2019; Wang *et al.* 2010). Determining a reasonable amount of irrigation is one of the main measures to solve the shortage of agricultural water resources. In addition, nitrogen plays an important role in crop growth. Nitrogen fertilizer has a significant impact on crop yield and is also an important measure to ensure food security (Li *et al.* 2020). In the past two decades, China's nitrogen fertilizer input has continued to increase, and the intensity of fertilizer use has been far greater than its ecological and economic suitability, exceeding the actual demand for crops (Liu & Pu 2019). A large

*Corresponding Author: 596088683@qq.com

DOI: 10.1201/9781003450818-8

number of studies at home and abroad have shown that there is an obvious coupling relationship between water and nitrogen. Appropriate irrigation amounts can promote the absorption and utilization of nitrogen by crops, reduce soil nitrogen loss, and improve water and fertilizer use efficiency (Cao *et al.* 2021; Nematpour *et al.* 2022; Wang *et al.* 2022).

As the fourth largest crop after maize, wheat, and rice, potato is a food and feed crop (Tang *et al.* 2018). Sensitive to water, water deficits reduce leaf number and leaf area, affect the accumulation and distribution of photosynthetic products, and ultimately affect potato stem yield. Nitrogen fertilizer can affect the chlorophyll content and photosynthetic system enzyme activity in potato leaves and also regulate the expansion time and size of potato stems. Nitrogen deficiency causes premature senescence of potatoes and reduces yield (Li *et al.* 2017; Šrek *et al.* 2010). The effects of water and nitrogen coupling on yield, quality, and water and fertilizer use efficiency of potatoes were summarized in this paper, which provided the theoretical reference for the water and nitrogen application rate of potatoes.

2 EFFECT OF WATER AND NITROGEN COUPLING ON POTATO TUBER YIELD

Water and nitrogen coupling can promote crop uptake and utilization of soil nutrients and water, reduce soil nutrient loss and soil pollution risk, and increase income and yield in agricultural production. Water and nitrogen are the keys to potato tuber yield formation. Appropriate soil moisture and fertilizer application can promote growth and development, as well as increase yield (Vashish *et al.* 2015), whereas excessive irrigation and nitrogen fertilizer application will reduce yield, water and fertilizer utilization, and ultimately wastewater and fertilizer resources. Yin *et al.* (Yin *et al.* 2020) found that the response of potato yield to different water and nitrogen regulations was different. Under the condition of low water (900 m^3/hm^2), the increase in nitrogen application rate would significantly increase potato yield, and the yield decreased with the increase in nitrogen application rate under high water conditions. Under the condition of a low nitrogen application rate (120 kg/hm^2), irrigation significantly promoted the formation of potato stems and increased the yield. Under the condition of a high nitrogen application rate (240 kg/hm^2), the yield increased first and then decreased with the irrigation amount. Zhou *et al.* (Zhou *et al.* 2004) found that excessive nitrogen application at the same irrigation amount would reduce potato yield, inhibit the formation of potato stem tubers, and reduce potato weight per plant and the potato commodity rate. Therefore, there is a certain threshold for irrigation and nitrogen application. Excessive nitrogen application will inhibit potato yield when water is sufficient, and excessive irrigation will also reduce yield when nitrogen fertilizer is sufficient. Optimizing water and fertilizer application schemes, as well as improving water and fertilizer utilization and potato tuber yield, are critical for sustainable potato development and food security in the face of a continuous shortage of agricultural irrigation water and an increase in fertilizer application.

3 EFFECTS OF WATER AND NITROGEN COUPLING ON WATER AND NITROGEN USE EFFICIENCY OF POTATO

Water and nitrogen coupling can improve crop water and fertilizer use efficiency to some extent while lowering agricultural production costs. It is an important index to measure the high yield and high efficiency of crops. Suitable water and nitrogen applications can obtain the best yield (Ma & Tian 2020). Li *et al.* (2016) conducted a water and nitrogen coupling experiment on potatoes using pots. The results showed that, under the same water treatment, the water use efficiency showed a downward parabolic trend with the increase in nitrogen application rate. The water use efficiency of moderate nitrogen treatment under severe water stress reached 22.4 g/kg. The nitrogen agronomic efficiency and nitrogen partial productivity were the best at the medium nitrogen level, and the nitrogen agronomic efficiency and nitrogen partial productivity were the largest under normal water conditions, which were 52.5 g/g and 143.9 g/g, respectively. Chen *et al.* found that increasing the nitrogen application rate under the same water condition would decrease nitrogen partial factor productivity, and increasing the nitrogen application rate under sufficient water conditions would also decrease nitrogen agronomic use efficiency and nitrogen physiological use efficiency. However, the agronomic efficiency of nitrogen fertilizer and the physiological efficiency of nitrogen fertilizer were different under low-water irrigation. From the overall level of water and nitrogen regulation, the irrigation amount of 2400 m^3/hm^2 and the nitrogen application rate of 150–225 kg/hm^2 were the nitrogen use efficiencies of potatoes. Jing *et al.* found that nitrogen application could promote potatoes to absorb soil moisture and increase the total water consumption during the whole growth period, but it reduced soil use efficiency and water use efficiency. Increasing the nitrogen application rate under the same water conditions reduced potato water consumption over the entire growth period, while soil water use efficiency and water use efficiency increased gradually. Among them, the soil water use efficiency and water use efficiency of irrigation amount 1350 and nitrogen application rate 300 were the highest, which were 160.14 kg/(mm·hm^2) and 115.87 kg/(mm·hm^2), respectively.

4 EFFECT OF WATER AND NITROGEN COUPLING ON POTATO QUALITY

Water and nitrogen coupling can achieve accurate regulation of water and nitrogen, and achieve sustainable development of a high-yield and high-quality potato. Shang *et al.* found that under the same nitrogen application level, the content of starch, vitamin C, and nitrate decreased with the increase in nitrogen application rate, and the content of soluble protein and soluble sugar showed a parabolic change rule. Under the same irrigation level, the content of soluble protein, nitrate, and vitamin C increased with the increase in nitrogen application rate. It was beneficial to the quality improvement and yield increase of potatoes under mulched drip irrigation when the irrigation amount was 900 m^3/hm^2 and the nitrogen application rate was 225 kg/hm^2. Irrigation and nitrogen application will affect the absorption of soil nutrients by potatoes, thereby affecting the growth status of plants and the assimilation and distribution of photosynthetic products, ultimately affecting the content of nutrients in potato stems. Zang *et al.* (2018) conducted a water-nitrogen coupling experiment on potatoes using sprinkler irrigation technology. The results showed that there was no significant difference in the content of starch and vitamin C in potato stems among different water-nitrogen combinations, and the crude protein content decreased first and then increased with the increase in nitrogen application rate. Irrigation and nitrogen application will affect the absorption of soil nutrients by potatoes, thereby affecting the growth status of plants and the assimilation and distribution of photosynthetic products, ultimately affecting the content of nutrients in potato stems. Therefore, reasonable regulation of water and nitrogen can promote potato stems to absorb soil nutrients, regulate photosynthetic products into the stem distribution of nutrients, and improve potato quality to achieve the purpose of saving water and fertilizer.

5 CONCLUSION

In recent years, frequent droughts have been expanding, and the shortage of agricultural irrigation water has seriously limited the production of potatoes. A reasonable application can promote the growth of potato stems and leaves, promote nutrient accumulation, and increase yield. The combination of the two can produce the best water and nitrogen coupling effect, improve crop yield and water and fertilizer use efficiency, improve crop quality, reduce soil nitrogen loss, and reduce harm to the farmland ecosystem (Nematpour *et al.* 2022; Wang *et al.* 2010, 2022). As one of the most important foods in the world, the high quality and high yield of potatoes are of great significance to the regional economy, national food security, and sustainable development of agronomy. The coupling technology of water and nitrogen can coordinate the absorption, accumulation, and distribution of potato nutrients to the greatest extent during the growth and development of potatoes. At the same time, it can improve the soil enzyme activity, promote the growth of potato plants and the formation of stems, improve the nutrients of potato stems, improve the water and nitrogen use efficiency of crops, and finally achieve the goals of high quality, high yield, and income increase. At the same time, the optimum amount of water and nitrogen varied in different agricultural climates, and the demand for potato varieties for water and nitrogen was also different. To deal with the difference between an arid climate and annual rainfall, reasonable water and nitrogen regulation should be carried out according to specific factors such as climate characteristics, soil fertility, and variety in the experimental site to give full play to the best interaction between water and nitrogen, improve the utilization of water and nitrogen in crops, improve the nutritional quality of potatoes, reduce production costs and soil environmental pollution, and realize the efficient utilization of water and nitrogen and the sustainable development of the potato industry.

ACKNOWLEDGMENTS

This work was financially supported by the Industrial Support Plan Project of the Gansu Provincial Department of Education (No. 2022CYZC-51), the National Natural Science Foundation of China (No. 52269008), and the Key Research and Planning Projects of Gansu Province (No. 18YF1NA073).

REFERENCES

Brian J B, Carl J R and David J M 2020. Impact of Variable Rate Nitrogen and Reduced Irrigation Management on Nitrate Leaching for Potato. *Journal of Environmental Quality*. 49 (2), 281–291.

Cao X C, Liu Z, Wu M Y, Guo X P and Wang W G 2019. Temporal-spatial Distribution and Driving Mechanism of Arable Land Water Scarcity Index in China from Water Footprint Perspective. *Transactions of the Chinese Society of Agricultural Engineering*. 35 (18), 94–100.

Cao X C, Wu L L, Zhu C H, Zhu L F, Kong Y L, Lu R H, Kong H M, Hu Z P, Dai F, Zhang J H and Jin Q Y 2021. Effects of Different Irrigation and Nitrogen Application Regimes on the Yield, Nitrogen Utilization of Rice and Nitrogen Transformation in Paddy Soil. *Scientia Agricultura Sinica*. 54 (07), 1482–1498.

Chen R Y, Meng M L, Liang H Q, Zhang J, Wang Y H and Wang Z X 2012. Effects of Different Treatments of Irrigation and Fertilization on the Yield and Nitrogen Utilization Characteristic of Potato. *Chinese Agricultural Science Bulletin*. 28(03), 196–201.

Jing T, Qin Y L, Fan M S and Zhou D B 2012.Effects of Water and Nitrogen Coupling on Wue of Potao in DRIP Irrigation Under MULCH. *Journal of Inner Mongolia Agricultural University(Natural Science Edition)*. 33(Z1), 41–45.

Li X Y, Leng X, Zhang J J, Guo Y, Ding Z J, Hu X D and Zhu K L 2020. Simulation and Optimization of Maize Growth and Nitrogen Utilization Under Degradation Film Mulching in Arid Areas of North China. *Transactions of the Chinese Society of Agricultural Engineering*. 36 (05), 113–121.

Li W T, Xiong B L, Wang S W, Deng X P,Yin L and Li H B 2017. Regulation Effects of Water and Nitrogen on the Source-Sink Relationship in Potato During the Tuber Bulking Stage. *PLoS ONE*. 11 (1), e0146877.

Li W T, Wang S W, Deng X P and Li H B 2016. Effects of Different Water and Nitrogen Levels on Yield and Water and Nitrogen Use Efficiency of Potato. *Agricultural Research in the Arid Areas*. 34(06), 191–196.

Liu Q P and Pu L J 2019. Spatiotemporal Variation of Fertilizer Utilization and its Eco-economic Rationality in Major Grain Production Areas of China. *Transactions of the Chinese Society of Agricultural Engineering*. 35 (23), 142–150.

Nematpour A, Eshghizadeh H R, Zahedi M and Mahdi G 2022. Interactive Effects of Sowing Date and Nitrogen Fertilizer on Water and Nitrogen Use Efficiency in Millet Cultivars Under Drought Stress. *Journal of Plant Nutrition*. 43 (1), 122–137.

Ma B and Tian J C 2020. Optimization of Irrigation Scheduling for Lycium barbarum L. Based on Water and Nitrogen Coupling. *Journal of Soil and Water Conservation*. 34 (06), 235–243.

Shang M S, Fang Z G, Liang B, Wang M and Li J L 2019. Effects of Different Water and Nitrogen Treatments on Potato Yield,Quality and Soil Nitrate Nitrogen Transport Under Drip Irrigation. *Acta Agriculturae Boreali-Sinica*. 34(06), 118–125.

Šrek P, Hejcman M and Kunzová E 2010. Multivariate Analysis of Relationship Between Potato (Solanum tuberosum L.) Yield, Amount of Applied Elements, their Concentrations in Tubers and Uptake in a Long-term Fertilizer Experiment. *Field Crops Research*. 118 (2), 183–193.

Tang J Z, Wang J, Fang Q X, Wang E L, Hong Y and Pan X B 2018. Optimizing Planting Date and Supplemental Irrigation for Potato Across the Agro-pastoral Ecotone in North China. *European Journal of Agronomy*. 98, 82–94.

Vashish, B B, Nigon T, Mulla D J, Rosen C, Xu H, Twine T and Jalota S K 2015. Adaptation of Water and Nitrogen Management to Future Climates for Sustaining Potato Yield in Minnesota: Field and Simulation Study. *Agricultural Water Management*. 152, 198–206.

Yin J, Zhang H J, Wang S, Wang C and Zhao Y B 2020. Effects of Different Water and Nitrogen Treatments on Photosynthetic Characteristics and Yield of Potato. *Water Saving Irrigation*. (06), 8–13.

Wang L M, Huang D F, Zhang B Y and Pan Z C 2022. Differences in Uptake, Utilization and Loss of Nitrogen and Phosphorus in a Chinese Double Rice Cropping System Under Different Irrigation and Fertilization Managements. *Chinese Journal of Applied Ecology*. 33 (04), 1037–1044.

Wang Y B, Wu P T, Zhao X N and Li J L 2010. Development Tendency of Agricultural Water Structure in China. *Chinese Journal of Eco-Agriculture*. 18 (02), 399–404.

Zhou N N, Zhang X J, Qin Y B and Xu Q 2004. Effect on Different Quantities of drip Irrigation and Nitrogen Fertilization for Yield and Quality of Potato. *Soil and Fertilizer Sciences in China*. (06), 11–12+16.

Zang W J, Li J J, Pei S S, Li Y J and Yan H J 2018. Effects of Different Water – Nitrogen Combinations on Potato Water Consumption,yield and Quality Under Sprinkler Irrigation. *J. Journal of Drainage and Irrigation Machinery Engineering*. 36(08):773–778.

Water Conservancy and Civil Construction – Oke & Ahmad (Eds)
© 2024 The Author(s), ISBN: 978-1-032-58614-4

Regulated deficit irrigation increasing water productivity of potato

Youshuai Bai
College of Agronomy and Agricultural Engineering, Liaocheng University, Liaocheng, China
College of Water Conservancy and Hydropower Engineering, Gansu Agricultural University, Lanzhou, Gansu, China

Shouchao Yu
College of Agronomy and Agricultural Engineering, Liaocheng University, Liaocheng, China

Hengjia Zhang*
College of Agronomy and Agricultural Engineering, Liaocheng University, Liaocheng, China
College of Water Conservancy and Hydropower Engineering, Gansu Agricultural University, Lanzhou, Gansu, China

Shenghai Jia, Zeyi Wang & Xietian Chen
College of Water Conservancy and Hydropower Engineering, Gansu Agricultural University, Lanzhou, Gansu, China

ABSTRACT: Water scarcity has become an important challenge to sustainable global agricultural development and food security. Regulated deficit irrigation is a novel, optimized water-saving measure based on traditional irrigation principles and methods. This paper briefly described the basic principles and practices of regulated deficit irrigation. The water-saving mechanism of regulated deficit irrigation and the research on the growth index, photosynthesis, yield, tube quality of potatoes, soil quality attributes, and nutrient utilization were reviewed. Finally, we summarized the problems of regulated deficit irrigation in practice and proposed the direction of its sustainable development to guide the further application of this irrigation technology and efficient agricultural water management.

1 INTRODUCTION

Freshwater plays an important role in the agriculture system worldwide (Gan *et al.* 2014; Kang *et al.* 2017). The increasing demand for water resources and constant climate change have further exacerbated water scarcity while increasing the complexity and difficulty of water resource management (Arnell 2004; Golfam *et al.* 2021). Water scarcity is a common problem in the 21st century and also an important factor affecting sustainable potato production. According to FAO statistics, about 702–928 million people will face hunger in 2021. As the fourth largest food crop in the world nowadays, the potato plays a highly essential role in food security and hunger reduction. China is the world's largest producer of potatoes, which are widely cultivated in arid and semi-arid regions.

The overexploitation of irrigation water and groundwater has now caused more serious ecological problems in arid regions, including declining groundwater levels, the degradation and death of vegetation, and sand and dust storms (Chai *et al.* 2014; Duda & El-Ashry 2000; Kourgialas *et al.* 2019). Therefore, the application of efficient water-saving irrigation

*Corresponding Author: 596088683@qq.com

 DOI: 10.1201/9781003450818-9

technology to improve crop water productivity is an important approach to the problem (Kang *et al.* 2017; Bai *et al.* 2022).

Regulated deficit irrigation provides another water-saving technology based on traditional irrigation principles and methods (Geerts & Raes 2009). Rationally regulated deficit irrigation helps to achieve a better compromise between yield and saving water (Badr *et al.* 2022). The farm management strategy of "regulated deficit irrigation" is to deliberately allow a significant irrigation reduction under reduced yield conditions (Costa *et al.* 2007). In recent years, water shortages have become more evident with the development of the economy and society. The problem of squeezed water for agriculture is prominent, which in turn leads to higher irrigation water charges. Regulated deficit irrigation saves irrigation water without significantly reducing crop production, and the comprehensive benefits are increased instead.

In this review, we focused on the water-saving mechanisms of regulated deficit irrigation and its effects on potato growth, photosynthetic characteristics, water productivity, and soil environmental aspects. Finally, the problems of regulated deficit irrigation in production practice and future development opportunities were explored.

2 BASIC PRINCIPLES AND METHODS OF REGULATED DEFICIT IRRIGATION

Crop species differ in their genetic and physiological growth characteristics. Therefore, there are also differences in the water requirements of different crops during the growth process. For example, the critical water requirement period of a potato is the tuber formation and tuber expansion period, and the critical water requirement period of the jujube tree is the blossoming and fruit expansion period. Based on the physiological growth characteristics of crops, regulated deficit irrigation needs to reduce the irrigation amount at non-water-demanding critical stages and artificially and actively increase a certain amount of water stress (Kriedemann & Goodwin 2003). With the purpose of regulating above- and below-ground growth dynamics, promoting reproductive growth, controlling nutritional growth, regulating the distribution of photosynthetic products to different tissues and organs, and achieving water savings, high yields, and better quality are needed.

Regulating the irrigation amount according to field capacity is one way of regulating deficit irrigation. Li *et al.* set three moisture deficit gradients (conventional irrigation, 65–75% field capacity; mild deficit, 55–65% field capacity; and moderate deficit, 45–55% field capacity) to investigate the effects of potato soil water deficit on growth, yield, and

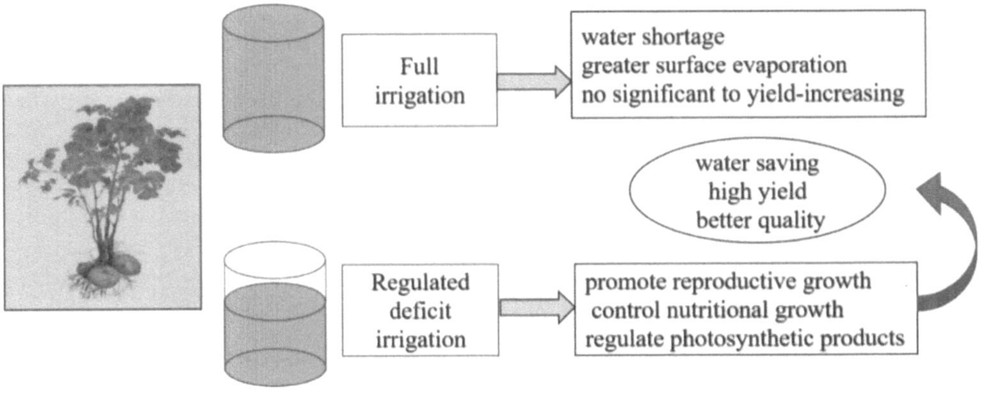

Figure 1. Basic principles of regulated deficit irrigation.

quality (Li *et al.* 2021). The results showed that mild water stress at the seedling stage improved yield and water use efficiency and could be used as an optimal irrigation strategy. Wang *et al.* demonstrated that applying mild water stress (65–75% field capacity) continuously during the growth of nutritive and fleshy roots improved water use efficiency and quality of Panax quinquefolium to some extent but did not significantly reduce yield (Wang *et al.* 2021). A more accurate irrigation technique is to provide water stress to the crop depending on the field capacity; this technique has several potential applications.

Regulated deficit irrigation is carried out by regulating the amount of irrigation based on crop evapotranspiration (ET). By testing the effects of different irrigation levels (90 mm, 60 mm, and 45 mm) on the yield and quality of pear jujube trees at different growth stages, Cui *et al.* showed that severe and moderate water deficits significantly increased jujube yield during the fruit maturation stage (Cui *et al.* 2008). Meanwhile, this study also showed that regulated deficit irrigation improved fruit quality to some extent. Studies have shown that stress irrigation using 75% of the crop water requirement improves water use efficiency compared with no stress (Kifle & Gebretsadikan 2016). This irrigation method saves water but does not reduce the corresponding yield, and it can be used as a better agricultural water management technique. In addition, the saved water can increase the irrigated area and expand production, increasing farmers' income. Mitchell *et al.* investigated the effect of deficit irrigation on peer tree growth and yield by regulating the amount of irrigation based on evaporation, and the results showed that regulated deficit irrigation could significantly increase the yield (Mitchell *et al.* 1984).

In summary, compared with full irrigation, regulated deficit irrigation reduces irrigation water at crop-specific growth stages and non-water-critical periods to optimize reproductive and nutritional growth, achieving water savings and yield maintenance.

3 WATER-SAVING MECHANISM OF REGULATED DEFICIT IRRIGATION AND THE RESEARCH ON THE GROWTH INDEX, YIELD, TUBE QUALITY AND SOIL ENVIRONMENT OF POTATOES

3.1 *Growth and photosynthesis*

Soil moisture conditions are one of the most significant indicators of crop growth. When crops are subjected to severe water stress, it may lead to growth restriction and yield reduction. Potatoes can be divided into seedling, tuber formation, tuber expansion, starch accumulation, and seed dormancy stages according to their physiological and growth characteristics. Several studies have shown that severe water deficits lead to lower photosynthetic rates in crops (Romero *et al.* 2013). Regulated deficit irrigation leads to root zone drying and causes water stress in the root systems. The response through hydraulic and chemical signals in the root system results in not only partial stomatal closure and increase photosynthetic water use efficiency, but a slight decrease in the top growth of potatoes (Jensen *et al.* 2010).

Studies have shown that deficit drip irrigation can reduce the photosynthetic rate, transpiration rate, and stomatal conductance in potatoes (Zhang & Li 2013). Li *et al.* found that net photosynthesis and stomatal conductance were significantly reduced by regulated deficit irrigation during the potato tuber formation and starch accumulation periods (Li *et al.* 2021). In conclusion, different water stresses applied at specific growth stages of potatoes affect their growth and photosynthetic characteristics.

3.2 *Yield, water use efficiency, and potato tube quality*

An important objective of using regulated deficit irrigation techniques is to increase water productivity. Soil hydrothermal status is an overwhelming influence on plant growth. Kifle and Gebretsadikan found that water deficit irrigation (water stress) significantly affected potato yield ($p < 0.05$) (Kifle & Gebretsadikan 2016). Zhang and Li concluded that deficit drip irrigation could reduce crop water consumption and promote water use efficiency, and that mild deficit irrigation during tuber formation does not reduce potato yield (Jensen *et al.* 2010). Related studies have also shown that the highest overall quality of potatoes was achieved with mild and moderate water stress at the seedling stage (Li *et al.* 2021). However, some studies concluded that mild water deficit treatment reduced potato yields by 12.7% ($p < 0.05$) and increased water use efficiency by 14.2% ($p < 0.05$) compared to full water supply (Li & Zhang 2013). El-Sawy *et al.* found that regulated deficit irrigation (60% and 40% of reference evapotranspiration) treatments significantly reduced nutritional growth, tuber yield, total chlorophyll content, and tuber quality parameters of potatoes (El-Sawy *et al.* 2022). In conclusion, crop yield reduction under regulated deficit irrigation conditions is based on improving water use efficiency and quality.

3.3 *Soil quality attributes and nutrient utilization*

The irrigation method is one of the most important factors affecting soil moisture. Soil moisture conditions and wetting characteristics affect the growth and yield formation of crops. For example, the irrigation regime for the same crop varies from one soil type to another. In turn, the soil-water-plant relationship has an impact on the soil environment. Studies have shown that regulated deficit irrigation reduces soil bulk density and water-holding pore content, contributing to improved nutrient uptake by plants (El Baroudy *et al.* 2014). Nitrogen application is important for potato production performance and dry matter accumulation (Darwish *et al.* 2006). In potato production, Xue *et al.* demonstrated that mulched drip irrigation with regulated deficit regulation effectively reduced soil available nutrient loss and increased soil available nutrient use efficiency (Xue *et al.* 2017). The above analysis shows that regulated deficit irrigation improves the soil environment and nutrient utilization for potato growth, resulting in increased water productivity.

4 PROBLEMS AND PROSPECTS

Several factors influence the application of regulated deficit irrigation technology, such as the genetic and physiological growth characteristics of the crop. Deficit irrigation regimes differ depending on crop species, soil conditions, meteorological and environmental conditions, and so on. Therefore, the application of deficit irrigation theory and technology should be based on years of continuous research before it has some reference value. In addition, the adverse effect of long-term deficit irrigation on soil ecology is also a concern. For example, long-term deficit irrigation may lead to a yearly reduction of water in the lower soil layer, resulting in a loss of water from the deeper soil layer, which in turn leads to soil desiccation. In addition, for regulated deficit irrigation, if the range of soil moisture is controlled based on the percentage of field water capacity, this narrow range of moisture is difficult to apply and operate in practice (Costa *et al.* 2007). Therefore, the application of regulated deficit irrigation may be limited by soil environment, meteorological conditions, and other factors.

In addition, the shortage of water resources leads to an accentuation of agricultural water conflicts. Water price increases cause an increase in agricultural inputs. However, fluctuations in potato prices can lead to a possible area reduction, which may endanger food security. In conclusion, deficit regulation irrigation is an optimal water-saving irrigation strategy that allows for yield reduction to improve water use efficiency and quality at certain crop growth stages and has broad application prospects. The application of this technique in some advanced water-saving irrigation technology combined with some agronomic measures is an effective way to improve agricultural water productivity and management.

ACKNOWLEDGMENTS

This work was financially supported by the National Natural Science Foundation of China (No. 52269008), the Industrial Support Plan Project of Gansu Provincial Department of Education (No. 2022CYZC-51), and the Key Research and Planning Projects of Gansu Province (No. 18YF1NA073).

REFERENCES

Arnell N. W. Climate Change and Global Water Resources: SRES Emissions and Socio-economic Scenarios. *Global environmental change*, 14 (1), 31–52 (2004).

Bai Y., Zhang H., Jia S., et al. Plastic Film Mulching Combined with Sand Tube Irrigation Improved Yield, Water use Efficiency, and Fruit Quality of Jujube in an Arid Desert Area of Northwest China. *Agricultural Water Management*, 271, 107809 (2022).

Badr M. A., El-Tohamy W. A., Salman S. R., et al. Yield and Water Use Relationships of Potato Under Different Timing and Severity of Water Stress. *Agricultural Water Management*, 271 (2022).

Chai Q., Gan Y., Turner N. C., et al. Water-saving Innovations in Chinese Agriculture. *Advances in Agronomy*, 126, 149–201 (2014).

Costa J. M., Ortuño M. F., Chaves M. M. Deficit Irrigation as a Strategy to Save Water: Physiology and Potential Application to Horticulture. *Journal of Integrative Plant Biology*, 49 (10), 1421–34 (2007).

Cui N., Du T., Kang S., et al. Regulated Deficit Irrigation Improved Fruit Quality and Water Use Efficiency of Pear-jujube Trees. *Agricultural Water Management*, 95 (4) (2008) 489–97.

Darwish T., Atallah T., Hajhasan S., et al. Nitrogen and Water Use Efficiency of Fertigated Processing Potato. *Agricultural water management*, 85 (1-2), 95–104 (2006).

Duda A. M., El-Ashry M. T. Addressing the Global Water and Environment Crises Through Integrated Approaches to the Management of Land, Water and Ecological Resources. *Water International*, 25 (1), 115–26 (2000).

El-Sawy S., Abd Elbaset M., El-Shafie A., et al. Effect of Irrigation Scheduling on Yield, Quality and Water Use Efficiency of Potato Plants Grown Under Deficit Irrigation Conditions. *Middle East Journal of Agriculture Research*, 11 (02) (2022) 693–711.

El Baroudy A., Ibrahim M., Mahmoud M. Effects of Deficit Irrigation and Transplanting Methods of Irrigated Rice on soil Physical Properties and Rice Yield. *Soil use and management*, 30 (1), 88–98 (2014).

Gan Y., Liang C., Chai Q., et al. Improving Farming Practices Reduces the Carbon Footprint of Spring Wheat Production. *Nature Communications*, 5 (1) 1–13 (2014).

Geerts S., Raes D. Deficit Irrigation as an On-farm Strategy to Maximize Crop Water Productivity in Dry Areas. *Agricultural water management*, 96 (9), 1275–84 (2009).

Golfam P., Ashofteh P. S., Loaiciga H. A. Modeling Adaptation Policies to Increase the Synergies of the Water-climate-agriculture Nexus Under Climate Change. *Environmental Development*, 37, 100612 (2021).

Jensen C. R., Battilani A., Plauborg F., et al. Deficit Irrigation Based on Drought Tolerance and Root Signalling in Potatoes and Tomatoes. *Agricultural Water Management*, 98 (3) (2010) 403–13.

Kang S., Hao X., Du T., et al. Improving Agricultural Water Productivity to Ensure Food Security in China Under Changing Environment: From Research to Practice. *Agricultural Water Management*, 179, 5–17 (2017).

Kifle M., Gebretsadikan T. Yield and Water use Efficiency of Furrow Irrigated Potato Under Regulated Deficit Irrigation, Atsibi-Wemberta, North Ethiopia. *Agricultural water management*, 170 133–9, (2016).

Kriedemann P. E., Goodwin I. Regulated Deficit Irrigation and partial Rootzone Drying: an Overview of Principles and Applications [M]. *Land & Water Australia* (2003).

Kourgialas N. N., Koubouris G. C., Dokou Z. Optimal Irrigation Planning for Addressing Current or Future Water Scarcity in Mediterranean Tree Crops. *Sci Total Environ*, 654, 616–32 (2019).

Li F., Deng H., Wang Y., et al. Potato Growth, Photosynthesis, Yield, and Quality Response to Regulated Deficit Drip Irrigation Under Film Mulching in a Cold and Arid Environment. *Sci Rep*, 11 (1), 15888 (2021).

Li J., Zhang h. Effect of Water Deficit Under Mulched Drip Irrigation on Photosynthetic Physiological Indexes and Yield of Potato During Tube Formation. *Journal of Irrigation and Drainage*, 32 (06), (2013)

Mitchell P., Jerie P., Chalmers D. The Effects of Regulated Water Deficits on Pear Tree Growth, Flowering, Fruit Growth, and Yield. *Journal of the American Society for Horticultural Science*, 109 (5) (1984) 604–6.

Romero P., Gil-Muñoz R., Del Amor F. M., et al. Regulated Deficit Irrigation Based Upon Optimum Water Status Improves Phenolic Composition in Monastrell Grapes and wines. *Agricultural water management*, 121, 85–101 (2013).

Vieira R. M. D. S. P., Tomasella J., Barbosa A. A., et al. Desertification Risk Assessment in Northeast Brazil: Current Trends and Future Scenarios. *Land Degradation & Development*, 32 (1), 224–40 (2021).

Wang Z., Zhang H., Wang Y., et al. Integrated Evaluation of the Water Deficit Irrigation Scheme of Indigowoad Root Under Mulched Drip Irrigation in Arid Regions of Northwest China Based on the Improved TOPSIS Method. *Water*, 13 (11) (2021)

Xue D., Zhang H. Z., Ba Y. B., et al. Effects of Regulated Deficit Irrigation on Soil Environment and Yield of Potato Under Drip Irrigation in Oasis Region. *Acta Agriculturae Boreali-Sinica*, 32 (3), 229–38 (2017).

Zhang H., Li J. Photosynthetic Physiological Characteristics and Water Use of Potato with Mulched Drip Irrigation under Water Deficit in Oasis Region. *Transactions of the Chinese Society for Agricultural Machinery*, 44 (10), (2013)

Water Conservancy and Civil Construction – Oke & Ahmad (Eds)

Effects of water and nitrogen regulation on physiological growth, yield, and quality of potato

Yong Wang
College of Agronomy and Agricultural Engineering, Liaocheng University, Liaocheng, China
College of Water Conservancy and Hydropower Engineering, Gansu Agricultural University, Lanzhou, Gansu, China

Shouchao Yu
College of Agronomy and Agricultural Engineering, Liaocheng University, Liaocheng, China

Hengjia Zhang*
College of Agronomy and Agricultural Engineering, Liaocheng University, Liaocheng, China
College of Water Conservancy and Hydropower Engineering, Gansu Agricultural University, Lanzhou, Gansu, China

ABSTRACT: The potato is a traditional food crop that is geographically adaptable, with rich nutrients, has high yield potential and good production efficiency, and ranks first in the world in terms of production, cultivation, and export in China. Ensuring high yields and quality development of the potato is an important way to effectively alleviate our food shortage. Irrigation and nitrogen addition are important means of influencing potato growth and development, quality, and yield formation. Too much or too little water and nitrogen can reduce potato water and nitrogen use efficiency, affecting crop yield and nutritional quality. Timely and appropriate water and nitrogen application facilitates the uptake of soil nutrients by potatoes, promotes the transport of photosynthetic products to the tubers, and improves potato yield and quality. The effects of water and nitrogen regulation on potato growth and physiology, yield and quality, water and nitrogen use efficiency, and water and nitrogen transport characteristics were reviewed. It is expected to provide theoretical references for water and nitrogen conservation, yield and quality improvement, and the sustainable development of potatoes.

1 INTRODUCTION

The potato (Solanum tuberosum L.), an annual herb of the Solanaceae family, is the fourth most important food crop in China after wheat, maize, and rice, and has nutritional, medicinal, and economic values. It is also effective in promoting intestinal peristalsis, tonifying the spleen, and detoxifying the body [1]. China is the country with the largest potato output in the world, and according to statistics, in 2020 China's potato cultivation area exceeded 4, 100, 000 ha, producing a total output of about 29, 874, 000 tonnes and a direct economic benefit of 41, 827 million yuan. With its main production areas widely distributed in the arid and semi-arid regions of China, it has become a pillar source for many provinces and cities [2, 3]. Potato growth and development are susceptible to external growth environments such as climate and soil conditions, and coordinating the coupling between soil components is an important way to improve the yield and quality of potato tubers [4].

*Corresponding Author: 596088683@qq.com

DOI: 10.1201/9781003450818-10

Water and nitrogen are key factors affecting potato quality and yield formation [5]. Drought stress is the main abiotic stress factor affecting potato growth and development [6]. The accumulation and distribution of photosynthetic products in the plant restrict material cycling, energy flow, and the rate of photosynthetic product accumulation and distribution, which in turn leads to lower tuber numbers and tuber deformation in potato plants, ultimately affecting yield and quality [7,8]. As a dense crop, the potato has a high potential for N losses, and exogenous N addition is an effective measure to improve soil N storage [9]. However, excessive nitrogen applications tend to cause redundant growth of potato nutrient organs, delay potato set, reduce tuber dry matter accumulation and starch content, and increase nitrate nitrogen accumulation in the root zone of the plant, affecting crop yield quality and soil environmental sustainability [10]. Studies have shown that there is an obvious reciprocal coupling effect between water and nitrogen and that appropriate soil water content is beneficial to potato uptake of soil nutrients, weakens nitrification, reduces nitrogen leaching losses, and improves nitrogen material use efficiency. As the demand for staple potato foods continues to expand, research on precise and efficient water and nitrogen regulation mechanisms is an important measure to promote potato quality, efficiency, and sustainable development [11,12].

2 RESEARCH PROGRESS

2.1 *Research on the physiological effects of water and nitrogen regulation on potato growth*

Reasonable water and nitrogen application is beneficial to enhance the photosynthetic physiological efficiency of plant leaves, delay leaf senescence, improve plant resistance performance, and promote crop growth and development [13]. Wang *et al.* [14] found that irrigation N application had significant effects on plant height, stem diameter, and leaf chlorophyll content, and water was the dominant factor affecting plant height and stem diameter of potatoes. At the same level of nitrogen application, potato plant height and stem thickness were positively correlated with the level of irrigation, and there was a significant positive correlation between potato stem thickness and urease activity in the 0–20 cm soil layer, while plant height was significantly positively correlated with peroxidase activity in the 0–20 cm soil layer. Water and nitrogen regulation promote the physiological processes of potato growth by influencing the soil environment for potato growth. Hamid *et al.* [4] discovered that under the conditions of combined water and nitrogen application, with the decrease of irrigation amount (more than 50% of field water capacity), the content of proline in potato leaves showed an increasing trend, and the high concentration of proline reduced the tissue water potential of potato leaves and avoided massive water loss. In a water and nitrogen control experiment between potatoes and oats, Wang *et al.* [15] found that the physiological indicators of potatoes were optimal when nitrogen was applied at 150 kg/ha to 225 kg/ha and that the peroxidase (POD) and catalase (CAT) contents of potato leaves at the seedling stage were lower than those at the tuber formation and maturity stages, and that with increasing nitrogen application, the chlorophyll content of potato leaves was higher than that at the tuber formation and maturity stages. The chlorophyll content, POD, and CAT of potato leaves showed an increasing trend, while malondialdehyde (MAD) and proline content showed a decreasing trend, indicating that water and nitrogen regulation in potato intercropping has the advantage of improving plant light permeability and stress resistance. Wang *et al.* [16] showed that under the same conditions of nitrogen application, potato plant height, stem thickness, and leaf area showed an increasing trend with increased irrigation, but high water and high nitrogen and low water and low nitrogen showed an obvious inhibitory effect on potato growth, and under high water and low fertilizer conditions, potato growth potential showed a continuous growth trend, which is a more suitable irrigation nitrogen application mode for potatoes. Yan *et al.* [17] showed that under normal water conditions, increasing nitrogen application significantly increased the net photosynthetic rate and stomatal conductance of potato leaves, and improved potato photosynthesis. However, under water stress conditions, excessive nitrogen application changed the osmotic pressure of leaf cells, inhibited the synthesis of photosynthesis, and significantly reduced photosynthetic efficiency.

2.2 *Study on the potato yield and quality effects of water and nitrogen regulation*

Soil water content directly affects the transport, assimilation, and nutrient effectiveness of soil nutrients, while nitrogen addition promotes plant adaptation to drought stress, both of which are important factors influencing potato quality and yield formation. Water and nitrogen influence potato yield by affecting the leaves of the potato plant and regulating the transport of photosynthetic products to the potato tuber [18]. According to Shang *et al.* [19], the potato yield showed a parabolic trend of "rising down" under the condition of irrigation quantity and nitrogen application, indicating that there was an optimal critical threshold for the potato irrigation nitrogen application. Too much or too little nitrogen in the water will affect the rate and yield of potato tuber expansion. When the nitrogen application quantity was 225 kg/ha and 900 m3/ha, respectively, the potato tuber yield reached 35299.9 kg/ha. Mai *et al.* [20] indicated that water and nitrogen control under drip irrigation had a significant effect on yield, with a 36.95% increase in potato yield compared to the no-nitrogen treatment at 450 m^3/ha and 225 kg/ha, respectively. Song *et al.* [21] discovered that under the same irrigation level, potato single plant tubers, vitamin C and starch content with increasing nitrogen parabolic trend, protein content with increasing nitrogen, in water (70% soil moisture ratio) nitrogen (180 kg/ha) yield and quality indicators are the best performances in the northwest arid region water-potato nitrogen coupling test. Furthermore, Yan *et al.* [17] discovered that normal water (65-75% of field water) and high nitrogen (11.11 g/plant) treatments, potato tuber starch, and crude fat content show that normal water and high nitrogen are conducive to potato starch and crude fat quality content accumulation, which could be attributed to the test potato varieties, regional climate, and soil physical and chemical traits.

2.3 *Effect of water nitrogen regulation on water nitrogen utilization efficiency in potatoes*

Soil nitrogen is a good source of nitrogen available to the potato plant for uptake and use. Nitrogen uptake efficiency is influenced by a number of factors, including irrigation water and soil heterogeneity [22]. Mohammad *et al.* [23] studied the response mechanism of potatoes to nitrogen in the field and showed that the high cation exchange capacity of fine-grained soil limited the migration of ammonium ions to the deep roots of plants, so the nitrogen use efficiency of coarse soil was higher than that of fine soil. The study also indicated that, due to the transformation and fixation of microorganisms, the nitrogen use efficiency of potatoes planted in the latter season was higher than that in the previous season, the nitrogen use efficiency of fertilization with water was higher than that of spreading application, and the nitrogen uptake of potato shoot plants increased with the increase of nitrogen application regardless of the nitrogen application method. Jiao *et al.* [24] studied the uptake, distribution, and transport of nitrogen in potato plants using the 15 N tracer technique and indicated that with increasing nitrogen application, nitrogen uptake in above-ground and below-ground potato plants (roots and tubers) showed an increasing trend, with nitrogen uptake and soil residual nitrogen content accounting for 17.29% and 71.83% of the total nitrogen content in potato plants, respectively; nitrogen uptake in above-ground potato plants, roots, and tubers was 0.19 g, 045 g, and 0.58 g, respectively, with a nitrogen use efficiency of 10.32%. Tang *et al.* [25] used the APSIM-Potato model to study water and nitrogen transport trials under different production objectives on typical farms in the north China agro-pastoral zone (APE, covering Ningxia, Gansu, Inner Mongolia, Hebei, Shanxi, Heilongjiang, Jilin, and Liaoning). The study indicated that appropriate water stress was conducive to improved water use efficiency in potatoes, with irrigation of 36.42. The combined water use efficiency of APE was the greatest with 36.42 mm of irrigation and 90 kg/ha of N fertilizer, and the amount of irrigation was less than that required for maximum yield. At the same time, water use efficiency was higher in central APE than in eastern APE and lowest in western APE due to the different transpiration and soil evaporation intensities of potatoes in different regions. It was also discovered that the higher the nitrogen application was, the greater the nitrogen loss was, and a corresponding increase in potato yield lacked. Nitrogen use efficiency was the highest at a 10 mm soil moisture deficit and 30 kg/ha of N fertilizer, but water use efficiency, yield, and net benefit were lower than in the other treatments. Therefore, a combination of irrigation of 53.42 mm and a fertilizer application of 30–120 kg/ha of nitrogen maintained higher levels of yield, water, nitrogen use efficiency, and profitability.

2.4 *Study of water and nitrogen transport mechanisms under ater and nitrogen regulation in potatoes*

Nitrate leaching and ammonium volatilization are the main forms of exogenous nitrogen loss. Carter *et al.* [26] indicated that nitrogen application had a direct impact on soil nitrate content by studying the migration law of nitrogen in potatoes and soil under irrigation, and the initial concentration of soil nitrate nitrogen was more severely affected by inter-annual climate change. Adding the same amount of organic fertilizer and inorganic nitrogen sources had no significant effect on soil nitrate content, and the nitrogen content of potato tubers with higher irrigation levels was lower than that of potato tubers with lower nitrogen application rates. Shang *et al.* [19] indicated that soil nitrate-nitrogen content showed a decreasing trend as fertility progressed, with the highest soil nitrate-nitrogen content at the seedling stage. Under the same conditions of nitrogen application, the nitrate-nitrogen content of the 0–40 cm soil layer increased with the increase in irrigation water, and the nitrate-nitrogen content of the 40–100 cm soil layer only showed the same trend at high nitrogen levels (300 kg/ha), but the 0–20 cm soil layer showed the same trend. The nitrate-nitrogen content of the 0–20 cm soil layer was negatively correlated, and with increasing irrigation, nitrate-nitrogen was transported from the potato root layer to the deeper layers of the soil with water, and this change in nitrogen leaching diminished as the fertility period progressed. According to Zhao *et al.*'s work [27], irrigation factors significantly influenced soil water distribution, with the upper and middle soil water contents fluctuating more than the lower and middle soil water contents, and appropriate nitrogen application expanded the radius of the effective water wetting front in the soil profile. In addition, the nitrate and ammonium nitrogen content of the soil showed a trend of increasing and then decreasing as

Table 1. Effects of coupling water and nitrogen on potato yield and utilization efficiency of water and nitrogen.

Study Area and Test Time		Nitrogen Application ($kg \cdot ha^{-1}$)	Water Consumption ($m^3 \cdot ha^{-1}$)	Yield ($t \cdot ha^{-1}$)	Water Use Efficiency ($kg \cdot m^{-3}$)	Nitrogen Partial Productivity ($kg \cdot kg^{-1}$)
Semi-Arid Area (2014–2015) [4]		180	5288.02	45.90	8.68	255.00
Arid and Semi-Arid Pastoral (1981–2010) [19]	Eastern region	30–120	2960.82–3906.25	25.70–35.00	8.96	291.67–856.00
	middle region	30–120	3082.71–4017.18	28.70–37.40	9.31	311.67–956.67
	Western region	30–120	3242.93–4115.57	27.50–34.90	8.48	290.83–916.67
Semi-Arid Area (2019) [6]		225	3869.52	35. 29	9.12	209.70
Arid Area (2013) [21]		135	4671.55	54.19	11.60	401.41
Arid Area (2019) [27]		180	3520.52	58.51	16.69	437.47

the fertility period progressed, with nitrate nitrogen generally accumulating at 0–20 cm and 40–60 cm soil depths, respectively, and ammonium nitrogen content at 0–20 cm being higher than the lower layer due to its volatility and difficulty in leaching. Wang *et al.* [28] indicated that there is a certain hysteresis in N application on soil fertility, that timely, split chasing of N is beneficial for soil nitrate N accumulation in the 0–80 cm soil layer, that multiple, small chasing of N fertilizer does not cause redundant growth of nutrient organs in the early stages of potato, nor does it affect nutrient translocation to the tubers in the late stages of greening, and that split application of N weakens the downward trend of nitrate N and N leaching, increasing potato yield and fertilizer use efficiency.

3 CONCLUSION

In recent years, China has faced important challenges to efficient potato production as water scarcity has increased and extreme weather events have occurred. Water and nitrogen are essential for potato growth and development, and precise water and nitrogen regulation is an important way to achieve high potato yields and quality, water conservation and weight loss, and a sustainable soil environment, based on the crop's water and fertilizer requirements. At the same time, the climate often varies greatly from region to region, so it is necessary to take into account the climatic conditions, soil conditions, and potato varieties at the trial site, and to combine the required cultivation patterns and agronomic practices to develop a scientific and efficient water and nitrogen control irrigation and fertilizer application system.

ACKNOWLEDGMENTS

This work was financially supported by the Industrial Support Plan Project of the Gansu Provincial Department of Education (No. 2022CYZC-51), the National Natural Science Foundation of China (No. 52269008), and the Key Research and Planning Projects of Gansu Province (No. 18YF1NA073).

REFERENCES

[1] Li F.Q., Zhang H.J., Li X., Deng H.L., Chen X.T., Liu L.T. (2022) Modelling and Evaluation of Potato Water Production Functions in a Cold and Arid Environment. *Water*, 14 (13): 2044.

[2] Peng J.W., Ding Z.L., Ding Q., Chai H.Q., Yan P.Y., Ren K.A. (2010) Effects of Soil Fertility and Nutrient Management on Rice Yield and Nitrogen Use Efficiency. *Chinese Agricultural Science Bulletin*, 26 (22): 190–195.

[3] Yu B., Yang H.Y., Wang L., Liu Y.H., Bai J.P., Wang P., Zhang J.L. (2018) Genetic Diversity Analysis and Comprehensive Evaluation of Phenotypic Traits of Introduced Potato Germplasm Resources in Arid and Semi-arid Regions. *Acta Agronomica Sinica*, 44 (01): 63–74.

[4] Hamid Reza B.F., Zohreh E.B., Barker A.V. (2022) Tuber Yield and Physiological Characteristics of Potato Under Irrigation and Fertilizer Application. *Communications in Soil Science and Plant Analysis*, 53 (11): 1432–1443.

[5] Jama-Rodzeńska A., Walczak A., Adamczewska-Sowińska K., Janik K., Pczkowski G. (2020) Influence of Variation in the Volumetric Moisture Content of the Substrate on Irrigation Efficiency in Early Potato Varieties. *PLoS ONE*, 15 (4): 0231831.

[6] Xing Y.Y., Zhang T., Jiang W.T., Li P., Shi P., Xu G.C., Cheng S.D., Cheng. Fan Y.T., Wang X.K. (2022) Effects of Irrigation and Fertilization on Different Potato Varieties Growth, Yield and Resources Use Efficiency in the Northwest China. *Agricultural Water Management*, 261.

[7] Elżbieta. Radzka, Tomasz. Lenartowicz. (2015) Rainfall Deficit and Excess Rainfall During Vegetation of Early Potatoes Varieties in Central-eastern Poland (1971-2005). *Nauka Przyroda Technologie*, 9.

[8] Schapendonk A., Spitters C., Groot P. (1989) Effects of Water Stress on Photosynthesis and Chlorophyll Fluorescence of Five Potato Cultivars. *Potato Res*, 32: 17–32.
[9] Jahanzad E., Barker A.V., Hashemi M., Sadeghpour A., Zandvakili O.R. (2016) Decomposition Rate and Release of Nitrogen from rye, Forage Radish, or Winter Peas Cover Crops Under Conventional or no-tilling Systems, C. Northeastern Branch Csa Meeting.
[10] I.Abdo. Ahmed, S.Elrys. Ahmed, K.Abdel-Fattah Mohamed, M.Desoky El-Sayed, H.T. Li, L.Q. Wang. (2020) Mitigating nitrate accumulation in potato tubers under optimum nitrogen fertilization with K-humate and calcium chloride. *Journal of Cleaner Production*, 259.
[11] Faradonbeh H.R.B., Bistgani Z.E., Barker A.V. (2022) Tuber Yield and Physiological Characteristics of Potato Under Irrigation and Fertilizer Application. *Communications in Soil Science and Plant Analysis*, 53 (11):1432–1443.
[12] Paff K., Fleisher D., Timlin D. (2022) Changes in the Effects of Water and Nitrogen Management for Potato Under Current and Future Climate Conditions in the U.S. *Computers and Electronics in Agriculture*, 197.
[13] Ju X.T & Gu B.J. (2014) Current Situation, Problems and Trends of Nitrogen Fertilizer Application in Farmland in China, *J. Plant Nutrition and Fertilizer Science*, 20 (04): 783–795.
[14] Wang C. (2020) Effects of Water and Nitrogen Regulation on Soil Enzyme Activities of Potato and its Response to Growth, Yield and Quality. Ningxia University.
[15] Wang X.G., Cai M., Wu N., Liu J.L., Ma Z.H. (2020) Effects of Intercropping and Nitrogen Application on the Physiological Characteristics and Quality of Potato in Ningnan arid Region. *Research on Agriculture in arid areas*, 40 (04): 69–76 + 98.
[16] Wang M.Q. (2019) Effect of Water-fertilizer Interaction on Physiological Characteristics, Growth Potential and Rhizospheric Soil Enzyme Activity of Potato, D. Yan'an University.
[17] Yan W.Y., Qin J.H., Duan S.G., Xu J.F., Jian Y.Q., Jing L.P., Li G.C. (2022) Effects of Water-nitrogen Coupling on Photosynthetic Properties, Tuber Formation, and Quality in Potato. *Journal of Horticulture*, 49 (07): 1491–1504.
[18] Zhang H.L., Smeal Dan., Arnold R.N., Gregory E.J. (1996) Potato Nitrogen Management by Monitoring Petiole Level, *J. Journal of Plant Nutrition*, 19 (10): 1405–1412.
[19] Shang M.X., Fang Z.G., Liang B., Wang M., Li J.L. (2019) Effects of Different Water and Nitrogen Treatments on Potato Yield and Quality and Soil Nitrate Transport Under Drip Irrigation. *Journal of North China Agriculture*, 34 (06): 118–125.
[20] Mai J.L, Yang C.L., Mi Z.M. (2016) Effect of Water and Nitrogen on Soil Moisture and Potato Yield in Potato Fields. *Water-saving irrigation*, (11): 28–31+35.
[21] Song N., Wang F.G., Yang C.F., Yang K.J. (2013) Effect of Water-nitrogen Coupling on the Yield, Quality and Water Utilization of Potato Under Membrane Drip Irrigation. *Journal of Agricultural Engineering*, 29 (13): 98–105.
[22] Hargert G.W., Frank K.D., Rehm G.W. (1978) Anhydrous Ammonia and N-serve for Irrigated Corn, *J. University of Nebraska-Lincoln Agronomy Dept Soil Sci Res Rep*, 31–34.
[23] Mohammad M.J., Zuraiqi S., Quasmeh W., Papadopoulos I. (1999) Yield Response and Nitrogen Utilization Efficiency by Drip-irrigated Potato. *Nutrient Cycling in Agroecosystems*, 54 (3), 243–249.
[24] Jiao F., Wu J., Yu L., Zhai R. (2013) ^{15}N tracer Technique Analysis of the Absorption and Utilisation of Nitrogen Fertiliser by Potatoes. *Nutrient Cycling in Agroecosystems*, 95 (3), 345–351.
[25] Tang J.Z., Xiao D.P., Wang J., Fang Q., Zhang J., Bai H. (2021) Optimizing Water and Nitrogen Managements for Potato Production in the Agro-pastoral Ecotone in North China. *Agricultural Water Management*, 253 (6):106945.
[26] Carter J.N. & Bosma S.M. (1974) Effect of Fertilizer and Irrigation on Nitrate-Nitrogen and Total Nitrogen in Potato Tubers. *Agronomy Journal*, 66 (2): 263–266.
[27] Zhao Y.B. (2019) Effects of Different Water Nitrogen Treatments on Potato Growth and the Distribution of Soil Water Nitrogen Transport, D. Ningxia University.
[28] Wang X.G. (2019) Effect of Nitrogen Chase Period and Frequency on Potato Growth and Soil Nitrogen Transport Under Film Drip Irrigation, D. Agricultural University of The Inner Mongol.

Water Conservancy and Civil Construction – Oke & Ahmad (Eds)
© 2024 The Author(s), ISBN: 978-1-032-58614-4

Application of blockchain in the quality control of concrete production in hydraulic engineering

Ciyin Chen
State Key Laboratory of Eco-hydraulics in Northwest Arid Region of China, Xi'an University of Technology, Xi'an, China

Peng Dong & Xiaofeng Song
Hanjiang-to-Weihe River Valley Water Diversion Project Construction Co. Ltd., Shaanxi Province, China

Yufei Zhao*
Department of Geotechnical Engineering, China Institute of Water Resources and Hydropower Research, Beijing, China

ABSTRACT: Currently, concrete materials are widely utilized in civil engineering and hydraulic engineering. To improve the many problems faced by concrete quality control in current hydraulic engineering, this paper achieves data effective aggregation and encrypted uploading and explores the feasibility of archival data information management by using the advantages of blockchain technology such as anti-tampering and traceability for both concrete raw material testing and finished concrete quality testing. It has achieved good application results in the Huangjinxia water project, which has a wide application prospect. It provides a feasible path for blockchain technology to solve similar file management work in hydraulic engineering.

1 INTRODUCTION

With the development of modernization in China, the number of infrastructure projects is increasing, especially the construction of some large projects such as hydraulic engineering and civil engineering, and the demand for concrete materials has increased. Concrete is a composite material in which multiple aggregates are cemented together. As the main construction material in infrastructure projects, it has the advantages of low cost, easy preparation, and simple structure. Although the emergence of new building materials and theories has caused some impact on concrete materials, it seems that concrete materials are still the most widely used building materials in the world. Different from civil engineering, hydraulic engineering is usually built in rugged mountainous terrain, and all of them are large-scale projects. They are large-scale, many types of hydraulic structures, damage with a catastrophic, wide range of impact, and other characteristics. Therefore, hydraulic engineering construction should be fully considered for its safety, so the quality of concrete is particularly critical. Because of the importance of concrete to hydraulic engineering, the quality control of concrete production has become a very important segment. Concrete production quality control is done through numerous forms to record some key quality data in real-time during the production process for subsequent quality certification and

*Corresponding Author: zhaoyf@iwhr.com

DOI: 10.1201/9781003450818-11

traceability. Traditional forms are usually paper-based and will face problems such as high resource consumption, storage access difficulties, and inconvenient transmission. That's why electronic forms are the future trend of archives. However, how to solve the current technical defects of centralization in electronic archives and achieve the anti-tampering of data is a problem that electronic archives need to solve urgently.

Since the emergence of Bitcoin designed by Satoshi Nakamoto in 2008, the global financial industry has entered the blockchain era first with Bitcoin as its representative. The essence of blockchain is a continuously growing list of records in blocks. Each block stores block information (block header) and transaction information (block body) for a certain period. The block header includes the hash value of the previous block to form a block-chain data structure. Because of its "block and chain" form, it is called blockchain (Nakamoto 2008), and its structure figure is shown in Figure 1. Blockchain is not a single technology, but a technology integration that combines encryption algorithm, timestamp, peer-to-peer network, consensus mechanism, smart contract, and other technologies to achieve decentralization of transactions and distributed storage of data (He 2017). Open degrees from low to high can be divided into a private blockchain, consortium blockchain, and public blockchain. With the further development of technology, the application of blockchain has gradually expanded from the financial industry to other industries and has been recognized globally. Due to the advantages of blockchain such as immutability and traceability, it has wide applications in logistics (Yu 2017; Zhang 2020), medicine (Wang 2019; Xue 2017; Zhang 2019), and other industries. General Secretary Xi made an important speech during the collective study of the Political Bureau of the Central Committee of the Chinese Communist Party on October 24, 2019. He proposed to strongly develop the independent innovation and application of blockchain from the national level. So far, blockchain has risen to a national strategy. As blockchain technology continues to mature, it has created opportunities for its application in the field of hydraulic engineering. Construction management architecture of hydraulic engineering based on blockchain was proposed by Hu Jing et al (Hu 2020), which provides a new idea for the application of blockchain in hydraulic engineering. Chen Genfa (Chen 2020) *et al.* solved the challenges in trust, management, and information interaction in water supply, water use, water rights trading, and government services in China by using blockchain. Chen Zuyu (Chen 2021) *et al.* explored the feasibility of securing electronic form entry and long-term preservation through blockchain based on

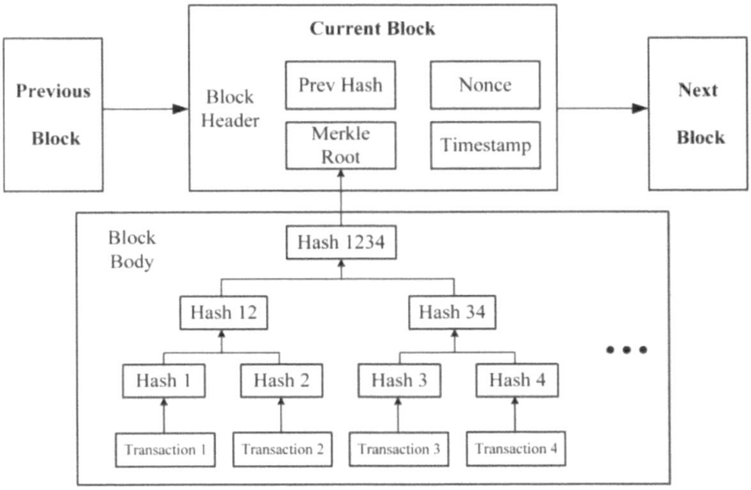

Figure 1. Block structure figure.

concrete production files. Jia Ningxiao (Jia 2021) *et al.* studied the application of blockchain in the field of intelligent construction and management of hydraulic engineering from the principle of blockchain technology and combined it with the characteristics of hydraulic engineering construction management.

The data generated during the production of concrete in hydraulic engineering is directly related to the security of hydraulic engineering, and these data will face serious security problems if they are circulated in the network in the form of informatization. According to the previous section, a Distributed Database System can be constructed using decentralization and trustless blockchain, and the data can be stored by encryption algorithms to ensure information security. It is a feasible scheme for the quality control of concrete production.

2 CONCRETE PRODUCTION QUALITY CONTROL POINTS

Concrete production quality control means the process supervision of concrete production in the type of form records, through various quality indicators testing and construction management to ensure that the final finished concrete meets the specified quality requirements. It can be divided into two separate categories according to the production process: concrete raw material testing and finished concrete quality testing.

2.1 *Concrete raw material testing*

Concrete raw materials generally include cement, fly ash, admixture, coarse aggregate, fine aggregate, etc. Its quality control according to different testing companies includes:

(1) Testing by raw materials manufacturers
The raw materials used in concrete production are generally provided by different raw material manufacturers, such as cement plants, fly ash plants, admixture plants, sand aggregate plants, etc. Raw material manufacturers need to strictly control the production process in the production of raw materials to ensure the production quality of raw materials and performance testing on the finished product of raw materials. Product testing is completed before delivery to customers, and quality documents such as the factory certificate of conformity of raw materials are issued.

(2) Testing by mixing plant
The mixing plant for incoming preliminary testing and laboratory performance testing is studied. The raw materials are sent to the mixing plant, and the staff of the mixing plant will check the quantity and appearance quality according to the raw material information issued by the raw material manufacturers, and the qualified raw materials can be put into the corresponding raw material storage room. The safety management of raw material storage room has corresponding system and specification requirements, such as various concrete raw materials should be stored separately according to varieties and quantities, and by their respective storage characteristics. Therefore, it is necessary to carry out preliminary checks such as quantity verification and appearance acceptance of raw materials entering the storage room, including species specifications, factory quality reports, product quantity, etc. After meeting the requirements, they are stored in the corresponding raw material warehouse, and samples are taken and sent for further detailed inspection to determine whether the quality indicators of the raw materials meet the actual production requirements.

The mixing plant laboratory staff samples raw materials for testing, by the specified sampling quantity and sampling range for sampling, while by the testing methods of different raw materials for their respective performance testing. The final raw material test report is obtained, and the raw material is judged to be qualified according to the specification requirements.

(3) Testing by third-party
The supervising company needs to sample raw materials for testing. Its testing is generally entrusted to a third-party testing center. The supervising company takes a sample from the raw material storage room of the mixing plant, entrusts the third-party testing center to carry out raw material performance testing, and judges whether the raw materials are qualified according to the raw material entrusted testing report issued by the third-party testing center.

2.2 *Finished concrete quality testing*

After the production of concrete is completed, its performance needs to be checked promptly, especially the compressive strength of concrete. Similarly, its quality testing according to different testing companies includes:

(1) Testing by mixing plant
A mixing plant, as a concrete production company, needs to be responsible for the quality of concrete produced. The concrete will be sampled at the mixer outlet after production and taken to the laboratory of the mixing plant for the slump, temperature, and other measurements. The concrete blocks are made and sent to the curing chamber for concrete curing, within a specified period for testing of compressive strength and other performance indexes. The concrete quality test report is issued to determine whether the concrete material is quality conformance.
(2) Testing by construction companies
The construction company is the unit that uses concrete. It conducts sampling and testing at both the concrete mixer outlet and the concrete pouring field to ensure that the concrete put into the pouring is quality conformance. The laboratory of the construction company issues concrete quality testing reports by the concrete testing process and relevant regulations to determine whether the concrete material is quality conformance.
(3) Testing by third-party
The supervising company also needs to sample the concrete for testing, which is usually entrusted to a third-party testing center. The supervisory company takes samples from the concrete mixer outlet and the concrete pouring field, and the third-party testing center is entrusted to conduct concrete performance testing. The concrete entrusted testing report issued by the third-party testing center determines whether the concrete material is quality conformance.

3 BLOCKCHAIN STRUCTURE

3.1 *Encryption algorithm*

The encryption algorithm is a very crucial link in blockchain. Using one encryption algorithm alone can have certain drawbacks. Therefore, blockchain combines multiple encryption algorithms to improve its security by combining their advantages. In concrete production, the quality control forms are stored in the form of information technology and the data is encrypted by a series of encryption algorithms, making data tampering only theoretically feasible.

(1) Hash Function
The hash function is a common one-way encryption algorithm that encrypts an input of arbitrary length to generate a fixed-length string, which is the hash value. Hash function usually has the following properties.
– Certainty: Always get the same hash value for a specific input;
– Unidirectionality: Impossible to find the input value for a given hash value in theory

- Randomness: A small change in the input value results in a huge change in the hash value;
- Anti-collision: There does not exist a fast and efficient way to find different input values but get the same hash value.

There are two main types of Hash functions, MDx and SHA. MDx includes MD4, MD5, etc. MD5 is improved by MD4 and has higher security. Table 1 shows the randomness of the MD5 function. By simply changing the input value, the output value becomes completely different. However, MD5 was confirmed unable to prevent collision attacks by China's academician Wang Xiaoyun (Wang 2005) in 2004, and the first example was given, so it is no longer applicable to security authentication. The main members of the SHA function are SHA-1, SHA-2, etc., which are widely used in data encryption and digital signature, and the well-known SHA-256 is one kind of SHA-2. Until February 2017, when the Dutch cryptography research group CWI and Google released the first instance of an SHA-1 hash collision, officially announcing that the SHA-1 function had been breached. On this basis, Liu Kun (Liu 2017) *et al.* analyzed its collision cause in depth to improve the anti-collision of the SHA-1 function.

Table 1. Example of MD5 function calculation.

Input	Output
Blockchain	3cc377f79bda308c750459a2caf7fc38
Blockchain	5510a843bc1b7acb9507a5f71de51b98

1. Symmetrical Encryption

   Symmetric encryption is the most basic encryption algorithm in cryptography. In symmetric encryption, the same key is used for encryption and decryption, which requires the sender and receiver to agree on a key privately before they can communicate securely. Such an encryption method has the advantages of an open algorithm, small calculation, fast encryption speed, and high encryption efficiency. However, it also faces many problems. The key confidentiality is especially important in symmetric encryption. It means the information is leaked if the key is lost, even if the key can be solved by directly calculating. In World War II, Germany used the most advanced symmetric encryption machine (Enigma) for encryption at that time, but it was still broken by the Turing machine researched by Turing, and even changed the course of World War II and the world development as a result. Therefore, it is necessary to change the key frequently to ensure the privacy of the key and thus secure the information.
2. Asymmetric Encryption

   Asymmetric encryption is the cornerstone of modern computer communication security. The public key and private key are used for encryption and decryption respectively, and the public key corresponds to the private key one by one. If the data is encrypted with the public key, only the corresponding private key can be decrypted, and vice versa, to achieve data encryption and digital signature. Asymmetric encryption, in contrast to symmetric encryption, the communicator does not need to disclose the private key, which is a good way to avoid the security of the key during transmission. But its complex encryption structure makes encryption and decryption slower. It can resist all cryptographic attacks known so far and is recommended by the International Organization for Standardization as the standard for public key data encryption (Zhang 2018).

3.2 *Merkle tree (aggregation structure of related statements)*

Merkle tree (Merkle 1980, 1987) is a tree data structure, also called a hash tree, proposed by Ralph Merkle in 1980, as shown in Figure 2. Merkle tree is the basic component of blockchain, which is mostly used in file systems and Peer-to-Peer transaction systems before the emergence of blockchain, and its generally binary tree. The Merkle tree is constructed from the bottom up. For every transaction, the hash operation is performed, and the computed hash value constitutes a leaf node of the Merkle tree. Based on this, two adjacent leaf nodes are aggregated, and the aggregated string generates a parent node by the hash operation again. Continue similar operations until only one root node, the Merkle root, remains at the top. From the structure of the Merkle tree, it is easy to see that a tampered transaction at any leaf node will result in a change in the Merkle root. Therefore, whether the hash value of the root node changes or not can be used as a criterion to assess whether a set of transactions has been tampered with. When the hash value of the root node is found to have changed, the child nodes whose hash values have changed are searched downwards for layer-by-layer traceability, and the tampered transactions are finally found. Because of its special structural form, traceability becomes very fast. Only seventeen verifications are needed for one hundred thousand transactions, which is extremely friendly for handling large amounts of data.

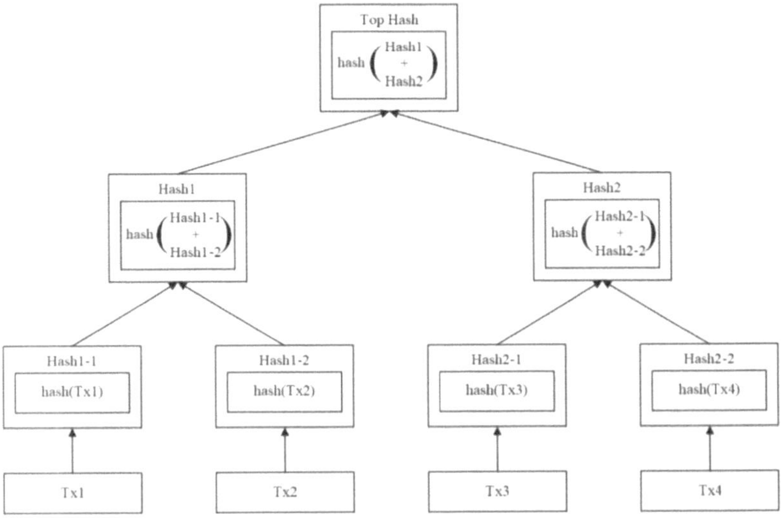

Figure 2. Merkle tree figure.

Theoretically, the Merkle tree is not a necessary component of blockchain, and the same effect can still be achieved by simply aggregating all the transactions simultaneously for hash operations. But this certainly makes traceability extremely difficult. With the data getting so large that it needs to be verified day and night, the adoption of the Merkle tree data structure was able to greatly reduce the amount of work required for traceability, which is what made the promotion of blockchain possible.

In concrete production, each form is a transaction. The forms are aggregated two by two after a hash operation in chronological order to regenerate a new hash value. The cycle goes on and on, and eventually, the unique string is obtained, which is the Merkle root hash. This hash value can easily determine whether the form data has been tampered with, which makes the security and traceability of the form well improved.

3.3 *Consensus structure (cochain and storage)*

In the blockchain, a decentralized distributed network structure, consensus mechanisms are needed to ensure that nodes can effectively agree on data without relying on trusted third parties. The consensus mechanism is the core component of blockchain, which is the norm and protocol based on consensus. Nodes can spontaneously and securely complete data exchange and verification without human intervention to achieve consistency of data from different nodes (Yu 2017). Consensus mechanisms are broadly classified into Byzantine Fault Tolerance and Crash Fault Tolerance based on whether malicious nodes are considered or not. The Crash Fault Tolerance consensus algorithm is mainly applied to the alliance blockchain and private blockchain where nodes trust each other, and its representative algorithms are Paxos, Raft, etc.; the Byzantine Fault Tolerance consensus algorithm is also applied to the public blockchain with complex nodes because it considers malicious node attacks, and the more applied algorithms are PoW, PBFT, etc. At present, there are a wide variety of mainstream consensus algorithms, which should reduce resource consumption and simplify the consensus process as much as possible under the premise of ensuring security.

The production process of concrete generally involves the production department, supervision department, testing department, and so on. The departments are close to each other and know each other, and the production of concrete is carried out according to a fixed process, which usually does not have malicious nodes. Therefore, a weakly-centralized alliance blockchain system can be used. In the choice of consensus algorithm, the Crash Fault Tolerance consensus algorithm with simple implementation and high speed is preferred. If the node is considered to become a malicious node by the attack, the algorithm can be improved and upgraded or the Byzantine Fault Tolerance algorithm can be directly selected.

The transactions agreed by the consensus algorithm will be recorded in blocks by cochain and stored in a distributed shared database. But the transactions already cochain cannot be changed again. For actual concrete production, it is neither practical nor necessary. Mistakes and omissions often occur in the actual form filling. It is possible to set up a summary and errata link before the cochain operation so that the cochain operation can be carried out again after the errata is completed to minimize the loss caused by human factors.

3.4 *Integrity verification, traceability*

Integrity verification is to check the transactions that have been cochain, determine whether the form data has changed, perform traceability operations and locate the specific form that has changed if the verification does not pass. The verification method adopts the Merkle tree structure, and the flow chart of verification is shown in Figure 3. The specific process of verification is as follows.

The Merkle root is computed for the individual block using the same hash algorithm and rule as when it is cochain, and the result of the computation is compared with the Merkle root of the corresponding block stored in the distributed shared database. If the result is the same, it means that all data in the block where the root node is located has not changed and the verification process is finished. On the contrary, it means that the data has changed, and further verification is needed to locate the changed form.

The hash value of the next-level child node is extracted and compared with the hash value of the corresponding node in the database. The nodes with different results are taken out separately and verified at the next level (it is not excluded that multiple nodes have different results in comparison, which indicates that there are multiple forms with changed data, and then they need to be verified separately). Loop like this, the loop ends if and only if the node is a leaf node. At this point, it indicates that the form data has changed, an alarm is issued, and the traceability process is completed. When integrality verification is completed for all existing blocks, the full process is finished.

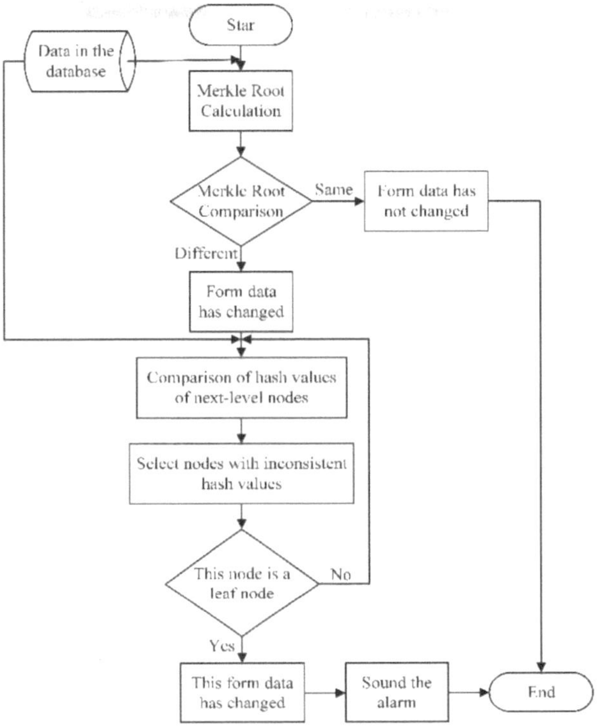

Figure 3. Integrity verification process figure.

4 APPLICATION OF BLOCKCHAIN IN HUANGJINXIA WATER CONTROL PROJECT

4.1 *Engineering background*

Hanjiang-to-Weihe River Valley Water Diversion Project consists of a water transfer project and a water diversion project which is one of the major national hydraulic engineerings during the 14th Five-Year Plan period. Huangjinxia Water Control Project is an important part of it, located in the mainstream of Han River. Its main project is a concrete gravity dam whose maximum dam height is 63m, with a 221 million m^3 total storage capacity. A large number of concrete materials was used in the construction process of the Huangjinxia Water Control Project, totaling about 1.3 million m^3. To achieve information supervision and quality control of the whole process of concrete production, the China Institute of Water Resources and Hydropower Research and Hanjiang-to-Weihe River Valley Water Diversion Project Construction Co. Ltd. jointly established the project "Development of Concrete Production Information Management System Based on Blockchain Technology for Huangjinxia Water Control Project of Hanjiang-to-Weihe River Valley Water Diversion Project". The project is based on the construction of the Huangjinxia Dam and combined with blockchain technology to achieve the whole process management of the concrete production process from raw materials to concrete production to concrete pouring, by using blockchain technology combined with electronic signature technology, to realize the safe transmission of form information and process management of concrete production. And the project acceptance was completed in February 2022.

4.2 *System completion*

At present, the Huangjinxia blockchain concrete production management system has been developed and applied for national software copyright, and the general functions include the whole process supervision of concrete production and the combination of blockchain technology in two parts. In total, the system has recorded 18,000 m^3 of actual concrete production data (about 3,500 electronic forms) from the end of November to the end of December 2020. It combines blockchain technology to achieve aggregation and cochain of forms, and regularly performs verification and traceability operations on the uploaded data to check whether the data has changed. Thus, it ensures the safety and reliability of data and improves the quality of concrete production.

After the system was completed, the National Center of Supervision & Inspection on Software Products Quality was commissioned to conduct software testing in November 2021. According to the national standard GB/T 25000.51-2016, the test was conducted including both user documentation and functionality. The test results showed that the Huangjinxia blockchain concrete production management system complied with the national standard, and two test reports on software testing and software defects were issued. The software defect reports proposed five S3-level (found problems affecting the correct implementation of the tested functions) software defects, which have been revised and verified.

5 CONCLUSIONS

Blockchain is well-known with the development of Bitcoin and is gradually becoming an important information technology to ensure the authenticity and security of data in national production and life. In this study, the application research of blockchain is conducted with the problem of concrete production quality control in hydraulic engineering. It proposes to choose the blockchain form of alliance chain, which combines various encryption algorithms and data storage structure of Merkle tree and sorts out the integrity checking process. It was practically applied in Huangjinxia Water Control Project and achieved good application results. The application of blockchain can effectively aggregate the information produced by different stages and different units in the concrete production process so far; it achieves the block structure design and data cochain according to the construction progress of the separated item project, which ensures the strict mapping relationship of the data; through the cochain, guarantees the safety and reliability of the data, and also provides an important technical mean for the rapid verification and positioning of data tampering.

This study is only a pilot in concrete production, but blockchain technology will not be limited to this. Because of its suitability with engineering files, it has a better universality in hydraulic engineering. The informatization of hydraulic engineering is the current problem faced, and blockchain is used to provide security for the information system, thus promoting the informatization process of hydraulic engineering.

ACKNOWLEDGMENT

This study was supported by the scientific research project of Hanjiang-to-Weihe River Valley Water Diversion Project Construction Co. Ltd. of Shaanxi Province (YHJW-D-71) and the China Institute of Water Resources and Hydropower Research Three Types of Talents Fund Project (GE0145B042022).

REFERENCES

Chen G. F. & NI H. Z. (2020). Prospect on the Application of Blockchain Technique in Water resource Management. *Water Resources and Hydropower Engineering*, 51(12):47–54.

Chen Z. Y. & Lei P. & Su Y. *et al.* (2021). Blockchain Supported Archive Management for Concrete Production: Framework and Safety Appraisals. *China Civil Engineering Journal*. 54(09):105–114.

He P. & YU G. & Zhang Y.F. *et al.* (2017). Survey on Blockchain Technology and its Application Prospect. *Computer Science*. 44(4), 1–7.

Hu J. & Chen Z. Y. & Wang Y. J. *et al.* (2020). The Architecture of the Construction Management Platform for Hydraulic Engineering is Based on Blockchain. *Journal of Hydroelectric Engineering*. 39(11), 40–48.

Jia N. X. & Feng M. & Huang B. H. (2021). Smart Construction and Management of Hydropower Projects Based on Blockchain Technology. *Yangtze River*. 52(S2):312–315.

Liu K. & Yang Z. X. (2017). Design and Research of the Improved SHA-1 Algorithm Based on the Local Collision Algorithm. *Software Engineering*. 20(11):27–29.

Merkle R. C. (1987). A Digital Signature Based on a Conventional Encryption Function. *In Conference on the Theory and Application of Cryptographic Techniques*. 369–378.

Merkle R. C. (2019). Protocols for Public Key Cryptosystems. *In Secure Communications and Asymmetric Cryptosystems*. 73–104.

Nakamoto S. (2008). Bitcoin: A Peer-to-Peer Electronic Cash System. *Decentralized Business Review*. 21260.

Wang R. J. & Yu S. Z. & Li Y. *et al.* (2019). Medical Blockchain of Privacy Data Sharing Model Based on Ring Signature. *Journal of the University of Electronic Science and Technology of China*. 48(6), 886–892.

Wang X. Y. & Yu H. (2005). How to break MD5 and Other Hash Functions. *In the Annual International Conference on the Theory and Applications of Cryptographic Techniques*. 19–35.

Xue T. F. &Fu Q. C. & Wang, C. *et al.* (2017). A Medical Data Sharing Model Via Blockchain. *Acta Automatica Sinica*. 43(9), 1555–1562.

Yu L. N. & ZHANG G.F. & JIA J.D. *et al.* (2017). Modern Agricultural Product Supply Chain Based on Blockchain Technology. *Transactions of the Chinese Society for Agricultural Machinery*. 48, 387–393.

Zhang, C. & Li, Q. & Chen, Z. H. *et al.* (2019). Medical Chain: Alliance Medical Blockchain System. *Acta Automatica Sinica*. 45(8), 1495–1510.

Zhang S. & Ye J. & Li G. (2020). Research and Implementation of Blockchain Technology Scheme for Cold chain Logistics. *Computer Engineering and Applications*. 56(3), 19–27.

Zhang Q. H. & Cao J. & Cao X. X. *et al.* (2018). Design and Implementation of an Asynchronous Low Power RSA Circuit Structure. *Acta Scientiarum Naturalium Universitatis Pekinensis*. 54(6), 1351–1354.

Water Conservancy and Civil Construction – Oke & Ahmad (Eds)
© 2024 The Author(s), ISBN: 978-1-032-58614-4

Response of potato growth, yield and quality to water deficit: A review

Xuan Li
College of Agronomy and Agricultural Engineering, Liaocheng University, Liaocheng, China
College of Water Conservancy and Hydropower Engineering, Gansu Agricultural University, Lanzhou, China

Shouchao Yu
College of Agronomy and Agricultural Engineering, Liaocheng University, Liaocheng, China

Hengjia Zhang*
College of Agronomy and Agricultural Engineering, Liaocheng University, Liaocheng, China
College of Water Conservancy and Hydropower Engineering, Gansu Agricultural University, Lanzhou, China

ABSTRACT: Deficit irrigation technology is crucial in the context of global warming because it helps farms achieve carbon neutrality by lowering the emissions and uptake of some essential greenhouse gases. This paper summarizes the current state of research on deficit irrigation technology for potatoes, focuses on the effects of deficit irrigation technology on its growth characteristics, yield, and water use efficiency and quality, and explains the fundamental workings of deficit irrigation technology systems to improve yield and quality. Finally, the shortcomings and drawbacks of the current development, demonstration, and promotion of deficit regulation irrigation technology are also analyzed, and the combination of deficit regulation irrigation technology with contemporary information technology is proposed, which will be an important idea for future water-saving agriculture and intelligent development, to provide theoretical support for sustainable development and system management of dryland agriculture.

1 INTRODUCTION

Water resources are a crucial vital resource that is crucial to human survival and growth. The supply-demand gap for water resources is growing increasingly pronounced as a result of economic globalization and rapid population expansion, and the issue of food security brought on by this gap has risen to the forefront of discussion in today's society. China is one of the 13 "water-poor countries," and its total freshwater resources are less than 2300 m^3, which is only 1/4 of the world's per capita freshwater resources (Huang 2007). China is a significant agricultural nation, and as such, agricultural water use makes up more than 80% of all national water use. Irrigation water use is also increasing quickly, although water use efficiency is low (Zheng & Yong 2022). Inadequate irrigation water supplies and infrequent rainfall, particularly in desert regions, severely limit agricultural development and obstruct long-term social growth. Therefore, it is vital to create water-saving agriculture.

*Corresponding Author: 596088683@qq.com

DOI: 10.1201/9781003450818-12

Because of its high nutritional value, high yield, and wide planting range, the potato, an annual herb, is one of China's four largest staple foods. It is grown in large quantities worldwide, with an annual output of about 18 million tons, and with the maturation of potato processing technology and planting structure optimization, has gradually changed from a food crop to a cash crop (FAO 2022; China Statistics Press 2020). During their growing phase, potatoes demand a lot of water and have high soil moisture requirements. Potato development and growth, as well as the transverse and longitudinal diameter of tubers, are directly impacted by water availability (Ma *et al.* 2005). Therefore, effective field management can result in greater financial gains. People expect higher-quality potatoes as their quality of living rises. Researchers both at home and abroad are very interested in finding the best way to cultivate potatoes in the arid northwest while using the least amount of water possible to assure the highest yield and quality (Huang *et al.* 2016; Liu *et al.* 2014; Reid & Gillespie 2017; Volschenk 2021). The research on potatoes grown under loss-regulating irrigation circumstances is reviewed in this publication along with its current status.

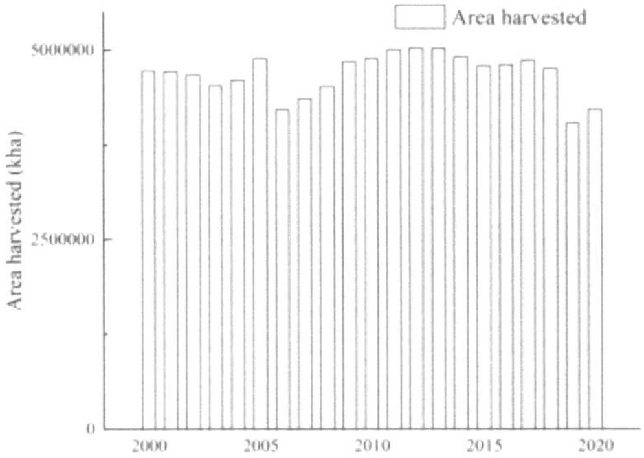

Figure 1. Area harvested of potato.

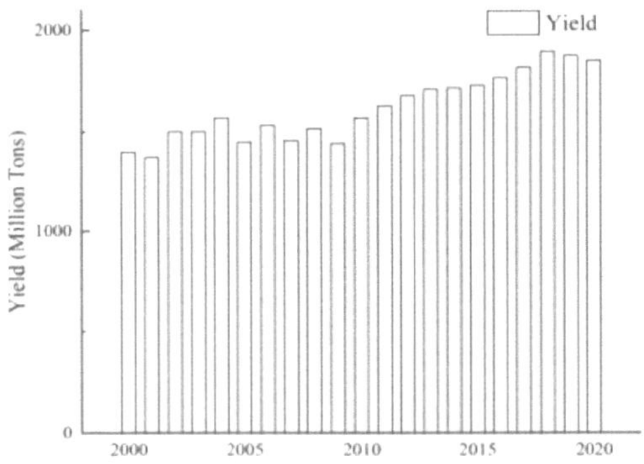

Figure 2. Yield of potato.

2 STATUS OF RDI MECHANISM RESEARCH

By regulating soil moisture during a specific reproductive period, crops are subjected to a certain level of water stress and are then treated with rehydration at a later reproductive stage. As a result, the treated crops not only improve their drought resistance at a later stage but also affect the redistribution of photosynthetic products to different tissues and organs and decrease the growth redundancy of nutrient organs, achieving the goal of water conservation and yield. When there is a water regulation shortage during a specific reproductive cycle, the crop uses its regulation to shift nutrients from other organs to the stressed organs. When crops are under water stress, their bodies experience physiological, biochemical, and other changes. By altering soil moisture promptly, circumstances can be created for activities like enhancing crop yield and quality and streamlining water consumption (Cai *et al.* 2004).

Under deficit irrigation, roots are the primary organ for communicating water stress information, and according to the deficit irrigation theory, roots are crucial to increasing the effectiveness of water consumption during crop reproduction. According to the root-crown functional balance theory, a crop's root and crown are mutually dependent and competing with one another. The crop can automatically transfer nutrients from other parts to the neediest organs through its regulation, minimizing the degree of damage to the crop itself, when the microclimate is in a turbulent state and the relationship between roots and crowns changes from interdependence to competition. The crop will redistribute the ratio of photosynthetic products between the roots and the crown when the roots sense a water deficit, and the roots will receive more of these products, which is better for the growth of the roots. However, the transfer of the generalized crop causes the growth of the crown to be inhibited, which reduces the leaf area and causes the crop to transpire less water, resulting in a decrease in water demand (Blackman, Davies 1985). Studies have shown that when deficit-regulating irrigation is used, the soil experiences a water deficit, the crop roots struggle to absorb water, and the supply of water to the above-ground parts is insufficient. This results in a decrease in the relative water content and water potential of the crop leaves and the production of a substance that can regulate the degree of leaf stomatal opening and closing, affecting physiological processes like photosynthesis and translocation (Chalmers *et al.* 1984; Cheng *et al.* 2003; Mills *et al.* 1996).

Deficit regulation irrigation has altered the distribution of photosynthetic products, which has promoted plant reproductive growth and improved fruit quality. Fruit Brix and storability have also improved. In recent years, research on the work of deficit regulation irrigation has shifted from the enhancement of crop yield to the improvement of crop quality. Numerous studies have demonstrated that fruits under deficit-regulated irrigation contain considerably more soluble sugars and organic acids than fruits under normal irrigation, increasing fruit quality. The ability of the fruit to compete for nutrients and water was improved, and the pace of fruit development after rehydration was accelerated by soluble solids, K^+, and organic acids in the fruit (Chang 2007; Failla *et al.* 1992; Khalili & Nejatzadeh 2001; Wang 2007; Yuan *et al.* 2007).

3 RESEARCH PROGRESS OF POTATO SUBJECTED TO RDI

Al-Mehmdy discovered that the use of surface drip irrigation at soil water availability depletion percent of 25% and 50% and partial soil surface drying reduced water waste and enhanced water-use efficiency without affecting potato growth and production (Al-Mehmdy *et al.* 2019). Wagg discovered that while short-term water deficit had a minor favorable impact on potato nutritional development, tuber yield, and quality (Wagg *et al.* 2021), it had overall detrimental impacts on potato nutritional growth and tuber yield and quality. According to the Kammoun study, under moderate water supply conditions, hybrid line plants had superior tuber quality than commercial line plants. Under a complete irrigation scheme, CN1 tubers had the highest levels of dry matter, starch, protein, and lipids. The starch, protein, and lipid contents of the various plant lines were somewhat affected by a 50% reduction in irrigation, but the CN1-treated tubers still contained more starch and reducing sugars than the other treatments (Karakas *et al.* 2021).

In contrast to quality indicators including starch, β-carotene, ascorbic acid, protein, and sucrose, Karakas discovered that both potatoes were severely affected by water scarcity to diminish growth indicators (Kammoun *et al.* 2018). The growth, yield, quality, and water productivity of potatoes were found to be significantly impacted by irrigation, fertilizer application, and potato variety in a study by Xing *et al.* The W2F3V1 treatment had the highest PFP, starch content, reducing sugar content, and vitamin C content (VC), while the IWUE and polyphenol oxidase activity (POA) were the most affected by irrigation. The W3F2V1 treatment had the highest IWUE and vitamin C levels as well as slightly greater starch content and POA (Xing *et al.* 2022).

4 PERSPECTIVES AND PROBLEMS

Deficit regulation irrigation technology is a low-cost, highly effective, and water-saving irrigation technology that has had a significant positive impact on both the economic and ecological spheres after being introduced. As a result, it has promising application prospects.

It aids in the efficient use of water resources for accurate irrigation and novel water-saving agricultural practices in contemporary farming. However, there are still certain issues with potato deficit-regulating irrigation technologies that require further study. (1) To achieve high yields and high quality to enhance economic efficiency under the presumption of obtaining efficient use of water and fertilizer, water and fertilizer saving effect strengthen the potato water-fertilizer coupling and water-heat coupling research. (2) Due to a scarcity of irrigation water resources in the area, it is possible to investigate a reasonable loss-adjusting irrigation theory by using brackish water, home sewage, or industrial effluent for irrigation. (3) Reducing manual input and enhancing management effectiveness through the integration of science, technology, and potato planting management, real-time monitoring of its growth dynamics, environmental conditions, and other elements are studied.

5 CONCLUSION

The wide planting area and great economic benefits of potatoes can significantly boost farmers' revenue. Numerous studies have demonstrated that potato tuber quality and water use efficiency may be efficiently increased without hurting output when using deficit adjustment irrigation technology. In conclusion, deficit adjustment irrigation in potato farming contributes to bettering fruit yield and quality, increasing the effectiveness of water resource allocation, and more. It also proposes an operational and simple-to-promote deficit adjustment irrigation technology system and forms a new model of intelligent, sustainable, and green agriculture by fusing the model with contemporary information technology.

ACKNOWLEDGMENTS

This work was financially supported by the National Natural Science Foundation of China (No. 52269008), the Industrial Support Plan Project of the Gansu Provincial Department of Education (No. 2022CYZC-51), and the Key Research and Planning Projects of Gansu Province (No. 18YF1NA073).

REFERENCES

Al-Mehmdy, S. M. H., Aldulaimy, S. E. H., & Aljanabi, M. A. A. Impact of Surface Drip Irrigation Manners and Allowed Moisture Ddepletion Percent on Potato Growth and Yield. *Int. J. Veg. Sci.*, 25(5), 503–510. (2019).

Blackman, P. G., & Davies, W. J. Root to Shoot Communication in Maize Plants of the Effects of Soil Drying. *J. Exp. Bot.*, 1: 39–48. (1985).

Cai D. X., Shen N. Z., Cui Z. C. Research Advance of Effects of Regulated Deficit Irrigation on Physiological and Ecological Characteristics of Crops. *J. Northeast Agric. Univ.*, 02: 239–243. (2004).

Chalmers, D. J., Mitchell, P. D., & Jerie, P. H. The Physiology of Growth Control of Peach and Pear Trees Using Reduced Irrigation. *Acta Hortic.*, 146: 143–150. (1984).

Chang L. F. *The Effects of RDI on the Growth, Production and Fruit Quality of Greenhouse Cucumber.* Northwest A&F University. (2007).

Cheng F. H., Li S. H., & Meng S. Q. Study on the Effect of Regulated Deficit Irrigation on the Vegetative Growth, Cropping, and Fruit Quality of Yali Pear Variety. *J. Fruit Sci.* 001: 22–26. (2003).

FAO. Retrieved August 27, 2022, Information on https://www.fao.org/faostat/zh/#data

Failla, O., Zoccffl, G., Treccani, C., *et al.* Growth, Development and Mineral Content of Apple Fruit in Different Water Status Conditions. *J. Hortic. Sci.*, 67(2), 265–271. (1992).

Huang H. B. *On China's Water Resource Utilization and Sustainable Development of the Economy Society.* Northwest A&F University (2007).

Huang Y., Wang W. J, Wang L. W., Xu Q. B., Kong Q S, Bie Z L. Effects of Regulated Deficit Irrigation on Photosynthetic Characteristics, Fruit Yield and Quality of Melon Under Plastic Greenhouse Conditions. *J. Huazhong Agric. Univ.*, 35(01): 31–35. (2016).

Kammoun, M., Bouallous, O., Ksouri, M. F., Gargouri-Bouzid, R., & Nouri-Ellouz, O. Agro-physiological and Growth Response to the Reduced Water Supply of Somatic Hybrid Potato Plants (*Solanum tuberosum* L.) Cultivated Under Greenhouse Conditions. *Agric. Water Manage.*, 203, 9–19. (2018).

Karakas, M. C., Kurunc, A., & Dincer, C. Effects of Water Deficit on Growth and Performance of Drip Irrigated Sweet Potato Varieties. *J. Sci. Food Agric.*, 101(7).2961–2973. (2021).

Khalili Mikaiel, Nejatzadeh Fatemeh. Effect of Deficit Irrigation and Kaolin Clay on Yield and Yield Components of Pumpkin (*Cucurbita pepo* L.). *SN Appl. Sci.*, 3(5). (2021).

Liu L. H., Mo Y. L., Yang X. Z., Li X. L., Wu M. M., Zhang X, Ma J. X., Zhang Y, Li H. Reasonable Drip Irrigation Frequency Improving Watermelon Yield and Quality Under Regulated Deficit Irrigation in Plastic Greenhouse. *Trans. Chin. Soc. Agric. Eng.*, 30(24): 95–104. (2014).

National Bureau of Statistics of China. *China Statistical Yearbook*. China Statistics Press, Beijing. (2020).

Ma F. S., Kang S. Z., Wang M. X. Research Advance and the Prospect of Regulated Deficit Irrigation on Fruit Trees. *Agric. Res. Arid Areas*, 04:225–228 (2005).

Mills, T. M., Behboudian, M. H., & Clothier, B. E. Water Relations, Growth, and the Composition of 'Braeburn' Apple Fruit Under Deficit Irrigation. *J. Amer. Soc. Hort*, 121.2, 286–291. (1996).

Reid, J. B., and R. N. Gillespie. Yield and Quality Responses of Carrots (*Daucus carota* L.) to Water Deficits. *N. Z. J. Crop Hortic. Sci.*, 45(4): 299–312. (2017).

Volschenk, Theresa. Effect of Water Deficits on Pomegranate Tree Performance and Fruit quality–A review. *Agric. Water Manage.*, 246: 106499. (2021).

Wagg, C., Hann, S., Kupriyanovich, Y., & Li, S. Timing of Short Period Water Stress Determines Potato Plant Growth, Yield, and Tuber Quality. *Agric. Water Manage.*, 247, 106731. (2021).

Wang F. (2007) *Study of Water Use and Quality Watermelon under Regulated Deficit Irrigation Condition in Arid-Hungriness Oasis Area.* Northwest A&F University. (2007).

Xing, Y., Zhang, T., Jiang, W., Li, P., Shi, P., Xu, G., ... & Wang, X. Effects of Irrigation and Fertilization on Different Potato Varieties Growth, Yield and Resources Use Efficiency in Northwest China. *Agric. Water Manage.*, 261, 107351. (2022).

Yuan B. Z., Zhang Q. Y., Bie Z. L. Effects of Regulated Deficit Irrigation on Growth of Drip-irrigated Muskmelon in the Greenhouse. *J. Drain. Irrig. Mach. Eng.* 33.7(2015): 611–617.

Zheng, J. W., and Yong H. "Research Progress on Waterlogging Tolerance of Cucurbit Crops." *Int. J. Hortic.*, 12 (2022).

Water Conservancy and Civil Construction – Oke & Ahmad (Eds)
© 2024 The Author(s), ISBN: 978-1-032-58614-4

The relationship between runoff and sediment load in the Malian river basin of Longdong loess plateau in Gansu, China

Qiao Yu, Ying Zhou & Jinzhu Ma*
College of Earth and Environmental Sciences, Lanzhou University, Lanzhou, China

Jijun Lv
Qingyang Soil and Water Conservation Ecological Environment Monitoring Substation, Qingyang, China

Gang Wang
Lanzhou Meteorological Bureau of Gansu Province, Lanzhou, China

ABSTRACT: The paper studied the relationship between runoff and sediment load and its long-term influencing factors in the Malian River Basin of the Longdong loess plateau of China. The results showed that the average annual runoff and the mean sediment runoff were 4.12×10^8 m^3 and 1.08×10^8t, respectively. The rainfall decreases with a weak trend, while the runoff and sediment load decrease obviously, with a reduced rate of 2.7 million m^3/yr and 1.24 million t/yr, respectively. The linear relationship between precipitation and runoff has abruptly changed since 1997, while the relationship between rainfall and sediment load has abruptly changed in 2003. The rate of runoff and sediment load has decreased sharply since 2003. The contribution rate of human activities to the benefit of sediment reduction is 61%. In the future, the principal measures of soil and water conservation should emphasize the principle of protecting and consolidating gully and cultivating gully.

1 INTRODUCTION

The formation and evolution of runoff and sediment are affected by many factors such as climate change and human activities, and the change in the water-sediment relationship is the most active one. Due to its special natural geographical conditions and long-term frequent human activities, the Loess Plateau has become one of the most serious areas of soil erosion and a key construction area for ecological construction and soil and water conservation in China. Soil erosion in the Loess Plateau directly affects the ecological security of the Yellow River and restricts the sustainable development of the regional economy and society for a long time. Under the influence of global large-scale climate change and regional human activities, the underlying surface conditions and river sections in the Yellow River basin have undergone drastic changes. Since the 1960s, terraced fields, afforestation, and grass planting have been built to increase forest and grass coverage of the loess plateau. While numerous silt dams have been built to reduce the amount of sediment entering the Yellow River, the water and sediment in the upper and middle reaches of the Yellow River have changed greatly (Wei *et al.* 2016). The research on the relationship between runoff and sediment load in 14 Loess Plateau sub-basins indicated that check dams and reservoirs significantly affected sediment load reduction, while vegetation restoration was the main factor since 2000 (Gao *et al.* 2017). Yao *et al.* (2017) observed a decreasing annual average runoff and sediment load trend in the Yellow River across 50 years of recent data, primarily due to anthropogenic activities. The long-term

*Corresponding Author: jzma@lzu.edu.cn

DOI: 10.1201/9781003450818-13

variation of runoff and sediment load in the changing environment of the Yellow River basin is of great scientific significance for basin management planning, soil and water loss benefit evaluation, disaster prevention, and reduction (Wang *et al.* 2017). At the same time, it can also better understand the benefits of current water and soil conservation measures, predict future runoff and sediment load trends, and provide a scientific basis for water and soil conservation work (Zuo *et al.* 2016). The long-term runoff variation of eight hydrological stations across the Yellow River mainstream during 1950–2013 was investigated and showed that the annual average streamflow exhibited an M-type spatial pattern and a parabolic annual average suspended sediment discharge curve (Wei *et al.* 2016).

The long time series of soil erosion and the relationship between water and sediment in small and medium-sized river basins are very important for the scientific assessment of the sediment change process in the Yellow River. On the other hand, a large number of studies have shown that human activities are the main cause of sediment reduction. However, the influence of human activities on sediment reduction is quite different in different small watersheds. At present, it is difficult to quantify the influence of soil and water conservation engineering measures such as vegetation restoration, terraced fields, and check dams on sediment change. The Malian River is the third tributary of the Yellow River and one of the most serious areas of soil and water loss in the middle reaches of the Yellow River, which contributes 8% of the total sediment to the Yellow River every year (Zhang *et al.* 2016). Although there is some research on rainfall, flood, and sediment discharge in the Malian river basin in recent years, there are few studies on erosion and sediment load in the Malian river basin, especially the analysis of long-time series sediment load data. The understanding of the long-term variation of runoff and sediment load in the Malian river basin is still insufficient.

The relationship between water and sediment in a river basin is a comprehensive reflection of regional natural conditions and human activities and has been a research hotspot in the fields of sediment erosion dynamics and river dynamics for many years. Many studies use rating curves to reflect the statistical and dynamic characteristics of runoff and sediment yield processes, including the characteristics of suspended sediment load, the calculation of sediment flux, and river regulation (Yang *et al.* 2017). Mu *et al.* (2012) conducted a systematic sediment change analysis of the Yellow River from 1919 to 2008 by using the anomaly accumulation method and double-mass curve. Hu *et al.* (2011) analyzed the characteristics of flood peaks and their water-sediment relationship in the Yangtze River basin by using the water-sediment rating curve. The special aim of the paper is to comprehensively analyze the long-term trend and stage characteristics of runoff and sediment load in the Malian river basin by using observed hydrological and meteorological data. The relationship between rainfall, runoff, sediment load, and the influencing factors was also studied by using various methods. It would provide a scientific basis for soil and water loss control in the gully region of the loess plateau.

2 METHODS

2.1 *Study area*

Malian River originates from Mahuang Mountain in Ningxia Province, flows to the Longdong loess plateau of Gansu Province, and flows into Jinghe River in Zhengping. Malian River has a length of 374.8 km and an area of 19,086 km^2. The catchment is rectangular with major tributaries such as the Dongchuan river, Rouyuan river, Heshui river, Guchengchuan river, and Jiulinghe river (Figure 1). The whole topography of the basin is high in the north and low in the south, with an elevation of 885–2082 m. The northern part of the basin is mainly composed of low mountains and hills, accounting for 60% of the basin area. The central and southwestern parts of the basin are flat and open Loess Plateau landforms (i.e. loess tableland), accounting for 30.5% of the total basin area (Wang *et al.* 2018). There are 12 large loess tablelands with an area of over 67 km^2. The Dongzhi Tableland is located between the Puhe River and Malian River within the Longdong Loess Basin and is the largest and most complete loessal tableland in the existing area with an area of 910 km^2. The Quaternary Loess Strata in the basin include the

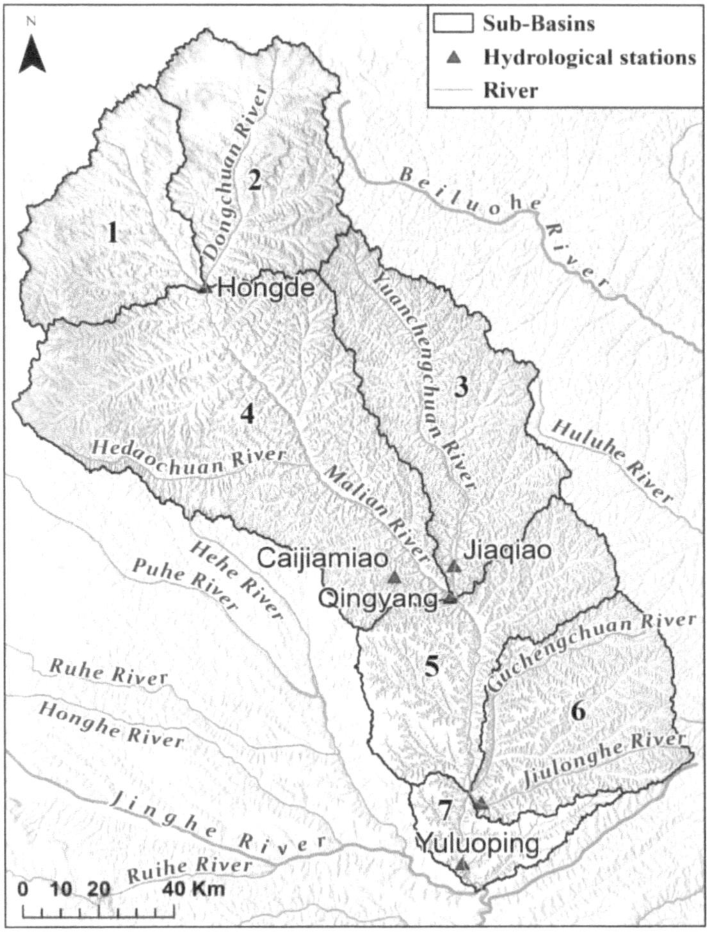

Figure 1. The location of the hydrological stations in the Malian river basin.

Lower Pleistocene Series Wucheng Loess of 40 – 60 m thick, the Middle Pleistocene Series Lishi Loess of 120 – 150 m thick, the Upper Pleistocene Series, and the Holocene Series Malan Loess with the thickness of around 10 m (Figure 2). There are alluvium strata at the triple terraces of the riverside, mainly consisting of loess silt with silty-fine sand and sandy gravel stratum, which are the main material sources of river sediment.

The southeast part is the mid-low Hill area of Ziwuling Mountain, which occupies 9.2% of the watershed area. There are more than 3100 km^2 of secondary forest with high vegetation coverage, which is the best area for soil and water conservation area in the basin. The main soil types in the basin are black loessal soil and yellow loamy soil, which are often interlaced and mainly distributed in the loess hilly land and tableland. The climate belongs to a temperate continental monsoonal climate with obvious seasonal variations. The annual average temperature is about 10.0°C, with the highest temperature of 38.4°C and the lowest of –22.0°C. The precipitation increases from northwest to southeast with an average annual precipitation of around 534 mm. However, nearly 60% of annual precipitation occurs as rainstorms from July to September. Since the 1980s, the local government has carried out large-scale soil and water conservation projects in the Malian river basin, such as closing off

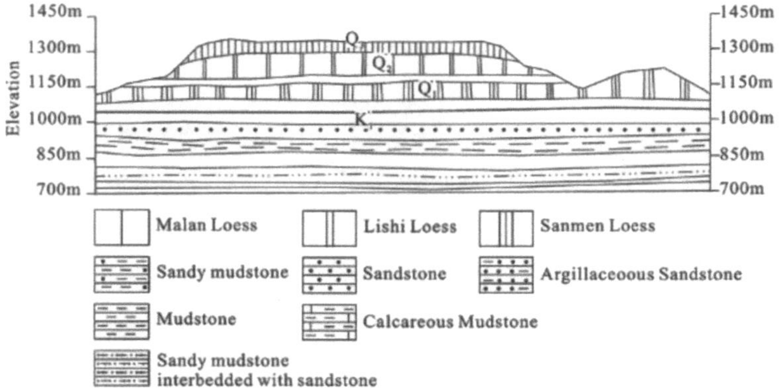

Figure 2. Geological cross-section of the Dongzhi Tableland.

the mountains to cultivate forests, planting trees and grasses, returning farmland to forests, returning pastures to grassland, and so on. In particular, the World Bank loan project for soil and Water Conservation in the Malian river basin began in February 1994. The ecological environment, agricultural production, and living conditions of the people have been greatly improved, and new rural construction has been promoted.

2.2 *Data sources analysis*

The runoff and sediment load data in six hydrological stations in the Malian River Basin from 1960 to 2020 were provided by the Yellow River Water Conservancy Commission (YRCC). There are three hydrological stations on the mainstream of the Malian River, namely Hongde, Qingyang, and Yuluoping. The hydrological stations on the tributaries are Yuele (Rouyuanchuan), Bangqiao (Heshuichuan), and Jiaqiao (Dongchuan). For some missing data, the method of near-site interpolation is used to ensure the integrity and continuity of the precipitation series. The rainfall data comes from 8 weather stations in Qingyang City from 1960–2018 were collected from eight meteorological stations (Chinese Climate Center; http://data.cma.cn/site/index.html).

2.3 *Analysis methods*

The changes and correlation of hydrological data over time series were mainly detected by linear regression tests of annual rainfall, runoff and sentiment load data, T-test method, double-mass curve method, and anomaly accumulation analysis. The anomaly accumulation method can accurately and intuitively analyze the stage characteristics of the interannual change of runoff and sediment load. The T-test is to test whether the average number of two samples and their respective representation of the overall difference is significant as follows:

$$\frac{t = \overline{x_1} - \overline{x_2}}{S\left(\frac{1}{n_1} - \frac{1}{n_2}\right)^{1/2}} \tag{1}$$

$$S = \left(\frac{n_1 S_1^2 + n_2 S_2^2}{n_1 + n_2 - 2}\right)^{1/2} \tag{2}$$

Eq. 1 obeys the distribution of degrees of freedom, $n_1 + n_2 - 2$, and n, $\overline{x}$, and S are sample number, mean value, and variance respectively. After giving the significance level a, the critical value t_a is obtained by checking the t-distribution table. If the sliding t value has not exceeded the critical value during the period, it means that there is no obvious abrupt change, and the year corresponding to the excess point is the year of abrupt change.

The double mass curve is a method for investigating the consistency and long-term trend of hydro-meteorological time series. The basic principle is that two variables are incrementally accumulated over the same length of time, one as a horizontal coordinate and the other as a vertical coordinate to describe the trend of both series. If the proportion between the two variables is constant, the relationship will be linear at the same time, and if the slope changes, the original relationship between the two variables will be changed. The slope is generally analyzed based on a double-mass curve of rainfall-runoff and rainfall-sediment load. If the slope of the straight line does not deviate significantly, it means that human activities have no significant effect on runoff and sediment load; otherwise, the results indicate the significance of anthropogenic activities (Gao *et al.* 2017).

The main source of runoff is rainfall. Soil erosion is caused by sediment carried by runoff, and the relationship between runoff and sediment load in the river basin directly determines the scouring and silting status of the river course. The runoff-sediment load synergy is of great significance to the study of runoff-sediment heterologous status and its control. In general, the concentration of suspended sediment increases with the increase of runoff. The increasing rate of suspended sediment concentration shows a great difference with the change in time conditions. The sediment rating curve is defined as the power exponent relationship between runoff Q and suspended sediment concentration S as follows:

$$S = aQ^b \tag{3}$$

$$\ln S = \ln a + b\ln Q \tag{4}$$

Where a is a coefficient; b is a power exponent. Parameter a indicates that the characteristics of runoff and sediment yield in rivers are mainly influenced by external factors, including dam and reservoir construction, soil and water conservation measures, conversion of cropland to forest and grassland, agricultural production and sand mining in rivers, etc. Parameter b represents the sediment load characteristics of the river, which are related to the shape of the river bed (channel shape, slope, and unit river power) or the soil erodibility and credibility of the river profile. It is greatly affected by internal factors such as flow velocity, discharge, sand gradation ratio, and so on. The values of a and b represent the supply of material sources and the corresponding change in the rate of growth of suspended sediment concentration. Zhao *et al.* (2012) derived the relationship between river runoff and sediment load based on Eq.3:

$$\mathrm{Ws} = k\frac{W^n}{(1 - S_{lv})^p} \tag{5}$$

$$S_{lv} = \frac{S_l}{\rho_s} \tag{6}$$

Where W_S represents the annual sediment load, W is the annual runoff; The coefficient k indicates the amount of sediment recharged from the river channel per unit along the way; the index n indicates the degree of influence of runoff on the amount of sediment load, and the index p indicates the possibility of the water flow obtaining sediment recharge from the river. S_l, the sediment content of upstream flow, and S_{lv}, volume-specific sediment content from upstream.

3 RESULTS AND DISCUSSIONS

3.1 *Temporal and spatial variation of runoff and sediment load*

The rainfall can directly reflect soil erosion. The annual rainfall in the Malian River basin ranged from 824 mm (1964) to 363 mm (1979), with an average value of 534 mm. The interannual variation of rainfall in the Malian river basin is relatively small, and the coefficient of variation is 0.13. The annual mean rainfall fluctuates periodically, and the increasing trend is very obvious during 2010–2020, with an average value of 570 mm.

The annual runoff in the Malian River basin ranged from 2.15×10^8 m^3 to 9.62×10^8 m^3, with an average value of 4.12×10^8 m^3, which showed a decreasing trend with a rate of 0.027×10^8 m^3/yr. The annual average sediment load is 1.08×10^8 t, with a decreasing trend (Table 1 and Figure 3). The maximum annual sediment load is 4.9×10^8 t and the minimum annual sediment load is 0.156×10^8 t. The average runoff and sediment load were the largest from 1990 to 1999, with an average mean value of 4.12×10^8 m^3 and 1.59×10^8 t, respectively. The average annual runoff and sediment load were the lowest from 2010 to 2020, with an average mean value of 3.3×10^8 m^3 and 0.54×10^8 t, respectively. The extreme value ratio was the largest from 1960 to 1969, the variation coefficient was the largest, and the fluctuation was the most violent from 1960 to 1969 and from 1980 to 1989. Both the annual runoff and sediment load exhibited an abrupt change in 2003. The average annual runoff and sediment load were 4.48×10^8 m^3 and 1.28×10^8 t during the period of 1960–2002, with decreasing to 3.27×10^8 m^3 and 0.61×10^8 t, respectively in 2003 – 2020.

Sediment load at different stations fluctuates sharply with time. The sediment load at Hongde station, Qingyang station, and Yuluoping station in the mainstream tend to decrease-increase-decrease in different eras, the sediment loads of Yuelue station and Banqiao station in different ages showed an increase-decrease-increase-decrease trend, while that of Jiaqiao station showed an increase-decrease trend. Except for Banqiao station, the other hydrologic stations reached their maximum value from 1990 to 1999 and showed a downward trend since 2003. The analysis results of the specific sediment yield of 6 hydrological stations show that the average annual specific sediment yield of the Malian River basin is 6000 t/km^2, and ~96% of the catchment area has an annual average specific sediment yield greater than 5000 t/km^2. The specific sediment yield has increased greatly up to 10,000 t/km^2 in the Malian river above Hongdae station and Rouyuanchuan above Yuele station since 1990. The specific sediment yield was decreasing to 5000 t/km^2 from 2000 to 2009, and less than ~2500 t/km^2 during 2010–2020.

Table 1. Summary of rainfall, runoff, and sediment load in the Malian river basin.

Duration	Rainfall (mm)	Runoff ($10^8 m^3$)	Sediment loads (10^8t)
1960–1969	583.1	4.59	1.29
1970–1979	520.3	4.54	1.29
1980–1989	523.6	4.28	0.99
1990–1999	505.7	4.75	1.59
2000–2009	504.1	3.33	0.85
2010–2020	570.1	3.30	0.54
1960–2002	531.9	4.48	1.28
2003–2020	539.2	3.27	0.61
1960–2020	533.9	4.12	1.08

3.2 *Trend analysis of runoff and sediment load*

The T-test was used to test the abrupt change of runoff and sediment load with Equations 1 and 2. The test results showed that the T value of annual rainfall only exceeded the confidence interval of 95% in 2009 and showed a decreasing trend. The runoff T value exceeded the 95% confidence interval between 1996 and 1997, indicating an abrupt change in runoff around 1996 and an increase in human-induced runoff since 1997. The T value of the annual sediment load in the basin exceeded the 95% confidence interval around 2003 and underwent an abrupt change. Accumulative anomaly curves of runoff and sediment load (Figure 4), and double-mass curves of rainfall-runoff and rainfall-sediment load in the Malian river basin during 1960–2020 (Figure 5) exhibited an abrupt change in 2003. During 1960–2002 and

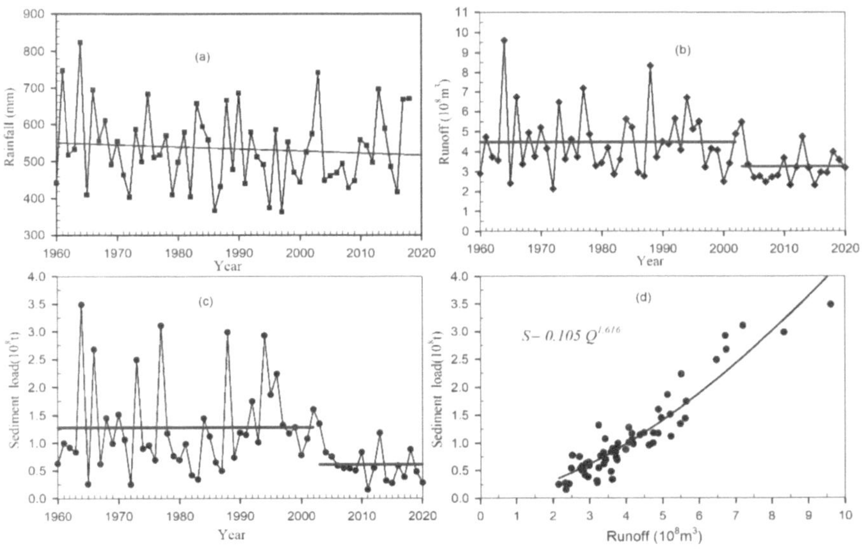

Figure 3. The trends of (a) rainfall, (b) runoff, (c) sediment load, and (d) sediment rating curve in the Malian River basin from 1960 to 2020.

2003–2020, the slope of the accumulative curve of rainfall-runoff was 0.0084 and 0.0058, respectively. It shows that the influence of climate on runoff is relatively weak since 2003. In 2003, the abrupt change of the double accumulative curve of rainfall-sediment load occurred with slopes of 24.1×10^8 t/mm and 113×10^8 t/mm, respectively. It can be seen that the sediment load in the basin increased with the increase of rainfall before 2002, while the slope of the curve tended to be smooth in 2003–2020, and the sediment output caused by the same rainfall was smaller, soil and water loss in the river basin is more moderate.

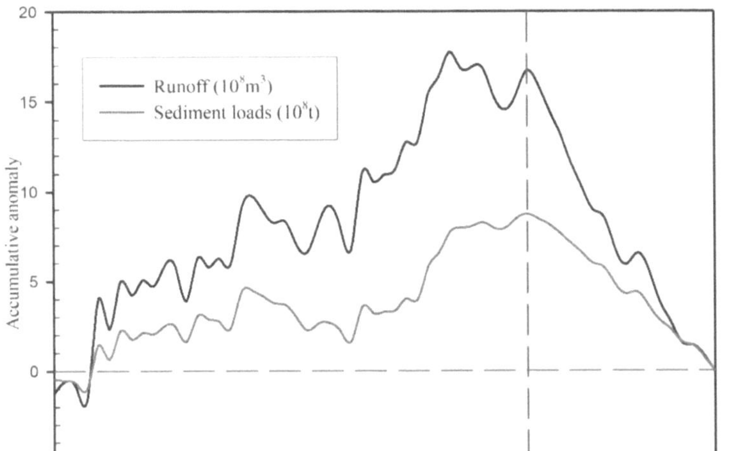

Figure 4. Accumulative anomaly curves of runoff and sediment load in the Malian river basin.

The sediment rating curve in the Malian river basin from 1960 to 2020 according to Eq. 3~Eq.6 is shown in Figure 3d and Table 2. Factor a = 0.105 represents the external influence, while Factor b = 1.616 represents the energy of sediment load in the river itself. The double mass curve of runoff and sediment load changed between 1993 and 2003 (Figure 6). The increasing rates of sediment load with runoff were 0.25, 0.34, and 0.18 during 1960–1992, 1993–2002, and 2003–2020, respectively. The linear relationship between accumulative runoff and accumulative sediment yield is stronger before 2003, and the sediment yield caused by the same runoff yield is larger. The decrease of sediment load with the increase of runoff may be due to the effects of human activities, including soil and water conservation measures and sediment control measures, which reduce the amount of soil erosion and the amount of sediment.

Table 2. The regression equations of sediment rating curves in the Malian river basin.

Rivers	Regression equations	Ws	R^2	k	n	p
Malian River	$y = 0.076x^{0.387}$	$Ws = 0.076*w^{1.210}/(1-S_v)^{6.672}$	0.977	0.076	1.210	6.672
Dongchuan	$y = 0.119x^{0.420}$	$WS = 0.119*W^{0.313}/(1-S_{v)}{}^{7.241}$	0.973	0.119	1.313	7.241
Rouyuanchuan	$y = 0.0.185x^{0.406}$	$Ws = 0.185*W1.269/(1-S_v)^{6.999}$	0.975	0.185	1.269	6.999
Heshuichuanr	$y = 0.475x^{0.235}$	$Ws = 0.475*w^{0.735}/(1-S_{v)}{}^{4.051}$	0.390	0.475	0.735	4.051

The relationship between runoff and sediment discharge of mainstream and tributaries of the Malian river, except the Heshuichuan river, is very significant (r^2 = 0.97) (Table 2, Figure 7). In the whole basin, the K value in the lower reaches is lower than that in the upper reaches. Because the K value reflects the amount of sediment supplement along the river, the amount of sediment supplement in the lower reaches is relatively small, the vegetation coverage rate in the upper reaches is relatively low, the ecological environment is relatively poor, and the soil erosion along the river is serious. The soil and water conservation measures in the lower reaches of the river are perfect and the sediment increase along the river is less. The n value is mainly related to the sediment concentration and increases with the increase of sediment concentration (Guo *et al.* 2015). The results show that the sediment concentration in the lower reaches is relatively lower, and the influence of runoff from the upper reaches to the lower reaches on sediment load is increasing. The P value reflects the sedimentation degree of the river, and the p-value of the upstream river is higher than that of the downstream, indicating that the velocity of the upstream river is relatively small, and the sedimentation along the river is more serious.

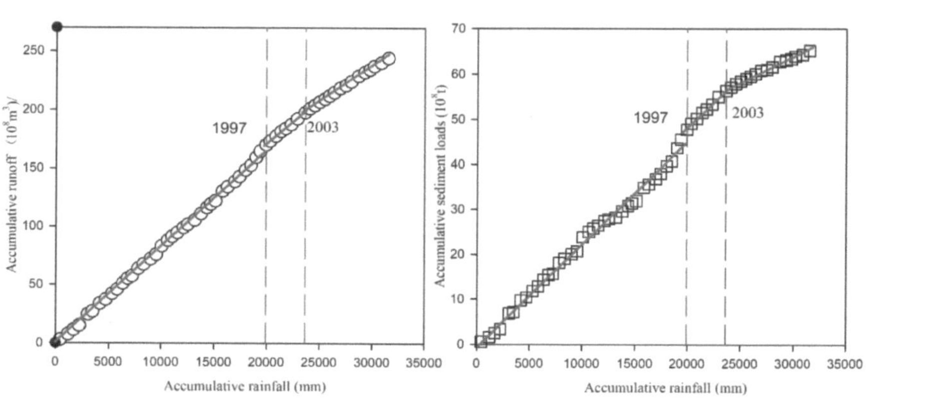

Figure 5. Double-mass curves of rainfall-runoff and rainfall-sediment load in the Malian river basin.

3.3 *Impacts of human activities on runoff and sediment load*

The linear relationship between precipitation and runoff is obvious, but the linear relationship between precipitation and sediment load is not obvious. The abrupt change year of runoff was 1997, and the linear relationship between precipitation and runoff was significant from 1997 to 2020. The size of runoff varies with the change in precipitation. The results show that the linear relationship between precipitation and runoff is more obvious since 1997 because of various soil and water conservation measures and human activities. The double mass curve also showed that the relationship between precipitation and runoff changed abruptly around 1997. The slope of the curve is smaller since 1997, and the runoff rate decreased by about 29% under the same precipitation condition. The impact of meteorological factors on runoff gradually weakened since 1997 due to the intensification of human activities. The poor linear relationship between precipitation and sediment load indicates that precipitation is not the most direct cause of the variation of sediment load. The double accumulation curve shows that the precipitation-sediment load relationship has experienced three stages, which are 1960–1993, 1994–2002, and 2003–2020 respectively. The slope of the curve has decreased by about 61% since 2003, indicating that human activities and meteorological factors contribute 61% and 39% to river sediment load, respectively.

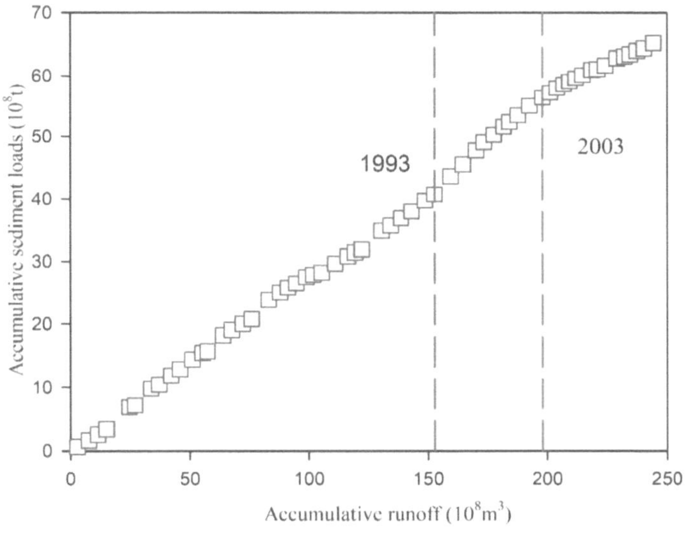

Figure 6. The double mass curve of runoff and sediment load in the Malian river basin.

The correlation coefficient between runoff and sediment load was 0.78 – 0.93 in 1960–2020, the strongest linear relationship was found before 1993, and then maintained at about 0.8. The year 2003 is not only an abrupt point in the sediment load but also an abrupt point in the relationship between runoff and sediment load (Figure 6). By using the double mass curve method, the contribution rate of human activities to the control of sediment load in the river basin reached 73.5% from 2003 to 2020. Many studies have also shown that human activities in the basin have a significant impact on the reduction of runoff and sediment load. A lot of local water conservation measures, including terraces, silt dams, afforestation, and other projects, have played an important role since 2003 and have become the dominant factor in runoff control and sediment load reduction (Zhang *et al.* 2017).

Climate change and human activities are the main drivers of the hydrological cycle. The relationship between runoff and sediment load changed due to the superposition of cyclical fluctuations of rainfall and human intervention. The above results show that since 2003, the relationship between runoff and sediment load in the Malian river basin has undergone great changes. The main factors affecting the amount of sediment load are no longer simply climatic, and the role of human activities and various water protection measures has gradually played a role and even dominated (Du *et al.* 2021; Zheng *et al.* 2021). From the above characteristics and the relationship between runoff and sediment load, it can be seen that runoff and sediment loads decreased significantly after the implementation of the World Bank-financed project in the Malian River Basin in 1994. It shows that the implementation of water and soil conservation in the Malian River Basin has achieved significant benefits in water and sediment reduction. The state attaches great importance to the control of soil erosion in the Malian River Basin, and in recent decades, large-scale soil erosion prevention and control projects have been carried out in the Malian River Basin, realizing comprehensive management from single measures and decentralized treatment to classified guidance of different types of areas with small watersheds as units. Especially in recent years, the government has implemented key soil and water conservation projects, such as the Malian River Key Soil and water conservation project, the Some Random Place Somewhere silt dam pilot project, and the comprehensive agricultural development project. After many years of construction, the soil and water conservation work in the Malian river basin has achieved remarkable results and effectively reduced the sediment entering the Yellow River.

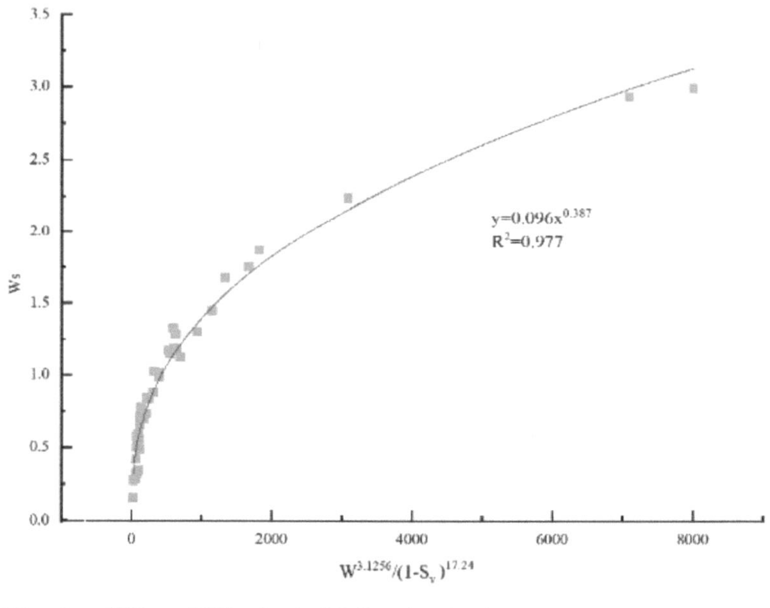

Figure 7. The plot of Ws vs $f(W)$ in the Malian river basin.

4 CONCLUSION

The paper studied the relationship between runoff and sediment load and its long-term influencing factors in the Malian River Basin of the Longdong loess plateau of China by using linear regression, the double-mass curve method, T-test, and the sediment rating curve method. The average annual runoff and the mean sediment runoff were 4.12×10^8 m^3 and 1.08×10^8 t, respectively. Significant reduction tendency was revealed with a reduction rate of 0.027×10^8 m^3/yr and 0.0124×10^8 t/yr, respectively. The linear relationship between precipitation and runoff and sediment load is not obvious. The linear relationship between precipitation and runoff abruptly changed in 1997 and the linear relationship between precipitation and sediment load in 2003. The contribution rate of human activities and meteorological factors to river desertification reduction was respectively 61% and 39%.

The relationship between runoff and sediment load fluctuated greatly from 2000 to 2010, with an abrupt change in 1993 and 2003. The rate of runoff sediment load decreased sharply since 2003. There is a significant downward trend in sand production. The contribution rate of human activities to the control of sediment load in the river basin reached 73.5% from 2003 to 2020. Under the background of large-scale conversion of cropland to forest and grassland projects and large-scale construction of silt dams since 2003, the vegetation coverage of the watershed is increasing in the Malian river basin, and the sediment yield of the watershed is decreasing. In the future, the principal measures of soil and water conservation and their collocation should emphasize the principle of protecting and consolidating gully and cultivating gully.

ACKNOWLEDGMENTS

The research work was supported by the Gansu Provincial Water Conservancy Scientific Research and Technology Promotion Program (22GSLK077). We would like to sincerely thank Dr. Du Min from the Institute of Soil and Water Conservation, Chinese Academy of Sciences & Ministry of Water Resources for her great help.

REFERENCES

Du, M., Mu X. M., Zhao, G., Gao, P., Sun, W. (2021). Changes in Runoff and Sediment Load and Potential Causes in the Malian River Basin on the Loess Plateau. *Sustainability* 13, 443.

Gao, P., Li, P., Zhao, B., Xu, R. R., Zhao, G. J., Sun, W., Mu, X. M. (2017). Use of Double Mass Curves in Hydrologic Benefit Evaluations. *Hydrol. Process* 31, 4639–4646.

Guo, A., Huang, Q., Chang, J. et al. (2015). Analysis of Evolution Characteristics of Water-Sediment Relationship in Jing River Basin based on Copula function. *J. Nat. Res.* 30(4), 673–683.

Hu, B., Wang, H., Yang, Z. et al. (2011). Temporal and Spatial Variations of Sediment Rating Curves in the Yangtze River Basin and Their Implications. *Quatern. Int.* 230(1), 34–43.

Mu, X. M., Zhang, X.Q., Shao, H. B., Gao. P., Wang, F., Jiao, J.Y., Zhu, J. L. (2012). Dynamic Changes of Sediment Discharge and the Influencing Factors in the Yellow River, China, for the Recent 90 years. *CLEAN–Soil Air Water* 40, 303–309.

Wang, S., Fu, B. J., Liang, W., Liu, Y., Wang, Y. F. (2017). Driving Forces of Changes in the Water and Sediment Relationship in the Yellow River. *Sci. Total Environ.* 576, 453–461.

Wang, Y.S., Cheng, X., Zhang, M., Qi, X. (2018). Hydrochemical Characteristics and Formation Mechanisms of Malian River in Yellow River Basin During the Dry Season. *Environ. Chem.* 37, 164–172.

Wei, Y. H., Jiao. J. Y., Zhao, G. J., Zhao, H. K., He, Z., Mu, X.M. (2016). Spatial-temporal Variation and Periodic Change in Streamflow and Suspended Sediment Discharge along the Mainstream of the Yellow River during 1950–2013. *Catena* 140, 105–115.

Yang, H., Li, E., Zhao, Y. (2017). Effect of Water-sediment Regulation and its Impact on Coastline and Suspended Sediment Concentration in Yellow River Estuary. *Water Sci. Eng.* 10(4), 311–319.

Yao, W. Y., Xiao, P. X., Shen, Z., Wang, J., Jiao, P. (2017). Analysis of the Contribution of Multiple Factors to the Recent Decrease in Discharge and Sediment Yield in the Yellow River Basin, *China. J. Geogr. Sci.* 26, 1289–1304.

Zhang, J. J., Zhang, X, P., Li, R. et al. (2017). Did Streamflow or Suspended Sediment Concentration Changes Reduce Sediment Load in the Middle Reaches of the Yellow River? *J. Hydrol.* 546, 357–369.

Zhang, Y. Z., Zhang, D.Y., Liu, Y. Y. (2016). Rainfall Variation in the Malianhe River Basin of the Loess Plateau in Recent 50 years. *Sci. Soil Water Conserv.* 14, 4.

Zhao, H., Hu, C., Chen, X. (2012). Study on the Relationship Between Water Transport and Sediment Load in the Mainstream of the Yellow River. *J. Hydraulic Eng.* 43(4), 379–385.

Zheng, H. Y., Miao, C. Y., Jiao, J. Y. et al. (2021). Complex Relationships between Water Discharge and Sediment Concentration Across the Loess Plateau, *China. J. Hydrol.* 596, 126078.

Zuo, D. P., Xu, Z. X., Yao, W. Y., Jin, S. Y., Xiao, P. Q., Ran, D. C. (2016). Assessing the Effects of Changes in Land Use and Climate on Runoff and Sediment Yields from a Watershed in the Loess Plateau of China. *Sci. Total Environ.* 544, 238–250.

Water Conservancy and Civil Construction – Oke & Ahmad (Eds)
© 2024 The Author(s), ISBN: 978-1-032-58614-4

Research progress on the effect of agronomic measures on water saving on potato quality

Jiandong Yu
College of Agronomy and Agricultural Engineering, Liaocheng University, Liaocheng, China
College of Water Conservancy and Hydropower Engineering, Gansu Agricultural University, Gansu Lanzhou, China

Shouchao Yu
College of Agronomy and Agricultural Engineering, Liaocheng University, Liaocheng, China

Hengjia Zhang*
College of Agronomy and Agricultural Engineering, Liaocheng University, Liaocheng, China
College of Water Conservancy and Hydropower Engineering, Gansu Agricultural University, Gansu Lanzhou, China

Jie Li, Zeyi Wang & Xietian Chen
College of Water Conservancy and Hydropower Engineering, Gansu Agricultural University, Gansu Lanzhou, China

ABSTRACT: As one of the four staple foods, potato plays an important role in ensuring global food security. While water scarcity and climate change-induced global acreage reduction are prominent, balancing the relationship between yield, water, and quality is a new challenge in recent years. This study introduces the unity of water regulation and quality improvement in potato cultivation in different regions, from potato quality connotation and use. This paper explores the link between water conservation and quality regulation under three agronomic measures: water-saving irrigation technology, on-farm moisture conservation technology, and irrigation system optimization, reveals the response mechanism of potato quality to different forms of agronomic measures and clarifies the significance of water conservation and quality regulation in its actual production.

1 INTRODUCTION

With global water scarcity and the intensification of the Russian-Ukrainian conflict, food security is under unprecedented impact (Chi & Guo 2022). As the world's fourth largest food crop after wheat, rice, and corn, the quality and safety of potato production is receiving increasing attention from scholars and government agencies (Hu et al. 2022). As the world's top potato producer in terms of acreage, China has achieved outstanding results in recent years in drought-tolerant potato breeding, pest monitoring and early warning technologies, and production mechanization. In 2015, China proposed a potato staple food strategy, after which China's potato production entered a new phase that emphasizes both yield and quality improvement.

*Corresponding Author: 596088683@qq.com

DOI: 10.1201/9781003450818-14

A series of factors such as severe shortage of water resources, backward potato industry technology, and serious imbalance in regional development restrict the rapid development of domestic potato water conservation and quality improvement, and water-saving agronomic measures such as the cultivation of drought-tolerant varieties, tillage cover, and optimization of irrigation system greatly improve the situation. This will play an important role in promoting the domestic agricultural industry structure layout, ensuring food security, and achieving synergistic water conservation and quality improvement of domestic potatoes (Hu 2021). In this study, we based on the current status of research on the effect of major water-saving agronomic measures on potato quality, review the connotation of potato quality and agronomic measures to save water, reveal the response mechanism of potato quality to alternate furrow irrigation, alternate root irrigation, under-film drip irrigation, moisture conservation by mulching, irrigation frequency, and irrigation quota, and clarify the practical production significance of potato water saving and quality adjustment. The aim was to provide a theoretical reference for China's potato water saving, high yield, high quality, and high-efficiency production.

2 QUALITY CONNOTATION AND APPLICATION OF POTATO

Potato quality is usually quantified in terms of elemental content or size and generally consists of nutritional quality, appearance quality, processing quality, and eating quality (Zhang 2011). Nutritional quality includes starch, vitamins, and mineral elements of the tuber, etc.; appearance quality includes potato shape, surface smoothness, tuber size, whether it turns green, etc.; processing quality mainly includes tuber appearance (skin thickness, number, and depth of bud eyes, color, size and potato shape, etc.), tuber interior (flesh color, no internal hollow, black heart, etc.) and reducing sugar content, etc.; eating quality mainly includes taste (no numb mouth) and palatability (Zhang 2011).

3 EFFECT OF AGRONOMIC MEASURES FOR WATER SAVING ON POTATO QUALITY

3.1 *Water-saving irrigation methods*

Water conservation by irrigation is one of the major technologies for efficient water use, and more advanced water-saving irrigation techniques such as alternate furrow irrigation, alternate root division, and under-film drip irrigation have been applied to potato water conservation and quality regulation studies. Sarker Khokan Kumer, *et al.* (2019) conducted a trial on the effect of Alternate Furrow Irrigation (AFI), Fixed Furrow Irrigation (FFI), and Every Furrow Irrigation (EFI) on potato quality in the central region of Bangladesh in South Asia., which showed that the yield remained stable, AFI saved 35% of irrigation water on average compared with EFI, and irrigation water productivity increased by 50%. The improvement of quality indexes such as soluble sugar was also highly significant. Wang Tengfei *et al.* (2020) in China also reached similar conclusions that AFI can save water significantly and ensure tuber quality improvement while keeping the harvest index stable. In addition, split-root alternate irrigation can also achieve water savings and quality improvement. Hu Chao *et al.* (2011) applied the split-root alternate irrigation technique to study potato quality response in Xinxiang City, Henan Province and showed that the technique could result in no significant yield reduction with a 30% reduction in irrigation water, which was not only improved quality but also reduced heavy metal accumulation in tubers.

Under-membrane drip irrigation technology, a combination of drip irrigation and mulching technology, was first applied to crops such as cotton in 1996 in the Xinjiang Production and Construction Corps and is also applicable to potato quality improvement. Liu Jinyang *et al.* (2018) showed in the western oasis region of the river that moderate water deficit (50–60%) during potato block formation and full irrigation (60–70%) during the rest of the reproductive period, based on saturated water content, could significantly improve water use efficiency, irrigation water use efficiency, and quality, although the yield was slightly reduced compared to full irrigation during the full reproductive period, which effectively reduced organic acid content and was effective in improving potato quality. Drip irrigation under the film of arch potatoes also contributed to quality improvement, and Liu Zhongliang *et al.* (2018) conducted a trial in Tai'an City, Shandong Province, which showed that deficit irrigation at 70% and 65% of full irrigation during tuber formation and starch formation, resulted in the highest soluble protein content (4.14 g/kg) and significant yield, quality and economic efficiency improvement, which was the best irrigation strategy.

3.2 *Agricultural mulching technology for moisture conservation*

With the rise of the plastic industry in the world, mulching technology has been popularized as an important agricultural water conservation technology in arid and semi-arid areas. China introduced mulching moisture conservation technology from abroad in 1978 and applied it to small-area vegetable trials with success, and now it has been applied to potato water conservation and quality control research on a large scale. Luo Lei *et al.* (2018) conducted a study on water conservation and quality control of potatoes by mulching monopoly in an arid area of central Gansu Province, and the results showed that the tuber starch content, crude protein content, vitamin C content, and Fe content were reduced to different degrees in both the black film duo-poly full mulching monopoly and black film monopoly mulching monopoly, and the white film duopoly full mulching monopoly had the best overall quality and all indexes increased. It is a more suitable mulching method in this area.

In addition, different types of mulch have different effects on tuber quality, and the results of a related study conducted by Zhang Shumin *et al.* (2017) in the Yellow-Huaihai region showed that compared to common polyethylene mulch, black and white color-matched mulch and biodegradable mulch not only achieved temperature and moisture conservation, water conservation and quality improvement (7.0%–37.6% increase in starch content and 1.0%–4.3% increase in vitamin C content) but also facilitated the reduction (26.7%–58.5%) of weed density. The results of Wang Dong *et al.* (2015) showed that the highest reducing sugar content was found in semi-membrane furrowed monopoly sown potato tubers (0.137%), while the highest amino acid content was found in flat beds without mulch (4.67mg/g), with no significant increase in protein and starch content.

Compared to plastic mulch for water and moisture conservation, straw return technology highlights its advantages of environmental protection and improvement of soil aggregation and microenvironment. The amount of maize straw returned to the field also has a promoting effect on potato yield, quality, and water use efficiency, Huang Kai *et al.* (2019) conducted a trial in the semi-arid region of central Gansu Province, and the results showed that at a maize straw application rate of 15 000 kg/hm^2, the starch content in tubers increased by 0.04% to 0.92%, the commercial potato rate increased by 5.2% to 21.5%, yield increased by 10.68% to 54.33% and WUE increased by 5.93% to 30.66%, achieving the high goal of water saving – yield increase – quality improvement.

3.3 *Optimizing the irrigation system*

Different fertility stages of crops have different sensitivity to soil moisture content, and their water consumption is also different. A large number of studies have shown that as the water carried by the seed potato itself can meet the physiological growth of water at the stage from sowing to seedling, the water demand at this stage is low, but if there is drought during the sowing period, it will lead to low seedling rate and other situations; after seedling, with the accumulation of dry matter and rising temperature, the water consumption of the plant and root system used to maintain normal life activities gradually increases. The water consumption of potato plants and roots peaks during the tuber expansion period, and this period is a critical period for the water demand of potatoes. Water deficit in this period in terms of an irrigation cycle, irrigation frequency, and irrigation quota can seriously affect the yield and quality of potato tubers. Geng Haojie *et al.* (2019) studied the water consumption pattern of potatoes in the Ningxia arid zone throughout the reproductive period and found that the best-combined value of reducing sugars (0.413%), starch (11.55%) and dry matter (18.93%) of potato tubers could be achieved when water is applied at the seedling (25%), tuber formation (25%) and tuber growth (50%) stages at the base of the irrigation quota, and potatoes could achieve water saving – yield increase – quality improvement.

Irrigation frequency affects the growth and development of the crop at each re-productive stage and is one of the main indicators of a crop irrigation system. Wang Ying *et al.* (2019) studied the effect of drip irrigation frequency and irrigation volume on potato yield, quality, and water use efficiency in the semi-arid region of Shaanxi, and the results showed that the optimal irrigation system with high yield and quality and high water use efficiency was obtained when the drip irrigation frequency was 8 d and the irrigation volume was the crop water requirement. Similar results were obtained by Ma Wei *et al.* (2011). The appropriate amount of deficit irrigation at the corresponding fertility stage could not only save water resources, and improve water use efficiency and yield, but also improve crop quality. Therefore, the optimal allocation of irrigation systems was also one of the important ways to achieve efficient crop growth and quality improvement.

4 CONCLUSION

In this study, we reviewed the research status of potato water-saving and quality control under different agronomic water-saving measures such as water-saving irrigation technology, on-farm mulching, and moisture conservation technology and irrigation system optimization, and took potato quality connotations, uses and agronomic water conservation measures as the starting point. The conclusions are as follows.

(1) Water-saving irrigation methods such as AFI, alternate root irrigation, and drip irrigation under film were applied to the potato growing period, which could all achieve the effect of water saving and quality improvement. Water use efficiency can be increased by about 50%, and AFI can significantly increase the content of soluble sugar. Alternate root irrigation can significantly reduce the content of heavy metals. Drip irrigation under the film can effectively reduce the content of organic acids, thereby increasing the content of soluble protein.
(2) The two agronomic measures of film mulching and straw returning have no significant effect on the water saving and conditioning of potatoes. The traditional black plastic film mulching will reduce the starch, crude protein, vitamin C, and iron content of tubers, while the black and white color matching plastic film and biodegradable plastic film mulching can not only increase the starch and vitamin C content but also significantly reduce the weed density. Water saving and starch quality improvement also improve the soil microenvironment and avoid “white pollution”.
(3) Irrigation cycle, irrigation frequency, and irrigation quota seriously affect the quality of potato tubers. Water (50% of the irrigation quota) is replenished during the tuber growth period to achieve the goal of saving water and improving quality. When the frequency of drip irrigation is 8d and the amount of irrigation water is the water demand of the crop, it is the optimal irrigation system with high quality, high yield, and water saving.

ACKNOWLEDGMENTS

This work was financially supported by the National Natural Science Foundation of China (No. 52269008), the Industrial Support Plan Project of the Gansu Provincial Department of

Education (No. 2022CYZC-51), and the Key Research and Planning Projects of Gansu Province (No. 18YF1NA073).

REFERENCES

Chi F.L. and Guo D. 2022 Research on the Overall Planning of Food Security and Development. *Academic J. of Zhongzhou* (07):38–43.

Geng H.J., Yin J., Wu J. and Liu Y.Z. 2019 Effects of Different Irrigation Amounts on Potato Growth, Water Consumption, Yield, and Quality. *Water Saving Irrig.* (03):43–47 + 58.

Hu C., Xu G.H., Qi X.B., Fan X.Y., Wu H.Q., Zhu D., Fan H.T. and Zhao Z.J. 2011 The Effect of APRI on Quality and Heavy Metal Accumulation of Potato by Reclaimed Water Irrigation. *J. of Irrig. and Drainage* 30(03):43–6.

Hu M.M. 2021 *Effects of Water Stress on Plant Growth, Physiological Characteristics, Tuber yield and Quality of Potatoes*. Hebei Agr. University.

Hu X.Y., Tan X.L., Sun X.H., Xie K.Z. and Liu Y.Q. 2022 Effects of Potato Continuous Cropping on Soil Fungal Community Characteristics. *Agr. Res. in the Arid Areas* 40(04):185–91.

Huang K., Wang J., He W.C., Tan W.J., He X.Q. and Han J.R. 2019 Effects of Amounts of Straw Mulching on Soil and Potato Yield and Water Use Efficiency. *Gansu Agr. Sci. and Technol.* (03):26–31.

Liu J.Y., Jia S.H. and Liang Z.G. 2018 Effects of Mulched Drip Irrigation Under Water Deficit on Potato Growth Index and Quality in Oasis Region, *Yellow River* 40(08):152–6.

Luo L., Li Y.J., Yao Y.H., Wang J., Zhang X.J. and Li D.M. 2018 Effect of Cover and Ridge Cultivation on Potato Growth, Yield, Quality and its Economic Benefit in a Dry Land. *Agr. Res. in the Arid Areas* 36 (01):194–9.

Liu Z.L., Gao X., Zhang Y.Y., Jiao J., Gu D.Y. and Gao J.J. 2018 Effects of Irrigation Amount on Growth, Yield, and Quality of Potato under Integrative Water and Fertilizer Planting. *J. of Heilongjiang Bayi Agr. University* 30(02):6–10.

Ma W. and Yin J. 2011 Effect of Different Treatments of Drip Irrigation on Quality and Yield of Potato. *Ningxia Eng. Technol.* 10(03):232–5.

Sarker K.K., Hossain A., Timsina J., Biswas S.K., Kundu B.C., Barman A., Murad K.F.I. and Akter F. 2019 Yield and Quality of Potato Tuber and its Water Productivity are Influenced by Alternate Furrow Irrigation in a Raised Bed System. *Agr. Water Manage* 224, 105750.

Wang D., Lu J., Qin S.H., Zhang J.L., Wang D. and Wang W.B. 2015 Effects of Film Mulch and Ridge-furrow Planting on Growth. *Yield and Quality of Potato in Continuous Cultivation* 31(07):28–32.

Wang T.F., Zhang R., Zhang M.H., Zhang Y.S., Lin B.J., Yang C.Y. and Wang C.H. 2020 The Effects of Different Furrow Irrigation on Potato Growth and Quality. *China Rural Water and Hydropower* (04):102–6.

Wang Y., Zhang F.C., Wang H.D., Bi L.F., Cheng M.H., Yan F.L., Fan J.L. and Xiang Y.Z. 2019 Effects of the Frequency and Amount of Drip Irrigation on Yield, Tuber Quality, and Water use Efficiency of Potato in the Sandy Soil of Yulin, Northern Shaanxi, China. *Chinese J. of Applied Ecol.* 30(12):4159–68.

Zhang S. 2011 *Effects of Genetic Factors and Environmental Conditions on Potato Tuber Yield, Quality and Nutrients Uptake*. Inner Mongolia Agr. University.

Zhang S.M., Ning T.Y., Liu Z., Wang B., Sun T., Zhang X.P., He Z.K., Yang Y. and Mi Q.H. 2017 Weed Infestation, Soil Moisture, and Temperature under Mulching Cultivation with Different Films and the Effects on Yield and Quality of Potato. *Acta Agronomica Sinica* 43(04):571–80.

Research progress on coupling effect of water and nitrogen in potatoes

Chenli Zhou
College of Agronomy and Agricultural Engineering, Liaocheng University, Liaocheng, China
College of Water Conservancy and Hydropower Engineering, Gansu Agricultural University, Lanzhou, China

Shouchao Yu
College of Agronomy and Agricultural Engineering, Liaocheng University, Liaocheng, China

Hengjia Zhang*
College of Agronomy and Agricultural Engineering, Liaocheng University, Liaocheng, China
College of Water Conservancy and Hydropower Engineering, Gansu Agricultural University, Lanzhou, China

Xietian Chen, Yingying Wang & Yong Wang
College of Water Conservancy and Hydropower Engineering, Gansu Agricultural University, Lanzhou, China

ABSTRACT: Water and nitrogen fertilizer are important factors affecting crop growth, yield, and quality. With the increasing water scarcity in agriculture and the increasing environmental problems caused by irrational nitrogen application, the research on the selection of appropriate irrigation and fertilization treatment to realize the sustainable development of potatoes with high quality, high efficiency, and stable yield has been widely paid attention to. Understanding the current research progress on the coupling effect of water and nitrogen in potatoes can provide a theoretical basis for further research on the mechanism of the coupling effect of water and nitrogen in potatoes. This paper summarizes the effects of the coupling of water and nitrogen on physiological characteristics, growth, yield, and quality of potato and soil environment in the tillage layer. At the same time, this paper pointed out the shortcomings of the current research on the coupling effect of water and nitrogen in potatoes and prospected future research work.

1 INTRODUCTION

The crop growth process was affected by many factors such as water, fertilizer, air, and heat, among which water and fertilizer play a decisive role. Coupling effects between water and fertilizers refer to the interaction of soil nutrients and water in agricultural systems to produce yield improvement and quality improvement (Cheng 2014). The coupling effect of water and fertilizer on plants can produce a synergistic effect, superposition effect, and antagonistic effect (Zhang *et al.* 2011). Potato (Solanum tuberosum L.) was the fourth most important food crop in the world after rice, wheat, and maize (Jackson 1999). Water and nitrogen affect the area and life span of leaves, and then affect the transport of

*Corresponding Author: *596088683@qq.com

DOI: 10.1201/9781003450818-15

photosynthates to tubers, thus affecting the yield (Zhang *et al.* 1996). Studies have found that the combined application of water and nitrogen was conducive to improving the water and fertilizer use for potatoes, thus improving the yield and quality (Mohammod & Zuraiqi 2003). However, when the allocation of water and nitrogen is unreasonable, antagonistic effects will be formed (Qi *et al.* 2010), which will reduce the utilization efficiency of water and fertilizer, affect plant growth and nutrient absorption, and even affect crop yield and quality (Li *et al.* 2019). Therefore, a reasonable combination of water and nitrogen is required to bring out the best interactive coupling for high quality and high yield of potatoes and efficient use of water and fertilizer resources (Darwish *et al.* 2006). This paper systematically summarized the effects of water-nitrogen coupling on potato physiological characteristics, growth and development, yield, quality, and soil environment in the tillage layer, pointed out the existing problems at the present stage, and prospected future research directions, to provide a theoretical basis for further research on water-nitrogen coupling in potato.

2 EFFECTS OF COUPLING OF WATER AND NITROGEN ON POTATO PHYSIOLOGICAL CHARACTERISTICS

Water was the most important raw material for photosynthesis in green plants, which can directly participate in the metabolic processes in plants (such as photosynthesis and respiration, etc.). Nitrogen is a component of many important compounds in crops such as proteins, nucleic acids, phospholipids, enzymes, and hormones. Therefore, water and nitrogen play a crucial role in plant life activities, which has a certain impact on the physiological characteristics of crops. The study found that the net photosynthetic rate (Pn), stomatal conductance (Gs), relative chlorophyll content, and plant height of potatoes improved with increasing nitrogen under normal or excess water (Yan *et al.* 2022). Under low water conditions, the Pn and Gs of potatoes gradually enhanced with the enhancement in nitrogen application rate (Yin *et al.* 2020). This was because, under low water conditions, the osmotic regulating capacity of the crop can be enhanced by moderate nitrogen application, which increased stomatal conductance and improves net photosynthetic rate. Too much or too little irrigation or fertilization can cause a decrease in relative leaf water content, SPAD, and root vigor, and an increase in relative conductivity, proline, malondialdehyde, soluble sugar, SOD activity, and CAT activity (Liu *et al.* 2015). Under appropriate drought stress, nitrogen fertilization can increase the photosynthetic rate of crop leaves, which was beneficial to plant growth, and alleviate the reduction of production caused by water shortage (Pinheiro *et al.* 2004).

3 EFFECTS OF COUPLING OF WATER AND NITROGEN ON POTATO GROWTH AND YIELD

Among the essential mineral elements of plants, nitrogen was the main limiting factor for crop growth. Water and nitrogen were two factors that both promote and constrain each other during crop growth. Sufficient water promotes plant root growth, while a strong root system facilitates plant absorption of nitrogen nutrients (Hu *et al.* 2016). Reasonable nitrogen fertilizer supply can coordinate the growth of the stem, leaf, and tuber of potato, and maintain the appropriate proportion of photosynthetic organs and storage organs, which was conducive to nutrient accumulation and yield improvement. If nitrogen fertilizer was insufficient, plant growth will be short and slow, leaves will be small and thin, premature senescence, and yield will be reduced. In contrast, excessive application of nitrogen fertilizer can cause early plant growth, late greying, and failure to shift growth centers at the right time, thus reducing yield. With the enhancement of soil moisture, the yield, whole plant biomass, and nitrogen use efficiency of potatoes increased obviously under the same nitrogen application rate (Li *et al.* 2016). Under arid climate conditions, potatoes can achieve favorable yield while reducing irrigation water and nitrogen supply (Badr & El-Tohamy 2012).

4 EFFECT OF COUPLING OF WATER AND NITROGEN ON POTATO QUALITY

Agricultural production processes are pursuing higher and higher requirements for the quality of agricultural products along with their yield. Irrigation and nitrogen application directly affect crop photosynthesis while affecting biomass accumulation, transfer, and distribution in various organs, thus affecting crop fruit yield and quality (Hou *et al.* 2018). The potato tuber mass, tuber starch content, tuber vitamin C content, and water consumption all showed parabolic trend changes with increasing nitrogen application under the same water conditions (Song *et al.* 2013). Excessive and insufficient irrigation amount and nitrogen application will have side effects on potato starch content (Sang *et al.* 2015). Under the irrigation method with a small amount of water and multiple irrigations, potatoes can achieve high yield and good quality at a low nitrogen level, which was an ideal drip irrigation method (Zhou *et al.* 2004). Appropriate irrigation and nitrogen application greatly improved the content of vitamin C, soluble protein, and starch in potato tubers (Gao *et al.* 2015).

5 EFFECTS OF COUPLING OF WATER AND NITROGEN ON THE SOIL PHYSICOCHEMICAL PROPERTIES

Improper irrigation and nitrogen application in agricultural production will lead to water leakage and nitrogen leaching, resulting in groundwater pollution and reducing the efficiency of crop water and nitrogen utilization (Zhang *et al.* 2018). Soil nitrate nitrogen is an effective form of nitrogen uptake by crops, which is not easy to be absorbed by soil particles and easy to be washed with water, which reduces the nitrogen use efficiency. In the potato seedling stage and tuber formation stage, the nitrate content increased with the enhancement of nitrogen application under the same irrigation condition (Shang *et al.* 2019). Soil enzymes participated in the decomposition of soil organic matter and humus formation, nutrient conversion, and cycling processes, providing a suitable growth environment for potatoes, thus increasing their yield (Yao *et al.* 2019). Soil urease, sucrase, and phosphatase activities can characterize soil fertility (Zhang *et al.* 2012). A moderate application of nitrogen fertilizer was beneficial to improve soil organic matter and nutrient content of nitrogen, phosphorus, and potassium, improve soil physical properties and enhance soil sucrase, urease, and phosphatase activities; while excessive nitrogen fertilizer can lead to a decrease in soil pH and soil enzyme activities (Wang *et al.* 2021). The different water and nitrogen treatments had a greater effect on soil urease activity in the 0–20 cm soil layer (Hu *et al.* 2022). Irrigation amount and nitrogenous fertilizer-based topdressing had remarkable influences on soil alkaline phosphatase and soil sucrase (Sun *et al.* 2022).

6 CONCLUSION AND PROSPECT

Compared with the traditional agricultural production mode, rational water and nitrogen management can maximize the advantages of water and nitrogen coupling, which was of great significance for the protection of the agricultural ecological environment, the realization of water fertilizer, fertilizer water transfer, and the full and efficient use of agricultural resources. At present, studies on the coupling of water and nitrogen in potatoes mainly focus on growth morphology and physiology, yield, quality, water, and nitrogen use efficiency, and soil environment. Therefore, the following aspects can be considered for further research in the future: (1) the responses of water and nitrogen coupling on carbon metabolism, transport, and allocation in potatoes; (2) the effect of water and nitrogen coupling on the microstructure, physiological mechanism, and molecular mechanism of potato cells.

ACKNOWLEDGMENTS

This work was financially supported by the Industrial Support Plan Project of the Gansu Provincial Department of Education (No. 2022CYZC-51), the National Natural Science Foundation of China (No. 52269008), and the Key Research and Planning Projects of Gansu Province (No. 18YF1NA073).

REFERENCES

Badr, M.A., El-Tohamy, W.A.: Yield and Water use Efficiency of Potato Grown under Different Irrigation and Nitrogen Levels in an Arid Region. *Agric. Water Manag.* 110, 9–15 (2012).

Cheng, J.H.: Research Progress on the Effect of Water-fertilizer Coupling on Crops. *Modern Agric. Sci. and Technol.* (5), 233–234, 236 (2014).

Darwish, T.M., Atallah, T.W., Hajhassn, S.: Nitrogen and Water use Efficiency of Fertigated Processing Potato. *Agric. Water Manag.* 85 (2), 95–104 (2006).

Gao, X., Li, C., Zhang, M.: Controlled Release Urea Improved the Nitrogen Use Efficiency, Yield, and Quality of Potato (solanum tuberous m l.) On Silt Loamy Soil. *Field Crop. Res.* 181, 60–68 (2015).

Hou, Y.S., Wang, Z.H., Li, W.H.: Effect of Water-fertilizer Coupling on Photosynthetic Characteristics and Relative Chlorophyll Content of Drip Irrigated Jujube in the Sandy Region of South Xinjiang. *J. Drain. and Irrig. Mach Eng.* 36 (9), 914–919, 924 (2018).

Hu, M.Y., Men, F.Y., Zhang, Y.J.: Research Progress on the Effects of Water-nitrogen Interactions on Physiological Characteristics and Nitrogen Utilization of Crops. *J. wheat crop.* 36 (3), 332–340 (2016).

Hu, P.C., Yin, J., Wei, X.D.: Effect of Different Water and Nitrogen Treatments on Potato Quality and Soil Urease Activity. *Jiangsu Agric. Sci.* 50 (06), 87–92 (2022).

Jackson, S.D.: Multiple Signaling Pathways Control Tuber Induction in Potato. *Plant Physiol.* 119, 1–8 (1999).

Li, H.H., Liu, H., Pang, J.: Effects of Water-nitrogen Interactions on Growth and Development and Nutrient Accumulation in Potted Tomatoes. *J. Agric. Mach.* 50 (09), 272–279 (2019).

Li, W.T., Wang, S.W., Deng, X.P.: Effect of Different Water and Nitrogen Levels on Potato Yield and Water and Nitrogen use Efficiency. *Agric. Res. Arid Regions.* 34 (06), 191–196 (2016).

Liu, S.J., Meng, L.L., Meng, M.L.: Study on the Physiological Response of Potato to Water Stress During Tuber Formation and Rehydration After Stress. *J. Irrig. Drain.* 34 (10), 45–51 (2015).

Mohammod, M.J., Zuraiqi, S.: Enhancement of Yield and Nitrogen and Water Use Efficiencies by Nitrogen Drip-fertigation of Garlic. *J. Plant Nutr.* 26 (9), 1749–1766 (2003).

Pinheiro, H.A., DaMatta, F.M., Chaves, A.R.M.: Drought Tolerance about Protection Against Oxidative Stress in clones of Coffea Canephora Subjected to Long-term Drought. *Plant Sci.* 167 (6), 1307–1314 (2004).

Qi, W., Li, F., Lin, Z.: Effects of Irrigation and Nitrogen Application Rates on Nitrate Nitrogen Distribution and Fertilizer Nitrogen Loss, Wheat Yield and Nitrogen Uptake on a Recently Reclaimed Sandy Farmland. *Plant Soil.* 337 (1), 325–339 (2010).

Sang, H.H., Qiu, X.C., Yin, J.: Effect of Water-fertilizer Coupling on the Starch Content of the Potato. *Water Conserv. Irrig.* (05), 5–8 (2015).

Shang, M.X., Fang, Z.G., Liang, B.: Effects of Different Water and Nitrogen Treatments on Yield, Quality and Soil Nitrate Nitrogen Transport in Potatoes under Sub-membrane Drip Irrigation. *North China J. Agric.* 34 (06), 118–125 (2019).

Song, N., Wang, F.X., Yang, C.F.: Effect of Water and Nitrogen Coupling on Yield, Quality and Water Use of Potato Under Drip Irrigation. *J. Agric. Eng.* 29 (13), 98–105 (2013).

Sun, F.B., Yin, J., Wei, X.D.: Evaluation of the Effects of Irrigation and N-fertilizer base Chasing Ratio on Potato Yield, Water, and Fertilizer use Efficiency and Enzyme Activity in Dryland Areas. *China Soil Fertil.* 1–15 (2022).

Wang, X.J., Sun, Y.Q., Yang, J.X.: Effect of long-term Crop Rotation and Fertilization on Soil Nutrients and Yield of Potato. *Chinese Melon Vegetable*. 34 (3), 42–46 (2021).
Yan, W.Y., Qin, J.H., Duan, S.G.: Effect of Water and Nitrogen Coupling on Photosynthetic Characteristics, Tuber Formation and Quality of Potato. *J. Hortic.* 49 (07), 1491–1504 (2022).
Yao, Z.S., Yan, G.X., Wang, R.: Drip Irrigation or Reduced N-fertilizer Rate can Mitigate the High Annual N_2O + NO fluxes from Chinese Intensive Greenhouse Vegetable Systems. *Atmos. environ.* 212, 183–193 (2019).
Yin, J., Zhang, N., Wang, S.: Effect of Different Water and Nitrogen Treatments on Photosynthetic Characteristics and Yield of Potato. *Water Conserv. Irrig.* (06), 8–13 (2020).
Zhang, L.Y., Wang, R.L., Zhang, J.S.: Effect of Water-fertilizer Coupling on Nitrogen Metabolism in Greenhouse Soilless Cucumber. *J. Hortic.* 38 (5), 893–902 (2011).
Zhang, H.L., Smeal, D., Amold, R.N.: Potato Nitrogen Management by Monitoring Petiole Level. *J. Plant Nutr.* 19 (10), 1405–1412 (1996).
Zhou, N.N., Zhang, X.J., Qin, Y.B.: Effect of Different Drip Irrigation and Nitrogen Application on Yield and Quality of Potato. *Soil Fertil.* (06), 11–12 + 16 (2004).
Zhang, Z.X., Chen, P., Nie, T.Z.: Effects of Different Water and Nitrogen Regulation Modes on Soil Nitrogen Distribution and Effectiveness in Rice Fields. *J. Agric. Mach.* 49 (11), 210–219 (2018).
Zhang, J.L., Gao, M.B., Wen, X.X.: Effects of Different Nitrogen Application Measures on Soil Enzyme Activity and CO_2 Emissions in Dry-crop Maize Fields. *J. Ecol.* 32 (19), 6147–6154 (2012).

Water Conservancy and Civil Construction – Oke & Ahmad (Eds)
© 2024 The Author(s), ISBN: 978-1-032-58614-4

Effects of regulated deficit irrigation on potato tuber quality

Lili Chen
College of Agronomy and Agricultural Engineering, Liaocheng University, Liaocheng, China
College of Water Conservancy and Hydropower Engineering, Gansu Agricultural University, Lanzhou, China

Shouchao Yu
College of Agronomy and Agricultural Engineering, Liaocheng University, Liaocheng, China

Hengjia Zhang*
College of Agronomy and Agricultural Engineering, Liaocheng University, Liaocheng, China
College of Water Conservancy and Hydropower Engineering, Gansu Agricultural University, Lanzhou, China

ABSTRACT: This paper introduces the basic situation of potatoes and the basic principle of regulated deficit irrigation. The main indexes of potato tuber quality include appearance quality and the bunching rate. The four indexes of nutritional quality, such as starch content, dry matter content, protein content, and reducing sugar content, were described respectively. It is found that a moderate water deficit in a certain growth period of potatoes can save water, stabilize production and improve the quality of potatoes. Moreover, according to the existing experimental results, the influence of regulated deficit irrigation on potato quality and its significance in water-saving and conditioning is demonstrated.

1 INSTRUCTION

1.1 *Basic overview of potato*

Like tomatoes, peppers, wolfberries, etc., potatoes also belong to the Solanaceae family of tubular flowers, which are extremely sensitive to moisture. Potatoes were born prematurely in South America and are rich in nutrients. Compared with some other food crops, potatoes have relatively high vitamin content and are very comprehensive, so they are soon widely planted in China, India, and other places. Potato is the fourth most important food crop in the world after wheat, corn, and rice, which is helpful to alleviate the global food crisis (Zheng 2014). All along, due to the poor soil environment and backward economic situation in northwest China, farmers' income cannot be increased. However, potatoes have strong adaptability to the land. In many areas, potatoes are used as raw materials and processed into instant products and snack foods, which sell well all over the world. For example, some potato starch processing industries and potato products processing industries in Dingxi City, Gansu Province, China, have quietly emerged and gradually become areas rich in potatoes. China is now a big potato producer in the world. In 2014, the country's total potato output reached 95.515 million tons (Yang 2017); it not only opened up a new way to increase grain production but also stimulated economic benefits, which is of great significance to solving

*Corresponding Author: 596088683@qq.com

 DOI: 10.1201/9781003450818-16

the poverty alleviation problem in poor areas in the northwest meaning. However, in recent years, due to the increase of the planting area of some other cash crops, farmers continue to adopt the traditional concept of extensive flood irrigation, and backward production technology and cultivation management problems, the potato planting area has decreased. The new technology of adjusted deficit irrigation will help to solve the problem of serious waste of water resources and standardize potato planting.

1.2 *Proposition and application of regulated deficit irrigation*

Regulated deficit irrigation technology was first proposed by the Australian Institute of Continuous Agricultural Irrigation in the mid-1970s. The technology is based on traditional irrigation principles and methods and integrates a variety of disciplines to achieve water saving, increase production, increasing production, and economic benefits. The principle is that in certain stages of crop growth and development, by actively applying certain water stress, photosynthetic products can be distributed to different tissues and organs, thereby increasing crop yield (Cui 2009).

In China, the research on regulated deficit irrigation began in the 1980s. After the regulated deficit irrigation was proposed, it was first applied to the cultivation of fruit trees such as pear trees and apple trees. Liang *et al.* used regulated deficit irrigation in the research of pear trees and found that regulated deficit irrigation can reduce irrigation water without affecting the growth and development of pear trees (Liang 2018). Huang *et al.* found that compared with full irrigation, the output of regulated deficit irrigation was not greatly affected, but the irrigation amount decreased by 17%~20%. These two experimental results show that regulated deficit irrigation has the advantage of saving water resources (Huang 2001). Later, China gradually carried out feasibility exploration research on corn, potato, and other food crops. Kang *et al.* found that the growth of corn above ground was reduced under the regulated deficit irrigation treatment, which increased the population density, which in turn was beneficial to increase its yield (Kang 1998). Pan *et al.* show that water deficit in the whole growth period of potatoes had different effects on potato quality. When the deficit was adjusted in the seedling stage, the content of protein and starch increased by 11.76% and 21.55% compared with conventional irrigation (Pan 2021), it can be seen that, from the application of fruit trees to field crops, adjusted deficit irrigation has great advantages in terms of increasing yield, saving water sources, and adjusting the quality.

2 QUALITY FACTORS OF POTATO TUBERS

Potato quality mainly includes three aspects: nutritional quality, appearance quality, and processing quality (Zhang 2020). The main influencing index of appearance quality is the rate of bunching potatoes. The measurement indicators of potato nutritional quality mainly include starch content, dry matter content, protein content, and reduced sugar content in potato tubers.

2.1 *Potato skewer rate*

Potato bunching is a common phenomenon in potato cultivation. The reason is that due to extremely high-temperature weather or irregular rainfall frequency, the tubers that had stopped growing started to grow for the second time. Repeated alternation of dry and wet causes potato deformity, which greatly reduces the yield and quality of potatoes. To reduce the occurrence of such phenomena, it is necessary to adopt the method of adjusted deficit irrigation in a certain growth period of potatoes. The study found that the potato tuber formation period will seriously affect the quality, and this period is also the formation stage of large potatoes. The growth of large potatoes is accelerated, but with the increasing degree of water deficit, the decline in the rate of large potatoes will be more obvious, which will affect the potato rate of its products. Zhang *et al.* found that there was no significant difference between the water deficit in the potato seedling stage and control, while mild and moderate water deficit in the tuber formation stage would increase the potato stringing rate by 12.80% and 16.06% respectively (Zhang 2019).

2.2 *Dry matter content of potato tubers*

Photosynthesis is a very important physiological process of crops, and it is also an important factor that determines the yield of potato tubers. The accumulation of dry matter is also positively correlated with the yield of tubers, that is, the greater the dry matter content of a single potato plant, the greater the yield of tubers big. Studies have shown that the dry matter accumulation rate of potatoes shows a fast-slow-fast curve trend, that is, the dry matter accumulation rate begins to decrease gradually after the maturity period, which was the same as that of Yan *et al.* which pointed out that the photosynthetic capacity of potato leaves began to decrease at the mature stage and the net photosynthetic rate of leaves was the highest at 75% irrigation amount at the ripening stage of potato, while there was no significant difference at 100%, 125%, and 150% irrigation amount, indicating that photosynthesis has a significant effect on the dry matter (Yan 2019). The accumulation of content plays a decisive role.

2.3 *Potato tuber starch content*

The starch content in potato tubers is generally 10%–20%, which is widely used in many fields because of its strong water absorption, convenience, and simple production. Cao *et al.* in Dingxi City studied the tuber formation stage-maturity stage of potatoes and found that the starch content of "Qingshu No. 9" was increased by 144.60% compared with the control when it was moderately deficient (Cao 2019). Ma *et al.* (Ma 2011) pointed out that the accumulation of starch is related to the irrigation quota and irrigation times. When the irrigation times are fixed, the irrigation quota is too small, which will affect the accumulation of starch content, and the water should not be too much during the starch accumulation period, otherwise, it will lead to the starch content of potato tubers was lower, which was consistent with the results of Li *et al.* (Li 2020) that the starch content of no irrigation was increased by 1.28–1.71% compared with other treatments.

2.4 *Protein content of potato tubers*

Protein is an indispensable nutrient for the human body. Common food crops have protein that can provide the human body. The protein content in potato tubers is about 2%. Wang *et al.* pointed out that the degree of water deficit was positively correlated with the protein content of corn in the effect of adjusted deficit irrigation on the physiological indicators of corn. With the increase in the degree of water deficit, the protein content of corn also increased (Wang 2004). In a two-year research experiment, Liu *et al.* found that the grain protein content was increased by 8.1% and 13.4% in the two-year average compared with the control experiment when the wheat heading-maturity period was heavily adjusted (Liu 2019). The most vigorous period of potato growth and development is the tuber expansion period, which is also a key stage that determines the level of potato yield. If the water is deficient at this stage, to maintain the growth of the tubers, a large number of leaves of the potato will dry up and die due to lack of water, thereby reducing the protein content; while in the starch accumulation period, the growth of the stems and leaves of the potato has stopped, and the leaves also begin to wither. Li *et al.* pointed out that the protein content of mild water deficit in the seedling stage was 18.95% higher than that of conventional irrigation (Li 2019). Liu *et al.* found that water deficit during the formation period of potato nuggets can increase the protein content by 21.89% compared with that under conventional irrigation, making the potato quality better (Liu 2018). This indicates that the protein content will increase to different degrees under water stress in different growth periods of potatoes.

2.5 *Reducing the sugar content of potato tubers*

With the rapid development of the food industry and people's strict requirements on dietary health, fried food has also attracted the attention of food processing enterprises. The reduced sugar content of potato tubers is an important detection indicator for fried processing enterprises (Wang 2005). Previous studies have found the human intervention of water stress during the potato growth period, from the tuber formation stage to the tuber expansion stage. After rehydration, the reduced sugar content in the potato tuber will increase with the rehydration. Geng *et al.* believe that the irrigation quota is inversely proportional to the accumulation of reduced sugar, and a smaller irrigation quota is more beneficial to the accumulation of reduced sugar (Geng 2019). However, the content of reduced sugar in potato tubers is not as high as possible. High sugar content makes potatoes sweeter and tastes better, but it is not easy to store. Since the potato itself is very sensitive to moisture and temperature, in a humid and high-temperature environment, it is easy to cause soil compaction, the soil permeability decreases, the respiration of potato tuber cells is weakened, the accumulation of CO_2 is accelerated, and the color of the fries eventually changes deep, affecting sales. Zhang *et al.* pointed out that during the crop growth period, the degree of deficit adjustment is proportional to the soil temperature (Zhang 2022). It can be seen that adjusted deficit irrigation can alleviate the damage caused by high temperatures to the soil by adjusting the temperature of a certain growth period of potatoes, and at the same time reducing the release of CO_2, so the production of reduced sugar in potato tuber is reduced, the color of potato chips is normal, and the quality of potato chips is improved.

3 CONCLUSIONS

China is a big agricultural country, but the shortage of water resources has always been a difficult problem in the process of crop cultivation. Therefore, the contradiction of agricultural water use has become increasingly prominent. Regulated deficit irrigation, as a technology, can exert a positive influence on the growth, yield, and quality of crops through artificial water stress. It is suitable not only for fruit trees and vegetables but also for field crops.

Using a regulated deficit irrigation experiment to study potato quality shows that moderate water deficit is beneficial to water-saving and conditioning. However, to improve potato quality, we should consider its genetic characteristics and other agronomic measures in addition to the factors affecting water. With the rapid development of science and technology, remote sensing technology, geographic information systems (GIS), and global positioning systems (GPS) play an increasingly important role in precision agriculture. In the future potato cultivation experiment, we should make full use of this advantage to realize the need for information acquisition technology in agricultural development.

ACKNOWLEDGMENTS

This work was financially supported by the National Natural Science Foundation of China (No. 52269008), the Industrial Support Plan Project of the Gansu Provincial Department of Education (No. 2022CYZC-51), and the Key Research and Planning Projects of Gansu Province (No. 18YF1NA073).

REFERENCES

Cao, Z. P.: *Effects of Deficit Irrigation on Potato Growth, Yield, Quality, and Water Use*. Gansu Agricultural University, MA thesis (2019).

Cui, N. B.: *Research on Water Efficient Utilization Mechanism and Optimal Deficit Irrigation Model of Pear and Jujube Trees in the Semi-arid Area of Northwest China* [D]. Shaanxi: Northwest A&F University (2009).

Geng, H. J., Yin, J., Wu, J., Liu, Y. Z.: Effects of Different Irrigation Amounts on Potato Growth, Water Consumption and Yield and Quality [J]. *Water Saving Irrigation* (03): 43–47 + 58 (2019).

Huang, X. F., Li, G. Y., Zeng, D. C.: Mechanism and Practice of Regulated Deficit Irrigation Technology for Fruit Trees [J]. *Chinese Journal of Agricultural Engineering* 17(4): 30–33 (2001).

Kang, S. Z., Shi, W. J., Hu, X. T., Liang, L. Y.: Effects of Adjusted Deficit Irrigation on Physiological Indexes and Water Use Efficiency of Maize [J]. *Chinese Journal of Agricultural Engineering* (04): 88–93 (1998).

Li, F. Q., Zhang, H. J., Deng, H. L., Ba, Y. C.: Effects of Drip Irrigation under Mulch on Water use Efficiency, Yield, and Quality of Potatoes [J]. *Water Conservancy Planning and Design* (06): 60–64 (2019).

Li, Y., Zhang, S., Hao, Y. F., Ma, J. C., Wang, C. M., Han, H. J., Gao, L., Gao, J. B.: Effects of Different Irrigation Period Combinations on Potato Yield and Water use Efficiency Under Drip Irrigation Under Film [J]. *North China Agricultural Journal* 35(03):160–167 (2020).

Liang, B. B., Wang, Q. J., Duan, P. W., Liu, T. C., Liu, B. Q., Cheng, F. H.: Effects of Adjusted Deficit Irrigation on Water Content and Chlorophyll Content of Huangguan Pear Leaves [J]. *Northern Gardening.* (14): 53–57 (2018).

Liu, J. Y., Jia, S. H., Liang, Z. G.: Effects of Drip Irrigation Under Oasis Film on the Growth and Quality of Potato [J]. *People's Yellow River* 40(08): 152–156 (2018).

Liu, X. F., Fei, L.J., Duan, A. W., Liu, Z. G., Meng, Z. J.: The Effect of Adjusted Deficit Irrigation on the Yield and Quality of Winter Wheat and its Relationship [J]. *Journal of Soil and Water Conservation* 33(03): 276–282 + 291 (2019).

Ma, W., Yin, J.: *Effects of Different Irrigation Treatments on Potato Tuber Quality and Yield* [J]. *Ningxia Engineering Technology* 10(03): 232–235 (2011).

Pan, X. F., Zhang, H.J., Deng, H. L., Li, F. Q.: Effects of Regulated Deficit Irrigation at Different Growth Periods in Hexi Oasis on Potato Growth, Yield and Quality [J]. *Agricultural Engineering* 11(2): 130–136 (2021).

Wang, M. X., Kang, S. Z., Cai, H. J., Ma, X. H.: Research on the Mechanism of Water-saving Regulation of Corn Deficit Irrigation [J]. *Journal of Northwest A&F University (Natural Science Edition)* (12): 87–90 (2004).

Wang, Y. P., Meng, M. L., Men, F. Y.: Research Progress on the Physiological Basis of Low Reducing Sugar Formation in Potato Tubers [J]. *Inner Mongolia Agricultural Science and Technology* (02): 10–12 (2005).

Yan, S. P., Jiao, R. A., Zhang, J. L., Li, J., Li, C. Z.: Effects of Irrigation Amount on Potato Physiological Characteristics and Tuber Yield and Quality [J]. *Agricultural Research in Arid Regions* 37(03): 41–51(2019).

Yang, K. J.: *Research on the Interaction Effect and Regulation Mechanism of Soil Water and Air in Potato Fields Under Drip Irrigation [D]*. China Agricultural University (2017).

Zhang, H. Z., Guan, Z. X., Li, J. Y.: Research on the Quality Evaluation System of Agricultural Products [J]. *Jiangxi Science* 20(3): 179–182 (2020).

Zhang, J. Y., Zhao, J. H., Yang, W. X., Jiang, Y. W., Liao, K. Halidanmu, T. E. D., Renaguli, K. R. B.: Effects of Regulated Deficit Irrigation on Soil Temperature and Yield of Walnut Trees Under Drip Irrigation [J]. *Xinjiang Agricultural Science* 59(01): 95–104 (2022).

Zhang, W. H., Zhang, H. J., Li, F. Q., Wang, Z. Y., Gao, J., Ba, Y. C.: Effects of Deficit-Adjusted Drip Irrigation at Different Growth Stages on Yield, Quality and Water use Efficiency of Oasis Potato [J]. *North China Agricultural Journal* 34(05): 145–152 (2019).

Zheng, N.: The Spread of Potatoes in China [J]. *Knowledge of Literature and History* 0(1) (2014).

Water Conservancy and Civil Construction – Oke & Ahmad (Eds)

Research progress on potato quality improvement and efficiency enhancement under water stress

Xietian Chen
College of Agronomy and Agricultural Engineering, Liaocheng University, Liaocheng, China
College of Water Conservancy and Hydropower Engineering, Gansu Agricultural University, Lanzhou, Gansu, China

Shouchao Yu
College of Agronomy and Agricultural Engineering, Liaocheng University, Liaocheng, China

Hengjia Zhang*
College of Agronomy and Agricultural Engineering, Liaocheng University, Liaocheng, China
College of Water Conservancy and Hydropower Engineering, Gansu Agricultural University, Lanzhou, Gansu, China

ABSTRACT: As an important crop to ensure national food security, the potato is grown in a high proportion of arid and semi-arid environments. Drought or water scarcity has become a key factor restricting the production of potatoes. This paper reviews the response of potato growth, physiological characteristics, water use efficiency, yield, and quality to water deficit, as well as provides an outlook on future research directions in this area. This review is expected to provide some scientific basis for research on water saving, quality, and efficiency improvement in potatoes in arid and semi-arid regions.

1 INTRODUCTION

Potato has the characteristics of drought resistance, barren tolerance, and wide adaptability. It is widely grown worldwide and is the fourth most important food crop globally, after rice, wheat, and maize. About 60% of the potato in China is grown in mountainous and arid areas, where water availability is a major constraint to potato production (Yin *et al.* 2017). Most potato varieties have relatively shallow roots, which makes them sensitive to water availability. It was found that potato leaves reduce water loss by closing stomata and reducing gas exchange between the cells of the leaves and the external environment under drought-stress conditions. However, this raises the temperature of the leaves and decreases the spread of CO_2 to the leaf cells, thereby decreasing photosynthesis in the leaves (Weisz *et al.* 1994). Although short-term water deficits at any stage of growth can have some effect on the development and yield of potatoes, studies have shown that the greatest effect is at the flowering stage and tuber formation (Lynch *et al.* 1995). Therefore, many studies have attempted to subject potatoes to some water stress at early stages to improve their tolerance to drought at later stages.

Irrigation is indispensable for growing potatoes in arid areas, but in practice, it is common for producers to over-irrigate in pursuit of high yields. Excessive irrigation not only increases water scarcity in arid zones but also has a detrimental impact on the growth and

*Corresponding Author: 596088683@qq.com

DOI: 10.1201/9781003450818-17

development of potatoes due to excess water. Excessive water content in the soil decreases the temperature of the cultivated layer and causes oxygen deprivation in the soil, which may cause early seed decay and delayed emergence of potato seedlings (Holder & Cary 1984). Furthermore, excessive irrigation water can increase the risk of nutrient leaching and reduce the efficiency of fertilizer use, leading not only to plant nutrient deficiencies but also to soil environmental degradation.

In conclusion, both insufficient and excessive water supply have significant effects on normal potato growth and development. Reasonable and scientific strategies for water deficit management may be more beneficial for potato production, and even potato quality and water use efficiency (WUE) may be improved. The purpose of this paper is to review the impacts of water stress on potato growth and development, physiological characteristics, yield, quality, and WUE at home and abroad in recent years, as well as to discuss future research directions in this area. The review is aimed at providing a scientific reference for efficient potato production and water-saving irrigation in arid regions.

2 IMPACTS OF WATER STRESS ON POTATO GROWTH, PHYSIOLOGICAL CHARACTERISTICS, YIELD, WUE, AND TUBER QUALITY

2.1 *Growth*

Under soil moisture stress conditions, potato growth and morphological characteristics are affected, and the magnitude of the impact depends on the period, degree, and duration of the moisture stress. Studies have shown that there is a significant relationship between soil moisture and potato seedling emergence rate, and appropriate soil moisture is beneficial to shorten the emergence time, which leads to increased potato emergence rate and vigorous seedling growth. Water stress after sowing can lead to delayed or no germination of seed potatoes, and sometimes even if germination occurs, it is not easy to break out through the soil (Zhao *et al.* 2014). The potato root system is most sensitive to water stress in the early period of stolon growth, and deficit irrigation at this stage will lead to a decrease in leaf expansion rate and leaf area index (Zhao *et al.* 2014). Soil water deficits between the seedling and tuber formation stages can lead to varying reductions in potato plant height, leaf area, and biomass, and the rate of reduction gradually increases with the extension of water stress time and the increase of stress intensity (Schapendonk *et al.* 1989).

2.2 *Physiological characteristics*

Photosynthetic physiological characteristics are important indicators of the sensitivity of crops to the response of physiological processes to adversity, and studying the response of photosynthetic parameters to soil moisture, it can help elucidate the physiological adaptations of crops in the changing drought environment. Previous studies have shown that the photosynthetic process of potato leaves is highly sensitive to soil water stress. The net photosynthetic rate, transpiration rate, and stomatal conductance of potato leaves decrease under water deficit conditions, and the decrease is greater as the water deficit increases (Rodríguez *et al.* 2016). Potato under soil water stress, the plant will start its own protection system, by reducing its own cell osmotic potential to adapt to the environment, the plant body of proline (Pro) and malondialdehyde (MDA) content increases, and the superoxide dismutase (SOD) activity declines (Kang *et al.* 2011). Studies have found that the smaller the change in Pro and MDA content under water stress, indicating that the drought resistance of potatoes is stronger. Therefore, Pro and MDA content can be used as physiological and biochemical indicators for evaluating the resistance of potato varieties to drought (Tian *et al.* 2009). In addition, under the condition of soil water stress, the contents of chlorophyll a, chlorophyll b, chlorophyll a/b, and total chlorophyll in potato leaves will decrease, and the more severe the extent of soil water stress, the greater the decrease in chlorophyll content values (Liu *et al.* 2004).

2.3 *Yield*

There are marked differences in the influence of different growth stages and different levels of soil water stress on potato yield. Previous related studies have found that the tuber formation stage appears to be the most sensitive stage of potatoes for soil water deficit (Wagg *et al.* 2011). Drought stress at this stage directly affects potato tuber formation, while increased irrigation facilitates tuber formation and later expansion, thereby increasing yield. Drought during the emergence-tuber formation stage significantly reduces the number of large potatoes and the weight of individual potatoes. Drought during flowering or after tuber formation has less impact on the number of potatoes produced. If drought occurs during late tuber expansion, the plant absorbs water from the potato tubers, which significantly reduces the water content in the tubers and leads to a decrease in yield (Iqbal *et al.* 1999). In general, the extent of the impact on potato yield when subjected to water stress at different stages of fertility is tuber formation > seedling > tuber expansion.

2.4 *Water use efficiency*

WUE is an important index reflecting agricultural water management. Studying the WUE of crops is important for the systematic evaluation of agricultural water resource utilization effects. Iqbal *et al.* (Li *et al.* 2021) reported that water stress applied at the early developmental stages of potatoes may result in a significant reduction in yield, but water stress imposed at the maturity stage can increase WUE without adversely impacting yield. Li *et al.* found that appropriate drought stress at the seedling and starch accumulation stages of potatoes was beneficial in increasing the WUE of potatoes. Additionally, it was shown that mild water deficit during tuber formation resulted in the highest WUE in potatoes, while yield reduction was not significant. Therefore, water stress may cause a decrease in potato yield, but appropriate water deficit can improve WUE.

2.5 *Tuber quality*

With economic development and the enhancement of people's material conditions, the quality of agricultural products has received more and more attention. Many studies have shown that appropriate water stress may be beneficial in improving fruit quality, which may be explained by the fact that water deficit avoids redundancy in the growth of crop nutrient organs, while assimilates allocated to fruits increase. Wang *et al.* (2021) showed that the starch and vitamin C content of potatoes tended to increase and then decrease with decreasing irrigation. Carli *et al.* (2014) found that reducing irrigation after tuber formation would have a non-significant effect on yield, but can dramatically increase WUE and starch content. Li *et al.* discovered that soil water stress at the seedling stage of potatoes had no apparent impact on total sugar content. Water deficit in seedlings reduces vitamin C content but favors higher protein and starch content in potatoes. However, Song *et al.* found that the decrease in irrigation reduced the starch content of potatoes. The reasons for this variation could be related to soil conditions, meteorological conditions, potato cultivars, cultivation schemes and methods, and management practices in the study area.

3 CONCLUSION

In potato water management, efficient irrigation modes (for example, drip irrigation and submembrane drip irrigation) have been widely used, and the determination of reasonable irrigation amounts in different growth stages has become the technical bottleneck to achieving efficient potato production. In addition to the different water requirements of different varieties, the water consumption of potatoes varies greatly due to light, temperature, wind speed, and soil texture. The current practice in production is usually to irrigate the soil until it is saturated. There is great blindness in irrigation amount, which often leads to a great waste of water and environmental pollution. Therefore, it is very necessary to fully exploit the water-saving potential of potatoes and work out the regulated deficit irrigation mode of water-efficient utilization.

In addition, as consumers increasingly value quality, efficient irrigation decision-making demands a balance between high yields and high quality. The most effective strategy is to understand the physiological mechanism of potatoes coping with water shortage and improving WUE, accurately measure and estimate potato water requirement, establish a yield-water-quality model under water stress, and actively seek an irrigation strategy that can achieve water saving, quality improvement, efficiency enhancement, and sustainable production.

ACKNOWLEDGMENTS

This work was financially supported by the National Natural Science Foundation of China (No. 52269008), the Industrial Support Plan Project of the Gansu Provincial Department of Education (No. 2022CYZC-51), and the Key Research and Planning Projects of Gansu Province (No. 18YF1NA073).

REFERENCES

Carli, C., Yuldashev, F., Khalikov, D., Condori, B., Mares, V., Monneveux, P.: Effect of Different Irrigation Regimes on Yield, Water use Efficiency and Quality of Potato (Solanum tuberosum L.) in the Lowlands of Tashkent, Uzbekistan: A Field and Modeling Perspective. *Field Crops Research* 163, 90–99 (2014).

Holder, C. B., Cary, J. W.: Soil Oxygen and Moisture in Relation to Russet Burbank Potato Yield and Quality. *American Potato Journal* 61(2), 67–75 (1984).

Iqbal, M. M., Shah, S. M., Mohammad, W., Nawaz, H.: Field Response of Potato Subjected to Water Stress at Different Growth Stages. *Crop Yield Response to Deficit Irrigation* 213–223 (1999).

Kang, Y., Gong, X., Zhao, H., Zhang, L., Zhang, W., Tian, Z., Qiao, H.: Physiological and Biochemical Response of Potato under the Drought Stress in Different Growth Period. *Chinese Agricultural Science Bulletin* (15),97–101 (2011).

Li, F., Deng, H., Wang, Y., Li, X., Chen, X., Liu, L., Zhang, H.: Potato Growth, Photosynthesis, Yield, and Quality Response to Regulated Deficit Drip Irrigation under Film Mulching in a Cold and Arid Environment. *Scientific Reports.* 11(1), 1–16 (2021).

Liu, L., Li, J., Li, C., Xia, P.: Evaluation of Resistance Level to Late Blight in the Populations Derived from the Horizontal Resistant Crosses of Potatoes. *Chinese Potato Journal* (04), 201–204 (2004).

Lynch, D. R., Foroud, N., Kozub, G. C.: Fames, B. C. The Effect of Moisture Stress at Three Growth Stages on the Yield, Components of Yield and Processing Quality of Eight Potato Varieties. *American Potato Journal* 72(6), 375–385 (1995).

Rodríguez, P. L., Sanjuanelo, C. D., Ñústez, L. C. E., Moreno-Fonseca, L. P.: Growth and Phenology of Three Andean Potato Varieties (Solanum tuberosum L.) under Water Stress. *Agronomia Colombiana* 34(2), 141–154 (2016).

Schapendonk, A. H. C. M., Spitters, C. J., Groot, T. P. J.: Effects of Water Stress on Photosynthesis and Chlorophyll Fluorescence of Five Potato Cultivars. *Potato Research.* 32(1) 17–32 (1989).

Song, N., Wang, F., Yang, C., Yang, K.: Coupling Effects of Water and Nitrogen on Yield, Quality and Water Use of Potato with Drip Irrigation under Plastic Film Mulch. *J. Transactions of the Chinese Society of Agricultural Engineering.* 29(13), 98–105 (2013).

Tian, F., Zhang, Y., Ma, J., Sun, D., Sun, Y.: Free Proline Content,Water Potential and Drought Resistance in Leaves of Different Potato Varieties. *Crops* (02), 73–76 (2009).

Treeby, M. T., Henriod, R. E., Bevington, K. B., Milne, D. J., Storey, R.: Irrigation Management and Rootstock Effects on Navel Orange [Citrus sinensis (L.) Osbeck] Fruit Quality. *Agricultural Water Management* 91(1–3), 24–32 (2007).

Weisz, R., Kaminski, J., Smilowitz, Z.: Water Deficit Effects on Potato Leaf Growth and Transpiration: Utilizing Fraction Extractable Soil Water for Comparison with Other Crops. *American Potato Journal* 71 (12), 829–840 (1994).

Wagg, C., Hann, S., Kupriyanovich, Y., Li, S.: Timing of Short Period Water Stress Determines Potato Plant Growth, Yield and Tuber Quality. *Agricultural Water Management* 247, 106731 (2021).

Wang, H., Cheng, M., Zhang, S., Fan, J., Feng, H., Zhang, F., Wang, X., Sun, L., Xiang, Y.: Optimization of Irrigation Amount and Fertilization Rate of Drip-fertigated Potato Based on Analytic Hierarchy Process and Fuzzy Comprehensive Evaluation Methods. *Agricultural Water Management*. 256, 107130 (2021).

Yin, Z., Guo, H., Feng, Y., Xiao G.: Research Progress of Potato Physiology Under Drought Tolerance. *Chinese Potato Journal* (04), 234–239 (2017).

Zhao, H., Wang, R., Ma, B., Xiong, Y., Qiang, S., Wang, C., Liu, C., Li, F.: Ridge-furrow with Full Plastic Film Mulching Improves Water use Efficiency and Tuber Yields of Potato in a Semiarid Rainfed Ecosystem. *Field Crops Research* 161, 137–148 (2014).

Water Conservancy and Civil Construction – Oke & Ahmad (Eds)
 ISBN: 978-1-032-58614-4

Advances in research on deficient irrigation of potatoes

Dan Wen
College of Agronomy and Agricultural Engineering, Liaocheng University, Liaocheng, China
College of Water Conservancy and Hydropower Engineering, Gansu Agricultural University, Gansu, China

Shouchao Yu
College of Agronomy and Agricultural Engineering, Liaocheng University, Liaocheng, China

Hengjia Zhang*
College of Agronomy and Agricultural Engineering, Liaocheng University, Liaocheng, China
College of Water Conservancy and Hydropower Engineering, Gansu Agricultural University, Gansu, China

Dandan Su
College of Water Conservancy and Hydropower Engineering, Gansu Agricultural University, Gansu, China

ABSTRACT: Regulated deficit irrigation is to make use of the characteristics of crop growth and development to properly treat the crop water deficit to achieve the purposes of ensuring production, increasing production, improving quality, and saving water. This paper summarized the effects of regulated deficit irrigation on potato growth and development, photosynthetic characteristics, water use efficiency, yield, and quality, and put forward solutions to the problems faced by regulated deficit irrigation, which provided a theoretical basis for water saving, high yield, high quality, and high efficiency in potatoes.

1 INTRODUCTION

The potato is one of the world's four major food crops, with high nutritional value, a short growth cycle, high adaptability, and high yield (Li 2020; Wei 2018). Potato has the properties of being used as a food, vegetable, and feed, and its processing uses are diverse, so it has a high potential for increasing production and income (Jiang 2015). The main producing areas of potatoes are widely distributed in China, Russia, Poland, the United States, and Mexico, while the main producing areas of potatoes in China are distributed in cooler climates such as Northeast China, Northwest China, Inner Mongolia, North China, and Yunnan-Guizhou (Zhan 2021).

Under the current climate change, urban, industrial, and agricultural water is being redistributed. Therefore, with the increasing shortage of water resources, water resources should be rationally planned and managed to save water and increase production (Ijaz-Ul-Hassan 2021). The concept of regulated deficit irrigation is well known to solve the contradiction between supply and demand for agricultural water. The so-called regulated deficit irrigation artificially imposes a certain water deficit during a certain period of crop growth

*Corresponding Author: 596088683@qq.com

 DOI: 10.1201/9781003450818-18

and development, thus affecting the redistribution of photosynthetic products to different tissues and organs and avoiding the growth redundancy of vegetative organs while increasing economic output (Chen 2021). At present, domestic and foreign scholars have carried out research on regulated deficit irrigation given the difficulties faced by potato planting, but the related research results are few, so its theoretical system is not perfect enough, resulting in less practical application, and most areas are still in the theoretical research stage. In this paper, the current regulated deficit irrigation theory of potatoes at home and abroad was summarized, mainly from the effects of regulated deficit irrigation on potato growth and development, photosynthetic characteristics, water use efficiency, yield, and quality. At present, the solutions to the problems faced by potato-regulated deficit irrigation are put forward.

2 GROWTH AND DEVELOPMENT

Appropriate water deficit can regulate the growth and development of crops, prevent excessive growth of their roots, stems, and leaves, control all parts of crops to maintain optimal growth, avoid growth redundancy, and improve water use efficiency (Wang 2015). Zhang (2019) showed moderate water deficits in potato tuber expansion periods will cause irreversible damage to the growth of leaf area and plant height. Pan (2021) studied the effects of regulated deficit irrigation at different growth stages on the growth, yield, and quality of potatoes in Hexi Oasis. It was found that the plant height, stem diameter, and biomass per plant were all reduced by water deficit treatment at different growth stages of the potato, especially at the tuber expansion stage. Liu (2018) found that the changes in plant height, stem diameter, and leaf area index of potatoes in different growth stages were similar, and the effects on the tuber formation stage, starch accumulation stage, and tuber expansion stage increased in turn. Li (2015) found that there were significant differences in leaf dry weight, stem dry weight, and tuber dry weight between mild and moderate water deficit adjustment treatments in the potato tuber expansion period, so the most sensitive water condition in the whole growth period of potatoes was the tuber expansion period. Yan (2022) used pot experiments to show that under all water deficit treatments, the average plant height of late-maturing varieties was lower than that of early-maturing varieties. Therefore, under mild water deficit treatment, both early and late potato varieties can be planted. However, under severe water deficit treatment, late-maturing potatoes are more suitable for planting.

3 PHOTOSYNTHESIS CHARACTERISTICS

The water condition in plants has an important influence on various physiological and biochemical metabolic processes. With the aggravation of the water deficit, the water condition in plants changes, which reduces the photosynthetic rate of leaves. Photosynthesis is an important metabolic process in plants, and its strength has an important influence on the growth and development of plants (Guo 2008; Shangguan 1990; Su 2020). Yan (2022) used pot experiments to show that the average decrease in transpiration rate, net photosynthetic rate, and stomatal conductance of late-maturing potato varieties were greater than that of early-maturing potato varieties after severe water deficit treatment. Xue (2018) showed that the leaf area per plant and leaf area index of potatoes decreased significantly after water deficit treatment at the seedling stage. Water deficit treatment in the potato starch accumulation period will significantly reduce its photosynthetic efficiency. Pan (2022) used a pot experiment to study that during potato tuber development, soil water deficit will significantly reduce chlorophyll content, and the photosynthetic characteristics index value will decrease with the increase of deficit degree. Li (2021) studied the response of potato growth, photosynthesis, yield, and quality to drip irrigation with film deficit under low temperatures and found that the net photosynthetic rate, stomatal conductance, and transpiration rate decreased significantly when water deficit treatment was applied in the tuber formation stage and starch accumulation stage. This is consistent with the results of Wen (2014). Liu (2018) showed that water stress has a significant effect on potato photosynthesis. After rehydration, photosynthesis has been restored, but it still has an effect compared with that before water stress. The main reason is that potato indirectly affects leaf Pn through changes in the relative content of chlorophyll (SPAD), transpiration rate (Tr), CO_2 concentration (Ci), and stomatal conductance (Gs), thereby affecting the photosynthesis mechanism, which is ultimately reflected in the yield. Hu (2021) studied the effects of water stress and rehydration on the growth and yield of potatoes and found that the number of leaves and population leaf area decreased significantly when water deficit treatment was carried out in the growth period of potatoes, and the significant degree increased with the increase in deficit degree.

4 WATER USE EFFICIENCY

It is found that regulated deficit irrigation technology can reduce the irrigation number of crops to a certain extent, improve water use efficiency, ensure stable economic output, and improve crop quality (Qi 2022). Zhang (2019) showed that the water consumption of potatoes was significantly reduced after water deficit treatment in the tuber formation stage and tuber expansion stage. Pan (2021) found that water deficit treatment at different growth stages of potatoes can improve their irrigation water use efficiency, especially in the seedling stage and tuber formation stage. Considering the growth and development, yield, quality, and water use efficiency of potatoes, the best water control treatment is mild water stress with soil relative water content of 55%~65% at the seedling stage and soil relative water content of 65%~75% at other growth stages. Xue (2018) showed that the yield, irrigation water use efficiency, and water use efficiency of potato tubers under water deficit treatment in the expanding stage decreased significantly, while mild water deficit treatment in other growth stages could improve crop water use efficiency and reduce daily water consumption intensity and water consumption modulus.

5 YIELD AND QUALITY

Water is the key factor that restricts the growth, yield, and quality of crops. Adjusting water deficits in some growth stages of crops can improve the total economic benefits of crops. Zhang (2019) found that the yield of potatoes was significantly increased by a slight water deficit during tuber formation. During the starch accumulation period, the quality of starch was significantly improved by a slight water deficit. The yield of potatoes decreased significantly after moderate water deficit treatment during tuber expansion. Pan (2021) showed that mild water deficit in the potato seedling stage could significantly increase the contents of protein, starch, amino acids, and Vc, while water deficit treatment in other growth stages could reduce the yield of potatoes, especially in the tuber expansion stage. Liu (2018) show that compared with a water deficit in other growth periods, a moderate water deficit in the potato block formation period promotes potato growth and development, improves potato quality, and thus significantly increases potato nutritional value and yield. When potatoes were treated with water deficit during tuber formation, the contents of total sugar, Vc, and starch were significantly increased, the contents of organic acids were significantly decreased, and the fruit quality was improved. Li (2015) found that when potato tubers were treated with water deficit adjustment during the expansion period, the yield decreased with the increase of deficit adjustment degree. Li (2019) found that moderate water deficit in the tuber formation stage, mild water deficit in the starch accumulation stage, water deficit treatment in the seedling stage, and tuber expansion stage had significant effects on tuber setting rate per potato plant. During the starch accumulation period, water deficit treatment significantly reduces its quality, which is not conducive to the accumulation of main nutrients.

6 EXISTING PROBLEMS AND SOLUTIONS

Through research on regulated deficit irrigation of potatoes at home and abroad, it is found that regulated deficit irrigation not only saves water and increases production but also faces many problems.

Most of the specific time of potato deficit adjustment depends on people's planting experience or some simple measuring methods such as a tensiometer. However, these methods have some disadvantages, such as inaccurate observation results and being time-consuming, etc. With the continuous development of electronic information technology in China, the combination of agricultural planting and information technology will be a new trend in agricultural development. We should make full use of science and technology to realize all-weather observation of farmland soil moisture in the whole growth stage and supplement crops with water in a timely and appropriate amount.

A single discipline to study the impact of regulated deficit irrigation on potatoes has certain limitations. We should cross-integrate multiple disciplines, not only to observe their macro impact but also to pay attention to micro changes. Further, the response relationship between water deficit regulation and crop growth should be studied and a more reasonable deficit regulation irrigation system should be formulated.

At present, regulated deficit irrigation of potatoes is mostly limited to the relationship between water use and yield, and there is little research on regulated deficit irrigation and quality. With the continuous development of China's economy, people's requirements for quality of life are constantly improving, which also inspires us not only to pay attention to output but also to focus on quality research. We should gradually transform the study of quantity into the pursuit of quality.

7 CONCLUSION

At present, furrow irrigation and border irrigation are the main irrigation methods for potatoes, but this traditional irrigation method will affect the soil structure and physical and chemical properties, thus affecting crop growth, causing waste of water resources, reducing water use efficiency, and restricting the development of the potato industry. Therefore, in future agricultural production, we should carry out water deficit irrigation according to the characteristics of potato growth and development to achieve yield increase, high quality, and high efficiency. The data monitoring method in traditional regulated deficit irrigation is an important factor that restricts its development. However, with the continuous development of science and technology, new monitoring methods and monitoring equipment will emerge as time requires, and the combination of science and technology with agriculture will be the new trend of agricultural development.

ACKNOWLEDGMENTS

This work was financially supported by the National Natural Science Foundation of China (No. 52269008), the Industrial Support Plan Project of the Gansu Provincial Department of Education (No. 2022CYZC-51), and the Key Research and Planning Projects of Gansu Province (No. 18YF1NA073).

REFERENCES

Chen X.T., Zhang H.J., Lei L., and Zhan U.K.: Research Progress of Regulated Deficit Irrigation of Sunflower. *Agr. Eng.* 11 (01), 69–72 (2021).

Guo C.F., Sun Y., and Zhang M.Q.: Effects of Soil Water Stress on Photosynthesis-light Response Characteristics of Tea Plants. *Chin. Ecol. Agr.* (06), 1413–1418 (2008).

Hu M.M., Zhang J.Z., Zhang L.F., Liu Y.H., and Huang P.J.: Effects of Water Stress and Rehydration on Potato Growth and Yield. *Agr. Res. Arid Areas.* 39(02), 95–101+121 (2021).

Ijaz-Ul-Hassan S., Khan A., and Erum S., F.: Effect of Deficit Drip Irrigation on Yield and Water Productivity of Potato Crop. *Agr. Extension.* 9, 239–244 (2021).

Jiang C.C., and Hu J.L.: Research on Potato Staple Food Strategy: A Review. *Shandong Agr. Univ.* 17(02), 52–58 (2015).

Li F., Deng H., Wang Y., Li X., Chen X., Liu L., and Zhang H.: Potato Growth, Photosynthesis, Yield, and Quality Response to Regulated Deficit Drip Irrigation under Film Mulching in a Cold and Arid Environment. *Sci. Rep.* 11(01), 15888 (2021).

Li F.Q., Zhang H.J., Deng H.L., and Ba Y.C.: Effects of Deficit Adjustment of Drip Irrigation under Mulch on Water use Efficiency, Yield and Quality of Potato. *Water Conservancy Planning Design.* (06), 60–64 (2019).

Li X.Z., Zhang H.J., Deng H.L., Yang X.T., Li J., and Ba Y.C.: Effects of Deficit Adjustment of Drip Irrigation under Mulch on Biomass Allocation, Yield and Water use Efficiency of Potato in an Oasis. *North China Agr.* 30(05), 223–231 (2015).

Li Y., Wang J., Tang J.Z., Zhang J., Hu Q., Pan Z.H., and Pan X.B.: Analysis of Production Characteristics, Limiting Factors and Countermeasures of Main Potato Producing Areas in China. *China Potato.* 34(06), 374–382 (2020).

Liu J.Y., Jia S.H., and Liang Z.K.: Effect of Drip Irrigation under Film on Potato Growth and Quality in an Oasis. *People's Yellow River.* 40(08), 152–156 (2018).

Liu S.J., Meng M.L., Chen Y.J., and Jiao R.Z.: Changes of Photosynthetic Characteristics of Potato Leaves Under Water Stress and its Response Mechanism. *Northwest A&F Univ.* 46(08), 29–38 (2018).

Pan N., Su W., Zhou Y., and Wang J.: Effects of Soil Water Stress on Photosynthetic Characteristics and Yield of Potato. *Hebei Agricultural Science.* 26(01), 70–75+94 (2022).

Pan X.F., Zhang H.J., Deng H.L., and Li F.Q.: Effects of Regulated Deficit Irrigation at Different Growth Stages on Potato Growth, Yield, and Quality in Hexi Oasis. *Agr. Eng.* 11(02), 130–136 (2021).

Qi Q., Sun Z.J., Zhu W.T., Yu Z., Li X.Q., and He J.: Effects of Regulated Deficit Irrigation on Yield, Quality and Water use Efficiency of Underground Infiltration Irrigation Cabbage. *Water-Saving Irrigation.* (01), 36–41 (2022).

Shangguan Z.P. and Chen P.Y.: Effects of Water Stress on Photosynthesis of Wheat Leaves and its Relationship with Drought Resistance. *Northwest Botany.* (01), 1–7 (1990).

Su Y., Fan M.S., Jia L.G., Qi M., Yang C.P., and Wu Y.: Real-time Monitoring of Potato Water Deficit Based on Hyperspectral. *China Potato.* 34(03), 180–186 (2020).

Wang J.Y., Li P.F., Cheng Z.G., Lu A., Batool G.C., Zhu Y., and Xiong Y.C.: Ideal Plant Type and Growth Redundancy of Dryland Wheat. *Ecol.* 35(08), 2428–2437 (2015).

Wei Q.H., Wang C., Fan C.M., Dai Q.L., and Peng E.R.: Review on Research Progress of Potato Planting Irrigation Methods and Fertilization. *Jiangsu Agr. Sci.* 46 (24), 20–23 (2018).

Wen A.C., Zhang H.J., and Zhang J.D.: Photosynthetic Physiology Characteristics of Potato (Solanum tuberosum) at Tuber Initiation Responses to Water Deficit Regulated with Mulched Drip Irrigation. *Adv. Mater. Res-Switz.* 838, 2370–2373 (2014).

Xue D.X., Zhang H.J., Ba Y.C., Wang Y.C., and Wang S.J.: Effects of Regulated Deficit Irrigation on Growth, Yield and Water use of Potato Under Film Drip Irrigation in a Desert Oasis. *Agr. Res. Arid Areas.* 36(04), 109116+132 (2018).

Yan W.Y., Qin J.H., Duan S.G., Xu J.F., Jian Y.Q., Jin L.P., and Li G.C.: Effects of Water Stress on Physiological Characteristics of Potatoes with Different Maturity. *Chin. Veget.* (05), 44–52 (2022).

Zhan B.C., Liang Y., Guo W.Z., Li Y.K., and Li L.: Research Progress and Development Trend of Water-saving Technology of Potato. *Agr. Technol.* 41 (16), 73–77 (2021).

Zhang W.H., Zhang H.J., Li F.Q., Wang Z.Y., Gao J., and Ba Y.C.: Effects of Regulated Deficit Drip Irrigation at Different Growth Stages on Potato Yield, Quality and Water use Efficiency in an Oasis. *North China Agr.* 34(05), 145–152 (2019).

Water Conservancy and Civil Construction – Oke & Ahmad (Eds)
© 2024 The Author(s), ISBN: 978-1-032-58614-4

Analysis of water diversion and water quality improvement based on a two-dimensional mathematical model

Yu Wang
School of Naval Architecture and Maritime Zhejiang Ocean University Zhoushan, China
School of Water Conservancy and Environmental Engineering Zhejiang University of Water Resources and Electric Power Hangzhou, China

Dongfeng Li*
School of Water Conservancy and Environmental Engineering Zhejiang University of Water Resources and Electric Power Hangzhou, China

Zihao Li
Qingtian County Organization Department, Qingtian, Zhejiang, China

Donghui Hu & Aijun Sun
Yuyao Water Conservancy Bureau, Yuyao, Zhejiang, China

Zhenghao Li
Department of Water Conservancy and Engineering Henan Vocational College of Water Conservancy and Environment, Zhengzhou, Henan, China

ABSTRACT: To improve the water environment of the central river, a two-dimensional water flow and water quality mathematical model is used to formulate a water diversion scheme to improve the water quality according to the water diversion gate and pumping station of Sanhui district planned by the river system in Jiangbin district, according to the water diversion flow of the new and existing gate stations. There are two diversion lines in the water diversion scheme in this paper. By controlling the gate and flow rate, the changing process of flow velocity, BOD concentration, and DO concentration with time and along the way is obtained, which shows that the method is effective.

1 INTRODUCTION

The riverside area of Shaoxing Binhai New City is an important part of Shaoxing Binhai New City. The riverside area of the coastal new town ranges from the Qiantang River in the north to the Cao 'e River in the southwest, and the Jiashao Expressway and Lihai Town boundary under construction in the east, including the entire town area of Lihai Street and its vast reclamation area in the north. Yang *et al.* (2010) calculated the economic benefits of water environment improvement, and Zhao *et al.* (2007) used a two-dimensional water quality mathematical model to optimize the water diversion scheme of the planned Jinshan Lake. Li *et al.* (2012) designed the water diversion and distribution scheme based on the improvement of hydraulic conditions; Zhang *et al.*'s (2020) research shows that the Qiputang water diversion and drainage project played an important role in improving water quality. Gu *et al.* (2019) established a one-dimensional water quality model in the east plain of the Dajiang River to analyze the water

*Corresponding Author: lidf@zjweu.edu.cn

DOI: 10.1201/9781003450818-19

diversion conditions and the improvement effect of water quality in one section. Jiang *et al.* (2013) used a two-dimensional hydrodynamic-water quality model to optimize the diversion scheme of the central lake and determined the optimal diversion scheme. Lu *et al.* (2019) constructed a one-dimensional unsteady hydrodynamic water quality model to simulate the effect of water diversion schemes and dispatching rules on water quality improvement in the study area. Gu *et al.* (2011) established a water quantity and quality model to analyze and calculate the water environment improvement scheme of Wujiang City. Zhu *et al.* (2021) used the MIKE11 water quality coupling model to simulate and analyze the influence of different water diversion schemes on the water quality of the water receiving area in the front of the eastern route of the South-to-North Water Diversion Project. Hua *et al.* (2017) used the Environmental Fluid Dynamics Code (EFDC) to establish a three-dimensional hydrodynamic and water quality model of cascade reservoirs to determine the main pollution sources. In the case of West Lake water diversion to improve water quality, Zhang *et al.* (2018) concluded that reasonable water diversion is conducive to the improvement of water quality in the case of West Lake water diversion. Based on the above research, a reasonable water diversion scheme can effectively improve water quality. The study area is the river network lake in the Sanhui district of the riverside planning of Binhai New City. Nanjiang Gate is the main water source of the Sanhui district. Based on the two-dimensional hydrodynamic model, this paper designs a separate water diversion scheme for the Sanhui District Nanjiang gate.

2 TWO-DIMENSIONAL MATHEMATICAL MODEL OF RIVER AND LAKE HYDRODYNAMIC ENVIRONMENT

2.1 *Basic theory*

The hydrodynamic water environment modeling theory of river network lakes includes the basic equations and definite conditions of water flow movement and water quality movement under its action.

The two-dimensional unsteady equations are:

$$\begin{aligned}
&\frac{\partial h}{\partial t}+\frac{\partial h\bar{u}}{\partial x}+\frac{\partial h\bar{v}}{\partial y}=hS\frac{\partial h\bar{u}}{\partial t}+\frac{\partial h\bar{u}^2}{\partial x}+\frac{\partial h\bar{u}\bar{v}}{\partial y}=f\bar{v}h-gh\frac{\partial \eta}{\partial x}-\frac{h}{\rho_0}\frac{\partial p_a}{\partial x}-\frac{gh^2}{2\rho_0}\frac{\partial \rho}{\partial x}\\
&+\frac{\tau_{sx}}{\rho_0}-\frac{\tau_{bx}}{\rho_0}-\frac{1}{\rho_0}\left(\frac{\partial S_{xx}}{\partial x}+\frac{\partial S_{xy}}{\partial y}\right)+\frac{\partial}{\partial x}\left(hT_{xx}\right)+\frac{\partial}{\partial x}\left(hT_{xy}\right)+hu_sS\frac{\partial h\bar{v}}{\partial t}+\frac{\partial h\bar{v}^2}{\partial x}\\
&+\frac{\partial h\bar{u}\bar{v}}{\partial y}=-f\bar{u}h-gh\frac{\partial \eta}{\partial y}-\frac{h}{\rho_0}\frac{\partial p_a}{\partial y}-\frac{gh^2}{2\rho_0}\frac{\partial \rho}{\partial y}+\frac{\tau_{sy}}{\rho_0}\frac{\tau_{by}}{\rho_0}-\frac{1}{\rho_0}\left(\frac{\partial S_{yx}}{\partial x}+\frac{\partial S_{yy}}{\partial y}\right)\\
&+\frac{\partial}{\partial x}\left(hT_{xy}\right)+\frac{\partial}{\partial y}\left(hT_{yy}\right)+hv_sS
\end{aligned} \tag{1}$$

In the formula: t is time (s); x and y are the three-axis directions of the Cartesian coordinate system; η is the water level (m); C was Xiecai coefficient ($m^{1/2}/s$); h is the total water depth (m); $h=\eta+d$, d is the static water depth; uandv are the velocity components (m/s) along the water depth in the x and y direction; μ is the viscous force coefficient (pa·s), g is the acceleration of gravity (m/s^2); ρ_0 is fluid density (kg/m^3); f is the Coriolis force coefficient, $f=2\omega\sin\varphi$; P_a is atmospheric pressure (pa); S_{yy}, S_{xx}, and S_{xy} are the radiation stress components; S is the source item; v_s and u_s are the flow rates. $\bar{u}\bar{v}$ is the average velocity along the water depth, defined by the following formula:

$$h\bar{u}=\int_{-d}^{\eta}u\,dz,\quad h\bar{v}=\int_{-d}^{\eta}v\,dz \tag{2}$$

T_{ij} is the horizontal viscous stress, including viscous force and horizontal convection, derived from the eddy current equation based on the velocity gradient averaged along the water depth:

$$T_{xx} = 2A\frac{\partial \overline{u}}{\partial x}, \quad T_{xy} = A\left(\frac{\partial \overline{u}}{\partial y} + \frac{\partial \overline{v}}{ox}\right), \quad T_{yy} = 2A\frac{\partial \overline{v}}{\partial y} \tag{3}$$

The water quality equation is:

$$\frac{\partial C}{\partial t} + u\frac{\partial C}{\partial x} + v\frac{\partial C}{\partial y} + w\frac{\partial C}{\partial z} = D_x\frac{\partial^2 C}{\partial z^2} + D_y\frac{\partial^2 C}{\partial z^2} + D_z\frac{\partial^2 C}{\partial z^2} + S_C + P_C \tag{4}$$

where C is the concentration of the state variable; D_x, D_y, and D_z are the diffusion coefficients; P_C is the process item of the ECO Lab; S_C are source sinks; $D_x\frac{\partial^2 C}{\partial z^2} + D_y\frac{\partial^2 C}{\partial z^2} + D_z\frac{\partial^2 C}{\partial z^2}$ is a diffusion term; $u\frac{\partial C}{\partial x} + v\frac{\partial C}{\partial y} + w\frac{\partial C}{\partial z}$ is a convection term. Each variable is coupled by P_C nonlinear or linear.

2.2 *Definite conditions*

The definite conditions include initial conditions and boundary conditions.

(1) Initial conditions
The initial value of the mathematical model, given the initial time of the river network water level, is 2.8 m, and the flow rate is 0; the initial condition of water quality is V class water, namely, BOD is 10 mg/l, DO is 2 mg/l.

(2) Boundary conditions
In the hydrodynamic mathematical model, there are six different boundary conditions, which namely the land boundary condition (zero vertical velocity), land boundary conditions (zero velocity), velocity boundary condition, flux boundary conditions, water level boundary conditions and flow boundary conditions. Each boundary must contain at least two nodes.

The initial boundary conditions use interpolation. Linear interpolation or piecewise cubic interpolation can be used when boundaries require time and space.

This calculation is the introduction of clean water for class III water, that is, BOD is 4mg/l, DO is 5mg / l, and the diversion port is given planning and design flow.

3 SANHUI NANJIANG GATE SEPARATE WATER DIVERSION SCHEME

Nanjiang Gate living water scheme is Nanjiang Gate diversion flow of $27m^3/s$ based on the opening of Yuwei Gate, improving the central river water environment.

3.1 *Calculating boundary conditions*

Table 1. Calculating boundary conditions.

Gate	Nanjiang Gate	Huayang Gate	Xinlian Gate	Lianyi Gate	Weimin Gate	Yuwei Gate
Flow rate (m^3/s)	27	5.13	5.13	5.13	5.13	Open

3.2 *Water diversion line 1*

The flow velocity vector diagram of the third diversion scheme of the Nanjiang gate is shown in Figure 1, and the diversion line 1 is shown in Figure 2. From Figure 3, it can be seen that the water level elevation is gradually increased within 1 day and 12 hours after the diversion, and the water level elevation is gradually decreased along the way. The water level changes along the way are similar in each period. As can be seen from Figure 4 velocity changes along the way, and the overall fluctuation of flow velocity within 12 hours of water diversion is relatively large.

From Figure 5 along the BOD concentration changes, it can be seen in the figure that from May 7 to May 8 at 12 o 'clock, the overall BOD concentration decreased significantly to 8 at 12 o 'clock. Most of the BOD concentration has dropped to 3.5 mg/l, reaching the three types of water standards. However, the BOD concentration did not change significantly within 12 hours at the beginning of 5–8 km from the water inlet, but the BOD concentration was greatly improved after 12 hours, reaching the standard of three types of water. The reason is that the opening of Yuwei Gate makes the living water of the whole line flow.

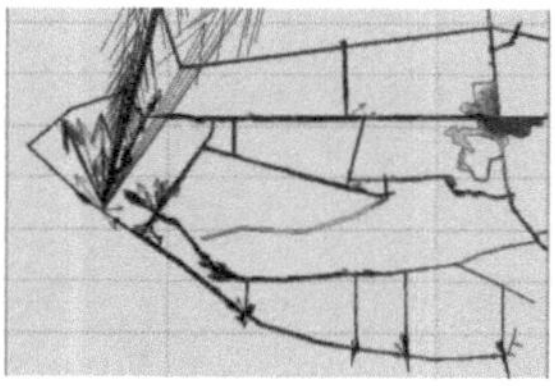

Figure 1. Velocity vector diagram.

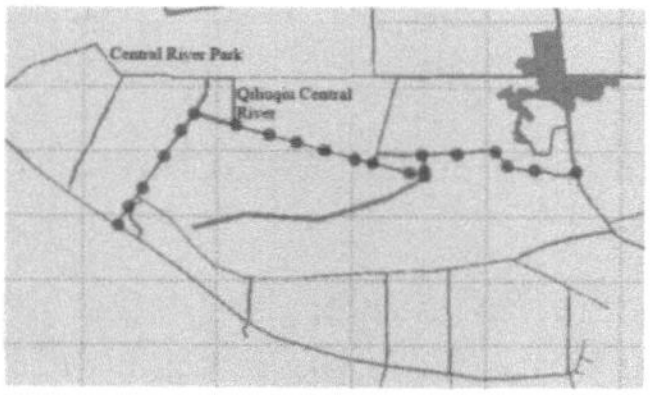

Figure 2. Line-location.

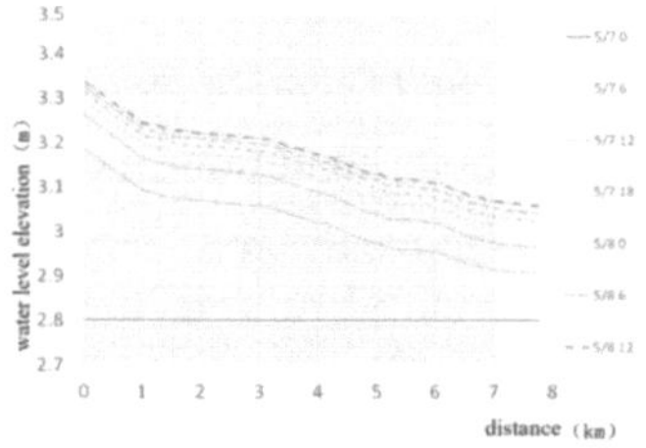

Figure 3. Variation of water level elevation.

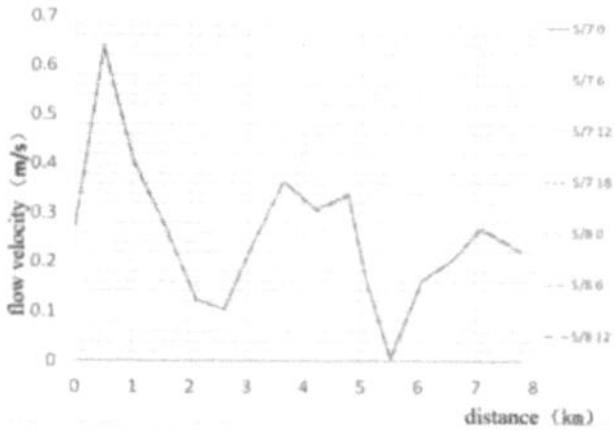

Figure 4. Velocity changes with flow.

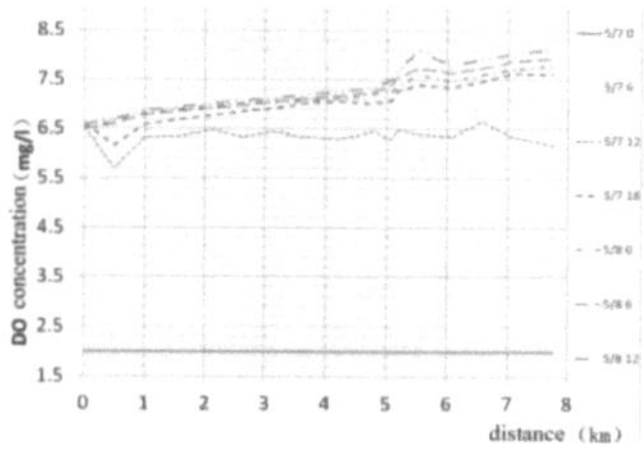

Figure 5. BOD concentration changes along the way.

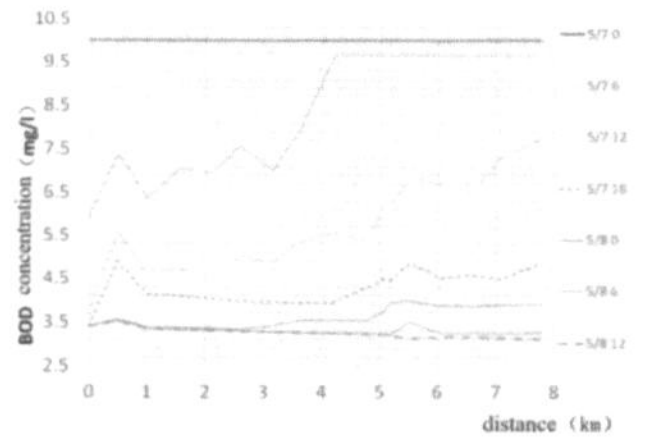

Figure 6. DO concentration changes the way.

3.3 *Water diversion line 2*

It can be seen in Figure 8 that water level elevation changes along the diagram in 1 day 12 hours after the diversion, the water level elevation generally gradually increased, but the water level elevation gradually decreased after 5 kilometers because the river network in this article has several bifurcations generated diversion. From Figure 9, it can be seen that the flow velocity varies greatly along the way. The flow velocity at 1–5 km, that is, the Nanhuantang River, is relatively low and stable but suddenly increases after 5 km. The reason is that the flow velocity increases after confluence, and the subsequent fluctuation is the decrease of flow velocity caused by the large loss of the local head.

From Figure 10 along with the BOD concentration changes in the figure, we can see that with the increase of water diversion time, 1–5 km away from the Nanjiang Gate BOD concentration is high in the water diversion within 12 hours the BOD concentration is high (8.5mg / 1 or more), but in 12 hours after the BOD concentration decreased, the water quality has been greatly improved, down to about 1mg/l, reached water standard, 6–8 km although the change is not stable, but after a day also dropped to about 2mg/l and reached a water standard. Figure 11 DO concentration changes along the way can be seen in the diagram, 1 to 5 km of DO concentration in 12 hours is low, but after 12 hours increased significantly and reached more than 9.5 mg/l, which shows that the water quality has been greatly improved. The reason is that the diversion flow of Nanjiang Gate is increased, and the water quality of the South Huantang River is improved. 6–8 km away from the water inlet DO concentration changes and Figure 10 BOD concentration changes are somewhat similar to the mirror, is also lower within 12 hours, 12 hours later become higher to a water standard.

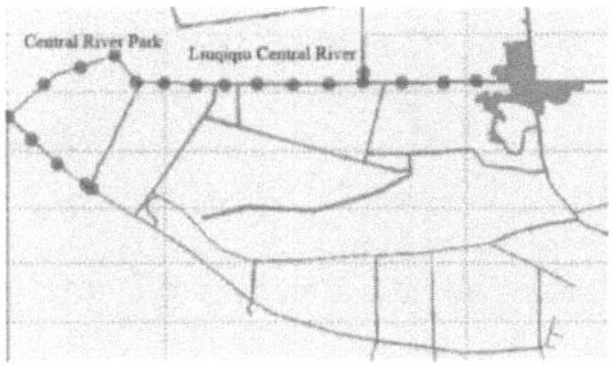

Figure 7. Line-location.

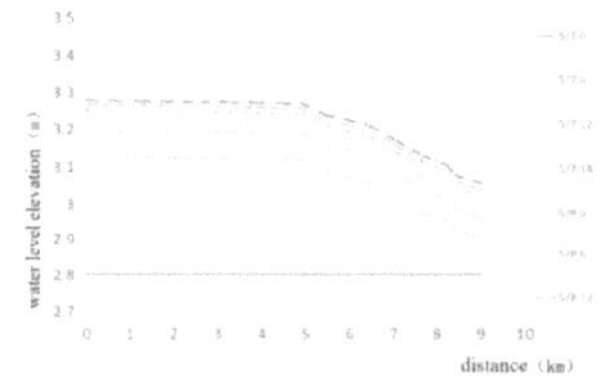

Figure 8. Variation of water level elevation.

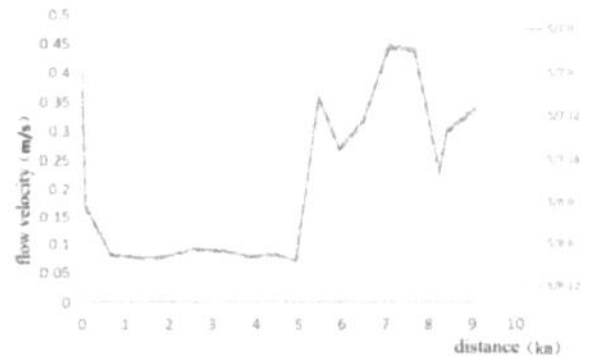

Figure 9. Velocity changes with flow.

Figure 10. BOD concentration variation along the way.

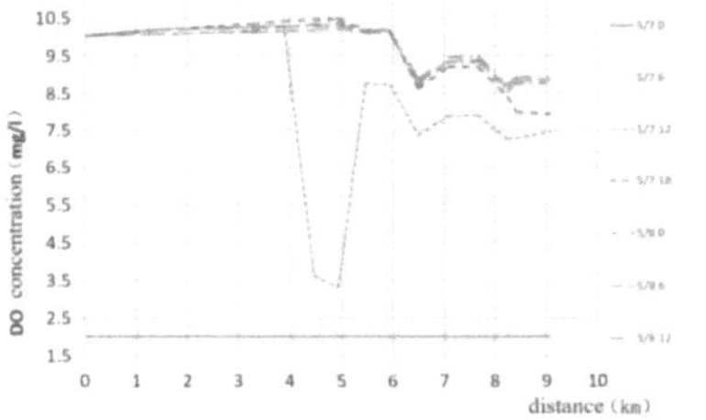

Figure 11. DO concentration variation along the way.

4 CONCLUSION

Through the above methods, the water quality of most river sections has reached the standard of Class III water. The BOD concentration of the river section connecting the Huantangnan River and the Qiliuqiu Central River reached 1.5 mg/l, which reached the standard of Class I water, and the water quality improved significantly. Therefore, the above water diversion scheme can improve the water environment of the Sanhui district. Using the calculation results can guide the gate dispatching, serve the improvement of the water environment through the opening and closing operation of the gate, and play the role of flood control and drainage.

ACKNOWLEDGMENT

This research was supported by the Funds Key Laboratory for Technology in Rural Water Management of Zhejiang Province (ZJWEU-RWM-202101), the Funds of Water Resources of Science and Technology of Zhejiang Provincial Water Resources Department, China (No. RC2239, No. RB2115, No. RC2040), the Joint Funds of the Zhejiang Provincial Natural Science Foundation of China (No. LZJWZ22C030001, No. LZJWZ22E090004), the National Key Research and Development Program of China (No. 2016YFC0402502), and the National Natural Science Foundation of China (51979249). The University Student Innovation and Entrepreneurship Project of Zhejiang University of Water Resources and Hydropower (S202211481017, S202111481001).

REFERENCES

Gu Jianzhong, Wang Fuyuan, Zhong Huimin & Lu Xuelin. (2011). Study on Water Diversion to Improve Water Environment in Wujiang City. *Jiangsu Water Conservancy* (01), 38–39 + 42.

Gu Xijun, Xia Dongmei & Li Can. (2019). A Preliminary Study on Improving the Water Quality of the East River Network in Hangzhou by Water Diversion from Qiantang River. *Zhejiang Hydromechanics* (04), 26–28.

Hua, R., & Zhang, Y. (2017). Assessment of Water Quality Improvements Using the Hydrodynamic Simulation Approach in Regulated Cascade Reservoirs: A Case Study of Drinking Water Sources of Shenzhen, China. *Water*, 9(11), 825.

Jiang Chao & Zhou Dandan. (2013). Study on Optimization of Central Lake Water Diversion Scheme in Xianju Shenxianju Tourist Resort. *Zhejiang Hydromechanics* (03), 84–86 + 89.

Lu Yiwei, Pang Yong & Zhou Ranran. (2019). Research on the Water Diversion Scheme for Water Environment Improvement in the Yundong District of Wuxi City in Taihu Basin. *Sichuan Environment* (01), 68–74.

Li Chaqing, He Wenxue & Chen Dongyun. (2012). The Scheme Design of Water Diversion and Distribution for Improving the Water Environment in the Flat River Network. *China Rural Water Conservancy and Hydropower* (07), 45–47 + 50.

Yang Tonghe, Yu Xuezhong & Luo Huihuang. (2010). Water Pollution Loss Model based on an Assessment of Environmental Benefit from Water Control and Diversion Project. *Water Conservancy and Hydropower Technology* (09), 20–23 + 27.

Zhang, M., Dolatshah, A., Zhu, W., & Yu, G. (2018). Case Study on Water Quality Improvement in Xihu Lake through Diversion and Water Distribution. *Water*, 10(3), 333.

Zhang Zhe, Fang Guohua, Cao Rong & Tan Qiaofeng. (2020). Analysis of the Effect of Water Diversion from Qiputang on Water Quality Improvement of Yangcheng Lake and the Surrounding River Network. *Jiangsu Water Conservancy* (05), 23–29.

Zhao Yanyan & Zheng Xiaoyu. (2007). Using Two-Dimensional Flow-Pollutant Mathematic Model to Optimize the Import Water Project of Jinshan Lake. *Henan Science* (02), 308–310.

Zhu Yuting, Chen Qihui, Zhang Yutian, Li Qiongfang, Chen Jing, Song Yun & Yàng Xuerui. (2021). Simulation Analysis of Water Quality Improvement Effect of the Yangtze River Diversion on the South-to-North Water Transfer Project. *Hydrology* (03), 25–31.

Water Conservancy and Civil Construction – Oke & Ahmad (Eds)
© 2024 The Author(s), ISBN: 978-1-032-58614-4

Response of potato water consumption characteristics to water deficit under film-mulched drip irrigation

Fuqaing Li
College of Agronomy and Agricultural Engineering, Liaocheng University, Liaocheng, China
College of Water Conservancy and Hydropower Engineering, Gansu Agricultural University, Lanzhou, Gansu, China

Shouchao Yu
College of Agronomy and Agricultural Engineering, Liaocheng University, Liaocheng, China

Hengjia Zhang*
College of Agronomy and Agricultural Engineering, Liaocheng University, Liaocheng, China
College of Water Conservancy and Hydropower Engineering, Gansu Agricultural University, Lanzhou, Gansu, China

ABSTRACT: The shortage of water resources and the traditional diffuse irrigation method in the cold and cool irrigation area of Hexi Oasis seriously restrict the sustainable development of agriculture in the area. To break the bottleneck, the goal of water saving, stable yield, and quality improvement of crops in this area can be achieved by introducing under-membrane drip irrigation to regulate deficit irrigation mode. This study was conducted at Yimin Irrigation Experiment Station, Flood River Management Office, Minle County, Zhangye City, Gansu Province, from April to October 2019. The test material variety was "Green potato 168", and eight deficit-regulating irrigation treatments and one adequate irrigation control were designed. The variation pattern of water consumption characteristics of potatoes under drip irrigation at different fertility stages was analyzed, and it was found that the maximum water consumption was 185.35~239.52 mm at the tuber expansion stage, followed by 100.02~132.30 mm and 82.48~112.36 mm at tuber formation stage and starch accumulation stage, respectively, and the least water consumption was 49.32~69.81 mm at the seedling stage. The intensity of water consumption showed a similar trend, with tuber expansion > tuber formation > starch accumulation > seedling stage.

1 INTRODUCTION

As a major potato-producing and consuming country, China's potato cultivation area is increasing year by year, from 2007 to 2018, the national potato cultivation area increased from 44.303 thousand hectares to 4758.1 thousand hectares, and the annual production increased from 12.958 million tons to 17.984 million tons. Meanwhile, with the rapid development of potato processing and the continuous optimization of crop planting structure, the potato has been gradually transformed from a major food crop to a cash crop with a high yield in the northwest (Xue *et al.* 2017).

At present, a large number of studies have shown that sub-membrane drip irrigation technology fully combines the advantages of ground cover and drip pipe water conservation,

*Corresponding Author: 596088683@qq.com

DOI: 10.1201/9781003450818-20

which can significantly improve irrigation water use efficiency and water use efficiency while providing thermal insulation and moisture conservation. Although this technology has been studied extensively in pepper, wood, cotton, and fruit trees, less research has been conducted on the application of this technology in potato cultivation (Xiao *et al.* 2007, 2008), and even less research has been seen on irrigated potatoes with different gradients of deficit regulation at different fertility stages (Hou *et al.* 2008). In this study, the proposed drip irrigation technology will break the conventional irrigation pattern of potato cultivation in this irrigation area. The local "Green Potato 168" variety was used as the research object, and under-membrane drip irrigation with loss-adjusting irrigation was carried out in each of the four fertility stages of potatoes to analyze the water consumption characteristics of potato at different fertility stages to provide a theoretical basis for achieving the goals of water saving, stable yield, and quality improvement in agriculture in this area (Wang *et al.* 2009; Wu *et al.* 2009; Xiao *et al.* 2011; Yang *et al.* 2012; Zhang *et al.* 2005).

2 EXPERIMENTAL DESIGN

The test material was "QingShu168", a potato variety provided by the Qinghai Provincial Institute of Agricultural Science. According to the fertility stage of potato, the experiment was conducted in four fertility stages of potato: seedling stage (from sowing to bud appearance), tuber formation stage (from bud appearance to first flowering), tuber expansion stage (from first flowering to final flowering), and starch accumulation stage (from final flowering to wilting) for water regulation loss treatment, namely WD1 (slight water deficit in seedling stage), WD2 (slight water deficit in tuber formation stage), WD3 (slight water deficit in tuber expansion stage), WD4 (slight water deficit in starch accumulation stage), WD5 (medium water deficit during seedling stage), WD6 (medium water deficit during tuber formation), WD7 (medium water deficit during tuber expansion), WD8 (medium water deficit during starch accumulation), and one control group CK.

Table 1. Experimental design.

Treatment	Deficit	Seeding	Tuber Initiation	Tuber Bulking	Starch Accumulation
CK	Conventional	65%~75%	65%~75%	65%~75%	65%~75%
WD1	Slight	55%~65%	65%~75%	65%~75%	65%~75%
WD2	Slight	65%~75%	55%~65%	65%~75%	65%~75%
WD3	Slight	65%~75%	65%~75%	55%~65%	65%~75%
WD4	Slight	65%~75%	65%~75%	65%~75%	55%~65%
WD5	Medium	45%~55%	65%~75%	65%~75%	65%~75%
WD6	Medium	65%~75%	45%~55%	65%~75%	65%~75%
WD7	Medium	65%~75%	65%~75%	45%~55%	65%~75%
WD8	Medium	65%~75%	65%~75%	65%~75%	45%~55%

3 RESULTS AND ANALYSIS

3.1 *Water consumption characteristics of potatoes at different stages*

As can be seen from Table 2, the water consumption at all stages of water stress affected by water stress was significantly lower ($P<0.05$) than that of the CK treatment for each water regulation loss treatment in potatoes in 2019 during the regulation fertility period. At the seedling stage, the water consumption of each treatment was between 58.86 and 77.59 mm, and its treatments WD1 and WD5 were 11.74% and 22.89% lower than the control CK, respectively, with significant differences; at the tuber formation stage, it's water consumption was between 109.42 and 137.43 mm, and its treatments WD2 and WD6 were 9.60% and 18.64% lower than CK, respectively, with significant differences. During the tuber expansion period, WD1, WD2, and WD6 increased by 0.59%, 1.87%, and 0.94%, respectively, compared with the control CK, with no significant difference, while the deficit treatment WD3 and WD7 decreased by 8.19% and 16.91%, respectively, compared with the control CK, with significant differences; during the starch accumulation period, WD1, WD2, WD3, WD5, WD6, and WD7 treatments were not significantly different from the control. In the starch accumulation period, WD1, WD2, WD3, WD5, WD6, and WD7 treatments were not significantly different from the control, while their deficit-regulating treatments WD4 and WD8 were reduced by 11.01% and 23.33%, respectively, compared with the control CK, with significant differences.

Table 2. Water consumption of potatoes at different growth stages.

Year	Treatment	Water consumption /mm				
		Seedling	Tuber Formation	Tuber Bulking	Starch Accumulation	Water consumption for the whole growth period
2019	CK	76.33ab	134.49ab	224.43ab	117.68ab	552.93a
	WD1	67.37b	134.16ab	225.76ab	116.55ab	543.83ab
	WD2	76.41ab	121.58b	228.62a	119.69ab	546.3ab
	WD3	74.92ab	135.19ab	206.05b	120.08ab	536.24ab
	WD4	75.83ab	135.88ab	225.82ab	104.72b	542.25ab
	WD5	58.86c	137.43a	223.39ab	115.37ab	535.05ab
	WD6	74.18ab	109.42c	226.53ab	119.13ab	529.26b
	WD7	77.59a	133.96ab	186.47c	120.81a	518.83c
	WD8	74.76ab	133.53ab	227.95ab	90.22c	526.46b

Note: Values are means of three replicates for each treatment. Data within one column followed by different letters are significantly different at $P<0.05$.

3.2 *Water consumption intensity of potatoes at different stages*

The intensity of water consumption is closely related to the fertility process, physiological characteristics, and environmental conditions in which the potato crop is located throughout the reproductive period. As can be seen from Table 3, in 2019, potato plants at the seedling stage were short and the atmospheric temperature was low, the intensity of water consumption was low, and the daily water consumption intensity of potatoes at this time ranged from 0.93 to 1.23 mm/d, which was lower than that of other fertility periods thereafter, while the daily water consumption intensity of WD1 and WD5 in the deficit adjustment treatment was 11.74% and 22.89%, lower than that of the control CK, respectively, with significant differences ($P < 0.05$). During the tuber formation period, with the rapid growth of potato plants, their daily water consumption intensity was much higher than that of the seedling period, ranging from 3.32 to 4.16 mm/d, and there was no significant difference ($P>0.05$) between the seedling period and the control after rehydration of the deficit treatments WD1 and WD5, while the deficit treatments WD2 (3.68 mm/d) and WD6 (3.32 mm/d) were 9.60% lower than the control CK (4.08 mm/d). /d), which were reduced by 9.60% and 18.64%, respectively, with significant differences. At the tuber expansion stage, the regulating deficit treatments WD1, WD2, WD5, and WD6 increased by 0.59%, 1.87%, 0.46%, and 0.94%, respectively, after rehydration compared with the control, with no significant difference, while the regulating deficit treatments WD3 (4.48mm/d) and WD7 (4.05mm/d) decreased by 8.19%, 16.91%, respectively, compared with the control CK (4.88mm/d), with significant differences of 16.91% with significant differences. During the starch accumulation period, no significant differences were observed in the loss-regulating treatments WD1, WD2, WD3, WD5, WD6, and WD7 compared to the control, while WD4 (2.83 mm/d) and WD8 (2.44 mm/d) were reduced by 11.01% and 23.33%, respectively, compared to the control CK (3.18 mm/d), with significant differences.

Table 3. Water consumption characteristics of potatoes at different growth stages.

		Seedling		Tuber Formation		Tuber Bulking		Starch Accumulation	
Year	Treatment	Intensity mm/d	Modulus /%	Intensity mm/d	Modulus /%	Intensity mm/d	Modulus /%	Intensity mm/d	Modulus /%
2019	CK	1.21ab	13.80b	4.08ab	24.32ab	4.88ab	40.59ab	3.18ab	21.28ab
	WD1	1.07b	12.39c	4.07ab	24.67ab	4.91ab	41.51ab	3.15ab	21.43ab
	WD2	1.21ab	13.99ab	3.68b	22.26b	4.97a	41.85ab	3.23ab	21.91ab
	WD3	1.19ab	13.97ab	4.10ab	25.21ab	4.48b	38.42b	3.25ab	22.39ab
	WD4	1.20ab	13.98ab	4.12ab	25.06ab	4.91ab	41.65ab	2.83b	19.31b
	WD5	0.93c	11.00d	4.16a	25.69ab	4.86ab	41.75ab	3.12ab	21.56ab
	WD6	1.18ab	14.02ab	3.32c	20.67c	4.92ab	42.80ab	3.22ab	22.51ab
	WD7	1.23a	14.95a	4.06ab	25.82a	4.05c	35.94bc	3.27a	23.29a
	WD8	1.19ab	14.20ab	4.05ab	25.36ab	4.96ab	43.30a	2.44c	17.14c

Note: Values are means of three replicates for each treatment. Data within one column followed by different letters are significantly different at $P<0.05$.

4 DISCUSSION

Since rainfall is scarce and spatial and temporal distribution is uneven in the west oasis irrigation area, artificial irrigation is mainly used to supply water to crops in this area. Therefore, it is important to study the water consumption characteristics of crops at different fertility stages to optimize the irrigation system afterward.

In this study, we found that the water consumption at seedling, tuber formation, tuber expansion, and starch accumulation stages ranged from 49.32 to 69.81 mm, 100.02 to 132.30 mm, 185.35 to 239.52 mm, and 82.48 to 112.36 mm, respectively, and the daily water consumption intensity at each fertility stage ranged from 0.77 to 1.09 mm·d-1, 3.08 to 4.07 mm·d-1, and 3.08 to 4.07 mm·d-1, respectively, 3.08∼4.07 mm·d-1, 4.39∼5.63 mm·d-1, and 2.66∼3.63 mm·d-1. Among them, the water consumption and daily water consumption intensity of the regulating deficit treatment were lower than those of the control treatment at all fertility stages affected by water stress, and the most significant effect of water stress on water consumption and water consumption intensity was observed at the tuber expansion stage, and all three showed a dynamic trend of maximum at the tuber expansion stage, followed by tuber formation stage and minimum at the seedling stage. Wang *et al.* (Wang, Zhang, Ba, *et al.*, 2018) showed that water consumption, intensity, and modulus of water consumption were the highest in the fruiting stage, the second highest in the flowering and fruiting stage, and the lowest in the seedling stage, and all treatments had different degrees of reduction compared with the control. The maximum water consumption was also observed during the berry expansion period when water stress was most significant, which is consistent with the findings of this study. In the seedling stage, the crop growth time is short, the number of leaves is small, photosynthesis and transpiration are the weakest, so water consumption and daily water intensity are the smallest, and water stress has the least effect on it; in the tuber formation stage, under the combined effect of rapid recovery of atmospheric temperature and rapid acceleration of crop nutritional growth, potato water consumption and daily water intensity are higher than in the seedling stage, and the effect of water stress also rises gradually; in the tuber expansion stage, external At the tuber expansion stage, the external temperature peaks, and the coexistence of nutritional and reproductive growth results in the highest water demand and the strongest sensitivity to water stress during this reproductive period; at the starch accumulation stage, as the external temperature falls and the nutritional growth slows down, the water consumption and daily water consumption intensity decrease, so the effect of water stress on the water consumption characteristics of potato at this stage decreases simultaneously.

5 CONCLUSIONS

As potato fertility progressed, its water consumption and water intensity showed a trend of increasing and then decreasing, with the largest water consumption and daily water intensity at the tuber expansion stage, followed by the tuber formation stage, and the smallest at the seedling stage. At the same time, all water stress treatments affected water consumption and water intensity of potato, with the greatest effect of water stress during tuber formation, the second greatest effect of water stress during tuber expansion, and the least effect of mild water stress on potato water consumption during starch accumulation.

ACKNOWLEDGMENTS

This work was financially supported by the National Natural Science Foundation of China (No. 52269008), the Industrial Support Plan Project of the Gansu Provincial Department of Education (No. 2022CYZC-51), and the Key Research and Planning Projects of Gansu Province (No. 18YF1NA073).

REFERENCES

Hou, X., Wang, F., Kang, S., et al.: Study on the Effect of Drip Irrigation on Potato Uncovering in Minqin Oasis, Northwest Arid Zone. *Research on Agriculture in Arid Areas* 04, 88–93 (2008).

Wang, Y., Zhang, Z., Fan, M.: Preliminary Study on Water Saving and Production Efficiency of Potato Under-membrane Drip Irrigation. *China Potato* 23(03):148–151 (2009).

Wu Z., Ren G., Li, J.: Experimental Study on Water Requirement and Irrigation System of Potato. *Journal of Irrigation and Drainage* 28(03):93–95 (2009).

Xiao, H., Sun, R., He J., et al.: Effect of Different Water Conditions on Water Consumption Characteristics and Yield of Potato. *Guizhou Agricultural Science* 39(01):73–75 (2011).

Xue D., Zhang, H., Ba, Y., et al.: Effects of Under-membrane Drip Irrigation and Deficit Regulation on Soil Environment and Yield of Potato in Oasis. *North China Journal of Agriculture* 32(03), 229–238 (2017).

Xiao, J., Liu, Z., Duan, A., et al.: Study on the Distribution of Winter Wheat Roots and Their Water Consumption Characteristics Under Different Soil Moisture Conditions. *China Rural Water Conservancy and Hydropower* 08, 18–21 (2007).

Xiao, J, Liu, Z., Duan, A., et al.: Relationship Between Water Consumption and Yield of Major Crops in China During the Whole Reproductive Period. *China Agronomy Bulletin* 03, 430–434 (2008).

Yang, H., Liu, H., Li, Y., et al.: Effect of Soil Substrate Potential on Yield and Water use Efficiency of Pepper Under Sub-membrane Drip Irrigation. *Agricultural Research in Arid Regions* 30(01):54–60 (2012).

Wang, S., Zhang, H., Ba, Y, et al.: Effect of Deficit Regulation Irrigation on the Growth and Water Use of Pepper Under Film Drip Irrigation. *Agricultural Research in Arid Regions*, 36(03):31–38 (2018).

Zhang, Z., Cai H., Yang R., et al.: Study on the Relationship Between Yield and Quality of Cotton Under Drip Irrigation and Crop Water Deficit Index. *Journal of Agricultural Engineering*, 06, 26–29 (2005).

Water Conservancy and Civil Construction – Oke & Ahmad (Eds)
© 2024 The Author(s), ISBN: 978-1-032-58614-4

A comprehensive evaluation of water quality of typical reservoirs in the Yellow River diversion plain

Qi-dong Liu
School of Water Conservancy and Environment, University of Jinan, Jinan, China

Xin Jiang*, Jian Liu & Ling-xiao Zhang
Water Research Institute of Shandong Province, Jinan, China

Gui-bin Pang
School of Water Conservancy and Environment, University of Jinan, Jinan, China

ABSTRACT: To ensure urban and rural water supply in the lower reaches of the Yellow River, the plain reservoirs were built to take the Yellow River water as the main water source for urban and rural water supply. To clarify the water environment of the Yellow River Diversion Reservoir, the most reasonable water quality evaluation method was selected based on the monitoring data of six reservoirs from 2018 to 2021. The single factor index evaluation, comprehensive pollution index method, and Nemerov pollution index method were used to analyze the water quality of the Yellow River Diversion reservoir, and principal component analysis was used to explore the main pollutants. The results showed that the evaluation results of the single-factor index method and the Nemerov pollution index method are similar. The Nemerov pollution index method is the most reasonable method to evaluate the water quality of the Yellow River reservoir, and its water quality evaluation is mostly between light pollution and moderate pollution. The main pollution factors of the reservoir were total nitrogen, dissolved oxygen, and fluoride by single factor index and principal component analysis. This study can provide guidance for the management of the water environment and the safety of water supply in urban and rural areas.

1 INTRODUCTION

The water quantity of groundwater and surface water in the plain lower reaches of the Yellow River is insufficient and the water quality is unstable, and the water supply guarantee rate of urban and rural areas is insufficient. To alleviate the water pressure, plain reservoirs were built in the lower reaches of the Yellow River to absorb the Yellow River water as the main water source for urban and rural water supply (Chen 2020).

The water source of the Yellow River Reservoir is mainly the Yellow River water with a lot of silt, which needs to be treated by the sedimentation basin before it can be discharged into the reservoir. The transportation process is relatively long and the water environment is relatively special. Therefore, this paper adopted a variety of water quality evaluation methods to evaluate the water quality of the Yellow River Reservoir, making the evaluation results more comprehensive and objective. Common water quality evaluation methods include the single factor evaluation method, comprehensive pollution index method,

*Corresponding Author: skyjiangxin@shandong.cn

DOI: 10.1201/9781003450818-21

Nemerov pollution index method, Canadian Water quality index method, principal component analysis method, and fuzzy comprehensive evaluation method (Ding 2022; Li 2012; Marselina 2022; Yan 2022).

2 MATERIALS AND METHODS

2.1 *Study area*

The study area is located in the Shandong section of the lower reaches of the Yellow River, at mid-latitude, with a warm-humid continental semi-humid monsoon climate, with hot and rainy summers and severe cold and dry winters. The six selected Yellow River diversion reservoirs are located in Jinan, Binzhou, and Zibo of Shandong Province, and the specific distribution of each reservoir is shown in Figure 1. The water quality of the Yellow River Reservoir is mainly influenced by the Yellow River water, which has a high content of cement sand and generally settles the sand through the front sedimentation basin (Hou 2016).

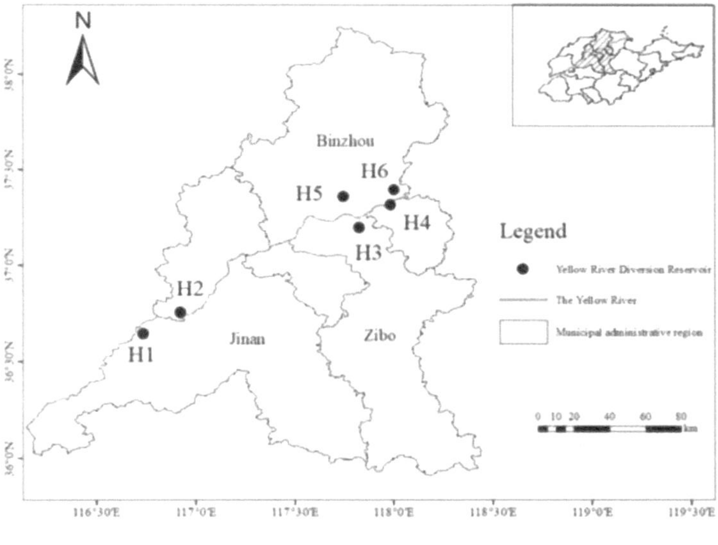

Figure 1. The Yellow River reservoir in the study area.

2.2 *Processing of data*

Samples were collected from the 6 reservoirs of Yellow River diversion during 2018 and 2021, as shown in Figure 1. Samples were collected once a month for each reservoir, except for H1 and H2 reservoirs, whose water surface froze in February 2018. Dissolved oxygen (DO), permanent index (COD_{Mn}), chemical oxygen demand (COD), 5-day biochemical oxygen demand (BOD_5), ammonia nitrogen (NH_3-N), total phosphorus (TP), total nitrogen (TN), nitrate (NO_3-N), sulfate (SO_4^{2-}), fluoride (F^-), arsenic (As), fecal coliform (FC), chloride (Cl^-), iron (Fe), manganese (Mn), and other 15 pollution indicators were evaluated.

The obtained experimental data were processed by Excel 2019, the Origin 2018 software was used for graph drawing and analysis, and the SPSS was used for principal component analysis calculation.

2.3 *Method of evaluation*

This paper used the single factor index method, comprehensive pollution index method, and Nemerov pollution index method to make a comprehensive comparative analysis of the water quality of the reservoir, used the variance of the evaluation results to study the variation range of the water quality of each reservoir and used the principal component analysis method to explore the main pollution indicators of the Yellow River reservoir. The water quality assessment standards were calculated according to the I-V water standards of GB3838-2002 "Surface Water Environmental Quality Standards" and referred to relevant literature, as shown in Table 1 (Guo 2019; Yang 2021). The evaluation methods are shown in Table 2.

Table 1. Water quality evaluation standards.

Degree of pollution	No pollution	Light pollution	Moderate pollution	Heavy pollution	Severe pollution
Single factor pollution index (p)	$p \leqq 1$	$1 < p \leqq 2$	$2 < p \leqq 3$	$3 < p \leqq 5$	$p > 5$
Composite Pollution Index (P)	$P \leqq 0.4$	$0.4 < P \leqq 0.7$	$0.7 < P \leqq 1$	$1 < P \leqq 2$	$P > 2$
Nemerov Pollution Index (P_n)	$P_n \leqq 1$	$1 < P_n \leqq 2$	$2 < P_n \leqq 3$	$3 < P_n \leqq 5$	$P_n > 5$
Overall score	$F \leqq -4.6$	$-4.6 < F \leqq -4.1$	$-4.1 < F \leqq -0.5$	$-0.5 < F \leqq 3.9$	$F > 3.9$

Table 2. Water quality assessment methods.

Method of evaluation	Principle of evaluation	Evaluation model
Single factor exponential method (Li 2017)	The measured value of the concentration of the participating index was compared with the standard value of the evaluation (Class III water), and the worst one was the result of the comprehensive evaluation of water quality.	$p = \frac{C_i}{C_j}$, in addition to DO $p_{DO,j} = \frac{\lvert c_{DO,f} - c_{DO,j} \rvert}{c_{DO,f} - c_{DO,s}}, c_{DO,j} \geq c_{DO,s}$ $p_{DO,j} = 10 - \frac{9c_{DO,j}}{c_{DO,s}}, c_{DO,j} < c_{DO,s}$
Comprehensive pollution index method (Lu 2022)	Based on the assumption that the contribution of each water quality assessment factor to water quality is the same, the value representing the water pollution degree is calculated by statistics considering all pollution factors.	$P = \frac{1}{n}\sum_{i=1} p_i, n = 15$
Nemerov pollution index method (Liu 2016)	Taking into account the extreme value and mean value of the evaluation index, the Nemerov pollution index and standard index were calculated for evaluation.	$P_n = \sqrt{\frac{p_{i,avg}^2 + p_{i,\max}^2}{2}}$
Principal component analysis (Liu 2015)	The original data was reduced dimensionality while the information of the original data was preserved to the maximum extent. Several comprehensive indexes were obtained for water quality evaluation.	$w_{ij} = \frac{\theta_j}{\sqrt{\lambda_i}}$ w_{ij}, Index weight; θ_j, Principal component coefficient; λ_i, Eigenvalue. $F_i = w_{i1}X_1 + w_{i2}X_2 + \ldots + w_{in}X_n$ $F = \alpha_1 F_1 + \alpha_2 F_2 + \ldots + \alpha_n F_n$ F_i, Principal component score; F, Comprehensive score.

In the principal component analysis, the original data should be standardized (dimensionality reduction), and SPSS should be used to calculate the correlation coefficient matrix, variance contribution rate (to determine the principal component), and the principal component load. Finally, the comprehensive score should be calculated through the evaluation model in Table 2.

3 RESULTS AND DISCUSSION

3.1 *Single factor index method evaluation*

The water quality evaluation results of the single-factor index method are shown in Figure 2. The results showed that the water quality of the H5 reservoir is the worst, 38% of the time is in a state of heavy pollution, which is the highest proportion of heavy pollution among all reservoirs. The second is H1 and H2 reservoirs, which are mainly moderately polluted, and their water quality has been severely polluted for a long time. H3 reservoir water quality status from no pollution to heavy pollution is distributed, of which the distribution of heavy pollution is relatively small. The water quality of H4 and H6 reservoirs is the best. The distribution of water quality of these two reservoirs is relatively concentrated, mainly with light pollution. The 4-year average single factor index of each reservoir was 2.54, 2.72, 1.74, 1.59, 2.76, and 1.43, respectively, indicating that the pollution degree of each reservoir was H5 > H2 > H1 > H3 > H4 > H6.

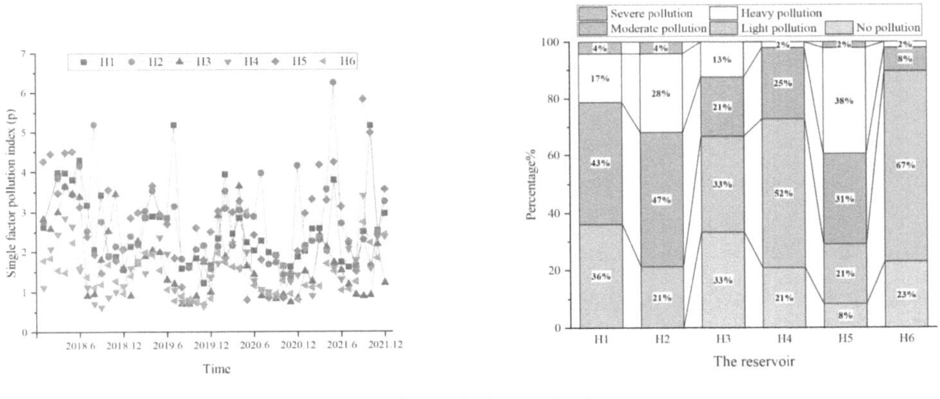

Figure 2. Evaluation results of the single factor index method.

3.2 *Comprehensive pollution index method evaluation*

The comprehensive pollution index of each reservoir was calculated according to the measured water quality index, and the water quality evaluation results are shown in Figure 3. The results showed that only in August 2018, the water quality of the H2 reservoir was moderately polluted, while the water quality of other reservoirs was non-polluted or moderately polluted. The water quality of H1, H2, and H5 reservoirs is mostly in a moderate pollution state, while the water quality of H3, H4, and H6 reservoirs is mostly in no pollution-free state. The non-polluted status of the H4 reservoir reached 85%, while the lightly polluted status of the H1 reservoir reached 72%. The average comprehensive pollution index of each reservoir in 4 years was 0.453, 0.446, 0.429, 0.348, 0.433, and 0.352, respectively, indicating that the pollution degree of each reservoir was H1 > H2 > H5 > H3 > H6 > H4.

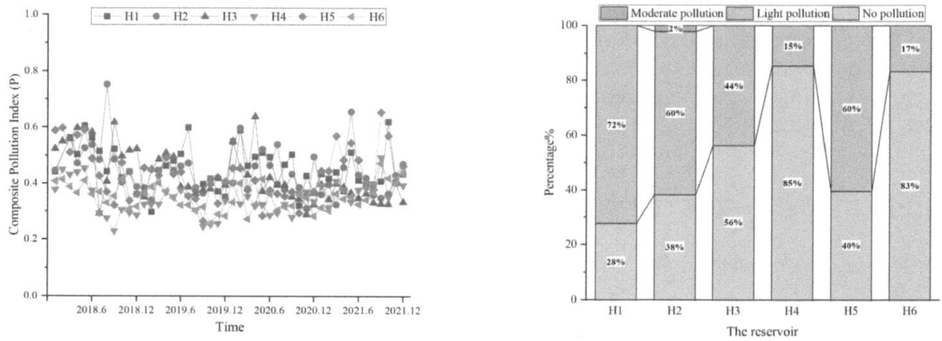

Figure 3. Evaluation results of the comprehensive pollution index method.

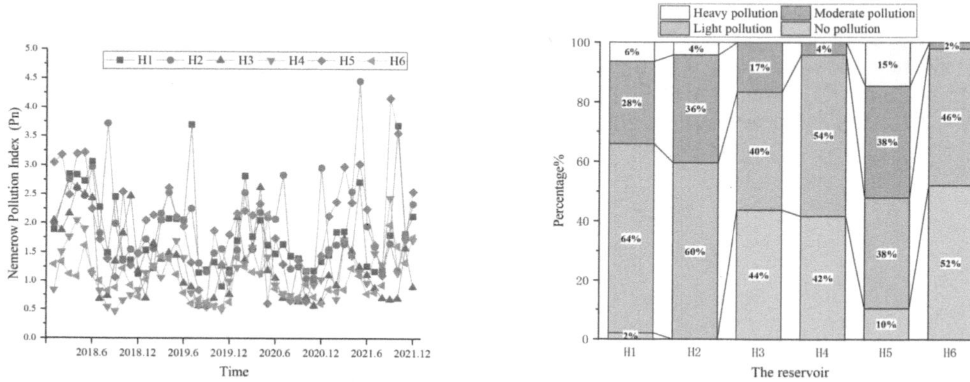

Figure 4. Evaluation results of the Nemerov pollution index method.

3.3 *Evaluation by Nemerov pollution index method*

Nemerov pollution index was calculated according to the measured concentration of each reservoir detection index, and the water quality evaluation results are shown in Figure 4. The results showed that the water quality of each reservoir was differentiated. The water quality of H1 and H2 reservoirs was poor, mainly with light and moderate pollution, and heavy pollution occurred in some months. The H3 reservoir was mainly pollution-free and lightly polluted, and the pollution index showed an obvious downward trend. The moderate pollution was mainly concentrated in 2018. H4 and H6 reservoirs have the best water quality, which is mainly pollution-free and lightly polluted, and their water quality changes gently. The distribution of water quality in the H5 reservoir is relatively dispersed, the pollution index has no obvious trend of decreasing or increasing, and the proportion of heavy pollution is the highest among all reservoirs. The average Nemerov pollution index of each reservoir in 4 years was 1.829, 1.947, 1.272, 1.154, 1.976, and 1.045, respectively, indicating that the pollution degree of each reservoir was H5 > H2 > H1 > H3 > H4 > H6.

3.4 *Evaluation results of principal component analysis*

SPSS was used to calculate the eigenvalue, variance contribution rate, and cumulative contribution rate according to the correlation coefficient matrix, and the evaluation results are shown in Table 3. The variance contribution rate of the first principal component was 62.8%,

Table 3. Evaluation results of principal component analysis.

The reservoir	Overall score: F	Ranking	Degree of pollution
H1	3.14	2	Heavy pollution
H2	3.49	1	Heavy pollution
H3	−4.29	6	Light pollution
H4	−1.06	5	Moderate pollution
H5	−0.25	3	Heavy pollution
H6	−1.05	4	Moderate pollution

and the contribution rate of the second principal component was 26.5%. The cumulative contribution reached 89.3%, which was sufficiently representative of most information from the original data. The main controlling factors of the first principal component are DO, F^-, As, FC, and Mn, and the main controlling factors of the second principal component are Fe, whose absolute values are all higher than 0.9.

According to the principal component load coefficient of the first principal component and the second principal component, the comprehensive index was constructed as follows:

$$F_1 = 0.315X_1 - 0.287X_2 - 0.265X_3 - 0.155X_4 + 0.268X_5 + 0.248X_6 + 0.242X_7 - 0.321X_8 - 0.304X_9 + 0.311X_{10} - 0.218X_{11} - 0.242X_{12} + 0.210X_{13} - 0.003X_{14} + 0.301X_{15}$$

$$F_1 = 0.315X_1 - 0.287X_2 - 0.265X_3 - 0.155X_4 + 0.268X_5 + 0.248X_6 + 0.242X_7 - 0.321X_8 - 0.304X_9 + 0.311X_{10} - 0.218X_{11} - 0.242X_{12} + 0.210X_{13} - 0.003X_{14} + 0.301X_{15}$$

$$F_2 = 0.062X_1 + 0.222X_2 + 0.279X_3 + 0.438X_4 + 0.262X_5 + 0.318X_6 - 0.154X_7 - 0.008X_8 + 0.142X_9 - 0.067X_{10} - 0.353X_{11} - 0.233X_{12} - 0.166X_{13} + 0.486X_{14} + 0.129X_{15}$$

$$F = 0.703F_1 + 0.297F_2$$

The comprehensive score value of each reservoir was calculated according to the above formula. The higher the comprehensive score is, the more serious the pollution is. The results are shown in Table 3. The H1, H2, and H5 reservoirs are heavily polluted, but the comprehensive scores of the H5 reservoirs are low and close to the critical value, while the H4 and H6 reservoirs are moderately polluted. The comprehensive scores of these two reservoirs are very close, and the water quality of the H3 reservoir is the best, and its comprehensive scores are far lower than other reservoirs.

3.5 *Comparison of evaluation results*

In terms of water quality evaluation results, the single factor index method has the worst evaluation results, the comprehensive pollution index method has the best evaluation results, the Nemerov pollution index method has the moderate evaluation results, and the principal component analysis method has many deviations from other methods. Most of the reservoirs' TN exceeds the standard seriously, which is the main pollutant of the Yellow River diversion reservoir. This is similar to the result of Wang *et al.* (Zhai 2021) evaluation of the water quality of the Yellow River diversion reservoir in Shandong Province, and the TN exceeds the standard is also a common problem faced by most lakes and reservoirs (Wang 2018).

Table 4. Variance of water quality evaluation results.

The reservoir	Single-factor exponential method	Ranking	The comprehensive pollution index method	Ranking	Nemerov pollution index method	Ranking
H1	0.894	3	0.0057	4	0.44	3
H2	0.972	2	0.007	3	0.49	2
H3	0.765	4	0.0085	1	0.38	4
H4	0.407	5	0.0034	5	0.2	5
H5	1.327	1	0.0079	2	0.66	1
H6	0.275	6	0.0022	6	0.14	6

The single-factor index method, comprehensive pollution index method, and Nemerov pollution index method evaluated the variation amplitude of water quality of each reservoir as shown in Table 4. The single-factor evaluation pollution index of H1-H6 reservoirs and the Nemerov pollution index ranked the variation amplitude of water quality as H5 > H2 > H1 > H3 > H4 > H6. This is consistent with the ranking of water pollution degree by the single factor index method and Nemerov pollution index method, indicating that the variation range of water quality is positively correlated with the degree of water pollution, that is, the reservoir with poor water quality also has a larger variation range of water quality. The range of water quality variation of the comprehensive pollution index of each reservoir is ranked as H3 > H5 > H2 > H1 > H4 > H6, which is inconsistent with the conclusions obtained by the other two methods.

4 CONCLUSION

1. Due to the special water environment of the Yellow River Diversion Reservoir, this paper adopted the single factor index method, the comprehensive pollution index method, the Nemerov pollution index method, and the principal component analysis method to evaluate the water quality of the six reservoirs in the lower reaches of the Yellow River. The comprehensive pollution index method has the best evaluation result, while the single-factor pollution index method has the worst evaluation result. The principal component analysis method is not suitable for the water quality evaluation of this reservoir. It is mainly used to analyze the main pollutants in the reservoir.
2. According to the common evaluation of the four evaluation methods on the Yellow River diversion reservoir, it can be seen that the water quality evaluation of H1, H2, and H5 reservoirs are similar and poor, the water quality of H3 reservoir has an obvious improvement trend, and the water quality of H4 and H6 reservoirs are similar and good. The main pollutant affecting the Yellow River reservoir is total nitrogen, and both the single factor index evaluation and Nemerov pollution index evaluation show that the reservoir with better water quality has a smaller range of water quality change. The principal component analysis shows that DO and fluoride are also important factors affecting water quality.

ACKNOWLEDGMENTS

This study was supported by the Open Research Fund of Henan Key Laboratory of Water Resources Conservation and Intensive Utilization in the Yellow River Basin (NO. HAKF202105) and the Optional Research Fund of Water Research Institute of Shandong Province (SDSKYZX202102).

REFERENCES

Chen, F., Liu, W., Pan, Z., *et al.* (2020). "Characteristics and Mechanism of Chitosan in Flocculation for Water Coagulation in the Yellow River diversion Reservoir." *Journal of Water Process Engineering* **34**: 101191.

Ding, F., Zhang, W., Chen, L., *et al.* (2022). "Water Quality Assessment using Optimized CWQII in Taihu Lake." *Environmental Research* **214**: 113713.

Guo, J., Wang, C. Huang, D., *et al.* (2019). "Pollution Characterization and Water Quality Assessment of Dongting Lake." *Environmental Chemistry* **38** (1): 152–160.

Hou, W., Sun, S., Jia, R. (2016). "Eutrophication and Water Characteristics of Mountain and Yellow River Reservoirs in Northern China." *Environmental Monitoring in China* **32** (2): 58–63.

Li, M.S., Zhang, J.H., Liang, N., *et al.* (2012). "Comparisons of Some Common Methods for Water Environmental Quality Assessment." *Progress in Geography* **31** (05): 617–624.

Li, X., Zhang, F., Wang, J., *et al.* (2017). "Analysis and Assessment of Water Quality of Ebinur Lake basin in Autumn." *Environmental Pollution & Control* **39** (6): 588–593.

Liu, M. & Chen, S. (2016). "Groundwater Quality Assessment of Honghu Area based on the Nemerov Index and Principal Component Analysis Method." *Journal of HuaZhong Normal University (Natural Sciences)* **50** (4): 633–640.

Liu, X., Xue, Y., Ji, Y., *et al.* (2015). "An Assessment of Water Quality in the Yellow River estuary and its Adjacent Waters based on Principal Component Analysis." *China Environmental Science* **35** (10): 3187–3192.

Lu, L., Wang, H., Hu, Y., *et al.* (2022). "Surface Water Quality in Second-phase Zhaokou Yellow River Irrigation District Project." *Journal of Irrigation and Drainage* **41** (9): 117–124.

Marselina, M., Wibowo, F., Mushfiroh, A. (2022). "Water Quality Index Assessment Methods for Surface Water: A Case Study of the Citarum River in Indonesia." *Heliyon* **8** (7): e09848.

Wang, L., Sun, Y. & Qi, F. (2018). "Analysis on Water Quality of Reservoirs Based on Improved Integrated Water Quality Index Method." *Bulletin of Soil and Water Conservation* **38** (4): 174–180.

Yan, T., Shen, S. & Zhou, A. (2022). "Indices and Models of Surface Water Quality Assessment: Review and perspectives." *Environmental Pollution* **308**: 119–611.

Yang, P., Tang, C., Lu, M., *et al.* (2021). "Pollution Characteristics of Water Body of Wenwusha Reservoir in Fuzhou City." *Wetland Science* **19** (5): 636–645.

Zhai, Z., Huang, T. & Chen, F. (2021). "Water Quality Evaluation and Pollution Source Analysis of Xikeng Reservoir." *Journal of Water Resources and Water Engineering* **32** (6): 57–64.

Deformation analysis of deep foundation pit in water-rich soft stratum of Foshan

Yunjun Qiu
China Construction Infrastructure Co., Ltd., Beijing, P.R. China
CSCEC Southern Investment Co., Ltd., Shenzhen, P.R. China

Shuang Zheng
China Construction Fourth Engineering Bureau Co., Ltd., Guangzhou, P.R. China
China Construction Fourth Engineering Bureau Civil Engineering Co., Ltd., Shenzhen, P.R. China

Jichao Li
China Construction Infrastructure Co., Ltd., Beijing, P.R. China
CSCEC Southern Investment Co., Ltd., Shenzhen, P.R. China

Houbing Xing & Dong Wei
China Construction Fourth Engineering Bureau Co., Ltd., Guangzhou, P.R. China
China Construction Fourth Engineering Bureau Civil Engineering Co., Ltd., Shenzhen, P.R. China

Zhanzhong Li*
School of Civil Engineering, Guangzhou University, Guangzhou, P.R. China

ABSTRACT: This study is based on the deep excavation of the water-rich soft stratum in Metro Line 3, Foshan. Three-dimensional finite element analysis (FEA) is adopted to investigate the characteristics and mechanism of the surrounding ground settlement and diaphragm wall deformation caused by deep and large excavation at different positions of the foundation pit. The study reveals that the surface ground settlement is in the shape of a groove, and the maximum settlement occurs within 10–20 m from the wall. The main factor affecting the surface settlement is the excavation depth. The lateral deflection of the diaphragm wall presents an arc shape, with small values at both ends and large in the middle. The maximum deflection occurs in the middle of the diaphragm wall, and the position will develop downward as the construction progresses. Reducing the span of the diaphragm wall can effectively increase the stiffness of the wall, thus effectively reducing the ground settlement and lateral deflection.

1 INTRODUCTION

With the continuous development of urbanization, the surface space is getting smaller and smaller and the utilization of underground space gradually becomes an important direction for urban sustainable development (Li *et al.* 2018). A large number of subway station excavations have been carried out in many cities (Feng *et al.* 2018). At present, the surrounding environmental conditions of the excavation of the subway stations are complex and usually located in the center of the prosperous city. Therefore, it can cause a large

*Corresponding Author: 2112116188@e.gzhu.edu.cn

DOI: 10.1201/9781003450818-22

number of excavation engineering problems, which have a large impact on the surrounding environment (Zheng *et al.* 2016).

Several pieces of research have been performed to provide some insight into the performance of deep excavation. By analyzing the monitoring data of the excavation, it was pointed out that the maximum deformation range of the retaining structure is from 0.35% to 1.44% H (H represents the excavation depth). And the maximum ground settlement is 1.27% H while the minimum is 0.28% H (Shen *et al.* 2022). Before the base slab of the foundation pit was poured, the deflection of the diaphragm wall varied slightly with the increase in the excavation depth at first and then increased gradually. When the base slab was poured, the deflection remained stable. The deflection values ranged from 0.08% H to 0.13% H (Elbaz *et al.* 2018). Several researchers have pointed out some factors affecting the performance of deep excavation. The excavation in the silt layer has the greatest impact on the surrounding ground settlement, and the precipitation in the pit can reduce the lateral deflection of the diaphragm wall (Yu *et al.* 2020). Taking the deformation of the retaining structure as the research object, different factors affecting the deformation at different depths are discussed. This study indicates that the foundation presents a high spatial and depth influence on the retaining structure and surroundings (Li *et al.* 2018). Besides, the depth at which the maximum value of lateral movement increment of the retaining wall is related to the hardness of the soil layer (Wang *et al.* 2015). Based on the research of ABAQUS, it shows that the Metro Jet System (MJS) technology can reduce the impact on the surrounding environment disturbed by deep excavation, (Yang *et al.* 2021). Based on the research results of Beijing in the fine silty sand stratum, the maximum surface settlement range can be preliminarily predicted according to the geological conditions, excavation depth, and construction method (Wu *et al.* 2020).

In summary, it can be seen that the construction process of excavation of deep and large foundation pits is complex and diverse, which has a great impact on the surrounding environment, especially in the soft stratum. In this study, the FEA is conducted to simulate the excavation process of the Zhongshan Park Station in a water-rich soft stratum of Foshan Metro. This paper aims to investigate the characteristics and mechanisms of the surrounding ground surface settlement and diaphragm wall deformation caused by deep excavation at different positions of the foundation pit. The results of the study can provide a reference for similar projects.

2 PROJECT OVERVIEW

2.1 *Overview of foundation pit engineering*

Zhongshan Park Station is the 26th station of the Foshan Metro Line. The station is an underground two-story double-column three-span station, which is constructed by the open-cut method. Figure 1 shows the planned view of the deep excavation. The total length of the station is 317 m, the outer package width of the standard section is 22.7 m, and the width of the expanded end is 28.8 m. The foundation pit is divided into three parts by the underground diaphragm wall, which are the East Pit (with a length of 95.7 m), Interchange Station Pit (with a length of 25.7 m), and West Pit (with a length of 195.6 m). The construction method of sectional excavation is adopted in the foundation pit engineering of the station. The main excavation depth is 16.7 m and the depth of the interchange station pit in the middle position is 24.8 m.

During the deep excavation, the ground surface settlement and diaphragm wall deformation are usually monitored to control the influence on the surrounding environment caused by excavation. Therefore, a set of monitor instruments is adopted. Among them, the monitoring points in the middle of the three parts are extracted for discussion in this study, respectively, which includes the lateral displacement monitoring points on the diaphragm

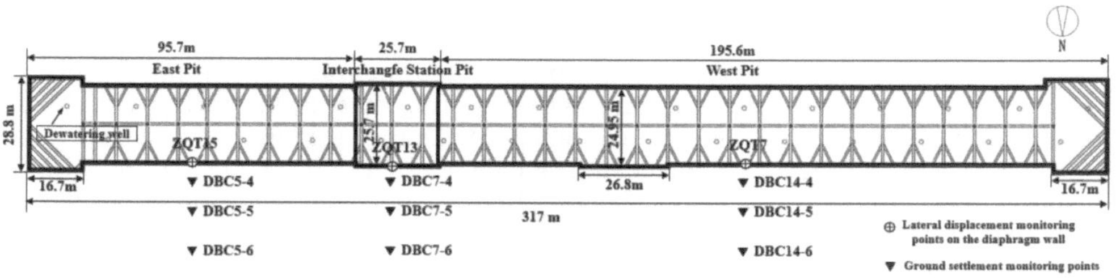

Figure 1. Plan view of the foundation pit.

wall (ZQT15, ZQT13, and ZQT7) and the ground surface settlement monitoring points (DBC5, DBC7, and DBC14). The locations of the extracted monitoring points are depicted in Figure 1.

2.2 *Parameters of supporting structure*

The Diaphragm walls with a thickness of 1 m and three internal supports are adopted as the retaining system of the foundation pit. The basic width of the diaphragm wall is 6 m as a waterproof curtain. The transfer station is supported by four sets of struts because of the deeper excavation, as shown in Figure 2. The first layer is supported by 700 × 900 mm reinforced concrete struts with a horizontal spacing of about 9 m, and the second one is supported by steel struts with the size of $\Phi = 609$, t = 16 with a horizontal spacing of about 3 m. Meanwhile, a 1000 × 1200 mm reinforced concrete strut is used as the third support, and the fourth support is an 800 × 1000 mm concrete strut. Both the standard section and the interchange station section are provided with neutral columns. A 150 mm-thick C20 concrete cushion is adopted at the bottom of the foundation pit.

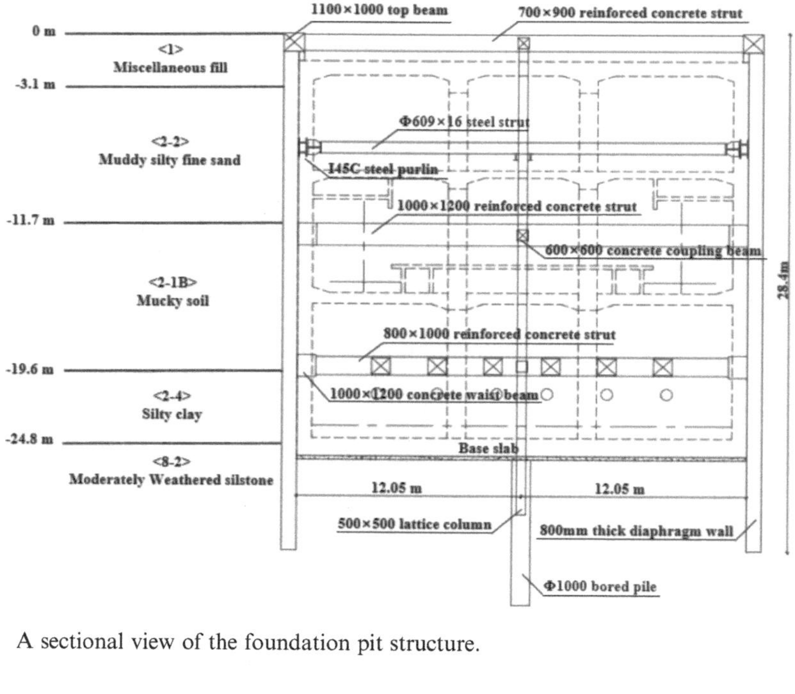

Figure 2. A sectional view of the foundation pit structure.

2.3 *Engineering geological conditions*

According to the geological survey report, the main distributed strata are divided into 6 sections, <1> miscellaneous fill, <2-2> muddy silty fine sand, <2-1B> mucky soil, <2-4> silty clay, <3-2> medium-coarse sand, and <8-2> moderately weathered siltstone, as shown in Figure 2. The position of the foundation pit floor in the East Pit and West Pit is mainly <2-1B> mucky soil and the bottom of the diaphragm wall is mainly located in the <8-2> moderately weathered siltstone stratum. The buried depth of the initial water level of groundwater is from 0.70 to 5.40 m, while the stable water level is from 1.30 to 5.70 m. The main aquifers are <2-2> muddy silty fine sand, <2-3> muddy medium-coarse sand, and <3-2> medium-coarse sand. These kinds of aquifers are mainly distributed in layers. The sand layer water is generally covered with a clayey soil layer, and the groundwater is slightly pressure-bearing. During the excavation, the groundwater level inside the foundation pit should be gradually lowered with the excavation so that the water level is maintained at 1 m below the excavation surface. Besides, 32 dewatering wells with a diameter of 1200 mm are arranged in the pit at a staggered interval of 20 m. The locations of the dewatering wells are depicted in Figure 1.

3 FINITE ELEMENT MODEL

3.1 *Establishment of finite element model*

The commercial FEA software ABAQUS is applied to carry out numerical calculations on the excavation construction of the foundation pit. The model diagram of the foundation pit project is shown in Figure 3. To reduce the influence of boundary conditions on the deformation of the foundation pit, this model is enlarged, and the influence width of the foundation pit excavation is about 3 to 4 times the excavation depth. Therefore, the geometrical dimensions of 443 m (length) × 173 m (width) × 80 m (depth) are adopted. A total of 324,106 elements and 359,567 nodes are adopted. 264,672 elements of type C3D8P are used to simulate the soil. The diaphragm wall and the base slab are simulated by 15,860 elements and 8,802 elements of type C3D8, respectively. Moreover, a total of 1,434 elements of type B31 are used for simulating the struts and pillars.

The interaction between both sides of the diaphragm wall and the surrounding soil is simulated by surface-to-surface contact. Besides, the bottom of the diaphragm wall is tied to the soil, while the supporting system is tied to the diaphragm wall. Uniformly distributed gravity is applied to the entire model. The simulation of the additional working load is realized by applying a uniform surcharge pressure of 20 kPa on the ground surface. The boundary conditions are applied to the entire model, in which the horizontal displacement of the four vertical boundary nodes around the model is set to zero. And the upper nodes of the model are defined as free boundaries, while the displacement constraints of the bottom nodes in all directions are restrained from displacement in all directions.

After the load and displacement boundary conditions are applied, the pore pressure elements are applied in the soil to realize the pore pressure/stress coupling analysis. The pore pressure boundary at the ground surface is restricted to zero.

3.2 *Material parameters*

As shown in Figure 3, the FEA model is simplified to a 6-layer homogeneous soil. Among them, the Modified-Cam-Clay (MCC) model and porous elastic model are adopted to simulate the constitutive behavior of the <2-1B> mucky soil and <2-4> silty clay. In contrast, the Mohr-Coulomb criterion and linear elastic model are used for the other four layers, which are <1> miscellaneous fill, <2-2> muddy silty fine sand, <3-2> medium-coarse sand, and <8-2> moderately weathered siltstone. The supporting structures, such as the

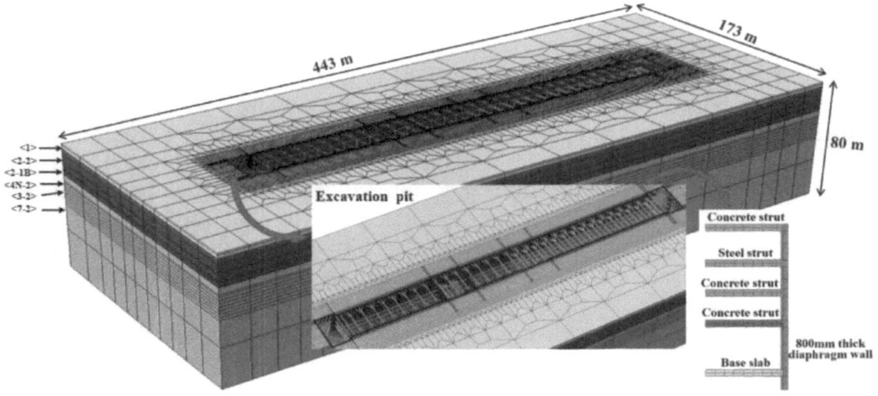

Figure 3. Overview of the FEA model.

diaphragm wall, bracing, and column piles, are simulated by linear elastic materials. According to the geological prospecting report, the friction coefficients of the soil layer from top to bottom are 0.25, 0.25, 0.25, 0.2, 0.25, 0.4, and 0.45, respectively. In the finite element analysis, it is considered that the permeability coefficient of soil does not change with the pore ratio. The dry density of the soil and the parameters of the triaxial shear test are adopted in the definition of material. The detailed material parameters adopted in the FEA model are listed in Table 1.

Theoretical research and experimental analysis show that the Mohr-Coulomb criterion belongs to the ideal elastic-plastic model, which is suitable for most soil properties and preliminary analysis of the foundation pit. The Modified-Cam-Clay (MCC) constitutive applies to both normally consolidated and overconsolidated clay soils and can also be used for investigations with high accuracy.

Table 1. Material parameters in FEA.

Component	γ/kN · m^{-3}	M	λ	κ	ν	e_1	k/m · s^{-1}	K	c$'$/kPa	φ'/°	E/MPa
<1>	18.72				0.30		2.31×10^{-5}		10	15	15
<2-2>	17.35				0.35		5.79×10^{-5}		5	25	18
<2-1B>	17.15	0.275	0.162	0.0148	0.25	1.205	1.16×10^{-8}	0.908			12
<2-4>	19.11	0.462	0.1047	0.0022	0.30	0.771	5.79×10^{-7}	0.846			20
<3-2>	19.11				0.25		1.16×10^{-4}		5	32	18
<8-2>	21.17				0.22		1.44×10^{-5}		40	28	2000
Diaphragm wall	24.5				0.22						3×10^4
Concrete struts	24.5				0.22						3×10^4
Steel struts	76.5				0.3						2.1×10^5
Bored piles	24.5				0.22						3×10^4

γ: Unit weight; M: Stress ratio; λ: Logarithmic plastic bulk modulus; κ: Logarithmic bulk modulus; ν: Poisson's ratio; e_1: Intercept of virgin consolidation line; k: Permeability coefficient; K: Flow stress ratio; c': Effective cohesion; φ': Effective friction angle; E: Young's modulus.

3.3 *Construction simulation steps*

The "element death" technology is adopted to simulate the excavation process of the foundation pit. Removal and addition of components are realized by "killing" or "activating" mesh units. The analysis steps and excavation times are divided according to the construction logs. Before the excavation of the foundation pit, the Geostatic analysis step is used to balance the initial ground stress. According to the construction on site, the foundation pit excavation is roughly divided into three parts, which include in East Pit, Interchange Station Pit, and West Pit. Since the excavation depth of the Interchange Station Pit in the middle is deeper than that on both sides, the side view of the whole foundation pit presents a "T shape". The project does not follow a fixed excavation sequence from left to right and from top to bottom. However, it is adjusted according to the actual construction. The detailed excavation sequence is shown in Figure 4.

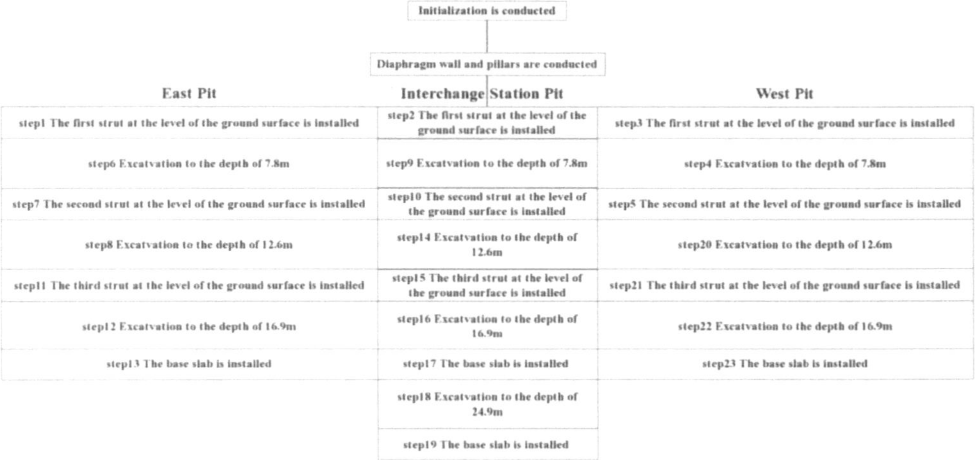

Figure 4. Flow chart of the simulation procedure.

4 ANALYSIS OF NUMERICAL CALCULATION RESULTS

4.1 *Model validation*

To verify the accuracy of the model, the daily monitoring value during the construction and the calculated values of the FE model are extracted to draw the deformation curves of each monitoring point with the construction. as shown in Figure 5. The monitoring points DBC4-

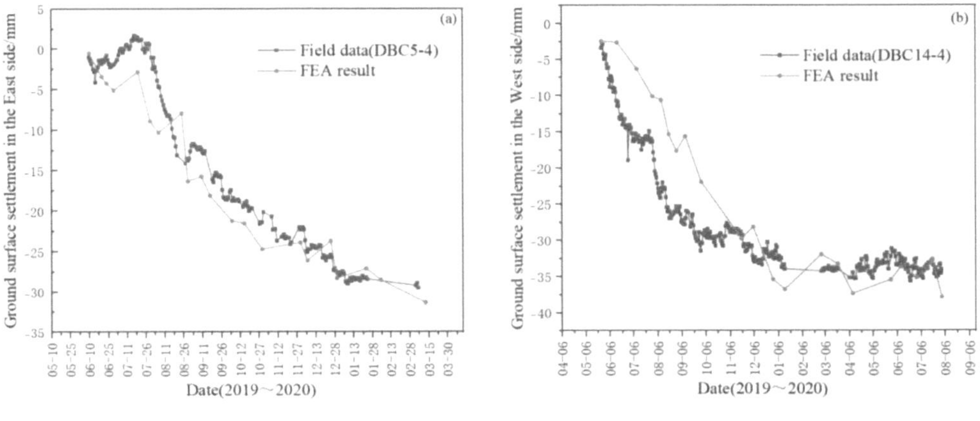

Figure 5. Comparison of FEA results and field data for ground surface settlement in the East and West sides of the foundation pit: (a) Results of ground settlement of DBC5-4; (b) Results of ground settlement of DBC14-4.

4 and ZQT15 in East Pit are extracted, as shown in Figure 1. In addition, the monitoring points DBC14-4 and ZQT7 in West Pit are extracted. As shown in Figure 5, the deformation development of the calculated values is similar to the field monitoring data. With the progress of construction on-site, the ground surface settlement outside the wall keeps increasing. At last, the settlement reaches the maximum value and stops increasing when the final construction is completed.

On the other hand, the monitoring data of lateral deflection of the diaphragm wall of ZQT7 and ZQT15 are compared with the calculated values of numerical simulation. As shown in Figure 6, the deformation development of FE simulation is similar to the monitoring data on-site. While the maximum and minimum values of the deformation and the location are similar. The underground diaphragm wall is installed as a support structure before excavation. It is worth noting that the horizontal lateral deflection of the diaphragm wall near the top is different. The monitoring value shifts in the opposite direction, which is the outer direction of the foundation pit. Among them, the deflection of the diaphragm wall is −9.62 mm of ZQT15 and −3.76 mm of ZQT4. The horizontal lateral of the numerical simulation value at this position is close to 0. This is because the plastic deformation of the first concrete strut is not considered in the numerical simulation, so the relative stiffness at the top of the diaphragm wall is larger and the stability of the structure is enhanced.

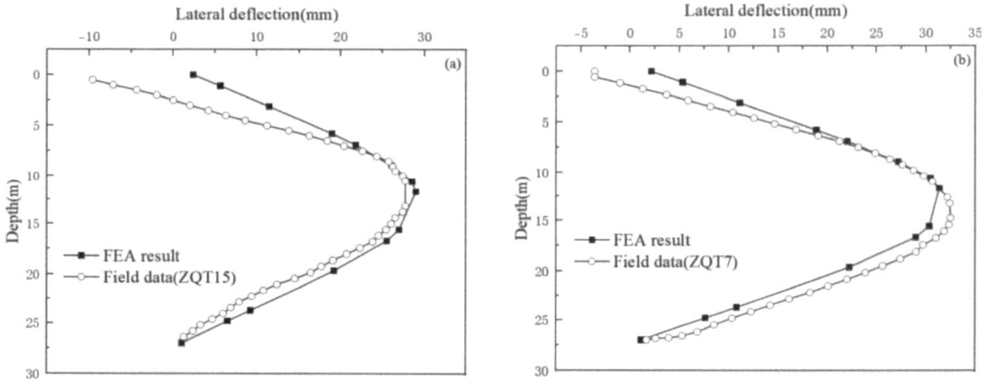

Figure 6. Comparison of FEA results and field data for wall lateral deflection in the East and West sides of foundation pit: (a) Results of wall deflection of ZQT15; (b) Results of wall deflection of ZQT7.

In summary, the accuracy of the FE model is verified by comparing the numerical simulation data with the monitoring data of ground surface settlement and the lateral deflection of the diaphragm wall. It can be considered that the FE model can simulate the deformation characteristics and mechanism of the actual excavation of the project.

4.2 *Analysis of ground surface settlement outside the wall*

Figure 7 shows the ground surface settlement outside the diaphragm wall at each construction stage in the middle of the West Pit. It can be seen that the surface deformation outside the wall is negative, which indicates that the longitudinal deformation is a settlement. The ground settlement increases with the increase in distance and reaches the maximum value in the range of 10–20 m. Subsequently, the settlement decreases slightly and eventually reaches a stable stage, forming a shape similar to a groove. This is because the original soil in the pit is removed at the beginning of the excavation so that the foundation pit forms a state of unloading. At that time, the soil outside the wall is disturbed, and the soil near the wall moves into the pit. The foundation pit reaches the maximum settlement of 32.36 mm when the excavation reaches the last excavation step of ex4.

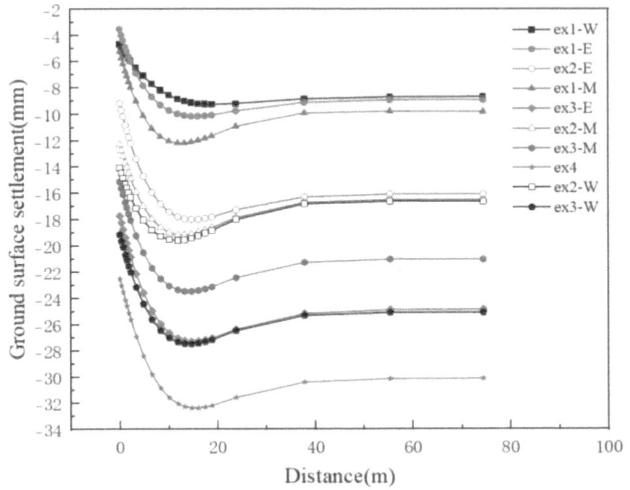

Figure 7. Results of ground surface settlement along line DBC16.

Since the main foundation pit is divided into three zones, East Pit, Interchange Station Pit, and West Pit. Among them, three layers of soil are excavated in East Pit and West Pit with an excavation depth of 16.7 m, while four layers of soil are excavated in Interchange Station Pit with an excavation depth of 24.8 m. Therefore, as shown in Figure 7, red represents the excavation stages of East Pit, green represents the excavation stages of Interchange Station Pit, and black represents the excavation stages of West Pit. Besides, different symbols represent the construction stages respectively. By comparing each construction stage, it can be found that the ground surface settlement increases with the progress of construction. However, the increments between the two construction stages are different. On the one hand, the increment of ground settlement is larger when a whole layer of soil is excavated along the length of the foundation pit, while the range of increment is about 7–10 mm. Taking the excavation in East Pit as an example, the maximum settlement of ex1-E (excavation of the first layer of soil) is 10.06 mm, while ex2-E is 17.92 mm and ex3-E is 27.19 mm. On the other hand, when the same layer of soil is excavated, the ground surface settlement induced by the later construction is larger than that of the first construction. And the increment range is within the range of 1 to 3 mm, which is smaller than the settlement caused by the increase in excavation depth. For example, ex1-M (the first layer of soil is excavated in the Interchange Station Pit) is constructed after ex1-E and ex2-E. Its settlement is larger than that of ex1-E but smaller than that of ex2-E.

In addition, the settlement of ex3-M is much smaller than that of ex3-E and ex3-W on the same layer. This is because the span of the diaphragm wall in the Interchange Station Pit is 25.7 m, which is smaller than that on both sides. It can be found that reducing the span of the diaphragm wall can effectively increase the stiffness and stability of the foundation pit, which leads to a decrease in the ground surface settlement.

4.3 *Deformation analysis of underground diaphragm wall*

The three diagrams in Figure 8 show the horizontal lateral deflection curves of the underground diaphragm wall in each construction stage of the East Pit, Interchange Station Pit, and West Pit. The curves present an arc shape with small values at both ends and large in the middle. With the excavation of the foundation pit, the lateral deflection of the diaphragm wall increases continuously. Then, the deformation reaches the maximum at the center of the

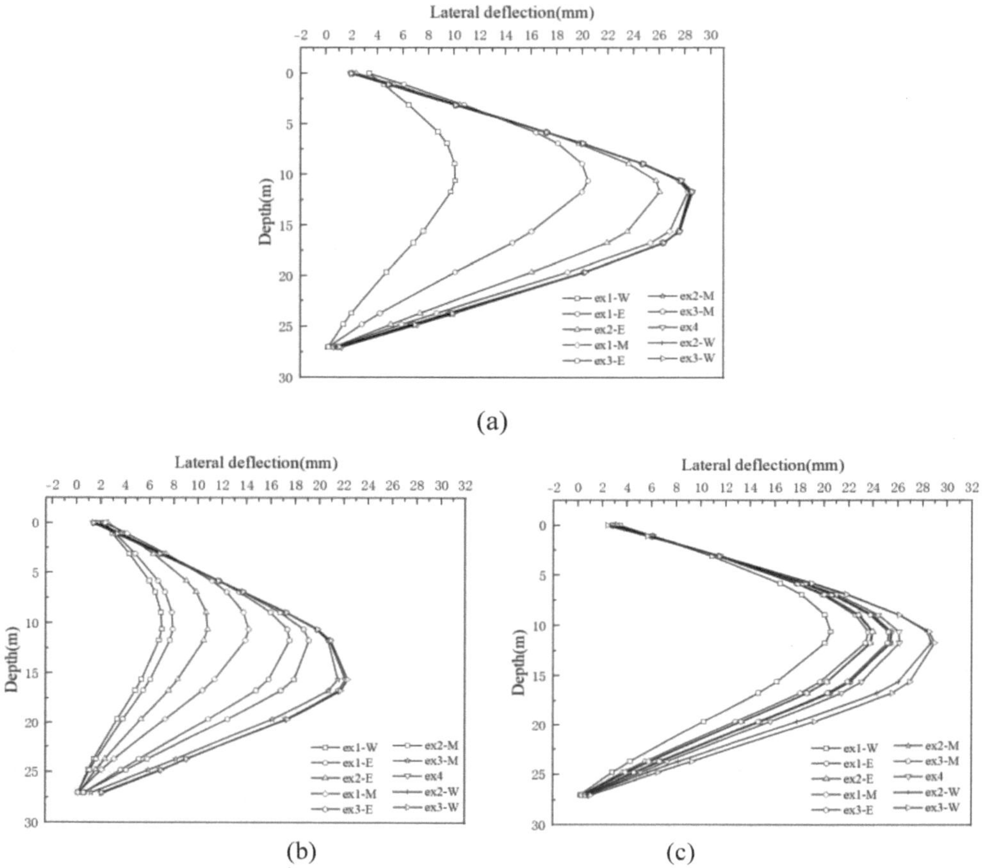

Figure 8. Lateral deflection of the diaphragm retaining wall: (a) Results of inclinometer ZQT15; (b) Results of inclinometer ZQT13; and (c) Results of inclinometer ZQT7.

wall and decreases after the peak value. Finally, the deformation reaches zero at the end of the diaphragm wall. Ex3-W (the third layer of the West Pit) is the last excavation step of the foundation pit, the horizontal lateral deflection at three locations reaches the maximum at that time, which are 28.49 mm in East Pit, 22.27 mm in Interchange Station Pit, and 29.36 mm in West Pit, respectively. There is a difference between the three values. The lateral deflection of the wall in the interchange station is the smallest, while the wall lateral deflection on the west side is the largest. This is because the span of the diaphragm wall in the interchange station is the smallest, while that in the west area is the largest. It shows that the stiffness of the wall increases effectively by reducing the span of the diaphragm wall, thus reducing the horizontal lateral deflection.

With the progress of construction, the maximum horizontal lateral displacement position of the wall develops downward. As shown in Figure 8(b), the maximum lateral deflection of ex1-W occurs at 10.59 m below the diaphragm wall, while the maximum horizontal lateral displacement of ex3-W appears at 15.59 m downward. Different from the analysis in the previous section, the surface settlement is mainly affected by the excavation depth, the horizontal deflection of the diaphragm wall is not greatly affected by the excavation depth, and the main influence comes from the construction step.

5 CONCLUSIONS

The characteristics of the surrounding ground settlement and wall deformation caused by deep and large excavation are investigated by the FEA software ABAQUS. The FEA model is validated by comparing it with the monitoring data. Based on the FEA model, the surrounding ground settlement and the diaphragm wall deformation caused by deep excavation at different positions of the foundation pit are researched. The following conclusions can be drawn:

1. The ground surface settlement outside the diaphragm wall caused by this foundation pit engineering is in the shape of a groove, and the maximum settlement occurs in the range of 10–20 m from the wall, and the maximum settlement is 32.36 mm.
2. In this study, the excavation sequence has little influence on the ground surface settlement, and the settlement value of the later construction is larger by 1–3 mm than that of the first construction. The main factor affecting the surface settlement is the excavation depth, which indicates that the ground surface settlement increases with the increase in excavation depth, and the increment is between 7–10 mm.
3. The lateral deflection of the diaphragm wall presents an arc shape with small values at both ends while large in the middle. The maximum horizontal lateral deflection appears in the central position of the diaphragm wall, which is 29.36 mm. With the progress of construction, the position of the maximum horizontal lateral displacement will develop downward.
4. Reducing the span of the diaphragm wall can increase the stiffness of the wall and enhance the effect of retaining structure, thus effectively reducing the ground surface settlement outside the wall and the horizontal lateral deflection of the wall.

REFERENCES

Elbaz, K. & Shen, S. L. & Tan, Y. & Cheng, W. C. (2018). Investigation into Performance of Deep Excavation in Sand Covered Karst: A Case Report. *J. Soils and Foundations*. 58, 1042–1058.

Feng, C. L. & Zhang, D. L. (2018). The General Deformation Mode and its Application of Subway Station Foundation Pit in Sandy Cobble Stratum. *J. Chinese Journal of Rock Mechanics and Engineering*. 37 (2), 4395–4405.

Li, J. P. & Chen, H. H. & Li, L. & Ma, J. S. (2018). Observation on Depth and Spatial Effects of Deep Excavation in Soft Clay. *J. China J. High. Transp.* 31 (2), 208–217.

Li, T. & Liu, B. & Chu, H. W. & Xuan, L. K. & Dai, J. D. (2018). Force Analysis and Control of Deep Excavation Lateral Supporting System in Shanghai Soft Clay Area. *J. Geotechnical Investigation & Surveying*. 1, 19–26.

Shen, H. J. & Jiang, Z. & Zhu, B. (2022). Deformation Analysis of Deep Foundation Pit of a Subway Station in Soft Soil Area. *J. Urban Geotechnical Investigation & Surveying*. 5, 184–189.

Wang, X. J. & Gong, J. & Zhao, X. H. (2015). Monitoring and Analysis of Lateral Deformation of Retaining Wall of a Super-Large Deep Excavation. *J. Chinese Journal of Underground Space and Engineering*. 11 (6), 1588–1595.

Wu, J. Y. & Ye, X. F. & Yu, P. & Tian, T. Y. (2020). Study on Settlement Law of PBA Station of Beijing Metro in Fine Silty Sand Stratum. *J. Tunnel Construction*. 40 (10), 1408–1416.

Yang, Y. B. & Li, J. S. & Liu, C. & Ma, J. J. & Zheng, S. & Chen, W. (2021). Influence of Deep Excavation on Adjacent Bridge Piles Considering Underlying Karst Cavern: a Case History and Numerical Investigation. *J. Acta Geotechnica*. 1–18.

Yu, W. & Chen, X. L. & Zhang, X. M. (2020). Analysis of the Influence of Deep Foundation Pit Construction in the Water-rich Soft Ground of Shenzhen on the Surrounding Environment. *J. Journal of Railway Science and Engineering*. 17 (9), 2251–2261.

Zheng, G. & Zhu, H. H. & Liu, X. R. & Yang, G. H. (2016). Control of Safety of Deep Excavations and Underground Engineering and its Impact on the Surrounding Environment. *J. China Civil Engineering Journal*. 49 (6), 1–24.

Water Conservancy and Civil Construction – Oke & Ahmad (Eds)
© 2024 The Author(s), ISBN: 978-1-032-58614-4

Review of study on the effects of regulated deficit irrigation on potato yield, quality, and water use efficiency

Yingying Wang
College of Agronomy and Agricultural Engineering, Liaocheng University, Liaocheng, China
College of Water Conservancy and Hydropower Engineering, Gansu Agricultural University, Gansu Lanzhou, China

Shouchao Yu
College of Agronomy and Agricultural Engineering, Liaocheng University, Liaocheng, China

Hengjia Zhang*
College of Agronomy and Agricultural Engineering, Liaocheng University, Liaocheng, China
College of Water Conservancy and Hydropower Engineering, Gansu Agricultural University, Gansu Lanzhou, China

ABSTRACT: With the development of the processing industry and the adjustment of the industrial structure, potatoes have become one of the advantageous crops with high yield, stable yield, and high economic benefits. However, under the traditional water management mode, water use is inefficient and water resources are scarce, so water has always been seen as a main factor limiting potato production. Regulated deficit irrigation cannot only improve the ability of crops to use soil water but also increase crop yield and improve crop quality, which is an effective way to promote the efficient development of the potato industry. This paper mainly reviews the effects of adjusted deficit irrigation on potato plant growth, yield, water use efficiency, and quality. The results show that timely and moderate water deficit can reduce the water consumption of potatoes throughout the growth period and enhance the absorption, utilization of water, and biomass yield, and improve tuber quality to some extent.

1 INTRODUCTION

China is a relatively water-poor country. The geographical distribution of domestic water resources is extremely uneven, especially in the northwest region, where drought and water shortage are important factors limiting agricultural development and national economic construction. Studies have shown [1,2] that the reduction of crop yield caused by water deficit has exceeded the sum of the effects of all other abiotic stresses, so irrigation is the main measure to alleviate drought stress in farmland. However, with the decreasing share of agricultural water in the country, the conflict between agriculture and other industries is becoming more and more prominent, while agricultural water is seriously wasted, with irrigation water use efficiency of less than 60% compared to developed countries [3]. Therefore, in the face of the increasing shortage of water resources and the prominent contradiction between supply and demand, how to improve the efficiency of farmland water use is the key to solving the current water shortage in agricultural production.

*Corresponding Author: 596088683@qq.com

DOI: 10.1201/9781003450818-23

Regulated deficit irrigation is based on the physiological characteristics of crops and artificially imposes certain water stress, which can maintain the normal growth of crops without fully meeting their water demand, thus improving the water use efficiency of crops [4]. Potato (*solanum tuberosum L.*) has high nutritional value, strong adaptability, and good economic benefits. Its roots are mainly distributed in the soil layer within 30 cm, and it is sensitive to changes in soil water and heats through the growing season [5]. Studies have revealed that adjusted deficit irrigation cannot only improve potato tuber yield and tuber quality [6,7] but also it has a significant impact on soil moisture changes and water consumption patterns during the whole growth period [8]. This paper summarizes and analyzes the effects of adjusted deficit irrigation on potato growth, water use efficiency, and quality, which provides a theoretical basis for efficient potato planting and cultivation.

2 OVERVIEW OF REGULATED DEFICIT IRRIGATION

In the mid-1970s, an Australian agricultural institution first proposed the concept of Regulated Deficit Irrigation (RDI) [9], in which water stress treatments are artificially applied at certain growth stages of crops, thereby changing the distribution ratio of photosynthetic products of crops among various tissues and organs, reducing redundant growth of vegetative organs of crops, and promoting the growth of reproductive organs to achieve water conservation, yield increase, and regulation of crops. The research on regulated deficit irrigation abroad was first applied to fruit trees [10]. After that, it was not until the mid-to-late 1990s that field crops were gradually studied by domestic and foreign scholars as experimental subjects. It was confirmed that regulated deficit irrigation was applicable and feasible for field crops [11,12]. Nowadays, research on regulated deficit irrigation has been widely applied to crops, such as wheat [13], maize [14], tomato [15], etc. Meanwhile, the scope of study on regulated deficit irrigation has become more and more extensive, such as other factors Simultaneous multi-factor experiments (water and fertilizer regulation, water-salt migration, intercropping, and other coupled controlled deficit irrigation experiments), the formulation and optimization of the regulated deficit irrigation system, and the design and application of the regulated deficit irrigation model.

The study has shown [16] that the reason why regulated deficit irrigation can improve the water use efficiency of crops mainly lies in the root system of crops. According to the theory of root-shoot function balance [17], the root-shoot ratio in the process of crop growth and development tends to have a stable value due to the role of inherent genetic factors in crops. When soil moisture is deficient at the root of the crop, the root system will develop preferentially, while the growth of the canopy of the plant is inhibited, the leaf area is reduced, the stomatal opening of the leaves is reduced, and the transpiration of the plant is weakened, thereby reducing the plant water consumption. However, when the water stress is relieved and rehydrated, the organic products accumulated by the crops will be used for their growth, and the previously developed root system will enhance the utilization of water and nutrients by the crops, so the final biomass and yield of the crops will not decrease or even appear a certain degree of increase in production [18].

3 YIELD

Moisture is the basis for the formation of potato yield. Studies have revealed [19] that the production of 1 kg of fresh potato tubers requires 200-300 kg of water, and the quality of soil moisture conditions is directly related to the level of potato yield and biomass. Soil water content is one of the important indicators to judge the degree of soil water deficit, which is affected by various factors, such as precipitation and irrigation [20]. Excessive irrigation will not only affect the normal growth of potato plants but also may cause tubers to rot and be intolerant of storage and transportation. Moderate deficit irrigation has effects on plant weight, transverse diameter, and tuber yield of potatoes. Potato tuber expansion is a critical period for dry matter accumulation and a water-sensitive period, and water deficit in this period will directly lead to potato yield reduction [21]. Li Xuanzhen et al. [22] found that mild and moderate water deficit adjustment during potato tuber formation had little effect on yield, while water deficit adjustment during tuber expansion had a significant impact on yield, and the greater the degree of deficit adjustment, the more severe the yield decline. The experimental results of Pan Xiaofan et al. [23] have shown that the water deficit adjustment at the seedling stage of potato cannot significantly reduce yield, but the yield of water deficit potato at the tuber formation stage, tuber expansion stage, and starch accumulation stage can be significantly reduced, with the most obvious reduction in tuber expansion.

4 WATER USE EFFICIENCY

Water use efficiency is the yield obtained by the crop consuming a unit mass of water, the greater the value, the better the water use of crops. Affected by the external environment and crop varieties, there are differences in the ecological characteristics of potato water. Xue Daoxin et al. [24] conducted a potato water deficit adjustment test and showed that since the tested potato was a late-maturing variety, the potato plants were luxuriant during the tuber formation period with high atmospheric temperature and evaporation intensity. The tuber expansion period showed a decreasing trend in atmospheric temperature and light intensity, plant transpiration and inter-tree evaporation decreased, and water consumption decreased accordingly. Therefore, the water consumption and daily water consumption intensity of potatoes in different growth stages are in the following order, tuber formation period, tuber expansion period stage, seedling stage, and starch accumulation stage. Water deficit adjustment has a great influence on the water consumption of potatoes in different growth stages and the water consumption decreases more significantly with the increase in the adjusted deficit level. Zhang Hengjia et al [25] indicated that compared with moderate deficit regulation during tuber expansion, mild deficit regulation during starch accumulation, and adequate water supply during the whole growth period, the water use efficiency of potatoes was the highest during mild water deficit regulation during tuber formation, which increased by 36.2%, 32.4%, and 14.2% respectively.

5 QUALITY

The quality of potatoes mainly includes three aspects, nutritional quality, commercial quality (appearance quality), and processing quality, of which tuber dry matter and starch content are the most important indicators affecting potato quality. The quality of potato pieces is affected by water. Moderate water deficit adjustment is beneficial to increase the starch content of potato tubers [26]. The starch accumulation period has the greatest impact on the nutritional quality of potatoes by water deficit. During the starch accumulation period, mild water deficit is beneficial to the growth of potato pieces and increases the protein and reduces sugar contents, respectively. However, a moderate water deficit will lead to reduced organic matter accumulation in potatoes [27]. Reducing sugar in potato tubers can undergo non-enzymatic browning Maillard reaction between the a-amino acids of nitrogen-containing compounds, which will deepen the surface color of French fries (chips) to tan, so the level of reducing sugar is the most stringent indicator of whether the potato can be used as a raw material for processing. Studies have shown [28] that deficit irrigation is not conducive to the formation of reducing sugars in potatoes. From sufficient water supply to severe water deficit, the reduced sugar content of 'Atlantic' decreased by 0.10% to 0.04%. The reduced sugar content of 'Qingshu 9' declined by 0.14% to 0.01%.

6 CONCLUSION

Regulated deficit irrigation uses the physiological response of crops to water stress to transfer photosynthetic products to specific organs, thereby affecting crop yield, water use, and quality. Studies have proved that the accumulation of potato above-ground biomass directly affects the formation of final yield. Tuber expansion stage is the water-sensitive period of potatoes and is the most important period for dry matter accumulation, so water loss should be avoided in the production. Potatoes consume the most water during the tuber formation period but their specific water ecological characteristics vary due to different varieties, regions, and other environmental factors. The starch content of potato tubers accounts for about 80% of its dry matter content, water deficit during starch accumulation has a great impact on potato nutritional quality, but moderate water deficit can improve soil air permeability, which is conducive to tuber respiration and metabolism to improve life vitality. With the rapid development of the potato processing industry and the continuous improvement of people's living standards, more and more attention is paid to potato quality and quantity. Therefore, appropriate water regulation and scientific and reasonable irrigation systems in the process of potato production cannot only improve crop water use efficiency, increase yield, and optimize quality but also improve soil structure and physical and chemical properties, thus effectively promoting the benign cycle and sustainable and healthy development of farmland system.

ACKNOWLEDGMENTS

This work was financially supported by the National Natural Science Foundation of China (No. 52269008), the Industrial Support Plan Project of the Gansu Provincial Department of Education (No. 2022CYZC-51), and the Key Research and Planning Projects of Gansu Province (No. 18YF1NA073).

REFERENCES

[1] Brdy, E. A., Hant. Responses to water deficit. *Trends Plant Science* 2:48–54 (1997).

[2] Xianzhen, Z., Zhengshu, Su. General Survey of Crop Water Deficiency Injury Physiological Research. *Journal of Shenyang Agricultural University* 01: 85–91 (1996).

[3] Huazhong, L. Water and Fertilizer Integration Technology and Its Application. *Agricultural Disaster Research* 4(8):50–52 (2014).

[4] Liguo, J., Yuzhen, C. Suyalaqiqige, Water efficient utilization of irrigated potato and its mechanism. *Soil Bulletin* 49(1):226–231 (2018).

[5] Schapendonk, A. H. C. M., Spitters, C. J. T., Grout, P. J. Effects of Water Stress on Photosynthesis and Chlorophyll Fluorescence of Five Potato Cultivars. *Potato Research* 32 (1):17–32 (1989).

[6] Lynch, D.R., Foroud, N., Kozub, G.C. The Effect of Moisture Stress at Three Growth Stages on the Yield, Components of Yield and Processing Quality of Eight Potato Cultivars. *American Journal of Potato Research* 72(6):375–386 (1995).

[7] Eldredge, E.P., Holmes, Z.A., Mosley, A.R. Effects of Transitory Water Stress on Potato Tube Stem-end Reducing Sugar and Fry Color. *American Potato Journal* 73 (11):517–530 (1996).

[8] Fengxin, W., Yuehu, K., Shiping, L. Study on the Law of Water Consumption and Water Demand of Potato Under Drip Irrigation. *Agricultural Research in Arid Areas* 23(1):9–15 (2005).

[9] Shaozhong, K., Huanjie, C. *Theory and Practice of Alternate Irrigation and Adjusted Deficit Irrigation of Crop Root Zones*. In: China Agricultural Press., Beijing (2002).

[10] Mitchell, P. D. Jerie, P. H. The Effects of Regulated Water Deficits on Pear Tree Growth, Flowering, Fruit Growth, and Yield. *Journal of the American Society for Horticultural Science* 109(5):604–606 (1984).

[11] Shaozhong, K., Wenjuan, S., Xiaotao, H. Effects of Adjusted Deficit Irrigation on Maize Physiological Ecology and Water Use Efficiency. *Chinese Journal of Agricultural Engineering* 14(4): 82–72 (1998).

[12] Xiying, Z., Xinyuan, Y., Wang. Preliminary Report on the Field Experiment of Winter Wheat Regulated Deficit Irrigation System. *Ecological Agriculture Research* 6(3):33–36 (1999).
[13] Panda, P. K., Behen, S. K., Kashyap, P. S. Effect Management of Irrigation Water for Wheat Under Stressed Condition. *Agricultural Water Management* 63: 37–56 (2003).
[14] Boyer, J.S. Differing Sensitivity of Photosynthesis to Low Leaf Water Potentials in Corn and Soybean. *Plant Physiology* 46(2): 236–239 (1970).
[15] Nangare, D.D, Singh, Y., Kumar, P. S. Growth, Fruit Yield and Quality of Tomato (Lycopersicon Esculentum Mill.) as Affected by Deficit Irrigation Regulated on Phenological Basis. *Agricultural Water Management* 171:73–79 (2016).
[16] Wenjuan, S., Xiaotao, Hu., Shaozhong, K. Research Status and the Prospect of Crop Deficit Irrigation Technology Under Drought and Water Shortage Conditions. *Agricultural Research in Arid Areas* (002):87–91 (1998).
[17] Anne-Maree, Boland. The Effect of Regulated Deficit Irrigation on Tree Water use and Growth of Peach. *Journal of Horticultural Science* 68(2):261–264 (1993).
[18] Baodi, D., ZhengBing, Z., Mengyu, L. Research Progress on the Compensation Effect of Crops Under Water Deficit. *Northwest Agricultural Journal* 13(3):31–34 (2004).
[19] Yuming, W. Effects of Drip Irrigation, Sprinkler Irrigation, and Pipe Irrigation on Potato Yield and Water Production Efficiency. *North China Agricultural Journal* 22 (z3):83–84 (2007).
[20] Yuehu, K., Fengxin, W., Shiping, L. Effects of Drip Irrigation on the Growth of Potato by Regulating Soil Moisture. *Journal of Agricultural Engineering* 20(2):66–72 (2004).
[21] Xuanzhen, L., Hengjia, Z., Haoliang, D., Xiaoting, Y., Yuchun, B. Effects of Sub-film Drip Irrigation on Biomass Allocation, Yield and Water use Efficiency of Oasis Potato. *North China Agricultural Journal* 30(05): 223–231 (2015).
[22] Daoxin, X. *Research on Potato Water Productivity of Deficient Potatoes with Drip Irrigation under Film in Desert Oasis*. Gansu Agricultural University (2017).
[23] Xiaofan, P., Hengjia, Z., Haoliang, D., Fuqiang L. Effects of Adjusted Deficit Irrigation at Different Growth Stages in Hexi Oasis on Potato Growth, Yield, and Quality. *Agricultural Engineering* 11 (02):130–136 (2021).
[24] Daoxin, X., Hengjia, Z., Yuchun, B., Shijie, W. Effects of Drip Irrigation Under Oasis Film on Soil Environment and Yield of Potato. *North China Agricultural Journal* 32(03):229–238 (2017).
[25] Hengjia, Z., Jing, L. Physiological Characteristics of Photosynthesis and Water Utilization of Potato Under Drip Irrigation Under Oasis Film. *Journal of Agricultural Machinery* 44(10):143–151 (2013).
[26] Jianyuan, Z., Ziyong, C., Huisheng, H. Experimental Research on Adjusted Deficit Irrigation Technology of Potatoes in No-tillage Cultivation with Wheat Straw Covering. *Agricultural Research in Arid Areas* 28 (6):7–11 (2010).
[27] Wanheng, Z. Effects of Water Deficit Adjustment in Different Growth Stages on Growth Characteristics, Yield and Quality of Potato Under Drip Irrigation Under Oasis Film. Gansu Agricultural University (2019).
[28] Zhengpeng, C. Effects of Deficit Irrigation on Potato Growth, Yield, Quality and Water Use. *Gansu Agricultural University* (2019).

Research progress on the effects of water-saving irrigation techniques and patterns on potato quality

Jie Li
College of Agronomy and Agricultural Engineering, Liaocheng University, Liaocheng, China
College of Water Conservancy and Hydropower Engineering, Gansu Agricultural University, Lanzhou, Gansu Province, China

Shouchao Yu
College of Agronomy and Agricultural Engineering, Liaocheng University, Liaocheng, China

Hengjia Zhang*
College of Agronomy and Agricultural Engineering, Liaocheng University, Liaocheng, China
College of Water Conservancy and Hydropower Engineering, Gansu Agricultural University, Lanzhou, Gansu province, China

Jiandong Yu
College of Water Conservancy and Hydropower Engineering, Gansu Agricultural University, Lanzhou, Gansu province, China

ABSTRACT: At present, the total water resources in China are insufficient and the amount of water used for agricultural irrigation is large. Therefore, it is particularly important to realize precision water-saving irrigation. Potato is one of the main food crops in China. Water deficit adjustment in different growth periods of potatoes has positive effects on improving the yield and quality of potatoes, which is significant to ensure food security in China. This paper summarizes the influence of different irrigation amounts, water-saving irrigation technology, and water-saving irrigation modes on potato quality in recent years. By discussing the response mechanism of potato quality to the water environment at different growth stages, this paper clarifies the significance of water-saving and quality adjustment and provides theoretical support for potato water-saving irrigation.

1 INTRODUCTION

The total water resource in China is insufficient. In 2021, the total water resource in China is 2,963.82 billion cubic meters and the total water consumption is 592.02 billion cubic meters. As a large agricultural-producing country, agriculture accounts for 61.5% of the total water consumption. In the face of decreasing freshwater resources and excessive water consumption in agriculture, it is particularly important to realize precise irrigation of crops while saving freshwater resources [1] and improving crop yield and quality.

Potato is an annual herb of Solanaceae [2] and its tuber has high nutritional value. With the continuous optimization of crop planting structure and the launch of potatoes as a staple food strategy, potatoes have become a crop that can be used for both food, vegetable, and feed [3]. In 2021, China's potato planting area was about 4,606 thousand hectares and its

*Corresponding Author: 596088683@qq.com

DOI: 10.1201/9781003450818-24

output was 18.309 million tons, both ranking first in the world. The growth and development of potatoes are light-loving, but the growing environment is not tolerant of high temperatures and requires loose soil structure, good drainage, ventilation, cooling, and moisture [4]. The important indexes to measure potato quality include starch content, protein content, total sugar content, amino acid content, vitamin C content, organic acid content, and so on. Potatoes are shallow-rooted crops and are sensitive to water requirements throughout the growth stage, and unreasonable irrigation and water supply are the main factors limiting potato yield and quality. Therefore, during its growth and development period, the exploration and research of efficient water-saving irrigation technology and mode can help to improve the water environment of potato planting and growth, thus ensuring stable yield and increasing yield and high quality of potato.

2 RESEARCH PROGRESS OF IRRIGATION AMOUNT ON POTATO QUALITY

Water is a necessity for crop growth and development and is an important raw material for photosynthesis. Water can regulate soil air, soil temperature, and soil fertility, improve farmland microclimate, and the quality and efficiency of farming. According to the difference in water requirements in different growth stages of crops, potato growth and development can be divided into four stages: seedling stage, tuber formation stage, tuber expansion stage, and starch accumulation stage [5]. The demands for irrigation amount of potato in each stage are different. Wang Ying *et al.* [6] studied the influence of drip irrigation frequency and irrigation amount on potato quality and found that starch and vitamin contents in potato tubers were the highest under medium-frequency drip irrigation, and potato quality increased with the increase in irrigation amount. Jin Jianxin *et al.* [7] set five gradient irrigation quotas for treatment. The results showed that each quality index of potato increased first and then decreased with the increase in irrigation amount.

Scientific and appropriate irrigation amount is the key to high yield and quality. Excessive irrigation amounts cannot only waste water resources and cause soil compaction but also can lead to the shallow distribution of potato roots. In this way, potato roots cannot absorb deep soil water, thus reducing the accumulation of photosynthetic assimilates in later stages. Too little irrigation water can cause crop growth retardation and reduce crop yield and quality. Therefore, the appropriate irrigation amount of different growth periods of potatoes can save water resources and ensure an increase in potato production and quality.

3 EFFECT OF WATER-SAVING IRRIGATION TECHNOLOGY ON POTATO QUALITY

3.1 *Regulated deficit irrigation technology*

Regulated Deficit Irrigation (RDI) refers to the artificial application of a certain degree of water stress at a certain stage of crop growth and development, especially at the vegetative growth stage [1], to promote the redistribution of photosynthates among different organs of crops and regulate the vegetative and reproductive growth rates of crops [8]. The main basic theories of RDI include root-shoot communication theory, growth redundancy theory, growth compensation effect of water deficit, stomatal regulation theory, and optimal allocation theory of limited crop water [9]. Root-cap communication theory refers to the process by which plant roots sense water stress and transmit information to the aboveground part, thereby inducing stomatal closure. The growth redundancy theory refers to the fact that the optimal growth of all parts of the crop can be regulated after the growth and development of the crop root system is properly regulated by irrigation [10], thus achieving the maximum yield of the crop. The growth compensation effect of water deficit refers to the application of a certain degree of water stress at the early stage of crop growth and development and the increase in crop yield and quality under the condition of rehydration in the later stage [9]. Stomatal regulation theory refers to the application of a certain degree of water deficit to crops, the root system of crops produces hormones and transmits them to leaves, thus causing stomatal regulation and changing the growth and yield of crops. The theory of the optimal allocation of limited water for crops refers to the rational allocation of limited irrigation water resources at different growth stages of crops.

In recent years, RDI is effective in improving potato quality. Zhang Wanheng *et al.* [11] studied the influence of water deficit on potato quality at different growth stages and showed that mild water shortage at the potato starch accumulation stage could increase the content of protein and reduce sugar in potato blocks by 3.11% and 15.63%, respectively, compared with full irrigation. Liu Jinyang *et al.* [12] set three deficit regulation treatments in the potato block formation stage, potato block expansion stage, and starch accumulation stage, respectively. They found that the influence degree of water deficit at different growth stages was as follows: potato block formation stage < starch accumulation stage < potato block expansion stage. Moderate water shortage in the formation period of potato blocks can significantly increase the content of amino acids, protein, and vitamin C in tubers and reduce the content of organic acids, resulting in better potato quality.

Many studies have shown that the appropriate adjustment period, degree, and duration of deficit cannot only improve the water use efficiency of crops but also control the growth of vegetative organs of crops and improve the quality of crops. At present, the research of deficit regulation technology in fruit trees, melons, and fruits has been basically mature. However, the research results of deficit regulation irrigation for field crops, such as potatoes, are still rare and need to be further explored.

3.2 *Controlled roots-divided alternative irrigation technology*

Controlled roots-divided alternative irrigation (CRAI) refers to artificially controlling or maintaining soil dryness in a part of the root zone, transmitting water stress signals generated by the soil to leaf stomata, forming an optimal stomatal opening, and reducing luxury transpiration and water consumption of crops. However, we should keep other parts of the root zone wet, maintain the normal water intake of crops, and ensure crop yield. The wet side and the dry sides need to be rotated and alternated repeatedly to avoid damage to part of the root system under a long-term drought environment, which can cause crop growth arrest [13]. Compared with other crops, the effects of controlled alternate root partitioning irrigation on potato quality are rarely studied. Wang Tengfei *et al.* [14] studied and set the alternate isolate ditch irrigation and conventional furrow irrigation, fixed infuse three research the growth and development and quality of potato processing. The results show that the protein and starch content of potatoes is significantly higher than the other two under alternate isolated furrow irrigation throughout the growth period and reduces the accumulation of reducing sugars in potato tubers, which improves the quality of potato tubers. According to the study of Hu Chao *et al.* [15], compared with the traditional irrigation method, alternate root irrigation not only reduces the content of reducing sugar in potato tubers but also increases the content of crude protein, starch, and titratable acid, thus significantly improving the quality of potato.

CRAI cannot only save irrigation water consumption and reduce the water loss caused by ineffective evaporation between plants but also can contribute to the compensatory growth of crop roots, improve soil permeability, make the roots better use of water and nutrients, and achieve water conservation and quality control while ensuring stable crop yield [16]. At present, the research on potato-CRAI irrigation is still at the exploratory stage, and the response of potato quality to alternate irrigation needs to be further studied. In addition, the appropriate control methods, alternate irrigation depth, sequence, and other indicators for different soil qualities and different crops need to be further studied.

4 EFFECT OF WATER-SAVING IRRIGATION MODE ON POTATO QUALITY

4.1 *Plastic film mulching*

Plastic film mulching can improve soil water and heat conditions, increase light and temperature, save water and moisture, collect rainwater flow, and improve water use efficiency. Plastic film mulching at the early stage of potato growth and development has played a role in increasing temperature and preserving soil moisture, inhibiting weeds, and improving water use efficiency, yield, and quality. According to the research results of Li Menglu *et al.* [17], compared with the traditional irrigation mode, plastic film mulching can significantly increase the content of starch, vitamin C, and dry matter in potato tubers, and significantly reduce the content of resistance indicators, such as soluble sugar, proline, and reducing sugar.

At the growth and development stage of potatoes, plastic film mulching technology not only shortens its growth cycle but also significantly improves its yield and quality. However, ordinary plastic film cannot be easily degraded, which can cause environmental pollution to a certain extent. Therefore, it is necessary to enhance the research, development, and application of degraded films while ensuring the film-coating effect.

4.2 *Drip irrigation under the membrane*

Drip irrigation under film is a new water-saving irrigation technology, which integrates drip irrigation technology and film mulching technology. Drip irrigation with film mulching cannot only save irrigation amount and improve irrigation efficiency but also can be beneficial to crop growth and development and increase crop yield, which is a major breakthrough in dry farming. Wei Fulong *et al.* [18] studied the influence of different topdressing treatments of drip irrigation under a membrane on potatoes and concluded that different fertilization treatments under drip irrigation under a membrane cannot only improve the utilization efficiency of fertilizer and the yield of potatoes but also increase the content of starch and crude protein in potato and make potato quality better compared with traditional furrow irrigation. Xue Daoxin *et al.* [19] showed that mild water deficit during tuber formation can further increase the starch content in potato tubers, thus improving potato quality.

At present, under-membrane drip irrigation has been widely used in the development of Chinese agriculture. Drip irrigation under a membrane can improve soil physical and chemical properties and increase soil temperature and water conservation, thus achieving water conservation and increasing yield. However, the problems of soil pollution and secondary soil salinization caused by long-term irrigation and the use of plastic film need to be solved.

5 CONCLUSION AND PROSPECT

In conclusion, the novel water-saving mode and efficient water-saving technology achieves scientific and appropriate water deficit regulation in the potato growth stage, plays a positive effect on the yield and quality of potatoes, and contributes to a large extent to the economic benefits of potatoes. Although researchers have promoted and applied potato water-saving irrigation in depth, there is still a lack of in-depth research on the impact of water-saving irrigation on potato quality compared with other crops.

In the actual production process, the impact of different geographical types, ecological environments, planting methods, irrigation methods, and different sewage sludge management measures on potato quality is different. It is important to understand the physiological mechanism of potato response to water shortage, to control field water rationally based on accurate monitoring and diagnosis of water regime, and to accurately measure and estimate the water requirements of potatoes at different stages. The establishment of the water-yield-quality model can further achieve water saving, high yield, and high quality of potatoes, which is significant to explore the change in potato quality.

ACKNOWLEDGMENTS

This work was financially supported by the Industrial Support Plan Project of the Gansu Provincial Department of Education (No. 2022CYZC-51), the National Natural Science Foundation of China (No. 52269008), and the Key Research and Planning Projects of the Gansu Province (No. 18YF1NA073).

REFERENCES

[1] Zhan, B. C., Liang, Y., Guo, W. Z., Li, Y. K., and Li, L. 2021. Research Progress and Development Trend of Potato Water-saving Technology. *Agric. Tech.* **9**(16): 73–7.

[2] Song, N., Wang, F. X., Yang, C. F., and Yang, K. J. 2013. Coupling Effects of Water and Nitrogen on Yield, Quality, and Water use of Potato with Drip Irrigation under Plastic Film Mulch. *Trans. Chin. Soc. Agric. Eng.* **29**(13): 98–105.

[3] Lin, Y. C., Zhang, D., and Xiao, Y. M. 2010. Development of Water-saving Cropping System on Potatoes in Northwest Regions in China. *Chin. Agric. Sci. Bull.* **26**(04): 99–103.

[4] Zhang, B. D., Liu, S. P., Song, X. P., and Feng, L. Y. 2018. Potato Dry Laminating Catchment Water Saving Cultivation Technique Analysis. *Chin. Agric. Abstract – Agric. Eng.* **30**(6): 70–2.

[5] Li, F. Q., Zhang, H. J., Deng, H. L., and Ba, Y. C. 2019. Effects of Drip Irrigation Deficit Under Mulch on Water Use Efficiency, Yield, and Quality of Potato. *Water Resour. Planning and Design* (06): 60–4.

[6] Wang, Y., Zhang, F. C., Wang, H. D., Bi, L. F., Cheng, M. H., Yan, F. L., Fan, J. L., and Xiang, Y. Z. 2019. Effects of the Frequency and Amount of Drip Irrigation on Yield, Tuber Quality, and Water Use Efficiency of Potato in the Sandy Soil of Yulin, Northern Shaanxi, China. *Chin. J. Appl. Ecol.* **30** (12): 4159–68.

[7] Jin, J. X., He, J. Q., Huang, J. C., and Gui, L. G. 2020. Effects of Different Irrigation Quotas on Growth, Yield, and Quality of Potatoes in the Arid Region of Central Ningxia. *Southwest Chin. J. Agric. Sci.* (5): 935–40.

[8] Guo, X. P. and Kang, S. Z. 1998. New Ideas of Regulated Deficit Irrigation and Water-Saving Irrigation. *Northwest Water Resour. Water Eng.* (04): 22–6.

[9] Yu, X. T., Cui, N. B., and Ma, Z. L. 2020. Research Progress on the Application of Regulated Deficit Irrigation. *Sichuan Water Resour.* **41**(01): 3–15.

[10] Wang, Y. D., Jiao, J., and Su, D. R. 2017. Effects of Deficit Degree and Deficit Duration on Biomass Allocation and Water Use Efficiency of Alfalfa (Medicago sativa L.). *Acta Agrestia Sin.* **25**(06): 1287–93.

[11] Zhang, W. H., Zhang, H. J., Li, F. Q., Wang, Z. Y., Gao, J., and Ba, Y. C. 2019. Effects of Regulated Drip Irrigation at Different Growth Stages on Yield, Quality and Water Use Efficiency of Potato in Oasis Region. *Acta Agric. Boreali-Sin.* **34**(05): 145–52.

[12] Liu, J. Y., Jia, S. H., and Liang, Z. G. 2018. Effects of Mulched Drip Irrigation under Water Deficit on Potato Growth Index and Quality in Oasis Region. *Yellow River* **40**(08): 152–6.

[13] Yang, Y., Guo, H. X., Zhu, J. R., Zhu, P. F., Mei, L. L., and Shu, L. Z. 2021. Research Progress of Alternate Partial Root-zone Irrigation. *J. Huaibei Normal Univ.* (*Nat. Sci.*) **42**(02): 35–42.

[14] Wang, T. F., Zhang, R., Zhang, M. H., Zhang, Y. S., Lin, B. J., Yang, C. Y., and Wang, C. H. 2020. The Effects of Different Furrow Irrigation on Potato Growth and Quality. *Chin. Rural Water Hydropower* (04): 102–6.

[15] Hu, C. 2011. Influence of Controlled Partial Root Drying Irrigation on Water Use Efficiency of Potato. *Nanjing Agric. Univ.*

[16] Kang, S. Z., Zhang, J. H., Liang, Z. S., Hu, X. T., and Cai, H. J. 1997. Controlled Alternate Irrigation – A New Idea of Farmland Water-saving Regulation. *Agric. Res. Arid Areas* (01): 4–9.

[17] Li, M. L. 2020. Effects of Drip Irrigation Under Film on Hydrothermal Effect and Yield and Quality of Potato in a Dry Land. *Ningxia Univ.*

[18] Wei, F. L., Zeng, L. S., Li, J. L., Fang, Z. G., Wu, X. H., and Shang, M. X. 2018. Effects of Different Topdressing Treatments on Commercial Potato Rate and Absorption Efficiency of Trace Elements Under Mulching Drip Irrigation. *Chin. Agric. Sci. Bull.* **34**(10): 28–34.

[19] Xue, D. X., Zhang, H. J., Ba, Y. C., Wang, Y. C., and Wang, S. J. 2018. Effects of Regulated Deficit Irrigation on Growth, Yield and Water use of Potato Under Mulched Drip Irrigation in Desert Oasis Region. *Agric. Res. Arid Areas* **36**(04): 109–116+132.

Water Conservancy and Civil Construction – Oke & Ahmad (Eds)
© 2024 The Author(s), ISBN: 978-1-032-58614-4

Design and application of water affairs early warning and collaborative disposal center system based on cloud architecture

Chaojun Yang
Agriculture, Forestry and Water Conservancy Bureau of Yuecheng, Shaoxing China

Xin Liu, Xudong Liu*, Yaping Liu & Dengbing Zhu
Hangzhou Dingchuan Information Technology Co. Ltd, Hangzhou China

ABSTRACT: There are many work contents in the form of early warnings in water conservancy work. For a long time, due to various reasons, such as the regulatory system of affairs, the early warning information is scattered, multi-sectoral involvement in water governance is prominent, coordination is weak, etc. At present, there is an opportunity to design and develop a B/S system architecture of water early warning and collaborative disposal center, integrate multiple areas of water-related early warning events, clear warning indicators, unify the warning classification, warning grade, and interface specifications, standardize disposal process, open up the information dissemination channels, make more timeliness of handling water-related events and cross-sectoral disposal of synergy. The actual application results show that through the system, early warning information can be quickly discerned and automatically distributed, and the events can be disposed of and linked so that the water governance capacity of the water administrative department has been comprehensively improved.

1 INTRODUCTION

The 19th National Congress of the Communist Party of China clearly proposed to build a cyber power, a digital China, and a smart society. To fully implement the spirit of the 19th National Congress of the Communist Party of China, the Zhejiang Provincial Government has comprehensively deployed and promoted the digital transformation of the government and promoted the modernization of the government governance system and governance capabilities (Bao 2020).

At present, the water-related operations in Zhejiang Province involve multiple departments. For example, the management and control of water garbage and water pollution involve the water sector, ecological environment sector, housing construction sector, and other departments, and water supply issues involve water conservancy departments and water supply companies. In general, there is no targeted system mechanism to meet the disposal feedback and online tracing of water-related incidents. The application of the Water Affairs Early Warning Collaboration Center integrates a unified classification and interface specification for early warning events, establishes standardized early warning disposal mechanisms and functions, provides overall early warning disposal control and supervision capabilities, and supports the retrospective display of disposal results from other related business systems.

*Corresponding Author: lxdtnn@126.com

 DOI: 10.1201/9781003450818-25

The application runs in the government cloud network environment to achieve multi-system access and loose coupling of multi-systems and meet the requirements of reliability, security, and accessibility, which can realize the important role of early warning triggering, supervision, and management of disposal, and auxiliary decision-making (Xie 2019).

2 SYSTEM DESIGN AND IMPLEMENTATION

2.1 *Demand analysis*

Comprehensive water resources sector on water-related business processing requirements and workflow, the application requirements are summarized as follows:

1. Early warning triggers. The monitoring data is integrated from the front-end sensing devices, analysis data from the system's embedded models, and early warning event information is generated by other business systems based on business classification specifications. For the accessed monitoring data, early warning trigger conditions can be set according to actual business work.
2. Early warning release. For the generated early warning information, the early warning level is set according to the grading specification and is automatically distributed according to the responsibility configuration and work requirements.
3. Early warning disposal. For the early warning disposal information of water-related events connected with other business systems, the output should be made by the standard disposal process. For early warnings issued based on monitoring data and embedded models, a standardized disposal process is designed, which forms a closed-loop control. For events that require collaborative disposal, the collaborative disposal requests are sent to relevant systems and handled according to the standard disposal process.
4. Information sharing. This part refers to docking with the water-related early warning platform of the higher authorities to receive early warning information and feedback on disposal, docking with the grassroots governance platform to share early warning information about water-related events and assist in event disposal, and docking with other water business systems to obtain early warning information and disposal information about water-related events. Water-related events can be configured in the disposal process of the responsible person and contact person, the relevant personnel can clearly know the coordination personnel during the disposal process.
5. Integrated display. Through the electronic map and display list of the Water Affairs Early Warning collaborative Center, the spatial distribution, details information, classification statistics, and disposal status of various early warnings are integrated and displayed by classification specifications.
6. Statistics query. Through the precipitation of early warning information of water-related events, the early-warning data of water-related events can be archived and traced and the historical disposal details can be queried.

2.2 *Early warning scheme design for water-related business*

According to the "116N" industry-wide unified construction framework principle of the digital reform of water conservancy in Zhejiang Province, the water-related business is preliminarily divided into several thematic application scenarios, such as flood, waterlogging, sewage, water supply, water resources, and water environment. We sort out a water-related early-warning business around thematic application scenarios, unify all early-warning event classification specifications, early-warning grading specifications, and interface specifications, fully access water-related early-warning events, and clarify early-warning indicators, collaborative departments, and disposal processes of water-related events as the institutional basis for the construction of water-related warning and collaborative disposal center system applications.

2.2.1 *Classification of water-related early warning events*

Thematic application scenarios in the early warning events include but are not limited to the following.

Flood: rainfall, river water level, reservoir water level, mountain torrents, reservoir rainfall storage, etc;

Waterlogging: waterlogging, etc;

Sewage: discharge of water pollutants, etc;

Water supply: insufficient water supply, substandard water quality, leakage of pipe network, abnormal water consumption by water users, etc;

Water resources: drought of water sources, excessive water intake, abnormal water intake, etc;

Water environment: water area 'Four Disorders' and water quality of river and lake Sections, etc;

Water-related early warning events that have been sorted out basically cover the main needs of water control work. In principle, all matters should be independent of each other, belong to a clear line of business, and meet the classification specifications of water management business and the daily needs of water management work. Subsequent additions must also meet the requirements of the classification specification.

2.2.2 *Early warning process design for water-related events*

The early warning and disposal process of water-related events mainly includes four links, early warning discovery, early warning release, linkage disposal, and early warning closure. Each link is concretely refined according to different types of water-related events. In the early warning discovery link, the early warning discovery link uses multiple technologies, such as water conservancy and water industry professional models, AI image recognition, and sensor monitoring, thus ensuring the reliability and automation of early warning. To realize early warning events information standardized by unifying all early warning event classification specifications, early warning grading specifications, and interface specifications, such as water problem, the AI intelligent recognition technology is used to interpret and capture the video surveillance, obtain the images of water area problem events, and generate standardized early warning information according to preset rules and classification and grading specifications. The links, such as early warning release, linkage disposal, early warning closure, and other links, are configured based on the standard process according to the territorial collaborative disposal system.

We take the water resources thematic scenario of the water source drought warning business process as an example.

The early warning findings are as follows: (1) The basic real-time data, such as rainfall forecast data, historical rainfall data from the meteorological department, agricultural irrigation water use data, water intake data, and real-time water level of regional water sources are collected and obtained. (2) The basic data are input into the water supply assurance model established by the integrated water supply unit division results, the available water supply, and the guaranteed supply days algorithm. (3) According to the model results, combined with the alert classification specification, the alert results are generated.

Early warning release: According to the division of responsibilities for water supply protection, the details of early warning, early warning disposal suggestions, and disposal feedback requirements are automatically released to the responsible person and contact person associated with the territory through Zhejiang government nail messages, SMS messages, etc. The responsible people include the personnel of the territorial water supply management department, the management personnel of the water supply enterprise, etc. The contact people include the actual management personnel of the water source area, the business operators of the water supply enterprise, etc.

Linkage disposal: The person in charge of the territory should take the lead in early warning disposal according to the level of the incident and the emergency plan measures. For example, the township (street) should strengthen the inspection of the water source area, restrict the water intake of the water source area, and restrict the water supply to the marginal water users, time-sharing water supply, and other measures to reduce the pressure on ensuring the supply. The control measures leave a trace on the business system or mobile terminal, and the water warning and collaborative disposal center can synchronize to capture the disposal actions. In addition to the water conservancy department, by the provisions of the emergency plan measures, an early warning message is sent to the water company and the information transmitted includes the person in charge of the water conservancy department's disposal work and the measures to be taken. The water company verifies the early warning and coordinates with the township to develop emergency measures.

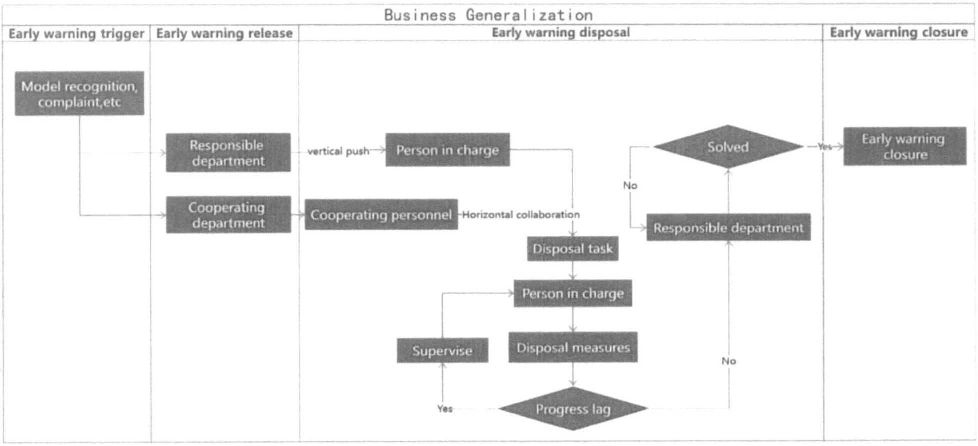

Figure 1. Process generalization.

2.3 *System architecture development*

2.3.1 *System overall architecture design*

To achieve this goal, the water affairs early warning and collaborative disposal center system is developed using the B/S mode with a layered architecture pattern designed using Service-Oriented Architecture (SOA) (Liu 2020). Unlike some other cloud infrastructure management systems (Li 2020; Liu 2021), we design the system architecture according to the characteristics of early warning and disposal of water affairs, the cloud architecture is designed as a 3-tier architecture and two sets of systems, and each tier architecture is independently deployed and maintained. The 3-tier architecture from top to bottom is the access layer, business layer, and data layer.

1. Access layer: We access the user system of Zhezhengding that a provincial general government affairs office service platform in Zhejiang province and use the unified user identity authentication service provided by the provincial water-related industry business system (Zhang 2022). At the same time, to ensure the security of the water affairs early warning collaborative disposal center system, the authorized access mode is adopted. The access layer also can be seen as the PaaS (platform layer). The access layer is mainly composed of users at the provincial, city, and county levels who use the system and users of the management unit. Different usage permissions are set for different users and application functions are assigned according to different permissions. In addition, to ensure data security and prevent the underlying data from being tampered with, unified

control is carried out at the data layer so that the database cannot be accessed directly. Only authorized users can access the interface and encrypt the data (Liu 2022).

2. Business layer: The business layer provides users with the function to interact with the system and provides an interactive display of graphics and text based on 3S technology and map base(Liu 2022). The business layer includes a PC platform and mobile terminal. The early warning collaborative disposal center system includes the cockpit of the early warning center, early warning disposal, early warning supervision, online collaboration, and background management modules. At the same time, the mobile terminal of the early warning collaborative disposal center is developed to provide a unified mobile platform for water-related early warning businesses for competent departments, operating units, and other users. The business layer also can be seen as the SaaS (application layer).
3. Data layer: The data layer relies on the existing local water conservancy data warehouse to call and maintain the real-time rainwater regime, water volume, water quality monitoring data, pipe network monitoring data, reservoir water level real-time monitoring data, and other water conservancy related data required by the system (Chang 2022). The data layer also can be seen as the DaaS (data service layer).

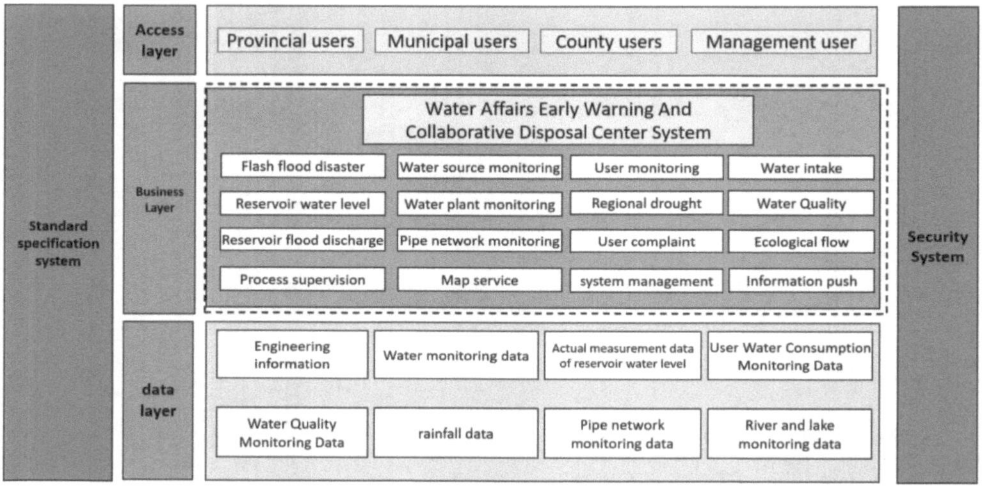

Figure 2. The overall framework of the system.

2.3.2 *System development and deployment*

The stable operation of the system requires a lot of computing and storage resources. At the same time, the server resources required are different due to different actual conditions in different regions. The system construction will use cloud service technology to realize the on-demand allocation of server resources, avoid idle resources, and maximize the utilization of system resources.

Based on the cloud environment, thematic data, thematic maps, analysis results, and data products are stored in a cloud data center in an integrated way to achieve results catalog and metadata management (Pan 2020).

Applications are developed in languages, such as JAVA and VUE, and cloud database resources are used to build basic support. Applications are deployed on the government cloud that meets safety applicability requirements. The cloud resources required for construction include ECS servers, relational databases, analysis databases, object databases, offline computing engines, and real-time computing engines. The open computing and storage service capabilities of elastic computing services and workspace services are built through the cloud computing engine. The open data service capabilities of data services and exchange services, including online map services and information query services, are built through the cloud service engine. The personalized application service capabilities of analytical services and customized services, such as online statistical services and online analytical services, can also be built through the cloud service engine.

2.4 *Design of main functional modules of the system*

Water early warning and collaborative disposal center is a comprehensive water-related early warning and disposal system built around the core water-related matters, such as water disaster prevention, water resource area protection, river, lake and reservoir protection, water development planning, water affairs supervision, and other core water-related affairs, and includes the function modules, such as home cockpit, early warning disposal, early warning supervision, and background management.

1. Cockpit: The cockpit serves as a comprehensive homepage of the Water Affairs Early Warning Collaborative Disposal Center. Relying on the electronic map, it superimposes various types of early warning risk element layers and early warning points and presents early warning thematic data intuitively, such as floods, waterlogging, sewage, water supply, water resources, and water environment. At the same time, it displays the current number of early warning events that users need to pay attention to, the type of early warning, the time of occurrence, the location distribution, the progress of disposal, etc.
2. Early warning treatment: Based on the early warning process design, the corresponding standardized disposal steps are formulated for each type of early warning and the early warning event disposal module is formed. As a secondary page of the water affairs early warning collaborative disposal center, we click on the specific early warning event on the cockpit home page to enter. The early warning event disposal module provides a view of the basic overview of early warning objects, event details, early warning discovery, early warning release, early warning disposal, and early warning closure.
3. Early warning supervision: According to the early warning process monitoring, the current disposal process lags behind the early warning events to provide supervision functions, including the issuance of supervision and execution of feedback modules. A comprehensive integrated, closed-loop management of water-related incident disposal supervision and operation mechanism is formed by the collaborative execution mechanism of water-related early-warning and disposal business.
4. Background management: It refers to the comprehensive management of various types of early warning information, providing data and technical support for the cockpit comprehensive home page data display and related statistical analysis, which mainly includes early warning list management, early warning process configuration, historical early warning records, and early warning briefings.

Table 1. Early warning events resolution rate.

Early warning events	Number of early warning events	Resolution rate
Flood	427	95.10%
Waterlogging	1	100.00%
Sewage	3	100.00%
Water supply	792	95.53%
Water resources	2	100.00%
Water environment	26	76.92%

2.5 *Application practical of the system*

At present, the system has been basically developed and applied in several counties (cities, districts), such as Jiande, Yiwu, and Yuecheng. Taking Yuecheng City as an example, the water warning and collaborative disposal center have been put into operation for 3 months, generating 1,251 warnings and completing 1,189, effectively improving the efficiency, standardization, and multi-span linkage of local water-related warning affairs.

3 CONCLUSIONS AND SUGGESTIONS

With the rapid development of the economy and society, the work of water conservancy has put forward higher requirements With the background of the "14th Five-Year Plan for Smart Water Conservancy Construction" and the purpose of improving the early warning and disposal system for water-related businesses, innovating the linkage management and control of early warning events of water-related businesses is the starting point (Zhao 2022). Through the establishment of a unified, standardized, multi-party coordination of water affairs early warning and collaborative disposal center, we unify all early warning events of event discovery access, classification norms, grading norms, and interface norms, standardizing the early warning disposal process, thus promoting the reshaping of water workflow, helping the new era of water work effectively, and providing strong support for the safe development of the economy and society.

REFERENCES

Bao Zhiyan, Jiang Xiaojun, Huang Kang, Tan Wei & Jin Xuanchen. (2020). Research on the Overall Framework and Key Technologies of Zhejiang Water Conservancy Digital Transformation. *J. Water Conservancy Informationization* (02), 1–8.

Chang Xing, Wang Mengqiang, Liu Gang & Shen Enshuai. (2022). Thinking and Design of Information Systems for Quality Supervision of Small Water Conservancy Projects. *J. Water Conservancy Technical Supervision* (10), 4–6+20.

Li Jialin, Yu Zhigang & Lin Yang. (2020). Research and Practice of "Natural Resource Cloud" Management Platform Based on Multi-cloud Architecture. *J. Land and Resources Informationization* (03), 15–21.

Liu Fafa, Xiong Xiaofeng, Liu Fei & Jiao Liwei. (2022). Design and Research of Dynamic Remote Sensing Monitoring and Supervision System for Natural Resources Based on Cloud Architecture. *J. Henan Science and Technology* (04), 13–16.

Liu Hengfei, Wang Hongchang & Liu Yuxin. (2020). Research and Construction of Key Technologies for Natural Resource Smart Supervision Platform. *J. Surveying and Mapping and Spatial Geographic Information* (S1), 76–79.

Liu Jiahong, Jiang Yunzhong, Mei Chao & Wang Jia. (2022). Research and Construction Progress of Digital Twin River Basin. *J. China Water Resources* (20), 23–24+44.

Liu Qingtao, Cai Siyu, Wang Yi, Shen Hongxia & Wang Yuanyuan. (2022). Design and Application of Ecological Flow Monitoring and Early Warning System for Key River and Lake Control Sections. *J. Water Conservancy Informationization* (02), 1–5+40.

Pan Liu. (2020). Design of Geological Disaster Monitoring and Early Warning Platform Based on Natural Resource Cloud. *J. Network Security and Informatization* (09), 65–67.

Xie Long, Wu Yan & Fang Chenliang. (2019). Design and Implementation of Safety Smart Management System for Water Conservancy Projects Under Construction. *J. Zhejiang Water Conservancy Technology* (04), 80–83.

Zhang Lujie. (2022). Design of Reservoir Water Regime SMS Forecasting and Early Warning System. *J. Hebei Water Conservancy* (02), 37–38+44.

Zhao Yongjun, Ma Songzeng, Luo Zhidong & Cheng Fu. (2022). Thoughts on Smart Soil and Water Conservation Construction during the "14th Five-Year Plan" Period. *J. China Soil and Water Conservation* (10), 74–78.

Water Conservancy and Civil Construction – Oke & Ahmad (Eds)
© 2024 The Author(s), ISBN: 978-1-032-58614-4

Research progress on the effects of water and nitrogen coupling on potato yield and quality

Xiaofan Pan
College of Agronomy and Agricultural Engineering, Liaocheng University, Liaocheng, China
College of Water Conservancy and Hydropower Engineering, Gansu Agricultural University, Lanzhou, China

Shouchao Yu
College of Agronomy and Agricultural Engineering, Liaocheng University, Liaocheng, China

Hengjia Zhang*
College of Agronomy and Agricultural Engineering, Liaocheng University, Liaocheng, China
College of Water Conservancy and Hydropower Engineering, Gansu Agricultural University, Lanzhou, China

ABSTRACT: Irrigation and nitrogen fertilizer are important factors affecting potato growth and development, stem yield, and quality. They are also two key factors for significant potato agricultural development and agricultural production efficiency. Therefore, research on technologies to improve crop water and fertilizer use efficiency, reduce crop irrigation and achieve high yields and quality has received widespread attention. Potatoes are sensitive to soil water and nitrogen. Excessive or small amounts of water and nitrogen can affect the growth and quality of its stems and ultimately reduce yield and quality. Water and nitrogen coupling can coordinate the absorption and utilization of water and nitrogen in potato stems and leaves, and promote the distribution of photosynthetic products to stems, ultimately leading to increased yield and improved quality. This paper reviews the effects of water and nitrogen coupling on potato yield and quality and provides a theoretical reference for the efficient utilization and sustainable development of potatoes.

1 INTRODUCTION

The Hexi Corridor irrigation area is located in the northwest arid region, with abundant light and heat resources but low precipitation. Sustainable high and stable agricultural production is greatly affected by water resources [1]. As the 'life element' of plants, nitrogen not only affects the vegetative and reproductive growth of plants but also affects crop yield [2]. The irrational application of nitrogen fertilizer causes a large loss of fertilizer and short expiration dates, which is not detrimental to nitrogen accumulation in crop organs [3]. In the end, the random input of water and nitrogen arbitrarily leads to the advantages of water and nitrogen resources not being fully exploited low crop production efficiency, and increasingly serious ecological problems of nitrogen deposition [4]. According to statistics, the fertilizer utilization rate in the irrigation area of the Hexi Corridor is only 20% to 30%, with great potential for improvement. Therefore, vigorously developing efficient water-saving irrigation technology and promoting the integration of water and fertilizer is one of the effective

*Corresponding Author: 596088683@qq.com

DOI: 10.1201/9781003450818-26

ways to solve the above constraints [5]. As one of the main economic crops in this area, the potato has the characteristics of high yield, good palatability, rich nutrition, and strong adaptability, which is widely used in the rural revitalization industry. Therefore, by collecting domestic and foreign literature, this paper systematically summarizes the relevant literature reported at home and abroad in recent years and clarifies the effects of different water and nitrogen treatments on potato quality, yield, and its components, to provide the technical basis and theoretical guidance for the scientific application of water and nitrogen coupling in production practice.

2 EFFECT OF WATER AND NITROGEN COUPLING ON POTATO YIELD

Water is the carrier of nutrient absorption by crops. Insufficient water can reduce the absorption of fertilizer by crop roots and excessive water can lead to the loss of nitrogen fertilizer. Too low or too high nitrogen fertilizer application limits the absorption of soil moisture by crop roots, thus limiting crop growth and reducing crop yield [6]. There is a close coupling relationship between water and nitrogen in the process of crop growth and development. Water and nitrogen coupling can achieve the goal of saving production costs, promoting crop growth, and increasing crop yield with optimal amounts of water and nitrogen. Potato is the fourth largest foodstuff and its stem yield is sensitive to water and nitrogen. Appropriate irrigation and nitrogen application are the main agronomic measures to increase potato yield [7]. Zang *et al* [8]used sprinkler irrigation technology to carry out water and nitrogen combination experiments on potatoes. The results show that different water and nitrogen combination treatments have a significant effect on potato yield. Among them, the total irrigation amount is 100 mm and the nitrogen application rate is 86.8 kg/hm^2, which is significantly higher than other treatments. The yield is 46,525 kg/hm^2. The application of nitrogen fertilizer at different frequencies is beneficial to the growth and weight gain of potato stems, which improves the utilization efficiency of potato nitrogen fertilizer and lays a foundation for improving potato yield. Yan Wenyuan *et al* [9]carried out water and nitrogen coupling experiments on potatoes through drought-resistant shed cultivation. The results show that under the condition of water deficit and water excess, the quality of potato stem tubers treated with medium nitrogen and high nitrogen is higher than that treated with low nitrogen and the insufficient nitrogen application rate can reduce the number of potato stem tubers. Under the same nitrogen application level, the quality of potato stem tubers treated with normal water and excessive water is significantly higher than that treated with water deficit and the high yield of potato appears in the medium and high levels of water and nitrogen, which indicates that water deficit and low nitrogen application can affect the growth and development of potato stem tubers and eventually lead to a decrease in potato yield. The study of Wang Shun *et al* [10] shows that under the same irrigation level or the same nitrogen application rate, the potato yield and commodity rate increase and then decrease with the increase in irrigation amount or nitrogen application rate. The potato yield is up to 53,698.95 kg/hm^2 and the potato commodity potato rate is up to 96.4% under the middle water nitrogen level (irrigation quota of 1500 m^3/hm^2 and nitrogen application rate of 210 kg/hm^2), which indicates that appropriate irrigation and appropriate nitrogen application can increase potato yield and potato commodity potato rate. It Makes potatoes more in line with market demand and also helps farmers to increase their income. Yinjuan *et al* [11] have shown that different water and nitrogen coupling treatments have a significant effect on potato yield. Under the condition of irrigation amount 900 m^3/hm^2, potato yield increases with the increase in nitrogen application rate. Under the condition of irrigation amount 1,350 m^3/hm^2 and 1,800 m^3/hm^2, the increase in nitrogen application rate shows a decreasing trend in potato yield followed by an increasing trend. Under the condition of low nitrogen (120 kg/hm^2), potato yield increases with the increase in irrigation amount. Under the condition of high nitrogen (240 kg/hm^2), potato yield increases with the increase in irrigation amount. Therefore, there is a certain threshold for both irrigation and nitrogen application. Excessive application can reduce potato yield and water and fertilizer use efficiency. A small amount of application can affect crop growth and nutrient distribution and ultimately reduce potato yield. Water and nitrogen regulation can be achieved by water to promote fertilizer to promote the purpose of crop uptake, utilization of nutrients, and the distribution of photosynthetic products to potato stem [12], thus achieving high yield, quality objectives, and efficient use of water and nitrogen, which contributes to the sustainable development of potato industry.

3 EFFECT OF WATER AND NITROGEN COUPLING ON POTATO QUALITY

Water and nitrogen coupling can increase potato yields and improve quality by determining the exact amount of water and nitrogen scientifically, avoiding waste of water and fertilizer resources while achieving water and fertilizer savings, high quality, and high yield. The coupling of water and nitrogen also has an important effect on the quality of potato stems. Shang Meixin *et al* [13] show that the coupling of water and nitrogen can significantly improve the quality of potato stems and reduce the nitrate content. The contents of vitamin C, starch, soluble protein, and soluble sugar are different under different water and nitrogen coupling treatments. The vitamin C content of high nitrogen and low water treatment reaches 256.6 mg/kg, which is 4.01%-26.85% higher than that of other treatments, while the starch content of low nitrogen and low water treatment reaches 157.2 g/kg. The soluble protein content of high nitrogen and high water treatment reaches 30.0 g/kg, which is 3.00%-29.00% higher than other treatments. The soluble sugar content of medium nitrogen and medium water treatment reaches 17.0 g/kg, which is 14.71%-32.35% higher than other treatments. The coupling experiment of water and nitrogen on potatoes is carried out by using sub-membrane drip irrigation in arid northwest China [14]. The results show that the content of vitamin C and starch in potato stems shows a parabolic trend with the increase in nitrogen application rate under the same water condition. Under the same nitrogen application rate, the content of vitamin C, starch, and protein increased with the increase in the wetting ratio. The research of wang chen [15] shows that different water and nitrogen treatments have significant effects on the content of starch, reducing sugar and vitamin C in potatoes. Compared with the control (no mulching, no irrigation, and no fertilization), the content of starch increases by 7.02%-41.23%, the content of reducing sugar increases by 11.76%-76.47%, and the content of vitamin C increases by 1.93%-9.53%. However, the content of reduced sugar and vitamin C also decreases under the high level of water and nitrogen, indicating that water and nitrogen coupling can improve the quality of potatoes. However, the excessive application produces negative effects and affects the quality of potato stems. The research of Zhou Nana *et al* [16] shows the starch content in potato stem increases and then decreases with the increase in nitrogen application rate under the same irrigation amount. However, the starch content increases with the decrease in irrigation amount under the same nitrogen application rate. Water and nitrogen can affect the distribution of nutrients and nutrients absorbed by crops, which in turn affects the growth status of potato plants, potato stem formation, and quality accumulation. Therefore, reasonable water and nitrogen application is an important way to high-yield and high-quality potatoes.

4 EFFECTS OF WATER AND NITROGEN COUPLING ON WATER AND NITROGEN USE EFFICIENCY OF POTATO

Potato is sensitive to water and fertilizer demand during its growth and development. Water and nitrogen are also important measures to increase crop yield and efficiency in arid areas [17]. Water and nitrogen use efficiency can reflect the utilization of water and nitrogen fertilizer by crops, which is an important index to measure the high yield and high efficiency of crops [18]. The research of Zhao Yanbo [19] shows that when the irrigation quota is 900 m^3/hm^2, the increase of nitrogen application rate significantly improves the irrigation water use efficiency and water use efficiency. However, with the increase in irrigation quota, the increase and decrease in irrigation water use efficiency and water use efficiency are different. Under the same irrigation amount, the partial productivity of nitrogen fertilizer decreases with the increase in nitrogen application rate, while the agronomic benefits of nitrogen fertilizer vary under different irrigation quotas. The studies of Mai Zizhen *et al* [20] have shown that the nitrogen application rate under the same irrigation level can affect crop water consumption, drip irrigation water production efficiency, and water production efficiency. Among them, the drip irrigation water production efficiency and water production efficiency can reach 1,143.66 kg/hm^2 and 65.48 $kg/(hm^2 \cdot mm)$ under the irrigation amount of 225 m^3/hm^2 and the nitrogen application rate of 225 kg/hm^2. However, with the increase in irrigation level, the drip irrigation water production efficiency and water production efficiency under the same nitrogen application rate show a decreasing trend, while the nitrogen production efficiency shows a decreasing trend except for the increase under the low nitrogen level. Song *et al* [14] show that the water use efficiency increases and then decreases with the increase in nitrogen application rate under the same soil moisture ratio. The water use efficiency reaches 12.86 kg/m^3 and 11.60 kg/m^3 when the soil moisture ratio is 40% and the soil moisture ratio is 70% and the nitrogen application rate is 135 kg/hm^2. Soil water status and fertilizer application rate affect crop yield, the utilization of water, and fertilizer by crops [21]. Therefore, appropriate water and nitrogen management is an important measure to improve potato yield and water and nitrogen use efficiency.

5 CONCLUSION

As the fourth largest food crop, it is significant to ensure its high yield and high quality for food security and agricultural development. Irrigation and nitrogen application are the main factors limiting the growth and development of potatoes. Insufficient water and nitrogen fertilizer can affect the growth of potatoes, which in turn affects the absorption of nutrients by stems and leaves, thus reducing crop yield. Excessive amounts of water and nitrogen can have a negative effect and wastewater and fertilizer resources. Water and nitrogen coupling can coordinate water and nitrogen absorption, exert the best water and nitrogen coupling effect, promote stem growth and improve quality, reduce soil nitrogen loss, and achieve the goal of efficient utilization of water and fertilizer resources and high yield and quality of potatoes.

ACKNOWLEDGMENTS

This work was financially supported by the Industrial Support Plan Project of the Gansu Provincial Department of Education (No. 2022CYZC-51), the National Natural Science Foundation of China (No. 52269008), and the Key Research and Planning Projects of Gansu Province (No. 18YF1NA073).

REFERENCES

[1] Aon, M. A. and Colaneri, A.C. II. Temporal and Spatial Evolution of Enzymatic Activities and Physico-chemical Properties in Agricultural Soil. *J. Applied Soil Ecology*, 18(3):255–270(2001).

[2] Chen, Y., Dai, X. Q., Yuan, L., Xi, X. Y., Ma, H. H., and Liu, M. Y. Advances in Effects of Water and Nitrogen Coupling on Soil Physicochemical Properties and Crop Growth. *Journal of Henan Agricultural Sciences.* (5):11–15(2009).

[3] Hu, P. C., Yin, J. Wei, X. D., and Wang, C. Effects of Different Water-nitrogen Treatments on Potato Quality and Soil Urease Activity. *J. Jiangsu Agricultural Sciences*.50(6):87–92(2022).
[4] Liu, N., Yang, W. X., Zhang, X. T., Yang, C. G., Wang, S. H., and Wang, X. R. Effects of Irrigation on Growth and Development and Yield of Different Wheat Varieties in Oasis Irrigation District of Hexi Corridor. *J. Water Saving Irrigation*. (8):1–7+17(2020).
[5] Mai, Z. Z., Yang, C. L., and Mi, Z. M. Affect Different Amounts of Drip Irrigation and Nitrogen Soil Moisture and Potato Production. *J. Water Saving Irrigation*, (11):28–31+35(2016).
[6] She, Y. J., Li, P., Du, Z. J., Bai, F. F., Guo, W., Liang, Z. J., Cui, J. X., Ma, C. C., and Qi, X. B. Effects of Different Groundwater Depth and Fertilizer Controlling on Nitrogen Uptake and Yield of Summer Maize. *J. Journal of Soil and Water Conservation*. 35(2):309–314(2021).
[7] Shang, M. S., Fang, Z. G., Liang, B., Wang, M., and Li, J. L. Effects of Different Water and Nitrogen Treatments on Potato Yield, Quality and Soil Nitrate Nitrogen Transport Under Drip Irrigation. *J. Acta Agriculturae Boreali-Sinica*. 34(6):118–125(2019).
[8] Song, N., Wang, F. X., Yang, C. F., and Yang, K. J. *Coupling Effects of Water and Nitrogen on Yield, Quality and Water Use of Potato with Drip Irrigation Under Plastic Film Mulch*. 29(13):98–105(2013).
[9] Shen, R. K., Wang, K., Zhang, Y. F., Yang, L.H., Mu, J.Y., and Zhao, L. X. Field Test and Study on Yield, Water Use and N Uptake Under Varied Irrigation and Fertilizer in Crops. *J. Transactions of the Chinese Society of Agricultural Engineering*, (5):35–38(2001).
[10] Vashisht, B. B., Nigon, T., Mulla, D. J., Rosen, C., Xu, H., Twine, T., and Jalota, S. K. Adaptation of Water and Nitrogen Management to Future Climates for Sustaining Potato Yield in Minnesota: Field and Simulation Study. *J. Agricultural Water Management*. 152, 198–206(2015).
[11] Wang, G.W., Li Y., and Wang, J. X. Effects of Different Fertilization Patterns on Growth Characteristic and Nitrogen Use Efficiency of Corn. *J. Southwest China Journal of Agricultural Sciences*. 32(9):2119–2125(2019).
[12] Wang, X. B., Gao, X. K., and Cai, D. X. Inter-reaction of Water and Fertilizer In Rainfed Farmland.S. *J. Agricultural Research in the Arid Areas*. (3):6–12(1997)..
[13] Wang, S., Yin, J. Zhang, H. J., and Wang, C. Effects of Different Water and Nitrogen Treatments on Soil Enzyme Activity and Yield of Potato. *J. Water Saving Irrigation*. (8):67–73(2021).
[14] Wang, C. *Effect of Water and Nitrogen Regulation on Soil Enzyme Activity and Its Response to Growth, Yield, and Quality of Potato*. D. Ningxia University, 2020.
[15] Yang, J. L., Ma, Z. M., Zhang, L. Q., Wang, Z. Q., Lian, C. Y., and Xue, L. Effects of Different Groundwater Depth and Fertilizer Controlling on Nitrogen Uptake and Yield of Summer Maize. *J. Journal of Soil and Water Conservation*. 35(9):1262–1268(2015).
[16] Yan, W. Y., Qin, J. H., Duan, S. F., Xu, J. F., Jian, Y. Q., Jin, L. P., and Li, G. C. The Effect of Water-nitrogen Coupling on Potato Photosynthesis. Tuber Formation and Quality. *J. Acta Horticulturae Sinica*. 49(7):1491–1504(2022).
[17] Yin, J., Zhang, H. J., Wang, S., Wang, C., and Zhao, Y. B. Effects of Different Water and Nitrogen Treatments on Photosynthetic Characteristics and Yield of Potato. *J. Water Saving Irrigation*. (6):8–13 (2020).
[18] Zang, W. J., Li, J. J., Pei, S. S., Li, Y. J., and Yan, H. J. Effects of Different Water –Nitrogen Combinations on Potato Water Consumption, Yield and Quality Under Sprinkler Irrigation. *J. Journal of Drainage and Irrigation Machinery Engineering*. 36(8):773–778(2018).
[19] Zhang, H. L., Smeal, D., Arnold, R. N., and Gregory, E. J. Potato Nitrogen Management by Monitoring Petiole Nitrate Level. *J. Journal of Plant Nutrition*. 19(10–11):1405–1412(1996).
[20] Zhou, N. N., Zhang, X. J., Qin, Y. B., and Xu, Q. Effect on Different Quantities of Drip Irrigation and Nitrogen Fertilization for Yield and Quality of Potato. *J. Soil and Fertilizer Sciences in China*. (6):11–12 +16(2004).
[21] Zhao, Y. B. *Effects of Different Water and Nitrogen Treatments on Potato Growth and Soil Water and Nitrogen Transport Distribution*. D. Ningxia University, 2019.

Water Conservancy and Civil Construction – Oke & Ahmad (Eds)
© 2024 The Author(s), ISBN: 978-1-032-58614-4

Characteristics of water quality changes in the Purdue River basin and the verification of the Kuznets curve from 2017–2021

Yadong Yu*
Yunnan University School of Ecology and Environment and Yunnan Basic Plateau Mountain Ecology and Restoration of the Restoration Environment, Yunnan University, Kunming University, China
The Ecological Environment Monitoring Station of the Yunnan Provincial Ecological and Environmental Monitoring Department in Kunming City, Kunming China

Mingshan Zhao
The Ecological Environment Monitoring Station of the Yunnan Provincial Ecological and Environmental Monitoring Department in Kunming City, Kunming China

Xiaoni Wu
Yunnan University School of Ecology and Environment and Yunnan Basic Plateau Mountain Ecology and Restoration of the Restoration Environment, Yunnan University, China

Fei Sun, Jin Liu & Liping Liu
The Ecological Environment Monitoring Station of the Yunnan Provincial Ecological and Environmental Monitoring Department in Kunming City, Kunming China

Sichen Wang
Yunnan University School of Ecology and Environment and Yunnan Basic Plateau Mountain Ecology and Restoration of the Restoration Environment, Yunnan University, Kunming University, China

ABSTRACT: The Purdue River Basin of Mantis Sichuan is located in the territory of Yunnan Province. The changes in water quality in this watershed are not only related to the good life of surrounding residents but also provide timely feedback on the governance measures of Dianchi Lake. From 2017 to 2021, it monitored the Mantis Sichuan-Purdue River Basin in southwest China. The five-day biochemical oxygen demand, ammonia nitrogen, fluorine ion, nitrate nitrogen, total phosphorus, high manganese, and high manganese in the basin characteristics of the basin are monitored. The characteristics of different time scale changes in the acid salt index exceed the situation. Environmental Kuz Sheets (EKC) believes that the level of environmental pollution increases with economic development and the increase in national income levels. When the economy develops to a certain degree, the pollution level will decrease with the rise of national income, which can be said to be better. The ground reflects the changes in economic development and environmental pollution. Based on the overall economic situation of the basin, it is found that the economy and environment of the river basin conform to the Kuznets curve. The main conclusion is that in 2021, the main pollution indicators are five days of biochemical oxygen demand, total phosphorus, and chemical oxygen demand. From the analysis of the monthly concentration of the main indicators, the water quality condition of the Mantis Sichuan-Purdue River is relatively stable after the rainy season (September-December) and decreases before and into the rainy season (January-August). The overall pollution of the basin is concentrated from

*Corresponding Author: Yu Yadong (b. 1991), Female, Engineer, Ph.D., Mainly engaged in the research of pollution and restoration ecology. e-mail: 13685311866@163.com
Grant project: Yunnan University Postgraduate Research Innovation Project (KC-22221078)

 DOI: 10.1201/9781003450818-27

May to August. The relationship between the economic and environmental quality of the basin conforms to the Kuznets curve, indicating that the watershed has stabilized the inverted U-shaped point of EKC. The development of the economy has more money at its disposal for water quality. Combining the seasonal and annual water quality changes in the basin and the upstream/downstream relationship, the impact of governance and human disturbance around Dianchi can be detected in time.

1 INTRODUCTION

The Purdue River Basin is one of the earliest regions to be developed in Yunnan Province. The central Kunming City of politics, economy and culture in the province is located on the banks of the Dianchi Lake in the upper reaches of the Purdue River. The Anning Industrial Zone is the main steel and chemical base of the province (Xu *et al.* 1992). Anning's economic development has a very important position in the province. The rich mineral resources in the basin and the development of large-scale mineral resources have caused serious soil erosion and surface pollution in the area, and the nutrient status of water bodies has become increasingly serious (Hui *et al.* 2020). In terms of ecology, Dianchi Lake's sewage function is undertaken by the Purdue River Basin of the Sichuan. Therefore, the evaluation of the water quality in the praying mantis Sichuan Basin is significant to maintain the water quality safety of the Jinsha River Basin.

Simon Kuznets, a Nobel laureate in economics, discovered the inverted curve of economy and environment (Kuznets S 1995). Environmental Kuz Sheets (EKC) believes that the level of environmental pollution will increase with economic development and the increase in national income levels. When the economy develops to a certain degree, the pollution level will decrease with the rise of national income, which can be said to be better. The ground reflects the changes in economic development and environmental pollution. Scientists have tried to verify the inversion relationship between income and environmental quality. Some studies have found that a single water environment index section follows the "Environment KUZNETS curve" (EKC) (Cai H. *et al.* 2020; Wang *et al.* 2017). However, many reports still lack evidence (Ekins P. 1997; Zhang & Ganggopadyay 2015). Due to data limitations, most research on the experience environment EKC focused on air quality and income (Hao Y. *et al.* 2018). In the study of KEC and environmental economy, only 14 % of research is related to water quality (Wong Y. L. *et al.* 2014). Moreover, many studies of EKC are concentrated on one or two specific pollutants. Different characteristics of different pollutants have led to different estimates. The comprehensive pollution index is an important way to evaluate the quality of the water environment. The Comprehensive Pollution Index (CPI) is an important way to evaluate the quality of the water environment. The selected indicators, pH, dissolving oxygen, permanganate index, biochemical oxygen demand quantity, ammonia nitrogen, and other indicators, can comprehensively represent the overall quality of the water body. Therefore, we use the comprehensive pollution index to represent water pollution and investigate the relationship between water quality changes in the Purdue River Basin of the Mantis Sichuan-Purdue River Basin.

2 RESEARCH AREA & METHOD

2.1 *Research area*

The Mantis River originates in the southwestern part of Dianchi Lake. It flows northward through the Xishan District of Anning City to Longquan Village of Fumin County, with a total length of about 110 kilometers and a river width of 25–40 meters. Crossing the river, it flows through Fumin County, Luquan County, and merges into the Jinsha River in northeastern Luquan County. The total length is about 213 kilometers, the river width is 25–47 meters, and the water depth is 1.3 to 5.4 meters. The sampling points are Zhongtanzhaomen (ZTZM), Wenquandaqiao (WQDQ), Qinglongxia (QLX), Fumindaqiao (FMDQ), Puduheqiao (PDHQ), Nigeshuiwenzhan (NGSWZ), Tongxianqiao (TXQ), Laomeishandaqiao (LMSDQ), Chengqidunxiaoqiao (CQDXQ), Yantang (YT), and Zhuanlong (ZL).

2.2 *Method*

Collection of samples: We collect water samples at the scheduled sampling point every month and determine the dissolved oxygen (DO) at the scene. When sampling ammonia nitrogen (NH_3–N) and nitric nitrogen (NO_3–N), it is necessary to add biological inhibitors to inhibit the effects of microorganisms. COD, TP, and hypertonic index samples need to be acidic to pH < 2 by adding sulfate. We reserve fluoride samples in polyethylene bottles, while the remaining samples are stored in glass bottles. After sampling, we store the sample at 0–4 ° C and send it to the laboratory for analysis.

Experimental method: We use the salicylic acid splitting optical meter to measure ammonia nitrogen and use an ionic electrode method to measure fluoride ions. The nitrate is determined by the phenol sulfuric acid splitting light method. the total phosphorus is measured by using the ammonium molybdenum luminosity method. The differences in dissolved oxygen are measured in a 5-day incubation after 5 days of oxygen. The high permanganate index is determined by using an acid method.

pH is the acidity of water quality, which is also known as the hydrogen ion concentration index, and is used to measure the strength of acidity and alkalinity of water quality as an indicator, usually a PH value of 0–14. The molecular state of oxygen dissolved in water is called dissolved oxygen, which is usually recorded as DO and expressed in milligrams of oxygen per liter of water. The Permanganate index is a common indicator reflecting the pollution of organic and inorganic oxidizable substances in water bodies. Chemical oxygen demand is the amount of oxidant consumed when treating water samples with a certain strong oxidant under certain conditions. BOD5 means 5-day biochemical demand, mainly representing the indicators of easy biochemical degradation in the sewage. Ammonia nitrogen is the nitrogen present in the water in the form of free ammonia and ammonium ions. Total phosphorus is the result of the measurement of the water sample after digestion to convert various forms of phosphorus into orthophosphate, measured in milligrams of phosphorus per liter of the water sample. Proper fluoride in the water body Fluorine is necessary for the human body. However, excess fluorine is harmful to humans.

The Comprehensive Pollution Index (CPI) is calculated based on parameters, including the needs of biological oxygen, nitrogen, total nitrogen, total phosphorus, dissolved oxygen, oxide, manganate index, and nitrate. We use the following formula to calculate the Comprehensive Pollution Index.

$$\mathrm{CPI} = \sum_{i=1}^{n} \frac{\mathrm{C_i}}{\mathrm{C_{ni}}} C_k \tag{1}$$

Where CPI is the comprehensive pollution index; C_K is the maximum allowable standard for uniformly allowed in the surface water; C_{Oi} is the highest allowable standard for various pollutants in the surface water; C_I is the concentration of various pollutants in the surface water (Yang W. *et al.* 2021).

3 RESULTS & DISCUSSION

3.1 *Basin 2021 overall analysis*

In 2021, the overall water quality status of 11 rivers was moderate pollution, the proportion of water quality categories (excellent water quality status) of Type I to III was 36.4%, and the proportion of inferior categories (severe water quality pollution) was 27.3%. The daily biochemical oxygen demand, total phosphorus, and chemical oxygen demand exceeded the standard rate of 65.38%, 61.11, and 60.53%. In the 11 rivers into the lake, 4 broken surfaces reached the water environmental protection target. 7 broken water quality did not meet the standard and the standard rate was 36.4%. The basin national examination indicators require that the assessment rate in the basin was 100%, the proportion of water quality categories I to III was 40.0%, and the proportion of inferior categories (severe pollution of water quality) was 20.0%. In 2021, the main pollution indicators of Mantis Sichuan-Purdue River were five days of biochemical oxygen demand, chemical oxygen demand, and total phosphorus, as shown in Figure 1. From the analysis of the monthly concentration of the main indicators, the water quality condition of the Mantis Sichuan-Purdue River was

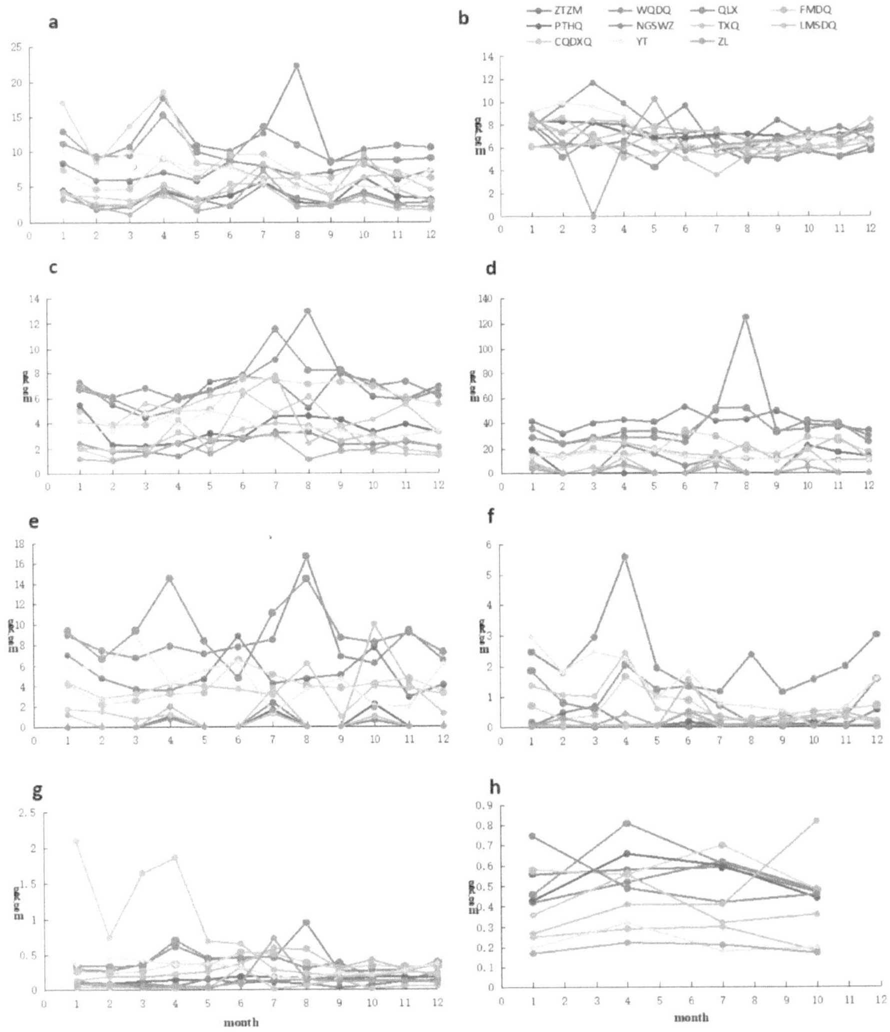

Figure 1. Monthly monitoring value of different pollutants in the basin. Note: a- Comprehensive pollution index; b- Discipline of oxygen value changes; c- Hanganate Index; d- Chemistry needs changes in the amount of oxygen; e- 5 days of biochemical oxygen demand; f- ammonia nitrogen; g- total phosphorus; h- fluoride.

relatively stable after the rainy season (September-December), and the water quality declined before and into the rainy season (January-August). The overall pollution of the basin is concentrated from May to August.

Analysis of the spatial distribution of the watershed shows that the level of water pollution upstream (middle beach gate, hot spring bridge, Qinglongxia) is heavy, and the pollution downstream (Purdue river bridge, nylon reservoir station) is light. The overall pollution in the middle reaches of the river is serious and the downstream water quality pollution is light. Among them, the comprehensive pollution index is the maximum monitoring section of the hot spring bridge. Among them, the two monitoring sections of the Mantisagchi tributary Salon River (the bridge of the adult pier) and the Purdue River tributary Hou River (Yantang) are more serious than the other sections (Figure 2).

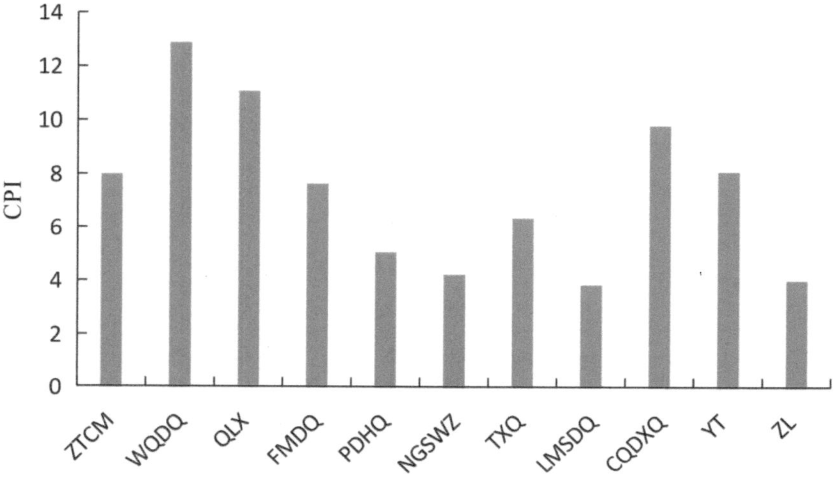

Figure 2. Mantis Sichuan-Purdue River comprehensive pollution index.

Water energy resources in the domain are rich in water and resources, which is conducive to the development of energy-saving and environmentally friendly hydropower industries. Among them, the water power stations that have been completed and put into use in the research area include Qinglong Power Station and Shilong Power Station.

The urban sewage collection capacity is insufficient, which results in pollution in the research area. The drainage system in the old town in the basin is a rainwater and sewage system. When the rainfall is strong on rainy days, the pollution capacity of the completed sewage collection and treatment facilities cannot effectively control the sewage in the rainy season and overflow pollution occurs. The domestic sewage generated by the third category of the service industry, such as the beauty salon industry, hotel, and catering industry in the Purdue River basin of Anning City Vocational Education Park, Yunkang Village area, Purdue River basin is not collected.

3.2 *Water quality changes in the past 5 years*

In the past 5 years, the comprehensive water quality pollution index of praying mantis Sichuan-Purdue River Basin has declined significantly. The comprehensive pollution index of the Fumin Bridge has declined significantly. Other sections are declining but not significant. The chemical oxygen demand for the overall watershed has decreased significantly (Figure 3).

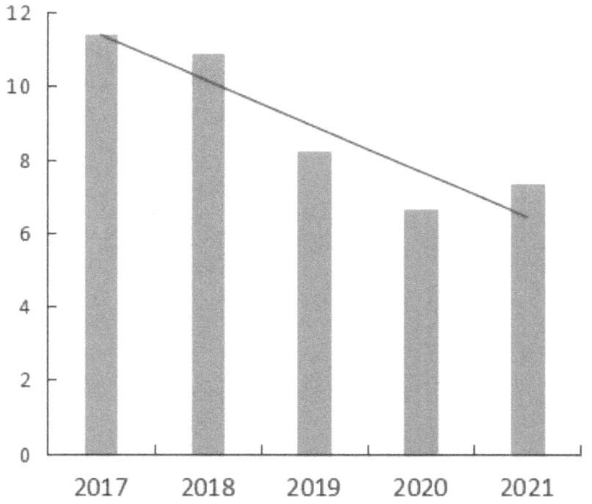

Figure 3. The trend of changes in the comprehensive pollution index of Mantis Sichuan-Purdue River.

Among the main pollution index indicators, the 5-day biochemical oxygen demand has decreased in all sections except Qinglongxia, where there is no significant increase. Chemical oxygen demand increases significantly only at the monitoring section of Qinglongxia and decreases at all other sections. Total phosphorus decreases significantly for all sections. The salt index has a significant increase in two sections of the Fumin Bridge and the Pier Pier Bridge and shows a downward trend in other sections. The monitoring section of ammonia nitrogen only rises in other monitoring sections. The trend is shown in Table 1.

Table 1. Statistics of rank correlation coefficient of water quality index of Mantis River.

Sample points	Coefficient	Do	ODMn	COD	BOD_5	NH_4-H	TP	Fluoride	CPI
overall	r_s	0.750	−0.600	−0.900*	−0.700	−0.800	−0.500	−0.600	−0.900*
ZTZM	r_s	0.250	−0.900*	−0.100	−1.000*	−0.300	−1.000*	−0.150	−0.700
WQDQ	r_s	−0.900*	0.800	0.800	−0.300	1.000*	0.000	0.450	−0.100
QLX	r_s	−0.800	0.850	0.900*	0.850	0.300	−0.900*	−0.450	−0.600
FMDQ	r_s	−0.900*	0.900*	−1.000*	−0.800	−0.200	−0.600	−0.500	−0.900*
PDHQ	r_s	0.100	0.000	−0.700	−0.450	−0.900*	−0.600	0.300	−0.600
TXQ	r_s	0.100	−0.400	0.300	0.350	−0.700	−1.000*	0.900*	−0.400
CQDXQ	r_s	0.150	0.900*	−0.800	−0.400	−0.700	0.000	−0.100	−0.800

Note: CPI- Comprehensive pollution index; Do- Discipline of oxygen value changes; ODMn- Hanganate Index; COD- Chemistry needs changes in the amount of oxygen; BOD_5–5 days of biochemical oxygen demand; NH_4-H- ammonia nitrogen; TP- total phosphorus

According to the statistical yearbook of Anning City, the GDP in 2017 was 31.761 billion yuan, the GDP in 2018 was 43.02 billion yuan, the GDP in 2019 was 57.514 billion yuan, the GDP in 2020 was 57.236 billion yuan, and the GDP in 2021 was 612,58 billion yuan.

The Environmental Kuzsnets Curve (EKC) model is used for correlation analysis to describe the evolutionary relationship between economic growth and environmental pollution. The original study of the correlation analysis between the economic status and water environmental quality (comprehensive pollution index) of the watershed in the basin is found to be consistent with the inverted U-shaped trend of the environmental Kuznets curve (EKC) model, as shown in Figure 4. The EKC model shows the "inverted U" curve, which reflects the deterioration of environmental quality is higher than economic growth at the beginning of economic development and slower than economic growth at a certain level of economic growth.

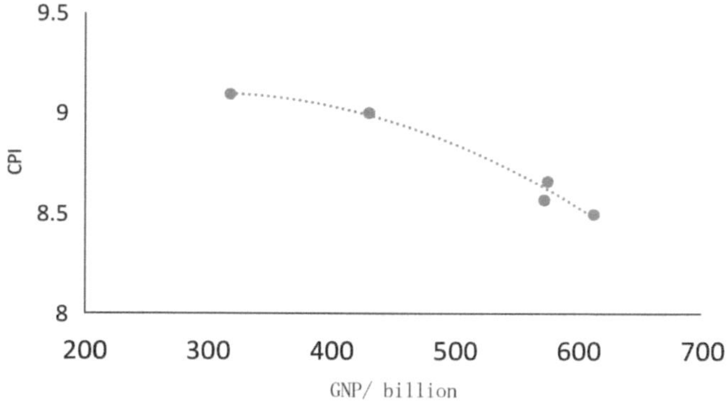

Figure 4. The relationship between the pollution index of Anning County Basin and GDP.

4 CONCLUSION

In the dry season (January-April, November-December), the water quality status is relatively stable and the water quality condition decreases after entering the rainy season (May-October). During the past five years, the water quality has improved significantly, except for the five-day biochemical and oxygen demand of the Qinglongxia monitoring section, and the other sections of the main pollution index indicators have decreased. The cross-section shows a significant upward trend and other monitoring sections show a downward trend. Chemical oxygen demand only increases significantly at Qinglongxia and decreases at other sections. Ammonia nitrogen only rises with a significant increase in other sections. The upstream of fluoride shows a downstream trend and the downstream is on the rise. The pollution indicators of the mainstream and tributary sections of Purdue River in praying mantis Sichuan within five days of biochemical oxygen demand, chemical oxygen, and total phosphorus, exceed the standard rate of 69.70%, 65.62%, and 56.25%, respectively. Combined with the local economic development, it is found that the pour U-shaped in this basin conforms to the environment of the Kuzshanes curve (EKC) model after passing the inflection point on the right. The watershed has stabilized the inverted U-shaped point of EKC. Economic development has made more funds available to treat water quality. The combination of seasonal and annual water quality changes in the watershed and upstream and downstream relationships allows the impacts of anthropogenic disturbances in and around Dianchi to be identified and addressed on time. The environment is precisely restored at the right time and point.

REFERENCES

Cai, H., Mei, Y. D., Chen, J. H., Wu, Z. H., Lan, L., Zhu, D. 2020. An Analysis of the Relationship Between Water Pollution and Economic Growth in China by Considering the Contemporaneous Correlation of Water Pollutants. *J. Cleaner Prod.* 276, 122783.

Ekins P. 1997. The Kuznets Curve for the Environment and Economic Growth: Examining the Evidence. *Environment and Planning A* 29(5), 805–830.

Hao, Y., Wu, Y., Wang, L., Huang, J. 2018. Re-examine Environmental Kuznets Curve in China: Spatial estimations Using an Environmental Quality Index. *Sustainable Cities and Society* 42, 498–511.

Wang, H. 2020. *Research on Spatial and Temporal Variation of Nitrogen and Phosphorus in Tanglang River Basin Based on the INVEST and CA-MAHKOV* model [D]. Yunnan University.

Kuznets, S. 1955. *Economic Growth and Income Inequality American Economic Review*. 49, 1–28.

Qing, F., Yan, Z., Shanjun, W. 2015. Water Quality and Spatial and Temporal Distribution Characteristics of Surface Drinking Water Sources in China. *Journal of Basic Science and Engineering*, 23(05), 886–894.

Wang, B., Liu, L., Huang, G. H. 2017. Retrospective and Prospective Analysis of Water Use and Point Source Pollution from an Economic Perspective-a Case Study of Urumqi, China. *Environ. Sci. Pollut. Res.* 24(33), 26016–26028.

Wong, Y. L., Lewis, L. 2014. The Disappearing Environmental Kuznets Curve: A Study of Water Quality in the Lower Mekong Basin (LMB). *J. Environ. Manage.* 131, 415–425.

Xu Z. M., Yan Z. S. 1992. Purdue River Basin Water Environment Features [J]. *Yunnan Environmental Science*, (01):29–32.

Yang W, Xia J. S., Wang C. 2021. Research on the Research Method of Mechanical Damage Facial Remote Sensing Information Extraction Method of Mantis Sichuan Basin [J]. *Yangtze River Basin Resources and Environment*, 30(12):2896–2904.

Zhang, J. J., Gangopadhyay, P. 2015. Dynamics of Environmental Quality and Economic Development: the Regional Experience from Yangtze River Delta of China. *Applied Economics*. 47(29), 3113–3123.

Water Conservancy and Civil Construction – Oke & Ahmad (Eds)
© 2024 The Author(s), ISBN: 978-1-032-58614-4

Optimization of Quay bridge scheduling of iron-water intermodal container terminal considering the time cost of unloading operation

Yihan An
School of Mechanic and Electronic Engineering, Wuhan University of Technology, Wuhan, China

Zihou Peng
Industrial Research Institute of Wuhan University of Technology, Suizhou, Wuhan, China

Cunrong Li*
School of Mechanic and Electronic Engineering, Wuhan University of Technology, Wuhan, China

ABSTRACT: Rail-water intermodal transport is a transport mode proposed to solve the large transport volume, and the high level of rail-water intermodal transport cannot be achieved without the efficient operation of container terminals. The shore bridge is usually considered the main factor limiting the loading and unloading efficiency of container terminals. To address the problems, such as the chaotic terminal unloading environment caused by the lack of a scheduling plan for the sequence of shore-bridge operations in the actual scenario, this paper establishes a model to minimize the operating time of the ship considering the shore-bridge collision constraint and improves the genetic algorithm for both the requirements of considering the sequence of operations and the operating time within the bays.

1 INTRODUCTION

1.1 *Background*

The container terminal is the connection point between water and land transport in rail-water intermodal transport, which is an important link to cargo transportation. With the unprecedented development of rail-water intermodal transport in the prosperous cargo trade, water transport and railroad transport have lower transport costs and larger transport volumes. Therefore, the rail-water intermodal transport itself on the container terminal has put forward higher hardware requirements and software requirements. Because of the convenience of water transport and railroad transport in the connection, rail-water transport can efficiently connect water transport and land transport to a certain extent and can effectively integrate the advantages of waterway freight transport and railroad cargo transport. In this case, improving logistics management and operation level, rational allocation of terminal resources, optimizing loading and unloading equipment scheduling, and improving container loading and unloading efficiency become the urgent need for further development of rail-water intermodal ports.

The level of operation of the container terminal has put forward high requirements. The operation efficiency of the container terminal depends largely on the operational efficiency of the terminal loading and unloading system. The reasonable scheduling and optimization of the existing loading and unloading resources of the terminal can effectively improve terminal operation efficiency.

*Corresponding Author: cunrong_li@163.com

 DOI: 10.1201/9781003450818-28

1.2 *Research status*

There is a large amount of literature studying the shore-bridge scheduling problem of container terminals. For example, Kim and Park (Kim & Park 2004) established a MILP model and proposed a branch-and-bound algorithm to solve the problem. Since then, Moccia *et al.* (2006), Sammara *et al.* (2007), and Bierwirth and Meisel (2009) have continuously improved the model and designed different heuristic algorithms to solve it. Chen, Li, and Goh (2014) proposed a new efficient one-way scheduling MILP model to directly solve the shore-bridge scheduling with performance exceeding previous literature. Qin and Unsal (Qin & Sha 2013a, 2013b) proposed constraint programming (CP) models to solve shore-bridge scheduling, respectively.

2 MODEL BUILDING

2.1 *Problem description*

From the current domestic development status, it can be seen that most of the current domestics take the water-yard-iron mode. The specific process can be roughly divided into 3 stages, Stage 1 is when the cargo ship arrives at the terminal, and the shore loading bridge (shore bridge) begins to unload containers from the ship. Stage 2 is for the container truck (collector truck) in the terminal front and the terminal to store containers between the yard to transport containers. Stage 3 is for the container yard, by the yard gantry crane (field bridge) completes the unloading of containers from the container truck as well as the yard. It is generally believed that the shore bridge is the main bottleneck that restricts the container terminal loading and unloading efficiency, which is one of the main factors that determine the container terminal throughput capacity. Therefore, to reduce the waiting time of the bridge and improve the efficiency of loading and unloading operations, the sequence of shore bridge operation needs to be optimally scheduled.

The optimal scheduling of the shore bridge is to solve the problem of deciding which task is to be handled by which specific shore bridge in which period so that multiple shore bridges can coordinate their operations in the process of serving the ship and making the loading and unloading time of the ship as short as possible.

2.2 *Description of the nature of shore bridge operations*

Shore bridge scheduling consists of assigning tasks to the shore bridge assigned to the vessel and arranging the starting time of each task. Ultimately, the ship loading and unloading operations can be completed before the ship's departure time to ensure the smooth operation of the terminal.

The number of the shore bridge scheduled to operate in a port is i. The shore bridge operation is constrained by the spatial location, that is, at the same time the shore bridge i cannot cross the neighboring bridges. For the sake of discussion, in the initial state, the default numbering principle for the shore bridges in this paper is to choose a certain direction to start numbering from small to large, and the shells on board are also numbered from the same direction according to this rule to ensure that the operation area of the smaller numbered shore bridge contains the same smaller numbered shells.

In the terminal loading and unloading, because the shore bridge resources compared to the container truck, field bridge resources are more nervous. Therefore, in the operation, we should ensure the rule of continuous shells and wait for the shore bridge to complete an operation area before starting the next operation area to minimize the operation switching time. An operation area generally includes multiple containers with similar locations or similar operation types.

If the shore bridge i of the sequence of inner shells is connected, it is concluded that i of the operating area is continuous. We suppose there are two shore bridges at a certain time, i and i + 1. According to the operation arrangement, the shore bridges i are assigned to two

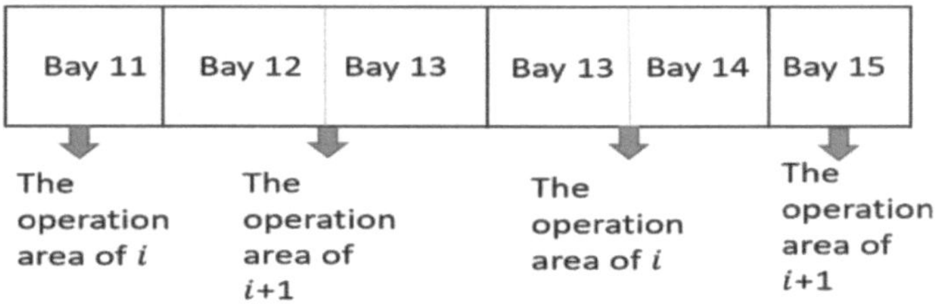

Figure 1. Schematic diagram of the shore bridge operation area.

operating areas, which are located on their adjacent shore bridges i + 1 on either side of the working area of i + 1. There is only an operating area. This gives rise to the space constraint problem, which defines an operation to swap the working areas of the two bridges: making the bridges i closer to i + 1 on one side of the bridge, which is operated by i + 1. The original operation area is operated by i. The operation is carried out. Before and after the operation, the operation sequences of i and i+ remain the same and the i operation area becomes continuous. In the case where there are several operating space constraints, the same operation can be experienced several times. Then the operating area of the shore bridge has no longer interfered. The adjustment time of the equipment position has been proportionally reduced compared to that before the adjustment. It approximates the ideal state.

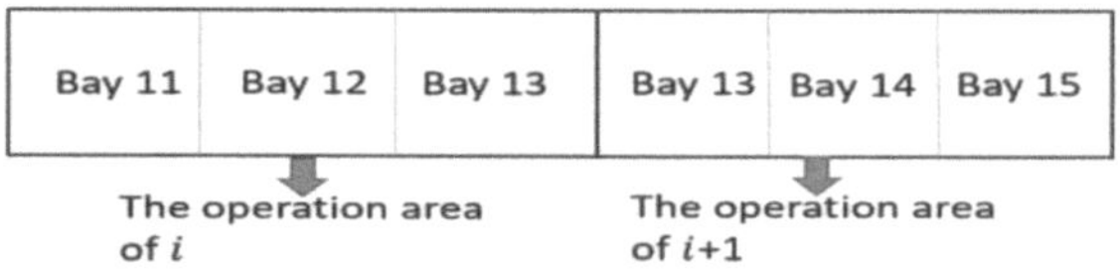

Figure 2. An example of an ideal state after job area swapping.

2.3 *Parameter description*

We define the Shore Bridge $\boldsymbol{i}\in\boldsymbol{I}$.for which the sequence of tasks is scheduled as $\boldsymbol{A_i}$. The number of tasks to be processed is $\boldsymbol{S_i}$ and $\boldsymbol{k_i}\in\boldsymbol{S_i}$, then we have $\{\boldsymbol{a_1},\boldsymbol{a_2},\ldots,\boldsymbol{a_{S_i}}\}\in\boldsymbol{A_i}$. We define the task is $\boldsymbol{a_{k_i}}$. The operation time required in the unloading process is $\boldsymbol{d_{k_i}}$. If the task $\boldsymbol{a_{k_i-1}}$ with $\boldsymbol{a_{k_i}}$ are two tasks that operate continuously on the shore bridge $\boldsymbol{i}$, the shore bridge $\boldsymbol{i}$ has an adjustment time at the end of the $\boldsymbol{a_{k_i-1}}$. After the operation, there is an adjustment time of $\boldsymbol{p_k}$.

We define $\boldsymbol{M}$ as a sufficiently large positive number, define $\boldsymbol{j}\in\boldsymbol{J}$ as the operational tasks to be handled by all shore bridges, and define $\boldsymbol{j=0,n}$ as the first and the last task of all shore bridges. The operation time is $\boldsymbol{d_j}$, the adjustment time for consecutive operations of the same shore bridge is $\boldsymbol{p_{jj'}}$ and $\boldsymbol{p_{0j'}}$, which is 0. The task completion time is $\boldsymbol{T_j}$. We define $\boldsymbol{D^S}$ as the pre-defined sequence of operations. $(\boldsymbol{j},\boldsymbol{j'})\in\boldsymbol{D^S}$ denotes $\boldsymbol{j}$ can be processed before $\boldsymbol{j'}$ is processed. We define $\boldsymbol{B_J}$ as the number of the bay.

When $\boldsymbol{j}\in\boldsymbol{A_i}$, the decision variable is defined $\boldsymbol{x_{ij}=1}$, otherwise, it is 0.

In all shore bridges, if two tasks are $\boldsymbol{j},\boldsymbol{j'}\in\boldsymbol{A_i}$, we will define the decision variable $\boldsymbol{y_{ijj'}=1}$, otherwise, it is 0.

When the task is $\boldsymbol{j},\boldsymbol{j'}\in(\boldsymbol{a_{k_i-b}},\boldsymbol{a_{k_i}})$, that is, although $\boldsymbol{j}$ is scheduled before $\boldsymbol{j'}$ and is processed on the same shore bridge, it is not a continuous task. We define the decision variable $\boldsymbol{z_{ijj'}=1}$, otherwise, it is 0.

When $\boldsymbol{T_j\geq d_j}$, that is, in all shore bridges, two tasks are processed consecutively before and after $\boldsymbol{j},\boldsymbol{j'}\in(\boldsymbol{a_{k_i-1}},\boldsymbol{a_{k_i}})$, the decision variable is defined $\boldsymbol{u_{ijj'}=1}$, otherwise, it is 0.

When two tasks are processed at the same time, we define the decision variable $\boldsymbol{v_{ii'jj'}=1}$, otherwise, it is 0.

2.4 *Mathematical model*

From the above, the present problem can be expressed as follows.

$$minT_{max} = max(T_j) \tag{1}$$

One and only one shore bridge per task is responsible for the operation.

$$\sum_{i\in I} x_{ij} = 1, \forall j \in J \tag{2}$$

When both tasks are handled on the same device, the following should be satisfied.

$$2y_{ijj'} \le x_{ij} + x_{ij'}, x_{ij} + x_{ij'} \le 2y_{ijj'} + 1, \forall j, j' \in J, \forall i \in I \tag{3}$$

$$y_{ijj'} = y_{ij'j}, \forall j, j' \in J, \forall i \in I \tag{4}$$

When two tasks are completed on the same shore bridge, there is always a sequence of these two task operations.

$$z_{ijj'} + z_{ij'j} = y_{ijj'}, \forall j, j' \in J, \forall i \in I \tag{5}$$

In addition, two tasks, j and j', on the same shore bridge i operate consecutively on (without distinguishing the sequence),

$$z_{ijj'} - u_{ijj'} \ge 0, \forall j, j' \in J, \forall i \in I \tag{6}$$

If j scheduled before j' and on the same shore bridge i is processed on, the following formula can be obtained.

$$T_j - \left(T_{j'} - d_j - \min\left(p_{j''j}\right)\right) + \left(1 - z_{ijj'}\right)M \ge 0, \forall j, j', j'' \in J, \forall i \in I \tag{7}$$

Further, the task j and j' are processed consecutively before and after, that is, $j, j' \in (a_{k_i-1}, a_{k_i})$ and $\forall p_{jj'} \in T$.

$$T_j - \left(T_{j'} + d_{j'} + p_{jj'}\right) + \left(1 - u_{ijj'}\right)M \ge 0, \forall j, j' \in J, \forall i \in I \tag{8}$$

If $v_{ii'jj'}$ meets the definition, then the following formulas can be obtained.

$$T_j - T_{j'} + \left(1 - v_{ii'jj'}\right)M \ge 0, \forall j, j' \in J, \forall i \in I \tag{9}$$

$$T_{j'} - (T_j - d_j) + \left(1 - v_{ii'jj'}\right)M \ge 0, \forall j, j' \in J, \forall i \in I \tag{10}$$

$$x_{ij} + x_{i'j'} \ge v_{ii'jj'} \tag{11}$$

Contrary to (2), a shore bridge cannot operate on two containers at the same time.

$$y_{ijj'} + v_{ii'jj'} \le 1, \forall j, j' \in J, \forall i \in I \tag{12}$$

The tasks that are processed consecutively before and after on the same shore bridge are called immediate pre-tasks and immediate post-tasks, it can be known from (12) that there is at most one immediate pre-task and immediate post-task per task.

$$\sum_{j\in J} u_{ijj'} \le 1, \forall j, j' \in J, \forall i \in I \tag{13}$$

$$\sum_{j\in J} u_{ij'j} \le 1, \forall j, j' \in J, \forall i \in I \tag{14}$$

From Formulas (13) and (14), it is clear that if the task is $j \in A_i$, it is impossible to form an immediately preceding and following relationship with the tasks assigned to other shore bridges.

$$2x_{ij} \geq \sum_{j \in J} u_{ijj'} + \sum_{j \in J} u_{ij'j}, \ \forall j, j' \in J, \forall i \in I \tag{15}$$

Similarly, if j, j' forms the immediately preceding and immediately following relation, there must be $j, j' \in A_i$, which is expressed as follows.

$$2x_{ij} - 1 \leq \sum_{j \in J} u_{ijj'} + \sum_{j \in J} u_{ij'j}, \text{ and } \forall j, j' \in J, \forall i \in I \tag{16}$$

Since the shore bridge movement is time-consuming and not easy to adjust, to make the shore bridge operation not fall into chaos due to the change of scheduling plan in the middle, the shore bridge operation always follows a predefined sequence.

$$T_j - d_j \leq T_{j'} - d_{j'}, \ \forall \left(j, j'\right) \in D^S \tag{17}$$

$$\left(1 - v_{ii'jj'}\right) M \geq B_j - B_{j'}, \quad \forall B_j, B_{j'} \in B_J, \forall j, j' \in J, \forall i \in I \tag{18}$$

3 ALGORITHM IMPLEMENTATION

Based on the excellent global search ability of the genetic algorithm, this paper selects a genetic algorithm to solve the model. Because there is only a single unloading process and the operation order of Bay and shore-bridge assignment are independent problems, that is, when the operation order of Bay is given, different shore-bridge assignments will not affect the operation time of Bay. Therefore, this study assumes that the minimum operation time of Bay is T_b and the container operation time is given by the empirical formula.

3.1 *Algorithmic framework*

We assume the current number of iterations is **t** and the maximum number of iterations is t_{max}. The first **t** generation population and the first **i** chromosomes are $P(t)$ and $P_i(t)$. The corresponding fitness function is $f(P_i(t))$. The population size is set to. The global optimal solution is found by the algorithm P_U, the genetic probability is p_i, the crossover probability is p_c, and the variation probability is p_m. From 2.2, it can be seen that the assignment of operation areas should be done according to the principle of consecutive operating regions in the order of numbering from smallest to largest. The order of macroscopic operations with the coded shore bridge is called a chromosome. The chromosome is of the form. $G_e = \{\ldots * b, T_b, b', T_b', b'', T_b'' \ldots * \ldots, 0, 0\}$, where * is the separator, the interval between two separators represents the operation order of one shore bridge, b is the number of the shellfish, and T_b is the time required to process the shells. To ensure that each chromosome is of the same length, chromosomes of insufficient length are used to make up the position with 0.

3.2 *Adaptation function*

The optimization objective in this paper is the minimization of time cost. For the minimization optimization problem, the general approach is $f(x_i) = 1/t_i$. Since this paper involves large data onshore bridge operation time, we use $f(x_i) = 1000/t_i$, where t_i is the solution of the chromosome.

The chromosome corresponding to the optimal solution of each generation is directly added to the next generation's population. The genetic probability is calculated based on fitness weighting according to this probability division from $P(t)$ from $N-1$. The new population is finally generated by selecting individuals from $P(t+1)$. A new population is generated.

3.3 *Lower bound*

To verify the efficiency of the genetic algorithm, a low bound for this problem is proposed under the condition of relaxing deck position and shore-bridge collision constraints.

It can be known that the adjustment time that moves the shore bridge from one shell to the adjacent shell is given as p_0, the number of shells is N_B, the number of shore bridges is N_Q, the total number of shore bridge moves is at least N_B-N_Q. Given that the total operation time of a shell is d_b, we can know that $T_{max}\geq(\sum_{b=1}^{N_B} d_b)/N_Q$. The final low bound for this problem is obtained as follows.

$$LowerBound = \frac{\left[\sum_{b=1}^{N_B} d_b + (N_B - N_Q)p_0\right]}{N_Q} \tag{19}$$

4 EXAMPLE VERIFICATION

In this paper, the genetic algorithm is set as $t_{max}=50$, the population size is $N=500$, the crossover probability is $p_c=0.7$, the variation probability is $p_m=0.3$. Numerical experiments are conducted on an Intel core i7–9700F processor with 16G RAM PC, implemented using MATLAB R2022b programming. Random data are generated with the number of shore bridges of 3, 4, 5, the number of bays of 24, and the number of containers of 3000–6000, and the arithmetic mean (AVEG) is calculated ten times for each case for comparison with the lower bound (LB). The results are shown in Table 1, which is called $K=\frac{AVEG-LB}{AVEG}*100\%$ as the effective degree of the genetic algorithm in this paper.

Table 1. Information on video and audio files that can accompany a manuscript submission.

Number of shore bridges	Total number of containers	AVEG (s)	LB (s)	K (%)
3	4000	7534.24	7218.00	4.20
3	4500	8458.85	8068.00	4.62
3	5000	9443.90	8918.00	5.57
4	4000	5703.92	5413.50	5.09
4	4500	6399.27	6051.00	5.44
4	5000	7126.60	6688.50	6.15
5	4000	4584.14	4330.80	4.87
5	4500	5123.98	4840.80	5.71
5	5000	5663.81	5350.80	6.09

5 CONCLUSIONS

5.1 *Analysis of the results*

In this case, with five shore bridges and 6000 boxes to process, the running time is 3.08 seconds, which proves the efficiency of the algorithm. It can be seen that the overall effective degree of the algorithm is around 5%. Considering that the low bound ignores the deck position and shore bridge collision constraints, it can be considered that the genetic algorithm plays a better optimization role.

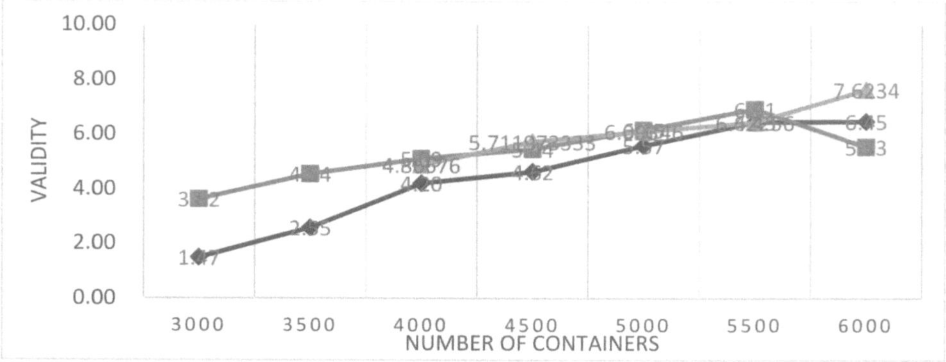

Figure 3. Effective degree distribution under different scenarios of a number of shore bridges and a number of containers in genetic algorithm.

5.2 *Summary and outlook*

At present, domestic rail-water transportation is facing the problem of "large volume and long waiting time". The problem has become an urgent task to improve terminal operation efficiency. Due to the characteristics of the shore bridge, constantly changing the scheduled operation plan can cause significant losses. In this paper, we propose a mathematical model. It considers the collision constraint and container location constraint of the shore bridge operation as the research direction to improve the efficiency of the shore bridge operation and the genetic algorithm for the optimization of the shore bridge unloading operation. The model proposed in this paper is practically significant and can be used as a reference for scheduling work. The shortcoming of this paper is that it does not consider the scheduling problem of shore-bridge operation when loading and unloading are carried out simultaneously. On the contrary, loading is the inverse process of unloading by default and the two processes do not affect each other.

REFERENCES

Bierwirth, C., Meisel, F. A Fast Heuristic for Quay Crane Scheduling with Interference Constraints [J]. *Journal of Scheduling*, 2009, 12(4): 345–60.

Chen, J. H., Li, D. H., Goh, M. An Effective Mathematical Formulation for the Unidirectional Cluster-based Quay Crane Scheduling Problem [J]. *European Journal of Operational Research*, 2014, 232(1): 198–208.

Kim, K. H., Park, Y. M. A Crane Scheduling Method for Port Container Terminal [J]. *European Journal of Operational Research*, 2004, 156(3): 752–768.

Moccia, L., Cordeau, J. F., Gaudioso, M., et al. A Branch-and-cut Algorithm for the Quay Crane Scheduling Problem in a Container Terminal [J]. *Naval Research Logistics*, 2006, 53(1): 45–59.

Qin Tianbao, Sha Mei. A CP Model for the Quay Crane Scheduling Problem Supporting Bidirectional Schedules [J]. *Systems Engineering*, 2013b, 31(4): 53–59.

Qin Tianbao, Sha Mei. Modeling and Solving Quay Crane Scheduling Problems Based on Constraint Programming [J]. *Computer Integrated Manufacturing Systems*, 2013a, 19(1):181–186.

Sammarra, M., Cordeau, J. F., Laporte, G., et al. A Tabu Search Heuristic for the Quay Crane Scheduling Problem [J]. *Journal of Scheduling*, 2007, 10(4–5): 327–336.

Water Conservancy and Civil Construction – Oke & Ahmad (Eds)

Variation of strength of cement-soil mixing column with depth in dredger fill site

Feng Cheng
CCCC (Tianjin) Rail Transit Engineering Construction Co., Ltd., Tianjin, China

Tao Liu*
Tianjin Transportation Research Institute, Tianjin, China

Zhiyuan Ma, Jiabin Chen, Shuo Li, Ruilong Liang, Yaoguang Zhang & Xin Zhao
CCCC (Tianjin) Rail Transit Engineering Construction Co., Ltd., Tianjin, China

ABSTRACT: The method of deep mixing has been widely used to improve the ground. However, the application of a deep mixing column in the dredger fill area has not been well studied. A series of laboratory tests and field column tests were carried out to investigate the strength of the cement-soil mixing column in the dredger fill area, in Nanjiang, Tianjin, China. The results show that the content of organic matter in the dredger fill in Nanjiang is not high with 1.5–2.0%. The PH value is 7.4 and the soil is neutral, which has little effect on the solidification and strength of cement soil. Laboratory tests show that the unconfined compressive strength of dredger fill in Nanjiang is not less than 2.0 MP in 28 days under the condition of 12% cement content. The core test results of the field column show that the strength of the test column core sample gradually decreases from top to bottom. There are two main reasons. First, the core sample of more than two meters is above the water level, but the core sample of fewer than two meters is below the water level. The strength of cement soil rises slowly below the water level. Second, silty clay is not easy to mix evenly and the slurry will return up, which has a lower amount of cement mixture in the middle and lower section, resulting in low strength or unable to form strength.

1 INTRODUCTION

In the face of increasingly severe land area pressure in the city, reclaiming land from the sea has become one of the common methods to alleviate the ground pressure, which is bound to reinforce the soft soil foundation of dredger fill soil. The method of cement-soil mixing column reinforcement is one of the most commonly used ground improvement methods, so the research on the column strength of cement-soil mixing columns is particularly important.

Oliveira *et al.* (2011) studied the effect of deep mixing column strengthening embankment on common consolidated soft soil foundation by the numerical model and analyzed the parameters for the numerical prediction results according to settlement and effective stress increment. It is found that using a deep mixing column to strengthen the embankment foundation is very effective, which can effectively control settlement and reduce uneven settlement. Because of the arch effect, the load exerted from the embankment is

*Corresponding Author: 36092342@qq.com

DOI: 10.1201/9781003450818-29

concentrated on the deep mixing column, and the effective stress increment of soil is negligible. Jiang *et al.* (2013) found that deep mixing columns can effectively improve the bearing capacity and stability of soft soil foundations and reduce the total settlement and uneven settlement of the foundation through three-dimensional finite element analysis. Liu (Liu *et al.* 2012) and Phutthananon (Phutthananon *et al.* 2020) introduced a new type of T-shaped cement-soil mixing column. Being Different from the traditional cement-soil mixing column, the cross-section of the new mixing column changes along the mixing depth. A large amount of cement slurry is injected, and a specially designed mixing blade is used to thoroughly mix with the foundation soil, which can not only better ensure the strength of the cement-soil mixing column, but also reduce the construction cost. Based on the engineering practice of several deep and large foundation pits in the Shanghai soft soil stratum, Huang *et al.* (2015) systematically studied the strength test results of TRD cement-soil mixing wall core samples with different depths, thicknesses, and application forms. The results show that there is no obvious difference in the appearance of wall core samples in different soil layers, and the integrity and uniformity of wall core samples are good in the range of wall depth. The 28-day unconfined compressive strength of wall core samples is generally higher than that of 1.0 MPa, and there is no significant difference in cement strength between shallow soft clay and deep sand. Yuan *et al.* (2018) carried out the laboratory unconfined compressive strength test of cement soil with cement content, comprehensive water content, and cement age as control variables to obtain its strength variation law. The test results show that the unconfined compressive strength of cement soil increases with the increase of cement content and age, and decreases with the increase of comprehensive water content. According to the reasonable and feasible construction process parameters, the comprehensive coordination parameter formula of the cement-soil mixing method is deduced. Through the test, Liang *et al.* (2009) found that the strength of cement soil is not only related to age, but also the properties of the soil layer, but there is no obvious difference between the strength of the beating area and non-beating area at the same elevation section. Compared with the core strength test results, the slurry strength test results are more uniform, less discrete, and closer to the actual strength of the mixing column. Because of the damage to the core sample in the coring process, the test strength value is lower. The ratio of the slurry strength value to the core strength value is between 1.3 and 1.6. Given some problems existing in the construction technology of cement-soil mixing columns in China, Zhao *et al.* (2014) have studied and developed a new technology for five-axis cement-soil mixing columns. It is found that in each soil layer, the strength value of the triaxial cement-soil mixing column is lower and its discreteness is larger. The strength of the two-axis cement-soil mixing column is better. The discreteness of the strength value of the five-axis cement-soil mixing column varies greatly at different ages. The strength value of the five-axis cement-soil mixing column is the highest and the discreteness is small, so it has a good prospect of popularization and application. Based on the on-site testing results of the cement-soil mixing column in Huanghua Port, Chen *et al.* (Chen & Li 2015) introduced a kind of cement mixing column bit with fan-shaped shotcreting in the middle position after turning the soil blade, which can improve the mixing uniformity of column body and reduce the amount of returning on the ground, so it can improve the strength and construction quality of cement-soil mixing column.

Most of the above research on the strength of cement-soil mixing columns is aimed at the soft clay foundation, but there are few studies on the column strength in the dredger fill foundation. And at present, there is no mature construction technology for the characteristics of high organic matter content, high salt content, high water content, large void ratio, and short settlement time in the dredger fill land area, which only depends on the construction enterprises to improve the cement content, the number of mixing times, and the column effect. The construction cost is high and the progress is slow. How to reasonably control the construction parameters in the process of foundation reinforcement columns in dredger fill land areas to maximize the quality of cement-soil mixing columns and reduce the settlement after construction is an urgent technical problem.

Based on the project of the dredger fill land areas in Nanjiang, Tianjin, this paper sampled and analyzed the physical and mechanical properties of the soil of the dredger fill site. The laboratory cement-soil strength test with the site soil was carried out. At the same time, the field column test is conducted, and the coring results were analyzed and discussed. Some suggestions were proposed to improve the reinforcement quality of cement-soil mixing columns in the future.

2 GENERAL SITUATION OF THE PROJECT

2.1 *Site location and stratigraphic structure*

The extension project line of Tianjin Nanjiang Port Ore Railway is located in Nanjiang Port, Binhai New Area, Tianjin. Tianjin Nanjiang Port Industrial Zone is located in the southeast of Tianjin Binhai New area. Cement-soil mixing column construction areas are alluvial marine plains and artificial reclamation areas (dredger fill-in sea areas). The overall topography is flat and the terrain is open, and most of the areas are industrial land.

The widely exposed strata are mainly the Quaternary Holocene artificial fill layer (artificial accumulation Qml), the I continental layer (Quaternary Holocene Upper formation riverbed ~ floodplain facies deposition Q43al), the I marine layer (Quaternary Holocene Middle formation shallow marine facies deposition Q42m), the II continental layer (Quaternary Holocene Lower formation swamp facies deposition Q41h and riverbed ~ floodplain facies deposition Q41al), the III continental layer (Quaternary Upper Pleistocene five formation riverbed ~ floodplain facies deposit Q3eal), the II marine layer (Quaternary Pleistocene four formation coastal ~ tidal zone facies deposition Q3dmc), and the IV continental facies layer (Quaternary Upper Pleistocene three formation riverbed ~ floodplain facies deposit Q3cal).

2.2 *Soil analysis*

Through the arrangement of drilling holes in the field, the site soil was obtained. The laboratory test was carried out to detect the content of organic matter in the soil and conduct the particle analysis.

Through the test, it is known that the content of organic matter in the dredger fill foundation of Nanjiang Port is 1.5–2.0%. Through experimental studies, Rao *et al.* (Rao & Huang 2009) believe that when the content of organic matter is less than 5%, the strength of cement soil is less affected by the content of organic matter. But when the content of organic matter is more than 5%, it has a great influence on the strength of cement soil. The above research shows that the organic matter content of 1.5–2.0% of the dredger fills foundation in Nanjiang Port is not high, which has little effect on the solidification and strength of cement soil.

Through the particle analysis of the site soil, the particle gradation curve is drawn from the results of the particle analysis, as shown in Figure 1. As can be seen from Figure 1, the soil particles are relatively small, and the particles less than 0.002 mm are more than 20%. And according to the test, the plasticity index of each layer of soil is 11–18, so the dredger fill soil of the Nanjiang Port has a greater viscosity. Through the research, Wu (2021) found that in the soft soil foundation, the cement slurry of the mixing column is not easy to stay in the soft soil and mix with the clay of the soil. Because of the characteristics of soft soil clay with electrostatic water film, such as soil structure, saturation, flow plastic shape, and high content of organic matter, it is not easy to mix. From the above results and research analysis, it can be seen that the viscosity of the dredger fills soil is relatively high. And according to its

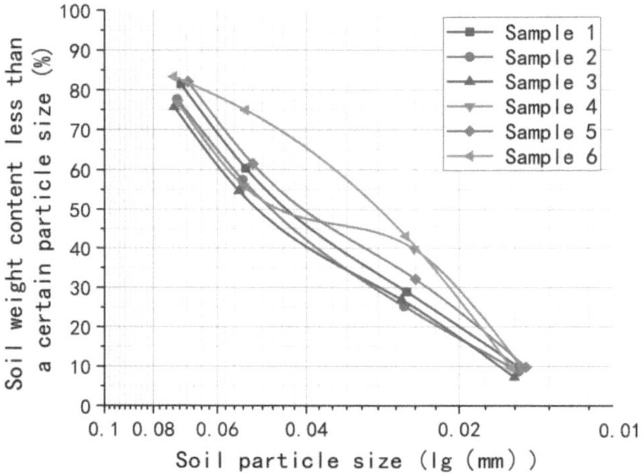

Figure 1. Particle gradation curve of site soil.

formation process, the source of the dredger fill soil is mostly muddy clay. In addition, the permeability coefficient is low and the water content is high. When adding cement slurry, more mixing times are needed to form uniform cement soil.

3 TEST RESULTS

3.1 *Laboratory test*

The laboratory strength tests were respectively carried out with different cement content and different foundation soil. According to the test results, the best cement mixing ratio which meets the requirements of design strength is selected. An unconfined compressive strength test is carried out with each mixing ratio for 7 days and 28 days. Among them, the cement content is 12% and 24% respectively, and the cement type is P.O 42.5 ordinary Portland cement, and the water-cement ratio is 0.6.

The final test results are shown in Table 1, which shows six kinds of cement soil samples made of different cement content and different foundation soil. All of them meet the design requirements (the 28-day unconfined compressive strength of cement soil is not less than 2.0 MPa).

Through the cement-soil test of different mix ratios, it can be known that the unconfined compressive strength of 28 days is not less than 2.0 Mpa and the strength quality is better under the condition of fully mixing. And the laboratory test verifies the conclusion that the

Table 1. Unconfined compressive strength (unit: MPa).

Batch	12% cement		24% cement		35% cement	
	Day 7	Day 28	Day 7	Day 28	Day 7	Day 28
The first soil was taken from the roadbed on April 11	2.2	4.1	4.9	10.7	9.7	13.9
The second soil was taken from the roadbed on June 29	1.9	3.3	4.3	6.0	9.4	12.8

content of organic matter in the site soil of Nanjiang Port does not affect the strength of cement soil.

3.2 *Field column test coring*

In the field column test, the cement adopts P.O 42.5 cement ordinary Portland cement; the construction water is tap water; the column diameter is 0.6 m; the water-cement ratio is 0.6; the natural water content is 32.2%; the wet density is 1.93 g/cm3; the grouting pressure is 0.5 Mpa; the drilling speed of the column mixer is 0.8 m/min; and the lifting speed is 0.6 m/min. The number of times of spraying and stirring follows the principle of two sprays and four stirrings. During the implementation of the test, the flow rate of the grouting pump is strictly guaranteed to be stable and sustained during the grouting process.

The test columns with different content and cementitious materials are selected in the test. Cement and curing agents were used as cementitious materials respectively. Through the test, the strength change law of the cement-soil mixing column under different parameters can be analyzed. The specific parameters of each test column construction are introduced in Table 2.

Table 2. Test column construction parameters.

Column number	Column diameter/m	Column length /m	Wet soil density	Mixing amount	Dosage per meter /kg	The total dosage of a single column /kg
GH-1, SN-1	0.6	10	1.93	12%	65.45	654.5
GH-2, SN-2	0.6	10	1.93	16%	87.27	872.67
GH-3, SN-3	0.6	10	1.93	20%	109.08	1090.84
GH-4, SN-4	0.6	10	1.93	24%	130.90	1309.00

(a) SN-2 (28-day)

(b) GH-2 (28-day)

Figure 2. On-site coring photo.

The unconfined compressive strength tests of the test column cores taken out on the spot are carried out at different depths respectively. According to the test results, the distribution map of the compressive strength of each test column along the column depth can be obtained, as shown in Figure 3.

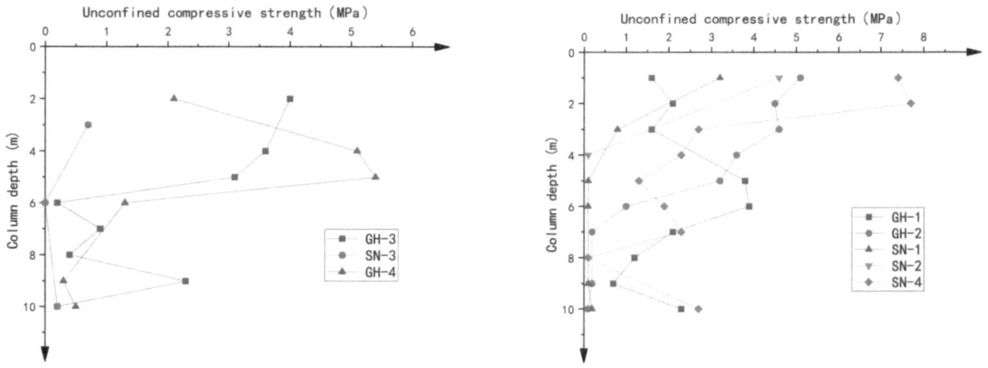

(a) 18-day

(b) 28-day

Figure 3. The distribution map of compressive strength of column sampling.

From Figure 3(a), it can be seen that the maximum strength of SN-3 of cement-soil core after 18 days of age is 0.7 MPa, while that of GH-3 and GH-4 of curing agent column core is 4.0 MPa and 5.4 MPa respectively, which appears at the depth of 2 m and 5 m respectively. That shows that the compressive strength of the curing agent column core is much higher than that of the cement-soil column core. Similarly, it can be seen from Figure 3(b) that the strength of the curing agent column core is significantly higher than that of the cement-soil core after 28 days of age. This shows that the curing agent can play an important role in improving the integrity of the column, and its performance is better than that of cement soil. And with the continuous increase in the amount of cementitious material, the strength of the test column core has been greatly improved. For example, in Figure 3(b), the compressive strength of SN-1 and SN-4 at the same column depth of 1 m is 3.2 MPa and 7.4 MPa respectively, increasing by about 131%.

Through the observation of all the distribution maps, it is found that these column cores have one thing in common: the compressive strength of the test column core shows a trend of gradual decline from top to bottom.

4 DISCUSSION

The above coring test results show that the compressive strength of the test column core shows a trend of gradual decline from top to bottom. The main reasons for this trend are as follows:

First, due to the influence of the groundwater level, the core sample of more than two meters is above the water level, while the core sample of fewer than two meters is below the water level. The strength of the cement soil below the water level rises slowly, so the strength of the upper core sample is greater than that of the lower core sample.

Second, as shown in Figure 4, double-layer straight blades are used to stir in the field. Its mixing capacity is lower than that of multi-layer blades. There are too few mixing times in the construction of the mixing column, so the cement and soil are not evenly mixed. At the same time, as shown in Figure 5, the formation pressure is small above and large below and the blocking ability of double-layer straight blades is not enough. So the unmixed cement slurry will return up in the soil layer, resulting in a lower cement content in the lower layer, which finally results in a lower strength in the lower layer of the column core.

Figure 4. Agitator blade.

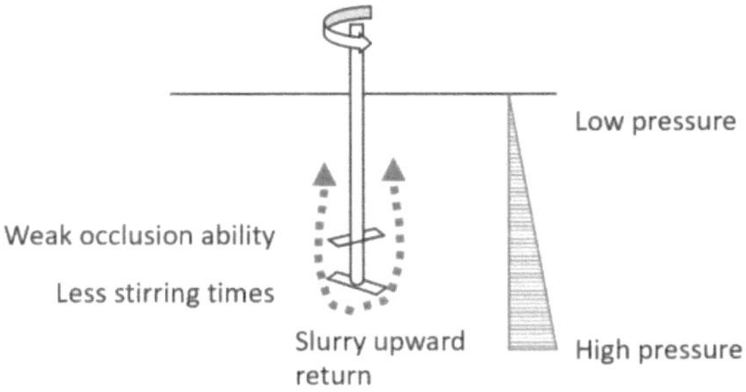

Figure 5. Schematic diagram of slurry returning upward.

5 CONCLUSION AND PROSPECT

In this paper, based on the project of Tianjin Binhai Nanjiang dredger fill land area, the physical and mechanical properties of the soil sampled in the dredger fill site are analyzed. The laboratory cement-soil strength test is carried out by using the site soil. At the same time, the field column test is carried out, and the coring results are analyzed and discussed. The main conclusions are as follows:

(1) Through field sampling, a series of laboratory tests were carried out, and the physical and mechanical properties of site soil in the dredger fill site were obtained. The content of organic matter in the foundation of the dredger fill in Nanjiang is not high, which is 1.5–2.0%. It has little effect on the solidification and strength of cement soil. But the site soil has a high viscosity, low permeability coefficient, and high water content. When adding cement slurry, more stirring times are needed to form uniform cement soil.
(2) Through the laboratory cement-soil strength test with different mix ratios, it can be known that the unconfined compressive strength of 28 days is not less than 2.0 MPa and the strength quality is better under the condition of full mixing.
(3) The strength of the curing agent column core sample is much larger than that of the cement column core sample, and the integrity of the curing agent column core sample is better than that of the cement column core sample. A curing agent plays an important role in improving the integrity of the column body.
(4) The strength of the test column core decreases gradually from top to bottom. First, the core sample of more than two meters is above the water level, and the core sample of fewer than two meters is below the water level. The strength of cement soil rises slowly below the water level. Second, silty clay is not easy to mix evenly, and the slurry returns up, which has a lower amount of cement mixture in the middle and lower section, resulting in low strength or unable to form strength.

In the reinforcement construction of the cement-soil mixing column in the dredger fill site in the future, it is suggested that the mixing equipment of the cement-soil mixing column should be improved to upgrade the mixing uniformity of cement and in-situ soil. How to better improve the construction equipment needs further research.

REFERENCES

Chen F. and Li H. "Study on Construction Technology of Deep Cement-soil Mixing Column in Huanghua Port area", *Chinese Journal of Geotechnical Engineering*, 37, 156–160 (2015).

Huang B., Wang W., and Di G. "Strength Test and Analysis of TRD Cement-soil Mixing Wall in Soft Soil Stratum in Shanghai", *Chinese Journal of Civil Engineering*, 48, 108–112 (2015).

Jiang Y., Han J., and Zheng G. "Numerical Analysis of Consolidation of Soft Soils Fully-Penetrated by Deep-mixed Columns", *KSCE Journal of Civil Engineering*, 17, 96–105 (2013).

Liang Z., Li Z., Liu J., Weng X., and Li W. "Strength Analysis and Experimental Study of Triaxial Cement Soil Mixing Column", *Chinese Journal of Underground Space and Engineering*, 5, 1562–1567 (2009).

Liu S. Y., Du Y. J., Yi Y. L., and Puppala A. J. "Field Investigations on Performance of T-Shaped Deep Mixed Soil Cement Column-Supported Embankments Over Soft Ground", *Journal of Geotechnical and Geoenvironmental Engineering*, 138, 718–727 (2012).

Oliveira P., Pinheiro J., and Correia A. "Numerical Analysis of an Embankment Built on Soft Soil Reinforced with Deep Mixing Columns: Parametric Study", *Computers and Geotechnics*, 38, 566–576 (2011).

Phutthananon C., Jongpradist P., Jongpradist P., Dias D., and Baroth J. "Parametric Analysis and Optimization of T-shaped and Conventional Deep Cement Mixing Column-Supported Embankments", *Computers and Geotechnics*, 122, (2020).

Rao C. and Huang H. "Study on Influencing Factors of Compressive Strength of Soft Soil Cement Soil in Shenzhen", *Journal of Huazhong University of Science and Technology (Urban Science `Edition)*, 26, 99–102 (2009).

Wu L. "Analysis of Factors Influencing the Quality of Concrete Mixing Column in Soft Soil Area of Zhuhai", *Building Technique Development*, 48, 140–141 (2021).

Yuan W., Cai Z., Xie S., Zhang J., and Pan J. "Study on Comprehensive Construction Parameters of Cement-soil Mixing Column Based on Strength Test", *Journal of Hunan University (Natural Science Edition)*, 45, 46–51 (2018).

Zhao C., Zou Y., Zhao C., Xie X., and Zhou Y. "Study on the New Technology of Five-axis Cement-soil Mixing Column Based on Strength Test", *Chinese Journal of Geotechnical Engineering*, 36, 376–381 (2014).

Water Conservancy and Civil Construction – Oke & Ahmad (Eds)
 ISBN: 978-1-032-58614-4

Optimization of rail-sea intermodal train organization scheme based on the new western land-sea corridor

Xin Qi*
School of Traffic and Transportation, Beijing Jiaotong University, Haidian District, Beijing, China

ABSTRACT: The current rail-sea intermodal train on the new western land-sea corridor has problems, such as the poor connection between rail and sea, and varying train frequency. It is urgent to optimize the train operation plan to realize the seamless connection between railway trains and ocean liners to create stable and efficient branded products. Based on comparing the train operation mode, the arrival time of the train, the number of grouping containers, etc. shall be carefully considered. With the shortest total time from the assembly point to the port and the lowest operating cost of the railway section as the goal, the optimization model of the train operation plan using multi-objective mixed integer nonlinear programming is constructed. A case facing the new western land-sea corridor is calculated to verify the validity of the model.

1 INTRODUCTION

China builds a new pattern of opening to the outside world with the construction of "the Belt and Road" as the leader, and the rapid development of economic and trade exchanges with countries along the route has led to the improvement of railroad interconnection and cross-border inter-regional multimodal transport corridors. China State Railway Group Co., Ltd. actively promotes the construction of multimodal transportation, and based on China Railway Express, actively builds the rail-sea intermodal train. The new western land-sea corridor is in the hinterland of the western region of China, which has an important strategic position in the pattern of regional coordinated development and is an essential area for the development of rail-sea intermodal transportation.

However, with the continuous growth of the new western land-sea corridor of rail-sea intermodal transportation, some bottlenecks that restrict its development have gradually come to the fore. There is an urgent need to coordinate the organization of trains from the overall level of the channel with a reasonable plan for the program of train operation.

2 LITERATURE REVIEW

At present, the domestic optimization of the operation scheme of the rail container train is mainly considered from the aspects of rail transportation efficiency, cargo owners' transportation cost, and transportation demand, and the corresponding mathematical model is constructed to minimize cargo transportation cost and optimize transportation time. Yan *et al* [1] optimized the operation scheme of container trains by using the linear programming model for the container assembly time and reorganization time along the route for

*Corresponding Author: 21120880@bjtu.edu.cn

DOI: 10.1201/9781003450818-30

inter-junction station operation. Zhang *et al* [2] established a mixed integer-based two-layer planning model by integrating container tariff and train operation scheme, aiming to achieve the highest profit for railroad transport enterprises as well as the lowest generalized cost for cargo owners. Yin *et al* [3] conducted a mixed integer planning model targeting minimum transportation cost and carbon emission based on the attractive regional division of inland railroad container hubs. In the studies related to container rail-sea intermodal transport, most of the studies were conducted from the perspective of cost, and fewer studies were conducted on how to effectively connect the railroad section with the port. Tang [4] designed a framework for operation schemes of container rail-sea intermodal liners and constructed a liner operation scheme model with the objective function of the shortest time consumption for rail-sea intermodal transportation. Mi *et al* [5] conducted an in-depth study on the train capacity and loading constraint problems and specified a planning model with the objective of the shortest total container truck transportation distance. In order to provide the whole process of the rail-sea intermodal transport service, it is necessary to fully consider the time problem in the process of connecting the rail section and the sea section of rail-sea intermodal transport and shorten the rail-sea connection time, improving the intermodal transport efficiency.

3 MODELING WORK

3.1 *Analysis of operating mode*

At present, the mode of operation of the rail-sea intermodal train can be divided into point-to-point mode and assembly mode. The point-to-point container train, that is, the train organization in the same direction of the goods from the origin of direct transport to a final station, midway without container loading, and unloading or dumping operations. The point-to-point mode mainly has a container "five fixed" and the current operation of some of the China Railway Express. The assembly mode is a supplementary mode of operation based on the point-to-point direct mode of operation, that is, the containers with different destinations are grouped into the same container train at the source and driven to the established assembly point. The assembly point will integrate the cargo flow from various sources according to the destination, and operate container trains with different destinations respectively. Two modes of operation are shown in Figure 1.

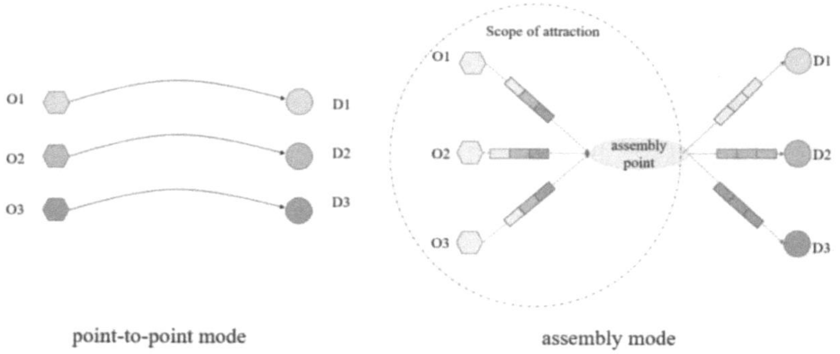

Figure 1. The figure of two modes of operation.

In the point-to-point mode, the routes of trains often overlap, and cannot be an effective use of resources and capacity. At present, the new western land-sea corridor is still in the early stage of development, and the channel coordination capacity is weak the quality of liner operation is not high, and cannot be on schedule, according to the formation of the opening of the situation from time to time. Therefore, because of the current situation of the new western land-sea corridor train operation, the assembly mode is more advantageous.

3.2 *Model establishment*

V denotes velocity, and T_0 denotes the time cycle. S denotes the set of assembly points; P denotes the cost set; T denotes the time set; R denotes the set of zones. j denotes the date in T_0, k denotes the number of trains operating, and f_s denotes the frequency of the train's operation in a time cycle at the assembly point. to^s_{jk} denotes the departure time of the kth train on a jth day at the assembly point; u^s_{jk} denotes the number of grouping containers of the kth train on a jth day at the assembly point. q_s denotes the volume of containers to be shipped to the assembly point, Cap_s denotes assembly point operation capacity in *the*T_0 cycle, and Cap_r denotes the passing capacity of segments during the T_0 cycle. $[c_1, c_2]$ denotes the number of grouping containers, based on 2 TEU per vehicle, and l_s denotes the transportation distance from the assembly point to destination D.

The train from the assembly point to the port process in the railroad section of the time required can be divided into four parts, respectively, the operation time at the assembly point, the train transit time, the train in the port unloading time, and the container storage time in the port.

The operation time in the assembly point is calculated in Equation (1). The train transit time is determined by the distance of the trajectory of the train and the average speed of the train, which is calculated in Equation (2), and the train in the port unloading time is calculated in Equation (3). The relationship between the arrival time of the train at the port and the collection time window is shown in Figure 2. If the train arrives within the collection time window of the day, there will be no need for storage, and can be loaded immediately without storage time; if the train arrives earlier than the beginning of the collection time of the day, the train will need to be stored at the port, generating storage time. The container storage time in the port is calculated in Equations (4)-(6).

$$T_1 = t^s_{1jk} = y_1 \times u^s_{jk} \tag{1}$$

$$T_2 = t^s_{2jk} = \frac{l_s}{V} \tag{2}$$

$$T_3 = t^s_{3jk} = y_2 \times u^s_{jk} \tag{3}$$

$$T_4 = t^s_{4jk} = \max\left[ET_j - td^s_{jk}{}', 0\right] \tag{4}$$

$$td^s_{jk}{}' = td^s_{jk} - Ltd^s_{jk} \div 24\mathrm{J} \times 24 \tag{5}$$

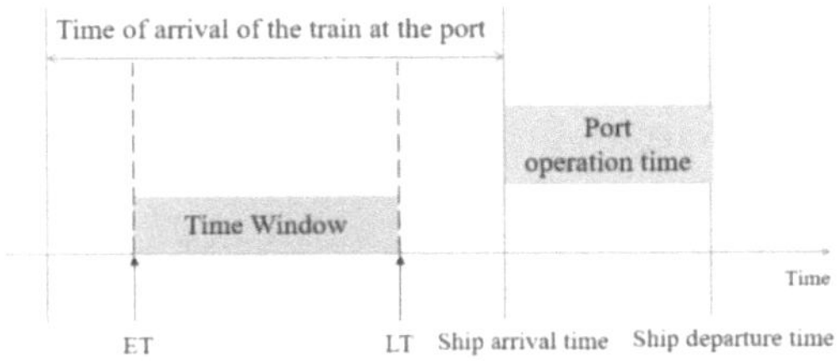

Figure 2. Relationship between the arrival time of the train and cargo collection time window.

$$td_{jk}^{s} = to_{jk}^{s} + t_{1jk}^{s} + t_{2jk}^{s} \tag{6}$$

where,

T_1: Assembly point operation time. Unit: hour;
T_2: Transit time. Unit: hour;
T_3: Port unloading operation time. Unit: hour;
T_4: Stacking time. Unit: hour;
td_{jk}^{s}: The arrival time of the kth train on a jth day at the s assembly point. Unit: hour;
$td_{jk}^{s}{}'$: The standardized arrival time of the kth train on a jth day at the assembly point, indicates the moment of arrival on a specific day in the time cycle. Unit: hour;
y_1: assembly point operation time factor, i.e., the average operating time per TEU for loading operations at the marshaling point. Unit: hour/TEU;
y_2: Port unloading operation time factor, i.e., the average operating time per TEU for unloading operations at the port. Unit: hour/TEU;
$[ET_j, LT_j]$: Port assembly time window, is the start of the collection time of the day and the cut-off time of the day. Unit: hour.

Under the "Measures for the settlement of incoming railroad cargo transportation (for trial implementation)", the operating costs of container rail-sea intermodal transport trains can be divided into the following five parts: locomotive traction fees, line usage fees, vehicle service fees, arrival service fees, and comprehensive service fees. The specific calculation formula is calculated in Equations (7)-(11).

$$P_1 = \mathrm{h}_1 \times w \times \frac{u_{jk}^{s}}{2} \times l_s \tag{7}$$

$$P_2 = h_2 \times l_s \times \frac{u_{jk}^{s}}{2} \tag{8}$$

$$P_3 = h_3 \times l_s \times \frac{u_{jk}^{s}}{2} \tag{9}$$

$$P_4 = h_4 \times u_{jk}^{s} \tag{10}$$

$$P_5 = h_5 \times u_{jk}^{s} \times b_1 + b_2 \times l_s \tag{11}$$

where,

P_1: Locomotive traction fee. Unit: Yuan;
P_2: Line usage fee. Unit: Yuan;
P_3: Vehicle service fee. Unit: Yuan;
P_4: Arrival service fee. Unit: Yuan;
P_5: Comprehensive service fee. Unit: Yuan;
h_1: Locomotive traction fee unit price. Unit: Yuan/ (10,000 gvt-km);
h_2: Line usage fee unit price. Unit: Yuan/ (vehicle-km);
h_3: Vehicle service fee unit price. Unit: Yuan/ (vehicle-km);
h_4: Arrival service fee unit price. Unit: Yuan/ TEU;
h_5: Comprehensive service fee unit price;
w: Average gross vehicle weight, i.e., the sum of the weight of the vehicle and the actual load. Unit: Ton/vehicle;
b_1: Fixed costs of operating rail-sea intermodal trains at the assembly point. Unit: Yuan/ TEU;

b_2: Variable costs of operating rail-sea intermodal trains at the assembly point. Unit: Yuan/(TEU*km).

In summary, the overall objective function and constraints of the model are expressed as:

$$Min\ Z_1 = T_1 + T_2 + T_3 + T_4 = \sum_s \sum_j \sum_k (t^s_{1jk} + t^s_{2jk} + t^s_{3jk} + t^s_{4jk}) \tag{12}$$

$$\begin{aligned} Min\ Z_2 &= P_1 + P_2 + P_3 + P_4 + P_5 \\ &= \sum_s \sum_j \sum_k \left(\begin{array}{c} \mathrm{h}_1 \times w \times u^s_{jk} \times l_s + h_2 \times l_s \times \frac{u^s_{jk}}{2} + h_3 \times l_s \times \frac{u^s_{jk}}{2} \\ + h_4 \times u^s_{jk} + h_5 \times u^s_{jk} \times b_1 + b_2 \times l_s \end{array} \right) \end{aligned} \tag{13}$$

$$s.t. \begin{cases} \sum_s \sum_j \sum_k u^s_{jk} \times x_r \le Cap_r \quad \forall r \in R \\ q_s \le Cap_s \quad s \in S \\ \sum_j \sum_k u^s_{jk} = q_s \quad \forall s \in S \\ c_1 \le u^s_{jk} \le c_2 \quad \forall s \in S \quad j = 1, 2 \ldots \frac{T_0}{24} \quad k = 1, 2, \ldots, f_s \\ td^s_{jk} \le LT_j \quad \forall s \in S \quad j = 1, 2 \ldots \frac{T_0}{24} \quad k = 1, 2, \ldots, f_s \\ u^s_{jk} \in N \quad \forall s \in S \quad j = 1, 2 \ldots \frac{T_0}{24} \quad k = 1, 2, \ldots, f_s \\ f_s \in N \quad \forall s \in S \\ to^s_{jk} \in T \quad \forall s \in S \quad j = 1, 2 \ldots \frac{T_0}{24} \quad k = 1, 2, \ldots, f_s \\ x_r \in \{0, 1\} \quad \forall r \in R \end{cases}$$

4 NUMERICAL EXPERIMENTS

Among the 15 alternative inland city nodes, the four cities are selected as assembly points in this paper, namely Chongqing, Lanzhou, Chengdu, and Guiyang. Combining the development targets in the Master Plan of the New Western Land-sea Corridor and the proportion of total exports to ASEAN countries in 2018 from the provinces where each city is located, the weekly container volume of each city in 2025 is discounted.

The relevant parameters concerning transportation costs are obtained according to the "Measures for the settlement of incoming railroad cargo transportation (for trial implementation)". The parameters related to container freight rates involved are determined according to the "Railway Freight Rate Table" and the actual data of the research. The decision period of this model is set to 1 week, i.e., 168 h, assuming that Qinzhou port realizes a daily frequency of ships with Hong Kong and Singapore, and assuming that the daily gathering time of the port in the decision period is 12:00 to 14:00. The rest of the parameters are derived from the research data.

In order to improve the efficiency of the solution, this paper converts the dual-objective model into a single-objective model based on the principal objective function method for

Table 1. Calculation results.

Assembly point	Lanzhou	Chengdu	Chongqing	Guiyang
Origin-destination	Lanzhou-Qinzhou Port Urumqi-Qinzhou Port Yinchuan-Qinzhou Port Xining-Qinzhou Port	Chengdu-Qinzhou Port Xi'an-Qinzhou Port	Chongqing-Qinzhou Port	Guiyang-Qinzhou Port Yibin-Qinzhou Port Zunyi-Qinzhou Port
Frequency (train/week)	14	28	28	56
The moment of departure	11:00 23:00	05:00 11:00 17:00 23:00	02:00 08:00 14:00 20:00	01:00 04:00 07:00 10:00 13:00 16:00 19:00 22:00
Number of containers (TEU)	72	76	58	58
The time required for the whole process (hour)	49	43	23.2375	20
Total time cost (million yuan)	9	13	10	18
Total operating cost (million yuan)	1,102	1,458	783	1,180
Total cost (million yuan)	1,111	1,471	793	1,198

solving. It is shown in Equation (14). The calculation results are shown in Table 1.

$$MinZ = \alpha MinZ_1 + MinZ_2 \tag{14}$$

where,

α: the time cost coefficient by which time is converted into time cost.

At the early stage of the development of the new western land-sea corridor for the rail-sea intermodal transport trains, the mode of operation based on the assembly point helps to organize scattered cargo sources, improving the support of cargo sources and making full use of the railroad capacity. We take Lanzhou as an example, up to the end of the research, Lanzhou does not have the rail-sea intermodal transport scheduled trains, and the start is applied each time according to the actual situation of cargo sources, and the train exists in route to make up the axle situation. In this case, not only the collection time is long, but more time is also consumed in transit, which isn't conducive to regular operation. If in the mode of assembly, Lanzhou as an assembly point can assemble cargo from Urumqi, Xining, and Yinchuan, a more significant amount of cargo can be mentioned from the Lanzhou assembly point to increase the frequency of the train, which is conducive to the formation of the regular transport of rail-sea intermodal transport, curing Lanzhou to Qinzhou port rail-sea intermodal transport lines and forming scale effect.

Under the assembly mode, the operation path of the liner revolves around the main assembly points, and the railroad lines between the assembly points generally have a higher passing capacity and more vital line capacity compared with other lines. The transport route chosen in this scheme mainly uses the current east and middle lines of the new western land-sea corridor, which are the two sub-channels with better line conditions. Compared with the current status of 50 TEU for a train, the average container capacity of the optimized line is 66 TEU, and the average capacity scale of the line is increased by 32%. In addition, the current operation of all rail-sea intermodal transport trains does not consider the matching problem between the rail section schedule and the sea section schedule. The model focuses on optimizing the matching problem between the rail schedule and the shipping schedule. By analyzing the research data and the model results, the average waiting time of the train is reduced by 44%. In terms of the total cost, it is effectively reduced due to the shortening of the transit time of the train and the increase of the capacity of the line, as there is no need to make up axles along the way and disperse into other direct trains. The total cost of the four lines is reduced by nearly 870,000 yuan.

5 CONCLUSIONS

In this paper, a multi-objective mixed integer nonlinear programming based on the optimization model for the operation scheme with the shortest total time from the assembly point to the port and the lowest operating cost of the railroad section is established. The validity of the model is verified through an example analysis oriented to the new western land-sea corridor. The decision on the departure time of the train included in the scheme helps to solve the problem of the poor connection between the railroad and sea from the overall perspective of the new western land-sea corridor, which can provide a reference for the actual cargo flow organization to a certain extent and provide decision support for the development of the western land-sea new corridor of rail-sea intermodal transport. In the future, based on this paper, the overall optimization of the departing and returning trains can be carried out, and the analysis of the influence of the shipping period on the schedule of the maritime section can be added to develop a more reasonable overall plan for the operation of the upstream and downstream trains.

REFERENCES

[1] Mi Weijian, Qin Zhao, Zhang Xiaohua, and Tian Ting. The Study on Stowage Problem of Container Train Loading with the Situation of Sea—rail Combined Transportation [J]. *Logistics Engineering and Management*, 2016, 38 (03): 113–116.

[2] Tang Yalong. *The Operation Plan Research of Container Sea—Rail Intermodal Transportation Regular Train [D]*. Beijing Jiaotong University, 2013.

[3] Yan Haifeng, Peng Qiyuan, and Yin Yong. Research on the Optimization Model of the Block Container Train Formation Plan [J]. *Journal of the China Railway Society*, 2003 (05): 14–18.

[4] Yin Chuanzhong, Ke Yuanding, Yan Yang, Lu Yu, and Xu Xingfang. Operation Plan of China Railway Express at Inland Railway Container Center Station [J]. *International Journal of Transportation Science and Technology*, 2020, 9 (3).

[5] Zhang Xiaoqiang, Liu Dan, Chen Bing, and Zhang Jin. Dynamic Pricing and Operation Planning of Container Train in Competitive Environment [J]. *Journal of the China Railway Society*, 2017,39 (02):17–23.

Water Conservancy and Civil Construction – Oke & Ahmad (Eds)
© 2024 The Author(s), ISBN: 978-1-032-58614-4

Saturated axisymmetric multilayered elastic system excess pore water pressure in asphalt pavement

Yanyang Li*
Heilongjiang Bayi Agricultural University, Daqing, China
Northeast Petroleum University, Daqing, China

Bin Zhang, Wei Guo & Guoliang Xie
Heilongjiang Bayi Agricultural University, Daqing, China

ABSTRACT: Water damage occurs in most asphalt pavements at the early stage of service. The coupling effect of pore-water and moving load has been focused on in the investigation. This study aimed to evaluate the explicit solution of Biot's consolidation equations through Laplace and Hankel transforms to derive stress, displacement, and excess pore water pressure of the saturated multi-layer elastic system. The accuracy and validity of the theory were verified by a computational example.

1 INTRODUCTION

Under the action of a moving load, pore water can not be discharged in time. Starting with growing of excess pore pressure and appearing of the flow of pore water, which accelerates the stripping of asphalt membrane and the loosening of aggregate, the bearing capacity of asphalt pavement is reduced, the water damage (Sha 2008) of most asphalt pavements arises at the early stage of use, simultaneously the traffic safety is greatly affected, and the service life of the pavement is tremendously reduced.

Therefore, a lot of research has been carried out by domestic and foreign scholars. It is generally believed that pore water is one of the main causes of asphalt pavement damage. Based on the Biot consolidation theory, many scholars assumed that asphalt concrete and subgrade as elastomers in the theoretical analysis of saturated asphalt pavement. The finite element method was used by most scholars (Dong *et al.* 2009; Eyad Mased & Chen 2007; Kettil 2005; Li *et al* 2003) in the theoretical analysis of saturated asphalt pavement. For example, Zhong (Zhong *et al.* 2006) *et al.* established the overall stiffness matrix according to the stiffness matrix of a single story to analyze the excess pore water pressure. Peng (Peng *et al.* 2004) *et al.* established the relation between the state vector of any depth and the state vector of $z = 0$ by using the Transfer-matrix method, hence the state vector of any depth was found. The above researches have certain theoretical value and practical significance. However, the explicit solutions of mechanical parameters such as stress and displacement are not given in most of the results, so it is very difficult to apply them in engineering.

In this research, we are interested in this work. Firstly, Hankel transform and Laplace transform (Guo & Ma 2001) were applied to solve the Biot consolidation equation and flow continuity equation. Explicit analytical expressions were obtained such as stress, strain, and displacement of multilayered elastic systems under axisymmetric conditions. The coefficient

*Corresponding Author: 191237636@qq.com

 DOI: 10.1201/9781003450818-31

expression was obtained based on the boundary condition and the contact condition between layers. Finally, a calculation example was itemized precisely, and the theoretical results were verified to be correct and efficient.

2 SOLUTION OF BIOT CONSOLIDATION AND THE FLOW CONTINUITY EQUATION

Regardless of gravity, the Biot consolidation equation and the flow continuity equation under Laplace transformation were defined as:

$$\frac{\partial \bar{\sigma}_r'}{\partial r} + \frac{\partial \bar{\tau}_{rz}}{\partial z} + \frac{\bar{\sigma}_r' - \bar{\sigma}_\theta'}{r} - \frac{\partial \bar{\sigma}}{\partial r} = 0 \tag{1}$$

$$\frac{\partial \bar{\tau}_{rz}}{\partial r} + \frac{\partial \bar{\sigma}_z'}{\partial z} + \frac{\bar{\tau}_{rz}}{r} - \frac{\partial \bar{\sigma}}{\partial z} = 0 \tag{2}$$

$$k'\nabla^2\bar{\sigma} = s\bar{e} - e(0) = s\bar{e} \tag{3}$$

Where $\bar{\sigma}_r'$ $\bar{\sigma}_\theta'$, $\bar{\sigma}_z'$ and $\bar{\tau}_{rz}$ are the radial, tangential, vertical effective stresses and shear stresses under the Laplace transformation; $\bar{\sigma}$ is the excess pore water pressure under the Laplace transform; $\nabla^2 = -\frac{\partial^2}{\partial r^2} + \frac{1}{r}\frac{\partial}{\partial r} + \frac{\partial^2}{\partial z^2} k' = \frac{k}{\gamma_w}$ and k is the coefficient of permeability, γ_w is the weight of water. In this paper, assuming $e(0) = 0$, the volume strain is equal to zero at the initial time.

The physical equations are substituted into the equilibrium Equations (1) and (2):

$$\frac{1}{1-2\mu}\frac{\partial \bar{e}}{\partial r} + \nabla^2\bar{u} - \frac{u}{r^2} - \frac{1}{G}\frac{\partial \bar{\sigma}}{\partial r} = 0 \tag{4}$$

$$\frac{1}{1-2\mu}\frac{\partial \bar{e}}{\partial z} + \nabla^2\bar{w} - \frac{1}{G}\frac{\partial \bar{\sigma}}{\partial z} = 0 \tag{5}$$

Where $\bar{u}$ and $\bar{w}$ are respectively radial and vertical displacements in Laplace space.

In Equation (4), it is multiplied by a differential operator $\frac{\partial}{\partial r} + \frac{1}{r}$. In Formula (5), it is taken a partial derivative from z, then added together:

$$\nabla^2\bar{\sigma} = M\nabla^2\bar{e} \tag{6}$$

In the formula: $M = \frac{2G(1-\mu)}{1-2\mu}$

By substituting Formula (6) into the seepage continuity Equation (3), the following results are obtained:

$$c\nabla^2\bar{e} = s\bar{e} \tag{7}$$

In the Formula: $c = k'M$

The first-order Hankel transformation of (4) and 0-order Hankel transform of (5) $\sim$ (7):

$$-\frac{\xi}{1-2\mu}\hat{\bar{e}}_0 + \frac{d^2\hat{\bar{u}}_1}{dz^2} - \xi^2\hat{\bar{u}}_1 + \frac{\xi}{G}\hat{\bar{\sigma}}_0 = 0 \tag{8}$$

$$\frac{1}{1-2\mu}\frac{d\hat{\bar{e}}_0}{dz} + \frac{d^2\hat{\bar{w}}_0}{dz^2} - \xi^2\hat{\bar{w}}_0 - \frac{1}{G}\frac{d\hat{\bar{\sigma}}_0}{dz} = 0 \tag{9}$$

$$\frac{d^2\widehat{\sigma}_0}{dz^2} - \xi^2\widehat{\sigma}_0 = \frac{\gamma_w s}{k}\widehat{e}_0 \tag{10}$$

$$\frac{d^2\widehat{e}_0}{dz^2} - q^2\widehat{e}_0 = 0 \tag{11}$$

Where 0 and 1 in subscript denote 0 and 1-order Hankel transformations, respectively $q^2 = \xi^2 + \frac{s}{c}$.

By solving the above four equations, the expressions can be obtained such as volume strain, excess pore water pressure, and displacement components in Laplace-Hankel space, as shown in Formulas (12) ~ (15):

$$\widehat{e}_0(\xi, z, s) = A_1 e^{qz} + B_1 e^{-qz} \tag{12}$$

$$\widehat{u}_1(\xi, z, s) = \frac{1}{\xi}\left(A_3 e^{\xi z} + B_3 e^{-\xi z}\right) - \frac{c\xi}{s}\left(A_1 e^{qz} + B_1 e^{-qz}\right) - z\left(A_2 e^{\xi z} - B_2 e^{-\xi z}\right) \tag{13}$$

$$\widehat{w}_0(\xi, z, s) = \frac{1}{\xi}\left(A_4 e^{\xi z} + B_4 e^{-\xi z}\right) + \frac{cq}{s}\left(A_1 e^{qz} - B_1 e^{-qz}\right) + z\left(A_2 e^{\xi z} + B_2 e^{-\xi z}\right) \tag{14}$$

$$\widehat{\sigma}_0(\xi, z, s) = 2G\left(A_2 e^{\xi z} + B_2 e^{-\xi z}\right) + M\left(A_1 e^{qz} + B_1 e^{-qz}\right) \tag{15}$$

The volume strain is $e = \frac{\partial u}{\partial r} + \frac{u}{r} + \frac{\partial w}{\partial z}$, and Laplace transform and 0-order Hankel transform are applied to the equation. By applying the properties of Hankel transformation, we obtain $\widehat{e}_0 = \xi\widehat{u}_1 + \frac{\partial\widehat{w}_0}{\partial z}$, and the expressions of volume strain, radial displacement, and vertical displacement are introduced into the equation, the following results can be obtained:

$$A_2 + A_3 + A_4 = 0 \tag{16}$$

$$B_2 + B_3 - B_4 = 0 \tag{17}$$

Such expressions for stress and strain at any layer showed as Formulas (18) ~ (23), and there are only six unknown coefficients $A_1 \sim A_3$ and $B_1 \sim B_3$.

$$\widehat{u}_1(\xi, z, s) = -\frac{c\xi}{s}\left(A_1 e^{qz} + B_1 e^{-qz}\right) - z\left(A_2 e^{\xi z} - B_2 e^{-\xi z}\right) + \frac{1}{\xi}\left(A_3 e^{\xi z} + B_3 e^{-\xi z}\right) \tag{18}$$

$$\widehat{w}_0(\xi, z, s) = \frac{cq}{s}\left(A_1 e^{qz} - B_1 e^{-qz}\right) + \left[\left(z - \frac{1}{\xi}\right)A_2 e^{\xi z} + \left(z + \frac{1}{\xi}\right)B_2 e^{-\xi z}\right] + \frac{1}{\xi}\left(-A_3 e^{\xi z} + B_3 e^{-\xi z}\right) \tag{19}$$

$$\widehat{\sigma}'_{z0}(\xi, z, s) = 2G\left[\left(\frac{\mu}{1-2\mu} + \frac{cq^2}{s}\right)\left(A_1 e^{qz} + B_1 e^{-qz}\right) + z\xi\left(A_2 e^{\xi z} - B_2 e^{-\xi z}\right) - \left(A_3 e^{\xi z} + B_3 e^{-\xi z}\right)\right] \tag{20}$$

$$\widehat{\tau}_{rz1}(\xi, z, s) = G\left[-\frac{2cq\xi}{s}\left(A_1 e^{qz} - B_1 e^{-qz}\right) - 2z\xi\left(A_2 e^{\xi z} + B_2 e^{-\xi z}\right) + 2\left(A_3 e^{\xi z} - B_3 e^{-\xi z}\right)\right] \tag{21}$$

$$\widehat{\sigma}_0(\xi, z, s) = M\left(A_1 e^{qz} + B_1 e^{-qz}\right) + 2\bar{G}\left(A_2 e^{\xi z} + B_2 e^{-\xi z}\right) \tag{22}$$

$$\widehat{\bar{v}}_0(\xi, z, s) = -k'\left[Mq(A_1 e^{qz} - B_1 e^{-qz}) + 2G\xi\left(A_2 e^{\xi z} - B_2 e^{-\xi z}\right)\right] \quad (23)$$

Where v is I layer porous water flow velocity (Gong 1992) of z direction; 0 and 1 in subscript denote 0 and 1-order Hankel transform respectively; ξ and s are the integral transform of Hankel transform and Laplace transform respectively; $A_1 \sim A_3$ and $B_1 \sim B_3$ are all functions of ξ and s, and its value is determined by the boundary condition.

3 BOUNDARY CONDITIONS AND INTERLAYER CONTACT CONDITIONS

We assume drainage conditions on the pavement surface, and the excess pore water pressure is equal to zero, therefore a uniformly distributed circular load is applied:

$$p(r,t) = \begin{cases} pH(t) & r \le \delta \\ 0 & r > \delta \end{cases} \quad (24)$$

Where $H\ (t)$ is the unit ladder function and $H\ (t) = 1\ (t \geq 0)$ or $0\ (t < 0)$. Then based on Laplace transform and Hankel transform, the boundary condition and the contact condition between layers can be expressed as:

$$\begin{cases} \widehat{\bar{\sigma}}'_{z0}(\xi, 0, s) = -\widehat{\bar{p}}(\xi, s) = \dfrac{-p_0 \delta J_1(\xi\delta)}{s\xi} \\ \widehat{\bar{\tau}}_{rz1}(\xi, 0, s) = 0 \\ \widehat{\bar{\sigma}}_0(\xi, 0, s) = 0 \end{cases} \quad (25)$$

and

$$\begin{cases} \widehat{\bar{\sigma}}'_{zi+10}(\xi, z, s)\big|_{z=H_i} = \widehat{\bar{\sigma}}'_{zi0}(\xi, z, s)\big|_{z=H_i} \\ \widehat{\bar{\tau}}_{rzi+11}(\xi, z, s)\big|_{z=H_i} = \widehat{\bar{\tau}}_{rzi1}(\xi, z, s)\big|_{z=H_i} \\ \widehat{\bar{u}}_{i+11}(\xi, z, s)\big|_{z=H_i} = \widehat{\bar{u}}_{i1}(\xi, z, s)\big|_{z=H_i} \\ \widehat{\bar{w}}_{i+10}(\xi, z, s)\big|_{z=H_i} = \widehat{\bar{w}}_{i0}(\xi, z, s)\big|_{z=H_i} \\ \widehat{\bar{\sigma}}_{i+10}(\xi, z, s)\big|_{z=H_i} = \widehat{\bar{\sigma}}_{i0}(\xi, z, s)\big|_{z=H_i} \\ \widehat{\bar{v}}_{i+10}(\xi, z, s)\big|_{z=H_i} = \widehat{\bar{v}}_{i0}(\xi, z, s)\big|_{z=H_i} \end{cases} \quad (26)$$

In the formula: the first number in the lower corner mark represents the number of layers; the second number represents the order of Hankel transformation; H_i is the depth of the contact surface between layer $i+1$ and layer i.

4 THE SOLUTION OF THE COEFFICIENT OF ANY LAYER

We should put Formulas (12) $\sim$ (17) into Formula (19) and write it in the form of a matrix:

$$[T_{i+1}(H_i)]\{C_{i+1}\} = [T_i(H_i)]\{C_i\} \quad (27)$$

In the formula:

$$\{C_{i+1}\} = \{\, A_{i+11} \quad A_{i+12} \quad A_{i+13} \quad B_{i+11} \quad B_{i+12} \quad B_{i+13} \,\}^T \quad (28)$$

$$\{C_i\} = \{\, A_{i1} \quad A_{i2} \quad A_{i3} \quad B_{i1} \quad B_{i2} \quad B_{i3} \,\}^T \quad (29)$$

$$[T_{i+1}(H_i)] =$$
$$\begin{bmatrix} -\frac{c_{i+1}\xi}{s}e^{q_{i+1}H_i} & -H_ie^{\xi H_i} & \frac{1}{\xi}e^{\xi H_i} & -\frac{c_{i+1}\xi}{s}e^{-q_{i+1}H_i} & H_ie^{-\xi H_i} & \frac{1}{\xi}e^{-\xi H_i} \\ \frac{c_{i+1}q_{i+1}}{s}e^{q_{i+1}H_i} & \left(H_i-\frac{1}{\xi}\right)e^{\xi H_i} & -\frac{1}{\xi}e^{\xi H_i} & -\frac{c_{i+1}q_{i+1}}{s}e^{-q_{i+1}H_i} & \left(H_i+\frac{1}{\xi}\right)e^{-\xi H_i} & \frac{1}{\xi}e^{-\xi H_i} \\ 2G_{i+1}\left(\frac{\mu_{i+1}}{1-2\mu_{i+1}}+\frac{c_{i+1}{q_{i+1}}^2}{s}\right)e^{q_{i+1}H_i} & 2G_{i+1}H_i\xi e^{\xi H_i} & -2G_{i+1}e^{\xi H_i} & 2G_{i+1}\left(\frac{\mu_{i+1}}{1-2\mu_{i+1}}+\frac{c_{i+1}{q_{i+1}}^2}{s}\right)e^{-q_{i+1}H_i} & -2G_{i+1}H_i\xi e^{-\xi H_i} & -2G_{i+1}e^{-\xi H_i} \\ -G_{i+1}\frac{2c_{i+1}q_{i+1}\xi}{s}e^{q_{i+1}H_i} & -2G_{i+1}H_i\xi e^{\xi H_i} & 2G_{i+1}e^{\xi H_i} & G_{i+1}\frac{2c_{i+1}q_{i+1}\xi}{s}e^{-q_{i+1}H_i} & -2G_{i+1}H_i\xi e^{-\xi H_i} & -2G_{i+1}e^{-\xi H_i} \\ M_{i+1}e^{q_{i+1}H_i} & 2G_{i+1}e^{\xi H_i} & 0 & M_{i+1}e^{-q_{i+1}H_i} & 2G_{i+1}e^{-\xi H_i} & 0 \\ -k'_{i+1}M_{i+1}q_{i+1}e^{q_{i+1}H_i} & -2k'_{i+1}G_{i+1}e^{\xi H_i}\xi & 0 & k'_{i+1}M_{i+1}q_{i+1}e^{-q_{i+1}H_i} & 2k'_{i+1}G_{i+1}\xi e^{-\xi H_i} & 0 \end{bmatrix} \tag{30}$$

$$[T_i(H_i)] = \begin{bmatrix} -\frac{c_i\xi}{s}e^{q_iH_i} & -H_ie^{\xi H_i} & \frac{1}{\xi}e^{\xi H_i} & -\frac{c_i\xi}{s}e^{-q_iH_i} & H_ie^{-\xi H_i} & \frac{1}{\xi}e^{-\xi H_i} \\ \frac{c_iq_i}{s}e^{q_iH_i} & \left(H_i-\frac{1}{\xi}\right)e^{\xi H_i} & -\frac{1}{\xi}e^{\xi H_i} & -\frac{c_iq_i}{s}e^{-q_iH_i} & \left(H_i+\frac{1}{\xi}\right)e^{-\xi H_i} & \frac{1}{\xi}e^{-\xi H_i} \\ 2G_i\left(\frac{\mu_i}{1-2\mu_i}+\frac{c_i{q_i}^2}{s}\right)e^{q_iH_i} & 2G_iH_i\xi e^{\xi H_i} & -2G_ie^{\xi H_i} & 2G_i\left(\frac{\mu_i}{1-2\mu_i}+\frac{c_i{q_i}^2}{s}\right)e^{-q_iH_i} & -2G_iH_i\xi e^{-\xi H_i} & -2G_ie^{-\xi H_i} \\ -G_i\frac{2c_iq_i\xi}{s}e^{q_iH_i} & -2G_iH_i\xi e^{\xi H_i} & 2G_ie^{\xi H_i} & G_i\frac{2c_iq_i\xi}{s}e^{-q_iH_i} & -2G_iH_i\xi e^{-\xi H_i} & -2G_ie^{-\xi H_i} \\ M_ie^{q_iH_i} & 2G_ie^{\xi H_i} & 0 & M_ie^{-q_iH_i} & 2G_ie^{-\xi H_i} & 0 \\ -k'_iM_iq_ie^{q_iH_i} & -2k'_iG_ie^{\xi H_i}\xi & 0 & k'_iM_iq_ie^{-q_iH_i} & 2k'_iG_i\xi e^{-\xi H_i} & 0 \end{bmatrix} \tag{31}$$

The coefficient of the $i + 1$ layer is obtained by the above formula:

$$\{C_{i+1}\} = [T_{i+1}(H_i)]^{-1}[T_i(H_i)]\{C_i\} \tag{32}$$

From the above formula, the coefficient of the lower layer is determined by the coefficient of the upper layer, namely:

$$\{C_{i+1}\} = [T_{i+1}(H_i)]^{-1}[T_i(H_i)]\{C_i\} \tag{33}$$

$$\{C_i\} = [T_i(H_{i-1})]^{-1}[T_{i-1}(H_{i-1})]\{C_{i-1}\} \tag{34}$$

$$\{C_{i-1}\} = [T_i(H_{i-2})]^{-1}[T_{i-2}(H_{i-2})]\{C_{i-2}\} \tag{35}$$

$$\cdots$$

$$\{C_2\} = [T_2(H_1)]^{-1}[T_1(H_1)]\{C_1\} \tag{36}$$

Then the recursive formula for $i + 1$ the layer coefficient of a multilayered viscoelastic system is obtained:

$$\{C_{i+1}\} = [T_{i+1}(H_i)]^{-1}[T_i(H_i)][T_i(H_{i-1})]^{-1}[T_{i-1}(H_{i-1})]\cdots\cdots[T_2(H_1)]^{-1}[T_1(H_1)]\{C_1\} \tag{37}$$

Through the above formulas, the coefficients of any layer can be expressed by the coefficients of the first layer, then the coefficients of other arbitrary layers can be obtained to require the coefficients of the first layer.

$$\{C_{i+1}\} = [D_{i+1}]\{C_1\} \tag{38}$$

In the formula:

$$[D_{i+1}] = [T_{i+1}(H_i)]^{-1}[T_i(H_i)][T_i(H_{i-1})]^{-1}[T_{i-1}(H_{i-1})]\cdots\cdots[T_2(H_1)]^{-1}[T_1(H_1)] \tag{39}$$

5 THE SOLUTION OF LAYER 1 COEFFICIENT

Then the 6 systems of layer 1 are solved by boundary conditions. By the boundary condition of Formulas (20) $\sim$ (22), (40) $\sim$ (42) is obtained:

$$2G_1\left[\left(\frac{\mu_1}{1-2\mu_1}+\frac{c_1q_1^2}{s}\right)(A_{11}+B_{11})-(A_{13}+B_{13})\right]=-\widehat{p}(\xi,s) \tag{40}$$

$$G_1\left[-\frac{2c_1q_1\xi}{s}(A_{11}-B_{11})+2(A_{13}-B_{13})\right]=0 \tag{41}$$

$$M_1(A_{11}+B_{11})+2G_1(A_{12}+B_{12})=0 \tag{42}$$

In the n layer, when $z\to\infty$, all the stress and displacement components tend to be zero at that time, i. e:

$$\lim_{z\to\infty}\langle\sigma_r',\sigma_\theta',\sigma_z',\tau_{zr},\sigma,u,w\rangle\to 0 \tag{43}$$

In Formulas (18) $\sim$ (23), e^{qz} and $e^{\xi z}$ are contained which was inconsistent with the boundary condition, so the coefficient A_{n1} A_{n2}, and A_{n2} were all equal to zero, that is to say:

$$\begin{Bmatrix}A_{n1}\\A_{n2}\\A_{n3}\end{Bmatrix}=\begin{Bmatrix}0\\0\\0\end{Bmatrix} \tag{44}$$

According to Formula (37), the coefficient n is expressed by the first layer as:

$$\{C_n\}=[D_n]\{C_1\} \tag{45}$$

the upper formula is expanded, according to Formula (26):

$$\begin{aligned}D_{n11}A_{11}+D_{n12}A_{12}+D_{n13}A_{13}+D_{n14}B_{11}+D_{n15}B_{12}+D_{n16}B_{13}=0\\D_{n21}A_{11}+D_{n22}A_{12}+D_{n23}A_{13}+D_{n24}B_{11}+D_{n25}B_{12}+D_{n26}B_{13}=0\\D_{n31}A_{11}+D_{n32}A_{12}+D_{n33}A_{13}+D_{n34}B_{11}+D_{n35}B_{12}+D_{n36}B_{13}=0\end{aligned} \tag{46}$$

Where D_{nij} represents i the row and j column elements of a matrix $[D_n]$.

The system of linear equations of the first-layer coefficients is established using the simultaneous Equations (40), (41), (42), and (46), and the form of the matrix was as follows:

$$[K]\{C_1\}=\{P\} \tag{47}$$

In the formula:

$$\{C_1\}=\{A_{11}\quad A_{12}\quad A_{13}\quad B_{11}\quad B_{12}\quad B_{13}\}^T \tag{48}$$

$$\{P\}=\left\{\frac{-\widehat{p}\,(\xi,s)}{2G_1}\quad 0\;0\;0\;0\;0\right\}^T \tag{49}$$

$$[K]=\begin{bmatrix}\frac{\mu_1}{1-2\mu_1}+\frac{c_1q_1^2}{s} & 0 & -1 & \frac{\mu_1}{1-2\mu_1}+\frac{c_1q_1^2}{s} & 0 & -1\\ -\frac{2c_1q_1\xi}{s} & 0 & 2 & \frac{2c_1q_1\xi}{s} & 0 & -2\\ M_1 & 2G_1 & 0 & M_1 & 2G_1 & 0\\ D_{n11} & D_{n12} & D_{n13} & D_{n14} & D_{n15} & D_{n16}\\ D_{n21} & D_{n22} & D_{n23} & D_{n24} & D_{n25} & D_{n26}\\ D_{n31} & D_{n32} & D_{n33} & D_{n34} & D_{n35} & D_{n36}\end{bmatrix} \tag{50}$$

According to Cramer's rule, the solution of the system of Equation (47) is obtained:

$$C_{1i} = \frac{|K_i|}{|K|}, \ i = 1, 2, \cdots, 6 \tag{51}$$

That is:

$$A_{11} = \frac{|K_1|}{|K|}, \ A_{12} = \frac{|K_2|}{|K|}, \ A_{13} = \frac{|K_3|}{|K|}, \ B_{11} = \frac{|K_4|}{|K|}, \ B_{12} = \frac{|K_5|}{|K|}, \ B_{13} = \frac{|K_6|}{|K|}.$$

Where C_{1i} i is an element of the vector $\{C_1\}$, $|K|$ is the determinant of a matrix $[K]$ of order 6, and $|K_i|$ is the determinant of the i column of the Matrix $[K]$ by substituting the elements of the i column with Vector $\{P\}$.

In the inverse Hankel transform, the integrand includes Terms $J_0(\xi)\,J_1(\xi)\,e^{-qz}$, and $e^{-\xi z}$, etc.

We can discover that the integrand tends to 0, which is very quickly with the increase of ξ.

Since the precision is satisfied, the part of the integrand function which is nearly zero can be omitted, and the integration interval is simplified from an infinite interval to a finite interval, the approximate solution of the Hankel inverse transform is obtained by using the complex Simpson quadrature formula. The inverse Laplace transform is based on the Durbin method (Durbin 1973).

6 CALCULATION CASE

The two-layer elastic system was taken as an example to verify the correctness of the formula. The elastic parameters of the upper layer and the lower layer were $G_1 = 550$ MPa, $\mu_1 = 0.3$, $G_2 = 400$ MPa, and $\mu_2 = 0.35$ respectively. The other parameters were $p = 0.7$ MPa, $\delta = 0.15$ m, $h = 0.2$ m. In this paper, the excess pore water pressure at different depths was calculated on the symmetrical axis of the load. Table 1 shows the results. In this paper, the surface was assumed to be a drainage condition, that is to say, the excess pore water pressure on the surface equals 0. We were surprised to detect that the excess pore water pressure increases first and then decreases with the depth, and the maximum depth is 0.075 m. With the increase of time, the dissipation rate of the excess pore water pressure gradually slows down. And the excess pore water pressure at 5 s is more than 50% of that at 1 s, hence the excess pore water pressure exists in the structure for a long time, and the stress state of the structure becomes complicated.

Table 1. The excess pore-water pressure (kPa).

	Depth								
Time/s	0.05 m	0.075 m	0.10 m	0.15 m	0.20 m	0.25 m	0.30 m	0.40 m	0.50 m
1	80.34	75.28	63.49	50.32	37.43	30.54	23.38	14.04	8.62
2	61.47	59.33	53.00	41.42	29.93	24.17	18.51	11.07	6.79
3	52.00	51.23	47.49	36.78	26.09	20.93	16.03	9.56	5.87
4	45.57	45.72	43.72	33.61	23.48	18.73	14.35	8.54	4.74
5	40.33	41.31	40.75	31.11	21.40	16.96	13.01	7.72	4.27

7 CONCLUSION

(1) By means of the Laplace transform and Hankel transform, the Biot consolidation equation and the flow continuity equation were solved, and the expressions under the integral transformation of strain, stress, and displacement in any layer were obtained. The explicit analytical expressions such as the strain, stress, and displacement of the multi-layer elastic system were obtained by the boundary condition and the contact condition between layers.
(2) According to the properties of the integrand function, the inverse Hankel transform was performed by using the complex Simpson quadrature formula, and the inverse Laplace transform was performed by the Durbin method.
(3) The accuracy and validity of the theory were verified by a computational example. The results showed that the excess pore water pressure increases first and then decreases with the increase in depth.

ACKNOWLEDGMENT

This paper was supported by the Graduated Talent Introduction Scientific Research Initiation Plan (XYB2014–06) and Three Horizontal and Vertical Support Plan (ZRCPY202225, ZRCPY202119) of Heilongjiang Bayi Agricultural University.

REFERENCES

Dong Zejiao, Cao Liping, Tan Yiqiu, et al. Analysis of the Dynamic Response of Three Directional Strains in Asphalt Pavement under Moving Vehicle Loads [J]. *China Civil Engineering Journal*, 2009, 42 (4): 133–139. (in Chinese)

Eyad Mased, Aslam Al Omari, and Hamn Ching Chen. Computations of Permeability Tensor Coefficients and Anisotropy of Asphalt Concrete Based on Microstructure Simulation of Fluid Flow [J]. *Computational materials science*. 2007, 40: 449–459.

Durbin F. The Numerical Inversion of Laplace Transforms an Efficient Improvement to Dubner and Abate's method [J]. *The Computer Journal*, 1973, 17 (4): 371–376.

Gong Xiaonan. *Advanced Soil Mechanics* [M]. Beijing: The People Communication Press, 1992. (in Chinese)

Gu Raozhang and Jin Bo. Biot Consolidation of Multilayer Subsoils Under 3D-Axisymmetric load [J]. *Engineering Mechanics*, 1992, 9 (3): 81–94. (in Chinese)

Guo Dazhi and Ma Songlin. *Engineering Mathematics in Pavement Mechanics* [M]. Harbin: Harbin Institute of Technology Press, 2001. (in Chinese)

Li Zhida, Shen Chengwu, Zhou Zengguo, et al. *The Influence of Super-void Water Pressure on Asphalt Concrete* [J]. Natural Science Journal of Xiangtan University, 2003, 25 (4):98–109. (in Chinese)

Marc E., Novak, and Bjorn Birgisson. Effects of Vehicle Speed and Permeability on Pore Pressure in Hot-mix Asphalt Pavements [C]. *Second MIT Conference on Computational Fluid and Solid Mechanics*: 532–536.

P. Kettil. Coupled Hydro-mechanical Wave Propagation in Road Structures [J]. *Computers and Structures*, 2005 (83): 1719–1729.

Peng Yongheng, Ren Ruibo, Song Fengli, et al. An Axisymmetric Solution of Multi-layered Elastic Body Super-pressure in Small Opening Water [J]. *Engineering Mechanics*, 2004, 21 (4): 204–208. (in Chinese)

Sha Qinglin. *Premature Damage and Its Preservation Measures of Bituminous Pavement on Expressway* [M]. Beijing: China Communications Press, 2008. (in Chinese)

Zhong Yang, Geng Litao, Zhou Fulin, et al. Computing the Express Pore Fluid Stress of Flexible Pavement by Stiffness Matrix Method [J]. *Journal of Shenyang Jianzhu University (Natural Science)*, 2006, 22 (1): 25–29. (in Chinese)

Water Conservancy and Civil Construction – Oke & Ahmad (Eds)
© 2024 The Author(s), ISBN: 978-1-032-58614-4

Study on alignment design of secondary highway in southeast humid and hot area: A case study from Xianshuitang village to Kengweitou village

Liangtao Deng*
School of Energy Science and Engineering, Henan Polytechnic University, Jiaozuo, Henan, China

ABSTRACT: The construction of secondary roads can not only improve the local transportation environment and promote economic development but also strengthen the connection between rural and urban areas, realizing the sharing and exchange of rural and urban resources. Combining the topography and landform of Zengcheng District in Guangzhou, the economic living conditions of the surrounding residents, and relevant norms, this paper studies the alignment design of the secondary highway from Xianshuitang Village to Kengweitou Village. The new road provides agricultural and fishery production along the route, as well as economic, cultural, and material exchange services, strengthening links with the outside world and thereby boosting the economic development of rural areas in the region.

1 INTRODUCTION

With the implementation of the strategy of transportation power, the State Council further proposed to comprehensively promote the construction of "four good rural roads" and accelerate the implementation of road hardening through villages. Now, the country faces the construction of rural roads, and reconstruction projects need special survey designs. In the plain and hilly areas, the linear shape is easier to arrange and the design index is higher, while in the mountainous, due to the influence of topography and geology, the design index is low and close to the limit value, and various factors need to be considered comprehensively, so the linear design is more complex. Thus, it can be seen that rural highway construction is an important measure for the country to achieve the goal of transportation power.

Tang [1] believed that rural roads in mountainous areas are a bridge connecting rural areas with the outside world and a way for local people to become rich. When designing routes, we should take care of the vast majority of the people as much as possible and make the roads close to their doorsteps to facilitate their passage. Wang [2] believed that the design of highway alignment should be integrated with the safety and comfort of vehicles, the economy of the project, and the aesthetics of the alignment. Li [3] believed that the roads in mountainous areas have complex terrain, changeable climate, and staggered distribution of horizontal and vertical planes, which requires relevant personnel to comprehensively consider external factors, climatic factors, surrounding vegetation planting factors, etc., and makes road alignment design according to actual local conditions. Yan [4] believed that the scientificness, safety, and reliability of linear design must be comprehensively considered in the design of mountain roads, and the visibility conditions, longitudinal slope rate, transverse slope rate, and safety design standards are all higher than those of plain roads. Tian [5]

*Corresponding Author: 212202020096@hpu.edu.cn

DOI: 10.1201/9781003450818-32

believed that rural roads are the link between urban and rural areas. Rural terrain is different and complex, so there will be many problems in the design. The designer always adheres to the purpose of keeping the alignment fit, continuous, and balanced, minimizing engineering diseases, reducing transportation, maintenance, and management costs, and ensuring smooth traffic and safe driving after the completion of the project.

2 GEOLOGICAL SURVEY OF HIGHWAY

Zengcheng District has low-lying terrain from north to south, mainly hills, occupying 8.3% of the land of Zengcheng District. It is an extension of the Jiulian Mountains, and from northeast to southwest, hills, and plains are successively distributed and there are the Dongjiang River and Zenghe River. The secondary highway passes through plains and hills and mostly passes through fish ponds and orchards.

Zengcheng District belongs to the monsoon maritime climate, characterized by short winters and long summers, wet and rainy. The average annual temperature in Zengcheng District is 21.9 degrees Celsius, with the highest temperature of 38.6 degrees Celsius. The annual precipitation is rich, but the change is not balanced in the four seasons. The annual precipitation is 2004.5 mm, among which 73% of the annual precipitation occurs from May to October, and the precipitation is up to 1479.9 mm.

3 ROUTE DESIGN

3.1 *Planar linear design*

This design section area is mostly plain and hilly, so the linear design adopts straight lines and circular curves. The maximum length of the straight line section should be selected according to the actual situation, to ensure the continuity of the straight line section and the adjacent curve section, according to the driving speed test.

The minimum length of the straight line is related to the circular curve shape. This design curve is the reverse circular curve. According to the specification, when the design speed is greater than or equal to 60 km/h, the minimum length of the straight line between the reverse circular curve (m) is not less than 2 times the design speed (km/h). When the two reverse curves are equipped with relaxation curves at both ends and the terrain is limited, the S-shaped curve can be used, and the radius of the relaxation curve and circle curve of the S-shaped curve should meet certain requirements [6]. Table 1 shows the minimum radius of circular curves on secondary highways.

Table 1. The minimum radius of a circular curve.

Design speed (km/h)	The minimum radius of the circular curve (general value) (m)	The minimum radius of the circular curve (limit value) (m)
60	200	125

According to the "Highway Engineering and Technical Standards", the radius of the circular curve is 270 m, 300 m, 260 m, 250 m, and 270 m, respectively.

In the design of the road route, a certain length of relaxation curve should be set, so that the vehicle can carry out different curvature transformations on the relaxation curve, the driver can turn smoothly and passengers feel comfortable. At the same time, it can not only make the lines beautiful and smooth but also make the ultra-high circle curve and the widening transition smooth [7].

Table 2. Minimum length of easing curves of highways at all levels.

Design speed (km/h)	120	100	80	60	40
Minimum length of relaxation curve (m)	100	85	70	50	35

In this design, the length of the relaxation curve is 70 m, which meets the requirements of the minimum relaxation curve.

This design plane linear design diagram is as follows:

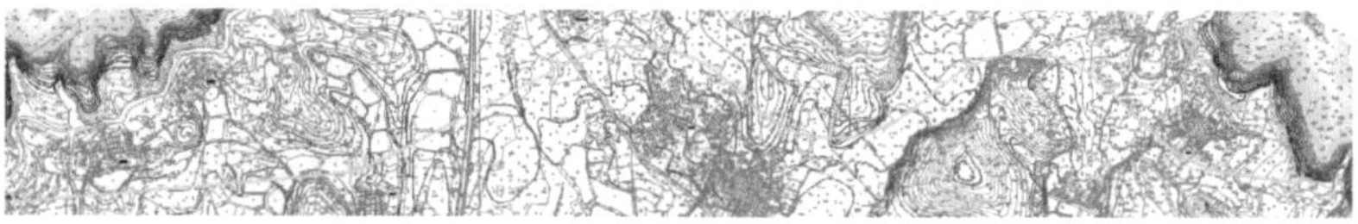

Figure 1. Planar linear design diagram.

3.2 *Longitudinal section linear design*

There are two main lines in the longitudinal section. One is the ground line, which is an irregular broken line dotted according to the elevation of each pile point on the middle line, reflecting the fluctuation of the ground along the middle line [8]. The other is the design line, which is determined from many aspects and has a regular geometric line shape, reflecting the ups and downs of the route, as well as the longitudinal design slope and vertical curve of the route [9].

The longitudinal section design should meet the different design requirements of longitudinal slopes, including maximum, and minimum longitudinal slope, slope length limit, average slope, composite slope, shortest vertical curve, shortest vertical curve, etc [10–12].

The maximum longitudinal slope allowed for each road should be considered according to the dynamic characteristics of the vehicle, road grade, natural conditions, engineering, operational economy, and other factors [13]. At the same time, the length of the longitudinal slope on the road must be limited, too long a slope will affect the driving speed and capacity [14]. The maximum longitudinal slope and slope length for different design speeds are specified in the following Table 3.

Table 3. Maximum longitudinal slope (%) and slope length (m) at different design speeds.

Design speed (km/h)		120	100	80	60	40
Degree of longitudinal slope (%)	3	900	1,000	1,100	1,200	
	4	700	800	900	1,000	1,100
	5		600	700	800	900
	6			500	600	700
	7					500

We all use not less than 0.3% design longitudinal slope as a minimum longitudinal slope to ensure drainage, and avoid water infiltrating to subgrade, thus affecting the stability of the subgrade [15]. On roads with poor or long horizontal drainage, where a horizontal slope (0%) or a longitudinal slope below 0.3% is used, the side ditch must be drained longitudinally.

It is generally believed that the minimum slope is to design an hour's drive of 9 to 15 seconds per hour [16]. On the expressway, the driving time of 9 seconds can meet the needs of driving and geometric layout, and the "Highway Engineering and Technical Standards" has made clear provisions on the minimum slope length of different grade highways, as seen in the table below.

Table 4. Minimum slope length.

Design speed (km/h)	120	100	80	60	40
Minimum slope length (m)	300	250	200	150	120

The minimum radius and minimum length of the vertical curve in the design are mainly affected by the requirements of cushioning impact, driving time, and visual range. The speed of this design is 60 km/h. The value specified in the "Highway Engineering and Technical Standards" is shown in the table below.

Table 5. Minimum radius and the minimum length of the vertical curve.

	Convex vertical curve (m)		Concave vertical curve (m)		Minimum length of a vertical curve (m)	
Design speed (km/h)	The general value	Limit value	The general value	Limit value	The general value	Limit value
60	2,000	1,400	1,500	1,000	120	50

The parameters of the longitudinal section in this design are summarized in the table below.

Table 6. Parameters of longitudinal section design.

Design speed (km/h)	Maximum longitudinal slope (%)	Minimum longitudinal slope (%)	Latest slope length (m)	Convex vertical curve		Concave vertical curve		Minimum length of a vertical curve (m)
				Maximum radius (m)	Minimum radius (m)	Maximum radius (m)	Minimum radius (m)	
60	2.46	0.47	160.21	10,000	9,800	8,000	5,000	97.85

We can take slope change point 1 as an example to calculate the vertical curve elements, the formula is as follows:

$$\omega = i_1 - i_2 \tag{1}$$

$$L = R\omega \tag{2}$$

$$T = L \div 2 \tag{3}$$

$$E = \frac{T\omega}{4} \tag{4}$$

where $i_{1,}i_2$ denote the longitudinal slopes of two adjacent straight slope lines at the slope change point, ϖ denotes slope difference, L denotes the length of the vertical curve, R denotes the radius of the vertical curve, T denotes tangent length, and E denotes outer distance.

The calculated elements of the vertical curve are as follows:

Table 7. Elements of vertical curve.

Slope difference (%)	Length of the vertical curve (m)	Tangent length (m)	From the outside (m)
2.32	116	58	0.3364

The longitudinal design drawing of this design is as follows:

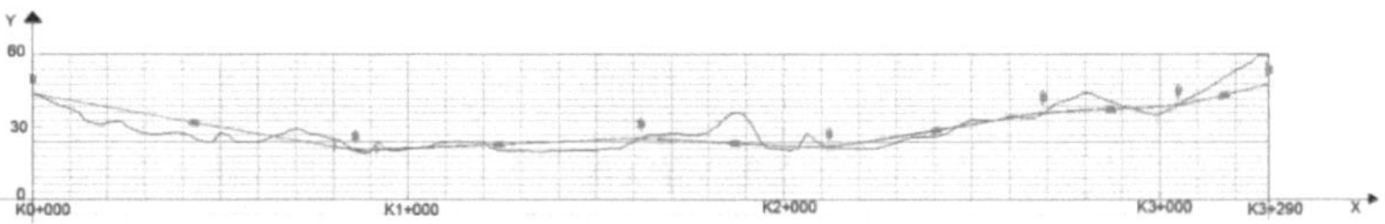

Figure 2. Longitudinal section design.

4 CONCLUSION

This paper mainly designs the alignment of the secondary road from Xianshuitang Village to Kengweitou Village. Considering the local natural conditions and economic conditions, the alignment is designed according to local conditions, to pass through as many villages as possible to promote the development of the local economy and the transportation of agricultural products. At the same time, on the basis of keeping the linear shape beautiful and direct as far as possible, we reduce the total cost of the project.

REFERENCES

[1] Tang Yang, Ren Rong, He Hongbin, et al. Analysis of Rural Highway Alignment Design in Mountainous Areas [J]. *Qinghai Transportation Science and Technology*. 2021, 33 (03): 69–72.

[2] Wang Kun. Analysis of Influencing Factors of Highway Alignment Design [J]. *Engineering Construction and Design*, 2020 (19): 134–135.

[3] Li Chengcheng. Analysis of Key Points of Highway Alignment Design Under Complex Conditions in Mountainous Areas [J]. *China New Technology and New Products*. 2020 (08): 111–112.

[4] Yan Xin. The Influence of Road Alignment Design on Road Alignment Design in Mountainous Areas [J]. *Transportation Research Part A*, 2019 (24): 116–117.

[5] Tian Hua. Analysis of Rural Highway Alignment Design [J]. *Science and Technology Innovation Review*, 2018, 15 (14): 63–65.

[6] Yan Jingxing. Analysis of Key Points of Highway Plane Alignment Design [J]. *Traffic World (Engineering Technology)*. 2015 (03): 68–69.

[7] Jiao Yinhe. Design of Highway Plane Alignment [J]. *Traffic World (Vehicle)*. 2008 (10): 85.

[8] Wang Lei. Key Points of Longitudinal Alignment Design of Mountain Highway [J]. *Technology & Market*, 2018, 25 (02): 150–151.

[9] Zhang Fenglan. Longitudinal Section Design of Highway Route [J]. *Heilongjiang Transportation Science and Technology*. 2011, 34 (10): 75–77.

[10] Li Niegui. Route Design and Analysis of Secondary Road Reconstruction and Expansion in Mountainous Areas. *Low-carbon World*. 2019, 9 (08): 273–274.

[11] Xu Youjun, Yang Huashi, and Zhang Zhiqing. Discussion on the Design Method of the Horizontal and Horizontal Lines of the Old Road Reconstruction and Expansion [J]. *Zhongwai Highway*. 2009, 29 (05): 9–12.

[12] Liang Deqiang. A brief Discussion on the Secondary Reconstruction Design of Local Highway [J]. *South Central Highway Engineering*. 2000 (04): 17–18.

[13] Hu Yanshan. Analysis of the Problems of Highway Alignment Design [J]. *Science and Technology Information*, 2010 (15): 48–49.

[14] Wu Xun. Discussion on the Limit of Maximum Longitudinal Slope Length of Highway [J]. *South Central Highway Engineering*. 1997 (02): 2–3.

[15] Liu Jingbo. Analysis of Key Points of Highway Longitudinal Section Design [J]. *Heilongjiang Transportation Science and Technology*. 2014, 37 (08): 4–6.

[16] Wang Yu Long. Discussion on Longitudinal Section Design of Low-grade Reconstructed Highway in Mountainous Area [J]. *Heilongjiang Transportation Science and Technology*. 2012, 35 (10): 22.

Water Conservancy and Civil Construction – Oke & Ahmad (Eds)
 ISBN: 978-1-032-58614-4

Cross-Section design of ultra-long underwater high-voltage cable tunnel by "double diamond model"

Tianli Song*, Yumeng Jiang*, Shihao Wang*, Yuchen Qi*, Ruanming Huang* & Haoen Li*
State Grid Shanghai Electric Power Company Economic and Technological Research Institute, Shanghai, China

Lin Cai*, Zihui Peng* & Ting Ni*
Shanghai Tunnel Engineering Rail Transit Design and Research Institute, Shanghai, China

ABSTRACT: With the national significant strategic goal of "carbon peak before 2030 and carbon neutrality before 2060", the power supply project from the western province to Shanghai has been placed in the foreground. The biggest challenge in this project is crossing the Yangtze River. Under the framework of using a special cable tunnel to cross the river, an ultra-long underwater high-voltage cable tunnel is proposed at a historic moment. In this paper, the cross-section scheme of an ultra-long underwater high-voltage cable tunnel is deduced through the double diamond design model, and the cross-section design elements such as double compartments and special accident air ducts are analyzed. Finally, a reasonable, feasible, and economic section scheme model is proposed.

1 INTRODUCTION

Under the premise of implementing the national significant strategic goal of "carbon peak by 2030 and carbon neutrality by 2060" (People's Daily 2021), and taking into account the economic growth, structural adjustment, and continuous improvement of terminal electrification rate in Shanghai, which has very limited electric power resources, so we need to advance layout and plan to promote the energy revolution. The biggest challenge of the power supply project from the western province to Shanghai lies in crossing the Yangtze River. Under the framework of the special cable tunnel across the river, the cross-section design of the ultra-long underwater high-voltage cable tunnel is the key to the project.

The double diamond design model proposed by British Design Association, is a divergent way of thinking in the design process, the core of which is to search for the problem and find the right solution. The double diamond design model divides the design process into four stages: problem discovery, problem definition, solution conception, and solution determination (Jonathan Ball 2019).

In this paper, the design and research method of the double diamond design model is used to analyze and deduce the cross-section design of an ultra-long underwater high-voltage cable tunnel, to realize the design objective of "designing the right thing" and "designing things right" in this project, which is the ethical aspects of design (EMAKINA 2021).

*Corresponding Authors: whppstl@163.com, 15021323909@163.com, 15021323909@163.com, dzbaxbb@163.com, 13761517479@163.com, 10885431@qq.com, 318597620@qq.com, 377039120@qq.com and 1040446299@qq.com

DOI: 10.1201/9781003450818-33

2 PROBLEM DISCOVERY: ULTRA-LONG CABLE TUNNEL INVESTIGATION

Through years of hard exploration and continuous innovation, cooperating with power grid planning, Shanghai Electric Power Co., Ltd. has implemented and built representative projects such as the Yanggao middle road cable tunnel, Xizang road cable tunnel, and expo cable tunnel. The representative domestic cable tunnels are analyzed one by one, in an attempt to find the common problems that need to be solved in the design of a cable tunnel.

2.1 *220 KV trans-river cable duct in Shanghai, Chongming, Jiangsu over Yangtze River tunnel*

The 220 KV cable duct project is built within the Shanghai Yangtze river tunnel, with a cross-river length of about 7.5 km. The inner and outer diameter of the shield tunnel is 13.7 m and 15 m respectively. The upper part of the cross-section is a three-lane highway, and the 220 KV cable is set in an independent cable compartment on the right side of the lower part. (Peng Zi Hui 2015)

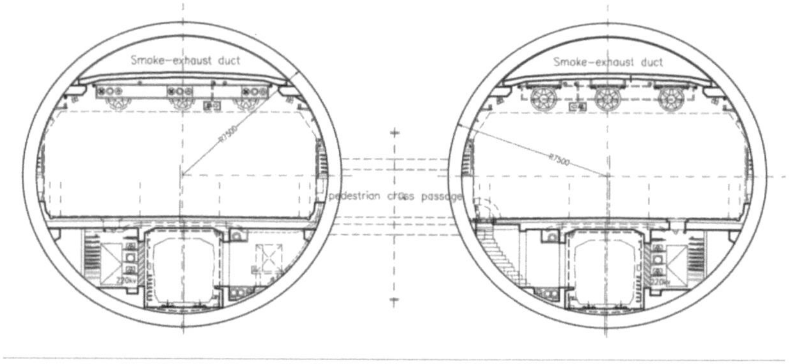

Figure 1. Cross section of Shanghai Yangtze river tunnel (STEDI 2015).

2.2 *The Sutong GIL pipeline corridor project (special purpose, underwater)*

The Sutong GIL Pipeline corridor project is the "Huainan ~ Nanjing ~ Shanghai 1,000 KV AC UHV Power Transmission and Transformation Project" constructed by the State Grid Corporation of China. The closed section of the project is about 5.7 km long, across the Yangtze River. The tunnel is a shield tunnel with an inner diameter of 10.5 m and an outer diameter of 11.6 m. Two 1,000 KV GIL pipelines are arranged at the upper part of the cross-section, which is distributed at both sides, and a set of SF6 gas exhaust systems and weak current channels are arranged at the lower part. (CPECC 2016)

2.3 *The expo cable tunnel (special purpose, including ground and underwater parts)*

The Shanghai World Expo cable tunnel project (Beijing West Road ~ Huaxia West Road) is the main power supply project of the 2010 Shanghai World Expo, connecting the downtown Expo 500 KV substation and Sanlin 500 KV substation. It is one of the backbone network projects for urban power transmission, and the total length of the project is about 15.3 km. There are two types of cross-section, i.e., a 3.5 m diameter for pipe jacking construction and a 5.5 m diameter for shield tunneling.

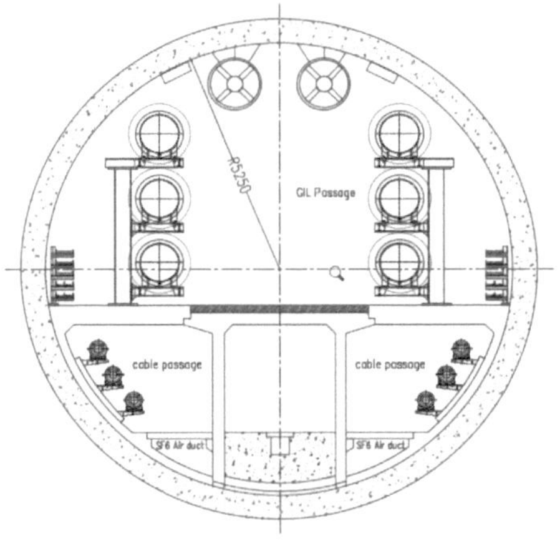

Figure 2. Cross section of the Sutong GIL pipeline corridor project (CPECC 2016).

The project is also a special cable tunnel, most of which is located underground in the city, so a total of 14 working shafts are set along the line. The working shafts are used for ventilation and gas exchange shafts, and the ventilation section is about 1,000-1,200 m (Xia Liang 2010).

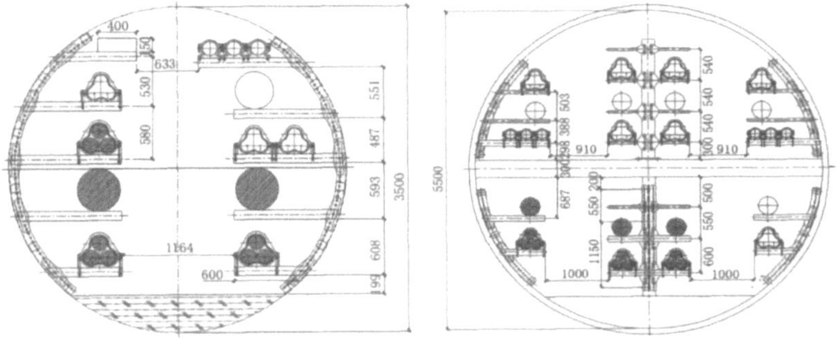

Figure 3. Cross section of the expo cable tunnel (SMEDI 2008).

2.4 *Summarization of common problems of the cable tunnel*

According to the analysis of the above three existing tunnels, it is not difficult to find that common issues that need attention in the design of cable tunnel cross sections are as follows: whether it is built with road tunnels and rail transit tunnels; cable layout form; ventilation scheme selection; fire protection design; cable operation and maintenance convenience; construction phases; convenience; project costs.

3 PROBLEM DEFINITION: ANALYSIS OF KEY PROBLEMS OF THE ULTRA-LONG UNDERWATER HIGH-VOLTAGE CABLE TUNNEL

3.1 *The ultra-long closed section*

According to the gap in Shanghai's power planning and the source of external power, Shanghai proposes to plan a cable tunnel from Chongming across the Yangtze River to

Shanghai. However, the Yangtze River is more than 15 km wide. Considering the construction conditions and environmental protection requirements, it is difficult to build a ventilation shaft and entrance through shield tunnels across the river. As the closed section is ultra-long at home and abroad, it is difficult to construct.

3.2 *Multi-loop high-voltage cables*

According to the current electricity consumption data of Shanghai, without large-scale external power from the city, and considering the local promotion of new energy development, there will still be a power gap of more than 7 million kilowatts and 35 billion kilowatt hours at the end of the "Fifteenth Five-Year Plan". Therefore, the new external power to the Shanghai DC project in the form of the "point-to-point" or "point-to-network" power supply to Shanghai, with peak power of no less than 8 million kilowatts, is an effective means to solve the power gap. Therefore, the cable scale of the proposed research project is in large demand, and 8 circuits of 500 KV cable (3 cables and single-layer arrangement each time) and 500 KV connector layer (single-layer arrangement) should be planned.

3.3 *Key problems*

Based on the above analysis of specific features, the author believes that the key problems to be solved for the cross-section scheme design of the ultra-long underwater high-voltage cable tunnel are mainly reflected in the following three aspects:

3.3.1 *We should speed up the fault repair function and ensure the stability of the power grid operation*

The foundation of cable tunnel construction is power supply, and undoubtedly its primary goal is to ensure the stability of power grid operation. In the process of cable operation, a fault will occur inevitably at a certain point. At this time, the partial fault of one cable should not affect the normal operation of other loop cables, and the quick repair of the faulty part is the way to ensure the operation stability of the power grid. In this way, laying cable in separate warehouses and perfect maintenance conditions are particularly important.

3.3.2 *We should optimize ventilation and fire protection design and ensure the safety of personnel evacuation*

The ultra-long closed section means that there is no condition to set the inlet & outlet and safe evacuation entrance with the ultra-long distance, causing great challenges to the ventilation design and fire protection design. The selection of a safe, comfortable, and economical design solution under such basic conditions is the top priority of section design.

3.3.3 *Scientific planning to realize engineering economy in the whole life cycle of the project*

The ultra-long underwater high-voltage cable tunnel is an infrastructure project that will be a huge investment in terms of both scale and construction difficulty, and the urgency is self-evident. At the premise of scientific and reasonable comfort, engineering economy should be taken into account in design accordingly. The economy involves both time and space dimensions, with time reflected in the construction cycle and space reflected in the project scale. The cross-sectional design directly affects the final investment in the project.

4 SOLUTION CONCEPTION: KEY POINTS OF SECTION DESIGN FOR THE ULTRA-LONG UNDERWATER HIGH-VOLTAGE CABLE TUNNEL

Based on the project conditions and the above core issues analysis, the section plan is compared and selected through the following points:

4.1 *Independence of cable arrangement*

This point is mainly for the cable operation stability of the above key problems.

Scheme 1: The cable is only laid on the upper layer (single-layer arrangement).
Scheme 2: Cables are laid on the upper and lower layers respectively (double-layer layout).

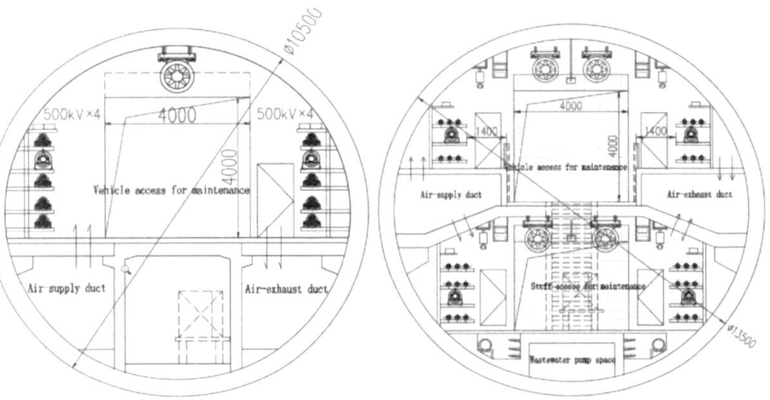

Scheme 1. Single-layer layout Scheme 2. Double-layer layout

Figure 4. Schemes for the cables space.

Comparing the two schemes, there is no doubt that the single-layer setting of the cable can reduce the section diameter relatively and the tunnel construction scale, thus directly reducing the project investment. However, considering the analysis of the problem definition stage of the double diamond design model, the power network operation stability is the primary goal of the construction of the cable tunnel, and the project should be built around this goal.

Therefore, the best way to improve the stability of power grid operation is to lay cables in two compartments.

4.2 *Disaster prevention design*

This point is mainly for the fire protection design of the key problems above. The disaster prevention design for the cable tunnel is based on fire prevention design, on the premise of the occurrence of a fire at a time.

4.2.1 A fire evacuation staircase is set every 200 m in the upper and lower layers of the tunnel section for excavation between the upper and lower single-hole cable channels. When a fire breaks out in the lower part of the tunnel, personnel can be evacuated to the upper safe area through the evacuation stairs.

4.2.2 The tunnel is used as a fire prevention unit every 400 m, and fire and smoke prevention partitions are set on both sides. In case of a fire in the tunnel, the fire and smoke are controlled within the 400 m fire prevention unit. The fireproof rolling curtain is used in the inspection path and passage of the tunnel, and normally open fire doors are provided on both sides. When a fire breaks out in the tunnel, the close command of the fire door and shutter can be triggered by the fire linkage system.

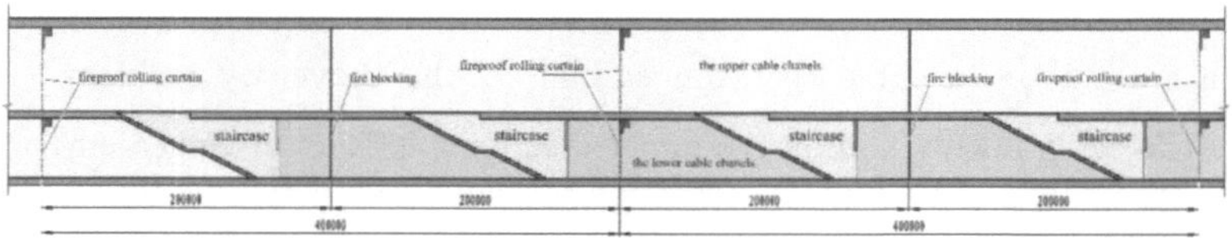

Figure 5. The separation diagram of the interval fire prevention unit.

4.2.3 Fire blocking is set at the cable support position of the tunnel at an interval of 200 m to strengthen the protection of the cable, improve safety, and reduce the impact on other facilities in the channel. (Cable Design Standard for Electric Power Engineering, GB50217-2018). The design scheme of fire blocking is as follows:

Scheme 1: Fire blocking extends out of the fire support 1 m wide to prevent the fire from spreading vertically. This scheme will occupy part of the personnel passage space;

Scheme 2: A fire blocking is made along the cable support perpendicular to the support, that is, a T-shaped fire blocking is formed to prevent the fire from spreading vertically. The scheme has a slight increase in the workload of fire blocking but does not occupy the outside space.

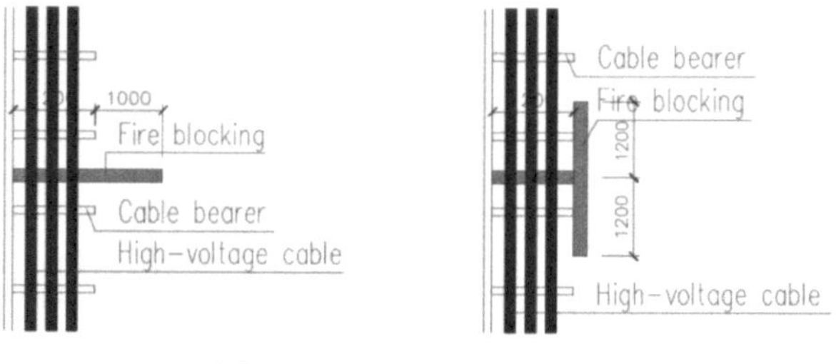

Scheme 1 Scheme 2

Figure 6. Schemes for the fire blocking.

Comparing these two schemes, it is believed that the corresponding schemes should be selected according to the specific situation. When the maintenance space is relatively tight, the second scheme is recommended first.

4.3 *Layout of ventilation and smoke exhaust duct*

This point is mainly considered for the ventilation design scheme in the key problems above.

The proposed research project spans the Yangtze River, 15 km across the river. It is an ultra-long cable tunnel, and ventilation shafts cannot be set in the middle as the inlet and outlet. Therefore, the cross-section design of the ventilation scheme is the difficulty of the project discussed in this paper.

Scheme 1: No special air supply and exhaust ducts are set up to save space, and the mid-plate forms a unified elevation plane, which is relatively convenient for transportation, installation, and even maintenance.

Scheme 2: The folded plate is arched into a special air supply and exhaust duct. In case of fire, the fire shutter doors in the fire area and adjacent sections should be closed. When it is confirmed that the fire is extinguished, and the flame is lower than the spontaneous combustion point and cannot be reignited, the exhaust fan can be opened to carry out the air behind the disaster area with the special exhaust duct, and the blower can be opened to inject fresh air into the area after the disaster area with the special fresh air duct.

Compared with the two schemes, and considering that the ventilation efficiency and air volume of the area behind the disaster area were much higher than those provided by the longitudinal ventilation, the smoke could be effectively controlled within 200~400 m, the fire area could be ventilated quickly, and a safe environment could be provided to meet the requirements of post-disaster recovery and maintenance in a short time. In the traditional Scheme 1, the post-disaster area can be exposed to outdoor fresh air for about 4~5 h after longitudinal ventilation to realize the ventilation in a real sense, while the post-disaster ventilation scheme provided by Scheme 2 for key areas only needs 8~10 min to expose the post-disaster area to outdoor fresh air. In addition, the traditional longitudinal post-disaster ventilation method needs to ventilate the whole tunnel to realize the ventilation of the post-disaster area, which consumes a lot of unnecessary transportation energy consumption, resulting in huge equipment configuration and increased operation energy consumption. However, post-disaster ventilation in this scheme only carries out ventilation for the post-disaster section and achieves the same air exchange for the post-disaster section. Compared with the traditional longitudinal ventilation scheme, the equipment scale of this scheme and the operational energy consumption is greatly reduced. As for the height difference of the mid-plate caused by the duct, it is also possible to accommodate the inconvenience caused by its mechanized design.

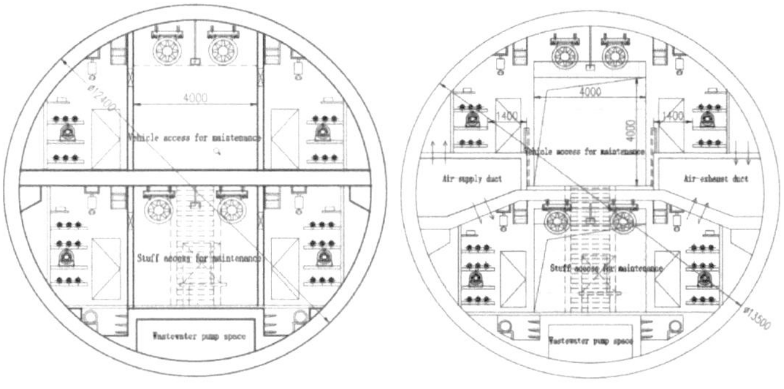

Scheme 1. Design without special air duct. Scheme 2. Design with special air duct.

Figure 7. Schemes for the ventilation design.

4.4 *Tunnel synchronous construction technology*

The tunnel inside is divided into two layers, and both the upper and lower layers are cable layers. The internal structure can be built by synchronous construction technology, making full use of characteristics of the round tunnel with the large diameter and long driving distance, and carrying on construction in the organization of the internal structure to achieve the flow operation. During the shield driving process, the prefabricated rectangle-shaped component is installed synchronously behind the frame to form a single lane for transporting the prefabricated internal components and segments. Then, the prefabricated walkway plate is installed, and the transport line forms two lanes in both directions. The cast-in-place middle plate and the bracket of the air duct plate are formed. The shield power supply line is moved to the bracket of the middle plate and hung. A trolley is set up on the bracket of the air duct plate to transport the medium plate to the installation place. The trolley will lift, rotate, and place the medium plate in the specified position. The prefabricated air duct plate is installed and bolted to the middle plate. At the same time, the mud conveying pipe, the shield water conveying pipe, and the air pipe are adjusted to the middle of the lower layer space to release the space on both sides of the lower layer.

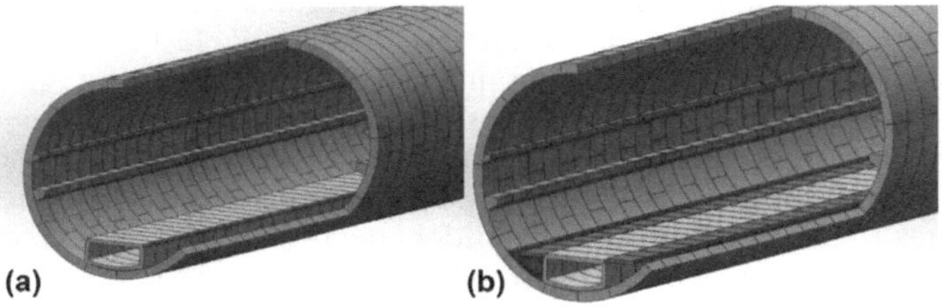

Figure 8. (a) Installation of the prefabricated mouthpiece (b) Installation of the prefabricated walkway.

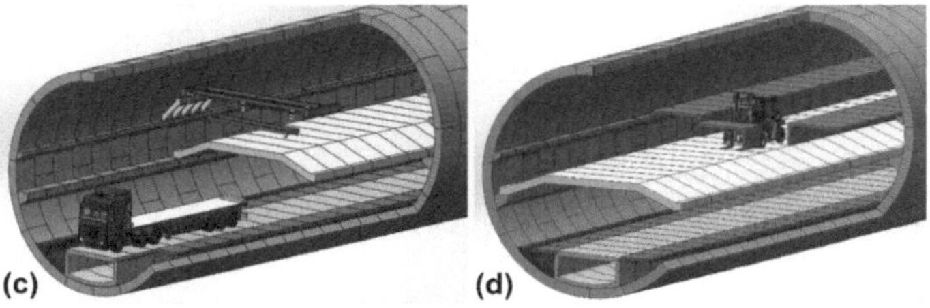

Figure 9. (c) Installation of the prefabricated mid-plate (d) Installation of air duct plate.

5 DELIVERY SCHEME: SECTION LAYOUT FOR THE ULTRA-LONG UNDERWATER HIGH-VOLTAGE CABLE TUNNEL

Through the analysis of the above section design points and the process of scheme conception, the section scheme with an outer diameter of 13.5 m is finally selected. The key factors of the scheme are mainly as follows:

1) The space is divided into compartments, especially for multi-loop high-voltage cables. The stratified and divided sides form four relatively independent spaces, which makes the independence of each loop cable optimal and effective and ensures the operation stability of the power grid.
2) The air duct for special accidents is set up. For the underwater cable tunnel without intermediate working well in the ultra-long closed section, the special accident air duct can effectively ensure the smoke efficiency after a disaster and the air exchange demand in the accident section after a disaster.
3) Considering the ventilation resistance comprehensively, the tunnel interval can be used as a fire prevention unit every 400 m. At the midpoint of the 400 m fire prevention unit, a fire prevention block can be added to the cable installation space on both sides.
4) For the long-distance double-decker tunnel, it is necessary to consider the entry of relevant operating vehicles into the tunnel and set 4 m×4 m vehicle access for maintenance. Considering the scale of the shield structure, we focus on a certain level of space for vehicle entry when designing the cross-section. Then it is connected with another level of space through stairs, which not only meets the convenience of operation mechanization but also takes into account the engineering economy.

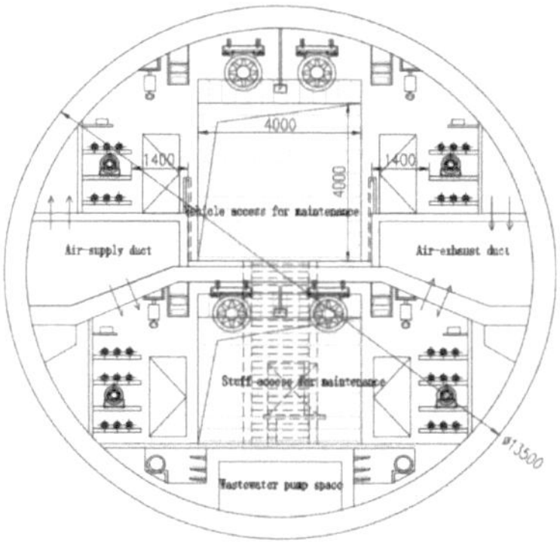

Figure 10. The delivery scheme for the section layout.

6 CONCLUSION

In this paper, we try to find out the problem of the ultra-long underwater high-voltage cable tunnel in the first stage by deducing the thinking frame of the "double diamond design model". In the second stage, on this basis, inspiration is constantly sought, divergent thinking is put forward, and the corresponding design scheme is finally refined and finalized, that is, the correct design is made.

With the continuous improvement of national space development, the rigid demand for terminal electrification rate is constantly increasing, and the ultra-long underwater tunnel across rivers, lakes, seas, and other spaces will spring up. We use the "double diamond design model" to deduce the design points of the long underwater cable tunnel and finally develop the ideal cross-sectional form. While ensuring the feasibility and optimizing the economy, it is essential to maximize the stability, security, and utilization of space during the operation of the cable tunnel, thus providing reference and reflection for the design of similar projects in the future.

REFERENCES

Cable Design Standard for Electric Power Engineering (GB50217-2018, 7.0.2).

CPECC (2016). Feasibility Study of Huainan~Nanjing~Shanghai 1000 KV UHV AC Sutong GIL Corridor Project.

CPECC (2016). Figures of the Feasibility Study of Huainan~Nanjing~Shanghai 1000 KV UHV AC Sutong GIL Corridor Project.

EMAKINA (2021). Ethical Design: How to Design the Right Thing Responsibly.

Jonathan Ball (2019). The Double Diamond: a Universally Accepted Depiction of the Design Process - Design Council. https://www.designcouncil.org.uk.

Peng Z. H. (2015). A Brief Review of Fire Evacuation and Rescue Design for the Yangtze River Tunnel in Shanghai Underground Engineering and Tunnels 2007 (4), 43~45+53+57

People's Daily (2021). "Aim for Carbon Peak by 2030 and Carbon Neutrality by 2060"—*Win the Tough Battle of Low-Carbon Transformation*. http://www.gov.cn/

SMEDI (2008). Preliminary Design Drawings of the Expo Cable Tunnel Project.

STEDI (2015). Construction Drawings of the Yangtze River Tunnel Project.

Xia L. (2010). *Study on the Planning and Alignment of Cable Tunnels in Central Urban Areas - an Example of Cable Tunnels into the Expo Station*, Shanghai Urban Planning, 2010 (04), 45~49

Water Conservancy and Civil Construction – Oke & Ahmad (Eds)
© 2024 The Author(s), ISBN: 978-1-032-58614-4

The genesis of ultra-deep overburden in the Qiaojia reach

Wanqiang Cheng
China Power Construction Group East China Survey, Design and Research Institute Co., Ltd., Hangzhou, China

Yunsheng Wang, Liang Song*, Zhengyou Li & Yuhang Zhou
State Key Laboratory of Geological Disaster Prevention and Protection, Chengdu University of Technology, Chengdu, China

ABSTRACT: Deep overburden is developed in most reaches of the Jinsha River, but the Qiaojia Reach has the largest thickness and abnormal buried depth of the base cover. It is of great theoretical and practical significance to study its sedimentary sequence and analyze its genesis. This paper redefines the fault depression boundary of the reach and reveals the filling sequence characteristics and fault depression mode of the deep overburden of the reach in detail based on the methods and means of radon gas measurement on boundary fracture of the reach, the detailed catalog of the core of the ultra-deep borehole in the deep overburden, the interpretation of the geophysical prospecting profiles across the river, the field record of large section excavation, and the 1:10000 engineering geological mapping methods. The results of the study show that: (1) The Qiaojia Reach fault depression takes the fault as the boundary, and there are obvious anomalies in the measured radon value; (2) The sedimentary sequence in the reach can be divided into six rhythms, representing six fault-depression-accumulation events, which is composed of coarse and fine rhythms of river and lake facies. The top is covered by proluvial and debris flow accumulation; (3) The Qiaojia deep overburden reach is formed by oblique right-handed strike-slip pull separation between the south section of the Zemuhe Fault and the north end of the Xiaojiang Fault. It is about 14 km long from north to south, about 5 km wide from east to west, and covers an area of about 70 km^2; (4) The Qiaojia Reach of the early Pleistocene began to slip and pull apart fault depression 1.59 million years ago, which lasted 1.15 million years and filled with more than 800 meters of deep overburden. 440,000 years ago, the fault depression activity ended after the complete connection between the Zemuhe Fault and the Xiaojiang Fault, and underwent the erosion and transformation of the Jinsha River; (5) The activity at the northern end of the Xiaojiang Fault has weakened for 440,000 years, and its upper overburden has been eroded by the torrents of the back mountains to form a nearly north-south trough landform.

1 INTRODUCTION

With the deepening of the exploration process of hydropower projects in western China, the thickness of the riverbed overburden has been continuously renewed, and the composition and origin of deep overburden have become important factors to be considered in dam site selection and reservoir resettlement projects (Shi 1986; Sun et al. 2008; Xu et al. 2008). Deep overburden refers to the riverbed accumulation layer with a depth of more than 30 m. Since

*Corresponding Author: songliang_92@163.com

DOI: 10.1201/9781003450818-34

the 1980s, there have been nearly 100 papers discussing the development characteristics, causes, engineering effects, and disposal measures of the deep overburden. The previously revealed deep overburden is mainly distributed in intermittent belts along the western valleys (Sun 2011; Wu 2009; Xu et al. 2010). Except that the overburden of Yele hydropower station in the upper reaches of Nanya River and the Milin-Zhibai reach, moraine, and moraine weir plug in the lower reaches of Yarlung Zangbo River is more than 200 m, the overburden depth of most landslide or debris flow weir plug riverbed in western rivers generally varies from dozens of meters to more than 100 m. However, the thickness of the riverbed overburden of the Qiaojia Reach of the Jinsha River reaches 728.2 m and the overburden elevation of the valley floor is much lower than that of the dam foundation of the downstream Baihetan Hydropower Station Dam. Its cause is still unknown.

Predecessors have done a lot of ground investigation on the supergene process of the Qiaojia Reach. For example, in terms of the origin and dating of the stratiform landforms in the valley. Jiang et al. (1999) and Chen et al. (2009) studied the grain size characteristics, age, and genesis of the loess-like deposits in the Qiaojia Reach of the Jinsha River. Zhang et al. (2009) measured radon at Dabakou and Liantang and believed that the starting point of the north end of the Xiaojiang Fault is connected with the Zemuhe Fault on the west of Huatan town on the opposite bank of the Jinsha River. He (2018) conducted a systematic dating of the nine-level terraces of the Qiaojia Reach, which provided valuable chronological data for the analysis of the valley evolution of this reach. In terms of the genesis of Qiaojia Reach, the 1:50000 geological map of Qiaojia County and the manual (1995) speculated that the deep overburden of Qiaojia Reach was caused by fault depression, but there was no exploration data on the structure and composition of overburden of this reach. He et al. (2008) and Yasutaka et al. (2008) believed that the quaternary connection (capture) between the newborn Daliangshan Fault and Xiaojiang Fault would become the eastern boundary of the Sichuan-Yunnan lozenge-shaped fault block. Yang Dayuan et al. [14] studied the valley geomorphology and evolution of the Baihetan Hydropower Station to reach the Jinsha River and put forward the viewpoint of river capture. Based on field investigation and shallow drilling in the reach, Pei et al. (2019) believed that the origin of the overburden of the Qiaojia Reach was the formation of weir based on the pull-apart depression of the tertiary Datong Fault and Xiaojiang Fault. Due to the lack of large-scale surveys in the study area, the deep structure of the overburden, and large section excavation data, it is difficult to have a comprehensive understanding of the boundary, filling sequence, and genesis of the deep overburden.

In this paper, the genesis of ultra-deep overburden in the Qiaojia Reach is discussed by means of on-site radon measurement, large-scale field investigation, ultra-deep drilling core logging, analysis, and geophysical prospecting profiles and dating across the river.

2 GEOLOGICAL BACKGROUND OF THE STUDY AREA

2.1 *Overview of the study area*

The Qiaojia Reach is located in Zhaotong City, Yunnan Province, and belongs to the intersection zone of the Yunnan-Guizhou Plateau, Western Sichuan Plateau, and Sichuan Basin. The whole reach is a long axis with a north-north-west direction, and the shape is an irregular ellipse, with an area of about 70 km^2. The overall terrain around the Qiaojia Reach is high in the east and low in the west. The mountain behind it belongs to the Yaoshan mountain range, and the mountain range is in the SN direction. The elevation of the mountain top is between 3,100-3,200 m. The slope elevation above 970 m is dominated by steep slopes with gullies on the slope surface (Figure 1). The exposed lithology of the Qiaojia Reach is composed of Permian Qixia-Maokou Formation limestone and Emeishan basalt, Devonian Yaopengzi Formation limestone, sandstone, mudstone, and quaternary loose deposits (Figure 2-a). The bottom of the reach has been eroded and reformed by the Jinsha River since the Middle Pleistocene. Except that the riverbed is flat (the slope is between 0° and 5°), there are quaternary accumulation scarps and branch ditches on both banks.

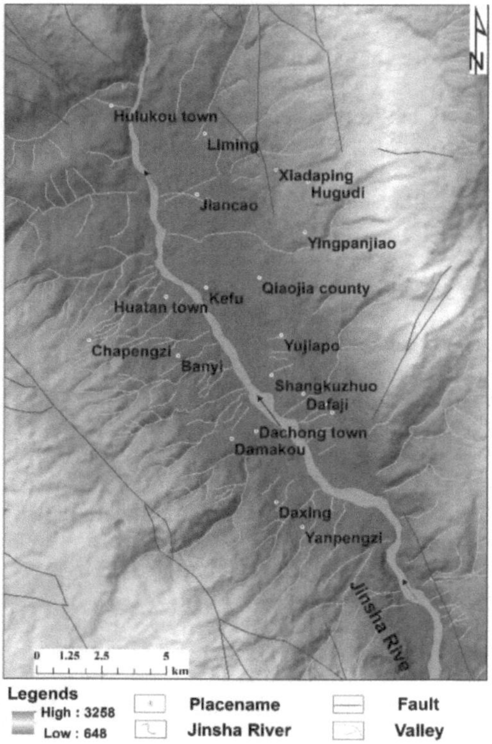

Figure 1. Geomorphic map of the Qiaojia Reach.

2.2 *Regional structural characteristics*

The Zemuhe Fault zone extending in a northwest direction and Xiaojiang Fault Zone in a nearly north-south direction are developed near the Qiaojia Reach. As the reach is located in the southeastern edge of the Sichuan-Yunnan rhombic block (Figure 2-b), the Sichuan-Yunnan rhombic block is strongly squeezed out in the south-east-east direction as a whole, making the faults in the reach mainly sinistral strike-slip (Song et al. 1998).

3 STRUCTURAL CHARACTERISTICS OF THE QIAOJIA REACH

The deep overburden of the Qiaojia Reach takes the fault as the boundary. In order to determine the spatial distribution of the boundary fault, 12 radon measurement profiles are arranged around the reach through remote sensing interpretation and on-site investigation. The location of the profile survey lines is shown in Figure 2. Radon measurement profiles arranged in the reach include Shuiniangou, Longjingwan, Liantang, and Honglu on the east boundary, with four profiles from 1-1' to 4-4' shown in Figure 2-a; the Shigaodi, Yanpeng Village, and Zihong Village on the south boundary, with three sections of 5-5' to 7-7' shown in Figure 2-a; the Shuitang Village, Xiaoyuanzi, and Dabaokou on the west boundary, with three sections of 8-8' to 10-10' shown in Figure 2-a; the Shanghulu and Lanyingpan on the north boundary, with two sections 11-11' and 12-12' shown in Figure 2-a. Statistics of radon measurement data are shown in Table 1.

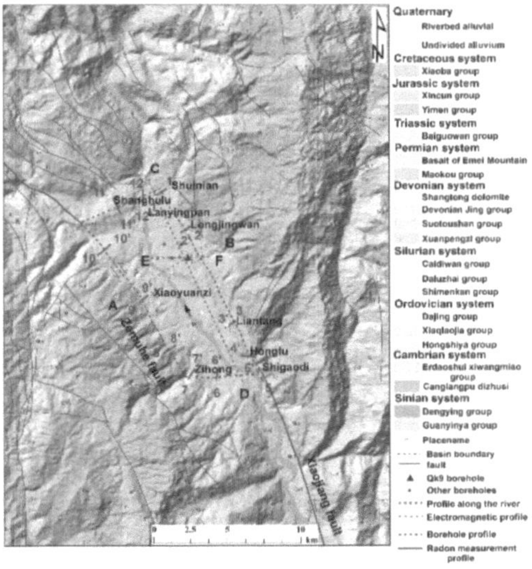

Figure 2. The geomorphic map of the Qiaojia Reach.

Table 1. Radon measurement data at the boundary of Qiaojia Reach.

Reach boundary	Profile	Abnormal value (Bq/m^3)	Maximum value (Bq/m^3)	Background value (Bq/m^3)	Maximum background value ratio
The East Boundary	Shuinian	26652.38	37594.05	1,500	25. 1
	Longjing Bay	4328.136	5374.2		3.6
	Liantang	7855.06	9176.7		6.1
	Honglu	13995.92	17111.25		11.4
The South Boundary	Shigaodi	16493.47	18328.05	1,500	12.2
	Yanpeng Village	14489.3	25451.4		16.9
	Zihong Village	10441.68	14196		9.5
The West Boundary	Damakou	18504.32	27104.01	1,500	18.1
	Xiaoyuanzi	7754.978	10723.05		7.1
	Shuitang Village	11578.54	12507.01		8.3
The North Boundary	Shanghulu	13704.05	22079.85	1,500	14.7
	Lanpajying	7255.89	9937.2		6.6

3.1 *Reach perimeter*

It can be seen from Table 1 that the maximum value of radon gas measured in each section exceeds the abnormal value (RNF), that is, there are faults passing under each profile. This shows that the east side of the reach is bounded by the north section of the Xiaojiang Fault.

Spatially, it starts from Liming Village in the north of Qiaojia county, passes through Qili Village, the pedestrian street, the passenger station in the east of Qiaojia county, and Liantang trough valley in the south, and ends near the gypsum land. The west boundary is the south extension of the Zemuhe Fault, starting from Daxing Village, Wuxing Village, Shuitang Village, Aotian, and Huapeng, ending at Zihong Village. The two are in a non-parallel oblique relationship in space. The northern boundary of the reach is located on the Laodukou-Hulukou-Liming Village line of the Heishui River, and the south boundary of the reach is the Xiaotian-Shigaodi area.

3.2 *Boundary fracture characteristics*

It has been agreed that the Xiaojiang Fault is the east boundary of the reach. According to the geomorphology and radon measurement, it starts from Liming Village in the north, passes through Fengshui ridge of the pedestrian street trough valley, and ends at Liantang trough valley in the south. Except for the high ratio between the north-south boundary of the reach and the background value, the main body ratio of the east boundary is 3.6-6.1 times, showing relatively weak activity.

West boundary fault: in recent years, we have conducted a 1:10000 geological survey on the Qiaojia Reach (Li 2009) and revealed that the south end of the Muhe Fault does not stop at Huatan, but continues to extend southward, and converges with the Xiaojiang Fault after passing Longtan Village, the west side of Luji Village, and Reshui Village, which are surface outcrops in Luji Village and Reshui Village. Radon measurement further reveals that the Zemuhe-Xiaojiang fault junction after the connection is significantly more active than the north end of the Xiaojiang Fault (Qiaojia-Liantang section). The thickness of overburden exposed by drilling at the resettlement point of Shuitang Village is more than 300 m. An on-site investigation also confirmed that there is no bedrock exposed from the west of Reshuitang Village to the Jinsha River. The radon measurement results show that the ratio of peak value to background value increases from north to south, from 7.1 to 18.1. The west boundary of the Qiaojia is the south extension of the Zemuhe Fault concealed under the quaternary system.

North boundary fault: there is a northwest-west fault on the east side of the Hulukou, which is near the Qiaojia Bridge of the Jinsha River. It hides eastward under the river bed overburden of the Jinsha River and its right bank. According to the occurrence of the fault, it is speculated that it intersects with the north end of the Xiaojiang Fault at Qiaojia Liming Village, whose measured radon ratio is 6.6-14.7.

South boundary fault: at present, there is no obvious fault corresponding to the surface, but there is a dense joint zone in the Shigaodi on the right bank of the Jinsha River. The measured radon ratio is 9.5-16.9, showing strong activity. According to Qiaojia County Chronicles (Tao et al. 2018), a high-speed remote landslide occurred on the high slope here in 1,800, which blocked the Jinsha River for 3 days and then burst. It can be seen that the south edge of the reach takes the dense joint intensive zone as the boundary.

4 CHARACTERISTICS OF FILLING SEQUENCE IN THE QIAOJIAHE REACH

4.1 *General characteristics of the deposits in the reach*

An on-site geological survey revealed that the deposit of the Qiaojia Reach is spatially divided into the Qiaojia platform on the right bank, the Huadan platform on the left bank, and the accumulation below the riverbed. Structurally, it can distinguish the accumulation in the fault depression period and the terrace accumulation of the river base (the cemented gravel layer in the fault depression period) after the end of the fault depression. Near-source alluvial deposits are developed at the top of the Qiaojia platform, which is mainly composed of limestone gravel sand mixed with giant boulders, and eight-level terraces are successively developed downward. The top of the huadan platform is red purplish red slope alluvium, composed of fine-grained loam with five-level terraces developing downwards. The accumulation below the riverbed is revealed by boreholes, mainly alluvial proluvial mixed with lacustrine.

4.2 *Internal characteristics of the reach*

To explore the internal characteristics of the Qiaojia Reach, a wide-area electromagnetic profile and a series of boreholes were arranged in the Qiaojia Reach (Figure 2), of which the borehole numbered QK9 reached a depth of 748 m and penetrated 19.8 m into the bedrock, exposing all the deep overburden of the Qiaojia County reach. Sections A-B, C-D, and E-F of the Qiaojia Reach as shown in Figures 3–5 were established through wide area electromagnetic method, borehole exploration, and comprehensive analysis across the river. It can be seen from Section 3 that the deepest fault depression of the Qiaojia Reach exceeds 728.2 m, which is composed of layered multi-stage proluvial facies, lacustrine facies, fluvial facies, and alluvial proluvial facies. The drilling sequence of bedrock at the bottom of the reach is the basalt of Emei Mountain and Qixia-Maokou Formation limestone. The overburden of the Qiaojia Reach is the deepest in the vertical direction near Qiaojia County, and the thickness of the overburden in the north-south section is relatively shallow, which also presents multi-stage sedimentary facies (Figure 4). The thickness of the shallow surface layer of the Chapengzi-Yingpanjiao profile is about 160 m, which is mainly composed of alluvial I-III-level terraces and proluvial (Figure 5).

Through logging and sedimentary facies analysis of all the boreholes, the filling sequence of the Qiaojia Reach was established, as shown in Figure 6. The sediments in the Qiaojia Reach are composed of gravel layers dominated by limestone, basalt, mud, and sand layers dominated by silt and fine sand, which are 6 quasi-cycles developed. According to the characteristics of sedimentary cycles, the sedimentary strata in the reach can be divided into 6 parasequences, which are briefly described from bottom to top as follows:

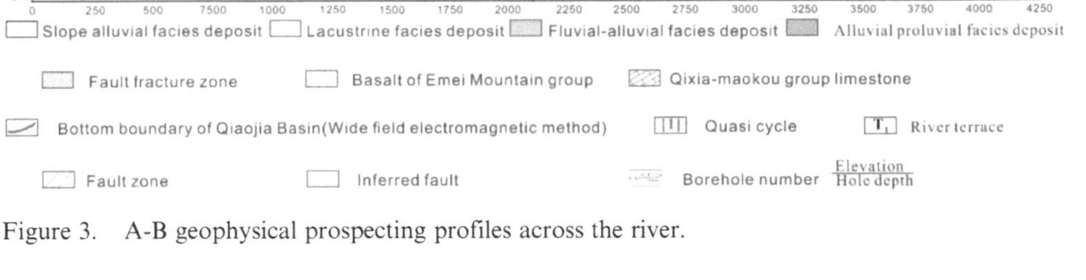

Figure 3. A-B geophysical prospecting profiles across the river.

The first layer (hole depth of 388.9-728.2 m): an alluvial proluvial-fluvial sedimentary layer. The gravel composition is mainly limestone and basalt, the gravel diameter is between 2-90 mm, the roundness is mostly angular to sub-circular, and the sorting is average. The muddy sand layer is mainly composed of coarse sand and mud.

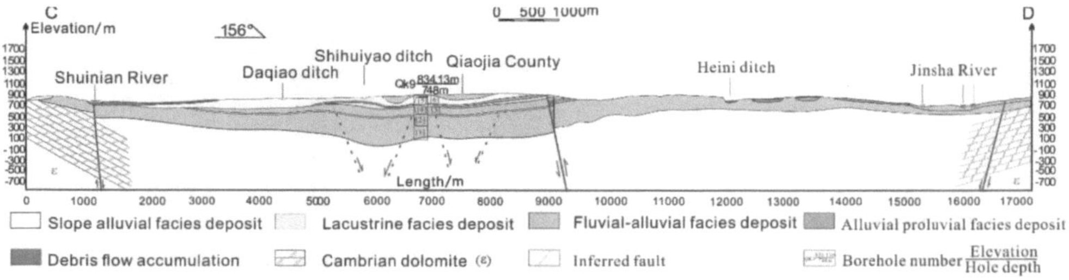

Figure 4. C-D profiles along the river.

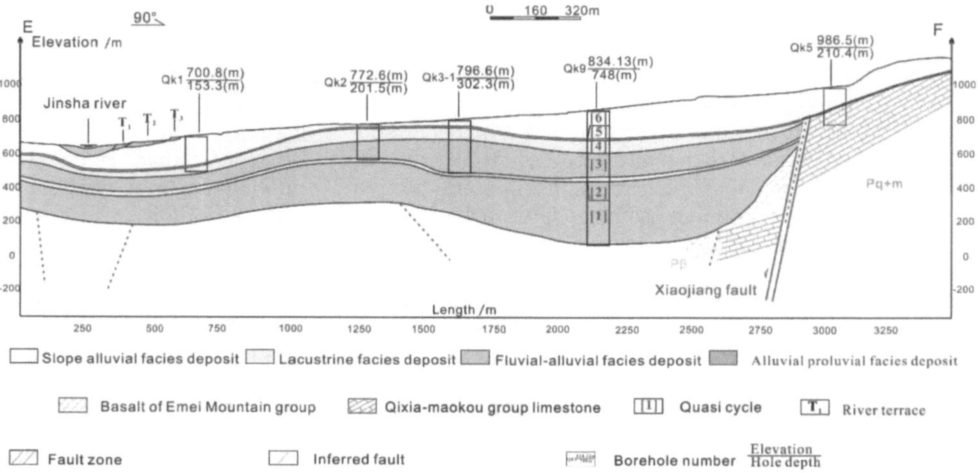

Figure 5. E-F shallow deposition structure characteristics.

The second layer (hole depth of 361.34-388.9 m): is a lacustrine sedimentary layer. It is inferred that it is a small fault depression period, which is a fine sand-argillaceous deposit of lacustrine facies. Due to the change in lake depth, the color of the sediment also changes to brownish yellow, yellowish brown, and grayish green. There is a 1.25 m thick gravel layer, mainly fine gravel with a particle size of 5-10 mm, and the sediment contains grayish-black carbon chips.

The third layer (hole depth of 220.4-361.34 m): an alluvial proluvial-fluvial sedimentary layer. The gravel is mainly composed of basalt and limestone, with general roundness, gravel diameter of 2-90 mm, and average sorting. The muddy sand layer is mainly composed of coarse sand and argillaceous silt.

The fourth layer (hole depth of 169.08-220.4 m): is a lacustrine sedimentary layer. It is inferred to be a large fault depression period, which is a sandy argillaceous deposit in the lacustrine facies. Due to the changes in lake depth, the color of the sediment also changes to brownish red, grayish brown, and grayish green. Horizontal bedding and cross-bedding can be seen.

The fifth layer (hole depth of 160.80-169.08 m): is an alluvial proluvial sedimentary layer. The gravel is composed of basalt, limestone, and sandstone, of which sandstone accounts for a small proportion and the sorting is poor. The gravel diameter is 2-80 mm, and the roundness is sub-angular to sub-circular. The muddy sand layer is mainly composed of fine sand and silt sand.

The sixth layer (hole depth of 0-160.8 m): is a proluvial slope sedimentary layer. The overall appearance is brownish yellow-brownish red. The gravels in this layer are seriously weathered, mostly moderately strongly weathered - completely weathered. The gravel is composed of near-source sediments such as basalt and limestone, with poor sorting. The gravel diameter is 2-90 mm and the roundness is angular to sub-angular. The muddy sand layer is mainly muddy with gravel.

The genetic types and combination models of sediments in the reach can usually reflect the activity of the fault zone in different periods and the specific sedimentary environment of the reach with deep overburden, which can also reflect the evolution stage of the reach. As shown in Figure 6, the evolution stage of the Qiaojia Reach can be divided into 6 stages. Among them, the second and fourth lacustrine facies belong to the period of strong fault period of the Qiaojia Reach overburden. It is inferred that the Xiaojiang Fault zone and the Zemuhe Fault zone have strong activities at this stage. The remaining parasequences are in the slow-filling stage of the reach, and the tectonic activity is relatively weak. The I-III-level river terraces can be seen on both banks of the Jinsha River, which are not completely overlying the above-old accumulation layer.

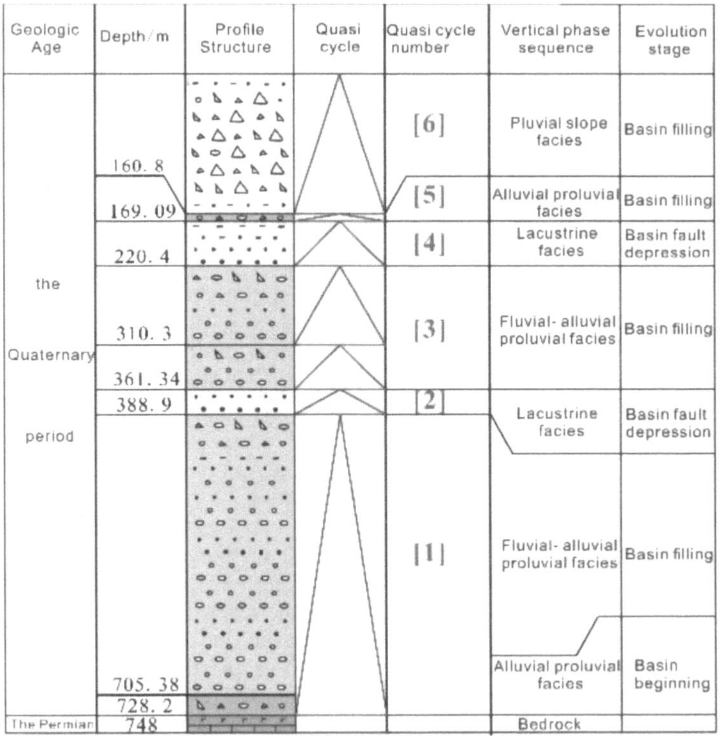

Figure 6. Filling sequence of the Qiaojia Reach.

5 THE FAULT DEPRESSION RATE AND EVOLUTION PROCESS OF THE QIAOJIAHE REACH

5.1 *The fault depression rate of the reach*

The QK9 borehole is a good time record carrier for the deep overburden of the Qiaojia Reach. Due to the exposure of the core, some dating cannot be realized. However, in recent years, geologists and geographers (He 2018) have provided the dating data of the Qiaojia stratified geomorphic system (Table 2), which laid a good foundation for the evolution analysis of the reach. The fault depression rates of the Qiaojia Reach in different periods were obtained.

Table 2. Basic parameters of terrace in the Qiaojia Reach.

Grade	The River elevation	Terrace type	Age (ka)
T1	15 (655)	Accumulation terrace	9-11 (Yang et al. 2009 and Tao et al. 2017, OSL)
T2	30 (670)	Accumulation terrace	15-20 (Yang et al. 2009 and Tao et al. 2017, OSL)
T3	65-75 (704)	Basement terrace	30 (He 2018)
T4	100-105 (733)	Basement terrace	80 (He 2018)
T5	135 (765)	Basement terrace	170 (He 2018)
T6	206 (836)	Basement terrace	220 (He 2018)
T7	238 (868)	Basement terrace	290 (He 2018)
T8	258 (888)	Basement terrace	440 (He 2018)
T9	298 (928)	Erosion terrace	1,200 (He 2018)
T10	382 (1,012)	Erosion terrace	1,290 (He 2018)
T11	415 (1,045-1,050)	Erosion terrace	1,590 (He 2018)

A survey on the right bank of the Qiaojia Reach by He (2018) revealed nine-level terraces. Through comparison, the river elevation of T1 is equivalent to our T3. Thus, there are eleven-level terraces in the Qiaojia Reach (Table 2). According to the site review, T1-T2 is the accumulation terrace, T3-T8 is the base terrace, and T9-T11 is the erosion terrace. The elevation of the front edge of the Daping platform (T11) is about 1,050 m.

According to the recovery of stratum distribution before the fault depression, the reach should start fault depression from the front edge of the Daping platform at 1,050 m, because the bedrock at the bottom of the reach is completely consistent with the lithology and horizon at the top of Daping platform, and corresponds exactly to the longitudinal extension space. In other words, the fault depression began on the T11 terrace, and the fault depression occurred 1.59 million years ago. At present, the elevation of the river alluvial component in the filling residual in the reach is about 890 m, which corresponds to the elevation of the T8 terrace (the highest elevation of alluvial residual in Kuigeliangzi area is 947 m, which is the debris flow and alluvial accumulation in back mountain). Therefore, the fault depression of the reach ended 440,000 years ago, the elevation of the reach foundation interface (QK9) is 106 m, the fault depression amplitude is 944 m, the fault depression rate is 0.82 mm/a, and the filling rate is 0.64 mm/a. During the past 440,000 years, the Qiaojia Reach entered into the normal river intermittent undercutting action, with an average undercutting rate of 0.59 mm/a, forming eight-level terraces. The field investigation shows that the IV-VII terraces are only partially preserved, the I-III terraces are well preserved, and terrace II is superimposed with a layer of weir plug silt fine sand.

The fault depression analysis of each parasequence is as follows:

The first layer corresponds to the 11th terrace to the 10th erosion terrace in time, which lasted for 300,000 years, and the average fault depression rate is only 0.1 mm/a. The fluvial facies superimposes the near-source diluvium of the primary tributary, and the near-source gravel has poor roundness, whose composition is consistent with the lithology (basalt and limestone) distributed in the tributaries on both banks of the Qiaojia Reach.

The second layer corresponds to the 10th terrace to the 9th terrace in time, which lasted for 90,000 years, with an average fault depression rate of 0.93 mm/a. A relatively clean water environment and lacustrine deposits were formed in the reach. There were obvious fluctuations in the fault depression. When the fault depression slowed down, a shallow water environment was formed, which accepted the fluvial facies deposition dominated by fine gravel, showing strong hydrodynamic conditions.

The third layer corresponds to the early stage of the formation period from the 9th terrace to the 8th terrace. The fault depression rate is relatively slow. The composition and diameter of the gravel are similar to those of the first layer. According to the sedimentary analogy, it lasted about 110,000 years.

The fourth layer corresponds in time to the middle of the formation period from the 9th terrace to the 8th terrace. The fault subsidence rate increases sharply, forming a relatively clean water environment and forming a large area of lacustrine deposit, which lasted for more than 500,000 years.

The fifth layer corresponds to the late stage of the formation period from the 9th terrace to the 8th terrace in time. The rate of fault depression sharply decreases, and the sedimentary riverbed facies and provides are estimated to last for 100,000 years by analogy.

The sixth layer corresponds to the end of the formation period of the 9th terrace to the 8th terrace in time. It is an accumulation of proluvial and debris flow. The thickness of the proluvial slope facies deposit of Qiaojia County is 200 m, with a distribution area of about 4×10^6 m^2 and a volume of 3×10^8 m^3. And the area of back mountain denudation is 1500×10^4 m^2. According to the "*Standards for Classification and Gradation of Soil Erosion*", the study area is a strongly eroded area, with an erosion modulus of 3.7~5.9 mm/a per year. Therefore, the alluvial proluvial layer with debris flow accumulation can be formed at the top in 3,000-5,400 years.

5.2 *Evolution of the reach*

The development of fault depression in the Qiaojia Reach is controlled by the staggered sinistral strike-slip pull-apart of the Xiaojiang Fault in the east and the Zemuhe Fault in the west. According to the spatial distribution change of the Xiaojiang Fault and the material composition of fault depression revealed by drilling in the Qiaojia Reach, the formation and evolution of fault depression reach can be divided into three stages: the embryonic stage of the formation of pull-apart fault depression reach, the pull-apart fault depression stage, and the post fault depression reformation stage.

The embryonic stage of the formation of pull-apart fault depression reach: when the Xiaojiang Fault and the Zemuhe Fault were in the late Pliocene, the strong crustal activity in the third episode of the Himalayan period resulted in the strong sinistral strike-slip activity. The Qiaojia Reach was just in the pull-apart area bounded by the initial section of the Xiaojiang Fault and the south end of the Zemuhe Fault (Figure 7a). The north-south boundary fault of the faulted reach is the transverse normal fault formed at the left front end by the strike-slip of the Zemuhe Fault and the Xiaojiang Fault, the east and west ends of which are strictly controlled by the Xiaojiang Fault and the Zemuhe Fault. When the normal fault continues to expand and connects with the adjacent strike-slip faults, the boundary of the faulted reach is fully connected and the faulted reach is formed (Figures 7a and 7b).

The pull-apart fault depression stage: a north-north-west to south-south-east stretch occurred within the pull-apart reach, forming the southern and northern boundaries of the north-south fault depression, and the internal blocks fall relatively. In the early stage, due to the slow fault depression rate, the filling and fault depression rates were close to balance, mainly alluvial and alluvial proluvial, which was followed by a rapid fault depression, forming a clean water environment and depositing silt fine sand layers. Then, the fault depression rate slowed down. The river pebble and gravel layer was deposited, and then entered the rapid fault depression period, where the lacustrine silt layer was deposited. With the strengthening of the strike-slip movement of the fault zone, intermittent secondary faults began to develop in Xiaojiang Fault Zone and Zemuhe fault zone (Figure 7c). Intermittent secondary faults continued to extend under the action of tectonic activities, connecting the

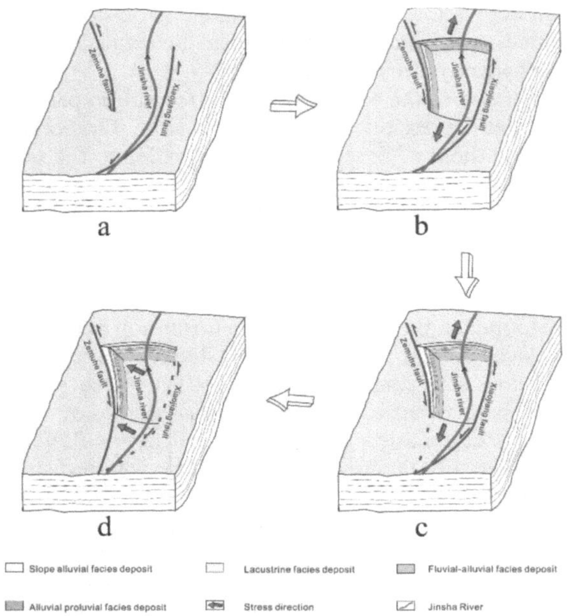

Figure 7. The schematic diagram of the evolution of the fault depression in Qiaojia Reach. a: In the early stage of Early Pleistocene, the south section of the Zemu River and the north end of the Xiaojiang Fault were distributed obliquely at the left boundary, forming a strike-slip pull-apart trend; b: In the middle stage of Early Pleistocene, the north-south boundary was formed and the strike-slip pull-apart stage was entered; c: In the early stage of the Middle Pleistocene, the southern segment of the Zemuhe Fault continued to extend southward; d: In the early stage of the Middle Pleistocene, the Zemuhe Fault was finally connected with the Xiaojiang Fault, and the fault depression activity ended.

Xiaojiang Fault and the Zemuhe fault on the opposite bank of Kugliangzi (Figure 7d). At this stage, with the connection of the Xiaojiang Fault and the Zemuhe Fault on the left bank of the Jinsha River, the Qiaojia-Liantang section at the north end of the Xiaojiang Fault Zone was replaced (the activity disappeared). At this time, the faulted reach changed from near north-south tensile stress to northwest tensile stress. Under the action of north-west tensile stress, the Qiaojia Reach continued to be faulted and deposited.

The post-fault depression reformation stage: after the 440,000 years of fault depression sliding ended, the Jinsha River entered the stage of normal intermittent river undercutting reconstruction, and successively formed eight-level terraces. These terraces are characterized by the development of round boulder gravel. The a-b planes tend to the upstream and the clay content is very low. Among them, the 8th terrace to the third terrace is the base terraces of the fault depression reach, and the first and second terraces are accumulation terraces. The second terrace is the most developed.

6 THE FAULT DEPRESSION RATE AND EVOLUTION PROCESS OF THE QIAOJIAHE REACH

According to the large-scale detailed survey on site, the exposure of ultra-deep boreholes, geophysical exploration, radon measurement verification, and large-section excavation profile recording, a new understanding of the boundary, boundary fracture characteristics, and filling sequence characteristics of the deep overburden in the Qiaojia Reach has been achieved. Furthermore, the reason for the ultra-deep overburden is explained, which provides a scientific basis for the evolution analysis of the Jinsha River valley and the risk assessment of Qiaojia resettlement sites. The main progress is as follows:

(1) The boundary of the Qiaojia Reach has been redefined: the west boundary is extended to the Zemuhe Fault, the north boundary is the Hulukou-Liming Village line, the south boundary is Zihong Village-Shigaodi line, and the east boundary is still the Qiaojia section of the Xiao Jiang Fault. The deep overburden of the Qiaojia Reach has a long radish shape with an area of about 70 km^2.
(2) There are six cycles in the riverbed cover of the deep overburden in Qiaojia Reach. The first five cycles are the filling sequence of faulted reach, of which there are two obvious rapid faulted lacustrine formation periods. The latest cycle is the erosion-deposition cycle of the reach following the upstream and downstream into normal rivers after the termination of fault depression activity.
(3) The deep overburden of Qiaojia Reach is a fault depression reach formed by sinistral strike-slip pull-apart of the south end of the Zemuhe Fault and the north end of the Qiaojia Fault on the basis of the early Pleistocene wide valley. The fault depression began on the basis of a 1050 m denudation surface in the middle of the Early Pleistocene, 1.59 million years ago. The fault depression rate was about 0.82 mm/a, and the filling rate reached 0.64 mm/a. 440,000 years ago, the river bottom fault depression reached 105.93 m above sea level (geophysical exploration shows that the deepest point may be lower than this elevation, which is subject to the disclosure of boreholes in this paper). At this time, the Muhe Fault and Xiaojiang Fault are connected in the south of each, the strike-slip pull-apart tectonic environment disappears, the fault depression activity ends, entering the normal river undercutting mode and forming eight-level terraces. The tendency of gravel on ab surface reveals that the Jinsha River has been connected at this time and flowed in the same direction as modern times.
(4) Since the Middle Pleistocene (440,000 years ago), the Xiaojiang Fault Zone has connected with the Zemuhe Fault in the Reshui Village area, so the activity of the east boundary of the reach to the north end of the Xiaojiang Fault has weakened, except that the west boundary fault to the Zemuhe Fault still has a strong strike-slip activity rate. Therefore, the tectonic stability of the Qiaojia County settlement is better because of the weakening of activity in the north end of the Xiao Jiang Fault.

REFERENCES

Chen J., Shi Z. T., and Su H. Grain Size Characteristics and Genesis of Loess in Qiaojia Basin [J]. *Study on Geographical Environment of Yunnan*. 2009: 03: 46–52.

He H. and Oguchi T. Late Quaternary Activity of the Zemuhe and Xiaojiang Faults in Southwest China from Geomorphological Mapping [J]. *Geomorphology*. 2008: 1–2: 62–85.

He R. *Study on the Formation Age of the Dry Hot Valley Based on River Terrace Records in the Qiaojia Section of Jinsha River* [D]. Yunnan Normal University, 2018.

Jiang F. C., Wu X. H., and Wang S. B. Age of Loess Like Deposits in Qiaojia Section of Jinsha River [J]. *Acta geomechanics*. 1999: 04: 35–40.

Li L., Yang D., and Huang D. Water System Evolution of Qiaojia Xinshizhen Reach of Jinshajiang River. *Quaternary Research*. 2009: 02: 327–333.

Pei X., Li T., Huang R., and Wang S. Structural Characteristics and Evolution of Qiaojia Pull apart Basin. *Journal of Southwest Jiaotong University*. 2019: 002: 278–286.

Shi J. Deep Overburden of Dadu River Bed and Its Engineering Geological Problems [J]. *Sichuan Hydropower*. 1986: 03: 14–19.

Sichuan Bureau of Geology and Mineral Resources (1995). *Description of the Geological Map of the People's Republic of China 1:50000 (Qiaojia Sheet g-48-38-b)*. Panxi geological printing house, Unknown place of publication.

Song F., Wang Y., and Yu W. *China Active Fault Research Album - Xiaojiang Active Fault Zone*. Beijing: Seismological Press, Beijing. 1998.

Sun Y. and Wang Q. Deep Overburden Characteristics and Engineering Significance of Hutiaoxia Reach of Jinsha River [J]. *Renmin Changjiang River.* 2011: 07: 1–4.

Sun Y., Wang Y., and Wu J. Characteristics and Genetic Mechanism of Deep Overburden and Unloading Relaxation Zone at the Valley Bottom of a Hydropower Station in Southwest China and Their Genetic Mechanism [J]. *Acta engineering geology*. 2008: 02: 169–172.

Tao Y., Chang H., and Qiang X. Characteristics and Chronology of Terraces in the First Bend of the Yangtze River. *Quaternary studies*. 2018: 1: 151–164.

Wu J. F. *Study on the Engineering Geological Property of Deep Overburden in the Qizong Hydroelectric Power Station, Jinsha River* [D]. Chengdu University of Technology, 2009.

Xu Q., Chen W., and Jin H. Characteristics and Development of Deep Overburden in Dadu River Valley [J]. *Quaternary research.* 2010: 1: 30–36.

Xu Q., Chen W., and Zhang Z. A New Understanding of the Genetic Mechanism of Deep Overburden in Southwest China [J]. *Progress in Geosciences.* 2008: 05: 448–456.

Yang D., Li L., and Ge Z. *Study on Valley Geomorphology and Geological Evolution of Baihetan Hydropower Station in Jinsha River* [R]. Nanjing University, 2009.

Yasutaka I., Masayoshi T., Masayoshi T., Tomoo E., and Shinsuke O. Newly-generated Daliangshan Fault Zone Shortcutting on the Central Section of Xianshuihe-Xiaojiang Fault System [J]. *Science China: Earth Sciences*. 2008: 51 (9): 1248–1258.

Zhang X. and Wang Y. S. (2017). Study on the Activity of Xiaojiang Fault Zone in Baihetan Hydropower Station Reservoir Area. *Journal of engineering geology*. 2017: 02: 531–540.

Zhang X. *Study on the Activity and Disaster-causing Effects of the Middle and Northern Section of Xiaojiang Fracture* [D]. Chengdu University of Technology, 2019.

Zou C. and Yang T. *Qiaojia County Annals*. Yunnan People's Publishing House, Kunming. 1997.

Water Conservancy and Civil Construction – Oke & Ahmad (Eds)
 ISBN: 978-1-032-58614-4

Characteristics of PH, turbidity, and nitrogen transport of shallow groundwater in river banks under rainfall

Yuyu Ji*, Huang Jian*, Jinhua Wen*, Lei Fu*, Aiju You* & Qiannan Jin*
Zhejiang Institute of Hydraulics & Estuary, Hangzhou, Zhejiang Province, China

ABSTRACT: We took the riverbank beach in Shanxi, the middle reaches of the Cao'e River, as the research object. By monitoring the change process of pH, turbidity, NH_4^+-N, and NO_3^--N concentration in shallow groundwater, whose change indicators in the riverbank beach under rainfall were studied. The results show that under the action of rainfall, the river bank beach has a good filtering effect on the shallow groundwater. The turbidity of the shallow groundwater at the river inlet section is only 55~106, which is lower than 159 of the river and has a certain impact on the pH in the water body. The pH in the shallow groundwater is significantly lower than that of the river. The rainfall caused the riverbank soil to be in a stagnant state, and NH_4^+-N accumulated. The average concentration of NH_4^+-N in # 1 reached 3.60 mg/L, which is 12 times that of surface water. As a result, the shallow groundwater NO_3^--N in the riparian zone decreases, and the average concentration of NO_3^--N in # 1 is 0.24 mg/L, which is only 1/8 of the surface water, indicating that the near water end of the riparian zone is the key zone for denitrification to remove nitrogen pollution, and the input of NH_4^+-N in the riparian zone needs to be focused.

1 INTRODUCTION

In recent decades, with the development of intensive agriculture and the massive application of chemical fertilizers and pesticides, nitrogen pollution in shallow groundwater has become increasingly serious, and the shortage of high-quality water resources caused by shallow groundwater pollution has become increasingly serious, which has become a worldwide concern [1,2]. Especially in China, where water resources are scarce and extremely unevenly distributed, about 20% of the water supply comes from groundwater [3]. The pollution of shallow groundwater will aggravate the eutrophication of surface water near the river and lake banks [4]. At the same time, it will bring certain risks to people's production, life, health, and safety. For example, drinking a certain amount of nitrate water can form methemoglobin, which can lead to hypoxia poisoning [5].

As an important zone for the hydrological cycle and material exchange between shallow groundwater and surface water [6], the nitrogen content of shallow groundwater directly affects the nitrogen level of rivers or lake surface water. At present, several scholars have observed and analyzed the nitrogen content of shallow groundwater, such as Yang Heng's analysis of nitrogen content in shallow groundwater around the lake [7], and Wang Jie's analysis of the temporal and spatial distribution of soil nitrogen in the riparian zone [8]. Chen Lihui et al. studied the influence of dry and wet environments on nitrogen content and spatial distribution characteristics of the riparian zone [9]. Nitrogen pollution in rivers and

*Corresponding Authors: jiyuyu2016@163.com, 877429264@qq.com, 422497597@qq.com, fulei8642@dingtalk.com, youaj@zjwater.gov.cn and 252102227@qq.com

DOI: 10.1201/9781003450818-35

lakes mainly comes from agricultural non-point source pollution input with surface and underground runoff during the rainy season [10]. The amount of nitrogen pollutants entering the river during the rainy season accounts for 81% of the whole year. The nitrogen pollutants carried by rainfall infiltrate into the shallow groundwater in the farmland area and then migrate to the riparian zone through the shallow groundwater runoff. This process is much slower than surface runoff, and the impact on the river is more lasting, However, the basic physical and chemical indexes and nitrogen transport laws of the riparian zone under the effect of rainfall in this process have not been clearly understood at present, which is difficult to guide the further development of river and management of lake water environment.

Based on this, this study selected a riverbank beach in Shanxi, the middle reaches of the Caoejiang River, as the research object, to dynamically observe the pH, turbidity, NH_4^+-N, and the NO_3^--N transport process of shallow groundwater in the riverbank beach during rainfall runoff. At the same time, we analyzed its laws, which provided a reliable theoretical basis and reference for the protection and restoration of ecological ecotones such as riverbank zone and beaches.

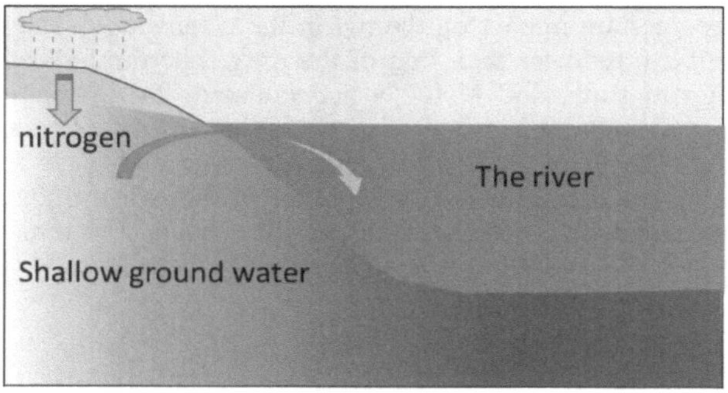

Figure 1. The schematic diagram of the river shoal.

2 MATERIALS AND METHODS

2.1 *Sample collection*

To reveal the nitrogen transport law in the process of water exchange between shallow groundwater on the river bank and the river, during the period when the river water level is relatively stable before rainfall, shallow groundwater sampling devices are buried along the direction from the river to the river bank. During rainfall, shallow groundwater is collected at intervals of 2 hours. A total of 75 water samples are collected. Specific collection and storage methods are as follows:

In the process of collecting the shallow groundwater in the river bank and shoal, the soil solution sampler (SQR100, Shenzhen Hengjinda Technology Co., Ltd., China) is used to place the sampling pump near the monitoring well. After it is stable, the shallow groundwater in the river bank and shoal is slowly collected at a sampling speed of 0~0.083 L/min to minimize the disturbance to the groundwater body during the sampling process. To avoid interference with relevant indicators of groundwater (pH, turbidity, etc.), the samples were collected and filtered by GF/F membrane (Whatman, Germany), then put into 500 ml polyethylene bottles cleaned by deionized water, and sealed with sealing membrane at 4 °C for cold storage.

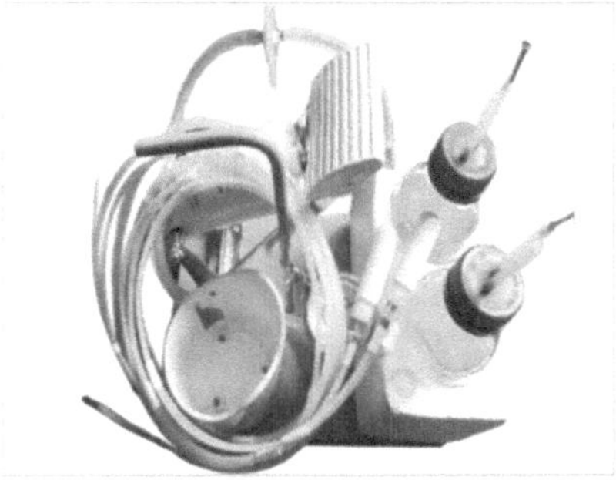

Figure 2. Soil solution sampler.

2.2 *Sample determination*

After the water sample is taken back to the laboratory, the concentrations of ammonia nitrogen (NH_4^+-N) and nitrate nitrogen (NO_3^--N) in the water body are measured respectively. The specific determination method is as follows: NH_4^+-N and NO_3^--N in the water sample are determined by 2 mol/L potassium chloride solution extraction spectrophotometry; the pH and turbidity in the water sample are determined by the potentiometric method and the Hach multi-parameter water quality analyzer (MS6100, Hach Company, USA).

3 RESULTS & DISCUSSION

The internal pH of the river beach is significantly lower than that of the river, and the pH of the river is 7.8. The maximum pH of # 1~# 4 in the river beach is 7.6, 7.7, 7.7, and 7.5, respectively. At 0 h, the pH of the shallow groundwater near Monitoring Well # 3 is the highest. At 2 h, the pH of the shallow groundwater near Monitoring Well # 3 decreases to some extent. At 4 h, the pH of the shallow groundwater near Monitoring Well # 3 increases again. At 6 h and 8 h, the pH of the shallow groundwater near Monitoring Well # 2 increases to some extent. We can see more details in Figure 3. The turbidity of shallow groundwater in the near water end of the river bank is lower than that of the river, while the turbidity of shallow groundwater in the far water end is higher than that of the river. The turbidity of shallow groundwater near Monitoring Well # 4 is the highest, reaching 258~336. The turbidity of shallow groundwater near Monitoring Well # 3 decreases to 24~62, and then gradually increases towards the river. At other times, the turbidity in the shallow groundwater near the four monitoring wells presents the same rule. At the same time, the turbidity in the shallow groundwater near each monitoring well is the lowest at 0 h, and the turbidity in the shallow groundwater near each monitoring well has increased in the later period, as seen in Figure 4 below for details.

The river bank beach has a good filtering effect on shallow groundwater, which has high turbidity and a certain impact on the pH of the water body. Under the effect of rainfall, the rainwater infiltrated into the groundwater will flow along the saturated water layer in the soil among the soil pores, forming a flow in the soil [11]. During the formation of the flow in the soil, plant roots and soil colloids have a strong interception and retention effect on suspended particles in the shallow groundwater, which can play a role in reducing the turbidity of shallow groundwater [12]. This study shows that after the rainfall generates runoff, the surface water with high turbidity infiltrates into the shallow groundwater. After a certain distance of transportation, the turbidity content decreases from the far water end to the near water end. The turbidity at the river inlet section is only 55~106, which is lower than 159 of

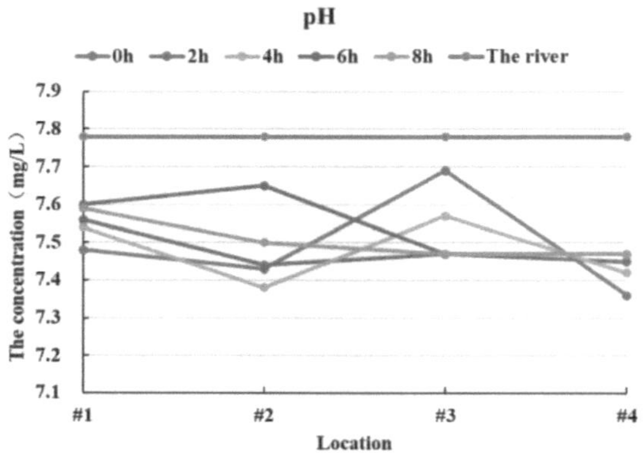

Figure 3. Change process of pH in the river bank.

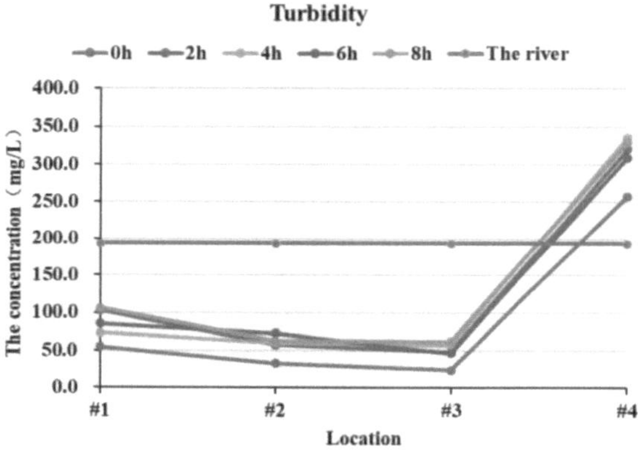

Figure 4. Change process of turbidity in the river bank.

the river. In addition, affected by rainfall, rainwater will dilute the shallow groundwater after infiltration, which may be the main reason why the pH of shallow groundwater in this study is significantly lower than that of the river.

The ammonia nitrogen content in the shallow groundwater inside the river bank is significantly higher than that in the river. The ammonia nitrogen content in the river water is 0.295 mg/L, while the lowest ammonia nitrogen content in the shallow groundwater near the monitoring wells # 1~# 4 are 3.19 mg/L, 2.09 mg/L, 2.90 mg/L, and 3.22 mg/L, respectively. At 0 h, the ammonia nitrogen content in the shallow groundwater near Monitoring Well # 4 was the highest. With the transportation of shallow groundwater, the ammonia nitrogen in the shallow groundwater near Monitoring Wells # 4~# 2 showed a downward trend, but there was a certain aggregation effect in the range of # 2~# 1, and the ammonia nitrogen concentration reached 3.19~4.22 mg/L respectively. The nitrate nitrogen content of the shallow groundwater on the river bank is significantly lower than that in the river. The nitrate nitrogen content in the river water is 1.82 mg/L, while the highest nitrate nitrogen content in the shallow groundwater near the monitoring wells # 1~# 4 is 0.88 mg/L, 0.12 mg/L, 0.67 mg/L, and 0.21 mg/L. At 0 h, the nitrate nitrogen content in the shallow groundwater near # 4~# 1 monitoring wells is higher than that at other times, and there are two low points at # 2 and # 4, and two high points at # 1 and # 3. At 2 h~8 h, the nitrate nitrogen trend of ammonia nitrogen in shallow groundwater is relatively consistent, and the nitrate nitrogen in shallow groundwater near Monitoring Well # 2 is still low, as seen in Figure 5 below for details.

Under the action of rainfall, NH_4^+-N accumulation and NO_3^--N decrease in shallow groundwater in the riparian zone. Soil nitrogen in riparian zone mainly comes from the input of nitrogen fertilizer, animal, and plant residues, biological nitrogen fixation, etc. Its migration and transformation process involves nitrogen mineralization, nitrification, and denitrification [13], the existing form and content of nitrogen mainly depend on the internal redox environment. In this study, rainfall caused the riverbank soil to be in a stagnant state, and NH_4^+-N accumulated. The average concentration of NH_4^+-N at # 1 reached 3.60 mg/L, which is 12 times that of surface water. The soil organic nitrogen generates NH_4^+-N under the action of microorganisms, and then generates NO_2^--N and NO_3^--N under the action of nitrification, which is further reduced under the action of denitrification, leading to the reduction of NO_3^--N. In this study, the average concentration of NO_3^--N at # 1 is 0.24 mg/L, which is only 1/8 of the surface water, as shown in Figure 6 below. In addition, some studies have shown that the soil is exposed to the air for a long time in the drought period, and its strong oxidation environment promotes the nitrification process, resulting in the reduction of NH_4^+-N content and the increase of NO_3^--N production. Therefore, the average content of NH_4^+-N in the riparian zone is low in the drought period, while the average content of NO_3^--N is high.

The riparian zone is the key zone for denitrification to remove nitrogen pollution. Denitrification is to convert NO_3^--N into N_2 and N_2O by microorganisms and release them into the atmosphere so that the part of the nitrogen is permanently removed from the soil. Therefore, denitrification is considered to be the best way to eliminate NO_3^--N and prevent it from entering the land water body. Many studies have confirmed that denitrification is the main mechanism of ammonia interception and transformation in the riparian zone. Denitrification is a process in which active denitrifying bacteria reduce nitrate or nitrite to gaseous nitrogen under sufficient NO_3^--N, organic carbon, and anaerobic conditions [14]. As the riparian zone is a typical high-productivity ecosystem, containing the amount of unstable organic substances, which is located at the edge of the river, it is often in a water-saturated state under the dual effects of rainfall runoff and river surface water, which is easy to form an anoxic environment [15]. While the continuous input of nitrogen in the shallow

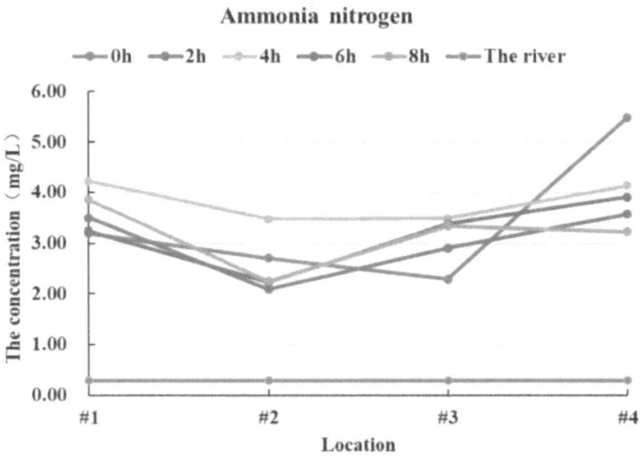

Figure 5. Change process of ammonia nitrogen in the river bank.

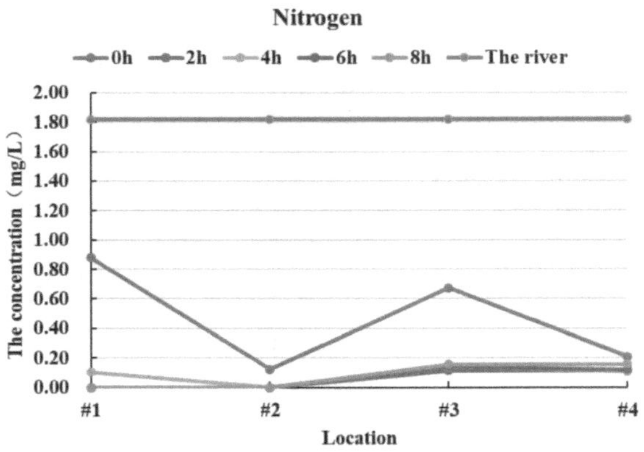

Figure 6. Change process of nitrogen in the river bank.

groundwater adjacent to the riparian zone highlands provides sufficient inorganic ammonia for denitrification, the riparian zone is the key zone for denitrification to remove nitrogen pollution. Generally, the NO_3^--N content in surface water accounts for 80%~90% of TN content [16], while the NO_3^--N content in groundwater in this study only accounts for 2%~20% of TN content, indicating that denitrification in riparian zone plays an important role in reducing nitrogen pollution input.

4 CONCLUSIONS

This chapter selects a riverbank beach in Shanxi, the middle reaches of the Cao'e River, as the research object. By monitoring the concentration change process of pH, turbidity, ammonia nitrogen, and nitrate nitrogen in shallow groundwater, the change characteristics of pH, turbidity, NH_4^+-N, and NO_3^--N indicators in the riverbank beach under rainfall are studied. The main conclusions are as follows:

(1) Under the effect of rainfall, the river bank beach has a good filtering effect on the shallow groundwater. The turbidity of the shallow groundwater at the river inlet section is only 55~106, which is lower than 159 of the river and has a certain impact on the pH of the water body. The pH of the shallow groundwater is significantly lower than the river pH.
(2) Under the effect of rainfall, the soil of the riverbank and shoal is in a stagnant state due to rainfall and accumulation of NH_4^+-N. The average concentration of NH_4^+-N in # 1 is 3.60 mg/L, which is 12 times that of surface water.
(3) Under the action of rainfall, NO_3^--N in shallow groundwater in the riparian zone will be reduced. The average concentration of NO_3^--N in # 1 is 0.24 mg/L, which is only 1/8 of the surface water. Moreover, 1.5 m near the water end of the riparian zone is the key zone for denitrification to remove nitrogen pollution.

ACKNOWLEDGMENT

This research was supported by the 2020 Science and Technology Plan Project of Zhejiang Provincial Water Resources Department (RC2045) and the Dean's Fund of Zhejiang Institute of Hydraulics & Estuary (ZH A20013).

REFERENCES

[1] Bechmann M. (2014). Long-term Monitoring of Nitrogen in Surface and Subsurface Runoff from Small Agricultural-dominated Catchments in Norway [J]. *Agriculture, Ecosystems Environment*, 198: 13–24.

[2] Cao S. W., Fei Y. H., Tian X., *et al.* (2021). Determining the Origin and Fate of Nitrate in the Nanyang Basin, Central China, Using Environmental Isotopes and the Bayesian Mixing Model[J]. *Environmental Science and Pollution Research*, 28 (35): 48343–48361.

[3] Chen L. H., Li H., Xiao J. W., *et al.* (2022). Effects of Dry and Wet Environment on Nitrogen Content and Spatial Distribution in the Riparian Zone [J]. *Journal of Shandong Agricultural University (Natural Science Edition)*, 53 (03): 401–405.

[4] Fowler D., Coyle M., Skiba U., *et al.* (2013). The Global I Nitrogen Cycle in the Twenty-first Century. *Philosophical Transactions of the Royal Society of London*, 368 (1621): 91–97.

[5] Li Feng, Wu Guangdong, Zhao Youlin, *et al.* (2022). Study on the Decoupling Trend and Spatial Aggregation of Groundwater Supply and Economic Development in China [J]. *Journal of Yangtze River Scientific Research Institute*, 39 (08): 159–166.

[6] Mahdizadeh K. M., Gholami S. M., Valipour M., *et al.* (2015). Simulation of Open and Closed-End Border Irrigation Systems Using SIRMOD [J]. *Arch. Agron. Soil Sci*, 61 (7): 929–94.

[7] Mayer P. M., Reynolds S. K., Mccutchen M. D., *et al.* (2007). Meta-analysis of Nitrogen Removal in Riparian Buffers [J]. *Journal of Environmental Quality*, 36 (4): 1172–1180.

[8] Pang M., Huang Z. L., and Zhang L. (2012). Experimental Study on the Effect of Nitrate Content on Denitrification in the Western Typical Area of Taihu Lake [J]. *Journal of Environmental Science*: 1–8.

[9] Tonian D. and Buffington J. M. (2007). Hyporheic Exchange in Gravel-bed Rivers with Pool-Riffle Morphology: Laboratory Experiments and Three-dimensional I Modeling [J]. *Water Resources Research*, 43: W01421

[10] Vavilin V. A. and Rytov S. V. (2015). The Non-linear Dynamic Model of Nitrogen Stable Isotope Fractionation with the Formation of Nitrous Oxide During Denitrification [J]. *Water Res*, 42 (2): 194–198.

[11] Wang J. (2022). Analysis of the Spatial and Temporal Distribution of Soil Nitrogen and Phosphorus and its Influencing Factors in the Riparian Zone [J]. *Applied Technology of Soil and Water Conservation*, 13–16.

[12] Xu Y., Xu Y. P., Wu L., *et al.* (2018). Dynamic Characteristics and Influencing Factors of Shallow Groundwater in the Water Network Area of Taihu Lake Basin [J]. *Journal of Lake Sciences*, 30 (02): 464–471.

[13] Yan Y., Ma T., Zhang J. W., *et al.* (2017). Experiment on Migration and Transformation of Nitrate Under the Interaction of Groundwater and Surface Water [J]. *Earth Science*, 42 (5): 783–792.

[14] Yang H., Li G. F., Ye Y. X., *et al.* Shallow Groundwater Around Plateau Lakes: Spatial-Temporal Distribution of Phosphorus and its Driving Factors [J]. (2022) *Environmental Science*, 43 (07): 3532–3542.

[15] Zhao S., Zhang B., Sun X., *et al.* (2021). Hot Spots and Hot Moments of Nitrogen Removal from Hyporheic and Riparian Zones: A Review [J]. *The Science of the Total Environment*, 762: 144–168.

[16] Zhu Q., Vries W. D., Liu X., *et al.* (2018). Enhanced Acidification in Chinese Croplands as Derived from Element Budgets in the Period 1980-2010 [J]. *Science of the Total Environment*, 618: 1497–1505.

Material control and structural repair and reinforcement

Water Conservancy and Civil Construction – Oke & Ahmad (Eds)
© 2024 The Author(s), ISBN: 978-1-032-58614-4

Overall dynamic analysis of FPSO ballast water pipeline support considering support function

Xiu Li*, Pei-Lin Dou & Shi-Fa Zhao
School of Naval Architecture and Ocean Engineering, Jiangsu University of Science and Technology, Zhenjiang, China

ABSTRACT: At present, the dynamic analysis of the pipeline system often simplifies the support into constraints, ignoring the influence of pipeline support deformation, which affects the correctness of the analysis results. Based on Ansys software, this paper took the pipeline section of a general Floating Production Storage and Offloading(FPSO) carrying ballast water as the research object and compared and analyzed the results of modal and harmonic response when considering the whole pipeline and simplifying the support into constraints. The results showed that, in modal analysis, simplifying the bracket into a constraint had a great influence on the results of modal analysis, which were higher than the actual value. In the harmonic response analysis, the way of simplifying the support into constraints also deviated from the actual situation, especially since the maximum amplitude along the Z axis was much smaller than the displacement response when the whole pipeline support was considered. The main reason for the above situation was that the bracket was simplified as a constraint, and the deformation influence of the bracket when excited was ignored, resulting in the stiffness value of the piping system in the calculation being greater than the actual stiffness of the piping system.

1 INTRODUCTION

As an important part of the ship, the pipeline system undertakes the task of conveying fluid, and it will vibrate under the action of pumps, internal fluids, and other random loads. As an important supporting component of the whole system, the support has a significant impact on the vibration of the pipeline system (Zhang *et al.* 2018). Dias (Dias *et al.* 2020) conducted a modal test on the gas-oil-water multiphase flow pipeline section through sensors. Al-Waily (Al-Waily *et al.* 2017) studied the effects of flow velocity and crack angle on pipeline vibration and fluid characteristics in the pipeline. Through ABAQUS finite element analysis software, Zhang *et al.* (2021) studied the vibration problems that may occur during the cleaning process of long-distance natural gas pipelines and put forward a pipe-soil coupling vibration model. Li (Li *et al.* 2021), North China University of Science and Technology, performed a modal analysis of oil pipelines under fluid-solid coupling by Ansys-Workbench but ignored the influence of supports in the system. Zhu (Zhu *et al.* 2018), based on the Timoshenko beam model, considered the interaction between fluid and the pipe wall, verified the correctness of the algorithm, and analyzed the influence of axial pretension and this boundary condition on the stability of the pipeline. Cao (Cao *et al.* 2018) and others modeled, simulated, and predicted the vibration characteristics of the pipeline in the aviation industry. In the above research, in the dynamic analysis of the pipeline system, the support was mostly

*Corresponding Author: 1512876905@qq.com

DOI: 10.1201/9781003450818-36

simplified as a constraint, while the vibration and deformation of the support under excitation were neglected. In this paper, the modal analysis and the harmonic response analysis results of two calculation methods, i.e., the whole pipeline support, and the simplified support as a constraint, were taken into account through the comparative analysis of some pipeline sections of a general FPSO ballast water system by Ansys software, so as to provide a reference for the dynamic numerical simulation analysis of the pipeline system in practice.

2 MODEL ESTABLISHMENT

2.1 *Numerical model*

Modal analysis methods include mathematical calculation, finite element method, and experimental analysis (Zhou *et al.* 2021). By discretizing the actual pipeline, a dynamic model for calculating the vibration characteristics of the pipeline can be obtained, and its general differential equation is expressed as follows (Fu & Hua 2000; Wang 1997):

$$[M][\ddot{x}] + [C][\dot{x}] + [K][x] = [F(t)] \tag{1}$$

The equation represents the mass matrix, the damping matrix, the stiffness matrix, the response vector, and the excitation vector. $[M]$ represents the quality matrix, $[C]$ represents the damping matrix, $[K]$ represents the stiffness matrix, $[x]$ represents the response vector, and $[F(t)]$ represents the excitation vector.

The coupled equations need to be decomposed into independent equations before being solved. Hence, the modal matrix and coordinates were introduced, and the parameters were specified. The Formula (1) was rewritten as the following steps: coordinate $[\varphi]$ and $[q]$, coordinate and specify parameters $[x] = [\varphi] + [q]$, and Equation (1) was rewritten into the following equation:

$$[\ddot{q}] + \begin{bmatrix} \ddots & & 0 \\ & 2\xi\omega_n & \\ 0 & & \ddots \end{bmatrix} [\ddot{q}] + \begin{bmatrix} \ddots & & 0 \\ & \omega_n^2 & \\ 0 & & \ddots \end{bmatrix} [q] = [\varphi]^T [f(t)] \tag{2}$$

The r-th modal of the coupled equations is stated as:

$$(k_r - \omega^2 M_r + j\omega C_\mathrm{r})q_r = F_r \tag{3}$$

In the formula, mass, stiffness, and damping were represented. In the equation $F_r = \Phi_r^T[\mathrm{f}(t)]$, M_r represents mass, K_r represents stiffness, and C_r represents damping.

$$q_r = \frac{F_r}{K_r - \omega^2 M_r + j\omega C_r} \tag{4}$$

The response of any point L on the structure can be written as:

$$\mathrm{x}_l(\omega) = \sum_r^N \varphi_{1r} q_r \tag{5}$$

2.2 *Pipeline system model*

According to the modal analysis of the pipeline support system, part of the pipeline section of a deepwater general FPSO was selected. The pipeline was made of FRP with good corrosion resistance. The length of the pipeline section was 7.328 m, the density was set at 8250 kg/m^3, the elastic modulus was 1.514 × 1011 Pa, and the Poisson's ratio was 0.33. The dimensions of each part of the structure are shown in Figure 1. The support selected by the system was the guide support, and the structural steel material was selected, which could be directly retrieved from the engineering database. The structural dimensions are shown in Table 1, and the structure is shown in Figure 2. Based on the three-dimensional model of the pipe section established by SolidWorks, after format conversion, it was imported into Ansys-Workbench, where the grid size was set to 0.05 m, the grid was divided, and the inlet and outlet of the fluid domain were set (only when considering the analysis of fluid-solid coupling; the other examples restrain the fluid domain) so that the overall numerical model of the pipe support (Figure 3) and the pipe model, when the support was simplified as a constraint (Figure 4), were obtained.

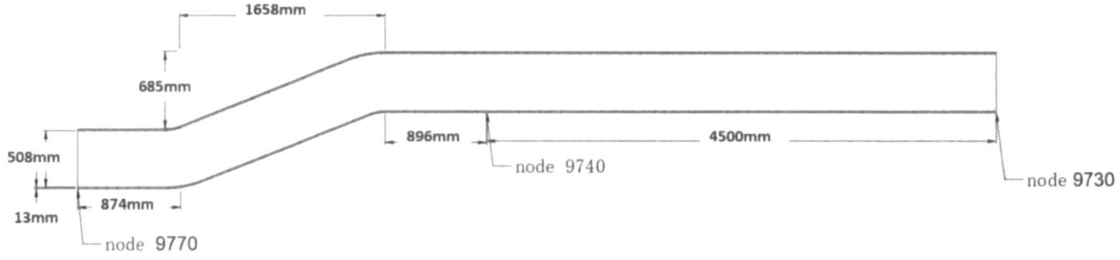

Figure 1. Structure diagram of the research pipe section.

Table 1. Parameter table of pipe support.

Overall Support Height /mm	697	Tube Height /mm	400
Clamp Ring Diameter /mm	508	Tube Thickness /mm	13
Thickness of Clamp Ring /mm	13	Tube Width /mm	150
Width of Clamp Ring /mm	150	Tube Bottom Plate Size /mm	634 × 150 × 13
Pipe Clamp Material	Structural Steel	Pipe Support Material	Structural Steel

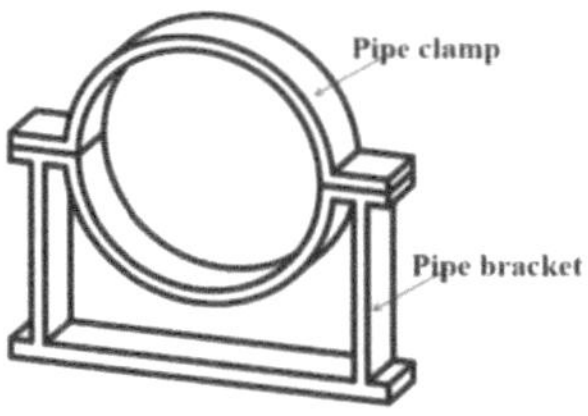

Figure 2. Structure diagram of pipeline support.

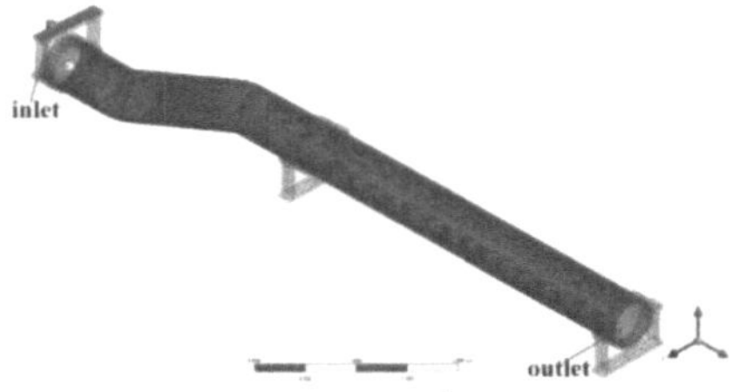

Figure 3. Overall analysis and calculation model of pipeline support.

Figure 4. Simplifying the support into a constraint calculation model.

3 MODAL ANALYSIS OF PIPELINE SYSTEM

At present, the finite element numerical simulation of modal analysis does not support nonlinear factors (Song & Ge 2020). The actual constraint between supports and pipes is between binding and frictionless. In order to better deal with this nonlinear problem, based on the model in Figure 3, the relative displacement between the pipe and the support under the action of fluid-solid coupling of water flow was analyzed so as to determine the real situation of its axial constraint. In the FPSO ballast water pipeline system, the pipeline was clamped by the bracket pipe and fixed by bolts. The water in the pipeline system was pumped by a water pump with a head of 6.7 m, so the pressure at the inlet was set at 67000 Pa, and the friction coefficient between the bracket and the pipeline was 0.2. Observe the deformation of the whole pipeline support system and the relative displacement between the pipeline and the support. The calculation results are shown in Figure 5. The maximum deformation of the pipeline system was 2.1491×10^{-6} m, which was negligible compared with the length of the whole pipeline of 7.328 m. Therefore, the binding contact method used in the calculation process was very close to the real constraint relationship. Under the studied working conditions, the stiffness of the pipeline was slightly improved by using the binding contact, and the modal analysis value was slightly larger, but the calculation model could be greatly simplified.

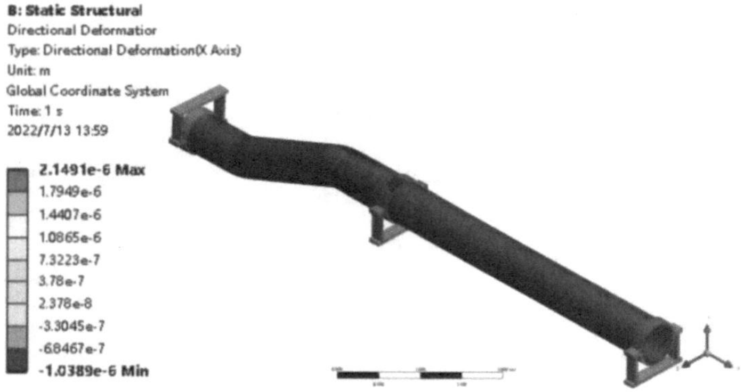

Figure 5. Deformation of the pipeline system.

(1) Mode of constraint: In the process of ship operation, especially the ballast water pipeline, when the ship is fully loaded, some pipelines are often empty, and only the support is constrained, so it is necessary to analyze such operating conditions. Based on the models shown in Figure 3 and Figure 4, the fluid domain was suppressed, and the first six-order

inherent characteristics of the pipe section, considering the whole pipe support and simplifying the support into a constraint mode, were analyzed by Workench's modal analysis module. The results are shown in Table 2.

Table 2. Results of the first six modal analysis of the system.

Project	First Order	Second Order	Third Order	Fourth Order	Fifth Order	Sixth Order
Pipe Support Whole	28.064	37.562	62.107	71.419	105.890	113.340
Pipes (Supports are Simplified as Constraints)	76.722	77.550	120.090	128.680	129.120	153.580

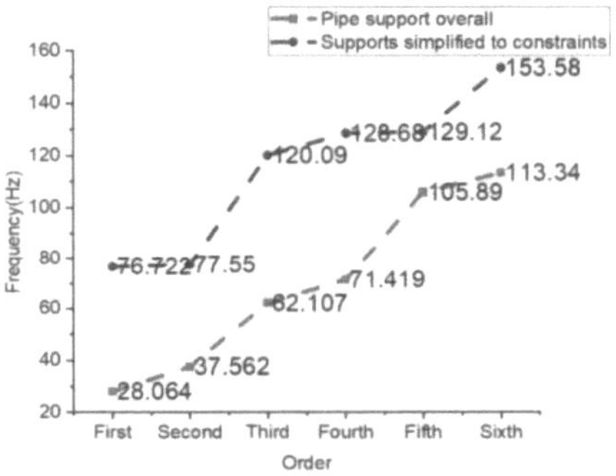

Figure 6. Results of the first six modal analysis of the system.

The above results were processed by the origin software, as shown in Figure 6. Through sorting out the first six modal analyses, it can be found that the modal frequency of simplified pipe support as a constraint was higher than that of the calculation form considering the whole pipe support.

(2) Considering the prestressed mode of fluid-solid coupling. The transportation of fluid is the main function of the pipeline system, so it is necessary to consider the interaction of fluid and structure in the modal analysis of the pipeline system. The pipe section selected in this paper was part of the general deep-water FPSO ballast water system, and its transport fluid was water. This time, the research object was the pipeline system, so the modal analysis of a single fluid-structure coupling could be done. The pump head used in the selected ballast water system was 6.7 m, so the fluid flow in the pipeline could be simulated in the form of a pressure inlet, and the inlet pressure was set at 67000 Pa. The models shown in Figures 3 and 4 were imported into Ansys, fluid analysis was carried out by the Fluent module, and the results were loaded on the static structure analysis module to further analyze the frequency of the pipeline system considering fluid-solid coupling. The first six frequencies are shown in Table 3.

The above results were processed by the original software. As shown in Figure 7, the pipeline system was analyzed by different calculation forms when considering the fluid-solid coupling condition, still showing that the modal frequency of the simplified pipeline support is higher than that of the calculation form considering the whole pipeline support.

Through the above analysis of two important working conditions of the pipeline system (constraint and considering fluid-solid coupling), it was found that the modal frequency of simplified pipeline support as a constraint was higher than that of the calculation form considering the whole pipeline support. The reason was that when the bracket was simplified as a

Table 3. Results of the first six modal analyses of the system.

Project	First Order	Second Order	Third Order	Fourth Order	Fifth Order	Sixth Order
Pipe Support Whole	27.985	37.341	61.956	71.410	105.830	113.320
Pipes (Supports are Simplified as Constraints)	65.077	65.686	101.270	110.770	111.060	130.530

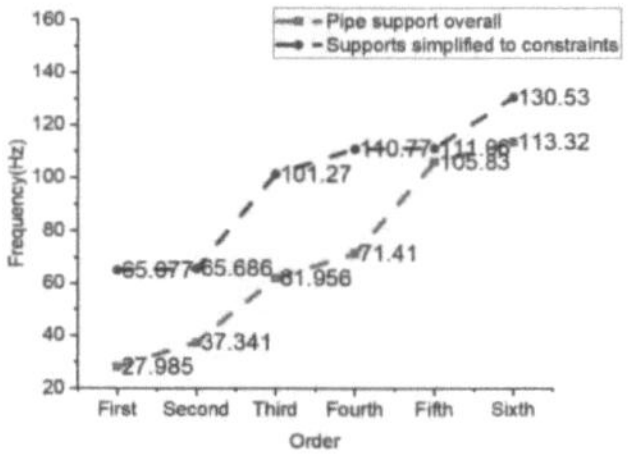

Figure 7. Results of the first six modal analyses of the system.

constraint calculation form, the flexibility of the bracket was ignored and the influence of the deformation of the bracket under external excitation was not fully considered, resulting in the calculated stiffness value being larger than the actual stiffness of the piping system. Therefore, when the bracket was simplified as a constraint, the calculated modal frequency was higher. To sum up, in the dynamic analysis of the pipeline system, the integrity of the pipeline and support should be fully considered to ensure the correctness of calculated values.

4 HARMONIC RESPONSE ANALYSIS OF PIPELINE SYSTEM

To further understand the response characteristics of the pipeline section in the frequency domain, frequency domain analysis was carried out based on modal analysis. This analysis adopted the simplest harmonic excitation to simplify the analysis. Although there are few cases of simple harmonic excitation in practical engineering, analyzing the response law of simple harmonic excitation of the pipeline system is the basis for understanding the response of the system to periodic excitation or more general excitation (Wang 2013). Therefore, it is necessary to analyze the harmonic response of the pipeline system correctly.

Through CAESAR II pipeline stress analysis software, the stress analysis of the deep-water general FPSO under the corresponding working conditions is carried out to obtain the stress state of the selected pipeline section, and the load value of each node of the pipeline section support is extracted as the load amplitude of the harmonic response analysis. The stress results in a cloud chart as shown in Figure 8, and the node loads are separated along the coordinate axis as shown in Table 4.

(1) Analysis of the harmonic response of the system under constraint only: Through modal analysis, the calculation range of the harmonic response analysis frequency was set to 0 Hz–200 Hz, and its loading condition was shown by the load of the corresponding nodes of the piping system extracted in Table 4.

According to the overall calculation method for pipeline support, the maximum displacements of the system along the X, Y, and Z axes under excitation were 1.484×10^{-4} m, 1.194×10^{-4} m, and 1.380×10^{-2} m respectively. As a result, the most dangerous vibration

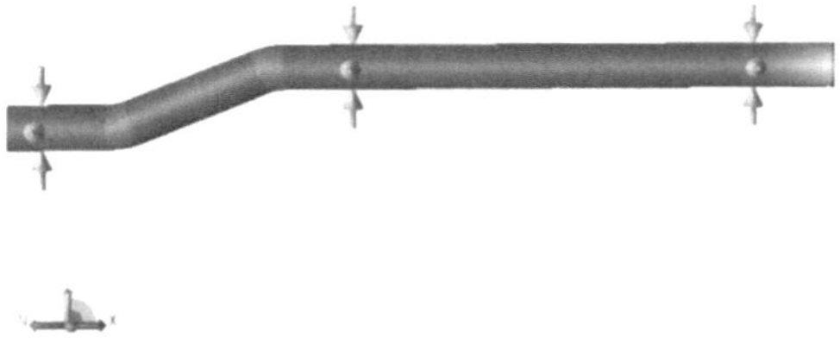

Figure 8. Nephogram of calculated stress of CAESAR II in the study section.

Table 4. Stress component table of each support node.

Support Node Number		9730	9740	9770
Force Condition /N	fx	0	0	0
	fy	−61559.79	−27965.51	27976.90
	fz	11382.86	12016.16	8268.63

situation when excited by the outside world was vibration deformation in the Z axis. The displacement response curve along the Z axis is shown in Figure 9, and the frequency that excites the maximum displacement along the Z axis is the first mode frequency (28.064 Hz). The modal shape before excitation was selected, as shown in Figure 10. The pipeline support system will have a large displacement along the Z-axis near the middle support. It also shows that the displacement of the bracket has a great influence on the overall displacement of the piping system when the piping system is excited.

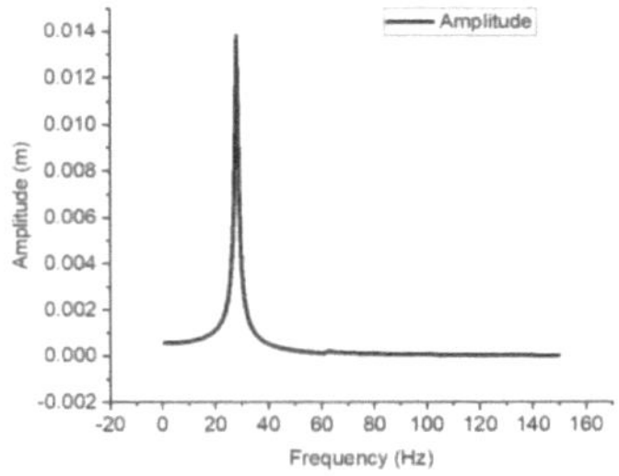

Figure 9. Displacement response curve of pipeline system along the Z axis.

Figure 10. First-order vibration mode of the pipeline system.

When the support was simplified into a constrained calculation form, the maximum displacements of the system along the X, Y, and Z axes under excitation were 4.943×10^{-5} m, 2.870×10^{-5} m, and 1.496×10^{-5} m, respectively. When excited, the displacement response along each coordinate axis was relatively small when the bracket was simplified as a constraint form, and the excitation modes were the third mode frequency of 120.090 Hz, the ninth mode frequency of 172.08 Hz, and the second mode frequency of 77.55 Hz, respectively. The modal shapes of each order before excitation are selected as shown in Figures 11–13, which shows that the maximum displacement of the pipeline in all directions along the coordinate axis occurs near the middle support. However, in the calculation method of simplifying the support into constraints, the constraints limited the displacement of the pipeline system in all directions along the coordinate axis, which could not reflect the flexible support characteristics of the support, so the displacement in all directions was not much different, which was quite different from the actual engineering phenomenon.

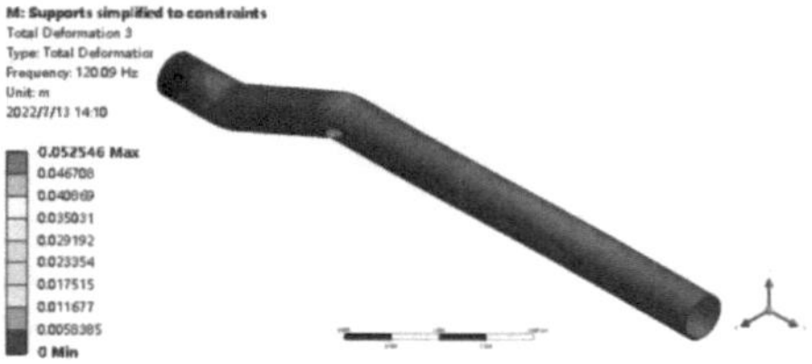

Figure 11. Third vibration mode of pipeline.

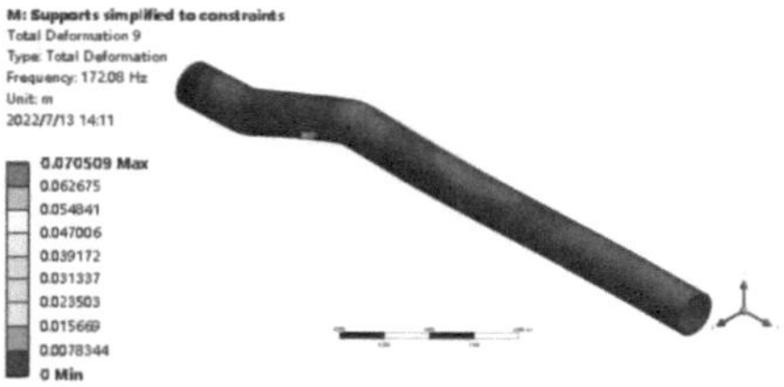

Figure 12. Ninth vibration mode of pipeline.

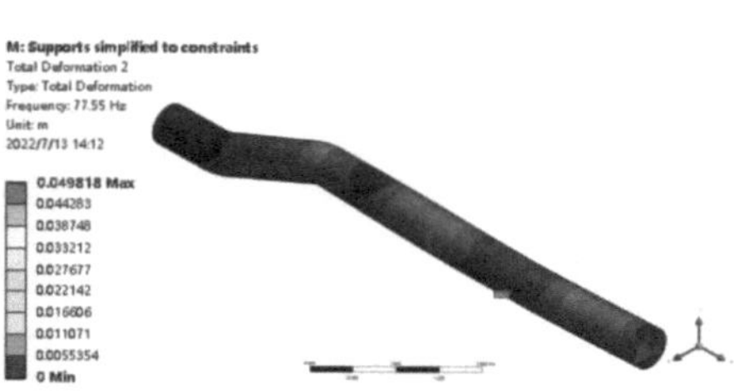

Figure 13. Second vibration mode of pipeline.

(2) Analysis of the harmonic response of the system under fluid-solid coupling: similar to the calculation settings for the harmonic response analysis of the system with only constraints, the calculation results are shown in Table 5. As shown in the table, it can be seen that the harmonic response analysis results of the system under fluid-solid coupling are

different from those of the support in the case of constraint only, but it still reflects the importance of the support in the dynamic analysis of the pipeline system. Considering the whole pipeline support along the Z axis, it has a large displacement response, and its displacement is 1.396×10^{-2}.

Table 5. Analysis and calculation results of harmonic response of system under fluid-solid coupling.

	Direction	X-Axis	Y-Axis	Z-Axis
Pipe Support Whole	Excitation Mode	Eight Order	Fourth Order	First Order
	Frequency (Hz)	115.400	71.410	27.985
	Maximum Displacement (m)	1.192×10^{-4}	1.196×10^{-4}	1.396×10^{-2}
The Support is Simplified as a Constraint.	Excitation Mode	Third Order	First Order	Second Order
	Frequency (Hz)	101.270	65.077	65.686
	Maximum Displacement (m)	1.832×10^{-6}	4.084×10^{-6}	7.782×10^{-7}

The two calculation methods in the above two sections were compared. In order to better reflect the importance of the support in the pipeline dynamics analysis, by sorting out the maximum amplitude of the harmonic response of the system along each coordinate axis, the comparison diagram only considering the constraint effect is shown in Figure 14, and the comparison diagram under the fluid-solid coupling effect is shown in Figure 15. From Figs. 14 and 15, it can be seen that there is a gap in the maximum amplitude of excitation along each coordinate axis during harmonic response analysis, especially in the direction of

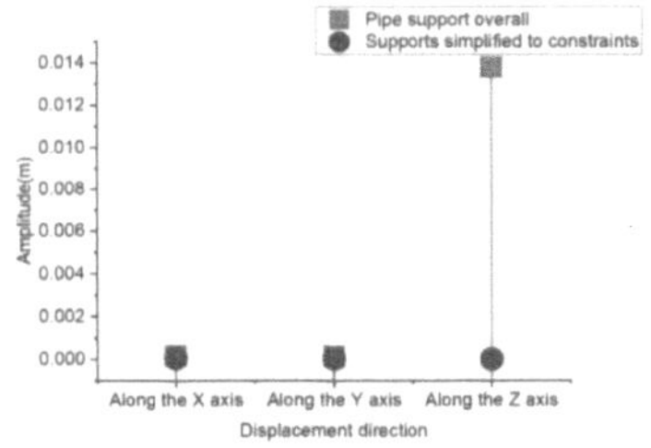

Figure 14. Scatter plot of maximum amplitude of harmonic response along the coordinate axis.

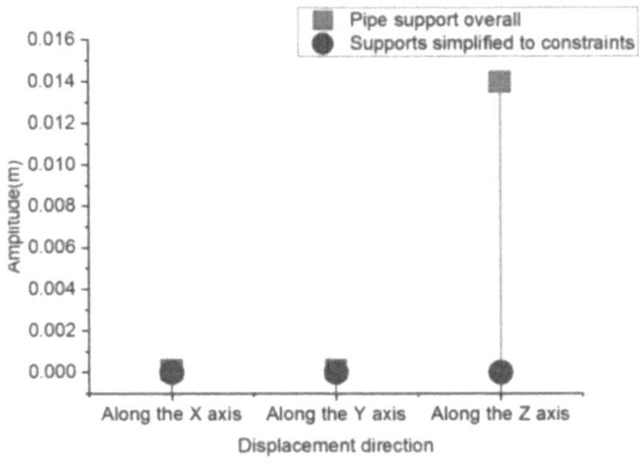

Figure 15. Scatter plot of maximum amplitude of harmonic response along the coordinate axis.

the Z axis. The maximum amplitude along the Z axis when the whole calculation method of pipeline support is adopted is much larger than that when the support is simplified as a constraint calculation method. The reason is that the vibration of the support itself in all directions under excitation is ignored. Therefore, it can be seen that the overall influence of pipeline support should be fully considered in frequency domain analysis.

5 CONCLUSIONS

In this paper, a part of the pipeline section of a general FPSO ballast water was taken as the research object, and the results of modal analysis and harmonic response analysis of the whole pipeline support and the two calculation methods of simplifying the support into constraints were compared and analyzed, and the following conclusions were drawn:

In modal analysis, simplifying the bracket into a constrained mode had a great influence on the results of modal analysis, and simplifying the bracket into a constrained calculation mode had a higher performance than the actual value.

In the harmonic response analysis, the way of simplifying the support into constraints also deviated from the actual situation, especially since the maximum amplitude along the Z axis was far less than the displacement response when considering the whole pipeline support.

In the dynamic analysis of the pipeline system, the overall influence of the pipeline support should be fully considered, and the vibration and deformation of the support in all directions cannot be ignored.

REFERENCES

Al-Waily M, Al-Baghdadi M, Al-Khayat R H. Flow Velocity and Crack Angle Effect on Vibration and Flow Characterization for Pipe Induce Vibration[J]. *International Journal of Mechanical & Mechatronics Engineering*, 2017, 17 (5): 19–27.

Cao Jianhua, Liu Yongshou, Liu Wei. Non-parametric Model Study of Aviation Pipeline Dynamics [J]. *Vibration and Shock*, 2018, 37 (18): 5.

Dias F, Santos E, Silva M, et al. New Algorithm to Discriminate Phase Distribution of Gas-Oil-Water Pipe Flow with Dual-Modality Wire-Mesh Sensor[J]. *IEEE Access*, 2020, PP (99): 1–1.

Fu Zhifang, Hua Hongxing. *Modal Analysis Theory and Application* [M]. Shanghai Jiaotong University Press, 2000.

Li Qing, Chen Weihua, Zhang Weijie, et al. Modal Analysis of Oil Pipeline Under Fluid-solid Coupling Based on ANSYS-Workbench [J]. *Journal of North China University of Science and Technology*, 2021, 18 (6): 9.

Song Xinyu, Ge Xinsheng. Unconstrained Modal Analysis of Dynamic Model of Flexible Spacecraft $ {\ \ BF 1) $ [J]. *Journal of Mechanics*, 2020, 52 (4): 954–964.

Wang Yongsheng. *Dynamic Study of Pipeline Support* [D]. China Naval Research Institute, 2013.

Wang Yucheng. *Basic Principle and Numerical Method of Finite Element Method* [M]. Tsinghua University Publishing House, 1997.

Zhang H, Qin M, Liao K, et al. Pipe-soil Vibration Characteristics of Natural Gas Pipelines During the Pigging Process[J]. *Journal of Natural Gas Science and Engineering*, 2021:104148.

Zhang Xiaofei, Li Liangbi, Gu Xiaomei. Effect of Connection Stiffness on Vibration Fatigue Life of Ship Vibration Damping Bracket [J]. *Journal of Jiangsu University of Science and Technology: Natural Science Edition*, 2018, 32 (4): 6.

Zhou Wei, Feng Zhongren, Wang Xiongjiang. Working Mode Analysis Based on Improved Empirical Fourier Decomposition [J]. *Vibration and shock*, 2021, 40 (9): 7.

Zhu Hongzhen, Wang Weibo, Yin Xuewen, et al. Vibration Modeling and Analysis of Marine Pipeline Based on Spectral Element Method [J]. *China Shipbuilding*, 2018, 59 (3): 15.

Water Conservancy and Civil Construction – Oke & Ahmad (Eds)
© 2024 The Author(s), ISBN: 978-1-032-58614-4

A temporary consolidation measure for continuous beam with cantilevers based on BIM

Shufeng Bai*, Qiyun Peng, Deqiang Yu, Wen Chen, Qiao Zhang, Qifei Wu & Huiying Sun
The Second Construction CO, LTD. of China Construction First Group, Daxing District, Beijing, China

ABSTRACT: With the development of bridge technologies, cantilever construction has been widely applied to large-span prestressed concrete bridges across China in recent years. Since beam piers are not consolidated during construction, it is difficult to maintain an absolute balance between the two sides. Temporary consolidation measures, as a crucial process, can ensure structural stability and the anti-overturning capability of beams during bridge construction. The consolidated structure will be destroyed using static crushing when the bridge satisfies design requirements. In this paper, we will introduce a consolidation measure for continuous beams with cantilevers that combines construction practices.

1 INTRODUCTION

Temporary consolidation, an important step in cantilever casting, mainly aims to connect continuous beams and bridge piers to improve construction safety. Most beams are designed with large spans for cantilever casting sections, while the top of each pier has a small area to meet requirements for use and appearance (Cheng 2021). In this case, the optimization design of the temporary consolidation is vital. The overall design and construction of temporary consolidation were hereby introduced through building information modeling (BIM) in the following aspects: the comparison and selection of temporary consolidation measures, collision detection, detailed design, and construction simulation.

2 OVERVIEW OF ENGINEERING EXAMPLE

The continuous beams of Tongren Bridge, Xining City, obliquely cross the Huangshui River, with a structural span of 56.5 m + 95 m + 66.5 m. The box beam is 12.2 m and 6.2 m wide at the top and bottom, respectively. The beam height is 5.8 m at the pivot of the main pier and 2.7 m in the linear cast-in-place segment. In addition, the thickness of the roof, baseboard, and web of the box beam is 0.3- 0.62 m, 0.3- 0.7 m, and 0.45- 0.9 m, respectively.

3 PRINCIPLE OF CONSTRUCTION TECHNOLOGY

In this study, the BIM technology was used throughout the whole design and construction process of temporary consolidation in this project. First, a model was created based on each consolidation strategy, the two-dimensional drawing was converted into a three-dimensional drawing, and the optimal technical measures were determined after analyzing their

*Corresponding Author: 515924922@qq.com

DOI: 10.1201/9781003450818-37

Figure 1. Layout of temporary consolidation.

Figure 2. Construction effect.

reasonability, safety, advantages, and disadvantages. Then, an embedded part model for pier top consolidation was constructed, followed by collision checks with the reinforcing steel bars in the beam body and bridge piers. To complete the detailed design, the reinforcing steel bar in bridge piers and the beam body were adjusted based on the detection report (Du 2020). A construction drawing was formed, and roaming visualization was performed as per the deepening model and drawing to improve the construction quality and express the construction procedures. Finally, the temporary consolidation construction was finished efficiently and with high quality.

In the cantilever casting segment, the span was large and the area at the pier top was small. The technical form was intuitively expressed based on the 3D visualization characteristic of BIM, and consolidating steel bars were reasonably arranged. On this basis, the optimal technical measure for temporary consolidation was determined (Hou 2021).

Temporary consolidated steel bars were arranged in detail to meet construction needs. With the range of the pier top fully utilized, the conflict point could be predicted through collision detection between models, followed by optimization to output the detailed design drawing.

The models were used repeatedly in the whole process with a high utilization rate, and the model value was fully utilized. Moreover, the application quality was improved in a total 3D visualized simulation form to guide and ensure accurate construction at this construction node (Liu 2020).

This technical measure can be applied to the construction of cantilever-cast continuous beams with small pier columns, projects with dense distribution of steel bars in 0# segments and pier columns, making temporary consolidating steel bars difficult to arrange, and projects with pier top consolidation and external consolidation combined construction.

4 FINITE ELEMENT SIMULATION ANALYSIS

4.1 *Modeling*

4.1.1 *Establishment of the major structural model*

The modeling standard was unified, and basic work contents like project information and the original point were determined. It was necessary to establish steel meshes in the 0# segment and at the pier top due to the inevitable collision problem with the steel bars in the major bridge structure and bridge piers, no matter what temporary consolidation method was adopted. Therefore, the major model of steel bars was established in advance as per the design drawing. During the establishment of major steel bars, some irregular steel bars were solved using the programmed nodes of Dynamo and directly transformed into structural steel bars using model lines.

4.1.2 *Establishment of an embedded part model*

The pier columns were specially vase-shaped, and the planar graph of the pier top consolidation was displayed in the following picture. Four temporary buttresses were set at each pier top, and for a single temporary buttress, its length in the transverse direction of the bridge, that in the longitudinal direction, and height were 1300, 400, and 820 mm, respectively. A total of 54 φ 32 deformed steel bars were arranged in temporary buttresses, and finished deformed steel bars were embedded, in two rows, in the pier body and the beam body by 1200 mm, respectively.

4.2 *Comparison and selection of technical measures*

The difficulties in the working procedures were determined as follows: First, because there was no design document to support temporary consolidation, the pier top consolidation did not meet construction requirements. Second, the external consolidation construction occupied the space at the beam bottom, and thus the construction supports for the 0# segment should be self-designed by combining the construction process. Based on the drawing design, the following two parts were compared and selected: the concrete form of temporary consolidation and the selection of cast-in-place supports for the 0# segment. Technical measures were completed in cooperation with BIM given its intuitively visualized display, thereby improving communication and picking efficiency during the determination of technical measures.

4.2.1 *Pier top consolidation + floor steel pipe supports*

The technical construction was simple and safe, and the foundation treatment outside the cushion cap and the strip foundation treatment were saved. The construction process was conventional,

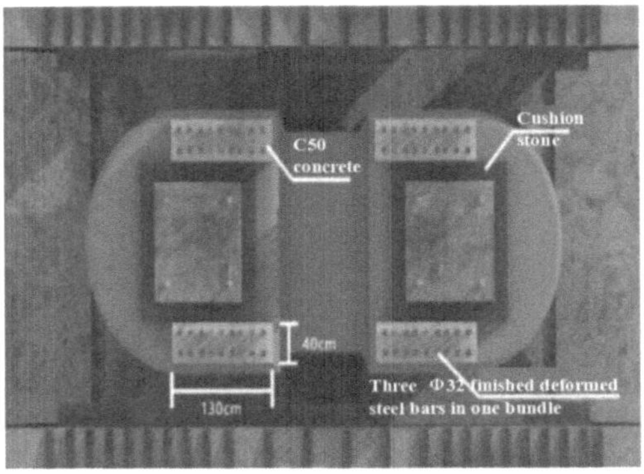

Figure 3. Overall preliminary design.

and floor supports could be consolidated externally. However, the pier top was small, which impeded temporary consolidation, so external consolidation should be added. There were many consolidating steel bars, and the emphasis should be on the concrete's compactness. Steel pipes needed processing with a long cycle. The quality requirement for brackets was high.

4.2.2 *Pier top consolidation + additional strip foundation for floor steel pipe supports*

This technical structure was mechanically stable. The construction process was relatively conventional. In addition, the safety risk of brackets was not considered. Floor supports could be externally consolidated. Nevertheless, the pier top was small, which discouraged temporary consolidation, so external consolidation should be added. There were many consolidating steel bars, and the emphasis should be on ensuring concrete compactness. The geological conditions were poor, which led to high foundation treatment costs (Mao 2020).

4.2.3 *Pier top consolidation + bracket construction*

This technology, which was commonly applied to high piers, was simple and convenient, with a conventional construction process and a short construction period. However, temporary consolidation failed to meet mechanical requirements, along with strict requirements for the pre-embedded position. Through comparative analysis and mechanical checking, the final temporary consolidation measure was presented as follows: A total of 216 PSB785 finished deformed steel bars were used for pier top consolidation, and a total of 4 φ 820 steel pipe columns were arranged for external consolidation.

4.2.4 *Analysis of calculated working conditions*

Condition I: unbalanced loads generated by normal construction;
Condition II: The hanging bracket and concrete at one side fell off under the maximum cantilever state;
Condition III: The left and right of the "T" – frame were subjected to wind loads from different directions and twisted under the maximum cantilever state.
Through calculation, construction needs were satisfied.

4.3 *Collision detection*

The models of different parts were integrated according to the temporary consolidation measure. Since the load checking was already done, pier-top consolidation had top priority without needing any adjustment. The steel bars in the beam body and bridge piers were taken as the adjustment objects (Wang 2021).

The integrated major structural model was imported into Navisworks to set collision groups in batches: 0# steel bars—consolidating steel bars; steel bars in the pier body—consolidating steel bars. Next, the real-time dynamic updating function of Navisworks and Revit was realized via the plug-in Navisworks Switch Back for the sake of collision adjustment. Finally, an analysis report was formed through the hard collision detection method. According to the report feedback, each collision point was positioned, and the collision image was checked and reproduced in the model segment for depth adjustment. In this way, the collision of steel bars at 340 positions was avoided to the greatest extent.

After the collision induced by the internal consolidating steel bars was solved, structural self-collision was performed to form the result.

4.4 *Construction simulation*

The construction sequence in each stage was determined to gradually simulate the installation of the overall temporary consolidating structure. Furthermore, key construction materials were introduced to enhance constructors' awareness and ability to master the construction process.

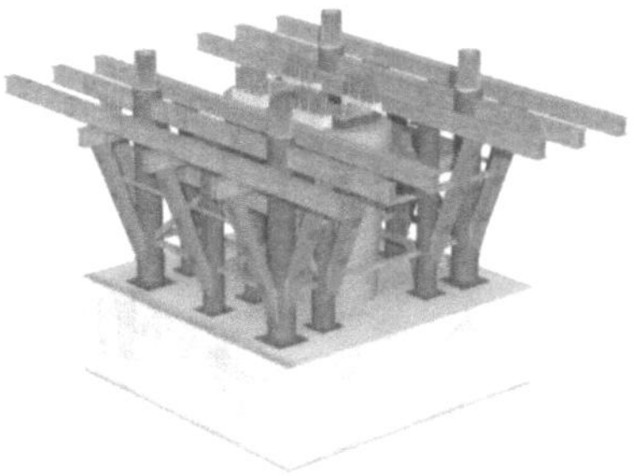

Figure 4. Final technical measure for temporary consolidation.

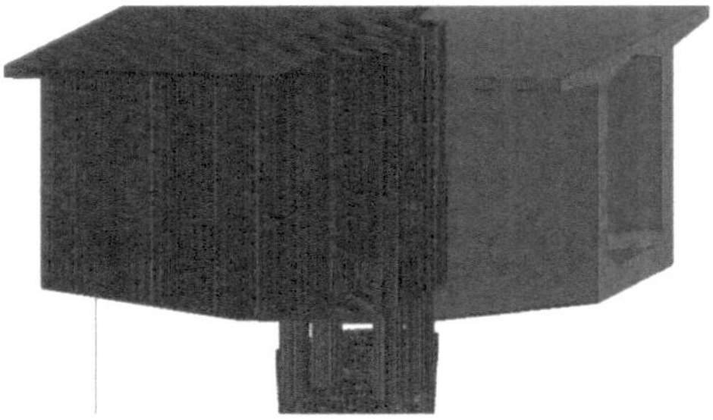

Figure 5. Preliminarily integrated model.

5 CONSTRUCTION QUALITY ASSURANCE MEASURES AND BENEFIT ANALYSIS

5.1 *Overall requirements*

After modeling was completed, the model was reviewed as per the drawing in contrast to the profile map, thus ensuring model-drawing consistency.

The temporary consolidating structure was strictly checked according to the drawing information, with all data strictly controlled during the construction.

Embedded parts were rechecked before concrete pouring.

5.2 *Quality assurance measures for embedded parts and reserved holes*

Delivery inspection reports of all raw materials and semi-finished products entering the construction site should be provided. Such raw materials and semi-finished products can be used for construction only after passing rechecking (Zeng 2020).

Surveying personnel should comprehensively master the code requirements for the design drawings and surveying of this project, check dimensions and coordinates in the drawings and timely handle any problem found.

The positioning points of embedded parts should be protected well and not be moved or destructed. The positions of embedded parts should be retested before concrete pouring.

Embedded parts should be fixed firmly and effectively connected to the existing structures.

During concrete pouring in the pier body, finished deformed anchoring steel bars should be embedded, and position-controlling supports should be set to ensure the accurate embedding position of such anchoring steel bars.

Finished deformed steel bars should be buried in the pier body and box beam with caps, without needing exposed anchorage.

Figure 6. Isolation of colliding steel bars.

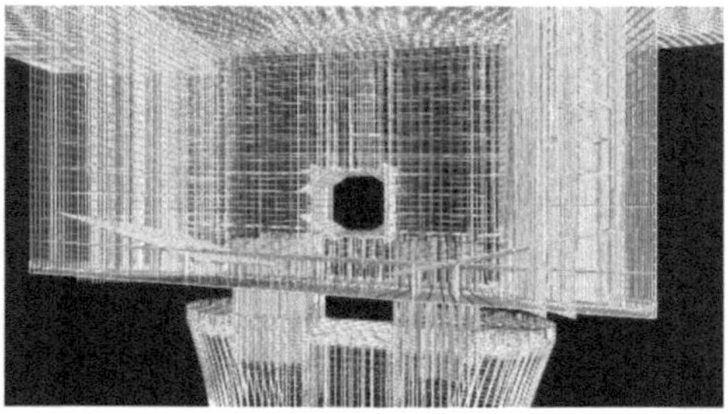

Figure 7. Overall conflict of the 0# segment with consolidating steel bars.

Figure 8. Simulation of support installation and demolition stage.

The top surface of temporary buttresses should be set according to the bottom elevation of the box beam. To facilitate demolition, isolation plates should be paved on the contact surface at the beam bottom or the surface should be smeared with oil.

5.3 *Quality assurance measures for reinforced concrete structures*

The supervision system at the construction site can be implemented for all important working procedures to ensure strict control of the construction process.

Concrete vibration control should be strengthened, particularly in parts with a dense distribution of steel bars.

5.4 *Quality assurance measures for steel structure*

Licensed welders must perform welding within the permissible scope specified in their welding qualification certificate, and welders without certificates are not allowed to do welding operations.

The welding process should be determined according to the evaluation report, and welding procedure specifications should be compiled to realize whole-process quality control.

The weld appearance should be controlled and non-destructive testing should be done. Other measures like repair welding should be taken at non-conforming positions.

5.5 *Benefit analysis*

In this project, the BIM technology was used to assist in the construction of embedded parts for temporary consolidation, thus solving the difficulties in embedding technologies and management of embedded parts, improving the embedding quality and efficiency, and saving costs. The efficient and high-quality embedding of embedded parts for temporary consolidation laid the foundation for the subsequent temporary consolidation construction. Meanwhile, the impact of BIM technology on bridge engineering was growing.

6 CONCLUSION

With the development of urban transportation, variable cross-section continuous beam overpass bridges will be increasingly built in the future, accompanied by an increasing overall mass, thus proposing higher safety requirements for temporary consolidation and bringing about certain challenges to the design and construction of temporary consolidation. In this study, BIM technology was used to run through the whole design and construction process of temporary consolidation. The construction scheme for the temporary consolidation of four T-frames in the bridge crossing Huangshui River with a half span in the east and west, respectively, on Tongren Road was optimized to guide the bridge construction. This study will provide certain reference values for the construction of similar bridges.

REFERENCES

Cheng J., Tian J., Yin X W. Stress Analysis and Study of Temporary Fixation of Cast-in-cantilever Continuous Box Beam [J]. *Chongqing Architecture*, 12 (10): 27–29 (2021).

Du B., Bao T., Huang C H., Wang Q M. Design of Temporary Fixing Structure for Construction of Large Span Prestressing Concrete Continuous Box Girder Bridge Cantilever [J]. *Journal of Guizhou University (Natural Sciences)*,30 (04): 125–127 (2020).

Hou J.J., Song W J., Li Y., Xu J.P., Gao W.J., Liu Z.B., Nie H.Y., Wu Z.Q., Huang Z.H., Gu J.M. *Construction Method of Temporary Consolidation and Post-anchorage System for Pier Beams* [Z]. (2021).

Liu G.Q. *Construction Technology of Temporary Consolidation System for Cast-in-cantilever Continuous Box Beams* [Z]. Guangzhou Municipal Group Co., Ltd, August 28, (2020).

Mao L.M., Xu L.W., Liao C.F. Discussions on Temporary Consolidation Design of Large-span Continuous Beam [J]. *Western China Communications Science & Technology*, 69–72 (2020).

Wang G.J. Temporary Consolidation Design and Analysis for Cantilever Construction of Continuous Box Beams [J]. *Traffic Engineering and Technology for National Defense*,12 (S1): 41–43 (2021).

Zeng Z.H., Xi A., Li S T. The Removal Order Research for Structure-temporary Consolidation of the Rigid Frame-continuous Girders [J]. *Highway Engineering*, 37 (02): 37–40 (2020).

Water Conservancy and Civil Construction – Oke & Ahmad (Eds)
© 2024 The Author(s), ISBN: 978-1-032-58614-4

An experimental study on strength parameters of unsaturated subgrade backfilled loess

Wei Wang*, Dongbo Cai, Zebin Wang, Zhen Zhang, Kunyang Yi, Qing Zhou & Lu Niu
The Seventh Engineering Co., Ltd. of CFHEC, Zhengzhou, China

ABSTRACT: The backfilled loess in engineering is easily affected by the change in water content caused by rainfall. In view of the limitations of the laboratory conditions of the construction company and the difficulties in analyzing the strength of unsaturated soil by using the volumetric pressure plate instrument, the strength characteristics of backfilled loess under different water contents were studied by using the traditional shear test in this paper. The results showed that the shear strength of loess decreased gradually with the increase in water content, which was mainly manifested by the decrease in internal friction angle. In addition, the larger the confining pressure was, the more obviously the shear strength would be affected by the shear rate.

1 INTRODUCTION

With the continuous development of China's infrastructure construction and the continuous improvement of the highway traffic network, high excavation, and deep filling are inevitable for highway construction. In general, the soil excavated in the project is piled up for backfilling the foundation pit, the roadbed, and other projects in the later period. Soil accumulation can easily form a hidden danger (Zhan 2010; Zhang 2019). Rainfall conditions have a significant effect on slope stability (Marin & Velásquez 2019; Teixeira *et al.* 2015). The excavation soil originally has a natural water content, but due to rainfall and other factors, the water content of the soil used for backfilling will change. The water content will have an important influence on the strength of the soil, especially for collapsible loess. The natural loess itself has a certain structural strength, but the original structural strength of the backfilled loess is destroyed during excavation. If the water content changes, the strength changes more (Hu *et al.* 2000). Therefore, it has certain engineering significance to study the influence of water content on the strength of backfilled soil.

It is generally considered that the shear strength of unsaturated soils consists of three parts: cohesion, friction, and matric suction. But the matrix suction is difficult to measure and needs to be determined by soil-water characteristic curves (SWCC) obtained through complex tests with expensive testing instruments such as volumetric pressure plate instruments (Fredlund & Morgenstern 1978). Thus, it is not practical in general engineering. However, there is a certain correspondence between matric suction and water content in unsaturated soil. Hence, the water content can be used to study the strength of unsaturated soil instead of matric suction (Emir *et al.* 2003).

In view of these, the mechanical parameters of backfilled soil were obtained by direct shear test and triaxial test on the background of collapsible loess excavation and piled soil slope engineering. The stability of slope filling and the landslide characteristics under the

*Corresponding Author: zhongjiao_wang@126.com

 DOI: 10.1201/9781003450818-38

action of rainfall were simulated by the numerical simulation method. The research results can provide references for similar projects.

2 OVERVIEW OF THE EXPERIMENTS

2.1 *Testing soil specimens*

The soil samples were taken from a construction site in Henan Province, all of which were 4–5 m deep. The basic physical property indices of each soil sample determined by indoor tests are shown in Table 1. Due to the close distance between each soil sampling point, the soil property parameters measured are similar. Table 1 shows the average value.

Remolded soil was used in all tests, and the preparation of soil samples was carried out in strict accordance with the *Standard for Geotechnical Test Methods* (GB/ T 50123- 2019). Soil samples with different water contents were wrapped with plastic wrap, numbered, and stored in a glass moisturizer after production. The water content of the soil sample was measured before the test, and the measured value was taken as the real water content of the reconstituted soil sample.

Table 1. Basic physical index of sample.

Nature Moisture Content w / (%)	Density ρ/(g/cm^3)	Specific Gravity of Soil Grain G_s	Void Ratio e	Liquid Limit w_L / (%)	Plastic Limit w_p / (%)	Plastic Index I_P
17-18	2.0-2.1	2.69	0.55-0.57	38	20	18

2.2 *Test scheme and method*

The specific test scheme is shown in Table 2. The direct shear test and triaxial shear test were carried out by the quadruple strain control direct shear tester and the triaxial shear tester, respectively. The size of the soil sample in the direct shear test was Φ 61.8 × 20 mm, while that in the triaxial test was Φ 39.0 × 81 mm. When analyzing the effect of water content by direct shear test, five soil samples with 8%, 10.8%, 15.3%, 17.3%, and 21% water contents were prepared, respectively, and the shear rates were all 0.8 mm/ min. In the analysis of the effect of shear rates, the tests were carried out at four shear rates for the soil samples with a water content of 15.3%. In the triaxial shear test, consolidated undrained shear was adopted, and the shear rates were 0.06, 0.08, 0.3, and 0.4 mm/min, respectively. Five kinds of confining pressures were applied to analyze the shearing process.

Table 2. Test scheme.

Test Type	Size of Soil Sample	Loads (kPa)	Water Content (%)	Shear Rate (mm/ min)
Direct Shear Test	Φ 61.8×20 mm	Normal Stress: 100, 200, 300, 400	8, 10.8, 15.3, 17.3, 21	2.4, 0.8, 0.1, 0.02
Triaxial Test ((CU))	(Φ 39.0×81 mm)	Confining Pressure: 100, 150, 200, 250, 300	Saturated Soil Sample	0.06, 0.08, 0.3. 0.4

3 ANALYSIS OF DIRECT SHEAR TEST RESULTS

3.1 *Shear strength parameters from direct shear test*

The strength parameters of soil samples under various conditions can be obtained by direct shear tests with the same shear rate and different water contents. Accordingly, the influence of water content on strength and parameters can be obtained by comparison. Similarly, the influence of shear rate on the strength and parameters of unsaturated soil can be obtained through shear tests with different shear rates for the same water content. Taking water content $w = 8\%$ and shear rate as an example, the shear stress-shear displacement curve can be drawn by sorting out the data of this group of direct shear experiments, as shown in Figure 1. The shear strength-normal stress relationship can be obtained by taking the peak values of each curve as the shear strength line. The strength parameters c and φ can be calculated by using linear fitting. Similarly, the test results of other water contents and shear rates can also be arranged according to the above method. Thus, the shear strength lines of direct shear tests of different water contents can be obtained. The c value is the intercept of the line with the vertical axis, and the slope is the tangent value of φ.

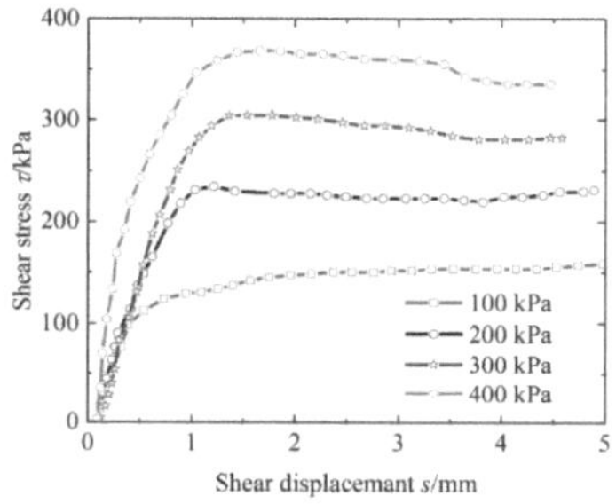

Figure 1. Shear stress and displacement relationship of the soil sample.

3.2 *Influence of different water content on shear strength*

By the shear strength lines of the soil samples with different water content, the corresponding strength parameters c and φ values can be obtained, as shown in Figure 2. It can be seen that water content has an obvious influence on the strength parameters of unsaturated soil. With the increase in water content, the cohesion increased first and then decreased. When the cohesion reaches the peak value, the corresponding water content is close to the plastic index of the soil. The internal friction angle decreases with the increase in water content, the trend is almost linear.

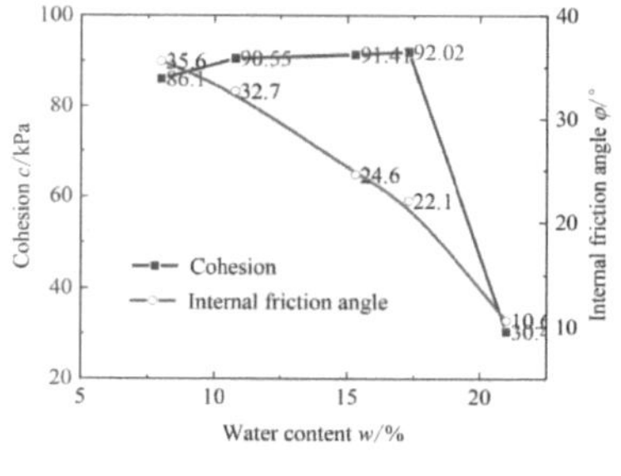

Figure 2. Influence of water content on strength parameters.

The shear strength is shown in Figure 3. Obviously, the shear strength decreases with the increase in water content. Moreover, within a certain water content range, e.g., 8%- 17.3%, the shear strength, and water content change linearly, the slope is negative, and the lines corresponding to different normal stresses are nearly parallel. The shear strength decreases sharply when the water content is close to the saturation value, i.e., 21%.

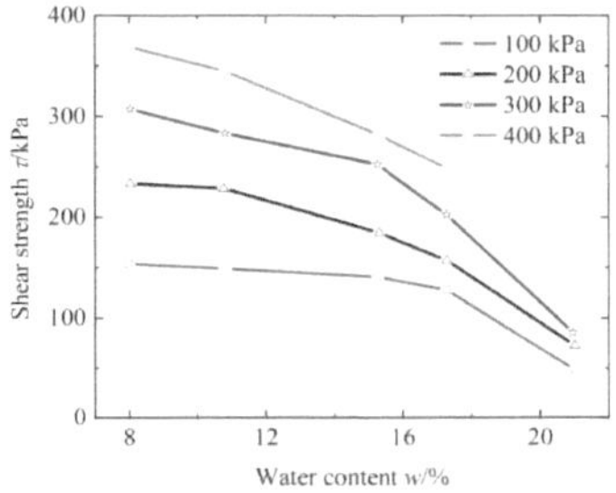

Figure 3. Influence of water content on strength parameters.

3.3 *Stress-strain relationship and failure characteristics from triaxial test*

Figure 4 shows the stress-strain relationship curve of soil samples during the triaxial test and a photo of soil samples during failure. At the beginning of loading, the elastic deformation is obvious. Then, the plastic limit deformation is obvious, and the stress change after the peak value is not obvious. The soil sample failure belongs to the weak softening type, which is the result of further development of plastic deformation after the soil sample has undergone obvious elastic-plastic deformation. In the process of plastic failure, large deformations occurred. Echelon shear fracture surfaces are formed during the failure of soil samples; some cracks were not connected and some plastic shear deformation bands were formed. Some shear failure surfaces or shear bands are not obvious; only plastic bending deformation occurs, and sometimes elongated failure occurs near the largest convex bending part. It can be seen from the figure that the test results of all samples have a relatively light strain softening phenomenon. At the initial stage of loading, the stress increases rapidly and is close to elastic deformation. The curve points near the peak value are more concentrated, and the strength drops slightly after the peak value. During the experiment, the peak strength is selected by whether there is a peak in the value of the curve. When the peak strength appeared when the strain did not reach 15%, the failure point was taken as the peak point. Otherwise, the point corresponding to the alternative strain at 15% is the failure point. Thus, the shear failure envelope of the triaxial test can be obtained.

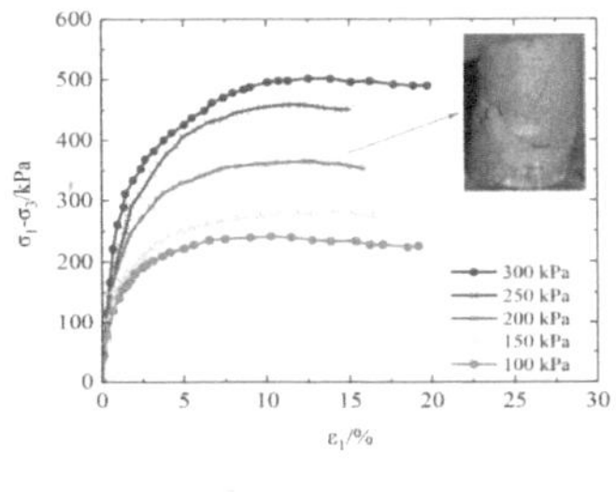

Figure 4. Stress-strain curves of soil samples during the triaxial test and a photo of soil samples during failure.

3.4 *Subsection heading*

Figure 5 shows the diagram of shear strength under different confining pressures, with the shear rate as the abscissa. At the same shear rate, the greater the confining pressure is, the greater the shear strength is. The difference in shear strength under different confining pressures gradually decreases as the shear rate increases. At a low rate, the difference in shear strength under different confining pressures is large, and the points are scattered. With the increase in rate, the points gradually concentrate. At the same time, it can be seen from the figure that the change in peak strength of different shear rates under the same confining pressure basically increases first and then decreases.

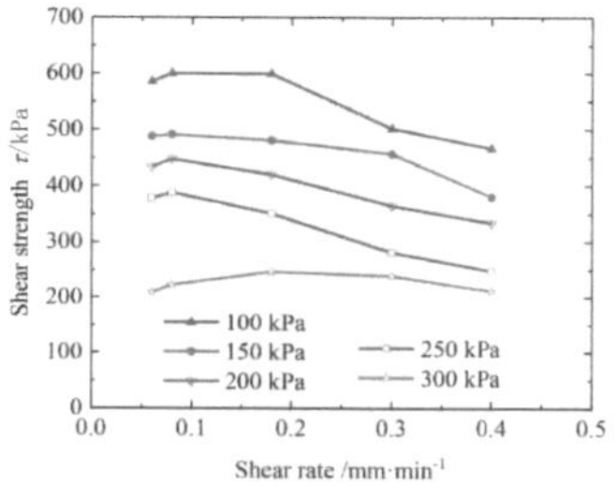

Figure 5. The relationship between shear strength and the shear rate under different confining pressures.

Table 3. Strength parameter values corresponding to different shear rates.

Shear Rate/ mm/min	*c'*/kPa	*φ'*/°	*c*/kPa	*φ*/°
0.06	2.26	35.68	19.82	28.04
0.08	5.74	35.62	24.06	27.99
0.18	6.31	35.24	24.50	27.14
0.3	6.67	32.29	27.99	24.43
0.4	5.84	30.91	23.12	23.09

Figure 6 shows the effective stress paths at different shear rates. The CSL line in Figure 6 is the critical state line. Previous studies on the CD and CU test results of normally consolidated clay found that the failure point of soil samples is a corresponding curve in the spatial coordinate p-q-e, which is the critical state line. In Figure 6, all CSL lines are in straight form, and the slope value of each failure line is close to 1.5. For different loading paths, the sample will fail as long as the stress state reaches the CSL line; that is to say, the sample failure is independent of the stress path. As can be seen from Figure 6, when the stress state is below the CSL, the soil sample is not damaged. The points of the stress path will be very dense when they are close to the CSL line. After the soil sample reaches its peak value, the soil strength decreases. As the confining pressure increases, the stress path becomes more curved, which reflects that the pore water pressure also increases with the confining pressure.

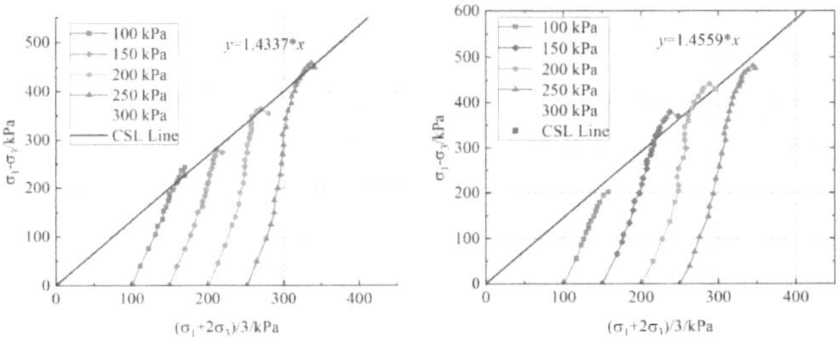

(a) Shear rate is 0.3 mm/min (b) Shear rate is 0. 06 mm/min

Figure 6. Stress path diagram for different shear rates (respectively).

4 CONCLUSIONS

(1) The moisture content has an obvious influence on the strength of unsaturated loess. Shear strength decreases with increasing water content and drops dramatically when the soil is close to saturation. Therefore, for projects requiring high backfill quality, the water content of the backfill soil should be properly controlled, and the accumulated backfill soil should be waterproof, drained, or sun-dried.
(2) The confining pressure has an important effect on the strength of saturated loess. Under the same shear rate, the greater the confining pressure is, the greater the shear strength is, and the smaller the radius of the Mohr circle is. At the same time, the confining pressure will affect the stress path of the soil, and the greater the confining pressure is, the more bending the stress path will be.
(3) The shear rate has an obvious effect on the strength parameters of soil. The friction angle decreases as the shear rate increases, while the cohesion increases first and then decreases.

REFERENCES

Emir, J. M., Laureano, R. H., Pedro, A., "Constitutive Modeling of Unsaturated Soil Behavior Under Axisymmetric Stress States Using a Stress/Suction-controlled Cubical Test Cell," *International Journal of Plasticity*, 19 (10): 1481–1515 (2003).

Fredlund, D. G., Morgenstern, N. R., Widger R. A., "The Shear Strength of Unsaturated Soils," *Canadian Geotechnical Journal.* 15 (3): 313–321 (1978).

Hu, Z. Q., Shen, Z. J., Xie, D. Y., "Research on Structure Behavior of Unsaturated Loess," *Chinese Journal of Rock Mechanics and Engineering*, 19 (06), 775–779 (2000).

Marin, R. J., Velásquez, M. F., "Influence of Hydraulic Properties on Physically Modelling Slope Stability and the Definition of Rainfall Thresholds for Shallow Landslides," *Geomorphology*, 106976 (2019).

Teixeira, E., Azevedo, R., Ribeiro, A., "*Influence of Rainfall Infiltration on the Stability of a Residual Soil Slope*," (2015).

Zhan, X. J., "Study on Stability Evaluation and Controlling Measure of a Highway Artificial Slope," *Soil Engineering and Foundation* (2010).

Zhang, X., "Evaluating Stability of Snisotropically Deposited Soil Slopes," *Canadian Geotechnical Journal*, 56 (5), 753–760 (2019).

Water Conservancy and Civil Construction – Oke & Ahmad (Eds)
© 2024 The Author(s), ISBN: 978-1-032-58614-4

Integral fabrication and hoisting technology of curved continuous steel box girder

Jisheng Hu* & Peihong Li
Guangzhou Second Municipal Engineering Company Limited, Guangzhou, China

ABSTRACT: The construction of the steel box girders of an urban bridge is difficult and risky, and the coordination of the construction process is complicated. In order to study the whole fabrication and hoisting technology of steel box girders, this paper conducted a comprehensive study from the aspects of bridge structure form, plate unit division, steel box girder fabrication, steel box girder hoisting planning, and hoisting safety analysis. In the research process, the standard beam segments were reasonably divided into different plate units along the transverse direction, and the manufacturing process of plate units was reasonably planned. The steel box girder was fabricated on the tire frame by adopting multi-beam continuous matching welding and a pre-assembly process. The installation sequence of steel box girder blocks and sections on site was to be erected from a large pile number to a small pile number. The hoisting and installation of the steel box girder were analyzed from the aspects of component size, crane selection, counterweight, installation radius, length of the lifting arm, and safety of the lifting weight. The research has made abundant achievements and has reference significance for the same type of bridges.

1 INTRODUCTION

With the rapid development of urban construction, more and more urban bridges have been constructed with steel box girders because of their large span, ease of construction, and short lifespan. However, the fabrication and hoisting of steel box girders are difficult and risky. In the process of making a steel box girder, it is difficult to ensure the precision of plate elements, reasonably divide the segments of the steel box girder, and weld the steel box girder. The hoisting of a steel box girder is subject to the lifting capacity of the crane, the hoisting site, the hoisting safety, and other problems, which lead to great risk in the hoisting process. In order to solve the above problems, this paper will focus on the overall production and hoisting technology of continuous steel box girders in curved sections and puts forward corresponding solutions, which can serve as a reference for similar projects (He & Lin 2021; Li *et al.* 2019; Li & Lin 2020; Yi & Fan 2021; Zhang 2021).

2 ENGINEERING BACKGROUND

The scope of the project route started from Songshan Lake Interchange in the east and ends at Tongsha Interchange in the west. The total length of the road transformation was about 3.5 km. Among them, the length of the newly built left-turn A ramp line was about 1882 m, and the length of the newly built straight B ramp line was about 703 m. The design speed was

*Corresponding Author: road-engineering@foxmail.com

 DOI: 10.1201/9781003450818-39

50 km/h. The standard section of the ramp A was a one-way dual carriageway with a road width of 8 m; ramp B was a single lane with a road width of 6.75 m. There were two ramp bridges, A and B, on the whole road. The total length of ramp bridge A was about 1303 m, and the total length of ramp bridge B was about 215 m.

3 STRUCTURAL FORMS OF BRIDGES

The road alignment of A ramp bridge had an easing curve and a circular curve. The minimum radius of the circular curve was 400 m. The bridge started at AK0 + 233 and ended at AK1 + 536. The seventh (45 + 52 + 45) m, the eighth (42 + 52 + 42) m, the tenth (35 + 40 + 35) m, and the eleventh (35 + 40 + 35) m superstructures of the ramp bridge were all continuous steel box beams with straight webs of equal height. The seventh and eighth beams were 2.3 m high, and the tenth and eleventh beams were 1.8 m high. A single box and chamber section were used in the steel box girder. The top of the box girder was 9 m wide, the cantilever length of the two flanks was 2.0 m, and the bottom plate was 5.0 m wide. These are shown in Figure 1. The middle line of the bridge was composed of circular curves. The steel beams were horizontally arranged by the bottom plate and vertically arranged by the web. The transverse slope of the bridge deck was formed by the different heights of the web. The main material of the steel box beam was a Q345qc steel plate, and the pressured concrete at the fulcrum position was C40 micro-expansion concrete. The vertical stiffening of the steel beam diaphragm and the web was arranged along the radial direction of the road design line, and it was arranged in a fan shape near the inclined steel beam.

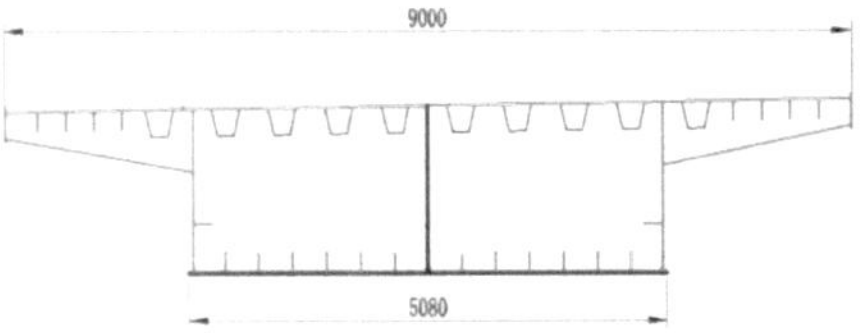

Figure 1. Typical steel box girder cross section.

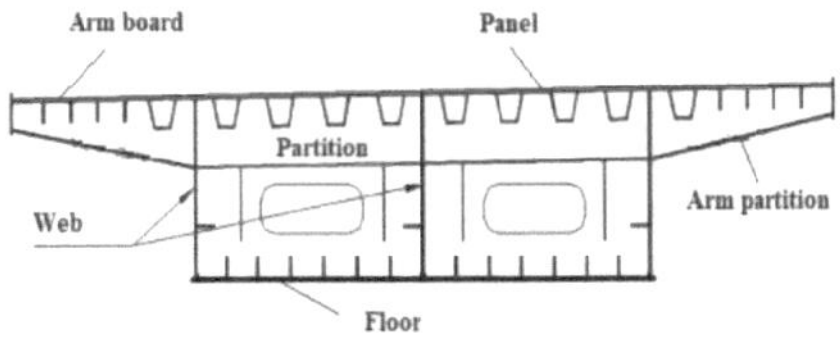

Figure 2. Schematic diagram of board unit division.

4 BOARD ELEMENT DIVISION

Each standard beam section was horizontally divided into 13 plate units, including 2 top plates, 2 bottom plates, 2 diaphragms, 3 webs, 2 arm units, and 2 arm baffles. This is shown in Figure 2. According to the classification of plate unit type, plate units can be formed in the workshop during special tire frame flow manufacturing. This was easy to achieve in terms of production standardization, product standardization, and quality stability.

5 SECTION DIVISION OF STEEL BOX GIRDER

Considering the requirements of manufacturing, transportation, and construction, the 7th and 8th steel box girders were divided into 7 manufacturing girder sections along the bridge direction: A, B, C, D, E, F, and G. The 10th and 11th steel boxes were divided into 8 manufacturing beam sections along the bridge: A, B, C, D, E, F, G, and H. Segments A, B, C, D, F, and G of the seventh link were divided into three transport blocks laterally. Segments A and F of the 8th joint were divided into three transport blocks laterally, and segments B, D,

E, and G were divided into two transport blocks laterally. Segments A and B of the 10th joint were horizontally divided into two transport blocks, and segments D, E, G, and H were horizontally divided into three transport blocks. The segments A, B, D, E, G, and H of the 11th joint were divided into three transport blocks horizontally. These are shown in Figures 3, 4, and Table 1.

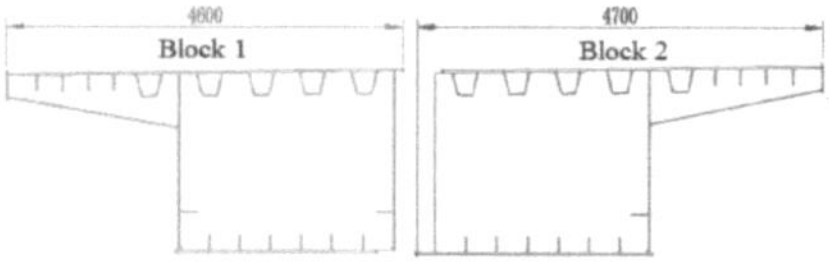

Figure 3. Segmental lateral block division (Two pieces).

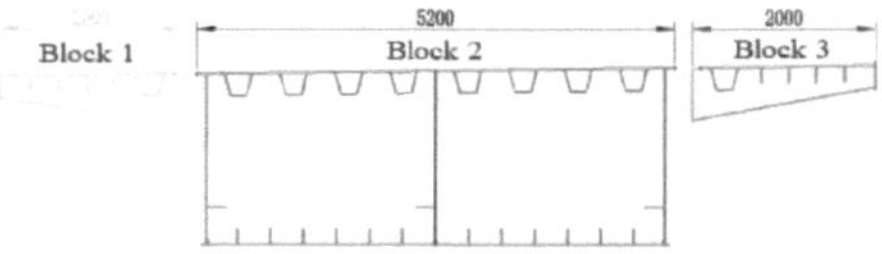

Figure 4. Segmental lateral block division (Three pieces).

Table 1. Breakdown of segments (Taking the seventh union for example).

Serial Number	Section	High	Wide	Long	Quantity	Weight	Transportation Method
The 7th Link	Section A	2300	5200	28460	1	112474	Double
	Section B (W)	2300	4700	23000	1	54957	Single
	Section B (N)	2300	4600	23000	1	61698	Single
	Section C	2300	5200	23000	1	90896	Single
	Section D	2300	5200	20000	1	79040	Single
	Horizontal Section E	2650	5000	18700	1	111832	Single
	Section F	2300	5200	23500	1	92872	Single
	Section G	2300	5200	18960	1	74930	Single
	Arm A	700	1800	28460	2	31875	Double
	Arm B	700	1800	23000	2	25760	Single
	Arm C	700	1800	23000	2	25760	Single
	Arm D	700	1800	20000	2	22400	Single
	Arm F	700	1800	23500	2	26320	Single
	Arm G	700	1800	18960	2	21235	Single

6 INTEGRAL ASSEMBLING METHOD OF STEEL BOX GIRDER

The beam assembly was carried out in the assembly frame, and the multi-beam continuous matching assembly welding and pre-assembly process were used to complete the beam assembly welding and pre-assembly at the same time. Box girder assembly adopts the formal method. The tire frame was used as the outer tire, and the diaphragm was used as the inner tire to control the shape and structure of the beam segment. The board units were placed on vertical and horizontal baselines and reinforced to ensure accuracy and safety. To make it easy to adjust the mutual position of each plate during beam docking, the longitudinal rib end welds of the top plate, the bottom plate, and the web were left 200 mm temporarily unwelded, waiting for the ring joint construction. Beam assembly was carried out according to the order of base plate, diaphragm plate, middle and inner web, diaphragm plate, outer web, roof plate, beam section pre-assembly, and assembly welding accessory structure. The assembly and welding of three-dimensional step-form propulsion were realized. The beam assembly focused on the alignment of the bridge, the geometric shape and dimensional accuracy of the steel box girder, and the exact matching of adjacent interfaces. Taking the seventh link, the A beam section, as an example, the continuous assembly process of the beam section is explained in detail as follows:

1. The base plate unit was assembled and welded: The base plate unit of the reference beam section was placed on the tire frame. Following the sequence from the small pile number to the large pile number, the horizontal and vertical baselines were precisely aligned with the baseline of the tire frame and fixed to the tire frame. Then two base plate elements were welded symmetrically. This is shown in Figure 5.
2. The inner diaphragm unit was assembled: Based on the horizontal and vertical baselines on the bottom plate, the ground mark line of the matching tire frame was assembled, positioned in the sequence from the small pile number to the large pile number, and fixed to the bottom plate unit through the flower basket screw. This is shown in Figure 6.
3. The inner web and middle web units were assembled: The vertical baseline of the bottom plate was used as the reference for scoring the web assembly line. According to the sequence from small pile number to large pile number, the inner web and middle web were assembled, positioned, and fixed with the bottom plate unit and the baffle unit. This is shown in Figure 7.
4. The outer diaphragm unit was assembled: Based on the horizontal and vertical baselines on the bottom plate, the positioning baffle unit was assembled in sequence from the small pile number to the large pile number and was fixed to the bottom plate unit through the flower basket screw. This is shown in Figure 8.
5. The outer web unit was assembled: Based on the longitudinal baseline of the bottom plate, the web assembly line was inscribed, and the outer web was assembled, positioned in the sequence from small pile number to large pile number, and fixed with the bottom plate unit and the baffle unit. The welds between the plate and the bottom plate, the plate and the web, and the web and the bottom plate were welded symmetrically. This is shown in Figure 9.
6. The roof unit is assembled: When assembling, the top plate and the diaphragm were closely attached to control the overall height of the steel box beam. The weld seams between the roof and diaphragm, the roof, and the web were symmetrical successively. This is shown in Figure 10.
7. Assembly welding boom unit: The overall width of the steel box girder should be controlled during assembly. The seams between the cantilever and the roof and between the cantilever and the web were symmetrically welded successively. This is shown in Figure 11.

Figure 5. Assemble and weld the base plate unit.

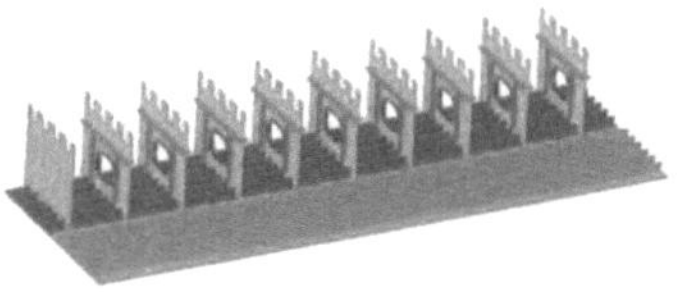

Figure 6. Assemble the inner diaphragm unit.

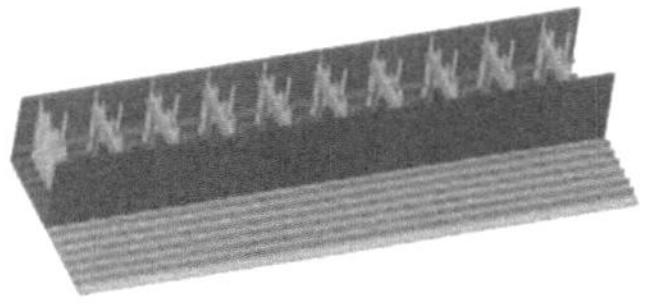

Figure 7. Assemble inner web and middle web units.

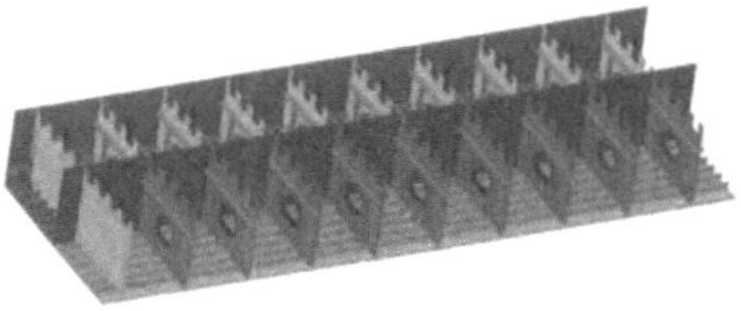

Figure 8. Assemble the outer diaphragm unit.

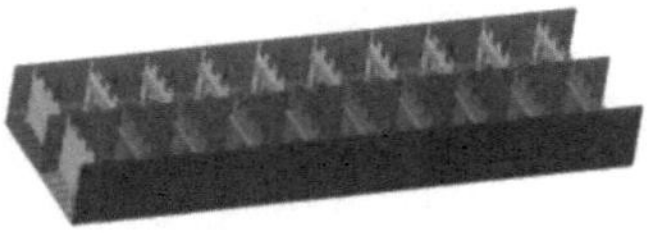

Figure 9. Assemble the outer web unit.

Figure 10. Assemble the roof unit.

Figure 11. Assembly welding boom unit.

7 SECTION LIFTING PLAN OF STEEL BOX GIRDER

Bridge installation entails lifting the pole into position and connecting the bridge. Bridge position connection mainly includes longitudinal block connection, inter-segment ring joints, welding of embedded parts, on-site installation and welding of bridge deck accessories, etc. To reduce the impact of welding on the bridge on passing vehicles and shorten the installation time of the steel beam, the box girder block was reassembled in the assembly area before installation. The installation sequence of steel box girder blocks and sections on site was planned to be erected from a large pile number to a small pile number: the 11th, 10th, 8th, and 7th links shall be successively hoisted and constructed.

1. Installation of block and section of the 11th joint steel box girder: Temporary support construction → hoisting section A→ hoisting section C→ hoisting section B→ hoisting section F→ hoisting section E→ hoisting section D→ hoisting section H→ hoisting section G→ removing temporary support → assembling and welding auxiliary structure.
2. Installation of block and section of the 10th joint steel box girder: Temporary support construction → hoisting segment C→ hoisting segment A + B (W) → hoisting segment A + B (N) → hoisting segment D→ hoisting segment H→ hoisting segment F→ hoisting segment G→ hoisting segment E→ removing temporary support → assembling and welding auxiliary structure.
3. Installation of block and section of the 8th steel box girder: Temporary support construction → hoisting segment A→ hoisting segment D + E + F1 (W) → hoisting segment D + E + F1 (N) → hoisting segment C1→ hoisting segment C2→ hoisting segment B (W) → hoisting segment G (W) → hoisting segment G (N) → hoisting segment F2→ removing temporary support → assembling and welding auxiliary structure.
4. Installation of block and section of the seventh steel box girder: Temporary support construction → hoisting section G→ hoisting section E→ hoisting section F→ hoisting section C→ hoisting section D→ hoisting section A→ hoisting section A (pick arm) → hoisting section B (W) → hoisting section B (N) → removing temporary support → assembling and welding auxiliary structure.

8 ANALYSIS OF LIFTING CAPACITY AND SAFETY OF STEEL BOX GIRDER

According to the section type of steel box girder, hoisting construction planning was carried out. The hoisting and installation of steel box girders were analyzed from the aspects of component size, crane selection, counterweight, installation radius, length of the lifting arm,

Table 2. Analysis of lifting capacity and safety of the 7th link.

Member Number	Section G	Section E	Section F	Section C	Section D
Component Size (m)	2.3 × 9.0 × 18.8	2.6 × 5.4 × 18.7	2.3 × 9.0 × 23.9	2.3 × 9.0 × 23	2.3 × 9.0 × 20
Selection of the Crane	500 t Crane	500 t Crane	500 t + 500 t Crane	500 t + 500 t Crane	500 t + 500 t Crane
Leg Size (m)	9.4 × 9.6 m	9.4 × 9.6 m	9.4 × 9.6 m	9.4 × 9.6 m	9.4 × 9.6 m
Weight of Component (t)	97	112	120	117	102
Counterweight (t)	142	142	142	142	142
Installation Radius (m)	9.5	9.0	15.6	15.7	16.0
Length of Outgoing Arm (m)	37	31.8	37	37	42.2
Rated Lifting Weight (t)	131	152.5	90.5	90.5	84.2
Reduction Factor 0.8	-	-	90.5*0.8 = 72.4	90.5*0.8 = 72.4	84.2*0.8 = 67.36
Comprehensive Lifting Weight X1.1 (t)	(97+2.0) *1.1 = 108.9	(112+2.0) *1.1 = 125.4	(120+2.0) *1.1/2 = 67.1	(117+2.0) *1.1/2 = 65.45	(102+2.0) *1.1/2 = 57.2
Safety Margin	16.87%	17.78%	7.32%	9.6%	15.08%
Conclusion	Security	Security	Security	Security	Security
Member Number	Section A	Arm A	Section B (W)	Section B (N)	
Component Size (m)	2.3 × 5.2 × 28.5	0.8 × 2.0 × 14.3	2.3 × 4.7 × 23.4	2.3 × 4.7 × 23.4 m	
Selection of the Crane	500 t Crane	500 t Crane	500 t Crane	500 t Crane	
Leg Size (m)	9.4 × 9.6 m	6.48 × 8.0 m	9.4 × 9.6 m	9.4 × 9.6 m	
Weight of Component (t)	113	8.0	55.0	62	
Counterweight (t)	142	13.8	142	142	
Installation Radius (m)	16.0	17.0	10.1	2	
Length of Outgoing Arm (m)	37	37.6	37	12	
Rated Lifting Weight (t)	90.5	11.4	115	37	
Comprehensive Lifting Weight X1.1 (t)	(113 + 2.0) *1.1/2 = 63.25	(8.0 + 2.0) *1.1 = 11	(55.0 + 2.0) *1.1 = 62.7	(62 + 2.0) *1.1 = 70.4	
Safety Margin	12.63%	3.51%	45.48.%	38.78.%	
Conclusion	Security	Security	Security	Security	

and safety of the lifting weight. According to the hoisting sequence of steel box girder members, the hoisting parameters are listed in Table 2 by taking the 7th link as an example.

9 CONCLUSIONS

This paper makes an in-depth study of the fabrication and hoisting of the curved continuous steel box girder. The main conclusions are as follows: (1) The superstructure of the seventh, eighth, tenth, and eleventh links of the ramp bridge of A adopted a continuous steel box girder with a straight web of equal height. The standard beam segments were reasonably divided into different plate elements along the transverse direction. All plate units could be manufactured in line with the special tire frame in the workshop according to their types to achieve standardization, standardization, and stability. (2) To ensure the manufacturing accuracy of plate units, the manufacturing of plate units was carried out in the order of "plate leveling, digital control precision cutting, parts processing, tire assembly, reverse deformation position welding, and local dressing." (3) The steel box girder was carried out in the assembling tire frame by adopting the multi-beam continuous matching welding and pre-assembling process. A formal assembly method was adopted. The frame was used as the outer tire, and the diaphragm was used as the inner tube to control the size of the girder. Each plate unit was placed in position according to the longitudinal and transverse baselines, supplemented with reinforcement facilities to ensure accuracy and safety. The beam section assembly shall be carried out in the following order: base plate, diaphragm plate, middle and inner web, diaphragm plate, outer web, roof, and beam section pre-assembly and assembly welding auxiliary structure. (4) The installation sequence of steel box girder blocks and sections on site was to be erected from a large pile number to a small pile number. Bridge position connection mainly included longitudinal block connection, inter-segment ring joints, welding of embedded parts, on-site installation and welding of bridge deck accessories, etc. The box girder block was reassembled in the assembly area and then installed. (5) The lifting steps of a steel box girder block were mainly divided into the steps of automobile crane position, hook, test lifting, lifting, rotation, falling beam and alignment adjustment, loosening hook, etc. The hoisting and installation of the steel box girder were analyzed from the aspects of component size, crane selection, counterweight, installation radius, length of the lifting arm, and safety of the lifting weight.

ACKNOWLEDGMENTS

This work was financially supported by the Applied Research Project of Guangzhou Second Municipal Engineering Company Limited.

REFERENCES

Aiping Zhang. Analysis of Mechanical Performance of Steel Box Girder Section Construction [J]. *Highway*, 2021, 66 (01): 157–161.

Chuanxi Li, You Li, Zhuoyi Chen. Fatigue Characteristics of Steel Box Girder Diaphragm Based on Measured Traffic Flow [J]. *Journal of Chang'an University (Natural Science Edition)*, 2019, 39 (05): 48–58.

Jiawu Li, Te Song, Lihua Lin. Study on Static and Aerodynamic Performance of Continuous Steel Box Girder Bridge with Variable Cross Section [J]. *Journal of Highway and Transportation Research and Development*, 2020, 37 (04): 88–95.

Youmiao Yi, Shaoche Fan. Key Technology of Steel Box Girder Construction of Qingzhou Waterway Bridge of Hong Kong-Zhuhai-Macao Bridge [J]. *Bridge Construction*, 2021, 51 (03): 138–144.

Zhigang He, Pengzhen Lin. Optimization of Girth Welding Sequence of Wide Steel Box Girder [J]. *Chinese Journal of Applied Mechanics*, 2021, 38 (01): 332–339.

Water Conservancy and Civil Construction – Oke & Ahmad (Eds)
 ISBN: 978-1-032-58614-4

Study on the failure mode of beam-plate structures of steel protective doors under different welding conditions

Ziye Liu, Xianxiang Zhou, Lan Xiao*, Xiao Li, Ce Tian & Fantong Lin
Institute of Defense Engineering, Academy of Military Science, PLA, Beijing, China

ABSTRACT: This paper aims to analyze the loadbearing capacity and failure characteristics of the door plates and I-beam steels under two welding techniques. We designed and crafted five groups of beam-slab structures and tested and measured their failure modes, ultimate bearing capacity, and strain curves under static pressure loads. The test results indicated that damage occurred at welding joints between the inner plate and the I-beam, where failure is easily caused. According to the displacement and strain change, the largest degree of deformation is found in the middle of the specimen. Before it reaches the yield point, the ultimate load capacity is equal to the ultimate strength of the weld plate. Combined with the test data, the calculation formula for the ultimate strength was obtained. Calculated errors and measurement errors are all within 10%, indicating that the formula can be used as a reference for designing protective doors.

1 INTRODUCTION

Protective doors have been widely used in military, commercial, and industrial applications. They are used as entrances to defensive bunkers or ammunition depots, to protect personnel and objects inside the bunker, and to control the effects of accidental explosions in specific areas. With the increase in terrorist activities, unexpected explosions threaten public safety (Hsieh 2008; Veeredhi 2015; Zhang 2019).

Commonly used protective doors can be divided into reinforced concrete and steel structure protective doors. Due to the large weight of concrete structures, they are ineffective in resisting weapons destruction. Steel structure protective doors have higher strength, easy processing, and are easy to use in large span and high strength projects, and thus are widely used (Koh 2003; Yelek 2021). The structural form of steel protective doors is mainly beam and plate type, i.e., the skeleton beam is the main force-bearing member, supplemented by the plate, as shown in Figure 1. The skeleton beam can be a flat plate, I-beam, etc., and the arrangement is cross-shaped, tic-tac-toe, etc. When the blast load is applied to the plate, the skeleton beam can carry most of the load. The beam structure is the most basic force unit in the steel explosion-proof door, and its performance directly affects the stability of the overall structure. Therefore, it is important to analyze the load-carrying capacity for beam-plate structures.

Currently, most researchers mainly study the performance of blast-resistant doors by changing door structures and materials used to improve mechanical properties. Guo Dong organically combined steel pipe, concrete and steel plate to design a steel-clad steel pipe and concrete structural form, which has the advantages of high resistance, high stiffness and good stability (Guo 2012, 2013). Chen *et al.* proposed a multi-arch double-layer plate blast

*Corresponding Author: festoon@sina.com

DOI: 10.1201/9781003450818-40

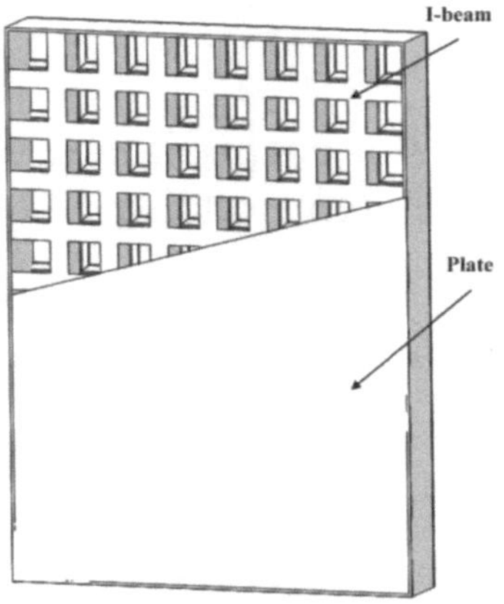

Figure 1. Structure of a steel protective door.

door. The effects of curvature, the thickness ratio of double-layer plates and material properties on the load-bearing capacity of blast doors were considered (Chen 2012). Dong Hui *et al.* designed a protective door with a cable film structure, which uses the steel plate's film force and the wire rope's suspension force to resist the blast load (Dong 2014). Thimmesh *et al.* compared the curved blast door with a rectangular door. They found that the peak displacement of the curved door is significantly reduced under the blast load, which improves the door's overall resistance performance 21).

Composites have low specific gravity and high specific strength and modulus compared to metal and concrete materials and are widely used in aerospace, automotive, electronics and electrical fields. With the development of composite materials, the original structural form is added to the composite materials with high blast resistance, lightweight and high strength in protection engineering, such as high damping rubber (Fang 2017; Xu 2018), aluminum foam (Zhang 2011, 2014), and polyurethane foam (Rashad 2012), to form a force-resisting system with high performance. Meng *et al.* used carbon fiber-reinforced plastics (CFRP) to design and fabricate a lightweight blast-resistant protective door. The results showed that CFRP could improve the stiffness and strength of the protective door and resist the pressure of the shock wave (0.45 MPa). Moreover, CFRP can enhance rebound effects under explosion loadings (Meng 2016). To test the performance of a new composite protective door, He *et al.* conducted an explosion test using a glass fiber-reinforced polymer (GFRP) sandwich plate combined with a steel frame. The test results showed that the hollow sandwich could reduce the door's mass and improve its blast resistance (He 2021).

The studies mentioned above only focus on the mechanical properties of the overall structure of the protective door, while few studies are on the characteristics of failure modes for protective doors. In a study by Luo *et al.*, the finite element method of the flat plate door revealed that the maximum stress is located at the point where the door plate meets the stiffener (Luo 2012). Nurick *et al.* studied the possible damage patterns of plates with stiffening ribs under blast pressure using experimental and numerical simulation methods (Nurick 1995). The protective door is an all-welded structure. There are some differences in the construction process between the upper and lower plates and the I-beam, where the upper plate is connected to the I-beam by intermittent welding.

In contrast, the lower plate is connected to the I-beam by plug welding. Under the action of external loads, due to the different welding methods, the degree of damage is caused by the difference. In addition, the explosion is a transient response, and the specific damage process inside the protective door cannot be well observed. Therefore, in response to the abovementioned problems, we studied the beam-plate structure in the internal component of the protective door under static loading. We then proposed an equation to describe the mechanical properties of this beam-plate structure based on the test results. This equation can improve the structural performance and structural design of protective doors.

2 TESTING SCHEMES

2.1 *Specimen design*

A total of five steel beam-plate specimens were designed and fabricated. The specimens were assembled and welded from type 12 I-shaped steel (120 mm in height, 74 mm in flange width, and 5 mm in web thickness) and 8-mm thick steel plates, with a total length of 1,000 mm. We have selected Q235 steel, which has a yield strength (fy) of 235 MPa and ultimate tensile strength (fu) of 370 MPa. A double-sided interrupted welding method was employed on the outer plate and I-beam to ensure that the process was similar to the actual construction, with a weld height of 4 mm and a weld length of 50 mm. Figure 2(b) shows that plug welds were used on the inner plate and I-beam.

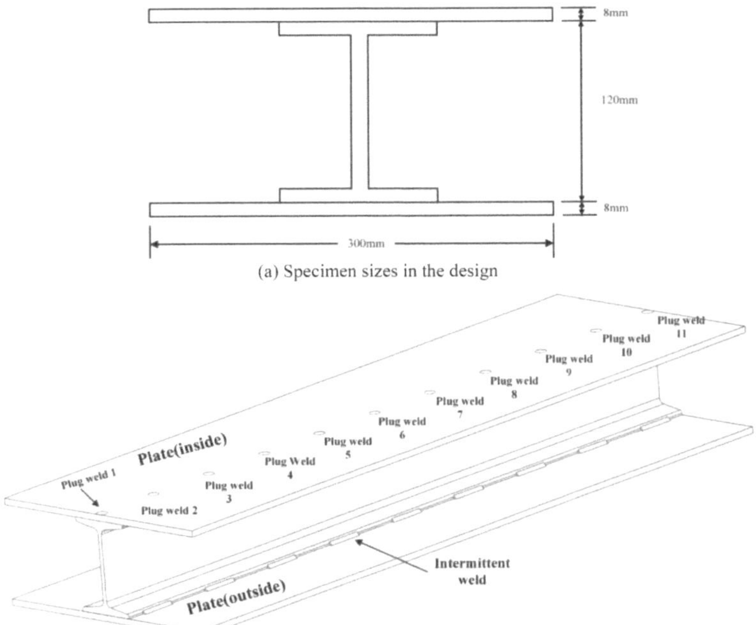

(a) Specimen sizes in the design

(b) Welding process

Figure 2. Schematic diagram of the specimen: size design and the welding process.

2.2 *Sensor arrangement*

Three strain gauges were pasted at the center of the specimen (see Figure 3). Two strain gauges were pasted at each location to reduce the error and prevent damage to the strain gauges during the test. The central strain gauge is located at plug weld 6, and the vertical deformation of this point is measured. Right, and left strain gauges were positioned 250 mm to the central strain gauge to measure vertical deformation.

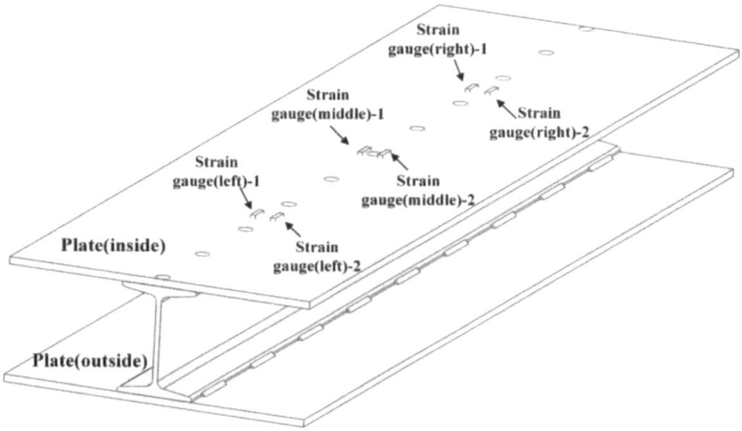

Figure 3. Arrangement for strain gauges.

2.3 *Loading solutions*

The specimen was supported on both sides and a concentrated load was applied from the middle. In the actual explosion, the blast shock wave was applied to the plate as a surface load, and to match the actual situation, the loading model (as shown in Figure 4) was adopted, and rectangular shims were placed above the outer plate to simulate the effect of the surface load applied in the middle. A universal material testing machine (WES-2000W) was used to apply vertical static loads at a loading rate of 5 mm/min (Figure 5). The specimen was first preloaded to a displacement of 2 mm to compress itself with the loading device. It was then unloaded to 0 mm and continuously loaded until damage occurred.

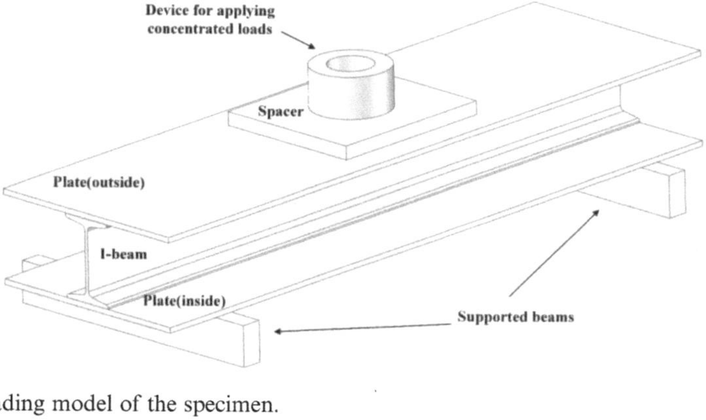

Figure 4. Loading model of the specimen.

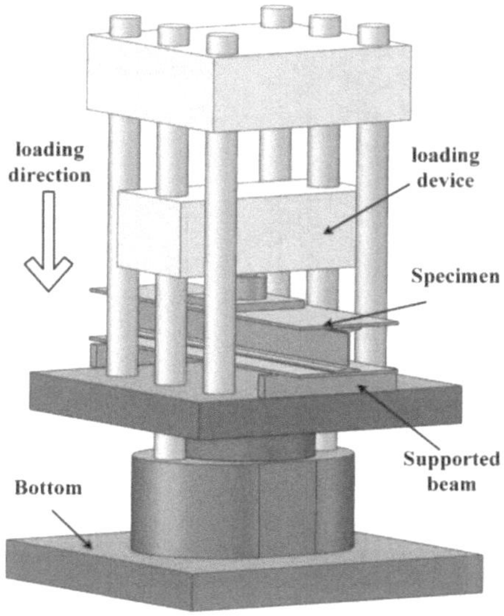

Figure 5. Loading device for testing.

3 TEST RESULTS

3.1 *Experimental observation*

The failure mode for specimen A is shown in Figure 6. In (a), it is apparent that the external plate becomes extruded and deformed in the middle region as the load increases, while both its left and right sides are warped. The welded joints are fractured and damaged due to excessive deformation in the middle. Under extrusion loads, the strength of the plug weld of the inner plate is less than that of the outer plate, which causes stress concentrations at plug welds, leading to fracture damage, as shown in (b). This specimen is primarily damaged on the left side, with all fractures occurring at plug welds 1 to 6.

Specimens B and C exhibit the same damage pattern as specimen A in Figures 7 and 8, respectively. Species B and C also exhibit the same damage pattern as specimen A, which is the right side fracture of the welding joints 6-11, whereas, for specimen C, only welding joints

(a) Failure types of the specimen during the test

(b) Failure occurred at the welding joint of the left side

Figure 6. Failure types of specimen A.

Figure 7. The failure mode of specimen B.

Figure 8. The failure mode of specimen C.

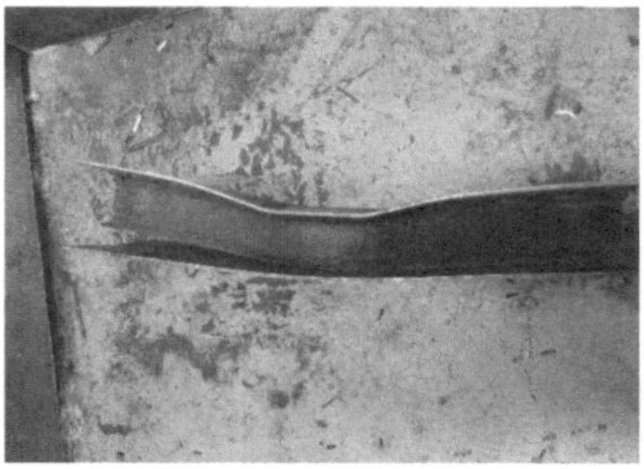

Figure 9. The failure mode of specimen D.

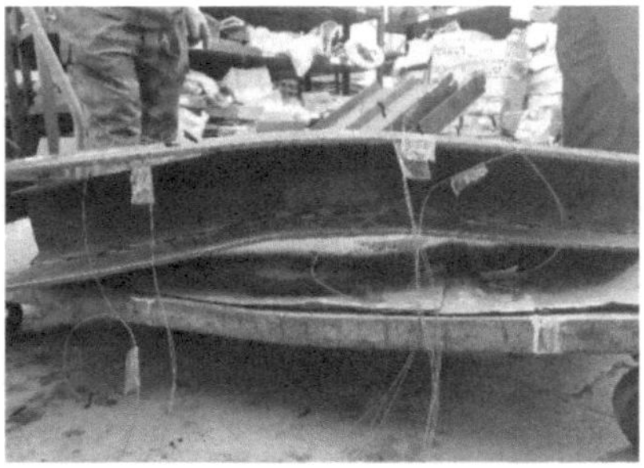

Figure 10. The failure mode of specimen E.

7 remain connected. In contrast, the remaining ten welded joints are fractured. As shown in Figures 9 and 10, specimens D and E exhibit different types of damage. In specimens D and E, there are fractures at I-beam junctions. The fractures are located at plug welds 1 to 5.

It can be observed from the above five specimens that failure modes appear at the welding joints between the inner plates and the I-beam when subjected to a vertical static load. As a result, the strength of plug welds is less than interrupted welding, leading to stress concentrations at plug welds. The connection between the inner plate and the I-beam must be strengthened in this case.

3.2 *Analysis of the force-displacement curves*

A universal material testing machine measures a specimen's displacement deformation. Because the displacement measured by the testing machine includes a component of the displacement of the loading device and a component of the specimen compression, a preloading process is applied to the specimen to minimize displacement errors caused by the loading device.

Figure 11 shows the force-displacement curve of specimen A. During the elastic stage AB, the specimen's bearing capacity shows a linear growth pattern, but the growth rate slows down when it reaches the plastic stage. When it reaches point C, the load reaches its maximum value of 326.6 kN, which is its maximum value. In stage CD, after continuing loading, a fracture appears between the inner plate and the I-beam, the bearing capacity starts to drop, the support of the inner plate fails, and the yield load is 296.3 kN. After passing point

D, the specimen completely yields. After point D, the specimen enters the plastic development stage (section DE). During this stage, the external load increases continuously, the cracks produced before are closed under the action of load, the bearing capacity of the specimen rises continuously, and the bearing capacity of the specimen is only supported by the I-beam and the outside plate together. At this point, the welding strength is insufficient to resist the external load, and the joints continue to fracture, resulting in the specimen's failure.

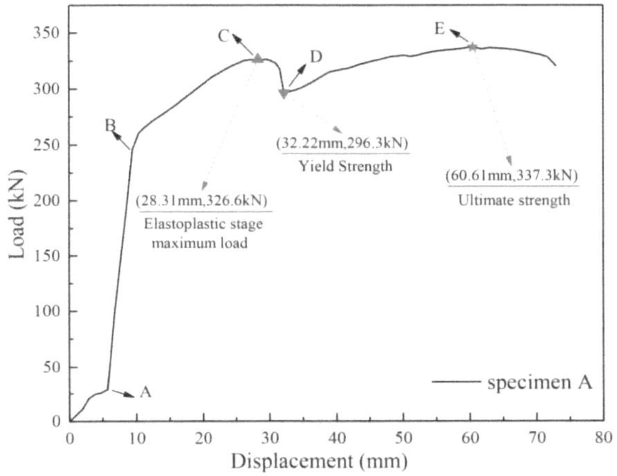

Figure 11. Force-displacement curve of specimen A.

Force-displacement curves for specimen B and specimen C are shown in Figures 12 and 13, respectively. We can determine that the force-displacement curves of specimens B and C have essentially the same trend. Species B and C have yield loads of 314.1 kN and 285.4 kN, yield displacements of 49.07 mm and 27.94 mm, ultimate bearing capacities of 331.38 kN and 329.9 kN, and ultimate displacements of 61.44 mm and 62.5 mm. During the elastic-plastic stage, the ductility of the specimen increased, prolonging the time it took for the specimen to reach the maximum load.

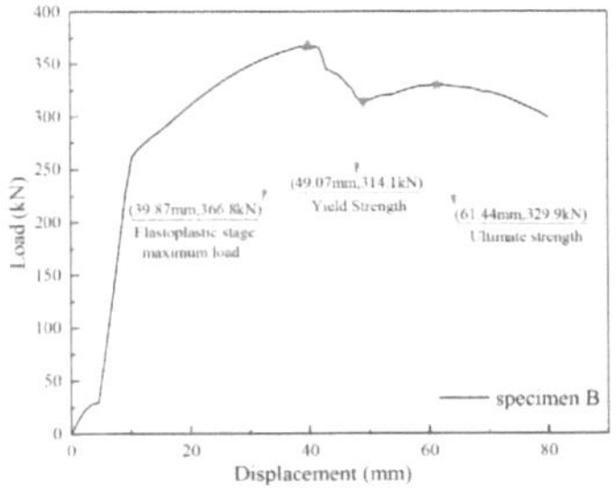

Figure 12. Force-displacement curve of specimen B.

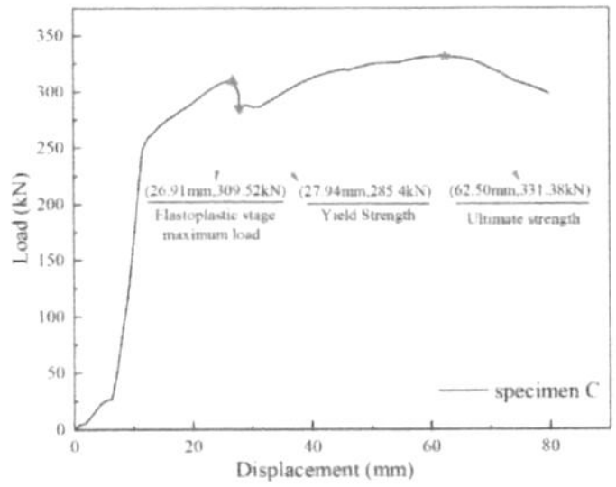

Figure 13. Force-displacement curve of specimen C.

Force-displacement curves for specimens D and E are illustrated in Figures 14 and 15. The figures indicate that specimens D and E possess yield strengths of 324.7 kN and 316.6 kN, respectively; Yield displacements of 33.38 mm and 33.66 mm, respectively; Their ultimate bearing capacities are 331.38 kN, 352.2 kN, and 353.78 kN, respectively; Their ultimate displacements are 62.5 mm, 57.45 mm and 62.01 mm, respectively. The yield strength, yield displacement, ultimate strength, and ultimate displacement data for the five specimens are summarized in Table 1. The average yield strength of the beam-slab structural members is 307.42 kN and the average ultimate strength is 340.91 kN.

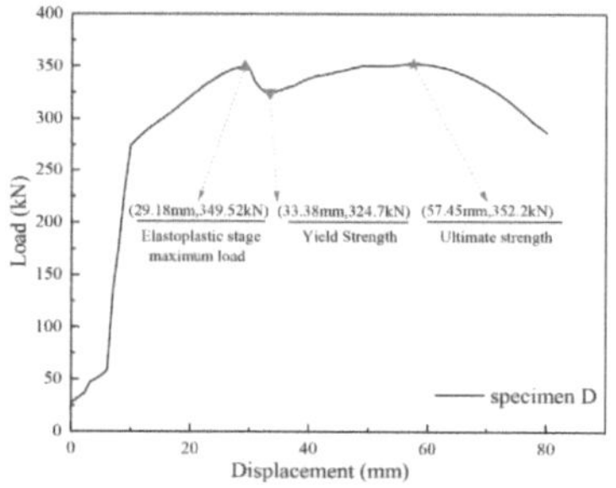

Figure 14. Force-displacement curve of specimen D.

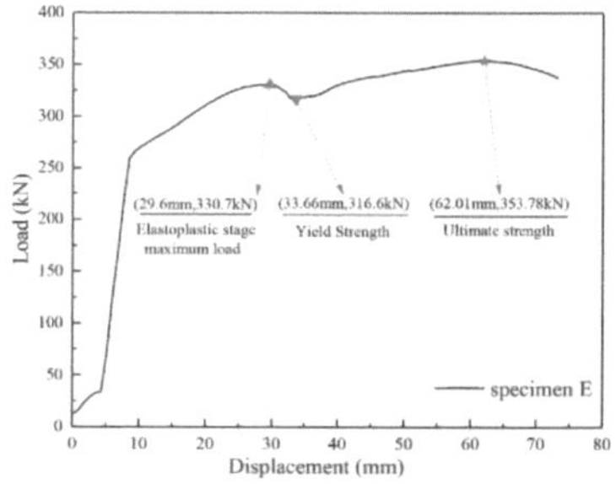

Figure 15. Force-displacement curve of specimen E.

Table 1. Data summaries for each specimen.

Number	Yield load (kN)	Yield displacement (mm)	Ultimate load (kN)	Ultimate displacement (mm)
Specimen A	296.3	32.22	337.3	60.61
Specimen B	314.1	49.07	329.9	61.44
Specimen C	285.4	27.94	331.38	62.50
Specimen D	324.7	33.38	352.2	57.45
Specimen E	316.6	33.66	353.78	62.01

Force-displacement curves of the five specimens indicate that as the specimen reaches the BC stage, the growth of load-bearing capacity slows. Displacement is increasing as the I-beam and the inner plate deformation is not synchronized, resulting in friction displacement between the two. Displacement in continuous increase occurs before the external load reaches the strength of the plug welding joint. The external load exceeds the welding strength, which causes the weld to break, reducing its load capacity. As a result of blast loading, the connection between the inner plate and the I-beam is a weak part of the entire structure. Therefore, the welding strength between the inner plate and the I-beam should be strengthened to make the structure more ductile.

An energy absorption capacity was calculated for the specimen in the AC phase according to the results of static loading tests. The energy absorption values were obtained by area calculation of the force-displacement curves obtained from the tests [4]. Table 2 provides information on the energy absorption of the five specimens in stage AC. Table 2 shows that

Table 2. Energy absorption values of specimens.

Number	Loadings at point C (kN)	Displacement at point C (mm)	Energy (J)
Specimen A	326.6	28.31	2946.27
Specimen B	366.8	39.87	3997.23
Specimen C	309.52	26.91	3132.31
Specimen D	349.52	29.18	3029.53
Specimen E	330.7	29.6	2622.05

the average value of the overall energy absorption of the specimen is 3145.48 J before the welding joint between the inner plate and the I-beam fails.

3.3 *Force-strain curve analysis*

Deformation data from Figure 3 were obtained using strain gauges pasted at the three locations of the specimen to measure strain data. The force-strain curve of specimen A illustrates the wide variation of central strain values under external loads, especially the central-2 curve, with a maximum strain value of 1278.4 $\mu\varepsilon$ under external load. Following is the left-2 strain curve, with a strain value of 713.26 $\mu\varepsilon$, consistent with the damaged position of the specimen.

Figures 16(b) and (c) show the force-strain curves of specimens B and C, respectively. The figures show that in the elastic stage, the strain change trend is basically the same at each position, and the strain curve grows almost linearly as the load keeps increasing. In the plastic development stage, the deformation in the middle of the specimen increases continuously, and the corresponding strain values continue to grow, with the maximum strains in the middle of specimen B and specimen C being 5602.9 $\mu\varepsilon$ and 5484.1 $\mu\varepsilon$, respectively.

Figures 16(d) and (e) show specimen D's force-strain curves and E's force-strain curves, respectively. The strain trends measured by the two strain gauges at each position of the specimen in the figures are the same, especially the force-strain curve of specimen E. With the increase of applied load, the deformation in the middle area of the specimen is the largest, and the strain value in the middle area is higher than the remaining four strain values. The maximum strain values in the middle area of specimens D and E are 6011.3 $\mu\varepsilon$ and 5186.1 $\mu\varepsilon$, respectively. The damage location of both specimens is located at the weld between the left inner plate and the I-beam, and the corresponding strain values are higher than the strain values on the right side, and the stretching degree on the left side is higher than that on the right side.

The force-strain curve of the specimen shows that the strain variation law at each point of the specimen coincides with the deformation of the specimen. Direct action of the load causes most deformation in the middle of the specimen, followed by strain on the cracked side, which is stretched more than the uncracked side.

Figure 17 shows the strain variation at each measurement point location for the five specimens under different loading conditions. Figures 17(a)-(c) show that the strain changes at each measurement point in stage AC of the selected specimens. Before the specimen reaches point C, the strain in the middle grows slowly, and the strain on the cracked side of the corresponding specimen becomes larger in this stage. Therefore, it can be determined that the inner plate of the specimen is under tension during stage AC, and the frictional deformation between the inner plate and the I-beam mainly generates its deformation. It can be seen from Figures 17(d) and (e) that when the external load exceeds the value of the specimen at point C, a sudden change in strain occurs in the middle of the specimen, indicating that the shear stress between the inner plate and the I-beam exceeds the strength of the

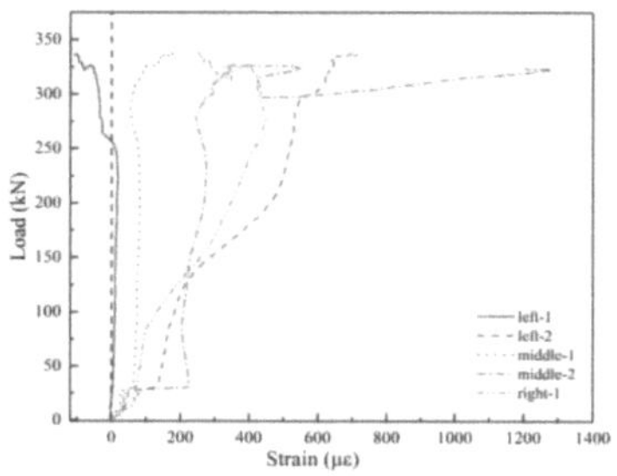

(a) Force-strain curve of specimen A

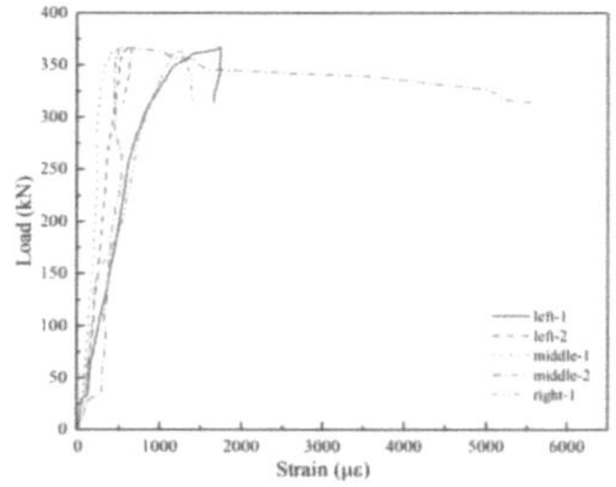

(b) Force-strain curve of specimen B

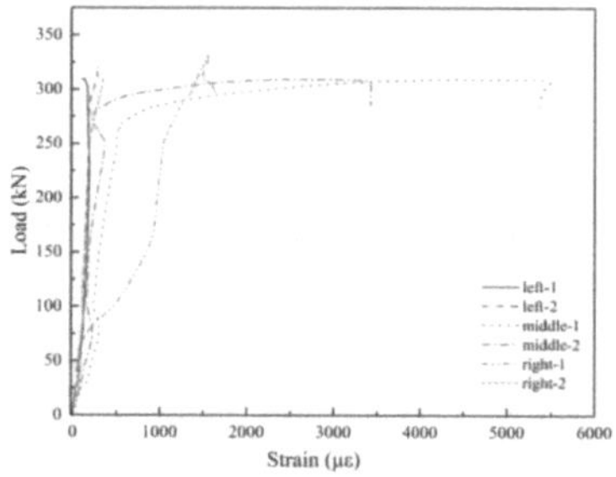

(c) Force-strain curve of specimen C

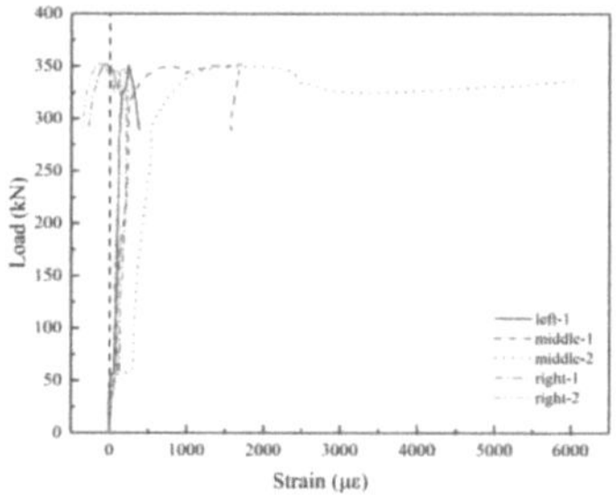

(d) Force-strain curve of specimen D

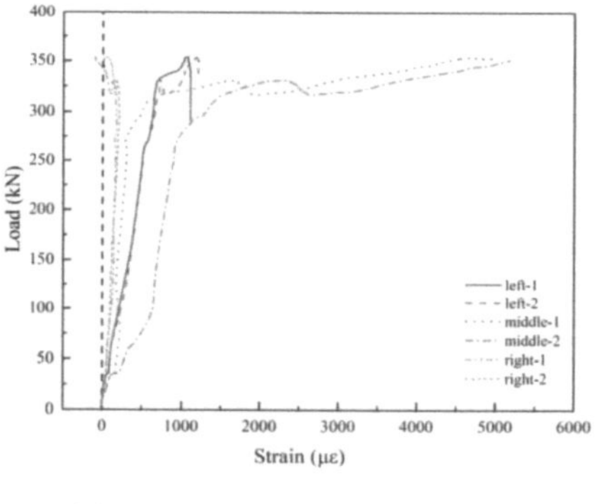

(e) Force-strain curve of specimen E

Figure 16. Force-strain curves of 5 specimens.

welding joint. As the weld joint cracks, the frictional displacement between the inner plate and the I-beam increases, causing the middle of the plate to deform rapidly. Therefore, point C is the critical value for the strength of the weld between the I-beam and the inner plate.

4 FITTING ANALYSIS

In light of the above analysis, it is clear that stage AC in Figure 11 constitutes the main force stage in the test piece as a whole. If there is damage to one side of the protective door under a

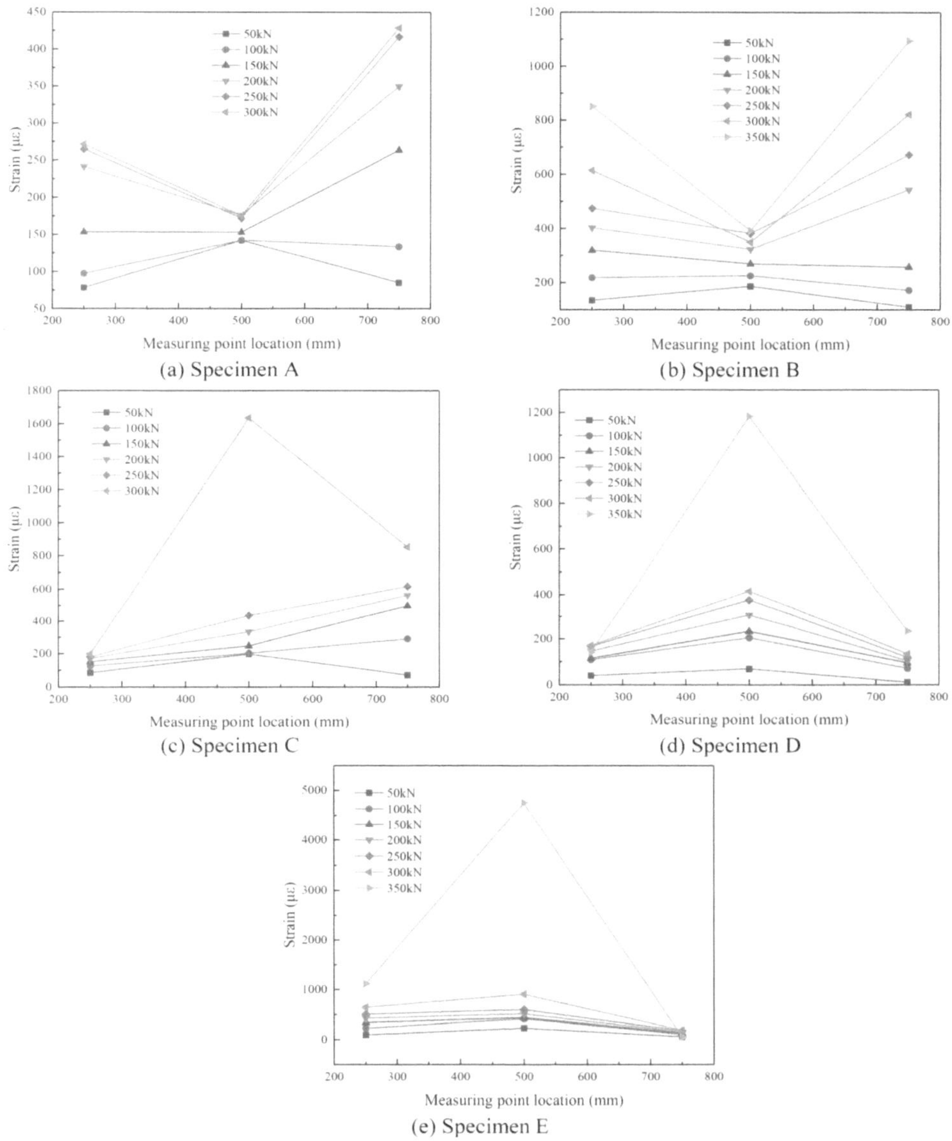

Figure 17. Strain variation at each measurement point under different loads.

real blast wave, the beam-plate structure loses functions. To enhance the protective performance of protective doors, it is necessary to accurately predict the variation in this phase.

As described in Figure 18, stage OA is the elastic stage of the specimen. From point A, the specimen enters plastic deformation, and the weld and the I-beam in section AB carry the load. Due to the friction between the I-beam and the inner plate, the growth rate of external load slows at this stage and displacement increases, and the weld undergoes shear deformation. When the shear stress exceeds the strength of the weld, the weld cracks, entering stage BC when the specimen begins to yield. Welds fail when they reach point C. During

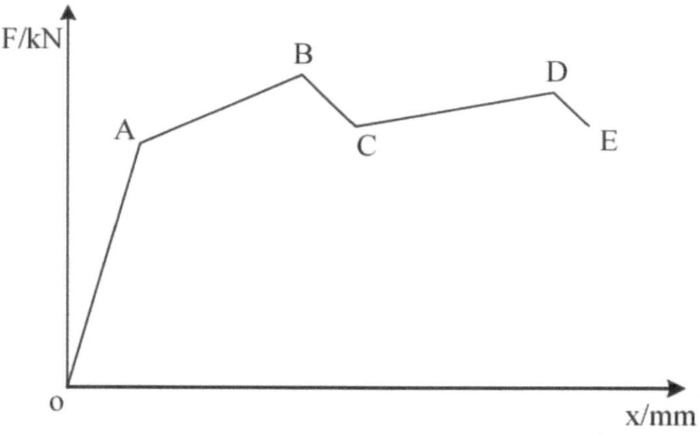

Figure 18. Schematic diagram of the force-displacement curve of specimens.

stage CD, the specimen is primarily supported by the I-beam and the outside plate, together with an external load. The specimen reaches ultimate strength at point D and then fails.

To accurately predict the trend of each stage, the four stages of section OD were fitted and the fitting equation for each stage was obtained as follows.

Elastic segment (OA): $F = (35.021 \pm 4.812)l + (-123.982 \pm 38.17)(3 < l \leq 11.36)$
Elastic segment (AB): $F = (3.644 \pm 0.135)l + (232.353 \pm 3.051)(11.36 < l \leq 28.21)$
Yield section (BC): $F = (-3.65 \pm 1.788)l + (438.09 \pm 56.643)(28.21 < l \leq 33.89)$
Plastic growth segment (CD): $F = (1.056 \pm 0.12)l + (278.612 \pm 5.896)(l > 33.89)$

where F is the load size of the beam-slab structure (Kn). l is the deformation size of the beam-plate structure (mm).

We have integrated the fitted 4-stage formula with the force-displacement curve measured by the test (Figure 19), and the fitted curve was found to match well with the test curve. It can accurately predict the trend for all stages, particularly OA and AB. According to Table 3, the maximum value of the elastic-plastic section at point B obtained from the fitted curve is 10% of the test values, showing that the fitted and test values are 10% of each other. Thus, the fitted formula can accurately predict the critical strength of weld damage in beam-slab structures of this type. Using the fitting formula, it is possible to determine the force stage of the specimen, which provides a reference for improving the overall performance of protective doors.

5 CONCLUSIONS

Due to different construction processes, steel protective doors of beam-plate structures are often equipped with different welding joints between the inner plate, outer plate, and I-beam. In this paper, five groups of beam-plate structures were tested with static pressure loading. The purpose is to determine the failure mode, bearing capacity, displacement, and characteristics of strain curves. Based on our analysis, a fitting formula for predicting the change of this type of structure was developed. The research findings are detailed below.

(1) When the inner plate of the specimen is pulled by static pressure, the inner plate and the I-beam deform uncoordinatedly, resulting in shear fracture damage between the two

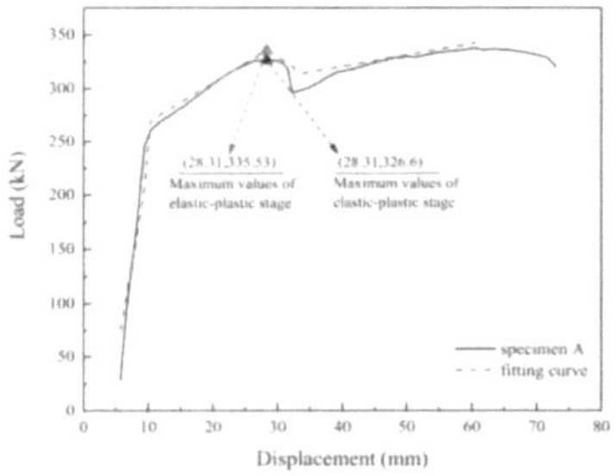

(a) Comparison of fitted curves of specimen A

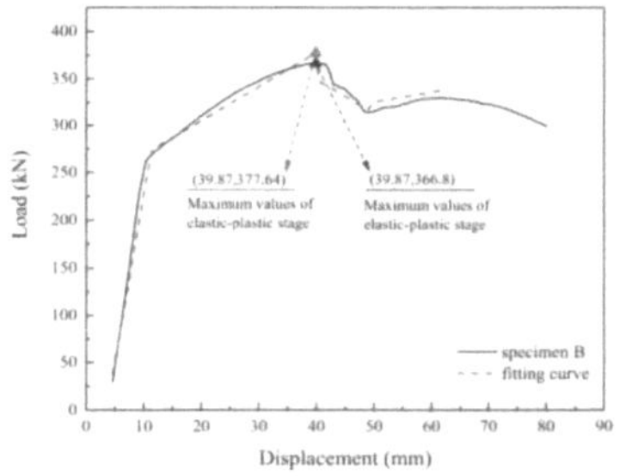

(b) Comparison of fitted curves of specimen B

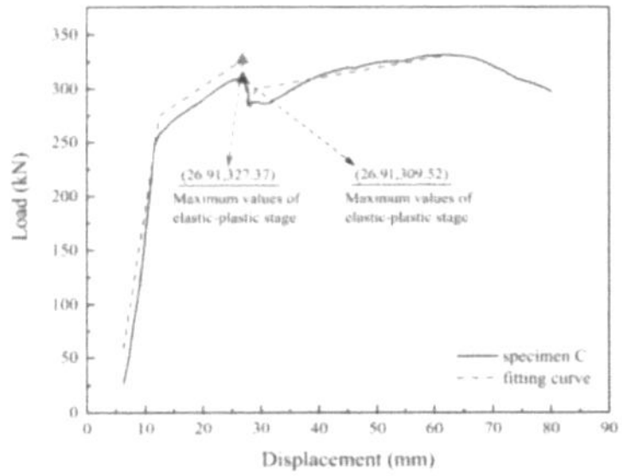

(c) Comparison of fitted curves of specimen C

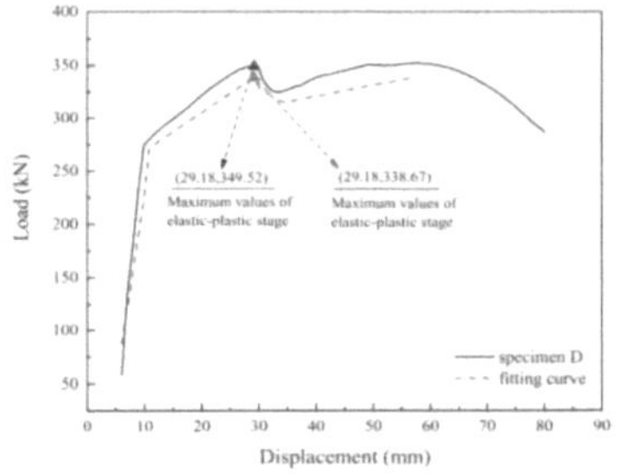

(d) Comparison of fitted curves of specimen D

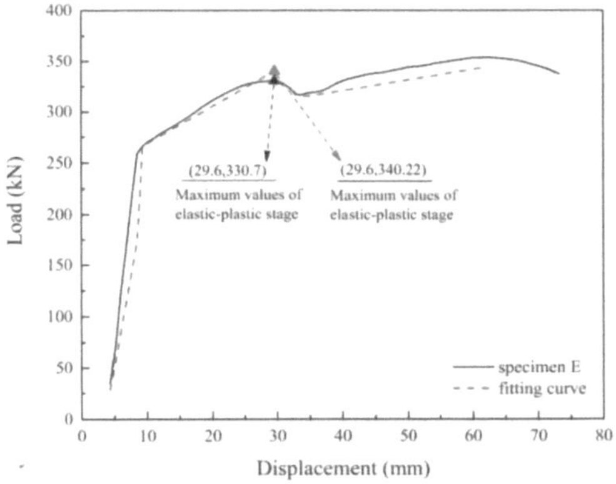

(e) Comparison of fitted curves of specimen E

Figure 19. Fitting equation vs. experimental results.

plug welding joints. To improve the overall deformation capacity of the protective door and delay damage, the welding strength at the connection between the inner plate and the I-beam should be strengthened when designing protective doors.

(2) When the specimen reaches point B in Figure 18, frictional deformation between the inner plate and the I-beam is the primary cause of deformation. When the load exceeds the B point, the specimen is damaged and its middle strain increases rapidly. Thus, point

B is the critical value of the strength of the welding joints between the inner plate and the I-beam of the specimen. Five groups of specimens at point B were found to have an average load-bearing capacity of 336.62 kN and an average displacement of 30.9 mm. The fitting curve matches the test curve well, with an error of 10% between the fitting values and the test values at point B.

Table 3. Comparison of experimental values and fitted values at point B.

Number	Test values (kN)	Fitted values (mm)	Error
Specimen A	326.6	335.53	2.73%
Specimen B	366.8	377.64	2.96%
Specimen C	309.52	327.37	5.77%
Specimen D	349.52	338.67	3.1%
Specimen E	330.7	340.22	2.88%

REFERENCES

Chen W., Hao H (2012). Numerical Study of a New Multi-arch Double-layered Blast-resistance Door Plate. *International Journal of Impact Engineering*. 43: 16–28.

Dong H., Chen L., Hong J (2014). Analysis of the Dynamic Response of New Flexible Cable Membrane Structure Protective Gate. *Industrial Architecture*. 44(S1): 269–273.

Fang H., Li X. D., Geng Z. G., Xu K (2017). Numerical Simulation of the Dynamic Response of High Damping Rubber-filled Steel Protective Doors. *Journal of Arms Equipment Engineering*. 38(09): 173–177.

Guo D (2012). *Dynamic Response Behavior and Bounce Mechanism of Protective Doors under Explosive Loads*. Tsinghua University.

Guo D., Li Z.P., Wang Q.S., Hou X.F (2013). Study on the Anti-burst Performance of Steel-clad Steel Pipe Concrete Protection Doors. *Protection Engineering*. (02): 38–43.

Hsieh M. W., Hung J. P., Chen D. J (2008). Investigation on the Blast Resistance of a Stiffened Door Structure. *Journal of Marine Science and Technology*. 16(2): 7.

He H., Zhang B., Zheng Q., et al. (2021). Anisotropic Dynamic Theory to Predict Blast Responses of Composite Fluted-core Sandwich Protective Door Plates. *Thin-Walled Structures*. 161: 107436.

Koh C. G., Ang K. K., Chan P. F (2003). Dynamic Analysis of Shell Structures with Application to Blast Resistant Doors. *Shock and Vibration*. 10(4): 269–279.

Luo X., Qian X., Zhao H., et al. (2012). Simulation Analysis on Structure Safety of Refuge Chamber Door Under Explosion Load. *Procedia Engineering*. 45: 923–929.

Meng F., Zhang B., Zhao Z., et al. (2016). A Novel All-composite Blast-resistant Door Structure with Hierarchical Stiffeners. *Composite Structures*. 148: 113–126.

Nurick G. N., Olson M. D., Fagnan J. R., et al (1995). Deformation and Tearing of Blast-loaded Stiffened Square Plates. *International Journal of Impact Engineering*. 16(2): 273–291.

Rashad M. A., Elwahab M., Mahfoz S. Y., et al. (2012). Numerical Study of Lightweight Sandwich Plates Under Explosion Using Rigid Polyurethane Foam and Vulcanized Rubber[C]//The International Conference on Civil and Architecture Engineering. Military Technical College. *9(9th International Conference on Civil and Architecture Engineering)*: 1–14.

Thimmesh T., Shirbhate P. A., Mandal J., et al (2021). Numerical Investigation on the Blast Resistance of a Door Plate. *Materials Today: Proceedings*. 44: 659–666.

Veeredhi L. S. B., Ramana Rao N. V (2015). Studies on the Impact of Explosion on Blast Resistant Stiffened Door Structures. *Journal of The Institution of Engineers (India): Series A*. 96(1): 11–20.

Xu K., Li X.D., Liu J.X., Mao H.Y (2018). Numerical Analysis of the Blast Resistance Mechanism of High Damping Rubber Reinforced Protective Doors. *Journal of Arms Equipment Engineering*. 39(10): 155–161.

Yelek İ., Ardalı R., Yılmaz İ (2021). Fracture Investigation of Different Welded Steel Beam-column Connections Under a Three-point Bending Test. *Journal of Constructional Steel Research*. 187: 106945.

Zhang B. Y., Zhai D. X., Sun J., et al. (2014). Numerical Simulation on the Dynamic Response of Blast-resistance Door with Built-in Aluminum Matrix Syntactic Foam. *Binggong Xuebao/Acta Armament*. 35: 263–267.

Zhang B., He H., Zhou J., et al. (2019). Construction and Failure Analysis of Ultra-light GFRP Fluted-core Sandwich Protective Structures. *Composites Science and Technology*. 173: 73–82.

Zhang D.X (2011). Research on the Blast Resistance of New Composite Aluminum Foam Sandwich Plate. *Harbin Institute of Technology*.

Water Conservancy and Civil Construction – Oke & Ahmad (Eds)
© 2024 The Author(s), ISBN: 978-1-032-58614-4

High-strength lightweight concrete preferred mix design

Xiangli Wang
China Highway Engineering Consulting Co., Ltd., China

Yunwu Wang
Hainan Transportation Engineering Construction Bureau, China

Shuirong Lin
Hainan CCCC Expressway Investment and Construction Co., Ltd., China

Dailiang Li
CCCC First Harbor Engineering Co., Ltd., China

Pengpeng Hou
CCCC-Shec Dong Meng Engineering Co., Ltd., China

Chenxu Li & Shanshan Zhang*
School of Architecture and Civil Engineering, Wuhan University of Technology, China

ABSTRACT: Lightweight concrete plays an important role in construction, especially for long-span bridges. This research aims at the national highway G360 Dingan to Lingao highway engineering. According to the actual needs of its research innovation and construction excellence, the preparation technology of lightweight high-strength concrete was studied. Four lightweight, high-strength concrete test blocks with different mix ratios were designed, and the compressive test was carried out. Finally, the optimal mix ratio was selected. FBG sensors adapted to strain detection. The results show that compared with ordinary concrete, lightweight high-strength concrete not only reduces the weight of concrete but also shows good mechanical properties and meets the requirements of use.

1 INTRODUCTION

Great progress has been made in the research and development of bridge deck pavement materials and structures in recent years (Chen *et al.* 2021; Fan & Luo 2021; Geng *et al.* 2017). However, the bridge deck pavement materials that are economical and can completely solve the problem and the matching process still need further research and development. The conventional bridge deck adopts conventional cement concrete pavement and asphalt pavement. The increase in pavement thickness will bring serious consequences of bridge deflection and cracking. Therefore, to ensure the bridge's stability, lightweight materials are used to increase the traffic volume of live load. In the long run, lightweight, high-strength concrete is preferred (Aslam *et al.* 2017; Sifan *et al.* 2023).

In the construction of long-span bridges and super high-rise buildings, the use of high-strength lightweight concrete can reduce structural weight by more than 30% (Bas *et al.* 2018; Li & Chan 2006; Xu *et al.* 2020). At the same time, it effectively reduces construction

*Corresponding Author: zhang_shanshan@whut.edu.cn

DOI: 10.1201/9781003450818-41

difficulty, material transportation costs, and overall project costs. The structural weight of the bridge is greatly reduced, which can reduce the density of the reinforcement so that the bridge foundation load will be reduced. High-strength, lightweight concrete brings a series of favorable factors to bridge structure design. This project is based on four design principles: high strength, lightweight, high durability, high crack resistance, lightweight, and high strength concrete mix optimization. In this study, FBG sensors were adopted to measure the strain state inside the concrete by using the matched fiber grating demodulator (Qin *et al.* 2022; Wei *et al.* 2018; Xu *et al.* 2013, 2020, 2014, 2018, 2022; Xu 2017). Finally, an optimal mix ratio of high-strength lightweight concrete is obtained for bridge construction.

2 MIX PROPORTION DESIGN OF HIGH-STRENGTH LIGHTWEIGHT CONCRETE

2.1 *Concrete raw materials and specifications*

Raw materials for preparing lightweight, high-strength concrete are: 1. Coarse aggregate using shale-broken ceramic, specifications for 5~9.5 mm ceramic, 9.5~19 mm ceramic, test to determine the blending ratio of 100% and 50%, synthesized for 5~20 mm continuous gradation. 2. Fine aggregate: using ordinary medium sand. P·O 42.5 cement is used as cement. Grade I fly ash is used, the technical index conforms to grade I, and the content is 10% of the cementitious material. 3. Silica fume: gray micro silica fume (96% content), 5% of cementitious materials dosage. 4. Additives: KXSP (KXPCA) polycarboxylate super-plasticizer, the dosage of cementitious materials 1.5%.

By calculating the selected benchmark mix ratio, the number of various materials in each concrete square is (Kg): cement: coal fly ash: silica fume: ceramic: sand: net water: water reducing agent powder =440:82.5:27.5:506:729.5:165:8.25.

2.2 *Mix proportion test design*

According to the calculated mix ratio as the benchmark mix ratio for trial, the calculated water-binder ratio remains unchanged, and by adjusting the mix ratio and other parameters, we make the slump, cohesion and water retention, strength, elastic modulus, and other properties of the concrete mixture meet the design objectives and construction requirements. Then we modify the calculation benchmark mix ratio, and put forward the trial mix ratio.

The concrete strength test is carried out based on the trial mix ratio and according to the following provisions: 1. Four different mix ratios should be adopted; 2. Concrete strength test, the performance of the mixture should meet the design objectives and construction requirements; 3. When the concrete strength test is carried out, at least one set of specimens should be prepared for each mixing ratio and cured for 28 days for testing. The specific trial scheme is shown in Table 1:

Table 1. Different mix proportion concrete quantity calculation table (Kg/m^3).

Test block	Water-binder ratio	Percentage of sand (%)	Cement	Coal fly ash	Silica fume	Sand	Ceramic	Gravel	Water	Additive
1	0.30	43	440	82.5	27.5	729.5	506	/	165	8.25
2	0.32	43	440	82.5	27.5	717.5	497.5	/	176	7.7
3	0.3	43	408	76.5	25.5	757.1	262.6	515	153	8.16
4	0.32	43	408	76.5	25.5	745.8	258.6	508	163	7.5

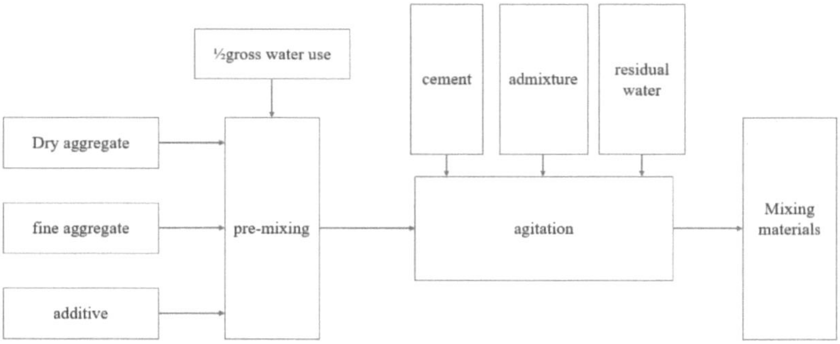

Figure 1. Feeding sequence when using lightweight aggregates not pre-wetted.

The concrete mixing on site adopts the self-falling mixer. When mixing high-strength lightweight concrete, the pre-wetted lightweight aggregate is used, and the following points should be noted: a. The net water consumption in the mix ratio is determined by dry sand, so the water content of sand must be determined before pre-mixing, and the water content of sand should be calculated. b. After determining the water content of the sand, the amount of other materials is calculated according to the mix; c. It needs to premix before formal mixing to avoid uneven mixing.

3 TEST RESULT

3.1 *Strength test results*

The experimental equipment for the compressive test is a 100 t universal testing machine. The test block's length, width, and height are 100 mm, the curing time is 7 days, and the loading rate is 0.5 MPa/s.

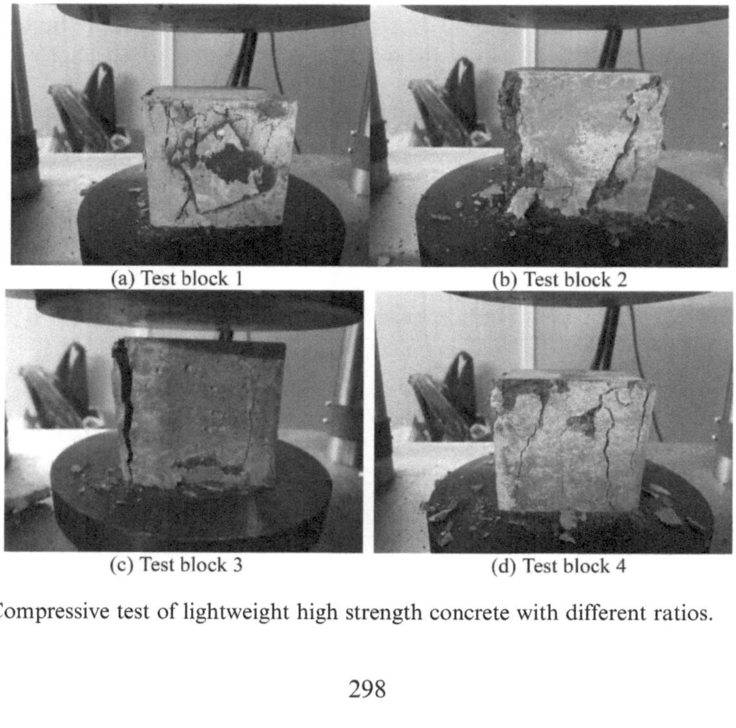

(a) Test block 1 (b) Test block 2 (c) Test block 3 (d) Test block 4

Figure 2. Compressive test of lightweight high strength concrete with different ratios.

The compressive strength of the test block 1 to 4 is 44.87 MPa, 41.87 MPa, 47.24 MPa, 45.26 MPa, and the average strength is 39.0 MPa. After multiplying the conversion coefficient of 0.95, the average strength of the test block for 7 days is 44.81 MPa. Adding silica fume and fly ash will reduce the early strength of concrete but will improve the late strength and durability. The test results are shown in Table 2. It can be seen from the test results that the seven-day strength value is 47.24 MPa when the slump values are similar under four different test conditions. Currently, the amount of gel material, cement, fly ash, and silica fume is less, so 3 is selected. Therefore, the final mix ratio is calculated according to the test strength test results, as shown in Table 3:

Table 2. Test results of concrete strength with the different mix proportion.

Test block	Cementitious materials/Kg	Cement/ Kg	Water-binder ratio	Percentage of sand	Coal fly ash /Kg	Silica fume/ Kg	Water reducer/ Kg	The proportion of ceramic	Slump/ mm	7-Day average intensity/ MPa
1	550	440	0.3	43%	82.5	27.5	8.25	100%	220	44.87
2	550	440	0.32	43%	82.5	27.5	8.25	100%	232	41.87
3	510	408	0.3	43%	76.5	25.5	8.16	50%	245	47.24
4	510	408	0.32	43%	76.5	25.5	7.5	50%	238	45.26

Table 3. Finalize construction mix ratio.

Cement Kg/m^3	Silica fume Kg/m^3	Coal fly ash Kg/m^3	Shale ceramic Kg/m^3	Crushed stone Kg/m^3	Medium sand Kg/m^3	Net water consumption Kg/m^3	Water reducer Kg/m^3
408	25.5	76.5	258.6	507.7	729	163.2	7.5

3.2 *Monitoring data*

Figure 3 is the layout plan of the optical fiber, and the strain change in Figure 4 is analyzed. The strain difference between the day of concrete pouring and one month after pouring reflects the shrinkage change of concrete within one month. It can be seen from the diagram that the micro-strain of lightweight concrete and ordinary concrete monitored by different sensors is not the same. This is due to the sensors' different locations and shrinkage stresses generated at different locations, which also has a certain impact on the installation process of the previous sensors. In addition, with lightweight concrete and ordinary concrete with an increase in age, there has been a certain degree of shrinkage. At the same scale, the phenomenon of lateral shrinkage than longitudinal shrinkage is more obvious. Lightweight concrete and ordinary concrete have a more obvious micro-strain change at the initial stage of pouring, which is due to the loss of plasticity of concrete. The analysis of monitoring data shows that compared with ordinary concrete, lightweight concrete not only reduces the weight of concrete but also shows good mechanical properties and meets the requirements of use.

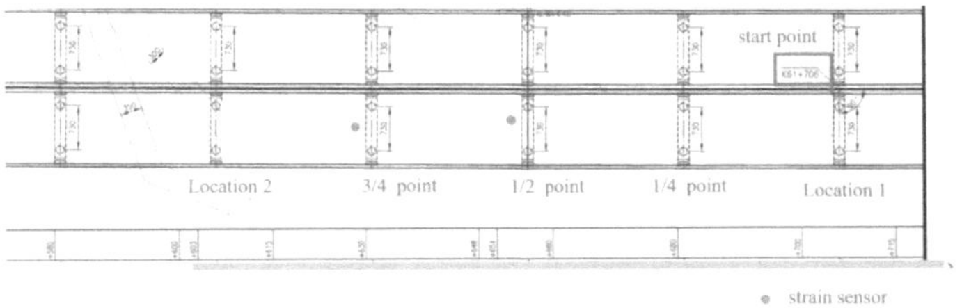

Figure 3. FBG sensor layout.

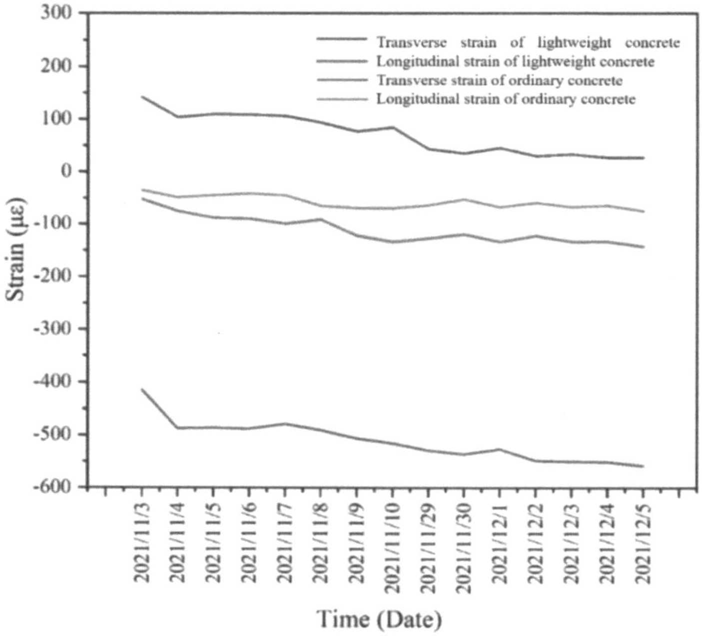

Figure 4. Micro strain data diagram of bridge deck concrete.

4 CONCLUSIONS

Different water-binder ratios studied the strength properties of high-strength lightweight concrete for bridge deck pavement, sand ratios, coarse aggregate ratios, fly ash, and silica fume. Cement: coal fly ash: silica fume: ceramic: sand: net water: water reducing agent powder =440:82.5:27.5:506:729.5:165:8.25 is liquidity and strength to meet the needs of the actual project site. At the same time, the analysis of monitoring data shows that compared with ordinary concrete, lightweight concrete not only reduces the weight of concrete but also shows good mechanical properties and meets the requirements of use.

REFERENCES

Aslam, M., P. Shafigh, & M. Jumaat (2017). High-strength Lightweight Aggregate Concrete Using Blended Coarse Lightweight Aggregate Origin from the Palm Oil Industry. *Sains Malays.* 46(4), 667–675.

Bas, S., N. Apaydin, A. Ilki, & F. Catbas, (2018). The Structural Health Monitoring System of the Long-span Bridges in Turkey. *Struct. Infrastructure. E.* 14(4), 425–444.

Chen, L., X. Zhang, W. Ma, & X. Zhang (2021). Development and Evaluation of a Pothole Patching Material for Steel Bridge Deck Pavement. *Constr. and Build. Mater.* 313125393.

Fan, X. & R. Luo (2021). Experimental Study of the Deformation Recovery Characteristics of Steel Deck Pavement Materials. *Constr. and Build. Mater.* 301124149.

Geng, L., Q. Xu, R. Ren, L. Wang, X. Yang, & X. Wang (2017). Performance Research of High-viscosity Asphalt Mixture as Deck-paving Materials for Steel Bridges. *Road Mater. Pavement* 18(1), 208–220.

Li, Z., & T. Chan (2006). Fatigue Criteria for Integrity Assessment of Long-span Steel Bridge with Health Monitoring. *Theoretical and Applied Fracture Mechanics*, 46(2), 114–127.

Qin Y., Q. Wang, D. Xu, J. Yan & S. Zhang (2022). A Fiber Bragg Grating-based Earth and Water Pressure Transducer with Three-dimensional Fused Deposition Modeling for Soil Mass. *J. Rock Mech. Geotech.* 14 (2), 663–669.

Sifan, M., B. Nagaratnam, J. Thamboo, K. Poologanathan, & M. Corradi (2023). Development and Prospectives of Lightweight High-strength Concrete Using Lightweight Aggregates. *Constr. and Build. Mater.* 362129628.

Wei, H., D. Xu, & Q. Meng (2018). A Newly Designed Fiber-optic-based Earth Pressure Transducer with an Adjustable Measurement Range. *Sensors* 18(4), 932.

Xu, D. (2017). A New Measurement Approach for Small Deformations of Soil Specimens Using Fiber Bragg Grating Sensors. *Sensors*, 17(5), 1016.

Xu, D., L. Borana, & J. Yin, (2014). Measurement of Small Strain Behavior of a Local Soil by Fiber Bragg Grating-based Local Displacement Transducers. *Acta Geotech.* 9(6), 935–943.

Xu, D., Liu H. & Luo W., (2018). Development of a Novel Settlement Monitoring System Using Fiber-optic Liquid-level Transducers with Automatic Temperature Compensation. *IEEE T. Instrum. Meas.* 67(9), 2214–2222.

Xu, D., Yin J., Cao Z., Wang Y., Zhu H. & Pei H. (2013). A New Flexible FBG Sensing Beam for Measuring Dynamic Lateral Displacements of Soil in a Shaking Table Test. *Measurement*, 46(1), 200–209.

Xu, D., Z. Su, B. Lalit, & Y. Qin, (2022). A Hybrid FBG-based Load and Vibration Transducer with a 3D Fused Deposition Modelling Approach. *Meas. Sci. Technol.* 33(60), 065106.

Xu, X., X. Yang, W. Yang, X. Guo, & Xiang, H. (2020). New Damage Evolution Law for Modeling Fatigue Life of Asphalt Concrete Surfacing of the Long-span Steel Bridge. *Constr. and Build. Mater*. 259119795.

Water Conservancy and Civil Construction – Oke & Ahmad (Eds)
 ISBN: 978-1-032-58614-4

Experimental study on ultrasonic properties of compressed concrete

Ping Fan*
Research Institute of Highway, Ministry of Transport, Haidian District, Beijing, P.R. China

Jinquan Zhang*
Research Professor, Research Institute of Highway, Ministry of Transport China, Haidian District, Beijing, P.R. China

Tao Wang* & Kanglin Zheng*
Research Institute of Highway, Ministry of Transport, Haidian District, Beijing, P.R. China

ABSTRACT: Coda wave interferometry (CWI) can detect small changes in the medium by using the wake waves formed by the repeated scattering of ultrasonic waves. CWI is widely used in seismic analysis due to its high sensitivity to subtle seismic disturbances. In this study, experiments were carried out on concrete specimens subjected to uniaxial compression loading to investigate the propagation behavior of ultrasonic waves in concrete. The variation of wave delay in concrete with uniaxial compressive stress is studied. The experimental results show that the ultrasonic coda wave velocity first increases with the increase of concrete compressive stress, reaches the maximum value, and then decreases at an accelerating rate with micro-cracks propagation. Therefore, ultrasonic coda wave velocity was sensitive to compressive stress changes and concrete cracking.

1 INTRODUCTION

Ultrasonic methods are widely used in civil engineering applications due to the ease of application, deep media penetration, and safety to humans (Kumar *et al.* 2013; Koch *et al.* 2015; Nobile *et al.* 2015; Popovics *et al.* 2004; Rens *et al.* 1999). In concrete structures, ultrasonic methods are mainly used to detect defects and evaluate concrete strength by measuring the velocity of direct ultrasonic waves propagating in concrete (Hannachi *et al.* 1999; Komlos 1996; Lee *et al.* 1998; Maierhofer *et al.* 2003; Ndagi *et al.* 2019). Ultrasonic methods are also used in bridge nondestructive testing (NDT) to detect internal damage or fatigue cracks in bridge structural elements (Khalifa *et al.* 2016; Korzeniowski *et al.* 2014; Nogueira *et al.* 2011; Zou *et al.* 2013).

Nevertheless, despite the early start of research on stress assessment, the application of ultrasonic methods for measuring stresses in concrete is relatively rare. Based on Murnaghan's theory of finite deformations (Murnaghan 1937), Hughes and Kelly (1953) developed the concept of acoustoelastic effects in stressed solids and presented expressions for the acoustoelastic theory of anisotropic materials. The practice has proved that the ultrasonic method can measure the stress of anisotropic materials such as concrete. Bergman R H *et al.* (1958) further discovered the double refraction phenomenon of acoustic waves in stressed materials, which laid the physical foundation for the acoustic, elastic stress analysis

*Corresponding Authors: fanping2008@ hotmail.com, jq.zhang@ rioh.cn, t.wang@ rioh.cn and kl.zheng@ rioh.cn

 DOI: 10.1201/9781003450818-42

method. Lin *et al.* (2011) studied how several wave parameters, including pulse velocity, amplitude, and spectrum area, varied with concrete stress and concluded that the wave velocity was insensitive to stress. Stähler *et al.* (2011) studied the relationship between the ultrasonic waves and stress in compressed concrete by measuring the direct wave velocity and found that in the elastic range, the wave velocity varies little with stress. However, due to the dispersion of the microstructure of concrete materials, it is difficult to establish a uniform standard of ultrasonic velocity for concrete of the same strength. In actual engineering, the ultrasonic method is mainly used to evaluate the residual stress of steel structural components. The accuracy of direct waves cannot meet the requirements of concrete stress measurement. Therefore, the stress measurement method using ultrasonic velocity in steel structures cannot be generalized in concrete structures.

Direct ultrasonic waves are mainly used for material quality, strength assessment, and defect detection. However, direct waves are rarely used for concrete stress assessment, mainly because this method usually fails to guarantee the accuracy required for stress measurements. Therefore, a more accurate method of measuring ultrasonic velocity is needed. Coda wave interferometry (CWI) is a Non-Destructive testing (NDT) method that exhibits high sensitivity to small (microstructural) changes in concrete by using multiple scattering of ultrasonic waves. CWI was originally used for seismic analysis and stress assessment of rocks (Aki & Chouet 1975; Gret 2006). Since concrete is a highly heterogeneous material similar to rock, the method can be extended to concrete stress measurements.

When the frequency is above 100 kHz, ultrasonic waves are mainly scattered in the medium. In this case, the wave propagation paths are extended, and the arrival time is delayed, similar to a long tail of the direct wave (the main mode, such as P, S, and surface waves), so they are called coda waves. Aki *et al.* (1975) proposed the concept of CWI and applied it to seismic research. Gret *et al.* (2006) systematically described the CWI theory and studied the effects of rock stress and temperature on the velocity of coda waves. Larose and Hall (2009) used CWI to perform a stress test on compressed concrete specimens and obtained an accurate estimate of the wave velocity change (0.001%), demonstrating the potential of CWI for NDT. Stähler *et al.* (2011) applied CWI to stress evaluation of a concrete bridge under construction. Niederleithinger *et al.* (2010) compared direct and coda waves' sensitivity to concrete stress, showing that the latter is much higher. Zhang *et al.* (2014) studied the effect of temperature on CWI in concrete and found that about 0.01% of a temperature change of about 1°C could cause about 0.01% of the velocity variation. It is very important to control the temperature for accurate CWI measurements.

In recent years, coda wave technology has been gradually applied to engineering practice, mainly in NDT (Hafiz *et al.* 2018; Hu *et al.* 2021; Legland *et al.* 2017; Spalvier *et al.* 2019).

CWI can be used not only for detecting internal defects in concrete but also for evaluating concrete stress due to its high accuracy. CWI has been applied to digital image processing techniques (Larose *et al.* 2010; Pacheco & Snieder 2005; Planès & Larose 2013; Rossetto *et al.* 2011; Zhang *et al.* 2016). This method can visually reveal defects (such as cracks and voids) and their location and visualize concrete stresses.

Despite these efforts, only limited studies using the CWI method to evaluate the mechanical properties of concrete have been reported, and the mechanisms by which ultrasonic parameters vary with stress have yet to be fully explored. In addition, the development principles, test conditions, and data analysis methods of CWI in concrete have yet to be well established.

This paper presents an experimental study of the propagation of ultrasonic waves in concrete under uniaxial compressive loads. The theoretical basis of CWI is first reviewed, and then the experimental procedures are described in detail. By analyzing the experimental results, the relationship between the time delay and stress of the ultrasonic coda waves and the propagation mechanism of the ultrasonic coda wave in concrete are discussed in detail. In this paper, the propagation speed of ultrasonic waves in concrete is quantitatively analyzed. Quantitative velocity data are closely related to the stress in concrete but different from the traditional stress-strain relationship. They can be used to evaluate the development of cracks in concrete.

2 THEORETICAL BASIS OF CWI

As mentioned above, CWI is used to detect subtle changes in a medium. After the direct wave passes through, the coda wave is formed by repeatedly scattering ultrasound waves. When the concrete specimen is slightly disturbed, the internal microstructure of the concrete changes and the propagation path of the ultrasonic wave in the concrete also changes. The theory of coda wave interferometry considers the scattering of waves in the medium, resulting in an infinite number of paths between the transmitting and receiving points. The coda wave is the superposition of waves propagating along these paths. The coda represents the sum of propagation paths and is invariant and repeatable when the receiving and transmitting positions are constant. According to Snider (2006), the field of ultrasonic waves scattered multiple times in the medium can be expressed as follows:

$$u(t) = \sum_{p} S_p(t) \tag{1}$$

where u is the ultrasonic wave field, S denotes the distance of the ultrasonic wave, and p ranges over all propagation paths of the ultrasonic wave in the medium. The propagation path p remains unchanged after perturbation of the medium caused by a change in stress, but the wave velocity and travel time vary. We suppose the change in wave propagation time is p after some disturbance in the medium and include it in Equation (1). Then, the wave field after perturbation, $u'(t)$, can be expressed as follows:

$$u'(t) = \sum_{p} S_p(t - \tau_p) \tag{2}$$

When the wave source remains unchanged before and after a small disturbance, the coda wave maintains its original form, but the travel time and amplitude of the wave will change. CWI senses small changes in the medium by analyzing the coda wave's time lag (or velocity). The time lag is calculated from the cross-correlation of the coda waves before and after the disturbance, defined as follows:

$$R^{t,t_w}(\tau_p) = \frac{\int_{t-t_w}^{t+t_w} u(t)u'(t+\tau_p)\mathrm{d}t}{\sqrt{\int_{t-t_w}^{t+t_w} u^2(t)\mathrm{d}t \int_{t-t_w}^{t+t_w} u'^2(t+\tau_p)\mathrm{d}t}} \tag{3}$$

where $2tw$ is the width of the time window and R is the cross-correlation function quantifying the similarity between the two waves before and after perturbation of the medium.

The time lag, τ_p, represents the time difference of the propagation of the coda wave in concrete before and after the disturbance. It corresponds to when the value of the cross-correlation function R reaches a maximum within a certain time window width.

During concrete stress testing, the time delay corresponding to each loading level through the passage of coda waves along a fixed propagation path can be evaluated. Then there would be a correspondence between the time lag and the stress level. If the relationship between time lag and stress level is readily available, it can be used to predict concrete stress levels quickly.

3 EXPERIMENTAL STUDY

This section explains the laboratory experiments to establish the relationship between coda waves' time lag (or velocity) and the stress level in a concrete specimen during mechanical loading.

3.1 *Specimen preparation*

Six concrete specimens (SP1-SP6) were prepared for testing. The concrete mix proportion adopted for the specimens is given in Table 1. Grade 42.5 ordinary Portland cement, mid-coarse sand, 5~25 mm gravel of good gradation and a high-range water-reducing admixture were used.

Table 1. Concrete mix proportion.

Component	Cement	Water	Sand	Aggregate	Water-reducing admixture
Weight ratio	1	0.5	1.6	2.9	0.01

The specimens were 150 mm × 150 mm × 300 mm prisms. They were compacted on a vibrating table for 90 seconds and removed from the formwork after 48 hours. The specimens were cured for three days in a standard curing room and dried indoors for 28 days.

3.2 *Test equipment*

The test rig comprises three main parts: the loading system, the ultrasonic transmitting, and receiving system, and the data acquisition and display system. As shown in Figure 2, a standard YAW-2000 compression testing machine of a capacity of 200 tons, RITEC RAM-5000 ultrasonic instrument, Tektronix MSO4104B-L oscilloscope, and two PAC R15a ultrasonic transducers were used. The main feature of RITEC RAM-5000 is that it can control the frequency, the number of cycles, and the amplitude of the input wave. The ultrasonic transducers had a resonant frequency of 150 kHz and an operating frequency range of 50~400 kHz. Tektronix MSO4104B-L was used for data acquisition and display. 3-m long double-shield cables were used for ultrasonic transducers to avoid signal attenuation. This ensured that the signal was strong enough without adding an amplifier. A high-power impedance matching of 50 Ω was used between the ultrasonic instrument and the transducers to condition the transmitted signal.

3.3 *Experimental method*

During the test, the ultrasonic transducers were fixed using cyanoacrylate adhesives on both sides of the concrete specimens perpendicular to the loading direction. The schematic diagram of specimen loading is shown in Figure 1.

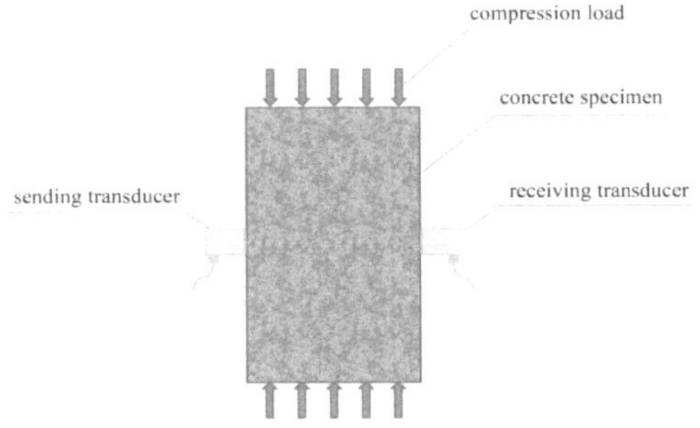

Figure 1. Schematic of specimen loading and ultrasound wave measurement.

Ultrasonic signals are affected by temperature. To reduce these influences, the testing rig was installed in a 140 cm × 140 cm × 210 cm insulated chamber with a door on one side for access, as shown in Figure 2. The temperature in the chamber can be controlled within ±1°C of the room temperature. After installing the specimen inside the chamber, the door was closed to maintain a constant temperature during testing. In addition, each test was performed in a relatively short period to avoid any further potential temperature variations.

Figure 2. Thermally insulated chamber and compression testing machine.

Loading occurs in several steps. Preloading was performed first, and the actual loading started after at least 3 minutes of preloading. At each step, the load was kept constant for at least 3 minutes before the next step. When the ultrasonic wave signals stabilized, the data were acquired, and the load was recorded simultaneously. This process was repeated for each concrete specimen until it failed. Figure 3 illustrates the failure of concrete specimens after reaching the ultimate load.

Figure 3. Failure of concrete specimen.

The frequency range of 150 KHz to 1 MHz in most practical cases defines multiple scattering mechanisms (Planes 2013). Therefore, the input signal was a sine wave with a frequency of 200 kHz, which corresponds to a wavelength similar to the aggregate size of 12 mm, which meets the requirements of multiple diffusion. The data acquisition frequency was 10 MHz, and the record length was 10,000 points, corresponding to 1 ms. To improve the signal-to-noise ratio, data acquisition was performed five times at each load step, with an interval of about 5-10 s between each acquisition. The coda wave signal for each load step was obtained by averaging over five measurements to remove noise. Before data processing, correlation analysis was conducted for the coda wave data for each load step. Data with a correlation of 0.90 were considered correct.

4 RESULTS AND DISCUSSION

In this experiment, the coda wave signals of six specimens under multiple loading steps and the ultimate compressive strength of each specimen at failure were obtained. The measured compressive strength of each specimen is listed in Table 2.

Table 2. Compressive strength test results.

Specimen index	Strength (MPa)
SP1	24.8
SP2	30.2
SP3	33.6
SP4	24.8
SP5	33.7
SP6	34.6

Figure 4 shows part of the ultrasonic input and received signals. Using Equation (3), 6,000 data points were analyzed for each load step, corresponding to a time length of 600 μs. For all concrete specimens, the propagation time of the direct wave was about 50 μs, but the time window of the coda wave was more than 10 times that of the direct wave. The precision of the coda waves was significantly higher.

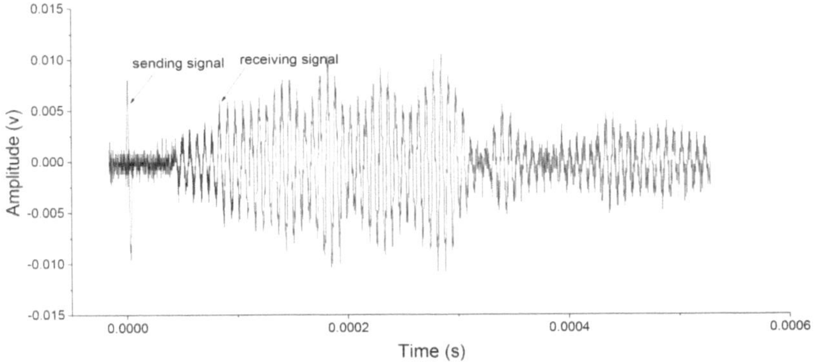

Figure 4. Transmitted and received ultrasonic signals.

Through the correlation analysis (Equation (3)) of the coda waves under different load levels, the relationship between the wave propagation time lag and the stress of each step of the concrete specimen under uniaxial compression load was obtained. The results are shown in Figure 5.

As shown in Figure 5, each stress-time lag curve is smooth. After using the load increment, the concrete stress increment corresponding to each load step was 0.88 MPa. Therefore, the ultrasonic coda wave resolution was also at least 0.88 MPa. Niederleithinger *et al.* (2010)

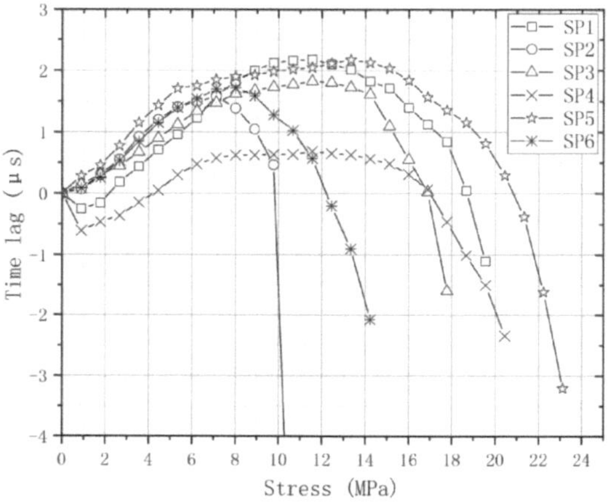

Figure 5. Measurement relationship between time lag and concrete stress for all specimens.

argued that the direct wave velocity is less sensitive to changes in concrete stress and that the wave velocity remains constant until cracks appear in the concrete. Therefore, compared with direct waves, the stress time lags generated by coda waves are much more accurate under the same experimental conditions.

The general trend of each curve in Figure 5 is similar and can be divided into four parts. Figure 6 illustrates the general relationship between load and time lag, indicating several characteristic points. The curve between points *a* and *b* is the initial stage, and as the load increases, the time lag gradually increases until it reaches the maximum at point *b*. Between points *b* and *c*, the time lag decreased slowly as the load increased, and after point *c*, the curve started to drop rapidly, reaching zero at point *d*. At this time, micro-cracks also appear in the concrete. After point *d*, the time lag changed from positive to negative and decreased sharply. At point *e*, the cracks widened significantly. As the load increased, the negative time lag became very large, and the cracks widened even more. Finally, the received ultrasound signal weakened until it disappeared completely, and the concrete sample collapsed.

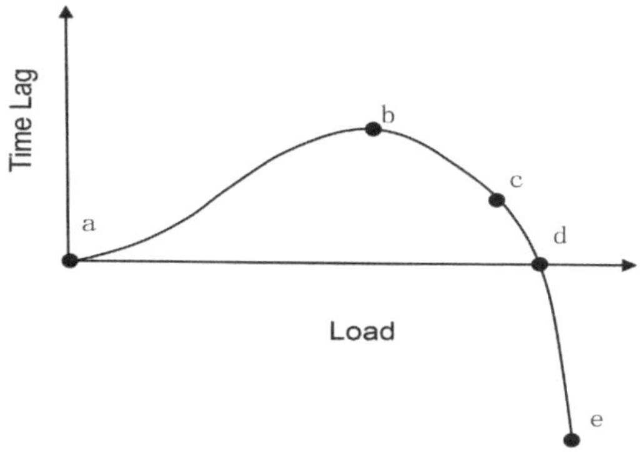

Figure 6. The general relationship between time lag and load with characteristic points is indicated.

As shown in Figure 5, the time lag increases with increasing stress, reaching a maximum value of about 2 μs. After reaching the maximum value and gradually reducing to 0, the time lag became negative when the concrete cracked and decreased rapidly, dropping to −2 μs.

The maximum stress for each specimen varied widely, ranging from 10 MPa to 22 MPa. Table 3 shows the stress values of each specimen at the four characteristic points *a-d* on the curve.

Table 3. Concrete stress at different stages (Unit: MPa).

Specimen number	Max. time lag (Point b)	0 μs (Point c)	< −2 μs (Point d)
SP1	11.5	18.7	19.5
SP2	8.0	12.2	14.2
SP3	7.1	9.5	9.7
SP4	11.5	17.0	17.7
SP5	11.5	17.0	20.4
SP6	13.3	20.8	23.1

Concrete cracking occurred when the time lag exceeded 2 μs for all specimens. Therefore, the relative stress was calculated concerning the stress corresponding to the time lag when concrete cracked. The results are shown in Figure 7.

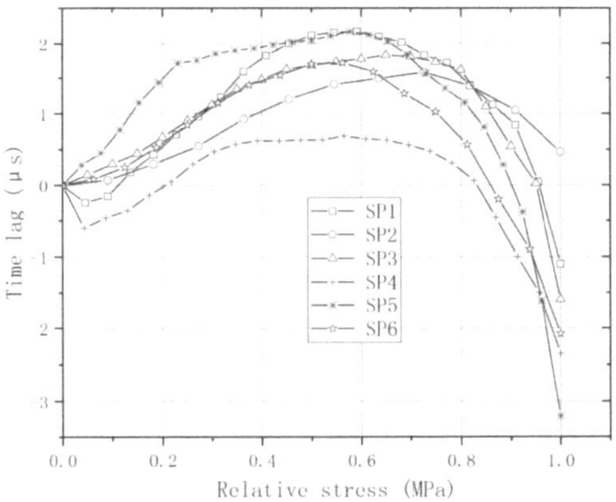

Figure 7. Measurement relationship between time lag and relative stress for all specimens.

It can be seen from Figure 7 that at the beginning of loading, the delay increases rapidly as the load increases. This is because the applied load compressed the concrete and increased its density. At the same time, as the load increased, cracks started to appear. These two factors simultaneously changed the time lag of ultrasonic waves, and the time delay-stress curves changed accordingly. Changes in the concrete density and internal structure are the main factors affecting the propagation of ultrasonic waves. As the load continued to increase, so did the concrete stress, but since micro-cracks already exist in the concrete, the delays caused by the micro-cracks and stress increases cancelled each other. Thus, the delay time–stress curves flattened out and gradually reached the maximum. When the load was further increased, micro-cracks in concrete grew, and damage continued to accumulate.

The damage gradually overcame the effect of the increased stress, and the time delay decreased. The time lag reached the maximum value when the relative stress was 0.5–0.7. This result is in good agreement with that reported by Niederleithinger *et al.* (2010). When the load was increased again, macroscopic cracks appeared in the concrete, the influence of cracks became dominant, and the time delay-relative stress curve dropped sharply. Based on the above results, the propagation of coda waves in concrete was divided into the following four stages corresponding to the internal state of concrete:

Stage 1: When the relative stress was $0 < \sigma < 0.7$, the concrete was compressed, and the density increased. At the same time, micro-cracks began to appear in the concrete, but their effect on the time-lag of the ultrasound waves was less than that of the increased concrete density, and the time-lag of the coda waves gradually increased.
Stage 2: When the relative stress was $0.7 < \sigma < 0.8$, as the stress increased, the density of concrete and the number of micro-cracks still increased, and the effect of crack propagation on the time lag of ultrasound waves was close to that of increasing concrete density. Therefore, the time lag-relative stress curve changed only slightly.
Stage 3: When the relative stress was $0.8 < \sigma < 1$, the micro-cracks gradually expanded and the time lag decreased. The effect of growing micro-cracks overcame the effect of increased concrete density.
Stage 4: When the relative stress was $\sigma > 1$ (not shown in Figure 7 due to scale limitations), the microcracks expanded rapidly and formed macro-cracks. On the other hand, the density of concrete barely increased. The time lag decreased rapidly. The amplitude of the ultrasonic waves also decreased rapidly, but the concrete stress did not reach the maximum value, although its failure was imminent. The time lags to failure varied significantly between different specimens, and the data obtained at this time were not of practical significance. It can be concluded that the proposed ultrasound stress measurement method failed at this stage.

5 CONCLUSIONS

The propagation of ultrasonic waves in concrete under uniaxial compressive stress was experimentally studied. CWI analyzed ultrasonic signals propagating in concrete. The main conclusions are as follows:

The coda waves are highly sensitive to stress changes in concrete. Using data interpolation, the resolution of stress changes was estimated to be at least 0.88 MPa. Since this accuracy is estimated from loading steps, the accuracy of the coda waves can be improved by using a more accurate compression testing machine. Since the propagation paths of coda waves are much longer than those of the direct waves, the accuracy of CWI is much higher than when using the direct waves. Therefore, CWI can be applied to stress testing and monitoring concrete structures under compression.

By analyzing the relationship between the ultrasonic wave time-lag and concrete stress, the entire process of concrete compression from the initial loading to failure can be divided into four stages, which can explain the changes in the ultrasonic coda waves in concrete as the stress increases. In the first stage, the time lag of the ultrasonic wave increased with the compressive load, and the density of concrete increased, leading to an increase in time lag. In the second stage, the time lag reached the maximum and then started decreasing rapidly. In the third stage, with the load increase, the influence of cracks became dominant, and the time lag decreased rapidly. In the fourth stage, the time lag decreased to zero and changed to negative. The cracks in the concrete widened, the negative time lag became very large, and the concrete specimen collapsed.

The coda waves were demonstrated to be very sensitive not only to concrete stress but also to the presence of cracks. Therefore, in this paper, the relative stress was defined concerning the stress at which macro-cracks appeared in concrete rather than the stress at which the concrete specimen failed. This is beneficial for discussing the transmission characteristics of ultrasonic waves in concrete. Compared with the traditional stress-strain relationship of concrete, the time lag of ultrasound waves can reflect not only the changes in stress but also the development of micro-cracks. This can monitor the formation and growth of cracks in concrete structures.

FUNDING

This study was funded by the Research Institute of Highways, Ministry of Transport China (grant numbers 2020-9031 and 2021-9082b).

REFERENCES

Aki, K., & Chouet, B.A. (1975). Origin of Coda Waves: Source, Attenuation, and Scattering Effects. *Journal of Geophysical Research*, 80, 3322–3342.

Bergman, R., & Shahbender, R. (1958). Effect of Statically Applied Stresses on the Velocity of Propagation of Ultrasonic Waves. *Journal of Applied Physics*, 29, 1736–1738.

Grêt, A., Snieder, R.K., & Scales, J.A. (2006). Time-lapse Monitoring of Rock Properties with Coda Wave Interferometry. *Journal of Geophysical Research*, 111.

Hafiz, A., & Schumacher, T. (2018). Monitoring of Stresses in Concrete Using Ultrasonic Coda Wave Comparison Technique. *Journal of Nondestructive Evaluation*, 37, 1–13.

Hu, H., Li, D., Wang, L., Chen, R., & Xu, X. (2021). An Improved Ultrasonic Coda Wave Method for Concrete Behavior Monitoring Under Various Loading Conditions. *Ultrasonics*, 116, 106498.

Hughes, D.S., & Kelly, J.L. (1953). Second-Order Elastic Deformation of Solids. *Physical Review*, 92, 1145–1149.

Jiang, H., Zhang, J., & Jiang, R. (2017). Stress Evaluation for Rocks and Structural Concrete Members through Ultrasonic Wave Analysis: Review. *Journal of Materials in Civil Engineering*, 29, 04017172.

Koch, C., Georgieva, K., Kasireddy, V., Akinci, B., & Fieguth, P.W. (2015). A Review of Computer Vision-based Defect Detection and Condition Assessment of Concrete and Asphalt Civil Infrastructure. *Adv. Eng. Informatics*, 29, 196–210.

Komloš, K., Popovics, S., Nürnbergerová, T., Babál, B., & Popovics, J.S. (1996). An Ultrasonic Pulse Velocity Test of Concrete Properties as Specified in Various Standards. *Cement & Concrete Composites*, 18, 357–364.

Korzeniowski, M., Piwowarczyk, T., & Maev, R.G. (2014). Application of the Ultrasonic Method for Quality Evaluation of Adhesive Layers. *Archives of Civil and Mechanical Engineering*, 14, 661–670.

Larose, É., & Hall, S.A. (2009). Monitoring Stress-related Velocity Variation in Concrete with a 2 x 10(-5) Relative Resolution Using Diffuse Ultrasound. *The Journal of the Acoustical Society of America*, 125 4, 1853–6.

Larose, É., de Rosny, J., Margerin, L., Anache, D., Gouédard, P., Campillo, M., & van Tiggelen, B.A. (2006). Observation of Multiple Scattering of kHz Vibrations in a Concrete Structure and Application to Monitoring Weak Changes. *Physical Review. E, Statistical, Nonlinear, and Soft Matter Physics*, 73 1 Pt 2, 016609.

Legland, J., Zhang, Y., Abraham, O., Durand, O., & Tournat, V. (2017). Evaluation of Crack Status in a Meter-size Concrete Structure Using the Ultrasonic Nonlinear Coda Wave Interferometry. *The Journal of the Acoustical Society of America*, 142 4, 2233.

Lin, J., Wu, D.L., Yang, H., & Zhao, M.J. (2011). Experimental Research on the Correlation of Acoustic Parameters and Stress of Concrete with Different Admixtures. *2011 International Conference on Remote Sensing*, Environment and Transportation Engineering, 1220–1223.

Maierhofer, C., Krause, M., Niederleithinger, E., & Wiggenhauser, H. (2003). *Non-destructive Testing Methods at BAM for Damage Assessment and Quality Assurance in Civil Engineering*.

Masera, D., Bocca, P.G., & Grazzini, A. (2011). Coda Wave Interferometry Method Applied in Structural Monitoring to Assess Damage Evolution in Masonry and Concrete Structures. *Journal of Physics: Conference Series*, 305, 012108.

Ndagi, A., Umar, A.A., Hejazi, F., & Jaafar, M.S. (2019). Non-destructive Assessment of Concrete Deterioration by Ultrasonic Pulse Velocity: A Review. *IOP Conference Series: Earth and Environmental Science*, 357.

Niederleithinger, E. (2017). *Detecting Subtle Changes in Concrete with Coda Wave Interferometry*.

Niederleithinger, E., Wang, X., Herbrand, M., & Müller, M. (2018). Processing Ultrasonic Data by Coda Wave Interferometry to Monitor Load Tests of Concrete Beams. *Sensors* (Basel, Switzerland), 18.

Nobile, L., & Nobile, S. (2015). Some Recent Advances in Ultrasonic Diagnostic Methods Applied to Materials and Structures (Including Biological Ones). *Physics Procedia*, 70, 681–685.

Nogueira, C.L. (2011). Wavelet-based Analysis of Ultrasonic Longitudinal and Transverse Pulses in Cement-based Materials. *Cement and Concrete Research*, 41, 1185–1195.

Pacheco, C., & Snieder, R.K. (2005). Time-lapse Travel Time Change of Multiple Scattered Acoustic Waves. *Journal of the Acoustical Society of America*, 118, 1300–1310.

Pao, Y.H., & Gamer, U. (1985). Acoustoelastic Waves in Orthotropic Media. *Journal of the Acoustical Society of America*, 77, 806–812.

Planès, T., & Larose, É. (2013). A Review of Ultrasonic Coda Wave Interferometry in concrete. *Cement and Concrete Research*, 53, 248–255.

Popovics, J.S. (2004). Non-Destructive Evaluation for Civil Engineering Structures and Materials. *Quantitative Nondestructive Evaluation*, 700, 32–42.

Rens, K. L., Wipf, T. J., Klaiber, F.W. (1999). Review of Nondestructive Evaluation Techniques of Civil Infrastructure. *Journal of Performance of Constructed Facilities*, 11(4), 152–160.

Schurr, D. (2010). *Monitoring Damage in Concrete Using Diffuse Ultrasonic Coda Wave Interferometry*.

Spalvier, A., Cetrangolo, G.P., Martinho, L.M., Kubrusly, A.C., Blasina, F., & Pérez, N. (2019). Monitoring of Compressive Stress Changes in Concrete Pillars Using Cross-correlation. *2019 IEEE International Ultrasonics Symposium (IUS)*, 2465–2468.

Stähler, S.C., Sens-Schönfelder, C., & Niederleithinger, E. (2011). Monitoring Stress Changes in a Concrete Bridge with Coda Wave Interferometry. *The Journal of the Acoustical Society of America*, 129 4,1945–52.

Zhang, Y., Abraham, O., Larose, É., Planes, T., Duff, A.L., Lascoup, B., Tournat, V., Guerjouma, R.E., Cottineau, L.M., & Durand, O. (2011). *Following Stress Level Modification of Real-size Concrete Structures with Coda wave Interferometry* (CWI).

Zhang, Y., Larose, É., Planès, T., Moreau, G., & Rospars, C. (2014). *Imaging of Early-Stage Cracking on Real-Size Concrete Structure from 4-Points Bending Test*.

Zou, Z., Wang, X., & Wang, Z. (2013). Application of Ultrasonic Testing in Concrete Filled Steel Tubular Arch Bridge. *Advanced Materials Research*, 639–640, 1025–1028.

Water Conservancy and Civil Construction – Oke & Ahmad (Eds)
© 2024 The Author(s), ISBN: 978-1-032-58614-4

Test method for bearing characteristics of deep rock mass based on force transfer pile foundation

ZhangTai Ke* & Zhu Peng*
Guangdong Highway Construction Co. Ltd, Guangzhou, China

Yang Ye*, ShiQi Long* & FuBai Yong*
CCCC Highway Bridge Construction National Engineering Research Center Co. Ltd, Beijing, China

ABSTRACT: Based on the project of the Guangdong Shiyang Bridge, a suitable basement elevation was selected for the anchoring foundation of the Shiyang suspension bridge. This project innovatively proposed and adopted the drilling and grouting pile technology of the double wall protection method to make the deep load plate test force transfer pile foundation, in which the inner steel cage was coated with a 0.8-meter-diameter steel sheath as the inner wall protection, so as to reduce the pile side friction resistance to the greatest extent. To increase the proportion of load transferred to the pile end, the outer layer uses a 1.2-meter-long steel casing as the borehole wall protection, and gravel is filled between the inner and outer walls to ensure the stability of the force transfer pile and the hole wall. In addition, by raising the mud weight control standard, the beginning of the first irrigation and the end of the second cleaning time must strictly ensure to be seamless in order to optimize and improve the process of pile tip sediment removal. The Q-S curve was obtained by using the method of monitoring pile tip stress with an embedded reinforcement meter and monitoring pile tip settlement with a preset reinforcement bar, and the characteristic value of the bearing capacity of the bearing layer was analyzed. The test results showed that 2461 kPa was the recommended bearing capacity characteristic value of the weathered bearing layer in the anchorage. It was effective and feasible to carry out the deep load plate test by using the double wall protection method of the bored pile, which can provide reference and experience for similar projects in the future.

1 INTRODUCTION

Since the 1950s, deep flat plate load tests have been carried out in China. It was done by using a reaction platform and a jack to put pressure on the force transfer column and push the force through the force transfer column into the deep load plate. The settlement between the jack load and the force transfer column was measured to obtain the load-settlement curve. As the technology was not mature in the last century and the application scenarios were few, it was no longer used in the 1980s. After the 1990s, with the increasing number of high-rise buildings in China, there were many application demands for the deep plate load test.

In recent years, new progress has been made in load tests. In 2001, Wu *et al.* (2001) from the Changchun Institute of Engineering developed the deep plate load test device

*Corresponding Authors: 3396411257@qq.com, 87359403@qq.com, 1769064254@qq.com, 775083789@qq.com and 42123043@qq.com

DOI: 10.1201/9781003450818-43

SP-1, which provided plate load test data for the deep foundation bearing layer over 20 m. Wu *et al.* (2015) improved the deep plate load test device by providing a reaction from the geotechnical layer above the working face, and the test depth was 30.9 m. Zheng *et al.* (Zheng 2014) studied the bearing capacity of fully weathered granite tested by a deep plate load test with an engineering example and introduced two methods for calculating the deformation modulus. Zhang *et al.* (Wang & Zhang 2018) designed an improved structure and test device for a deep loading plate test chamber, which has the advantages of high safety and can support a higher test reaction force. Combined with engineering examples, Zhang *et al.* (Ding *et al.* 2007) found that the bearing capacity of soft rock foundations tested in situ was greater than the conclusion obtained from the uniaxial compressive strength test. Lun *et al.* (Lun 2019) and Liu *et al.* (Liu 2022) found that PHC pipe pile bearing capacity and other aspects of performance could better meet the requirements of large tonnage deep plate load tests. Chen *et al.* (Chen & Xie 2021) introduced a new deep load plate test method. The test was carried out on the surface, and the double casing scheme was adopted. The inner casing pipe was filled with concrete as the experimental force transfer column, and the outer casing and inner casing were filled with sand.

For the bearing layer under the geological conditions of large burial depth and high-water level, the manual borehole test method has great safety risks and a high cost, and if the bearing layer is deep or hard rock, the device is easy to lose stability because the force column is too long. In addition, with the conventional deep load plate test, the settlement value of the bearing plate is not measured by the actual bearing plate but is indirectly measured by the surface of the column top. The column will be compressed under the load, and the friction resistance between the column and the surrounding soil in the settlement process will lead to a reduction of the load transmitted to the bearing plate, especially in the case of a deep bearing layer. The impact of measurement errors cannot be ignored.

To that end, the project proposed and used double-walled drilling and grouting pile technology to create the deep rock bearing capacity test force transfer component, as well as a pre-buried pile end settlement observation system and a pile axial force monitoring system to measure the end settlement and end pile end stress of the hole bottom force transfer pile foundation. Compared with the manual hole-digging method, the cost was greatly saved, and the data was safe, controllable, and reliable.

2 PROJECT OVERVIEW

The Shiyang Bridge is a double-layer steel truss-beam suspension bridge with a main span of 2180 m. The mooring foundation is planned to be a Figure-8 gravity mooring foundation with a length of 150 m along the bridge direction and a large scale. With a thickness of 1.5 m, the underground diaphragm wall is used as a water-stopping and supporting measure. It is recommended to use medium-weathered argillaceous sandstone as the holding layer. According to geological prospecting, the medium-weathered argillaceous sandstone is argillaceous cement, mainly composed of clay and sandy minerals, with sandy contents of about 50–60% and 45–65%. The core is columnar or massive (RQD > 65%), the hammer sound is clear, and the uniaxial compressive strength is 9.67 MPa. The recommended bearing capacity characteristic value is fa 0 = 1200 kPa.

It is critical to determine the plane scale of the anchoring base based on the correct value of the foundation bearing capacity, which can serve as a foundation and a reference for design. Based on the rationality of the test, the bearing capacity characteristic test of the bearing layer is carried out, and the bearing capacity characteristic value of the anchor basement can be determined based on the geological prospecting situation.

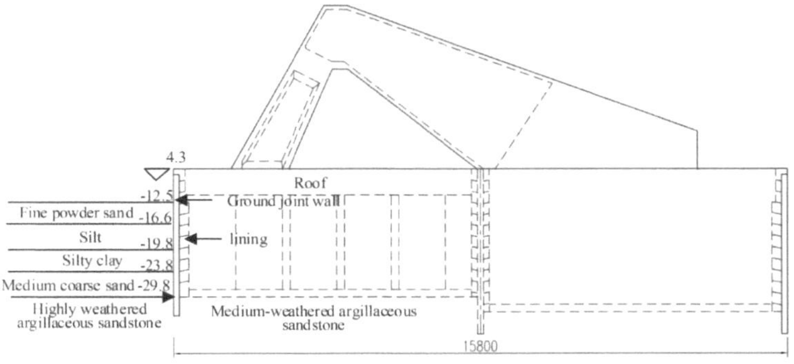

Figure 1. Schematic diagram of anchor foundation and bottom section.

3 TEST SCHEME

3.1 *Test principle*

This project proposed to test the bearing capacity characteristics of rock mass in the deep bearing layer based on the technology of bored piles and obtain the bearing capacity characteristic value of rock mass in the deep bearing layer by monitoring the load-displacement settlement curve of the pile end. The bored pile was constructed using the mud wall protection method. The diameter of the drilled hole was 1.2 m, and the steel guard tube of 0.8 m was set inside and poured together with the pile foundation to reduce the side friction resistance of the pile and increase the load sharing at the pile end. The gap between the drilled hole and the inner guard tube was filled with gravel to ensure the stability of the pile foundation under vertical load. The test principle for the bearing capacity of the bearing layer is shown in Figure 2.

In order to obtain settlement and deformation of the pile end, a settlement observation system was arranged. The pile end was pre-embedded with two casing sleeves; the casing end was sealed at the bottom; and the through-length steel bar was placed inside the pipe. The steel bar was completely separated from the casing, and the casing was sealed without water leakage, so as to avoid infiltration into the casing of settlement monitoring when concrete was poured.

In addition, in order to monitor the pile end stress, an axial force observation system was set up to measure the axial force at the corresponding section position of the pile through the reinforcement meter embedded at the top and the pile end. During the test, the pile load on the ground provided a reaction, and the pile top arranged jacks and reaction beams. As the pile top reaction loaded step by step, the load was transferred to the pile end and produced a settlement, thus indirectly loading the deep foundation rock mass. The values of each steel bar meter under different graded loads were obtained, and the corresponding "load-settlement curve" was drawn.

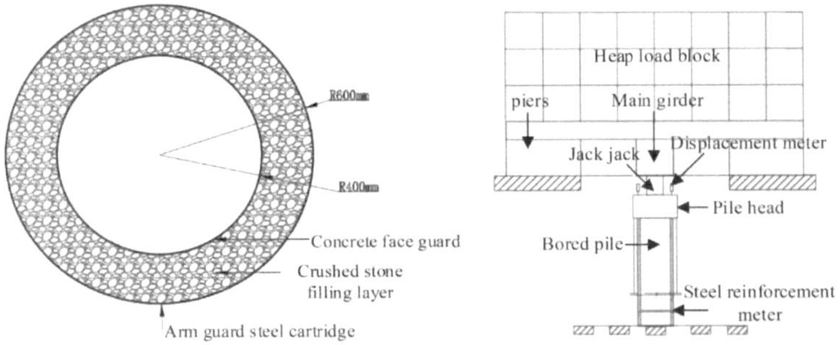

Figure 2. Deep plate load test system by the surcharge method.

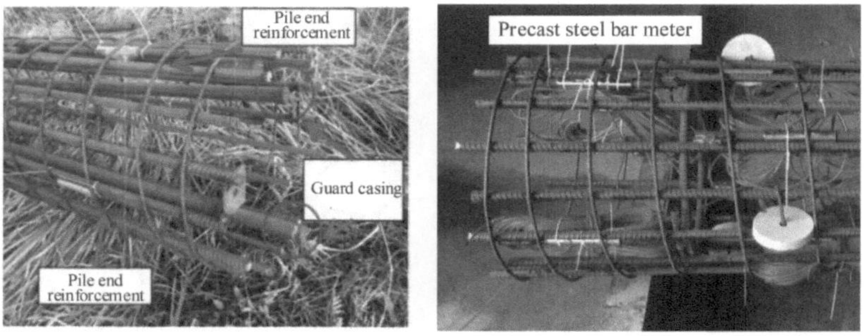

Figure 3. Stress monitoring system.

3.2 *Content detected*

According to the geological conditions of the anchorage, three groups of bearing capacity tests were designed, and the test points were 46#, 47#, and 49#, respectively. The bored pile was used as the force transfer pile foundation; the pile diameter was 0.8 m, the pile length was 34.1 m, 39 m, and 33.1 m, and the pile end was inserted into the medium-weathered mudstone. The test groups were shown in Table 1. Figure 4 shows the position of the mooring holes in Figure 8.

Table 1. Test groups of deep rock bearing capacity.

Group	The Pile Diameter /m	Basement Soil Layer	Reference Drill Hole	Elevation of Medium-Weathered Argillaceous Sandstone /m	Pile Length /m	Elevation /m	
						The Top	At the End of
46#	0.8 m	Medium-Weathered Argillaceous Sand-stone	cszk-46	-30.10	34.1	3.00	-31.10
47#	0.8 m	Medium-Weathered Argillaceous Sand-stone	cszk-47	-35.90	39	2.10	-36.90
49#	0.8 m	Medium-Weathered Argillaceous Sand-stone	cszk-49	-29.71	33.1	2.39	-30.71

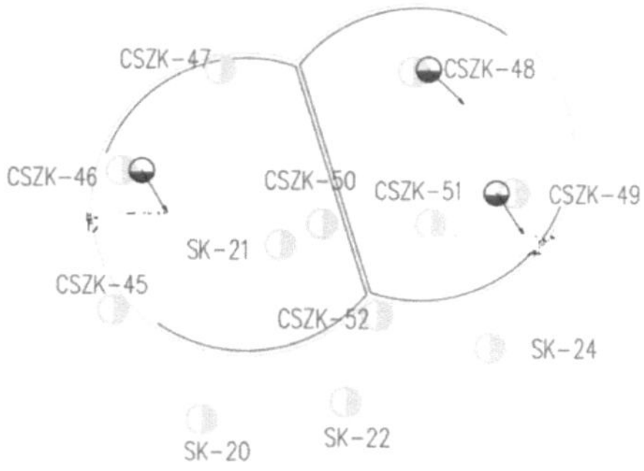

Figure 4. Figure-8 location of mooring holes.

3.3 *Loading scheme*

In this test, the test scheme was designed and operated in accordance with the relevant provisions of the test technology for the characteristic value of bearing capacity in the "Code for Design of Foundation and Foundation of Highway Bridges and Culverts" (JTG3363-2019) (JTG3363-2019, 2019) and the "Engineering Geology Manual" (fifth edition) (Chang & Zhang 2007).

Considering the design average stress of the base of 1200 kPa and the maximum loading stress of 3 times, the load stress of the pile end should reach 3600 kPa, the base area is 0.5 m^2, and the vertical pressure required by the test is 1800 kN. Furthermore, given the influence of pile friction resistance, the load sharing ratio of the pile end should be set at 30%. Therefore, the maximum load on the pile top should be 6000 kN, and the load reaction should be safe by 1.2 times. The load weight of the pile should be 7200 kN. The loading conditions can be terminated according to the relevant provisions of the deep load test in the Code for Design of Foundations and Foundations of Highway Bridges and Culverts (JTG3363-2019) or until the maximum design load of 6000 kN is reached.

The test was divided into 10 stages of slow and equal loading step by step, and the load increment of each stage was 600 kN. When the settlement was less than 0.25 mm for two consecutive hours, it was deemed to meet the standard of stability, and the next level of load could be applied. During unloading, the next level of load can be unloaded in an equal quantity step by step at five stages of slow speed.

Table 2. Load and load design scheme.

Estimated Characteristic Value of Bearing Capacity /kPa	Estimate the Maximum Loading Stress /kPa	Pile Tip Area /m^2	Pile End Load /kN	Pile End Load Sharing ratio /%	Expected Load /kN	Load Capacity /kN
1200	3600	0.5	1800	30	6000	7200

Figure 5. Site of load test.

3.4 *The characteristic value of bearing capacity*

For the selection of the characteristic value of carrying capacity, according to the relevant specifications, combined with the importance of the project, after comprehensive research and judgment with the design, owners, supervisors, and other units, the final determination of the value basis is as follows:

(1) For those with obvious scale limits, the load corresponding to the scale limits on the Q-S curve is taken as the characteristic value of the bearing capacity;
(2) When the termination loading condition is met, the first stage load at the time of failure is taken as the ultimate load. When the ultimate load is less than 2 times the corresponding proportional limit, the limit load is taken as half of the limit load;
(3) For the slow curve, the relative settlement method is used to determine the characteristic value of the bearing capacity. In this test, combined with the importance of the project, through communication with the test parties, the load corresponding to s/d = 0.006 d is actually selected as the characteristic value of the bearing capacity, and its value should not be greater than half of the maximum load;

Finally, a statistical method is adopted to determine the characteristic value of soil foundation-bearing capacity. The number of test points from the same soil layer participating in statistics should not be less than 3 points. If the range of the three groups of data does not exceed 30% of the average value, the average value is taken as the characteristic value of the bearing capacity; otherwise, the minimum value is taken as the characteristic value of the bearing capacity.

3.5 *Construction technology of pile foundations by force transfer*

The pile position shall be determined according to the coordinates, and the base and top shall be kept stable after the drill is in position. In the first alignment, the drill bit shall be aligned with the pile site, and after drilling to the depth of the outer steel guard cylinder, the crosshair shall be pulled for the second alignment, and the center of the guard cylinder shall be moved to make sure that the center of the guard cylinder coincides with the center of the drilling hole, and the outer steel guard cylinder shall be lowered. After drilling to the bottom elevation of the hole, a flat-bottom scraper bit was used to smooth the bottom of the hole and clean the residue, and then the lower pipe was cleaned by the gas lift reverse circulation method. After the inner steel sheath welded in advance is lowered to the design elevation at the bottom of the hole by the crane, gravel was filled between the inner and outer steel sheaths. After ensuring the stability of the inner steel sheath, the gas lift reverse circulation process was adopted for the second cleaning, so that the sediment was in suspension and ejected from the nozzle at the top of the pipe. The hole was cleaned continuously until the relative density of the ejected mud was less than 1.05 and the slagging standard was reached. Concrete pouring should be carried out immediately, and the pile foundation construction process with force transfer is shown in Figure 6 (a)–(f).

In order to ensure the construction quality of the force-transferring pile foundation, the pile integrity test was carried out to judge the quality of the pile concrete after the strength of the pile foundation met the requirements. In this test, the three pile foundations were all tested as Class I piles, then the pile heads were broken and pile caps were poured as bearing platforms.

4 TEST RESULTS

4.1 *Load sharing and pile end stress analysis*

The embedded steel bar meter was used to monitor the pile axial force and draw the comparison curve of the pile end and pile side load sharing ratio, as shown in Figure 7. The analysis shows that under different loads, the load sharing of the pile side and pile end was relatively stable, among which the load sharing of the pile side is about 70% and the load sharing of the pile end is about 30%, which can basically meet the loading requirements of the pile end. In addition, given the pile diameter of 0.8 m and the section area of the pile end of 0.5 m^2, the pile end stress under each level of load can be converted by combining the load

(a) Under the inner steel sheath　(b) Gravel filling　(c) Gas lift reverse circulation clearance

(d) Air lift reverse circulation hole clearing　(e) Reinforcement cage and reinforcement meter lowering　(f) Pile integrity detection

Figure 6. Construction technology of pile foundation by force transfer.

sharing value and section area of the pile end. The change of pile end stress under graded load is shown in Figure 8, which shows that the pile end stress increases linearly, step by step, with the increase in load. Among them, the pile tip stress of 46 and 47 pounds is basically similar under a graded load. Under the final load of 6000 kN, the pile tip stress is about 8000 kPa, while the extreme stress of 49# is 4392 kPa, which is significantly lower than the stress levels of 46# and 47#.

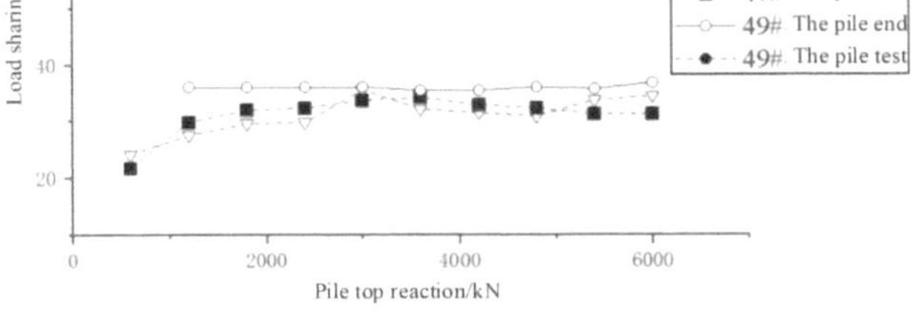

Figure 7. Pile end pile side load sharing curve.

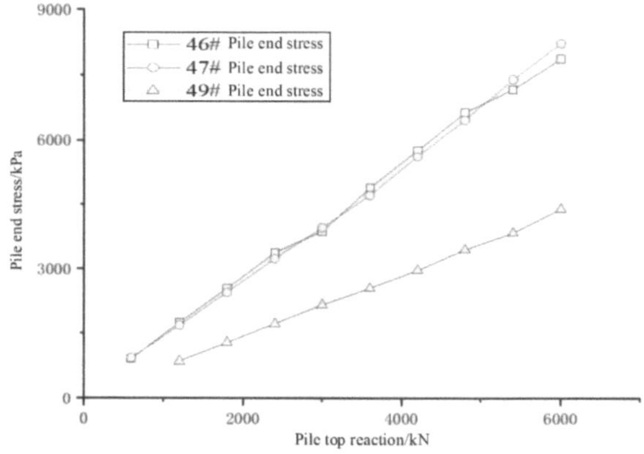

Figure 8. Pile tip stress curve.

4.2 *Analysis of foundation bearing capacity*

The load-settlement relationship curves of the three groups of tests are shown in Figure 9 (a)–(c). The pile tip loads of the three test groups were all applied to 6000 kN, and the load-settlement relationship curves of the pile tip showed slow deformation characteristics; there was no obvious proportion limit and no obvious abnormal phenomenon. Under the maximum load, there was no obvious failure phenomenon in the pile tip-bearing layer.

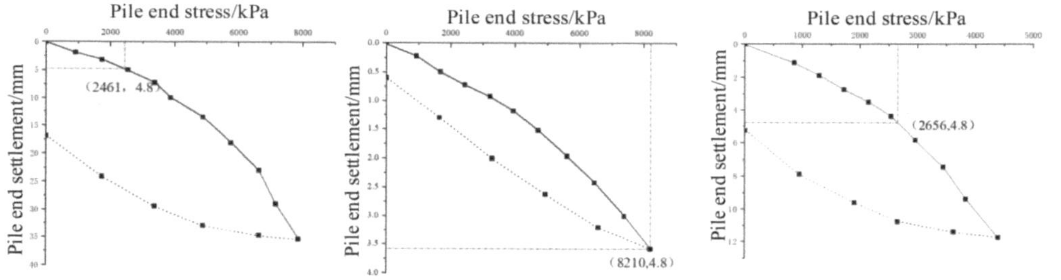

(a) 46# load-settlement curve (b) 47# load-settlement curve (c) 49# load-settlement curve

Figure 9. Load-settlement relation curve of the pile tip.

Figures 9(a) and (c) show that curves 46# and 49# show slow deformation characteristics, and the ultimate load method and proportional limit method cannot be used to determine the characteristic value of the bearing capacity. In this test, the relative settlement method of s = 0.006 d was used to evaluate the bearing capacity. When the maximum pile tip was loaded at 7864 kPa, the maximum cumulative settlement of the pile tip was 35.31 mm. The characteristic value of bearing capacity calculated according to the difference of the relative settlement method (s = 0.006 d = 4.8 mm) was 2461 kPa, which was no more than 3932 kPa, half of the maximum pile tip load. Therefore, the bearing capacity characteristic value of the 46# pile tip rock mass was 2461 kPa; for the 49# pile tip, when the maximum pile tip was loaded at 4392 kPa, the maximum cumulative settlement of the pile tip was 11.74 mm. The characteristic value of bearing capacity calculated according to the difference of the relative settlement method (s = 0.006 d = 4.8 mm) was 2656 kPa, which was more than 2466 kPa, half of the maximum pile tip load. As a result, the characteristic value of the bearing capacity of 49# pile tip rock mass was 2466 kPa.

Figure 9(b) shows the load settlement relation curve of the 47# test pile. It can be seen from Figure 9(b) that when the top load increases to 8210 kPa, the pile tip's cumulative settlement is only 3.58 mm, which is very small, but it has reached the maximum loading requirement of this test. The characteristic value of foundation bearing capacity was taken as half of the maximum loading amount, namely 4105 kPa. Compared with the test data of 46# and 49#, the characteristic value of the bearing capacity measured by 47# was significantly higher. The research group has analyzed the reasons for this phenomenon. According to the geological survey report, the stratum of the 47# test pile was interbedded with moderately weathered sandstone, and the actual rock stratum at the end of the pile might be moderately weathered sandstone. The saturated uniaxial compressive strength of the moderately weathered mudstone was 9.67 MPa, and the saturated uniaxial compressive strength of the moderately weathered sandstone was 27.2 MPa. Therefore, the characteristic value of the bearing capacity of pile 47# was larger than that of pile 46# and pile 49#. The bearing capacity characteristic values of the three groups of tests are shown in Table 3.

Combined with engineering practice and design conditions and other factors, the characteristic values of the three groups of anchorage tests were respectively 2461 kPa, 4105 kPa, and 2466 kPa. The range of the three groups of data was more than 30% of the average value, and the minimum value was taken as the recommended value of the characteristic value of the bearing capacity of the bearing layer, namely 2461 kPa.

Table 3. Characteristic value of bearing capacity.

Test Pile Number	Maximum Settlement /mm	Half of the Maximum Pile End Stress /kPa	S = 0.006 d Corresponding to the Load /kPa	Characteristic Value of Bearing Capacity /kPa
46#	35.31	3932	2461	2461
47#	3.58	4105	>8210	4105
49#	11.74	2466	2656	2466

5 CONCLUSIONS

Based on the technology of bored piles, this paper carried out three groups of bearing capacity tests of rock mass in the deep bearing layer. Through monitoring and analysis of pile tip settlement and pile tip stress, the bearing characteristics of the rock mass in the pile tip bearing layer were obtained, and the following conclusions were drawn:

(1) Three groups of test studies have been carried out to prove that the scheme of using a steel guard with a pass length set in the inner layer to reduce the side friction resistance, using the outer steel guard as the borehole wall protection, filling gravel between the inner and outer walls to ensure the stability of the pile foundation and the hole wall, and setting the axial force monitoring system and pile end settlement monitoring system were feasible and could be applied to the follow-up similar deep rock bearing capacity tests.
(2) Through the analysis of pile axial force, the pile tip reaction increased linearly with the pile tip reaction. As can be seen from the load sharing ratio of pile end to pile side, the ratio of pile side load of the three groups of test piles was 70%, 70%, and 65%, respectively, and the load transferred to the pile end met the requirements of pile end test loading and testing.
(3) There was no obvious scale limit in the pile curves of the three groups of tests, and no obvious failure occurred in the pile tip-bearing layer. Taking 0.006d as the standard

value of the bearing capacity characteristic value and not exceeding half of the ultimate loading value, the three groups of tests obtained the characteristic values of 2461 kPa, 4105 kPa, and 2466 kPa, respectively, and the data range of the three groups exceeded 30% of the average value. After discussion with the design team, owners, and supervisors, the minimum value was recommended as the characteristic value of the bearing capacity of the bearing layer, which was 2, 461 kPa.

REFERENCES

Chang Shipiao, Zhang Sumin. *Engineering Geology Manual* [M]. China Architecture and Building Press, 2007.

Chen Xinkui, Xie Lifei. *Test Design and Analysis of New Deep Load Plate* [A]. Digital Publishing Center of China Architecture and Building Press, 2021 202: 3. 203–205.

Code for Design of Foundation and Foundation of Highway Bridges and culverts: JTG3363-2019. [S]. Beijing: People's Communications Press, 2019.

Ding Youliang, Gao Wenhua, Zhang Zhimin. Study on the Bearing Capacity of Soft Rock End of Artificially Dug Pile by Deep Plate Load Test [J]. *Building Science*, 2007 (07): 75–77.

Lun Yuning. Improvement and Application of Force Transfer Device for Deep Plate Load Test [J]. *Construction Supervision, Inspection and Cost*, 2019, 12 (03): 29–32.

Liu Yasheng. The Shear Bearing Capacity of Large Diameter PHC Pipe Piles for Numerical Simulation Studies [J]. *Industrial construction*, 2022 (12): 1–9.

Wang Xiao, Zhang Li. Research on Structure and Device of Deep Load Plate Test [J]. *Building Structure*, 2018, 48 (S1): 830–832.

Wu Liping, Wu Yinzhu, Wang Wenchen, Yang Guochun. Research on Deep Plate Load Test Device [J]. *Engineering Survey*, 2001 (06):4–7.

Wu Dandan, Wang Jiangtao, Yang Chengbin. Application of Deep Plate Load Test in a Project [J]. *Building Structure*, 2015, 45 (01): 91–93.

Zheng Shengjie. Application of Deep Plate Loading Test to Fully Weathered Granite [J]. *Fujian Construction Science and Technology*, 2014 (04): 26–28.

Water Conservancy and Civil Construction – Oke & Ahmad (Eds)
© 2024 The Author(s), ISBN: 978-1-032-58614-4

Stress analysis and reinforcement design of bottom orifice of Baihetan arch dam

Jianrong Xu
Powerchina Huadong Engineering Corporation Limited, Hangzhou, China

Ruiqi Niu, Tongchun Li & Lanhao Zhao*
College of Water Conservancy and Hydropower Engineering, Hohai University, Nanjing, China

Yu Peng & Jianxin Wang
Powerchina Huadong Engineering Corporation Limited, Hangzhou, China

Jiayu Qian
College of Water Conservancy and Hydropower Engineering, Hohai University, Nanjing, China

ABSTRACT: The stress distribution around the orifice of a high arch dam is complex, and the proper reinforcement is necessary to improve the stress state of the orifice structure and ensure the safe operation of the arch dam. Taking the No. 3 bottom orifice of the Baihetan Arch Dam as an example, a 3-D finite element calculation was first made on the orifice structure under construction conditions for analyzing the stress distribution and mechanism around the orifice. Then, the stress graph method was used to obtain a reinforcement design for the orifice. Finally, the nonlinear calculation of reinforced concrete was carried out for the orifice with reinforcement. The results showed that the Poisson's ratio effect of the concrete was the main cause of the tensile stress along the river at the top and bottom of the orifice under the action of gravity on the dam body. The overall stress of the reinforcement was not high, and the reinforcement design scheme had a great safety margin, which meant the stress graph method was conservative. However, the stress graph method should still be adopted for the reinforcement design in terms of safety.

1 RESEARCH BACKGROUND

To meet the requirements of flood discharge, water delivery, and sediment ejection, various orifices should be set up in the arch dam body. Previous studies have shown that orifices have little influence on the overall stress of the dam, but the stress near the orifice changes significantly, resulting in the generation of cracks (Pan 1987). To improve the stress state of the orifice and limit the development of concrete cracks, it is necessary to properly configure reinforcements to ensure the safety of the orifice structure and the normal operation of the arch dam (Pan 2014).

The stress distribution around the orifice of the high arch dam is very complicated. It was traditionally believed (Tan 2015; Zhang 1999) that stress concentration was the main reason for the large tensile stress around the orifice, and the stress distribution around the orifice of the high arch dam had its own unique causes and characteristics. According to Guo (2014), the sub-model method was adopted to carry out the finite element analysis of the arch dam

*Corresponding Author: zhaolanhao@hhu.edu.cn

DOI: 10.1201/9781003450818-44

bottom outlet. The results showed that under the arch dam and the action of the arch thrust, more significant bending occurred on two sides, and the orifice was subjected to internal water pressure, resulting in a possible large vertical tensile stress in the side walls of the orifice. According to Li (2010), the stress distribution around the orifice was studied and analyzed by changing the structural form of the arch dam. It was found that when the cantilever structure was arranged at the orifice, its gravity would be transferred to the dam body through the concrete so that the orifice near the inlet and outlet positions would produce a large tensile stress along the river.

According to the stress distribution around the orifice, the stress graph method (Han 2011; Li 2016) is used to allocate reinforcement to the orifice in engineering. The amount of reinforcement required for the bearing capacity can be determined by the elastic stress graph area obtained from the elastic theoretical analysis method to obtain the corresponding reinforcement design scheme. For the high arch dam orifices, an important non-bar system structure, the stress states around the orifices before and after concrete cracking are quite different, so it is appropriate to use the nonlinear finite element method (Yin 2017; Zhao 2018) of reinforced concrete to analyze and adjust the reinforcement scheme.

In this paper, the No. 3 bottom outlet of the Baihetan arch dam was taken as an example. First, the stress distribution of the orifice under the working conditions during the construction period was obtained by using the sub-model method, and the generation mechanism of the tensile stress in the top and bottom of the orifice along the river was studied and analyzed. Then, according to the stress graph method to determine the amount of reinforcement in different parts of the orifice, the specific reinforcement design scheme was obtained. Finally, the reinforced concrete finite element nonlinear analysis was performed on the completed reinforced orifice, and the reinforcement design scheme was evaluated according to the stress level, which provided a basis for the orifice optimization design and reinforcement research.

2 CALCULATION MODEL AND SCHEME

2.1 *Engineering situation*

Baihetan Arch Dam has a crest elevation of 834.0 m and a maximum dam height of 289.0 m. To meet the requirements of flood discharge and water transfer, the arch dam adopted an arrangement of 6 surface outlets, 7 deep outlets, and 6 bottom outlets. Among the 6 bottom outlets, the size of No. 1–No. 5 bottom outlets was 6 m × 10 m, and the outlet elevation was 630 m, with a blocking gate on the upstream side and a free flow downstream. The No. 6 bottom outlet had an orifice size of 5 m × 7 m and an outlet elevation of 665 m, with a working radial gate downstream to stop the water.

2.2 *Grid model*

Based on the form of the arch dam and the corresponding detailed structure, such as various orifices, piers, and so on, the whole 3-D finite element model is established in Figure 1. In this paper, the focus was on the detailed structure of the dam body, so the mesh was appropriately encrypted in the orifice and the surrounding area, and the mesh profiles in other parts were relatively sparse. The paper applied the local incompatible mesh interpolation method to calculate (Li 2003), which was coordinated by the displacement interpolation method on the cut boundary. In this method, the node displacement on the contact surface was applied to the detached body as a constraint condition, and the stress of the sub-model was calculated together with other loads.

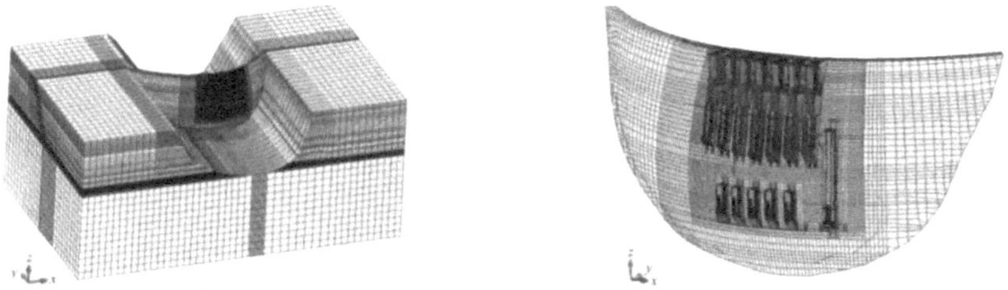

(a) Dam and foundation model (b) Dam body model

Figure 1. The whole 3-D finite element model of the dam.

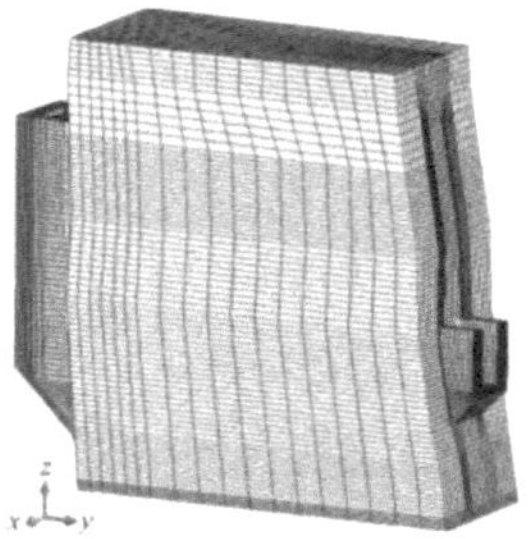

Figure 2. No. 3 bottom outlet sub-model.

A typical orifice was selected from the whole model as a sub-model for secondary encryption, at which time the grid size could guarantee the accuracy requirements of orifice stress and reinforcement. To facilitate the interpolation operation, the coordinate system and cell type of the sub-model should be the same as the whole model when modeling. According to the calculation results of the whole model, the nodal displacement values on the contact surface between the sub-model and the rest of the whole model were extracted and applied to the boundary of the sub-model as constraints, and then the load corresponding to the working condition was applied to the sub-model for the finite element calculation of the sub-model. The encrypted No. 3 bottom outlet sub-model is shown in Figure 2.

2.3 *Working conditions and material parameters*

The dam's working conditions were classified into two types. Condition 1 (anhydrous condition): during the construction period, the dam was poured to the top. The upstream water was not stored (cofferdams block water), and the calculated load was only self-weight. Condition 2: For the construction period's water storage conditions, the dam was filled to the top. The calculated load was mainly self-weight, water pressure, and gate thrust, where the upstream water level was 750.00 m and the corresponding downstream water level was 606.00 m. The No. 1–No. 5 bottom outlets were blocked by the upstream flat door, and the No. 6 bottom outlet was blocked by the downstream radial gate. The positive thrust of the radial gate was 64360 kN, and the side thrust was 3220 kN. The specific parameters of the No. 3 bottom outlet materials are shown in Table 1.

Table 1. Physical parameters of the bottom outlet material.

Material	Unit Weight/ $kN{\cdot}m^{-3}$	Elasticity Modulus /Gpa	Boisson's Ratio	Standard Compressive Strength/ $N{\cdot}mm^{-2}$	Standard Tensile Strength/ $N{\cdot}mm^{-2}$	Thermal Conductivity/ $10^{-3}(m^2{\cdot}h^{-1})$	Coefficient of Linear Expansion $/10^{-6}/°C$
C_{90} 40 Concrete	24	24	0.167	40	3.2	3.0	6.5
C_{10} 35 Concrete				35	2.8		
HRB400 Rebar	78	200	0.230		360		

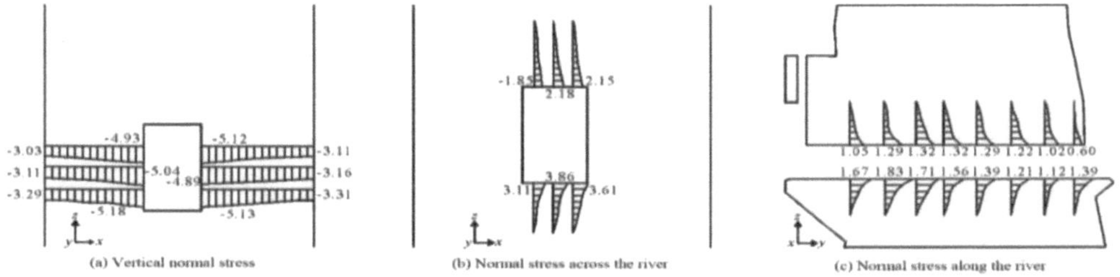

Figure 3. Normal stress distribution of the bottom outlet under working condition 1 (unit: MPa).

3 STRESS RESULT AND CAUSE ANALYSIS

3.1 *Calculation*

The sub-model was used to carry out the finite element calculation of the No. 3 bottom outlet under the working conditions during the construction period. The normal stress distribution of the bottom outlet under the anhydrous condition during the construction period is shown in Figure 3. According to the stress results, the lateral wall of the No. 3 bottom outlet was mainly subjected to compressive stress. In working condition 1 (an anhydrous condition during construction), there was a large tensile stress in the x and y directions at the top and bottom of the orifice. The maximum tensile stress in the x direction was 3.86 MPa, which occurred in the middle of the bottom plate of the orifice section. The maximum tensile stress in the y direction was 1.83 MPa, which occurred in the bottom plate of the upstream section of the orifice. When the water pressure upstream of the dam was included, the maximum tensile stress in the x-direction decreased to 1.46 MPa and the maximum in the y-direction decreased to 1.28 MPa.

The above calculation results showed that the gravity of the structure was the main load that caused the tensile stress along the river and across the river at the top and bottom plates of the orifice. However, when the reservoir was filled, the tensile stress of the dam body was partially offset by the action of water pressure, and the stress state was improved.

3.2 *Analysis of tensile stress along the river of the orifice*

The above stress calculation results show that the vertical and transversal stress distribution of the No. 3 bottom outlet was consistent with the general law. However, the top and bottom of the bottom outlet had large tensile stresses along the river, which contradicted the traditional view that the tensile stress along the river is small. Through the comparative analysis of multiple schemes, it can be seen that the dead weight applying method and the inlet cantilever structure had a certain influence on the range and value of the tensile stress along the river of the top and bottom of the orifice, but they were not the main causes of this tensile stress. From the perspective of elastic mechanics, the stress state of a point around the orifice under the action of gravity on the dam body was solved to explore its mechanical properties, to further analyze the causes of the tensile stress along the river, and to verify it with the finite element calculation results.

According to Hou (2011), the stress on the orifice perimeter of the infinite region with a square orifice can be written as follows:

$$\sigma_\theta = \frac{4(A \cdot C + B \cdot D)}{C^2 + D^2} P_x + \frac{4(A' \cdot C' + B' \cdot D')}{C'2 + D'2} P_z \tag{1}$$

$$\sigma_\rho = \tau_{\rho\theta} = 0 \tag{2}$$

where $A = 14 - 24\cos 2\theta - 7\cos 4\theta$, $A' = 14 + 24\cos 2\theta - 7\cos 4\theta$, $B = -24\cos 2\theta - 7\sin 4\theta$, $B' = -24\cos 2\theta - 7\sin 4\theta$, $C = C' = 56 + 28\cos 4\theta$, $D = D' = 28\sin 4\theta$, $P_x = \sigma'_x = \mu\sigma'_z = \mu\sigma'_{z0}/(1-\mu^2)$, $P_z = \sigma'_z = \sigma'_{z0} = \mu\sigma'_{z0}/(1-\mu^2)$, P_x and P_z are x and z-direction uniform distribution load; σ'_{z0} is the vertical stress at infinity when Poisson's ratio is equal to 0.

The center point of the orifice roof ($\theta = \pi/2$) was taken as an example, the transverse stress $\sigma_x = \sigma_\theta$, the vertical stress $\sigma_z = 0$. For the plane problem, we can know from the physical equation as follows:

$$\varepsilon_y = \frac{1}{\mathrm{E}}\left[\sigma_y - \mu(\sigma_x + \sigma_z)\right] \tag{3}$$

Assuming that the strain in the y-direction of the orifice boundary on the section was the same as that at an infinite distance, it can be obtained by Formula (3):

$$\varepsilon_y = -\frac{\mu}{\mathrm{E}(1-\mu)}\sigma'_{z0} \tag{4}$$

Finally, Formulas (1), (2), and (4) were substituted into Formula (3) to obtain the stress along the river at the center point of the orifice roof:

$$\sigma_y = \sigma'_{z0}\left(\frac{1}{21}\frac{17\mu - 31\mu^2}{1-\mu^2} + \frac{\mu}{1-\mu}\right) \tag{5}$$

where E is the elastic modulus (Gpa), and μ is the Poisson's ratio.

When the orifice structure form is certain and the load is constant, σ'_{z0} is constant. Therefore, the stress along the river is only related to Poisson's ratio, and the top and bottom of the orifice are always strained along the river under Poisson's ratio effect. When $\mu = 0$, $\sigma y = 0$. Therefore, Poisson's ratio effect is the main reason for the tensile stress along the river.

An infinite thick plate model with a square orifice was established to test the correctness of the tensile stress formula along the river of the top and bottom of the orifice. It was assumed that the thick plate was a common material, and the Poisson's ratios were 0.10, 0.15, 0.18, and 0.20. The y-direction tensile stress at the top of the orifice was calculated by the finite element method and compared with the analytical solution of elasticity. The results are shown in Table 2, which shows that the finite element calculation results are consistent with the analytical solution of elasticity. When the orifice's thick plate material was concrete, its Poisson's ratio must be greater than 0, resulting in a larger value of tensile stress along the river, which further showed that the concrete Poisson's ratio effect was the main reason for the tensile stress along the river of the orifice.

Table 2. The tensile stress values along the river under different poisson's ratios (MPa).

Poisson's Ratio	The Finite Element Solution	Analytical Solution of the Elastic Mechanics
0.10	1.72	1.71
0.15	2.57	2.56
0.167	2.86	2.85
0.18	3.08	3.08
0.20	3.44	3.43

4 REINFORCEMENT DESIGN AND NONLINEAR ANALYSIS

4.1 *Orifice reinforcement design*

The reinforcement layout at the orifice of the high arch dam is very complicated. Although the forms of reinforcement are different, they are mainly composed of reinforcement across the river, reinforcement along the river, and vertical reinforcement. In this paper, an appropriate number of sections and lines were intercepted based on the stress distribution of the bottom outlet, the principal tensile stress patterns of different parts of the orifice were obtained, and the amount of reinforcement in each part was determined by using the stress graph method (Zhu 2016). The design scheme of the reinforcement for the bottom outlet is shown in Table 3.

Table 3. The design scheme of reinforcement for the bottom outlet.

Position		Direction	Control Tensile Stress (MPa)/ condition	Calculate the Among of Reinforcement/mm^2	Reinforcement Scheme	Design the Among of Reinforcement/ mm^2
Entrance Section of Gate Piers		Along the River	0.621 / Condition1	18914.67	4-layer C36@200	20358.0
		Vertical	Compressive Stress	Constructional Reinforcement	1-layer C36@200	5089.5
Entrance Section of Brackets		Across the River	2.882 / Condition1	11548.33	3-layer C36@200	15268.5
		Along the River	1.124 / Condition1	10598.54	3-layer C36@200	15268.5
Orifice Body	The Topand Bottom	Across the River	3.861 / Condition1	7822.83	2-layer C36@200	10179.0
		Along the River	1.828 / Condition1	10026.78	2-layer C36@200	10179.0
	Side Wall	Along the River	0.246 / Condition1	4808.83	1-layer C36@200	5089.5
		Vertical	Compressive Stress	Constructional Reinforcement	1-layer C36@200	5089.5
Outlet Section of		Vertical	0.187 / Con-	592.17	1-layer	5089.5

(continued)

Table 3. Continued

Position	Direction	Control Tensile Stress (MPa)/ condition	Calculate the Among of Reinforcement/mm^2	Reinforcement Scheme	Design the Among of Reinforcement/ mm^2
Gate Piers		dition1		C36@200	
	Along the River	0.377 / Condition1	4853.52	1-layer C36@200	5089.5
Outlet Section of Brackets	Across the River	1.326 / Condition1	1809.40	1-layer C36@200	5089.5
	Along the River	1.341 / Condition1	4736.89	1-layer C36@200	5089.5

Through the calculation and analysis of the reinforcement of the orifice, it was known that the control tensile stress of the entrance section bracket and the top and bottom of the orifice was larger, and the calculated amount of reinforcement was also larger, while the tensile stress along the river at the entrance pier was small, but the calculated amount of reinforcement was very large. It was mainly due to the deep extension of tensile stress in this part, resulting in a large stress graph area (Figure 4).

According to the needs of the actual project, the concrete cover thickness was 200 mm, the steel bar was C36, and the spacing was 200 mm, determining the specific reinforcement scheme of each part of the bottom outlet (Table 3). According to the above reinforcement layout scheme, the 3-D finite element model of the bottom outlet reinforcement was established, as shown in Figure 5.

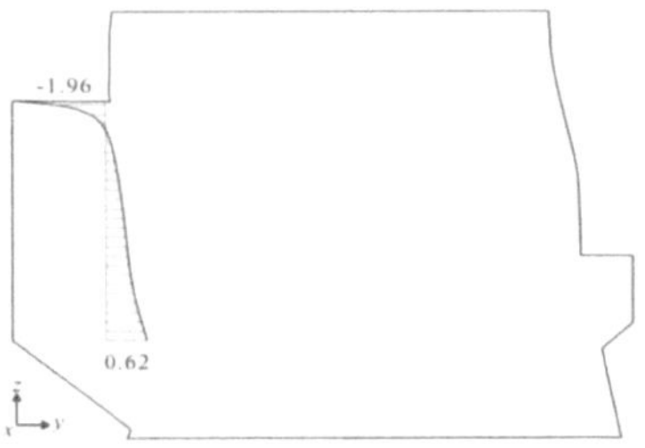

Figure 4. Stress of entrance section of piers.

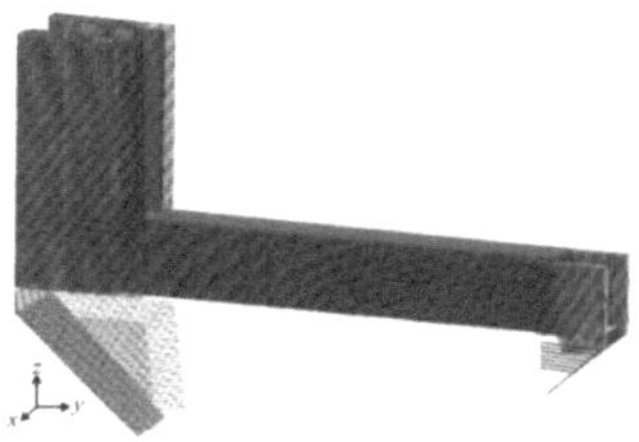
Figure 5. Model of bottom outlet reinforcement.

4.2 *Nonlinear analysis of reinforced concrete*

From the above reinforcement results, it can be seen that the reinforcement amount for each part of the orifice was the largest under condition 1. Therefore, based on the existing 4-parameter damage model (Wei 2004), the single spring coupling element method (Yan 2011; Zhao 2015) was used to simulate the interaction between reinforcement and concrete. The reinforced concrete nonlinear calculation of the No. 3 bottom outlet under this condition was carried out, and according to the results of reinforcement stress, it provided a reference basis for perfecting and optimizing the design scheme of orifice reinforcement.

Table 4 shows the maximum reinforcement stress in various parts of the No. 3 bottom outlet under condition 1. It can be seen from Table 4 that the tensile stress of the reinforcement mainly appears at the top and bottom of the channel and the entrance section and the maximum value is 51.581 MPa. On the whole, the stress level of reinforcement is not high, which is much less than the tensile strength of reinforcement. It was mainly due to the coordinated deformation of reinforcement and concrete. When there was no damage or cracking of the concrete, the strain on the reinforcement was the same, and the stress on the reinforcement was only about 10 times that of the concrete.

Table 4. Maximum stress of reinforcement in different parts of bottom outlet.

Position		Direction	Maximum Stress
Entrance Section of Gate Piers		Along the River	1.623
		Vertical	-5.421
Entrance Section of Brackets		Across the River	51.581
		Along the River	31.547
Orifice Body	The Top and Bottom	Across the River	47.942
		Along the River	28.426
	Side Wall	Along the River	7.817
		Vertical	-3.549
Outlet Sectionof Gate Piers		Vertical	1.211
		Along the River	0.871
Outlet Section of Brackets		Across the River	10.472
		Along the River	11.783

The reinforcement bore less tensile stress, and its performance was not fully played, which indicated that the reinforcement design scheme adopted in this paper had a large safety margin and the stress graph method was conservative. But in the actual project, due to the influence of construction, temperature, and other external factors, concrete would often crack. Once the crack appears, most of the concrete in the tensile area of the crack section exited the work, and the tensile stress was almost all borne by the reinforcement, and the stress of the reinforcement would suddenly increase. Therefore, the stress graph method should still be adopted for the reinforcement design in terms of safety.

5 CONCLUSION

(1) The stress state around the orifice of a high arch dam was relatively complicated. The traditional thought was that the stress concentration led to the large tensile stress around the orifice, but the arch thrust and cantilever structure of the arch dam were the main reasons for the large tensile stress around the orifice.
(2) Only under the action of dead weight were there larger cross-river and downstream tensile stresses at the top and bottom of the No. 3 bottom outlet. When the upstream portion of the dam was impounded, the tensile stress would be reduced, and the stress state would be improved with the existence of internal water pressure.
(3) From the point of view of elasticity, the analytical solution of the tensile stress along the river of the orifice top and bottom was derived and verified by the finite element method, which showed that the concrete Poisson's ratio effect was the main reason for the tensile stress along the river.
(4) The overall stress of the reinforcement was not high, and the reinforcement design scheme had a great safety margin, which meant the stress graph method was conservative. However, the stress graph method should still be adopted for the reinforcement design in terms of safety in practical engineering.

REFERENCES

Guo Xiaojing and Xu Jinjin. Application of Submodel Method in Stress Analysis of Arch Dam Discharge Bottom Outlet [J]. *Journal of Water Resources and Architectural Engineering*, 2014, 12 (1): 205–208+212.

Han Yulian, Zhao Jixun, Gao Yang, Wang Lei and Zhang Xiaolei. Based on the Study on the Method of Finite Unit Method of Reinforcement Concrete Structure [J]. *Water Science and Engineering*, 2011 (4): 87–89. The DOI: 10.19733/j.carolcarrollnki.1672-9900.2011.04.034.

Hou Huaqiang, Wang Lianguo, Lu Yinlong and Zhang Bei. Study on Surrounding Rock Stress Distribution and Failure Mechanism of Rectangular Roadway [J]. *Chinese Journal of Underground Space and Engineering*, 2011, 7 (S2): 1625–1629.

Li Pengchong. Design of Sand-flushing Meso-hole Reinforcement for Wanjiakouzi Arch Dam [J]. *Hongshuihe River*, 2016, 35 (5): 30–33.

Li Shouyi, Yang Sheng and Gao Jumei. Analysis on Influencing Factors of Stress in Arch Dam Discharge Hole [J]. *Chinese Journal of Basic Science and Engineering*,2010,18(1):20–27.

Li Tongchun, Li Miao, Wen Zhao-Wang and Shen Hong-jun. Application of Local Uncoordinated Mesh in Stress Analysis of High Arch Dam [J]. *Journal of Hohai University (Natural Science Edition)*, 2003, 31 (1): 42–45.

Pan Yanfang and Li Manlin. DaGang Mountain Arch Dam Flood Discharge Deep Orifice Reinforcement Design Study [J]. *The People of the Yangtze River*, 2014, 45 (22): 65–68. The DOI: 10.16232/j.carolcarrollnki.1001-4179.2014.22.020.

Pan Yongbao. Views on Arch Dam Opening Some [J]. *Journal of Northeast Water Conservancy and Hydropower*, 1987 (3): 28–32. DOI: 10.14124/j.carolcarrollnkidbslsd22-1097.1987.03.005.

Tan Lin and Guo Yuan. Finite Element Analysis of Stress Concentration in a Finite Plate with Open Orifices [J]. *Journal of Chongqing University of Technology*, 2015, 29 (7): 35–39.

Wei Wei and Li Tongchun. Isotropic Damage Model Based on Four-parameter Equivalent Strain [J]. *Journal of Hohai University (Natural Science Edition)*, 2004, 32 (4): 425–429.

Yan Tianyou, Li Tongchun and Zhao Lanhao. Improved Algorithm of Single Spring Coupling Element Method [J]. *Journal of China Three Gorges University (Natural Science Edition)*, 2011, 33 (06): 23–26+32.

Yin Ming, Li Tongchun, Zhao Lanhao and Zhang Wei. Non-linear Finite Element Analysis of Pore Reinforcement of Baihetan Arch Dam Based on Single spring Coupling Element Method [J]. *Water conservancy and hydropower technology*, 2017 (01): 1352–58. DOI: 10.13928/j.carolcarrollnkiwrahe.2017.01.010..

Zhang Jianhai, He Jiangda, Fan Jingwei, Xiao Baiyun, Ding Yutong and Zhao Wenguang. Substructural Element Method for Stress Analysis of Pore groups in Dam Body [J]. *Design of Hydropower Station*, 1999, 15 (2): 29–35.

Zhao Lanhao, Zhang Mengdi and Zhang Yunfeng. Non-linear Finite Element Analysis of Reinforced Concrete on the Neck of Peokong Sluice Pier [J]. *Journal of Hydropower Energy Science*, 2018, 36 (1): 120–123+115.

Zhao Lanhao, Zhang Wei, Bai Xin, et al. Single Spring Joint Element based on the Mixed Coordinate System [J]. *Mathematical Problems in Engineering*, 2015 (18): 1–16.

Zhu Fenglin and Li Tongchun. Reinforcement Design of Surface Orifice of a Concrete Arch Dam under Earthquake [J]. *Hydropower Energy Science*, 2016, 34 (6): 75–79.

Water Conservancy and Civil Construction – Oke & Ahmad (Eds)
© 2024 The Author(s), ISBN: 978-1-032-58614-4

Numerical simulation research of compressive and flexural mechanical properties of CFRP-strengthened reinforced concrete

Junyu Chen, Yichen Wang, Zhenyu Feng* & Jiaqi He
Chang'An Dublin International College of Transportation, Chang'An University, Xi'an, China

ABSTRACT: Carbon fiber-reinforced polymer (CFRP) has the advantages of being lightweight and having high strength, good corrosion resistance, strong adaptability, and no additional defects (Ali *et al.* 2012), and can be used to reinforce various materials. As a new reinforcement method, CFRP reinforcement technology has been widely studied and applied in the field of civil engineering in recent years. In order to study the axial compression and flexural mechanical properties of reinforced concrete reinforced with CFRP, this paper used the finite element software ABAQUS to carry out the axial compression test and the numerical simulation and mechanical properties analysis of the three-point bending test of reinforced concrete beams and CFRP-strengthened reinforced concrete beams. The analysis of the displacement-load curves of two kinds of reinforced concrete beams that came from the test showed that CFRP reinforcement could effectively inhibit the crack propagation of beams and improve their flexural performance.

1 INTRODUCTION

Reinforced concrete has become a common building component in building structures due to its good economic effect, excellent compressive performance, and simple construction. It has a wide range of applications in tunnels and housing construction projects (Zhao *et al.* 2021). At present, most of China's building structure systems are still reinforced concrete structures, but the concrete itself has certain defects, such as low tensile strength and poor crack resistance. At the same time, unreasonable design, improper construction, and other natural or human factors may result in the insufficient bearing capacity of reinforced concrete columns and prevent them from meeting the requirements of normal service (Yang *et al.* 2021). Civil engineering design standards have been constantly updated during the National 14th Five-Year Plan, and the requirements for environmental protection and building safety are constantly increasing. It has become the focus of scholars' work to figure out how to repair and strengthen the buildings that have been designed (Liu *et al.* 2006). For reinforced concrete columns, although increasing section, outsourcing angle steel, cement grouting reinforcement, replacement reinforcement, and other traditional reinforcement methods can improve the stiffness and strength of reinforced concrete to a certain extent, these methods have the problems of long construction times and poor chemical corrosion resistance (Zhang & Liu 2003). Therefore, novel materials and reinforcement methods are needed in the field of reinforced concrete to improve the current issues. Carbon fiber-reinforced polymer (CFRP) has gradually become a hot spot in the field of material reinforcement in recent years. Compared with the traditional reinforcement methods, CFRP reinforcement has the advantages of being lightweight and having high strength, a wide

*Corresponding Author: 2019903302@chd.edu.cn

DOI: 10.1201/9781003450818-45

application, and low environmental contamination. At the same time, CFRP has an extremely high tensile strength, and almost no plastic deformation occurs before reaching its ultimate tensile strength. Strengthening reinforced concrete with CFRP can fully combine the merits of both and effectively improve the mechanical properties of reinforced concrete (Sha *et al.* 2021). In order to analyze the strengthening effect of CFRP on reinforced concrete beams, this paper used ABAQUS to simulate the axial compression and three-point bending tests of reinforced concrete beams, expecting to get economical and effective CFRP reinforcement methods to improve the fatigue performance of reinforced concrete.

2 METHODOLOGY

2.1 *Progress and application of CFRP-strengthened reinforced concrete beams*

At present, world researchers have achieved abundant achievements in the study of CFRP-strengthened reinforced concrete beams. Zhao *et al.* (Zhao *et al.* 2000) used non-prestressed CFRP reinforcement technology to carry out an axial compression test on reinforced concrete columns. In the experiment, it was observed that some zigzag cracks appeared on the carbon fiber cloth. The outer part of the reinforced concrete column was partially desquamated, while the inner core was slightly cracked but still integrated. By comparing the formulas of other FRP-strengthened reinforced concrete members, the equation of the bearing capacity of CFRP-strengthened reinforced concrete columns was obtained, and the research results were successfully applied to the reinforcement of reinforced concrete columns in a station.

Maaddawy (El Maaddawy 2008) studied the eccentric compression test of non-prestressed reinforced concrete columns reinforced with CFRP in a highly corrosive environment and the anti-corrosion protection effect of carbon fiber cloth on reinforced concrete columns under different wrapping methods. In this test, reinforced concrete columns were eroded for 30 days under corrosion conditions and then divided into two groups of tests with carbon fiber cloth in full package and half package, followed by corrosion for 60 days. The research results have manifested that the full package form of reinforced concrete columns can efficaciously prevent corrosion and maintain the reinforcement effect. In contrast, although the half-packaged method can maintain the reinforcement effect of reinforced concrete columns, it cannot sufficiently prevent the erosion of concrete, so the outcome of improving the bearing capacity of concrete was weakened.

Ghanim *et al.* (Ghanim & Al-Abbas 2018) used the method of installing CFRP sheets near the surface to study the performance of strengthened reinforced concrete beams. The test results showed that the stiffness of reinforced concrete increases significantly after the installation of CFRP on the surface, and its ultimate load can reach twice that of ordinarily reinforced concrete beams without reinforcement.

Mortazavi (Mortazavi *et al.* 2003) studied the improvement in compressive strength of reinforced concrete columns after strengthening by using carbon fiber cloth tubes to inject expanding materials. The research showed that the compressive strength of the strengthened reinforced concrete column was 35% higher than that of the reinforced concrete column without reinforcement, and the compressive strength was more than 4 times higher than that of the generally reinforced concrete column.

2.2 *ABAQUS plastic damage model (concrete damaged plasticity, CDP)*

In this paper, the concrete damage plasticity model, which was widely used in ABAQUS, was used as the constitutive model (Zhou *et al.* 2022). The proposed model was developed based on the elastoplastic theoretical framework and introduced damage mechanics parameters. The determination method of constitutive relation referred to GB 50010-2010,

"Code for Design of Concrete Structures." The stress-strain relationship of concrete under uniaxial loading is given as follows:

$$\sigma = (1 - d_k)E_C\epsilon \quad (k = t, c)$$

where d_k is the damage evolution parameter of concrete under uniaxial load; E_C is the elastic modulus of concrete.

The yield surface function in the CDP structures model mainly includes hardening variables ε^{pl} and ε^{el} which represent compressive and tensile equivalent plastic strain, respectively. As shown in Figure 1, the CDP model uses the damage factor to describe the stress-strain curve of the concrete material.

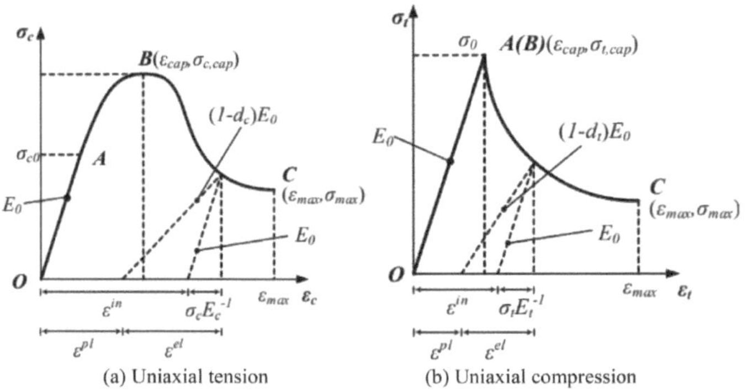

Figure 1. Uniaxial stress-strain curve of CDP.

Under uniaxial compression, the concrete material first hardened and then softened after reaching the initial yield stress σ_{c0}, until the crushing failure. Under uniaxial tension, the concrete material exhibited softening after tensile yield until cracking. The strain input by the user in the CDP model was the compressive inelastic strain ε_c^{pl} and the tensile cracking strain ε_t^{pl}. The ABAQUS program was automatically converted into plastic strain. The relationship between compressive plastic strain ε_c^{pl} and compressive inelastic strain, tensile plastic strain ε_t^{pl}, and tensile cracking strain can be described by the following formulas (Zhang *et al.* 2022).

$$\begin{cases} \varepsilon_{0c}^{el} = \dfrac{\sigma_c}{E_0} \\ \varepsilon_c^{in} = \varepsilon_c - \varepsilon_{0c}^{el} \\ \varepsilon_c^{pl} = \varepsilon_c^{in} - \dfrac{d_c}{(1-d_c)}\dfrac{\sigma_c}{E_0} \end{cases}$$

where σ_t is the compressive stress; ε_t is the compressive strain; E_0 is the initial elastic modulus; d_c is the compressive damage factor; ε_{0c}^{pl} is the elastic compressive strain.

$$\begin{cases} \varepsilon_{0t}^{el} = \dfrac{\sigma_t}{E_0} \\ \varepsilon_t^{ck} = \varepsilon_t - \varepsilon_{0t}^{el} \\ \varepsilon_t^{pl} = \varepsilon_t^{ck} - \dfrac{d_t}{(1-d_t)}\dfrac{\sigma_t}{E_0} \end{cases}$$

where σ_t is the tensile stress; ε_t is the tensile strain; E_0 is the initial elastic modulus; d_t is the tensile damage factor; ε_{0t}^{pl} is the elastic tensile strain.

3 FINITE ELEMENT SIMULATION ANALYSIS

3.1 *Establishment of the finite element model of the axial compression test*

The model employed a reinforced concrete rectangular section simply supported beam structure. The beam span was 2000 mm, the section size was 300 mm × 300 mm, the strength grade of concrete was C30, the thickness of concrete cover was 25mm, the longitudinal tensile steel bar was 4∅20, and the stirrup was 20∅6. The size of the beam and the configuration of the rebar are shown in Figure 2.

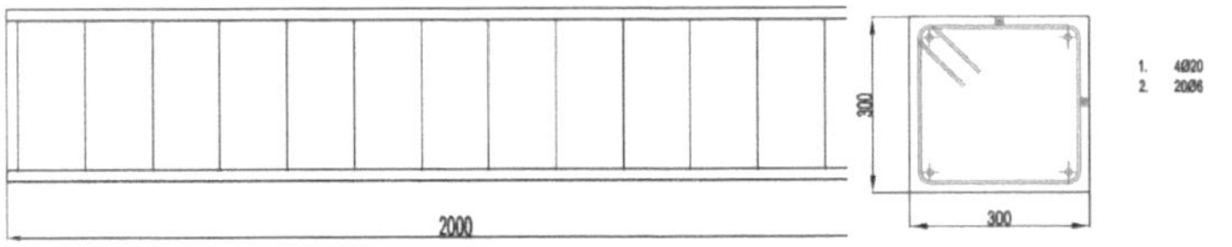

Figure 2. The size of specimen and reinforcement (unit: mm).

3.2 *Related parameters and test results of axial compression specimens*

The elastic modulus and Poisson's ratio of concrete are $E = 3 \times 10^4$MPa and $\mu = 0.2$, respectively. The elastic modulus and Poisson's ratio of Q345 rebar are $E = 2 \times 10^5$MPa and $\mu = 0.3$, respectively. The CFRP cloth was created by Composite Layup. The number of layers was 8, the thickness of each layer was 0.146mm, and the spread angle was 90°.

Table 1. Material Properties of CFRP Strip.

Model	Monolayer Thickness (mm)	Tensile Strength (MPa)	Elasticity Modulus for Tension (MPa)	Interlaminar Shear Strength (MPa)	Elongation (%)
UT70-30 Carbon Fibre Cloth	0.146	4216	25200	46.9	1.76

Figure 3 shows the stress nephogram of reinforced concrete beams without CFRP reinforcement, and the extremum of the stress is 20.92MPa. Figure 4 shows the stress nephogram of CFRP-strengthened reinforced concrete beams, and the extremum of the stress is 22.15MPa. The finite element simulation results show that the carbon fiber cloth can share some of the bearing capacity, resulting in an increase of 5.9 % in the bearing capacity of reinforced concrete beams in the loading position.

Figure 3. Equivalent stress distribution of reinforced concrete beams without CFRP reinforcement.

Figure 4. Equivalent stress distribution of reinforced concrete beams with CFRP reinforcement.

The displacement-load curves of reinforced concrete beams without reinforced CFRP (Figure 5) and with reinforced CFRP (Figure 6) are made using Origin. It can be discerned from the comparison of the curves that both types of beams break when the load reaches 3.0×10^6N. The reinforcement of CFRP has little effect on the axial maximum pressure of reinforced concrete beams because CFRP is a tensional material, and it usually works for the tension and also the flexural force.

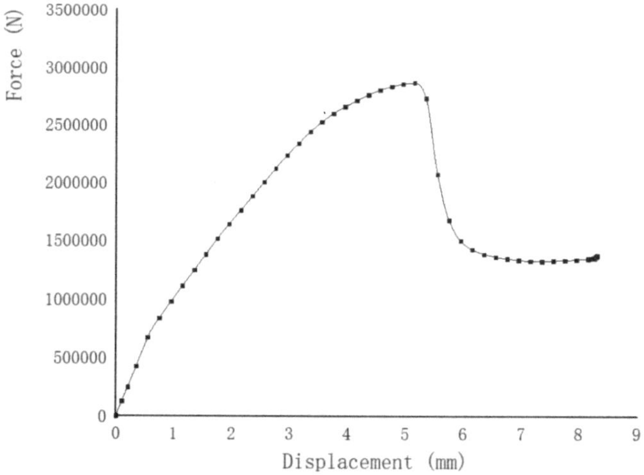

Figure 5. The displacement-load curve of reinforced concrete beams without reinforced CFRP.

3.3 *Establishment of finite element model of three-point bending loading test*

In order to study the flexural mechanical properties of reinforced concrete strengthened with CFRP, a three-point bending loading test was carried out by using finite element software. The model adopted a reinforced concrete rectangular section simply supported beam structure, the beam span was 2000 mm, and the section size of the beam was 300 mm × 200 mm. The grade of the concrete strength was C30, the thickness of the concrete protective layer

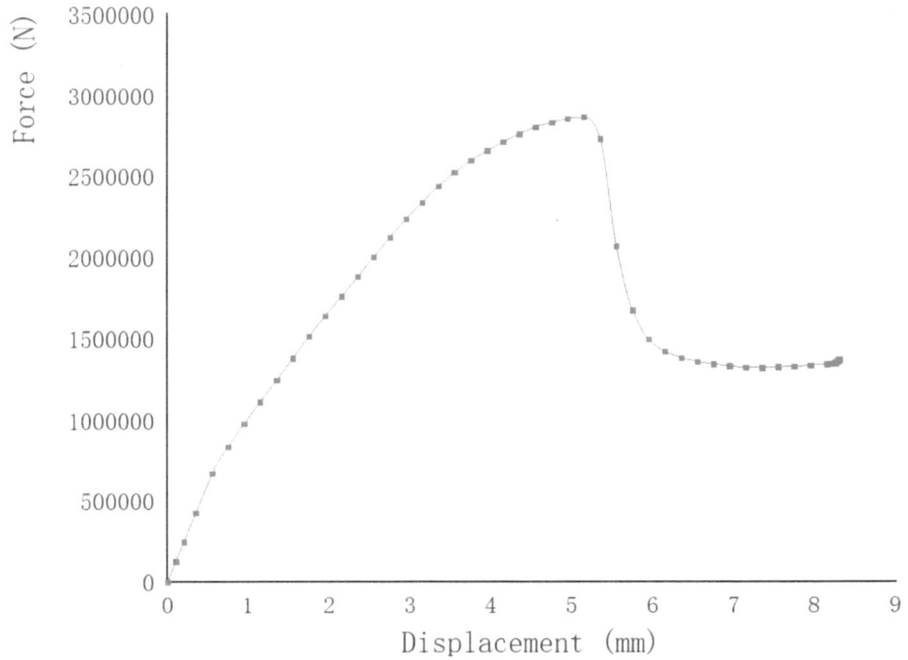

Figure 6. The displacement-load curve of reinforced concrete beams with reinforced CFRP.

was 25 mm, and the longitudinal tensile bar was 4∅16. The size of the test beam and the configuration of the rebar are shown in Figure 7.

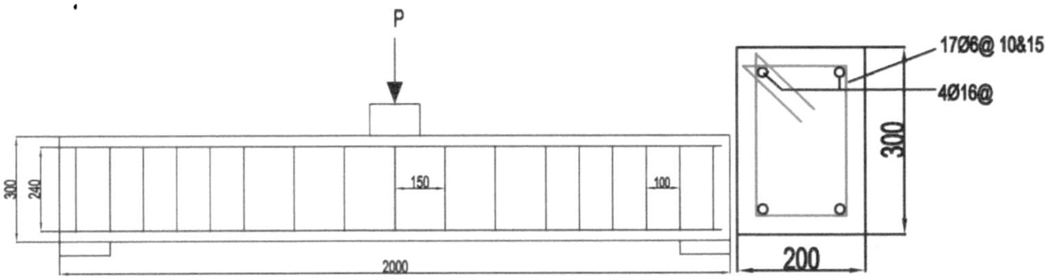

Figure 7. The specimen size and reinforcement drawing of the three-point bending loading test.

3.4 *Related parameters and test results of three-point bending loading test specimens*

The specimen concrete was C30, the type of tebar was Q345, the cloth of CFRP was UT70-30 carbon fibre cloth, and the relevant property parameters were consistent with the parameters of the axial compression test. Figure 8 is the equivalent stress distribution map of reinforced concrete beams without reinforced CFRP, and its extremum stress is 17.06MPa. Figure 9 is the equivalent stress distribution map of CFRP-strengthened reinforced concrete beams, and the extremum stress is 22.63MPa. It can be seen from the figures that in the case of single-point mid-span loading, when loaded to the ultimate load, the stress at the loading

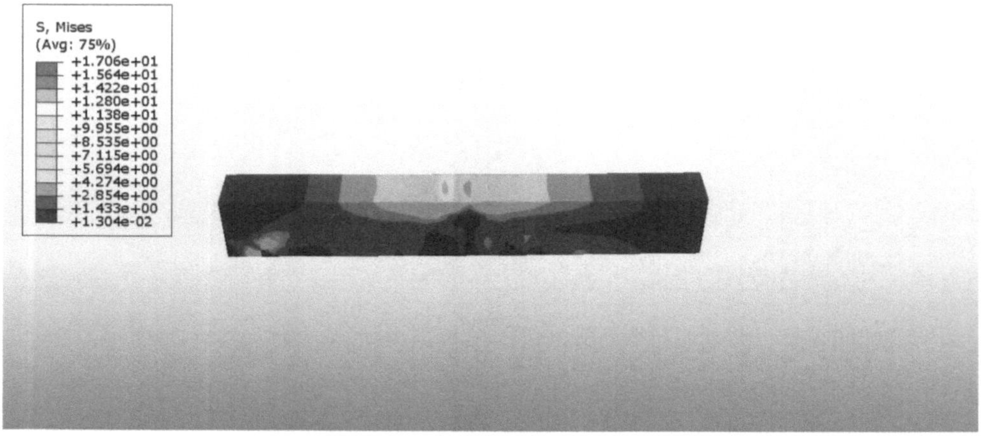

Figure 8. Equivalent stress distribution of reinforced concrete beams without CFRP reinforcement.

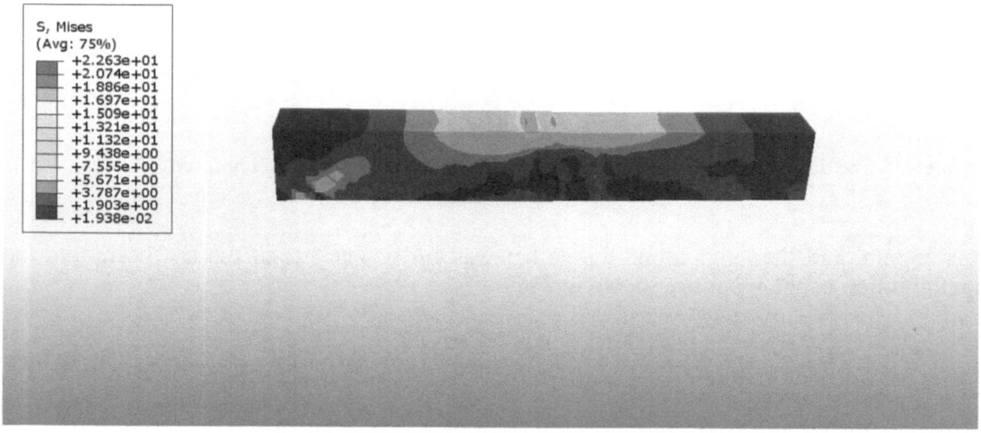

Figure 9. Equivalent stress distribution of CFRP-strengthened reinforced concrete beams.

point area of the reinforced concrete beam reaches the design compressive strength, and the concrete at the lower edge of the beam near the mid-span also reaches the tensile strength.

The origin is used to make the displacement-load curves of reinforced concrete beams without and with CFRP reinforcement, respectively, and the curves are merged into a diagram (Figure 10). It can be seen from the curve that the ultimate flexural bearing capacity of CFRP-strengthened reinforced concrete beams is 74.5 % higher than that of reinforced concrete beams without CFRP reinforcement. The finite element simulation results explain that carbon fibre cloth can share part of the bearing capacity and effectively improve the strength and stiffness of the beam.

4 CONCLUSION

In this paper, the axial compression test and three-point bending loading test numerical simulations of reinforced concrete beams without and with CFRP reinforcement were

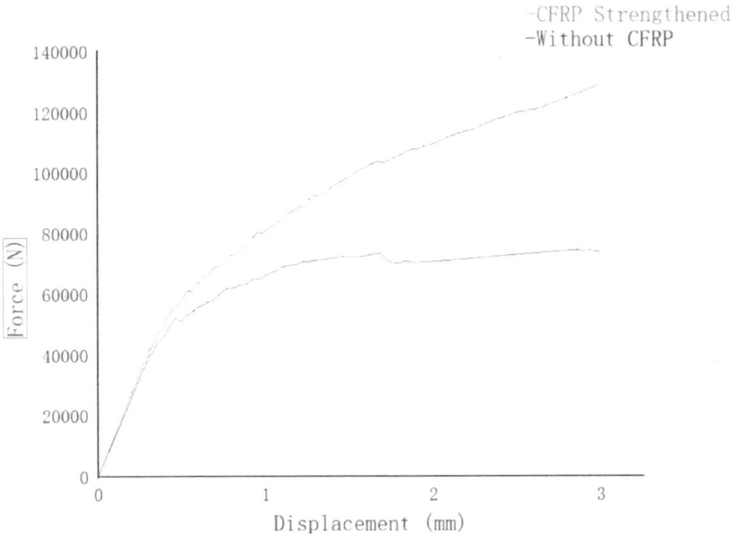

Figure 10. The combined displacement-load curve of reinforced concrete beams without and with CFRP reinforcement.

carried out, respectively, and the test results were contrasted and analyzed. The improvement effects of compressive bearing capacity and flexural bearing capacity of CFRP-strengthened reinforced concrete beams were discussed, and the following conclusions were obtained as follows:

(1) The introduction of the concrete damaged plasticity model (CDP) and Hashin failure criterion made the simulation results of CFRP-strengthened reinforced concrete more realistic. This approach was also applicable to the simulation of other CFRP-strengthened reinforced concrete structures.
(2) The extremum stress and flexural capacity of CFRP-strengthened reinforced concrete beams have been significantly improved, indicating that the bearing capacity of strengthened reinforced concrete has been obviously lifted.
(3) The crack distribution of concrete beams strengthened with CFRP was relatively dense. Circumferentially wound CFRP can availably inhibit the expansion of the interface, make the crack expand along the bonding surface, delay the cracking of the concrete, and promote the bearing capacity of reinforced concrete beams.

ACKNOWLEDGEMENTS

Thanks to the project tutor, Dr. Cheng Gao, for his guidance and the funding and support of Chang'an University. This project has been funded by Chang'an University's 2022 College Students' Innovative Entrepreneurial Training Plan Program (provincial-level project), and the project number is S202210710285.

REFERENCES

Ali O., Bigaud D. and Ferrier E. 2012 Comparative Durability Analysis of CFRP-strengthened RC Highway Bridges *Constr. Build. Mater.* **30** 629–42

El Maaddawy T. 2008 Post-repair Performance of Eccentrically Loaded RC Columns Wrapped with CFRP Composites *Cem. Concr. Compos.* **30** 822–30

Ghanim A., Al-Abbas B. 2018 Experimental Study on the Flexural Strengthening of Reinforced Concrete Beams using NSM CFRP Bars *Proceedings of the 5th national and 1st International Conference on Modern Materials and Structures in Civil Engineering* (Teheran: Amirkabir University of Technology)

Liu T., Feng W., Zhang Z.M. and Wei G.F. 2006 A Study on the Compressive Performance of Rectangular Concrete Columns Confined with CFRP Sheets *China Civil Engineering Journal* **12** 41–7

Mortazavi A.A., Pilakoutas K. and Son K.S. 2003 RC Column Strengthening by Lateral Pre-tensioning of FRP *Constr. Build. Mater.* **17** 491–7

Sha L.R., Chen W.L. and Wang X.L. 2021 Experimental Study on the Mechanical Performance of RC Beam Strengthened with CFRP *Journal of Jilin Jianzhu University* **38** 12–6

Yang G., Yan Y.K., Zhuo Y.H., Deng J.L., Liu X.D. and Bi W.B. 2021 Review of Research on Concrete Columns Strengthened with ‘ *Shandong Chemical Industry* **50** 103–5

Zhao H.D., Zhao M. and Zhang Y. 2000 Experimental Study on Concrete Circular-columns Wrapped with CFRP Under Axial Compression *Building Structure* **7** 26–30

Zhao J.H., Chen Y.X. and Zhang H.Q. 2021 Load-bearing Capacity of RC Square Short Column Strengthened with CFRP and Circular CFST *Journal of Guangxi University. Natural Science Edition* **46** 1130–8

Zhang Y.D. and Liu R.G. 2003 Survey on Research and Application of Strengthening Techniques in Reinforced and Pre-stressed Concrete Structures *Journal of Jiangsu University. Natural Science Edition* **6** 91–4

Zhang Y.J., Chen L., Xie P.C., Tang B.J. and Shen H.J. 2022 Rate Correlation of the ABAQUS Damage Parameter in the Concrete Damage Plasticity Model and its Realization Method *Explosion and Shock Waves* 1001–1455

Zhou J.N., Kong X.L., Wang P., Wang X.P., Chen X.S. and Jin F.N. 2022 Compressive Behavior of Plain Concrete Arches Reinforced by CFRP: Numerical Study *Journal of Army Engineering University of PLA* **1** 80–8

Water Conservancy and Civil Construction – Oke & Ahmad (Eds)
 ISBN: 978-1-032-58614-4

Creep model of compacted loess considering parameter randomness and engineering application

Changming Hu, Minghui Tian*, Yili Yuan, Fangfang Wang, Tingting Hu & Xuhui Hou
China College of Civil Engineering, Xi'an University of Architecture and Technology, Xi'an Shaanxi, China

ABSTRACT: The randomness of creep deformation will occur due to the variability of geotechnical materials and the contingency of the test evaluation process. In this paper, through repeated one-dimensional consolidation creep experiments, a creep model that can describe the one-dimensional consolidation creep characteristics of compacting loess was proposed, and a multivariate normal distribution model of creep parameters was established. The random analysis of creep deformation can be realized by combining it with Matlab software. The creep model was developed in Fortran, and the ABAQUS finite element software was used to analyze and verify the random deformation of a slope. The creep model and application method proposed in this paper can provide a new idea and a new channel for the analysis of creep deformation.

1 INTRODUCTION

A large number of loess high-fill projects have emerged in the northwest of China, and the engineering problems caused by the post-construction settlement of high-fill foundations caused by the creep of compacted loess frequently appear. Accurate evaluation of the creep behavior of fill is the basis for predicting post-construction settlement of fill foundations and the basic premise of disaster prevention.

As early as 1925, Terzaki mentioned the creep characteristics of soil in his book Principles of Soil Mechanics. Up until now, there have been many achievements in the study of creep. In the study of the creep test, the one-dimensional consolidation test of loess with fiber yarn reinforcement could effectively restrain its creep deformation (Chu *et al.* 2022). Direct shear creep tests on silty clay showed that the creep characteristics of soft soil samples were more significant at high shear stress (Liang *et al.* 2021). Li *et al.* (2021) studied the microstructure of the triaxial creep test samples of Malan loess and found that the pore size, particle size, and shape were the three microstructure parameters with the most obvious changes in the creep process. The purpose of the creep test was to analyze the creep characteristics of rock and soil masses on the one hand and to find a suitable creep model for quantitative analysis on the other hand. Li *et al.* (2021) introduced time, water content, and stress state into the Burgers model to better describe the one-dimensional consolidation creep characteristics of undisturbed loess. Many scholars introduced the theory of fractional calculus to establish a creep model with small parameters that can describe the accelerated creep stage of rock mass (Su 2022; Wei *et al.* 2022). Based on the Nishihara model and considering the coupling effects of stress and time on the model elements, Luo *et al.* (2020) proposed a model that can describe the decay, steady state, and accelerated creep

*Corresponding Author: 1939137520@qq.com

DOI: 10.1201/9781003450818-46

characteristics of frozen sand. Based on the strain energy theory, Shen *et al.* (2014) could effectively judge the occurrence time of accelerated creep by defining the critical strain energy density value of accelerated creep, providing a new idea for the study of the rock creep fracture process. The ultimate purpose of establishing the creep model is to better apply it in engineering. However, due to the particularity of loess, the variability of rock and soil material itself, and the randomness of test evaluation, the creep deformation and creep model parameters of loess appear random. As a result, a creep model capable of describing the creep characteristics and creep randomness of compacted loess should be developed.

In this paper, repetitive one-dimensional consolidation creep tests were carried out on a high-sticking slope project of compacted loess. A nonlinear creep model that can describe the one-dimensional consolidation creep characteristics and creep randomness of compacted loess was proposed, and the secondary development of the creep model in ABAQUS software was realized by using Fortran. The creep model was applied to realize the random analysis of the deformation of the high-sticking slope.

2 ONE-DIMENSIONAL CONSOLIDATION CREEP TEST

2.1 *Soil sample testing*

In this one-dimensional consolidation creep test, the soil sample was a highly filled slope of compacted loess. The soil was mainly silty, with a small amount of white calcareous film and sporadic calcareous nodules. The basic physical parameters of soil samples are shown in Table 1.

Table 1. Physical properties of soil samples.

Optimal Water Content $\omega_0/\%$	Maximum Dry Density $\rho_d/(g * cm^{-3})$	Plastic Limit $W_P/\%$	Liquid Limit $W_L/\%$	Index of Plasticity I_P
13.45	1.78	18.4	29.7	11.3

2.2 *Test scheme*

The sample with a dry density of 1.75 g/cm^3 and moisture content of 10% was prepared, and a one-dimensional consolidation creep test was carried out by a WG type single lever consolidation instrument. The sample size was cylindrical 50 cm^2 × 2 cm, the loading method was to preload at 25 kPa pressure and stabilize, according to the five loads of 100 kPa, 200 kPa, 400 kPa, 800 kPa, and 1600 kPa step-by-step loading, and nine samples were made under each pressure for the test. When the cumulative vertical deformation over two consecutive 24 hours was less than or equal to 0.005 mm, an electronic micrometer was used to measure it. During the test, the indoor temperature should be kept as constant as possible.

2.3 *Analysis of test results*

The Boltzmann linear superposition method was used to process the one-dimensional consolidation creep test curve under graded loading, and the strain-time curve under various loads was obtained, as shown in Figure 1.

Instantaneous deformation will appear in the test at the moment the load was applied. Over time, the deformation rate of the sample gradually decreased and finally tended to be stable. When the load was small, the instantaneous deformation of the sample was small. Since it is easier for the large load to overcome the gravity between soil particles and the pore structure formed, the instantaneous deformation of the sample will increase when the load is increased, and the time to reach the stable creep stage is also relatively longer, so the creep effect is more obvious. According to the analysis of nine samples under each grade of load, although the conditions were the same, the variability of rock and soil material and the accidental error in the test process made the stable strain of creep appear to different degrees, so it was reasonable to consider the randomness of creep deformation.

3 ESTABLISHMENT AND ANALYSIS OF THE CREEP MODEL

3.1 *The establishment of a nonlinear creep model*

The commonly used creep models of soil include the component combination model and the empirical model. The concept and principle of the element model are clear, but the description of nonlinear attenuation creep is limited. Compared with the component model, the empirical model has a higher degree of fit, but it lacks a rigorous theoretical basis and a definite physical meaning. Therefore, many scholars used the element model to simulate the linear rheological part of the soil and an empirical formula to simulate the nonlinear rheological part of the soil. In this way, a semi-empirical and semi-theoretical model was established, which can be well used in engineering practice (Zhang *et al.* 2009; Fan *et al.* 2007).

The nonlinearity of the creep curve was observed. First, a nonlinear function was defined as follows:

$$f(t) = \frac{mt^n}{1 + mt^n} \tag{1}$$

where m and n = shape parameters, whose values are greater than zero; and t = time.

$f(t)$ is a continuous increasing function. When the soil creep increases to a certain time, it is close to 1.

The Kelvin model in the Merchant model was replaced with a nonlinear softening spring. Under constant stress, its nonlinear attenuation creep will satisfy.

$$E(t) = E_1/f(t) = E_1(1 + mt^n)/mt^n \tag{2}$$

where E_1 = the initial value of elastic coefficient of nonlinear element.

At zero, constant stress is applied to the creep model, and the corresponding one-dimensional creep constitutive equation can be written as follows:

$$\varepsilon(t) = \frac{\sigma_0}{E_0} + \frac{\sigma_0}{E(t)} = \frac{\sigma_0}{E_0} + \frac{\sigma_0}{E_1}\frac{mt^n}{1 + mt^n} \tag{3}$$

where E_0 = Elastic modulus of instantaneous deformation under loading; E_1 = Elastic modulus of nonlinear softening spring, and the parameter m and n together control the decay rate of the nonlinear decay creep.

Taking the first set of data for each stage of pressure as an example, the improved nonlinear creep model of this paper, the Merchant model, and the Burgers model were compared, as shown in Figure 2. The Merchant model cannot reflect well the process from decay creep to stable creep and the development of the later stage of creep, while the Burgers model has poor convergence in the later stage of creep. The fitting effect of instantaneous deformation is not high. The nonlinear creep model proposed in this paper can well describe the instantaneous deformation, decay creep, and stable creep states of compacted loess, and its correlation coefficient of fit degree is very high, above 0.99. It is more accurate to use this model to analyze the creep characteristics of compacted loess.

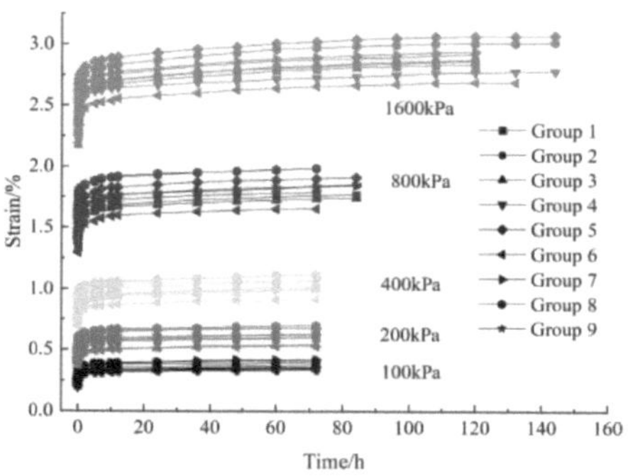

Figure 1. Strain-time curves.

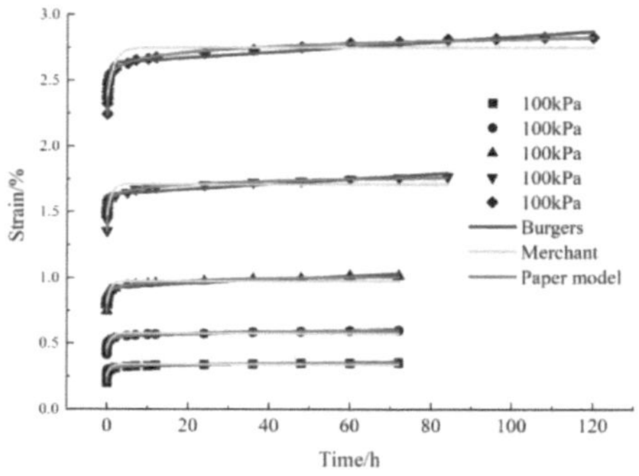

Figure 2. Comparison chart of creep model fitting.

3.2 *Analysis of the creep parameters*

The Levenberg-Marquardt optimization algorithm has the advantages of both the gradient method and the Newton method. Creep parameters are obtained by using the fitting test curve of the Levenberg-Marquardt optimization algorithm, and the statistical data are shown in Table 2.

Table 2. Statistics of the creep parameters.

	Mean					Coefficient of Variation				
Vertical Pressure/kPa	100	200	400	800	1600	100	200	400	800	1600
E_0/MPa	45.55	45.66	51.63	56.01	70.06	9.31%	11.42%	7.89%	7.05%	3.85%
E_1/MPa	46.66	74.24	96.65	93.04	104.06	10.91%	9.79%	12.47%	14.26%	10.14%
m	0.90	0.68	0.66	0.40	0.27	24.44%	27.50%	29.31%	20.05%	19.19%
n	0.23	0.25	0.23	0.18	0.17	11.21%	8.10%	7.33%	9.55%	7.47%

The inherent randomness of geotechnical materials and the external randomness caused by the parameter evaluation process are manifested in the deformation of each group of samples. Whether it is an instantaneous or stable deformation, there is a certain difference. When reflected in the creep model, it is manifested in the variability of the creep parameters, and the variability of the other four parameters is also different. The variation coefficient of the creep parameter m is much larger than the other three parameters, with a maximum value of 29.27% and a minimum value of 19.1%. The remaining three coefficients of variation are all below 15%, indicating that the variability of each group of deformation data has a greater impact on the creep parameter and a relatively small impact on the remaining three parameters.

3.3 *Application method of the creep model considering parameter randomness*

The above statistics indicate that creep parameters had randomness, and it was more reasonable to treat them as random variables rather than definite values when applied. Since there is a certain relation between the parameters, if each parameter is regarded as a separate random variable, the relation between the creep parameters will be ignored, resulting in a large deviation between the solution results and the real results. According to the study of the statistical model of creep parameters by Ma and Wang (2013), the parameters obtained from the test were put together, and the four parameters constituted a four-dimensional population random sample:

$$\boldsymbol{X} = [\boldsymbol{E}_0\ \boldsymbol{E}_1\ \boldsymbol{mn}] \tag{4}$$

Assuming that the four parameters obey the normal distribution, then the random matrix $\boldsymbol{X}$ obeys the quaternion normal distribution, and the joint distribution function of its quaternion function is:

$$f(x) = \frac{1}{(\sqrt{2\pi})^4 |\sum|1/2} \exp\left[-\frac{1}{2}(x-\boldsymbol{\mu})'\sum\nolimits^{-1}(x-\boldsymbol{\mu})\right] \tag{5}$$

The mean vector $\boldsymbol{\mu}$ and covariance matrix Σ of the random vector $\boldsymbol{X}$ are:

$$\boldsymbol{\mu} = \boldsymbol{E}(\boldsymbol{X}) = [E(E_0)E(E_1)E(\mathrm{m})E(n)]^{\mathrm{T}} \tag{6}$$

$$\sum = E\left[(\boldsymbol{X}-\boldsymbol{E}(\boldsymbol{X}))(\boldsymbol{X}-\boldsymbol{E}(\boldsymbol{X}))^{\mathrm{T}}\right] \tag{7}$$

In the random analysis of creep deformation, the mvnrnd function in Matlab software can be used to sample the multivariate normal model and obtain the number of parameter groups that meet the calculation requirements.

4 SECONDARY DEVELOPMENT OF THE CREEP MODEL

The material constitutive model is the mathematical description of the material stress-strain relationship, which is the basis of finite element calculation. Abaqus software provides a wealth of material models, but because of the complexity of the actual situation, Abaqus cannot include all the problems, so it provides several subroutines to support users in using the model for secondary custom development of custom. UMAT (user-defined mechanical and thermal material behavior) is a user subroutine interface provided by Abaqus for users to carry out secondary development of the material constitutive model. Using this interface, users can easily define the material constitutive model they need. In this paper, the UMAT subroutine of the nonlinear compacted loess creep model is written in Fortran using the secondary development platform.

To verify the correctness of the subroutine written, a 3D cylindrical model with a diameter of 79.8 mm and a height of 20 mm was established, which was divided into 7792 units and 9225 nodes. The unit type was C3D8. The bottom of the model limited the displacement in the x, y, and z directions, while the side limited the displacement in the x and y directions and imposed a vertical load on the top. The finite element model of the sample is shown in Figure 3. The calculation process was divided into two analysis steps: the instantaneous analysis step and the creep analysis step. The first step was to simulate the instantaneous deformation under loading, and the second step was to simulate the creep process under constant load.

The subroutine was called to conduct a numerical simulation with the first set of creep parameters at each pressure level, and the strain-time curve and test strain-time curve obtained are shown in Figure 4. The numerical simulation results were in good agreement with the experimental results under different pressures. The maximum error of the instantaneous strain was 0.33% and the minimum error was 0.04%; the maximum error of the final stable strain was 1.073% and the minimum error was 0.028%. They proved that subroutine development was reasonable.

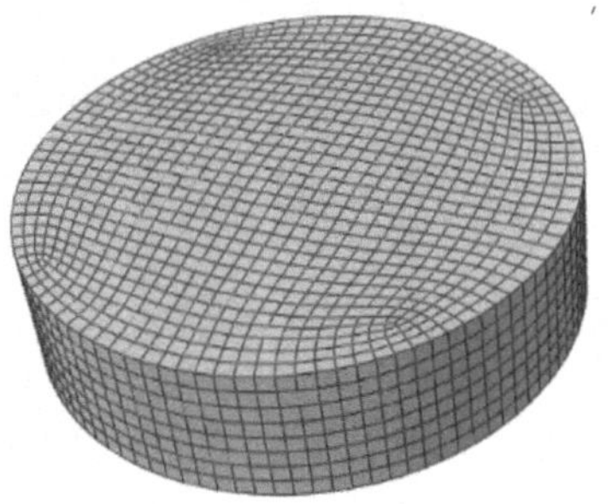

Figure 3. One dimensional consolidation test finite element model.

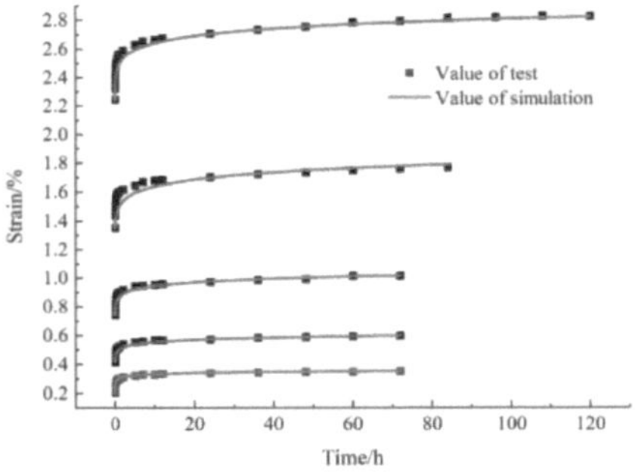

Figure 4. Comparison between numerical simulation and test results.

5 ENGINEERING APPLICATIONS OF THE CREEP MODELS

5.1 *Numerical analysis model*

With the high-sticking slope as an example, according to the geological survey report of the high-filling project, the undisturbed soil slope was generalized into 4 layers, which were respectively loess (Q_3^{eol}), paleosol (Q_2^{el}), red clay (N_2), and sandstone (J_1^{yl}) from top to bottom. The majority of the slope sticking is made of loess filling. The total length of the high sticking slope was 301 m and the height was 127 m, of which the body length was 222 m and the height was 82 m. Its geometric model is shown in Figure 5.

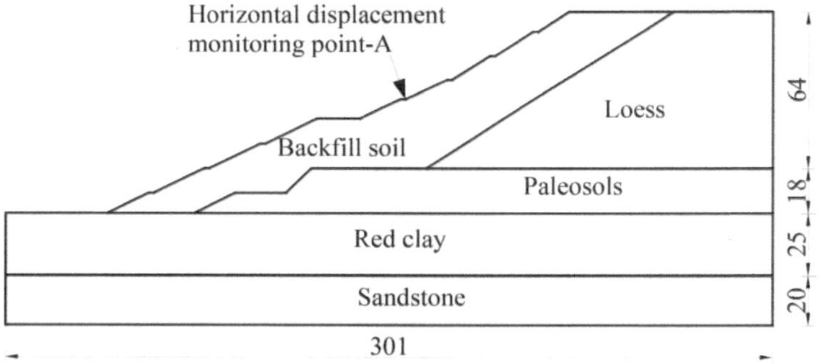

Figure 5. Slope geometry.

The plane strain model was established by ABAQUS software. There were 8614 units and 26254 nodes in the high sticking slope, among which 1745 units were in the sticking slope. The unit type was CPE4. The boundary conditions were set as fixed constraints at the bottom, limited horizontal displacement at the left and right sides, and free boundaries at the top. The model applied a self-weight load. As this paper focused on the study of creep deformation on a sticking slope based on the deformation analysis of a high sticking slope, the elastic constitutive model was adopted for the undisturbed soil slope and the nonlinear creep constitutive model was adopted for the sticking slope. The physical and mechanical parameters of each layer of the original slope are shown in Table 3. The sticking slope was divided into 18 layers. The function of the life and death unit was used to simulate the stratified filling process. After sampling the four-element normal distribution model under a fifth-order load with the mvnrnd function in MATLAB, the creep parameters of each layer were obtained by linear interpolation according to the thickness and overlying load.

Table 3. Physical and mechanical parameters of rock and soil layer.

	Modulus of Elasticity E/MPa	Poisson's Ratio	Bulk Density γ/kN·m^{-3}
J_1^{yl}	300	0.25	26
N_2	40.9	0.35	21.9
Q_2^{el}	34.77	0.35	19.6
Q_3^{eol}	35.8	0.36	20

5.2 *Analysis of results*

According to the slope displacement monitoring data, the deformation quantity and deformation rate of the horizontal displacement of a high sticking slope were much larger than those of the vertical displacement. Using horizontal displacement monitoring point A as an example, the mvnrnd function sampled 500 random parameters. The comparison between the numerical simulation results at point A and the monitoring data is shown in Figure 6. The trend of the numerical simulation results was roughly the same as that of the monitoring displacement, and the monitoring displacement was within the range of the numerical simulation results, which proved that the random variable creep model was reasonable for the random analysis of slope deformation.

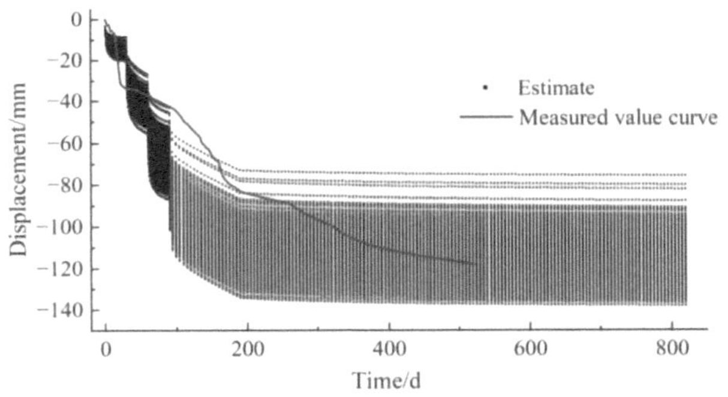

Figure 6. Comparison between numerical simulation and test results.

6 CONCLUSIONS

(1) When the load was small, the creep deformation of compacted loess was small, and the stability time was relatively short. When the load was large, the creep deformation of compacted loess was large, the stability time was relatively long, and the creep effect was more obvious.

(2) The creep of compacted loess was random due to the variability of rock and soil material and the contingency of the test process. In this paper, a nonlinear creep model was proposed to describe the one-dimensional consolidation creep characteristics of compacting loess. By establishing a multivariate normal distribution model of creep parameters, the stochastic analysis of creep deformation can be realized.

(3) Through the secondary development, the application of the nonlinear creep model of compacted loess in ABAQUS software was realized, which widened the engineering application channel of the proposed creep model, and the numerical simulation of an example slope was carried out to verify the rationality of the model in the random analysis of slope deformation.

REFERENCES

Chu, F., Shao, S.J., Deng, G.H., Chen, C.L., 2022. Experimental Study on one Dimensional Creep Behavior of Loess Reinforced with Fiber Yarn [J]. *Chinese Journal of Rock Mechanics and Engineering*, 41 (5): 1054–1066.

Fan, Q.Z., Gao, Y.F., Cui, X.H., Fu, Z.L., 2007. Study on Nonlinear Creep Model of Soft Rock [J]. *Chinese Journal of Geotechnical Engineering*, (4): 505–509.

Li, A., Chen, J.B., Sun, X.H., Ding, X.S., Ji, B.N., 2021. One Dimensional Creep Test and Creep Model of Undisturbed Loess [J]. *Science Technology and Engineering*, 21 (21): 8789–8796.

Li, Z.X., Wang, J.D., Yang, S., Liu, S.H., Li, Y.W., 2022.Characteristics of Microstructural Changes of Malan Loess in Yan'an Area During Creep Test [J]. *Water*, 14 (3).

Liang, R.K., Zhang, Q.J., Zhang, C., Li, D., Su, D., Li, Q., 2021. Creep Characteristics and Model Parameters of Qianhai Soft Soil [J]. *Chinese Journal of Geotechnical Engineering*, 43 (S2): 133–136.

Luo, F., Zhang, Y., Zhu, Z.Y., Zhang, D.J., He, J.L., 2020. Creep Constitutive Model for Frozen Sand of Qinghai-Tibet Plateau [J]. *Journal of Harbin Institute of Technology*, 52 (2): 26–32.

Ma, J.W., Wang, Z.Y., 2013. Probability Distribution of Rock Salt Creep Parameters Based on Multivariate Normal Distribution [J]. *Science & Technology Review*, 31 (36): 41–45.

Shen, C.H., Zhang, B., Wang, W.W., 2014. A new Visco-elastoplastic Creep Constitutive Model Based on Strain Energy Theory [J]. *Rock and Soil Mechanics*, 35 (12): 3430–3436.

Su, Y.J., 2022. Fractional Creep Constitutive Model of Rock in Consideration of Aging Damage [J]. *Journal of Yangtze River Scientific Research Institute*, 39 (3): 92–97, 103.

Wei, E.R., Hu, B., Li, J., Cui, K., Zhang, Z., Cui, A.E., Ma, L.Y., 2022. Nonlinear Viscoelastic-plastic Creep Model of Rock Based on Fractional Calculus [J]. *Advances in Civil Engineering*, 2022.

Zhang, Y.P., Cao, P., Zhao, Y.L., 2009. Visco-Plastic Rheological Properties and a Nonlinear Creep Model of Soft Rocks [J]. *Journal of China University of Mining & Technology*, 38 (1): 34–40.

Water Conservancy and Civil Construction – Oke & Ahmad (Eds)
© 2024 The Author(s), ISBN: 978-1-032-58614-4

Study on the joint influence of slope gradient and P_5 on the stability of a reinforced high-fill slope

Yan Bin* & Ma Jie
CAAC Central Southern Airport Design & Research Institute (Guangzhou) Co., Ltd., Guangzhou

ABSTRACT: To study the joint influence law of slope and filler's P_5 (particle size greater than 5mm content in filler) on the stability of a reinforced high fill slope, based on the No. 2 ditch of Shennvfeng airport in Wushan, Chongqing, this study analyzed the stress distribution in the slope using PLAXIS numerical simulation. The law of deformation and stability changing with slope and filler's P_5 shows that the slope stability coefficient decreases linearly with the increase of slope; with the increase of P_5, the slope stability shows a trend of "first decrease and then increase", in which the P_5 of filler is 55%, and the overall stability of slope is the worst. Therefore, this graded filler is not suitable for reinforced fill slopes, which provides a reference for practical projects in the future.

1 INTRODUCTION

With the development of Chinese western mountainous area construction, the treatment of cut and fill slopes has gradually become a hot issue in the field of geotechnical engineering. For excavation slopes, retaining walls, anti-slide piles, or anchor cables are often used in practical engineering to restrain the instability of the slope. For the filling slope, due to the limitation of space and cost caused by land acquisition scope, undisturbed terrain, and other reasons, the design slope of the filling slope is often steep in the construction of the mountainous area. At the same time, the influencing factors of the internal stress state of the filling slope are complex. Therefore, the overall and local stability analysis and slope treatment of high and steep filling slopes in mountainous areas are also important geotechnical engineering problems. Due to the low tensile strength of rock and soil, reinforcement is often used in the filling process to increase the strength of composite filler. Peng Xianqing has systematically explored the application of geotextile cells in the filling engineering of mountain highway construction (Peng 2020). This method is to lay geogrid and other reinforced materials in the filler layer by layer in the filling process, to increase the tensile and shear strength of the composite filler on the one hand. Emma Burak has systematically expounded on the influence of reinforcement on the shear strength index of soil mass (Burak 2021). On the other hand, the use of reinforcement soil interaction gives full play to the self-supporting capacity of rock and soil, so it has the characteristics of energy conservation and environmental protection. Niu Xiaodi has proved that the reinforced soil has good engineering applicability and environmentally friendly characteristics through field tests (Niu 2021).

Reinforced fill has many advantages in the process of filling slope treatment and is widely used in practical projects. As an influencing factor of reinforced soil slope stability, it has always been a frontier hotspot in engineering research. Taking the reinforced high fill slope

*Corresponding Author: yanbincisco@qq.com

DOI: 10.1201/9781003450818-47

of the No. 2 ditch of Shennvfeng Airport in Wushan, Chongqing as the research prototype, through the PLAXIS numerical simulation of the reinforced fill slope under different variables, this paper explores the joint influence law of the slope of the reinforced fill slope and the P5 in the filler on the stability of the reinforced high fill slope.

2 BASIC CONDITIONS

2.1 *Prototype overview*

The prototype of the plan and plane figure of the reinforced fill slope of ditch 2 of Shennvfeng airport studied in this paper is shown in Figure 1.

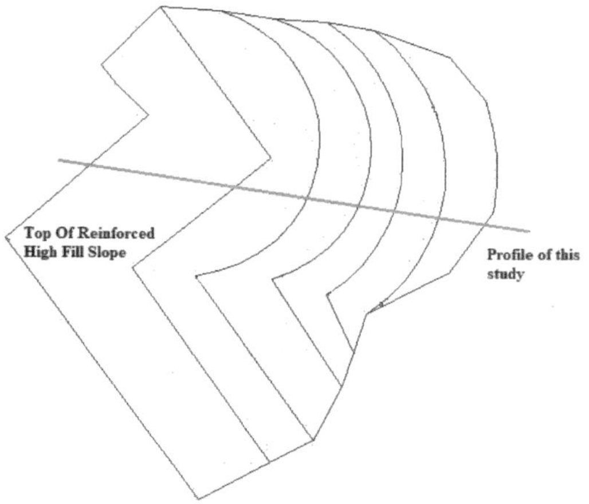

Figure 1. Schematic diagram of the slope receiving plan.

The reinforced fill slope is 50 m high. It is divided into five levels. Each level is provided with a berm with a slope of 45°, and the compactness of the backfill is not less than 93%. There are 4m unreinforced in the upper part. The reinforced soil adopts the back-wrapped geogrid reinforcement method, with a reinforcement spacing of 50 cm. The filler is taken from the site excavation, which is weakly weathered limestone, with a maximum particle size of 15 cm. The sectional drawing is shown in Figure 2.

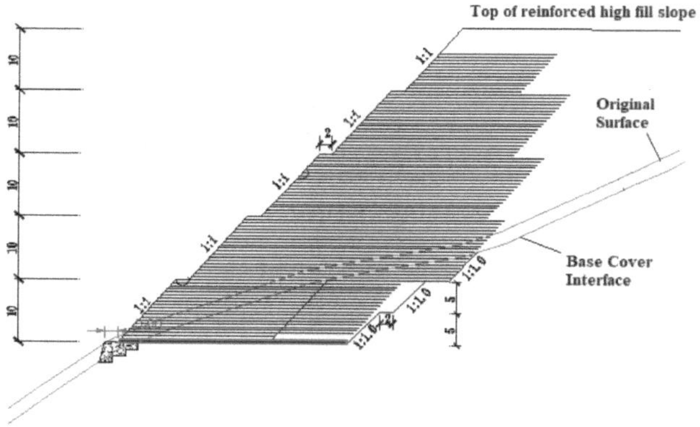

Figure 2. The sectional drawing of the No. 2 ditch filling slope of Wushan Shennvfeng Airport.

Three types of geogrids are used in the project. According to the design documents and the tensile test of the three types of geogrids on site, the geogrid laying of the reinforced high-fill slope is summarized in Table 1.

Table 1. Grid strength of filling slope of No. 2 ditch of Shennvfeng Airport.

Location of laying	Model of Grid	Grid strength (kN/m)	Length of laying
4 m-10 m below the slope top elevation	HDPE-130	137.91	18
10 m-20 m below the slope top elevation	HDPE-130	137.91	28
20 m-30 m below the slope top elevation	HDPE-200	212.17	35
30 m-40 m below the slope top elevation	GSJ300-300	318.25	40
40 m-50 m below the slope top elevation	GSJ300-300	318.25	35

2.2 *Numerical simulation parameters*

The geogrid unit built in PLAXIS software is slender with axial tensile stiffness but no bending stiffness. In this simulation, a 15-node triangular element grid is selected as the soil element, and the geogrid is defined with 5 nodes. This setting can simulate the interaction between reinforcement and soil (Zuo 2022).

In this simulation, the constitutive model of the grid is an elastic-plastic model, and the default grid is only subject to tension and not pressure. In the actual process, the exertion degree of grid tensile capacity is often far less than its design tensile strength (Yan 2021). Therefore, in PLAXIS software, the calculation formula of grid tensile force is as follows:

$$F = EA\frac{\Delta l}{l} \tag{1}$$

where: F = tension of geogrid (kN/m); EA = axial stiffness of geogrid; Δl = elongation of geogrid (m); l = original length of geogrid (m).

As the strength and layout of geogrid are non-interference variables in this study, they are controlled in numerical simulation. To facilitate the generation of a numerical model grid, the layout of the engineering prototype grid is simplified. The laying density of the prototype grid in the proportion of 1:6 (and increase the grid strength in proportion) is reduced. The specific definition, parameter meaning, and value can be seen in Table 2, and the specific layout of the model can be seen in Figure 3.

Table 2. Grid parameter setting.

Parameter	Parameter meaning	Type A Grid	Type B Grid	Type C Grid
EA (kN/m)	Axial stiffness	420	650	1000
N_p (kN/m)	Maximum axial tension	780	1200	1800

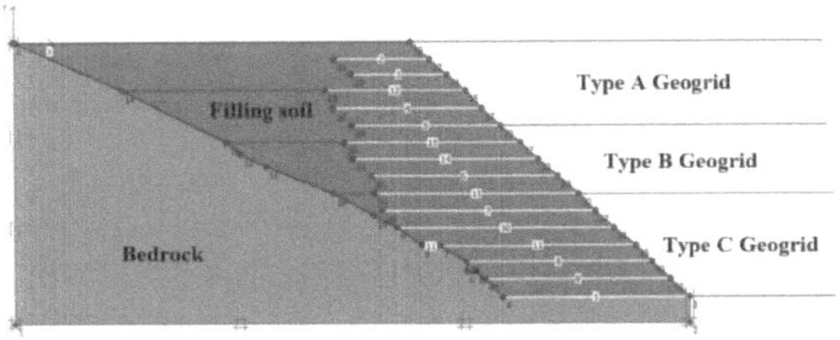

Figure 3. Layout of the numerical model geogrid.

To systematically study the influence of filler P_5 content and slope angle on the stability of reinforced high fill slope under natural conditions, this study cited the research results of the sub-topic ***"Research on long-term deformation laws and optimization design methods of airport high fill" in the "National Key Basic Research and Development Program (973 Program) funded project"*** personally participated by the author of this paper Ma Jie (Ma 2019), and selected the "Mohr-Coulomb" constitutive model built in PLAXIS. The physical and mechanical properties of 9 kinds of limestone fillers with different P_5 (particle size greater than 5mm) are tested. The specific property parameters are shown in Table 3.

Table 3. Statistical table of grading properties of fillers with different P_5.

Parameter	P_5 (%)	γ_d (kN/m^3)	γ_{sat} (kN/m^3)	k (m/d)	E_{ref} (MPa)	μ	c (kPa)	φ (°)
Parameter meaning	Content of particles with particle size greater than 5 mm	Dry unit weight	Saturated unit weight	Permeability coefficient	Compression modulus	Poisson's ratio	Cohesion	Internal friction angle
Test result	35	21.8	28.3	5.61	52	0.25	83.4	34
	45	21.9	28.4	7.26	71	0.23	62.8	37
	55	22	28.6	7.56	87	0.21	16.8	37
	60	22.1	28.8	7.78	85	0.2	30.5	39
	65	22.2	28.9	47.52	115	0.19	31.3	41
	70	22.3	29	66.01	190	0.18	36.5	42
	75	22.4	29.1	73.35	237	0.17	37.5	43
	80	21.5	28.7	354.9	225	0.19	78.9	42
	85	21.1	28.4	1973.4	212	0.21	73.8	40

The bedrock of the filling project is mainly weakly weathered limestone. The parameters are obtained by in-situ test and controlled as non-interference variables in numerical simulation.

When using PLAXIS software to calculate slope stability, the analysis method used when using the finite element method to calculate slope stability is the strength reduction method. The law was proposed by Bishop in 1955 (Yang 2022). Its definition formula is:

$$\tau = \frac{c + \sigma \tan\varphi}{F_s} \tag{2}$$

where: τ = shear strength under reduction; c = cohesion; φ = internal friction angle; σ = principal stress; Fs = stability safety factor calculated by strength reduction method.

For F_S under the Mohr-Coulomb model, it is defined by the following formula (Hu 2021):

$$F_s = \frac{c - \sigma_n \tan\varphi}{c_r - \sigma_n \tan\varphi_r} \tag{3}$$

where: c = cohesion input value; φ = internal friction angle input value; σ_n = normal stress component in failure critical state; c_r = Failure critical state cohesion; φ_r = Internal friction angle in critical failure state.

PLAXIS numerical simulation adopts a *ph-i* algorithm to realize the strength reduction. The basic principle of this method is to calculate the cohesion c and internal friction angle of soil shear strength parameters φ. And the parameters of the reinforcement soil interface are reduced to make the strength parameters c and $tan\varphi$ reduced in proportion. The reduced proportion is the stability safety factor, which is calculated by the total multiplier in the numerical simulation process of PLAXIS software ΣMsf indicates. Its definition is as follows:

$$\sum M_{sf} = \frac{c}{c_r} = \frac{\tan\varphi}{\tan\varphi_r} \tag{3}$$

As the number of loading steps increases, ΣMsf increases continuously until the soil reaches the critical failure state. After the soil reaches the critical state, ΣMsf will remain constant, and the value at this time can be regarded as the stability safety factor F_S.

3 STUDY ON THE INFLUENCE OF DIFFERENT SLOPES AND THE FILLER WITH DIFFERENT P_5 ON THE STABILITY OF A REINFORCED HIGH-FILL SLOPE

To systematically study the influence of filler gradation and slope on the stability of reinforced high-fill slope under natural working conditions, we take the data shown in Tables 2 and 3 as calculation parameters and strictly control other variables. Taking the reinforced high-fill slope of the No. 2 ditch of Wushan Shennvfeng Airport as the prototype through PLAXIS software, a total of 72 numerical simulations were carried out for 9 fillers with different P_5 and 8 numerical models with different slopes and 2 cross variables. The statistical stability coefficients are shown in Table 4.

Table 4. P_5 statistics of stability coefficient F_S of each model under different fillers and different slopes.

Slope \ P_5	35%	45%	55%	60%	65%	70%	75%	80%	85%
45°	1.496	1.418	1.002	1.398	1.334	1.433	1.495	1.738	1.658
48°	1.313	1.234	0.882	1.099	1.021	1.176	1.221	1.412	1.438
54°	1.185	1.054	0.706	0.802	0.841	0.960	0.982	1.191	1.288
60°	1.095	0.968	0.656	0.776	0.768	0.804	0.856	0.109	1.126
66°	0.937	0.863	0.537	0.697	0.707	0.688	0.716	0.909	1.010
72°	0.929	0.824	0.505	0.554	0.557	0.657	0.670	0.814	0.907
78°	0.75	0.640	0.307	0.496	0.521	0.542	0.551	0.698	0760
84°	0.7	0.590	0.262	0.424	0.488	0.530	0.570	0.616	0.720

Note: the underlined data in the table do not meet the requirements of safety factor for the stability of medium and high fill slopes of civil airport engineering under natural conditions specified in the code for Chinese Geotechnical Design of Civil Airport (MH/ t5027-2013).

To more intuitively display the variation law of stability coefficient with the content of coarse particles of filler and slope, the "Fs-P_5" curve under each slope (as shown in Figure 4), and the "F_S- *Slope*" curve under each P_5 (as shown in Figure 5) are drawn respectively.

Under different slopes, with the increase of filler's P_5, the stability coefficient of reinforced high fill slope shows a nonlinear trend of "first decreasing and then increasing", in which the lowest point is P_5, which is about 55%. Therefore, under natural conditions and other conditions unchanged, the filler with P_5 of about 55% is the least suitable for reinforced fill slope.

It can be seen from Figure 5 that the variation trend of the stability coefficient of the reinforced slope with the slope under different fillers is the same, which is similar to the linear

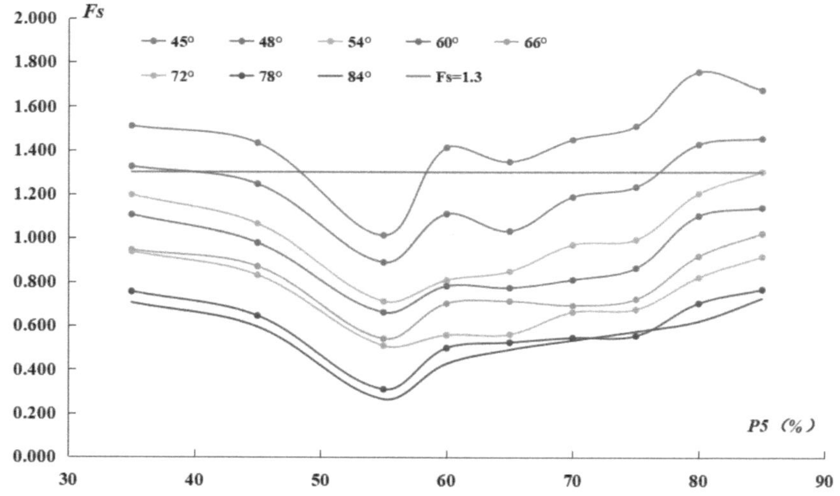

Figure 4. Stability coefficient P_5 curve of a reinforced slope with different slopes.

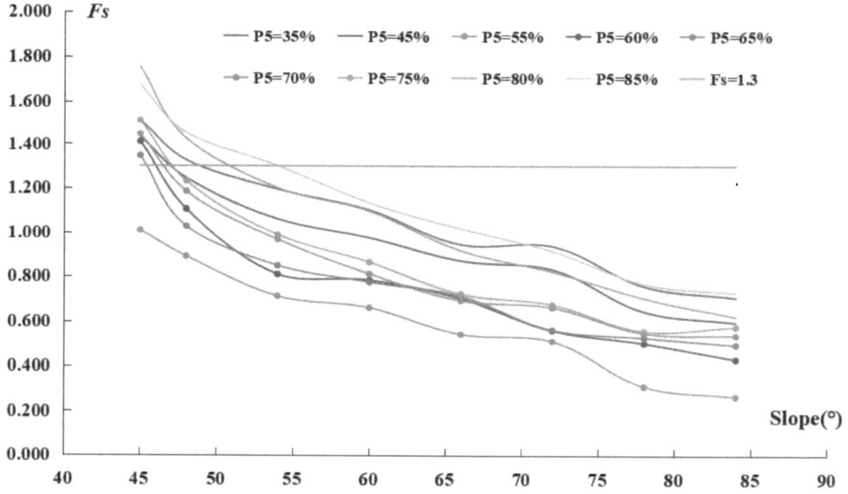

Figure 5. Stability coefficient slope curve of a reinforced slope with different P5 filler.

downward trend. Now the perform linear fitting on the curves is shown in Figure 5, and the fitting results are summarized in Table 5.

It can be seen from Table 5 that the linear fitting degree of the variation curve of the slope stability coefficient of fillers with different P_5 with the slope is very high, and the correlation coefficient generally reaches more than 0.88. At the same time, it can be found that when P_5 is 55%, the slope of the curve linear fitting equation is the smallest, and when P_5 is 80%, the slope of the curve linear fitting equation is the largest. It shows that when P_5 is in the range of 35% ~ 85%, and P_5 is 55%, the sensitivity of the stability coefficient with slope is the lowest. And by the same token, it is inferred that when P_5 is 80%, the sensitivity of the stability coefficient with the slope is the highest.

Table 5. Linear fitting results of slope stability coefficient of fillers with different P5 with slope.

P_5 (%)	Linear fitting equation	Fitting coefficient (R^2)
35	$y = -0.0193x + 2.2816$	0.96
45	$y = -0.0198x + 2.2098$	0.95
55	$y = -0.0182x + 1.7671$	0.97
60	$y = -0.0218x + 2.1705$	0.89
65	$y = -0.0191x + 1.9961$	0.88
70	$y = -0.0216x + 2.2244$	0.90
75	$y = -0.0224x + 2.3077$	0.92
80	$y = -0.0262x + 2.7287$	0.98
85	$y = -0.0232x + 2.5932$	0.97

Note: In the linear fitting equations, y represents the stability coefficient F_S, and x represents the *slope* (°).

At the same time, as shown in Figure 4 and Figure 5, by comparing the data points in the figure with the F_S limit of 1.3 required by the specification, it can be found that for ditch 2 of Wushan Shennvfeng Airport, under natural conditions and 45° slope, the filler with P_5 = 55% should not be used, and the filling slope should not be greater than 48°.

4 CONCLUSION

(1) Under different slopes, with the increase of filler's P_5, the stability coefficient of the reinforced slope shows a nonlinear trend of first decreasing and then increasing. Among them, the lowest point of stability coefficient of reinforced high fill slope corresponds to filler P_5 = 55%. It can be seen that under natural conditions, the filler with P_5 of 55% is the least suitable for reinforced fill slopes.

(2) Under the same other conditions, the slope stability coefficient of fillers with different P_5 decreases approximately linearly with the increase of slope, and when the filler P_5 is in the range of 35% ~ 85%, the reinforced high fill slope with filler P_5 = 55% has the lowest sensitivity to the change of slope. The stability coefficient of reinforced fill slope with filler P_5 = 80% is the most sensitive to the change of slope.

(3) For the No. 2 ditch of Wushan Shennvfeng Airport, under natural conditions and a 45° slope, the filler with P_5 = 55% shall not be used, and the filling slope shall not be greater than 48°.

(4) At present, the coefficient of nonuniformity (C_u) and coefficient of curvature (C_c) are often used as reference indexes in the research on the gradation of high-fill materials, and the research variables are often single. This paper explores the stability law of reinforced high-fill slopes with different P_5 and slope gradients, aiming to explore a new direction to control the stability of the reinforced high-fill slope and provide a new idea of design and construction for the quality control of reinforced high-fill slope in practical projects.

REFERENCES

Emma Burak, Ian C. Dodd, John N. Quinton. Do Root Hairs of Barley and Maize Roots Reinforce Soil Under Shear Stress? [J]. *Geoderma*, 2021, 383.

Hu Jiaju, Xu Ming, Liu Xianshan, Luo Bin. Simplified Bishop Method Analysis Based on the Upper Bound Principle of Limit Analysis [J]. *Journal of Lanzhou University (Natural Science Edition)*, 2021,57 (05): 622–626. DOI: 10.13885/j.issn.0455-2059.2021.055.007

Ma Jie, Han Wenxi, Liu Zhongxuan, Liang Chunxiao. Study on the Influence of P_5 Content on the Compressibility of Coarse-grained Soil [J]. *Renmin Changjiang River*, 2019, 50 (S2): 179–184
Ma Jie, Han Wenxi, Li Baocheng, Liang Chunxiao. Study on the Influence of Coarse Particle Content on the Shear Strength Index of Coarse-grained Soil [J]. *People's Pearl River*, 2019, 40 (12): 25–30
Ma Jie, Han Wenxi, Liu Zhongxuan. Study on the Influence of Coarse Particle Content on the Compactness of Coarse-grained Soil [J]. *Henan Science*, 2019, 37 (07): 1099–1103
Niu Xiaodi, Yang Guangqing, Wang He, Ding Shuo, Feng Fan. Field Test Study on Structural Characteristics of Reinforced Earth Retaining Walls with Different Panel Forms [J/OL]. *Geotechnical Mechanics*, 2021 (01): 1–11 [2020-11-01] https://doi.org/10.16285/j.rsm.2020.0578.
Peng Xianqing, Xia Shuai Shuai. Study on the Influence of the Selection of Geocell on the Stability of Retaining Wall Slope [J/OL]. *Highway*, 2020 (10): 11–17 [2020-11-02] http://kns.cnki.net/kcms/detail/11.1668.U.20201013.1506.006.html.
Yang Bo. Comparison of Slope Stability Analysis by Strength Reduction and Limit Equilibrium Methods [J]. *Western Exploration Engineering*, 2022, 34 (07): 22–25
Yan Han Research and Numerical Analysis on Earth Pressure of Geogrid Reinforced Retaining Wall [D]. *Hubei University of Technology*, 2021. DOI: 10.27131/d.cnki.ghugc.2021.000098
Zuo Jianzhong, Jiang Xin, Fu Youguo, Chen Xinni, Huang Cenyi, Qiu Yanjun. Numerical Simulation of Embankment Reinforcement Based on PLAXIS and OPTUM [J]. *Transportation Science and Technology*, 2022 (01): 5–10;

Water Conservancy and Civil Construction – Oke & Ahmad (Eds)
© 2024 The Author(s), ISBN: 978-1-032-58614-4

Reinforcement scheme of EPB shield tunneling through dense buildings optimal analysis

Hao Li & Bowei Wen*
CCCC Second Harbor Engineering Company Ltd., Wuhan, Hubei Province, China

ABSTRACT: Based on the example of Guangzhou Metro Line 14 EPB shield tunnel construction, this paper brings forward the reinforcement scheme of the EPB shield tunnel under dense buildings in a water-rich sand layer. After comprehensively considering the feasibility of the scheme implementation, technical difficulty, construction period and cost, etc., the technically feasible and economically reasonable MJS pile foundation isolation reinforcement scheme is then preferred in this paper. Practice showed that MJS piles were used to isolate and reinforce dense buildings, which effectively reduced the foundation settlement in the process of double-track shield tunneling.

1 INTRODUCTION

The shield method has become the most commonly used method in urban rail transit construction in the soft soil area of Guangdong, and the EPB shield has become the main construction machinery in shield construction due to its applicability and economic advantages, Guo (2013) and Zhao (2010) (Guo 2013; Zhao & Zhang 2010). However, the supporting pressure of the excavation face is difficult to keep stable and shield tunneling is a process of continuously maintaining dynamic balance, and using EPB shield construction will inevitably cause stratum loss, which would lead to settlement and deformation, Han (2007), Wei (2013), and Xia (2016) (Han et al. 2007; Wei et al. 2013; Xia et al. 2016), especially in areas with dense ground buildings. The excessive settlement will endanger the regular service of surrounding buildings and even lead to engineering accidents, Chen (2010) (Chen et al. 2010). Therefore, it is necessary to study how to adopt auxiliary construction methods for ground reinforcement to reduce the impact of shield construction on the surrounding environment, Wei (2010), Han (2011), Gao (2021), and Huang (2021) (Gao et al. 2021; Han et al. 2011; Huang et al. 2001; Wei 2010).

2 PROJECT OVERVIEW

The left and right lines of shield tunnels between Dengcun Station and Jiangpu Station in Section 1 of Civil Engineering of Guangzhou Rail Transit Line 14 pass through Zhongwei Architectural complex at mileage Y (Z) CK60 + 800 ~ Y (Z) CK61 + 100, with a length range of about 405 m and a buried depth of 10 ~ 18 m. The number of houses directly penetrated underground by shield tunnels is 15, and the number of houses within the influence range of tunnels is 38. The soil layer of the tunnel excavation section is mainly

*Corresponding Authors: wenbowei2200@163.com and wenbowei@cccltd.cn

DOI: 10.1201/9781003450818-48

water-rich gravel stratum, including silty fine sand, medium coarse sand, gravel sand, and strongly weathered sandy conglomerate. The floor height of the underpass building varies from 3 to 7 floors. Generally, strip foundations and brick concrete structures are adopted, and the spacing between houses is 2 ~ 15m. The plane and section of the positional relationship between the tunnel and the housing group named Zhongwei can be seen in Figures 1 and 2.

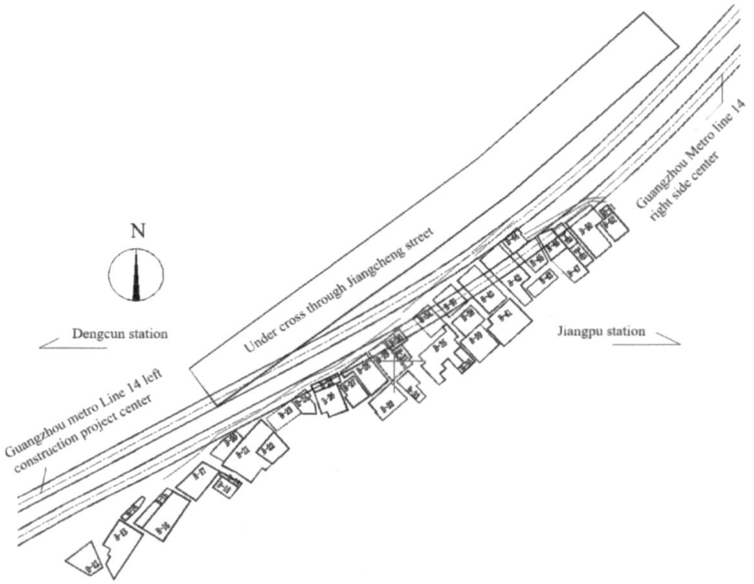

Figure 1. Plan the relationship between the tunnel and Xing Zhong Wei house group.

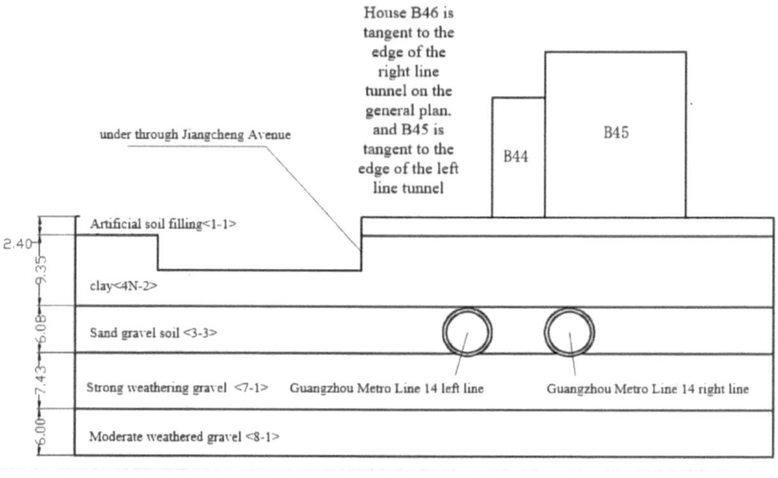

Figure 2. Typical cross-sectional relationship between tunnel and Xing Zhong Wei house group.

3 REINFORCEMENT SCHEME OF SHIELD TUNNEL PASSING THROUGH THE BUILDING

The shield tunnel between Dengcun Station and Jiangpu Station passes through Zhongwei architectural complex for a long distance, and the layer distribution changes little in the crossing range. The shield on the right line is close to the house, which causes great disturbance to the house due to the earth's surface settlement. Therefore, the reinforcement scheme mainly focuses on the foundation of the house and the shield tunnel on the right line. The following five reinforcement schemes are mainly considered in the construction process.

1) Sleeve Valve pile Grouting Reinforcement

Sleeve valve pile grouting is made of a sleeve valve tube, grouting device, shell material, grouting slurry, and grouting pipe composed of a relatively perfect construction method. This construction method has the characteristics of strong adaptability and repeatable lifting and grouting. Sleeve valve tube grouting reinforcement is carried out before shield tunneling to Zhongwei architectural complex, and pre-grouting reinforcement can be carried out on the sand layer at the bottom of 43 houses in Zhongwei House Estate. Two rows of sleeve valve tube grouting holes shall be drilled along the 1.5 m and 3.5 m lines outside the building structure, with a distance of 1.5 m between the holes in the same row. Inclined holes shall be drilled below the projection of the building, and the depth of the hole opening invading the foundation shall not be less than 3 m, to reinforce the bearing layer of the foundation under the building to the greatest extent. Then, before the shield tunnel passes through, cement slurry is injected to change the compression characteristics of the foundation soil and reduce the settlement caused by stratum loss. However, the sleeve valve tube grouting scheme is greatly affected by the ground buildings, and the implementation difficulty of sleeve valve tube construction is greatly increased when there are many pipelines around the buildings and the distance between houses is narrow. The schematic diagram of the grouting reinforcement of the sleeve valve tube can be seen in Figure 3.

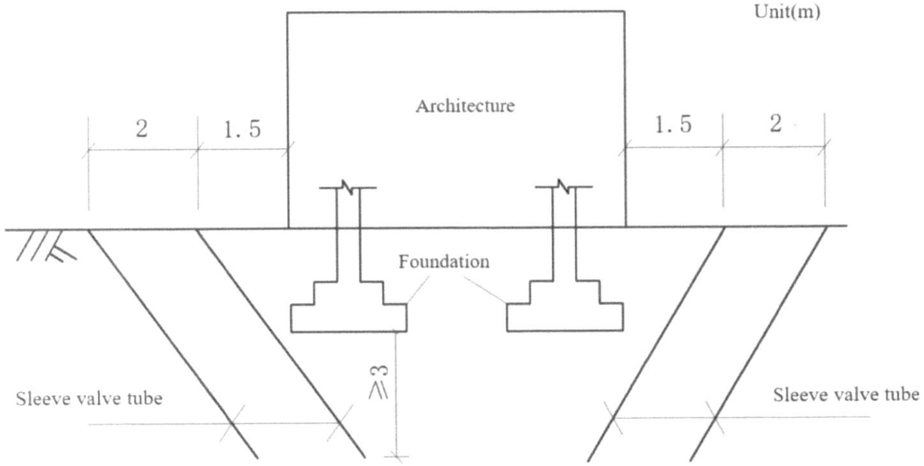

Figure 3. Sleeve valve tube grouting reinforcement section.

2) Horizontal directional drilling with grouting reinforcement technology

Horizontal directional drilling with grouting reinforcement technology includes two engineering skills, which means this technology needs two steps to achieve the purpose of reinforcement. This reinforcement technology is to use directional drilling to drill holes, then place prefabricated φ 108 steel valve pipes, seal the annular space with low-strength rapid-setting materials, and then realize fixed-point and quantitative grouting with hydraulic slurry packer. According to the situation of the project, the control pre-grouting reinforcement is carried out. After the shield machine enters the reinforcement range, supplementary tracking grouting is carried out according to the monitoring data of land subsidence. Horizontal directional drilling with grouting reinforcement can realize long-distance drilling and grouting reinforcement engineering, which has little impact on the structural environment such as buildings.

The implementation method is as follows: the holes are formed in the two ends of the area of the Zhongwei house group, and a row of steel valve pipes are set up parallel to the tunnel above the left and right line to form a pipe shed, with a length of about 360 m (including 120 m inclined section). Steel valve pipe spacing of 3 m, a total of 16 sets. The steel valve pipe and the surrounding sand layer are consolidated by grouting, to reduce the loss of the vault and in the process of shield tunneling and reduce the risk of ground subsidence.

3) Isolation and reinforcement of MJS pile

When the shield side passes through buildings and structures, it can be isolated between buildings and tunnels by rotary grouting piles. At present, the main construction methods are single (multi) heavy pipe rotary grouting reinforcement and derivative high-pressure rotary grouting pile construction methods MJS and RJP, etc. The main purpose of reinforcement is to block or weaken the transmission of stratum deformation, reduce the influence of soil erosion caused by over-excavation in the shield tunnel construction stage on buildings, and play a role in protecting adjacent buildings. Ordinary jet grouting pile has a great influence on houses in the construction process, which often causes surface subsidence or bulge. MJS pile could help control the internal pressure of the foundation and reduce the deformation of the foundation, meanwhile, it could reduce foundation deformation. MJS piles can reinforce the soil in a small space, and the construction site requirements are not strict.

According to the reinforcement range, the diameter of the pile body is designed to be 1 m and 1.3 m. In the scheme that the occlusal parts between adjacent piles are 0.2 m and 0.3m, it is proposed to reinforce the soil at the vault of the tunnel. According to the different relative positions between the tunnel and the residential building, it is divided into five reinforcement zones. The inclination angle of MJS piles in each reinforcement zone varies with the position relationship between the tunnel and the building, ranging from 28 to 66 degrees. The diameter of MJS piles is 1 m and 1.3 m respectively, and the width of occlusion between adjacent piles is 0.2 m and 0.3 m respectively. The reinforcement section of the MJS pile is shown in Figure 4.

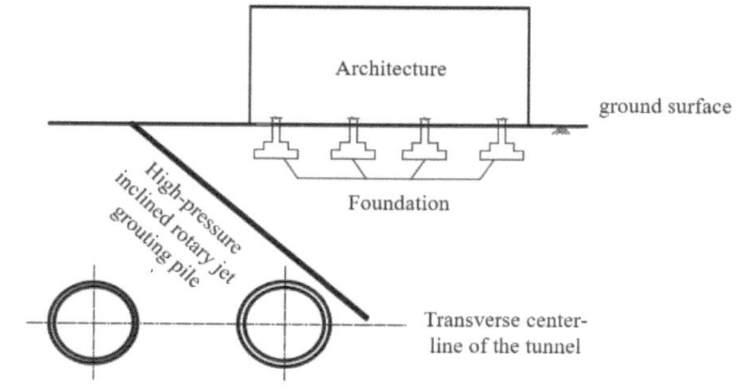

Figure 4. MJS pile reinforcement section in key housing section.

4) Advance reinforcement of large pipe shed

Given the complexity and particularity of the surrounding environment, and considering that the outer side of the left line tunnel is a completed highway underpass tunnel, which has the conditions for large pipe shed construction, the stratum can be reinforced by setting a super-long large pipe shed in advance. A row of large pipe sheds is set up above the left and right lines of shield tunnels in the built highway tunnel of Conghua Avenue, with a setting range of about 140 m. The top of the tunnel is covered with soil about 3 ~ 5 m, and a large pipe shed with a diameter of 108 mm is adopted, with a distance of 300 mm and a single length of 20 m. The length of the pipe shed is 3 m longer than the sideline of the house within the influence range of the right tunnel, and the total length of the single pipe is about 50 ~ 60 m. The construction stratum of the pipe shed is mainly a silty clay layer. The pipe joint needs to be drilled to form a slurry hole, and the pipe shed and the surrounding rock and soil are consolidated into a whole by grouting in the pipe shed to increase the strength of the surrounding rock. In an ideal state, a thin shell and solid can be formed to improve the bearing capacity of the foundation. Due to the limited length of the highway tunnel, the construction of a large pipe shed cannot be carried out under the remaining 100 m of houses, and other auxiliary construction methods are needed for reinforcement. The reinforcement section of the large pipe shed is shown in Figure 5.

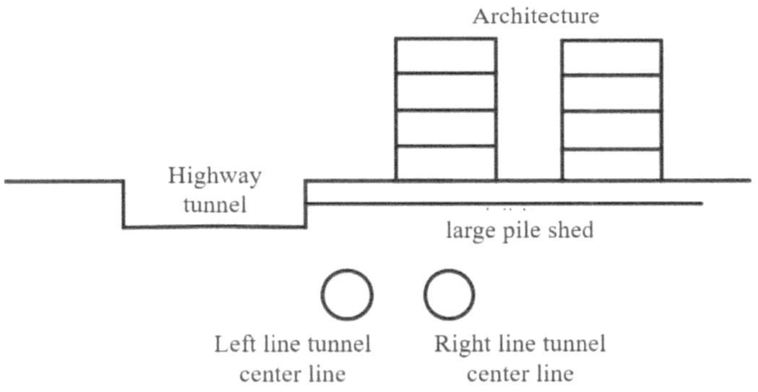

Figure 5. Cross-section plan of pipe shed reinforcement.

5) Grouting reinforcement inside the tunnel

When the ground buildings are dense and cannot meet the construction conditions of grouting reinforcement, underground grouting reinforcement can also be adopted. By grouting in the left tunnel in advance to reinforce the soil layer at the top of the right tunnel, the right tunnel passes can be reduced. Cement slurry or cement-water glass double-liquid slurry is used for reinforcement grouting in the tunnel. Under the condition of not changing stratum composition, water between soil particles is squeezed out during grouting, so that the gaps between particles are filled with slurry and consolidated, thus achieving the purpose of improving soil properties. The specific implementation method is as follows: grouting reinforcement is carried out within the length range of 6 m above the center of the tunnel to be excavated, 5 m on both sides of the tunnel center, and 240m in the longitudinal direction. The bottom spacing of grouting holes is 0.9 ~ 1.5 m and 6 grouting holes are designed for each ring segment in the tunnel. A total of 954 grouting holes are arranged in the 240 m reinforcement length of the tunnel. The grouting reinforcement section can be seen in Figure 6.

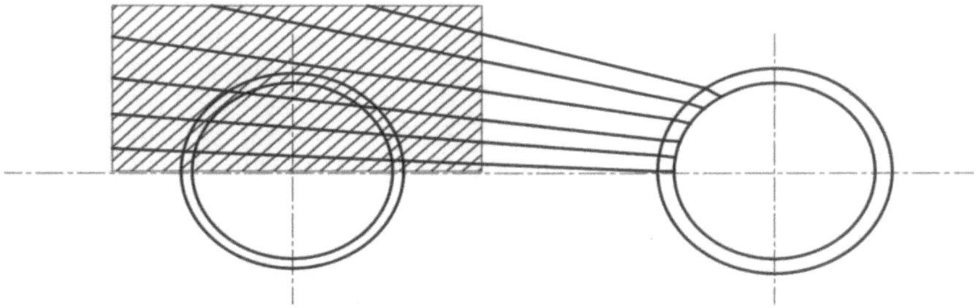

Figure 6. Schematic diagram of grouting reinforcement in the tunnel.

4 COMPARISON AND COMPREHENSIVE ANALYSIS OF 3 REINFORCEMENT SCHEMES

According to the situation of the site, considering the reliability, practicability, and maturity of each reinforcement scheme, taking into account the economy and rationality, it is also necessary to consider the time and cost saved by shortening the construction period, and draw the following conclusions:

1) All kinds of auxiliary construction methods are affected by site restrictions and environmental restrictions to varying degrees in the implementation process, especially the conventional sleeve valve tube grouting construction, which requires large-scale temporary relocation of personnel and has the lowest practicability. Horizontal directional drilling with grouting reinforcement technology has the least impact on the environment and the highest practicability, but its application in municipal reinforcement projects is less, and the reinforcement effect remains to be verified.
2) Among all kinds of auxiliary construction methods, the large pipe shed has the best reinforcement effect, and has been successfully applied to municipal projects with poor geological conditions and high requirements for land subsidence control, such as shallow burial and underground excavation, and has mature construction experience. However, it needs to carry out working well construction before construction. Although the existing highway tunnel structure can be used for working well construction in this project, the cost is high.
3) The grouting scheme in the tunnel can be carried out in the tunnel, which has little impact on the ground environment. However, it must be in an existing tunnel as the reinforcement construction area, which has a great impact on the shield construction organization. At present, the grouting reinforcement in the tunnel is mainly aimed at the construction and reinforcement of the connecting passage, and there is no experience for reference in large-scale use, so the impact on the regional segment needs further analysis and research.
4) MJS high-pressure jet grouting pile method can effectively control the pressure through the active slurry discharge monitoring system and has little influence on the ground houses. At the same time, high-pressure jet grouting pile, pile radius, and pile strength can be controlled, and the reinforcement effect is second only to the pipe shed method. The construction equipment of the MJS method has been made in China, and the construction cost is slightly higher than that of sleeve valve tube grouting technology but far lower than that of the pipe shed method, and the economy is better.

Table 1. Comparison of different reinforcement schemes.

Compare content	Grouting reinforcement of sleeve valve tube	Horizontal directional drilling with grouting reinforcement	Isolation and reinforcement of MJS pile	Advanced large pipe shed reinforcement	Grouting reinforcement in the tunnel
Stratum reinforcement effect	General	Better	Better	Better	General
Applicable stratum	Silty clay, silty mudstone, broken rock mass	Sand, clay, pebble soil	Silty fine sand layer, clay layer, and gravel sand layer	Gravel sand layer, clay layer, and weak rock layer with low strength	Round gravel layer, full-section gravel sand layer
Risk prevention	Improper pressure control in the pre-grouting process is easy to cause house cracking	The diffusion radius of slurry is difficult to control	The grout injection pressure, pile forming position, and angle can be controlled	It is easy to cause damage and leakage of segments in tunnels	It is difficult to control the injection volume and diffusion radius of slurry
Construction accuracy	Higher	Higher	Higher	Poor	Difficult to control
Difficulty of implementation	Less difficult	The difficulty is average	The difficulty is average	Difficult	Difficult
Reliability	Reliable	Reliable	Reliable	Very reliable	Reliable
Scope of application	Extensive	Less	More	Extensive	Less
Construction Equipment	Relatively simple	Relatively simple	More complex	Simple	Complex
Construction period	3 months	3 months	6 months	8 months	3 months
Comprehensive unit price	22,000/tunnel linear meter	20,000/tunnel linear meter	29,000/tunnel linear meter	54,000/tunnel linear meter	19,000/tunnel linear meter

To sum up, under the condition of similar complex geological conditions and dense ground houses without relocation, the isolation, and reinforcement scheme of MJS pile for houses is the most favorable under comprehensive conditions such as reinforcement effect, economy, and environmental impact.

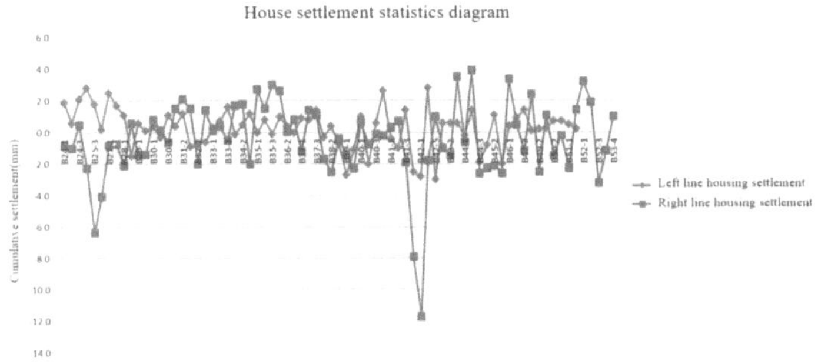

Figure 7. Housing settlement statistics.

5 REINFORCEMENT EFFECT OF 4 MJS PILE

After MJS pile reinforcement, the shield successfully passes through the dense group of houses, and the statistical values of house settlement during shield tunneling are shown in Figure 7. In the process of shield tunneling, the settlement curve of the house fluctuates greatly after the right tunnel passes under, which has a great influence on the house. Local settlement of individual houses is large, and the maximum accumulated settlement value is 12mm, which is mainly due to the continuous setting of occlusal piles during the construction process, and the construction cannot be carried out according to the established construction sequence of "one jump and two jumps" due to the influence of the site. Comprehensive analysis shows that in the double-line shield passing project, the settlement of the house is controlled within 4 mm, the differential settlement is controlled within 1 mm, and the inclination of the house meets the specification requirements. This shows that the use of high-pressure micro-disturbance jet grouting pile (MJS) occlusion reinforcement has a good protective effect on the building.

6 CONCLUSIONS

In this paper, five stratum reinforcement schemes are proposed based on the shield tunneling of Guangzhou Metro Line 14, and the optimization analysis of the five schemes is carried out. The main conclusions are as follows:

1) All five reinforcement schemes have certain reinforcement effects. According to the comprehensive comparison of various auxiliary reinforcement methods for this project, the construction period and economy are unreasonable.
2) Using MJS inclined pile reinforcement effect can effectively eliminate the double-track tunnel construction caused by the superposition effect of ground subsidence, and effectively reduce the settlement of the foundation. The practice has proved that the maximum uplift and settlement of the building foundation strengthened by the MJS pile are controlled within 4mm.

REFERENCES

Chen Jianfeng, Wang Xin, Shi Zhenming. Analysis of Construction Interaction of Highway Shield Tunnels with Short-distance and Large Diameter[J]. *Rock and Soil Mechanics*, 2010; 31(S2): 242–253.

Gao Wenshan, Wang Lichuan, Zhang Huijian, Zhang Lela, Zhao Wei, Hou Feng. Analysis of the Contribution of Face Support and Primary Support to the Surface Settlement Control of Shallow Tunnel [J]. *Journal of Railway Science and Engineering* 2021, 13(3): 720–727.

Guo Yuhai. Study on Ground Surface Movement Induced by Large-diameter Earth Pressure Balance Shield Tunneling [J]. *China Civil Engineering Journal*, 2013(11): 128–137.

Han Xuan, Li Ning, Standing J R. An Adaptability Study of Gaussian Equation Applied to Predicting Ground Settlements Induced by Tunneling in China[J]. *Rock and Soil Mechanics*, 2007, 28(1): 23–28, 35.

Han Changrui, He Guangzong, Wang Guibin. Analysis of Surface Settlement Induced by Some Factors in Parallel Dual-tunnel Construction[J]. *Rock and Soil Mechanics*, 2011, 32 (S2): 484–487.

Huang Changfu, Long Wen, Song Qilong, Li Dong, Su Dong, Chen Xiangsheng. Influence of Reinforcement Range on Ground Settlement of Shallow-buried Super-large-diameter Shield Tunneling in Soft Soils[J]. *Chinese Journal of Geotechnical Engineering*, 2021, 43 (Supp.2): 76–79.

Wei Xingjiang, Zhou Yang, Wei Gang. Research of EPB Shield Tunneling Parameter Relations and Their Influence on Stratum Displacement[J]. *Rock and Soil Mechanics*, 2013, 34(1): 73–79.

Wei Gang. Selection and Distribution of Ground Loss Ratio Induced by Shield Tunnel Construction[J]. *Chinese Journal of Geotechnical Engineering*, 2010, 32(9): 1354–1361.

Xia Guanghui, Zhu Fengbin, Zhang Haiya. Study on the Influence of Dynamic Shield Construction on the Enclosure Structure of the Adjacent Underground Continuous Wall. *Science Technology and Engineering*, 2016, 16(23): 70–74.

Zhao Yubo, Zhang Zhongmiao. The Additional Stress of Surrounding Soil is Caused by Propelling of Shield Tunneling [J]. *Chinese Journal of Geotechnical Engineering*, 2010, 32(9): 1386–1391.

Water Conservancy and Civil Construction – Oke & Ahmad (Eds)
© 2024 The Author(s), ISBN: 978-1-032-58614-4

Experimental study on uniaxial dynamic performance of concrete core samples of Xiluodu dam

Lijun Zhao* & Haibo Wang
State Key Laboratory of Simulation and Regulation of Water Cycle in River Basin, China Institute of Water Resources and Hydropower Research (IWHR), Beijing, China
Earthquake Engineering Research Center, China Institute of Water Resources and Hydropower Research (IWHR), Beijing, China

Hailong Huang
China Three Gorges Construction Engineering Corporation, Chengdu, China

ABSTRACT: The dynamic performance of dam concrete is a critical factor in the seismic design and safety verification of dams. To fully understand the dynamic characteristics of the dam concrete, a 15MN large MTS material testing machine was utilized to carry out the uniaxial compression test, uniaxial compression elastic modulus test, and splitting tensile test of a core specimen of concrete under different strain rates. The test results show that the uniaxial compressive strength and splitting tensile strength of the core specimen of concrete is significantly affected by the rate of strain. The energy absorption capacity of concrete increased with the rise of strain rate and the two showed a strong correlation. The compressive strength and splitting tensile strength of concrete core samples are different from those of full-graded concrete large specimens and wet-screened small specimens. The reasons are analyzed, including the influence of the growth of the compressive strength of dam concrete in the late stage, the differences between the site mixing, pouring, and curing conditions of dam concrete, and the molding curing conditions of a laboratory.

1 INTRODUCTION

The dam of Xiluodu Hydropower Station on Jinsha River is a concrete double-curvature arch dam with a maximum height of 285.5 m. According to the Code for seismic design of hydraulic structures of hydropower project (NB 35047-2015), the seismic fortification classification is Class A. After the Wenchuan earthquake, the earthquake department determined that the horizontal peak acceleration of the design bedrock corresponding to the exceedance probability of 0.02 in the 100-year base period was 0.357g. The current standard for seismic design of hydraulic structures (GB 51247-2018) stipulates that the dynamic performance of concrete should be determined based on specialized material tests for concrete dams of Class A.

In 1917, Abrams (Abrams D.A. 1917) proposed the property of the rate sensitivity of concrete compressive strength. Since then, many scholars at home and abroad have studied the relationship between peak strength and the loading rate of concrete. The results show

*Corresponding Author: zhaolj@iwhr.com

DOI: 10.1201/9781003450818-49

that the mechanical properties of concrete have obvious side effects and rate effects (Caverzan A. 2016; Dong Yan 2006; Gang Peng 2015; Qu Fujin 1997; Malvar L.J. 1998; Solomos G. 2015). In this paper, the dynamic performance of Xiluodu dam concrete core samples under different loading rates is studied, and the test results are analyzed and compared with the full-graded large specimens and wet-screened small specimens. The relationship and difference between the concrete strength indexes of different specimen sizes and different test types are comprehensively analyzed, which provides a scientific basis for further research on the seismic safety of Xiluodu dam during operation.

2 TEST SCHEME

The test equipment is a 15MN large-scale material testing machine built by the China Institute of Water Resources and Hydropower Research (Figure 1). The specimen is $C_{180}40$ concrete core sample drilled into Xiluodu dam, and the diameter of the core sample is 220mm. According to the test code for hydraulic concrete (SL/T 352-2020), the length-diameter ratio of the compressive strength specimen and splitting tensile strength specimen is 1.0, and the length-diameter ratio of the dynamic axial compressive elastic modulus test specimen is 2.0.

Figure 1. The 15MN large-scale material testing machine.

The core concrete test of Xiluodu dam includes static and dynamic compressive strength tests, axial compressive elastic modulus tests, and splitting tensile strength tests. The strain rate of tensile failure of concrete is about 5×10^{-4}/s, and the strain rate of compressive failure is about 5×10^{-3}/s. To study the variation of dynamic characteristics of concrete with loading strain rate, quasi-static (10^{-6}/s) and three different dynamic monotonic loading strain rates were selected for the three types of tests. The loading conditions for each group of compressive and splitting tensile tests are shown in Table 1, with the loading strain rate or cyclic loading frequency in parentheses.

3 TEST RESULTS

3.1 *Results and analysis of compression test*

The quasi-static compression makes the specimen produce more vertical small cracks. The concrete on the surface of the specimen is partially peeled off. The specimen maintains the initial shape after the test is stopped, and no burst or large block fragmentation occurs. With

Table 1. The loading conditions for each group.

No.	Group	Loading Conditions
1	Compressive strength test (Specimen C)	Quasi-static (10^{-6}/s)
2		Dynamic strain rate (5×10^{-4}/s)
3		Dynamic strain rate (5×10^{-3}/s)
4		Dynamic strain rate (3×10^{-2}/s)
5	Axial compressive elastic modulus test (Specimen D)	Quasi-static (10^{-6}/s)
6		Dynamic strain rate (5×10^{-4}/s)
7		Dynamic strain rate (5×10^{-3}/s)
8		Dynamic strain rate (3×10^{-2}/s)
9	Splitting tensile strength test (Specimen B)	Quasi-static (10^{-6}/s)
10		Dynamic strain rate (10^{-4}/s)
11		Dynamic strain rate (5×10^{-4}/s)
12		Dynamic strain rate (2×10^{-3}/s)

the increase of loading rate, the degree of fragmentation of the specimen increased (Figure 2). The compressive strength of the concrete core sample increases with the increase of loading rate (Figure 3). The maximum dynamic increase factor of compressive strength is 1.26, and the dynamic increase factor of strength corresponding to the strain rate of the fundamental frequency compression failure of the dam is 1.18, which is slightly smaller than 1.20 recommended by the seismic design code of hydraulic structures.

(a) Quasi-static (10-6/s) (b) Dynamic strain rate (5×10-4/s) (c) Dynamic strain rate (5×10-3/s) (d) Dynamic strain rate (3×10-2/s)

Figure 2. Compressive failure modes of concrete core under different loading rates.

Different length-to-diameter ratios lead to different effects of end friction constraints on test results. The average compressive strength of the quasi-static concrete core specimen D (the length-diameter ratio is 2.0) is 48.5 MPa, which is about 74.9 % of the quasi-static compressive strength of specimen C. And the average compressive strength of 48.5 MPa is lower than the reference value given by the test code for hydraulic concrete. The average dynamic compressive strength values of the second group (5×10^{-4}/s), the third group (5×10^{-3}/s), and the fourth group (3×10^{-2}/s) were 59.28 MPa, 67.15 MPa, and 79.18 MPa, respectively. The compressive strength of the concrete core sample increases with the increase in loading rate, and the maximum dynamic increase factor is 1.63 (Figure 3).

The results of four groups of axial compressive elastic modulus tests show that the average quasi-static compressive elastic modulus of concrete core samples is 46.2 GPa. And all the dynamic compressive elastic moduli are greater than the quasi-static compressive elastic modulus. However, unlike the dynamic compressive strength, it does not show a monotonic increase with the loading rate (Figure 3). The compressive elastic modulus increases by 5.9 %

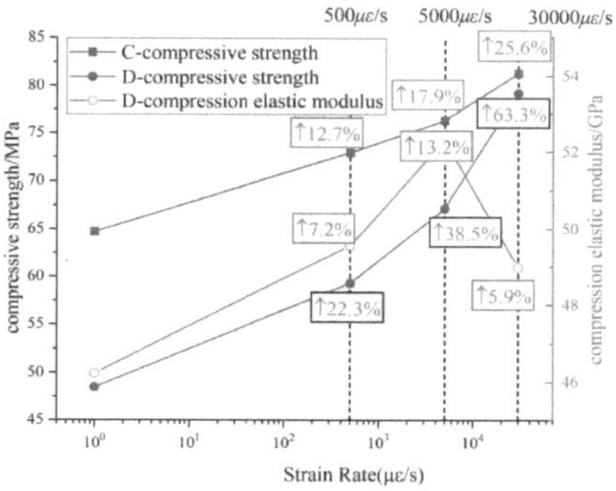

Figure 3. Variation trend of compressive strength of concrete core under different loading conditions.

~ 13.2 %. It should be pointed out that the quasi-static compressive modulus obtained in this test is a transient modulus that does not reflect the influence of creep under long-term load. The standard value of static elastic modulus in seismic design code includes the influence of long-term creep, so the dynamic modulus is 1.5 times the static standard value. According to the code for the design of concrete structures (GB50010-2010), the static elastic modulus of C40 concrete is 32.5GPa, and the dynamic modulus used in the seismic design is 48.75GPa, which is only about 5% higher than that measured by Xiluodu core sample.

Compared with the test results of fully-graded concrete and wet-screened concrete of Xiluodu dam laboratory specimens in 2014 (Yanhong Zhang 2014), the quasi-static compressive strength of fully-graded cubic specimens is 50.34 MPa, and the dynamic strength improvement factor corresponding to the fundamental frequency loading rate of the dam is 1.19. The quasi-static compressive *strength* of the wet-screened cube specimen is 51.00 MPa, and the average value of the dynamic improvement factor of the dam base frequency loading rate strength is 1.19. The dynamic compressive strength improvement factor of the concrete core sample specimen (the length-diameter ratio is 1.0) is 1.18, and the difference is not significant.

After conversion, the quasi-static compressive strength of the cylinder specimen corresponding to the wet sieve cube specimen (Cube with a side length of 150 mm) is about 43.22 MPa, which is only 66.77 % of the average quasi-static compressive strength of the core concrete (64.73 MPa). The reason for the difference in compressive strength between the two specimens may contain two aspects. One is the influence of the growth of the compressive strength of the dam concrete in the later period, and the other one is the difference between the conditions of mixing, pouring, and curing of the dam concrete on site and the conditions of molding and curing in the laboratory.

3.2 *Results and analysis of splitting test*

Quasi-static splitting produces macroscopic cracks at the edge of the spacer and extends until the specimen loses its bearing capacity. The section of the splitting surface of the core specimen is not smooth and the proportion of interface damage is not high. With the increase in loading rate, the range of damage around the splitting surface of the specimen increases under high-speed impact load, and some specimens have multiple failure surfaces (Figure 4). When the loading rate increases, the specimens with a larger proportion of interface failure usually have lower splitting tensile strength.

The average quasi-static splitting tensile strength of the specimens was 4.05 MPa, which was 6.26 % of the quasi-static compressive strength (64.73 MPa). The average dynamic splitting tensile strength values of the second group (10^{-4}/s), the third group (5×10^{-4}/s), and the fourth group (3×10^{-3}/s) were 4.41 MPa, 5.33 MPa, and 5.36 MPa, respectively.

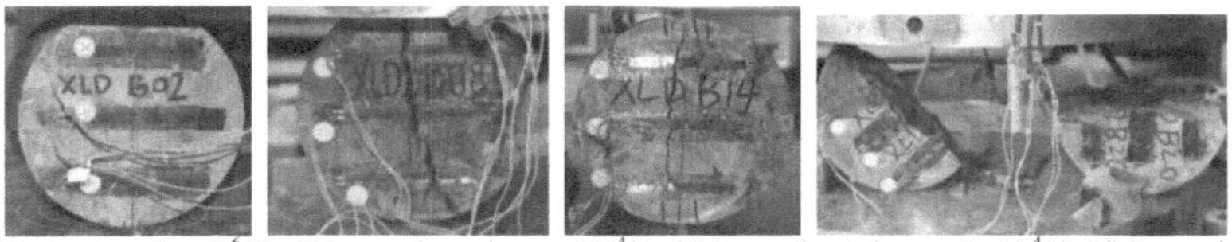

(a) Quasi-static (10^{-6}/s); (b) Dynamic strain rate (10^{-4}/s); (c) Dynamic strain rate (5×10^{-4}/s); (d) Dynamic strain rate (2×10^{-3}/s)

Figure 4. Failure modes of split tensile specimens under different strain rates.

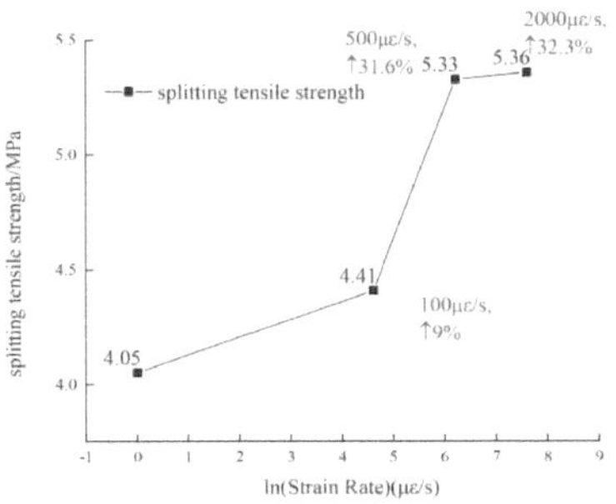

Figure 5. Variation trend of splitting tensile strength of concrete core under different loading conditions.

The variation of splitting tensile strength of the concrete core sample with loading rate is shown in Figure 5. The dynamic splitting tensile strength increases with the increase of loading rate, and the maximum strength increase factor is 1.32. The dynamic strength increase factor corresponding to the strain rate of fundamental frequency splitting tensile failure is about 1.32.

Comparing the test results of fully-graded concrete and corresponding wet-screened concrete of Xiluodu dam laboratory formed and maintained specimens in 2014 (Yanhong Zhang 2014), the quasi-static splitting tensile strengths of fully-graded and wet-screened cubic specimens are 2.33 MPa and 3.42 MPa respectively. And the corresponding dynamic increase factors of dam fundamental frequency loading rate strength are 1.39 and 1.14 respectively. The quasi-static splitting tensile strength of the concrete core sample is 4.05 MPa, which is significantly higher than the previous test results. It is not only the influence of the growth of concrete strength in the later period but also the difference between the conditions of mixing, pouring, and curing of dam concrete on site and the forming and curing conditions of a laboratory. The dynamic strength improvement factor corresponds to the splitting tensile failure strain rate of the fundamental frequency of the dam, the test results of the fully-graded cube are consistent with those of the dam core concrete specimens, and the difference is about 5%.

4 CONCLUSIONS

In this paper, the compressive, axial compressive, and splitting tensile tests of Xiluodu concrete core samples were carried out and compared with the test results of fully-graded concrete specimens and wet-screened concrete specimens prepared in the laboratory. The research results reveal the static tensile and compressive characteristics of the Xiluodu dam and the influence of initial continuous static load on dynamic tensile and compressive properties. The main conclusions are as follows.

The static axial compressive strength of the concrete core sample of the Xiluodu dam is 64.73 MPa. The dynamic axial compressive strength is significantly affected by the loading rate, and the maximum strength dynamic increase factor is 1.26. After conversion, the corresponding cylinder static compressive strength of the wet sieve cube concrete specimen is only 66.77 % of the static compressive strength of the concrete core. The difference in compressive strength between the two specimens is not only influenced by the growth of compressive strength of dam concrete in the later stage but also the difference between the conditions of mixing, pouring, and curing of dam concrete on site and the forming and curing conditions of a laboratory. The static splitting tensile strength of the dam concrete core sample is 4.05 MPa, which is significantly higher than the static splitting tensile strength of the fully graded large specimen and the wet sieve small specimen prepared in the laboratory. The dynamic splitting tensile strength of the dam concrete core sample is significantly affected by the loading rate, and the maximum dynamic increase factor is 1.32. Corresponding to the strong dynamic increase factor of fundamental frequency splitting tensile failure strain rate of the dam, the test results of the fully-graded cube are consistent with those of dam core concrete specimens.

ACKNOWLEDGMENT

Project funding: China Three Gorges Group Co., Ltd. Research funding (Contract No. XLD / 2170).

REFERENCES

Abrams D.A. Effect of Rate of Application of Load on the Compressive Strength of Concrete[A]. *Proc. 20th Annu. Meeting*, ASTM[C]. West Conshohoeken, Pa., 1917:366–374.

Caverzan A. Compressive Behavior of Dam Concrete at Higher Strain Rates[J]. *The European Physical Journal Special Topics*, 2016, 225(2): 283–293.

Dongming Yan. *Experimental and Theoretical Study on the Dynamic Properties of Concrete [D]*. The Dalian University of Technology. 2006.

Gang Peng. Study on Dynamic Compressive Properties of Concrete Under Pore Water Pressure Environment [J]. *China Civil Engineering Journal*, 2015, 48(1):11–18.

Fujin Qu. Effect of Strain Rate on the Dynamic Behavior of High-performance Fiber Reinforced Concrete[J]. *Construction Technology*, 1997(05):5–7.

GB 50010-2010 *Code for Design of Concrete Structures* [S]. 2010.

GB 51247-2018 *Standard for Seismic Design of Hydraulic Structures*[S]. 2018.

Malvar L.J. and Crawford J. *Dynamic Increase Factors for Concrete*, Twenty-Eighth DDESB Seminar, Orlando, FL, August 1998

NB 35047-2015 *Code for Seismic Design of Hydraulic Structures of Hydropower Project:* [S]. 2015.

SL/T 352-2020 Test Code for Hydraulic Concrete: [S]. 2020.

Solomos G, BERRA M. Compressive Behavior of Recycled Aggregate Concrete under Impact Loading[J]. *Cement and Concrete Research*, 2015(71):46–55.

Weihua Hu. Energy Absorption Characteristics and Size Effect of Concrete under Different Strain Rates[J]. *China Academic Journal Electronic Publishing House*. 2015,32(05):132–136.

Yanhong Zhang. Statistical Analysis of Concrete Strength of Xiluodu Dam[C]. *The Technological Progress of Construction and Operation Management of High Dam-Academic Annual Meeting of China Dam Association*, 2014:138–146.

Water Conservancy and Civil Construction – Oke & Ahmad (Eds)
© 2024 The Author(s), ISBN: 978-1-032-58614-4

Performance monitoring of new lightweight high-strength concrete

Qi Song
Hainan CCCC Expressway Investment and Construction Co., Ltd., China

Chao Pan & Huijie Jia
CCCC Fourth Highway Engineering Co., Ltd., China

Shuirong Lin
Hainan CCCC Expressway Investment and Construction Co., Ltd., China

Lijun Xiao
Guangdong Guanyue Highway & Bridge Co., Ltd., China

Chenxu Li & Shanshan Zhang*
School of Architecture and Civil Engineering, Wuhan University of Technology, China

ABSTRACT: Lightweight high-strength concrete combines the characteristics of lightweight and high-strength concrete and it is more and more popular because of its increased strength and reduced self-weight, easy handling, and economic benefits. In this study, a new lightweight high-strength concrete material suitable for bridge deck pavement is prepared, and the strain state inside the concrete is measured by using the fiber Bragg grating (FBG) sensor. The results show that the transverse shrinkage is more obvious than the longitudinal shrinkage in the same testing period and scale. Compared with ordinary concrete, lightweight high-strength concrete has more internal voids, which reduces the self-weight of concrete and also shows good mechanical properties.

1 INTRODUCTION

Compared with conventional concrete (Payam et al. 2011; Tanyildizi & Coskun 2008; Tanyildizi 2013), due to the relatively high volume of voids in LWC, it provides lower density, excellent thermal insulation, improved fire resistance, and reduced production cost, and can provide lower deadweight and higher efficiency (Qi & Fourie 2019; Shafigh et al. 2012; Wu et al., 2015). Therefore, it has been used for structural and non-structural elements, such as long-span bridges, high-rise buildings, and floating and offshore structures (Domagała 2020; Meyer & Kahn 2002; Sengul et al. 2011).

With the continuous application of continuous rigid frame long-span concrete bridges, it has been found that this bridge type has common diseases, such as web cracks, transverse cracks in the top slab of the bridge deck, main beam transition down disturbance, etc. These diseases can be attributed to the material problems such as the large dead weight of the concrete used, large shrinkage, and creep of high-performance concrete. The weight of concrete is one of the most important parameters for manufacturing economic structures. According to the design principles of high strength, lightweight, high durability, and strong

*Corresponding Author: zhang_shanshan@whut.edu.cn

DOI: 10.1201/9781003450818-50

crack resistance, the mix proportion of lightweight high-strength concrete is optimized. The workability, homogeneity, durability, and crack resistance of lightweight high-strength concrete for bridge deck pavement have been improved. A new lightweight high-strength concrete material suitable for bridge deck pavement is prepared, and the strain state inside the concrete is measured by using the FBG grating (Qin et al., 2022; Xu et al., 2013, 2014, 2017, 2018, 2022; Wei *et al.* 2018) with the supporting fiber grating demodulator, giving the development trend of the internal strain of the concrete under natural conditions.

2 MATERIALS AND METHODS

2.1 *Materials and mix proportion*

The raw materials of the new lightweight high-strength concrete include cement, sand, ceramsite, fly ash, silica fume, water reducer, and mixed water. According to Table 5.2.1 in a technical standard for the application of lightweight aggregate concrete (JGJ/T 12-2019), issued by the Ministry of Construction, brand 42.5 ordinary Portland cement with a true density of 3100 kg/m^3 is required for preparing lightweight high-strength concrete. Fine aggregate sand is ordinary medium sand with an apparent density of 2580 kg/m^3, the particle size is 0.25-0.5 mm, and its moisture content needs to be measured before mixing. 800-grade shale broken ceramsite is used to improve the overall strength of the test block, with a particle size range of 5-20mm and bulk density of 783 kg/m^3, the cylinder pressure strength is 6.8 MPa, the water absorption rate is 3% in 1h, pre-wet for 1h before mixing, and drain the surface water. Grade I fly ash is used as fly ash, and grey micro silica fume (96% content) is used as silica fume. KXSP (KXPCA) polycarboxylic acid series high-performance water-reducing agent produced by Guizhou Kaixiang New Materials Co., Ltd. is used, with a dosage of 1.5% of the cementitious material. The designed lightweight high-strength concrete mix is shown in Table 1. The slump of the mix proportion test block is 245mm, and the average strength in seven days is 47.24 MPa.

Table 1. Lightweight and high-strength concrete mix.

Cement (kg/m^3)	Silica fume (kg/m^3)	Fly ash (kg/m^3)	Shale ceramsite (kg/m^3)	Spall (kg/m^3)	Medium sand (kg/m^3)	Net water consumption (kg/m^3)	Water reducer (kg/m^3)
408	25.5	76.5	258.6	507.7	729	163.2	7.5

2.2 *Data acquisition equipment and materials*

The data acquisition system is arranged in the monitoring center. The monitoring center is temporarily set in the temporary room near the project construction. The acquisition frequency of optical fiber monitoring is 1HZ. The data acquisition system will record, store, and display the test data in real-time and continuously within a working period of about one month. As shown in Figure 1, the data acquisition material is mainly composed of a fiber grating demodulator and a BOTDA strain sensor. The BOTDA strain sensor collects the real-time strain information generated during the use of the bridge and then transmits it to the fiber grating demodulator. FBG belongs to short-period fiber grating. When a section of broadband light waves incident into the fiber grating, it is divided into two parts: reflected light and transmitted light. If the wavelength of the beam meets the Bragg condition, it will be reflected at the grating, and the light of other wavelengths will be transmitted. The reflected light can be superimposed into a reflection peak. It can be considered that fiber

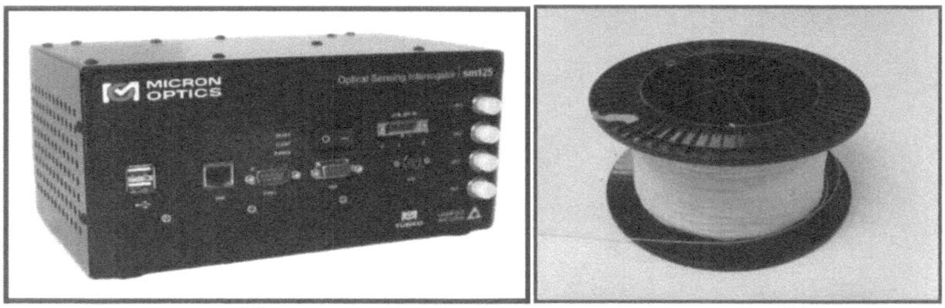

(a) (b)

Figure 1. Data acquisition materials: (a) FBG demodulator; (b) BOTDA strain sensor.

grating plays a role in filtering and screening light waves. The central peak value of the wavelength of this specific light wave is the central wavelength of the fiber Bragg grating. The wavelength expression of the grating reflection is:

$$\lambda_B = 2n_{eff}\Lambda \tag{1}$$

where, λ_B is the central wavelength of fiber grating, n_{eff} is the effective refractive index of the fiber, and Λ is the period of the grating.

2.3 *Optical fiber sensor layout*

In this study, the sensor is arranged on the right side of the K61706 Canjunxi Bridge. Figure 2 mainly shows the engineering layout of the strain sensor. The strain sensor is arranged at the joint of each part of the box girder and placed in the middle of the pavement cross-section. According to the actual situation of the project, the sensor is placed in the box girder, the strain sensor is placed at 3/4 and 1/2, the sensor is fixed, and then concrete is poured.

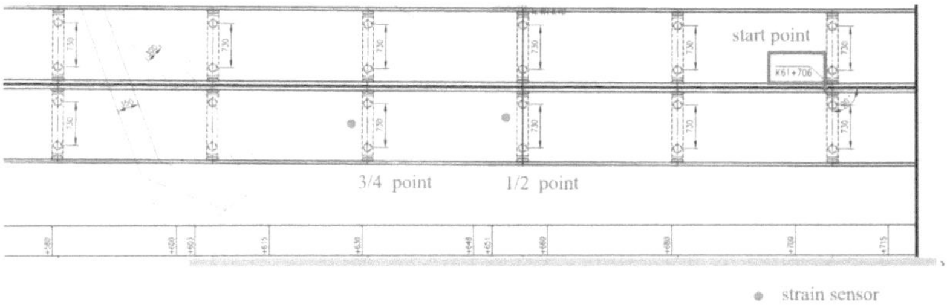

Figure 2. The layout of bridge body sensors.

3 RESULTS

3.1 *Strain at 1/2 of the bridge deck*

Figure 3 shows the monitored strain change of the concrete pouring near the completion of pouring at 1/2 of ordinary concrete and lightweight concrete. It can be seen from the figure that the microstrain of lightweight concrete and ordinary concrete monitored by different sensors is not the same. This is due to the different positions of sensors, and the

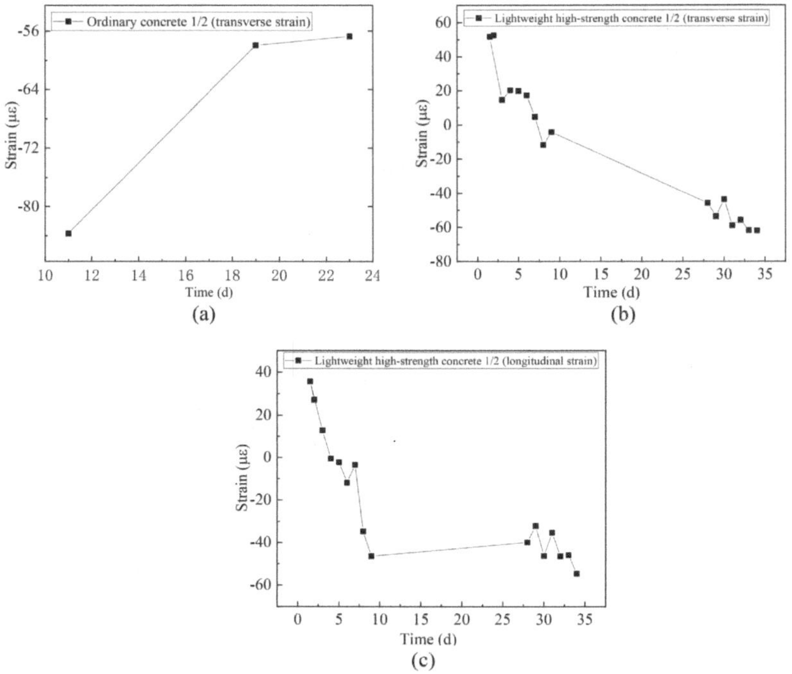

Figure 3. Strain at 1/2 of ordinary concrete and lightweight high-strength concrete: (a) transverse strain of ordinary concrete; (b) Transverse strain of lightweight concrete; (c) Longitudinal strain of lightweight concrete.

different shrinkage stress generated at different positions, which also has a certain impact on the previous sensor installation process. Lightweight concrete and ordinary concrete have a relatively obvious micro strain change at the initial stage of pouring, which is a large change caused by the loss of plasticity of concrete and conforms to the normal phenomenon. Figure 3 shows that the strain of the two kinds of concrete increases with time, which may be caused by two reasons: (1) The hydration reaction of concrete and the change of ambient temperature cause the concrete deformation; (2) The development of concrete strength makes the interior of concrete more compact. Figure 3(b) and (c) show the transverse and longitudinal strain changes at 1/2 of the lightweight concrete, which can well reflect the self-compacting process in the concrete. The longitudinal strain and transverse strain of light-weight concrete are similar. Through the analysis of the monitoring data, it is shown that lightweight concrete, compared with ordinary concrete, not only reduces the self-weight of the concrete but also shows good mechanical properties, meeting the requirements of use.

3.2 *Strain at 3/4 of the bridge deck*

Figure 4 shows the comparison of transverse strain at the 3/4 position between ordinary concrete and lightweight concrete. It can be seen from the figure that the self-compacting process of ordinary concrete is faster than that of lightweight concrete. The hydration reaction of ordinary concrete in the later stage is relatively large, resulting in large deformation fluctuation of concrete. Lightweight concrete has a “micropump” effect due to its large number of internal voids, which makes its later hydration reaction less volatile, and its internal micro-cracks are less, with good durability.

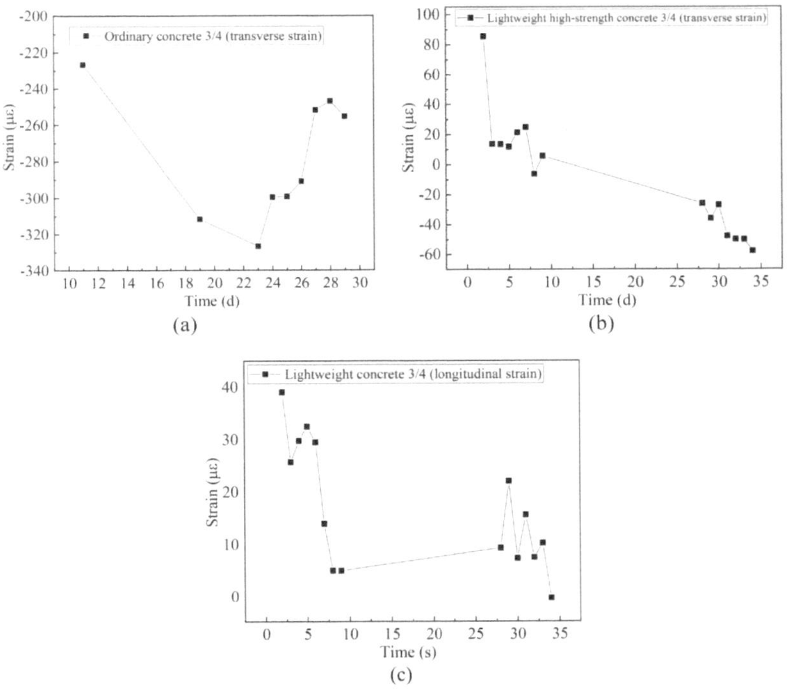

Figure 4. Strain at 4/3 of ordinary concrete and lightweight high-strength concrete: (a) transverse strain of ordinary concrete; (b) Transverse strain of lightweight high-strength concrete; (c) Longitudinal strain of lightweight high-strength concrete.

4 CONCLUSIONS

Field installation and monitoring of sensors shall be carried out according to the G360 project. After the sensor is installed, the matched demodulator to monitor is used, and the measured data are collected and analyzed. The FBG sensor has stable performance, high accuracy, and long life, which can realize long-term stable monitoring of target objects. The monitoring system has the characteristics of usability, reproducibility, convenience, stability, etc. The following conclusions are drawn:

1) Both lightweight concrete and ordinary concrete have a certain shrinkage phenomenon with the increase of age, and the transverse shrinkage phenomenon is more obvious than the longitudinal shrinkage phenomenon at the same scale.
2) The changing trend of the micro strain of lightweight concrete and ordinary concrete is the same. No matter the transverse strain or the longitudinal strain, they all show the same shrinkage trend, which indicates that the volume reduction of lightweight concrete and ordinary concrete in the curing process is roughly the same in the initial setting and hardening process
3) Lightweight concrete and ordinary concrete have a relatively obvious micro strain change at the initial stage of pouring, which is a large change caused by the loss of plasticity of concrete and conforms to the normal phenomenon.
4) Through the analysis of the monitoring data, it is shown that lightweight concrete, compared with ordinary concrete, not only reduces the self-weight of the concrete but also shows good mechanical properties, meeting the requirements of use.

REFERENCES

Domagała L. (2020) Durability of Structural Lightweight Concrete with Sintered Fly Ash Aggregate. *Materials*, 13(20), 4565.

Meyer K. & Kahn L. (2002) Lightweight Concrete Reduces Weight and Increases the Span Length of Pre-tensioned Concrete Bridge Girders. *PCI J.* 47(1).

Payam S., Mahmoud H. & Mohsen K. (2011) An Investigation of the Flexural Behavior of Reinforced Lightweight Concrete Beams. *Int. J. Phys. Sci.* 6(10), 2414–2421.

Qi C. & Fourie A. (2019). Cemented Paste Backfill for Mineral Tailings Management: Review and Future Perspectives. *Miner. Eng.* 144, 106025.

Qin, Y., Wang, Q., Xu, D., Yan, J., & Zhang, S. (2022). A Fiber Bragg Grating-based Earth and Water Pressure Transducer with Three-dimensional Fused Deposition Modeling for Soil Mass. *J. Rock Mech. Geotech.* 14(2), 663–669.

Sengul O., Azizi S., Karaosmanoglu F. & Tasdemir M. (2011). Effect of Expanded Perlite on the Mechanical Properties and Thermal Conductivity of Lightweight Concrete. *Energy. Buildings* 43(2–3), 671–676.

Shafigh P., Jumaat M. & Mahmud H. (2012). Effect of Replacement of Normal Weight Coarse Aggregate with Oil Palm Shell on Properties of Concrete. *Arab. J. Sci. Eng.* 37(4), 955–964.

Tanyildizi, H., & Coskun, A. (2008). The Effect of High Temperature on Compressive Strength and Splitting Tensile Strength of Structural Lightweight Concrete Containing Fly Ash. *Constr. Build. Mater.* 22(11), 2269–2275.

Tanyildizi H. (2013) Variance Analysis of Crack Characteristics of Structural Lightweight Concrete Containing Silica Fume Exposed to High Temperature. *Constr. Build. Mater.* 47, 1154–1159.

Wu Y., Wang J., Monteiro P. & Zhang M. (2015). Development of Ultra-lightweight Cement Composites with Low Thermal Conductivity and High Specific Strength for Energy-efficient Buildings. *Constr. Build. Mater.* 87, 100–112.

Xu, D., Yin, J., Cao, Z., Wang, Y. L., Zhu, H., & Pei, H. (2013). A New Flexible FBG Sensing Beam for Measuring Dynamic Lateral Displacements of Soil in a Shaking Table *Test, Measurement*, 46, 200–209.

Xu, D., Borana, L., & Yin, J. (2014). Measurement of Small Strain Behavior of a Local Soil by Fiber Bragg Grating-based Local Displacement Transducers, *Acta Geotech.* 9(6), 935–943.

Xu, D. (2017), A New Measurement Approach for Small Deformations of Soil Specimens using Fiber Bragg Grating Sensors, *Sensors* 17(5), 1016.

Xu, D. S., Liu, H., & Luo, W. (2018). Development of a Novel Settlement Monitoring System using Fiber-optic Liquid-level Transducers with Automatic Temperature Compensation, *IEEE T. Instrum. Meas.* 67 (9), 2214–2222.

Xu, D. S., Su, Z. Q., Lalit, B., & Qin, Y. (2022). A Hybrid FBG-based Load and Vibration Transducer with a 3D Fused Deposition Modeling Approach, *Meas. Sci. Technol.* 33(6), 065106.

Xu, X. Q., Yang, X., Yang, W., Guo, X., & Xiang, H. L. (2020). New Damage Evolution Law for Modeling Fatigue Life of Asphalt Concrete Surfacing of a Long-span Steel Bridge, *Constr. Build. Mater.* 259, 119795.

Water Conservancy and Civil Construction – Oke & Ahmad (Eds)
 ISBN: 978-1-032-58614-4

Force analysis of pile foundation by the thickness of bearing platform under flat turning bridge girder and length of pile foundation

Xingbang Chen & Richen Ji*
College of Civil Engineering, Lanzhou Jiaotong University, Lanzhou, Gansu, China

ABSTRACT: To study the force distribution of the pile foundation of a flat turning bridge and the influence of lower bearing thickness and pile foundation length on the force of pile foundation, taking Taichengxi (94.8+125) m bridge of new Fuxia passenger dedicated line as engineering background, the model of rotary cap and pile foundation with software and numerical analysis is established. Four groups of cap thickness and four groups of pile foundation length are selected to compare and analyze the influence of different cap thicknesses and different pile foundation lengths on pile foundation stress. The results show that the pile top reaction force decreases as the center distance from the ball hinge increases for the pile foundation of a flat-turned bridge girder. The increase in the thickness of the bearing platform or the length of the pile foundation will significantly reduce the pile top reaction force of the central pile foundation, resulting in a more uniform force distribution in the pile group foundation. A new pile-top reaction force inhomogeneity coefficient is proposed, and the relationship between pile-top reaction force inhomogeneity coefficient and pile-bearing platform relative stiffness is established, which solves the problem of discriminating the force inhomogeneity of rotating pile foundations.

1 INTRODUCTION

To reduce the impact of new railroads on existing lines, flat turnaround construction is widely used in the construction of cross-line bridges (Gu 2017). The flat turn is divided into the pier top turn (Gao 2017) and the pier bottom turn, and the pier bottom turn is commonly used at present (Chen 2017). The pier bottom flat turn bridge girder bearing is divided into upper and lower bearings by a ball hinge, which is located between the upper and lower bearing so that the upper and lower bearing cannot form a whole. The lower bearing carries the weight of the bridge in the process of level rotation and the entire weight of the bridge superstructure is transferred by the ball hinge. Because the size of the ball-hinged contact area is much smaller than the size of the lower bearing, the lower bearing has less stiffness, resulting in uneven force distribution on the bearing and pile foundation. With the continuous improvement of construction technology, the weight of the rotating bridge is increasing (Feng 2020), and the bearing pile foundation is more complicated in terms of force, so it is necessary to conduct an in-depth study on the bearing pile foundation to improve the safety in the construction process (Liu 2019). At present, there are many pieces of research on rotating bridges at home and abroad, mainly using friction mechanics to study the force performance of ball hinges in rotating construction, and using the pile foundation tension bar model as the basis to study the force calculation method of pile foundation bearing, etc. The friction factor of the ball hinge during the rotation of the bridge and the calculation method of the frictional moment of

*Corresponding Author: 1165251533@qq.com

DOI: 10.1201/9781003450818-51

horizontal rotation has also attracted the attention of many researchers (Luo 2020); Jingquan Wang *et al.* (Wang 2013) established a conical surface space tension bar model and proposed a method of bearing platform damage mode and load capacity prediction.

The existing code for bearing design approximates rigid bearing. The force on the pile foundation is uniformly distributed. However, due to the specificity of the structure of the bearing platform of the level turn bridge, the existing design is in error with the actual situation. This paper takes the Tai Cheng Xi (94.8+125) m special bridge of the new railroad Fuxia passenger dedicated line as the engineering background. ABAQUS finite element software was used to analyze the effects of different pile lengths and bearing thickness on the forces on the pile foundation of the rotating bearing, to provide a reference for the design and optimization of bearing thickness and pile length of a turn-around bridge.

2 PROJECT OVERVIEW

The new Fuxia Railway Tai Chengxi Special Bridge uses concrete T-structure cast in side supports at Shenhai Highway, then the beam is turned flat in place with the main pier as the center. The rotating bearing platform is divided into two parts: the upper and lower bearing platforms. The lower bearing platform is 22.9 m long, 20 m wide, and 6 m thick. The diameter of the ball-hinged projection is 6 m, and the length of the pile foundation is 25 m. The lower bearing is surrounded by chamfers. The pile foundation is arranged in a plum pattern. The rotating system consists of a lower turntable, ball hinge, upper turntable, bracing legs, rotating traction system, boosting system, and axial trim system (Yu 2020). The weight of the beam is 38,000 tons. The profile of the rotating system is shown in Figure 1.

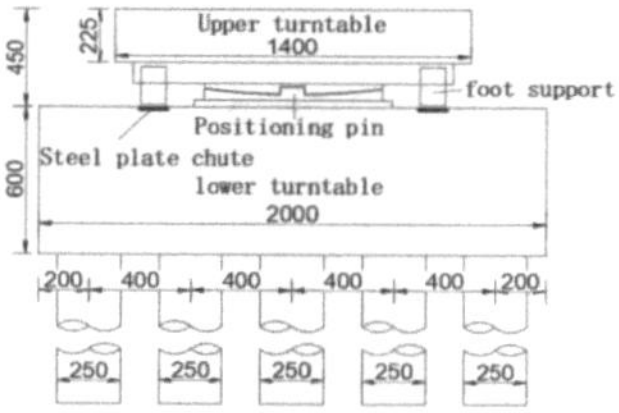

Figure 1. Rotation system profile unit (cm).

3 FINITE ELEMENT NUMERICAL SIMULATION

3.1 *Cell and mesh division*

In this paper, the finite element analysis software ABAQUS is used to simulate the rotation system numerically. The model is divided into six parts: pier body, upper turntable, upper ball hinge, lower ball hinge, lower turntable, and pile group foundation. All six parts use 3D solid linear reduced integration cells (C3D8R cells). C3D8R cell calculation is fast and cell density is high, which is convenient for parts to mesh. During the mesh delineation process, the node has localized the mesh refinement for the parts of stress concentration and concern, such as the upper ball hinge, lower ball hinge, and pile group foundation.

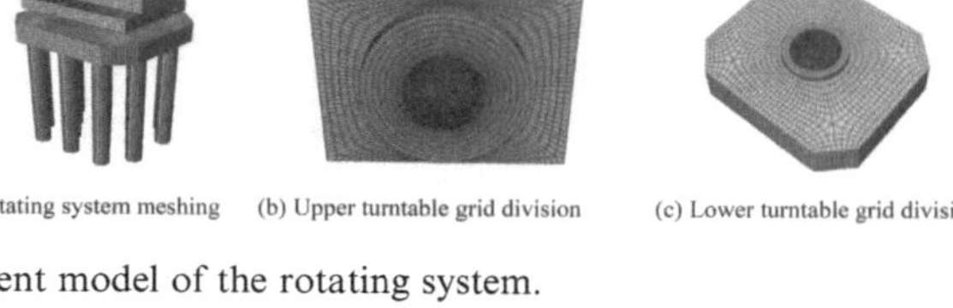

(a)Rotating system meshing (b) Upper turntable grid division (c) Lower turntable grid division

Figure 2. Finite element model of the rotating system.

3.2 *Material property definition*

According to the engineering design requirements, Q235 steel was selected as the material for the upper and lower ball hinges, and C55 was selected as the material for the upper and lower bearing. In the finite element model of the rotating body construction, the components are considered as uniform continuous materials (Shipping 2018). The upper ball hinge and lower ball hinge contact surface tangential to take the ball hinge friction coefficient design value of 0.06. The specific parameters of the materials are shown in Table 1.

Table 1. Model material parameters.

Materials	Poisson's ratio	Elastic modulus (MPa)	Friction coefficient
Q235	0.3	2.1e5	0.06
C55	0.2	3.55e4	–

3.3 *Contact simulation*

In the rotating system, the upper ball hinge and the lower ball hinge belong to contact friction. The ball hinge is defined as face-to-face contact between the hinges, the normal behavior is defined as a hard contact, which is unrestricted for transferring contact pressure, and the contact pressure will be 0 when there is no contact or contact separation [14], which can better reflect the actual situation. Since the penalty function-n formulation allows for elastic slip on the contact surface (Shipping 2018) and applies to most contact problems, the penalty friction is used for the tangential behavior. The rest of the components are connected by "tie" binding relationships, making the model easier to converge.

3.4 *Load boundary conditions*

The bearing platform bears the weight of the beam mainly in the vertical force, and the simulation process applies pressure on the pier body instead of the vertical force. The pile foundation is a column pile, and according to the actual situation, only the bottom of the pile is fully fixed and restrained. The model is solved with a static and general analysis step.

4 FINITE ELEMENT ANALYSIS RESULTS

4.1 *Pile top force distribution*

The pile arrangement is divided into symmetrical, plum-shaped, and ring-shaped. In this paper, the pile arrangement is plum-shaped and divided into 5 groups according to the distance from the pile to the center of the bearing, and the pile arrangement is shown in Figure 3.

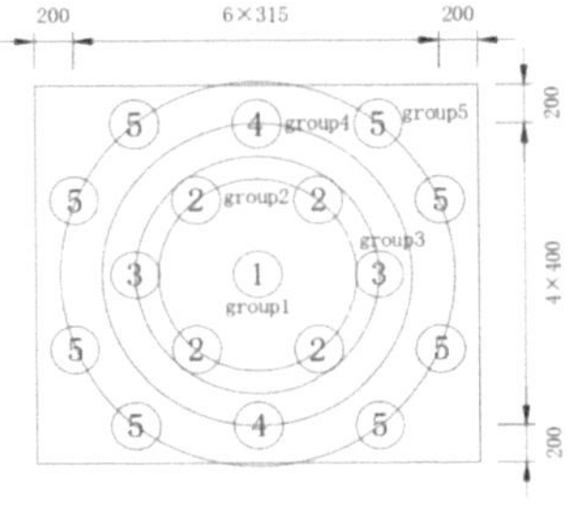

Figure 3. Grouping of piles (cm).

The vertical stress and vertical displacement of the pile top of the pile foundation during the bridge leveling process and the one-half profile of the pile foundation and the bearing are shown in Figures 4 and 5.

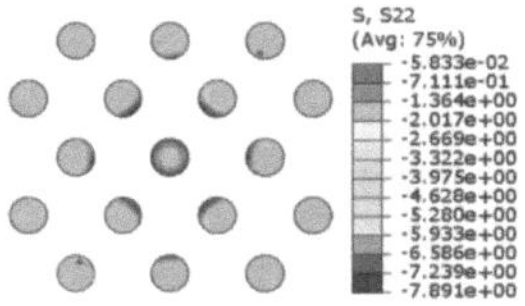

Figure 4. Pile top vertical stress (MPa).

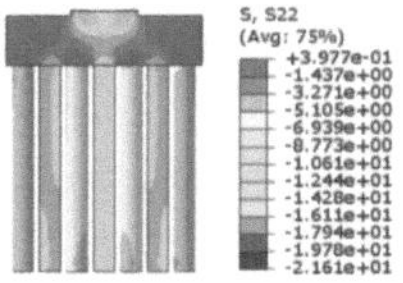

Figure 5. Section of bearing and pile foundation (MPa).

Extracting the pile top reaction force of the pile foundation can be obtained: the maximum pile top reaction force of group 1 is 27655.1KN, which is 1.24 times the average pile top reaction force; the minimum stress of group 5 pile is 17160.6KN, which is 0.77 times the average value of pile top reaction force, and the difference between the maximum and minimum top force is 37.9%; the maximum displacement of group 1 pile is 4.42 mm, which is 1.36 times the average displacement; the minimum displacement of group 5 pile is 2.78 mm, which is 0.85 times the average displacement; the difference between the maximum and minimum values of the displacement is 37.1%. From Figures 4 and 5, it can be seen that the pile stress is larger in the lower part of the ball hinge, and the stress gradually decreases the farther the horizontal distance from the center of the ball hinge, and the stress distribution in the bearing and pile base is not uniform.

4.2 *Effect of bearing thickness on pile top reaction force*

To study the effect of bearing thickness on the force of pile foundation, the thickness of 4m, 5 m, 6 m, and 7 m bearing was selected as the object of analysis for calculation, and the calculation results are shown in Figure 6, Figure 7, and Table 2.

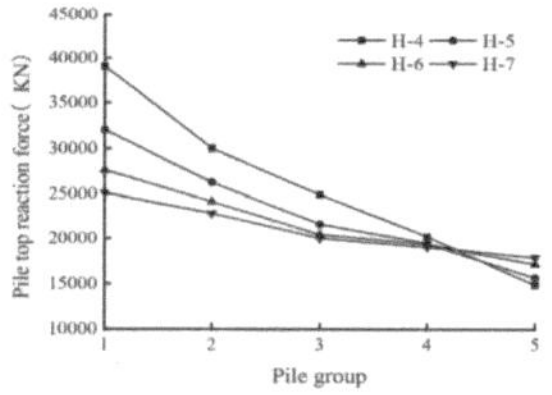

Figure 6. Vertical stress on top of the pile with different bearing thickness.

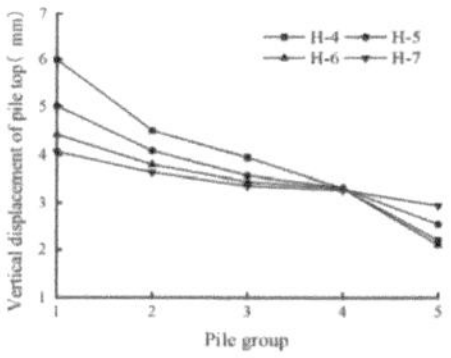

Figure 7. Vertical displacement of pile tops with different thicknesses of bearing.

Table 2. Pile top reaction force and displacement concerning the mean value as a percentage.

	Pile top reaction force		Vertical displacement of pile top	
Ratio	Bearing Table 4 m	Bearing Table 7 m	Bearing Table 4 m	Bearing Table 7 m
Group 1	1.74	1.12	1.81	1.24
Group 5	0.66	0.80	0.66	0.90

From Figure 6, it can be seen that: the pile top reaction force from group 1 to Group 3 decreases as the thickness of the bearing increases, Group 4 piles remained unchanged, while Group 5 piles increased with the increase of bearing thickness, with a decrease of 35.6% for Group 1 piles and an increase of 20.3% for Group 5 piles. From Figure 7, it can be seen that: the vertical displacement of Group 1 to Group 3 piles decreases as the thickness of the bearing increases, and Group 4 piles remain unchanged, Group 5 piles increased with the increase of bearing thickness, the decrease of Group 1 piles was 32.1% and the increase of Group 5 piles was 33.9%. From the above: it can be obtained that as the thickness of the bearing increases, the stiffness of the bearing increases, the force coordination capacity of the bearing increases, and the force transmitted by the lower ball hinge contact surface is more uniformly distributed.

4.3 *Pile length to pile top stress analysis*

To study the influence of pile length on the force of pile foundation, pile foundations with pile lengths of 15 m, 25 m, 35 m, and 50 m were selected as the objects of analysis for calculation, and the calculation results are shown in Figure 8, Figure 9, and Table 3.

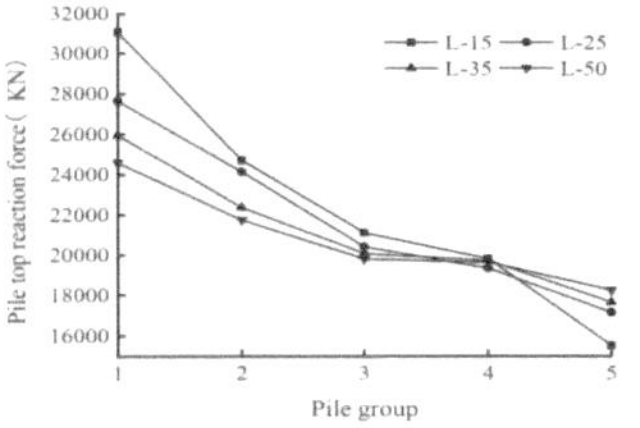

Figure 8. Vertical stress at the top of the pile for different pile lengths.

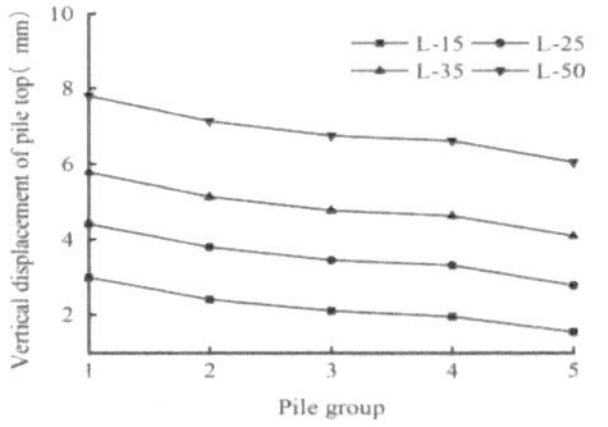

Figure 9. Vertical displacement of pile tops with different pile lengths.

Table 3. Pile top reaction force and displacement concerning the mean value as a percentage.

	Pile top reaction force		Vertical displacement of pile top	
Ratio	Length of pile foundation 15 m	Length of pile foundation 50 m	Length of pile foundation 15 m	Length of pile foundation 50 m
Group 1	1.39	1.05	1.53	1.19
Group 5	0.67	0.81	0.74	0.90

From Figure 8, it can be seen that: Group 1 to Group 3 pile top reaction force decreases with increasing pile length, Group 4 pile remains the same, and Group 5 pile increases with increasing pile length. The reduction of pile top reaction force in Group 1 is 24.4%, and the increase of pile top reaction force in Group 5 is 20.3%. From Figure 9, it can be seen that the vertical displacement of the center pile increases with the increase of pile length, the increase of vertical displacement of the center pile is 1.61 times, and the increase of vertical displacement of the corner pile is 3.08 times. From the above, it can be obtained that as the pile length increases, the pile top reaction force of the central pile base decreases, and the pile top reaction force of the corner pile increases. The displacement increase of the center pile is smaller and the displacement increase of the corner pile is larger, which leads to the partial force transfer from the center pile to the side pile and the corner pile.

5 UNEVEN COEFFICIENT OF PILE TOP REACTION FORCE

5.1 *Relative stiffness factor of the pile-bearing platform*

Based on the existing design codes, the design of pile foundations is carried out by considering the bearing as a rigid block. The force on the pile foundation is uniformly distributed, and the reaction force on the top of each pile is shown in the following equation.

$$N_i = \frac{N + G}{n} \tag{1}$$

Where N_i is the monopile pile top reaction force, N is the vertical force acting on the top surface of the bearing, G is the self-weight of the rotating bearing, and n is the number of pile bases in the pile foundation.

Guo Chao *et al* (Guo 2010) considered that the pile reaction force distribution is related to the pile support stiffness and the thickness of the bearing platform, etc. The pile-bearing platform relative stiffness coefficient β can reflect the distribution law of pile reaction force.

$$\beta = \frac{\sqrt{(B/L)E_R H a D_P}}{\sqrt[3]{K_P}\mathrm{D}\mathrm{s_a}} \tag{2}$$

L and B are the length and width of the bearing; E_R is the modulus of elasticity of concrete of the bearing; H is the thickness of the bearing, a is taken as 2.5 in the bridge design, D_P is the pile diameter, K_P is the vertical support stiffness of the pile, D is the closest distance from the calculated perimeter of the load action to the center of the farthest pile, and s_a is the center distance of the pile.

It was shown (Wang 2013) that when $\beta \leq 1$, the top reaction force of the central pile is greater than that of the side piles and corner piles; When $1 < \beta \leq 4$, the reaction force at the top of the central pile decreases and the reaction force at the edge and corner piles increases. This type of bearing is called a "semi-rigid bearing". When $\beta \geq 4$, the pile top reaction forces of corner piles, side piles, and center piles are gradually close to each other and the distribution tends to be uniform, and this kind of bearing is called "rigid bearing". This shows that the force effects on the pile foundation by the thickness of the bearing and the length of the pile foundation are consistent with the force characteristics expressed by the relative pile-bearing stiffness, indicating that the model simulation of the rotation system is accurate.

5.2 *Uneven coefficient of pile top reaction force*

From the above, it can be seen that the magnitude of the pile-bearing table relative stiffness coefficient β directly determines the distribution of the pile top reaction force and also affects the deformation of the pile foundation. In this section, the pile top reaction force inhomogeneity coefficient will be introduced to reflect the distribution of pile top reaction force, which is used to quantify the forced uniformity of the pile foundation. From the analysis results of finite element software, the most unfavorable location of the pile foundation of the flat turning bridge girder is the pile foundation on the lower side of the ball-hinged contact surface. Therefore, the force in the center pile can best reflect the distribution of the force at the top of the pile. Parameter λ is introduced to establish the relationship between the central pile-top reaction force and the average pile-top reaction force.

$$\lambda = \frac{nN_{中}}{N + G} \tag{3}$$

where $N_{中}$ is the pile top reaction force of the central pile. The pile-bearing relative stiffness coefficient and pile-top reaction force inhomogeneity coefficient for different bearing thicknesses and pile lengths are calculated, to investigate the influence of bearing thickness and pile length on the parameters. The calculation results are shown in Table 4.

Table 4. Unevenness coefficient vs. relative stiffness of pile-bearings.

	Thickness of the bearing platform				Length of pile foundation			
Parameters	4m	5m	6m	7m	15m	25m	35m	50m
Unevenness factor	1.74	1.43	1.24	1.12	1.39	1.24	1.16	1.05
Relative stiffness factor	2.15	2.69	3.23	3.77	2.72	3.23	3.70	4.07

The relationship between the pile-top reaction force inhomogeneity coefficient and pile-bearing relative stiffness coefficient is established in the following equation.

$$\lambda = 0.1398\beta^2 - 1.2048\beta + 3.6664 \quad (4)$$

Equation (4) shows the relationship between the magnitude of a and the consistency of the pile foundation bearing rigidity and flexibility, solving the problem of discriminating the uneven force of the rotating pile foundation.

6 CONCLUSIONS

(1) The setting of ball hinges on the bearing deck of the rotating bridge leads to the gravity of the superstructure being concentrated on the bearing deck on the lower side of the ball hinges and transferred to the pile foundation, resulting in uneven forces on the pile foundation of the rotating bridge. The concept of uniform distribution of pile top reaction force in the current code does not apply to the rotating bridge, and it is unreasonable to design the rotating bearing platform accordingly.
(2) By analyzing the force of rotating pile foundations through ABAQUS finite element software, it presents center pile > side pile > corner pile. Increasing the thickness of the bearing platform or the length of the pile foundation can effectively reduce the top reaction force of the center pile and increase the top reaction force of the corner pile. The thickness of the bearing or the length of the pile foundation can effectively improve the reasonableness of the force on the pile foundation.
(3) The pile-top reaction force inhomogeneity coefficient is established, reflecting the distribution of the pile-top reaction force of the rotating bearing platform. From the relationship between the pile-top reaction force inhomogeneity coefficient and the pile-bearing table relative stiffness coefficient, obtaining the thickness of the bearing is required to satisfy the rigidity condition. It can be used as the basis for the design of the thickness of the rotating bearing.

REFERENCES

China Academy of Building Science. Technical Specification for Building Pile Foundations JGJ 94-2008 [M]. *Building Pile Foundation Technical Specification JGJ 94-2008*, 2008

Cheng Fei, Zhang Qifeng, Wang Jingquan. The Development Status and the Prospect of Bridge-turning Construction Technology in China[J]. *Railway Standard Design*, 2011(06):67–71.

Feng Y., Qi J., Wang J., et al. Rotation Construction of Heavy Swivel Arch Bridge for High-speed Railway[J]. *Structures*, 2020, 26(4): 755–764.

Gu Lanyu. *Research on Construction Technology and Construction Plan Comparison Method for Bridges Over Railroad Business Lines* [D]. Tsinghua University, 2017.

Guo Chao, Lu Bo, Gong Weiming, et al. Experimental Study on the Bearing Performance of Nine-pile Thick Bearing Platform Under Column[J]. *Journal of Civil Engineering*, 2010, 43(1): 95–102.

Gao Gengyuan. Construction Technology of High-level Turning of Gusaoshu Road Over Railroad Overpass in Wuhan[J]. *Railway Construction*, 2017(06):44–48.

Luo Lijun. Accurate Calculation Method for the Moment of Vertical Rotation of Spherical Hinges in Bridge Transit Construction[J]. *Railway Construction*, 2020, 60(01):27–30.

Liu Wanquan. Key Points of Bridge Superstructure Turning Construction Technology[J]. *Traffic World*, 2019 (11):132–133.

Shiping H., Mengyu H., Yonghui H., et al. A New Model for Optimal Mechanical and Thermal Performance of Cement-Based Partition Wall[J]. *Materials*, 2018,11(4): 615.

Wang Jingquan, Zhang Qifeng, Qian Guifeng, et al. Design and Analysis Method of 16,800-ton Rotating Arch Bridge Bearing for Shanghai-Hangzhou High-speed Railway[J]. *Bridge Construction*, 2013, 43(05): 99–105.

Yu Yanxia, Wang Dezhi. Design of (95+125) M Asymmetric Single-tower Partial Cable-stayed Bridge for Fuxia High-speed Railway[J]. *Railway Standard Design*, 2020, 64(S1): 147–151.

Water Conservancy and Civil Construction – Oke & Ahmad (Eds)
© 2024 The Author(s), ISBN: 978-1-032-58614-4

Optimization inversion of material parameters of arch dam based on PSO-LSTM

Dongyan Jia* & Jie Yang
State Key Laboratory of Eco-hydraulics in Northwest Arid Region, Xi'an University of Technology, Xi'an, China

ABSTRACT: Particle Swarm Optimization (PSO) and Long Short-term Memory (LSTM) are combined to optimize the inverse of arch dam material parameters to improve the accuracy and efficiency of inverse analysis of arch dam material parameters. Firstly, the Latin Hypercube Sampling (LHS) is used to construct the mechanical parameter samples. The Finite Element Method (FEM) calculates the radial deformation of the arch dam under different working conditions. The mechanical parameter samples and the calculated hydraulic component deformation are taken as the training samples of the LSTM network, and the proxy model is established. The least square method is used to establish the statistical model of the measured deformation of the arch dam and separate the water pressure component. The measured water pressure component is combined with the calculated value of the proxy model to establish the objective fitness function. The inverse analysis of the material parameters is carried out through the PSO algorithm. To verify the method's reliability, the inverse analysis of the comprehensive elastic modulus of the dam body and foundation is carried out for a high arch dam. The results show that the algorithm has high accuracy and can meet the application requirements of practical projects.

1 INTRODUCTION

During the FEM of the arch dam, the accuracy of material parameters is crucial to the reliability of simulation analysis results. However, the material parameters in the design stage often cannot accurately reflect the actual structural characteristics of the arch dam (Giovanni 2013); the finite element calculation based on the design value of material parameters will produce large errors, so it is necessary to carry out inverse analysis on the parameters of arch dam materials and foundation materials.

In recent years, the research on the inversion of dam material parameters based on the measured data has gradually increased. For example, Wang (2021) used the inversion method based on the mixed model to inverse the elastic and viscoelastic parameters of the dam body concrete of Jinping First Stage Arch Dam; Cao (2016) and Jia (2015) respectively used the quantum genetic algorithm to optimize inversion of the dam body and dam foundation material parameters of different concrete gravity dams; Zhao (2012) combined the PSO algorithm with ADINA to optimize the back analysis of the material parameters of earth rock dam; as well as many inversion methods based on the mixed model and intelligent optimization algorithm (Sun 2018) using geotechnical mechanics theory, the inversion research of dam material parameters have been developed rapidly.

However, many finite element calculations are required in parameter optimization and inversion. Support vector machines (Nui 2020), response surface models, and multi-output support vector machines (Yuan 2017) are widely used as proxy models to improve calculation efficiency. In recent years, the LSTM (Hochreiter 1997), which is based on the improvement of recurrent neural network (RNN), has been widely used in many industries

*Corresponding Author: 2093105625@qq.com

 DOI: 10.1201/9781003450818-52

because it solves the problem of gradient explosion and gradient disappearance in RNN (Ma 2015) and relies on its excellent prediction ability and simple usage (Ren 2021; Song 2021).

Therefore, this paper proposes a material parameter optimization inversion method based on PSO for parameter optimization and the LSTM as the surrogate model. Firstly, the required data set is obtained through finite element calculation, and a reliable LSTM surrogate model is established through the training set. PSO is used to optimize the parameters and obtain reliable material parameter values. Taking a concrete arch dam in northwest China as an example, the inversion calculation of the comprehensive elastic modulus of dam concrete and bedrock is carried out, which verifies the reliability and effectiveness of this method, and provides a new method for the inversion of the concrete arch dam body and dam foundation material parameters; Moreover, LSTM surrogate model can significantly improve the computational efficiency of parameter inversion compared with time-consuming finite element calculation.

2 BASIC THEORY

2.1 *Statistical model theory of dam deformation*

During the long-term operation of an arch dam, its deformation is mainly affected by water pressure, temperature, and aging factors (Wu 2003). According to the influencing factors, the deformation expression is:

$$\delta = \delta_H + \delta_T + \delta_\theta \tag{1}$$

Where δ_H is the water pressure component of deformation; δ_T is the temperature component of deformation; δ_θ is the aging component of deformation.

2.2 *LSTM network theory*

LSTM network is an improved RNN network. Based on the RNN network, the unit of the hidden layer is converted into a hidden state and three gate structures (input gate, forgetting gate, and output gate). The control information is updated in the hidden state by controlling each memory unit's state. The calculation formula of the LSTM network is as follows:

$$\begin{cases} i_t = \sigma(W_{xi}x_i + W_{hi}h_{i-1} + W_{ci}c_{i-1} + b_i) \\ f_t = \sigma\left(W_{xf}x_t + W_{hf}h_{t-1} + W_{hc}c_{t-1} + b_f\right) \\ c_t = f_t h_{t-1} + i_t \tanh\left(W_{xc}x_t + W_{hc}h_{t-1} + b_c\right) \\ o_t = \sigma(W_{x0}x_t + W_{h0}h_{t-1} + W_{c0}c_t + b_0) \\ h_t = o_t \tanh\left(c_t\right) \end{cases} \tag{2}$$

Where i_t, f, c_t, and o_t represent input gate, forgetting gate, updated cell state, and output gate, respectively; x_t is the input information; h_t is the acquired input information; W is the weight coefficient; b_i, b_f, b_c, b_0 are offset; t is the accumulation phase.

2.3 *PSO optimization algorithm theory*

PSO algorithm assumes that in an N-dimensional space, there is a population composed of n particles $X = (X_1, X_2, \ldots, X_n)$, where the N-dimensional vector $X_i = (X_{i1}, X_{i2}, \ldots, X_{in})$ of the ith particle represents the position of the ith particle in space. This paper uses the comprehensive elastic modulus of concrete and bedrock in finite element calculation. The velocity of the ith particle is $V = [V_{i1}, V_{i2}, \ldots, V_{iN}]^T$, its extreme individual value is $P_i = [P_{i1}, P_{i2}, \ldots, P_{iN}]^T$, and the extreme global value of the population is $P_g = [P_{g1}, P_{g2}, \ldots, P_{gN}]^T$. In each optimization iteration, the velocity and position of particles passing through individual extremum to the global extremum finer particles are updated as follows:

$$V_{id}^{k+1} = \omega V_{id}^k + c_1 r_1\left(P_{id}^k - X_{id}^k\right) + c_2 r_2\left(P_{gd}^k - X_{id}^k\right) \tag{3}$$

$$X_{id}^{k+1} = X_{id}^{k} + V_{id}^{k+1} \tag{4}$$

Where ω is the inertia weight, taken as 0.7; $d = 1,2, \ldots, N$; $i = 1,2 \ldots ,n$; r_1 and r_2 are random numbers distributed between (0, 1); c_1 and c_2 are acceleration factors, which are taken between (0, 2). This paper sets them as 1.5 and 1.7; k is the current iteration number; V_{id} is the particle velocity.

2.4 *Optimized inversion model based on PSO-LSTM*

The sample set of material parameters is obtained by Latin hypercube sampling (LHS), the database for training the LSTM network is obtained by finite element calculation, and the agent model is established. The PSO algorithm optimizes the parameters, and the objective fitness function determines the optimal material parameter values. The inversion analysis steps established in this paper are as follows:

1. The distribution range of material parameters is determined according to the engineering data, and the LHS method is used for sampling to establish the value set of material parameters.
2. The finite element calculation model of the arch dam is established, the corresponding calculation conditions are selected, and the calculated values of water pressure components corresponding to the material parameters are obtained through finite element calculation.
3. The calculated values of water pressure components and material parameters are used as the training set and test set of the LSTM to establish the agent model.
4. The PSO is used for parameter optimization. The optimization parameters are substituted into the LSTM network to determine the optimal inversion result by whether the moderate objective function is satisfied.
5. The parameter values after inversion are used for finite element calculation, and a hybrid model for arch dam deformation monitoring is established. The calculated values of the hybrid model are compared with the measured values to verify the accuracy of the inversion results.

3 ENGINEERING EXAMPLE

3.1 *Project introduction*

A concrete double curvature arch dam is located northwest of China. Its maximum dam height is 250 m. There are five vertical monitoring points PL4-1~PL4-5 in the arch crown beam dam section, which are located at the elevations of 2405 m, 2350 m, 2295 m, and 2250 m, respectively; Three inverted vertical IP4-1~IP4-3 anchor points are located at the elevation of 2180 m, 2160 m, and 2130 m respectively. The dam body's three-dimensional finite element calculation

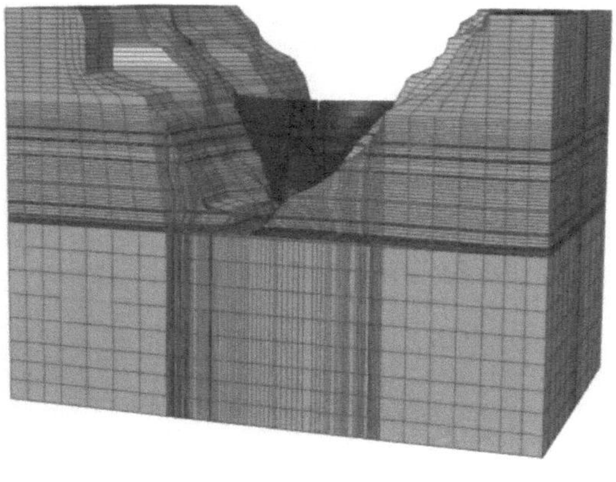

Figure 1. Finite element calculation model.

model selects 1.5 times the dam height to establish the foundation range. The dam concrete and foundation rock mass are calculated using the linear elastic model. The calculation model is shown in Figure 1. The distribution of material parameters obtained according to the self-inspection report of the design unit of the Project is shown in Table 1.

Table 1. Numerical calculation parameters.

Position	Comprehensive elastic modulus/ (GPa)	Standard deviation/ (GPa)	Density/ (kg/m^3)	Poisson's ratio
Dam concrete	[16, 24]	2.50	2400	0.167
Rock mass	[18, 25]	2.25	2600	0.250

3.2 *Establishment of statistical model and LSTM surrogate model*

In the deformation of the arch dam, the deformation analysis of the arch crown beam dam crest is more important than that of other dam parts. In this paper, the monitoring data from January 2012 to July 2021 at the measuring point PL4-1 at the 2460 m elevation (arch crown beam crest) of the arch dam during the operation period are selected as the basis for inversion research. The least square regression is used to analyze the data processed by gross error, and the statistical model of dam deformation is established. The fitting effect is shown in Figure 2:

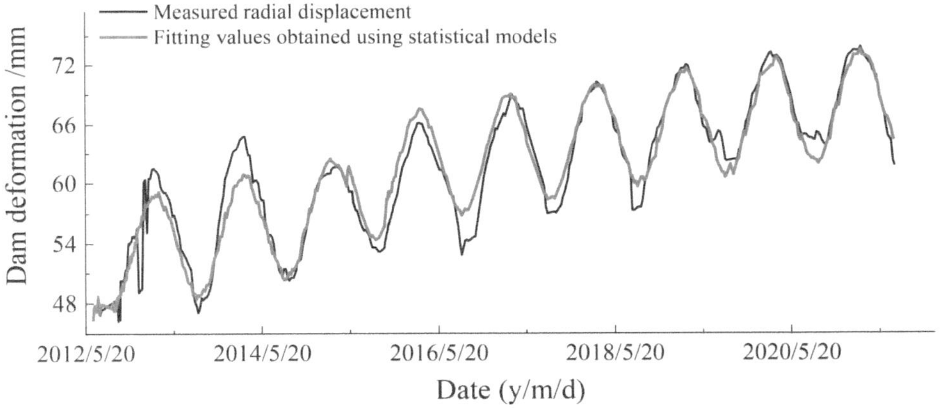

Figure 2. Fitting diagram of deformation statistical model of the PL4-1 measuring point.

The determination coefficient R^2 of the fitted dam deformation statistical model is 0.981, which shows that the overall accuracy of the established statistical model meets the requirements, the fitting effect is good, and it can truly reflect the variation law of the dam deformation measured. The results of separating water pressure components from the statistical model are shown in Figure 3.

From Figure 3, the water pressure component separated by the statistical model is consistent with the change rule of the upstream reservoir water level. The obtained water pressure component can accurately describe the deformation change trend of the arch crown beam dam crest with the change of the upstream water level.

The LHS method is used to sample the data in Table 1. To meet the training and testing needs of the LSTM surrogate model, 100 groups of material parameters are selected to generate 100 groups of data for finite element calculation. One hundred groups of data were

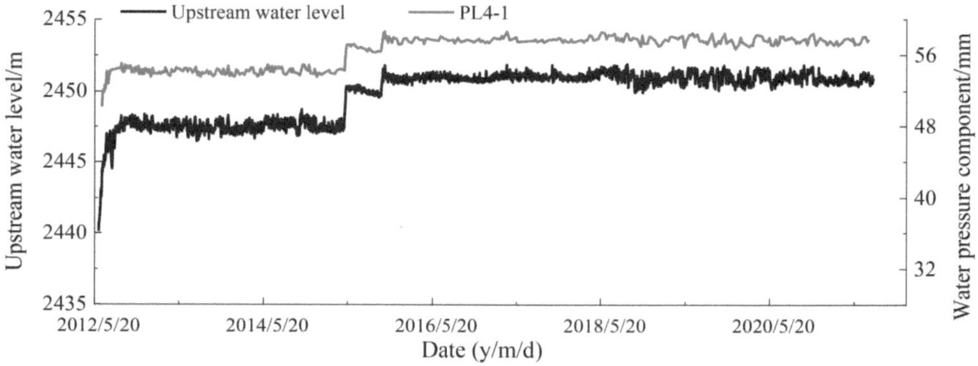

Figure 3. Water pressure component separation results.

randomly divided into the training set and test set according to 7:2. The initial learning rate of the LSTM network is set to 0.7, the learning rate decline cycle to 100, the learning rate decline coefficient to 0.6, and the maximum number of iterations to 4000. The LSTM network training effect is shown in Figure 4:

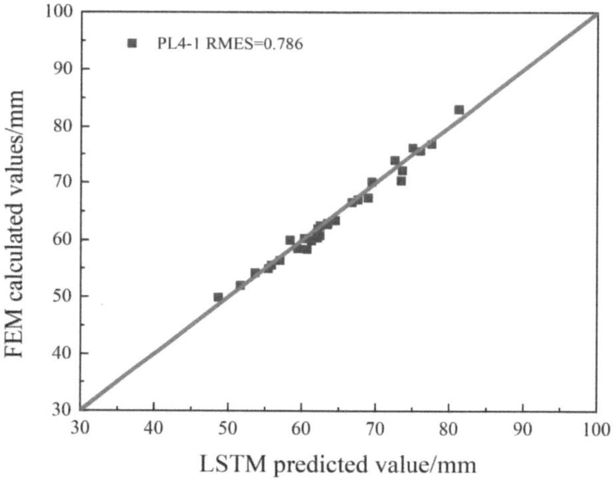

Figure 4. Rendering of LSTM surrogate model.

Figure 4 shows that the difference between the predicted value of the LSTM surrogate model and the calculated value of finite element is slight, which proves that the LSTM surrogate model trained by 70 groups of data has a good test effect and can better reflect the nonlinear relationship between the comprehensive elastic modulus of the dam body and dam foundation and the radial deformation of the PL4-1 measuring point.

3.3 *Analysis of parameter inversion results*

The PSO optimization algorithm is used for material parameter inversion, and the particle swarm size is set as 20. Referring to Table 1, the optimization range of the comprehensive elastic modulus of dam concrete is 16~24 GPa, and the optimization range of the comprehensive elastic modulus of dam foundation rock mass is 18~25 GPa; The termination condition is set to 200 optimizations. Through PSO-LSTM inversion, the elastic modulus of dam concrete is 22.65GPa, and the elastic modulus of dam foundation rock mass is 19.55GPa. At the same time, based on the Wu Zhongru inversion method (Wu 2003), the elastic modulus of dam concrete is 23.23 GPa, and the elastic modulus of dam foundation rock mass is 19.61 GPa.

From the inversion results, it can be concluded that the parameters obtained by the PSO-LSTM inversion method are close to those obtained based on the Wu Zhongru inversion method. To further verify the reliability of the inversion results, the inversion results are used to establish a hybrid model for dam deformation monitoring and to predict the horizontal radial deformation of the vertical line PL4-1 on June 16, 24, July 7, 15, and 22, 2021. Figure 5 shows the comparison results between the calculated values of the mixed model and the measured values.

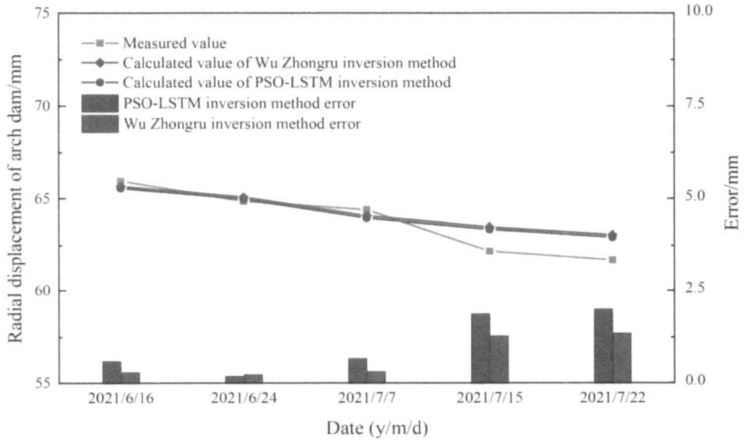

Figure 5. Comparison diagram of calculated deformation value of the hybrid model and measured deformation value.

In Figure 5, the deformation calculation value of the PSO-LSTM inversion method is close to that of the inversion method based on the mixed model; The difference with the measured deformation value is small, and the maximum absolute error is about 2.5 mm. It can be explained that the method of inversion of mechanical parameters of the dam body and dam foundation materials proposed in this paper is feasible.

4 CONCLUSION

(1) The LSTM network as the surrogate model can better reflect the nonlinear relationship between dam material parameters and deformation. Using the PSO optimization inversion method to optimize the inversion parameters requires a lot of numerical calculations. Using an ordinary computer for a numerical simulation takes about six minutes. Using the LSTM surrogate model can obtain the dam deformation values corresponding to the material parameters in just a few seconds, which greatly improves the calculation efficiency of the optimization inversion. The entire inversion process only takes about 10 minutes.

(2) In this paper, the particle swarm optimization algorithm is combined with the long-term and short-term memory network to establish the objective fitness function to inverse the elastic modulus of the dam body and bedrock based on the water pressure component of

the measured value of the arch dam deformation and the long-term and short-term memory network calculation value, which is close to the accuracy of the inversion method based on the hybrid model.

(3) The hybrid model is established based on the comprehensive elastic modulus of the dam body and dam foundation obtained by the PSO-LSTM inversion method, and the calculated value of the hybrid model is compared with the measured value. The results show that the PSO-LSTM optimization inversion method is practical and can provide a reference for high arch dam material parameters inversion.

ACKNOWLEDGMENTS

This work was supported by the National Natural Science Foundation of China (Grant Nos. 52279140), the National Natural Science Foundation of China (Grant Nos. 51409205), and the State Key Program of National Natural Science of China (Grant Nos. 52039008).

REFERENCES

Cao M.J., Cao X. and Xu Z.Z. (2016). Application of Quantum Genetic Algorithm to InverseCalculation of Comprehensive Elastic Modulus of Concrete Gravity Dam. *Journal of Yangtze River Scientific Research Institute* 33(04):111–114.

Giovanni B. and Paolo P. (2013). The 1963 Vajont Landslide:50th Anniversary. *Rock Mech Rock Eng* 46 (6):1267–1270.

Hochreiter S. and Schmidhuber J. (1997). Long Short-term Memory *Neural Comput* 9(8):1735–1780.

Jia C.L. and Zhu K. (2015). The Application of Classified Quantum Genetic Algorithm in the Reversion Calculation of Comprehensive Elastic Modulus of Concrete Gravity Dam. *Journal of Water Resources and Architectural Engineering* 13(04):91–95.

Nui J.T. (2020). Dam Deformation Monitoring Model based on Singular Spectrum Analysis and SVM Optimized by PSO. *Advances in Science and Technology of Water Resources* 40(06):60–65.

Ma X.L., Tao Z.M., Wang Y.H., Yu H.Y. and Wang Y.P. (2015). Long Short-term Memory Neural Network for Traffic Speed Prediction using Remote Microwave Sensor Data. *Journal of Transportation.* 54:187–197.

Ren X.Q., Liu S.L., Yu X.D. and Dong X. (2021). A Method for State-of-charge Estimation of Lithium-ion Batteries based on PSO-LSTM. *Energy* 234:121236.

Song Y., Yang J., Song J.T. and Chen L. (2021). Concrete Dam Deformation Prediction based onCEEMDAN-PE-LSTM model. *Hydro-Science and Engineering* 3:41–49.

Sun F.T., Zhang X.L., Wang Y.J. and Shen H.Y. (2018). Research and Application of Quick DamParameter Inversion Instrument Based on Improved GA. *Water Power* 44(01):55–58.

Wang S.W., Xu C. and Liu Y. (2021). Mixed-coefficient Panel Model for Evaluating the Overall Deformation Behavior of High Arch Dams Using Spatial Clustering. *Struct Control Hlth* 28(10): e2809.

Wu Z.R. (2003). *Safety Monitoring Theory and Application of Hydraulic Structures.* Beijing: Higher Education Press.

Yuan Y.L., Guo Q.Q., Zhou Z.J., Wu Z.Y., and Chen J.K. (2017). Back Analysis of Material Parameters of High Core Rockfill Dam Considering Parameters Correlation. *Rock and Soil Mechanics* 38(S1):463–470.

Zhao D., Zhang Z.L. and Chen J.S. (2012). A Combined Application of Particle Swarm Optimization Algorithm and ADINA for Parametric Inversion of Earth-rock Dams. *Advances in Science and Technology of Water Resources* 32(3):43–47.

Water Conservancy and Civil Construction – Oke & Ahmad (Eds)
© 2024 The Author(s), ISBN: 978-1-032-58614-4

Study on the influence of silica fume on sulfate attack resistance of concrete

Ganggang Xu*
College of Architecture and Civil Engineering, Xi'an University of Science and Technology, Xi'an, Shaanxi, China
Xi'an Research Institute of China Coal Technology & Engineering Group, Xi an, Shaanxi, China

ABSTRACT: In the construction of underground concrete engineering facilities, the ability of concrete to resist sulfate attack was the basic performance index to guarantee the durability of concrete under a sulfate attack environment. In this paper, the physical and mechanical properties of cement mortar specimens with 0, 5%, 10%, 15%, and 20% silica fume (SF) were studied by full immersion test for 19 months. The results showed that the flexural strength and compressive strength of mortar with different contents of SF increased significantly with the increase of curing age and the content of silica fume. When the concentration of Na_2SO_4 is less than 3%, adding 0~10% SF could significantly improve the sulfate resistance of cement-based materials; for 15% Na_2SO_4 sulfate attack environment, to improve the anti-corrosion ability of cement slurry, the optimal amount of SF was 15~20%.

1 INTRODUCTION

There were a lot of strong corrosive media such as sulfate in the soil in some areas of Western China; there were a lot of sulfates in the coastal saline soil and seawater in the eastern coastal area; there were also a lot of sulfate ions in the wastewater discharged from various petrochemical, iron, and steel metallurgical plants. These sulfate ions in the soil, groundwater, seawater, and industrial wastewater infiltrated into the concrete and reacted with hydration products, resulting in expansion, cracking, spalling, and other phenomena, so the strength and viscosity of concrete were reduced and lost (Zhu 2020).

The main inclined shaft of a mine in Gansu Province passed through a quicksand layer. The water was rich in $SO4^{2-}$, the concentration was as high as 4001 mg/L, the total salinity was 12179 mg/L, and the pH value was 7.6. Due to the corrosion of sulfate ions, the concrete in the main inclined shaft was corroded and peeled in different degrees, resulting in water gushing, watering, and dripping in different degrees at the top, side, and bottom of the shaft roadway, especially in the quicksand section, which would seriously threaten the shaft safety and coal mine production. Silica fume (SF) was one of the industrial wastes with a large discharge capacity in China (Li 2017), and its large amount of storage had a great impact on the environment. Therefore, it was of great practical significance to study the utilization of silica fume. To reduce the influence of sulfate ions on concrete, the influence of silica fume on the sulfate resistance of concrete from the aspect of mineral admixtures was studied.

Sulfate corrosion was an important factor affecting the durability of cement-based materials. Domestic and foreign scholars had carried out a lot of research on its erosion mechanism and anti-corrosion methods (Liu 2013, 2014), mainly focusing on improving the impermeability of cement-based materials by reducing the content of tricalcium aluminate in cement, and adding fly ash,

*Corresponding Author: xuganggang@cctegxian.com

DOI: 10.1201/9781003450818-53

calcium aluminate, and calcium aluminate in cement Slag and other mineral admixtures were used to improve the corrosion resistance of cement-based materials in sulfate attack environment (Ding 2018; Li 2012). SF was a by-product of the industrial production of silicon or silicon steel. It was composed of ultra-fine amorphous SiO_2 microspheres with high pozzolanic activity and micro aggregate effect. Adding SF into concrete could improve the fluidity, filling, stability, mechanical properties, and durability of concrete (He 2013; Yao 2022). However, there were few studies on improving the corrosion resistance of cement-based materials by adding SF in a sulfate attack environment (Huang 2020; Yan 2015); therefore, the influence of SF on the mechanical properties of cement slurry and the effectiveness of sulfate resistance were analyzed through the full immersion test of cement mortar in Na_2SO_4 solution, which provided the basis for grouting reinforcement construction in this special environment.

2 RAW MATERIALS AND TEST METHODS

2.1 *Raw materials*

In this experiment, P·O42.5 cement produced by Gezhouba Shimen Special Cement Co., Ltd. was used; SF was micro SF produced by Henan Fengkai refractory Co., Ltd.; sand for mixing was Xi'an Weihe River sand with fineness modulus of 2.1, which belonged to medium sand. Na_2SO_4 solution was prepared with a chemical analysis reagent, and its mass fraction was 0.6%, 3%, and 15%. The water came from the tap water of the laboratory.

2.2 *Test method*

In the specifications of GB/T749-2008 "Test Method for Sulfate Resistance of Cement" and GBT50082-2009 "Test Method Standard for Long-term Performance and Durability of Ordinary Concrete", the erosion test of cement or concrete was conducted only after cement mortar specimens and concrete specimens were cured to the strength value of 28. However, in the sulfate attack environment, the grouting project was attacked by sulfate ions when the slurry entered the sulfate attack stratum, and the whole process of slurry coagulation was attacked by sulfate. Therefore, to truly simulate the actual working conditions, and referring to the above specifications of cement sulfate attack resistance, the specimen size was 40mm × 40mm × 160mm, and the designed water-cement ratio was 1. To speed up the erosion test process, the cement and sand ratio was designed to be 1:4.5, and the dosage of SF was 0 and 20%, respectively. After the test piece was formed, it was put together with the test mold into the water and the corresponding prepared Na_2SO_4 solution for soaking and curing. The container was then closed. After the slurry was finally set, the mold was dismantled, and then the test piece was put into the container for further maintenance. Considering the flow of groundwater in the field, the pH value and sulfate ion concentration in the solution remains unchanged, so the erosion solution should be replaced every two months, and after reaching the predetermined soaking age (1M, 3M, 5M, 10M, 12M, and 19M), the flexural strength and compressive strength tests should be carried out respectively.

Table 1. Mortar mixing ratio and curing conditions.

Test number	Cement content/%	Silica fume content/%	Water/%	Medium sand/%	Curing conditions			
PO	100	0	100	450	tap water	0.6% Na_2SO_4 solution	3% Na_2SO_4 solution	15% Na_2SO_4 solution
SF05	95	5						
SF10	90	10						
SF15	85	15						
SF20	80	20						

The sulfate corrosion resistance of the specimen was expressed by the flexural and corrosion resistance coefficient (FCRC).

$$K = R_s/R_w \tag{1}$$

where k was the corrosion resistance coefficient of the specimen; R_s was the flexural strength of the specimen soaked in the solution for a predetermined age, MPa; R_w was the flexural strength of the specimen soaked in water for a predetermined age, MPa.

3 TEST RESULTS AND DISCUSSION

3.1 *Influence of SF content on mortar strength*

The compressive and flexural strength of mortar specimens of different ages for 1–19 months in water is measured, as shown in Figures 1 and 2. It can be seen from Figure 2 that, as a whole, the compressive strength of mortar specimens with different SF content increased with the prolongation of curing time in tap water and increased with the increase of SF content. At 0–5M, the compressive strength of PO, SF05, SF10, SF15, and SF20 grout stones increased linearly and rapidly, and at 6–19M, the compressive strength increased slowly and tends to be stable.

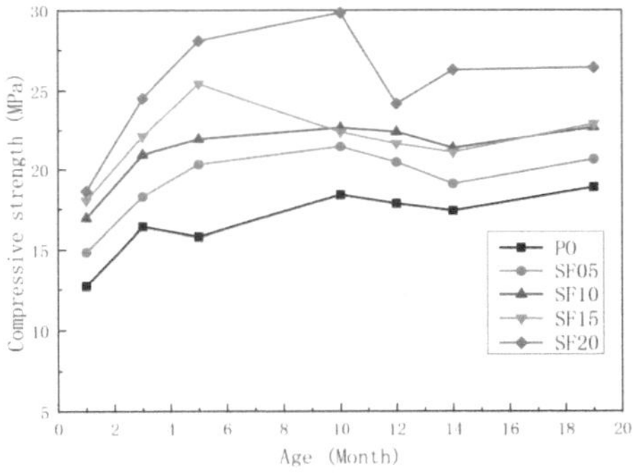

Figure 1. Compressive strength of mortar specimens at different water curing ages.

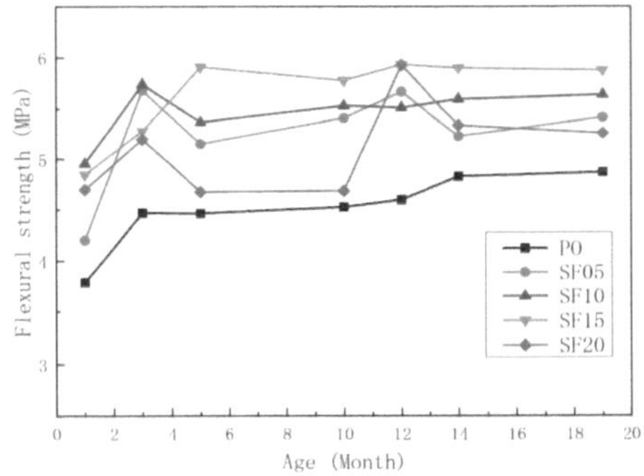

Figure 2. Flexural strength of mortar specimens at different water curing ages.

It can be seen from Figure 2 that the flexural strength of mortar specimens with different SF content increased with the prolongation of curing time in the water. The flexural strength of mortar specimens added with SF increased with the increase of SF content and decreased with the decrease of SF content, only the change trend of SF20 was abnormal. At 0–3M, the flexural strength of PO, SF05, SF10, SF15, and SF20 mortar specimens increased rapidly. At 3–5M, PO remained unchanged, and the flexural strength of SF05, SF10, and SF20 decreased rapidly, while SF15 showed an increasing trend. At 5–10M, the flexural strength increased slowly or remains unchanged; at 10–19M, the flexural strength of PO, SF05, SF10, SF15, and SF20 increased slowly and gradually became stable.

3.2 *Influence of SF content on sulfate attack resistance of mortar specimens*

By measuring the flexural strength and mass loss of mortar specimens in different ages for 1–19 months under the conditions of water culture and Na_2SO_4 solution, the flexural and corrosion resistance coefficient (hereinafter referred to as corrosion resistance coefficient) was calculated according to Formula 1, and the influence of SF content on sulfate attack resistance of mortar specimens was analyzed, respectively.

The variation curve of the corrosion resistance coefficient of mortar specimens in 0.6% Na_2SO_4 solution was shown in Figure 3. At 0–5M, the corrosion resistance coefficients of PO, SF05, and SF10 increased with the prolonging of curing time, while the corrosion resistance coefficients of SF15, SF20, and PO decreased with the prolonging of curing time, all of which were greater than 1. At 5–19M, the corrosion resistance coefficient of PO increased slightly and then tended to decrease. The corrosion resistance coefficients of SF05 and SF10 decreased rapidly with the increase of erosion time, but the corrosion resistance coefficients were all higher than PO, while the corrosion resistance coefficients of SF15 and SF20 increased slightly and then gradually decreased, and the final values were all greater than 1.

At 1M erosion time, the corrosion resistance coefficient of mortar specimens with SF was higher than that of PO specimens without silica fume. Within 1–19M erosion time, the corrosion resistance coefficients of SF05 and SF10 mortar specimens were all greater than PO, while the corrosion resistance coefficients of SF15 and SF20 mortar specimens were less than PO, which indicated that adding 0–10% SF could significantly improve the sulfate corrosion resistance of cement-based materials in a low sulfate ion erosion environment.

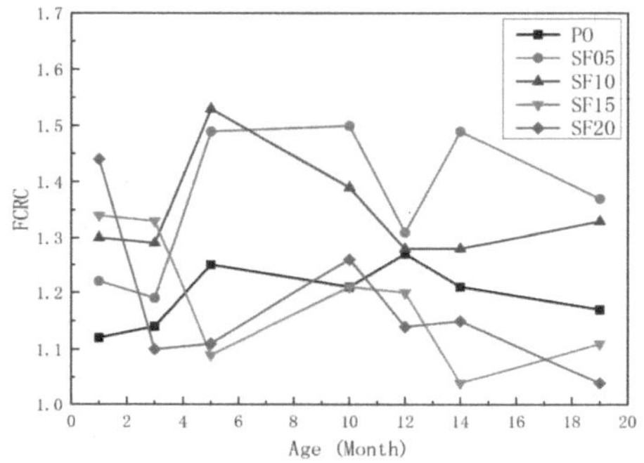

Figure 3. Variation curve of FCRC in 0.6% Na_2SO_4 solution.

The variation curve of the corrosion resistance coefficient of mortar specimens in 3% Na_2SO_4 solution was shown in Figure 4. At 0–3M, the corrosion resistance coefficients of SF05, SF15, and SF20 decreased with the prolongation of curing time, while the corrosion resistance coefficients of PO and SF10 increased with the prolongation of curing time. At 3–5M, the corrosion resistance coefficients of PO, SF05, SF10, and SF20 increased with the extension of curing time, while the corrosion resistance coefficient of SF15 decreased with the extension of curing time. When it was 5–19M, the corrosion resistance coefficients of PO, SF15, and SF20 decreased with the prolonging of curing time, while the corrosion resistance coefficients of SF05 and SF10 mortar specimens decreased first and then increased with the increasing of erosion time, and tended to be stable. At the same time, in the erosion age, the corrosion resistance coefficients of PO mortar specimens and mortar specimens with different amounts of SF were all greater than 1.

At 1M erosion time, the corrosion resistance coefficient of mortar specimens with SF was higher than that of PO specimens without silica fume. At 0–3M, on the contrary, the corrosion resistance coefficient of SF05, SF10, SF15, and SF20 was smaller than those of PO. At 3–5M, the corrosion resistance coefficient of SF05 and SF10 were larger than those of PO, while the corrosion resistance coefficient of SF15 and SF20 were smaller than those of PO. At 5–14M, the corrosion resistance of SF05, SF10, SF05, and SF20 with SF was larger than that of PO. The anti-corrosion coefficients of 14–19M, SF05, and SF10 mortars were all greater than PO, while the anti-corrosion coefficients of SF15 and SF20 gradually decreased but were less than PO. It showed that in a 3%Na_2SO_4 corrosive environment, adding 0∼10% SF can significantly improve the sulfate resistance of cement-based materials.

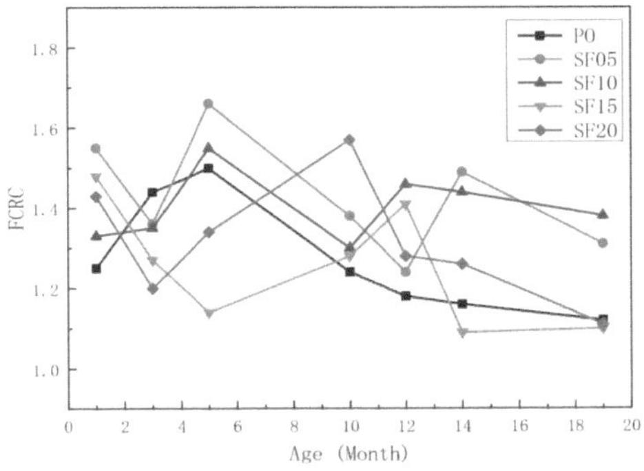

Figure 4. Variation curve of FCRC in 3% Na_2SO_4 solution.

The variation curve of the corrosion resistance coefficient of mortar specimens in 15% Na_2SO_4 solution was shown in Figure 5. At 0–3M, the corrosion resistance coefficient of SF05, SF15, and SF20 mortar specimens decreased with the prolonging of curing time, while that of PO and SF 10 mortar specimens increased slightly with the prolonging of curing time. At 3–5M, the corrosion resistance coefficient of PO, SF10, SF15, and SF20 mortar specimens increased with the increase in erosion time, but the corrosion resistance coefficient of SF05 decreased with the increase in erosion time. At 5–19M, the corrosion resistance coefficient of PO, SF05, and SF10 mortar specimens decreased gradually with the increase of erosion time, while the corrosion resistance coefficient of SF15 and SF20 mortar specimens decreased first and then increased slowly with the increase of erosion time, and became stable.

At 0–3M, the corrosion resistance of SF5, SF10, SF15, and SF20 mortar specimens was larger than that of PO, while at 3–19M, the corrosion resistance of SFO5 and SF10 mortar specimens was smaller than that of PO, while the corrosion resistance of SF15 and SF20 mortar specimens was larger than that of PO. It showed that in the environment of 15%Na_2SO_4, adding 15∼20% SF can significantly improve the sulfate resistance of cement-based materials.

Using SF to replace part of cement (0∼20%) could significantly improve the resistance of mortar or concrete to sulfate attack. This was because, on the one hand, the content of the initial tricalcium aluminate in the mortar was reduced, thus reducing the content of hydrated

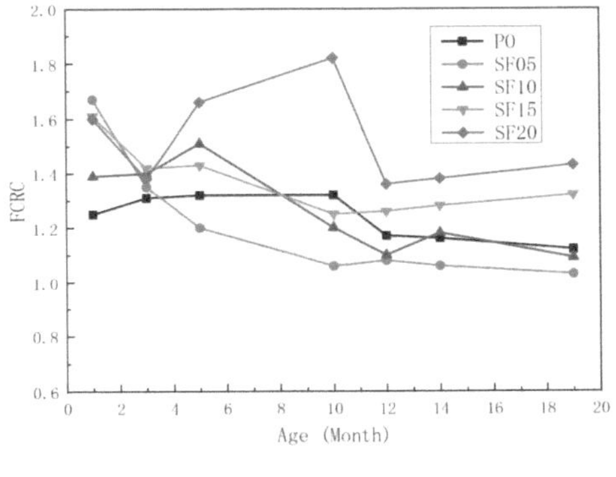

Figure 5. Variation curve of FCRC in 15% Na_2SO_4 solution.

calcium aluminate and the subsequent ettringite. On the other hand, the active substances in SF reacted with calcium hydroxide, the hydration product of cement, to produce hydrated calcium silicate, which not only reduced the content of calcium hydroxide in the cement paste but also the newly formed cement paste gel could fill the pores in the structure, improve the compactness of mortar specimens, and prevent SO_4^{2-} ions from invading into the mortar specimens.

4 CONCLUSIONS

By simulating the sulfate attack environment in mines, the influence of SF on the sulfate attack resistance of cement mortar was studied, and the following conclusions were drawn:

(1) When curing in water, the flexural strength and compressive strength of mortar specimens with different amounts of SF increased significantly with the increase of curing age, and with the increase of SF content.
(2) Adding 5%–20% SF to water-injected mud in a sulfate attack environment could significantly improve its sulfate attack resistance. For a sulfate attack environment with a Na_2SO_4 concentration of less than 3%, adding 0-10% SF could significantly improve the sulfate attack resistance of cement-based materials. However, for a 15% Na_2SO_4 sulfate attack environment, to improve the corrosion resistance of cement slurry, the optimal SF content was 15~20%.

REFERENCES

Ding Y.F., Zeng L., Zhang J.D., et al (2018). Anti-corrosion Test of Tower Foundation Concrete for Changji-Guquan ±1100 kV UHVDC Power Transmission Project. *New Building Materials* 45(8), 92–95.
He X.F., Lu J.T., Li X.N., et al (2013). Progress in Research on the Effect of Silica Fume on the Performance of Cement Concrete. *Journal of the Chinese Ceramic Society*. 32(3), 423–428.
Huang F.L., Zhao Q.J., Tao J.Q., et al (2020). Gel Material Component of Tunnel Lining Concrete under the Environment of Sulfate Attack. *China Railway*. 3, 120–125.
Li H., Sun W., Zuo X.B. (2012). Effect of Mineral Admixtures on Sulfate Attack Resistance of Cement-based Materials. *Journal of the Chinese Ceramic Society*. 40(8), 1119–1126.
Li Y.X., Cao Y.D., Zhang J.S., et al (2017). Current Situation of Comprehensive Utilization of Silica Fume in China and Analysis of Existing Problems. *Applied Chemical Industry*. 46(10), 2031–2034
Liu C., Ma Z.C., Liu H.Y. (2013). A Review on Sulfate Corrosion of Cement Concrete. *Material Guide a Review*. 27(4), 67–71.
Liu J., Niu D.T., Song H. (2014). Influences Brought by Admixtures to Sulfate Corrosion of Concrete. *Concrete*. 3(293), 79–82.
Yan B.G., Lu W.S., Yang P. (2015). Experimental Study on the Application of Silica Fume in a Mine Filled with Seawater. *Mining Research and Development*. 35(10), 6–9.
Yao L.H., Wang Y.P., Chen C.X., et al (2022). Effect of Silica Fume on Properties of Fly Ash-Slag based Geopolymer. *New Building Materials*. 49(05), 48–52.
Zhu M.M., Liu S.M., Ren Z.G., et al (2020). Experimental Study on the Improvement of Sulfate Resistance of High-strength Concrete with Mineral Admixtures. *Journal of North China University of Water Resources and Electric Power (Natural Science Edition)*. 41(6), p 67–72.

Water Conservancy and Civil Construction – Oke & Ahmad (Eds)
© 2024 The Author(s), ISBN: 978-1-032-58614-4

Research on bonded steel plates of hydraulic tunnel lining structure

Meng Zhou
China South-to-North Water Diversion Middle Route Co., Ltd., Beijing, China

Zhiyong Zhou*
Wuda Jucheng Structure Co., Ltd., Wuhan, China

ABSTRACT: Aiming at the technical problem of lining reinforcement of hydraulic tunnels, this paper conducts tunnel reinforcement research on a hydraulic engineering tunnel by bonded steel plates. The causes of cracks were analyzed by finite elements, and then in-depth finite element analysis was carried out before and after the reinforcement of the lining structure in combination with the reinforcement scheme, and the conclusion of the effectiveness of bonded steel plates was obtained, which could provide an engineering basis for the reinforcement treatment of similar tunnels.

1 INTRODUCTION

Cracks are the main diseases of concrete lining structures of hydraulic tunnels, which have different degrees of impact on the normal use and safe operation of concrete structures [Cyclic 2007; Zhu 1999]. Cracks reduce the service life of the lining structure [Fan 2011; Ye 2012], and also accelerate the carbonization of concrete and rust of rebar, further affecting the durability and safety of the structure. Therefore, it is necessary to reinforce the tunnel lining structure promptly.

2 COMMON REINFORCEMENT SCHEMES

The commonly used methods for reinforcement and reinforcement of lining structures mainly include the viscous fiber sheet reinforcement method, bonded steel plates, lining steel structure arch reinforcement method, and enlarged section reinforcement method [Qiu 2022; Wang 2020; Xu 2017]. The viscous fiber sheet reinforcement method is to paste the fiber sheet on the inner surface of the lining member through a structural adhesive, which is more suitable for the reinforcement of the tensile bending member and can withstand the tensile stress of the surface. The bonded steel plates can enhance the bearing capacity of the lining structure, effectively limit the development of cracks and cracking on the inside of the concrete structure, and strengthen the anchorage between the steel plate and the concrete. The focus on the reinforcement of the lining structure and the reinforcement method of lined steel structure arch can also better enhance the bearing capacity of the lining structure, and the overall rigidity of the structure is also obvious, but this method has strict requirements for the height of tunnel water. The method of increasing cross-section reinforcement is to increase the cross-sectional area of the original lining structural components and add steel bars through the same material to improve their bearing capacity and improve structural rigidity.

*Corresponding Author: 124411319@qq.com

DOI: 10.1201/9781003450818-54

3 RESEARCH ON REINFORCEMENT SCHEME BY BONDED STEEL PLATES

Based on the advantages of convenient construction, small self-weight, and good reinforcement effect of bonded steel plates, this method is widely used for reinforcement of hydraulic tunnel lining structure, and this paper takes the tunnel lining bonded steel of an urban water conservancy project as an example for in-depth research.

3.1 *Project overview*

The tunnel has been in use for more than ten years and is located in the surrounding rock of III Class, and IV class, using the new Austrian construction, after the tunnel excavation and spray anchor support. The tunnel passes through the hill, the maximum depth is 30m, the total length of the cave is more than 200 meters, the section (Figure 1) is in the form of a "city gate cave", the upper semi-circular arch, the lower straight side wall, flat bottom, is a pressureless tunnel.

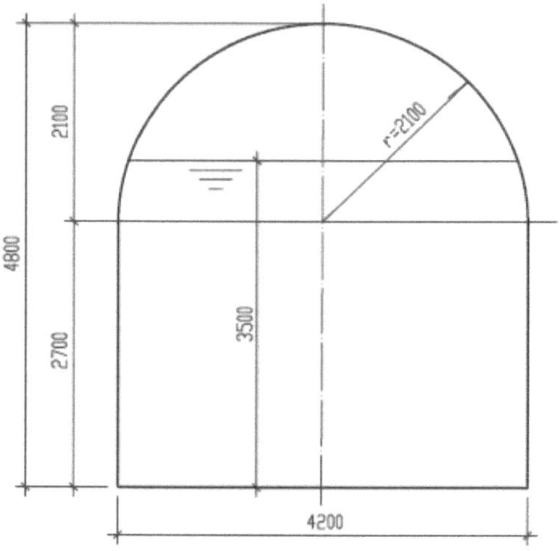

Figure 1. Cross-sectional view of the tunnel.

During a water shutdown for maintenance, an obvious longitudinal crack appeared on the left and right straight side walls of the tunnel lining, and support treatment was carried out here; one year later, the water was stopped again for maintenance, and it was found that the tunnel produced more cracks, and the proportion of the crack length of each section to the total length of the tunnel was between 10%~30%. Geological radar image detection found that there was a phenomenon of filling water between the surrounding rock and the lining in many cave sections, and the preliminary analysis was that the lining caused longitudinal cracks in the left and right walls under the action of hydrostatic pressure.

3.2 *Fracture genesis analysis*

To further analyze the cause of lining cracks, the lining was taken as the analysis object and considered according to the plane strain problem. The lining is longitudinally per unit length, the thickness is 250mm, and the concrete grade is C25 [SL 279 2016, SL191 2008]. According to the empirical formula for calculating the surrounding rock pressure of thin layer and fragmentation structure, the vertical uniform surrounding rock pressure of

31.8~47.8 kN/m^2 and the horizontal uniform surrounding rock pressure was obtained at 8.8~17.6 kN/m^2. To be considered unfavorably, the surrounding rock pressure is divided into two load combinations:

(1) The vertical uniform surrounding rock pressure is 31.8 kN/m^2, and the corresponding horizontal uniform surrounding rock pressure is 17.6 kN/m^2;
(2) The vertical uniform surrounding rock pressure is 47.8 kN/m^2, and the corresponding horizontal uniform surrounding rock pressure is 8.8 kN/m^2.

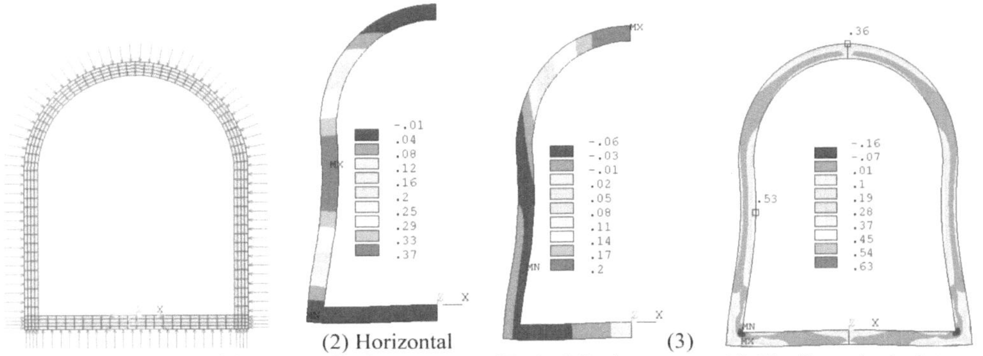

(1) Calculation model (2) Horizontal displacement (3) Vertical displacement (4) The first principal stress

Figure 2. Lining calculation model and calculation results.

It is modeled and analyzed with ABAQUS finite element software. The lining calculation model is shown in Figure 2(a). The displacement unit is mm, and the stress unit is MPa. The horizontal displacement of the lining is shown in Figure 2(b), there is a displacement in the middle of the side wall pointing to the center of the tunnel, and the maximum horizontal displacement is 0.37mm. The vertical displacement of the lining is shown in Figure 2(c), the middle part of the top arch is moved upward, and the maximum vertical displacement is 0.2 mm. The main stress of the first lining is shown in Figure 2(d), and there is a large tensile stress of 0.53 MPa on the tunnel side at a height of 2.4 m in the middle of the side wall. According to the tensile strength of 1.78MPa, it is equivalent to cracking when the head is 1.78/0.53 = 3.4m. The upper part of the middle of the top arch has a large tensile stress of 0.36MPa, which is calculated according to the tensile strength of 1.78MPa, which is equivalent to cracking when the head is 1.78/0.36 = 4.9 m.

The above calculation results show that under the action of external water pressure, the plain concrete lining of the city gate hole is easy to produce horizontal cracks in the inner middle of the straight-side wall.

3.3 *Scheme design of bonded steel plates*

The tunnel is affected by the surrounding rock pressure and groundwater osmosis pressure, which causes greater tensile stress on the cave-shaped lining of the city gate. The low tensile strength of plain concrete structures leads to the cracking of lined side walls, floor slabs, and top arches, causing serious safety hazards, and effective reinforcement measures must be taken to ensure structural safety. Therefore, for the lining structure of the hydraulic tunnel, the scheme design of bonded steel plates is proposed:

Given the low tensile strength of plain concrete lining, pasting a layer of 10mm thick steel plate on the inner surface of the lining is equivalent to transforming the lining into a single-layer reinforced concrete structure. To ensure that the steel plate and the original concrete structure can form a whole and work together, structural glue is poured between the steel plate and concrete, and the connection is strengthened by planting bars.

At the same time, the following reinforcement measures shall be taken:

(1) Grouting treatment is adopted for the hollow part of the surrounding rock lining. The elastic binding of the surrounding rock on the lining by grouting is increased and rock fractures are plugged to reduce the adverse effects of groundwater on the lining.
(2) Grouting the lining cracks to restore the integrity and strength of the lining.
(3) Rock bolts are driven into the side walls and top arches to apply to prestress to improve the tensile stress state of the outer surface of the lining.
(4) Fill drainage pipes to reduce the influence of groundwater osmotic pressure on the lining.

Bonded steel plates are shown in Figure 3.

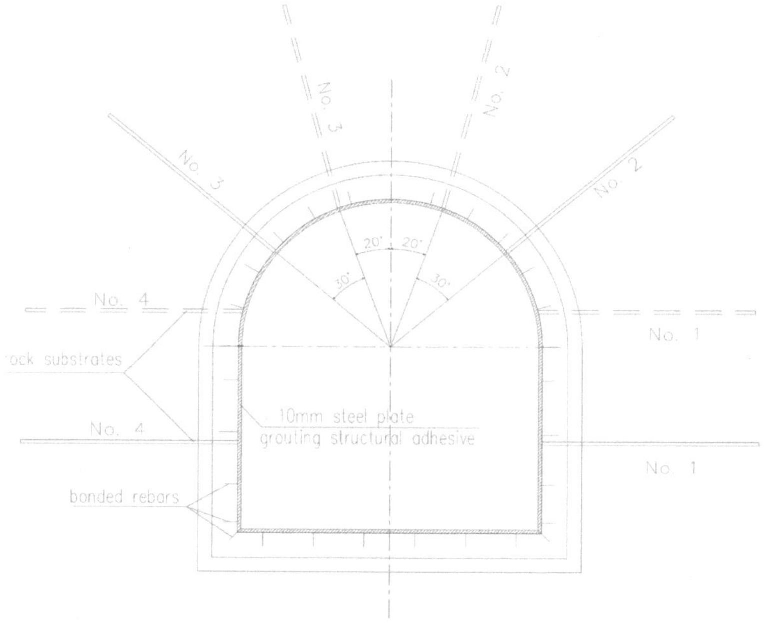

Figure 3. Reinforcement and reinforcement design.

3.4 *Finite element calculation and analysis*

3.4.1 *Computational model*

The finite element model of the reinforced pre-lining structure and the overall cofferdam is shown in Figure 4, and the tunnel lining and the surrounding rock are defined as frictional contact, which constrains the two directions of freedom on the left and right sides of the surrounding rock and at the bottom. The tunnel is excavated using the new Austrian method, and the surrounding rock outside the tunnel is self-balancing due to the new Austrian construction process. Therefore, the self-weight of the outer surrounding rock was not taken into account when performing the finite element analysis. The load applied to the inner lining is shown in Figure 5.

After the tunnel lining is treated by the reinforcement design scheme, the same method is used to model the reinforced lining and surrounding rock, and the reinforcement effect is analyzed. The hardened model should reflect the following:

(1) A 10 mm-thick steel plate is pasted on the inner surface of the tunnel lining, and the steel plate and the concrete lining are connected to work together as a whole;

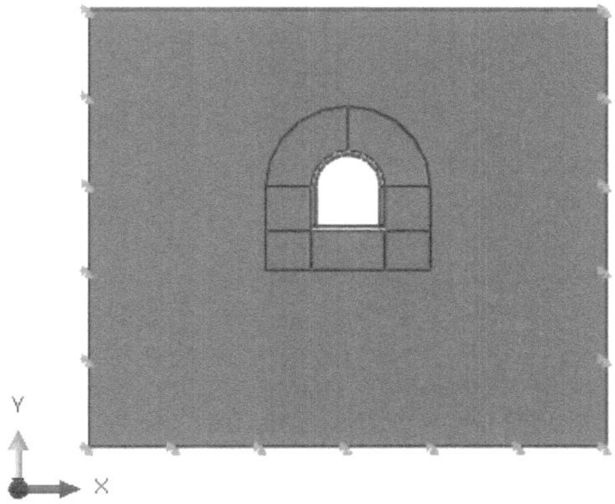

Figure 4. Overall finite element model.

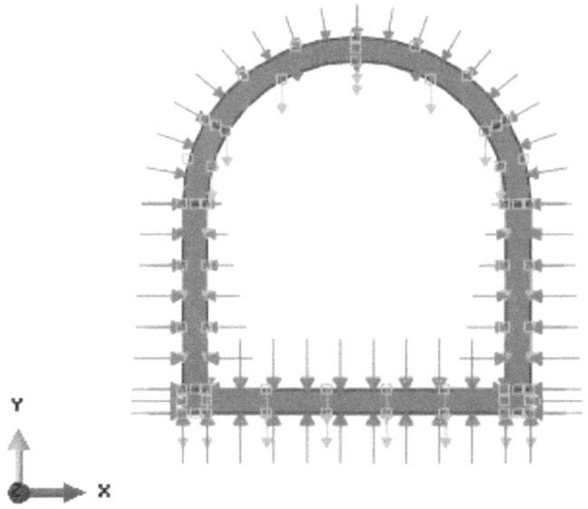

Figure 5. Schematic diagram of lining load.

(2) A 30kN prestressed anchor is provided on the side wall and top arch lined in the tunnel, and the anchor is buried at a depth of 3m. Four bolts are arranged along the side walls and top arches of the tunnel cross-section, and there are two ways to arrange them (Figure 6). The bolts are arranged in a group with a longitudinal interval of 1.5m along the hole, each group adopts one of the arrangements, and the two arrangements pass through to form a tunnel plum blossom arrangement.

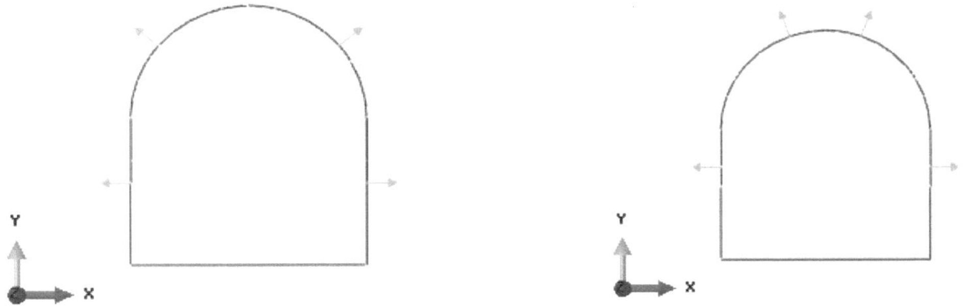

(a) Way 1 (b) Way 2

Figure 6. Two arrangements for prestressed bolts.

The prestressed bolt acts on the steel plate at one end and on the surrounding rock at the other. To simulate the reinforcement effect more accurately, the cross-sections of these two arrangements are calculated separately. The reinforcement effect of the bolt longitudinally converted into a length of 1 m along the tunnel is equivalent to applying 20kN prestress to a single bolt of the calculation model.

3.4.2 *Calculation of working conditions*

According to the calculated load specified in the specification, considering the two working conditions of maintenance and operation, and considering the most unfavorable surrounding rock pressure, there are four calculation conditions:

Maintenance conditions: surrounding rock pressure (1) + groundwater 20m + lining weight;
Maintenance conditions: surrounding rock pressure (2) + groundwater 20m + lining weight;
Operating conditions: surrounding rock pressure (1) + groundwater 20m + lining weight + internal water pressure;

Operating conditions: surrounding rock pressure (2) + groundwater 20m + lining weight + internal water pressure.

3.4.3 *Calculation results*

Considering that the tunnel concrete lining is cracked due to tension, the calculation results of the plane problem give the first principal tensile stress. The lined side walls and floor plates are horizontal and vertical, and the calculation results also give horizontal positive stress and vertical positive stress, as well as horizontal and vertical displacement. The typical cross-sectional positions of the lining structure are shown in Figure 7(a), and 9 groups of typical cross-section positions are selected to extract the stress values; The typical cross-sectional position of the reinforced steel plate is shown in Figure 7(b), and nine groups of positions are selected to extract the Mises stress value. The statistical results of maximum main tensile stress before and after lining reinforcement under maintenance conditions are shown in Table 1, and the statistical results of maximum main tensile stress before and after lining reinforcement under operating conditions are shown in Table 2. The statistical results of Mises stress when the steel plate is calculated with the surrounding rock pressure (1) are shown in Table 3, and the surrounding rock pressure (2). The statistical results of Mises stress when participating in the calculation are shown in Table 4.

(a) Typical cross-sectional position of the lining (b) Typical cross-sectional position of steel plate

Figure 7. Schematic diagram of the typical cross-sectional position.

Table 1. List of maximum main tensile stresses before and after lining reinforcement under maintenance conditions.

Place Working conditions		Before Reinforcement (MPa).		After reinforcement (MPa).			
				Surrounding rock pressure (1).		Surrounding rock pressure (2).	
		Surrounding rock pressure (1).	Surrounding rock pressure (2).	Way 1	Way 2	Way 1	Way 2
Top arch	1	−0.02	−0.02	−0.05	−0.05	−0.05	−0.05
	1'	−0.13	−0.12	−0.12	−0.12	−0.13	−0.13
	2	−0.03	−0.04	−0.08	−0.09	−0.09	−0.09
	2'	−0.08	−0.08	0.32	0.34	0.36	0.22
	3	−0.03	−0.03	−0.02	−0.02	−0.02	−0.03
	3'	−0.05	−0.09	−0.10	−0.10	−0.11	−0.11

(*continued*)

Table 1. Continued

Place Working conditions		Before Reinforcement (MPa).		After reinforcement (MPa).			
				Surrounding rock pressure (1).		Surrounding rock pressure (2).	
		Surrounding rock pressure (1).	Surrounding rock pressure (2).	Way 1	Way 2	Way 1	Way 2
Side walls	4	**2.65**	**2.33**	0.82	0.72	0.67	0.54
	4'	-0.11	-0.10	-0.12	-0.12	-0.11	-0.11
	5	**4.70**	**4.11**	1.19	1.25	0.92	0.95
	5'	-0.12	-0.11	-0.13	-0.12	-0.12	-0.12
	6	0.03	0.03	0.13	0.14	0.12	0.12
	6'	**5.53**	**5.32**	**3.91**	**3.88**	**3.79**	**3.73**
Motherboard	7	0.11	0.11	0.11	0.11	0.12	0.12
	7'	**10.95**	**10.90**	**7.74**	**7.75**	**7.68**	**7.69**
	8	**1.36**	**1.58**	0.71	0.70	0.80	0.79
	8'	-0.08	-0.08	-0.09	-0.09	-0.09	-0.09
	9	**8.13**	**8.36**	**3.21**	**3.19**	**3.34**	**3.32**
	9'	-0.11	-0.11	-0.12	-0.12	-0.12	-0.12

Table 2. List of maximum principal tensile stresses before and after lining reinforcement under operating conditions.

Place Working condition		Before Reinforcement (MPa).		After reinforcement (MPa).			
				Surrounding rock pressure (1).		Surrounding rock pressure (2).	
		Surrounding rock pressure (1).	Surrounding rock pressure (2).	Way 1	Way 2	Way 1	Way 2
Top arch	1	−0.02	−0.02	−0.05	−0.05	−0.04	−0.05
	1'	−0.11	−0.12	−0.12	−0.12	−0.13	−0.13
	2	−0.03	−0.04	−0.08	−0.09	−0.09	−0.09
	2'	−0.08	−0.07	0.38	0.26	0.30	0.13
	3	−0.03	−0.03	−0.02	−0.03	−0.03	−0.04
	3'	−0.08	−0.09	−0.10	−0.10	−0.11	−0.11
Side walls	4	**2.25**	**1.92**	0.68	0.55	0.50	0.37
	4'	−0.11	−0.11	−0.12	−0.12	−0.11	−0.11
	5	**3.94**	**3.39**	0.93	0.96	0.67	0.69
	5'	−0.12	−0.12	−0.13	−0.13	−0.12	−0.12
	6	−0.01	−0.01	0.07	0.07	0.06	0.06
	6'	**4.14**	**3.82**	**2.80**	**2.77**	**2.55**	**2.50**
Motherboard	7	0.06	0.05	0.04	0.03	0.04	0.04
	7'	**7.53**	**7.52**	**5.42**	**5.43**	**5.38**	**5.39**
	8	0.54	0.68	0.32	0.32	0.37	0.36
	8'	−0.10	−0.10	−0.10	−0.10	−0.10	−0.10
	9	**5.14**	**5.31**	**1.93**	**1.92**	**2.03**	**2.02**
	9'	−0.12	−0.12	−0.12	−0.12	−0.12	−0.12

Table 3. Surrounding rock pressure (1) participates in the calculation of Mass stress of steel plate.

	Overhaul (MPa).		Operating conditions (MPa).	
Typical location	Way 1	Way 2	Way 1	Way 2
1	7.91	7.88	7.22	7.24
2	15.77	16.20	14.91	14.71
3	18.32	18.18	15.84	15.85
4	8.24	7.36	6.87	5.77
5	8.12	8.79	6.40	6.89
6	35.63	35.62	28.98	29.09
7	35.94	**36.10**	26.58	26.65
8	7.65	7.51	3.45	3.38
9	23.50	23.35	14.33	14.26

Table 4. Steel plate mises's stress.

	Overhaul (MPa).		Operating conditions (MPa).	
Typical location	Way 1	Way 2	Way 1	Way 2
1	7.10	7.28	5.88	6.52
2	15.32	15.05	13.96	13.53
3	16.85	16.92	14.70	14.76
4	6.82	5.72	5.29	4.14
5	6.13	6.64	4.48	4.93
6	35.32	35.33	28.22	28.32
7	34.91	35.06	25.88	25.98
8	8.61	8.48	4.12	4.02
9	24.46	24.32	15.00	14.91

The analysis led to the following conclusions:

1. The tunnel is supported by prestressed anchors, and the steel plate pasted on the inner side of the lining is reinforced, which significantly improves the tensile stress state of the lining concrete. The tensile strength of the concrete of the side wall and floor part of the lining before reinforcement exceeds the standard design value (data is shown in bold), and the tensile stress of these parts is greatly reduced after reinforcement, and some even meet the requirements of the standard design value.
2. The parts that fail to meet the strength requirements after reinforcement are concentrated in the foot and floor area under the side wall of the lining, which is mainly caused by the osmotic pressure of groundwater. Compared with the crack development of the real structure, it shows that the actual groundwater permeation pressure of the floor is much smaller than the given calculated load, so the calculation result is quite different from the actual load. The addition of drainage pipes to the reinforcement program is necessary to artificially reduce groundwater infiltration pressure.
3. The arrangement of the two prestressed bolts has little effect on the tensile stress of the lining and has little effect on the stress distribution of the steel plate Mises, and the staggered arrangement during actual construction is conducive to improving the stress concentration of the surrounding rock and lining.
4. Although the use of a 10mm thick steel plate cannot give full play to its strength, the stiffness of the thick steel plate is very necessary to effectively reduce the tensile stress of

concrete. After the reinforced concrete structure, if the concrete cracks, the tensile stress released will be transferred and borne by the steel plate. Based on the above calculation results comparison and causal analysis, the safety of tunnel use can be ensured by reinforcement.

4 CONCLUSIONS

(1) Under the action of hydrostatic pressure, the gate hole-shaped plain concrete lining is prone to horizontal cracks on the straight side wall.
(2) The theoretical calculation of the maximum tensile stress on the inside of the lining edge wall is consistent with the crack unfolding position of the side wall on site. The tensile stress calculation result of the floor is large, which does not match the crack on site. The reason is that the concrete and the surrounding rock in this part are closely bonded and do not form an external water pressure surface, and the actual permeable external water pressure of the floor is not calculated and set by the external water pressure, and the external water pressure has an obvious effect on the tensile stress of the floor.
(3) The tunnel lining bonded steel plates was proposed, and the finite element model analysis was carried out before and after the reinforcement, which showed the effectiveness of the reinforcement scheme; At the same time, it is necessary to artificially reduce the osmotic pressure of groundwater by adding drainage pipes to the reinforcement plan. After being treated with this reinforcement scheme, the safe operation of the reinforced tunnel can be ensured.
(4) The research process and conclusions that were given in this paper can provide a reference for the reinforcement treatment of similar tunnels, and can provide an engineering basis for further improving the durability of tunnel safety operation.

REFERENCES

Cyclic Freeze-thaw Damage Prediction of Concrete Structures based on Response Surface Method[J]. *Architecture and Building Materials*, 2007, 21(12): 2031–2040.

Fan Siyi. Discussion on the Diseases and Prevention and Control of Reservoir Leakage and Coagulation Crack Diseases[J]. *Pearl River Water Transport*, 2011, (22): 76–77.

Qiu Zenghao. Analysis of Reinforcement Scheme of Lining Structure of Hydraulic Tunnel[J]. *Heilongjiang Water Conservancy Science and Technology*, 2022, 50(07): 160–162.

SL 279, *Code for Design of Hydraulic Tunnels[S]*. Beijing: China Water Resources and Hydropower Press, 2016.

SL191, *Code for Design of Hydraulic Concrete Structures[S]*. Beijing: China Water Resources and Hydropower Press, 2008.

Wang Xiaoyan. Research on Emergency Reinforcement Treatment Method of Billing Tunnel in Dongjiang Water Source Project[J]. *Shaanxi Water Resources*, 2020, (03): 164–165+168.

Xu Lan. Discussion on Reinforcement Scheme of Lining Structure of Hydraulic Tunnel[J]. *Northeast Water Resources and Hydropower*, 2017, 35(07): 3–5,18,71.

Ye Chunxiu. Review of Treatment Measures for Cracks in Hydraulic Concrete[J]. *Building Materials Development Orientation*, 2012, (1): 117–118.

Zhu Bofang. *Temperature Stress and Temperature Control of Large-volume Concrete[M]*. Beijing: China Electric Power Press, 1999.

Water Conservancy and Civil Construction – Oke & Ahmad (Eds)
© 2024 The Author(s), ISBN: 978-1-032-58614-4

Recent progress and development trends of acoustic emission detection technology for concrete structures

Qi Shi & Jingxian Zhang
CCCC Road & Bridge South China Engineering Co., Ltd. Zhongshan, China

Yuhui Jin*
School of Resources and Safety Engineering, Central South University, Changsha, China

ABSTRACT: Concrete damage research is an important part of building and infrastructure management, and acoustic emission (AE) monitoring has many advantages. However, there are some challenges in the research and application of AE in concrete damage. This paper reviews the existing research on the types and locations of concrete damage sources using qualitative and quantitative AE analysis. This paper uses the literature analysis method to review and summarize the analysis methods and specific applications. In addition, this paper evaluated and prospected the application of AE in the field of concrete by comparing the application of different analytical means, to provide a reference for future research on damage sources.

1 INTRODUCTION

As the most widely used material in the field of construction, the physical and mechanical properties of concrete and its durability have become the subject of numerous scholars' research. Meanwhile, its detection approaches and methods are also improving. In current concrete nondestructive testing, the accuracy and validity of the data are of great concern. Among the current nondestructive testing techniques (NDT), the effectiveness of AE detection has been proved precisely and aroused attention extensively. AE is the phenomenon of transient elastic waves generated by local damage to an object or material under the loading condition (Yang 2016). In 1950, Josef Kaiser discovered that the metal deformation process was accompanied by the AE phenomenon. In 1959, Riisch first applied AE technology in concrete (Li 2017). Currently, the research focuses on the attributes of the injury source and the precision determination of localization. 2455 papers on the use of acoustic emission technology for concrete research are now available on China Knowledge. Among these studies, the focus is on the precise determination of the properties and localization of the damage source. In addition, many scholars have used AE techniques to study concrete damage under corrosion and freeze-thaw cycling conditions. Among them, there are 188 articles on corrosion and 56 articles on freeze-thaw cycles.

2 DAMAGE SOURCE STUDY

2.1 *Damage source generation type study*

The characteristics of the AE signal and parameters are closely related to the nature of the damage source, so the feature of the damage source can be characterized. Studies have

*Corresponding Author: 225511014@csu.edu.cn

DOI: 10.1201/9781003450818-55

shown that microcracks generated during concrete damage can be divided into tensile cracks and shear cracks. Existing studies can describe the damage mechanism of concrete materials not only by analyzing the microcrack types but also by predicting the extent of damage and possible macroscopic fracture patterns (Farhidzadeh 2014). Two types of AE analysis methods, parameter-based and signal-based, are currently the main methods for microcrack determination (Liu 2015; Ohno 2014).

2.1.1 *Parameter-based AE analysis method*

The parameter-based AE analysis method is also known as the qualitative analysis method. This method analyzes concrete members by relating certain parameters in their AE activity to the damage mechanism of the specimen. This analysis method mainly determines the microcrack type by calculating the relative relationship (AF/RA) between the rise angle (RA = rise time/peak amplitude) and the average frequency (AF = ringing count/duration) (Han 2019; Lacidogna 2018; Ranjan 2018). When the AF/RA value is large, the cracks producing AE sources are tension cracks, and vice versa for shear cracks. This method is referred to as the RA-AF method. In the current open literature, different scholars have proposed different methods of AF/RA taking for concrete materials (Abdur Rasheed 2018, 2019; Gan 2020; Li 2017, 2018; Ohno 2010). However, a convincing method for taking AF/RA has not been concluded, and there will be a small error in determining the type of cracks with little accuracy. Therefore, existing studies of concrete damage based on AE parameters mostly use Gaussian mixture models and other aids for analysis.

Table 1. The part about the FA/RA values for distinguishing injury types.

Name	Ohno	Biao Li	Rasheed	Banjara	Yixiong Gan
Recommended value	200	14/25	200	1	90
Time	2010	2017/2018	2018/2019	2019	2020

Xudong Chen *et al.* used GMM (Gaussian Mixture Model) to analyze the acoustic emission parameters of a rock-concrete composite beam with four-point bending, and concluded that the percentage of tensile damage in the whole loading rupture phase is greater than 87.9% (Chen 2022). Das *et al.* studied uniaxial compression damage of concrete and uniaxial tensile damage of SHCC using the Gaussian mixture model and support vector machine (SVM) to give microcracking patterns that dominate concrete damage (Das 2019). Shuizhou Song *et al.* combined the AE technique to carry out experimental studies on the damage characteristics of steel fiber concrete with prefabricated central cracked Brazilian discs (BDCN) under the action of mixed type I-II loading (Song 2022). In the study, a Gaussian mixture clustering method and SVM theory are combined to analyze the acoustic emission parameters. Among them, a hyperplane is given for RA-AF map data, which has an accuracy of over 95.6%. The tensile-type microcracks in all stages of the test exceeded 64%, revealing the damage mechanism of steel fiber concrete and its crack development process.

2.1.2 *Signal-based AE analysis methods*

Signal-based AE analysis methods, i.e., quantitative analysis methods, are also called waveform analysis methods. In this method, the waveform parameters are used as computational data for computational analysis to find out the relevant values of the damage mechanism. To determine the moment tensor of the AE source, Ohtsu developed and implemented SiGMA (simplified green function for moment tensor analysis) (Ohtsu 1998). This analysis includes a three-dimensional (3D) AE source localization procedure and moment tensor analysis of the AE source. The location of the AE source is determined by the difference in arrival time (Yun 2010). Then, the components of the moment tensor are determined from the amplitude of the first motion at the AE channel (Kawasaki 2013). In this regard, SiGMA is a sophisticated method for estimating the size, direction, crack type, location, and fracture mode of individual microcracks (Shigeishi 2001). According to the results of SiGMA analysis, it is known that tensile-type cracks are obtained mainly at the early stage, followed by mixed-mode cracks of tensile and shear types, and finally, shear-type cracks are produced mainly during concrete fracture (Yuyama 1999).

Although the signal-based AE analysis method started later, it is more advantageous in determining the crack mechanism of the member, etc. Parameter-based AE analysis can only characterize a trend of variation. And while signal-based AE analysis methods, such as the moment tensor inversion method, can infer the quantitative value of each point source cracking mechanism generated when the member is damaged. Zonglian Wang *et al.* based on the wavelet transform noise reduction AE source localization method and moment tensor theory to investigate the cracking of concrete specimens containing cutouts on both sides during uniaxial compression and revealed the fine view crack expansion mechanism (2022).

2.2 *Damage source localization*

When a crack occurs in a material, energy is released and the resulting elastic wave is detected by the AE sensor and the location of the crack can be detected by analyzing the data. In the AE test, it is significant how to accurately locate the source location. At the same time, because concrete exhibits anisotropy inside, its damage sources are not as easy to locate as those of metallic materials. Therefore, improving its accuracy and its use in the field of locating large components is critical in the future. As early as 2005, Kurz *et al.* proposed an improved algorithm based on the red pool information criterion (AIC) to improve the accuracy of AE signals to eliminate false or inaccurate information (Kurz 2005). The result shows that the maximum deviation of the localization with onset times determined manually and the AIC-picker values is 11%. In 2012, Aljets *et al.* developed a certain sensor array to identify the location of AE sources and used a new source localization algorithm that facilitated the accuracy of damage source localization for large plate-like structures (Aljets 2012).

However, the most traditional AE source localization method uses constant wave velocity without considering the attenuation of the AE signal. Dongxu Li *et al.* analyzed the effects of water-cement ratio, sand ratio, and maximum aggregate size on the attenuation of standard AE wave velocity (Li 2020). They also developed a distance attenuation model of AE wave velocity according to the degree of influence of each factor on amplitude and wave velocity. Combining the exhaustive and regional localization methods, a regional exhaustive localization method was established based on the modified wave velocity which completely improved the accuracy of localization. Jie Chen *et al.* combined signal processing and deep learning techniques to propose a new method of rock acoustic emission signal localization based on wavelet spectrum analysis and convolutional neural networks. This method can accurately obtain the 3D coordinates of known acoustic emission signal sources (Chen 2022). Shuaijie Miao *et al.* proposed a high-precision fiber optic laser (RFLs) using designed RFLs and an adaptive arrival time extraction strategy for concrete fiber optic AE localization system to localize the weak damage sources. In this system, the discrete layout of 16 PLB points and 4 comparison tests for each point allows a more reasonable assessment of positioning accuracy and system reliability (Miao 2022).

3 THE APPLICATION OF AE IN NON-DESTRUCTIVE TESTING OF CONCRETE

3.1 *Research on AE in reinforced concrete corrosion*

In 2016, Lei Wang *et al.* performed three-point bending loading on reinforced concrete with varying degrees of corrosion and compared the resulting acoustic emission signals (Lei

2016). The study shows that it is feasible to monitor the defect sources during the damage of rusted reinforced concrete beams using acoustic emission technology. However, the modification process is very lengthy in practical engineering, and the specific use of acoustic emission techniques still needs to be studied.

In 2019, Charlotte Van Steen *et al.* proposed a high-precision RFL-AE system for weak source locations on concrete structures (Steen 2019). Combined with a high resolution and adaptive trigger threshold search strategy, the system can extract TDOA with high precision and accurately locate the pseudo-damage source. In their study, a correlation-based aggregated hierarchical clustering algorithm was developed to successfully distinguish signals from corrosion and concrete cracking. The results show that for the small-scale samples, the combination of clustered AE events and micro-CT can help to understand the different damage processes caused by corrosion.

In 2019, Gang Xu *et al.* believed that reinforced concrete materials also produce acoustic emission signals in the early stages of reinforcement corrosion and have an impact on reinforced concrete life (Xu 2019). The experiment was conducted by using the AE technique to monitor the reinforced concrete material in the curing stage and the wet and dry cycles, and the damage was explored by comparing the data from different stages. The results show that the mortar has a large number of acoustic emission signals during the maintenance process, and the number of signals gradually decreases with the development of the hydration process. Among them, the peak frequencies were mainly concentrated in the 2 frequency bands from 20 to 60 kHz and 130 to 200 kHz.

In 2022, Kawasaki Yuma *et al.* used electrochemical noise and AE parameters to analyze the corrosion of reinforcing steel in concrete under galvanic corrosion tests (Yuma 2022). In the paper, oxide film damage corresponds to tensile damage, and oxide film peeling corresponds to shear damage in the assessment of reinforcement corrosion. Later, the oxide film peeling is detected and the subsequent corrosion state is evaluated by combining the electrochemical noise-related parameters. The results show that the compressive strength of the steam-reinforced concrete decreases with the increase in the number of freeze-thaw cycles. Meanwhile, the activity of tension rupture also decreases with the increase in the number of freeze-thaw cycles.

3.2 *Damage study of concrete under the action of temperature fields using AE*

Concrete has been used in construction practically since it was discovered as cement. During their service life, concrete elements are subject to various degradation factors, such as mechanical and chemical influences, and rapid temperature changes. From a scientific point of view, especially from the point of view of practical applications in construction, it is crucial to understand the influence of temperature changes on the quality of concrete during its service life. In addition to the degradation effects of very high temperatures, the alternation of positive and negative temperatures is one of the most damaging operational factors for many concrete products. Freeze-thaw (F-T) cycles can adversely affect the durability of concrete structures in a very rapid way.

In 2020, Hernán Xargay *et al.* tested self-compacting high-strength concrete (SCHSC) and fiber-reinforced concrete (SCHSFRC) samples by exposing them to high temperatures of 300°C and 600°C, then testing them under three-point bending and recording AE data (Xargay 2020). The test results show that the peak flexural strength of concrete is significantly affected by temperature, but the presence of fibers changes the magnitude of residual flexural strength, and their bridging effect inhibits the crack extension. In addition, the increase in thermal damage and the incorporation of fibers transformed the cracking mode into a mixed mode, which is characterized by a gradual decrease in the average frequency and an increase in the RA value. Meanwhile, the article also confirms the potential of AE in examining the damage level of cementitious composites.

In 2021, based on signal analysis, Topolá⊠ Libor *et al.* evaluated the resistance of concrete to the effects of freeze-thaw (Libor 2021). In the article, concrete specimens of different sizes and shapes are tested in freeze-thaw cycles, and their mechanical and acoustic emission parameters are collected. The results show that the principal frequencies are significantly shifted toward lower values by fast Fourier transform analysis of acoustic emission signals when concrete is subjected to freeze-thaw cycles. In addition, there was no significant change in concrete quality after 100 cycles. The decrease in dynamic modulus of elasticity was only 1.6 percentage points. The tensile splitting strength did not decrease after freeze-thaw. The only standard parameter that exhibited some decrease was the flexural strength, which was reduced to 85.3% of the original value.

In 2022, Bo Chen *et al.* investigated the rupture forms and crack development of evaporated concrete under a freeze-thaw environment (Chen 2022). The analysis was carried out by performing 200 freeze-thaw cycles on the steamed concrete specimens in combination with the steamed concrete acoustic emission parameters (including acoustic emission source location, RA-AF value, and b value) during the freeze-thaw cycles. The results showed that the compressive strength of the autoclaved concrete decreased with an increasing number of freeze-thaw cycles. After 0, 50, 100, 150, and 200 freeze-thaw cycles, the strength loss of the autoclaved concrete was 4.23%, 10.2%, 24.7%, and 41.94%, respectively. In addition, the activity of tension rupture also decreased, with a reduced range of 13.54% to 80.23%.

4 CONCLUSION AND EXPECTATION

AE technique is an important method for the damage assessment of concrete structures. With the development of AE technology, the application of AE technology in concrete structure damage assessment is becoming more and more widespread. In this process, more and more AE parameters are used for concrete damage assessment, and the accuracy of this method is improving with the progress of research. Studies using moment tensor theory can effectively obtain initial crack information (crack location, number of cracks, and crack type) and conform well with test results, and even capture internal cracks in concrete in advance. It has an irreplaceable role, especially in freeze-thaw and corrosion.

There are still some problems to be solved in the use of AE in concrete nondestructive testing:

(I) To avoid noise or unwanted AE signals, laboratory tests and field monitoring usually use a two-step method. However, during the filtering process, the entire AE signal needs to be acquired, so the pre-trigger portion of the transient waveform and transient signal needs to be stored. Therefore, in addition to careful benchmarking and virtual sensors, appropriate big data analysis methods are needed in subsequent studies. For example, automatic clustering and (un)supervised machine learning are applied to enhance the processing of acoustic emission data from noisy environments and to improve the accuracy of the collected acoustic emission data.

(II) The accuracy of AE location is not enough. Because concrete is composed of different materials, its interior presents anisotropy. This means that the reflection, refraction, and transmission of waves need to be considered more in future AE localization.

(III) It is still challenging to conduct a comprehensive coupling of AE data and the full range of structures. The advantage of AE technology is to obtain information about the damage process at the local scale. However, the expanded access to information will facilitate the enhancement of knowledge related to structural performance under corrosive conditions of existing members and thus determine the urgency of interventions on concrete structures. Therefore, there is a need to develop a multi-scale monitoring approach. For example, detection is performed using a combination of local acoustic emission techniques and global vibration-based monitoring methods. These two techniques are complementary to each other and can more accurately quantify the structural capacity of deteriorated reinforced concrete structures.

REFERENCES

Abdur Rasheed M., Suriya Prakash S. & Gangadharan R. (2019). Acoustic Emission Characterization of Hybrid Fiber Reinforced Cellular Concrete Under Direct Shear Loads. *Journal of Nondestructive Evaluation* (1).

Abdur Rasheed M., Suriya Prakash S., Gangadharan Raju & Yuma Kawasaki. (2018). Fracture Studies on Synthetic Fiber Reinforced Cellular Concrete using Acoustic Emission Technique. *Construction and Building Materials.*

Alireza Farhidzadeh, Anastasios C. Mpalaskas, Theodore E. Matikas & Dimitrios G. Aggelis. (2014). Fracture Mode Identification in Cementitious Materials using Supervised Pattern Recognition of Acoustic Emission Features. *Construction and Building Materials.*

Avik Kumar Das, Deepak Suthar & Christopher K.Y. Leung. (2019). Machine Learning-based Crack Mode Classification from Unlabeled Acoustic Emission Waveform Features. *Cement and Concrete Research.*

Biao Li, Lihua Xu, Yin Chi. & Changning Li. (2017). Experimental Investigation on the Stress-strain Behavior of Steel Fiber Reinforced Concrete Subjected to Uniaxial Cyclic Compression. *Construction and Building Materials.*

Biao Li, Lihua Xu, Yuchuan Shi. & Changning Li. (2018). Effects of Fiber Type, Volume Fraction, and Aspect Ratio on the Flexural and Acoustic Emission Behaviors of Steel Fiber Reinforced Concrete. *Construction and Building Materials.*

Charlotte Van Steen, Lotfollah Pahlavan, Martine Wevers & Els Verstrynge. (2019). Localization and Characterization of Corrosion Damage in Reinforced Concrete using Acoustic Emission and X-ray Computed Tomography. *Construction and Building Materials.*

Chen Bo, Chen Jialin, Qiang Sheng & Zheng Yongjie. (2022) Experimental Study on the Acoustic Emission of Steam-cured Concrete in the Freeze-thaw Environment. *Journal of Huazhong University of Science and Technology (Natural Science Edition).*

Chen Jie, Chen Ziyang & Pu Yuanyuan. (2022). Acoustic Emission Source Localization in Rocks based on Spectral Analysis and Convolutional Neural Network. *Journal of Rock Mechanics and Engineering*(S2), 3271–3281.

Chen Xudong, Guo Yuzhu, Hu Linagpeng, Bai Ying & Ning Yingjie. (2022). Cluster Identification of Acoustic Emission Parameters for Bending-tensile Damage of Rock-concrete Composite Beams. *Vibration and Shock* (19),274–281.

Dirk Aljets, Alex Chong, Steve Wilcox & Karen Holford. (2012). Acoustic Emission Source Location on Large Plate-like Structures using a Local Triangular Sensor Array. *Mechanical Systems and Signal Processing.*

Gan Yixiong, Wu Shunchuan, Ren Yi & Zhang Guang. (2020). Evaluation Indexes of Granite Splitting Failure based on RA and AF of AE Parameters. *Rock and Soil Mechanics* (07), 2324–2332.

Hernán Xargay, Paula Folino, Nicolás Nuñez. & Enzo Martinelli. (2018). Acoustic Emission Behavior of Thermally Damaged Self-Compacting High Strength Fiber Reinforced Concrete. *Construction and Building Materials.*

Hyun-Do Yun, Won-Chang Choi & Soo-Yeon Seo. (2010). Acoustic Emission Activities and Damage Evaluation of Reinforced Concrete Beams Strengthened with CFRP Sheets. *NDT and E International* (7).

Kawasaki Yuma, Fukui Shinya & Fukuyama Tomoko. (2022). Phenomenological Process of Rebar Corrosion in Reinforced Concrete Evaluated by Acoustic Emission and Electrochemical Noise. *Construction and Building Materials.*

Kentaro Ohno & Masayasu Ohtsu. (2010). Crack Classification in Concrete based on Acoustic Emission. *Construction and Building Materials* (12).

Kentaro Ohno, Kimitaka Uji, Atsushi Ueno & Masayasu Ohtsu. (2014). Fracture Process Zone in a Notched Concrete Beam Under Three-point bending by Acoustic Emission. *Construction and Building Materials.*

Kurz Jochen H, Grosse Christian U & Reinhardt Hans-Wolf. (2005). Strategies for Reliable Automatic Onset Time Picking of Acoustic Emissions and Ultrasound Signals in Concrete. *Ultrasonics* (7).

Lacidogna, Giuseppe; Piana, Gianfranco; Carpinteri, Alberto (2018). Damage Monitoring of Three-point Bending Concrete Specimens by Acoustic Emission and Natural Frequency Analysis. *Engineering Fracture Mechanics.*

Li Dongsheng, Yang Wei, Yu Yan. (2017). *Acoustic Emission Monitoring and Evaluation of Structural Damage in Civil Engineering: Theory, Method, and Application.* Beijing: Science Press.

Li Dongxue, Yang Kang, He Zhaoyi, Zhou Hanlin & Li Jiaqi. (2020). Acoustic Emission Wave Velocity Attenuation of Concrete and Its Application in Crack Localization. *Sustainability* (18).

Liu Jian-po, Li Yuan-hui, Xu Shi-da. & Jin Chang-yu. (2015). Cracking Mechanisms in Granite Rocks Subjected to Uniaxial Compression by Moment Tensor Analysis of Acoustic Emission. *Theoretical and Applied Fracture Mechanics.*

Masayasu Ohtsu, Takahisa Okamoto. & Shigenori Yuyama. (1998). Moment Tensor Analysis of Acoustic Emission for Cracking Mechanisms in Concrete. *Structural Journal* (2).

Miao Shuaijie, Gao Liang, Tong Fengzhuang & Zhong Yanglong. (2022). Research on High Precision Optical Fiber Acoustic Emission System for Weak Damage Location on Concrete. *Construction and Building Materials.*

Mitsuhiro Shigeishi & Masayasu Ohtsu. (2001). Acoustic Emission Moment Tensor Analysis: Development for Crack Identification in Concrete Materials. *Construction and Building Materials* (5).

Nawal Kishor Banjara, Saptarshi Sasmal & V. Srinivas. (2019). Investigations on Acoustic Emission Parameters during Damage Progression in Shear Deficient and GFRP-strengthened Reinforced Concrete Components. *Measurement.*

Prabhat Ranjan Prem, A. Ramachandra Murthy & Mohit Verma. (2018). Theoretical Modeling and Acoustic Emission Monitoring of RC beams Strengthened with UHPC. *Construction and Building Materials.*

Qinghua Han, Guang Yang, Jie Xu. & Alberto Carpinteri. (2019). Acoustic Emission Data Analyses based on Crumb Rubber Concrete Beam Bending Tests. *Engineering Fracture Mechanics.*

Shigenori Yuyama, Zheng-wang Li, Yoshihiro Ito & Masaki Arazoe. (1999). Quantitative Analysis of Fracture Process in RC Column Foundation by Moment Tensor Analysis of Acoustic Emission. *Construction and Building Materials* (1).

Song Shuizhou, Ren Huilzn & Ning Jianguo. (2022). Acoustic Emission Parameters in the Damage Process of Steel Fiber Reinforced Concrete under Mixed Loading. *Journal of Military Engineering* (08), 1881–1891.

Topolá⊠ Libor, Kocáb Dalibor, Pazdera Luboš & Vymazal Tomáš. (2021). Analysis of Acoustic Emission Signals Recorded during Freeze-Thaw Cycling of Concrete. *Materials* (5).

Wang Lei, Zhong Liheng, Xia Hailong & Zhang Xin. (2016). Acoustic Emission (AE) Characteristics of Corroded Reinforced Concrete Beams in Loading Process. *Journal of Building Materials* (04),682–687.

Wang Zonglian, Wang Huaiwei, Ren Huilan, Zhao Mingyan & Luo Zhiqiang. (2022). Inversion of Crack Mechanism in Concrete Materials Based on Moment Tensor Theory of Acoustic Emission. *Journal of Military Engineering* (01),181–189.

Xu Gang, Zeng Zhen, Zhang Rui, Peng Yanzhou & Yang Zewen. (2019). Characteristics of Acoustic Emission Signals from Initial Corrosion of Steel Bar in Cement-Based Materials. *Journal of Building Materials* (03), 385–393+423.

Yang Nengjun, Yao Chunjiang, Yuan Xiaojing. (2016). *Acoustic Emission-based Material Damage Detection Technology.* Beijing: Beijing University of Aeronautics and Astronautics.

Yuma Kawasaki, Tomoyo Wakuda, Tomoe Kobarai & Masayasu Ohtsu. (2013). Corrosion Mechanisms in Reinforced Concrete by Acoustic Emission. *Construction and Building Materials.*

Water Conservancy and Civil Construction – Oke & Ahmad (Eds)
© 2024 The Author(s), ISBN: 978-1-032-58614-4

Experimental and numerical study on the bending behavior of cup joints of prefabricated utility tunnel

Jianqiu Wu*, Lei Han & Xingjie Fang
China Construction Eighth Engineering Division Corp., Ltd.

ABSTRACT: To study the mechanical performance of the cup joint of the prefabricated utility tunnel, three specimens were designed and made for the plane static loading test. The test results show that the embedment depth of the cup joint has a great influence on the failure form of the structure. Failure of cup-groove joint specimens can be divided into three stages: elastic stage, plastic stage, and failure stage. Based on experimental research, the numerical simulation analysis of the loading process of the specimens was conducted using ANSYS and the numerical results are in good agreement with the experimental results, which verifies the validity of the numerical simulation.

1 INTRODUCTION

The prefabrication technology of a utility tunnel is a construction method of a utility tunnel that assembles the components of the utility tunnel into the whole structure on site after prefabrication. Compared with the cast-in-place utility tunnel, the prefabricated utility tunnel has the characteristics of a short construction period, easy quality control, and remarkable comprehensive economic benefits. The common assembly forms of prefabricated utility tunnels include full prefabricated assembly, half prefabricated assembly, and composite prefabricated assembly (Qian & Chen 2007; Tan *et al.* 2016).

Full precast utility tunnel construction speed and construction environment are good, with less manual input, and is suitable for single-room or double-room utility tunnels; however, for multi-room utility tunnels, the prefabrication is too heavy, lifting and transportation difficulties arise, the assembling precision demand is high, and the problem of one-time investment is too large. Furthermore, in full precast utility tunnels, there are too many joints, and waterproof quality is difficult to guarantee. The half-prefabricated utility tunnel combines the advantages of whole cast-in-place and full precast, but there are still some problems such as wet operation and low construction efficiency. The composite precast utility tunnel reduces the on-site formwork of reinforced concrete and avoids large hoisting equipment, but the size and position of the vertical inserts connecting the wall panel to the bottom plate are strictly required. These problems restrict the popularization and application of prefabricated technology in utility tunnel construction (Xue *et al.* 2018; You 2017; You 2018).

Based on the above reasons, this study made a structural improvement on the original prefabricated utility tunnel connection node form by referring to the basic form of the cup foundation commonly used in China and proposed a cup joint suitable for the assembly of the prefabricated utility tunnel. To study the mechanical performance of this cup joint, three specimens were designed and made for a plane static loading test (Hu & Xue 2010; Tian 2016). Based on the test, ANSYS software was used for numerical simulation analysis to further study the stress process, weak parts, and failure mechanism of the specimens, and to verify the effectiveness of the numerical simulation.

*Corresponding Author: Jianqiuw1987@163.com

DOI: 10.1201/9781003450818-56

2 OVERVIEW OF THE TEST

2.1 *Test design*

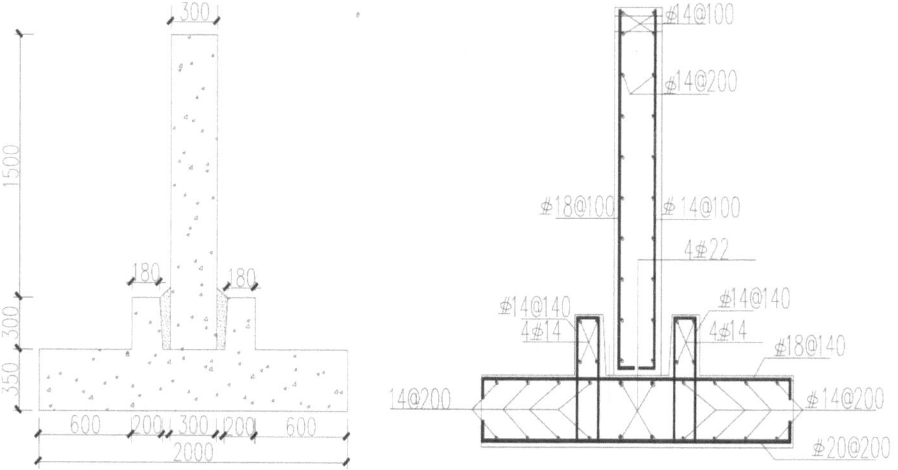

(a) Size of specimen 1 (b) Reinforcement of specimen 1

Figure 1. Diagram of size and reinforcement of Specimen 1. (a) Size of specimen 1 (b) Reinforcement of specimen 1.

Table 1. Geometrical parameters of specimens.

No.	Depth (mm)	Thicknesses of the top (mm)	Thicknesses of the bottom (mm)	Height of the foundation (mm)
1	300	180	200	650
2	300	120	140	650
3	150	180	200	450

Three specimens were designed and made for the test. The structural form, size, and reinforcement of Specimen 1 are shown in Figure 1. Specimens 1 and 2 were selected with different wall thicknesses of cup joints, and Specimens 1 and 3 were selected with different embedment depths to study the influences of different cup wall thicknesses and embedment depths on the forces of cup joints. The specific parameters of the specimen are shown in Table 1. The thickness of the three specimens' prefabricated wall panels is 300 mm, the wall width is 800mm, and the wall height above the top of the foundation is 1500 mm. The precast and cast-in-place parts of the test are made of fine stone concrete, the Concrete strength grade is C30, and the steel bar is HRB400.

2.2 *Test plan*

The loading device of the test is shown in Figure 2. The test uses the electro-hydraulic servo loading system to carry out horizontal loading and process control of the specimen. The single-point loading on the top of the wall is adopted, and the horizontal load is measured and controlled by tension and pressure sensors. The base of the foundation of the specimen is fixed on the ground of the laboratory by the foot bolt to simulate the fixed end constraint condition to limit the joint rotation of the foundation and the wall. Force-controlled loading method was used for horizontal loading in the test, which was divided into two stages: preloading and force-controlled loading. In the formal horizontal loading, 1 kN was loaded in each stage until the specimen was

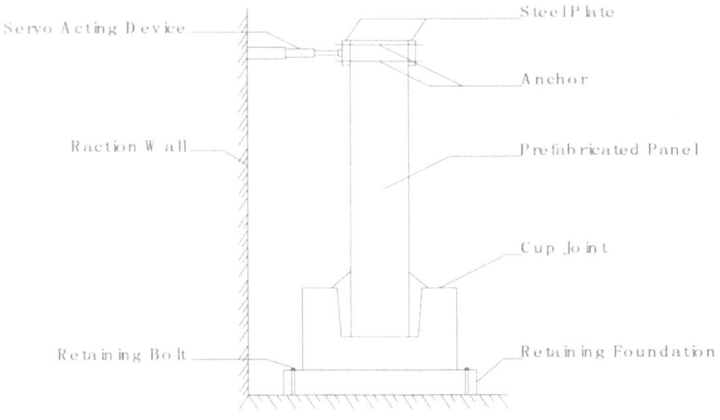

Figure 2. Diagram of the test device.

damaged. During the test, the load, wall plate top displacement, support displacement, and cup joint vertical bar strain were measured, and the crack development was monitored.

3 RESULT OF TEST

3.1 *Failure process of Specimen 1*

At the beginning of loading, each section of the cup joint remained elastic, and no cracks appeared on the surface of the specimen. When the loading was up to 8.5 kN, the first crack appeared along the contact surface between the wall panel and the cup joint on the loading side of the foundation of the 1 cup joint of the specimen. As the loading continued, the crack gradually extended vertically downward from the top of the foundation of the cup and groove. When the loading reached 20.3 kN and 22 kN, the crack expanded to the bottom of the cup and continued to extend rapidly from the outer edge of the loading side along the horizontal direction. In the process of loading to 28.2 kN, the vertical cracks at the bottom of the cup joint on the loading side developed continuously to the bottom of the foundation, while the main cracks widened continuously and appeared with multiple micro-cracks. Continuing to load to 33.1 kN, the horizontal crack continues to extend outward, and the vertical crack continues to extend upward from the bottom of the cup joint along the contact surface between the cup joint and the wall panel. When the load reaches 39.3 kN, the main crack on the loading side of the cup joint expands and widens continuously. The vertical crack extending from the bottom of the cup joint to the bottom of the foundation develops slowly, and the horizontal crack extending to the outside of the cup wall appears. The horizontal cracks in the inner side of the cup joint develop slowly, and the vertical cracks extend to the bottom of the foundation. When the loading was continued to 52.1 kN, the inclined crack appeared at the bottom of the wallboard, the main crack continued to expand and the micro-crack near the main crack increased and widened, and the wallboard and the cup joint gradually separated. When the loading is continued to 60.2 kN, the bottom of the prefabricated wall plate is an inclined crack through, the outer concrete protective layer of the reinforcement is crushed, the prefabricated wall plate is peeled off from the cup joint, and the loading device cannot continue to load, resulting in the specimen being finally damaged. The final failure state is shown in Figure 3.

3.2 *Failure process of Specimen 2*

The failure process of Specimen 2 with the same embedding depth is similar to that of Specimen 1. At the beginning of loading, no surface cracks were found. When the load is up to 6.4kN, the first crack appears along the interface between the wall panel and the cup joint on the loading side of the cup joint. When loading continued, the crack gradually extended vertically from the top of the foundation of the cup joint. When the load reached 12.9 kN, the crack extended to the bottom of the cup joint and extended laterally from the loading side along the horizontal direction. In the process of loading to 21.6 kN, the horizontal cracks at the bottom of the cup developed continuously. Due to the thinning of the cup wall of Specimen 2, vertical cracks extending downward and horizontal cracks extending to the unloaded side appeared at the bottom of the cup at the same time. When the load was continued to 29.4 kN, the main crack of the specimen continued to widen, and multiple fine cracks appeared. At the same time, new vertical cracks continued to extend upward from the bottom of the cup wall. When loading to 37.4 kN, inclined cracks continuously extending from the bottom of the cup joint to the unloaded side of the cup will appear at the bottom of the prefabricated wall. When loading to 39 kN, the horizontal crack extending from the outside of the cup joint to the inside of the reinforced concrete protective layer at the bottom of the precast wallboard appears. After the crack appears, it develops violently and quickly goes through. At the same time, the crack on the outside of the cup develops rapidly and widens, the wall panel is separated from the cup, and the inside of the cup is crushed. The loading device could not continue loading, and the specimen was finally damaged. The final failure state is shown in Figure 4.

3.3 *Failure process of Specimen 3*

Due to the small embedment depth of the prefabricated wall panel in Specimen 3, the failure process of Specimen 3 developed rapidly. When the prefabricated wall panel was loaded to 4.7kN, the first crack extending downward along the contact face between the wall panel and the cup joint appeared. When the loading was continued to 10 kN, a horizontal crack appeared at the bottom of the cup joint and extended from the outside to the inside along the bottom of the cup joint. When the load was loaded to 13kN, horizontal cracks developed laterally at the bottom of the cup joint, and vertical cracks extending downward appeared in the middle of the bottom of the cup joint. In the process of loading to 21.4 kN, cracks developed continuously at the bottom of the cup, and vertical cracks continued to extend downward at the bottom of the cup. Different from Specimen 1 and Specimen 2, the horizontal cracks and vertical cracks at the bottom of the cup and groove of Specimen 3 did not continue to extend to the cup wall and the bottom of the foundation. Continue to load to 24 kN, the bottom of the cup joint appears to have vertical cracks extending upward along the inner side of the cup wall, while the reinforced concrete protective layer at the bottom of the wall plate appears to have bending cracks extending from the

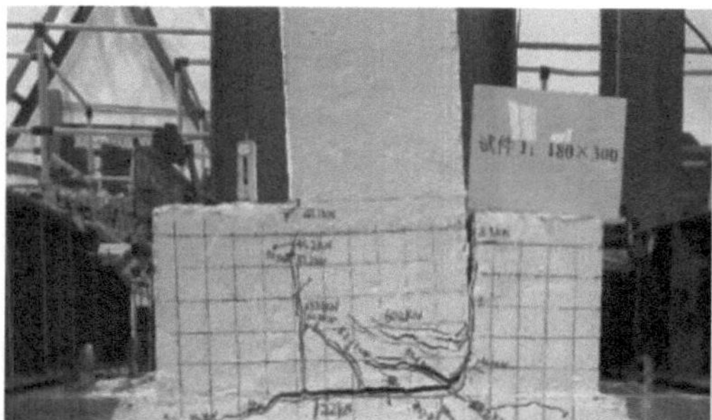

Figure 3. Final failure state of Specimen 1.

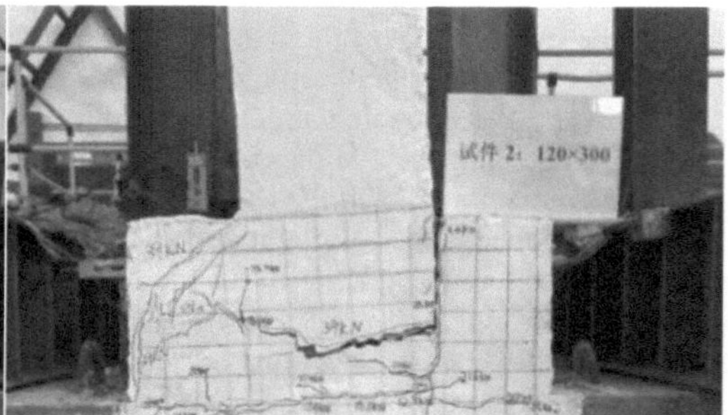

Figure 4. Final failure state of Specimen 2.

Figure 5. Final failure state of Specimen 3.

outside to the inside. Continue to load to 26 kN, the curve crack rapidly extended and widened through, the concrete protective layer was crushed, and there were bending cracks with uneven thickness. Also, bending cracks developed rapidly, and the surrounding concrete continued to peel. At the same time, the cracks at the bottom of the cup continued to expand to the outside of the cup wall and the bottom of the foundation, and the wall panel was separated from the cup joint. The loading device could not continue loading, and the specimen was finally damaged. The final failure state is shown in Figure 5.

4 ANALYSIS OF THE RESULT

The load–displacement curves at the vertices of the three specimens of prefabricated wallboard are shown in the figure. It can be seen from the figure that the curves can be roughly divided into three stages: the first stage is the elastic stage, and the load and displacement show a linear relationship; the second phase is the elastic–plastic stage or the plastic stage, and the load–displacement curve shows the first obvious turning point, displacement of Specimen 1 and Specimen 2 growth speed, and structure into the elastic-plastic state. Due to the specimen, 2 cups of tank wall thickness are less than 1 and the specimen of wallboard lateral restraint is weak, so Specimen 2 first turning point corresponding to the load is less than Specimen 1. For Specimen 3, due to the shallow embedding depth, the load on the top of the wall panel increased a little while the horizontal displacement of the wall increased rapidly, and the structure entered a plastic state. The third stage is the breaking phase, due to cracks at the bottom of the wallboard concrete, Specimen 1 is well versed in, being the basis and crushing the inside of the cup slot section to achieve ultimate shear capacity and failure; due to the wallboard root fracture penetration and the inside of the cup tank wall, Specimen 2 is crush and damage; because of the wallboard root crack and the concrete cover, Specimen 3 is crush and destroy it. The load–displacement curves of the three groups of specimens all entered the declining stage.

5 NUMERICAL ANALYSIS

ANSYS was used to establish a three-dimensional numerical model for numerical simulation and analysis of the stress test process (Zhong *et al.* 2015, 2017). The interaction between the cup joint and precast and cup joint caulked concrete is considered, and the bonding slip

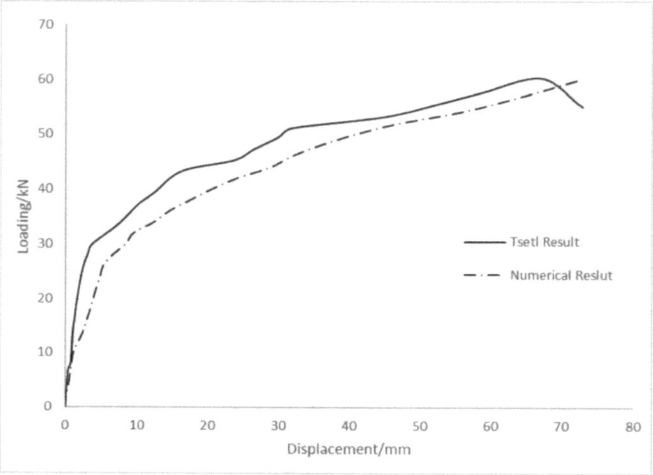

a) Specimen 1

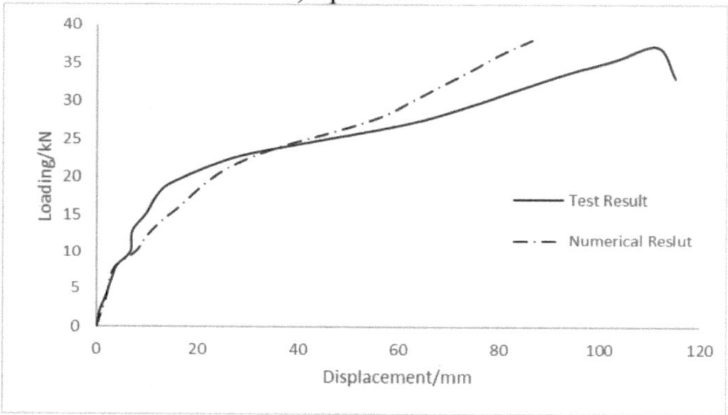

b) Specimen 2

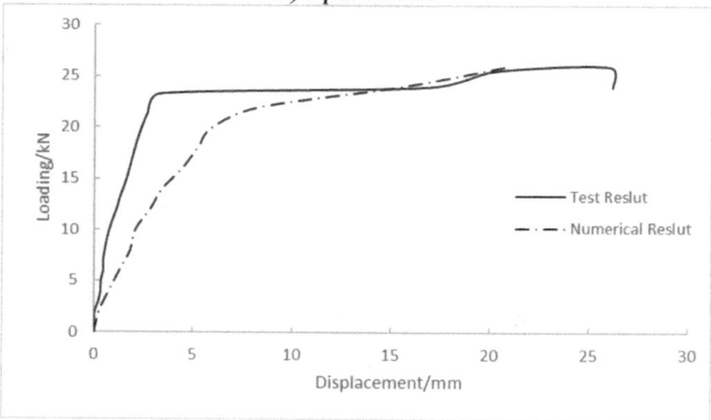

c) Specimen 3

Figure 6. Comparison of load–displacement curves at the top of specimens. a) Specimen 1 b) Specimen 2 c) Specimen 3.

between reinforcement and concrete is not considered. The specimen model is modeled by the separation method. Figure 7 shows the overall grid element division of Specimen 1, and Figure 8 shows the reinforcement element division of Specimen 1. The steel pad is set up at the loading point of the model to prevent local crushing. The freedom of three directions is restricted at the bottom of the cup joint model. The uniform load applied to the pad at the top of the wall plate is shown in Figure 9.

5.1 *Material and element types*

The concrete element is simulated by the Solid65 element, the reinforcement element is simulated by the Link8 element, and the steel pad is simulated by the Solid45 element. The axial compressive strength of concrete is 0.76 times the experimental compressive strength of a concrete cube, and the falling section of the concrete constitutive curve is not considered. The stress–strain curve of the ascending section under uniaxial compression is determined according to the Code: Design of Concrete Structures (GB50010-2010), and the multi-linear follow-up strengthening model (MKIN) is adopted for the rebar (Huang 2014).

Figure 7. Meshed model.

Figure 8. Meshed reinforcement.

Figure 9. Model constraints and loading.

5.2 *Contact setting*

Two contact surfaces are set in the cup joint foundation to simulate the stress of the new and old concrete connection parts, as shown in Figure 10. An external contact surface is the contact surface between the cup foundation groove and the post-poured caulked concrete in the foundation, wherein the surface of the post-poured concrete is set as the contact surface, and the surface of the groove of the cup joint is set as the target surface. The inner contact surface is the contact surface between the precast wall panel and the post-poured caulked concrete in the foundation, wherein, the surface of the precast wall panel is set as the target surface, and the post-poured caulked concrete in the foundation is set as the contact surface.

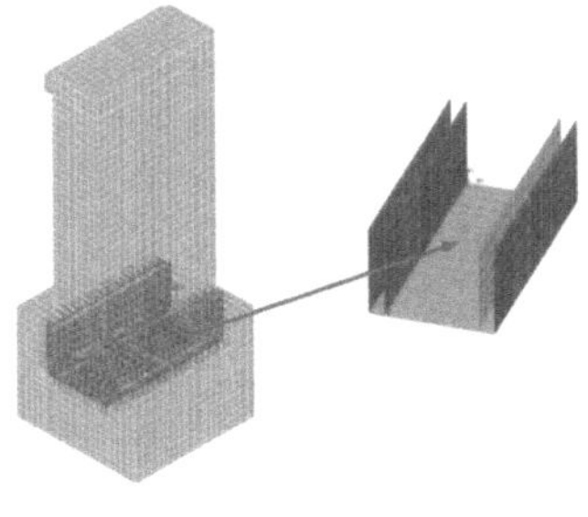

Figure 10. The contact surface of the model.

The contact surface and target surface are simulated by Conta174 and Targe170, respectively. The maximum penetration (FTOLN) of the contact element is set to 0.1 to achieve higher calculation accuracy, and the contact stiffness (FKN) is set to 1 to avoid too many iterations and ensure convergence. No rough treatment measures were taken for the contact surface between the new and old concrete, so the friction coefficient of the contact surface was 0.6.

5.3 *Analysis of the result*

The comparison of load–displacement curves obtained from the analysis and test is shown in Figure 6. As can be seen from the figure, the numerical simulation results are in good agreement with the test results, indicating that the analysis results have a certain reliability. To ensure the convergence of calculation, the descending section of the concrete constitutive curve was not considered in the calculation, so the load–displacement curve of the numerical simulation did not have a descending section, which was inconsistent with the test results.

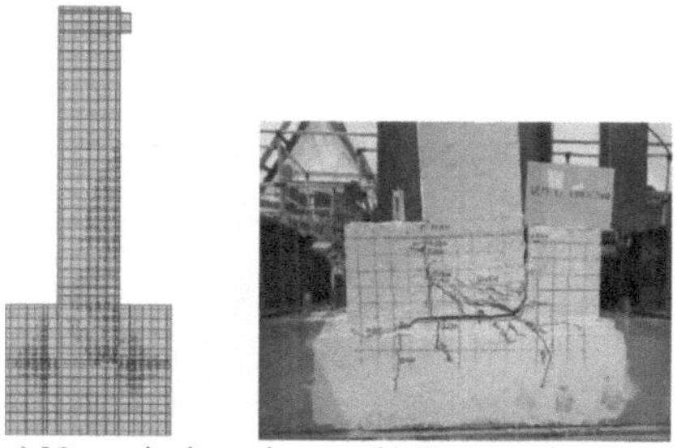

a) Numerical result b) Test result

Figure 11. Comparison between numerical and test of crack distribution in Specimen 1. a) Numerical result b) Test result.

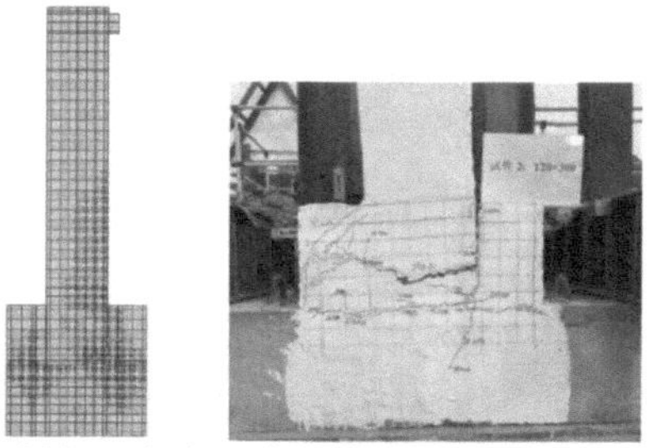

a) Numerical result b) Test result

Figure 12. Comparison between numerical and test of crack distribution in Specimen 2. a) Numerical result b) Test result.

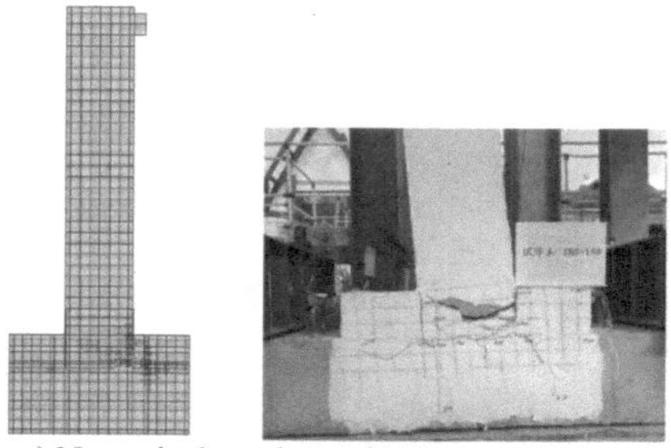

a) Numerical result b) Test result

Figure 13. Comparison between numerical and test of crack distribution in Specimen 3. a) Numerical result b) Test result.

Figures 11–13 show the numerical simulation and test comparison of crack distribution during the final failure of the specimen. It can be seen from the figure that the numerical simulation results of the crack distribution of the three groups of specimens are consistent with the test results. The oblique crack at the bottom of the wall panel of Specimen 1 was connected, and the cup joint crack was distributed along the wall and bottom of the cup joint due to the embedment of the foundation, which was consistent with the development of the final crack in Specimen 1 in the test. Due to the thin cup wall of Specimen 2, cracks were densely distributed in the cup wall and bottom of the loading side when the specimen finally failed, which was also consistent with the test results. Due to the shallow embedment depth of Specimen 3, the cup foundation had a weak embedment effect on the wall panel, and the cracks of the wall panel were mainly distributed in the attachment of the concrete protective layer. When the wall panel was pulled out under a small load, there were fewer cracks in the inner cup wall of the specimen, which was also consistent with the test results.

6 CONCLUSION

In this paper, the static loading test was carried out on three cup joints of the prefabricated integrated utility tunnel, and the numerical simulation analysis was carried out on the stress process of test specimens by using ANSYS software. The main conclusions were drawn as follows:

(1) Embedment depth of the cup joint has a great influence on structural bearing capacity and failure form. When the embedment depth of the cup joint is designed reasonably, the anchoring function of the cup joint is firm and reliable, which can be equivalent to the solid connection between the prefabricated wallboard and the foundation. When the depth of the cup joint is shallow, the cup joint does not play the anchoring role, and the main failure mode of the specimen is the extrusion failure of the protective concrete of the wall panel.
(2) Failure of cup joint specimens can be divided into three stages: elastic stage, plastic stage, and failure stage.
(3) The established finite element model of the cup joint considering the contact surface of old and new concrete shows that the load–displacement curves of the specimens are in good agreement with the test results, which verifies the validity of the numerical simulation.

REFERENCES

Hu Xiang, Xue Weichen. Experimental Study on Mechanical Performance of Prestressed Composite Pipe Corridor [J]. *Chinese Journal of Civil Engineering*, 2010, 43(05):29–37.

Huang Li. *Test and ANSYS Parameter Analysis of External Bending Behavior of Laminated Underground Outdoor Wall Panel with Embedded Foundation* [D]. Hefei University of Technology, 2014.

Qian Qihu, Chen Xiaoqiang. The Current Situation, Problems, and Countermeasures of Underground Integrated Pipeline Corridor Development at Home and Abroad [J]. *Chinese Journal of Underground Space and Engineering*, 2007(02):191–194.

Tan Zhongsheng, Chen Xueying, Wang Xiuying, Huang Mingli. Management Mode and Key Technologies of Urban Underground Integrated Pipeline Corridor Construction [J]. *Tunnel Construction*, 2016, 36 (10):1177–1189.

Tian Zixuan. *Experimental Study on Mechanical Performance of Assembled Composite Concrete Underground Composite Pipe Corridor* [D]. Harbin Institute of Technology, 2016.

You Xinhua. Current Situation and Development Trend of Urban Integrated Pipeline Corridor [J]. *Urban Housing*, 2017, 24(03):6–9. (In Chinese)

You Xinhua. Current Situation and Future Development Trend of Urban Integrated Pipeline Corridor Construction in China [J]. *Tunnel Construction (Chinese and English)*, 2018, 38(10):1603–1611.

Xue Weichen, Wang Hengdong, You Xinhua, Hu Xiang. Development Status and Prospect of Prefabricated Integrated Pipe Corridor Structure System [J]. *Construction Technology*, 2018, 47(12):6–9. (In Chinese)

Zhong Xun, Fang Yicheng, Jiang Qing, Ye Xianguo, Fan Hua, Xing Wei, Shao Huibin. Study on Mechanical Behavior and Embedment Depth of Laminated Underground Outdoor Wall Panel with Embedded Foundation [J]. *Chinese Journal of Civil Engineering*, 2017, 50(03):44–53.

Zhong Xun, Huang Junqi, Jiang Qing, Ye Xianguo, Zhou Bowen. Experimental Research and Numerical Simulation Analysis on External Flexural Behavior of Embedded Base Laminated Wallboard [J]. *Engineering Mechanics*, 2015, 32(05):131–137.

Water Conservancy and Civil Construction – Oke & Ahmad (Eds)
© 2024 The Author(s), ISBN: 978-1-032-58614-4

The effect of fly ash admixture in concrete on the performance of concrete

She Yuhao* & Meng Xiangxi
Liaoning Technical University School of Civil Engineering, Liaoning, China

ABSTRACT: Fly ash is a fine solid particle produced by the combustion of coal and its main source is generated by thermal power generation. It is a threat that cannot be ignored, both in terms of the environmental impact of fly ash itself and in consideration of the environmental impact of greenhouse gases produced during concrete production and power generation. Therefore, how to deal with the use of fly ash has become a challenge for researchers to overcome in recent years. With the demand for high-performance concrete construction materials, researchers began to notice the special properties of fly ash and its activation characteristics. In the early 20th century, many researchers and scholars began to research the impact of fly ash on the performance of concrete, and their research found that fly ash can optimize the mechanical properties of concrete to a certain extent, and since then there has been the practice of applying fly ash in concrete, which began to be widely used. Qian Jueshi and other scholars established a life prediction model for the resistance to carbonation of concrete with high fly ash dosing (Zhou & Xie 2022). This has led to a study on the optimization of the properties of fly ash added to concrete. In this paper, the effect of the amount of fly ash addition to concrete on the properties of concrete is investigated about each other, and the effect of adding different amounts of fly ash on the strength properties of concrete, such as frost resistance, compression resistance, and flexural resistance, is investigated through a controlled variable design. In this experiment, a numerical prediction model was designed and constructed based on the experimental data obtained from the actual calculations and measurements using the software Stata. The best admixture ratios of fly ash are 23.19% and 28.06%, respectively, and its corresponding best compressive strength and flexural strength are 26.2 MPa and 5.7 MPa, respectively, in the form of data analysis and prediction for engineering data analysis, which has certain research significance and development prospect.

1 INTRODUCTION

Concrete is the most widely used material in civil engineering construction today, and with the increasing social and economic development, the requirements for construction are gradually increasing, and the strength of the concrete applied and the various flexural, compressive, frost-resistant, and durable aspects have higher demands. Therefore, scholars and researchers have begun to focus on the research and production of high-performance concrete. In the context of the rapid development of the concrete industry, the initial method of concrete production used was a mixture of cement, crushed stone, and sand. As for the research on high-performance concrete, according to the relevant journal literature, it is known that internationally, the construction material production industry in various countries considers durability as the main indicator for the design of high-performance concrete, followed by considerations of durability,

*Corresponding Author: syh15515000812@163.com

 DOI: 10.1201/9781003450818-57

strength, and economy (Wang & Cheng 2011). For the study of high-performance concrete, Europe, the United States, and other countries in this area of research earlier and faster development, there are many technical breakthroughs, at the forefront of this industry. In the United States, the University of California at Berkeley in the use of construction materials, the addition of fly ash in concrete greatly reduces the amount of cement in concrete (Li et al. 2016). The strength of the concrete used in the Esch-born office building in Germany has also been significantly increased with a concrete strength of C105 (Sun & Wang 2015).

In this paper, the effect of fly ash on the properties of concrete is investigated through a controlled variable design to study the effect of adding different amounts of fly ash on the strength properties of concrete, such as frost resistance, compression resistance, and flexural resistance. In this experiment, a numerical prediction model NPMFA is designed and constructed based on the experimental data obtained from the actual calculations and measurements, and the data is processed and predicted using the software Stata. The model is finally used to derive the optimum value of fly ash admixture in concrete under the condition of optimization of all properties for engineering data analysis.

Fly ash is the fine solid particles emitted with the flue gas after coal combustion and collected by dust collection equipment. The main source of fly ash is the flue gas emitted from thermal power plants. In all countries of the world, electricity is an important and indispensable source of energy. Therefore, the number of power plants is naturally innumerable. Fly ash is the biggest by-product of power plants in each country at present, adding fly ash as an admixture in concrete is an efficient and reasonable way to utilize fly ash at present.

This cannot only effectively improve the comprehensive performance of concrete in all aspects but also improve the utilization of resources and reduce the amount of cement in concrete while achieving the environmental purpose of saving energy and protecting the environment. The properties of fly ash mainly refer to physical and chemical activity. As an efficient active admixture, fly ash admixed concrete is particularly utilized in water conservancy projects due to its unique advantages and improved performance.

As the scale of construction infrastructure grows, the amount of cement used also increases significantly. According to the survey, the amount of carbon dioxide produced by the decomposition of cement stone alone is very large. When added to the greenhouse gases and harmful gases produced by power plants, the impact on the environment is unpredictable. In some regions, to promote the rational use of fly ash, relevant announcements and policies have been issued to improve the utilization of fly ash. By reviewing the relevant literature, it is clear that the main application areas of fly ash are construction and infrastructure, which is currently the most effective way to improve the utilization rate of fly ash.

Fly ash is a major waste material of thermal power plants and its main components include silica, iron oxide, and alumina, which mostly exist in unstable chemical forms. Similar to the nature of volcanic ash, fly ash can also undergo hydration reactions under alkaline conditions. The products of its hydration reaction, calcium oxide, and silicate gel can enhance the strength properties of concrete to some extent. In the early stages of considering the use of fly ash in concrete, related studies have incorporated fly ash as an active admixture to enhance the strength of concrete. The incorporation of fly ash can improve certain properties of cement to a certain extent, making it more widely used. At the same time, it reduces cost, improves resource utilization, reduces greenhouse gas emissions, saves energy, and protects the environment while ensuring the strength and yield of concrete. By understanding the relevant fly ash structure analysis report, it is found that fly ash has the advantages of higher adhesion and chemical stability at room temperature, while the same genus also has the deficiency of slow hydration reaction rate and long setting time due to lower chemical activity.

The first part of this paper introduces the selection of experimental materials and relevant performance parameters, including concrete, fly ash, aggregates, etc. The second part is a reproduction of the experimental process, including the experimental design and experimental implementation process. The third part constructs a numerical prediction model through experimental data, uses Stata software for data analysis and prediction to achieve the purpose of engineering analysis data, and derives the optimal value of fly ash in concrete admixture. In the fourth part, the experimental results are discussed and finally, the conclusion is drawn and summarized throughout the paper.

2 EXPERIMENTAL MATERIALS

The sources and physical properties of the experimental materials used in this experiment, including cement, fly ash, aggregates, etc., are described in detail in this section and the properties of these raw materials are outlined in the following sections.

2.1 *Cement*

Ordinary silicate cement is a hydraulic cementitious material made of silicate cement clinker, 6%-20% of mixed materials, and an appropriate amount of gypsum ground finely. P.O42.5 cement produced by Yangchun Cement Co., Ltd. is chosen in this experiment and its detailed physical properties are shown in Table 1:

Table 1. Details of the physical properties of the cement selected for the experiments.

Number	Content	P.O42.5Standard Indicators
1	Burning loss (%)	4
2	Specific surface area (m2/kg)	358
3	Coagulation time Initial condensation(min)	172
	Final condensation (min)	234
4	Flexural strength (MPa) 3d	5.5
	28d	7.8
5	Compressive strength (MPa) 3d	27.2
	28d	47.5
6	Adequacy	Qualified

2.2 *Fly ash*

Fly ash is a fine solid particulate matter discharged during the combustion process and its particle size is generally between 1 and 100 μm. Due to the existence of surface tension, fly ash is mostly spherical, with a small internal surface area, good permeability, and a strong capillary phenomenon. The surface is smooth and the micropores are small. According to relevant research data, burning 1t of coal can produce about 300 kg of fly ash. If fly ash is not controlled and treated, it can cause air pollution and even harm to living creatures and the human body.

The next is an overview of the chemical activity of fly ash: fly ash is a volcanic ash material and the main components and content variations are shown in Table 2:

Table 2. Fly ash main components and content table.

Ingredients	Silicon Dioxide	Aluminum oxide	Calcium oxide	Iron oxide	Burning loss
Scope(%)	31~58.9	14.5~38.0	0.8~8.8	0.5~1.9	2.1~25.3

The activity of fly ash mainly comes from the product of the hydration of SiO_2 and Al_2O_3 in alkaline conditions, calcium silicate, and calcium aluminate hydrate, which can change the porosity of concrete (Peng et al. 2022). At room temperature, the calcium hydroxide crystals in concrete are unstable and its secondary hydration reaction can occur with fly ash, filling the capillary space of concrete (Zhang & Li 2017). It enhances the bonding of concrete and improves its performance to some extent. After the discovery of the physical and chemical activity of fly ash, it began to be added more and more widely as an admixture to replace the amount of cement in concrete. In modern concrete, it has become an important factor in the performance of concrete and is an admixture with good safety properties.

The detailed physical properties of Class II fly ash produced by Henan Plutonium Casting Materials Co, Ltd. selected for this experiment are shown in Table 3:

Table 3. Details of the physical properties of Class II fly ash selected for the experiment.

Inspection		Requirement	Result
Fineness	Sieve residue (%)	5um ≤ 18%	16
Burning loss	Scorch difference subtraction	≤ 5%	2.8
	Aluminum oxide (%)	≤ 30%	24.2
	Silicon Dioxide (%)	≤ 50%	45.1
	Water content (%)	≤ 1.0%	0.85
	Chloride ion (%)	≤ 0.02%	0.015
	Sulfur trioxide (%)	≤ 3%	2.1
	Alkali content (%)	≤ 1.5%	1.2
	Free calcium oxide (%)	≤ 1.0%	0.85
	Density (g/cm3)	2.1-3.2	2.55
	Stacking density (g/cm3)	0.63-1.38	1.12

2.3 *Aggregate*

(a) Fine aggregates

The fine aggregate produced by Shandong Jialin Energy Technology Co., Ltd. is selected for this experiment and the indicators of the fine aggregate are shown in Table 4:

Table 4. Details of the fine aggregates selected for the experiment.

Number	Inspection	Specification(mm)	Result%
1	Sieving	9.5	95.8
		4.75	3.2
		1.18	0.2
		16	58.8
		13.2	27.5
		31.5	37.4
2	Mud content	4.75-31.5	0.7
3	Crushing value	9.2-14.5	15.2

(a) Coarse and large aggregates

The crushed stone aggregate produced by Shandong Jialin Energy Technology Co., Ltd. is selected for this experiment and the indices of the coarse aggregate are shown in Table 5:

Table 5. Details of the coarse and large aggregates selected for the experiment.

Number	Inspection	Specification(mm)	Result%
1	Sieving	9.5	99.8
		4.75	99.4
		2.36	85.6
		1.18	63.2
		0.6	33.2
		0.3	15.1
2	Mud content		2.3
3	Stacking density(g/cm3)		1.254

2.4 *Water*

The water used in this experiment for the plates and concrete is tap water from Fuxin.

3 EXPERIMENTAL DESIGN AND EXPERIMENTAL PROCEDURE

3.1 *Experimental design*

This paper investigates the effect of adding different amounts of fly ash to several commonly used concretes on the strength properties of concrete, such as frost resistance, compression resistance, and flexural resistance through a controlled variable design, and correlates the effect of the amount of fly ash added in concrete on the various properties of concrete. In this experiment, a numerical prediction model of fly ash, NPMFA (Nkomo et al. 2022), is designed and constructed based on the experimental data obtained from the actual calculations and measurements. The data is processed and predicted using the software Stata. The model is used to derive the optimum value of fly ash admixture in concrete with the optimization of all properties.

3.2 *Experimental implementation process*

(a) Concrete mixture mixing
In this experiment, concrete specimens are prepared according to the test protocol, in which the groups of fly ash admixture of 0, 5%, 10%, 15%, and 20% are designed. In this experiment, the raw concrete material is maintained at 22-25 degrees Celsius at all times.

The newly made concrete in this experiment is by the National Standard of the People's Republic of China: Ready-mixed Concrete (GB/T14902-2012) (Tan 2013). The specimens are first demolded and placed at a constant temperature of 23 degrees Celsius for 24d and 28d.

(b) Compression resistance test
Concrete specimens of 150 mm * 150 mm * 150 mm are tested for compressive strength of specimens after reaching the curing age of 28d using a microcomputer-controlled electro-hydraulic servo static and dynamic material and structure testing machine. The experimental data is measured and recorded (Niu et al. 2018).

(c) Bending resistance test

Concrete specimens of 150 mm * 150 mm * 550 mm are tested by a hydraulic universal testing machine for flexural strength of specimens after reaching the curing age of 28d. The experimental data is measured and recorded.

4 PREDICTIVE MODEL CONSTRUCTION

To dramatize engineering production, this experiment develops a univariate prediction model by modeling and parameter adjustment. The modeling is done with Stata software and both modeling and statistical analysis are performed with Stata software. The data entry model is optimized for the slab strength parameters during the entry of the experimental data and the continuous factor optimized in the experiment is the proportion of fly ash blending.

4.1 *Construction of NPMFA, a numerical prediction model for fly ash*

For this experiment, the univariate linear regression model is as follows.

$$y_t = \beta_1 + \beta_2 x_t + \beta_3 x_t + \ldots + \beta_k x_{kt} + \varepsilon_t \quad (1)$$

It is assumed that Equation (1) is the actual data fit image model with sample iT and bi is the estimate of the least squares βi so that the fitted values become in the constructed model.

$$y_t^* = b_1 + b_2 x_t \rightarrow b_3 x_t + \cdots + b_k x_{kt} \quad (2)$$

In Equation (2), the in-sample predictions in the model are constructed based on the point estimates using the regression coefficients and the actual observations of the regressors, that is, when t = 1,2,3...T, the actual values are constructed.

Model revision and optimization the model is revised and optimized by comparing the predicted output of the model with the actual results of the experiment. The changes between the best model output results and the actual data response are compared by statistical methods.

5 RESULT

5.1 *Experimental results and predicted results of compressive performance.*

The compressive strength table for concrete specimens with a maintenance age of 28d is shown in Table 6:

From the experimental results, it can be obtained that the admixture of fly ash in concrete

Table 6. 28d compressive strength of concrete with different fly ash admixture.

Fly ash mixing amount (%)	0	5	10	15	20
Compressive strength (Mpa)	25.1	25.5	25.7	22.5	21

can adversely affect the compressive strength of concrete at 28 d. The compressive strength of concrete is increased by 0.4%, 5%, 9%, and 15% for fly ash admixture of 5%, 10%, 15%, and 20%, respectively. With the increase in fly ash admixture, the number of cement increases, and the strength of the early stage increases. The cementitious materials, such as hydrated calcium aluminate and hydrated calcium silicate generated in the later stage, make the concrete structure dense and the strength of the later stage increases. However, after replacing cement with a large amount of fly ash, the early strength of concrete develops slowly, the late strength increases with time, and its compressive strength generally increases more in 28-180 d (Malhotra 2002).

It can be seen from Figure 2 that the measured data and the predicted data have a reference value within a certain range.

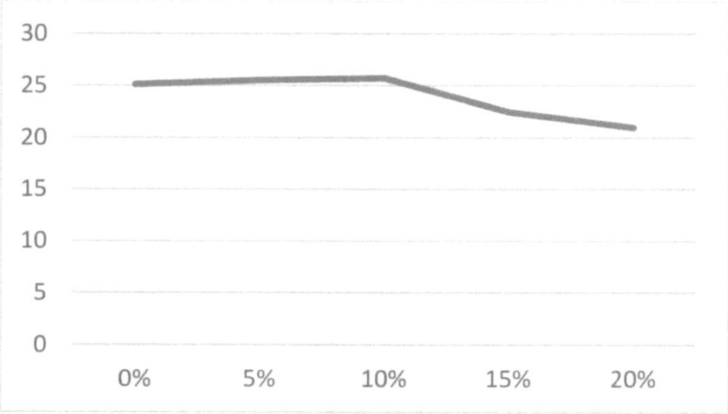

Figure 1. 28 d Compressive strength of concrete with different fly ash admixture.

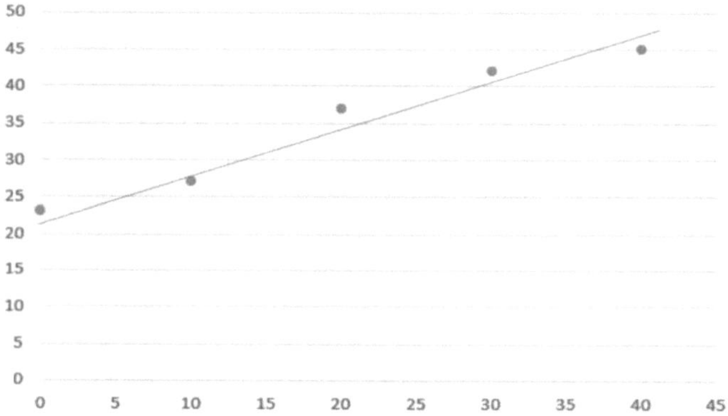

Figure 2. Reference chart of concrete compressive strength measured results and NPMFA results.

5.2 *Experimental results and predicted results of flexural properties*

The table of flexural strength of concrete specimens with a maintenance age of 28 d is shown in Table 7 below:

Table 7. Table of flexural strength of concrete with different fly ash admixture at 28 d.

Fly ash mixing amount (%)	0	5	10	15	20
Flexural strength (Mpa)	5.6	5.5	5.4	5.4	5.1

From the experimental results, it can be obtained that the admixture of fly ash in concrete can adversely affect the compressive strength of concrete at 28 d. The compressive strength of concrete is reduced by 0%, 2.1%, 2.1%, and 4.1% for fly ash admixture of 5%, 10%, 15%, and 20%, respectively. With the increase in fly ash admixture, the amount of cement decreases, and the early strength decreases. The flexural strength and compressive strength of fly ash concrete both decrease with the increase in fly ash admixture.

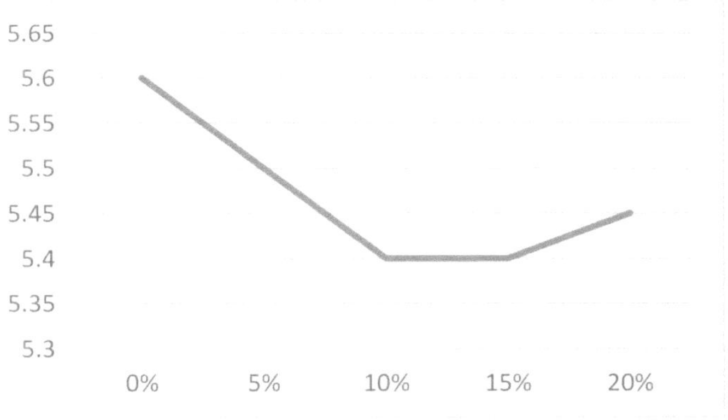

Figure 3. 28 d Flexural strength of concrete with different fly ash admixture.

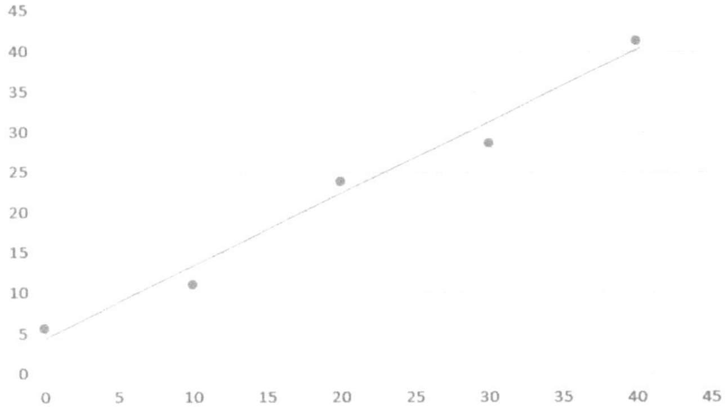

Figure 4. Reference chart of concrete flexural strength measured results and NPMFA results.

It can be seen from Figure 4 that the measured data and the predicted data have a reference value within a certain range.

6 CONCLUSION

Fly ash has excellent activation characteristics and application prospects and can be mixed into concrete to enhance and optimize the overall mechanical properties and working durability of concrete to different degrees. Fly ash concrete has been widely used in engineering fields due to its good performance. In this paper, the following conclusions are obtained by testing the mechanical properties of concrete with different fly ash dosages in the experiments, mainly the compressive strength and flexural strength of concrete specimens with fly ash admixture.

(1) Compared to normal concrete, the compressive strength of the specimens tested gradually increases from 0 to the predicted optimum value of 23.19% for fly ash. However, when the amount of fly ash admixture in concrete continues to increase, the compressive properties of the concrete specimens decrease significantly.
(2) Compared to normal concrete, the compressive strength of the specimens tested gradually increases from 0 to the predicted optimum value of 28.06% for fly ash, and gradually increases to 5.7 MPa. However, the flexural properties of the concrete specimens decrease significantly when the fly ash admixture in the concrete continues to increase.

(3) By observing the prediction results of the NPMFA model and the actual measured data, it can be found that the results obtained from the numerical prediction model are quite accurate and have some reference value within a certain permissible range. The numerical model construction and prediction proposed in this experiment achieves the purpose of engineering data-based analysis in the form of data analysis and prediction, which has certain research significance and development prospects.

The research on fly ash is significant to improve the environment and increase production efficiency. In terms of the application prospect of fly ash, fly ash mainly comes from the production of coal power. However, due to the damage to the environment, such as the outbreak of widespread haze, the development path of coal-based power has gradually shifted to the path of green development, so the production of fly ash will be somewhat restricted. Therefore, fly ash will be more clear in the application direction. First, in the construction industry, fly ash can replace a large number of traditional building materials products, such as cement. Second, in agriculture, fly ash can help agricultural production by improving the physical and chemical properties of soil and improving clayey and acidic soil.

Considering that the amount of cement decreases with the increase in fly ash, the strength of concrete mixed with fly ash decreases in the early stage. The cementitious materials, such as hydrated calcium aluminate and hydrated calcium silicate generated in the later stage, make the concrete structure dense, and the strength in the later stage increases. After, replacing cement with a large amount of fly ash, the early strength of concrete develops slowly and the later strength increases with time, and generally, its compressive strength in 28-180 d increases more. Therefore, high fly ash concrete is suitable for concrete projects with low early strength requirements.

The numerical model construction and prediction proposed in this experiment achieves the purpose of engineering data-based analysis in the form of data analysis and prediction, which has certain research significance and development prospects.

REFERENCES

Li Yebin, Zhu Chaoyan, Wang Jianhua, *et al.* Research and Development Atatus of High-Performance Concrete [J]. *Shanxi Construction*, 2016, 42(23): 117–118.

Nkomo, N. Z., Masu, L. M., Nziu, P. K. Optimisation of Mechanical Properties of Polyethylene Terephthalate Fiber/fly Ash Hybrid Concrete Composite [J]. *Case Studies in Construction Materials*, 2022,17.

Niu Yinsheng, Gao Caizhen, Wang Liyong, Wu Weilong. Research on the Anti-Freezing Performance of Fly Ash-doped Concrete [C]. *Progress of Concrete Admixture Research and Application in China - 2018 Cologne Cup Paper Compilation (below)*. [publisher unknown], 2018:163–172.

Malhotra, V. M. High-performance High Volume Fly Ash Concrete [J]. *ConcreteInternational*, 2002, 24 (7):30–34.

Peng Yufa,Zhu Yunhua,Li Xun,Wang Qingshan, Zhang Manni. Experimental Study on the Effect of Fly Ash on the Performance of Concrete [J]. *Chongqing Construction*, 2022, 21(09):70–72.

Sun, G., Wang, Q. Introduction to the Prospect of High-performance Concrete Development [J]. *Journal of Hebei Higher Institute of Engineering and Technology*, 2015 (3): 39–41, 61.

Tan Hongguang. The Relevance of the New Concrete Standard - Analysis of Ready Mixed Concrete GB/ T14902 [J]. *Building Materials Development Orientation*, 2013, 11(05):34–38.

Wang Guoqing, Cheng Liping. Development History and Research Status of Self-Compacting Concrete [J]. *China Water Transport (second half of the month)*, 2011, *11*(1): 240, 243.

Zhang, M., Li, M. Effect of Fly Ash Admixture on the Durability of Concrete [J]. *Housing and Real Estate*, 2017(3):105–107.

Zhou, T., Xie, L. Current Status and Prospect of Fly Ash Application in Concrete [J]. *Jiangxi Building Materials*, 2022(3):9–11.

Water Conservancy and Civil Construction – Oke & Ahmad (Eds)
© 2024 The Author(s), ISBN: 978-1-032-58614-4

Study on ground settlement of EPB shield in upper-soft and lower-hard composite stratum

Dan Miao*, Fucheng Wu*, Jianxin Ye*, Qiqi Zeng*, Fengzhi Wang*, Quan Liu* & Yang Chen*
Guangzhou Guangjian Construction Engineering Testing Center Co., Ltd., Guangzhou, China

ABSTRACT: Tunneling in the upper-soft and lower-hard composite stratum is a difficult problem in shield construction. Inadvertent control and monitoring can lead to engineering accidents such as ground subsidence and collapse. To analyze the causes of excessive ground settlement, two shield lines crossing the upper-soft and lower-hard composite stratum in a cross-sea tunnel project are used. Through a statistical analysis of the shield parameters, the principle of shield parameter selection for earth pressure balance shield tunneling in the upper-soft and lower-hard composite stratum is proposed, which can provide some reference for similar projects.

1 INTRODUCTION

With the rapid development of urban metro tunnel engineering, the corresponding geotechnical engineering problems have emerged in large numbers (Hou & Liao 1993). When the construction of a metro tunnel is carried out in the interior of a rock and soil mass, the rock and soil can create disturbances that lead to formation deformation. When this deformation develops to a certain extent, it poses a threat to the normal use of surface buildings, roads, and bridges. Therefore, safety monitoring and relevant analysis of the tunnel shield are necessary.

Various factors can lead to ground displacement. The main factor in the soil pressure balancing shield process is the geological conditions, especially the complex stratum (Liu 2009). With the development of urban underground space and urban rail transit, the diameter and burial depth of the shield tunnels are required to be higher. However, shield tunnel construction will inevitably encounter more varied composite strata. A composite stratum refers to the stratum combination composed of two or more kinds of different strata within the excavation section and in the excavation extension direction (Rankin 1998). The features of these formations are highly diverse, such as rock mechanics, engineering geology, and hydrogeology. There are three categories for the combination of composite strata: (1) different combinations of strata in the vertical direction of the section, (2) different combinations of strata in the horizontal direction, and (3) both of the above (Attewell *et al.* 1985; Chen *et al.* 2011; Mair & Talor 1999). The upper-soft and lower-hard composite stratum is the most typical type of the different combinations of strata in the vertical direction of the section. It has soft soil on the top and hard rock on the bottom. It can also be a soft rock on top and a hard rock on the bottom. Occasionally, soft rock is between the hard rock. Nowadays, shield construction in the upper-soft and lower-hard composite stratum is prevalent in tunnel construction at home and abroad. There are numerous problems in the actual construction, such as serious wear and low driving efficiency (Chen 2010; Wang *et al.* 2012; Zhou & Pu 2002; Zi *et al.* 2010). Additionally, there is no well-established theoretical design and construction

*Corresponding Authors: 61790367@qq.com, 2545504274@qq.com, 58109650@qq.com, 1195400307@qq.com, 2545504274@qq.com, 766167808@qq.com and 18079910562@163.com

DOI: 10.1201/9781003450818-58

system theory for shield machines in the upper-soft and lower-hard composite stratum for reference. The particularities of the strata also make special demands on the shield machine. If the chosen design of the shield is not adapted to the geological conditions of this stratum, it will bring great losses to the project and cause adverse social effects.

Based on the Gongbei to Hengqin section of the Zhuhai-Zhuhai Airport Intercity Railway Project, safety monitoring was carried out for the left and right tunnels crossing the upper soft and lower hard composite strata. The data on land subsidence during the shield tunnel is studied, where the settlement of the two sides of the tunnel is compared and analyzed. Lastly, the methods and principles of soil pressure balance shield tunnel in the upper-soft and lower-hard composite stratum are summarized.

2 PROJECT OVERVIEW

To study the ground settlement problem of the earth pressure balance shield in the upper-soft and lower-hard composite stratum, ensure the shield construction safety, and reveal the deformation characteristics of the site, the settlement data of the left and right tunnels of Gongbei to Hengqin Railway Station of Zhuhai-Zhuhai Airport Intercity Railway Project is monitored and analyzed.

2.1 *Geological environment*

The Gongbei to Hengqin line of the Zhuhai-Zhuhai Airport Intercity Railway under the project is 16.86 km from Zhuhai Railway Station to Zhuhaichanglong Railway Station. The main components of the project include the Qianshan Bridge, the Contact Line Bridge, the Subgrade, the Hengqin Tunnel, and the station. Among them, Zhuhai Station is an elevated station, while Wanzaibei Railway Station, Wanzai Railway Station, Shizimen Railway Station, Hengqinbei Railway Station, Hengqin Railway Station, and Hengqin Chimelong Station are underground stations. A new earth pressure balancing shield machine is put into the tunnel on the left and right lines. The outer diameter of the shield machine is 8.78 m and the excavation diameter is 8.82 m. The project consists of 7 sections of upper-soft and lower-hard soil layers and all these upper-soft and lower-hard composite strata are mainly concentrated in the tunnel section between Hengqinbei Railway Station and Hengqin Railway Station. The upper section and the vault located in the tunnel are mainly on the upper soft and lower hard soil layers. The layers contain soft soil, residual soil, and heavily weathered rock. And the lower part of the tunnel is located in granite rock. As a result, these 7 sections of upper-soft and lower-hard soil layers in the Jinheng section construction of the Zhuhai-Zhuhai Airport Intercity Railway Project are given high priority. The parameters of the upper soil layer and the lower hard rock involved in the composite strata of this project are shown in Tables 1 and 2.

Table 1. Parameters of upper-soft soil.

Stratum Code	Stratum Name	Density (ρ, g/cm^3)	Internal Friction Angle(φ, °)	Internal Cohesion (c, kPa)	Modulus of Compressutility (Es, MPa)
(1) 1-1	Mucky	1.65	4.16	5.15	2.44
(1) 1-2	Mucky clay soil	1.77	7.1	14.5	4.08
(1) 1-3	Mucky clay soil	1.79	7.2	15.1	4.2
(1) 2-2	Clay	1.79	6.39	16.87	4.09
(1) 2-3	Clay	1.89	11.3	21	4.4

Table 2. Hard rock compressive strength below.

Stratum Code	Stratum Name	The Standard Value of Compression Strength (MPa)
(5) 3	light-weathered granite	73.02
(5) 4	slightly-weathered granite	85.4

2.2 *Field monitoring data*

To investigate the relationship between excavation parameters and formation deformation in the upper-soft and lower-hard composite strata, the two typical strata sections with similar geological conditions on the left and right lines are selected for study and analysis in the supporting project. The two sections are from 368 to 468 rings on the right and 851 to 949 rings on the left. During the construction, there is a large settlement on the left line. After the occurrence of this condition, some pressure preservation measures are taken. However, due to the ground settlement, the through-crack and the pressure cannot be maintained. It eventually leads to the collapse of the ground in front of the shield and causes the adjacent sunken roadside wall to crack, as shown in Figures 1 and 2. Compared to the left line, the right line is constructed relatively smoothly. The ground settlement control on the right line is better controlled and no similar situation occurs.

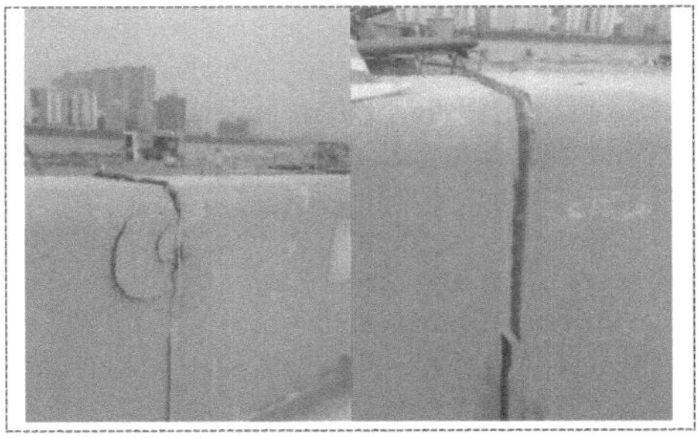

Figure 1. The left-line shield causes the side wall of the road to crack.

Figure 2. The left-line shield causes the ground cracking and air leakage.

The two sections are constructed by two different shield machines (defined as Shield Machine 1 corresponding to the right line and Shield Machine 2 corresponding to the left line, respectively). When the two shields pass through similar ground sections, two different types of ground settlement occur. The ground settlement in the section passed by Shield Machine 1 is mainly controlled below the alarm value, while the settlement of the other section is large. Most of the settlement measuring points passed by Shield Machine 2 are above the alarm value. Figure 3 shows the axial settlement of the two machines in the selected sections. The number of segment rings is normalized for the same length section for a clear comparison of the two sections' tunnel settlement. The monitoring cross section is set to a large one every 50 m and a small one every 10 m. In the cross small section, the settlement points are arranged only on the shield axis. As for the large cross-sections, the settlement points are arranged not only on the axis but also at 30m to the left and right of the axis. A small number of large cross-sections and the settlement of only the shield axis are taken out for comparison.

3 INFLUENCE OF SHIELD PARAMETERS ON SITE DEFORMATION

The above field monitoring data show that the ground settlements measured at the two selected monitoring sections are different in the shield tunnel. As shown in Figure 3, the

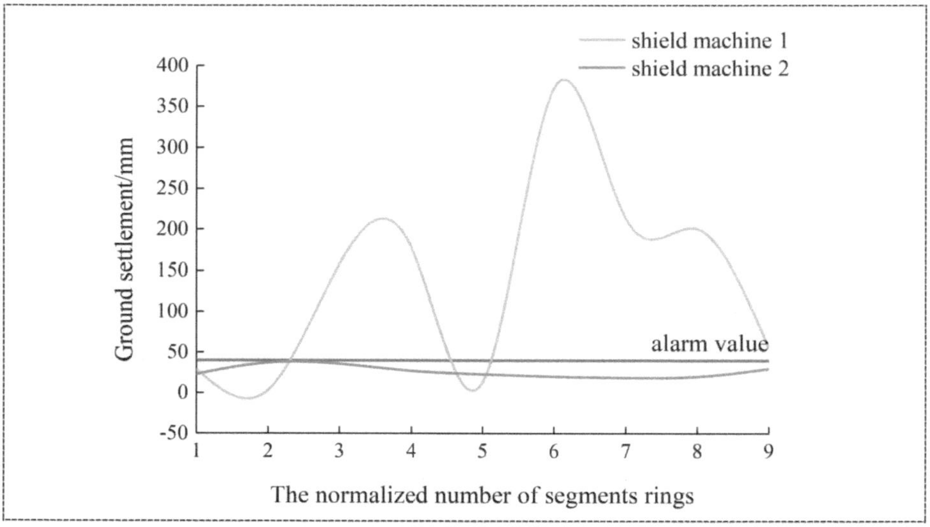

Figure 3. Comparison of ground settlement caused by the shield.

settlement of Shield Machine 2 through the section is significantly larger than that of Shield Machine 1. The settlement is fluid. To compare the different shield parameters between the two sections and the reasons for the differences in the settlement, five main shield parameters including cutter head speed, torque, thrust force, advance velocity, and penetration are selected for analysis. The number of segment rings is normalized in the same length sections. Rings 1-30 of Shield Machine 1 are the soft soil section. Rings 30-62 are the upper-soft lower-hard stratum transition section and rings 62-99 are the hard rock section. For Shield Machine 2, rings 1-20 are the soft soil section, rings 20-89 are the upper-soft lower-hard stratum transition section, and rings 89-99 are the hard rock section.

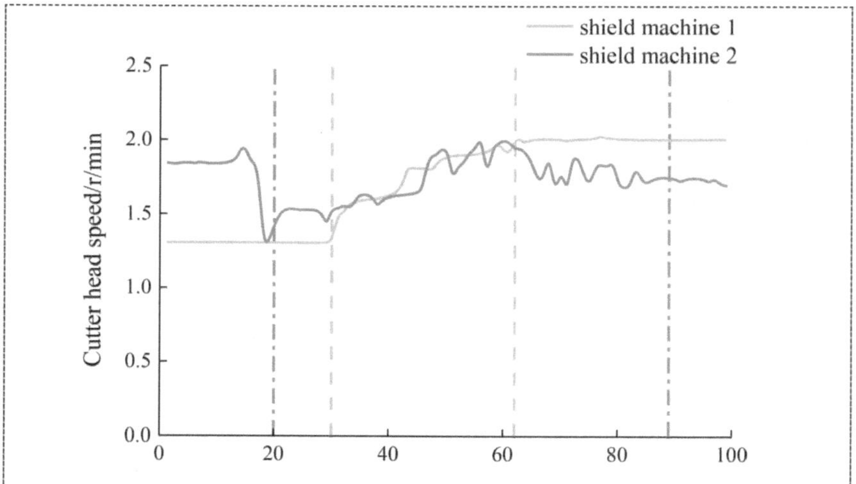

Figure 4. Comparison of cutter head speed.

Figure 4 shows a gradual increase in the transition section of the upper soft and lower hard strata for both shield machines. The difference between the two machines' cutter head speeds is that Shield Machine 1 is more stable than Shield Machine 2. In the soft soil section and the hard rock section, the cutter head speed of Shield Machine 1 remains essentially constant. In the soft soil section and hard rock section, the cutter head speed of Shield Machine 1 remains the same. The cutter head speed of Shield Machine 2 fluctuates greatly, especially when the soft soil section enters the upper-soft lower-hard stratum transition section. There are large oscillations in the transition part from the upper-soft lower-hard stratum transition section to the hard rock section corresponding to huge settlements.

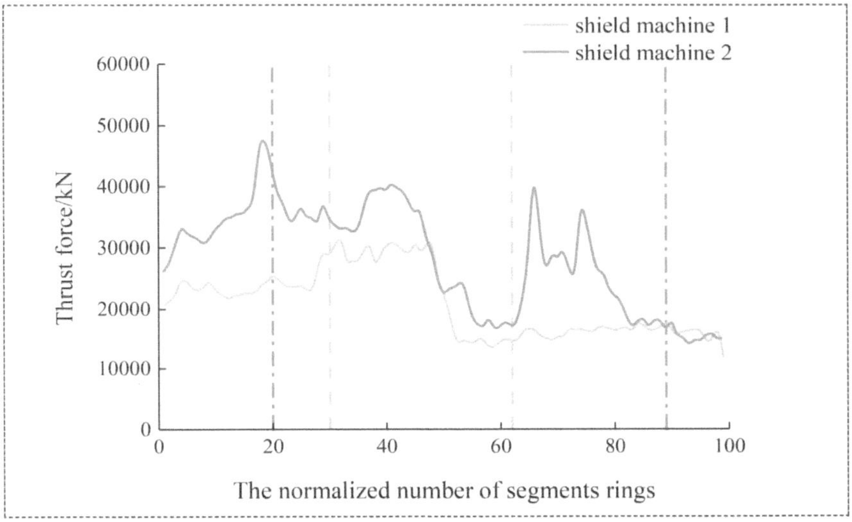

Figure 5. Comparison of thrust force.

Figure 5 shows the thrust comparison. The trend of the thrust force is the same for the two shield machines before entering the hard rock section from the soft soil section and the upper-soft lower-hard stratum transition section. In the soft soil section, the thrust force of Shield Machine 2 is significantly greater than that of Shield Machine 1. During the first half of the upper-soft lower-hard stratum transition section, the thrust forces of the two shield machines are the same and they both start to decrease. In the second half of the transition section, the thrust increases sharply. At this time, the settlement of Shield Machine 2 starts to increase. After entering the hard rock section, the thrust force of the two shield machines maintains at a small level. Shield Machine 2 maintains a high thrust force during the whole process but the cutter head speed and penetration are low. It is indicated that the cutting tool can be damaged as the tunneling efficiency is reduced. When the right line ring 385 and ring 386 (normalized ring 19 and ring 20) enter the upper-soft lower-hard stratum transition section, the cutting tool damage leads to the reduction of driving efficiency. Tables 3 and 4 show the cutting tool damage condition.

Table 3. Damage of cutters from construction to ring 385.

S/N	Tool Number	Wear Condition	S/N	Tool Number	Wear Condition
1	9#	Partial wear	18	39#	Normal wear 20mm
2	12#	Partial wear	19	40#	Normal wear 25mm
3	17#	Cutter ring broken	20	41#	Normal wear
4	18#	Partial wear	21	42#	Cutter ring falling

(*continued*)

Table 3. Continued

S/N	Tool Number	Wear Condition	S/N	Tool Number	Wear Condition
5	21#	Partial wear	22	43#	Cutter ring falling
6	24#	Partial wear	23	44#	Cutter ring falling
7	25#	Partial wear	24	45#	Cutter ring falling
8	26#	Normal wear 10mm	25	46#	Cutter ring falling
9	27#	Normal wear 8mm	26	47#	Blade hub damage
10	29#	Normal wear 13mm	27	48#	Normal wear 27mm
11	30#	Normal wear 10mm	28	49#	Normal wear 28mm
12	31#	Partial wear	29	50#	Normal wear 25mm
13	32#	Normal wear 13mm	30	51#	Partial wear
14	34#	Normal wear 13mm	31	53#	Normal wear 15mm
15	35#	Partial wear	32	55-3#	Normal wear 5mm
16	36#	Partial wear	33	56-1#	Partial wear
17	38#	Normal wear 15mm	34	56-3#	Normal wear 5mm

Table 4. Damage of cutters from construction to ring 386.

S/N	Tool Number	Wear Condition	S/N	Tool Number	Wear Condition
1	1#	Cutter ring broken	10	42#	Cutter ring falling
2	2#	Partial wear	11	43#	Cutter ring falling
3	3#	Partial wear	12	44#	Cutter ring falling
4	4#	Cutter ring broken	13	45#	Cutter ring falling
5	5#	Cutter ring broken	14	46#	Cutter ring falling
6	6#	Cutter ring broken	15	47#	Cutter ring falling
7	7#	Cutter ring broken	16	48#	Cutter ring falling
8	8#	Cutter ring broken	17	50#	Cutter ring broken
9	13#	Partial wear	18	51#	Partial wear

Figure 6. One-sided tool wear and broken cutting ring.

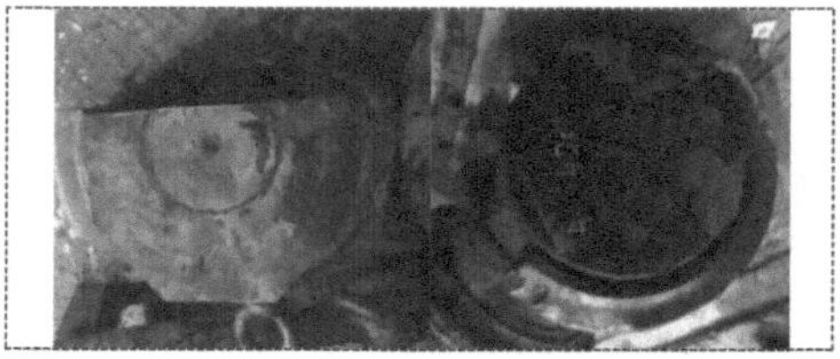

Figure 7. Knife hub wear.

Figure 8. Hob cover clearance and bearing entering sand.

After entering the upper-soft lower-hard stratum transition section, the rock face begins to intrude from the bottom of the tunnel. The shield construction parameters are a forward speed of 15 mm/min and a cutter head speed of 1.8–2.0 r/min. The cutting tool is cooled by adding water to the soil bin. The soil bin is full of bin muck and the muck is a mixture of stone slag and slurry. First, the slag temperature is too steep in the advancing process. On the left line of the second opening inspection, partial wear of the cutter is found. This phenomenon accounts for about 30% of the entire tool change situation. In the third abnormal drive parameter, a large number of hobbing knives with partial wear and broken knife rings are found (Figure 6) and even the knife hub is damaged (Figure 7). The abnormal damage rate reaches 81% and the shield construction is adjusted. The advance velocity is 4–5 mm/min. The penetration is 2–3 mm/r and the cutter head speed is 1.5–1.6 r/min. After adjusting the parameters, the subsequent abnormal wear on the left line is reduced. In the whole upper-soft lower-hard stratum transition section, the normal wear rate of the left line hob is 31% and that of the right line is 62%. It is suitable for the upper-soft lower-hard stratum transition section to use a 19-inch hob. As the rock surface rises, the working temperature of the cutter head surface rises, causing the cutter ring seal to fail. Water and soil can enter the bearings (Figure 8), causing the tool oil leakage and sand to enter the tool body damaging the tool bearings and the tool to stop turning.

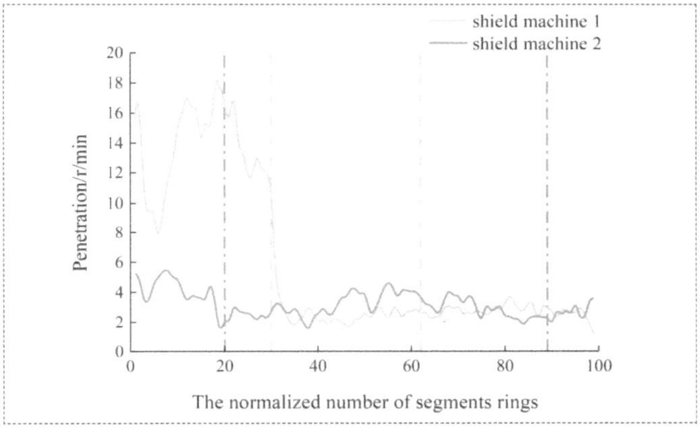

Figure 9. Comparison of penetration.

As shown in Figure 9, the trend of the advance velocity and penetration is essentially the same. These parameters of Shield Machine 1 in the soft soil section are much larger than that of Shield Machine 2. After entering the upper-soft lower-hard stratum transition section and the hard rock section, the two shield machines show little change in these parameters and the changing trend and magnitude are the same.

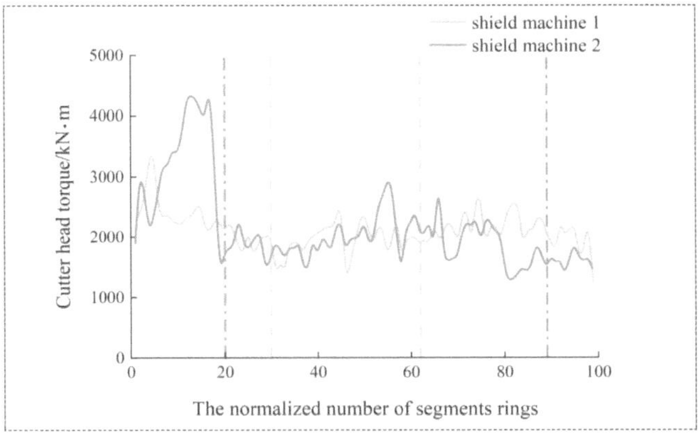

Figure 10. Comparison of cutter head torque.

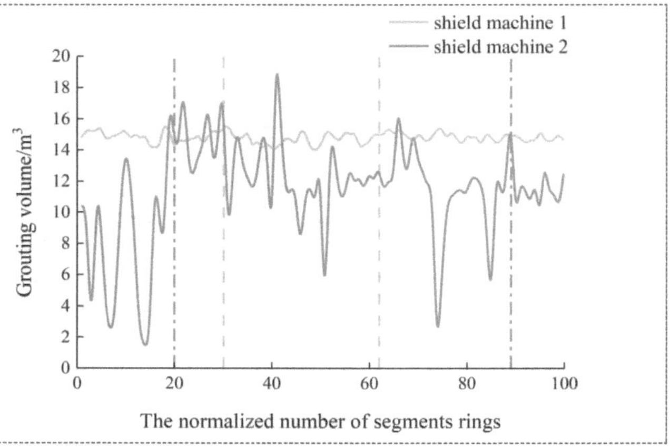

Figure 11. Comparison of grouting volume.

In the soft soil section, the cutter head torque of Shield Machine 2 is unstable and changes greatly (Figure 10). The cutter head torque of Shield Machine 1 does not change much in the whole process, which remains at about 2000 kN·m. The soil loss caused by shield construction is balanced by synchronous grouting. The grouting volume is mainly determined based on the pore volume of the shield tail, and factors, such as the penetration, compaction of the slurry, and over-excavation, are also taken into consideration. The formula for calculating the grouting volume is as follows:

$$Q = \lambda\pi(D^2 - d^2)L/4 \tag{1}$$

where Q is the grouting volume; λ is the grouting ratio; D is the shield machine cutting outer diameter; d and L are the segment outer diameter and segment width, respectively.

In this project, the shield machine has an outer cut diameter of 8.780 m and an inner cut diameter of 8.820 m. The segment diameter is 8.1 m. The thickness is 400 mm and the width is 1.6 m. Considering that after excavation, the upper soil of the shield has a certain displacement due to its weight, which has a certain offset effect on the loss of grouting. Therefore, λ is 1 and the theoretical grouting volume is 15.3 m^3 calculated by Formula 1. The grouting volume of the two shield machines in the selected section is shown in Figure 11. Figure 11 shows that the grouting amount of Shield Machine 2 is generally less than that of Shield Machine 1. The grouting volume of Shield Machine 1 is concentrated around 15 m^3. Shield Machine 2 is all below 15 m^3 except for some rings. Meanwhile, the fluctuation of Shield Machine 2 grouting is large and some of them are even below 5 m^3, indicating that the grouting volume of Shield Machine 2 is insufficient.

According to the above analysis, the reasons for the excessive settlement of this section in terms of shield parameters are as follows:

- In the upper-soft lower-hard stratum transition section, large thrust, high cutter head speed, and large fluctuation of speed can cause excessive disturbance to the soil;
- The survey report of the section of Shield Machine 2 shows that the normalized 60 ring is a hard rock. Additional drilling is subsequently carried out. The actual situation is that a considerable part of soft soil is sandwiched in the hard rock and the shield parameters in the hard rock are still used, which leads to excessive settlement;
- In the sort soil section, the penetration is too small and the torque fluctuation is too large;
- Generally, large thrust forces can result in large penetration, tool wear acceleration, and abnormal wear. There is a large thrust force and no corresponding increase in the

penetration of Shield < Machine 2, showing that the tool has been damaged, which is also confirmed by the actual opening inspection later;
- The excessive amount of earth excavation can cause over-excavation of the excavation surface, resulting in a large settlement.

4 CONCLUSIONS

Based on the Gongbei to Hengqin section of the Zhuhai-Zhuhai Airport Intercity Railway Project, the data on land subsidence during the construction of the shield tunnel are studied and the settlement on both sides of the tunnel is compared and analyzed. The conclusions are as follows:

- It is easy to make an excessive disturbance to the soil when using large thrust, high cutter head speed, and large fluctuation of speed in the upper-soft and lower-hard composite stratum;
- Generally, large thrust force can result in large penetration, tool wear acceleration, and easy abnormal wear. Therefore, when the thrust force is large and the penetration does not increase correspondingly, the cutter tool has a high probability of damage;
- The simultaneous grouting volume has an obvious effect on the ground settlement in the upper-soft and lower-hard composite stratum;
- It is necessary to pay attention to the shield parameters when entering the upper-soft and lower-hard composite stratum. The parameters should be adjusted in real-time according to the changes in the strata.

REFERENCES

Attewell, P. B. et al. (1985). *Soil Movements Induced by Tunneling and Their Effects on Pipelines and Structures. Methuen Inc New York Ny* 25(4), 36–42.

Chen Qiang. *Research on the Supporting Pressure of the Excavation Face of the Shield Tunnel in the Upper Soft and Lower Hard Strata.* (2010). Huazhong University of Science and Technology.

Chen Renpeng et al. (2011). Model Test Study on the Stability of the Excavation Surface of the Dry Sand Shield. *Chinese Journal of Geotechnical Engineering* 33(1), 117–122.

Hou Xueyuan & Liao Shaoming. (1993). Prediction of Shield Tunnel Settlement. *Underground Engineering and Tunnels* (4), 24–32.

Liu Lianwei. (2009). *Study on the Ground Subsidence Law of Subway Construction by Shield Method in the Composite Stratum.* China University of Mining and Technology.

Mair, R.J. & Taylor, R.N. (1999). Bored Tunneling in Urban Environments. *Proceedings International Society for Soil Mechanics and Foundation Engineering.* 4.

Rankin, W. J. (1988). *Ground Movements Resulting from Urban Tunneling: Predictions and Effects.* Geological Society London Engineering Geology Special Publications 5(1), 79–92.

Wang Heng et al. (2012). Effect of Earth Pressure on Earth Settlement Caused by Composite Stratum Shield Construction. *Anhui Architecture* 4(03), 96–97.

Zhou Xiaowen & Pu Jialuo. (2002). Experimental Study on Ground Settlement Caused by Tunnel Excavation in the Sand. *Rock and Soil Mechanics* 23(5), 559–563.

Zi Yi et al. (2010). The Effect of the Supporting Pressure of the Tunnel Excavation Surface on the Stratum Deformation in the Composite Stratum. *Journal of Civil Engineering and Management* 27(4), 36–41.

Water Conservancy and Civil Construction – Oke & Ahmad (Eds)
© 2024 The Author(s), ISBN: 978-1-032-58614-4

Fatigue life distribution characteristics and reliability of epoxy resin pavement mixture

Xiaoqing Wang & Biao Ma*
Chang'an University, Xi'an, China

ABSTRACT: Epoxy resin (ER) mixtures used as pavement materials are subjected to the repeated vehicle and temperature loads. Therefore, their fatigue performance should be considered in the design of mixtures and pavement structures. In this study, the fatigue characteristics of ER mixtures are analyzed using flexure fatigue tests under different stress levels in the stress control mode. The probability distribution of the fatigue life obeys a two-parameter Weibull distribution. Therefore, the two-parameter fatigue equations for the average and equivalent fatigue lives are established based on the S-N equation and the probabilistic stress life curve is obtained. The semi-logarithm and dual-logarithm fatigue equations demonstrate a high linear fitting correlation for the average and effective fatigue life and the single logarithmic model provides relatively better and more conservative results. Based on the Miner damage theory, the fatigue damage reliability of ER mixtures is analyzed. The results indicate that fatigue reliability decreases with the increase in the load cycle number and stress level. The study findings provide a reference for the fatigue life prediction of ER mixtures and pavement structure design and lay the foundation for further study on fatigue damage characteristics and failure mechanism of ER mixtures.

1 INTRODUCTION

With the development of road traffic, new requirements for green operation, environmental protection, and intelligence of road construction and maintenance have been put forward. In recent years, resins have been widely used in road engineering, including epoxy asphalt, polyurethane asphalt, epoxy emulsified asphalt, epoxy concrete, and polyurethane mixtures, due to their excellent strength, durability, adhesion, corrosion resistance, and good compatibility (Xiang & Xiao 2020; Yu *et al.* 2020; Zhang *et al.* 2021).

Epoxy resin concrete is a type of composite material produced by mixing sand, stone, and other aggregates with ER, where ER replaces total or partial cement as a binder. This concrete exhibits good compressive, tensile, and flexural strengths, as well as good wear resistance. In addition, ER concrete exhibits waterproofing characteristics, frost, and corrosion resistance (Wang *et al.* 2004). The successful application of ER concrete has stimulated the use of ER as a substitute for asphalt in asphalt mixtures to produce a composite material, here called ER mix, for pavement construction and maintenance. The main differences between the ER mixture and ER concrete are the grain size of the stone and the amount of ER present. In general, there are significant differences in performance between them due to the different amounts of ER (Ferdous *et al.* 2020; Zhou *et al.* 2016).

Researchers have conducted several studies on the mechanical properties, durability, and mix design, among other aspects of ER concrete (Ferdous *et al.* 2020; Gagandeep 2020; Ribeiro *et al.* 2002; Marinela & Daniel 2008), thus promoting its application. However, relatively few studies

*Corresponding Author: mabiao@chd.edu.cn

 DOI: 10.1201/9781003450818-59

have been conducted on the properties of ER mixtures. Liu (2009) studied the road performance of ER mixtures and indicated that these mixtures exhibited high compressive strength and strong low-temperature deformation ability, thus enabling their use in pavement layers of bridge decks or pavement surface layers. Xu *et al.* (2021) studied the strength and formation law of ER mixtures and established a prediction model for their strength and deformation characteristics. The existing research has shown that ER mixtures have the potential for use as pavement materials. Their static mechanical strength is greater than that of asphalt mixtures and their deformation performance is better than that of cement concrete and asphalt mixtures, indicating good toughness. In addition, as pavement surface materials, ER mixtures are subjected to alternating tensile and compressive stresses under the continuous action of vehicle axle loads, and their properties gradually decay, leading to fatigue damage (Tigdemir *et al.* 2002; Li *et al.* 2007). Therefore, it is vital to characterize and evaluate their fatigue performance. However, few relevant studies have focused on this topic of research.

In this study, the flexural fatigue test of an ER mixture with different stress levels under the stress control mode is performed and its static flexure and tensile properties are tested. In addition, the probability distribution of the fatigue life test data is analyzed. Based on the *S-N* equation, the two-parameter fatigue equation of the average and equivalent fatigue life is regressed. The reliability of the fatigue damage of the ER mixture is analyzed based on the Miner damage theory. The study findings provide a reference for the fatigue life prediction of ER mixtures and pavement structure design, which lays the foundation for further study of the fatigue damage characteristics and failure mechanism of ER mixtures.

2 MATERIALS AND METHODS

2.1 *Materials*

2.1.1 *Raw materials and composition of ER binder*

The raw materials of the ER binder in a mass ratio of 1:0.8:0.7 are hydrogenated bisphenol A ER, curing agent polyamide 650, and a toughening agent, respectively. The properties of the curing material are listed in Table 1.

Table 1. Characteristics of curing ER binder.

Sample	Tensile strength, 25 °C (MPa)	Elongation at break, 25 °C (%)	Glass transition temperature (°C)
Curing ER binder	3.2	199	–10

2.1.2 *Aggregates of ER mixture*

The coarse and fine aggregates and mineral powder are limestones. The gradation is shown in Table 2.

Table 2. Aggregate gradation.

Size (mm)	13.2	9.5	4.75	2.36	1.18	0.6	0.3	0.15	0.075
Passing (%)	91.51	66.24	40.10	32.00	25.79	20.90	16.66	12.05	8.00

2.1.3 *Preparation of ER mixtures*

Three types of ER mixtures with binder-aggregate ratios of 4%, 5%, and 6% are prepared. The preparation steps are as follows. First, coarse and fine aggregates and ER binder are stirred for 90 s and the mineral powder is added and stirred for another 90 s to obtain an ER mixture. All processes are performed at room temperature.

2.2 *Experimental method*

2.2.1 *Static flexural test*

The static flexural test of the ER mixture is performed using a universal testing machine. The test temperature is 25 °C and the loading rate is 50 mm/min. The test sample is a 42 × forty-two × 250 mm beam, cured for 28 days before the static bending test.

2.2.2 *Flexural fatigue test*

The bending fatigue test adopts the stress control mode using a three-point load and the number of load cycles for the beam to rupture is considered as the fatigue life. The test temperature is 25 ± 1 °C and the load is a sine wave load with a frequency of 10 Hz. The test is divided into five groups, with five samples per group. The stress levels were 0.3, 0.4, 0.5, 0.6, and 0.7.

3 RESULTS AND DISCUSSION

3.1 *Static flexural and tensile properties of ER mixtures*

The measured static flexural and tensile properties of the ER mixtures are listed in Table 3.

Table 3. Flexural and tensile properties of ER mixtures.

Binder-aggregate ratio (%)	Flexural-tensile strength (MPa)	Flexural-tensile strength strain (με)
4	10.8	4985
5	12.4	5889
6	11.8	8123

As can be seen from Table 3, the flexural strength of the ER mixture is not significantly different under the three binder-aggregate ratios. When the cement-aggregate ratio is 5%, flexural strength is the highest. The flexural strain increases with an increase in the binder-aggregate ratio. Thus, increasing the amount of ER binder can enhance the flexibility of the mixture without significantly affecting or reducing its strength.

Table 4. Fatigue test results of ER mixtures with different binder-aggregate ratios.

Binder-aggregate ratio (%)	Stress level	Nf (times)					
		Sample 1	Sample 2	Sample 3	Sample 4	Sample 5	Average
4	0.5	51879	38012	33015	20086	32115	35021
5		113957	125042	95337	141993	70815	109428
6		40287	33257	26879	30957	30198	32316

3.2 *Effect of binder-aggregate ratio on fatigue life of ER mixtures*

As shown in Table 4, the binder-aggregate ratio has a significant impact on the fatigue performance of the mixture. The specimen with a binder-aggregate ratio of 5% exhibits the largest fatigue life and the best fatigue resistance. This indicates that this ER mixture has the optimum ER content in terms of fatigue durability and that better results are not necessarily related to the increase in ER content. The fatigue life of the epoxy mixture with a binder-aggregate ratio of 5% is further analyzed in the following sections.

3.3 *Statistical distribution of fatigue life of ER mixtures*

3.3.1 *Fatigue test results*

The fatigue life of the ER mixture with a 5% binder-aggregate ratio at different stress levels is shown in Table 5.

Table 5. Fatigue life of ER mixture under different stress levels.

	Nf (times)							
Stress level	Sample 1	Sample 2	Sample 3	Sample 4	Sample 5	Average	Stdev	Stdev /Average
0.3	752918	797450	995489	1202987	1248142	999397	226314	0.226
0.4	175875	192987	236854	268872	382594	251436	81921	0.325
0.5	70815	95337	113957	125042	141993	109429	27451	0.251
0.6	11041	20187	27587	31547	41894	26451	11645	0.440
0.7	3594	3869	5687	10014	14187	7470	4548	0.609

As shown in Table 5, the fatigue life of the ER mixture decreases with an increase in the stress level. The higher the stress level, the more sensitive the fatigue life test results. This is because voids and initial defects are distributed in the structure of the ER mixture and are amplified at high-stress levels, thus increasing the dispersion of the test results.

3.3.2 *Fatigue life distribution test based on two-parameter Weibull distribution*

Under the same load level cycle, the distribution law of the fatigue life of specimens can be expressed by the Weibull function as follows (Weibull 1951):

$$\mathrm{f}(N_\mathrm{f}) = \frac{\mathrm{b}}{N_\mathrm{a} - N_0}\left(\frac{N_\mathrm{f} - N_0}{N_\mathrm{a} - N_0}\right)^{\mathrm{b}-1} \exp\left[-\left(\frac{N_\mathrm{f} - N_0}{N_\mathrm{a} - N_0}\right)^{\mathrm{b}}\right] \quad N_0 \leq N_\mathrm{f} < \infty \tag{1}$$

Where N_f is fatigue life; N_a is the characteristic fatigue life parameter; N_0 is the minimum fatigue life. Considering the discrete nature of the ER mixture material, this parameter is taken as 0 and b is the Weibull shape parameter.

The Weibull probability parameter, which is expressed as N_p, is built according to the Weibull distribution function. Then, the distribution function $\mathrm{F}(N_\mathrm{p})$ of the Weibull parameter N_p can be obtained from Equation (1) as follows:

$$\mathrm{F}(N_\mathrm{p}) = \mathrm{P}(N_\mathrm{f} < N_\mathrm{p}) = 1 - \exp\left[-\left(\frac{N_\mathrm{p}}{N_\mathrm{a}}\right)^{\mathrm{b}}\right] \tag{2}$$

Then,

$$\mathrm{P}(N_\mathrm{f} > N_\mathrm{p}) = 1 - \mathrm{P}(N_\mathrm{f} < N_\mathrm{p}) = \exp\left[-\left(\frac{N_\mathrm{p}}{N_\mathrm{a}}\right)^{\mathrm{b}}\right] \tag{3}$$

Noting $\mathrm{P}(N_\mathrm{f} > N_\mathrm{p})$ as P, Equation (3) can be written as follows.

$$\frac{1}{P} = \exp\left(\frac{\mathrm{N_p}}{\mathrm{N_a}}\right)^{\mathrm{b}} \tag{4}$$

Taking the natural logarithm twice on both sides, we obtain the following equation.

$$\ln\left[\ln\left(\frac{1}{P}\right)\right] = \mathrm{b}\ln N_\mathrm{p} - \mathrm{b}\ln N_\mathrm{a} \tag{5}$$

The k fatigue life test results under the same stress level are named as i from small to large and the reliability P is calculated according to Equation (6) (k is set as 5).

$$P = 1 - \frac{i - 0.3}{k + 0.4} \tag{6}$$

We assume $y = \ln\left[\ln\left(\frac{1}{P}\right)\right]$, $x = \ln N_p$, and $c = b\ln N_a$, Equation (5) can be written as follows.

$$y = bx + c \tag{7}$$

Equation (7) can be used to check whether the fatigue test data obey a two-parameter Weibull distribution. The fatigue loading test results of the ER mixture with a binder ratio of 5% (Table 3) at different stress levels are processed and fitted according to the Weibull distribution and the results are depicted in Figure 1. The linear fitting correlation coefficients R^2 are all above 0.85, indicating that $\ln\left[\ln\left(\frac{1}{P}\right)\right]$ has a good linear correlation with $\ln N_f$, which confirms that the flexural fatigue life of the ER mixture obeys the two-parameter Weibull density distribution function.

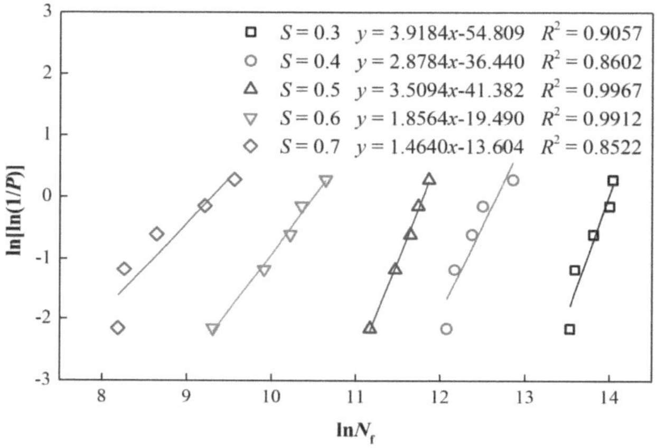

Figure 1. Linear regression fitting results of two-parameter weibull distribution.

3.4 *Flexural fatigue strength analysis of ER mixture*

3.4.1 *Average S-N curve*

In the fatigue test, a set of corresponding fatigue failure cycles is obtained by applying cyclic loads under different stress levels to the same group of specimens. With the stress level S as the abscissa and the number of fatigue cycles N_f as the ordinate, the S-N curve can be drawn according to the test data. Semi-logarithm ($S - \lg N_f$) and dual-logarithm ($\lg S - \lg N_f$) fatigue equations are commonly used in engineering to fit the fatigue life. In general, the survival rate of the semi-logarithm and dual-logarithm fatigue equations is only 50%. The semi-logarithmic fatigue equation is given by Equation (8).

$$\lg N_f = \alpha S + \beta \tag{8}$$

The dual-logarithm fatigue equation is given by Equation (9).

$$\lg N_f = \gamma \lg S + \lg \eta \tag{9}$$

The linear regression between the average fatigue life and the corresponding stress level in Table 3 is performed using the semi-logarithm and dual-logarithm fatigue equations. The regression curves are depicted in Figure 2.

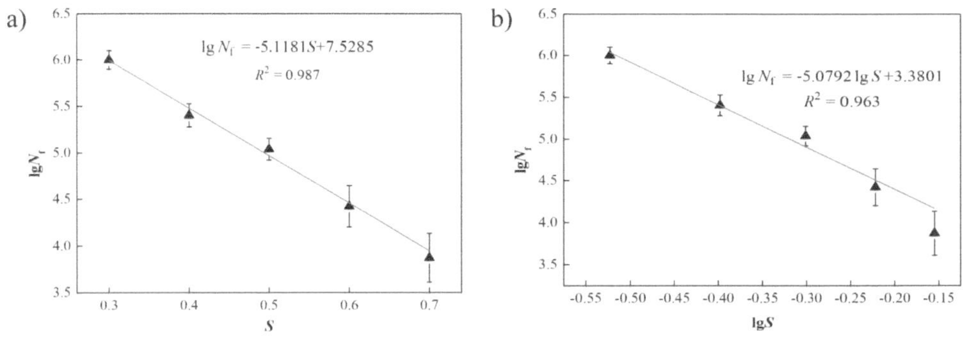

Figure 2. S-N curve: (a) Semi-logarithm; (b) Dual-logarithm.

As shown in Figure 2, the linear correlation coefficients R^2 are all above 0.96, indicating that both S and lg$S$ exhibit a good linear relationship with lgN_f. The fitting model based on the semi-logarithm and dual-logarithm fatigue equations can be used to roughly estimate the stress level when the fatigue life of the mixture tends to be infinite (2 million times). For the semi-logarithm and dual-logarithm fatigue equations, with N_f equal to 2 million times, we determine S is 0.24 and 0.28, respectively, and the stress level obtained is low. In addition, the corresponding load amplitudes are 3.06 MPa and 3.47 MPa, respectively, indicating that the semi-logarithmic fatigue equation is more conservative in the prediction of the fatigue performance of the ER mixture.

3.4.2 *Fatigue equation and P-S-N curve of two-parameter Weibull distribution*

As described in Section 3.2.2, the fatigue life N_f of the epoxy mixture obeys the two-parameter Weibull distribution function. Thus, the fatigue life of the ER mixture under different reliabilities P can be explored on this basis. The reliability P can be calculated using Equation (3) and the corresponding equivalent fatigue life N is obtained using Equation (10):

$$N = N_a \left[\ln\left(\frac{1}{P}\right) \right]^{\frac{1}{b}} \tag{10}$$

Substituting the regression parameters b and c obtained in Section 3.2.2 into Equation (3), the equivalent fatigue life N under different stress levels and reliabilities can be obtained. The calculation results are presented in Table 6. In addition, the linear regression on the equivalent fatigue life N is performed according to Equations (8) and (9) to obtain the semi-logarithm and dual-logarithm fatigue equations under different reliabilities P, as shown in Figure 3.

Table 6. Equivalent fatigue life under different reliabilities and stress levels.

	Stress level						Stress level				
Reliability	0.3	0.4	0.5	0.6	0.7	Reliability	0.3	0.4	0.5	0.6	0.7
0.95	556590	112187	56693	7323	1427	0.65	958069	234975	103963	23044	6107
0.90	668833	144062	69600	10792	2334	0.60	1000658	249308	109136	25260	6860
0.85	747057	167472	78749	13630	3138	0.55	1041643	263310	114139	27493	7639
0.80	810011	186971	86194	16169	3896	0.50	1081717	277196	119053	29774	8451
0.75	864266	204223	92665	18540	4635	0.65	958069	234975	103963	23044	6107
0.70	913005	220059	98518	20816	5368						

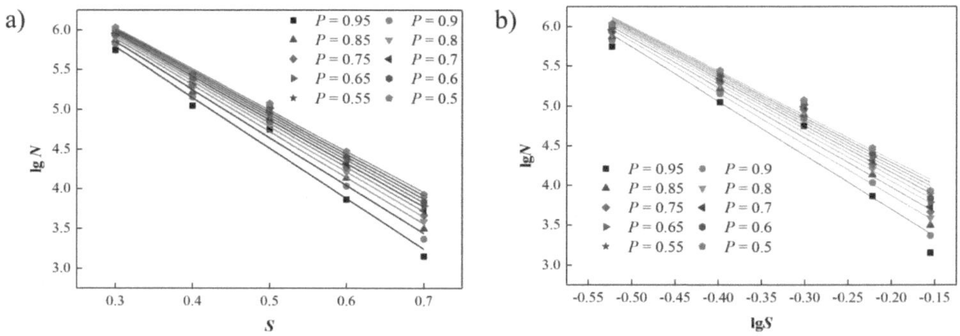

Figure 3. P-S-N curve of equivalent fatigue life: (a) Semi-logarithm; (b) Dual-logarithm.

As shown in Figure 3, the linear correlation coefficients R^2 of the equivalent fatigue life regression equations under different reliabilities of the two parameters are above 0.9, indicating that the equivalent fatigue life of the epoxy mixtures under the two-parameter Weibull distribution is equally adapted to the semi-logarithm and dual-logarithm fatigue equations. As the reliability increases, the absolute value of the slope of the corresponding *S-N* curve increases, indicating that the fatigue life is more sensitive to the stress level at larger reliability requirements. From the regression parameters of the equivalent fatigue life regression equation under different reliability degrees, there is a certain correlation between the degree of reliability and slope. However, comparing the intercepts of the semi-logarithm and dual-logarithm regression equations with reliability, it is found that the intercepts of the semi-logarithm fatigue equations vary slightly and monotonically with reliability, whereas the intercepts of the dual-logarithm fatigue equations show irregular variations. The intercept of the logarithmic fatigue life equation reflects the strength of the material bending fatigue performance. Thus, the change rule of the semi-logarithm fatigue equation intercept is by its physical meaning. Although the semi-logarithm fatigue equation cannot satisfy the boundary condition of $S \to 0$ when $N_f \to \infty$. The fitting degree of the single logarithmic fatigue equation to the average and equivalent fatigue lives is higher in the stress level range of [0.3, 0.7] selected in this study. Using the *P-S-N* curve or the equivalent fatigue life equation in Table 7, the fatigue life under different reliabilities can be roughly predicted in practical engineering.

3.5 *Fatigue damage reliability analysis of ER mixtures based on Miner damage theory*

Although according to the *P-S-N* curve, the fatigue reliability of the ER mixture gradually decreases with the increase in load cycle times, the specific details of this change are unclear. Therefore, the distribution function of the fatigue damage model of the ER mixture is further obtained based on the two-parameter Weibull distribution and the Miner damage model.

Assuming that the probability function of random variable fatigue damage is F(D) and $D_p = \frac{n}{N_p}$, according to Equation (2), we can determine F(D) as follows.

$$F(D) = \exp\left[-\left(\frac{n}{DN_a}\right)^b\right] \tag{11}$$

By substituting the regression parameters b and c from Section 3.2.2 into Equation (11), the distribution function of random variable fatigue damage under different stress levels can be determined as follows.

$F_{0.3}(D) = \exp\left[-\left(\frac{n}{1187780D}\right)^{3.9814}\right]$, $R^2 = 0.9057$; $F_{0.4}(D) = \exp\left[-\left(\frac{n}{314837D}\right)^{2.8784}\right]$, R^2=0.8602;

$F_{0.5}(D) = \exp\left[-\left(\frac{n}{132158D}\right)^{3.5094}\right]$, $R^2 = 0.9967$; $F_{0.6}(D) = \exp\left[-\left(\frac{n}{36272D}\right)^{1.8564}\right]$, $R^2 = 0.9912$;
$F_{0.7}(D) = \exp\left[-\left(\frac{n}{10854D}\right)^{1.4640}\right]$, $R^2 = 0.8522$

As the ER mixture has different critical values of fatigue damage under different stress levels, the residual strength of the mixture is calculated using the ratio of the average modulus E_{min} obtained from the last five cycles before fatigue failure to the average modulus E of the 101st loading. That is, the stiffness is used to define the damage and the critical value of fatigue damage is calculated according to Equation (12). The calculation results are listed in Table 7.

$$D_c = 1 - \frac{E_{min}}{E} \tag{12}$$

Where E_{min} is the stiffness before fatigue failure and E is the initial stiffness.

Table 7. Critical fatigue damage under different stress levels.

Stress level	0.3	0.4	0.5	0.6	0.7
Critical fatigue damage	0.468	0.314	0.258	0.184	0.163

The critical value of fatigue damage in Table 7 is substituted into the distribution function of the random variable fatigue damage under different stress levels and the fatigue reliability of the ER mixture can be obtained by calculating F (D) after a given number of load cycles n. Assuming a set of load cycles n (100, 200, 400, ... , 819200, 1638400), the fatigue reliability of the ER mixtures based on the Miner damage model is calculated, and the results are depicted in Figure 4.

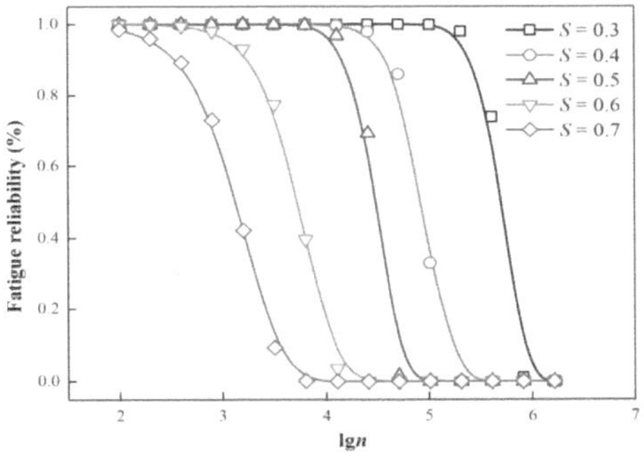

Figure 4. Fatigue reliability curve of ER mixtures.

As shown in Figure 4, with an increase in the logarithmic load, the reliability of the ER mixture gradually decreases from almost 100% to 0% at the same stress level. Regardless of the stress level, the fatigue reliability is relatively stable at the beginning of the load cycle. However, with the increase in the stress level, the duration of this stage gradually decreases and the number of logarithmic load cycles corresponding to the fatigue reliability attenuation of 0% also decreases, thus resulting in lower fatigue reliability of the ER mixture in the reliability attenuation stage.

4 CONCLUSION

(1) The amount of ER has a significant impact on the fatigue life of ER mixture. The flexural-tensile strength and flexural-tensile strain exhibit better values when the binder-aggregate ratio is 6%, but the fatigue performance is poor.

(2) The fatigue life of the mixture decreases with an increase in the stress level. A larger stress level magnifies the internal defects of the mixture and increases the dispersion of the test results. The fatigue life test results of the ER mixture are in good agreement with the Weibull distribution function.
(3) The semi-logarithm and dual-logarithm fatigue equations have a high linear fitting correlation for the average and effective fatigue life. The semi-logarithmic model provides more conservative results. The two-parameter equivalent fatigue life regression equation and *P-S-N* curve obtained from the regression of the *S-N* curve can be used to predict the flexural-tensile fatigue performance of the ER mixture and provide a reference for the design of pavement structures.
(4) The increase in the number of loading cycles and stress levels gradually reduces fatigue reliability. To maintain the fatigue reliability of the ER mixture at a high level, the overload should be maximally limited, layer thickness should be appropriately increased, and defects of the pavement structure layer should be regularly detected.

REFERENCES

Beeldens, A., Monteny, J., Vincke, E., et al. Resistance to Biogenic Sulphuric Acid Corrosion of Polymer-modified Mortars. *Cement Concrete Comp.* 2001; 23(1):47–56.

Ferdous, W., Manalo, A., Wong, H. S., et al. Optimal Design for Epoxy Polymer Concrete Based on Mechanical Properties and Durability Aspects. *Constr Build Mater.* 2020; 232:117229.

Gagandeep. Experimental Study on Strength Characteristics of Polymer Concrete with Epoxy resin. *Materials Today: Proceedings.* 2020; 37(2):2886–9.

Li, H., Zhang, M., Ou, J. Flexural Fatigue Performance of Concrete Containing Nano-particles for Pavement. *Int J Fatigue.* 2007; 29(7):1292–301.

Liu, K. Study on Pavement Performance of Epoxy Resin Concrete. *Journal of Southeast University (Natural Science Edition).* 2009.

Marinela, B., Daniel, L. Mechanical Characteristics Investigation of Polymer Concrete Using Mixture Design of Experiments and Response Surface Method. *Journal of Applied Sciences.* 2008; 8(12).

Ribeiro, M. C. S., Tavares, C. M. L., Ferreira AJM. Chemical Resistance of Epoxy and Polyester Polymer Concrete to Acids and Salts. *J Polym Eng.* 2002; 22(1):27–44.

Tigdemir, M., Karasahin, M., Sen, Z. Investigation of Fatigue Behavior of Asphalt Concrete Pavements with the Fuzzy-logic Approach. *Int J Fatigue.* 2002; 24(8):903–10.

Wang, G., Liu, W., Zhang, C., Xun, Z., et al. Adhesive Properties of Modified Epoxy Resin Used for Reinforced Concrete Structure. *Journal OF Harbin Institute OF Technology.* 2004; 36(2):187–90.

Weibull, W. A Statistical Distribution of Wide Applicability. *J Appl Mech.* 1951; 18(2):293–7.

Xiang, Q., Xiao, F. Applications of Epoxy Materials in Pavement Engineering. *Constr Build Mater.* 2020; 235:117529.

Xu, J., Ma, B., Mao, W., et al. Strength Characteristics and Prediction of Epoxy Resin Pavement Mixture. *Constr Build Mater.* 2021; 283:122682.

Yu, H., Ma, T., Wang, D., et al. Review on China's Pavement Engineering Research–2020. *China Journal of Highway and Trasport.* 2020; 33(10):1–66.

Zhang, Z., Sun, J., Huang, Z., et al. A Laboratory Study of Epoxy/Polyurethane Modified Asphalt Binders and Mixtures Suitable for Flexible Bridge Deck Pavement. *Constr Build Mater.* 2021; 274:122084.

Zhou, J., Zheng, M., Wang, Q., et al. Flexural Fatigue Behavior of Polymer-Modified Pervious Concrete with Single-sized Aggregates. *Constr Build Mater.* 2016; 124:897–905.

Water Conservancy and Civil Construction – Oke & Ahmad (Eds)

Study on alignment design of secondary highway in southeast gumid and hot area—take Xinxi village to Huangtian village as an example

Jiahao Wang*
School of Energy Science and Engineering, Henan Polytechnic University, Jiaozuo, Henan, China

ABSTRACT: Highway construction plays a vital role in regional economic development. As a widely used secondary highway, its overall investment and cost are low and its comprehensive use value is high. In highway design, route design is the key point of highway design. In the process of route design, the index of highway routes is mainly determined according to the tasks, requirements, and various factors of highway construction. In this paper, we mainly analyze the principles of plane alignment selection briefly and discuss the design points and design process with cases.

1 INTRODUCTION

Alignment is the basis of the design of secondary highways. The influence of topography and geomorphology should be considered comprehensively in the selection of routes, and mutual coordination should be achieved. The ratios of straight lines, gentle curves, and circular curves are between 1:1:1 and 1:2:1 in strict accordance with the highway design specification. The line shape of the design should be continuous. A good line shape cannot only provide drivers with a comfortable driving process but also reduce the project cost, and more importantly, improve the safety performance of road traffic.

The rationality of the cross-section should be ensured when selecting the line, and the balance of the longitudinal section should be improved to enhance the effect of the homeostatic operation. The actual requirements of the design of secondary roads according to the terrain, the relevant designers should establish effective data analysis and research of the life and work to ensure the cleverness of the layout, solve the problem of quantity control and reduce the extent of damage on the surrounding environment of the project. High performance is used to establish the corresponding processing work and improve the project cost management effectiveness.

Li Chengcheng believes (Li 2020) that route design is an important framework of highway engineering and that the quality of route design directly affects the safety and performance of the whole highway engineering. Route selection is the focus of route design of highway engineering, which has a very important influence on the cost of the highway and the service life of the highway.

2 GEOLOGICAL SURVEY OF HIGHWAY

As an autonomous prefecture under the jurisdiction of Hubei Province, Enshi Prefecture has 29 ethnic groups, including Tujia, Miao, and Han. It has a high forest coverage rate and is wet all year round with abundant rainfall, which is suitable for planting economy, such as tea and fruit trees. And it has rich coal resources. The construction of secondary roads here can

*Corresponding Author: 18839505751@163.com

DOI: 10.1201/9781003450818-60

enhance transportation capacity, transport crops, and mineral resources, greatly promote local economic development, and provide more convenient services for residents.

3 ROUTE DESIGN

3.1 *Principles of linear design*

(1) Multiple schemes Mandatory (Liu 2016): in the design of secondary highways, several linear design schemes are made when selecting routes. Based on comprehensive consideration of terrain, geology, economy, safety, and other factors, the best scheme is proved by multiple options.
(2) Principle of alignment: in the process of highway alignment design, it is necessary to enhance the consistency between longitudinal section and plane alignment, thus reducing the hidden danger of driving on the highway.
(3) Principles of environmental protection (Lin 2013; Zhu 2015): to fundamentally improve the practical value of secondary highway survey and route design, it is necessary to construct an effective treatment mechanism for complex terrain and hydrological conditions, reduce the damage to the ecological environment, and take the regional ecological environment protection as the key in the specific operation process.
(4) The principle of the economy: in the selection of lines, we minimize the occupation of farmland, reduce the demolition of houses, and reduce the construction of Bridges, tunnels, and other structures. Selecting a line in the range of smaller height differences can avoid deep filling and deep excavation, thus improving the economy of the project.
(5) Security principles (Xu 2007): the continuity index based on running speed is used as the parameter to test and evaluate the safety of the linear shape.

3.2 *Planar linear design*

In the design of the line shape, the height difference of the terrain in the selected area should be fully considered as much as possible. In this design, we choose the line shape of the same curve and the reverse curve to cooperate, thus achieving the equilibrium and continuity of the line shape and maximizing the economic benefits (Hu & Zhang 2014).

Not all roads are straight. However, there is a need for flat curves at the corners. Highway Route Design Standard (Zhao 2013) makes relevant provisions on the minimum radius of circular curves, as shown in Table 1.

Table 1. The minimum radius of the circular curve.

Design speed (km/h)	Minimum radius of the circular curve (general value) (m)	Minimum radius of the circular curve (limit value) (m)			
		Imax = 4%	Imax = 6%	Imax = 8%	Imax = 10%
60	200	150	135	125	115

There are 5 intersection points in this plane design. Intersection point 1 and intersection point 2 form the same direction curve with a radius of 350 m and 310 m, respectively. Intersection point 2 and intersection point 3 form a reverse curve with a radius of 310 m and 280 m, respectively. In the same direction and the reverse curve, the radius of the easing curve is 120 m, 150 m, and 160 m.

In this design, the minimum radius of the circular curve is 200 m. The radius of the circular curve should be as large as possible because the larger the radius of the circular curve, the smaller the centripetal force, thus avoiding traffic accidents, such as planning and rollover. However, the larger the radius of the circular curve, the better it is. According to the specification, the maximum radius of the circular curve should not exceed 10,000 m.

In the design of plane lines, we insert a moderate curve of a certain length between the straight line and the circular curve (Miu 2022) to transition the ultra-high and widening of the circular curve (Zhao 2013). Since the radius is larger than 250 meters, no widening is needed. The minimum length of the relaxation curve is calculated using the following formula:

$$L_{s(\min)} = \frac{v}{1.2}(m) \tag{1}$$

Where $L_{s(\min)}$ is the minimum length of the transition curve; v is the design speed (km / h).

The length of the relaxation curve is greater than the minimum length of the relaxation curve, which meets the requirements.

3.3 *Longitudinal section linear design*

The longitudinal slope degree is more than 0.3% to ensure the balance of filling and digging and the total amount of filling and digging is moderate. The line shape should meet the design principles. Under the premise of ensuring the sight distance requirement, it can make the sight line more continuous and avoid the phenomenon of "dark concave" and "broken back".

When the design speed is 60 km/h according to the specification, the minimum synthetic slope should not be less than 0.5% to ensure the drainage of the road surface.

The calculation formula of the synthetic longitudinal slope is as follows.

$$\mathrm{I} = \sqrt{\mathrm{i}_{\mathrm{h}}^2 + \mathrm{i}^2} \tag{2}$$

Where I is the synthetic slope (%); i_h is the ultra-high transverse slope (%); i is the route design longitudinal slope (%).

The minimum combined slope of this highway is I = 8.01%, which meets the requirements (Miu 2022).

Table 2. Synthetic slope.

Highway grade	Highway, first-class highway				Two, three, or four highways				
Design speed (Km/h)	120	100	80	60	80	60	40	30	20
Composite slope (%)	10	10	10.5	10，5	9	9.5	10	10	10

To prevent the increase in the number of slope points caused by the short length of the slope, the frequency of weight changes frequently during the vehicle driving process, which leads to the inconvenience of vehicle driving (Liu 2014; Zhao 2022).

The longitudinal slope in a relatively steep situation (Wu 1997) can cause the water tank to "boil" during the exercise. The frequent braking of the automobile is easy to cause the failure of the heater, which affects the safety of driving and causes the reduction of capacity and service level. According to the test, the maximum slope length of the straight slope section is 395.78 m, which meets the limit requirement of the maximum slope length.

Table 3. Position table of slope turning point (part).

Change pile point pile number	Elevation	Slopei$_1$	Slopei$_2$	Radius of vertical curve
K0+485.005	44.0971	-0.65%	1.49%	8335 (concave)
K0+662.880	46.75	1.49%	5.54%	1236 (concave)
K0+874.180	58.45	5.54%%	0.998%	1400 (convex)
K1+336.964	63.07	0.998%	-6%	1400 (convex)
K1+490.554	53.85	-6%	5.99%	16445 (concave)

The location of each slope point is determined and the data parameters of each slope point are used to design the longitudinal cross-section.

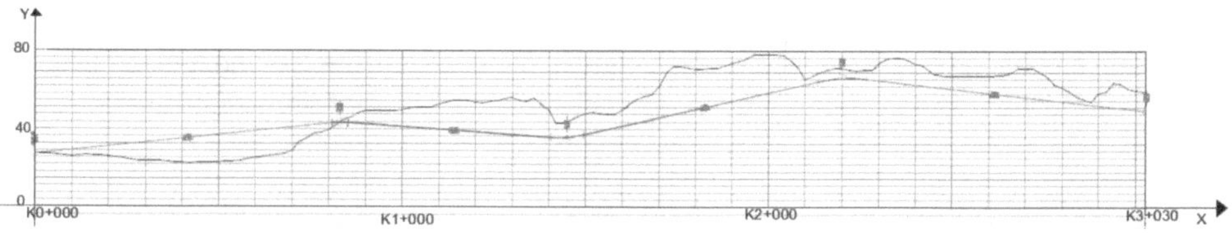

Figure 1. Design drawing of La slope.

4 CONCLUSION

This paper mainly designs the alignment of the Enshi secondary highway. The local terrain conditions are taken into account comprehensively to choose a suitable alignment and the local economic development is combined with people-oriented road transportation to improve the local economic development. In the specific design, the occupation of arable land should be reduced as much as possible, and the scheme with less construction difficulty should be selected. The area through which this route is designed is mostly plain, which prevents deep filling and digging, reduces the difficulty of construction, and saves construction costs. When designing secondary highways in the future, we should pay attention to route selection, maintain the original natural state of the road through the natural landscape and scenic spots and historic sites, and coordinate with it. In the selection of routes, we can choose a variety of schemes to demonstrate and compare the methods of selection to determine the best route scheme.

REFERENCES

Hu Hui, Zhang Bowen. [J]. *Wireless Internet Technology*. 2014(07): 110.

Li Chengcheng. Analysis of Key Points of Highway Alignment Design Under Complex Conditions in Mountainous Areas [J]. *China New Technology and New Products*. 2020(08): 111–112.

Lin Yongming. Study on Coordinated Design of Secondary Highway Alignment and Ecological Landscape in the Mountainous Area of Guizhou Province [J]. *Highway Transportation Science and Technology (Applied Technology Edition)*. 2013, 9(03): 294–300.

Liu Bin. Comparison Selection and Optimization of Route Scheme of Secondary Highway [J]. *Transportation Science and Technology*, 2016(04): 80–82.

Liu Jingbo. Analysis of Key Points of Highway Longitudinal Section Design [J]. *Heilongjiang Transportation Science and Technology*. 2014, 37(08): 4–6.

Miu Guoping. Analysis of Highway Route Longitudinal Section Design Method [J]. *Transportation Manager World*. 2022(05): 22–24.

Wu Xun. Discussion on the Limit of Maximum Longitudinal Slope Length of Highway [J]. *South Central Highway Engineering*. 1997(02): 2–3.

Xu Dihui. Safety Evaluation of Highway Design [J]. *Chinese and Foreign Highways*, 2007(03): 15–17.

Zhao Xin. Discussion on Route Design Experience of Secondary Highway in the Mountainous Area [J]. *Shanxi Architecture*, 2013, 39(34): 170–171.

Zhao Fang. Design and Safety Evaluation Method of Highway Longitudinal Section Alignment [J]. *Engineering Technology Research*, 2022, 7(16): 173–175.

Zhu Yunqiang. Analysis of Route Selection Principles and Key Points of Secondary Highway in Mountainous Areas [J]. *Science and Technology Innovation and Application*. 2015(33): 228.

Water Conservancy and Civil Construction – Oke & Ahmad (Eds)
© 2024 The Author(s), ISBN: 978-1-032-58614-4

Viscoelastic dynamic analysis of saturated asphalt pavement under semi-sinusoidal harmonic load

Bin Zhang
Heilongjiang Bayi Agricultural University, Daqing, China

Yanyang Li
Heilongjiang Bayi Agricultural University, Daqing, China
Northeast Petroleum University, Daqing, China

Wei Guo & Guoliang Xie*
Heilongjiang Bayi Agricultural University, Daqing, China

ABSTRACT: For the water damage from the asphalt pavement, the porous medium theory can be used to analyze the damage mechanism. In this paper, the asphalt pavement is considered a saturated layered viscoelastic axisymmetric system. The semi-sinusoidal harmonic load is employed to simulate the vehicle load. By using the integral transformation method, dynamic balance equations of the asphalt pavement and fluid and the seepage continuity equation are solved. General solutions of stresses, displacements, strains and the excess pore water pressure in the integral space are obtained. By introducing boundary conditions, analytical solutions of physical quantities, such as stresses, displacements, and the excess pore water pressure of the saturated semi-infinite space body are derived. The F. Durbin method and the Simpson quadrature formula are used to inverse Laplace and Hankel transformation to calculate the excess pore water pressure. Results indicate that with the decrease in the permeability coefficient, the permeability of the asphalt pavement decreases, which leads to an increase in the excess pore water pressure. And the peak value of the excess pore water pressure appears earlier than that of the vehicle load. The change rule of the excess pore water pressure is revealed.

1 INTRODUCTION

The asphalt pavement is a porous medium, which is full of water because of precipitation and groundwater. Under the action of the vehicle load, the excess pore water pressure and flow of pore water are produced. The pore water is subjected to scouring, pumping, and squeezing effects on the asphalt pavement. The pore water directly acts on the asphalt membrane, causing the asphalt membrane to peel off from the surface of the aggregate and resulting in water damage to the asphalt pavement. Water damage to the asphalt pavement shortens the service life of asphalt pavement, increases the maintenance cost, and reduces driving comfort and safety. Therefore, the damage mechanism of water damage to the asphalt pavement should be studied. In studying the damage mechanism of water damage to the asphalt pavement, the effects of pore water and the vehicle load must be taken into account simultaneously. The asphalt pavement can be regarded as a saturated layered

*Corresponding Author: xieguoliang1219@163.com

DOI: 10.1201/9781003450818-61

system. We assume that the pores of the asphalt pavement are full of water. Therefore, the porous medium theory is used for analysis and research.

Most scholars regard the asphalt pavement as a saturated layered elastic system, establish mechanical models of the saturated asphalt pavement, and solve them by using the transfer matrix method (Peng *et al.* 2004), the stiffness matrix method (Zhong *et al.* 2006), the integral transformation method (Li & Deng 2008), the discrete method in wave-number domain (Cong *et al.* 2007), the Fourier series expansion and the Fourier transformation (Luo *et al.* 2012), the finite element method (Dong *et al.* 2007; Fu *et al.* 2006; Ren *et al.* 2011), and other methods. Bin Zhang and Yinghua Zhao (Zhang & Zhao 2012) regard the asphalt pavement as a saturated axisymmetric layered viscoelastic system and solve the Biot consolidation equation by using the integral transformation method. In the current research, most scholars regard asphalt pavement as a saturated layered elastic system. Even if the layered viscoelastic system is used, static analysis is carried out without considering the dynamic response of the asphalt pavement. This is inconsistent with the viscoelastic and dynamic characteristics of asphalt pavement. Therefore, the asphalt pavement is regarded as a saturated layered axisymmetric viscoelastic system in this paper. Dynamic equilibrium equations of solids and liquids and the seepage continuity equation are solved by using the integral transformation method. And general solutions of physical quantities, such as stresses, displacements, and the excess pore water pressure in the integral space, are obtained. By introducing boundary conditions, analytical solutions of physical quantities, such as stresses, displacements, and the excess pore water pressure of the semi-infinite space body, are solved, and the excess pore water pressure is calculated and analyzed.

2 THE SOLUTION OF THE DYNAMIC EQUILIBRIUM EQUATION

In the axisymmetric problem, the inertia term of fluid on the asphalt pavement is ignored and the dynamic balance equations of the saturated asphalt pavement are as follows.

$$\frac{\partial \sigma_r'}{\partial r} + \frac{\partial \tau_{rz}}{\partial z} + \frac{\sigma_r' - \sigma_\theta'}{r} - \frac{\partial \sigma}{\partial r} = \rho \ddot{u}_r \tag{1}$$

$$\frac{\partial \tau_{rz}}{\partial r} + \frac{\partial \sigma_z'}{\partial z} + \frac{\tau_{rz}}{r} - \frac{\partial \sigma}{\partial z} = \rho \ddot{u}_z \tag{2}$$

Dynamic equilibrium equations of fluid are as follows.

$$-\frac{\partial \sigma}{\partial r} = \frac{1}{k_d'} \dot{w}_r + \rho_f \ddot{u}_r \tag{3}$$

$$-\frac{\partial \sigma}{\partial z} = \frac{1}{k_d'} \dot{w}_z + \rho_f \ddot{u}_z \tag{4}$$

Assuming that both the asphalt pavement and fluids are incompressible, the seepage continuity equation is as follows.

$$\frac{\partial \dot{u}_r}{\partial r} + \frac{\dot{u}_r}{r} + \frac{\partial \dot{u}_z}{\partial z} = -\left(\frac{\partial \dot{w}_r}{\partial r} + \frac{\dot{w}_r}{r} + \frac{\partial \dot{w}_z}{\partial z}\right) \tag{5}$$

Carrying out Laplace transformation on time t for Equations (1) and (2) and assuming that in the initial state $u_r = \dot{u}_r = 0$ and $u_z = \dot{u}_z = 0$, the following equations can be obtained.

$$\frac{\partial \bar{\sigma}_r'}{\partial r} + \frac{\partial \bar{\tau}_{rz}}{\partial z} + \frac{\bar{\sigma}_r' - \bar{\sigma}_\theta'}{r} - \frac{\partial \bar{\sigma}}{\partial r} = \rho s^2 \bar{u}_r \tag{6}$$

$$\frac{\partial \bar{\tau}_{rz}}{\partial r} + \frac{\partial \bar{\sigma}'_z}{\partial z} + \frac{\bar{\tau}_{rz}}{r} - \frac{\partial \bar{\sigma}}{\partial z} = \rho s^2 \bar{u}_z \tag{7}$$

Carrying out Laplace transformation on time t for Equations (3) and (4) and assuming that in the initial state, $w_r = w_z = 0$, the following equations can be obtained.

$$-\frac{\partial \bar{\sigma}}{\partial r} = \frac{1}{k'_d} s \bar{w}_r + \rho_f s^2 \bar{u}_r \tag{8}$$

$$-\frac{\partial \bar{\sigma}}{\partial z} = \frac{1}{k'_d} s \bar{w}_z + \rho_f s^2 \bar{u}_z \tag{9}$$

By substituting viscoelastic constitutive equations in the Laplace space into Equations (6) and (7), Equations (10) and (11) can be obtained as follows.

$$\frac{\overline{G}}{1-2\bar{\mu}} \frac{\partial \bar{e}}{\partial r} + \overline{G}\left(\nabla^2 - \frac{1}{r^2}\right)\bar{u}_r - \frac{\partial \bar{\sigma}}{\partial r} = \rho s^2 \bar{u}_r \tag{10}$$

$$\frac{\overline{G}}{1-2\bar{\mu}} \frac{\partial \bar{e}}{\partial z} + \overline{G}\nabla^2 \bar{u}_z - \frac{\partial \bar{\sigma}}{\partial z} = \rho s^2 \bar{u}_z \tag{11}$$

By performing operations $(\frac{\partial}{\partial r} + \frac{1}{r})$ (10) + $\frac{\partial}{\partial z}$ (11), Equation (12) can be obtained as follows.

$$\nabla^2 \bar{\sigma} = \frac{2\overline{G}(1-\bar{\mu})}{1-2\bar{\mu}} \nabla^2 \bar{e} - \rho s^2 \bar{e} \tag{12}$$

By performing operations $(\frac{\partial}{\partial r} + \frac{1}{r})$(8) + $\frac{\partial}{\partial z}$ (9), Equation (13) can be obtained as follows.

$$-\nabla^2 \bar{\sigma} = \frac{1}{k'_d} s \left(\frac{\partial \bar{w}_r}{\partial r} + \frac{\bar{w}_r}{r} + \frac{\partial \bar{w}_z}{\partial z}\right) + \rho_f s^2 \bar{e} \tag{13}$$

Carrying out Laplace transformation on time t for Equation (5), Equation (14) can be obtained as follows.

$$s\bar{e} = -\left(\frac{\partial \bar{w}_r}{\partial r} + \frac{\bar{w}_r}{r} + \frac{\partial \bar{w}_z}{\partial z}\right) \tag{14}$$

Substituting Equation (14) into Equation (13), Equation (15) can be obtained as follows.

$$\nabla^2 \bar{\sigma} = \frac{1}{k'_d} s\bar{e} - \rho_f s^2 \bar{e} \tag{15}$$

According to Equations (12) and (15), Equation (16) can be obtained as follows.

$$\nabla^2 \bar{e} - \frac{a_2}{a_1} \bar{e} = 0 \tag{16}$$

Where $a_1 = \frac{2\overline{G}(1-\bar{\mu})}{1-2\bar{\mu}}$ and $a_2 = \frac{1}{k'_d} s + \left(\rho - \rho_f\right) s^2$.

Carrying out 0-order Hankel transformation on the coordinate r for Equation (16), Equation (17) can be obtained as follows.

$$\frac{d^2 \tilde{\bar{e}}_0}{dz^2} - \alpha^2 \tilde{\bar{e}}_0 = 0 \tag{17}$$

Where $\alpha^2 = \frac{a_2}{a_1} + \xi^2$.

According to Equation (15), a similar formula can be obtained as follows.

$$\nabla^2\bar{\sigma} = a_3\bar{e}$$

Where $a_3 = \frac{s}{k_d'} - \rho_f s^2 = a_2 - \rho s^2$.

Carrying out the 0-order Hankel transformation of the above equation, Equation (18) can be obtained as follows.

$$\frac{d^2\tilde{\bar{\sigma}}_0}{dz^2} - \xi^2\tilde{\bar{\sigma}}_0 = a_3\tilde{\bar{e}}_0 \tag{18}$$

Carrying out 0-order Hankel transformation for Equation (11), Equation (19) can be obtained as follows.

$$\frac{d^2\tilde{\bar{u}}_{z0}}{dz^2} - \beta^2\tilde{\bar{u}}_{z0} = \frac{1}{\overline{G}}\frac{d\tilde{\bar{\sigma}}_0}{dz} - \frac{1}{1-2\bar{\mu}}\frac{d\tilde{\bar{e}}_0}{dz} \tag{19}$$

Where $\beta^2 = \frac{\rho s^2}{\overline{G}} + \xi^2$.

Carrying out the first-order Hankel transformation of Equation (10), Equation (20) can be obtained as follows.

$$\frac{d^2}{dz^2}\tilde{\bar{u}}_{r1} - \beta^2\tilde{\bar{u}}_{r1} = -\frac{\xi}{\overline{G}}\tilde{\bar{\sigma}}_0 + \frac{\xi}{1-2\bar{\mu}}\tilde{\bar{e}}_0 \tag{20}$$

We can solve Equations (17)–(20) to get the following equations.

$$\tilde{\bar{e}}_0 = A_1e^{\alpha z} + B_1e^{-\alpha z} \tag{21}$$

$$\tilde{\bar{\sigma}}_0 = \frac{a_1a_3}{a_2}(A_1e^{\alpha z} + B_1e^{-\alpha z}) + \left(A_2e^{\xi z} + B_2e^{-\xi z}\right) \tag{22}$$

$$\tilde{\bar{u}}_{z0} = \frac{a_1\alpha}{a_2}(A_1e^{\alpha z} - B_1e^{-\alpha z}) - \frac{\xi}{\rho s^2}\left(A_2e^{\xi z} - B_2e^{-\xi z}\right) + \left(A_3e^{\beta z} + B_3e^{-\beta z}\right) \tag{23}$$

$$\tilde{\bar{u}}_{r1} = -\frac{a_1\xi}{a_2}(A_1e^{\alpha z} + B_1e^{-\alpha z}) + \frac{\xi}{\rho s^2}\left(A_2e^{\xi z} + B_2e^{-\xi z}\right) + \left(A_4e^{\beta z} + B_4e^{-\beta z}\right) \tag{24}$$

Laplace transform and Hankel transform of the volume strain can be obtained: $\tilde{\bar{e}}_0 = \xi\tilde{\bar{u}}_{r1} + \frac{\partial\tilde{\bar{u}}_{z0}}{\partial z}$. Substituting the analytical solutions of $\tilde{\bar{e}}_0$, $\tilde{\bar{u}}_{r1}$, and $\tilde{\bar{u}}_{z0}$ into the above equation, we can get $A_4 = -\frac{\beta}{\xi}A_3$, $B_4 = \frac{\beta}{\xi}B_3$.

According to constitutive equations, other physical quantities are obtained as follows.

$$\tilde{\bar{\tau}}_{rz1} = -\frac{2a_1\overline{G}\alpha\xi}{a_2}(A_1e^{\alpha z} - B_1e^{-\alpha z}) + \frac{2\overline{G}\xi^2}{\rho s^2}\left(A_2e^{\xi z} - B_2e^{-\xi z}\right) - \frac{\overline{G}(\beta^2+\xi^2)}{\xi}\left(A_3e^{\beta z} + B_3e^{-\beta z}\right) \tag{25}$$

$$\tilde{\bar{\sigma}}'_{z0} = 2\overline{G}\left(\frac{\bar{\mu}}{1-2\bar{\mu}} + \frac{a_1\alpha^2}{a_2}\right)(A_1e^{\alpha z} + B_1e^{-\alpha z}) - \frac{2\overline{G}\xi^2}{\rho s^2}\left(A_2e^{\xi z} + B_2e^{-\xi z}\right) + 2\overline{G}\beta\left(A_3e^{\beta z} - B_3e^{-\beta z}\right) \tag{26}$$

3 BOUNDARY CONDITIONS

For boundary conditions at infinity, physical quantities, such as stresses, strains and displacements are equal to 0. These expressions contain $e^{\alpha z}$, $e^{\beta z}$, and $e^{\xi z}$, which are inconsistent with boundary conditions at infinity. Therefore, we can get $A_1 = A_2 = A_3 = 0$.

Boundary conditions on the upper surface are as follows.

(1) The vertical stress is equal to the vehicle load, then $\sigma_z(r, 0, t) = \begin{cases} -p(t) & r \leq \delta \\ 0 & r > \delta \end{cases}$;

(2) The action of the horizontal load is not considered, then $\tau_{rz}(r, 0, t) = 0$;

(3) The upper surface is in drainage condition, then $\sigma(r, 0, t) = 0$.

Laplace transformation and Hankel transformation are carried out for the above boundary conditions to obtain the following equation.

$$\tilde{\bar{\sigma}}_{z0}(\xi, 0, s) = -\bar{p}(s)\int_0^\delta rJ_0(\xi r)dr = -\bar{p}(s)\frac{\delta J_1(\xi\delta)}{\xi},\ \tilde{\bar{\sigma}}_0(\xi, 0, s) = 0,\ \tilde{\bar{\tau}}_{rz1}(\xi, 0, s) = 0 \tag{27}$$

The vehicle load is as follows.

$$p(t) = \begin{cases} p_{\max}\sin\left|\frac{\pi t}{T}\right| & 0 \leq t \leq T \\ 0 & T \leq t \leq T_a \end{cases} \tag{28}$$

By Laplace transformation of Equation (28), Equation (29) can be obtained as follows.

$$\bar{p}(s) = \frac{p_{\max}\pi}{T\left(\frac{\pi^2}{T^2} + s^2\right)} \tag{29}$$

4 AXIAL SYMMETRY VISCOELASTIC SEMI-INFINITE BODY

According to boundary conditions, Equations (30)–(32) can be obtained as follows.

$$\tilde{\bar{\sigma}}_0(\xi, 0, s) = \frac{a_1 a_3}{a_2}B_1 + B_2 = 0 \tag{30}$$

$$\tilde{\bar{\tau}}_{rz1}(\xi, 0, s) = \frac{2a_1\overline{G}\alpha\xi}{a_2}B_1 - \frac{2\overline{G}\xi^2}{\rho s^2}B_2 - \frac{\overline{G}(\beta^2 + \xi^2)}{\xi}B_3 = 0 \tag{31}$$

$$\tilde{\bar{\sigma}}_{z0}(\xi, 0, s) = 2\overline{G}\left(\frac{\bar{\mu}}{1 - 2\bar{\mu}} + \frac{a_1\alpha^2}{a_2}\right)B_1 - \frac{2\overline{G}\xi^2}{\rho s^2}B_2 - 2\overline{G}\beta B_3 = -\bar{p}(s)\frac{\delta J_1(\xi\delta)}{\xi} \tag{32}$$

We can get the following equations.

$$B_1 = -\frac{1}{2}a_2 s^2\rho(\beta^2 + \xi^2)B \tag{33}$$

$$B_2 = \frac{1}{2}a_1 a_3\rho s^2(\beta^2 + \xi^2)B \tag{34}$$

$$B_3 = -a_1\xi^2(\alpha\rho s^2 + a_3\xi)B \tag{35}$$

Where

$$B = \frac{\left(\bar{\mu} - \frac{1}{2}\right)\bar{p}(s)\dfrac{\delta J_1(\xi\delta)}{\xi}}{\overline{G}\left(\left(a_3\xi^4 - 2a_3\beta\xi^3 + \left(\alpha^2\rho s^2 - 2\alpha\beta\rho s^2 + a_3\beta^2\right)\xi^2 + \alpha^2\beta^2\rho s^2\right)\left(\bar{\mu} - \frac{1}{2}\right)a_1 - \frac{1}{2}s^2 a_2\bar{\mu}\rho\left(\beta^2 + \xi^2\right)\right)}$$

By bringing in the equation, similar equations can be obtained as follows.

$$\widetilde{\bar{e}}_0 = -\frac{1}{2}a_2 s^2\rho\left(\beta^2 + \xi^2\right)Be^{-\alpha z}$$

$$\widetilde{\bar{\sigma}}_0 = \frac{1}{2}a_1 a_3 s^2\rho B\left(\beta^2 + \xi^2\right)\left(-e^{-\alpha z} + e^{-\xi z}\right)$$

$$\widetilde{\bar{u}}_{z0} = \frac{a_1 B}{2}\left[\alpha s^2\rho\left(\beta^2 + \xi^2\right)e^{-\alpha z} + a_3\xi\left(\beta^2 + \xi^2\right)e^{-\xi z} - 2\xi^2\left(\alpha\rho s^2 + a_3\xi\right)e^{-\beta z}\right]$$

$$\widetilde{\bar{u}}_{r1} = \frac{1}{2}a_1\xi B\left[s^2\rho\left(\beta^2 + \xi^2\right)e^{-\alpha z} + a_3\left(\beta^2 + \xi^2\right)e^{-\xi z} - 2\beta\left(\alpha\rho s^2 + a_3\xi\right)e^{-\beta z}\right]$$

$$\widetilde{\bar{\tau}}_{rz1} = \frac{2a_1\overline{G}\alpha\xi}{a_2}B_1 e^{-\alpha z} - \frac{2\overline{G}\xi^2}{\rho s^2}B_2 e^{-\xi z} - \frac{\overline{G}\left(\beta^2 + \xi^2\right)}{\xi}B_3 e^{-\beta z}$$

$$\widetilde{\bar{\sigma}}_{z0} = 2\overline{G}\left(\frac{\bar{\mu}}{1 - 2\bar{\mu}} + \frac{a_1\alpha^2}{a_2}\right)B_1 e^{-\alpha z} - \frac{2\overline{G}\xi^2}{\rho s^2}B_2 e^{-\xi z} - 2\overline{G}\beta B_3 e^{-\beta z}$$

5 CALCULATION OF THE EXCESS PORE WATER PRESSURE

In this paper, the Burgers model is used to simulate the asphalt pavement and material parameters are as follows.

$$\overline{G}(s) = \frac{4.25 \times 10^{11}s + 5.74 \times 10^{13}s^2}{2\left(1 + 1.16 \times 10^4 s + 4.40 \times 10^5 s^2\right)},\ \bar{\mu}(s) = 0.3$$

Other parameters are ρ_s = 2400 kg/m^3, n = 0.08, 0.042[10]; k_d = 13.41 × 10^{-6} m/s, 2.641 × 10^{-6} m/s; ρ_f = 1000 kg/m^3; $p_{\max}$ = 0.7 MPa; δ = 0.151 m; v = 20 m/s; T = 0.0906 s.

The excess pore water pressure is solved and the calculation results are shown in Figure 1. Through comparative analysis, it can be found that since the asphalt pavement is a drainage condition, pore water can be discharged from the surface of the asphalt pavement and the excess pore water pressure is equal to zero. With the increase in depth, the excess pore water pressure first rapidly increases to the maximum value, and then slowly decreases. For example, in Figure 1, the maximum excess pore water pressure of 0.0091 s is located at 0.05 m, while at other times it is located at 0.1 m.

The permeability coefficient is 13.41×10^{-6} m/s and 2.641×10^{-6} m/s, respectively, and the maximum excess pore water pressures are 158 kPa and 339 kPa, respectively. It is obvious that with the decrease in the permeability coefficient, the permeability of the asphalt pavement decreases and the pore water is not easy to discharge, which increases the excess pore water pressure. Therefore, on the premise of ensuring the bearing capacity of the asphalt pavement, the porosity of the asphalt pavement should be increased and the drainage capacity of the asphalt pavement should be improved. And the discharge of pore water and the dissipation of the excess pore water pressure should be accelerated. The pressure of pore water directly acting on asphalt membranes should be reduced.

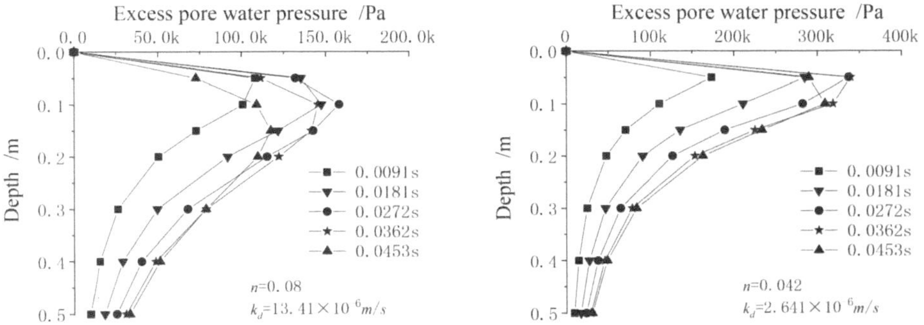

Figure 1. The curve of the excess pore water pressure changes with depth.

The vehicle load is simulated by the semi-sinusoidal load. The load acting time is 0.0906 s and the maximum load is 0.7 MPa at 0.0453 s. It can be found in Figure 2 that the peak value of the excess pore water pressure is not at 0.0453 s. With the increase in depth, the arrival of the peak value of the excess pore water pressure is delayed. The peak value at 0.05 m depth is 0.0181 s, 0.0272 s at 0.10 m, 0.15 depth, and 0.0362 s at 0.20 m and 0.30 m depth. With the increase in the vehicle load, the asphalt pavement will deform and effective stresses will increase. The pore water flows out of the pores of the asphalt pavement, leading to the dissipation of the excess pore water pressure. The load originally borne by the pore water can be transferred to the asphalt pavement. When the excess pore water pressure generated by the increase in vehicle load is less than the dissipation value of the excess pore water pressure, the excess pore water pressure is generally reduced. Therefore, the peak value of the excess pore water pressure will appear earlier than the peak value of the vehicle load.

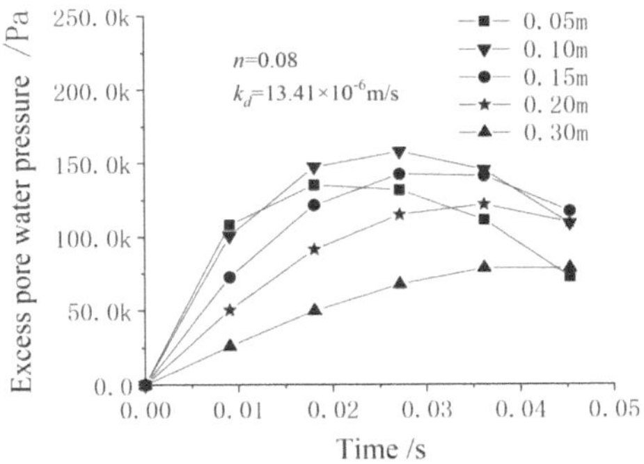

Figure 2. The curve of the excess pore water pressure changes with time.

6 CONCLUSION

In this paper, the damage mechanism of asphalt pavement water damage is analyzed. The asphalt pavement is regarded as a saturated layered viscoelastic system. The dynamic equilibrium equations and the seepage continuity equation are solved by using the integral transformation method. General solutions of physical quantities, such as excess pore water pressure, displacements, and strains in the integral space, are obtained. By introducing the boundary condition of the semi-infinite body, the dynamic analytical solutions of physical quantities in the integral space are obtained and the excess pore water pressure is calculated.

Through calculation and analysis, the following conclusions are drawn. With the increase in depth, excess pore water pressure first rapidly increases to the maximum value and then slowly decreases. With the decrease in permeability coefficient, the excess pore water pressure increases. The peak value of the excess pore water pressure occurs earlier than the peak value of the load.

ACKNOWLEDGEMENTS

This paper is very grateful for the Heilongjiang Bayi Agricultural University Graduated Talent Introduction Scientific Research Initiation Plan (XYB2014-06) and the Heilongjiang Bayi Agricultural University Three Horizontal and Three Vertical Support Plan (ZRCPY202225, ZRCPY202119).

REFERENCES

Cao Jian. *Research on the Influence of Voidage and Permeability On-road Performance of Asphalt Pavement in Seasonal Frost Areas* [D]. Jilin University, 2007.

Cong Bori, Zhang Xiaochun, Li Cunyong. Express Pore Fluid Stress Analysis of Asphalt Pavement under Dynamic Loads[J]. *Transportation Science & Technology*, 2007, (224):51–53.

Dong Ze-jiao, Tan Yi-qiu, Cao Li-ping, et al. Research on Pore Pressure Within Asphalt Pavement Under the Coupled Moisture-loading Action [J]. *Journal of Harbin Institute of Technology*, 2007, 39(10): 1614–1617.

Fu Bo-feng, Zhou Zhi-gang, Chen Xiao-hong, et al. The Numerical Simulation Analysis of Asphalt Pavement Moisture Damage Fatigue Failure Process [J]. *Journal of Zhengzhou University (Engineering Science)*, 2006, 27(1): 51–58.

Li Zhigang, Deng Xiaoyong. Axial Symmetric Elastic Solution of Pore Water Pressure in Asphalt Pavement Under Mobile Load[J]. *Journal of Southeast university (Natural Science Edition)*, 2008, 38(5):804–810.

Luo Su-ping, Dan Han-cheng, Li Liang, et al. Coupled Hydro-Mechanical Analysis of Saturated Asphalt Pavement Under Moving Traffic Loads [J]. *Journal of South China University of Technology (Natural Science Edition)*, 2012, 40(2):104–111.

Peng Yong-heng, Ren Rui-bo, Song Feng-li, et al. An Axisymmetric Solution of Multi-layered Elastic Body Super-pressure in Small Opening Water [J]. *Engineering Mechanics*, 2004, 21(4):204–208.

Ren Ruibo, Qi Wenyang, Li Meiling. Analysis of Dynamic Response of Saturated Asphalt Pavement Under Moving Vehicle Loads by 3D Finite Element Method [J]. *Journal of Highway and Transportation Research and Development*, 2011, 28(9):11–16.

Zhang Bin, Zhao Ying-hua. A Solution of the Saturated Axial Symmetrical Viscoelastic Body [J]. *Chinese Journal of Computational Mechanics*, 2012, 29(6):978–982.

Zhong Yang, Geng Litao, Zhou Fulin, et al. Computing the Express Pore Fluid Stress of FlexiblePavement by Stiffness Matrix Method [J]. *Journal of Shenyang Jianzhu University*, 2006, 22(1):25–29.

Study on mechanical and thermal properties of green and environmentally friendly three-doped concrete self-insulating blocks

Zhenhui Xu & Zhengchao Jin*
College of Engineering, Yanbian University, China

ABSTRACT: In this paper, according to the three-doping method of replacing cement with 6% straw powder, 10% silica fume, and 20% fly ash, the green environmentally friendly three-doped concrete self-insulating block is trialed and the structure of the block is designed according to the requirements of the current national standard "ordinary concrete small block". Then, according to the mixing ratio, the concrete block with strength grade MU7.5 is made and the compressive strength of the block is tested and recorded through the mechanical property test. The failure forms and mechanisms are analyzed, all of which reach the standard of strength grade MU7.5. Finally, the thermal performance of the block is simulated by using the Abaqus finite element analysis software, the stress distribution and crack development law of each part of the block are recorded, and the thermal properties, such as temperature gradient distribution and heat flux density of the block, are analyzed, and the thermal performance of the block exceeds the national standard for building energy saving, thus ensuring that the self-insulating block meets both the mechanical thermal performance of the cold regions in the north and the green environmental protection requirements.

1 INTRODUCTION

Silica fume is a filter powder generated from the reduction of high purity quartz - ferrosilicon metals or silicon metals. Silica fume consists mainly of very fine, smooth, spherical silicon oxide particles with an extremely high surface area. The typical chemical and physical composition of silica fume is shown in Figure 1, in comparison with OPC cement, fly ash, and slag.

Fly ash is a heterogeneous by-product material produced during the combustion of coal for power stations. It is a fine gray powder with spherical glassy particles that rise with the flue gas. Since fly ash contains volcanic ash material components, these components form cementitious materials together with lime. Therefore, fly ash is used in concrete, mines, landfills, and dams.

According to statistics, the total output of crop straw in China has reached 865 million tons in 2021, while burning and landfilling straw can cause air pollution, environmental damage, and waste of land resources. In 2019, China's total coal output was 3.75 billion tons, accounting for 47.3% of the global total coal output and 51.7% of the global consumption. At present, China still takes coal as the main energy source and the energy structure will not change for a long time in the future. A large number of dust and harmful

*Corresponding Author: jinzhengchao@ybu.edu.cn

DOI: 10.1201/9781003450818-62

	Unit	Portland Cement	Battle	Flyash	Silica Fume
Tall	%	60 - 67	32 - 48	1 - 5	0.1 - 0.6
SiO_2	%	17 - 25	28 - 40	40 - 55	86 - 95
To the_2Or_3	%	2 - 8	10 - 22	20 - 30	0.2 - 2.2
Fe_2Or_3	%	0 - 6	3 - 5	5 - 10	0.2 - 2.2
MgO	%	0.1 - 4.0	2 - 16	1 - 4	0.3 - 3.5
SO_3	%	1 - 4	2 - 4	0.4 - 2.0	0.4 - 1.5
On_2Or	%	0.1 - 1.0	0.2 - 1.2	0.1 - 2.0	0.2 - 1.8
K_2Or	%	-	-	1 - 6	

Figure 1. Chemical composition.

gases will be emitted during coal combustion, which can cause serious harm to human health and the ecological environment (Wang 2018).

Straw concrete is a mixture of crop straw, cement, aggregate, sand, and water in a certain mix proportion. It makes use of the characteristics of lightweight, high strength, and good crack resistance of crop straw (Chang 2019), so the produced concrete has lightweight, high strength, strong crack resistance, and tensile resistance, and is easy to obtain materials, cheap in cost and simple in production. Su Youwen, Li Chaofei, and others calculated the heat transfer coefficient and thermal resistance of hollow blocks by using the building heat transfer theory through experimental research on the thermal performance of ordinary concrete and straw fiber concrete hollow blocks with different shapes. The results show that the straw fiber concrete hollow block has a better thermal insulation effect than the ordinary concrete hollow block. With the increase in straw fiber content, the thermal insulation effect of straw fiber concrete hollow blocks is getting better. The optimal straw fiber content is 6%. As a new green building wall material, straw fiber concrete has certain application prospects (Su 2016).

Green building materials, known as green building materials, ecological building materials, environmental protection building materials, and healthy building materials, refer to healthy, environmentally friendly, and safe building materials. Green building materials do not refer to individual building materials products but to the evaluation of the "health, environmental protection, safety" character of building materials. It is concerned with the impact of building materials on human health and environmental protection, as well as safety and fire performance. It has the properties of demagnetization, sound reduction, dimming, temperature regulation, heat insulation, fire prevention, antistatic, and special new functional building materials that regulate human body functions.

Green concrete as a new environmentally friendly material, not only has good mechanical properties, volume stability, and durability but also can save resources, protect the environment, and improve the comprehensive utilization of industrial wastes, and other solid wastes, which is the direction of concrete development. The straw of agricultural waste (wheat straw, corn straw, finely milled rice husk powder, and dry straw), and industrial waste silica fume, fly ash as a product of secondary use, are indispensable parts of green concrete. In this paper, silica fume, rice husk powder, and fly ash are used as concrete admixtures to prepare green and environment-friendly concrete load-bearing building blocks. The best matching ratio in the paper (Tuo 2021) is further investigated, the strength of the concrete blocks produced is further discussed, and their mechanical properties and thermal performance are further explored and analyzed, providing a broad development prospect for the development and utilization of self-insulating blocks with triple admixture concrete.

2 STUDY ON THE DESIGN AND MATCHING RATIO OF THREE-DOPED CONCRETE BLOCKS

2.1 *Test materials and design*

This experiment is a study of three-doped concrete insulation blocks based on the paper (Tuo 2021), so the matching scheme of 1 m^3 ordinary concrete with a ratio of 215.6: 201.6: 684.26: 1272.73 (cement: water: sand: crushed stone) is still selected. The origin of straw powder (rice husk powder), silica fume and fly ash, and other three-doped concrete materials are the same as in this paper.

Table 1. Three-doped concrete test mix ratio table.

Cement/kg	Water/kg	Sand/kg	Natural crushed stone/kg	Silica fume/kg	Straw powder/kg	Fly ash/kg
6.69	6.65	22.58	42.00	1.02	0.43	2.03

(1) Block design

By analyzing the advantages and disadvantages of masonry construction forms in the summary literature, this paper makes design innovations for the three-doped concrete block structure and grants a practical emerging patent (Su 2016) (Jin 2022).

(2) Design strength

This experimental design is mainly applicable to load-bearing blocks rated MU7.5. According to the relationship between block strength and concrete cube strength (Tuo 2021), according to the empirical formula (1) of Yongyu Li [8], the following formula can be obtained.

$$RK/RL = 0.9577 - 1.129K \tag{1}$$

Where RK represents the 28-day strength of concrete hollow blocks (MPa); RL represents the concrete cube test block's 28-day strength (MPa); K represents the hollow ratio of hollow blocks of 0.22.

The design strength of concrete blocks is as follows. $RK = (0.9577 - 1.129 \times 0.22) \times 13.8 = 9.79MPa$

3 STUDY ON THE MECHANICAL PROPERTIES OF THREE-DOPED CONCRETE BLOCKS

3.1 *Block mold making*

The green and environmentally friendly three-doped concrete self-insulating block is composed of the concrete outer wall, rib wall, three-layer B1 grade EPS board, and foamed mortar concrete. The main material of the mold is a wood stencil and B1 grade EPS board, the template base is 420 mm × 240 mm, and the main size of the mold is 390 mm × 190 mm × 190 mm. To avoid the formwork from the side expansion when pouring concrete, the base and the side of the formwork are fixed with wood or air nail guns around the mold. The B1 grade EPS board is fixed in the designated position with glue inside the mold, the mortar filling cavity is fixed with a formwork, the wooden formwork is removed after the initial setting of the concrete, and the foamed concrete mortar is poured.

Figure 2. Concrete block formwork making.

Figure 3. Concrete block mixing and pouring.

Figure 4. Concrete forming blocks.

3.2 *Block pouring and mold removal*

After the completion of concrete mixing, the measured slump is 38 mm and the slump that meets the construction requirements is T = 30-50 mm, and then it is poured according to the homemade mold. After the pouring is completed, the blocks are put into the standard curing room (the temperature is controlled within 18-22 °C and the relative humidity is controlled above 80%). Watering is required for curing and the mold is removed after 3D curing.

3.3 *Block compressive strength experiment*

According to the test method for concrete blocks and bricks (GB/T4111-2013) (Song 2021), it is necessary to cut the special-shaped concrete blocks into flat hexahedral blocks, as shown in Figures 5 and 6. The uneven surface of the cured concrete block is closed and leveled, the thickness of the plastering mortar cannot exceed 3 mm, and the plastered block needs to be cured in the curing room for 3d before the compressive strength test of the block.

Figure 5. Compression block diagram.

Figure 6. Masonry damage diagram.

The block is placed on the pressure plate of the testing machine, the loading speed is set to 3 KN/s, and the uniform load is applied so that it can reach the limit load at about 1.5 min. at this time, the compressive strength F of the block at the time of failure is recorded.

Table 2. Compressive strength of concrete blocks.

Numbering	Failure load P (kN)	Compressive strength F (MPa)	Average compressive strength F (MPa)
1	576.21	9.19	9.04
2	568.69	9.07	
3	573.08	9.14	
4	565.55	9.02	
5	550.51	8.78	

From Table 2, it can be concluded that the average value of the compressive strength F of the block is 9.04 MPa and the minimum value of compressive strength Fmin is 8.74 MPa, which meets the design standard of compressive strength of non-load-bearing block with strength grade MU7.5. According to the comparison and calculation of the measured value of the block and the theoretical value obtained by the empirical formula, it can be concluded that the correction coefficient $\alpha = 0.92$, and the relationship between the compressive strength of double-doped straw and gangue self-insulating concrete block and the concrete hollow block is as follows.

$$RK/RL = (0.9577 - 1.129K)\cdot\alpha \tag{2}$$

4 FINITE ELEMENT ANALYSIS OF THREE-DOPED CONCRETE SELF-INSULATION BLOCKS

4.1 *Finite element analysis of mechanical properties of blocks*

4.1.1 *Block model diagram*

In this paper, the isotropic method is adopted for simplification. The load on the block is mainly the dead weight of the masonry structure above the block (Wang 2019). We simulate the stress-strain relationship of the block and analyze the stress state of the block when building the wall material, as shown in Figures 7 and 8.

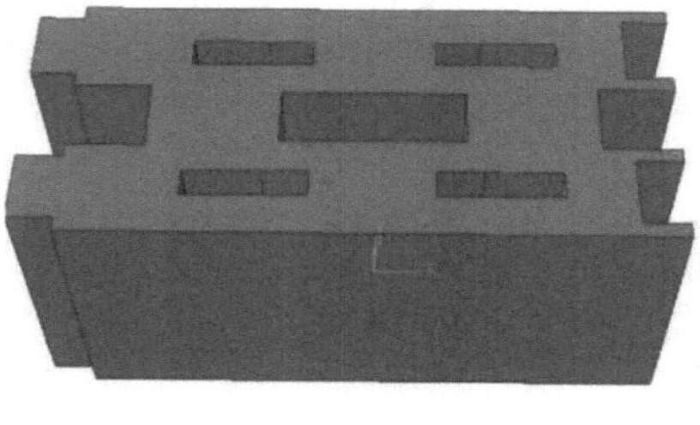

Figure 7. Block model diagram.

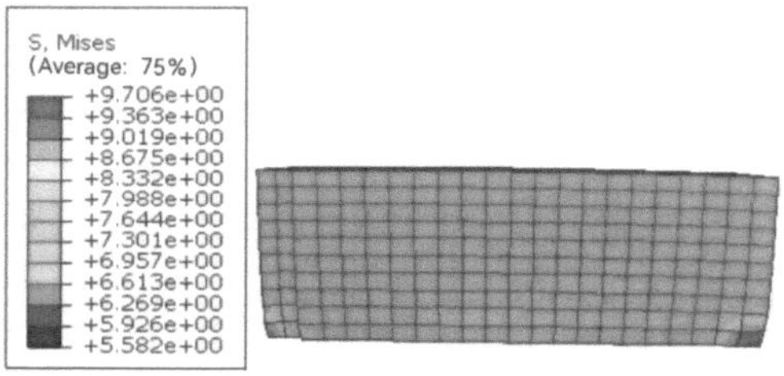

Figure 8. The positive view of stress distribution.

4.1.2 *Isometric force cloud diagram*

The maximum stress value in the figure is 9.706 MPa, which is mainly distributed at the corner of the block. The stress at the end and middle of the block is greater than that at other parts and the minimum stress value is 5.582 MPa. The stress distribution of the block is more uniform from the four corners to the middle. The stress around the cavity and in the corners of the blocks is more variable, which has a significant effect on the strength of the blocks.

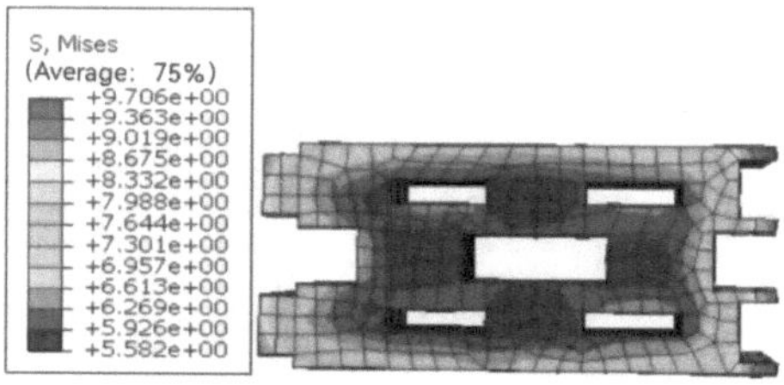

Figure 9. A positive view of stress distribution.

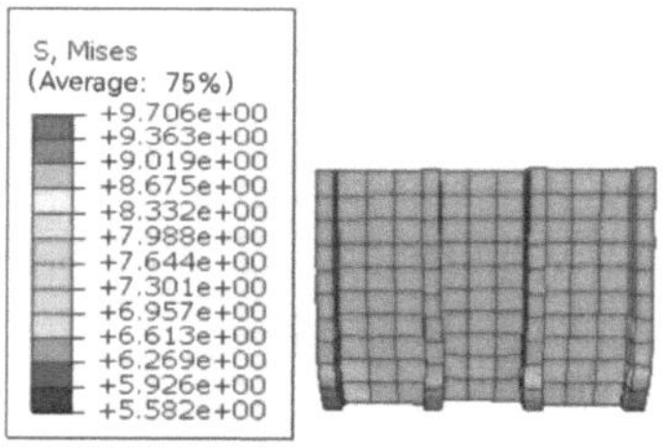

Figure 10. Bottom view of stress distribution.

4.2 *Finite element analysis of thermal properties of blocks*

According to the measurement results of the thermal performance of green and environment-friendly three mixed concrete measured in Literature (Song 2021), the thermal performance of the block is analyzed by using the temperature field under thermal load. The heat transfer effect applicable to different internal and external temperatures is obtained.

The grid division diagram, temperature gradient diagram, and heat flow density diagram are shown in Figures 11 to 14 below.

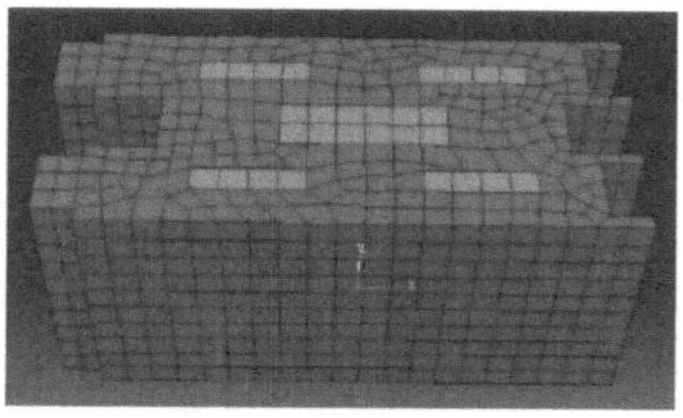

Figure 11. Meshing diagram.

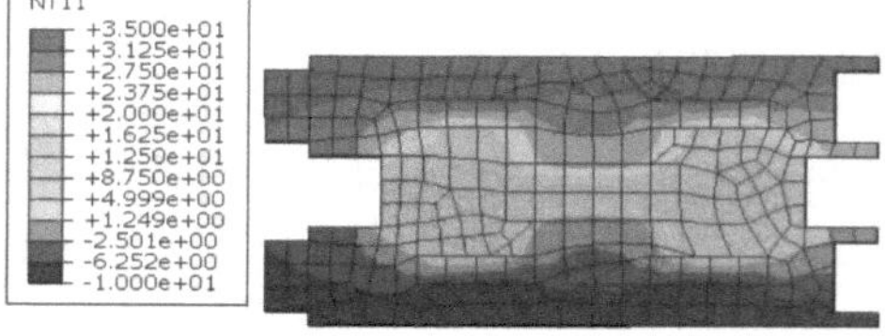

Figure 12. Temperature distribution plot.

Since the materials of each part of the block are different, the block should be divided according to the material properties, and the grid size should be assigned. This is the model after grid division. The analysis type is two-dimensional steady-state simulation analysis. The constraints are applied and temperature loads are applied.

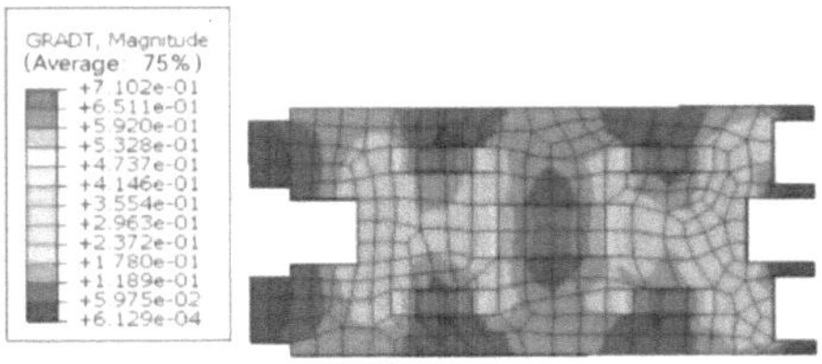

Figure 13. Temperature gradient plot.

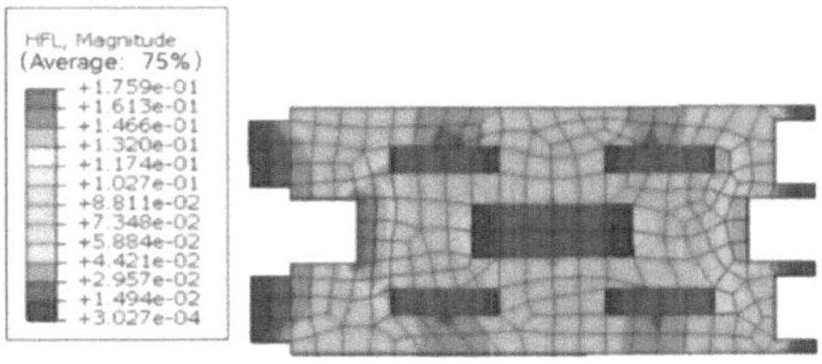

Figure 14. Heat flux density diagram.

Where the temperature changes greatly, the material change and block type change are mainly caused by the material change and the change of thermal conductivity at the corner. The heat is mainly concentrated in the thermal insulation cavity because the thermal resistance is greater here than that of other materials. The heat flow density distribution in the figure is uniform and stable and the heat flow transmission at both sides and the middle of the block are relatively concentrated. When studying the integral masonry structure, the integral heat flow of the connecting blocks can be transferred between the blocks, thus improving the thermal insulation performance and making the thermal performance of the blocks better. In the process of steady heat analysis, it is concluded that the heat transfer mode of the block belongs to steady heat transfer.

5 CONCLUSION

In response to the national development strategy of green, environmental protection, lightweight, energy conservation, and sustainability, this paper prepares three-doped concrete self-insulation blocks (rice husk powder, silica fume, fly ash) based on the matching ratio designed by the paper (Tuo 2021), designs the block structure, and uses this ratio of concrete to pour concrete blocks. The thermal insulation chamber is filled with a B1-grade EPS board and thermal insulation mortar. By summarizing and analyzing the compressive properties, mechanical properties, thermal properties, and other test results of the block, the following conclusions are drawn:

(1) According to the designed self-insulating concrete hollow block and the selected paper matching ratio, the average compressive strength F of the concrete block made is 9.04 MPa, which meets the design standard of non-load-bearing block compressive strength with strength grade MU7.5, and the modified formula $RK/RL = (0.9577 - 1.129K)\cdot\alpha \quad (\alpha = 0.92)$ is proposed.

(2) In this paper, through Abaqus finite element analysis, the mechanical and thermal properties of three mixed concrete self-insulating blocks are analyzed, respectively. From the results of finite element analysis, it can be concluded that the maximum compressive strength is 9.706 MPa and the compressive strength of the block with this mix ratio can meet the design requirements. Secondly, the thermal steady state analysis is carried out through Abaqus finite element software and the block heat flow density cloud diagram is obtained. It can be seen that the block heat flow density distribution is relatively uniform and stable and the heat flow transfer at the end and middle of the block is relatively concentrated.

Therefore, the green and environment-friendly three-mixed concrete block has strength up to the standard, has a good thermal insulation effect, and can be applied to building reconstruction and new construction projects in cold regions.

ACKNOWLEDGMENTS

This paper was supported by the Science and Technology Research Project (JJKH20220539KJ) of the Jilin Provincial Department of Education.

REFERENCES

GB/T 4111-2013. *Test Method for Concrete Blocks and Bricks* [S]. Beijing: China Building Industry Press, 2013.

Jianxin Wang, Li Jing, Shibao Zhao. Research Progress and the Prospect of Resource Utilization of Fly Ash in China [J]. *Bulletin of the Chinese Ceramics*, 2018, 37(12):3834–3841.

Liping Wang, Li Chao. Research Progress on the Development and Utilization of Fly Ash Resource Technology [J]. *Mineral Protection and Utilization*, 2018, 39(4):38–45.

Qiuying Chang. Application of Straw Resources in Civil Engineering [J]. *Housing and Real Estate*, 2019 (28):104.

Wang Zheng. *Study on New Cinder Concrete Self-insulating Block and its Masonry Strength* [D]. Yanbian University, 2019.

Youwen Su, Chaofei Li, Tinghui Yang, Yang Kai, Zhao Pu. Experimental Study on Thermal properties of Hollow Blocks of Straw Fiber Concrete [J]. *Concrete*, 2016(10):131–134,138.

Yuhao Song, Zhengchao Jin. Experimental Study on Freeze-Thaw Properties and Thermal Conductivity of Green and Environment-Friendly Three-Admixture Concrete [J]. *IOP Conference Series: Earth and Environmental Science*, Volume 787 (2021012128).

Yuli Yong, Xiping Jiang. Research on New Composite Self-insulating Blocks [J]. *Concrete*, 2012(01):109–112.

Zhengchao Jin, Zhongyu Chen, Li Chao, Ouyang Song, Zijie Tuo, Yuhao Song, Yiming Wu. *A Kind of Green and Environmentally Friendly Self-insulating Concrete Block* [P]. Jilin Province: CN216810590U, 2022-06-24.

Zijie Tuo, Zhengchao Jin. Mix Design and Compressive Strength Test of Green and Environment-friendly Concrete with Three Admixtures [J]. *IOP Conference Series: Earth and Environmental Science*, Volume 787 (2021012069).

Water Conservancy and Civil Construction – Oke & Ahmad (Eds)

Study on bending behavior of shape memory alloy (SMA) smart concrete beams

Li Xu & Xian Cui*
Civil Engineering, Yanbian University, Yanji, China

ABSTRACT: To study the influence of shape memory alloy (SMA) on the flexural performance of concrete beams, Ni-Ti SMA bars and wires with different materials and diameters were used to replace steel bars in the tensile zone of concrete beams by using the characteristics of shape memory effect (SME) and superelasticity (SE) of SMA. Firstly, 12 concrete beams were designed and tested by static loading. The cracks of SMA concrete beams were observed by heating, and then the self-repaired SMA concrete beams were loaded until they were destroyed, to study the flexural performance of SMA intelligent concrete beams. The test results show that, compared with ordinary reinforced concrete beams, the yield load and ultimate load of super-elastic SMA concrete beams with 4 mm main reinforcement material are increased by 35.2% and 53.8% respectively, while when the diameter is increased to 8 mm, the yield load and ultimate load are only increased by 4.6% and 18.4% respectively. Small-diameter super-elastic SMA has a good prospect for market application. When the main reinforcement material is 8 mm, the yield load and ultimate load of the specimens laid with one-way memory SMA wire at the bottom are increased by 38.2% and 19.7% respectively. The one-way memory SMA wire can repair tiny cracks and effectively improve the flexural performance of cracked reinforced concrete beams.

1 INTRODUCTION

Shape Memory Alloy (SMA) is a new functional material developed in recent decades. It will deform at a low temperature. When it is heated, its original shape and volume will be restored as the temperature reaches a certain value. Because of its unique shape memory effect and super-elasticity, academic and engineering circles at home and abroad have done a lot of experimental research (Cicekli *et al.* 2007; Cervenka & Papanikolaou 2008; Cui *et al.* 2010; El-Tawil & Ortega-Rosales 2004; Effendy *et al.* 2006; Kuang & Ou 2008; Li *et al.* 2014) on the deformation control of concrete members by using its driving force. Di Shengkui (Di *et al.* 2010) *et al.* studied the relationship between the resistance change rate of SMA wire and the crack width of the concrete beam during loading and found that increasing the cross-sectional area of the alloy wire appropriately can enhance its driving recovery effect. Shi Yan (Shi *et al.* 2010) etc. studied the self-healing characteristics of SMA intelligent concrete beams and their influencing factors, and found that increasing the reinforcement ratio of SMA wire can improve the driving effect of alloy on concrete beams. Sun Li (Sun *et al.* 2015) *et al.* studied the influence of the pre-strain value of SMA wire on the self-repairing performance of concrete beams and found that the bearing capacity of beams increased with the increase of pre-strain.

At present, the research on shape memory alloys in China mainly focuses on auxiliary embedding, while the research on concrete members that completely replace steel bars

*Corresponding Author: cxian@ybu.edu.cn

DOI: 10.1201/9781003450818-63

becomes less. In this experiment, three kinds of main reinforcements made of different materials were set in the tensile zone of concrete, and SMA wires with the same number and diameter were placed respectively, and their self-repairing performance was tested and studied. The bearing capacity and crack development of concrete beams are tested by single-point loading in the middle of the span, and the mechanical properties of concrete beams under different conditions are discussed, which can provide some reference for the engineering application of SMA.

2 EXPERIMENTAL DESIGN

2.1 *Specimen design*

Four types of concrete beams are designed and manufactured: contrast beam, ordinary reinforced concrete beam, the concrete beam of SMA bar with one-way memory effect in the tension zone, and concrete beam with super-elastic SMA bar in the tension zone. The geometric dimension of the specimen is 400 mm × 100 mm × 100 mm (length × width × height), and the thickness of the protective layer is 10 mm. The diameter of the main reinforcement is 4 mm, 6 mm, and 8 mm, respectively. The upper part is equipped with two $\varphi 6$ steel bars as stud bars, in which the stirrup is $\varphi 6@60$, and six one-way memory SMA wires with a diameter of 1 mm are buried 20 mm away from the bottom, as shown in Figure 1. The SMA bar was designed to completely replace the steel bar in the tensile zone of the concrete beam 10 mm away from the bottom. The specific design parameters of the concrete beam specimen are shown in Table 1.

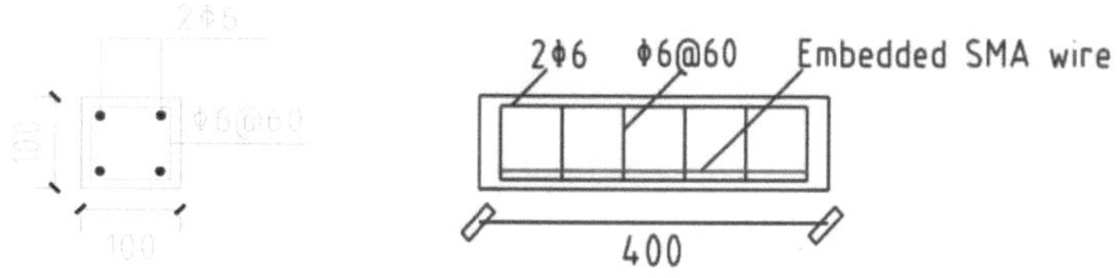

Figure 1. Geometric dimensions and structure of the specimen.

Table 1. Design parameters of concrete beam specimen.

Test piece number	Reinforcement in the tension zone		Single-pass SMA wire configuration
	Main reinforcement material	Diameter (mm)	
RC-1	Reinforcing steel bar	4	0
RC-2	Reinforcing steel bar	6	0
RC-3	Reinforcing steel bar	8	0
RC (S)-1	Reinforcing steel bar	4	$6\varphi 1$
RC (S)-2	Reinforcing steel bar	6	$6\varphi 1$
RC (S)-3	Reinforcing steel bar	8	$6\varphi 1$
CT (S)-1	Super-elastic SMA	4	$6\varphi 1$
CT (S)-2	Super-elastic SMA	6	$6\varphi 1$
CT (S)-3	Super-elastic SMA	8	$6\varphi 1$
DC (S)-1	One-way SMA	4	$6\varphi 1$
DC (S)-2	One-way SMA	6	$6\varphi 1$
DC (S)-3	One-way SMA	8	$6\varphi 1$

2.2 *Test materials*

In this experiment, Ni-Ti shape memory alloy and Q235 steel bar produced by Dongguan Yongshengda Special Steel Co., Ltd. were used. The shape memory alloy was processed into the wire with a diameter of 1 mm and rod with a diameter of 4 mm, 6 mm, and 8 mm respectively. The yield strength of SMA was 195-690 MPa; the ultimate strength was 850 MPa; the elongation was 25%-50%; the pseudo-elasticity of phase transformation was 8%. Among them, the martensitic transformation starting temperature M_s and finishing temperature M_f of SMA with shape memory effect are 14 °C, 25 °C, 32 °C, and 49 °C respectively. The test mixture ratio is water: cement: sand: stone =0.175: 0.343: 0.621: 1.261; the cement strength grade is 32.5 MPa; the concrete strength grade of the specimen is C20. The 28-day measured compressive strength of the concrete standard cube is 27.9 MPa.

2.3 *Test device and loading system*

The test device is the PWS-500 hydraulic servo fatigue tester, in which the actuator has its force sensor and displacement sensor, and the DH2817 dynamic and static strain test system is used to collect the strain data of concrete beams. Firstly, the concrete specimen cured for 28 days is subjected to a three-point bending test to simulate the damage, and strain gauges are set on both sides of the main reinforcement in the tension zone and the top surface of the concrete compression zone in the middle of the beam span. The test adopts the graded static loading mode, and the loading rate is 0.1 kN/s. When the test force is loaded to 1 KN, we shall hold the load for 1 min, then continue to load to 10 KN, and unload.

The self-repairing process test device is the LAREYE TSD-20 industrial heater, which can control a constant temperature of 0-90 °C. In the test, a TSD-20 industrial heater is placed on the left and right sides of the specimen after initial loading, and a thermometer is placed on the concrete specimen to measure the heating temperature. We shall observe and record the crack changes of the specimen, then measure and record the strain of the main reinforcement, next crack width, and alloy wire temperature every 30 s, and stop heating after heating for 10 min.

The Re-load concrete beam after heating and self-repairing, with the loading rate of 0.1 KN/S, until the specimen is damaged or the maximum load displacement reaches 20 mm.

3 LB TEST RESULTS AND ANALYSIS

3.1 *Test phenomenon*

In the process of applying load, the mid-span deflection of the specimen increases gradually with the increase of load, and tiny cracks appear at the bottom of the specimen. After unloading, the cracks were repaired to different degrees. In the process of reloading, the tiny cracks appeared before continuing to develop. After the concrete in the tension zone quit working, the tensile force of the section was borne by the main reinforcement in the tension zone, and then the cracks widened and quickly spread along the beam height to the top of the beam. When the specimen reaches the ultimate load or the maximum displacement reaches 20 mm, it will be regarded as a beam failure and stop loading.

Table 2 shows the yield load, ultimate load, yield-strength ratio, maximum deflection, and failure mode of each specimen. As can be seen from the table, when the diameter of the main reinforcement material is 4 mm, the specimen CT-1 with super-elastic SMA bar as the main reinforcement shows obvious advantages in both yield strength and ultimate load, which are increased by 35.2% and 53.8% respectively compared with ordinary reinforced concrete beams, and its yield-strength ratio is also in a reasonable range. The failure form is shown in Figure 2, which is an ideal bending failure. Analysis shows that compared with other materials, super-elastic SMA has a large elastic limit and a small elastic modulus. After the

Table 2. Yield load, ultimate load, yield-strength ratio, maximum deflection, and the failure form of TABLE specimen.

Test piece number	Yield load P_y/KN	Ultimate load P_u/KN	Yield ratio	Maximum deflection f_u/mm	Destruction forms
RC-1	14.5	18.4	0.788	20.0	Bending failure
RC-2	23.8	33.9	0.702	19.0	Oblique compression failure
RC-3	29.6	42.2	0.701	17.9	Bending failure
RC (S)-1	12.0	14.7	0.816	20.0	Bending failure
RC (S)-2	19.2	30.3	0.634	20.0	Bending failure
RC (S)-3	40.9	50.5	0.810	20.0	Oblique compression failure
CT (S)-1	19.6	28.3	0.693	19.9	Bending failure
CT (S)-2	17.7	28.8	0.615	19.8	Bending failure
CT (S)-3	20.5	33.5	0.612	20.0	Bending failure
DC (S)-1	13.9	18.3	0.760	19.7	Bending failure
DC (S)-2	19.4	25.2	0.770	19.2	Oblique compression failure
DC (S)-3	16.9	41.0	0.412	17.7	Oblique compression failure

Figure 2. CT-1 damage.

Figure 3. RC-2 damage.

Figure 4. RC (S)-3 damage.

concrete cracks in the tensile zone, the super-elastic SMA begins to bear the tensile force at the bottom of the beam. After unloading, the driving force of the rebound of the super-elastic SMA makes the cracks close, further improving the ultimate bearing capacity of the concrete beam in the process of reloading. When the diameter of the main reinforcement is 6 mm, the yield load and ultimate load of the RC-2 specimen are larger than those of the other three kinds of beams, but it has oblique compression failure that needs to be avoided in practical engineering, as shown in Figure 3. The analysis may be that the artificial binding stirrup of this beam in the manufacturing process is unqualified, which leads to the unsatisfactory failure form. When the diameter of the main reinforcement is 8 mm, the RC(S)-3 with one-way memory SMA wire laid at the bottom has both a larger yield load and ultimate load, which are 38.2% and 19.7% higher than those of RC-3. The analysis shows that when the bar with the larger size is used as the main reinforcement, the superiority of the larger rigidity of the steel bar itself begins to manifest. At this time, however, the specimen also suffered from baroclinic failure, as shown in Figure 4. Through calculation and analysis, because the shear span ratio is small, it is influenced by the bearing reaction force and the unidirectional direct pressure caused by the load. When the oblique crack appears in the belly of the beam, the stirrup does not yield. With the increase of the load, the belly of the beam is divided into inclined compression columns, and finally, the oblique compression failure occurs.

3.2 *Load-deflection curve*

Figure 5 shows the load-deflection curves of specimens with different main reinforcement materials when the diameter is 4 mm. It shows that when the diameter of the main reinforcement is 4 mm, the ultimate bearing capacity of the specimen with the super-elastic SMA bar as the main reinforcement is the largest, and the bearing capacity will be further improved after it cracks and yields. Because the strain of the super-elastic SMA is greater than that of concrete under the action of a large load, the reverse driving force increases the ultimate load of the specimen. Figure 6 shows the load-deflection curves of specimens with different main reinforcement materials when the diameter is 6 mm. It shows that the common reinforced RC-2 has an obvious yield point and yield platform, and the deflection rises rapidly after reaching the ultimate bearing capacity. Compared with the other three beams, it has a larger safety reserve. Figure 7 shows the load-deflection curves of specimens with different main reinforcement materials when the diameter is 8 mm. As can be seen from Figure 7, the two specimens with steel bars as the main reinforcement have greater advantages in bearing capacity than the other two beams, indicating that in the large-section members, steel bars still have their outstanding advantages because they have the higher rigidity and can work well with concrete. After reaching the yield load, the bearing capacity of the specimens with super-elastic SMA and one-way memory effect SMA as the main reinforcement is still improved to a certain extent. The analysis suggests that SMA has a large elastic limit, and when it is used as the main reinforcement, it can increase the ductility of the specimens to a certain extent.

The load-deflection curve when the main reinforcement is the super-elastic SMA bar is shown in Figure 8. It can be seen that the mechanical properties, such as ductility and stiffness of each beam, do not improve with the increase of the diameter of the main reinforcement. When the diameter of the main reinforcement increases from 4 mm to 8 mm, the yield load and ultimate load of the beam only increase by 4.6% and 18.4% respectively. The

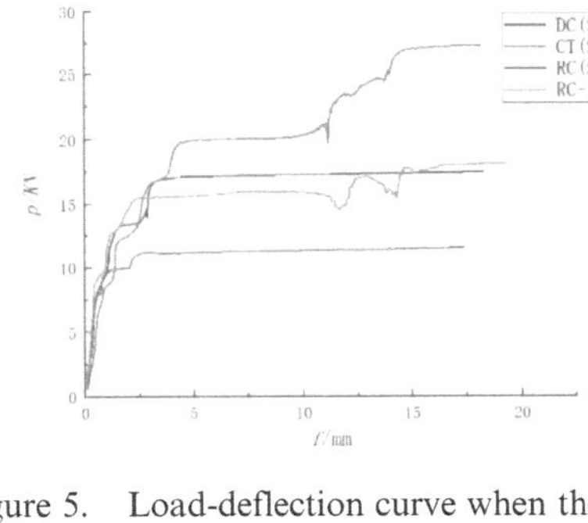

Figure 5. Load-deflection curve when the diameter of the main reinforcement is 4 mm.

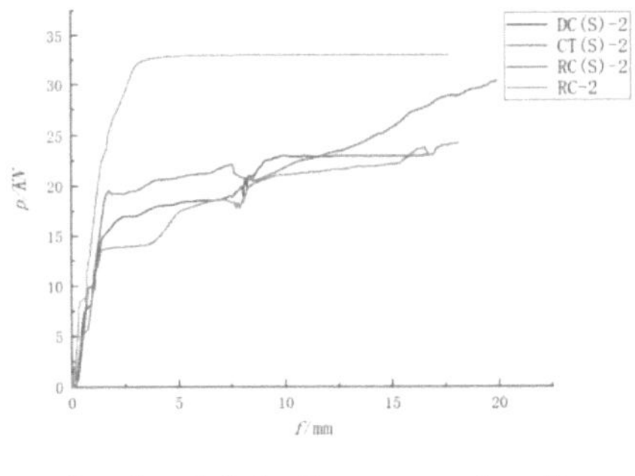

Figure 6. Load-deflection curve when the diameter of the main reinforcement is 6 mm.

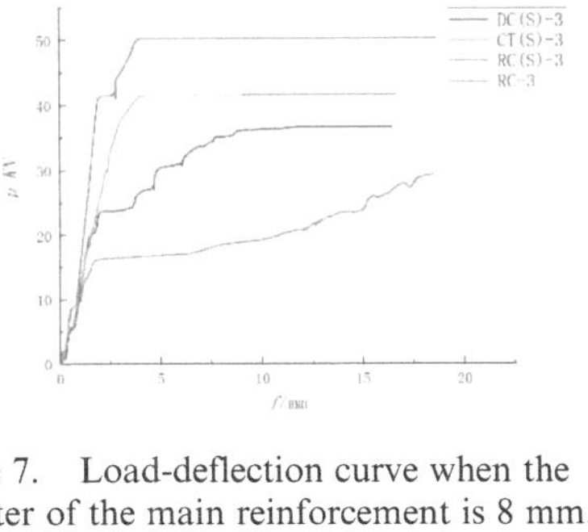

Figure 7. Load-deflection curve when the diameter of the main reinforcement is 8 mm.

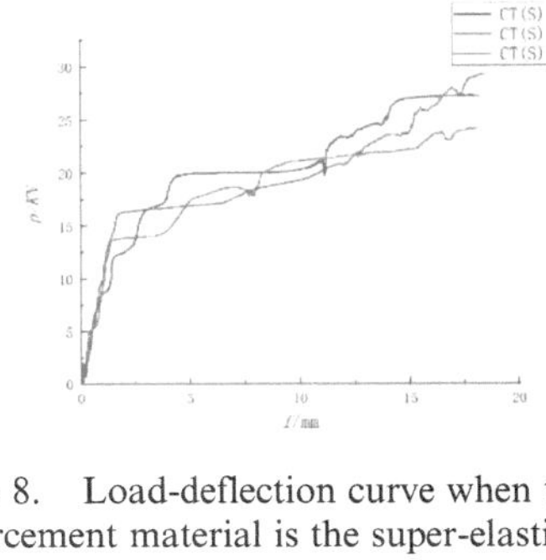

Figure 8. Load-deflection curve when the main reinforcement material is the super-elastic SMA.

calculation results show that super-elastic SMA is very feasible in the practical application of small-sized components.

4 INFLUENCE OF THE SMA WIRE ON THE BENDING PERFORMANCE OF THE BEAM

As shown in Figures 9 and 10, the load-deflection curve of the intelligent concrete beam with SMA wire at the bottom of the beam is drawn when the main reinforcement material is the same by comparing it with the previous experiment of our research group.

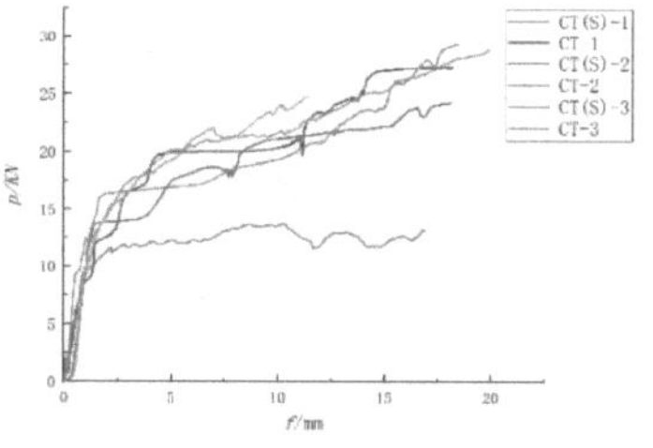

Figure 9. Load-deflection curve when the main the reinforcement material is super-elastic SMA.

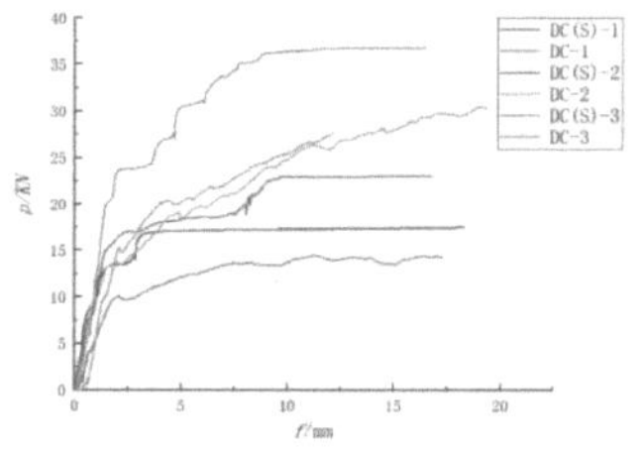

Figure 10. Load-deflection curve when the main reinforcement material is one-way memory SMA.

When the main reinforcement material is a super-elastic SMA bar, the load-deflection curve of each beam is shown in Figure 9. It can be seen that the SMA wire has the most significant reinforcement effect on CT-1, and the ultimate load of CT (S)-1 has increased by 105.97%, which makes its bearing capacity close to CT (S)-2 and CT (S)-3, further improving the application value of small-diameter super-elastic SMA bars in practical components.

The load-deflection curve of each beam is shown in Figure 10 when the main reinforcement material is the single-pass memory SMA bar. Figure 10 shows that when the reinforcement ratio is large, the SMA wire has the best reinforcement effect on the beam. Compared with DC-3, the ultimate load of DC (S)-3 increases by 44.88%, and the ductility of the beam is greatly improved.

When the main reinforcement material is steel, the load-deflection curve of each beam is shown in Figure 11. As can be seen from Figure 11, only when the diameter of the steel bar reaches 8 mm, the SMA wire has a reinforcing effect on the concrete beam. The analysis thinks that when the reinforcement ratio is small, the steel bar plays a leading role in intelligent concrete. With the increase of reinforcement ratio, the driving force of SMA wire after destruction plays a synergistic effect on the improvement of beam bearing capacity, and the exploration of its mechanism provides a good verification direction for subsequent tests.

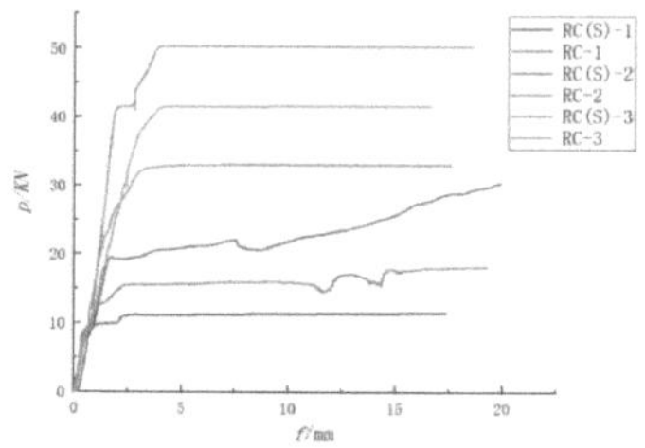

Figure 11. Load-deflection curve when the main reinforcement material is steel bar.

5 CONCLUSION

In this experiment, Ni-Ti SMA bars and wires of different materials and diameters are used to replace steel bars in the tensile area of concrete beams to make SMA intelligent concrete beams, and loads are applied to them. Then the SMA intelligent concrete beams are heated to self-repair and then loaded to destroy them. The bending performance of SMA intelligent concrete beams is studied. By drawing and comparing the load-deflection curves, the following conclusions are drawn:

(1) When applied to large-scale structural systems, SMA bars are inferior to steel bars in bearing capacity because of their low rigidity. However, it can be considered to replace a certain number of steel bars by configuring a certain number of SMA bars in the case of a high reinforcement ratio, which not only meets the requirements of strength, but also meets the requirements of durability, repair, and reinforcement. It provides a new idea for future research.

(2) Compared with ordinary reinforced concrete beams, the yield load and ultimate load of super-elastic SMA beams with 4 mm main reinforcement are increased by 35.2% and 53.8% respectively. When the diameter of the main reinforcement is increased from 4 mm to 8 mm, the yield load and ultimate load of super-elastic SMA beams are only increased by 4.6% and 18.4% respectively. The test results show that the super-elastic SMA has a good prospect for market application among the members with a low reinforcement ratio.

(3) When the main reinforcement material is 8 mm, the yield load and ultimate load of ordinary reinforced concrete beams laid with one-way memory SMA wire at the bottom are increased by 38.2% and 19.7% respectively. The test results show that one-way memory SMA wire can effectively repair tiny cracks and effectively improve the flexural performance of cracked reinforced concrete beams.

REFERENCES

Cervenka J., Papanikolaou V. K. Three Dimensional Combined Fracture–the Plastic Material Model for Concrete [J]. *International Journal of Plasticity*, 2008, 24 (12): 2192–2220.

Cicekli U., Voyiadjis G. Z., and Al-Rub R. A Plasticity and Anisotropic Damage Model for Plain Concrete [J]. *International Journal of Plasticity*, 2007, 23 (10–11): 1874–1900.

Cui Di, Li Hongnan and Song Gangbing. Experimental Study on Mechanical Properties of Shape Memory Alloy Concrete Beams [J]. *Engineering Mechanics*, 2010, 27 (2): 117–123.

Di Shengkui, Hua Weipan, Ji Shengwei, et al. Crack Monitoring and Self-repair of Restrained SMA Concrete Beams [J]. *Journal of Building Materials*, 2010 (02): 237–242.

Effendy E., Liao W. I., Song G., et al. Seismic Behavior of Low-Rise Shear Walls with SMA Bars [C]. Workshop on Biennial International Conference on Engineering. 2006.

El-Tawil S., Ortega-Rosales J. Prestressing Concrete Using Shape Memory Alloy Tendons [J]. *ACI Structural Journal*, 2004, 101 (6): 846–851.

Kuang Yachuan and Ou Jinping. Study on Deformation Characteristics of Shape Memory Alloy Intelligent Concrete Beams [J]. *China Railway Science*, 2008, 29 (4): 41–41.

Li Shuangbei, Liang Qingguo, Jiang Linjie, et al. Double Spline QR Method for Crack Self-healing Performance Analysis of SMA Concrete Beams [J]. *Journal of Guangxi University (Natural Science Edition)*, 2014, 39 (1): 180–186.

Shi Yan, Sun Jing, and Wang Wei. Study on Deformation Characteristics of Shape Memory Alloy Intelligent Concrete Continuous Beam [J]. *Concrete*, 2010 (03): 8–12.

Sun Li, Chen Xiaodan, and Gao Qianqian. Experimental Study on Repair Performance of Concrete Beams with Prestressed Shape Memory Alloy Wires [J]. *Journal of Architectural Structure*, 2015, 36 (S2): 265–269.

Water Conservancy and Civil Construction – Oke & Ahmad (Eds)
© 2024 The Author(s), ISBN: 978-1-032-58614-4

Research on mix ratio of track slab and self-compacting concrete for low-temperature environment in a laboratory

You Xueqi*
Faculty of Architecture, Civil and Transportation Engineering, Beijing University of Technology, Beijing, China

ABSTRACT: In this paper, the mix proportion of track slab concrete and self-compacting concrete for CRTS III slab track for the high-speed railway was taken as the research object. The binder content, rate, and water-binder ratio were comprehensively considered. Then the functions and control standards of main components are analyzed, such as cement, coarse aggregate, fine aggregate, admixture, and other main components. Finally, the laboratory mix proportion of track slab and self-compacting concrete for CRTS III slab track in a low-temperature environment was put forward. The proposed mix proportion was verified by the method of performance on ordinary fresh concrete and concrete cubic compressive strength test. The results show that the concrete prepared with the laboratory mix ratio proposed in this paper has a good working performance, and its cubic compressive strength and workability meet the requirements.

1 INTRODUCTION

The independently developed CRTSIII type of ballastless track has been widely used in the high-speed railway passenger line[1]. CRTS III-type slab ballastless track is mainly composed of the steel rail, fastener system, prefabricated track slab, SCC filling layer, isolation layer, base plate, etc. The composite-composed track slab and self-compacting concrete are one of the main load-bearing structures of CRTS III slab ballastless track. The working performance of the track slab and self-compacting concrete composite layer affects the function and durability of the ballastless track and even affects the running safety of the high-speed railway. To facilitate the systematic study of the track slab and self-compacting concrete layer, it is necessary to clarify the standardized production process of the track slab and self-compacting concrete layer, which requires the determination of the laboratory mix ratio of both. In the actual operation process, a low-temperature environment will be used, so it is of great significance to study the mix ratio of CRTSIII plate ballast track board and self-compacting concrete in a low-temperature environment.

Among the existing studies on C60 concrete and C40 self-compacting concrete, there are studies on the mixture ratio of C60 concrete and C40 self-compacting concrete under normal conditions[2-7]and application of C60 concrete and C40 self-compacting concrete in high-cold or low-temperature environment[8-13]. For the current studies of the mix ratio under normal conditions, except for the mix ratios that do not meet the requirements of the railway concrete code, there is no mixing ratio suitable for the low-temperature environment among the remaining mix ratios. At present, the concrete mix ratio under a specific low-temperature environment is either not in line with the requirements of railway concrete specifications, or

*Corresponding Author: youxueqi687095@163.com

 DOI: 10.1201/9781003450818-64

the quantified mix ratio is not given directly. Therefore, it is particularly important to study the mix ratio of track slabs and self-compacting concrete in a low-temperature environment.

In this paper, the laboratory mix ratio of the track plate and self-compacting concrete suitable for the low-temperature environment is given. Firstly, the design principles of two kinds of concrete mix ratios were summarized based on the literature research. According to the role of various materials in concrete, the standard prescribed by screening controlled the quality of raw materials. Then, according to the standard mix ratio, it was adjusted by repeated test mixing in the laboratory. Finally, the mix ratio of the track slab and self-compacting concrete suitable for the low-temperature environment was determined. To control the quality of concrete engineering, we shall verify the correctness of the mix ratio proposed in this paper and check the quality of the selected materials, and it is necessary to test the performance of concrete mix and the compressive strength of concrete cube for these two kinds of concrete. The experimental results show that the mixture ratio proposed in this paper is suitable for a low-temperature environment.

2 DESIGN PRINCIPLES FOR THE MIX OF TRACK PLATE AND SELF-COMPACTING CONCRETE

The track slab is made of high-strength concrete with a graded strength of C60. Portland cement of 42.5, higher strength grade, or ordinary Portland cement should be selected for the preparation of high-strength concrete C60. The amount of cementite material is generally controlled at no more than 500 kg/m^3. The water-binder ratio should not be greater than 0.35[14]. The slump of the high-performance concrete used in the track plate is usually controlled at 100~140 mm[15]. In this paper, the laboratory mix ratio was studied in the cold winter of northern China, and it was considered to increase the number of cementitious materials to reduce the water-binder ratio, which could not only improve the fluidity, gap passing ability, and filling ability of the concrete mix, but also ensure the overall performance of the track plate concrete mix[16-17]. Under the condition of meeting the concrete design requirements, moderately increasing the content of mineral admixture in cementing materials[18-21] cannot only reduce the hydration heat of concrete and reduce the cracking phenomenon of concrete. It can also improve the viscosity and fluidity of concrete. At the same time, the sand rate of concrete should be controlled. Too much sand rate will affect the gap passage of the concrete mixture, but too small a sand rate will affect the actual effect of the concrete. Since the water-binder ratio of the track plate is controlled at 0.3 and the maximum particle size of the track plate aggregate is 20 mm, the sand rate should be controlled at 34%-38%.

Self-compacting concrete is a kind of high-performance concrete with high fluidity, clearance passing rate, and segregation resistance. It can be filled and formed evenly by its weight without vibration during pouring. Self-compacting concrete should increase the volume of slurry by increasing the powder material, and the viscosity and fluidity of slurry can also be improved by adding admixture. An appropriate amount of high-quality fly ash, ground slag powder, and other mineral admixture as well as water-reducing agent, air entraining agent, expansion agent, viscosity modification material, and other admixture should be added to self-compacting concrete to ensure that self-compacting concrete has a good working performance. The dosage of cementing material for self-compacting concrete should not be more than 580 kg/m^3; the water consumption should not be more than 180 kg/m^3; the total amount of slurry per unit volume should not be more than 0.40 m^3; the extension time should be controlled within 3~7 s; the slump extension should not be more than 680 mm. If the slump extension is too large, the bond performance between the self-compacting concrete and the track plate will decrease [22].

3 EXPERIMENTAL MATERIALS

The tracking board and self-compacting concrete have added admixtures, especially the self-compacting concrete. Therefore, the selection of concrete raw materials (including cement varieties and properties as well as material specifications and gradation of sand and stone), the dosage, and the type of admixture should be selected reasonably according to the actual engineering needs and specifications requirements.

3.1 *Cement*

Portland cement or ordinary Portland cement should be used for the track board and self-compacting concrete. The cement strength grade of the track board in this paper should not be lower than 42.5 grade; early strength cement should not be used; cement alkali content should not be greater than 0.60%; sulfur trioxide content should not be greater than 3.0%; chloride ion content should not be greater than 0.06%; C_3A content in clinker should not be greater than 8.0%. Other performance requirements and self-compacting concrete shall comply with the provisions of "Railway Concrete" TB/T 3275.

Ordinary Portland cement is adopted in this paper. The hardening time of ordinary Portland cement is short and other cement may be used, therefore the performance advantage of the material will be weakened under the influence of other mineral admixtures such as fly ash and mineral powder, leading to the reduction of concrete strength[23]. Considering that the construction is in winter, 52.5 high-strength cement is properly used to prepare the concrete to ensure the overall performance of concrete [24].

Cement, P.O 52.5 ordinary Portland cement produced by Yangchun Cement Co., Ltd., whose chemical composition is shown in Table 1 and physical properties are shown in Table 2.

Table 1. Chemical composition of cement (%).

CaO	SiO_2	Al_2O_3	Fe_2O_3	MgO	SO_3	CL	LOSS
56.77	20.86	590	3.61	3.50	2.43	0.021	1.16

Table 2. Physical properties of cement.

	Setting time (min)			Flexural strength (MPa)		Compressive strength (MPa)	
Specific surface area (m^2/kg)	Initial set	Final set	Stability	3d	28d	3d	28d
381	115	184	Qualified	6.2	10.8	33.8	61.2

3.2 *Mineral admixture*

According to experience, if only cement is used when making up concrete, the early hydration heat of concrete will increase, resulting in greatly weakened stability and durability of concrete. Fly ash can react with Ca $(OH)_2$ generated after the heat of hydration of cement, and the resulting calcium silicate hydrate gel has a filling effect. It can also inhibit the alkali-aggregate reaction, reduce the erosion of chloride ions, increase the compactness of concrete, and reduce the output of heat of hydration. Fly ash can also increase the workability of concrete. The performance of fly ash and slag powder shall meet TJ/GW 112-2013 "Interim

Technical Conditions for Self-Compacting Concrete of high-speed Railway CRTSIII Type Plate-type Ballastless Track and TJ/GW", "Interim Technical Conditions for Prestressing Track Plate of High-speed Railway CRTSIII Type Plate-type Ballastless Track" 156-2017 and "Concrete for Railway" TB/T 3275. The performance of fly ash and slag powder shall meet "Interim Technical Conditions for Self-compacting Concrete of High-speed Railway CRTSIII Type Plate-type Ballastless Track" TJ/GW 112-2013 and "Interim Technical Conditions for Prestressing Track Plate of High-speed Railway CRTSIII Type Plate-type Ballastless Track" TJ/GW 156-2017 and "Concrete for Railway" TB/T 3275.

Keywords: Fly ash, grade I fly ash, mineral powder, and S95 ground fine mineral powder.

3.3 *Fine aggregate*

As one of the main materials, the fluidity of fine aggregate affects the performance of the whole concrete mix. Natural river sand with hard material, a clean surface, and reasonable gradation is selected. If the sand is too coarse, it will lead to poor cohesion and fluidity of the concrete mixture. If the sand is too fine, the elastic modulus of the concrete mixture will decrease greatly and the strength will be difficult to meet the requirements. For track board sand requirements, the fineness modulus should be 2.3~3.0; mud content is not more than 1.5%; chloride content is not more than 0.02%; and self-compacting concrete requires a fineness modulus of no more than 2.7; mud content is not more than 2.0%. Other properties should conform to the provisions of "Railway Concrete" TB/T 3275.

Keywords: Fine aggregate, natural river sand, medium sand, and fineness modulus 2.6.

3.4 *Coarse aggregate*

Like fine aggregate, the fluidity of coarse aggregate will have a great influence on the performance of the concrete mix. The tracking board should be made of hard material, surface clean secondary or multi-level single-grain gravel. The maximum particle size is 20 mm, and self-compacting concrete stone should be selected with good grain shape, solid texture of clean broken, and pebble, which should be used with two grade aggregate mixed. The maximum particle size should not be greater than 16 mm, and other performance should conform to the "Railway concrete" TB/T 3275 provisions. We should give priority to round stones to reduce the content of needle-like and flaky particles and friction and to enhance the fluidity of concrete.

Keywords: Coarse aggregate; two kinds of gravel size of 5~10 mm and 10~16 mm.

3.5 *Water-reducing agent*

Water reducing agent, as the main component of admixture, is to prevent the condensation of dispersed particles and plays the role of plastic. We shall reduce the viscous resistance of the mixture and the incidence of shrinkage to improve the compactness of the concrete. But we shall also ensure the dosage, plays an effective role. Selecting a high-performance superplasticizer with stable quality can significantly improve the durability of self-compacting concrete. The superplasticizer should have good compatibility with cement and mineral admixture, and other properties should conform to the provisions of "Railway Concrete" TB/T3275.

Keywords: Water-reducing agent; a water-reducing agent with a high-performance polycarboxylic acid.

3.6 *Viscosity-modified materials*

Viscosity-modified materials are mainly used in the production of self-compacting concrete. The use of viscosity modifiers is to increase the segregation resistance and viscosity of self-compacting concrete so that the concrete mix can better wrap the steel bar. The materials can improve the working performance of self-compacting concrete, such as anti-isolation ability and viscosity, without reducing the mechanical performance and durability of self-compacting concrete. And their performance shall comply with the provisions of "Provisional Technical Conditions for Self-compacting Concrete for High-speed Railway CRTSIII Slab Ballast Track" TJ/GW 112-2013.

Keywords: Viscosity-modified material; REW-V06.

3.7 *Expansion agent*

Self-compacting concrete has a shrinkage phenomenon, which can be compensated by using an expansion agent. The shrinkage of concrete can be reduced by adding an appropriate amount of expansion agent, and the cohesion of concrete can be improved by adding an appropriate amount of expansion agent. All indexes of the expansion agent shall meet the requirements of the Concrete Expansion Agent (GB 23439).

Keywords: Expansion agent; type II expansion agent; REW-E014.

3.8 *Mixing water*

Because the development period was in the northern winter, the influence of temperature must be considered. In addition to the consideration of increasing the amount of cementitious material and the proportion of mineral admixture, it is also necessary to use heated mixing water during configuration. The 52.5 ordinary Portland cement shall be used, and the mixing water temperature is controlled at 60°C. If the water temperature is not enough to meet the requirements of thermal calculation, we can also increase the water temperature to 100°C, but it should be noted that the cement shall not be in direct contact with water of more than 80°C.

The mixed water shall comply with the provisions of "Railway Concrete" TB/T 3275.

4 MIX RATIO OF CONCRETE

In the process of studying the mix ratio, based on the standard mix ratio, after several trials and mixing in the laboratory, the following laboratory mix ratios meeting the requirements were finally obtained.

The mix ratio of track board concrete is shown in Table 3. The water-binder ratio is 0.30; the sand ratio is 36%; the amount of cementitious material is 492.

The mix ratio of self-compacting concrete is shown in Table 4. The water-binder ratio is 0.36; the sand ratio is 50%; the amount of cementitious material is 474.4.

Table 3. A mix ratio of track board concrete.

Raw material	Cement	Fly ash	Sand	Gravel (5~10)	Gravel (10~16)	Water reducing agent	Water
Mix ratio	452	40	600	352	706	19.2	130.7

Table 4. A mix ratio of self-compacting concrete.

Raw material	Cement	Fly ash	Mineral powder	Expansive agent	Viscosity modified materials	Sand	Gravel (5~10)	Gravel (10~16)	Water reducing agent	Water
Mix ratio	325	19	93	37	0.4	775	451	301	18	171.8

5 EXPERIMENTAL METHODS

5.1 *Performance test of concrete mixture*

The performance of the concrete mix of the track board is evaluated by the slump index, and the slump requirement should meet 100~140 mm. The performance of the self-compacting concrete mix is evaluated by the spread and expansion time T500 of the slump. The slump spread should be less than or equal to 680 mm, and the expansion time T500 should meet 3~7 s.

The first step of the slump test: we should ensure that there is no obvious watermark on the inner wall and bottom plate of the slump cylinder. The bottom plate is made of steel plate and placed on a solid horizontal surface. We should put the slump cylinder in the center of the bottom plate and step on the pedal with our feet. The second step is feeding, and the concrete mix is divided into three layers evenly to join the slump cylinder to ensure each layer of concrete with a tamper solid. The third step: after the top layer is inserted and tamped, the top layer of excess concrete mix is scraped off and smoothed along the hole. The fourth step: we should lift the slump cylinder vertically and smoothly after removing the concrete from the bottom plate on the side of the slump cylinder. When the specimen does not continue to slump or the slump time reaches 30 s, we can use a steel ruler to measure the height difference between the height of the cylinder and the highest point of the concrete after the slump as the slump value of the concrete mixture.

The extending degree of self-compacting concrete slump refers to the concrete mixture slump behind the diameter of the extension. The preliminary preparation steps for this test are not different from the slump test, except that after lifting the slump cylinder vertically and smoothly when the sample no longer diffuses or the diffusion duration reaches 50 s. We should use a steel ruler to measure the maximum diameter of the spread surface of the concrete mixture and the diameter perpendicular to the maximum diameter. When the difference between the two diameters is less than 50 mm, the average value of the two diameters is taken as the slump spread value of the concrete mix. When the difference between the two diameters is greater than 50 mm, resampling should be adopted for determination. When measuring the extension time T500, it is slightly different. It is necessary to fill the slump cylinder with the concrete mixture at one time, without compaction or vibration. Then we can lift the slump cylinder to a height of 250±50 mm, and the extension time is from the time when the slump cylinder is lifted off the ground to the circumference of 500 mm where the outer edge of the extended concrete mix reaches.

5.2 *Compressive strength test of concrete*

In the preparation of the specimen, the standard cube test block was selected, and the side length was 150 mm. After maintenance for 1 day, the specimen was removed from the mold and cured with a standard curing box (the temperature at 20±2 °C and humidity above 95%). The machine of electro-hydraulic pressure testing was used to uniformly load at a loading speed of 0.8 MPa/s, and the cubic compressive strength of 7 d, 14 d, and 28 d was measured, respectively.

6 TEST RESULTS AND ANALYSIS

6.1 *Experimental results*

To ensure that the obtained mixture ratio is suitable for a low-temperature environment, the whole test conditions from concrete preparation to later curing are carried out in the outdoor low-temperature environment.

The performance test results of track slab concrete and self-compacting concrete are shown in Table 5.

Table 5. Performance of concrete.

Test content	Slump (mm)	Slump spread (mm)	Extension time T500 (s)	Compressive strength (MPa)		
				7 d	14 d	28 d
Track plate	120			46.5	59.6	66.3
Self-compacting		650	5	32.9	37.7	44.5

6.2 *Result analysis*

The working and mechanical properties of the mix of track plate and self-compacting concrete presented in this paper meet the requirements of relevant codes. It can realize the standardized configuration of CRTS type III rail plates and self-compacting concrete in a low-temperature environment. It is specifically reflected in:

(1) The measured slump of the track slab concrete mix is 120 mm, which meets the requirement of a slump from 100 to 140 mm, and the compressive strength of the 28 d cube is 66.3 MPa.
(2) The measured slump extension of self-compacting concrete mix is 650 mm, which meets the requirement that the slump extension of self-compacting concrete is less than or equal to ($\leq$) 680 mm, the extension time T500 is 5 s, and the extension time is 3 s-7 s. The compressive strength of the 28 d cube is 44.5 MPa.

REFERENCES

[1] Ge C., Zhang J., *et al.* Study on the Influence of Mineral Admixture on Hydration Heat of Cement, *Architectural Engineering Technology Design*, 2016 (23): 2365–2365.

[2] Wu C., *et al.* Preparation of CRTSIII type Self-compacting Concrete and Analysis of Quality Control Points in Winter Construction, *Communications Science and Technology Heilongjiang*, 2020, 43 (12): 182–183.

[3] Leng F., Ding W., Wei Q., *et al.* Experimental Research on C60~C100 High Strength and Performance Concrete, *Building Structure*, 2011, 41 (11): 155–158+40.

[4] Peng G., Gao R., Liu Y., *et al.*, Experimental Investigation of Concrete Characterized by High Frost Resistance of F300 and Strength Grade of C50 or C60, *Journal of Railway Science and Engineering*, 2004 (02): 14–18.

[5] You G., *et al.* Study on Performance of Self-compacting Concrete for C40 Ballastless Track, *Proceedings of the 7th National Academic Exchange Conference on Concrete Expansion Agents*, 2018: 125–130.

[6] Dong H., He C., *et al.* Mix Proportion Design of C60 High-strength Concrete, *Northern Communications*, 2010 (06): 82–84.

[7] Li H., Ding Z., Xing F., *et al.* Effect of Fly Ash and Slag on Hydration Heat Evolution of Cement, *Concrete*, 2008 (10): 54–57.

[8] Hong J., Li J., Li Y., *et al.* Experimental Study on High-performance Frost Resistant Concrete and its Application in Engineering, *Development Guide to Building Materials*, 2010, 8 (01): 53–56.

[9] Huang J., *et al.*, Construction Technology of CRTSIII Track Slab Concrete for High-Speed Railway, *Architecture and Decorations*, 2020(8):138.

[10] Pei J., *et al.* Preparation of C60 High-strength Ultra-high Performance Concrete, *Tianjin Construction Science and Technology*, 2014, 24 (02): 29–30.

[11] Wang J., *et al.* *Study on Mechanical Properties and Frost Resistance of High-Performance Self-compacting Concrete*, Jilin University, 2019.

[12] Wang J., Wang M., Liu W., Zhao Y., *et al.*, Technology of CRTSIII Ballastless Track System, *China Railway*, 2017 (08): 11–15.

[13] Wei J., *et al.* Factors and Key Technology of Quality of Construction in Winter (to pingdu region as an example), *Journal of Qingdao University of Technology*, 2015.
[14] J. Zhang, *et al.* Experimental Study on the Effect of Mineral Admixture on Cement Hydration Heat, *Architectural Engineering Technology Design*, 2017 (1): 897, 902.
[15] Li L., Tan Y., Li H., *et al.* Preparation and Performance of Self-compacting Concrete with Low Gel Content in CRTSIII Slab-type Ballastless Track, *Railway Engineering*, 2016 (01): 84–88.
[16] Li L., Tan Y., Li H., *et al.* Research on Bond Performance of Self-compacted Concrete for CRTSIII Slab-type Ballastless Track, *Railway Engineering*, 2015 (01): 137–140.
[17] Luo Q., *et al.* Effect of Fly Ash on Frost Resistance of C40 Self-compacting Concrete, *Sichuan Cement*, 2021 (06): 55–56.
[18] Ji W., *et al.* Experimental Research and Analysis of Hydration Heat of Cement, *Technology of Highway and Transport*, 2016, 32 (1): 13–16.
[19] Ding X., Zhao X., Xu X., *et al.* Effect of Admixtures on Properties of Sulphoaluminate Cement-common Portland Cement Composite System, *New Building Materials*, 2020, 47 (3): 40–44.
[20] Liu X., Yu Z., Jin C., *et al.* Experimental Study on Composite Plate of CRTSIII Slab Track under Transverse Bending Moment, *Journal of the China Railway Society*, 2018,40 (12): 153–160.
[21] Chen Y., *et al.* Freezing Mechanism of Concrete and Winter Construction Control, *Shanxi Architecture*, 2005 (04): 93–94
[22] Liu Y., *et al.* Experimental Study on Mix Proportion of Roller Compacted Concrete in Winter at Low Temperature, *Shaanxi Water Resources*, 2021 (08): 248–250.
[23] Pang Z., Yin H., Zhang L., *et al.*, Experimental Research of the Effect of Fly Ash on the Frost Resistance of Concrete, *Concrete*, 2015 (08): 53–55+62.
[24] Zhao Z., Wang H., Nie Z., *et al.* Research and Application of CRTSIII Type Self-Compacting Concrete for Ballastless Slab Track, *Architecture*, 2019 (16): 68–69.

Water Conservancy and Civil Construction – Oke & Ahmad (Eds)
© 2024 The Author(s), ISBN: 978-1-032-58614-4

Identification of bridge surface roughness based on displacement influence line of contact points of two single axle vehicles

MingHua Wang*
School of Civil Engineering, Chongqing University, Chongqing, China

ABSTRACT: A new method is proposed for the identification of bridge surface roughness based on the responses of two single-axle vehicles. The key of the study is to introduce the displacement influence line to represent the relationship between the dynamic responses of two contact points. Based on the relationship, the roughness formula only related to vehicle responses can be deduced. At the same time, after numerical verification, it is found that the method has high accuracy.

1 INTRODUCTION

The surface roughness of a bridge is closely related to its health but with the increase in service time, the bridge pavement will become rougher and rougher, which will lead to an increase in vehicle responses. The increase in vehicle responses will make the pavement rougher, forming a vicious circle. The maintenance cost of the bridge deck is one of the main expenditure sources of the bridge maintenance budget. Therefore, if the degradation of surface roughness can be identified in time and corresponding measures can be taken, the maintenance cost of the bridge can be effectively reduced.

Given this, a new method for identifying the surface roughness of the bridge based on the responses of two single-axle vehicles is proposed in this paper. The method is applied to simply supported beam bridges, and its reliability is verified by numerical analysis.

2 FORMULATION OF THE CONCERNED PROBLEM

When the vehicle moves on the bridge, both roughness and bridge vibration can cause the vehicle to vibrate. This means that the signals recorded by the sensors on the moving vehicle will contain information about both the roughness and the bridge, which is not conducive to roughness identification. To solve this problem, a method to identify the surface roughness of the bridge is proposed in the study, in which the influence of the bridge vibration on the vehicle response is eliminated by using the displacement influence line.

2.1 *Vehicle-bridge interaction (VBI) model*

The vehicle model used in this paper is shown in Figure 1, where subscripts “1” and “2” denote the moving and stagnant vehicles, respectively. It is assumed that the wheels of the two vehicles always keep good contact with the bridge deck during driving, and the contact points between the vehicle and the bridge are P_1 and P_2 respectively. The bridge is assumed

*Corresponding Author: MinghuaW@cqu.edu.cn

DOI: 10.1201/9781003450818-65

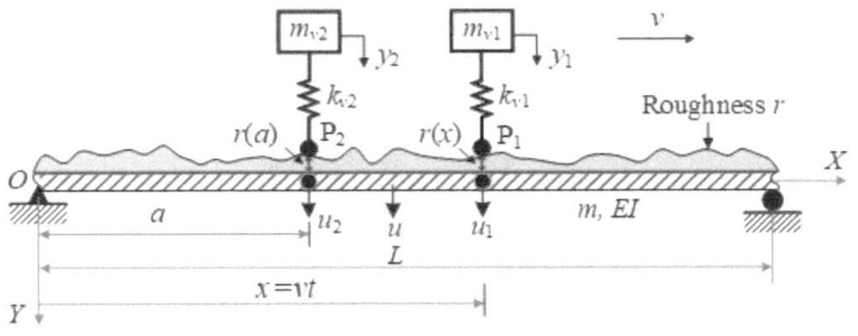

Figure 1. The VBI model with surface roughness.

to be the Bernoulli–Euler type with constant bending stiffness EI and constant mass m per unit length. To improve the fidelity of the vibration transmitted to the vehicle, damping is not considered in the design of both vehicles.

When vehicle "1" is driving on the bridge, the positions of contact points P_1 and P_2 are x and a respectively. The vertical displacement of the two contact points is composed of two parts: the deflection u of the bridge and the surface roughness r at the point. Both vehicles will be excited by the vibration transmitted from the contact point, and their motion equations are as follows:

$$m_{v1}\ddot{y}_1(t) + k_{v1}\{y_1(t) - [u_1(x) - r(x)]|_{x=vt}\} = 0 \tag{1a}$$

$$m_{v2}\ddot{y}_2(t) + k_{v2}\{y_2(t) - u_2(a)\} = 0 \tag{1b}$$

where y_1 and y_2 represent the vertical displacement of vehicles "1" and "2" respectively; $u_1(x)$ and $u_2(a)$ are the vertical displacements of the bridge at the two contact points respectively; and $r(x)$ is the corresponding surface roughness value.

The acceleration responses, i.e., $\ddot{y}_1$ and $\ddot{y}_2$ in Equation (1) can be measured by the accelerometer installed on the vehicle. According to the method given in References (González 2012; Han 2010; Hong *et al.* 2013; Park *et al.* 2005), the displacement responses, i.e., y_1 and y_2 can be obtained by numerical integration of the acceleration responses. At this time, there are still three unknowns, i.e. $u_1(x)$, $u_2(a)$, and $r(x)$ in Equation (1), which means that the equation still cannot be solved. Meanwhile, we notice that the displacement responses, i.e., $u_1(x)$ and $u_2(a)$ of the bridge at the two contact points are both caused by the movement of the vehicle "1", so there will be a relationship between them. As long as the relationship is found, the theoretical solution of roughness can be obtained.

2.2 *Introduction of displacement influence lines*

Because the mass of the vehicles is far lower than that of the bridge, the dynamic impact caused by the vehicle passing through the bridge at a relatively low speed is very small. Then the dynamic deflection of the bridge caused by the vehicle can be taken equal to the one caused by the static weight (Yang *et al.* 2004, 1995). In this study, the dynamic deflections, i.e., $u_1(x)$ and $u_2(a)$ of the two contact points will be approximated by the displacement influence line:

$$u_1(x) = \delta_{11}(x)m_{v1}g + \delta_{12}(x)m_{v2}g \tag{2a}$$

$$u_2(a) = \delta_{21}(a)m_{v1}g + \delta_{22}(a)m_{v2}g \tag{2b}$$

where g is the acceleration of gravity, and δ_{ij} is the displacement influence line value of the bridge at the contact point P_i caused by a unit load acting at the contact point P_j. According to the Maxwell–Betti reciprocal theorem, we can know that the values of $\delta_{12}(x)$ and $\delta_{21}(a)$ are the same.

By dividing Equation (2b) by Equation (2a), the formula expressing the relationship between the displacements of the two contact points can be obtained:

$$u_1(x) = \frac{\delta_{11}(x)m_{v1} + \delta_{12}(x)m_{v2}}{\delta_{21}(a)m_{v1} + \delta_{22}(a)m_{v2}} \cdot u_2(a) \tag{3}$$

2.3 *Theoretical solution of bridge surface roughness*

By Substituting Equation (3) into Equation (1), after some operations, we can get:

$$r(x) = \frac{\delta_{11}(x)m_{v1} + \delta_{12}(x)m_{v2}}{\delta_{21}(a)m_{v1} + \delta_{22}(a)m_{v2}} \cdot \frac{m_{v2}\ddot{y}_2(t) + k_{v2}y_2(t)}{k_{v2}} - \frac{m_{v1}\ddot{y}_1(t) + k_{v1}y_1(t)}{k_{v1}} \tag{4}$$

From Equation (4), it can be found that the theoretical solution of bridge surface roughness proposed in this section only depends on the responses of the vehicle. Therefore, as long as the displacement influence line values δ_{ij} at the contact point can be obtained, the corresponding surface roughness can be obtained in Equation (4).

2.4 *Application to a simply-supported beam*

Figure 2 shows displacement influence lines of a simply supported beam of length L at the two contact points. The virtual work principle is used to calculate the displacement influence line values, and the results are as follows:

$$\delta_{11}(x) = \frac{x^2(L-x)^2}{3EIL} \tag{5a}$$

$$\delta_{22}(x) = \frac{a^2(L-a)^2}{3EIL} \tag{5b}$$

$$x \le \mathrm{a}, \delta_{12}(x) = \delta_{21}(a) = \frac{x(L-a)}{6EIL}(2aL - a^2 - x^2) \tag{5c}$$

$$x > \mathrm{a}, \delta_{12}(x) = \delta_{21}(a) = \frac{a(L-x)}{6EIL}(2xL - a^2 - x^2) \tag{5d}$$

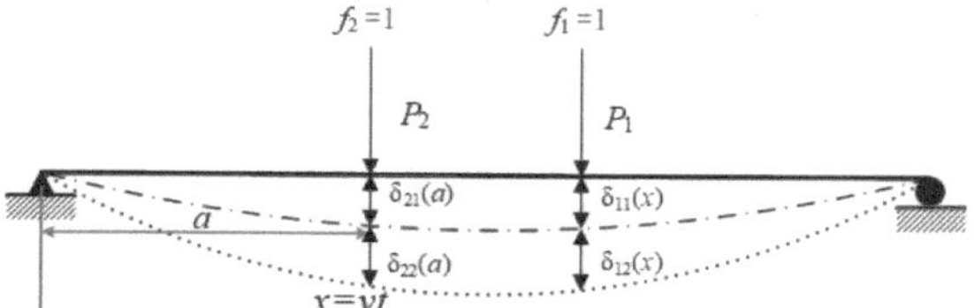

Figure 2. Displacement of influence lines at the two contact points.

By substituting Equation (5) with Equation (4), the surface roughness of the bridge can be obtained.

3 METHODOLOGY OF NUMERICAL SIMULATION

In this study, the finite element simulation (FEM) is used to verify the proposed theory. The following study will briefly introduce VBI elements and the generation of roughness.

3.1 *Finite element simulation for the VBI system*

Figure 3 shows the main features of VBI elements. For vehicles with damping of zero, the motion equation of VBI elements considering surface roughness is as follows (Yang *et al.* 2012):

$$\begin{bmatrix} m_{vi} & 0 \\ 0 & [m_b] \end{bmatrix} \begin{Bmatrix} \ddot{y}_i \\ \{\ddot{u}_b\} \end{Bmatrix} + \begin{bmatrix} 0 & 0 \\ 0 & [c_b] \end{bmatrix} \begin{Bmatrix} \dot{y}_i \\ \{\dot{u}_b\} \end{Bmatrix} + \begin{bmatrix} k_{vi} & -k_{vi}\{N\}_i^T \\ k_{vi}\{N\}_i & [k_b] - k_{vi}\{N\}_i\{N\}_i^T \end{bmatrix} \begin{Bmatrix} y_i \\ \{u_b\} \end{Bmatrix}$$
$$= \begin{Bmatrix} -k_{vi}r_i \\ -m_{vi}g\{N\}_i - k_{vi}r_i\{N\}_i \end{Bmatrix} \tag{6}$$

where $\{u_b\}$ denotes the nodal displacement vector of 4 DOFS, and $[m_b]$, $[c_b]$, and $[k_b]$ denote the mass, damping, and stiffness matrices of the beam element, respectively; for the vehicle, y_i is the vertical displacement of the ith vehicle (i = 1, 2), r_i is the surface roughness value at the contact point P_i under the ith vehicle, and $\{N\}_i$ is the cubic Hermitian interpolation value evaluated at the coordinate x_i of element i.

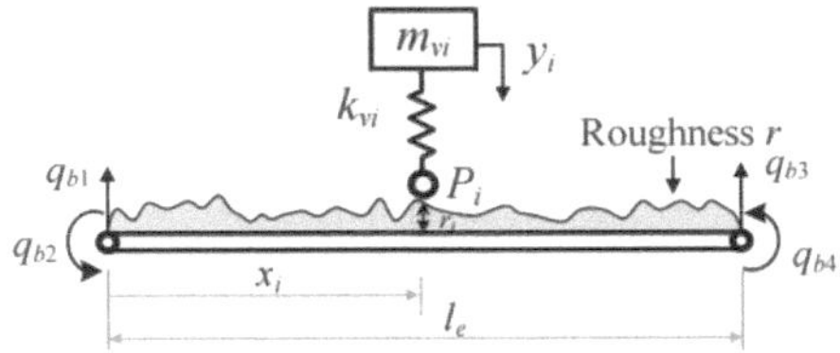

Figure 3. Vehicle-bridge interaction element.

For conventional beam elements that are not in direct contact with the vehicle, the equations of motion are:

$$m_b\{\ddot{u}_b\} + c_b\{\dot{u}_b\} + k_b\{u_b\} = 0 \tag{7}$$

By assembling VBI elements and conventional beam elements, the overall motion equation can be established, and then the dynamic response of the whole VBI system is obtained by using the Newmark-β method.

3.2 *Generation of surface roughness*

To generate the surface roughness in the numerical simulation, the power spectral density (PSD) function proposed by ISO 8608 (2016) is adopted. According to ISO 8608, the PSD function for the roughness profile is defined as:

$$G_d(n) = G_d(n_0)\left(\frac{n}{n_0}\right)^{-\omega} \tag{8}$$

where n denotes the spatial frequency, n_0 denotes the reference spatial frequency taken as 0.1 cycles/m, and ω is a constant equal to 2. The roughness coefficient $G_d(n_0)$ is determined according to the roughness classes listed in Table 1.

Table 1. Roughness coefficient $G_d(n_0)$ for different roughness classes.

Road class	A	B	C	D	E	F	G	H
$G_d(n_0)/10^{-6}$ m^3	16	64	256	1,024	4,096	16,384	65,536	262,144

Then, the road surface roughness can be represented by a normal zero-mean real-valued stationary Gaussian process as:

$$r(x) = \sum_{i=0}^{N} \sqrt{2G_{\text{d}}(n_i)\Delta n}\cos(2\pi n_i x + \theta_i) \tag{9}$$

where N denotes the total number of harmonic waves used to construct the roughness profile, n_i is the ith spatial frequency, Δn denotes the sampling interval of the spatial frequency, and d_i and θ_i denote the spatial frequency amplitude and random phase angle of the ith cosine function, respectively.

4 NUMERICAL VERIFICATION

In this section, the theoretical solution of surface roughness will be verified by numerical simulation. The parameters of bridges and vehicles are listed in Table 2 and the vehicle speed is v=18 km/h. In the simulation, each element length is taken as 0.5 m, the time step is 0.001 s and the roughness class is B.

Table 2. Properties of the vehicle and bridge.

Vehicle	Mass	$m_{\text{v}} = 1{,}200$ kg
	Stiffness	$k_{\text{v}} = 50$ kN/m
	Damping	$c_{\text{v}} = 0$
Bridge	Length	$L = 25$ m
	Young's modulus	$E = 27.5$ GPa
	Moment of inertia	$I = 0.12$ m^4
	Mass per unit length	$m = 4{,}800$ kg/m

With the vehicle responses, in Equation (4), the surface roughness can be calculated and the result is shown in Figure 4. At the same time, the real roughness curve and the deviation between the two curves are also given in the figure. It can be observed from the figure that the calculated roughness is highly consistent with the real roughness, and the deviation between them is close to zero, which confirms the feasibility of Equation (4).

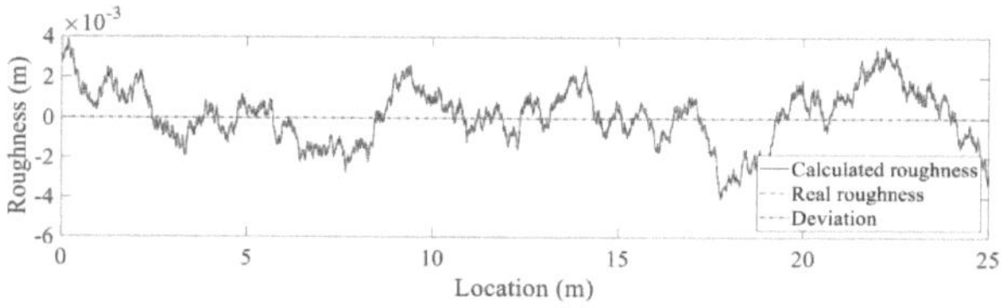

Figure 4. Identification results of class B roughness.

5 PARAMETER ANALYSIS

This section shows the effects of different vehicle speeds and vehicle/bridge mass ratio (Yang *et al.* 2013) on the recognition effect, and the error indicator used is the root mean square error (RMSE) (Kang *et al.* 2019). The results are shown in Table 3.

Table 3. The effects of different vehicle speeds and vehicle/bridge mass ratio.

Vehicle speed (km/h)	RMSE (*10^{-4})	$2m_v/m_L$ (%)	RMSE (*10^{-4})
30	7.59	1	4.64
40	7.63	1.5	6.50
50	8.08	2	8.79

$$\mathrm{RMSE} = \frac{1}{r_{R,\max}}\sqrt{\frac{1}{N}\sum_{i=1}^{N}\left(r_{c,i} - r_{R,i}\right)^2} \tag{10}$$

where $r_{c,i}$ and $r_{R,i}$ denote the ith data of the two kinds of roughness, respectively, $r_{R,\max} = |r_{R,i}|_{\max}$, and N is the number of sampling points.

6 CONCLUDING REMARKS

A new formula for identifying bridge surface roughness based on vehicle response is proposed. The core is to establish the relationship between the displacement response of two contact points by using the displacement influence line. Based on the paper, we can conclude: (1) It has high reliability to use the displacement influence line to express the relationship between the responses of different contact points. (2) The measurement error increases with the increase of the speed. (3) The measurement error increases with the increase in mass ratio.

REFERENCES

González A., Brien E. J. O. and McGetrick P. J. Identification of Damping in a Bridge Using a Moving Instrumented Vehicle[J], *J. Sound Vib.*, 331: 4115–4131 (2012).

Han. S. Measuring Displacement Signal with an Accelerometer, *J. Mech. Sci. Technol.* 24 (6) 1329–35 (2010).

ISO-8608, the International Organization for Standardization, Mechanical Vibration-Road Surface Profiles-Reporting of Measured Data (2016).

Park K. T., Kim S. H., Park H. S. and Lee K. W. The Determination of Bridge Displacement Using Measured Acceleration, *Eng. Struct.* 27 (3) 371–78 (2005).

Kang S. W., Kim J. S. and Kim G. W. Road Roughness Estimation Based on Discrete Kalman Filter with an Unknown Input, *Vehicle Syst. Dyn.* 57 (10) 1530–1544 (2019).

Yang Y. B., Chen W. F., Yu H. W. and Chan C. S. Experimental Study of a Hand-drawn Cart for Measuring the Bridge Frequencies, *Eng. Struct.* 57 (2013) 222–231.

Hong Y. H., Lee S. G. and Lee H. S. Design of the FEM-FIR Filter for Displacement Reconstruction Using Accelerations and Displacements Measured at Different Sampling Rates [J], *Mech. Syst. Signal. Pr.*, 38: 460–481 (2013).

Yang Y. B., Li Y. C., and Chang K. C. Using Two Connected Vehicles to Measure the Frequencies of Bridges with the Rough Surface: a Theoretical Study, *Acta Mech.* 223 (8) 1851–61 (2012).

Yang Y. B., Liao S. S. and Lin B. H. Impact Formulas for Vehicles Moving Over Simple and Continuous Beams *J. Struct. Eng.*, 121(11) 1644–50 (1995).

Yang Y. B., Yau J. D., Yao Z., and Wu Y. S. *Vehicle-Bridge Interaction Dynamics: with Applications to High-Speed Railways* (Singapore: World Scientific Publishing Company), p 530 (2004).

Water Conservancy and Civil Construction – Oke & Ahmad (Eds)
© 2024 The Author(s), ISBN: 978-1-032-58614-4

Study on stability of Tongjiaping landslide

Quanyi Li* & Huaxi Gao
Zhejiang Ocean University, Zhoushan, China

ABSTRACT: In the three major geological disasters, collapse, landslide, and debris flow, the frequency of landslide are high and their damage is severe, several of which on the construction of the city, mountain living environment, and people's life, transportation, water conservancy, and hydropower construction caused a huge impact and harm. To ensure the safety of people's life and property, this study takes the Tongjiaping landslide as the research object, finds out the basic characteristics, formation conditions, stable state, and harm degree of Tongjiaping landslide, and puts forward the comprehensive prevention and control project plan of geological disasters.

1 INTRODUCTION

Landslides cause huge economic losses every year. Therefore, analysis of slope failure mechanism and stability has always been a key topic in slope research. Many scholars have put forward a lot of methods, such as the Swedish method, simplified Bishop method, Janbu method transfer coefficient method, wedge method, Fellenius method finite element strength reduction method, and more than a dozen other calculation methods of slope stability. Based on the study of numerical analysis, more scholars began to develop many different calculation methods. The limit equilibrium method is a common method for slope stability analysis. According to different assumptions of limit equilibrium, more than a dozen calculation methods have been extended. It was first pointed out by Morgenstern and Price that the solution obtained by the limit equilibrium bar fraction method must satisfy two rationality conditions: 1. No tensile stress can be generated between the strips, and the points must fall within the side of the strips. 2. The inter-bar shear force acting on the bars shall not exceed the ultimate shear strength provided by the Mohrcoulomb strength criterion.

The unbalanced thrust method in the strip division is an original analytical method of slope stability research, which is simple in calculation and can provide design thrust for practical projects. Zheng Yingren *et al.*, Zhang Luyu *et al.*, and Shi Weimin *et al.* discussed the problems existing in the calculation of the unbalanced thrust method. The Morgenstern-Prince method is recognized by geotechnical academia as having high accuracy. As a reference, by comparing the calculation and analysis of multiple slopes, it is found that the unbalanced thrust method has a large error in calculating the curved slope, and the error is less than 3% compared with the M-P method in calculating the circular slope, in which may be an unreasonable phenomenon that the shear force between the strips exceeds the shear strength between the strips. Based on previous studies, this paper calculates the stability of the Tongjiaping landslide. The unbalanced thrust method is used to determine the stability coefficient of the landslide better, and the landslide prediction is further promoted by combining it with the practice.

*Corresponding Author: 1248609980@qq.com

 DOI: 10.1201/9781003450818-66

Tongjiaping is located in the transition zone between Wuxia and Xiling Gorge of the Yangtze River, belonging to the structural erosion of the middle and low mountain gorge geomorphology unit. The highest point is 424 m outside the landslide, and the lowest point is 67 m along the riverfront of the Tongjiaping landslide, with a relative elevation difference of 357 m. The flow direction of the Yangtze River in the test area is N40°E, and the cross-section of the valley is V-shaped. The river surface is 350–500 m wide and the water level is 66–86 m. In the survey area, the overall strike is N40°E, and several nearly EW and SN trending gullies are developed on the bank slope. The gully cutting depth varies from 3 to 15 m, which makes the gullies crisscross in the area. The overall terrain is high in the south, low in the north, high in the west, and low in the east. The landslide is a combination of bedding rock and soil. The front shear zone is exposed at the lower part of the slope, and the boundary on both sides is mainly controlled by the ridge of chessboard outlying bedrock extending in the direction of 250°~290° and the Zichang groove extending in the direction of 310°~315°. The boundary is clear with typical landslide landform characteristics. The sliding body material is (T_2b_2) block fissure, block stone soil, and silty clay crushed stone. The block fissure is mainly distributed in the middle and lower part of the sliding body, with the outcropping in the front and north Zichang groove. The rocky soil is mainly distributed in the posterior margin and the middle and upper layers of the sliding body. Silty clay crushed stone is mainly distributed in front and surface of the sliding body. The landslide mainly slides along the bedrock layer, and the soil structure of the slip zone is dense too dense, and the breccias trapped are mostly in the form of sub-angular to sub-circular shape. The slip bed consists of (T_2b_2) purplish red silty mudstone and gray-green calcareous mudstone, which are soft and medium-thick layered clastic rocks, belonging to easy slip strata.

According to the drilling data and ground mapping data, Tongjiaping landslide material can be divided into three layers: the silty clay gravel layer, the stone soil layer, and the block cracked rock layer. The silty clay crushed stone layer on the surface is mostly distributed on the gentle slope of the landslide front platform, with an elevation of 100–160 m, and the central axis can be extended upward to 220 m. In addition, the gentle slope angle of 260–280 m is scattered on the slope. The rock layer is mostly distributed in an elevation of more than 160 m, and the elevation of more than 220 m on the auxiliary line, is mostly distributed in the slope with a relatively steep slope. The fragmentation rocks are found in the steep slope section 100 m below the front elevation and Zichang gully at the northern boundary of the landslide. The material composition and structure of the slip zone are controlled by the location of the slip body and the strata in which the slip zone develops. Due to the short moving distance of the landslide front, the extrusion milling effect is weaker than that of the middle and rear area, and the content of gravel is significantly increased. The material composition of the sliding zone is mainly gravel breccia, and the ratio of soil to rock is 3:7–4:6. The grinding degree of the gravel in the upper part of the sliding zone is worse than that in the middle and rear area. The ratio of soil to rock in the middle and rear slip zone is 7:3–6:4, and the content of clay is higher. According to the borehole data, due to the difference in lithology in the development area of the slip zone, the composition and thickness of slip zone soil are different, the lithology is relatively weak and the weathering degree is high. (T_2b_2) Purple silty mudstone and mudstone stratum slip zone soil material is mainly silty clay, followed by gravel. The ratio of soil to rock is generally 7:3–6:4 with a thickness of 1.36–2.1 m, while the lithology is relatively hard, and the degree of weathering is relatively weak gray-green calcareous mudstone stratum in the slip zone soil gravel content increased significantly. The ratio of soil to rock in the slip zone soil is 4:6–5:5, and the thickness of 0.24–0.85 m.

The soil material of the Tongjiaping landslide is mainly silty clay mixed with silty clay, and a few parts are gravel mixed with silty clay. The slip-zone soil is silty clay with a soil-rock ratio of 7:3–6:4. The soil is purplish-pink silty clay, which is extruded and grounded. The structure is dense and delicate, with a small scratch smooth surface, and the gravel is sub-angular and sub-rounded. The slip-zone soil is composed of gray-green calcareous mudstone with a ratio of 3:7–5:5. The silty clay is gray-green and viscous, and the gravel and blithe composition is (T_2b_2) gray-green calcareous mudstone. The physical and mechanical properties of slip soil are shown in Table 1.

Table 1. List of test results of physical and mechanical properties of Tongjiaping slip soil.

Sampling position		Depth of sampling (m)	Water content (%)	Severe (KN/m^3) Natural	Dry	The proportion	Degree of saturation (%)	Index of plasticity (%)	Fluid property index
Eastern trailing edge of the landslide	ZK1	22.65–22.78	15.4	22.2	19.2	2.77	97.0	7.6	0.197
		23.36–23.60	6.4	21.6	20.3	2.73	50.7	7.8	<0
Midline front of the landslide	ZK2	45.14–45.34	8.6	23.1	21.3	2.75	80.8	6.4	<0
		46.37–46.63	16.7	22.0	18.9	2.74	100	8.9	<0
	ZK4	33.29–33.56	18.3	21.6	18.3	2.73	100	8.6	<0
	ZK6	35.47–35.67	10.4	21.5	19.5	2.77	68.2	17.1	<0
	ZK3	34.96–35.13	8.1	20.8	19.2	2.77	51.0	7.4	<0
West of landslide front	ZK7	21.22–22.45						5.9	<0
	ZK5	22.10–22.30	11.9	22.0	19.7	2.77	80.6	7.0	<0

2 FORMATION MECHANISM AND EVOLUTION PROCESS

According to Table 1, in terms of the regional structure, the test area has experienced several periods of tectonic movement with strong fractures. In the Yanshan period, linear anticline and syncline were formed. The anticline was tight and the syncline was wide and gentle. In the middle and late period of the middle Pleistocene, the western Hubei area was strongly uplifted and the river was cut down, forming a high and steep open surface near the modern Yangtze River. The sloped body produced elastic-plastic deformation under the action of self-weight and formed a tensile surface at the rear edge. With the gradual accumulation of elastic-plastic strain of the slope body, the tensile surface at the rear edge was connected with the shear surface at the front edge, resulting in sudden failure and the formation of a landslide.

2.1 *Master control conditions and influence factors of landslide formation*

Formation lithology: formation lithology is the material condition of landslide formation, and TongJiaPing landslide develops at the second paragraph of the badong group (T_2b_2) in a thick layer of purple pigments sandy mudstone, argillaceous siltstone, siltstone, celadon calcareous mudstone, on both sides of the occurrence of rock mass 350 ° to 10 ° < 26 ° to 60°, trailing occurrence 26° 320°. Lithology is relatively weak, and the rock has weak resistance to weathering, low mechanical strength, and is easy to soften or muddy when exposed to water. It is a typical slippery stratum, which provides material conditions for the generation of Tongjiaping landslide.

Geological structure: structure is often the weak part of the rock mass strength, and the growth of structural fracture damages the integrity of the rock mass, which provides channels for groundwater flow, making the deep development of weathering. There are several groups of structural fractures in the Tongjiaping landslide area, among which the bedrock in the middle and rear of the test area is consistent with the slope inclination, which provides favorable conditions for the formation of the slip zone. The Sn-trending and EW-trending cracks control the boundary of both sides of the slide body, and the NE-trending cracks provide the dominant cutting surface for the formation of the landslide trailing edge.

Conditions of terrain: refer to terrain conditions in the late Pleistocene to the end of the western Hubei mountain uplift stage. The Yangtze river formed deep canyon terrain in, Changjiang badong area during this period, which formed the multilevel bedrock and eroded terrace for the formation of the international airport.

Climatic conditions: refer to climate conditions in the western Hubei area late Pleistocene to the mid and late Pleistocene in the late quaternary third interglacial and interglacial stage 4. Warm and humid climate, abundant rainfall, and various strong natural weathering battalion force formed to endanger the stability of the slope of the environmental conditions.

2.2 *Analysis of formation mechanism of landslide*

Tongjiaping landslide area is located at the top of the Yangtze River coming out of Wu Gorge. The continuous downward cutting of the Yangtze River forms a high and steep free surface, which releases the accumulated stress in the slope and causes the rock mass to expand and rebound toward the free surface. Meanwhile, new stress concentration is generated near the free surface, especially at the shoulder and foot of the slope. Under the long-term action of stress dominated by the dead weight of the slope, a shear fracture surface is formed along the shear stress concentration zone at the foot of the slope along the slope body. It is exposed at the foot of the slope, showing continuous peristaltic deformation. Under the combined action of groundwater and weathering, the cracks in the shallow part of the slope (surface) along the stress concentration zone continue to develop to the deep and both sides. With the continuous development of deformation, the rock and soil mass in the upper part of the slope gradually relax, and the slope deformation enters the progressive failure stage. The continuous infiltration of surface water, the continuous intensification of groundwater runoff activities, the continuous expansion of the tracking plane, and the surface deformation tend to be obvious. Under the action of prolonged rain or heavy rain, the trailing fissure fills with water, generating hydrostatic pressure. Meanwhile, the rock mass in the slip zone softens, and its shear strength decreases rapidly. Due to the self-weight of the slope body, it is concentrated on the unclipped convex body, which causes its failure and deformation, and finally leads to the overall landslide.

3 MATERIALS AND METHODS

Considering the current stable state of the landslide, the design of the backwater level combined with the Three Gorges reservoir and the geological structure location of the landslide, and various working conditions were formulated as follows: natural state (earthquake), natural state (135 m water level), natural state (135 m water level; earthquake), natural state (175 m water level), natural state (175 m water level; earthquake [10]). In Figure 1, it shows the noodles:

The unbalanced thrust method is the main calculation method, and the general strip method is used for comparison [7]. We can see Figure 1 for the calculation of section strips. The calculation of the unbalanced thrust Method [11] is as Equation (1) and Equation (2):

$$P_i = W_i \sin \alpha_i - \frac{1}{K}\left\{ {}_iL + \begin{bmatrix} W_i \cos \alpha_i + P_{i-1} \sin(\alpha_{i-1} - \alpha_i) \\ -U_i - PD_i \sin \alpha_i \end{bmatrix} tg\varphi_i \right\} + P_{i-1} \cos(\alpha_{i-1} - \alpha_i) \tag{1}$$

$$F.O.S = \frac{C_n L_n + [W_n \cos \alpha_n + P_{n-1} \sin(\alpha_{n-1} - \alpha_n) - U_n - PD_n \sin \alpha_n] tg\varphi_n}{W_n \sin \alpha_n + P_{n-1} \cos(\alpha_{n-1} - \alpha_n)} \tag{2}$$

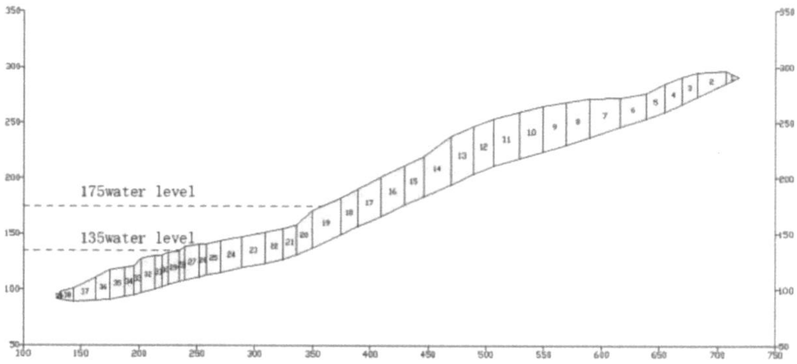

Figure1. Section bar graph.

Where P_i denotes the thrust force of the i block (KN), W_i denotes the dead weight of block i (KN/m^3), U_i denotes the water pressure of the sliding surface in Section i (KN), PD_i denotes the seismic horizontal force of block i (KN), C_i denotes the cohesive force of sliding surface in Section i (KPa), Φ_i denotes internal friction Angle (°) of sliding surface in Section i, L_i denotes the length of sliding surface in Section i (m), α_i denotes slope (°) of the sliding surface in block i.

Table 2. List of calculation parameters of Tongjiaping landslide.

	Overall calculation parameters and parameter values	Overall calculation parameters and parameter values	Overall calculation parameters and parameter values
State of nature	C: 55 φ: 18.5	C: 55 φ: 18.5	C: 55 φ: 18.5
State of fullness	C: 47.3 φ: 15.91	C: 47.6 φ: 16.0	C: 47.6 φ:16.0

According to Table 2, it is calculated in Equation (1) and Equation (2). The stability calculation results are shown in Table 3:

Table 3. Calculation result table.

Condition of construction	Factor of safety	Factor of safety	Factor of safety
Natural	1.27	1.25	1.21
Natural + Earthquake	1.13	1.11	1.00
Natural +135 m	1.16	1.12	1.18
Natural +135 m+ earthquake	1.01	0.99	1.04
Natural +175 m	1.17	1.15	1.15
Natural +175 m+ earthquake	1.00	1.00	1.00

4 CONCLUSIONS

4.1 *Comprehensive evaluation of landslide stability*

According to Table 3, the stability evaluation under various working conditions is as follows: under the natural state (the Yangtze River water level is 67 m), the stability coefficient of the Tongjiaping landslide is 1.21–1.27, indicating good overall stability [11]. Under the natural condition (the water level of the Yangtze River is 67 m) and considering the influence of earthquakes (according to VI), the stability coefficient of Tongjiaping landslide is 1.00–1.13, indicating that the overall stability of the landslide is good. Under the natural condition, considering the influence of 135 backwaters, the stability coefficient of the Tongjiaping landslide is 1.12–1.18, indicating the overall stability of the landslide is good. Under the natural condition, the stability coefficient of the Tongjiaping landslide is 1.00–1.04, considering the influence of 135 backwaters and earthquakes (according to VI). The whole landslide is stable. Under the natural condition, considering the influence of backwater 175, the stability coefficient of the Tongjiaping landslide is 1.15–1.17, indicating the overall stability of the landslide is good. Under the natural condition, the stability coefficient of the Tongjiaping landslide is 1.00–1.05, which is in the limit equilibrium state, after considering 175 backwaters and adding the earthquake factor at the same time.

Tongjiaping landslide accumulation body has been formed for a long time. Through the comprehensive analysis of the landform, material composition, and stability calculation results of the accumulation body, the results show that the landslide body is in a stable state in the natural state. Although the local section has a small range of deformation, the overall slip is less likely to occur again in the natural state. After the backwater of 135 and 175 of the three Gorges project, the overall stability of the landslide is good without considering the superposition earthquake factor. The landslide stability is better when earthquake factors are superimposed. The test results under different working conditions show that landslide is sensitive to water levels and earthquakes, and their sensitivity to an earthquake is higher than that of water. With the rise of groundwater level and the superposition of earthquake conditions, the landslide safety factor decreases rapidly, but the overall stability of the landslide is good, and the overall slip possibility is small.

Tongjiaping in Badong County is 75 km away from the Sandou Ping dam site. The sliding body is currently in a stable state. After the impoundment of the reservoir, part of the sliding body is underwater, and the slip distance is short. As the overall slip probability is small, manifested as shallow and local slip, it has little influence on the reservoir capacity. Landslides and bank collapse will not constitute the main source of reservoir sediment and have no substantial influence on the storage capacity and life of the reservoir. After the formation of the reservoir, the width and depth of the water surface are greatly increased, the channel is greatly improved, and the landslide and bank collapse will not cause serious obstruction to navigation.

Tongjiaping landslide is a rock and soil mixed bedding sliding landslide with a large thickness of loose soil at the surface. After impoundment of the reservoir, under the action of immersion and scouring of the reservoir water, it has a high probability of inducing shallow landslide. Once sliding occurs, the affected objects are mostly permanent buildings. Tongjiaping landslide is located at the exit of Wuxia. The bank slope is facing the current and rushing, and the surface of the slope is loose. The slope is high and steep at the height of 135–200 m, and the natural conditions of the reservoir bank are poor. It is not difficult to see from the prediction of bank reconstruction that the bank reconstruction is strong in the Tongjiaping landslide area. Under the condition of water levels 135 and 175, there is strong reservoir bank scour transformation, and the sliding body material will be carried away by the wave reservoir water in large quantities, thus forming wave erosion concave, steepening the slope, and inducing shallow slope deformation and failure.

The new town of Badong County was first relocated to the loess slope. With the discovery of the landslide of the Loess slope and the fact that the whole county is located on the slope of the Yangtze River, the construction land is in shortage. The new town is planned to develop westward to the west soil slope, forming five groups: Huangsoil slope, Daling slope, Baitu slope, and Yuntuo and Xilang Slope. West to East power transmission Wan (county) Yi (Chang) of 500,000 volts high voltage transmission lines passed above the landslide. Once the landslide occurs, it will directly endanger the life and property safety of the residents on the slide and the leading edge of the tourist wharf, and it is estimated that the direct economic loss will be 4 million yuan. The surface of the eastern side of the landslide rear boundary is covered by landslide deposits, the underlying bedrock is strongly weathered and the rock structure is broken. After the landslide occurs, traction and deformation may also occur, endangering the 500,000–volt high-voltage transmission tower. In addition, the luxury cruise ship docked at the wharf will have huge potential harm and cause huge economic losses and social impact. Once the Tongjiaping landslide declines, it will not only seriously affect the construction and development of Xinling Town and the process of immigration, but also adversely affect the reform and opening up of Badong County and the sustainable economic development. Therefore, the Tongjiaping landslide comprehensive treatment is imminent and necessary, with important social significance and realistic economic significance.

4.2 *Suggestions*

Since the Tongjiaping landslide is in a stable state at present, the control project should be based on the principle of strengthening the stability of the landslide. The prevention and control project plan for the landslide area should be unified consideration and comprehensive treatment. According to the different topographic conditions, formation mechanism, the material composition of the sliding body and its potential influence degree, and the emphasis of the prevention plan should be different. At the same time, the prevention and control project should be based on the principle of combining town planning, environmental protection, bank protection, economic and practical, as far as possible in the implementation of the prevention and control project, which can improve the environment around the new city. According to the field investigation data and indoor stability calculation results, the specific suggestions for the prevention and control project are as follows:

(1) The feasibility study of the prevention and control project should fully consider the harm and severity of the landslide, and determine a reliable prevention and control plan based on a comprehensive and comprehensive analysis of the causes of the landslide and geological characteristics.
(2) Since the overall stability of Tongjiaping landslide is good, considering the loose material structure of the middle and rear sliding body of Tongjiaping landslide, the large slope and sliding surface, the easy occurrence of reservoir bank reconstruction and shallow soil slide, and other characteristics, Tongjiaping landslide should focus on reservoir bank protection.
(3) The reservoir bank protection project is mainly arranged in the 130–180 m zone to prevent the large-scale collapse of the reservoir bank after the backwater of the reservoir area and to reduce the adverse impact of the reservoir bank reconstruction on the rear of the sliding body. It is suggested that the reservoir bank protection should be mainly reinforced concrete lattice, anchor cable anchoring, slurry slope protection, and surface drainage.

REFERENCES

[1] Bishop A. W. The Use of the Slip Circle in the Stability Analysis of Slopes [J]. *Géotechnique*, 1955 (1): 7–17.

[2] Chen Zuyu. General Section Method for Soil Slope Stability Analysis and its Improvement [J]. *Journal of Rock and Soil Engineering*, 1983, 5(4): 11–27.

[3] Fellenius W., Erdstatische Berechnungen, Mit Reibung, Kohsion (Adhsion), and Unter Annahme Kreiszy Lindrischer Gleitflchen. 1927.
[4] Guo M. W., Ge X. R., and Wang S. I. Slope Stability Analysis Under Seismic Load by Vector Sum Analysis Method [J]. *Journal of Rock Mechanics and Geotechnical Engineering*, 2011, 3 (3): 282–288. J.
[5] Janbu N. Slope Stability Computations [J]. *Embankment Dam Engineering Casagrande* Volume, 1973.
[6] Lee W.A., Lee T. S., Sharma S., et al. *Slope Stability and Stabilization Method* [M]. New York: Wiley-Interscience Publication, 1996.
[7] Lin Feng and Huang Runqiu. Discussion on the Limit Equilibrium Strip Division Method for Slope Stability. *Geological Hazards and Environmental Protection*, 1997, 8(4): 9–13.
[8] Shi Weimin, Zheng Yingren, Tang Boming, et al. Precision of Unbalanced Thrust Method for Slope Stability Analysis and Application Conditions [J]. *Chinese Journal of Geotechnical Engineering*, 2004, 26(3): 313–317. (in Chinese)
[9] Zhang Luyu, Zheng Ying, and Shi Weimin. On Unbalance Push in Slope Stability Analysis Discussion on force method [J]. *Chinese Journal of Rock Mechanics and Engineering*, 2005, 24(1): 177–182.
[10] Zheng Ying was Born to Wei People at the Time. Problems Needing Attention in the Application of Unbalanced Thrust Method [J]. *Heavy Qing Architecture and Architecture*, 2004, 3(2): 6 – 8.
[11] Zheng Yingren, Shi Weimin, and Yang Mingcheng. Discussion on Unbalanced Thrust Method and Sarma Method [J]. *Chinese Journal of Rock Mechanics and Engineering*, 2004, 23(17): 3030–3036.

Structural seismic design and safety assessment

Water Conservancy and Civil Construction – Oke & Ahmad (Eds)

Mechanical characteristics and slope stability of deposit slope excavated by highway construction under rainfall

Yong Chen, Wei Wang*, Dongbo Cai, Zirui Li, Qihao Chen, Guoliang Zhang & Kun Li
The Seventh Engineering Co., Ltd. of CFHEC, Zhengzhou, China

ABSTRACT: Earthworks will regularly accumulate excavated soil sites to form deposit slopes, which is easily induced slope instability disaster under the action of rainfall, causing safety risks and affecting construction progress. In this paper, the shear strength parameters and deformation characteristics of the soil were obtained through the triaxial test by taking the deposit slope soil of a road project. Then, based on finite-state software (ABAQUS), the deposit slope was analyzed using the strength reduction method. The stability of the deposit slope under rainfall action was carried out based on the local rainfall characteristics. The research results showed that the deposit slope stability coefficient of this construction section was high under the action of non-rainfall, but the slope instability was easily generated under the action of rainfall. In addition, the mechanical mechanism of slope instability of the deposit slope was accordingly conducted. The research results will have a positive role in the safety control of earthwork excavation in similar works.

1 INTRODUCTION

In general, the soil obtained from an excavation in the project is stockpiled and used for backfilling of later foundation pits, subgrade, and other projects. The excavated soil originally has a natural water content, and the water content of the soil used for backfill will change due to rainfall accumulation and other factors. The water content will have an important effect on the strength of the soil, especially for collapsible loess. The natural loess itself has a certain structural strength, but the original structural strength of the backfilled loess is destroyed during excavation. If the water content changes, its strength will change more. Therefore, it has some engineering significance to study the effect of water content on the strength of backfilled soil.

The construction of highway projects will inevitably involve high excavation and deep filling, thus making them inevitably prone to the formation of deposit slopes and the creation of safety hazards (Yu & Zheng 2018; Zheng *et al.* 2009). Rainfall conditions have a significant contribution to slope stability (Alonso *et al.* 2003; Ng & Shi 1998; Pan *et al.* 2000). At present, much research has been conducted on the influence of rainfall on the slopes of deposit bodies. For example, Ng *et al.* (2001) studied the slope stability of the drainage field slopes under different rainfall intensities from the perspective of reliability. Herrada *et al.* (2014) studied the effect of different volumes of water content on hydraulic heads under rainfall conditions. Muntohar *et al.* (2014) studied the infiltration law of slope under the action of short-term strong rainfall and long-term weak rainfall. Most of the research mainly focuses on open-air residue-deposited soils, but research on the slope of collapsible loess excavation deposits is rare.

*Corresponding Author: zhongjiao_wang@126.com

DOI: 10.1201/9781003450818-67

In view of this, this study conducted some research by using a project of collapsible loess excavation and deposited slope as the background. The triaxial tests were conducted to obtain the mechanical parameters of the backfilled soil. And the numerical simulation was conducted to analyze the stability of the slope under the action of rainfall. The research results can provide references for similar projects.

2 ENGINEERING BACKGROUND

2.1 *Project introduction*

A road engineering project in Henan Province was taken as the project background in this study. The soil layer of this road project was mainly loess-like silt, silt, and silty clay and was uniformly distributed from top to bottom. No groundwater was found in the subgrade filling area. The loess-like soil had collapsibility with a grade of II. The soil quality was uniform, the upper surface layer was mostly loose due to human disturbance, and the bearing capacity was low. There were two high excavation sections in this road engineering project. The excavation and filling height of the road section, as shown in Figure 1, was the highest. The maximum height of the soil-filled slope was 18.6 m. In the research process of this paper, the three most unfavorable sections were taken as the slope stability analysis surface for the convenience of the modeling calculation.

Figure 1. Actual view and the calculated sections of a road embankment excavation and filling slope.

2.2 *Shear mechanical properties of the backfilled soils*

The newly deposited loess in this project had not reached the state of consolidation, and the whole presented a loose state; the pores were relatively large, and the strength parameters were susceptible to the influence of water. As a result, the triaxial test was used to evaluate the saturated remodeled soil. In the test, the consolidated undrained shear, i.e., the CU test, was used, and the shear rate was set to 0.08 mm/min. Figure 2 shows photos of the soil samples after the damage of this consolidated, undrained triaxial test of saturated loess. The test results showed that the soil sample damage described in this paper belonged to weakly softening type damage, and this form of damage generally occurred after the soil sample had experienced obvious elastic-plastic deformation, which was the result of further development of plastic deformation. The shear expansion effect during the test was greater than the shear contraction effect; thus, the stress–strain curve of the soil sample showed a weakly softening type.

In addition, the shear strength envelope of the soil sample was obtained by collating the test results, as shown in Figure 3. The strength parameters of the backfilled soil layer can be obtained by fitting as c = 30 kPa and φ = 26°.

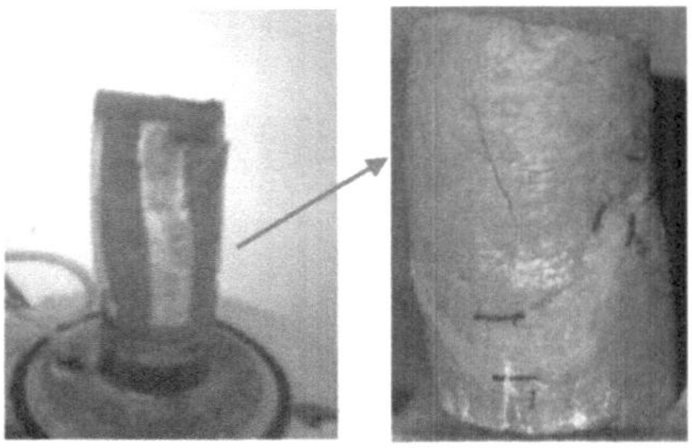

Figure 2. Soil sample after triaxial test damage.

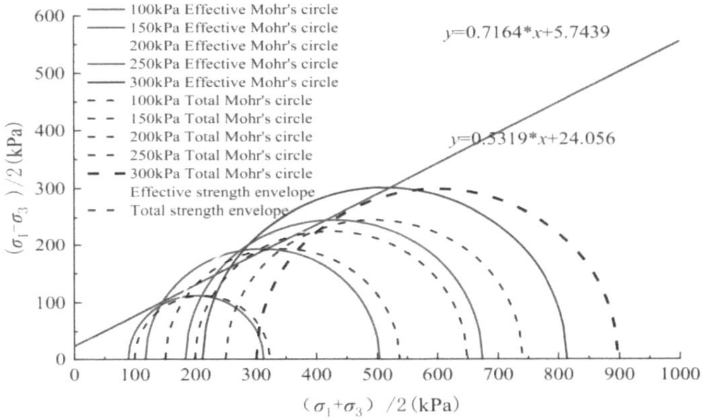

Figure 3. Molar circle and strength wrap.

3 NUMERICAL SIMULATION ANALYSIS

In this numerical simulation, the two-dimensional model was used to analyze the stability of the actual engineering slope. The size of the model was taken as 120 m × 30 m (length × width); the total height of the backfilled soil was 38 m; the slope ratio was 1:1.5. The height of the lower layer was 23 m, and the height of the upper layer was 15 m. The backfill platform in Section 1-1 was 33 m away from the foundation, and its width was 2 m. The backfill platform in Section 2-2 was 28 m away from the foundation. There was no backfill platform in Section 3-3. Figure 4 shows the finite element model of Section 1-1. The other two sections are similar to the model of this section. The Mohr–Coulomb model was used in numerical simulations. The soil parameters are shown in Table 1. In the simulation of a realistic rainfall situation, the rainfall intensity amplitude curve was selected for the free boundary. Therefore, the infiltration intensity was set to 0.02 m/h for the horizontal free surface and 0.0167 m/h for the slope surface, and the rainfall intensity amplitude curve is shown in Figure 5.

The CPE4 unit type was used in this model. The boundary conditions are set as follows: (1) The horizontal movement of the left and right sides is restricted, while the vertical movement is not limited, i.e., $u_1 = 0$; (2) the bottom boundary is completely fixed, i.e., $u_1 = 0$, $u_2 = 0$; (3) the top and slope surfaces are set as a free boundary. In the actual project, each soil layer is not completely horizontal, and the depth of each layer does not stay the same. However, for the convenience of calculation, the average thickness of each soil layer was used as the thickness of the soil layer in the simulation process of this paper.

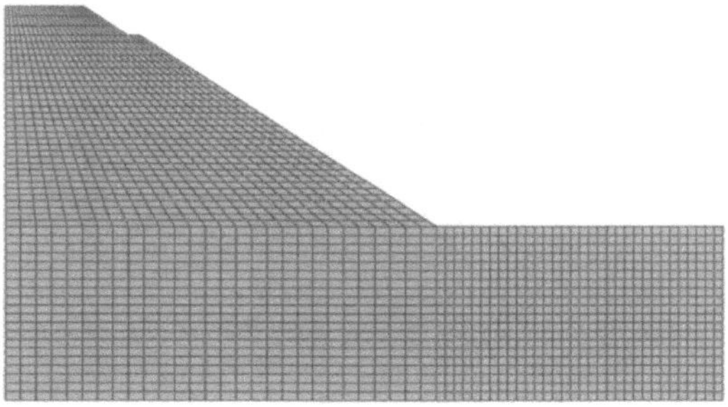

Figure 4. Finite element model diagram.

Table 1. Table of soil parameters.

Soil Type	Density g/m^3	Modulus of Elasticity MPa	Poisson's Ratio μ	Cohesion c/kPa	Internal Friction Angle φ/°
Backfilled Soil	1.62	80	0.26	30	26
Original Loess	1.87	90	0.31	32	28

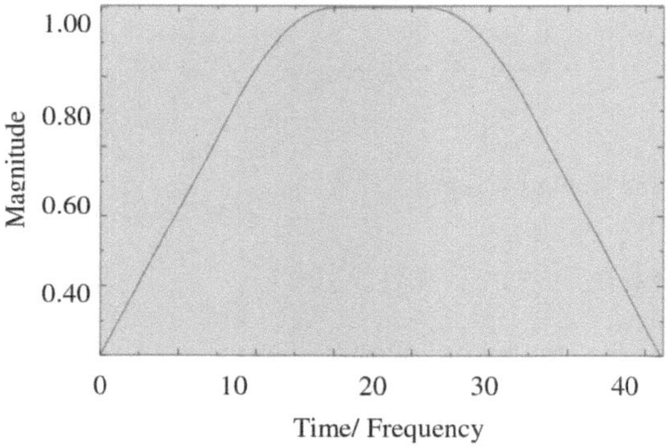

Figure 5. 72-h rainfall intensity amplitude curve.

In this paper, the stability of real engineering slopes was simulated by the finite element strength reduction method. The specific steps of the simulation included the following: (1) A field variable is defined, which is usually taken as the strength reduction factor F_r. (2) The material model parameters that vary with the field variable are defined. (3) The magnitude of the field variables is specified and a gravity load is applied to the model to establish the equilibrium stress state. To avoid damage at this point, F_r can be obtained small, e.g., $F_r < 1$, which amplifies the strength. (4) In the subsequent analysis step, the field variable F_r is linearly increased, and the results are processed after the calculation is stopped, i.e., the value does not converge, and the safety factor is determined according to the instability evaluation standard. In the analysis, two analysis steps were set separately. In the first analysis step, a static general analysis step was set to make the slope produce a small displacement under the action of dead weight and establish the equilibrium stress state. In the second analysis step, the strength reduction was started to reduce the material cohesion and internal friction angle, respectively, to make the slope unstable, and then the stability coefficient of the slope is analyzed.

4 ANALYSIS OF NUMERICAL RESULTS

4.1 *Stability of soil slope in the absence of rainfall*

Figures 6 and 7 show the stress clouds, strain clouds, and displacement clouds of the slope under the strength discounting method. The convergence of the numerical results and the inflection point of the displacement of the characteristic parts were used as the evaluation criteria to judge the instability of the slope. From Figure 7, it is clear that the slope safety factor Fs = 1.3925 for Section 1-1, Fs = 1.3964 for Section 2-2, and Fs = 1.36646 for Section 3-3. All the model safety factors of the three sections are greater than 1, so the slope is in a stable state. As Figure 8 shows, the displacements of all three sections occur mainly at the top of the slope, mainly in the vertical direction, i.e., the consolidation settlement caused by gravity.

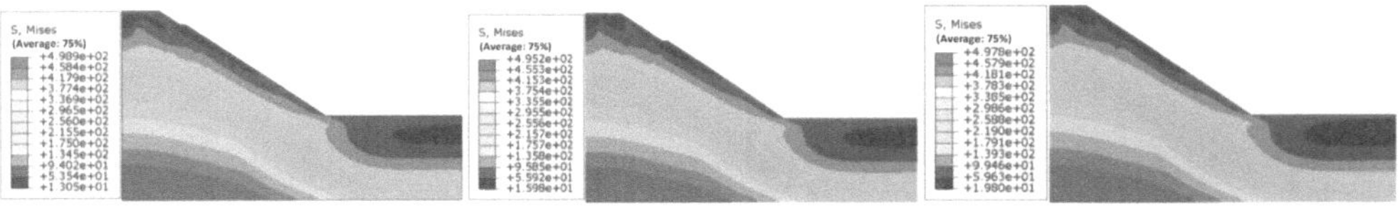

Figure 6. The stress cloud map of the calculated slope.

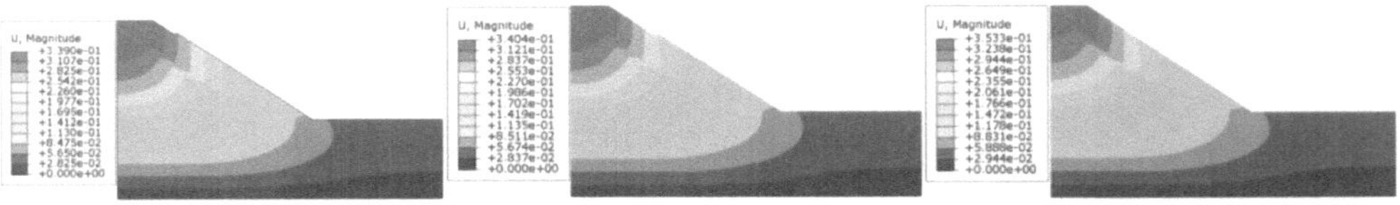

Figure 7. The displacement cloud map of the calculated slope.

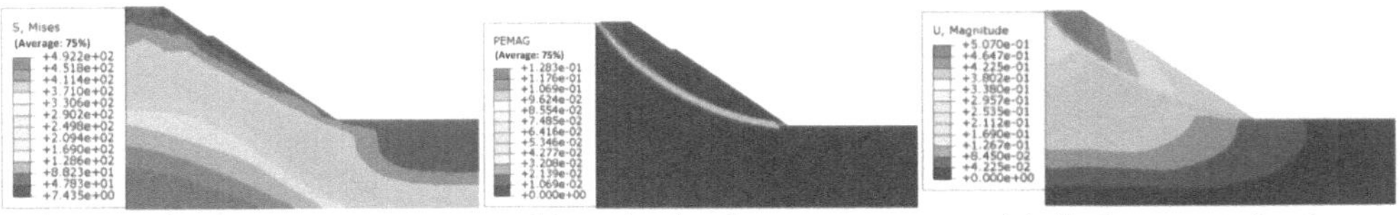

(a) stress cloud map (b) strain cloud map (c) displacement cloud map

Figure 8. Section 2-2 cross-sectional calculation results chart. (a) stress cloud map (b) strain cloud map (c) displacement cloud map.

4.2 *Stability of soil slope under rainfall conditions*

Figure 8 shows the stress cloud diagram, strain cloud diagram, and displacement cloud diagram of the slope (Section 2-2) after 72 h of rainfall. It can be seen from the figure that the stress in the middle and lower parts of the slope is still elevated to a certain extent due to internal rainwater transport. Besides, the stress concentration area at the foot of the slope is expanding with the passage of time, resulting in a larger stress concentration. The section's strain cloud diagram and displacement cloud diagram show an obvious horizontal displacement at the sliding surface, indicating that the slope was unstable. It can be clearly seen that the slope after rainfall appears to have an obvious sliding surface at a certain depth to form a circular arc area with a heterogeneous distribution of shear stress.

Figure 9 shows the plot of the reduction factor and horizontal displacement under this working condition. Similarly, the convergence of the numerical results and the inflection point of the displacement of the characteristic parts are used as the evaluation criteria to judge the instability of the slope. The calculation model does not converge when t = 1.47484, and the reduction coefficient *Fs* = 1.21225. The left side of the free surface of the slope is taken as the characteristic point, and the reduction coefficient and horizontal displacement are plotted at this point. It can be seen from Figure 9 that there is an inflection point near t = 1.47484, which means the slope was unstable.

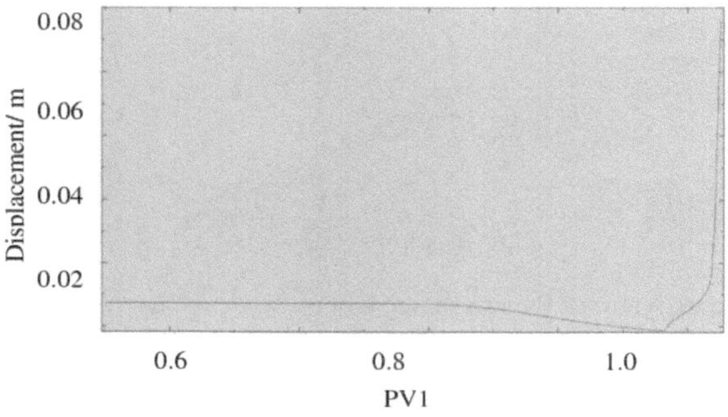

Figure 9. Section 2-2 discount factor and horizontal displacement diagram.

The calculation results of the plastic zone under the above conditions are shown in Figure 10. It is obvious to see the distribution of the plastic zone along the foot of the slope upward through and with a large area. This indicates that the slope was in a stable or less stable state without rainfall but was not destabilized; when experiencing drought conditions, the slope finally became destabilized. According to the set rainfall duration, the whole foot of the slope basically yielded at the end of the rainfall, and the plastic zone still developed continuously after the rainfall ended and finally penetrated, forming a landslide.

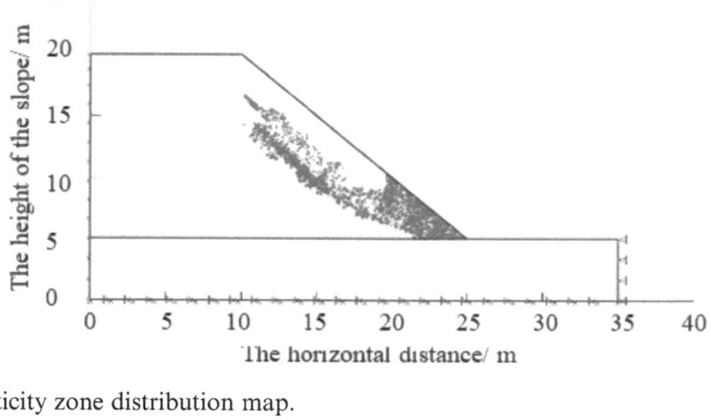

Figure 10. Plasticity zone distribution map.

5 CONCLUSIONS

Based on an excavation filling project, the stability of the backfilled slope was studied based on field investigation, indoor testing, and numerical simulation analysis. The following conclusions are obtained:

(1) The compression results of the backfilled loess showed that the deformation damage was weakening and was the result of the further development of plastic deformation. The strength of this type of soil obviously decreased with the increase in water content. Therefore, attention should be paid to rainfall in practical engineering to avoid its influence on soil stability.
(2) Under the non-rainfall effect, the slope stability coefficient of this construction section of the mound was high, but the slope instability was easy to occur under the rainfall effect.
(3) Under the action of rainfall, the plastic zone of the slope started to appear from the foot of the slope, and the whole foot part of the slope basically yielded at the end of the rainfall.

REFERENCES

Alonso, E. E., Gens, A., Delahaye, C. H., "Influence of Rainfall on the Deformation and Stability of a Slope in Overconsolidated Clays: a Case Study," *Hydrogeology Journal*, 11(1): 174–192, (2003).

Herrada, M. A., Gutiérrez-Martin, A., Montanero, J. M., "Modeling Infiltration Rates in a Saturated/Unsaturated Soil Under the Free Draining Condition," *Journal of Hydrology*, 515: 10–15 (2014).

Muntohar, A., "Effect of Rainfall Intensity and Initial Matric Suction on the Stability of Residuals Soils Slope," 18th Annual Conference on Geotechnical Engineering (PIT HATTI ke-18) (2014).

Ng, C., Shi, Q., "Influence of Rainfall Intensity and Duration on Slope Stability in Unsaturated Soils," *Quarterly Journal of Engineering Geology & Hydrogeology*, 131(2): 105–113 (1998).

Ng, C.W., Wang, B., Tung, Y. K., "Three-dimensional Numerical Investigations of Groundwater Responses in an Unsaturated Slope Subjected to Various Rainfall Patterns," *Canadian Geotechnical Journal*, 38(5): 1049–1062 (2001).

Pan, Y., Wu, G., Zhao, ZM., "Analysis of Rock Slope Stability Under Rainfall Conditions Considering the Water-induced Weakening of Rock," *Computers and Geotechnics*, 128(2): 103806 (2000).

Yu, Z. L., Zheng, X.G., "Research on Evaluation of Operation Safety of Expressway Tunnel in Mountainous Areas Based on Fuzzy Synthetic Evaluation Method," *Technology of Highway and Transport* (2018).

Zheng, J.L., Rui, Z., Yang, H.P., "Highway Subgrade Construction in Expansive Soil Areas," *Journal of Materials in Civil Engineering*, 21(4): 154–162 (2009).

Water Conservancy and Civil Construction – Oke & Ahmad (Eds)
© 2024 The Author(s), ISBN: 978-1-032-58614-4

Analysis of track-ground interactions

Zhanjun Niu & Changting Li*
Vehicle Application Department, Army Academy of Armored Forces, School of Noncommissioned Officers, Changchun, Jinlin, China

ABSTRACT: In view of the interaction between the track and the ground of special vehicles, this paper systematically combs the different models of the interaction between the track and the ground, introduces the classification and characteristics of the soil, analyzes the research methods of the interaction between the track and the ground, and describes the characteristics of the interaction between the track and the ground, laying the foundation for the next step of research on the high mobility of tracked vehicles.

1 INTRODUCTION

Tracked vehicles have good off-road mobility and are suitable for driving on soft ground in the wild, as well as on special ground such as sand, snow, and mud. Therefore, they are widely used in the modern military, engineering machinery, construction, and other fields and play an important role. In the modern military field, tracked vehicles need to have high mobility and be able to drive at high speed on complex road conditions such as roads, earth roads, gravel roads, and even Gobi roads. In order to deeply analyze the high trafficability of tracked vehicles and give full play to their high mobility, it is necessary to systematically study the interaction between tracks and the ground. In view of this, researchers have conducted a multi-directional study on the interaction between the track and the ground. Jiang Wenfeng designed the ground adhesion coefficient test program and determined the relationship between the ground adhesion coefficient and the pavement slip rate by using theoretical analysis; Li Jun et al. developed a three-dimensional model of the interaction between the track and the ground by using the Mckyes-Ali three-dimensional model and combining the research methods of Baker and Nobutaka; and Yang Congbin investigated the adhesion characteristics of tracked vehicles. After a series of tests and model derivations, he finally proposed the optimization scheme of track shoes on different typical surfaces; Hu Jiyong built a virtual prototype model of a heavy-duty track by using the dynamics simulation software Recurdyn and obtained the support reaction force of the track under different working conditions.

The above researchers mainly conducted in-depth research on one aspect of different models of track-ground interaction. On the basis of the existing research, this paper will systematically discuss and analyze the relationship between the track and the ground, which will lay the foundation for the next round of research on the high mobility of tracked vehicles.

*Corresponding Author: leect1991@163.com

DOI: 10.1201/9781003450818-68

2 SOIL CLASSIFICATION AND CHARACTERISTICS

The traction performance of tracked vehicles is affected by many factors, among which different types of ground soil are important factors affecting its traction performance, so it is necessary to classify the soil according to its characteristics. In addition, soil bearing characteristics and shear characteristics are the main indicators of soil performance evaluation and the main soil characteristic factors affecting the driving performance of tracked vehicles. Therefore, the study of soil classification and soil characteristics is a theoretical basis for the study of the interaction between tracks and the ground.

2.1 *Soil classification*

In the field of vehicle engineering, the study of soil mechanics will be involved, in which the constant characteristics of soil are defined, which means that no matter what state the vehicle will be in during driving, those characteristics of soil will not change significantly. In a short period of time, environmental factors have little impact on the particle diameter of soil, which is generally considered to be unchanged. Therefore, the particle diameter with qualitative change will be taken as the boundary size of soil particles in the project, and the soil will be divided into several particle groups according to this size, as shown in Table 1.

Table 1. Classification of granular groups.

Granule group name	Gravel	Coarse Sand	Fine Sand	loam	Clay	Colloidal Particles
Particle Diameter/ mm	>2	2–0.5	0.5–0.05	0.05–0.005	<0.005	<0.001

According to the different diameters of soil particles, the soil can be divided into different types, and the properties of each type of soil are also different. For example, the mineral composition of soil within the range of gravel and sand has little influence on soil properties, but the particle diameter has a greater influence on soil properties. The smaller the particle diameter is, the greater the binding force between soil particles is.

2.2 *Soil-Bearing characteristics*

When a tracked vehicle travels on the soil, its gravity and the driving effect of the driving wheel will make the track contact the soil and generate a certain pressure on the soil. Because of this pressure, the soil will have a certain amount of subsidence, and the soil will have a supporting force on the track, which is the performance of the soil-bearing characteristics. Many researchers have studied the bearing characteristics of soil, among which the bearing model proposed by American scholar M.G. Bekker is widely used, and the relationship between pressure and subsidence is shown in Formula 1.

$$\sigma = \left(\frac{k_c}{b} + k_\phi\right) \cdot z^n \tag{1}$$

Where σ is the soil normal stress, kPa; K_c is the cohesive modulus of soil, kN/ m^{n+1}; k_φ is the soil friction modulus, kN/ m^{n+2}; b is the width of the short side of the track shoe, m; z is the soil subsidence, mm; n is the soil deformation index.

This mathematical model is based on the semi-empirical formula obtained from a large number of press plate tests. In practice, the mathematical model can be well matched with the test data, indicating that the mathematical model has a very important application value.

Where kc, k_φ, and N are the three parameters that affect the relationship between soil pressure and subsidence, representing the bearing capacity of the soil.

2.3 *Soil shear properties*

When the tracked vehicle is traveling on the soil, the surface of the track and the track teeth will squeeze and shear the soil, causing the soil to shear displacement. At the same time, shear stress is generated. Shear stress is an important factor affecting the tracked vehicle's ability to pass through the soft ground. The soil shear strength is characterized by the relationship between the soil shear stress and shear displacement.

The common soft ground is divided into two typical situations. One is fragile ground, including compacted sand, silt, soil, frozen snow, and marsh; the other is plastic ground, including loose sand, saturated dry soil, and dry snow. The relationship between shear stress and the deformation characteristic curve of these two kinds of ground is shown in Figure 1.

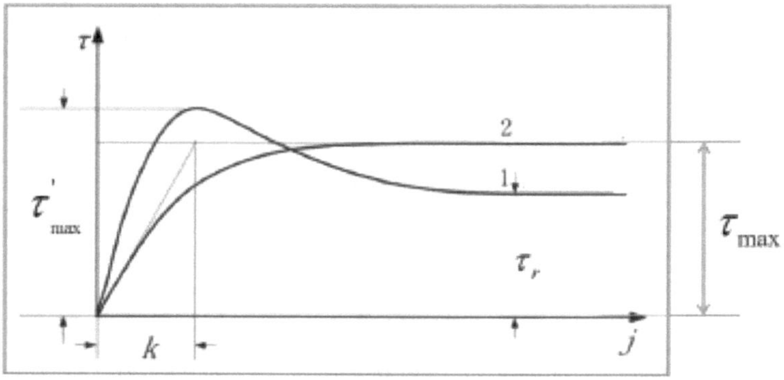

Figure 1. Relation curve between ground shear stress and shear deformation.

It can be seen from Figure 1 that the brittle ground will have the maximum shear stress τ_{max} as well as an obvious "hump," and residual stress will appear after the ultimate yield, making the curve tend to be flat; however, the plastic ground's shear stress will increase at first, and then the curve will tend to be gentle, without the "hump" phenomenon.

Generally speaking, Janosi believed that the most common ground was plastic, which can be expressed by Formula 2.

$$\tau = \tau_{max}\left(1 - e^{-\frac{j}{k}}\right) \tag{2}$$

Where τ is the shear stress, kPa; τ_{max} is the maximum shear stress, kPa; j is the shear displacement, m; k is the shear modulus, m.

3 RESEARCH METHODS OF TRACK GROUND INTERACTION

3.1 *Semi-empirical model analysis method*

Bekker is representative of the semi-empirical model analysis method. Through the mechanical analysis of the interaction between the track and the ground and supported by certain test data, an approximate simplified formula for the interaction between the track and the ground is established.

In the later period, Beck combined the semi-empirical formula with the relevant formula of civil engineering on the basis of previous studies. After long-term verification and reasoning, he found that this semi-empirical method can well represent the physical characteristics of the ground, and the relationship between the track and the ground can be represented by an appropriate model, which provides a certain theoretical basis for studying the interaction between the tracked vehicle and the ground, and is likely to be widely used by future generations.

3.2 *Simulation test method*

The simulation test method is based on the relevant theory of ground mechanics and uses simulation software to establish and analyze the interaction model between the track and the ground.

With the rapid development of computer technology and algorithms, many problems that are difficult to solve by theoretical or experimental methods can be solved by numerical simulation. At present, finite element simulation software is widely used in structural stress analysis, model numerical simulation, numerical simulation calculation, and other aspects. Multi-body dynamics, track, and ground dynamics models are among the key technologies of tracked vehicle dynamics simulation. Using the finite element simulation software to establish a suitable model of the interaction between the track and the ground can more comprehensively analyze the stress situation, providing strong support for an in-depth study of the interaction between the track and the ground.

3.3 *Empirical methods*

The cone index method of the US Army Engineering Corps' waterway test station is a typical representative of the empirical method (WES method for short). The conical index method is mainly used to conduct a large number of tests on existing vehicles and establish a certain relationship with soil strength characteristics, thus obtaining a limited empirical relationship, so this method cannot be applied to new vehicles. Because this method has no theoretical support and is only applicable to some specific environments, the results are not always consistent with the test.

For the research methods of track-ground interaction, the above three methods are relatively common, and the simulation test method is the most widely used at present. In addition to the above three methods, there are other rare research methods, such as basic theoretical methods and model methods.

4 INTERACTION BETWEEN TRACK AND GROUND

4.1 *Characteristics of interaction between track and ground*

4.1.1 *Similarities and differences between track and wheel and ground*

Compared with the wheels of wheeled vehicles, the ground area of the track rarely changes with the sinking of the vehicle and is not sensitive to the change of load or interface shear stress. Therefore, the soil response under the track is not necessarily subject to the distribution of the soil effectiveness conditions subject to the plastic theory. Even if the soil under the track is in an effective state, the sag of the track cannot be determined by the balance conditions alone, which is very different from the rigid wheel.

The action of interface shear stress in the interaction between track and ground is similar to that between wheel and ground. Along the direction of possible failure, the acting shear stress reduces the possible bearing capacity provided by the soil. If the ground pressure under the track is smaller than the support stress corresponding to the shear force applied by the track, the soil under the track is not in a failure state. The track sag under the action of the applied stress system depends on the degree to which the stress state in the soil is close to the plastic state.

4.1.2 *Ground pressure of track*

The nominal grounding pressure (also called average grounding pressure, defined as the weight of the whole vehicle divided by the track grounding area) has been used to measure the overall performance of tracked vehicles for many years. This indicator is applicable to tracks with small track widths, small load wheels, and long pitches. When the pressure is within the range of the track grounding length, the pressure on the surface soil can be approximately regarded as having a uniform distribution, so the nominal grounding pressure can be used to roughly estimate the effect of tracked vehicles on the road surface. With the increase in vehicle speed, the design of tracked vehicles tends to adopt large load wheels and short pitch tracks, so that significant pressure changes occur within the range of track grounding length. The corresponding peak pressure at the load wheel position can reach 2–3 times the nominal grounding pressure, while the pressure between the two load wheels can be very low or even close to zero. B. G. Schreiner's test research showed that tracked vehicles with the same nominal grounding pressure could have different maneuverability on soft ground, indicating the necessity to propose new indicators to correctly evaluate the maneuverability of high-speed tracked vehicles on soft ground.

4.1.3 *Core area of the track grounding plane*

The core area of the track grounding plane is an area around the geometric center of the two-track contact areas of the tracking device. As long as the center of gravity of the machine acts within this area, the track contact area can bear a certain load along the length direction. However, when the center of gravity of the machine exceeds this area, only part of the ground area of the track contact area bears the load along the length direction. In this case, the maximum ground-specific voltage must increase significantly. Therefore, it is of practical value for the design of crawler-type construction machinery to use the theory of the core region of the crawler grounding plane to study the law that the specific pressure of the crawler grounding varies with the change of the longitudinal eccentricity e of the machine.

4.2 *Types of forces acting on the interaction between track and ground*

When the tracked vehicle runs on soft ground, it should overcome the plastic deformation of the ground and be subject to the earth's moving resistance and settlement resistance from the ground. In addition, it also needs to overcome the same external resistance and internal resistance as hard pavement. External resistance mainly includes inertial resistance, ramp resistance, steering resistance, wind resistance, etc. The internal resistance includes the friction resistance between the drive wheel axle, the guide wheel axle, the roller axle, and the bushing; the friction resistance generated when the tracking link is wound in and out of the driving wheel and guide wheel; the rolling resistance of the roller and track chain; the friction resistance between the upper track chain and carrier roller; and the resistance caused by the polygon effect of the track chain.

5 SUMMARY

(1) The influence of soil characteristics on the traction performance of tracked vehicles is analyzed; the soil particle group is divided according to the soil particle diameter; furthermore, the bearing and shear characteristics of soil are analyzed, and its mathematical model is introduced. It can serve as a theoretical foundation for analyzing the interaction of tracks and ground.

(2) Three research methods of track-ground interaction are introduced in detail, and the characteristics of each research method are analyzed, which can provide better technical means for the research of track-ground interaction.
(3) The characteristics of the interaction between tracked vehicles and wheeled vehicles and the ground are analyzed and compared; the concepts of track grounding pressure and plane core region are introduced, and the relationship between them and the traction performance of tracked vehicles is analyzed; the types of forces acting on the interaction between the track and the ground are described.
(4) Through the systematic discussion and analysis in this paper, a good foundation for further research can be laid on the interaction between tracks and the ground and also pave the way for follow-up research on the high mobility of tracked vehicles.

REFERENCES

Hu Jiyong. *Research on Soft Ground Driving Performance of Heavy Tracked Vehicles* [D]. Changchun: Jilin University, 2012, 6.

Jiang Wenfeng. *Evaluation of Adhesion Between Tire and Road Surface During Driving* [D]. Xi'an: Chang'an University, June 2008.

Lei Xueyuan. *Optimal Design of Track Shoes for Tracked Vehicles* [D]. Beijing: Beijing University of Technology, January 2016.

Li Jun, Zhang Yu, Li Qiang, *et al.* Establishment of Three-dimensional Model for Analyzing Track Ground Interaction [J]. *Agricultural Equipment and Vehicle Engineering*, 2013, 51(6): 38–42.

Liu Honghai, Zhang Chunyan, Feng Zhongxu. Adhesion of Caterpillar Pavers Running on Geotextiles [J]. *Journal of Chang'an University (Natural Science Edition)*, 2006, (05): 94–98.

Ni Na. *Research on Topology Optimization of Track Shoe of Large Crawler Running Device* [D] Changchun: Jilin University, June, 2011.

Yang Congbin. *Research on Adhesion Characteristics and Optimization of High-speed Track and Soft Ground* [D]. Beijing: Beijing University of Technology, January 2015.

Zhang Kejian. Discussion on Mechanical Model in Finite Element Analysis of Shoe Prick Effect [J] *Journal of Agricultural Machinery*, 1983, (2).

Zhang Kejian. *Vehicle Ground Mechanics* [M]. Beijing: National Defense Industry Press, 2002.

Water Conservancy and Civil Construction – Oke & Ahmad (Eds)
© 2024 The Author(s), ISBN: 978-1-032-58614-4

Wind resistance safety of oil derrick based on reliability index

Dongying Han
School of Vehicles and Energy, Yanshan University, Qinhuangdao, Hebei, China

Nian Liu
China North Industries Group Jiangshan Heavy Industry Research Institute Co., Ltd, Xiangyang, Hubei, China

Yan Huang*, Guoqing Zhu & Xujia Li
School of Vehicles and Energy, Yanshan University, Qinhuangdao, Hebei, China

Rongrong Fu
School of Electrical Engineering, Yanshan University, Qinhuangdao, Hebei, China

ABSTRACT: Aiming at the problem of wind resistance performance evaluation of oil derricks, a wind resistance safety analysis method based on the reliability index is proposed. Based on the stress response data and wind-induced response analysis, the reliability index under design conditions is calculated using random reliability theory, and then the safety under wind load is evaluated. The results show that: 1. the structural safety analysis method based on the reliability index could evaluate wind safety; 2. under the wind load required, the safety does not meet the requirements; and 3. under the backwind load, the reliability of the middle diagonal member is lower than the safety threshold, so the strength of this part should be optimized.

1 INTRODUCTION

With the development of modern industrial technology, the height of much operating equipment is constantly increasing. As important as bearing equipment is in the process of oilfield development, if the structure of the oil derrick is defective, it may lead to insufficient wind resistance, affecting the safety of drilling and production operations (Liu *et al.* 2016). As a result, analyzing wind resistance safety is critical for improving safety.

Stochastic reliability theory calculates the probability that a structure can maintain a reliable state. This method can quantitatively describe the safety of the structure, and it has been widely used in the safety assessment of the structure in recent years (Tao 2021). Jia *et al.* studied the wind resistance characteristics of cooling tower structures through wind tunnel experiments and finite element calculation methods (Jia *et al.* 2019). Karadeniz studied the parameters that will affect the safety of the derrick and concluded that the reliability of the derrick will be lower than the condition of the safety threshold (Karadenz 2001). Jia *et al.* introduced the subset simulation method in the structural nonlinear dynamic reliability solution problem and improved the efficiency of the bridge structural reliability solution under seismic (Jia *et al.* 2021).

In this paper, the results of wind vibration response analysis are used as sample data to design a wind resistance safety analysis method for oil derricks based on the reliability index. The safety is evaluated and the weak position is analyzed based on the reliability index analysis results.

*Corresponding Author: 292788857@qq.com

 DOI: 10.1201/9781003450818-69

2 RESPONSE ANALYSIS OF OIL DERRICK UNDER WIND LOAD EXCITATION

2.1 *Design conditions*

According to the requirements in APISPEC·4F, when the design condition is the back wind with a wind speed of 38.6 m/ s, the drilling rig stops drilling and all the vertical roots are put down (Han & Wu 2013).

2.2 *Establishment of a finite element model of the oil derrick*

The object of the study is the ZJ70 oil derrick, which is 52.8 meters high and made of Q345 steel. The second-floor platform and base are omitted. The influence of the omitted parts on the whole derrick is equivalent to the corresponding binding force. The finite element model of the simplified ZJ70 oil derrick is established, as shown in Figure 1. The positive direction of the X-axis in the figure is the backward wind direction of the oil derrick. The rods on both sides are classified into horizontal rods, inclined rods, and vertical rods according to the different bearing directions.

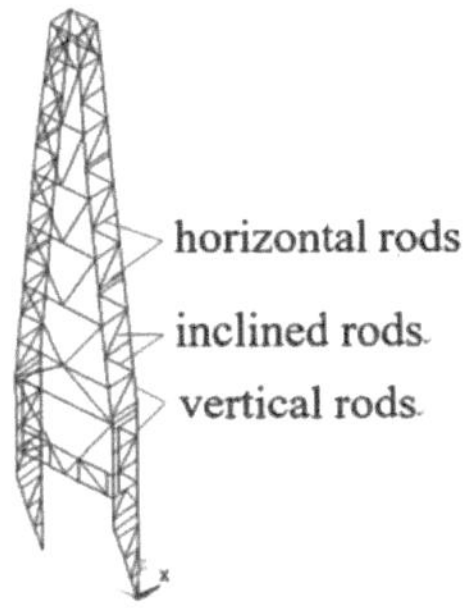

Figure 1. Schematic diagram of a ZJ70 oil derrick.

2.3 *Wind load simulation of the oil derrick*

Natural wind can be divided into the average wind and pulsating wind based on the change period of wind speed. The period of the average wind far exceeds the natural vibration period of the structure, so it can be treated as a static load. The period of pulsating wind is short, which can be regarded as a stationary random process. The wind speed is broken down as follows:

$$\mathrm{v} = \overline{V} + V_1 \tag{1}$$

where v is the instantaneous wind speed; $\overline{V}$ is the average wind speed at the z height; V_s is the average wind speed at the reference height (generally the average wind speed at a height of 10 meters); α is the ground roughness; V_1 is the pulsating wind speed at the z height.

The wind speed in the design condition is the average wind speed within 10 min, while Equation (1) is the average wind speed within 2 minutes. The calculated average wind speed at the reference height under the design condition is 42.9 m/s.

The fluctuating wind speed can be simulated using the linear filtering method according to the fluctuating wind power spectral density. The formula is as follows:

$$
\begin{cases} S_v(n) = \dfrac{4K\bar{v}_{10}x^2}{n(1+x^2)^{\frac{4}{3}}} \\ x = \dfrac{1200n}{\bar{v}_{10}} \end{cases} \tag{2}
$$

where n is the frequency (Hz); K is the roughness coefficient and the oil field is taken as 0.03.

The linear filtering method is also called the AR method. It can directly simulate the random process from the time domain. It is widely used in the linear prediction of stationary random sequences.

The combined wind speed power spectrum formula and AR model can obtain the time-history value of the backwind wind speed that the oil derrick bears at a height of 10 meters under the design conditions, as shown in Figure 2. At a height of 10 m, the wind speed fluctuates around 42.9 m/s and ranges from −6 to 103 m/s within 200 seconds.

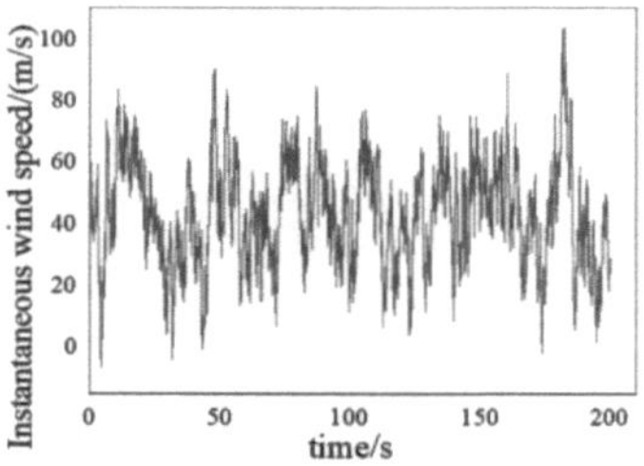

Figure 2. Instantaneous wind speed at 10 m height.

2.4 *Wind vibration response results of the oil derrick under design conditions*

The wind load mainly acts on the surface of the oil derrick structure in the form of pressure. Taking the wind pressure at a height of 10 meters as an example, the time-history value of the instantaneous wind speed at a height of 10 meters in Figure 2 is substituted into the wind pressure time-history value calculated in the wind pressure formula, and the image is drawn as shown in Figure 3.

The same method can be used to calculate the time-history value of wind load and pressure on other heights and extract all axial stress data for side members. Statistical processing and calculation of the response data can be obtained: the mean value of the axial stress response is −24.5 MPa, the peak value is −123.9 MPa, and the mean square error is 19.6 MPa.

Draw the distribution frequency diagram of the axial stress response of the E23 member under the design conditions, as shown in Figure 4. It can be seen that because the frequency of the 10 MPa response value is too high, the stress response data does not meet the normal

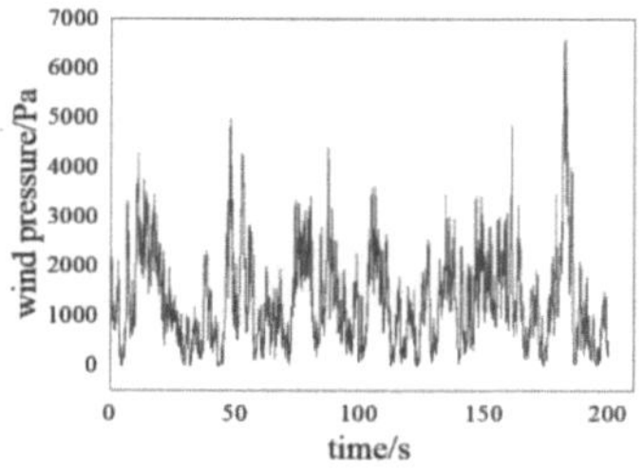

Figure 3. Wind pressure at 10 m height.

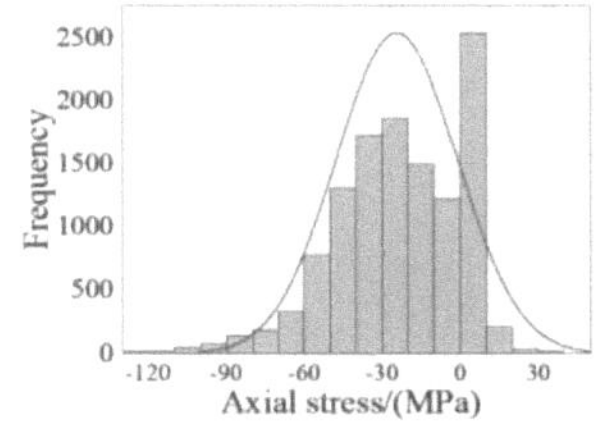

Figure 4. Frequency distribution of axial stress response.

distribution characteristics. It cannot be calculated with the relevant properties of the normal distribution in the subsequent reliability analysis.

3 BASIC PRINCIPLES OF RELIABILITY ANALYSIS

3.1 *Stochastic reliability theory*

As a natural load with strong randomness, it is unreasonable to simply analyze its influence on the structure from the point of view of statics. Stochastic reliability theory was proposed to analyze the influence of uncertain factors in engineering analysis on the structure. Suppose there are n random variables related to the structure function, written as $X = (X_1, X_2, \ldots \ldots, X_n)^T$, and the function

$$Z = g(X) \tag{3}$$

is called the state function of the structure. When Z = 0, the structure reaches the limit state; when Z > 0, the structure remains in a reliable state; when Z < 0, the structure is in a failure state.

The probability that the structure fails to achieve its intended purpose is called the failure probability, denoted by p_f. Assuming that the probability density function of the state function Z is $f_z(z)$ and the distribution function is $F_Z(z)$. When $f_Z(Z)$ is plotted as a graph, as shown in Figure 5, the area of the shaded portion of the graph represents the probability of failure p_f.

Z = 0 means that the structure reaches the limit state and then the failure probability is:

$$p_f = F_N(0|\mu_Z, \sigma_Z) = \phi\left(-\frac{\mu_Z}{\sigma_Z}\right) \tag{4}$$

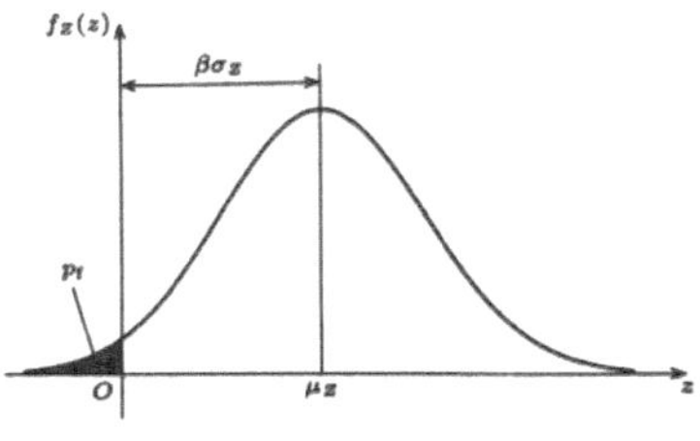

Figure 5. Probability density function diagram.

The variable β can be used to describe the ratio relationship between the mean and the standard deviation, as shown in Figure 5. β is the reliability index of the structure.

$$p_f = \phi(-\beta) = 1 - \phi(\beta) \tag{5}$$

3.2 *Normalization of the state function*

When the state function does not obey the normal distribution, an approximation algorithm is needed to calculate the reliability index. The method of Taylor series expansion of the state function at a point is called the design point method in stochastic reliability theory, with the following state function:

$$Z = g_X(x^*) = 0 \tag{6}$$

After calculation, the structural reliability index based on the design point method can be obtained by:

$$\beta = \frac{\mu_{Z_L}}{\sigma_{Z_L}} = \frac{g_X(x^*) + (\mu_X - x^*)^T \nabla g_X(x^*)}{\|diag[\sigma_X]\nabla g_X(x^*)\|} \tag{7}$$

3.3 *Variable data normalization*

Many structural variables in the project require normalization processing. Rackwitz-Fiessler method is used in the normalization of variables.

A variable X_i is assumed to be non-normally distributed, with a mean of u_{Xi}, a probability density spectrum of $F_{xi}(x_i)$, and a cumulative distribution function of $F_{xi}(x_i)$. If the coordinate x^* is stored so that the probability density spectrum of X_i and the normal distribution variable Y_i at this coordinate are equal to the cumulative distribution function, then the JC method can be normalized. After normalization, it can be obtained:

$$\mu_{Y_i} = x^* - \phi^{-1}[F_{Y_i}(x^*)]\sigma_{Y_i} \tag{8}$$

$$\sigma_{Y_i} = \frac{\varphi\{\Phi^{-1}[F_{Y_i}(x^*)]\}}{f_{Y_i}(x^*)} \tag{9}$$

4 RELIABILITY ANALYSIS RESULTS

4.1 *The establishment of the state equation*

For oil derricks, safety can be judged from the relationship between the bearing capacity and the load. When the bearing capacity is less than the load, the structure is in failure; its state equation is:

$$Z = R - Q \tag{10}$$

where R is the resistance of the member; Q is the bearing capacity of the member.

4.2 *Rod resistance and bearing capacity*

The material used for the rod is Q345 steel. According to the test data, the average yield strength is 371.8 MPa, and the root mean square is 29.1 MPa. According to the calculation, the allowable compressive stress value of the rod is lower than the allowable compressive stress value. Therefore, the allowable compressive stress value is taken as the resistance value of the rod. Taking the E23 element member as an example, after calculation, the average resistance of the member is 220.5 MPa, and the mean square error is 17.3 MPa. If the axial stress response of the rod in the wind-vibration response analysis can be taken as its bearing capacity, then the average bearing capacity of the E23 element member is 24.5 MPa and the mean square error is 19.6 MPa.

4.3 *Reliability calculation process of the ZJ70 oil derrick*

Figure 4 shows that the stress response value does not strictly obey the normal distribution, and the reliability cannot be calculated directly by the mean and mean square error. Before carrying out the reliability analysis, the resistance R and bearing capacity Q of the member have been normalized, and then the Taylor series expansion on the state function is performed.

Formula (5) shows that the reliability index β is negatively correlated with the failure probability p_f. The unified standard stipulated that the reliability index β of structural components should not be lower than 3.2. The resistance and bearing capacity data of the E23 member are substituted into the calculation, and the reliability index β of the member under the design condition is 3.45, and the failure probability is 0.028%, which meets the requirements of the specification.

4.4 *Reliability analysis of the ZJ70 oil derrick under design conditions*

The side members are classified by centroid height and bearing form, and the reliability under the design conditions is calculated. The calculation results are shown in Figures 6 and 7:

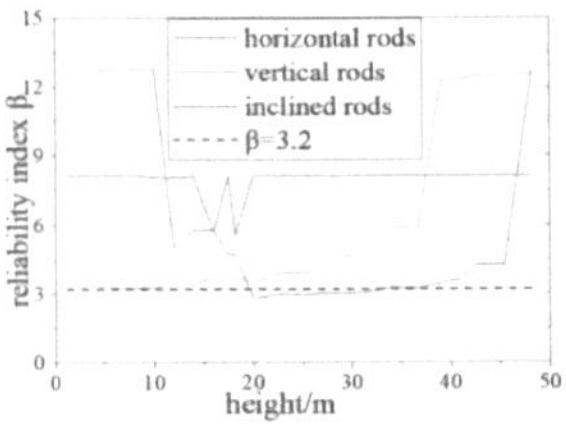

Figure 6. Reliability index of each member under design condition.

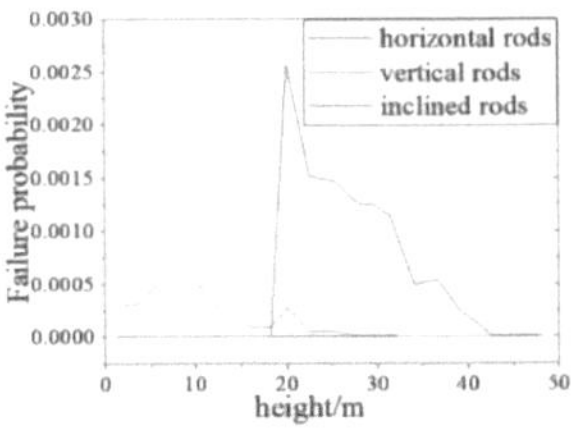

Figure 7. Failure probability of each member under design condition.

The calculation results of the reliability index β under the design conditions are shown in Figure 6. The horizontal axis is the centroid height, the vertical axis is the calculation result, and the dotted line is $\beta = 3.2$. Under the design conditions, the lowest reliability index β of the cross bar in the side members is 5.54, which is higher than the required value of the specification. For the vertical rod that mainly bears the gravity, the reliability index β of the bottom rod is low, and the minimum value of the reliability index β of the vertical rod at the centroid height of 10 m is 3.26; the reliability index β of the inclined rod between the centroid heights of 20–30 m is lower than 3.2. It is unsafe.

The calculation results of the failure probability under the design conditions are shown in Figure 7. The crossbar has a failure probability of 1.5×10^{-8}; that of the vertical rod at the centroid height of 10 m is the highest, at 2.8×10^{-4}; that of the inclined rod at the centroid height of 20 m is the highest, reaching 0.26%.

If the reliability index $\beta = 3.2$ is used for safety evaluation, the ZJ70 oil derrick is not safe under the wind load required by the specification, especially the diagonal members in the middle. The structure of this part should be optimized to ensure the wind resistance and safety of the ZJ70 oil derrick.

5 CONCLUSION

Based on the wind vibration analysis response data of oil derricks, this paper designs a wind resistance safety analysis method based on reliability indexes and studies the safety under wind load:

(1) The structural safety analysis method based on the reliability index fits the randomness of wind load and could effectively evaluate the safety of oil derricks under wind load.
(2) Under the action of the wind load required by the APISPEC·4F, the reliability is lower than the safety threshold allowed by the specification, and the oil derrick is in an unsafe state.
(3) The safety of the diagonal member in the middle of the ZJ70 oil derrick under the reverse wind load in the design condition is the lowest and lower than the safety threshold. This part of the member should be optimized to ensure wind resistance and safety during the service of the oil derrick.

ACKNOWLEDGMENTS

The studies are funded by the National Natural Science Foundation of China (Grant number 51875500), the Natural Science Foundation of Hebei Province (Grant number E2020203147), and other projects of the Hebei Province Department of Science and Technology (Grant number 216Z4301G).

REFERENCES

Han Dong-ying, Wu Wei. Safety Analysis of Derrick Steel Structures in Wind Loads[J]. *Journal of Guangxi University: Nat Sei Ed*, 2013, 38(06): 1292–1299.

Jia Ming-ming, Li Zhi-ping, Lyu Da-gang, et al. Analysis of Wind Pressure Time-history Response and Wind Resisting Performance for Super Large Cooling Tower [J]. *Engineering mechanics*, 2019, 36(S1): 118–124.

Jia Shao-min, Wang Zi-qi, Chen Hua-ting, et al. Hamiltonian Monte Carlo based Subset Simulation for Reliability Analysis of Energy-dissipation Bridge Structures with Viscous Dampers under Multi-support Seismic Excitations [J]. *Journal of Engineering Mechanics*, 2021, 34(02):

Karadenz H. Uncertainty Modeling in the Fatigue Reliability Calculation of Offshore Structures [J]. *Reliability Engineering and System Safety*, 2001, 74(3): 323–335.

Liu Hong-bing, Chen Guo-ming, Lyu Tao, et al. Wind-induced Response of Large Offshore Oil Platform [J]. *Petroleum Exploration and Development*, 2016, 43(04): 647–655.

Song Zi-ling. Dynamic Time History Analysis of Elevator Test Tower Under Wind Action[J]. *Engineering and Construction*, 2017, 31(02): 162–164.

Tao Ning-yan, Optimal Design of Bridge Structure based on Rebliability [J]. *Construction & Design for Project*, 2021(04): 99–101.

Water Conservancy and Civil Construction – Oke & Ahmad (Eds)
© 2024 The Author(s), ISBN: 978-1-032-58614-4

The cost of anchorage in the sea of Lingdingyang bridge on Shenzhen-Zhongshan link

Jun Mo*
Shenzhen-Zhongshan Link Administration Centre, Zhongshan, China

Hou-qing Huang
CCCC Second Harbor Engineering Company Ltd, Wuhan, China

ABSTRACT: The Lingdingyang bridge of Shenzhen-Zhongshan Link adopts a (580 + 1666 + 580) meter three-span steel box girder suspension bridge. The East and West anchorages are built in the sea (River), with an offshore distance of about 15 km. The construction of anchorage includes the formation of the land area of the island cofferdam, the construction of a diaphragm wall and foundation, anchor body and anchorage system, etc. The work content, temporary facilities, and construction organization of anchorage in the sea are quite different from those on the land. Based on the study of the construction characteristics and construction organization of the offshore anchorage, this paper analyzes the main aspects of the increased cost of offshore anchorage construction; puts forward the key factors to be considered in the cost compilation of the offshore anchorage, such as the secondary transfer fee of materials and equipment, the use of large ship machinery, construction monitoring, monitoring costs; and put forward opinions on the relevant contract terms and the floating rate of bidding. The research results have important guiding significance for offshore construction or marine (water) engineering construction project approval, investment calculation in the design stage, and cost management during implementation.

1 INTRODUCTION

The Shenzhen-Zhongshan Link is a system cluster project integrating "bridge, island, tunnel, and underground interworking", with a total route length of about 24 km, a design speed of 100 km/h, and a two-way 8-lane motorway technical standard. The Lingdingyang Bridge is a three-span steel box girder suspension bridge with a main span of 580 + 1666 + 580 meters. The east and west anchorages are both large gravity anchorages in the sea. The anchor foundation is constructed using an island cofferdam, and the anchor foundation uses an 8-way underground diaphragm wall.

The construction site of anchorages in the sea is far from land conditions, and there are significant differences in the temporary facilities, construction organization, selection of mechanical equipment, and other anchor projects compared with other land construction. Among the existing research literature, Chen *et al.* (2018) analyzed the influence of marine climate characteristics, hydrological environment characteristics, complex geological conditions, and the marine corrosive environment on the construction cost of railway bridges in the marine environment and proposed the necessity of conducting research on the

*Corresponding Author: 1784646584@qq.com

DOI: 10.1201/9781003450818-70

engineering environment of cross-sea bridges and supplementing the relevant engineering estimate (budget) quotas. Zhong *et al.* (Wang 2006; Wang & Yan 2003; Zhong 2017; Zhao 2013) relied on engineering examples to analyze the main factors affecting the cost of cross-sea bridges from the perspective of the preparation of cost documents such as estimates, budget estimates, and budgets. Lin took a cross-sea bridge as an example to analyze the key points of cost management in the bidding and contract signing stage, construction stage, and settlement stage of a highway (Lin 2018).

This paper studies the construction characteristics of anchorages in the sea and analyzes the main factors affecting the cost of anchorages, such as the layout of temporary construction facilities, the design of construction organizations, and the selection of large mechanical ships. On this basis, the key aspects of the cost preparation of anchorages at sea and related cost management issues are proposed, which will provide a reference for the management of similar projects and the formulation and revision of relevant policies in the future.

2 CHARACTERISTICS OF ANCHOR CONSTRUCTION IN THE SEA

Anchor bridge construction in the sea has the following distinctive features compared to conventional bridge construction on land:

2.1 *Harsh construction conditions and complex construction environment*

The anchorage is located at the junction of salt and fresh water at the mouth of the Pearl River, adjacent to the Lingdingyang Channel, the Longxuan South Channel, and the Alum Rock Channel, one of the busiest waters in the world. The construction area is mostly hydroplastic silt, with thick layers of silt and very poor lateral stability, and typhoons are frequent in the area. The climate is warm and humid, with high rainfall and intensity, and the average tidal difference is in the range of 0.85 m to 1.7 m. The construction operation environment is characterized by high humidity, high temperature, high corrosiveness, complex geological conditions, high winds, heavy rainfall, rapid waves, and fast-flowing water in the natural environment (Yao 2020). In addition to the need to face more complex and changing natural conditions and construction environments, it also requires communication and coordination with relevant management departments such as waterways, water conservancy, maritime, aviation and defense, and the military. The complex construction conditions place greater demands on design standards and technology, as well as on project management levels, construction organization capabilities, and the large scale of ship and machine equipment.

Figure 1. Anchor construction of Lingdingyang bridge in Shenzhen-Zhongshan link.

2.2 *Constrained construction site area and more complex organization*

The construction of an island or water platform requires permission from the relevant waterway, marine, maritime, and flood control authorities, and the economic comparison of the scale and workload of secondary transshipment of personnel, materials, and equipment is an important factor in the development of the island construction program. The construction organization scheme adopted in the land formation stage of the offshore anchorage construction project is mainly based on large ship machinery equipment, supplemented by a small amount of manual work. Therefore, compared with the anchorage construction project on the land, the construction organization, construction technology, and work efficiency of the anchorage project in the sea are quite different.

2.3 *Low labor efficiency and low equipment utilization*

Offshore operations are characterized by humid environments, high temperatures, and changing weather conditions. Long-term isolation on islands can easily lead to mental stress, irritability, physical discomfort, etc., and reduce the production efficiency of construction workers. In addition, the transfer cost of equipment to and from the offshore operation platform is high, and the conversion of the process easily leads to the inability of the equipment to be transferred to the site, resulting in short-term equipment idleness and a low utilization rate.

3 CONSTRUCTION ORGANIZATION OF ANCHORAGE IN THE SEA

3.1 *The main construction contents*

The east and west anchorages of the Shenzhen-Zhongshan link include a land cofferdam, a foundation, an anchor body, and auxiliary facilities. Except for the differences in cofferdam and foundation treatment methods, other work contents were the same (limited by space, this paper was only based on the study of East Anchorage Development). Among them, the East Anchorage adopted the cofferdam structure combining the processing of locking steel pipe piles and the external underwater pipe bag sand surface covered with membrane bag concrete. The main construction contents were as follows: cofferdam construction → site leveling → diaphragm wall → foundation pit excavation and construction lining → construction blind ditches and bottom sealing concrete → construction bottom plates, core filling, and top plates → construction anchor block Splay saddle pier → main cable construction (not included in the scope of anchor construction) → construction of front anchor chamber roof and front wall → construction of auxiliary facilities → removal of the anchor cofferdam.

3.2 *The main construction procedure*

3.2.1 *Centralized processing, transportation and installation of reinforcement parts*

The reinforcement was processed and manufactured in the centralized distribution center. After the prefabrication of the reinforcement cage was completed, the cage was hoisted to the storage area by the girder crane, transferred to the wharf by the flatbed truck, and then transported to the anchorage platform by the barge. The reinforcement cage was assembled into two sections on the special jig frame of the anchor platform, and the lifting and lowering were completed by one 260-ton crawler crane and one 150-ton crawler crane.

3.2.2 *Construction of the island cofferdam*

The island cofferdam was a circular structure with a diameter of 150 m, and the combination scheme of locking steel pipe piles and I-shaped sheet piles plus steel wire rope was adopted. The diameter of the construction area was about 200 m. The main construction contents were as follows: desilting construction in the anchorage area → construction of a sand cushion → construction of a plastic drainage board → construction of a locking steel pipe pile → construction of steel box hoop → construction of cofferdam backfilling sand → sandbag throwing outside the cofferdam → block stone scouring protection → triaxial mixing pile.

3.3 *Main labor and ship engine input*

The total construction period for anchorage construction was about 34 months. The details of the main ship machinery equipment and input time are shown in Table 1. During the peak period, the number of on-site laborers in construction was about 400, and the total labor force was about 173, 000 people per day (30 days per month). The main labor input is shown in Table 2. Intelligent detection of road diseases was essential to realize two-dimensional or three-dimensional imaging and recognition of diseases. The imaging conditions were random, and image recognition depended on the algorithm. Measurements such as potholes require good imaging results in a reasonable space, while assessments such as cracks require very accurate imaging in a local area (Huang *et al.* 2020). These two extreme cases posed challenges for imaging techniques and recognition algorithms. Through the comprehensive application of relevant technologies, common diseases such as pavement flatness, skid resistance, rutting, and cracks were intelligently detected.

Table 1. Input of main ship machinery equipment for underwater anchorage construction.

Serial Number	Equipment Name and Model	Quantity and Time	Serial Number	Equipment Name and Model	Quantity and Time
1	1000 t Split Barge	1 3 months	15	YZ300 Vibrating Hammer	3 6 months
2	1000 t Plastic Drain Board Ship	1 4 months	16	DZJ120 Vibrating Hammer	3 6 months
3	300 t Floating Crane	1 6 months	17	1000 t.m Boom Tower Crane	1 18 months
4	300 t Floating Crane	1 6 months	18	BC32-1.5 Hydraulic Tank Washing Machine	2 6 months
5	1600 t Floating Crane	1 3 months	19	GB46 Hydraulic Grab	1 9 months
6	4000 HP Anchor Boat	1 24 months	20	150 t Crawler Crane	1 9 months
7	Tender	1 34 months	21	260 t Crawler Crane	1 9 months
8	1000 t Material Carrier	1 34 months	22	Three Shaft Mixer	1 3 months
9	Guard Ship	2 34 months	23	HGY18 Concrete Distributor	4 12 months
10	2000 HP Tug	1 34 months	24	135 t Tower Crane	3 12 months
11	90m^{3}X2 Mixing Vessel	1 24 months	25	HZS180Q Mixing Station	2 24 months
12	Sand Blowing Boat	2 5 months	26	Grab Boat	1 1 month
13	Belt Carrier	2 7 months	27	Bulk Freighter	2 24 months
14	Concrete Pump Truck	3 15 months	28	90 t Crawler Crane	1 12 months

Table 2. Main labor input for underwater anchorage construction.

Main Project		Labor Quantity (person)	Time (month)
Anchor Body	Peak	200	22
	Off Peak	80	3
Diaphragm Wall		170	6
Anchorage System		20	6

Note: This table does not include the sub-divisional works such as plastic drainage board, dewatering well, sand dumping and filling, dewatering well, etc. with less labor.

4 COMPARATIVE ANALYSIS OF BUDGETING RESULTS WITH ACTUAL INPUTS

4.1 *Labor costs*

The number of labor days in the budget was about 447, 000, including the salary, regulation fee, welfare, main and non-staple food subsidy, and various allowances for all construction workers. The labor cost was about 41 million yuan. According to the actual site input for the construction, the total labor input was about 173, 000 people * days, the labor wage standard was based on the current average wage level of water construction workers and was 450 yuan per day, and the estimated labor cost was about 78 million yuan. As a result, the labor cost increased by 37 million yuan (90 percent) compared to the budgeted labor cost.

4.2 *Costs of ship machinery equipment*

In addition to labor costs, ship machinery, and equipment costs were major contributors to the budget's deviation from actual input. The main machinery and equipment costs for budget preparation were about 110 million yuan. According to the actual investment, the estimated cost of machinery and equipment was about 180 million yuan, an increase of about 70 million yuan (64%) compared to the budget. There were two main reasons why the actual input cost of ship machinery and equipment was higher than the budgeted one. For one thing, the actual input equipment model and specification were different from the fixed equipment model and specification. For example, although the actual engineering consumption (2.99 shifts per 100 m^3) was almost the same as the budgeted fixed consumption (3.02 shifts per 100 m^3), the actual unit price of imported equipment was relatively high, resulting in a higher input cost than the budget. For another thing, the market rental ship machinery equipment price was higher.

4.3 *Costs of auxiliary construction measure*

The material costs were not significantly different in land environment construction and water environment construction, but in the auxiliary construction measures cost, the actual investment was higher than the budget, which was primarily reflected in the high quality of the project, the use of imported formwork, the project application being small (high amortization cost), as well as a variety of testing, monitoring, evaluation, and other costs compared to the conventional method. The actual investment of about 33 million yuan, compared to the budget of about 20 million yuan, increased by 13 million yuan.

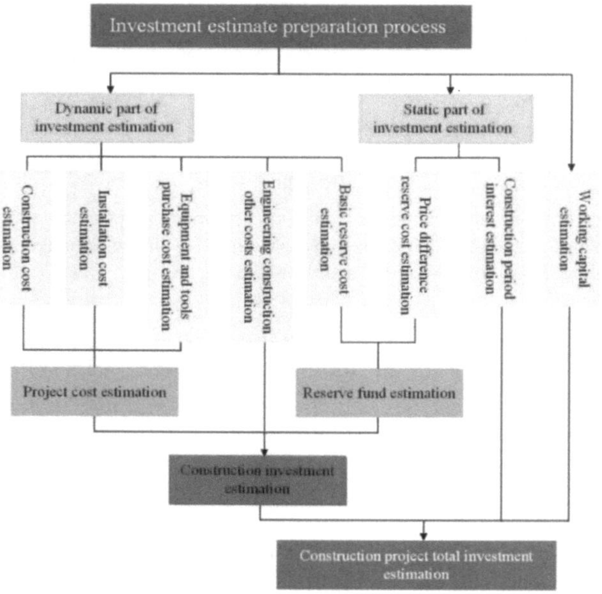

Figure 2. Investment estimate preparation process.

5 THE KEY FACTORS OF THE COST MANAGEMENT OF SEA ANCHORAGE

5.1 *Budgeting*

In terms of budget, it focuses on the secondary transfer costs of materials and equipment, the installation and demolition costs of large temporary facilities, the auxiliary costs of construction monitoring, evaluation, etc., and the cost of a large ship and machinery equipment.

5.1.1 *The secondary transfer costs of materials and equipment*

Personnel, materials, equipment, and machinery need to be transferred from the shore to the water platform through the ship, involving the transfer of the back yard, the ship transport, the installation equipment of the front yard transfer, and the labor costs of the transfer. The material information price used in the cost preparation refers to the transportation cost of the conventional land construction site, excluding the transshipment cost. Therefore, the increased secondary transfer fee for materials and equipment needs to be combined with the construction organization and input situation, and the key consideration is given in the cost preparation.

5.1.2 *Costs for the installation and demolition of large temporary facilities*

Due to the limitations of offshore construction sites, it is usually necessary to build steel trestle bridges, underwater platforms, various island cofferdams, and island roads for construction platforms or living platforms for workers. These costs are outside the scope of regular project rates and require special design and budgeting.

5.1.3 *Construction monitoring costs*

The safety of temporary structures, especially in the open ocean, is a critical issue, so special additional construction monitoring and testing are often needed, and construction organizations need to be prepared and considered in the budget.

5.2 *Bid float rate*

The labor costs and accommodation costs of anchoring construction in the sea are higher than those on land. For example, with the costs of cooling, keeping warm, emergency materials, water and electricity bills, and household garbage disposal in the construction environment, the increase in labor costs is often difficult to solve in the budget. According to the above comparison of labor costs, the increase in labor costs for sea anchorage accounts for about 6% of the budget for sea anchorage, so it is suggested that it be considered in the range of a floating rate of bid.

6 CONCLUSION

This paper analyzed the construction characteristics and construction organization of the anchorage in the sea of the Lingdingyang Bridge in the Shenzhen-Zhongshan Link. The specific conclusions are as follows:

Based on the current cost pricing system, this paper put forward three factors in the need to be considered in strengthening the cost management of anchors in the sea, including materials to be considered in cost preparation, equipment secondary transfer costs, the use of large ship machinery, construction monitoring, monitoring costs, etc., and the impact of the degree of analysis.

Adjustment clauses were set for aspects that have a large impact on the cost (such as large ship machinery and equipment), and the shortage of budget preparation was compensated by reducing the bid float rate.

The research in this paper can provide a reference for the preparation of the cost of similar projects and the formulation of related policies, which are also applicable to other projects such as bridge works, artificial islands, and tunnels implemented in the sea.

REFERENCES

Chen Hongjun, Peng Weili. Analysis of Influence Factors of Marine Environment on Railway Bridge Engineering Cost [J]. *Railway Engineering Technology and Economy*, 2018(9): 16–19.

Huang houqing, Li Mian, Liu Jianbo. Optimization of Construction Scheme of Island Cofferdam for East Anchorage of Lingdingyang Bridge of Shenzhen China Channel [J]. *Bridge Construction*, 2020, 50(06): 104–109.

Lin Jiezhen. Analysis of Key Points of Cost Management in the Implementation Stage of Expressway Project – Taking a Sea Crossing Bridge as an Example [J]. *Fujian Architecture*, 2018(11): 82–84.

Wang Jiufu. Characteristics and Cost Analysis and Determination of Cross Sea Bridge [J]. *Railway Engineering Technology and Economy*, 2006(5): 11–15.

Wang Yupeng, Yan Yu. Cost Analysis and Control in the Feasibility Study and Preliminary Design Stage of Hangzhou Bay Sea Crossing Bridge [J]. *Highway*, 2003(10): 121–127.

Yao Zhian. Key Technology of Island Cofferdam Construction for Lingdingyang Bridge of Deep Medium Channel [J]. *World Bridge*, 2020, 48(02): 15–19.

Zhao Ying. Thoughts on Cost Compilation of Offshore Projects [J]. *Highway*, 2013(10):155–158.

Zhong Qiong. Preparation and Analysis of Preliminary Design Estimate of Xiamen Zhangzhou Sea Crossing Bridge [J]. *Highway Traffic Technology*, 2017(2): 137–142.

Water Conservancy and Civil Construction – Oke & Ahmad (Eds)

Analysis and optimization of a structural seat plate

Yi Yao* & Lebin Tan
Nanjing University of Science and Technology, Nanjing, China

ABSTRACT: To solve the stress concentration problem of the seat plate when shooting on hard ground, the structural design of the mortar seat plate was redesigned. On the premise of ensuring shooting form and stability under soft soil, surface support was used to reduce the maximum stress under hard soil. For the three-dimensional modeling of the new structure's seat plate, a suitable soil constitutive model was selected to draw more accurate conclusions, and the chamber pressure curve of the internal ballistic of the mortar was calculated. The finite element analysis software was used to analyze the Stress distribution in the state. To further reduce the weight of the seat plate and improve its strength of the seat plate, a multi-objective optimization based on a genetic algorithm was carried out. The optimized seat plate significantly reduces mass with reduced maximum stress.

1 GENERAL INSTRUCTIONS

Since its appearance, the mortar has been an effective suppressive weapon accompanying infantry combat. It has a simple structure, convenient operation, and flexible maneuvering. The seat plate is an important part of the mortar. When the mortar is fired, the recoil force is transmitted to the soil or other supports through the seat plate. Therefore, the first thing to consider when designing the seat plate is the demand for the strength and stiffness of the seat plate. Both the traditional circular window seat and the trapezoidal seat have excellent performance in launching in soft soil. Still, in hard soil (such as cement ground, frozen soil, and other rigid ground), due to the three-point support, the stress situation is more difficult. Some positions have a large stress concentration (Zhou 2016). This paper analyzes and optimizes a self-designed mortar seat plate, conducts finite element analysis of the design attitude under different launch conditions, samples multiple sets of data to establish a proxy model, and uses a genetic algorithm to analyze the size parameters of the seat plate multi-objective optimization, lightweight design of the seat panel.

2 SEAT STRUCTURE DESIGN

Figure 1a is a schematic diagram of the trapezoidal seat plate structure. The seat plate adopts a trapezoidal pyramid structure. The plane of the seat plate is approximately triangular. Three large vertical ribs and three small vertical ribs are symmetrically distributed on the back of the seat plate (Wang 2012). Compared with the circular seat plate in the same area, the distance from the support point to the center of the abutment is increased, which benefits shooting stability (Ge 2020). The trapezoidal seat plate must be constructed to build a position on soft soil. On hard soil (such as cement pavement or frozen soil pavement), it is placed flat on the ground for three-point support shooting. Stress concentration is easily generated during the shooting process, and even Plastic deformation occurs locally. Given

*Corresponding Author: yy18851199306@163.com

 DOI: 10.1201/9781003450818-71

this problem, combined with the design experience of the mortar seat plate, a new seat plate model structure is designed, as shown in Figure 1b.

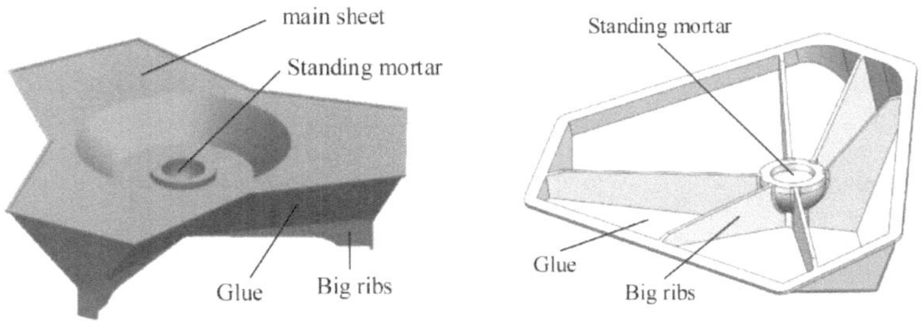

a. Trapezoidal seat plate b. new structure

Figure 1. Seat plate.

This seat plate has different shooting forms on soft ground and hard ground. Shooting on soft ground is similar to the traditional trapezoidal seat plate. It contacts with the hard ground to reduce stress concentration and other conditions. Figure 2 shows the launch situation of the two terrain seat plates.

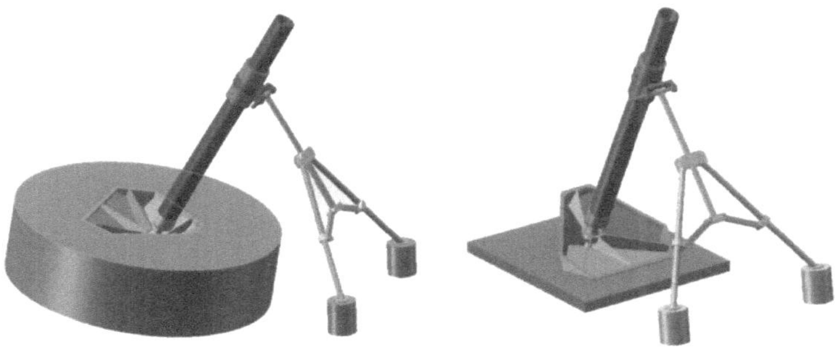

a. Shooting in soft soil b. Shootingin hard soil

Figure 2. Shooting form.

3 FINITE ELEMENT ANALYSIS

3.1 *Finite element modeling*

We carry out three-dimensional modeling of the seat plate and the initial assignment to each dimension parameter of the seat plate. We obtain the dimension parameters of each key part of the seat plate, as shown in the table below. The volume of the seat plate after modeling is 2.50 dm^3, and the mass is 11.55 kg.

Table 1. Values of main dimensions.

Parameter	Bottom thickness (d_1)	Rib Thickness (d_2)	The radius of the base column (r_1)	Chamfering of key parts (r_2)
Value (mm)	5	6	20	70

Zhang Yuyi concluded through experiments and analysis that when the diameter of the finite element area of the soil is taken to be more than 1.5 times the width of the structure, the finite element calculation can meet the accuracy requirements, and the diameter of the finite element area of the soil can no longer be increased (Zhang 2006, Hu 2014). To ensure the calculation accuracy, this calculation takes three times the maximum section width of the seat plate for soil modeling. It constructs a depression according to the shape of the seat plate to ensure that the soil and the seat plate are fully bonded.

3.2 *Material properties*

For this seat plate, the material is TC4 titanium alloy, the elastic modulus is 96 GPa, the strength is 1.012 GPa, the Poisson's ratio is 0.34, and the density is 4620 kg/m^2. There are two types of launch sites: hard soil and soft soil. In hard soil, cement materials with higher overall hardness are used. In soft soil, the soil will be compacted during the construction of the site and launch, and it will fit the seat plate and be compacted (Xu 2014). Based on this situation, hard and compact sandy soil was selected for this analysis. The stress-strain relationship of soil is relatively complex, with special cases such as nonlinearity, elastoplasticity, anisotropy, and shear tension (Ristic 2009, Wang 2014). In this paper, the Ducekr-Prager nonlinear constitutive model is used to simulate the stress and deformation of soil. The general expression of the constitutive model is:

$$^{t+\Delta t}f^{DP} = \alpha^{t+\Delta t}I_1 + \sqrt{^{t+\Delta t}J_2} - k \tag{1}$$

Among them, α and k are the material characteristic parameters; f^{DP} is the stress on the soil; I_1 is the first invariant of the stress tensor; J_2 is the second invariant of the stress tensor; $t + \Delta t$ is the time t and its increment.

The soil parameters are shown in Table 2:

Table 2. Parameters of the hard soil model.

Elastic Modulus (E/Mpa)	Poisson's ratio	Density ($\rho/(\mathrm{kg/m^2})$)	Friction angle (°)	Expansion angle (°)
150	0.21	2000	30	10

3.3 *Set contact*

On the contact surface between the breech and the seat plate, there is only the displacement change in the normal direction, and the tangential displacement is very small. It will have little influence on the calculation process. To simplify the calculation model, the contact between the breech and the abutment surfaces uses binding constraints. The contact model between the seat plate and hard soil adopts the Coulomb friction model, the number of contact surfaces is 1, and the friction factor is 0.6. The Coulomb friction model is the same as that of soft soil, the number of contact surfaces is 10, and the friction coefficient is 0.2.

3.4 *Mesh*

The seat plate is mainly processed and produced by welding, so the base and surrounding parts of the seat plate are divided by solid elements, the bottom plate and ribs are divided by plate and shell elements, and the soil model is divided by solid models (Wang 2019). The finite element model under the hard soil condition is divided into 188,001 units, and the finite element model under the soft soil condition is divided into 489,221 units. The preliminary estimate meets the accuracy requirements. The results of mesh division under the two working conditions are as follows:

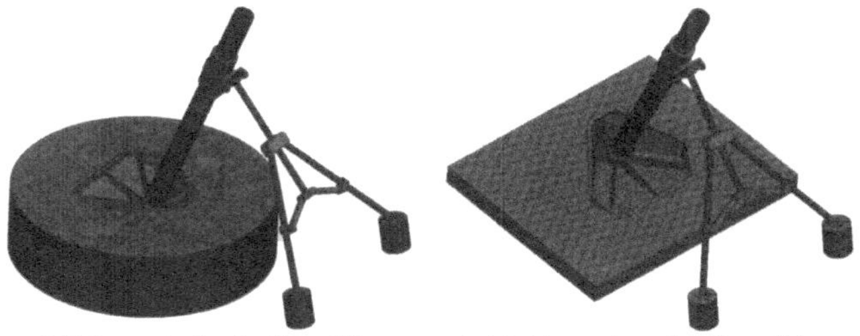

a. Gridding under hard soil conditions b. Meshing under soft soil conditions

Figure 3. Meshing.

3.5 *Boundary conditions*

The basic equation of the internal ballistic of the mortar is as follows (Liu 2011):

$$\begin{cases} \varphi = xZ + x\lambda Z \\ \sigma = 1 + 2\lambda Z \\ \dfrac{dZ}{dt} = \dfrac{P}{I_k} \\ Spdt = \varphi m dv \\ y = \varphi_2 A S_\Delta I_K \\ SP(l_\varphi + l) = \overline{f}\omega\left[\dfrac{x_o}{1+x_0} + \dfrac{\varphi}{1+x_0} - \overline{\eta_K}Z - \dfrac{\theta\varphi m}{2\overline{f}\omega}v^2\right] \end{cases} \tag{2}$$

Among them, S and S_Δ are the cross-sectional area of the gun bore and the gap area; y is the outflow of gas; $\overline{f}$ is the force of the powder; φ is the calculation coefficient of the secondary work; Z is the relative thickness of the powder; λ and$x$ are characteristic quantities of gunpowder shape; I_k is the full impulse of pressure.

Combined with the initial parameters of the mortar's internal ballistics, the pressure curve of the mortar's internal ballistics bore bottom is drawn by Matlab software, as shown below, and the pressure in the figure changes with time evenly applied to the bottom of the barrel.

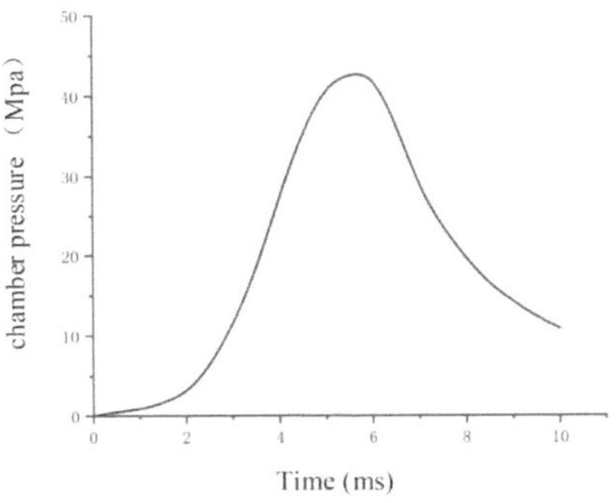

Figure 4. Variation curve of bore pressure with time.

Taking the most commonly used shooting conditions: the direction angle is 0°, the height angle is 70°, and the dynamic analysis is carried out on the seat plate. It can be seen from the results that the maximum stress point of the seat plate appears at about 6 ms, the maximum stress value is 338.19 Mpa, and the maximum stress value of the hard soil condition is 142.24 Mpa. The stress cloud diagrams under the two working conditions are shown in the figure below:

a. Soft soil

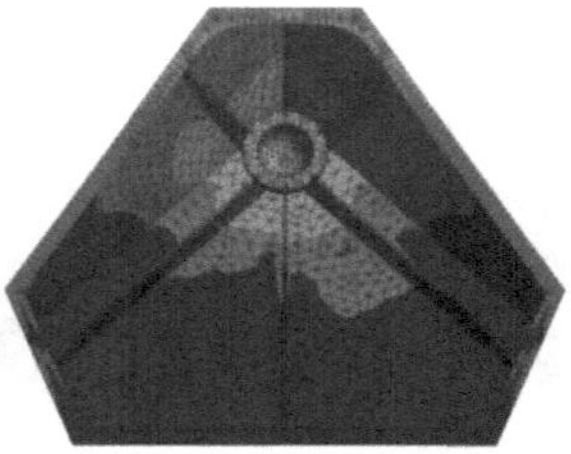

b. Hard soil

Figure 5. Stress cloud diagram.

It can be seen from the stress cloud diagram under the soft soil condition that the overall stress distribution of the seat plate is relatively uniform, and there is a slight stress concentration at the front edge of the seat plate. By adjusting the chamfer of the edge part, the stress concentration can be significantly reduced. Still, it will also Correspondingly increase the mass of the seat plate, so the chamfer here will be set as a design variable in the multi-objective optimization process. The reinforcement around the mortar still has a large stress distribution. Because most of the resultant bore force on the mortar is transmitted from the rib to the whole seat plate, the stress at the connection point between the rib and the mortar is significantly greater than that of other parts of the mortar and the rib. It can be seen from the stress cloud diagram under hard soil conditions that the stress of the dock and below the dock is relatively large, and the stress of the structures around the dock decreases with the distance from the dock. The stress on most areas of the front end of the seat plate is relatively small. The reason is that when shooting on hard soil, the main stress point and direction are the rear of the seat plate, and the front ribs and contact surfaces are not stressed. The maximum stress occurs at the reinforcing rib behind the mortar, close to the bore axis. The reinforcing rib is one of the key load-bearing parts, so it is easy to generate large stress.

The finite element analysis results show the structural seat plate's maximum stress position and magnitude under different shooting conditions. Parameters such as the thickness of the rib will have a great influence on the maximum stress. The soft ground will have a certain degree of stress concentration at the bottom of the seat plate and the upper edge of the seat plate. Under the two working conditions, the stress of each part of the seat plate can be further improved through parameter optimization, and the quality of the seat plate also has room for further optimization, so further optimization design of the seat plate is required.

4 OPTIMAL DESIGN OF THE SEAT PLATE

To obtain a more reasonable structural size for the seat plate, it is necessary to optimize the design of each size parameter of the seat plate based on structural improvement. The optimal size of the seat plate is sought through multi-objective optimization. Firstly, parametric modeling is carried out on the seat plate, and the four dimensions of the seat plate include the thickness of the main plate, the thickness of the ribs, the diameter of the base and below the base, and the size of the chamfer in the part of the stress concentration area. We design variables during optimization. The objective function is defined as the minimum stress at the maximum stress position under soft and hard soil conditions and to make the overall mass of the seat plate smaller. At the same time, the absolute displacement of the seat plate in the seat is also kept within a certain limit. The four design variables of the seat plate and the range of values in the optimization process are shown in the table below.

Table 3. Initial values and ranges of design variables.

Location	Bottom thickness d_1	Rib thickness d_2	The radius of the base column (r_1)	Chamfering of key parts (r_2)
Upper limit	7	7	25	100
Initial value	5	6	20	70
Lower limit	3	3	15	50

Each finite element analysis will calculate the maximum stress in hard and soft soil. To simplify the calculation, the optimization objective function of the seat plate is set as follows:

$$P_0 = \alpha P_1 + \beta P_1 \tag{3}$$

Among them: P_0 is the optimization objective function; P_1 is the maximum stress under hard soil; P_2 is the maximum stress under soft soil; α and β are the weight coefficients of the corresponding working conditions, and $\alpha = \beta = 0.5$ is taken for this analysis. After determining the objective function, combined with the above design variables and constraints, the optimization model is established as follows:

$$\begin{cases} \min\ (P_1, m) \\ X = (\mathrm{d_1 d_2 r_1 r_2})^T \\ s.t. \quad 3 \le \mathrm{d_1} \le 7,\ 3 \le \mathrm{d_2} \le 7 \\ \ 15 \le \mathrm{r_1} \le 25,\ 50 \le \mathrm{r_2} \le 100 \end{cases} \tag{4}$$

The fast non-dominated sorting genetic algorithm (NSGA-II) with elite strategy is used for multi-objective optimization solution, and the initial sample is 100 groups. Find the optimal solution set after reaching the maximum number of iterations. Otherwise, we solve the result directly. After iterative calculation, the most suitable solution with relatively small quality is selected from the solution set, and the obtained parameters are shown in the following table:

Table 4. Parameters after optimization.

Parameter	Bottom thickness (d_1)	Rib Thickness (d_2)	The radius of the base column (r_1)	Chamfering of key parts (r_2)
Value (mm)	4.48	4.74	18.8	75.6

The mass of the optimized rear seat panel is 10.24 kg, compared with the mass of 11.55 kg before optimization. The overall weight has been reduced by 11.3%. Finally, the model of the seat panel before and after optimization is imported into the finite element analysis software, and its performance in soft and hard soil is calculated. The stress distribution under the condition. The obtained stress cloud diagram is as follows:

a. Soft soil

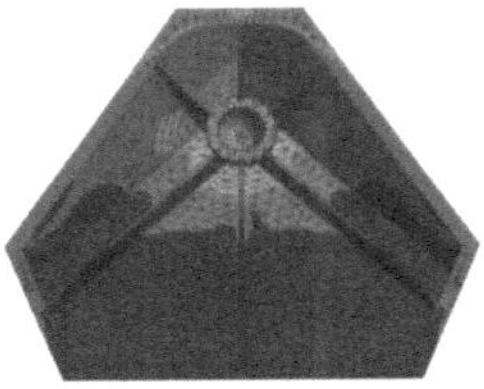

b. Hard soil

Figure 6. Stress contour after optimization.

The following table shows the maximum stress value and seat sheet quality before and after optimization:

Table 5. Comparison before and after optimization.

Seat panel	Soft soil maximum stress (MPa)	Hard soil maximum stress (MPa)	Quality (kg)
Before optimization	338.19	142.24	11.55
After optimization	326.84	130.00	10.24

Comparing the finite element analysis results before and after optimization, it can be seen that the maximum stress of the two working conditions after optimization is reduced, and the overall mass of the seat plate is reduced by 11.3%. It can be seen that the optimized seat panel not only achieves the purpose of lightweight design but also improves the strength of the seat panel to a certain extent.

5 CONCLUSION

(1) In this paper, a new structure of the mortar seat plate is studied, and the finite element simulation analysis is carried out to explore the stress distribution of this structure under different working conditions, laying the groundwork for optimizing the structure of the seat plate.
(2) The genetic algorithm performs multi-objective optimization on the parametrically modeled seat panel. After optimizing the size, the strength of the seat panel is slightly improved, and the mass is reduced by 11.3%, which improves portability and achieves the design goal of lightweight.

REFERENCES

GE Jian, Xie Xinyu, Liu Guozhi, Yang Guolai. Dynamic Response Analysis and Structure Optimization of Composite Seat Plate [J]. *Journal of Ballistics*, 2020, 32 (04): 83–90. (In Chinese)

Hu Guimei, Zhang Yajuan. Dynamic Performance Analysis of Soil Under Impact Load of Launcher [J]. *Value Engineering*, 2014. 33 (31): 23–25. (In Chinese)

Liu Yunfeng. Transient Dynamic Analysis of Mortar Seat Plate Based on AWE[J]. *Mechanical Engineering and Automation*, 2011 (03): 41–42. (In Chinese)

Ristic Z, Kari A, Bajevic M. Dynamic Analysis of a Model of the Mortar Base Plate Applying the Proengineer Software[J]. *Military Technical Courier*, 2009, 57(1): 81–91.

Toivola J, Moilanen S, Jussila H R. Force, Pressure and Strain Measurements for Traditional Heavy Mortar Launch Cycle[J]. *Rakenteiden Mekaniikka*, 2012, 44(4): 309–329.

Wang Fengfeng, Yang Guolai, GE Jian. Optimization of Mortar Seat Plate Structure based on HyperWorks [J]. *Journal of Ballistics*, 2019.

Wang Wei, Gao Yuefei. Finite Element Analysis and Optimization of Mortar Seat Plate [J]. *Mechanical Engineering and Automation*, 2014: 72–73. (In Chinese)

Xu Xiaole, Liu Shuhua. Dynamic Analysis of the Influence of Different Soils on Mortar Firing Stability [J]. *Mechanical Engineering and Automation*, 2014 (005): 49–50. (In Chinese)

Zhang Xiaoming, Liu Shuhua, Peng Kexia, Liu Xingguo. Topology Optimization Design of a Mortar Seat [J]. *Journal of Ordnance and Equipment Engineering*, 2016 Magi 37(04): 33–35. (In Chinese)

Zhang Yue, Zhang Fengqing, Zhu Yinglei. Influence of Analysis Range on Infinite Element Dynamic Response of Underground Structures [J]. *Railway Construction*, 2006 (08): 42–45. (In Chinese)

Zhou Zhongjia, Bu Huatao, he Yong. Study on Launch Dynamics of Mortar Pedestal Plate-soil Coupling [J]. *Journal of Ordnance Equipment Engineering*, 2016: 18–21. (In Chinese)

Water Conservancy and Civil Construction – Oke & Ahmad (Eds)
© 2024 The Author(s), ISBN: 978-1-032-58614-4

Influence of filling coefficient of long spiral drilling pressure grouting pile on pile quality (WG22027)**

RongJun Ding*
Wanjiang University of Technology, Ma'anshan City, Anhui Province, China

ABSTRACT: Aiming at the problem of the concrete filling coefficient of the long spiral bored pile being too large in the Yunlu project of Yunlu, Dali, this paper examined the possible influencing factors in detail and, by using fractal theory, it examines the filling coefficient and the diffusion mechanism of concrete in the sandy soil layer from a microscopic perspective, as well as its influencing factors and causes. According to the diffusion mechanism, targeted solutions were proposed, which will play an important role in reducing the filling coefficient, controlling the quality of the pile foundation, and ensuring the economic benefits of the project in its later construction.

1 INTRODUCTION

With the acceleration of urbanization and infrastructure construction in China, pile foundations have been widely used in various industrial and civil buildings and have become a common form of deep foundation in engineering. The cast-in-place pile is the most commonly used type of pile foundation. The long auger-bore pile has high construction efficiency due to the simultaneous completion of hole formation and concrete pouring. The long spiral drill press grouting pile shall be drilled with a long spiral drill. After reaching the design depth, the drill shall be lifted, the concrete shall be poured, and the reinforcement cage shall be vibrated to form the pile. The existing hole and pile shall be completed by one machine at a time (Wang & Pan 2018).

The pressure grouting concrete pile developed based on the long spiral dry drilling method has been widely used throughout the country in recent years. The long spiral dry drilling pressure grouting pile has the following advantages: the concrete has good continuity, and there is no sediment at the pile bottom; no vibration, low noise, and little impact on the environment during construction; there is no need for mud wall protection and no need for sewage discharge and for soil squeezing, and pose no impact on groundwater level; compared with other construction methods, the construction cost is relatively low, and the comprehensive benefit is high. In addition, the application scope of a long spiral bored pile is very wide, which applies to sand, silt, clay, mucky soil, miscellaneous fill, etc., and it is also applicable to the composite foundation, independent foundation, strip foundation, and raft foundation. The filling coefficient is one of the most important indexes in cast-in-place pile construction. The profit coefficient refers to the ratio of the actual amount of concrete poured into a single pile to the pile volume calculated according to the designed effective pile

*Corresponding Author: 1344630087@qq.com
**This paper is one of the achievements of the school level project of Wanjiang University of Technology "The influence of the construction filling coefficient of bored pile based on fractal theory on the pile quality" (project approval No.: WG22027)

DOI: 10.1201/9781003450818-72

length and diameter. It reflects the ratio of the actual amount of concrete to the designed amount, which is both a quality control indicator and a cost control indicator for pile foundation engineering. According to the current pile foundation construction specifications, the filling coefficient of cast-in-place piles shall not be less than 1 or greater than 1.3. If the filling coefficient is less than 1, it indicates the actual amount of concrete poured. It is less than the theoretical calculation, indicating that there are some defects in the pile quality. If the filling coefficient is greater than 1.3, it indicates that there may be hole collapse and diameter expansion. In actual construction, if the filling coefficient is greater than the quota filling coefficient, the project benefit will be seriously affected. If the construction filling coefficient is too small, the pile quality cannot be guaranteed, and the problem of broken piles and empty piles may occur. The relationship between the detailed filling coefficient and the pile quality is conducive to intuitive management and control of the construction site (Wang *et al.* 2017).

2 PROJECT OVERVIEW

Dali • Erhai Qingcheng Project is located in Dali City, south of Heping Road and west of Hongshan Road. The land area of the project is 98209.4 m^2, and the total building area is 488283.2 m^2. The building is mainly composed of 7 high-rise buildings (8#–13#), 1 commercial building (20#), an auxiliary podium building, and a pure basement. A basement is set on the site as a whole, and the excavation depth is about 3.4–4.8 m below the ground. The structure type of the proposed building is a shearing wall and frame, and the proposed foundation type is a pile raft foundation.

According to the geological exploration report, the maximum exposed depth of the drilling is 75.60 m. The foundation soil on the site from top to bottom, consists of Quaternary artificial accumulation (Qml) layer distributed on the surface, with a plain fill lithology; an upper part being Quaternary alluvial and pluvial (Qal + pl) layer, with the clay and clay lithology; the lower part being Quaternary alluvial and lacustrine (Qal + l) layer, with silt, silt, gravelly sand, and clay lithology (Xiong 2020).

3 SITE PROBLEMS AND CAUSE ANALYSIS

According to the statistics and calculation of daily construction records, the preliminary filling coefficient of cast-in-place piles constructed in March and April 2020 has been calculated. It was found that the minimum filling coefficient of the pile foundation was 1.12, the maximum filling coefficient was 1.6, and the average filling coefficient was 1.35. Too large a filling coefficient, on the one hand, means that the number of concrete increases, and the cost greatly exceeds the expectation of the project department; on the other hand, there may be defects in the pile quality, which will affect the service life of the pile foundation. Therefore, it is of great importance to solve the problem of the excessive filling coefficient of the long spiral-bore pile.

Many factors may affect the excessive filling coefficient of the long spiral-bore pile. Generally speaking, the reasons for the excessive filling coefficient of concrete are mainly four main aspects: local hole enlargement, full hole section enlargement, local formation leakage, and excessive grouting. The most likely reason for the abnormal filling coefficient of conventionally bored piles is that a pebble or fluid soil layer is encountered during drilling. If the drilling speed is too fast, it is very easy to form a hole collapse in the local stratum, which will increase the local pile diameter and increase the concrete pouring amount. The long spiral bored cast-in-place pile is adopted in the project, and the stability of the hole wall is beyond the conventional bored cast-in-place pile. Therefore, the analysis of possible situations shows that one of the situations that may lead to an excessive fill coefficient is subjective, that is, due to human factors. However, the excessive filling coefficient caused by human factors often occurs in individual piles, and it is rare for the overall filling coefficient of hundreds of piles to be excessive. Therefore, it can be judged that the large overall filling coefficient of hundreds of piles should be caused by the stratum (Yu 2003).

During drilling and grouting, if the pump creates too much pressure or is too slow, it will result in an excessive pressure of concrete in the pile hole. Under the action of high pressure, the poured concrete is likely to cause grout leakage, resulting in an excessive actual filling coefficient of concrete.

After rechecking the survey report and the field trench inspection results, it was found that there was indeed a layer of quicksand at the base of the long spiral pile, and the quicksand layer might leak under the pressure of the long spiral drilling and pressure grouting pile, so the sand was obtained and the permeability of the sand was studied using the fractal theory.

4 RESEARCH ON CONCRETE DIFFUSION IN SAND

4.1 *Research on the relationship between sand pore structure and permeability*

The penetration of cement slurry into the water-rich sand layer is a problem in which a fluid composed of two components (water and cement) flows through a porous medium composed of sand.

A porous medium is a heterogeneous multiphase medium composed of soil particles and filled with liquid gas that is generally characterized by heterogeneity and anisotropy. This makes people pay attention to the porous media seepage problem. Because of the complexity of the porous media microstructure, it is difficult to simulate the microstructure of porous media when a theoretical analysis of the porous media seepage problem is carried out, which can only be carried out under the ideal conditions of the continuous media method. As a heterogeneous porous medium, this paper has the characteristics of heterogeneity and anisotropy. Generally speaking, all permeability characteristics of heterogeneous porous media, such as permeability and permeability coefficient, are affected by pore structure in varying degrees. However, because of its complex pore structure, it is difficult to describe and calculate it with classical European geometry. At the same time, in the process of studying it, it was found that the pore microstructure of a heterogeneous hollow ring had a certain degree of self-similarity, which made many properties difficult to explain at the macro level but well explained at the micro level using fractal theory.

Pore distribution in the plane and a curved pore pipe in space represent the pore structure. Therefore, no matter whether the pore structure of sand or the permeability of sand is studied, the study of the bending degree of the pore pipe is unavoidable. Because the pore pipeline of gravelly soil also has a fractal feature, the fractal dimension of pore tortuosity was introduced here to express this feature. The fractal dimension of pore tortuosity is a reflection of the degree of curvature of the pore pipeline. The apparent length of the pore is assumed as L0, and the total length of the pore pipeline is assumed as Lt(λ). Because the pore pipeline is straight or zigzag, the curve length is always greater than or equal to the pore appearance length, that is, Lt(λ) L0. If the pore channel is straight, then Lt(λ) = L0; the smaller the pore diameter is, the longer the pore channel is. The pipe length of a straight pore channel is equal to the apparent length of the pore and the calculated pore tortuous fractal dimension (DT = 1), and when the degree of pore bending is extremely complex, even the entire pore channel will be filled in some extreme cases, then the pore channel will change from one-dimensional to two-dimensional (DT = 2).

The relationship between pore fractal dimension and pore permeability is as follows:

$$K = \frac{\mu L0Q}{\Delta p A0} = \frac{\pi}{128}\frac{A}{A0}\frac{D}{3 + DT - D}L_0 - D_T\lambda_{\max}^{3+DT-D}$$

The following conclusions can be drawn:

(1) In a certain range, when the fractal dimension of pore tortuosity is fixed, the permeability of gravelly soil increases with the increase of fractal dimension D.
(2) When the fractal dimension of the pore structure of gravelly soil is fixed, the permeability decreases with the increase in the fractal dimension DT of the pore tortuosity and the tortuosity of the pore pipeline.
(3) The permeability of gravelly soil varies with the maximum pore diameter increasing with the increase of λ (Ding 2018).

4.2 *Research on the diffusion range of concrete in sand*

Permeation grouting of cement slurry in a water-rich sand layer can be simplified as Bingham fluid, considering the starting pressure gradient.

Seepage flow occurs in porous media. The length of the small grouting duct is h, the radius is pr, the grouting pressure is 1 p, and the semi-infinite formation water pressure is 0 p. Considering these factors, the following assumptions are adopted:

(1) A sand stratum is homogeneous, isotropic, and isothermal.
(2) The slurry flows horizontally in a single phase in an infinite formation, resulting in radial seepage.
(3) The slurry is incompressible, and the influence of temperature is not considered during the penetration process.
(4) The influence of gravity on the seepage model is ignored.

According to the theory of seepage mechanics, the seepage grouting of Bingham fluid under constant pressure belongs to passive unsteady seepage, so the general form of the Bingham fluid seepage differential equation can be deduced.

$$\left(1-\frac{4}{3}\lambda H+\frac{1}{3}\lambda^4 H^4\right)\nabla^2 \mathrm{p}-\frac{4}{3}(\lambda+\lambda^4 H^3)\nabla \mathrm{H}\cdot\nabla \mathrm{p}=\frac{\varphi u c_1}{K}\frac{qp}{dt}$$

where ϕ is the porosity of porous media, K is the permeability of porous media, and μ is the dynamic viscosity of Bingham fluid (Zheng & Li 2015).

At this point, we can draw the following conclusions: The dimensionless dynamic boundary diffusion distance of Bingham fluid is related to the permeability of porous media and the dynamic viscosity of Bingham fluid. The greater the permeability is, the greater the diffusion distance is, and the greater the dynamic viscosity is, the smaller the diffusion distance is.

According to the above deduction, for the same kind of sand, its dynamic viscosity is a fixed value, and its permeability is related to the pore fractal dimension; that is, the diffusion range of concrete in the sand can be qualitatively or quantitatively judged by determining the pore fractal dimension (Zhou *et al.* 2020).

5 CONTROL MEASURES FOR EXCESSIVE FILLING COEFFICIENT

According to the analysis of the above-mentioned influencing factors, the reason for the excessive filling coefficient of the bored pile in this project was mainly due to the stratum and the excessive permeability of the sandy soil layer of the project. However, since the modified stratum was located below the pile foundation surface, it was impossible to directly improve the underlying conditions through conventional foundation treatment methods. Therefore, we can improve the permeability of the concrete paste in the sand by improving its initial dynamic conditions and its dynamic viscosity, so as to reduce its filling coefficient. Therefore, the following methods were selected:

1. The drilling and pipe pulling speeds play a key role in the drilling quality and the filling coefficient of concrete. The hole-forming test can also be carried out before construction. During drilling, the drill bit shall rise and fall at a uniform speed, without excessive rise and fall or sudden acceleration. The drilling speed shall be adjusted according to the rock stratum distribution of the geotechnical investigation data according to the specifications, especially when encountering sudden resistance. The situation must be investigated and handled carefully. Through the trial hole-forming test, the optimal drilling speed under specific engineering and geological conditions is sought to avoid the occurrence of hole collapse and slurry leakage in the local stratum and better control the hole-forming quality and concrete filling coefficient.
2. As the penetration range of concrete in the sand is related to its initial grouting pressure, it may cause grout leakage under high pressure, leading to an excessive actual filling coefficient of concrete, while too small grouting pressure may cause insufficient pile density and poor body quality, which may affect project safety. According to the current technical specifications for building pile foundations, the end grouting pressure of pile end grouting shall be determined according to the nature of the soil layer and the depth of the grouting point. For the saturated sand layer at the site, the grouting pressure should be 1.2–4.0 MPa. At the same time, it should be noted that the grouting flow should not exceed 75 L/min. Therefore, before construction, it is also necessary to carry out different grouting pressure contrast experiments for different formations to select the optimal grouting pressure.
3. Although subjective personnel factors are not the main reason for the excessive concrete filling coefficient of the project, the quality of the concrete pile can be improved by avoiding human factors. The cast-in-place pile project has high requirements for professional technology, which requires not only the construction personnel to have sufficient professional knowledge and theory, but also sufficient on-site practical experience. It can be seen from the above analysis that improper operation of construction personnel during construction is also one of the reasons for the excessive filling coefficient of cast-in-place pile concrete. Therefore, it is necessary to strengthen the skill training of construction personnel (Yang *et al.* 2021).

6 SUBSEQUENT FILLING COEFFICIENT AND PILE QUALITY INSPECTION

After using the above methods, the filling coefficient of the subsequent construction of long auger-bore pressure grouting piles in this project has been effectively reduced. During the subsequent construction of more than 1000 piles, the filling coefficient of 30-meter-long piles did not exceed 1.3, and the average filling coefficient was only 1.14. However, the small concrete consumption of 20-meter-short piles led to more residual concrete on site, resulting in more concrete than the actual pile. The calculated filling coefficient was large, but the maximum value did not exceed 1.3, and the average value was 1.22. When the construction was completed and accepted, the qualified rate of pile quality inspection was 91%, which was excellent.

7 CONCLUSIONS

Aiming at the problem that the concrete filling coefficient of the long spiral bored pile was too large in the Qingcheng project in Yunlu, Dali, this paper analyzed the possible influencing factors in detail. This paper examined the filling coefficient and diffusion mechanism of concrete in the sandy soil layer from a microscopic perspective, as well as the influencing factors and causes, by using fractal theory. According to the diffusion mechanism, targeted solutions were proposed, which played an important role in reducing the filling coefficient, controlling the quality of the pile foundation, and ensuring the economic benefits of the project in its later construction. In the construction of a long auger-bore pile, many factors affected the construction quality. It can also be said that in the whole geotechnical engineering field, the problems encountered by different engineering geological conditions had both common and unique aspects. Only by fully understanding the geological conditions and taking targeted improvement measures, we can ensure that the project quality meets the design and specification requirements.

REFERENCES

Bin Zheng, Juhua Li Fractal Permeability Model Based on Kozeny Carman equation [J] *Natural Gas Geoscience*, 2015,26 (01): 193–198

China Academy of Building Sciences. *Technical Code for Building Foundation Treatment* [M] Beijing: China Planning Press, 2012

Fei Yang, Hailiang Wang, Lu Yonghong, *et al* Cause Analysis and Control of Excessive Filling Coefficient of Long Spiral Bored Pile [J] *Building Engineering Technology and Design*, 2021 (6): 2547–2548

Feng Zhou, Yong Xu, Rui Zhu, *et al* Diffusion Characteristics of Grouting Slurry in Sandy Formation [J] *Journal of Building Science and Engineering*, 2020, 37 (5): 182–192

Leijin Xiong *Mechanism and Experimental Study of Seepage Grouting in Water Rich Sand Layer Considering Seepage Effect* [D] Beijing: Beijing Jiaotong University, 2020

Rongjun Ding *Study on Permeability of Gravelly Soil based on Fractal Theory* [D] Kunming: Kunming University of Science and Technology, 2018.

Wei Wang, Wei Ding, Nie Qingke. Analysis and Research on Control of Concrete Filling Coefficient of Bored Pile [J] *Survey Science and Technology*, 2017 (S1): 22 – 26 + 54

Xiuzhu Yang *Theoretical and Experimental Study on Slurry Diffusion under Static and Dynamic Action* [D] Hunan: Central South University, 2005

Ying Wang, Guotai Pan, Key Points and Problem Analysis of Long Spiral Bored Pile Construction Technology [J] *Journal of Chifeng University (Natural Science Edition)*, 2018, 34 (5): 82–83

Yu Boming. Progress in Fractal Analysis of Transport Properties in Porous Media Advances in Mechanics, 2003, 33 (3): 333–346

Water Conservancy and Civil Construction – Oke & Ahmad (Eds)
 ISBN: 978-1-032-58614-4

Influence of umbrella arch systems on stability of soft surrounding rock and safety of support structure

Zhanbiao Li & Da Hu
Central Yunnan Water Diversion Project Co., Ltd. of Yunnan Province, Kunming Yunnan, China

Jiangrong Pei*
China State Key Laboratory of Simulation and Regulation of Water Cycle in River Basin, China Institute of Water Resources and Hydropower Research, Beijing, China

Zhengwei Zhang
Central Yunnan Water Diversion Project Co., Ltd. of Yunnan Province, Kunming, Yunnan, China

Jibin Jiang
Kunming Branch of Central Yunnan Provincial Water Diversion Project Construction Administration Branch, Kunming, Yunnan, China

Hongtao Miao
Central Yunnan Water Diversion Project Co., Ltd. of Yunnan Province, Kunming Yunnan, China

ABSTRACT: In order to realize the full-section construction of the drilling and blasting method in the weak surrounding rock section of a deep-buried tunnel, the reasonable design of an advanced pipe shed is a key step. Taking a hydraulic tunnel with weak surrounding rock as an example, the effects of an advanced pipe shed and the external insertion angle of the pipe shed on surrounding rock displacement, the plastic zone, the anchor, and the internal force of the steel arch were studied by means of numerical simulation. The results showed that the advanced pipe shed support can reduce the plastic yield depth range, volume, surrounding rock convergence value of tunnel excavated rock mass, tension, steel arch stress, and bending moment. For the roof position of the pipe shed, the anchor tension and the bending moment of the steel arch could be reduced by 33.48% and 8.94%, respectively, but the stress of the steel arch could be reduced by only 1.22%. When the extrapolation angle of the pipe shed was 5°–7°, the radial displacement of the top of the hole increased noticeably, and it was almost unchanged when the extrapolation angle was 8°–15°. The stability control of the surrounding rock could be achieved by selecting a 10° extrapolation angle after thorough consideration of the project. The research results could provide not only technical support for the pipe shed design of the project but also a reference for the advanced support measures and parameter determination of similar soft rock projects at home and abroad.

1 INTRODUCTION

In order to prevent the risk of weak rock mass collapse or surface settlement of shallow caverns during the excavation of deep underground caverns, the geotechnical engineering field often uses advanced pipe-shed technology to reinforce the surrounding rock and stimulate the self-stabilization of the rock mass (Volkmann 2007; Oke 2012; Oke 2014; Oke

*Corresponding Author: 969943845@qq.com

DOI: 10.1201/9781003450818-73

2015). Some scholars have carried out related research on the interaction between the pipe shed and rock and soil masses, the control effect of the pipe shed bearing load on surrounding rock deformation, and its joint action with an anchor, a steel arch frame, etc. Some scholars have carried out related research by using the physical model test, field monitoring, and other techniques (Chen 2009; Hong 2014; Li 2009). With the development of numerical simulation technology, simulation software is gradually being used in the study of pipe shed support to control the yield deformation of surrounding rock. According to different engineering cases, a large number of scholars used FLAC3D numerical simulation software to adopt the fine model. From the point of view of surrounding rock deformation, plastic zone depth change, and so on, pipe shed technology has carried out in-depth research on many aspects, such as surrounding rock reinforcement, pipe shed and other support combined bearing, palm face reinforcement, and so on (Lunardi 2000; Li 2014; Ocak 2008; Pashaye 2021; Zhang 2020; Zhou 2009). Many useful conclusions have been obtained. However, due to the differences in underground engineering occurrence environments, design parameters, and support parameters, it is difficult to directly apply the relevant research results from other projects, which need to be analyzed according to the specific engineering conditions.

The excavation diameter of the tunnel of a certain water diversion project was 10 m, the maximum buried depth was 1450 m, and the length of the tunnel section greater than 600 m accumulated 42.18 km, accounting for 67.38% of the total length of the tunnel. The problem of surrounding rock stability was prominent, with a Type IV-V surrounding rock length of 41.99 km, accounting for 67.08% of the total. To solve the existing problems such as collapse and large deformation, the advanced pipe shed technology was proposed to strengthen the surrounding rock at the top of the tunnel in front of the face. In order to study and determine the influence of pipe-shed measures on the stability of the surrounding rock and the stress of the supporting structure, a fine three-dimensional model of a typical section was established by using FLAC3D numerical simulation software in this paper. Considering the actual step-by-step excavation length and support sequence, the plastic zone and deformation of surrounding rock under the pipe shed and the stress changes of bolts and steel arch frames were studied. At the same time, the effects of different extrapolation angles of the pipe shed were compared and analyzed. The research results can provide technical support for the design of the pipe shed for the project. At the same time, it can be used as a reference for the determination of advanced support measures and parameters for similar soft rock projects at home and abroad.

2 NUMERICAL ANALYSIS

2.1 *Design parameters*

The diversion tunnel was excavated by the full-section method, and the initial support of the soft rock mass tunnel section was composed of anchor rods, steel mesh, steel arch frames, and sprayed polypropylene coarse fiber concrete. The specific design parameters are shown in Table 1.

Table 1. Support parameters.

Support Type	Layout	Size (mm)	Spacing (m)	Length (m)
Advance Pipe Shed	The Vault is Distributed at 120 Degrees.	ϕ108	0.9	12
Anchor Rods	Full Cross-Section Distribution	ϕ25	1.0 × 1.0	8
40U-Shaped Steel	Full Cross-Section Distribution	171 × 141.9	0.5	–
I22a-Shaped Steel	Full Cross-Section Distribution	73 × 200	0.5	–
Concrete	Full Cross-Section Distribution	250	–	–

The research was mainly carried out on the V-type surrounding rock, and the physical and mechanical parameters of this type of rock mass are shown in Table 2.

Table 2. Physical and mechanical parameters of surrounding rock.

Young's Modulus (Gpa)	Poisson's Ratio	Cohesion (MPa)	Friction Angle (°)	Dilation Angle (°)	Density (kg/m³)
2	0.3	0.5	28.81	8.81	2200

2.2 *Model building*

The refined three-dimensional model established is shown in Figure 1. The overall size of the model was 70 m × 70 m × 60 m, including initial support such as shotcrete, an advanced pipe shed, and anchor rods. Normal constraints were applied around and at the bottom of the model, and stress boundaries were applied to the upper boundary of the model to simulate the self-weight stress of the overlying rock mass (H = 775 m). Assuming that the surrounding rock was a homogeneous and isotropic continuum, the influence of the creep characteristics of the rock mass was not considered, and the Mohr-Coulomb yield criterion was followed.

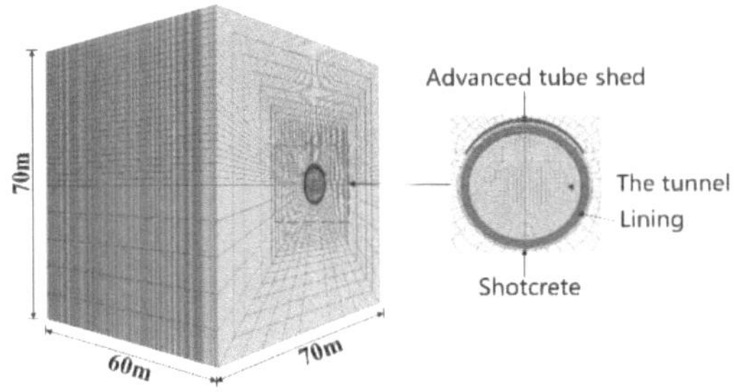

Figure 1. Three-dimensional numerical analysis model.

The top arch is 120° with an advanced grouting pipe shed, using 12 m of ϕ108@40 × 900 cm hot-rolled seamless steel pipe. The joints were 15 cm threaded butt joint connections, each pipe shed was inserted at an angle of 10°, and the end was welded to the steel arch frame's outer edge (Figure 2). The pure cement slurry with a water-cement ratio of 1:1 for grouting was used. The initial pressure was 0.5–1.0 MPa, the final pressure was 2.0 MPa, and the diffusion radius was not less than 50 cm. Considering the strengthening effect of grouting on the surrounding rock after diffusion, and referring to other engineering experiences, a surrounding rock reinforcement ring was set around the pipe shed in the simulation analysis to simulate the effect of grouting reinforcement (Figure 2).

(a) Umbrella arch (b) Reinforcement range

Figure 2. Umbrella arch and its reinforcement range.

2.3 *Support parameters*

The advanced tube shed and steel arch frame were simulated by a beam unit, the anchor rod was a cable unit, and the others were solid units. The elastic modulus of the pipe shed and the steel arch was converted by the equivalent section method, and the specific parameters are shown in Table 3.

Table 3. Support parameters in numerical simulation.

Support Type	Young's Modulus (Gpa)	Poisson's Ratio	Density (kg/m^3)	Cohesion (MPa)	Friction Angle (°)
Advance Pipe Shed	92	0.3	3386	–	–
Pipe Shed Reinforcement Area	2.9	0.3	2200	0.59	32.62
C25 Concrete	28	0.3	2500	–	–
C30 Concrete	30	0.3	2700	–	–
40U-Shaped Steel	60	0.3	7850	–	–
I22a-Shaped Steel	63.1	0.3	7850	–	–
Anchor Rods	130	0.3	7850	–	–

3 UTILITY ANALYSIS OF PIPE SHED

3.1 *Monitoring section*

To analyze the influence of pipe shed support on the stability of surrounding rock, a monitoring section was set at Y = 24 m, and monitoring points were set on the roof, side wall, arch waist, and bottom of the cave (Figure 3) to obtain the surrounding rock during excavation.

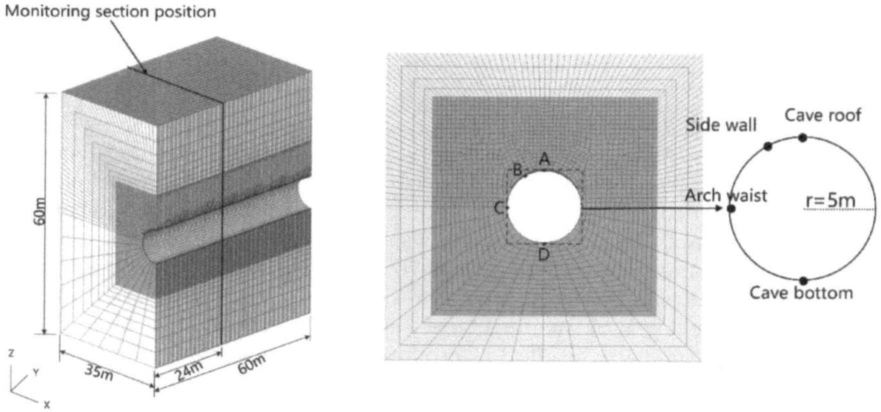

(a) Monitoring section (b) Monitoring point layout

Figure 3. Tunnel monitoring section and layout.

3.2 *Surrounding rock stability*

After step-by-step excavation with a step length of 3.0 m and additional supports such as bolts, shotcrete, and steel arches, the plastic zone of the surrounding rock with or without a pipe shed is shown in Figures 4 and 5.

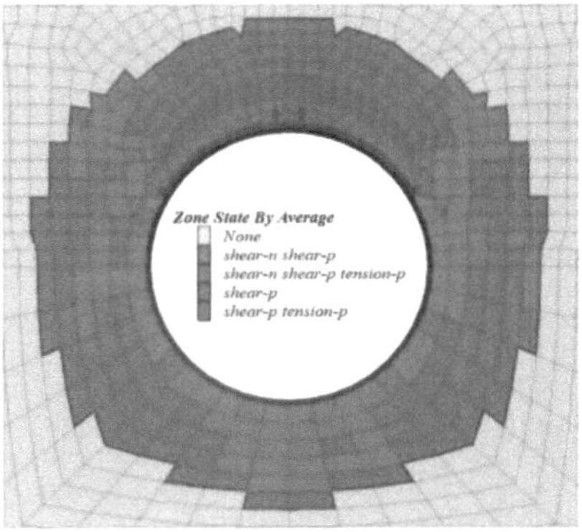

(a) Front view

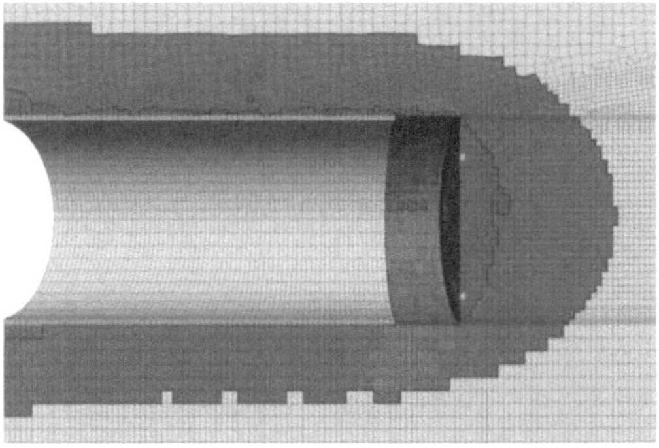

(b) Sectional view

Figure 4. Plastic zone of surrounding rock without pipe shed support.

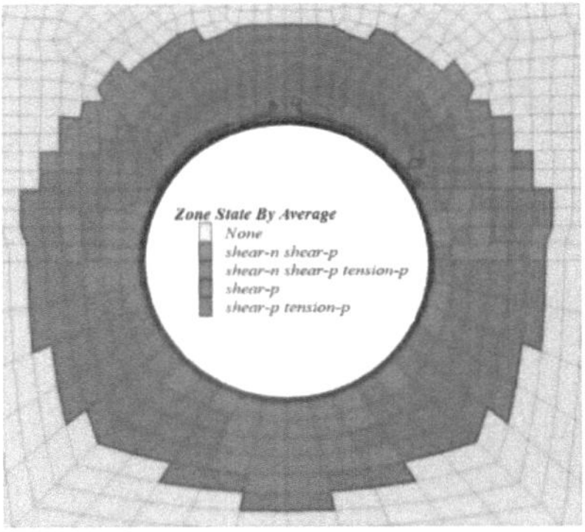

(a) Front view

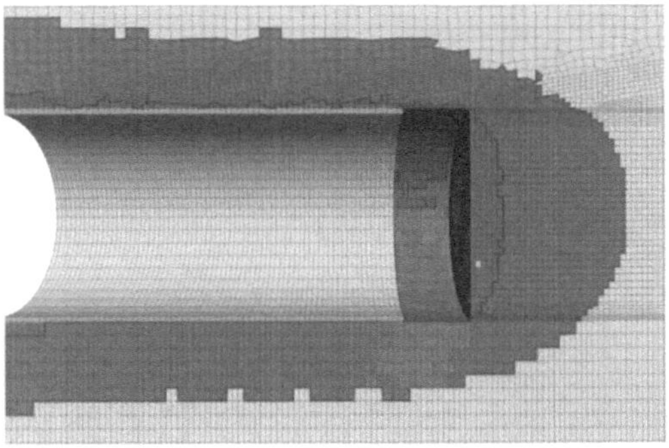

(b) Sectional view

Figure 5. Plastic zone of surrounding rock with pipe shed support.

It can be seen from the figure that the plastic zone of the surrounding rock, with or without the pipe shed, was mainly distributed in the range of 0.5 times the depth of the tunnel diameter in the radial direction and 1 time the depth of the tunnel diameter in front of the tunnel face. It was cracked and yielded, but the depth of the plastic zone changed slightly. The maximum depth of the plastic zone was 5.01 m when there was no tube shed, and it was 4.68 m when there was a tube shed, which was correspondingly reduced by 0.33 m. At the same time, after analyzing statistics on the volume of the plastic zone of the model under the same tunnel face excavation pile number with or without the tube shed, it was found that the volume of the plastic zone was 11202.2 m^3 when there was no tube shed and 10783.1 m^3 when there was a tube shed. At Y = 24 m, the four monitoring points A, B, C, and D at the monitoring section changed with the progress of the tunnel face, as shown in Figure 6.

Figure 6 shows that each monitoring point basically had a certain displacement when the tunnel surface did not reach the position of the monitoring surface, and the displacement when the excavation surface reached the monitoring surface was basically 1/3 of the final displacement. Gradually advancing away, the displacement of the surrounding rock gradually increased and finally stabilized. The maximum radial displacement of the cave roof (monitoring point A) was 14.80 cm when there was no pipe shed support and 14.36 cm when

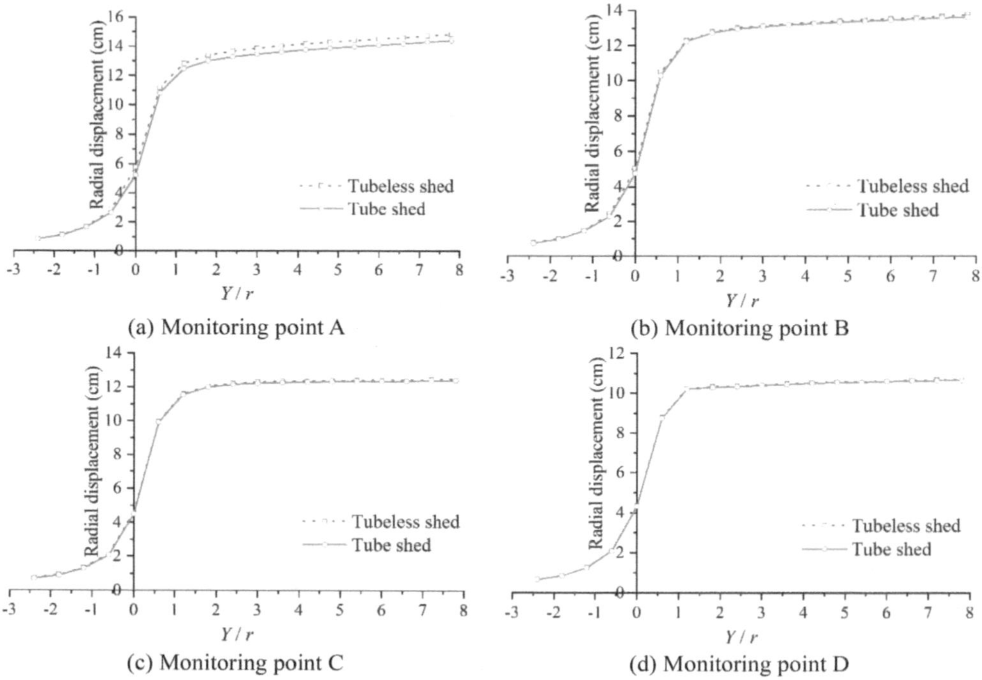

Figure 6. The displacement of each monitoring point varies with the advance of the palm face.

there was pipe shed support. Under the action, the radial displacement of the cave roof was relatively reduced, and the advanced support of the pipe shed could effectively restrain the deformation of the surrounding rock within its range. At the same time, it could be seen from the figure that the displacement of monitoring point A relative to the roof of the cave changes significantly, and the displacement of the side wall, arch waist, and cave bottom (monitoring points B–D) did not change significantly under the pipe shed. It could strengthen and limit the deformation of the surrounding rock at the support position of the pipe shed, but it could not increase the stability of the surrounding rock of the entire section, and the surrounding rock still needed to be supported by shotcrete, anchor rods, and steel arches.

3.3 *Support security*

Due to the deformation and yielding of the surrounding rock, the continuous advancement of the working face will cause stress changes on the bolts and steel arches. After the deformation of the surrounding rock was stable, the force on the bolts at the monitoring section is shown in Figure 7.

It can be seen from the figure that the internal force of the anchor with or without the tube shed basically showed that the value of the arch waist was the largest, and the position of the bottom and the vault was relatively low. For a single anchor, the maximum internal force generally appeared near the middle. The maximum axial force of the anchor rod was 186.41 kN when there was no tube shed, and the maximum axial force was 178.29 kN when there was an anchor rod. At the same time, at the top of the cave, the internal force of the anchor rod was 64.51 kN when there was a pipe shed and 98.47 kN when there was no pipe shed,

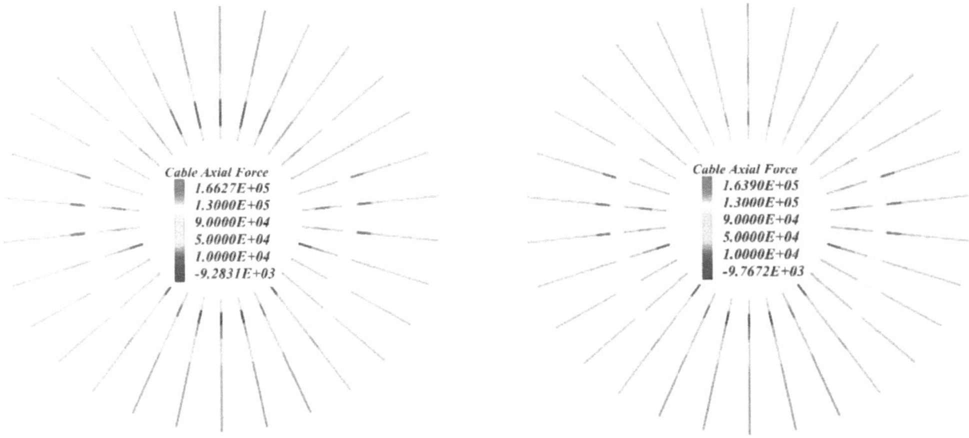

(a) Tubeless shed (b) Tube shed

Figure 7. With or without the change of axial force of anchor under pipe shed (N).

which meant that the pipe shed could significantly reduce the tension of the anchor rod at the position where the pipe shed was used. The axial stress of the steel arch at the corresponding position is shown in Figure 8.

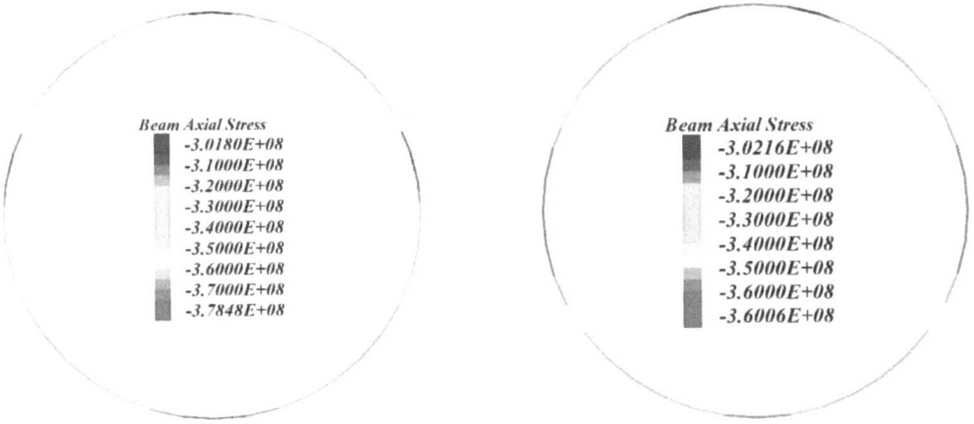

(a) Tubeless shed (b) Tube shed

Figure 8. With or without axial stress of steel arch under pipe shed (Pa).

It can be seen from the figure that the axial stress distribution of the steel arch with or without the tube roof was basically the same. For the cave roof position, the maximum compressive stress of the steel arch was 302.16 MPa without the tube roof, which was basically the same as 301.79 MPa with the tube roof. There were some differences in overall stress. When there was no tube shed, the maximum axial compressive stress was located at the upper part of the arch waist, with a value of 378.48 MPa. With the existence of the pipe shed, the maximum axial compressive stress was basically located at the arch waist. The value was 360.06 MPa, indicating that after the surrounding rock was strengthened by the pipe shed, the force on the steel arch could be reduced. The bending moment of the steel arch is shown in Figure 9.

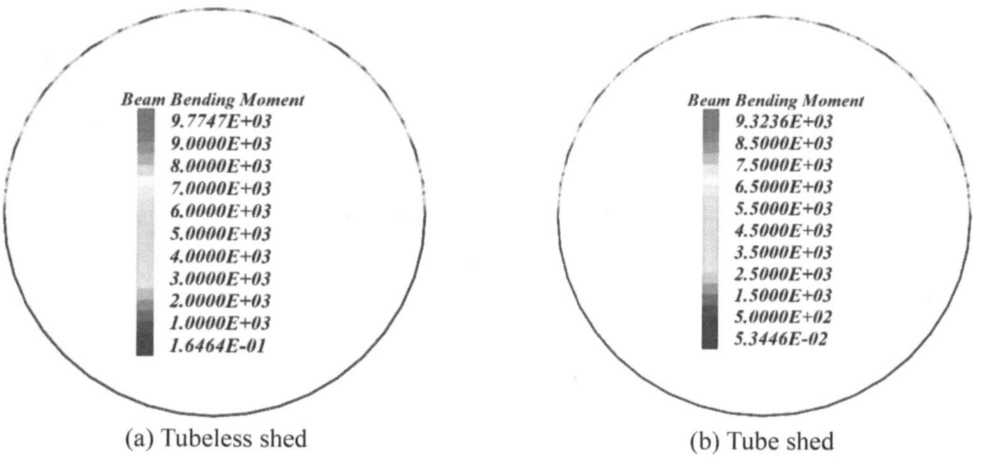

Figure 9. Bending moment of steel arch frame with or without tube shed (N·m).

Figure 9 shows that the bending moment distribution form of the steel arch frame was the same, and the larger value was distributed above the arch waist. The maximum bending moment was 9.77 kN·m without a pipe shed and 9.32 kN·m when there was a pipe shed. The time was 7.33 kN·m, and the pipe shed could reduce the bending moment value of the steel arch.

The results of the comprehensive analysis showed that after the top arch 120° advanced grouting pipe shed support measures were applied to this project, the deformation of the surrounding rock at the top of the cave, the depth and volume of the plastic zone, the tensile force of the anchor rod, and the axial direction of the steel arch frame could be effectively reduced. The construction of a steel arch may increase the stability of the surrounding rock and the safety of the support structure as compared to no advanced pipe shed in terms of compressive stress, steel arch bending moment, etc.

4 EFFECT OF EXTRAPOLATION ANGLE

The external insertion angle of the pipe shed is different, and its force and supporting effect on the surrounding rock will change accordingly. Using the same rock mass and support structure parameters, the pipe shed external insertion angle was studied from 5°–15° equidistant. The influence of the change on the deformation of the surrounding rock, considering that the pipe shed has little effect on the deformation of the side wall, the arch waist, and the position of the floor, caused the deformation of the surrounding rock at the top of the cave to change with the advancement of the tunnel face under different inset angles of the pipe shed, which are shown in Figure 10.

It can be seen from the figure that the deformation of the surrounding rock changes with the advancement of the face under different pipe shed angles, and the changing trend is the same, but the control effect of the pipe shed on the displacement of the cave roof is slightly different. It is large, but when it exceeds a certain value, the change slows down. For

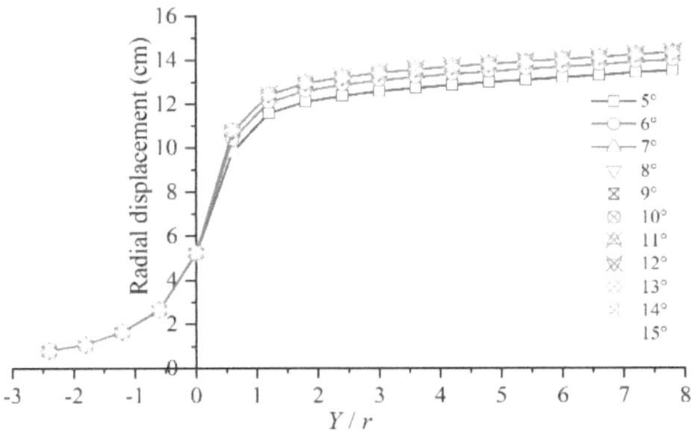

Figure 10. Changes in the displacement of the roof of the tunnel under different interpolation angles of the pipe shed (monitoring point A).

example, in this project, when the outer insertion angle of the leading pipe shed was 5°–7°, with the increase in the outer insertion angle of the leading pipe shed, the diameter of the cave roof increased toward displacement, but when the extrapolation angle was 8°–15°, there was almost no change in the radial displacement of the roof at the extrapolation angle. It should be noted that if the external insertion angle were reduced, on the one hand, it would make it difficult for the drilling rig to make holes, and if the pipe shed's external insertion angle was too large, it would cause an increase in the corresponding cost. This project considered comprehensive factors and selected a 10° external insertion angle to meet the stability control requirements of the surrounding rock.

5 CONCLUSIONS

In this paper, the numerical simulation method was used to study and compare the surrounding rock stability and support safety of a deep-buried hydraulic tunnel with or without pipe shed measures in the soft rock mass section and the deformation of the surrounding rock when the pipe shed's external insertion angle changed. The values were analyzed, and the specific conclusions are as follows:

(1) After the excavation of the project, the plastic zone of the surrounding rock was mainly distributed in the range of 0.5 times the depth of the tunnel diameter in the radial direction and 1 time the depth of the tunnel face in front of the tunnel face. It was generally U-shaped, and the yield type was mainly shearing yield and partial tension cracking.
(2) When bolts, steel arches, and shotcrete support were used comprehensively, the depth of the plastic zone under the tubeless shed was 5.01 m, the maximum deformation of the surrounding rock reached 14.80 cm, the maximum tensile force of the anchors was 186.41 kN, the maximum stress of the steel arch was 378.48 MPa, the maximum bending moment was 9.77 kN·m, the depth of the plastic zone was 4.68 m in the case of a pipe shed, the maximum deformation of the surrounding rock was 14.36 cm, the maximum tensile force of the anchor was 178.29 kN, the maximum stress of the steel arch was 360.06 MPa, and the maximum bending moment was 9.32 kN·m.
(3) The pipe shed could significantly reduce the tensile force of the anchor rod and the bending moment of the steel arch at the position of the pipe shed, which could be reduced by 33.48% and 8.94%, respectively, but had little effect on the stress of the steel arch, only 1.22%.

(4) When the external insertion angle of the pipe shed in this project was 5°–7°, the radial displacement of the roof of the cave increased significantly with the increase of the external insertion angle, and it was almost unchanged when the external insertion angle was 8°–15°. The project was given careful consideration, and a 10° extrapolation angle was chosen.

ACKNOWLEDGEMENTS

This work was financially supported by grants from the Major Science and Technology Projects in Yunnan Province (Grant No.202102AF080001).

REFERENCES

Chen Weitao, Wang Mingnian, Zhang Lei,Wei Longhai, He Yulong (2009). Influence of Pre-reinforcement Measures on Excavation Stability of Tunnelling, *J. Chinese Journal of Rock Mechanics and Engineering*, 2009 (8): 1640–1645.

Hong Kairong, Yang Chaoshuai, Li Jianhua (2014). Analysis on Impact of Advanced Support on Space Deformation of Tunnel in Soft Rock Mass, *J. Chinese Journal of Underground Space and Engineering*, 10 (2): 429–433.

Li Bin, Qi Taiyue, Kuang Wentao, *etc* (2009). Application of Nitm in the Design of Liu Yang River Tunnel, *J. Modern Tunnelling Technology*, (4): 83–88.

Li Ning (2014). *Numerical Simulation of the Pre-consolidation and the Excavation of Large Cross-section Tunnel(D)*, Beijing, Beijing Jiaotong University.

Lunardi P. (2000). The Design and Construction of Tunnels Using the Approach Based on the Analysis of Controlled Deformation in Rocks and Soils, *J. Tunnels and Tunnelling International*, 3–30.

Ocak I. (2008). Control of Surface Settlements with Umbrella Arch Method in Second Stage Excavations of Istanbul Metro, *J. Tunnelling and Underground Space Technology*, 23 (6): 674–681.

Oke J., Vlachopoulos N., Diederichs M.S. (2012). Improved Input Parameters and Numerical Analysis Techniques for Temporary Support of Underground Excavations in Weak Rock, *J. RockEng.* Edmonton.

Oke J., Vlachopoulos N., Diederichs M.S. (2014). Numerical Analyses in the Design of Umbrella Arch Systems, *J. Journal of Rock Mechanics and Geotechnical Engineering*, 6 (6): 546–564.

Oke J., Vlachopoulos N., Diederichs M.S. (2015). *Recent Advances in the Design and Understanding of Umbrella Arch Systems*, C//13th ISRM International Congress of Rock Mechanics. OnePetro.

Pashaye N., Fard Moradinia S., Ferdousi A. (2021). Using Umbrella Arch Method in Design of Tunnel Lining, Case Study: Water Transfer Tunnel of Kani-sib, Urmia Lake, *J. Journal of Numerical Methods in Civil Engineering*, 6 (2): 1–13.

Volkmann G.M., Schubert W. (2007). Geotechnical Model for Pipe Roof Supports in Tunneling, C//Proc. of the 33rd ITA-AITES World Tunneling Congress, Underground Space–the 4th Dimension of Metropolises, 1: 755–760.

Zhang Yiteng, Wang Mingnian, Yu Li, *etc* (2020). Study on the Influence of Advanced Support on the Stability of Tunnel Face in Soft Surrounding Tocks, *Modern Tunnelling Technology* (S1), 57 (S1): 119–128.

Zhou Jie, Qi Taiyue, Kuang Wentao, *etc* (2009). Numerical Simulation on Stability of Tunnel with Large Cross-section in Process of Pre-reinforcement, Excavation and Support, *J. Tunnel Construction*, 29 (2): 185–188.

Water Conservancy and Civil Construction – Oke & Ahmad (Eds)
© 2024 The Author(s), ISBN: 978-1-032-58614-4

Seismic performance level of a framed underground structure

Weishen Li* & Wenting Li*
School of Civil Engineering, Shanghai Normal University, Shanghai, China

ABSTRACT: Underground structures such as subway stations, which are operated at a high capacity and frequency, are prone to damage by large earthquakes, as seen in recent times. The seismic performance level classifications of an underground structure are very important in its seismic design. However, the present classifications of seismic performance levels for underground structures have not reached an agreement. The paper reviews the present classifications and classifies the seismic performance level of a one-story and two-span frame subway station. The seismic performance of the underground structure was simulated by the three-dimensional (3D) nonlinear finite element model. The structural damage under various seismic intensities was discussed and compared. The research results showed that structural damage was highly related to structural inter-story displacement. When the underground structure in this study was nearly undamaged after an earthquake, the structural inter-story displacement ratio was 1:1011; when the structure was slightly damaged, the structural inter-story displacement ratio was 1:415; when the structure was heavily damaged, the structural inter-story displacement ratio was 1:210; and when the structure was almost collapsed, the structural inter-story displacement ratio was 1:126.

1 INTRODUCTION

Subway networks are an important component of the planning and expansion of cities. The damage to the subway station in Japan during the 1995 Hanshin earthquake prompted research on the seismic performance of underground structures (Du 2018; Lin 2009). It is necessary to study the seismic performance of underground structures (Han 2021; Lu 2022). At present, the classification standards for the seismic performance level of underground structures have not reached an agreement. Wang et al. (2011) proposed a method for evaluating the deformation capacity and strength of structural members. Using the pushover method, Du et al. (2019) investigated the seismic performance level of subway stations. Zhuang (2019) investigated the classification of the seismic performance level of a framed subway underground station structure under seven different site types and gave the structural inter-story displacement ratio concerning the four performance levels. Dong et al. (2018) examined 25 cases and calculated the limited structural inter-story displacement ratio. It is not difficult to find that most of the present research was conducted to study the seismic performance level of underground structures by using the method of ground structures as a reference or simply analyzed the underground structures by using two-dimensional models. However, there are few types of research on dynamic time history analysis using a three-dimensional model. This method can reflect the seismic performance of underground structures more comprehensively and accurately.

*Corresponding Authors: lws5974@163.com and lwt@shnu.edu.cn

DOI: 10.1201/9781003450818-74

The paper established a three-dimensional finite element model based on a one-story and two-span frame subway station structure. A dynamic time-history analysis was conducted. The structural performance under various seismic intensities was compared and discussed.

2 OVERVIEWS OF THE PROJECT

This paper employed a single-story and two-span station (Figure 1) with a total height of 7.2 m, a total width of 15.7 m, and a burial depth of 8 m. The cross-sectional dimensions of the center column were 0.7 m × 1.1 m, and the distance between columns was 8 m. The thicknesses of the upper slab, lower slab, and side walls were 0.7 m, 0.9 m, and 0.7 m, respectively. The concrete of the middle column was C50, and the other parts were C30. The structural longitudinal reinforcement was HRB400, and the hoop reinforcement was HRB300.

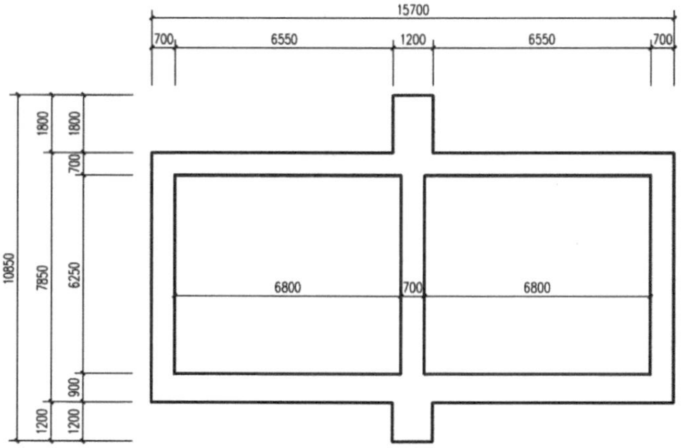

Figure 1. Section of the underground structure (mm).

He (2018) simplified the site into eight horizontal soil layers. The properties of the site are listed in Table 1. According to the Code for Seismic Design of Buildings (MOHURD 2010), the site in this study was classified as Grade II.

Table 1. Site properties.

Number	Type	Thickness /(m)	Unit weight / (kg/m^3)	Shear Wave Velocity (m/s)	Poisson Ratio
1	Clay	4.0	1850	87	0.35
2	Clay	4.0	1830	110	0.38
3	Clay	2.0	2000	244	0.40
4	Sand	12.0	1935	225	0.30
5	Clay	16.0	2000	400	0.30
6	Sand	10.0	2000	420	0.30
7	Clay	8.0	2000	400	0.30
8	Rock	4.0	2000	550	0.30

The Northridge wave, also known as the input seismic wave, was measured in 1994 at the LA-Wonderland Ave station with a peak acceleration of 0.103 g. The acceleration time history and acceleration response spectrum are shown in Figure 2. The seismic wave was input at the bottom of the model along the X direction. The peak acceleration was first adjusted to 0–0.5 g when inputting seismic waves, and the incremental step was set to 0.1 g to simply estimate the peak acceleration interval corresponding to each performance level limit. Then, to obtain more accurate limits, further smaller intervals were set for 0–0.2 g and 0.2–0.4 g, with incremental steps of 0.025 g and 0.05 g, respectively.

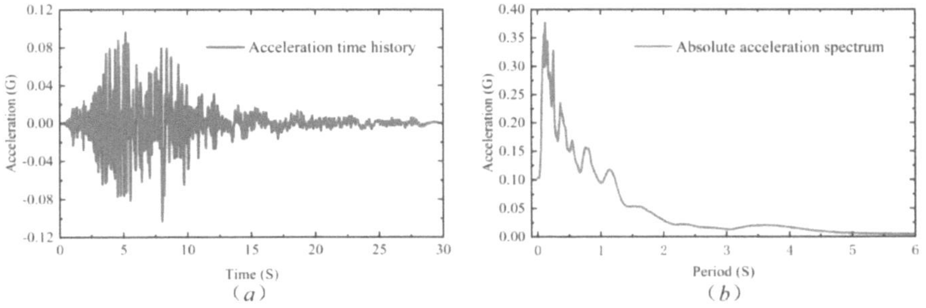

Figure 2. Acceleration time history and acceleration response spectrum of Northridge wave: (a) Acceleration time history; (b) Acceleration response spectrum.

3 FINITE ELEMENTS MODEL

The general finite element software ABAQUS was used to establish the 3D nonlinear finite element model of the subway station-soil interaction system, as shown in Figure 3. The concrete-plastic damage model was employed to model the behavior of concrete material, and the bilinear model was used to model the steel bars' material. The interaction of concrete and bars was modeled by using embedded techniques. The interface between the structure and the ground was modeled as a frictional surface with a friction coefficient of 0.4 (Ma 2019). Soil nonlinearity was simulated by equivalent linearization (Bardet 2000) by using the parameters in Seed et al.'s work (1970). The curves of dynamic shear modulus and equivalent damping ratio concerning strain are shown in Figure 4.

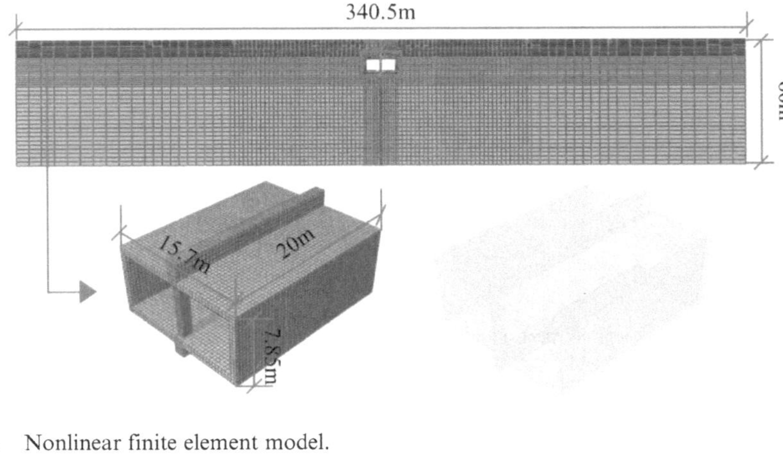

Figure 3. Nonlinear finite element model.

The damping was simulated by using the full-form Rayleigh damping (Xu 2019), with a stiffness-dependent damping coefficient and a mass-dependent damping coefficient β expressed as

$$\alpha = 2\zeta \frac{\omega_1 \omega_2}{\omega_1 + \omega_2} \tag{1}$$

$$\beta = 2\zeta \frac{1}{\omega_1 + \omega_2} \tag{2}$$

where ζ is the equivalent damping ratio; ω_1 and ω_2 are the first and second natural frequencies.

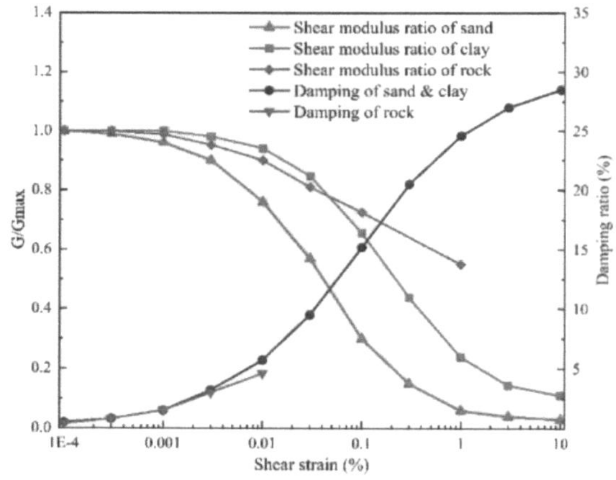

Figure 4. Soil properties (Seed 1970).

A repeatable boundary condition was employed, which was important to prevent waves generated at the model boundary from re-entering the FE model. The artificial lateral boundary was set at 162.4 m from the structure. The ground stress was first balanced before introducing the seismic load (Wang 2017).

4 PERFORMANCE LEVEL CLASSIFICATION

According to Seismic Design Standards for Underground Structures (MOHURD 2018) and other research (Du 2019), the seismic performance levels of underground structures are classified in Table 2.

Table 2. Classification of seismic performance level of underground structures.

Performance Level	Status	Description
I	Undamaged	The structure is in an elastic state. Only side walls or floor slabs locally appear to have slightly visible cracks. The middle column is not cracked. The structure does not need to be repaired or only needs minor repairs. The damage does not affect normal operation.

(continued)

Table 2. Continued

Performance Level	Status	Description
II	Minor Undamaged	The structure is in an elastic state. The floor slab and side wall have visible cracks but no plastic hinges at the joints. The middle column has microscopic cracks, but the structure as a whole is intact. The structure can be restored to normal use after a short period of minor repair.
III	Heavy Undamaged	The structure is in the elastic-plastic stage. Macro cracks appear in the floor slab, side walls, and their connections. Visible cracks appear at the end of the middle column. Plastic hinges initially appear in the connection between the floor slab and the side walls. It takes a long time to repair before it can be used.
IV	Collapse	Cracks in structural members are serious. The structure as a whole is close to collapsing. It is difficult to repair.

5 RESULTS

Existing research (Zhou 2016) indicated that the tensile damage of underground structures during an earthquake was much greater than the comprehensive damage. The concrete possessed visible cracks when the concrete tensile damage factor was greater than 0.75 (Zhou 2016). The seismic damage distribution of the subway station structure is shown in Figure 5.

When the peak acceleration of the input seismic wave was as small as 0.075 g, the inter-story displacement angle of the subway station was 1:1011. Figure 5(a) depicts its tensile damage. Only tensile damage and minor visible cracks occurred in the side walls and floor slabs of the subway stations, which only required quick repair. Therefore, the subway station was in the "linear elastic working stage," also called the "functional normal stage."

When the peak acceleration of the input seismic wave was 0.15 g, the inter-story displacement angle of the subway station was 1:415. Its tensile damage is shown in Figure 5(b). The floor slab and side walls of the subway station showed a large area of visible cracks, but the floor slab and side walls had not formed plastic hinges at this time. In addition, the center column also showed slight damage. The station structure as a whole was intact, and only short-term repairs to the visible cracks in the floor slab and side walls were required to restore normalcy. Therefore, the subway station as a whole was basically in the flexible working stage, also called the "minor repairs" stage.

When the peak acceleration of the input seismic wave was 0.25 g, the inter-story displacement angle of the subway station is 1:210. Figure 5(c) depicts its tensile damage. The visible cracks in the floor slab and side walls of the subway station developed more obviously, and the floor slab and side walls formed preliminary plastic hinges. What's more, visible cracks appeared in the center column. The subway station as a whole was in good condition, but the floor slab, sidewalls, and the joints between them and the center column needed to be reinforced, repaired, and waterproofed before they can be restored to service. Therefore, the subway station as a whole was in the elastic-plastic working stage at this time, also called the available stage for major repairs.

When the peak acceleration of the input seismic wave was 0.35 g, the inter-story displacement angle of the subway station was 1:126. Figure 5(d) depicts its tensile damage. The plastic hinge had been fully formed at the connection between the floor slab and the side wall of the subway station, and the overall crack penetration of the subway station was obvious. Furthermore, the floor slab and the side walls were damaged, and the center column was severely damaged. Therefore, at this time, the subway station as a whole was close to collapse and difficult to repair and was in the stage of collapse prevention.

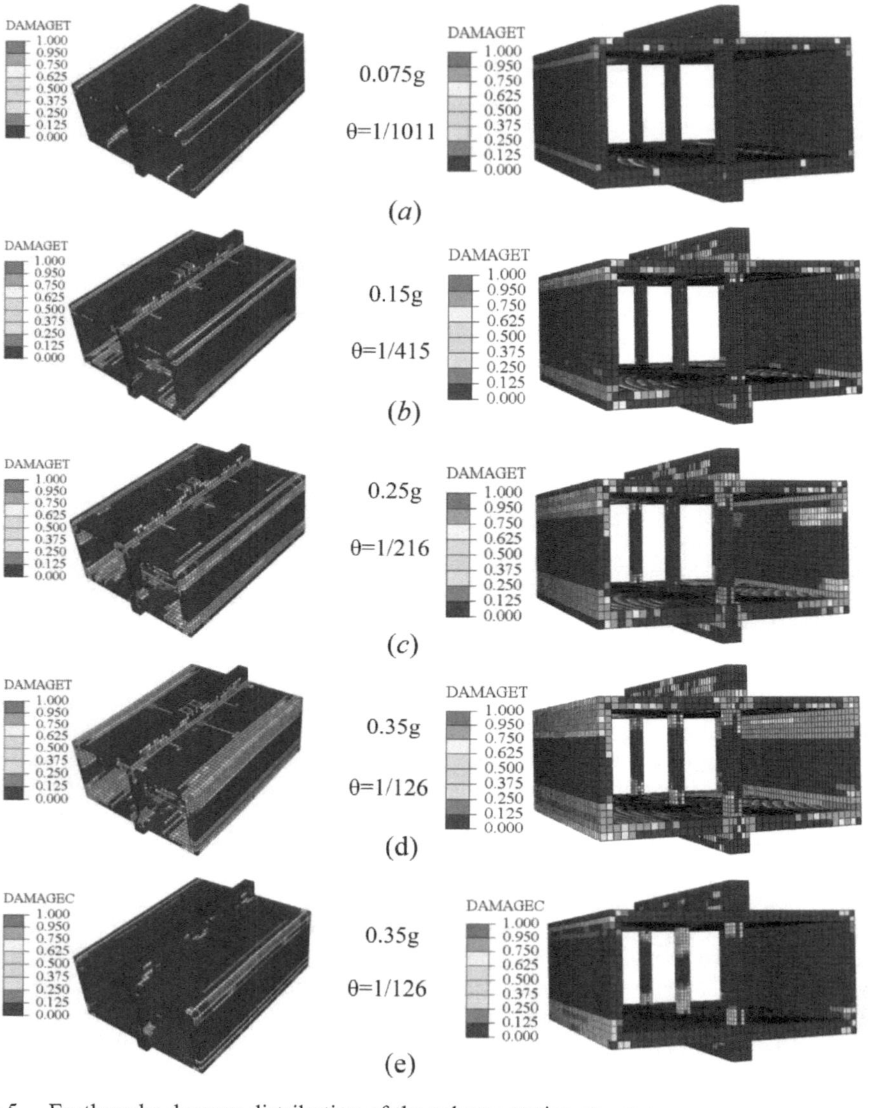

Figure 5. Earthquake damage distribution of the subway station structure.

In addition, when the peak acceleration of the seismic wave was 0.35 g, the compressive damage to the subway station is shown in Figure 5(e). The overall subway station only showed slight damage, which was much smaller than the tensile damage of the subway station under the peak acceleration of the same seismic wave.

6 CONCLUSIONS

In this paper, a nonlinear refined finite element 3D model of the subway station-soil interaction system was established for a single-story and two-span subway station, and the peak ground vibration acceleration was adjusted for dynamic time analysis. Based on the reference to existing codes and relevant studies, the seismic performance level of the underground structure was defined, and the corresponding functional states were described. The following conclusions can be drawn from the analysis of the results of the subway station structure:

(1) Performance levels I, II, III, and IV corresponded to inter-story displacement ratios of 1:1011, 1:415, 1:210, and 1:126, respectively. All of them were within the range of existing studies, indicating that the performance level limits of underground structures obtained in this paper were reasonable.
(2) The plastic damage occurred at the connections of structural members. The structure's plastic damage first appears at beam-to-roof slab connections, followed by damage to beam-to-roof slab connections, columns' tops, walls' tops, roof slab edges, and walls' bottom.

ACKNOWLEDGMENTS

This work is funded by the National Natural Science Foundation of China (Grant No. 52008248).

REFERENCES

Bardet J P & Ichii K. (2000). *EERA: A Computer Program for Equivalent-linear Earthquake Site Response Analyses of Layered Soil Deposits*. R. California, Department of Civil Engineering, University of Southern California.

Dong Z & Zeng F. (2018). Research on the Seismic Performance Limits of Rectangular Underground Stations Based on IDA Method for Inter-story Displacement Ratio. *J. Modern Tunnel Technology*. 55 (S2), 441–449.

Du X & Jiang J. (2019). Study on Calibration of Seismic Performance Index of Shallowly Buried Rectangular Frame Subway Station Structure. *J. Journal of Civil Engineering*. 52 (10), 111–119.

Du X & Li Y. (2018). Progress of Research on the Causes of Earthquake Damage and Mechanism Analysis of Disaster Formation in the Daikai Subway Station of the 1995 Hanshin earthquake in Japan. *J. Journal of Geotechnical Engineering*. 40 (02), 223–236.

Han B & Yang Z. (2021). A Review of Statistics and Analysis of Urban Rail Transit Operations in the World in 2020. *J. Metropolitan Express Transportation*. 34 (01), 5–11.

He Z & Chen Q. (2018). Exploration on the Performance Index of Vulnerability of Typical Underground Space Structures Considering Vertical Seismic Effects. *J. Quarterly Journal of Mechanics*. 39 (01), 117–125.

Lin G & Luo S. (2009). Earthquake Damage of Subway Structures and Measures to Deal with it. *J. Modern Tunnel Technology*. 46 (04), 36–41.

Lu D & Ma C. (2022). Progress of Research on Seismic Toughness of Urban Underground Structures. *J. Chinese Science: Technical Science*. 52 (10), 1469–1483.

Ma C & Lu D. (2019). Structural Components Functionalities and Failure Mechanism of Rectangular Underground Structures During Earthquakes. *J. Soil Dynamics and Earthquake Engineering*. 119, 265–280.

MOHURD. (2010). *Code for Seismic Design of Buildings*. S. Beijing.

MOHURD. (2018). *Seismic Design Standards for Underground Structures*. S. Beijing.

Seed H B & Idriss I M. (1970). Seismic Response of Soil Deposits. *J. Journal of the Soil Mechanics & Foundations Division*. (96), 631–638.

Wang G & Xie W. (2011). Research on the Evaluation Method of Seismic Performance of Underground Frame Structure. *J. Journal of Geotechnical Engineering*. 33 (04), 593–598.

Wang X & Zhuang H. (2017). Study on the Effect of Underground Diaphragm Wall on the Seismic Response of Stacked Wall Subway Station Structure. *J. Journal of Geotechnical Engineering*. 39 (08), 1435–1443.

Xu Z & Du X. (2019). A Comparative Study of Site Rayleigh Damping Construction Methods in Seismic Response Analysis of Underground Structures. *J. Geotechnical Mechanics*. 40 (12), 4838–4847.

Zhou H & Zheng G. (2016). Analysis of Nonlinear Seismic Response of Soft Ground Iron Structures. *J. Journal of Tianjin University (Natural Science and Engineering Technology Edition)*. 49 (04), 361–368.

Water Conservancy and Civil Construction – Oke & Ahmad (Eds)
© 2024 The Author(s), ISBN: 978-1-032-58614-4

Seismic vulnerability analysis of existing frame shear wall structure based on layered shell element

Yi Zhang
CEA Key Laboratory of Earthquake Monitoring and Disaster Mitigation Technology, Guangdong Earthquake Agency, Guangzhou, China

Shiqian Ding*
Shenzhen Academy of Disaster Prevention and Reduction, Shenzhen, China

Jin Li
CEA Key Laboratory of Earthquake Monitoring and Disaster Mitigation Technology, Guangdong Earthquake Agency, Guangzhou, China

Hui Jiang
Shenzhen Academy of Disaster Prevention and Reduction, Shenzhen, China

Jiajian Zhu
CEA Key Laboratory of Earthquake Monitoring and Disaster Mitigation Technology, Guangdong Earthquake Agency, Guangzhou, China

ABSTRACT: Frame shear wall structure is widely distributed in China because of its flexibility and ease of use, and the study of its seismic performance is of great importance. In this paper, an elastoplastic analysis model of a frame shear wall structure was established based on the layered shell elements, and its IDA analysis was carried out to evaluate the seismic performance of the structure based on the seismic susceptibility probability of each limit state under different intensity earthquakes. The results showed that the seismic response of this structure was minimal due to the existence of a "cylinder"-like structure in the second layer alone, while the seismic response of the remaining layers increased gradually from the bottom to the top layer. The probability of the performance level of this structure met the requirements of seismic protection under multiple earthquakes, fortified earthquakes, and rare earthquakes.

1 INTRODUCTION

The frame-shear wall structure is a structural system formed by combining two structures, the frame, and the shear wall, together. The vertical load of the house is shared by the frame and shear wall, respectively, while the horizontal action is mainly borne by the shear wall with higher lateral stiffness. This structure has the characteristics of a flexible layout and convenient use of a frame structure, with greater stiffness and a strong seismic capacity, and is widely used in public buildings and civil buildings. China is an earthquake-prone country, and almost all provinces have areas with a fortification intensity of 7 degrees or higher. Once the structure collapses under the force of an earthquake, the damage caused will be huge.

*Corresponding Author: movezhangsx@163.com

DOI: 10.1201/9781003450818-75

Therefore, it has important engineering value and theoretical significance to study the seismic resistance of frame shear wall structures.

In the numerical simulation analysis of frame shear wall structure, the beam and column elements can be accurately simulated by fiber units, while the shear wall forces are more complex and are the hot spot of research. The most used ones are the multi-spring unit (SFI-MVLEM) (Kolozvari 2013; Orakcal 2012) and the layered shell elements developed by Lu at Tsinghua University based on the OpenSEES platform (Lu 2015). Wu (2019) compared the numerical simulation effects of the two models and found that for shear wall structures, the layered shell elements could predict the structural load capacity better, while SFI-MVLEM underestimated the structural load capacity, and the layered shell elements converged better. Therefore, the layered shell elements were used to simulate the shear wall elements in the structure.

In this paper, an established frame shear wall structure was used as the research object by using the IDA analysis method. Firstly, an elastoplastic analysis model was established based on the layered shell elements, and the seismic vulnerability analysis was carried out to obtain the exceedance probability curves for different performance levels of the structure. Then the damage risk was analyzed to comprehensively evaluate the seismic performance of the structure.

2 MODEL BUILDING

2.1 *Project overview*

In this study, a seven-story frame shear wall structure was used as the object of study, and the height of the structure was 4.8 m, 4.5 m, 3.3 m, 3.3 m, 4.2 m, 3.6 m, and 2.8 m from the bottom to the top, respectively. The basic design seismic acceleration value of the project was 0.20 g, the design seismic group was Group II, the site category was Class II, and the characteristic period of the site was 0.4 s.

2.2 *Analytical modeling*

In this study, the elastoplastic analysis of the structure was modeled based on OpenSEES, which is a software framework for analyzing and simulating the response of structural and geotechnical systems to seismic effects. It has become one of the most influential open research platforms in the field of earthquake engineering with its powerful nonlinear numerical simulation functions, rich material, and cell libraries, various efficient algorithms, open program architecture, and advanced concept of continuous integration of the latest research results.

For the beam and column elements of the structure, the fiber section unit was used to simulate the cross-section of the rod unit into several fibers, and each fiber could be set as a uniaxial tension-compression unit of different material instants. The deformation coordination of the cross-section was considered based on the theory of flat section assumptions, and the shear force between fibers was ignored. Therefore, the unit could accurately reflect the coupling effect of tensile and compression bending of the elements with high simulation accuracy. To consider the restraining effect of hoop reinforcement on the concrete in the core area, the full-section concrete of beam and column members was considered constrained concrete constitutive, and the uniaxial concrete constitutive material Concrete01 in OpenSEES was used, as shown in Figure 1. The material constitution was derived from the Kent-Scott-Park uniaxial concrete constitutive model, and the Karsan-Jirsa addition and removal criterion determined the linear unloading stiffness without considering the tensile strength of the concrete. The reinforcement constitution was the uniaxial reinforcement constitution Steel02 in OpenSEES, which was derived from the uniaxial isotropically strengthened Giuffre-Menegotto-Pinto reinforcement constitution.

For shear wall members, the layered shell elements were used based on the principle of composite mechanics, and a shell element was divided into several layers along the thickness

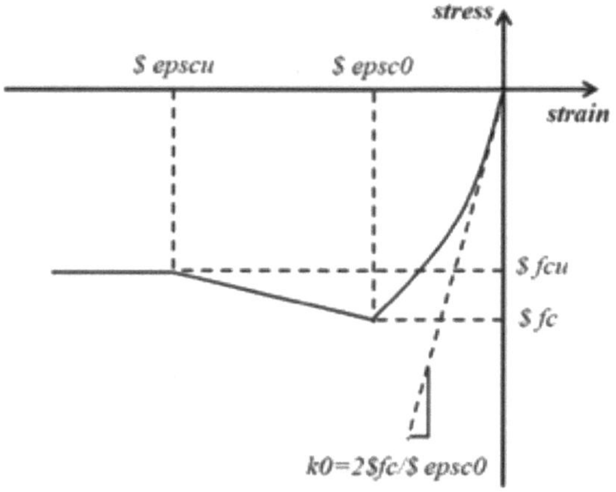

Figure 1. Concrete 01 constitution curve.

direction, each of which could be given the corresponding material and thickness according to the actual size and reinforcement of the member, as shown in Figure 2.

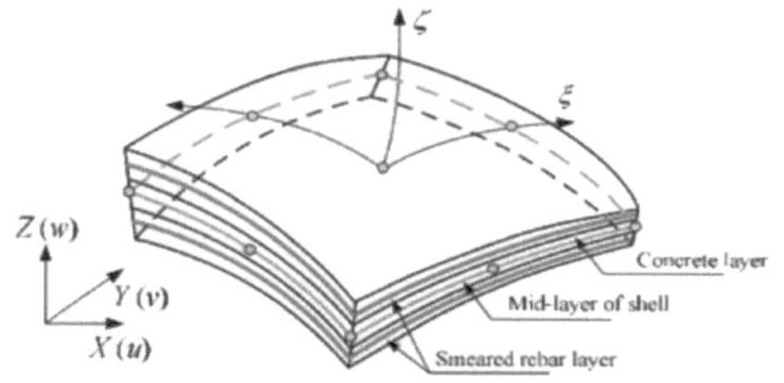

Figure 2(a). Schematic diagram of the layered shell element.

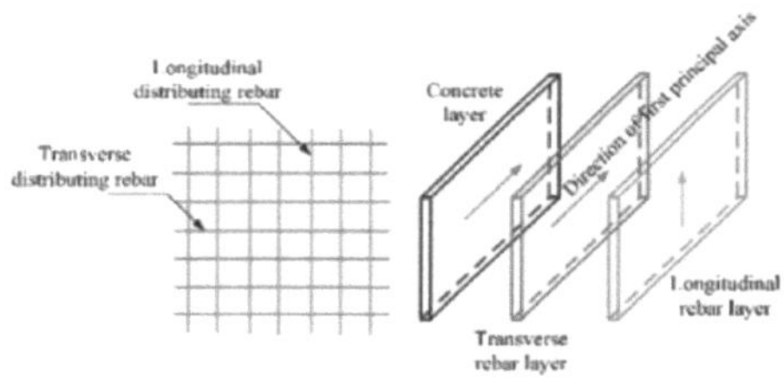

Figure 2(b). Schematic diagram of reinforcement layer distribution.

The distributed longitudinal reinforcement and hoop reinforcement in the middle wall of the layered shell model was simulated by using a diffuse reinforcement layer, and the thickness of the layer could be calculated according to the reinforcement rate on one side. The longitudinal reinforcement in the edge-constrained member was simulated by a discrete steel truss unit, and the deformation coordination between the two was achieved by co-nodes with the shell unit. In the unit analysis, the strain and curvature of the central layer of the shell unit were first obtained, and the strains of the other layers were calculated according to the assumption of a flat section between the layers. The stresses at the integration points of each layer were then obtained from the material constitutive equation of each layer, and finally, the internal forces of the shell unit were obtained by numerical integration. The layered shell elements took into account the coupling effect among in-plane bending, in-plane shear, and out-plane bending, which could reflect the spatial mechanical properties of the shell structure more comprehensively.

A rigid floor slab was assumed for the structure, and the established finite element model was shown in Figure 3. The first five revitalization periods of the structure were 0.56 s, 0.51 s, 0.44 s, 0.19 s, and 0.15 s.

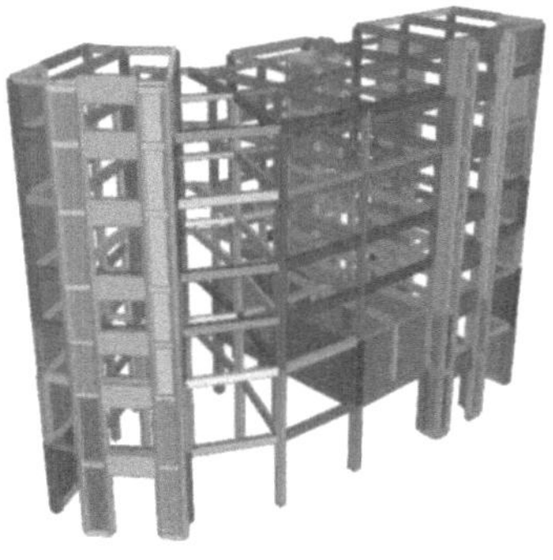

Figure 3. Structural finite element model.

3 INCREMENTAL DYNAMICS ANALYSIS (IDA)

3.1 *Basic principles*

Incremental Dynamic Analysis (IDA) is a parametric analysis method based on dynamic elastoplastic time analysis, and its analysis results can reflect the seismic performance of the whole structural system under different earthquake strengths more intuitively, which can make a more comprehensive and realistic evaluation of the seismic capacity of the structure (Lv 2012). The basic principle of the IDA method is to select a certain number of earthquake records and adjust the intensity index of the records according to certain rules. In this way, multiple-intensity earthquake records are obtained, which can be used to analyze the structure in elastoplastic time by using the amplitude-modulated earthquake records. Based on the analysis results, the IM-DM relationship curve (FEMA 2008) was established by using the intensity measure (IM) and damage measure (DM) as the coordinate parameters. In the selection of intensity measure (IM) and damage measure (DM), the maximum inter-story displacement angle can not only characterize the deformation performance of the structure but also serve as a basis for the damage degree of the structure (Sun 2021). Therefore, the peak ground acceleration (PGA), which can characterize the strength of earthquakes, was selected as the intensity measure (IM), and the maximum inter-story displacement angle was used as the damage measure (DM).

3.2 *Earthquake selection*

Due to the unstable vibration characteristics of the earthquakes, there are many uncertain influences in the processes of vibration and propagation, and in the process of structural elastoplastic dynamic time analysis, the different choices of the earthquake records can make the seismic response of the structure vary greatly. Therefore, choosing a reasonable earthquake record is a prerequisite for obtaining accurate analysis results. Han's team at the South China University of Technology statistically analyzed a large number of strong earthquake records collected from the Japanese Strong Earthquake Observatory Network and the seismic wave library of the Pacific Earthquake Research Center and selected strong earthquake records that were statistically consistent with the canonical response spectrum (Han 2021). The set of strong vibration data records for elastoplastic analysis with different site records and different basic period structures was given. Referring to the characteristic period of 0.4 s and the basic period of 0.56 s of the structure at the site where the project was analyzed in this paper, the corresponding recommended record group was selected for this study. The seismic information is shown in Table 1, and the response spectrum information is shown in Figure 4. From the response spectrum curves, it can be seen that the average response spectrum of the selected ground shaking was slightly larger than the normative spectrum of the site where the structure was located at the platform end, and the average spectrum was very close to the normative spectrum at the value of the fundamental period (0.56 s) of the structure.

Table 1. Seismic information for structural analysis.

Serial No.	Record Name	Record No.	Location of Occurrence	Time of Occurrence	Earthquake Magnitude	Duration (s)
1	A-SON033	RSN40	Borrego Mtn	1968/4/9	6.63	45.205
2	H-NIL360	RSN186	Imperial Valley-06	1979/10/15	6.53	40
3	H-VC6090	RSN366	Coalinga-01	1983/5/2	6.36	59.99
4	DAP000	RSN517	N.Palm Springs	1986/7/8	6.06	59.99
5	A-EUC292	RSN618	Whittier Narrow-01	1987/10/1	5.99	29.655
6	B-POE360	RSN725	Superstition Hills-02	1987/11/24	6.54	22.3
7	0655-292	RSN983	Northridge-01	1994/1/17	6.69	28.635
8	TCU119-E	RSN2418	ChiChi Taiwan-02	1999/9/20	5.9	81
9	CHY057-N	RSN2721	ChiChi Taiwan-04	1999/9/20	6.2	65
10	TTN018-N	RSN2914	ChiChi Taiwan-04	1999/9/20	6.2	58
11	CHY024-N	RSN3264	ChiChi Taiwan-06	1999/9/20	6.3	81.02
12	CHY075-E	RSN3301	ChiChi Taiwan-06	1999/9/25	6.3	67.995
13	TCU046-N	RSN3454	ChiChi Taiwan-06	1999/9/25	6.3	59.95

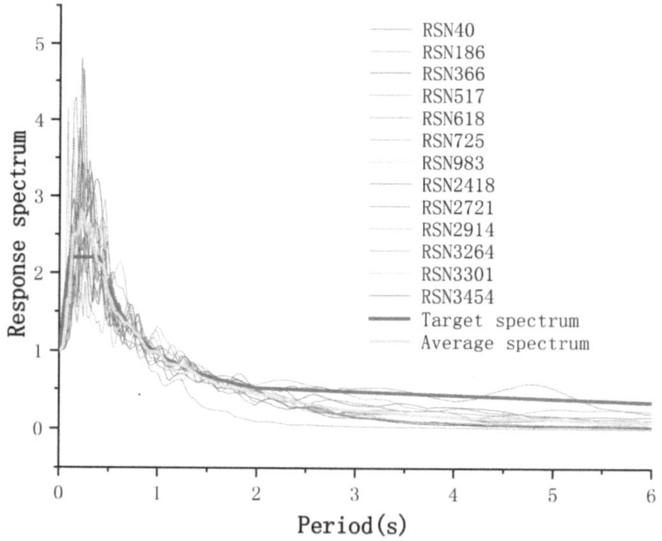

Figure 4. Seismic response spectrum curve.

3.3 *Results analysis*

The selected seismic groups were amplitude-modulated to 0.07 g, 0.2 g, 0.3 g, 0.4 g, 0.5 g, 0.6 g, 0.7 g, 0.8 g, 0.9 g, and 1.0 g and input into the structure one by one for elastoplastic analysis. The maximum inter-story displacement angular response structures of the collated structures under multiple earthquakes (0.07 g), fortified earthquakes (0.2 g), and rare earthquakes (0.4 g) are shown in Figure 5.

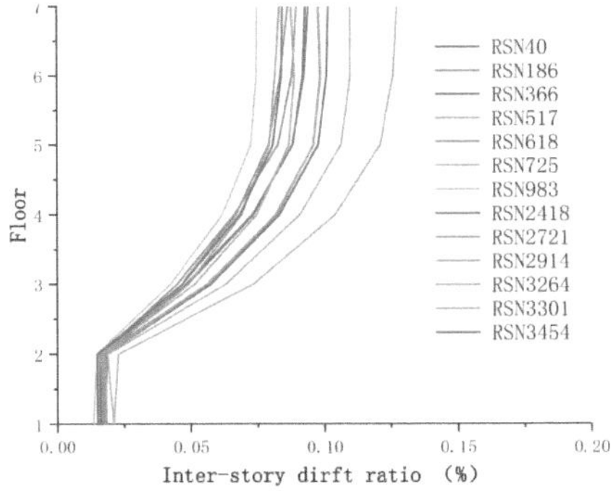

Figure 5(a). Maximum inter-story displacement angle of each layer of the structure under multiple encounter earthquakes.

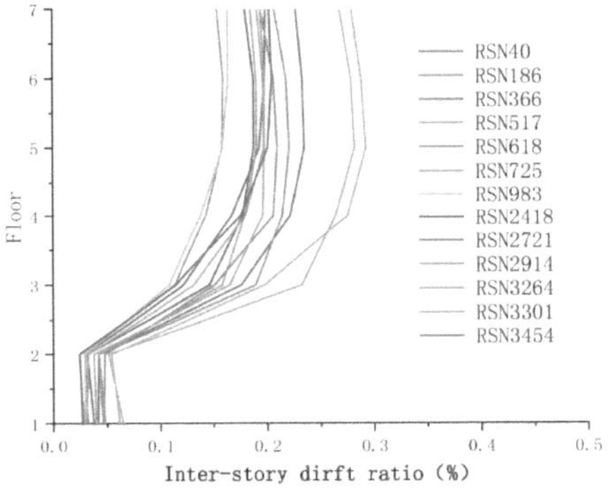

Figure 5(b). Maximum inter-story displacement angle of the structure at each layer under fortified earthquakes.

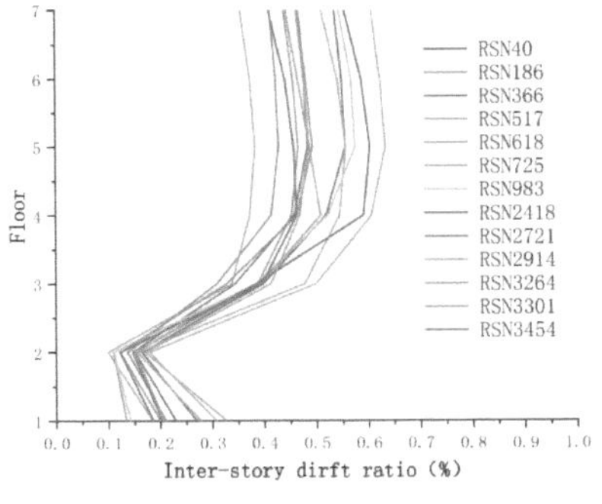

Figure 5(c). Maximum inter-story displacement angle of each layer of the structure under rare earthquakes.

Figure 5 shows that the distribution curves of the inter-story displacement angles of the structure show a similar trend under the three earthquake intensities. The inter-story displacement angles of the first and second floors are similar and much smaller than those of the other floors. In rare earthquakes, the inter-story displacement angle of the second floor is even smaller than that of the first floor, probably due to the existence of a separate "cylinder"-like space enclosed by shear walls on the second floor of the structure, which increases the lateral stiffness of the second floor. From the second floor to the fifth floor, the inter-story displacement angle gradually increases, while the inter-story displacement angle from the fifth floor to the seventh floor does not differ much.

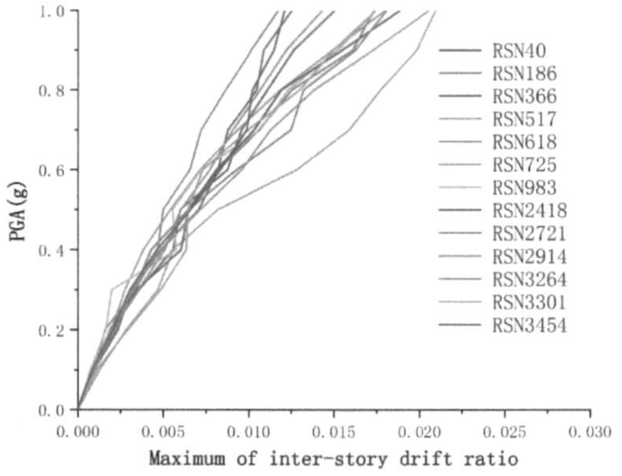

Figure 6. IDA Curve Cluster.

Using the maximum inter-story displacement angle of the structure as the horizontal coordinate and the seismic intensity PGA as the vertical coordinate, the IDA curve of each seismic wave can be obtained (see Figure 6). From the curves, it can be seen that the growth trend of the IDA curves of each earthquake is the same, and with the increase in PGA, the value of the inter-story displacement angle increases. Even for the seismic records that follow the same site response spectrum selection, some of them have more than double the difference under the same PGA, reflecting the large influence of earthquakes' randomness on the structure's seismic response. Therefore, selecting multiple seismic records for calculation is necessary when performing time-course analysis.

4 SEISMIC VULNERABILITY ANALYSIS

4.1 *Performance level classification and quantification of index limits*

The expected maximum damage state of the structure under a certain intensity of earthquake load is the structural performance level, so the structural performance level can be divided into several classes according to different damage states. According to schedule M.1.3-4 of GB50011-2010 "Seismic Design Code for Buildings" (GB 50011-2010), there are five performance levels in this paper, and their corresponding inter-story displacement angle intervals are shown in Table 2.

Table 2. Structural performance levels and corresponding inter-story displacement angle ranges.

Performance Status	Intact	Slight damage	Moderate damage	Severe damage	Close to collapse
Maximum Inter-Story Displacement Angle Ranges	<1:800	1:800–1/:400	1:400–1:200	1:200–1:100	>1:100

4.2 *Probabilistic seismic demand model*

As an important part of seismic vulnerability analysis and seismic risk analysis, the core of probabilistic seismic demand analysis is to establish a probabilistic seismic demand model, which characterizes the probabilistic relationship between seismic intensity and seismic demand. Under the effect of certain earthquake intensities, the seismic demand for erection usually follows a log-normal distribution. The median seismic demand and the intensity of ground shaking are assumed to obey the logarithmic prior relationship, while the logarithmic standard deviation of seismic demand is assumed to remain constant (Yu 2013). The median seismic demand SD of the structure and the intensity measure IM satisfy the following relationship (see Equation 1).

$$S_D = a(IM)^b \tag{1}$$

where S_D is the median seismic demand; a and b can be obtained by means of statistical regression to establish the probabilistic relationship between the seismic demand and the seismic intensity of the structure. Using the maximum inter-story displacement angle as the structural seismic demands D and the peak ground acceleration (PGA) as the intensity index, IDA analysis was performed on the structural model to obtain a large number of PGA and maximum inter-story displacement angle data samples. The median value of $ln(\theta\_(max))$ for different seismic excitations was selected and linearly regressed with $ln(PGA)$ (see Equation 2) to find the relationship (Figure 7).

$$\ln(\theta_{\max}) = lna + bln(PGA) \tag{2}$$

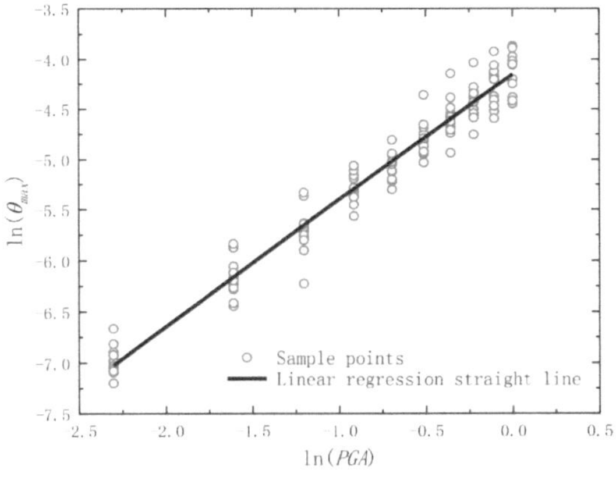

Figure 7. Seismic demand probability model.

4.3 *Vulnerability analysis*

Combining the range of inter-story displacement angles corresponding to the performance level of the structure in Section 3.1 and the logarithm and standard deviation of the seismic demand probability model in Section 3.2, the probability that the inter-story displacement angle of the structure exceeds the range of values for each performance level, $P_E\ (DS \geq DS_i/IM)$, can be obtained according to the statistical principle of probability as follows (see Equation 3).

$$P_E(DS \geq DS_i/IM) = \Phi\left[\frac{1}{\sqrt{\beta_C{}^2 + \beta_D{}^2}} \ln\left(\frac{S_D}{S_C}\right)\right] = \Phi\left[\frac{\ln a + b\ln(PGA) - \ln(\theta_i)}{\sqrt{\beta_C{}^2 + \beta_D{}^2}}\right] \tag{3}$$

where $\beta_C = lna$, $\beta_D = b$, and β_D is the logarithmic standard deviation.

The probability of exceeding the performance level for different PGA values was calculated, as the X-axis to obtain the probability of exceeding the performance level curve for different seismic strengths (Figure 8). Based on the curves, the probability of the structure being at each performance level under multiple (0.07 g), fortified (0.2 g), and rare (0.4 g) earthquakes was calculated (Table 3).

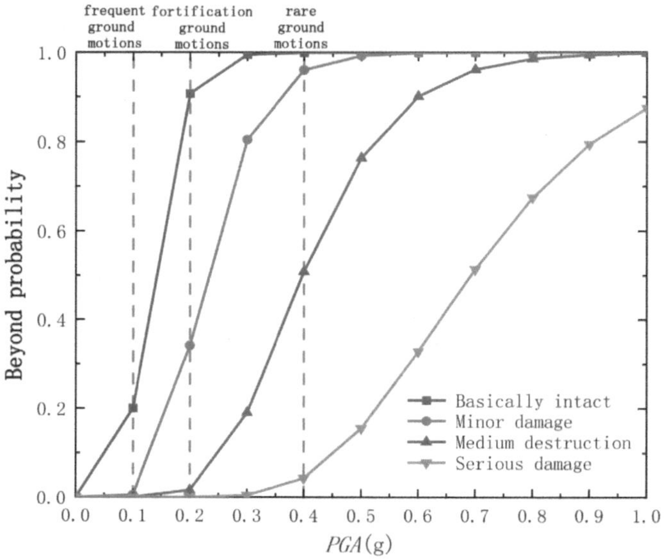

Figure 8. Exceedance probability curves for each limit state of the structure.

Table 3. Seismic vulnerability matrix of the structure at each limit state.

		Probability of Each Limit State of the Structure/%				
Seismic Level	A_{pg}/g	Intact	Slight Damage	Moderate Damage	Severe Damage	Close to Collapse
Multiple Earthquakes (Magnitude 8)	0.07	79.94	19.56	0.50	0	0
Fortified Earthquakes (Magnitude 8)	0.2	9.27	56.58	32.54	1.61	0
Rare Earthquakes (Magnitude 8)	0.4	0.02	3.94	45.14	46.54	4.36

From the table, the probability that the structure is intact under multiple earthquakes is 79.94%, the probability of slight damage is 19.56%, and the probability of moderate damage is only 0.5%. The probability that the structure is intact under the fortified earthquake is 9.27%, the probability of slight damage is 56.58%, the probability of moderate damage is 32.54%, and the probability of severe damage is 1.61%. The probability that the structure is intact under a rare earthquake is 0.02%; the probability of slight damage is 3.94%; the probability of moderate damage is 45.14%; the probability of severe damage is 46.54%; and the probability of near collapse is 4.36%.

5 CONCLUSION

This paper adopted a layered shell unit to simulate shear wall elements and a fiber model to simulate beam-column elements, and established an elastoplastic analysis model for an existing frame shear wall mechanism. 13 seismic records were selected, and a seismic vulnerability analysis based on IDA analysis was conducted. The following conclusions were obtained:

(1) The second floor of the structure had the smallest seismic response due to the "cylinder"-like structure enclosed by shear walls alone, while the response of the rest of the structure gradually increased from the bottom to the top floors.
(2) There were considerable differences in the structural IDA curves for different earthquakes, which proved that the randomness of the seismic records had a large influence on the structural response and that sufficient seismic records should be selected for analysis in the time-course analysis.
(3) In the case of multiple earthquakes, the structure had a higher probability of being at the performance level of basically intact. In the case of fortified earthquakes, the structure was more likely to be in the performance level of minor damage or moderate damage. In the case of rare earthquakes, the structure had a higher probability of being at the performance level of moderate damage or severe damage. It met the seismic design requirements of "no damage in minor earthquakes, repairable in moderate earthquakes, and no collapse in severe earthquakes."

ACKNOWLEDGMENTS

This work was financially supported by the National Key Research and Development Program (2019YFC1509401-01) and the National Natural Science Foundation of China (U1901602-05).

REFERENCES

FEMA P695(ATC-63 90%). *Quantification of Building Seismic Performance Factors* [S]. Federal Emergency Management Agency, Washington, D.C., 2008.

GB 50011-2010,Code for Seismic Design of Buildings [S].

Han Xiaolei, Ji Jing. *Performance-based Seismic Design of Reinforced Concrete Structure* [M]. Beijing. China Architecture and Architecture Press, 2011: 288, 310.

Kolozvari K. *Analytical Modeling of Cyclic Shear-Flexure Interaction in Reinforced Concrete Structural Walls* [D]. State of California: Department of Civil and Environmental Engineering, University of California, 2013.

Lu X, Xie L, Guan H, *et al.* A Shear Wall Element for Nonlinear Seismic Analysis of Super-tall Buildings Using OpenSees [J]. *Finite Elements in Analysis and Design*, 2015, 98: 14–25.

Lu Xilin, Su Ningfen, Zhou Ying. IDA-based Seismic Fragility Analysis of a Complex High-rise Structure [J] *Journal of Earthquake Engineering and Engineering Vibration*, 2012, 32 (5): 20–25.

Orakcal K., Massone L.M., Ulugtekin D. *Constitutive Modeling of Reinforced Concrete Panel Behavior Under Cyclic Loading:15th World Conference on Earthquake Engineering* [C]. Lisbon, Portugal, 2012.

Sun Xiaojing, Yang Feng, Zhang Haitao. Seismic Vulnerability Study of Concrete Filled Steel Tubular Frame Based on IDA [J]. *Structural Engineers*, 2021, 37(1): 75–81.

Wu Zinan, Luo Yu, He Ruibo. Nonlinear Finite Element Analysis of Shear Wall under Shear-Flexural Interaction [J]. *Guangdong Architecture Civil Engineering*, 2019, 26 (01): 15–19 + 58. DOI: 10.19731/j.gdtmyjz.2019.01.004.

Yu Xiaohui, Lü Dagang, Wang Guangyuan. Discuss on Probabilistic Seismic Demand Models [J]. *Engineering Mechanics*, 2013, 30 (08): 172–179.

Water Conservancy and Civil Construction – Oke & Ahmad (Eds)
© 2024 The Author(s), ISBN: 978-1-032-58614-4

Influence of structural parameters on erosion characteristics of solid-liquid flow in U-shaped combined elbows

Xi Shi*, Li Gong, Hu Tao, Guoming Wu & Chunbin Tan
College of Civil Engineering, Lanzhou Jiaotong University, Lanzhou, China

ABSTRACT: To study the erosion characteristics of solid-liquid flow in U-shaped combined elbows, a numerical model for erosion calculation was established based on computational fluid dynamics (CFD) methods. Then, three erosion models, including the McLaury model, the Oka *et al.* model, and the Generic model, were verified with literature data. The results showed that the calculated results of the generic model were the closest to the experimental results. On this basis, the effects of structural parameters such as combination spacing (*L*/*D*), curvature radius to diameter (*R*/*D*) ratio, and pipe diameter on erosion rate were discussed. The results showed that the erosion rate was lower when the combination spacing was $L/D = 0$, higher when $L/D = 1.0$, and then decreased as the combination spacing increased. When the combination spacing L/D = zero, the elbow 2 suffered the most severe erosion. It also commonly appeared in both elbows as $L/D = 1.0$ and $L/D = 4.0$. It occurred in elbow 1 when $L/D = 7.0$ and $L/D = 10.0$ and the erosion variations in elbow 1 were consistent with the single elbow. When L/D was = 0, the average and maximum erosion rates decreased as the *R*/*D* ratio increased. As the *R*/*D* ratio increased, the bending angle decreased, and the number and strength of particles directly impacting the wall decreased. The average erosion rate and the maximum erosion rate decreased with an increase in pipe diameter. With the increase in pipe diameter, the Stokes number of particles decreased, the following property of water flow becomes better, and the effect of inertia force weakened.

1 INTRODUCTION

Erosion caused by solid particles on the pipe wall is a serious problem that frequently occurs in hydraulic transmission pipelines. Some large particles or small particles impact the pipe wall and the deformation parts for a long time, causing the pipe wall to thin or even wear through, which affects transmission safety. The erosion mainly occurs at deformation parts such as elbows, tees, and valves; therefore, the erosion of these parts comes to be the focus of our research. Recent years have seen a large number of research reports apply methods of experimentation and CFD simulation to study the erosion of deformation parts such as elbows. Sedrez *et al.* (2019) researched the erosion characteristics of a multi-phase flow elbow composed of water, air, and sand by adopting experiments and CFD methods. Zeng *et al.* (2014) analyzed the erosion behavior of the X65 pipeline elbow and established a numerical model, which showed that numerical simulation could work well in studying the erosion of the elbow. Peng and Cao (2016), employing the numerical simulation method, discussed the erosion characteristics of gas-solid two-phase flow in 90° elbows and obtained the prediction model of the maximum erosion location in the elbow. Gabriel *et al.* (2014) studied the model of particle erosion in a 90-degree elbow and found that the calculated results of the Oka *et al.* model were in good agreement with the experimental results. Hu *et al.* (2017) analyzed the erosion characteristics of a throttle valve using the Ahlert erosion

*Corresponding Author: shixi103@163.com

DOI: 10.1201/9781003450818-76

model. Laín and Sommerfeld (2019) calculated the erosion characteristics and particle distribution in the pneumatic conveying elbow with the model of Oka *et al.* It can be seen that a lot of experimental attention has been attributed to the single elbow, and CFD numerical simulation has become the main method to study erosion problems, whose reliability has been well verified. There are also a few achievements in the study of particle erosion in combined elbows. For instance, Zhang *et al.* (2018) studied the erosion of two Z-shaped combined elbows by using the CFD model; however, the influence of the change in combination spacing was not discussed. Wang *et al.* (2018) used the E/CRC erosion model to predict the erosion characteristics of gas-solid flow in combined elbows. Zhao *et al.* (2021) analyzed the erosion behavior of gas-solid flow in two elbows mounted in series, and the erosion morphology was also discussed. Notably, previous research on the erosion characteristics of solid-liquid flow in combined elbows is inadequate. In this paper, the effects of structural parameters such as combination spacing (*L*/*D*), curvature radius to diameter (*R*/*D*) ratio, and pipe diameter on the erosion characteristics of solid-liquid flow in U-shaped combined elbows were discussed by numerical simulation.

2 NUMERICAL MODEL

2.1 *The continuous phase equation*

The continuous phase flow of solid-liquid flow satisfies the Navier-Stokes (N-S) equation and features turbulence characteristics that are three-dimensional and incompressible. Its basic governing equations are as follows:

$$\frac{\partial \rho}{\partial t} + \frac{\partial}{\partial x_i}(\rho \boldsymbol{u}_i) = 0 \tag{1}$$

$$\frac{\partial}{\partial t}(\rho \boldsymbol{u}_i) + \frac{\partial}{\partial x_i}(\rho \boldsymbol{u}_i \boldsymbol{u}_j) = -\frac{\partial p}{\partial x_i} + \frac{\partial}{\partial x_j}\left(\mu \frac{\partial \boldsymbol{u}_i}{\partial x_j} - \rho \overline{\boldsymbol{u}'_i \boldsymbol{u}'_j}\right) + \boldsymbol{S}_i \tag{2}$$

where ρ is the density of the continuous phase, $\boldsymbol{u}_i$ and $\boldsymbol{u}_j$ are the velocity components in the directions of x_i and x_j respectively, p is the pressure on the fluid micro components, μ is the coefficient of dynamic viscosity, $-\rho u'_i u'_j$ is the Reynolds stress, $\boldsymbol{S}_i$ is the generalized source phase.

When calculating the flow of the continuous phase, a new turbulence model is needed to close the equations. Here in this study, the standard $k - \varepsilon$ turbulence model was used, in which turbulent kinetic energy k and dissipation rate ε are two basic unknowns. For incompressible fluids, the corresponding transmission equations are as follows:

$$\frac{\partial(\rho k)}{\partial t} + \frac{\partial(\rho k \boldsymbol{u}_i)}{\partial x_i} = \frac{\partial}{\partial x_j}\left[\left(\mu + \frac{\mu_t}{\sigma_k}\right)\frac{\partial k}{\partial x_j}\right] + G_k - \rho\varepsilon \tag{3}$$

$$\frac{\partial(\rho \varepsilon)}{\partial t} + \frac{\partial(\rho \varepsilon \boldsymbol{u}_i)}{\partial x_i} = \frac{\partial}{\partial x_j}\left[\left(\mu + \frac{\mu_t}{\sigma_\varepsilon}\right)\frac{\partial \varepsilon}{\partial x_j}\right] + \frac{C_{1\varepsilon}\varepsilon}{k} G_k - C_{2\varepsilon}\rho\frac{\varepsilon^2}{k} \tag{4}$$

where σ_k and σ_ε are the Prandtl numbers corresponding to turbulent kinetic energy k and dissipation rate ε respectively, which are 1.0 and 1.3; μ_t is turbulent viscosity; G_k is the generated term of turbulence kinetic energy caused by the mean velocity gradient; $C_{1\varepsilon}$ and $C_{2\varepsilon}$ are empirical constants, which are 1.44 and 1.92 respectively.

2.2 *The dispersed phase equation*

To obtain the particle trajectories, the motion equation of the particles was integrated into the Lagrangian coordinates. And the equilibrium equation of forced particles is expressed in

Cartesian coordinates as follows (Yang & Zhu, 2019):

$$\frac{d\boldsymbol{u}_p}{dt} = \boldsymbol{F}_D(\boldsymbol{u} - \boldsymbol{u}_p) + \frac{\boldsymbol{g}_x(\rho_p - \rho)}{\rho_p} + \boldsymbol{F}_x \tag{5}$$

where $F_D(u - u_p)$ is the mass force per unit of particles, u and u_p are the velocity of the continuous phase and particle phase respectively, g_x is the gravity acceleration in x direction, ρ_p is the particle density, $\boldsymbol{F}_x$ are other forces in x direction, including pressure gradient force, virtual mass force, Saffman force, and so on.

2.3 *Erosion model*

Three common erosion models, the Generic model, the McLaury model, and the Oka *et al.* model, were used for calculation. Then, to verify, the erosion test data of a single elbow from the literature are used. Finally, the best erosion model was selected for analyzing erosion at combined elbows.

Generic Erosion Model (Wang *et al.* 2019):

$$R_e = \sum_1^N \frac{m_p C(d_p) f(\theta) {u_p}^{b(v)}}{A} \tag{6}$$

where R_e is the erosion rate, kg·m^{-2}·s^{-1}, N is the number of colliding particles per unit area, $C(d_p)$ is the function of particle diameter, set as 1.8×10^{-9}, u_p is the relative velocity of particles, $b(v)$ is the relative function of particle velocity, set as 2.6, θ is the particle impact angle, $f(\theta)$ is the impact angle function, calculated according to the following formula (Huser & Kvernvold 1998):

$$f(\theta) = 2.69\theta + 1.61\theta^2 - 8.84\theta^3 + 7.33\theta^4 - 1.85\theta^5 \tag{7}$$

Mclaury Erosion Model (Mclaury 1993):

$$ER = AV^n f(\theta) \tag{8}$$

$$A = FB_h{}^k \tag{9}$$

$$\begin{cases} f(\theta) = b\theta^2 + c\theta(\theta \le \theta_{\lim}) \\ f(\theta) = x\cos^2(\omega\theta) + y\sin^2\theta + z(\theta > \theta_{\lim}) \end{cases} \tag{10}$$

where ER is the mass erosion rate, kg/kg, B_h is the Brinell hardness of the pipe material, $A = 1.99 \times 10^{-7}$, $n = 1.73$, $\theta_{\lim} = 15°$, $b = 13.3$, $c = 7.85$, $\omega = 1$, $x = 1.09$, $y = 0.125$, $z = 0.872$.

Oka Erosion Model (Oka & Okamura 2005; Oka & Yoshida 2005):

$$E(\theta) = f(\alpha)E_{90} \tag{11}$$

$$f(\theta) = (\sin\theta)^{n_1}[1 + H_v(1 - \sin\theta)]^{n_2} \tag{12}$$

$$E_{90} = K(H_v)^{k_1}\left(\frac{v}{v'}\right)^{k_2}\left(\frac{d_p}{d'}\right)^{k_3} \tag{13}$$

where $E(\theta)$ is the volume erosion rate, mm^3/kg, E_{90} is the reference erosion rate under the positive impact angle, mm^3/kg, reference erosion rate, v' is the reference speed, set as 104 m/s, d' is the reference particle diameter, set as 326 μm, H_v is the Vickers hardness of the pipe material, set as 1.61 GPa, k_3 is the index, set as 0.19, $n_1 = 0.71(H_v)^{0.14}$, $n_2 = 0.71(H_v)^{-0.94}$, $k_2 = 2.3(H_v)^{0.038}$.

2.4 *Particle-wall rebound model*

There is energy transformation and loss after a particle-wall collision, which is mainly reflected in the change of the velocity component before and after the collision (Parsi *et al.* 2014).

The energy loss after the collision is usually described with the particle-wall rebound coefficient, which is the ratio of velocity components before and after the collision. In this study, the Grant and Tabakoff model was used for the wall rebound coefficient (Grant & Tabakoff 2012), which is written as:

$$\begin{cases} \varepsilon_N = 0.993 - 1.76\theta + 1.56\theta^2 - 0.49\theta^3 \\ \varepsilon_T = 0.988 - 1.66\theta + 2.11\theta^2 - 0.67\theta^3 \end{cases} \tag{14}$$

where N and T represent normal and tangential directions, respectively.

3 GEOMETRIC MODEL AND NUMERICAL METHOD

3.1 *Geometric model and relative parameters*

The combined elbows were calculated as common seamless steel pipes, and a three-dimensional geometric model was established with CAD software. Figure 1 presents the geometric model under the conditions of combination spacing $L/D = 0$, curvature radius to diameter ratio $R/D = 1.5$, and pipe diameter $D = 65$ mm. As shown in Figure 1, the combined elbows consist of three parts: an inlet section, a connection section, and an outlet section. To ensure the full development of turbulence in the pipeline, the length of the inlet section was set at $L_1 = 14$ d and the length of the exit section at $L_2 = 16$ d for the calculation area (d was the pipe diameter).

The continuous phase in the elbows was liquid water, the dynamic viscosity was 1.003×10^{-3} Pa·s, and the density was 998.2 kg/m^3; the dispersed phase was solid particles; the density was 2650 kg/m^3, the particle diameter was 0.2 mm, and the mass flow rate was 0.2 kg/s. The velocities of the continuous phase and particle phase were set to the same, which were both 8 m/s; considering the influence of gravity, the gravity direction was set to be negative to the Z axis.

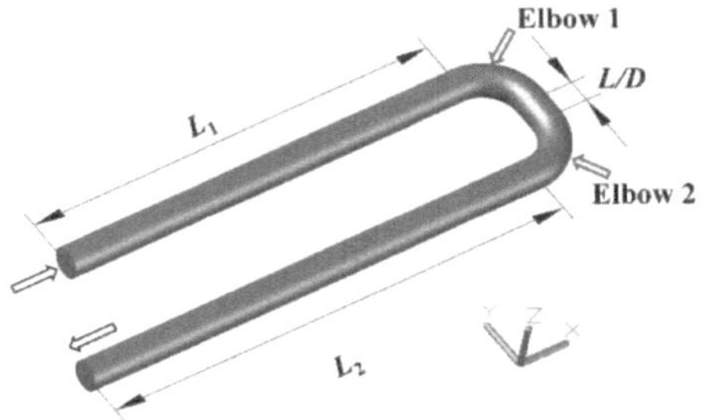

Figure 1. Geometric model of combined elbows.

3.2 *Mesh generation and independence test*

A tetrahedral grid with good adaptability was used to encrypt the local part and connection section of combined elbows; the hexahedral grid was used in other parts; and six layers of the boundary layer grid were set at the pipe wall.

To ensure that the number of grids has little influence on the calculation results, an independence validation was carried out on the average erosion rate of the combined elbows as an example, with the pipe diameter D = 40 mm, curvature radius to diameter ratio R/D = 1.5, and combination spacing L/D = 0. The results are shown in Figure 2. It can be known that when the number of grids divided was less than 400, 000, and the grid had a great influence on the calculation results, while the influence was little when the number of grids increased beyond 400, 000. The number of grids used in this case was 696, 000, and Figure 3 shows the divided grids.

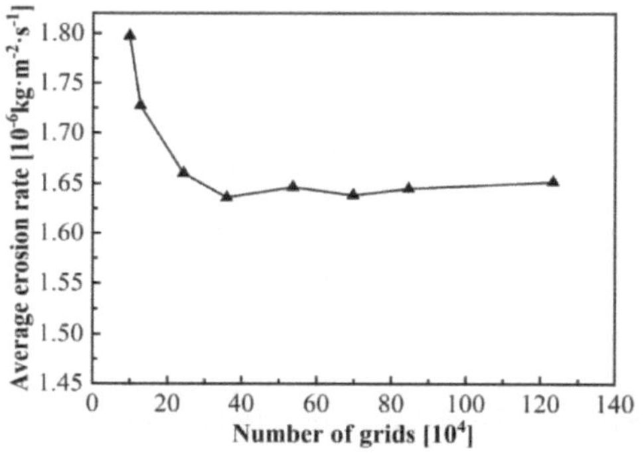

Figure 2. Grid dependence test.

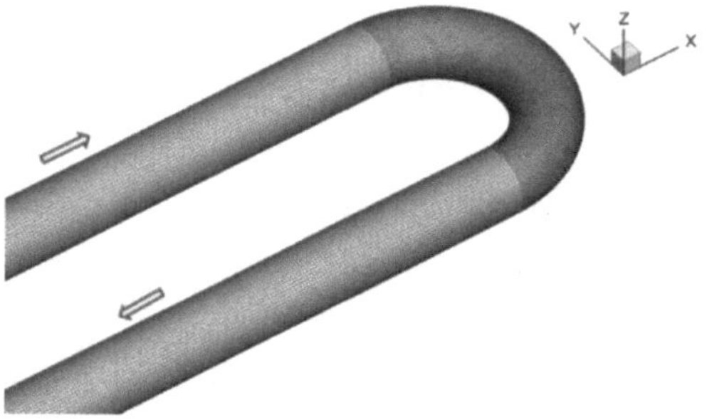

Figure 3. Mesh generation.

3.3 *Boundary conditions and numerical method*

The standard turbulence model was used to calculate the continuous phase, the inlet boundary was the velocity inlet, and the outlet adopted a free outflow boundary. The flow near the wall surface was calculated by the standard wall function method.

The combined elbows' inlet and outlet were set to the escape condition, while the wall surface was set to the rebound condition. The particles were assumed to be spherical with uniform size and to move in an irrotational motion, and the damage to the particles after the collision was ignored. The pressure gradient force, virtual mass force, Saffman force, Magnus force, and other forces in the flow field were also neglected (Lin *et al.* 2014).

The coupling between pressure and velocity was iterated with the SIMPLE algorithm, and a two-order discrete scheme was used for the pressure term and momentum term. The continuous phase and dispersed phase were coupled bidirectionally.

4 RESULTS AND ANALYSIS

4.1 *Numerical validation of different erosion models*

To figure out the best erosion model for numerical calculation, Zeng *et al.* (2014) carried out a test based on the experimental results of erosion of solid-liquid flow in a single 90° elbow. The relevant parameters were: pipe diameter of 50 mm, elbow radius of 76.9 mm, particle velocity of 4 m/s, particle density of 2650 kg/m^3, mass flow rate of 0.235 kg/s, and particle diameter of 400–500 μm. The calculated results of different erosion models are shown in Figure 4, which shows that the calculated results of the McLaury model are quite different from the experimental results, while the calculated results of the Oka *et al.* model and the Generic model were close to the experimental results, and the calculated results of the Generic model are the closest. Since the combined elbows were composed of two identical single elbows, the erosion calculation of the combined elbows was carried out by the generic model.

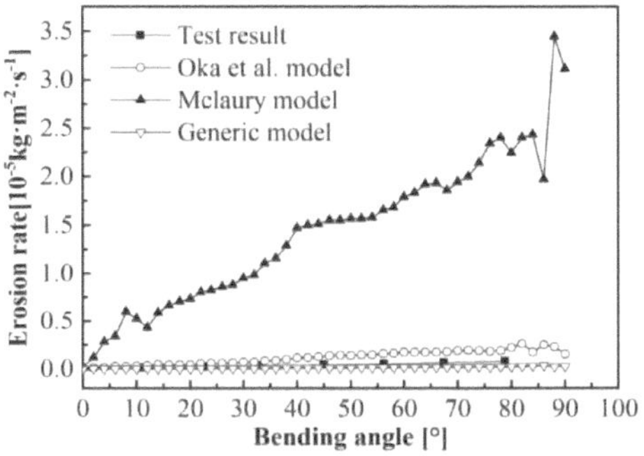

Figure 4. Validation of different elbow erosion models.

4.2 *Variation with different combination of spacings*

Figure 5 shows the erosion rate of combined elbows with combination spacing when D = 80 mm and R/D = 2.0. Except for the combination spacing L/D = 0, the average and maximum erosion rates generally decrease with increasing combination spacing. The value of the erosion rate was small when L/D = 0, and the reason may be as follows: when L/D = 0, the erosion area decreases, which makes the erosion rate smaller; when L/D = 1.0, the erosion area increases and the erosion rate increases; when L/D = 1.0 and increased again, the extent of the erosion area is limited and the adjacent effect between two elbows is weakened, which makes the erosion rate decrease.

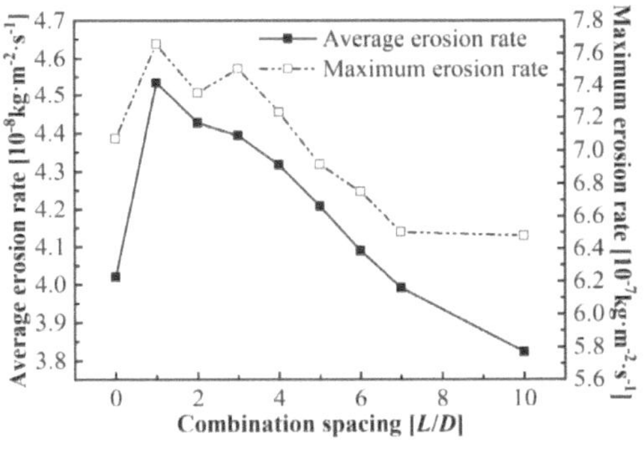

Figure 5. Erosion rate under different combination spacing.

Figure 6 shows the erosion contours under different combination spacings. It can be seen that with the increase in combination spacing, the most serious erosion location gradually transits from elbow 2 to elbow 1. When combination spacing L/D = 0, the most serious erosion occurs near the outer wall of the outlet of elbow 2; when L/D = 1.0 *and* L/D = 4.0, the most serious erosion appears at the outlet side wall of elbow 1 and the outlet outer wall of elbow 2, and the serious area in elbow 1 is relatively less; when L/D = 7.0 and L/D = 10.0, the most serious erosion appears at the outlet side wall of elbow 1, when the erosion change in elbow 1 is consistent with that of the single elbow and is immune from the influence of elbow 2.

Pipeline erosion is caused by particle movement, and the change of flow field directly affects the particle movement trajectory. Figure 7 shows the velocity distribution of the center section under different combination spacings when Z = 0. According to the figure, the

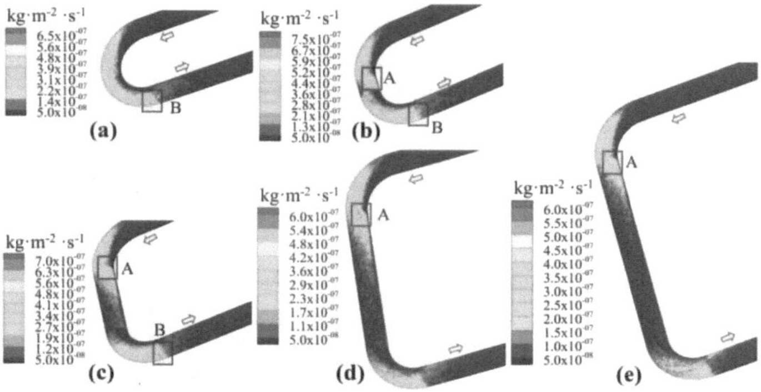

Figure 6. Erosion contours under different combination spacings (a) L/D = 0, (b) L/D = 1.0 (c) L/D = 4.0 (d) L/D = 7.0 (e) L/D = 10.0.

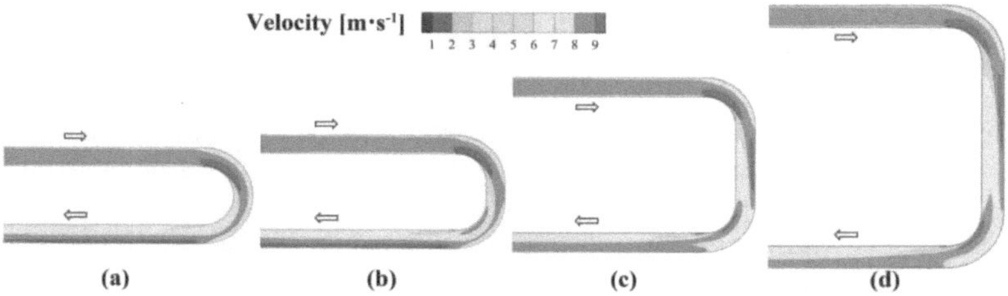

Figure 7. Velocity distribution under different combination spacings (a) L/D = 0, (b) L/D = 1.0 (c) L/D = 4.0 (d) L/D = 7.0.

velocity distribution of elbow 1 remains similar when the combination spacing is changed. And the distribution at the front section of the inlet is relatively uniform. After turning, the main stream deviates to the inner wall under the action of centrifugal force, resulting in a velocity gradient between the inner and outer walls; however, the velocity distribution in elbow 2 varies a lot with different combination spacing. When $L/D = 0$, the main stream flows out of elbow 1 and then directly enters elbow 2. Under the influence of elbow 1, the mainstream is biased towards the outer wall of elbow 2, and the flow velocity near the outer wall of elbow 1 is small. At this time, the most serious erosion location is near the outer wall of elbow 2; when $L/D > 1.0$, since a straight pipe is connected between the two elbows, the mainstream flowing out of elbow 1 first deviates to the outer wall of connected straight pipe and then enters elbow 2; when $L/D = 1.0$, because the connecting straight pipe is short, the mainstream in elbow 2 is biased towards the outer wall, and the erosions at the outlet of elbow 1 and near the outlet of elbow 2 are both serious; when $L/D > 4.0$, the influence of elbow 2 on elbow 1 gradually decreases due to the increase of combination spacing, and the flow velocity at the inner wall of elbow 2 increases and the velocity at the outer wall decreases, while the most serious erosion location gradually transits to elbow 1. The above analysis shows that the change in combination spacing has a great influence on the velocity distribution in elbow 2, thus affecting the erosion change.

4.3 *Variation with different curvature to diameter ratios*

Under the condition of pipe diameter $D = 100$ mm and combination spacing $L/D = 0$, the change in erosion rate of combined elbows with a curvature radius to diameter (R/D) ratio is shown in Figure 10. It can be seen that both the average erosion rate and the maximum erosion rate tend to decrease with an increase in the *R*/*D* ratio. The average erosion rate is 2.83×10^{-8} kg·m^{-2}·s^{-1} when *R*/*D*=1.5, which reduces to 2.31×10^{-8} kg·m^{-2}·s^{-1} when *R*/*D* = 4.0; the maximum erosion rate is 5.80×10^{-7} kg·m^{-2}·s^{-1} when *R*/*D* = 1.5, which reduces to 2.50×10^{-7} kg·m^{-2}·s^{-1} when *R*/*D* = 4.0. Overall, the variation range is restricted.

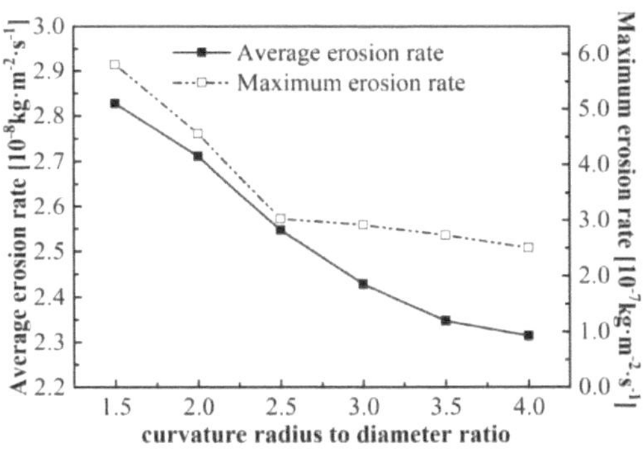

Figure 8. Erosion rate under different curvature radius to diameter ratios.

Figure 9 shows the erosion contours under different *R*/*D* ratios. When *L*/*D* = 0, erosion under different *R*/*D* ratios mainly occurs in the part of the outer pipe wall of elbow 1 and the whole outer pipe wall of elbow 2, and the most serious erosion appears at the outer pipe wall toward the outlet of elbow 2. With the increase in the *R*/*D* ratio, the erosion rates in all erosion regions tend to decrease.

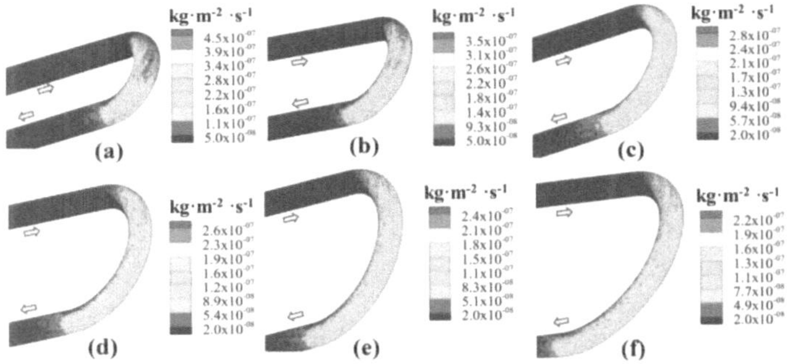

Figure 9. Erosion contours under different curvature to diameter ratios (a) R/D = 1.5 (b) R/D = 2.0 (c) R/D = 2.5 (d) R/D = 3.0 (e) R/D = 3.5 (f) R/D = 4.0.

Figure 10 shows the velocity distribution of the central section and the velocity vector diagrams of different sections when Z = 0, under the conditions of *R*/*D* = 1.5, *R*/*D* = 2.5, and *R*/*D* = 4.0, respectively, which are observed along the horizontal direction of the flow. The flow velocity distribution is relatively uniform in the front section of the elbow inlet, and the flow velocity gradient appears after entering elbow 1. Under the influence of elbow 1, the flow velocities in elbow 2 and near the outer wall after flowing out of elbow 2 are relatively large. It can be found that with the increase of the *R*/*D* ratio, the curvature radius increases, the bending angle becomes lower, and gradually the mainstream fits better to the outer pipe wall.

In addition to the influence of velocity distribution, the change in erosion is also related to the effect of secondary flow. As can be seen from Figure 10, there is no obvious secondary flow

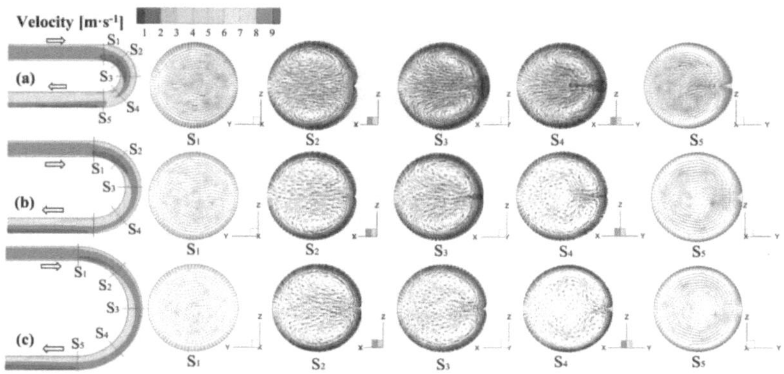

Figure 10. Velocity distribution and velocity vector under different R/D ratios (a) R/D = 1.5 (b) R/D = 2.5 (c) R/D = 4.0.

at the inlet S1 section, while there are obvious secondary flow phenomena in the S2, S3, and S4 sections, among which the secondary flow in the S3 section is the severest, and the secondary flow in the S5 section practically disappears. It is found that with the increase of the *R/D* ratio, the phenomenon of secondary flow in different sections tends to weaken. From the erosion contours in Figure 9, it can also be seen that the erosion rate at the side wall is relatively small at different *R/D* ratios, which shows that the secondary flow exerts less influence on erosion.

4.4 *Variation with different pipe diameters*

Under the conditions of $R/D = 1.5$ and combination spacing $L/D = 0$, the change in erosion rate of combined elbows with pipe diameter is shown in Figure 11. It shows that with the increase in pipe diameter, the average erosion rate and the maximum erosion rate decreased. The average erosion rate is 1.64×10^{-7} kg·m^{-2}·s^{-1} when $D = 40$ mm, which reduces to 1.33×10^{-8} kg·m^{-2}·s^{-1} when $D = 150$ mm. The maximum erosion rate is 5.25×10^{-6} kg·m^{-2}·s^{-1} when $D = 40$ mm, which reduces to 2.40×10^{-7} kg·m^{-2}·s^{-1} at $D = 150$ mm.

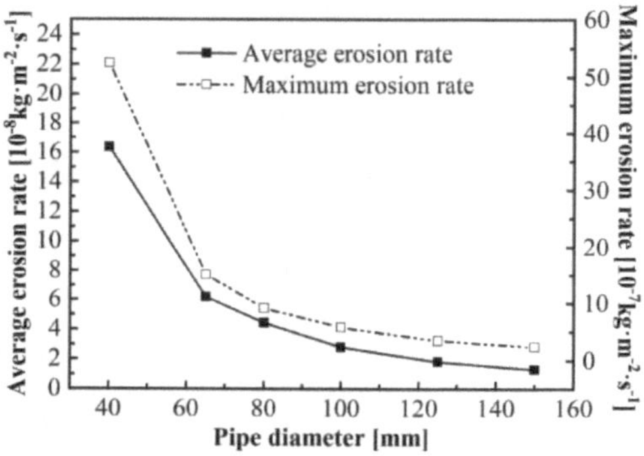

Figure 11. Erosion rate under different pipe diameters.

Figure 12 illustrates the variations in the erosion contour under different pipe diameters. The erosion regions remain broadly the same when pipe diameter changes, and the most serious erosion location still appears at the outer wall of elbow 2 near the outlet, and the erosion rate decreases with the increase in pipe diameter. According to related research (Wang *et al.* 2014), the motion characteristics of particles in water flow can be described by a change in the Stokes number. When other parameters are kept constant, with the increase of pipe diameter, the Stokes number of particles decreases, the following properties of particles with water flow become better, the inertia force weakens, and the erosion degree weakens.

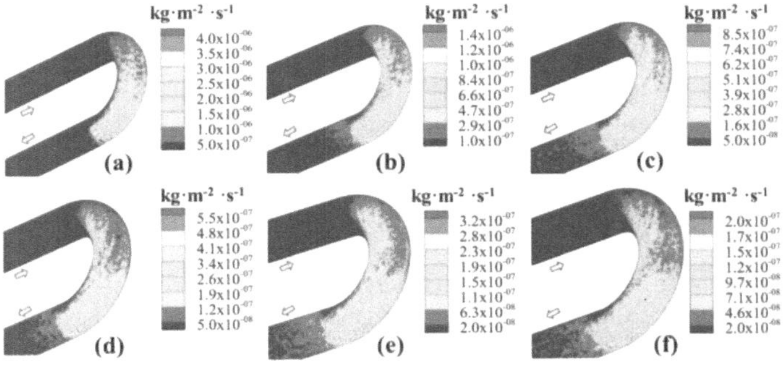

Figure 12. Erosion contours under different pipe diameters (a) D = 40 mm (b) D = 65 mm (c) D = 80 mm (d) D = 100 mm (e) D = 125 mm (f) D = 150 mm.

Figure 13 shows the flow field distribution under different pipe diameters. According to Figure 13, the velocity distribution of the central section remains unchanged with different pipe diameters when Z = 0. According to the velocity vector diagram of different sections, there are still obvious secondary flow phenomena in the S2, S3, and S4 sections, which recover in the S5 section. And the severity of secondary flow gradually decreases as pipe diameter increases. It can also be seen from the erosion contours in Figure 12 that the erosion rate at the side wall of the elbow is relatively smaller under different pipe diameters, which shows that the influence of secondary flow on erosion characteristics is still limited.

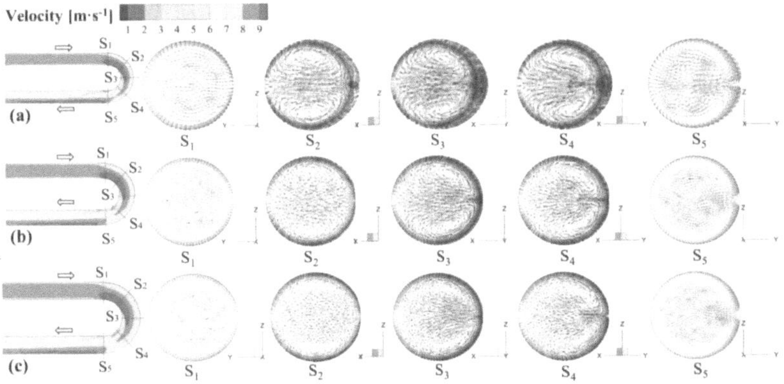

Figure 13. Velocity distribution and velocity vector under different pipe diameters (a) D = 40 mm (b) D = 100 mm (c) D = 150 mm.

5 CONCLUSIONS

Taking the horizontal U-shaped combined elbows as the research object, this paper discussed the influence of different structural parameters such as combination spacing (*L/D*), curvature radius to diameter (*R/D*) ratio, and pipe diameter on the particle erosion characteristics of the combined elbows by using the CFD method. Through analysis, the following conclusions are drawn:

(1) The erosion rate changes with combination spacing: when $L/D = 0$, the erosion rate was smaller; when $L/D = 1$, the erosion rate was higher and then decreases as the combination spacing increased. This change may lay in the size change and mutual influence of erosion regions between two elbows: when $L/D = 0$, the erosion region decreased, which led to a smaller erosion rate; when $L/D = 1.0$, the erosion area increased, which led to an increase in the erosion rate; when $L/D = 1.0$ and increased, the increasing range of the erosion region was limited, and the adjacent effect between two elbows was weakened, causing the erosion rate to decrease.
(2) The most serious erosion was located in the outer wall of elbow 2 near the outlet when the combination spacing $L/D = 0.2$; when $L/D = 1.0$ and $L/D = 4.0$, the most serious erosion appeared at the outlet side wall of elbow 1 and the outer wall of the outlet of elbow 2; when $L/D = 7.0$ and $L/D = 10.0$, the most serious erosion appeared at the outlet side wall of elbow 1, and the erosion change in elbow 1 was consistent with that of a single elbow. This change may depend on the mutual influence, the secondary flow, and the inertial force, which compete with each other.
(3) When the combination spacing $L/D = 0$, the average and maximum erosion rates decreased as the R/D ratio increased. With the increase of the R/D ratio, the bending angle decreased, the number of particles directly impacting the wall decreased, and the impact force decreased, thus, the erosion rate decreased.
(4) When $L/D = 0$, the average erosion rate and the maximum erosion rate of combined elbows decreased with the increase in pipe diameter. After analyzing the Stokes number of particles, with the increase in pipe diameter, the Stokes number decreased, the following property of particles with water flow became better, and the action of inertia force was weakened. Therefore, the erosion degree was weakened.

ACKNOWLEDGMENTS

This study is supported by the National Natural Science Foundation of China (Grant No. 51969011), the Natural Science Foundation of Gansu Province of China (Grant No. 20JR10RA239), and the Higher Education Industry Support Plan Project of Gansu Province of China (Grant No. 2022CYZC-32).

REFERENCES

Gabriel, C. P., Francisco, J. D. S. & Diego, A. D. M. M. (2014). Numerical Prediction of the Erosion Due to Particles in Elbows. *Powder Technol.* 261, 105–117.
Grant, G. & Tabakoff, W. (2012). Erosion Prediction in Turbomachinery Resulting from Environmental Solid Particles. *J. Aircraft.* 12 (5), 471–478.
Hu, G., Zhang, P., Wang, G. R., et al. (2017). Performance Study of Erosion Resistance on Throttle Valve of Managed Pressure Drilling. *J. Petrol. Sci. Eng.* 156, 29–40.
Huser, A. & Kvernvold, O. (1998). Prediction of Sand Erosion in Process and Pipe Components. Pro 1st North American Conference on Multiphase Technology, *Banff*, 217–227.
Lin, Z., Ruan, X. D., Zhu, Z. C., et al. (2014). Numerical Study of Solid Particle Erosion in a Cavity with Different Wall Heights. *Powder Technol.* 254, 150–159.
Laín, S. & Sommerfeld, M. (2019). Numerical Prediction of Particle Erosion of Pipe Bends. *Adv. Powder Technol.* 30 (2), 366–383.
Mclaury, B. S. (1993). *A Model to Predict Solid Particle Erosion in Oilfield Geometries.* The University of Tulsa.
Oka, Y. I., Okamura, K. & Yoshida, T. (2005). Practical Estimation of Erosion Damage Caused by Solid Particle Impact: Part 1: Effects of Impact Parameters on a Predictive Equation. *Wear.* 259 (1), 95–101.

Oka, Y. I. & Yoshida, T. (2005). Practical Estimation of Erosion Damage Caused by Solid Particle Impact: Part 2: Mechanical Properties of Materials Directly Associated with Erosion Damage. *Wear*. 259 (1), 102–109.

Peng, W. S. & Cao, X. W. (2016). Numerical Prediction of Erosion Distributions and Solid Particle Trajectories in Elbows for Gas-solid Flow. *J. Nat. Gas. Sci. Eng.* 30, 455–470.

Parsi, M., Najmi, K., Najafifard, F., et al. (2014). A Comprehensive Review of Solid Particle Erosion Modeling for Oil and Gas Wells and Pipelines Applications. *J. Nat. Gas. Sci. Eng.* 21, 850–873.

Sedrez, T. A., Shirazi, S. A., Rajkumar, Y. R., et al. (2019). Experiments and CFD Simulations of Erosion of a 90° Elbow in Liquid-dominated Liquid-solid and Dispersed-bubble-Solid Flows. *Wear*. 426–427, 570–580.

Wang, B., Kang, K., Chu, K. T., et al. (2019). Numerical Simulation of the Erosion Wear of Pipes with a Low Concentration of Particles and Multiphase Flow. *J. B. Univ. Chem. Technol. (Nat. Sci. Ed.)* 46 (2), 22–30.

Wang, K., Li, X. F., Wang, Y. S., et al. (2014). Numerical Prediction of the Maximum Erosion Location in Liquid-solid Two-phase Flow of the Elbow. *J. Eng. Thermophys-Rus.* 35 (4), 691–694.

Wang, Y., He, Q., Yu. F., et al. (2018). Numerical Simulation of the Erosion Characteristics and Structure Optimization of Elbows Connection for Gas-solid flow. *Prog. CSEE.* 38 (3), 832–839.

Yang, D. C. & Zhu, H. W. (2019). Erosion Wear Analysis of Natural Gas-sand Two-phase Flow at Bend Pipe. *China Pet. Mach.* 47 (10), 125–132.

Zeng, L. Zhang, G. A. & Guo, X. P. (2014). Erosion-corrosion at Different Locations of X65 Carbon Steel Elbow. *Corros. Sci.* 85, 318–330.

Zhang, J., McLaury, B. S. & Shirazi, S. A. (2018). Modeling Sand Fines Erosion in Elbows Mounted in Series. *Wear*. 402–403, 196–206.

Zhao, X. Y., Cao, X. W., Cao, H. G., et al. (2021). Numerical Study on Gas-solid Erosion of Two Elbows Mounted in Series. *J. Chin. Univ. Pet.* 45 (6), 152–160.

Water Conservancy and Civil Construction – Oke & Ahmad (Eds)
 ISBN: 978-1-032-58614-4

Quasi-Static test and finite element analysis of U-shaped concrete shear wall

Jiuyang Li, Xinmei Fan, Yuepeng Zhu, Jingwei Luo & Xiaoyu Wang*
School of Civil Engineering, Changchun Institute of Technology, Changchun, China

ABSTRACT: In order to study the seismic performance of the U-shaped reinforced concrete (RC) cross-section shear wall, the U-shaped cross-section shear wall with a 1/5 scale was tested under low cyclic loading based on the subway operation library in Changchun. The skeleton curve, bearing capacity, ductility, and stiffness degradation were analyzed. Based on the experimental study, the finite element simulation of the specimen was carried out by ABAQUS software, and the calculation results were compared with the experimental results. The results show that the U-shaped shear wall has good seismic performance and should be applied to more projects through practical design.

1 INTRODUCTION

In recent years, specially shaped shear wall structures have been widely developed in China. Short limb special-shaped shear wall refers to the length of the wall limb being 5–8 times the thickness of the shear wall structure, commonly known as "T" shaped, "L" shaped, "cross-shaped," "Z" shaped, broken linear, or "one" shaped (China Construction Industry Press 2010). With the vigorous development of China's construction industry and the diversification of building structure forms, people's requirements for layout have also improved, not only to increase the indoor use area but also in comfort, safety, rationality, and other aspects.

Zhou *et al.* investigated the force transmission characteristics, failure mechanism, ductility analysis, material strain, and other properties of the zigzag short-leg shear wall as part of their research on the influencing factors of the seismic performance of the zigzag short-leg shear wall (Zhou *et al.* 2018). Li *et al.* of Xi'an University of Architecture and Technology comprehensively analyzed the stress and deformation characteristics of a short-leg special-shaped shear wall structure, broke through the limitation of the plane section assumption in the bar model, and improved the existing multi-vertical bar model to the section displacement model (Li *et al.* 2004, 2010). Zhang of Kunming University of Science and Technology studied the seismic performance of a T-shaped short-leg shear wall under axial compression ratio, load direction, and section height-thickness ratio and obtained the restoring force model of the T-shaped short-leg shear wall to be "trilinear" (Zhang & Liu 2019; Zhang 2011). Zhang *et al.* from Guangxi University of Science and Technology found that the load-displacement curves of T-shaped and L-shaped short-leg shear walls were not uniform, and the hysteresis loop was asymmetric in the low-cycle horizontal load test. They proposed the local slotted short-leg shear wall to improve the seismic performance of T-shaped and L-shaped short-leg shear walls (Zhang 2015; Zhang *et al.* 2015). Ding and Ji studied the bending-shear deformation characteristics of a cross-shaped shear wall structure intensely and deduced the parameter formulas of interlayer displacement angle, bearing capacity limit,

*Corresponding Author: wxy18339168120@163.com

 DOI: 10.1201/9781003450818-77

and stiffness degradation considering the rigid domain factors. The cross-shaped shear wall structure has good seismic performance and meets the deformation requirements under strong earthquakes (Ding & Ji 2000).

At present, the research on U-shaped shear walls is mainly focused on the common special-shaped shear wall at home and abroad, but the research on U-shaped shear walls is less, and the corresponding seismic performance and design suggestions are lacking. Therefore, it is indispensable to research its hysteretic performance. ABAQUS finite element software was used to simulate and analyze it.

2 TEST SURVEY

2.1 *Design and fabrication of specimens*

The cantilever-reinforced concrete U-shaped shear wall with a section size of 870 mm × 730 mm × 1600 mm was designed in the experiment. The specimen design, size, and reinforcement are shown in Figure 1. HRB400 was used as reinforcement, the concrete strength grade was C35, and the measured average cube compressive strength was 35.6 MPa. The design parameters of each specimen and the material properties of reinforcement are shown in Table 1.

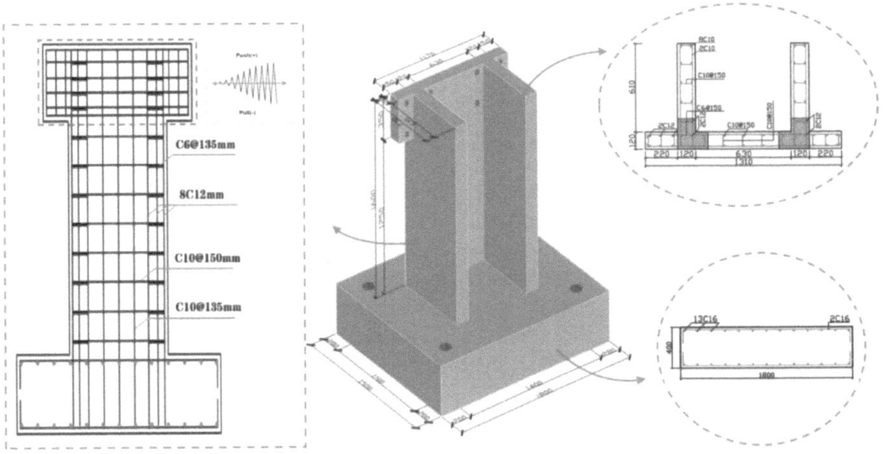

Figure 1. Sample design schematic diagram.

Table 1. Design parameters of specimens.

Wall Limb Size/mm	Web Size/mm	Overhang Wing Size/mm	Stirrup	Horizontal tendon	Longitudinal bar	Overhanging wings Longitudinal tendons
730 × 1600 × 160	870 × 1600 × 160	220 × 400 × 160	C6@150	C10@200	16C12 22C10 (1.41%)	12C10 (1.51%)

2.2 *Loading device and loading system*

A double electrohydraulic servo actuator was designed to apply load when combined with the laboratory's existing equipment conditions. The loading device and loading program are shown in Figure 2. The horizontal thrust and tension were applied to the TWD actuator, which stipulated that the flange was pulled in by the horizontal load, which was a positive load, and the pressure was a negative load. The loading program was controlled by displacement. Before steel yielding, each stage was loaded once at 20 KN to approach the yield load. After yielding, the steel bars were loaded three times, with integral times of yield displacement at each stage. When the load was reduced to 85% of the peak load, the specimen was identified as a failure, and the load was stopped.

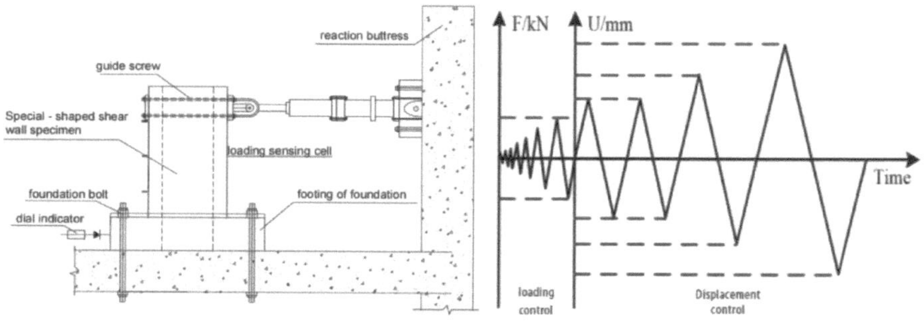

A) Charger b) Loading program

Figure 2. Loading diagram.

2.3 *Creating specimens*

The specimens were constructed in two stages with concrete pouring. When pouring the foundation base, the holes for the anchor bolts were reserved in the corners of the foundation base to secure the foundation base and the laboratory floor (Figure 3(a)). The laboratory ground was smooth and rigid, and there was no need for a bottom mold during the experiment. Plastic cloth was laid on the ground to prevent pollution and facilitate demolding. The side die was fixed with steel formwork and steel wire. Before the foundation base was poured, to ensure the spacing and location of longitudinal reinforcement, the shear wall longitudinal reinforcement was inserted into the foundation for anchoring (Figure 3(b)).

(a) Base reinforcement skeleton (b) The bottom supporting mold

Figure 3. Manufacturing process of the base skeleton.

To ensure that the shear wall and the foundation base are integrated after pouring, the chiseling process was adopted at the connection between the wall limb and the base after the initial concrete solidification. (Figure 4(a), (b)): When the foundation strength reaches 20 MPa (GB 50204-2015 "Concrete Structure Engineering Construction Quality Acceptance Specification"). The "concave" shear wall formwork was supported after the wall reinforcement was tied. Scaffolds were supported on the outside of the formwork, while tension bolts were used between the inside and outside of the wall limb formwork to ensure the stability of the formwork during concrete pouring. (Figure 4(c), (d)).

(a) Pre-embedded thread sleeve (b) Chiseling connection surface

(c) Formwork Support (d) Specimen Shaping

Figure 4. Template support for preliminary work.

3 EXPERIMENT RESULTS AND ANALYSES

The failure mode of the specimen is shown in Figure 5.

At the beginning of loading, the specimen was in the elastic state, and there was no obvious crack in the component. When the load reached + 165 kN, the cracks first appeared at the interface between the two wall limbs and the base, and then the horizontal cracks with a length of about 50 mm appeared at the lower abdominal end and the ends of the wall limbs. With the reciprocating loading of the actuator, the horizontal cracks on both sides developed alternately to the center of the wall limbs, forming a "V-shaped crack" as a whole.

The first crack of the web appeared at about + 220 kN, located at the junction of the web and wall limb 1. The lateral load of the web continued to increase, the original horizontal cracks at the bottom of the web became longer and wider, and a number of cracks of about 30° were added at the middle and lower parts of the web. When the load reached + 370 kN, the inclined cracks connected with the original horizontal cracks to form the main cracks, and the web cracks penetrated. After the wall cracks penetrate, the continuous load caused the existing crack width to increase but basically did not produce new cracks. When the lateral load of the web reached + 460 kN, the concrete at the bottom of the wall limb fell off, the steel bars yield, the horizontal load did not increase or decrease, and the displacement continued to increase, and then the displacement control was adopted. When loading to + 460 kN, the concrete at the bottom of the wall limb fell off, the steel bars yield, and the horizontal load did not increase or decrease. When the displacement increased to 67 mm, the longitudinal reinforcement broke suddenly, and the test ended.

a) Failure of webs b) Failure of wall limbs c) Concrete spalling off d) Steel fracture

Figure 5. Breakdown phenomenon.

4 TEST RESULTS AND ANALYSIS

4.1 *Skeleton curve*

It can be seen from the test skeleton curve that the specimen is basically in a linear development before the descending section, indicating that the specimen can use its own structure to dissipate the gradually increasing web sideload, and the structural rigidity is almost unchanged when the web site load reaches 458 kN, the main reinforcement yields, and the curve decreases. The skeleton curves in both positive and negative directions have better symmetry, indicating that the integrity of the U-shaped shear wall under the web side load is better. It can be seen from the test skeleton curve that the specimen is basically in a linear development before the descending section, indicating that the specimen can use its own structure to dissipate the gradually increasing web sideload, and the structural rigidity is almost unchanged when the web site load reaches 458 kN, the main reinforcement yields, and the curve decreases. The skeleton curves in both positive and negative directions have better symmetry, indicating that the integrity of the U-shaped shear wall under the web side load is better.

It can be seen from Figure 6 that the skeleton curves derived from the test results are generally consistent with the finite element results. Especially in the early stage of loading, both of them are in the elastic stage, and the slope of the curve is almost coincident. When loaded to the cracking load, the two curves begin to deviate, but the curve trend is still consistent, mainly due to the significant difference in bonding forces between the concrete and the steel bar.

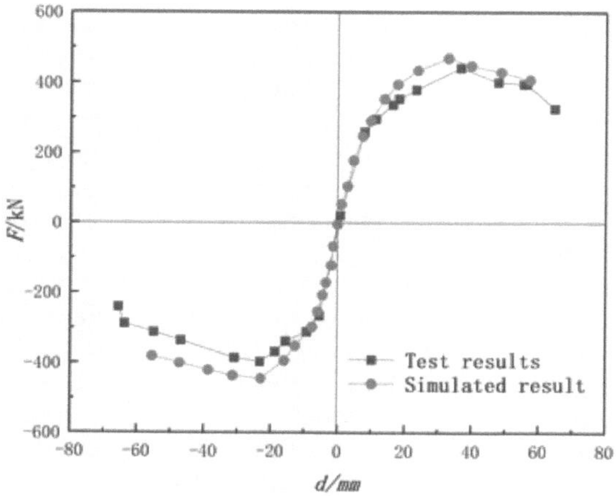

Figure 6. Skeleton curve of the specimen.

4.2 *Analysis of bearing capacity and Deformation capacity*

The main failure process was divided into three stages: the cracking stage, the yield stage, and the ultimate stage. From Table 2, we can obtain the horizontal load, displacement, and ductility coefficient of each stage of the specimen. We can see that the ductility coefficient of the test value of the specimen is close to 4.0, indicating that the U-shaped shear wall can meet the seismic requirements with a certain reasonable design.

Table 2. Comparison of the finite element analysis value and test value of each load point.

Specimen Number	Method	Δ_y/mm	P_{max}/kN	Δ_{max}/mm	Δ_u/mm
URC-2	Test	19.19	458.16	36.56	64.56
	Simulation	18.75	464.92	32.88	57.18

4.3 *Rigidity degeneration*

The stiffness degradation curve of the U-shaped shear wall to web side load is shown in Figure 7. As a whole, it can be seen that the stiffness degradation curves in both positive and negative directions are highly symmetrical, which further illustrates that the U-shaped section shear wall presents better integrity to the lateral load of the web. The stiffness degradation rate in both positive and negative directions is equal, and the initial stiffness is the same. It shows that the development of cracks and the fatigue degree of steel bars in U-shaped cross-section shear walls are linear in the website reciprocating load test, and there is no apparent defect.

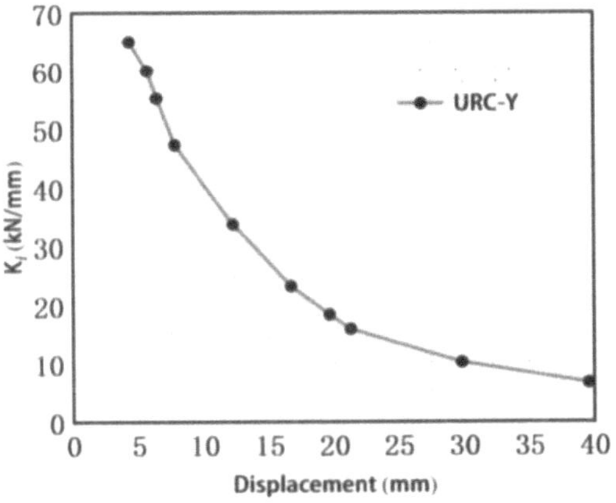

Figure 7. Stiffness degradation curve of a U-shaped shear wall.

5 CONCLUSION

Through the low-cyclic loading test and finite element analysis of the URC shear wall, we can get the following conclusions:

(1) The free end of the two sides of the U-shaped cross-section shear wall was independent and lacked connection. Faced with the lateral load of the web, it belonged to the weak area, which was mainly manifested as bending cracks, bending shear cracks, concrete

crushing of the protective layer, and longitudinal reinforcement bonding failure, showing a typical bending shear failure mode.

(2) The failure process analysis showed that the overall ductility and energy dissipation capacity were reduced but still met the requirements of seismic design; the website load was in the weak area, and the end of the wall limb was often destroyed first. In practical application, some strengthening measures should be taken at the free end of the wall limb.

REFERENCES

China Ministry of Construction Industry Standards: High-rise Building Concrete Structure Technical Specification GJ3-2010 [S]. Beijing: China Construction Industry Press, 2010.

Ding Yongjun, Ji Gang. Calculation of Inter-story Ultimate Deformation Capacity of Short-leg Shear Wall Structure [J]. *Journal of Tianjin University*, 2000, 33 (3): 363–366.

Li Qingning, Wang Ning, Cai Weining. Elastoplastic Model and Seismic Response Analysis of T-shaped short-leg Shear Wall [J]. *Seismic Engineering and Engineering Vibration*, 2004, 24 (2): 69–74.

Li Qingning, Li Xiaolei, Lei Weining. Failure law and Stiffness Ductility of T-shaped and L-shaped Short-leg Shear Walls [J]. *Architectural structure*, 2010, 40 (2): 37–40.

Zhang Min. *Seismic Performance Test and Calculation Method of Recycled Heat Preservation Concrete Shear Wall* [D]. Taiyuan: Taiyuan University of Technology, 2015.

Zhang Min, Yi Qi, Wang Zhulin, et al. Experimental Study on Torsional Behavior of Local Slotted Short-leg Shear Walls with T-shaped and L-shaped Sections [J]. *Structural engineer*, 2015, 31 (2): 185–193.

Zhang Pingle, Liu Junxiong. Experimental Study on Restoring Force Model of T-shaped Short-leg Shear Wall [J]. *Sichuan Architectural Science Research*, 2019, 45 (5): 37–41.

Zhang Pinle. Experimental Study on Seismic Performance and Damage Analysis of Short-leg Shear Wall [D], 2011.

Zhou Yun, Liu Zili, Liu Feng, et al. Model Test Research on Seismic Performance of Zigzag Short-leg Shear Wall [J]. *Journal of Building Structures*, 2008, 29 (4): 81–88.

Water Conservancy and Civil Construction – Oke & Ahmad (Eds)
© 2024 The Author(s), ISBN: 978-1-032-58614-4

Study on the improvement of anti sliding bearing capacity of rock socketed gravity anchorage foundation

Lingzheng Wu & Fengchao Guo
Guang Dong Bay Area Traffic Construction Investment Co., Ltd, Guangzhou, China

Wei Li*, Ye Yang, Haiyang Shi & BaiYong Fu
China Communications Highway Long Bridge Construction National Engineering Research Center Co., Ltd., China

ABSTRACT: In the code of specifications for the design of the foundation of highway bridges and culverts, the bearing capacity of traditional gravity anchorage foundation is calculated only by considering the friction effect of anchorage foundations, without considering the effect of rock socketed effect of the bottom plate on the improvement of bearing capacity. To study the influence of rock socketed effect on the bearing capacity of gravity anchorage, two groups of scale model tests of non-rock socketed and rock socketed gravity anchorage with a scale of 1:50 were designed and carried out to study the bearing characteristics and load distribution law of different models under vertical and horizontal combined loads, to obtain the effectiveness of rock socketed effect. The test results show that the ultimate failure loads of the two groups of models are 300kN and 350kN respectively, and the deflection values under the ultimate load are 5.57cm and 2.14cm respectively. Therefore, under the same horizontal cable force, the displacement value of gravity anchor considering the rock socketed effect is not only smaller but also the horizontal ultimate failure load is increased by about 17.7%. In addition, after the floor is embedded in the moderately weathered layer, the rock stratum provides the lateral friction resistance for the anchor structure, thereby reducing the stress level of the floor, and changing the ultimate failure state of the gravity anchor from "sliding friction failure mode" to "overturning failure mode", providing technical support for the design and construction of the gravity anchor foundation on site.

1 INTRODUCTION

At present, a gravity anchorage is a common form in domestic suspension bridge construction projects. As the span of the suspension bridge increases, its size becomes larger and larger. In some shallow geological conditions, anchorage foundations will generally be embedded in moderately weathered or slightly weathered rock layers to a certain depth, but the calculation of anti-overturning and anti-slip only considers the friction between foundation and foundation, balancing the cable horizontal force with the friction produced by anchorage self-weight, and does not consider the anti-slip bearing capacity improvement function exerted by anchorage embedded rock effect. As a result, the gravity anchorage foundation has a large design scale, high construction cost, and large optimization space.

Industry scholars have carried out extensive research on gravity anchorage foundations. Caichu Xia (1997) carried out a 1:50 analogy field structural model test on the tunnel

*Corresponding Author: 1349080105@qq.com

DOI: 10.1201/9781003450818-78

anchorage of Humen Bridge and studied the deformation mechanism and failure mode of the anchorage structure and rock mass. According to the numerical study on the anchorage foundation of the suspension bridge of the second Humen Bridge, Gang Ya (2017) pointed out that the anti-sliding ability of the foundation will be greatly improved and the failure mode of the foundation will change from sliding failure to overturning failure when the compound effect of diaphragm wall is considered. Yongsheng Li (1995) carried out a 1:100 similar material model test on the North Anchorage of Jiangyin Yangtze River Bridge, and proposed measures for foundation reinforcement, structure selection, and embedment depth to strengthen the stability of the anchorage structure and adjacent strata. Rongan Lin (2008) used the finite element method to analyze the gravity anchorage of the Yangluo Yangtze River Bridge in Wuhan. The bearing capacity, sliding stability, overturning stability, and settlement characteristics of the anchorage foundation are studied. Xiaowen Zhu (2005) studied that the horizontal displacement of anchorage is not sensitive to the variation of friction coefficient due to its huge mass, while the depth of bedrock has a greater influence on the foundation displacement. Enjie Zou (2018) carried out field tests on the semi-diagenetic shear strength and bearing capacity of the anchorage foundation and obtained the friction resistance and bearing capacity of the contact surface between the bedrock and concrete. Xin Tan (2015) proposed to comprehensively determine the friction coefficient of gravity anchorage by shearing resistance, shearing resistance, and single point friction test and giving an engineering example for verification. Linge Luo (2019) studied the influence of the embedding action of the diaphragm wall below the base plate of the foundation of the diaphragm-gravity composite anchorage on its bearing capacity and determined that when the local diaphragm wall structure is embedded in the bedrock with high deep strength. The contribution of the embedding action of the diaphragm wall to its bearing capacity can be considered in the anchorage foundation design of the foundation pit enclosure structure.

2 PROJECT OVERVIEW

According to the actual geological survey and structural design scheme, the anchorage of a suspension bridge for a river crossing passage is a gravity-type anchorage, which bears a huge cable force and has certain rock strata under geological conditions. The foundation elevation of the diaphragm wall is – 43m, which is located in moderately weathered argillaceous sandstone. To give better play to the effectiveness of the rock stratum and reduce the scale of the anchorage, it is proposed to explore the changes in the displacement, stress distribution characteristics, and failure patterns of the anchorage foundation under the action of vertical and horizontal loads with or without rock socketed conditions through a physical scale model.

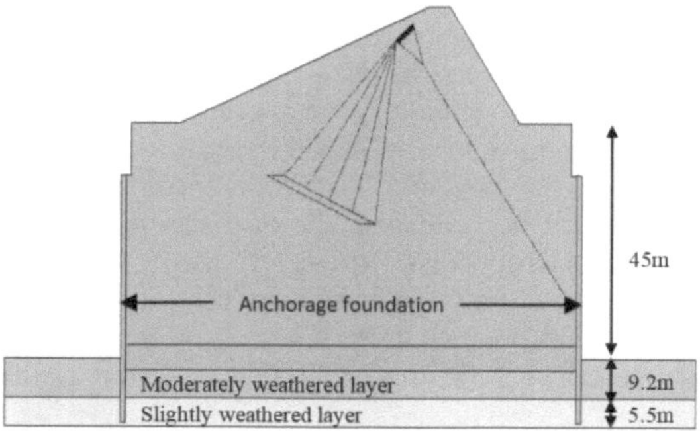

Figure 1. Anchorage style and rock stratum.

3 MODEL DESIGN

3.1 *Model scheme*

According to the space of the laboratory and the capacity of loading equipment, the scale of the selected model is 1:50. The simulation is conducted according to the size of the preliminarily formed anchor foundation scheme. The actual outer diameter of the diaphragm wall is 60m, and the outer diameter of the model is 1.2m. All geometric dimensions shall be considered according to 1:50 to ensure the accuracy of the load distribution relationship. The comparison of geometric scales of different models is shown in the table.

Table 1. Anchor model size.

Content	Prototype size/m	Model size/m
Anchorage height	45	0.9
Anchorage diameter	60	1.2
Moderately weathered rock stratum thickness	9.2	0.185
Embedded in moderately weathered rock stratum depth	2.76	0.055
Moderately weathered thickness under the base plate	6.44	0.13

Based on a similar theory, the interaction between the anchorage foundation and surrounding rock and soil mass is simulated. Without considering the influence of soil mass on anchorage, the only variable – the influence of embedded rock on anchorage is determined. The soil mass on the upper part of the anchorage is not considered in this test. On this basis, the comparison test between the non-embedded gravity anchorage foundation and the embedded gravity anchorage foundation is designed, which are:

S1 model — a physical model of gravity anchor foundation without rock socketed, that is, the depth of the moderately weathered layer embedded in the bottom plate of 5.5 cm is not considered, and the vertical and horizontal loads are offset by its gravity and base friction.

S2 model — a physical model of a rock socketed gravity anchor foundation, that is, in addition to the contact between the foundation floor and the rock stratum, the lower part of the gravity anchor also contacts the rock stratum with a certain depth, the contact depth is 5.5 cm, and the anchor base friction and the embedded moderately weathered rock stratum share the cable force load.

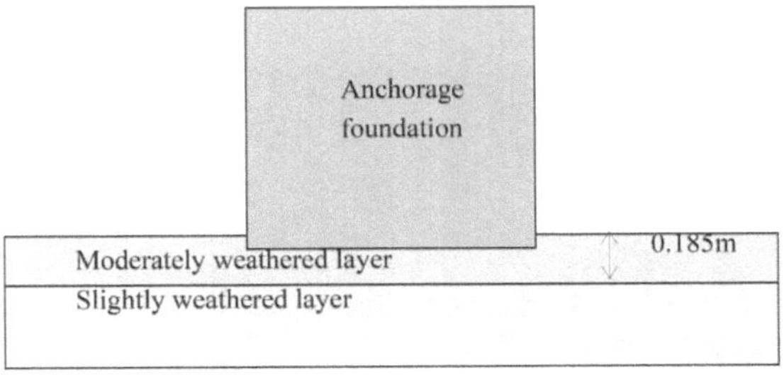

Figure 2. Rock socketed gravity anchorage.

3.2 *Model materials*

The anchorage consists of three parts: top plate, core filling material inside the anchorage, and steel hoop. Among them, the anchor top plate is made of a 30 mm thick steel plate according to the equivalent conversion of bending stiffness. The steel hoop is a 6 mm Q235B steel plate as the internal filling template, and the filling material is C40 micro-expansion concrete.

Due to the high strength of moderately weathered mudstone and slightly weathered mudstone at the bottom of the anchorage, mortar is used to simulate the weathered rock stratum. The two rock layers are simulated by controlling the elastic modulus and axial compressive strength. In the test, cement mortar is used to simulate moderately weathered rock and slightly weathered rock, and the mortar mix proportion is debugged with two control indexes of axial compressive strength and elastic modulus. The following table shows the comparison between mortar and geological exploration parameters.

Table 2. Mortar test index and geological exploration parameter index to be used in the test.

Rock type	Index	Geological exploration parameters	Mortar parameters
Moderately weathered	Compressive strength/MPa	21.2	18.8
	Elastic modulus/GPa	16.3	16.6
Slightly weathered	Compressive strength/MPa	56	44.5
	Elastic modulus/GPa	66	35

3.3 *Fabrication of test model*

Before the anchor model is made, the bottom shall be poured to make a rigid foundation to simulate the fresh bedrock. Then, the 18.5 cm thick mortar shall be poured according to the design proportion in advance to simulate the moderately weathered rock mortar. The fabrication and installation of the anchor model are divided into four key steps: 1) Bedrock simulation layer mortar pouring. The location and elevation of the anchor model are determined on the weathered rock according to the test plan. 2) Anchor model processing and manufacturing. The top plate and the inner side of the steel wall are welded with connecting reinforcement. After the welding is completed, it is moved to the set position, and then laid the soil pressure cell monitoring data line. 3) Pour core filling concrete. The vibration shall be strengthened during pouring, and the top plate shall be welded after pouring. 4) Concrete curing. During curing, displacement sensor support, horizontal and vertical actuators, commissioning test equipment, etc. can be installed. The fabrication process of the anchor model is shown in Figure 3.

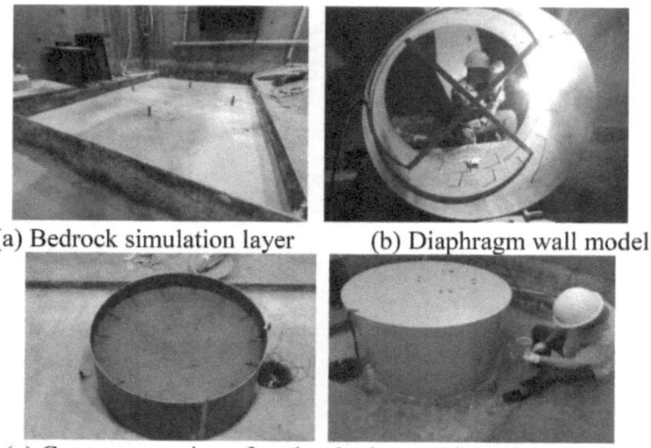
(a) Bedrock simulation layer (b) Diaphragm wall model
(c) Concrete pouring of anchor body (d) surface coating

Figure 3. Anchorage fabrication process.

3.4 *Monitoring measurement and loading scheme*

(1) Loading scheme

The vertical load is 500 kN, which is loaded by V1~V5 five levels. Each level is 100 kN, and the vertical load is 15.5 cm away from the center of the circle; The horizontal step load is 50 kN, which is loaded step by step from H1~H4 until the ultimate failure. The horizontal load is 110cm away from the anchor base plate.

(2) Monitoring program

It is necessary to arrange a micro soil pressure cell at the corresponding position to measure the stress and earth pressure change of the anchor under the load. The soil pressure cell is arranged at the center, edge, and between the center and edge of the model base plate. Considering the boundary effect of the soil pressure cell, it is necessary to ensure that the distance between the soil pressure cell and the inner edge of the model is more than one time the diameter of the soil pressure cell. The layout of the soil pressure cell is shown in Figure 4.

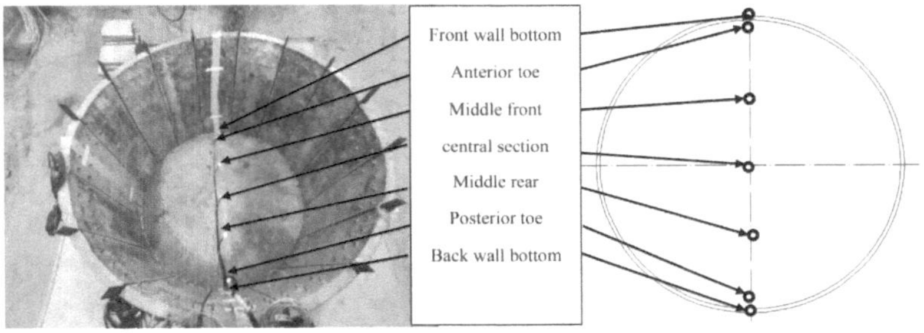

Figure 4. Layout of soil pressure cell.

To monitor the structural displacement characteristics and differences of each group of anchor models under the action of structural load, four vertical settlement monitoring displacement meters are arranged respectively in the direction before and after the anchor roof is loaded and in the orthogonal direction to monitor the vertical displacement of the anchor roof under the action of vertical load and horizontal load; At the same time, a horizontal displacement meter is arranged horizontally at the upper and lower positions of the front toe area of the anchorage to monitor the horizontal displacement law of the top plate and front toe of the anchorage. The layout of displacement monitoring points and test loading diagram are shown in Figure 5.

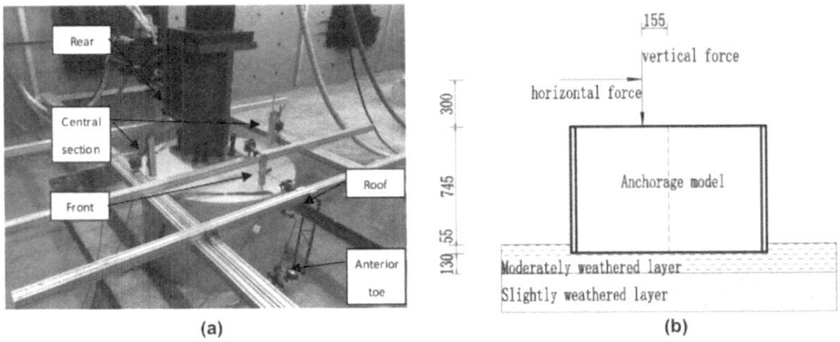

Figure 5. Schematic diagram of model displacement monitoring point layout and test loading. (a) Layout of Anchorage Model Monitoring Points (b) Schematic Diagram of Model Loading (Unit: mm).

4 ANALYSIS OF TEST RESULTS

4.1 *Analysis of anchorage settlement under vertical load*

The vertical load-settlement curve and average settlement curve of gravity anchorage in groups S1 and S2 are shown in the figure. According to the curve, due to a certain depth of regolith embedded in the model of group S2, the front, middle, and rear of the anchorage of group S2 and the average settlement of the anchorage structure under the same vertical load are smaller than those of group S1. Taking the final vertical load of 450kN and 500kN as an example, the front, middle, and rear settlement of the anchoring in group S2 decreased by about 20% compared with that in group S1. After the floor is embedded into the middle regolith at a certain depth, the rock layer provides a certain lateral friction resistance to the anchorage structure, and at the same time can share and disperse the basement stress to a certain extent. Therefore, the relative settlement of gravity anchorage in Group S2 is relatively low. The vertical displacement of the rock-socketed group S2 is significantly lower than that of the un-socketed group S1, and the average floor stress of the two models in the comparison group S2 is also significantly lower than that in group S1.

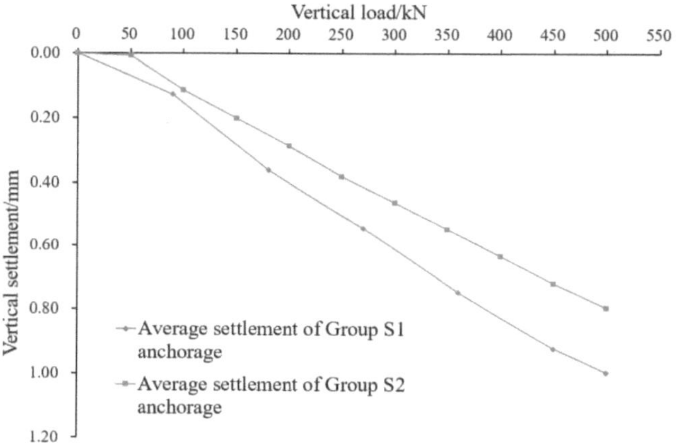

Figure 6. Comparison of load settlement curves of anchorage.

Table 3. Percentage decrease of settlement of two groups of anchorage models.

Classification	Vertical load /kN					
Group number	450	500	450	500	450	500
S1 Group	0.80	0.87	0.94	1.02	1.02	1.1
S2 Group	0.66	0.72	0.72	0.80	0.78	0.86
Percentage reduction	0.18	0.17	0.23	0.22	0.24	0.22

4.2 *Analysis of displacement of anchorage under horizontal load*

(1) Horizontal Loads — Analysis of horizontal displacement

After the vertical load is applied to the design value, the vertical load is kept unchanged and the horizontal load is applied step by step until the ultimate failure. In the process of horizontal load application, the horizontal displacement of the roof and floor of S1 and

S2 mooring models with load application was monitored. The contrast curves of horizontal load-horizontal displacement of the S1 and S2 models are shown in the figure.

Definition of ultimate failure load: In the process of applying the same load, the displacement control standard of the two groups of models is 50 mm. When the anchorage exceeds this standard instantaneously, it is considered that ultimate failure has occurred and the upper stage is the ultimate failure load.

The ultimate failure load was analyzed. According to the curve analysis, the mean horizontal displacement of the anchorage structure in group S2 was smaller than that in group S1, due to the embedment effect of 5.5cm regolith on the fore toe of the model in group S2 under all levels of horizontal loads. In the analysis, the first stage of failure is taken as the ultimate failure load. The ultimate failure load of the group S1 model is 300kN and the group S2 model is 350kN, and the ultimate failure load is increased by about 17.7%.

The mean value of the horizontal displacement of the top plate and fore toe of the anchorage in group S2 is lower than that in group S1, and the horizontal displacement of the top plate and fore toe of the anchorage in group S1 increases with the increase of the horizontal load. However, before the horizontal load of 300kN, the fore toe of the anchorage model in group S2 did not displace, and the horizontal displacement was all 0mm. With the continuous increase of the horizontal load, the fore toe of the anchorage was slightly dislocated, then the overturning failure occurred, and the test was terminated.

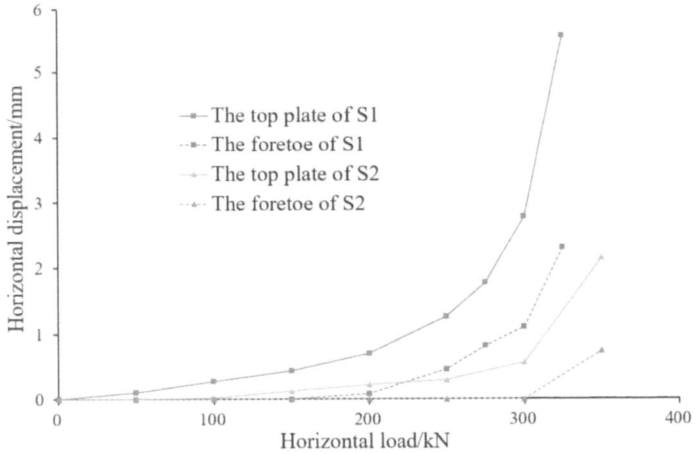

Figure 7. Comparison of two groups of horizontal displacement.

In addition, to obtain the shear strength and friction coefficient of the anchor base, based on the S1 group of anchor models, the measurement tests under two groups of different vertical loads, including 500kN and 700kN, were carried out respectively. The load of the previous level of failure was taken as the horizontal ultimate load $\mu = Fh/Fv$,

Table 4. Measurement statistics of base shear strength and friction coefficient.

Working condition	Vertical force/kN	Vertical force + dead weight/kN	Horizontal force/kN	Shear strength/ kPa	Friction coefficient
Working Condition I	500	520.81	300	265	0.624
Working Condition II	700	720.81	425	376	0.590

the calculated friction coefficient values are 0.624 and 0.590 respectively, and the base shear strength is 265kPa and 376kPa respectively.

(2) Horizontal loads-analysis of vertical displacement

By analyzing the vertical displacement rule of the roof of the anchorage under horizontal load, it is concluded that:

Both S1 and S2 mooring models showed the displacement phenomenon of mooring roofs from front to bottom to top. The difference between the two models is that the S2 model presents overturning failure mode with the front toe as the axis, the S1 model has no overturning trend, and the horizontal displacement of the front toe and roof is larger than the S2 model, showing a sliding failure mode, which further verifies the two different failure modes of gravity anchorage foundation due to the play of the front toe rock socking effect of the floor. With the increase of the horizontal load, the horizontal displacement of the top plate and the front toe of the Group S1 anchorage model increase, resulting in sliding failure. However, before the horizontal load of 300kN, the fore toe of the group S2 mooring model did not displace. As the horizontal load continued to increase, the fore toe of the mooring model would overturn and fail immediately after a small displacement.

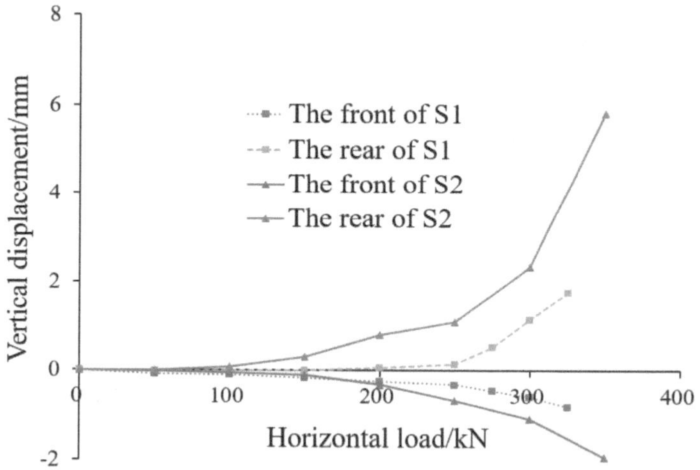

Figure 8. Roof displacement of group S1 under horizontal load.

4.3 *Comparative analysis of floor earth pressure*

Five earth pressure boxes are arranged at the bottom plates to analyze the distribution regularity of foundation stress along the stress direction of two groups of anchorages by using the earth pressure box as monitoring means and are analyzed according to the data of the earth pressure box of the anchorage bottom plates. The distribution of earth pressure under the bottom plate of two groups of anchorage models, S1 and S2, was calculated. The vertical load was defined as five values from V-1 to V-5 according to the loading classification stage, and the horizontal load was similarly defined as four stages from H-1 to H-4. The comparative analysis showed that:

(1) At the stage of vertical load application, the distribution curve of floor load along the front and rear toes presents a concave shape, and the stress level at the edge of the anchor base is higher than that in the inner area of the base. Due to the vertical load eccentricity, the base's stress level at the anchorage's back toe is higher than that at the front toe.

(2) In the stage of horizontal load application, the stress at the rear toe of the anchorage floor gradually decreases due to the bending moment load from the back toe to the toe, and the stress at the front toe increases with the increase of horizontal load. Finally, the stress concentration in the fore toe area was significant when the failure state approached.
(3) Due to the lateral friction resistance provided by medium-weathered rock socketed, under vertical load, part of the vertical load is transferred to rock strata through the rock-socketed structure, thus reducing the load sharing of the floor. Compared with the non-rock-socketed gravity anchorage of group S1, the floor stress of group S2 is relatively lower under the same vertical load. For example, under vertical loads of 300 kN, 400 kN, and 500 kN, the average stress of the bottom plate of group S1 is −342 kPa, −453 kPa, and −554 kPa, respectively, and that of group S2 is −251 kPa, −334 kPa, and −423 kPa, respectively. Stress reduction amplitudes were 27%, 26%, and 24%, respectively.

In general, the load distribution of the foundation floor has the following characteristics: 1) Under vertical load, part of the vertical load is transferred to the rock stratum through the rock-socking structure, thus reducing the load sharing of the bottom floor. Therefore, the average stress of the bottom floor of group S2 is relatively low, and under the same vertical load and eccentricity, both of them show "the front and back edge stress is large, and the front and middle stress is the least. In between" distribution characteristics. This indicates that the stress of the anchorage foundation floor is dumbbell-like, and the stress of the rock-socked gravity anchorage foundation is lower. 2) Under the horizontal load, the stress in the front toe gradually increases, while the stress in the back toe gradually decreases, and the

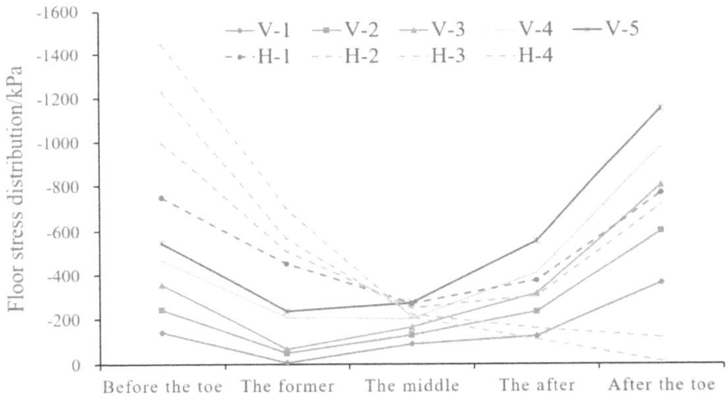

Figure 9. Load base stress distribution curve of group S1.

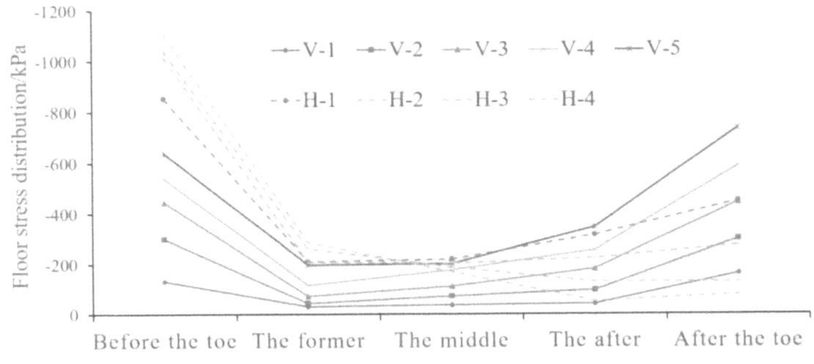

Figure 10. Load base stress distribution curve of group S2.

final stress redistribution gradually decreases from the front toe to the back toe, indicating that the stress distribution under the horizontal load gradually becomes higher in the front and lower in the back.

5 CONCLUSION OF THE TEST

In this experiment, model tests of the un-socketed gravity anchorage S1 group and socketed gravity anchorage S2 group were carried out. The ultimate failure load, ultimate failure state, and stress distribution law of the foundation floor of the two groups of models were compared, and the rock-socking effect of gravity anchorage was revealed. The following conclusions were drawn:

(1) In the vertical load stage, the rock socking effect can improve the vertical bearing capacity of the gravity anchorage, and because the floor is embedded in a certain depth of the middle regolith, the rock layer provides a certain lateral friction resistance to the anchorage structure, and also has a certain sharing and dispersing effect on the basement stress. The gravity anchorage presents the characteristics of low floor stress levels and small vertical settlements.
(2) In the stage of horizontal load, due to the constraint of the front toe, the horizontal displacement is limited and the failure mode is changed. Thus, the rock-socked gravity anchorage presents greater horizontal bearing capacity, and the vertical deformation of the rear toe area is greater than that of the front toe area.
(3) In the ultimate failure stage, the anchorage's front toe plays a resistance role. When the load exceeds the allowable value, the ultimate failure will occur. Thus, the sudden failure of the rock-socketed gravity anchorage appears, and the failure style changes from slip failure mode to overturning failure mode.

REFERENCES

Caichu Xia & Hongxin Cheng & Rongqiang Li. Field Model Test on the East Anchorage of Humen Bridge in Guangdong Province [J]. *Journal of Rock Mechanics and Engineering*, 1997, 16 (6): 571–576. (In Chinese)

Enjie Zou & Bin Tian & Jiang Xu & Kaiyu Jiang & Yuchen Liu Semi-diagenetic In-situ Experimental Study on the Foundation of the Side Anchorage of Maputo Bridge [J]. *Construction Technology*, 2018,47 (07): 31–34. (In Chinese)

Gang Ya & Dongdong Han & Yang Shihai, *et al.* Study on the Stability of Gravity Anchor Foundation of the Time Second Bridge Taizhou Waterway Bridge [J]. *Highway*, 2017 (4): 129–134. (In Chinese)

Linge Luo & Lichuan Cui & Haiyang Shi & Chao Guo & Shaoping Yi. Experimental Study on Bearing Capacity of Diaphragm Wall Gravity Composite Anchorage Foundation [J]. *Geotechnical Mechanics*, 2019, 40 (03): 1049–1058. (In Chinese)

Rong'an Lin. *Force Analysis and Optimization of Gravity Anchorage System of Suspension Bridge* [D]. Xi'an: Chang'an University, 2008. (In Chinese)

Xiaowen Zhu & Qilin Zhao & Kai Zhu. *et al.* Three-dimensional Numerical Simulation Analysis of North Anchorage Foundation of Runyang Bridge [J]. *Journal of Southeast University (NATURAL SCIENCE EDITION)*, 2005, 35 (2): 293–297. (In Chinese)

Xin Tan & Houqing Xu & Wei Peng. Friction Coefficient Test and Value Method of Gravity Anchor of Suspension Bridge [J]. *Journal of Underground Space and Engineering*, 2015, 11 (S2): 549–552 + 584. (In Chinese)

Yongsheng Li. Experimental study on North Anchorage Model of Jiangyin Yangtze River Highway Bridge [J]. *Journal of Tongji University*, 1995, 23 (2): 134–140. (In Chinese)

Water Conservancy and Civil Construction – Oke & Ahmad (Eds)
 ISBN: 978-1-032-58614-4

Selection of support parameters and rationality verification of a deep-buried soft rock hydraulic tunnel

Yajie Liu*
School of Engineering and Technology, China University of Geosciences (Beijing), Beijing, China
State Key Laboratory of Simulation and Regulation of Water Cycle in River Basin, China Institute of Water Resources and Hydropower Research, Beijing, China

Yongming Zhang, Wei Huang, Chao He, Xiyang Li & Zhao He
Central Yunnan Water Diversion Project Co., Ltd. of Yunnan Province, Kunming, Yunnan, China

ABSTRACT: The selection of support types and the determination of parameters under large deformation of surrounding rock in deep hydraulic tunnels have always been the focus and difficulty of research in the industry. A hydraulic tunnel is deeply buried, mainly in soft rock, and there is a risk of large deformation. In this paper, the relationship between the physical and mechanical parameters of the surrounding rock and the ground stress is used to determine the magnitude of deformation of the surrounding rock and the supporting force required under the control target. According to the support force that different support types can provide, combined with engineering experience, the corresponding support types and parameters are determined, and the rationality of support parameters is simulated and verified by numerical simulation analysis. The results show that the support type and parameters determined by this method can meet the code requirements, and the force of the support structure exceeds the ultimate strength.

1 INTRODUCTION

The problem of extrusion and large deformation has always been a major problem in the field of geotechnical engineering (Wang 2020), such as Tauern Tunnel in Austria, Arlberg Tunnel, Ena Mountain Tunnel in Japan, and other projects (He 2002). Due to the large magnitude of deformation and long deformation time has brought great threats to excavation support design, surrounding rock stability, and permanent lining safety, the final construction period, construction cost, and subsequent maintenance frequency and cost far exceeded expectations (Barla 1995; Steiner 1996). Scholars at home and abroad have carried out correlative research on support measures for large deformation disasters in surrounding rocks (Guo 2009; Barla 2011; Tran 2015; Li 2016; Karampinos 2016; Zheng 2021). The current support measures mainly use steel arches, anchor rods, shotcrete, etc. The prediction of extrusion value and the determination of the required support force is of great significance to the stability of surrounding rock and support safety in soft rock tunnels (Singh 2007). A deep-buried soft rock hydraulic tunnel has a total length of 62.596 km, a maximum buried depth of 1450 m, and sections with a buried depth greater than 600m account for 67.38% of the total tunnel length. The tunnel passes through many soft rock formations, including several large-scale regional fracture zones, and the total length of soft rock is 13.11 km,

*Corresponding Author: 1255691700@qq.com

DOI: 10.1201/9781003450818-79

accounting for 20.94% of the total length of the tunnel. The problem of large deformation of soft rock is more prominent.

Based on this, this paper studies the estimation of the soft rock deformation magnitude of the project based on the empirical formula of surrounding rock deformation, and the required support force is determined according to the requirements of controlling deformation. After determining the support type and parameters in combination with the code and engineering experience, numerical simulation tools are used to consider the actual excavation process to verify and analyze the rationality of the support. The research results can provide a certain reference for the design of similar projects at home and abroad.

2 SOFT ROCK DEFORMATION AND REQUIRED SUPPORT FORCE SELECTION

Hoek *et al.* used the Monte Carlo method (Hoke 2000) to simulate and analyze the relationship between the deformation of the surrounding rock and the depth of the yield zone in the plastic zone under a large number of different tunnel diameters, and obtained the formula for the deformation of the surrounding rock and the depth of the plastic zone, as shown in Formula (1):

$$\varepsilon_u = \frac{u_i}{R_0} \times 100 = \left(0.2 - 0.25\frac{p_i}{p_0}\right)\frac{\sigma_{cm}}{p_0}^{\left(2.4\frac{p_i}{p_0}-2\right)} \tag{1}$$

where ε_u = radial displacement strain of surrounding rock; u_i = radial displacement of surrounding rock (m); R_0 = surrounding rock excavation diameter (m); ε_p = strain in plastic zone of surrounding rock; R_p = surrounding rock plastic zone diameter (m).

The buried depth of the selected typical analysis section tunnel is about 775m, the rock mass of V type, the initial ground stress P_0 = 15.19MPa, the elastic modulus of the surrounding rock is 2.0GPa, the Poisson's ratio is 0.30, the internal friction angle is 28.81°, the cohesion force is 0.50MPa, and the excavation hole diameter is 10m. According to Formula (1), it can be seen that without support, the radial displacement deformation rate of the surrounding rock is 16.5%, which belongs to a typical large deformation tunnel section, and a reasonable support design is required to control the stability of the surrounding rock and ensure the safety of the support. According to Formula (1), the relationship between support force and surrounding rock deformation is drawn, as shown in Figure 1. According to the requirement of 1.0%~3.0% for the allowable deformation rate of V-type surrounding rock in tunnels with a depth greater than 300 m in the "Technical specification of shotcrete and rock bolt for water resources and hydropower project" (SL377-2007), the support force provided by the support system needs to be 3.7~6.1MPa.

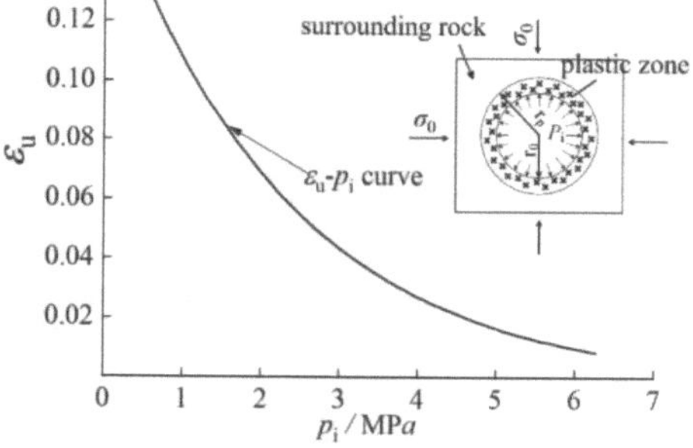

Figure 1. Relationship curve between support force and surrounding rock deformation.

Hoek *et al.* gave the support force that different support measures can provide under different spacing and sizes (Hoke 1998). Based on this, combined with the existing engineering experience, the selected support is mainly composed of an advanced pipe shed, anchor rod, steel arch frame, and C25 concrete sprayed with polypropylene coarse fiber for this project. 120° symmetrical arrangement of leading pipe shed vault, the model is L12mφ108@40 × 900 cm; Anchor rods are arranged in a ring with a length of 8m, the model is φ25@1.0 × 1.0m; Circular layout of steel arch frame, 40U-shaped steel primary support, spacing 50cm, I22 type steel secondary support, spacing 50cm; C40 shotcrete thickness 25cm. The supporting force they can provide is 5.2MPa. At this time, the corresponding surrounding rock deformation is 0.12 m, and the deformation rate is 2.4%, which meets the requirements of the "Technical specification of shotcrete and rock bolt for water resources and hydropower project" (SL377-2007).

3 RATIONALITY VERIFICATION

To verify the rationality of the determined support parameters, a refined 3D model of the initial support including the excavation section, shotcrete, advanced pipe shed, anchor rod, etc. was established. The model size is 70 m × 70 m × 60 m (Figure 2), the step-by-step excavation method is adopted, and the excavation step length is 3.0 m. Normal constraints are applied around and at the bottom of the calculation model, and stress boundaries are applied to the upper boundary of the model. Assuming that the surrounding rock is a homogeneous and isotropic continuum, the influence of the creep characteristics of the rock mass is not considered, and the Mohr-Coulomb yield criterion is followed.

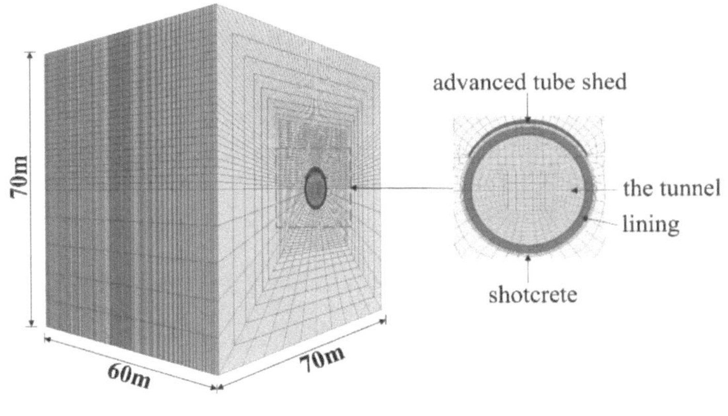

Figure 2. Three-dimensional finite element numerical model.

The surrounding rock parameters and support parameters in the numerical simulation are shown in Table 1.

Table 1. Surrounding rock parameters and initial partial support parameters.

Support type	E (GPa)	μ	ρ (kg·m^{-3})	c (MPa)	ϕ (°)
Advance pipe shed	92	0.3	3386	–	–
Pipe shed reinforcement area	2.9	0.3	2200	0.59	32.62
C25 concrete	28	0.3	2500	–	–

(*continued*)

Table 1. Continued

Support type	E (GPa)	μ	ρ (kg·m^{-3})	c (MPa)	φ (°)
C30 concrete	30	0.3	2700	–	–
40U-shaped steel	60	0.3	7850	–	–
I22a-shaped steel	63.1	0.3	7850	–	–
Anchor rods	130	0.3	7850	–	–

After the tunnel is excavated and supported, the plastic zone and deformation of the surrounding rock are shown in Figures 3 and 4.

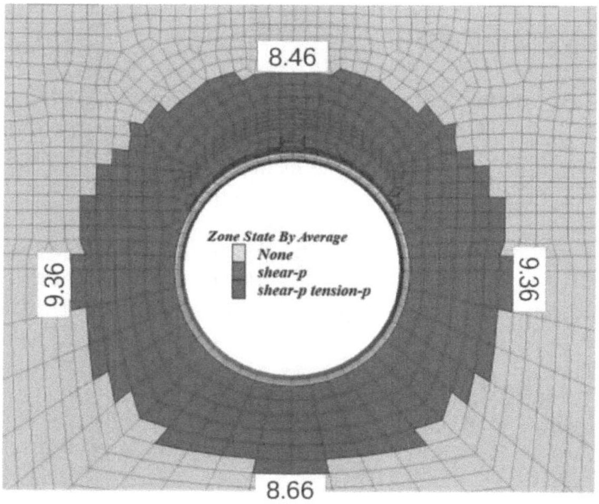

Figure 3. Distribution of plastic zone (cm).

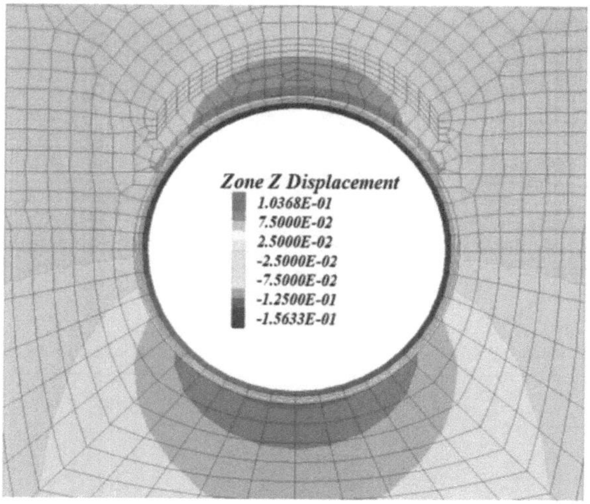

Figure 4. Surrounding rock deformation cloud map (cm).

It can be seen from the figure that under the designed support parameters, the maximum thickness of the surrounding rock plastic zone is 4.36 m, while the maximum displacement of the surrounding rock is 10.36 cm, and the deformation rate is 2.07%. The design length of the anchor bolt of this project is 8.0 m, and the length of the anchor bolt and the controlled deformation all meet the relevant requirements of the "Technical specification of shotcrete and rock bolt for water resources and hydropower project" (SL377-2007). The stress conditions of anchor rods, steel arches, and shotcrete are shown in Figures 5 to 7.

Figure 5. The compressive stress of shotcrete (Pa).

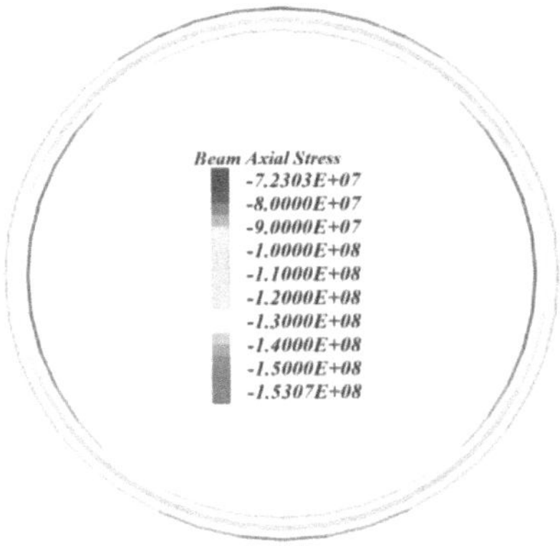

Figure 6. Steel arch axial force (Pa).

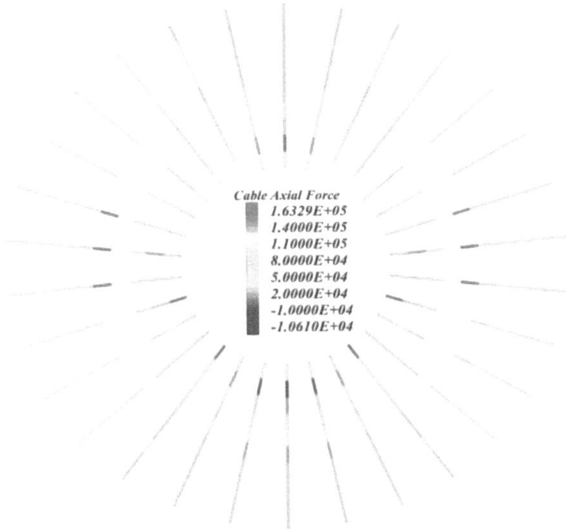

Figure 7. Anchor force (N).

It can be seen from the figure that the force of the anchor is 10.61 kN~163.29 kN, the compressive stress of the steel arch is 72.30 MPa~153.07 MPa, and the compressive stress of the shotcrete is 5.40 MPa~10.94 MPa. The ultimate bearing capacity of the anchor is used in this project. The tensile force is 180 kN, the ultimate strength of the steel arch is 235 MPa, and the C25 shotcrete is 11.9 MPa. None of the above stress values exceed their ultimate loads, and they are all in a safe state.

4 CONCLUSIONS

In this paper, the relationship between the physical and mechanical parameters of the surrounding rock and the in-situ stress is used to determine the magnitude of deformation of the surrounding rock and the support force required under the control target. Types and parameters, and numerical simulation analysis were used to simulate and verify the rationality of support parameters, and the following conclusions were drawn:

(1) In the unsupported state, the deformation rate of the surrounding rock after excavation of the typical section of the project is 16.5%, which is a typical large deformation tunnel section, and a reasonable support design is required to control the stability of the surrounding rock and ensure the safety of the support.
(2) Based on the allowable deformation rate of 1.0%~3.0% for tunnels with a buried depth greater than 300 m, the support force provided by the support system needs to be between 3.7 and 6.3 MPa, and the selected support type is mainly determined by the advanced pipe shed, anchor rod, steel arch frame, and sprayed polypropylene coarse fiber C25 concrete.
(3) Under the design parameters, the maximum deformation of the tunnel is 10.36 cm, and the thickness of the plastic zone is 4.36m. The length of the bolt and the controlled deformation all meet the relevant requirements of the "Technical specification of shotcrete and rock bolt for water resources and hydropower project" (SL377-2007). Anchor rods, steel arches, and shotcrete are stressed beyond the ultimate strength.

ACKNOWLEDGMENTS

This work was financially supported by grants from the Major Science and Technology Projects in Yunnan Province (Grant No.202102AF080001).

REFERENCES

Barla G (1995). Squeezing Rocks in Tunnels, *J. ISRM News Journal*, 4: 44–49.

Barla G, Bonini M, Semeraro M (2011). Analysis of the Behavior of a Yield-control Support System in Squeezing Rock, *J. Tunnelling and Underground Space Technology*, 26(1): 146–154.

Guo Zhibiao, Li Qian, Wang Jiong (2009). Coupled Bolt-mesh-anchor-truss Supporting Technology and its Engineering Application to Deep Soft Rock Roadway. *J Chinese Journal of Rock Mechanics and Engineering*, 28(A02): 3914–3919.

He Manchao (2014). Progress and Challenges of Soft Rock Engineering in Depth, *J. Journal of China Coal Society*, 39(08): 1409–1417.

Hoek E (1998). *Tunnel Support in Weak Rock*, C//Keynote address, Symposium of Sedimentary Rock Engineering, Taipei, Taiwan. 12.

Hoek, E P.M. (2000). Predicting Tunnel Squeezing Problems in Weak Heterogeneous Rock Masses. *Tunnels and Tunnelling International.*

Karampinos E, Hadjigeorgiou J, Turcotte P (2016). Discrete Element Modeling of the Influence of Reinforcement in Structurally Controlled Squeezing Mechanisms in a Hard Rock Mine, *J. Rock Mechanics and Rock Engineering*, 49(12): 4869–4892.

Li Shucai, Xu Fei, Li Liping, *et al* (2016). State of the Art: Challenge and Methods on Large Deformation in Tunnel Engineering and Introduction of a New Type Supporting System, *J. Chinese Journal of Rock Mechanics and Engineering*, 35(07): 1366–1376.

Singh M, Singh B, Choudhari J (2007). Critical Strain and Squeezing of Rock Mass in Tunnels, *J. Tunnelling & Underground Space Technology Incorporating Trenchless Technology Research*, 22(3):343–350.
Steiner W (1996). Tunneling in Squeezing Rocks: Case Histories, *J. Rock Mechanics and Rock Engineering*, 29 (4): 211–246.
Tran Manh H, Sulem J, Subrin D, *et al* (2015). Anisotropic Time-dependent Modeling of Tunnel Excavation in Squeezing Ground, *J. Rock Mechanics and Rock Engineering*, 48(6): 2301–2317.
Wang Jianyu (2020). The Key Way is to Release the Genuine Rock Pressure—discussion on Problems of Tunneling in Squeezing Ground. *J Modern Tunnelling Technology*, 57(4):11.
Zhang Guangze, Deng Jianhui, Wang Dong (2021). Mechanism and Classification of Tectonic-induced Large Deformation of Soft Rock Tunnels, *J. Advanced Engineering Sciences*, 53(01): 1–12.

Water Conservancy and Civil Construction – Oke & Ahmad (Eds)
© 2024 The Author(s), ISBN: 978-1-032-58614-4

Research on bending resistance performance of a modular steel construction innovative connection with installed bolts in the columns

Junwu Xia & Hang Xu*
Jiangsu Key Laboratory Environment Impact and Structural Safety in Engineering, Chin University of Mining and Technology, Xuzhou, Jiangsu, China

Yongrui Wang
School of Naval Architecture, Ocean and Civil Engineering, Shanghai Jiao Tong University, Shanghai, China

Nianxu Yang
Jiangsu Key Laboratory Environment Impact and Structural Safety in Engineering, Chin University of Mining and Technology, Xuzhou, Jiangsu, China

ABSTRACT: To solve the problem that most of the existing connections between modules lack construction space and it is difficult to connect eight modules, an innovative modular steel construction connection with Installed bolts in the columns is proposed. Through a full-scale column-column bending test, force characteristics, failure process, strain distribution, and the moment-rotation relationship under pure bending moment are investigated. The results show that its failure characteristics are the tensile failure of the connection weld between the bottom-end plate and the column; the upper and lower columns are independently stressed, and the neutral axis is located in the middle of the web.

1 INTRODUCTION

A modular building is a new type of prefabricated building. Its prefabrication ratio can reach 90%. Most of the work is completed in the factory. Only the foundation pouring and the connection between modules need to be completed on the site, which can realize rapid construction, intelligent construction, and green construction. Modular buildings are connected as a whole at the construction site through inter-module connections, and the connection between modular units ensures the safety, stability, and applicability of the entire building. Therefore, many scholars have conducted a lot of research and put forward a variety of connection forms.

The first type is bolted connection. R.M. Lawson (Lawson 2008, 2010) proposed a bolt connection mode with a connection plate; R.W. Ma (2021a, 2021b) developed a new type of square steel tube column-channel steel beam in-build components connection, which realizes the connection between modules and the natural fit of ceiling beam and floor beam through the in-build components and stay bolts. The second type is the cross-plate bolted joint. Keum-Sung Park (2016) proposed a cross-plate bolted connection, in which cross-plates are installed on the contact surfaces of adjacent modular columns, and then fixed on the webs of horizontal beams by high-strength bolts, to realize the beam-to-beam connection between the module units. The third type is a welded connection. Annan (2008, 2009) proposed an all-welded connection suitable for the box-type module, where the upper and lower modules are welded

*Corresponding Author: xh874351832@163.com

DOI: 10.1201/9781003450818-80

together through the cover plate at the end of the module column. Some other connection methods have also been proposed (Chen Z. 2017; Dai Z. 2020, 2021), including new grouting sleeve connections and prestressed connections between angular support steel frame modules.

At present, there are many forms of modular steel structure building connections at home and abroad. However, most of them lack sufficient construction space, which makes it difficult to connect eight modules and forms "eight columns and sixteen beams". To overcome the above difficulties, this paper proposes a modular steel construction innovative connection with installed bolts in the columns. It has good construction space and will not damage indoor and outdoor decoration. A column-column bending test is designed in this paper. The four columns connected by the connection are loaded. Through the four-point bending test method, the bending performance and failure mode of the connection under horizontal load is studied.

2 EXPERIMENTAL STUDY

2.1 *Design and assembly method of the innovation connection*

The modular steel construction innovative connection with installed bolts in the columns proposed in this paper is shown in Figure 1. The connection is mainly composed of five parts: locating plate, locating tube, bolt, upper-end plate, and bottom-end plate. The connection connects the upper and lower columns with bolts, which is a new type of column-column connection between modules. The locating plate, the locating tube, and the bottom-end plate are connected to the lower end of the square steel tube column by welding to fix the approximate position of the bolt and ensure that the bolt will not tilt or even reverse. The threaded upper-end plate is divided into three parts and welded with the upper end of the column in turn.

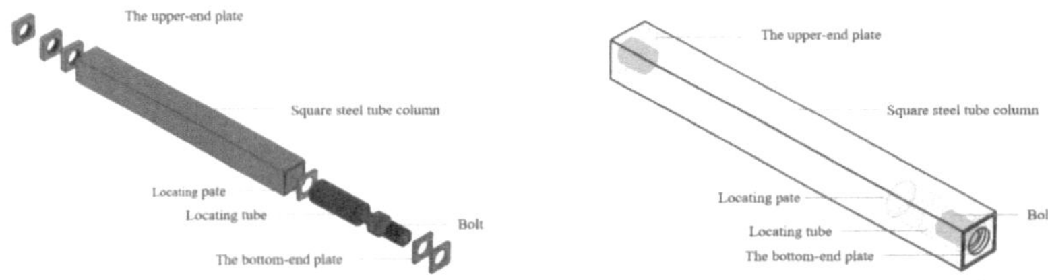

(a) Schematic diagram of node decomposition (b)Overall schematic diagram of nodes

Figure 1. Schematic diagram of the new connection structure.

After the lower module is hoisted on the construction site, the upper module is hoisted. The special twisting tool is used to extend and tighten the bolts from the upper part of the square steel tube column of the upper module, to realize the connection between the vertical modules, and then the modules in the horizontal direction are also connected through the base plate to form "eight columns and sixteen beams". The twisting tool and the base plate are shown in Figures 2 and 3 respectively. Figure 4 is a schematic diagram of the connection in 8 modules.

2.2 *Designing and preparation of the experimental specimen*

Four columns of modular structure connected by the innovative connection were taken to conduct a column-column bending test. The bolts were made of NO.45 steel, and other elements were made of Q235B. The upper and lower module columns were made by cold-bending 150 mm × 150mm × 8mm square steel, and the columns were 1500 mm long. The thickness of the base plate is 30mm, the thickness of the upper-end plate of the column is 30mm,

Figure 2. Base plate.

Figure 3. Twisting tool.

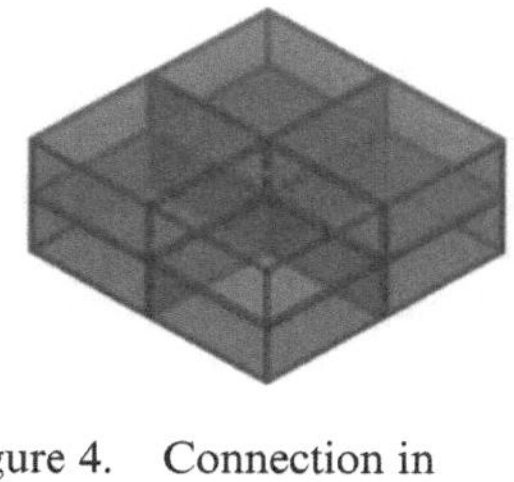

Figure 4. Connection in modular building.

and the thickness of the bottom-end plate is 15mm. The physical and dimensional drawings of each component are shown in Figure 5. Three groups of column steels and bolt steels were taken for the material property test. The results of the material property test are shown in Table 1. Three upper-end plates and two bottom-end plates were respectively welded with the upper and lower ends of the module column. They were polished with a polishing machine after the welding was completed after the welding is completed. To ensure that the bolts could be tightened smoothly, there was no gap between the end plates. E43 welding rods were adopted for all of the welding.

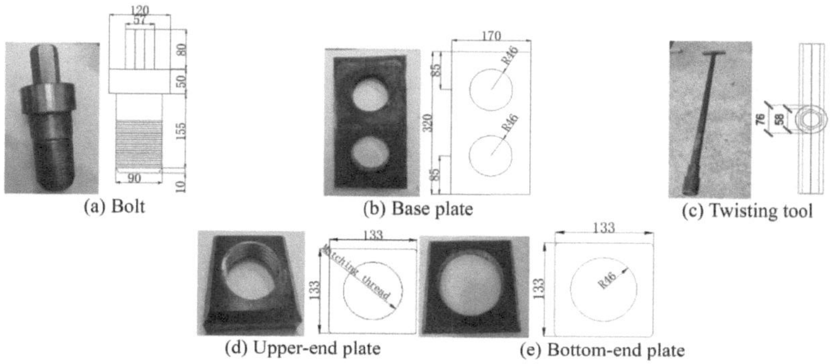

Figure 5. Physical and dimensional drawings of components.

Table 1. Results of the material test.

Type	Number	f_y(MPa)	f_u(MPa)	(f_u/f_y)	$E(\times 10^5$MPa)
Column	3	309.87	415.73	1.34	1.77
Bolt	3	352.13	643.33	1.82	1.95

2.3 *Test scheme*

The column-column bending test device in this paper is shown in Figure 6. To avoid out-of-plane instability of the specimen under concentrated force, lateral constraints were added. To avoid local buckling of the specimen during loading, the loading plate is welded at the loading position of the specimen and the support. Three-point loading mode is adopted in the test. During loading, the jack applies a load to 1/3 of the specimen through a distribution beam. Before the formal loading, 30kN preloading was conducted to check whether the loading device and all instruments were normal, and ensure that all parts of the specimen were fully contracted, which were unloaded after confirming that the loading system and the test system were normal. The formal loading adopted the loading method of force control and the load increment of each level was 10kN. After the abnormal sound was heard, the load increment of each level was changed to 5kN. When obvious deformation or damage of the specimen is observed, the test was terminated.

Figure 6. Test setup.

2.4 *Measurement scheme*

Three electronic displacement meters were arranged under the midspan of the specimen and the loading point. The specific locations were shown in Figure 7a. The strain of the control point was collected by a DH3816N static strain tester. A certain number of strain gauges were arranged at the corresponding positions of the web and flange of the column to obtain strain development in the pure bending section. The specific locations are shown in Figure 7b, where ABCD area and X1-X4 are the frontal strain gauge arrangements of the connection area; Zone Z is the arrangement of strain gauges on the upper part of the column and base plate; Zone Y, X5, and X6 are the arrangement of strain gauges on the lower part of the column and the base plate.

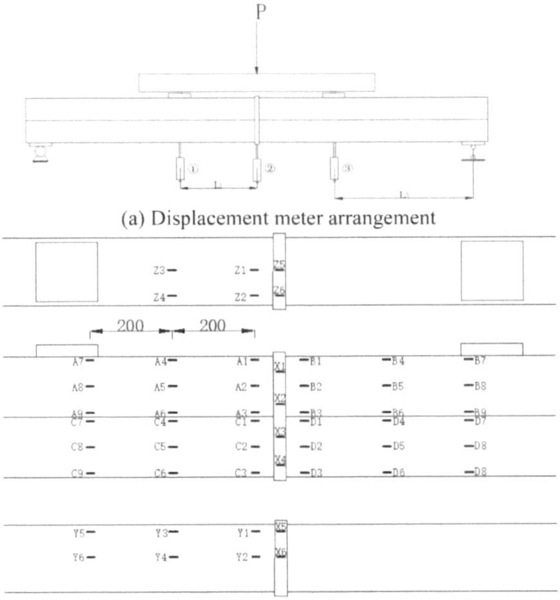

(a) Displacement meter arrangement

(b) Strain gauge arrangement

Figure 7. Strain gauge and displacement meter arrangement.

3 TEST PHENOMENA AND FAILURE CHARACTERISTICS

When the vertical load was less than 50kN, all parts of the specimen were not abnormal and the moment-rotation curve of the specimen also showed a good linear relationship, indicating that the specimen was in the elastic stage at this time. When the load was about 60kN, it could be observed that the initial relative sliding of the upper and lower columns was accompanied by sound, and the load was changed to 5 kN per stage. When the load was up to 150kN, there was a gap between the end plate and the base plate, and the gap on the side of the bottom-end plate was larger than that on the side of the upper-end plate, as shown in Figure 9a. When the load reached 250 kN, the specimen had a large bending deformation, and as shown in Figure 8, the loading device and the specimen had a more severe squeezing and sliding sound. When the load was up to 295 kN, the tensile weld between the lower side of the bottom-end plate and the lower column broke with the sound of a "bang", and the load dropped rapidly. Because the end plates were embedded in the column after the weld was cracked, the failure phenomenon is difficult to detect. Therefore, the specimen continues to be loaded. After the load is continued for 20 mm, the end plate and the column were separated to a certain extent, as shown in Figure 9b.

Figure 8. Deformation under bending.

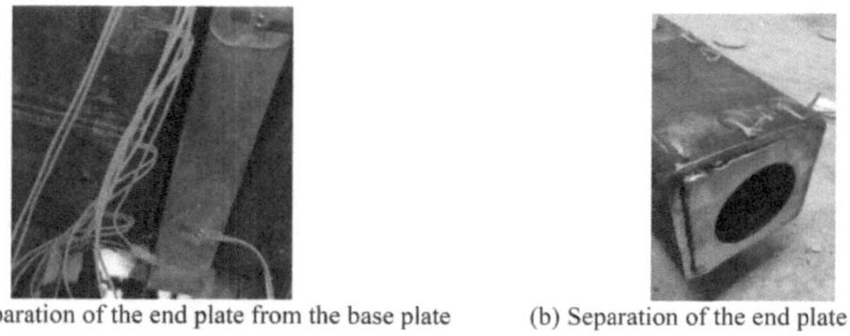

(a) Separation of the end plate from the base plate (b) Separation of the end plate from the column

Figure 9. Experimental phenomenon.

4 TEST RESULTS AND ANALYSIS

4.1 *Moment-rotation curves*

The moment-rotation curve is an important index to evaluate the flexural performance of connections. In this paper, the bending moment and rotation are defined as:

$$M = \frac{1}{2} P L_2 \tag{1}$$

$$\theta = \arctan \frac{2\Delta_2 - \Delta_1 - \Delta_3}{L_1} \tag{2}$$

Where M, P, and θ are the bending moments, concentrated load, and rotation respectively; Δ_1, Δ_2, Δ_3, L_1, and L_2 are the measured displacements of the displacement meters ①, ②, and ③, the distance between the midpoint of the loading plate and the midspan of the specimen, and the distance between the midpoint of the loading plate and the support (Figure 7a).

The moment-rotation curve of the column-column bending test in this paper is shown in Figure 10. It can be seen from the figure that the load in the early stage increases steadily, and the connection can bear the load normally. When the moment reaches 135.7kN·m, the specimen is damaged by welding, and the load begins to drop after the fracture of the weld. With the further increase of the deflection, the end plate continues to move outwards, the load further falls back, and the specimen loses its bearing capacity completely.

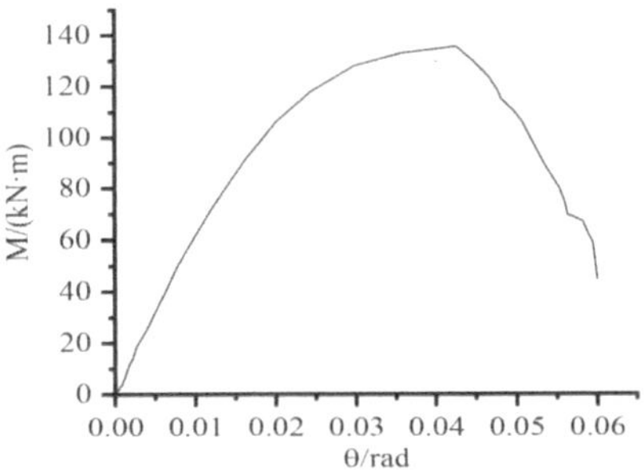

Figure 10. Moment-rotation curve.

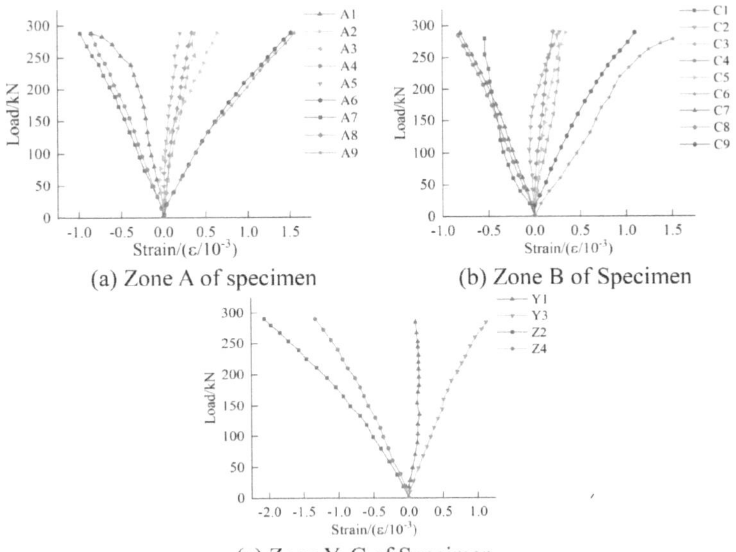

Figure 11. Strain-load curves of each region of the specimen.

4.2 *The strain development in the core area*

Figure 11(a) shows the strain-load curve of the specimen in Zone A. For the upper column, in the vertical direction, the force on the edge of the column is large, the force on the center is small, the upper edge is under pressure, and the lower edge is under tension. In the horizontal direction, the stress at both ends is large, and the central stress is small. The strain level of test point A5 is always low, indicating that the neutral axis of the upper column is near the test point when the specimen is loaded.

Figure 11(b) shows the strain-load curve of the specimen in Zone C, and the strain development law is similar to that in Zone A. The lower side of the lower column is tensioned while the upper side is compressed, and the upper and lower columns show independent bending resistance. The strain level of measuring point C8 is always small, indicating that the neutral axis of the lower column is near the measuring point when the specimen is loaded.

It can be seen from Figure 11(c) that the strain of Z4 is less than Z2, indicating that the maximum compressive stress occurs in the area near the base plate at the upper flange of the column. No strain change was collected in the X zone of the specimen, indicating that no obvious deformation occurred on the outer ring of the backing plate.

5 CONCLUSIONS

In this paper, a full-scale column-column bending test is carried out to load four columns connected by the new type of connection. The test results show that the failure mode of the specimen is the tensile failure of the connecting weld between the lower side of the bottom-end plate and the lower column. Each column is compressed at the upper part and tensioned at the lower part. The neutral axis of the upper and lower columns is located in the middle of the web, showing an independent bending phenomenon.

REFERENCES

Annan C.D., Youssef M.A., El Naggar M.H. (2008). Seismic Overstrength in Braced Frames of Modular Steel Buildings. *Journal of Earthquake Engineering*. 13(1), 1–21.

Annan C.D., Youssef M.A., El Naggar M.H. (2009). Experimental Evaluation of the Seismic Performance of Modular Steel-braced Frames. *Engineering Structures*. 31(7), 1435–1446.

Chen Z., Li H., Chen A., Yu Y., Wang H. Research on Pre-tensioned Modular Frame Tests and Simulations. *Engineering Structures*, 151, 774–787.

Dai Z., Pang S.D., Liew J.R. (2020). Axial Load Resistance of Grouted Sleeve Connection for Modular Construction. *Thin-walled Structures*, 154, 106883.

Dai Z., Cheong T.Y.C., Pang S.D. Experimental Study of Grouted Sleeve Connections under Bending for Steel Modular Buildings. *Engineering Structures*, 243, 112614.

Lawson R.M., Ogden R.G. (2008). 'Hybrid' Light Steel Panel and Modular Systems. *Thin-Walled Structures*. 46(7–9), 720–730.

Lawson R.M, Richards J. Modular Design for High-rise Buildings. (2010). *Proceedings of the Institution of Civil Engineers – Structures and Buildings*. 163(3), 151–164.

Ma R., Xia J., Chang H. (2021a). Experimental and Numerical Investigation of Mechanical Properties on Novel Modular Connections with Superimposed Beams. *Engineering Structures*, 2021, 232: 111858.

Ma R., Xia J., Chang H. (2021b). A Component-Based Model for Novel Modular Connections with Inbuild Component. *Applied Sciences*. 11(21), 10503.

Park K., Moon J., Lee S. (2016). Embedded Steel Column-to-foundation Connection for a Modular Structural System. *Engineering Structures*. 110, 244–257.

Water Conservancy and Civil Construction – Oke & Ahmad (Eds)
© 2024 The Author(s), ISBN: 978-1-032-58614-4

Research on frequency-magnitude relationships for Ryukyu subduction zone seismicity and the geological implications

Wangqi Li*
Shenzhen Vanke Meisha Academy, Shenzhen, China

ABSTRACT: The occurrence of natural earthquakes has a power-law frequency-magnitude relationship. The b-value in the frequency-magnitude relationship is indicative of the occurrence frequency of large earthquakes and their distribution in time and space. Therefore, the b-value is critical for understanding the nature of faults, seism genic patterns, and earthquake hazard prediction and warning. The Ryukyu-Philippine subduction zone is one of the major origins of megathrust earthquakes. The hazard potential in and around the Ryukyu subduction zone needs to be studied by analyzing the b-value. Besides, the study of b-values in different regions of the subduction zone is indicative of subduction rates, fault properties, etc. Here, we conduct sub-regional statistics using the well-documented Ryukyu-Philippine subduction zone earthquakes in the time range of 2008–2018 to estimate the b-values in different regions, depths, and periods. Our preliminary result demonstrates that the b-values have a clear along-strike variation feature and are correlative to the subduction rate and depth. Compared with previous studies, we used the latest seismic observations in the Ryukyu Islands to calculate b-values for different regions and discussed the geological significance.

1 INTRODUCTION

Natural earthquake occurrence frequency and magnitude have a power-law relationship, which is first reported by Gutenberg and Richter (1944):

$$\log N = a - bM, \tag{1}$$

where N is the occurrence frequency for earthquakes with a magnitude larger than M. a and b are parameters describing the intercept and slope of the frequency-magnitude relationship. The b-value is a key parameter because it indicates the potential of large earthquakes in the fault zone, which is related to the fault properties including stress state, fault rigidity, and the size of asperities. Besides, b-value analysis is important to the earthquake hazard potential investigation and prevention. The Ryukyu-Philippine subduction zone has occurred megathrust earthquakes. Both earthquakes and tsunamis triggered by megathrust earthquakes are threatening major cities around the Philippine Sea (Fan & Zhao 2021). Therefore, in this paper, we focus on the spatial distribution of the b-value in the Ryukyu-Philippine subduction zone and surrounding regions. Based on the distribution between the magnitude and occurrence frequency of natural earthquakes, the b value can be optimally estimated using the maximum likelihood method (Wu *et al.* 2019). The maximum likelihood method is also successfully applied in the estimation of b-values of multiple

*Corresponding Author: zzz030709@126.com

DOI: 10.1201/9781003450818-81

geologic regions including the eastern part of the Okhotsk microplate (Wang *et al.* 2020), the Western East Sea, Korea (Hong *et al.* 2020), and the southeastern part of Sichuan, China (Xie & Meng 2021). Here, we examine the b-value using a well-documented earthquake catalog of the Ryukyu subduction zone, published by the International Seismology Center (ISC) from 2008 to 2018. After estimating the b-value and the completeness magnitude, we obtain a well-constrained spatial distribution of b-values and observe a clear along-strike and along-depth variation of the b-values. For the study of subduction rate and fault properties, it is particularly pertinent to examine the b values in different regions of the subduction zone. It is also possible to determine the stress state of the fault zone from the change in the b value over time. We present inferences based on the combination of previous literature and the analysis of the calculated data, including the stress accumulation rate, earthquake depth, and earthquake pre- and post-earthquake effects.

2 METHODS AND DATA

We study the Ryukyu subduction zone and surrounding regions, i.e., in the latitude range of 20–50W and longitude range of 118–135E, including Fujian and Taiwan Provinces, China; Kyushu, and Shikoku, Japan (Figure 1). A high-quality earthquake catalog is critical for studying the b-value. We use the ISC catalog from the international earthquake center (International Seismological Centre (2008–2018)). The earthquake catalog in the study region is within the time frame from 2008 to 2018, with a magnitude between M1.4 and M6.6, a depth between 0 and 600 km, and a total of 940559 earthquakes. Figure 1 also shows the 2D distribution of the earthquakes. We observe that most of the earthquakes have occurred along the Ryukyu subduction zone.

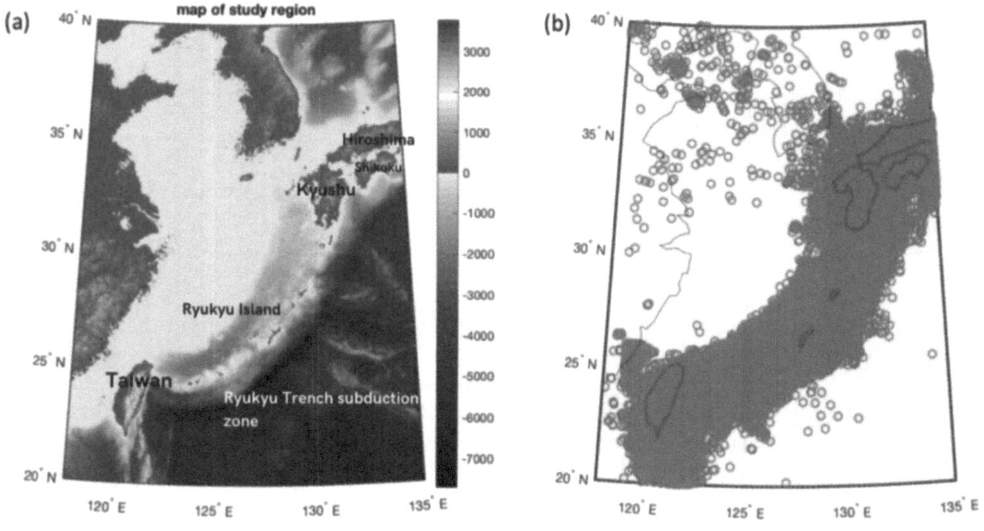

Figure 1. Ryukyu subduction zone and surrounding regions. (a) Ryukyu subduction zone study region and geological units. (b) Earthquake distribution. Red circles are the earthquake epicenters from the ISC catalog in the time range of 2008–2018.

There are different methods for estimating the b-value, including the least-squares method and the maximum likelihood method. In this study, the maximum likelihood method (Wiemer & Wyss 2002) was used to determine the values of the model parameters. Using this method, the parameter values are determined in such a way as to maximize the probability that the process described by the model will produce the actual observed data.

In detail, first, we divide the magnitudes into several bins. The bin size is selected to be 0.1. The b value is estimated by the following equation:

$$b = \log_{10}(\exp)/(\text{Mmean} - \text{Mmin}) \tag{2}$$

where Mmean is the average magnitude and Mmin is the minimum magnitude in the predefined magnitude bin.

After calculating the b-value, the standard deviation of the b-value is estimated by:

$$\delta b = 2.3b^2\sqrt{\frac{\sum_i (M_i - M)^2}{n(n-1)}} \tag{3}$$

The earthquake catalog might not be complete, especially, since many small-magnitude earthquakes are possible to be undocumented due to current seismic station distributions, sensor sensitivity, etc. The frequency-magnitude relationship will be valid above a certain completeness magnitude Mc for a certain region. To estimate the completeness magnitude of Mc, we apply a grid-search method (Wiemer & Wyss 2002). The first step in this research was to estimate the b- and a-values of the frequency-magnitude distribution (FMD) as a function of minimum magnitude using the maximum likelihood estimation method. To test the fitness of the real value and the calculation result, this study uses a misfit function (Equation 4) to measure the misfit between the estimated FMD and the real data. As a result, if Mi is smaller or larger than the correct Mc, the misfit will not reach the minimum, then the synthetic distribution cannot be used to model FMD appropriately. We choose the Mi value that minimizes the misfit function, which means that the synthetic FMD can best model the real FMD.

$$R(a, b, M_i) = 100 - \left(\frac{\sum_{M_i}^{M_{max}} |B_i - S_i|}{\sum_i B_i}\right) \tag{4}$$

To get the spatial distribution of the b-values, we estimate b-values in local spatial windows of $1^\circ \times 1^\circ$, centered at each grid point. The spacing between grid points is 0.1°. To keep the b-value estimation stable, for each local window, the b-value and Mc are estimated using Equations 2 and 4 if the total number of earthquakes is greater than 100. If the total number of earthquakes is less than 100, we think the earthquake catalog is incomplete to perform a stable b-value estimation. The method is similar to Wang *et al.* (2020).

In comparison with the maximum likelihood method, the ordinary least squares method is generally considered to have fewer desirable optimization properties than the maximum likelihood method. Further, due to the non-normal distribution of earthquake data, it would be more accurate for researchers to use the maximum likelihood method to reduce error (Wu *et al.* 2019). The maximum likelihood method can ensure that the distribution parameters are estimated more accurately, and the estimated variance is less, to calculate the confidence intervals and tests of the model parameters more precisely (Genschel & Meeker 2010).

There are several advantages to the method used in this study, which is the maximum likelihood method. This method has the advantage of providing an overall consistent approach to parameter estimation. In other words, maximum likelihood estimation can be applied when a large number of parameters are available. Moreover, the study aims to test earthquake data so that it can be better incorporated into the model.

3 RESULTS

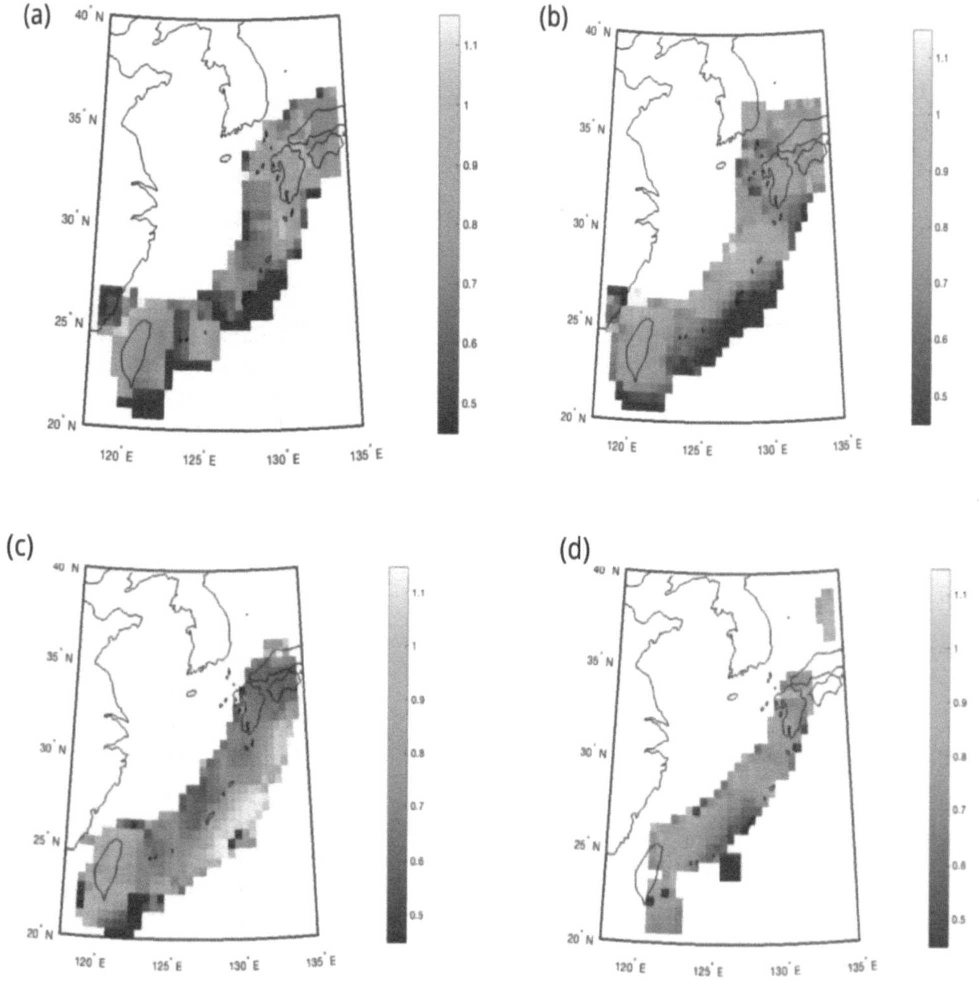

Figure 2. b-Value distributions with depth. (a) 0–10 km; (b) 10–30 km; (c) 30–100 km; (d) 100–600 km; the Color map shows the b-value

We first investigate the spatial distribution of b-values in our study region and its relation to the depth of the earthquakes. Figure 2 shows the distribution of b-values at 0–10 km, 10–30 km, 30–100 km, and 100–600 km depths.

First, at 0–10 km depth, the b-value in Fujian Province, China, located at 25–27°N and 118–120°E, is low with a value of about 0.5 and the image is dark purple; while in the whole of Taiwan Province, the b-value is around 0.8 throughout the region. The b-values in the Kyushu and Shikoku regions of Japan between 32–36 degrees north latitude and 128–135 degrees east longitude are mostly around 0.8, while the b-values in the Kitakyushu region are slightly higher, at around 1.05 (Nishimura *et al.* 2004).

Looking next at the depth of 10–30 km (Figure 2(b)), the b-value data for the Fujian and Taiwan regions of China are the same as those at 0–10 km. While in the northern part of the Ryukyu Islands region of Japan, i.e., between 26–29°N and 122–129°E b-values increase from 0.7 to between 1.05 and 1.1. In the Kyushu, Shikoku, and Tsushima Strait regions of Japan, there is no significant difference except that the b-value in the Kitakyushu region decreases from 1.05 to about 0.65. From Figures 2(a) and (b), the areas with large differences in b-values for depths of 0–10 km and 10–30 km are the northern Ryukyu Islands and the Kitakyushu region of Japan.

The area with large differences in b-values between 30–100 km and 10–30 km (Figure 2(b) and (C)) is the southeastern Okinawa region to the southern Shikoku region of Japan between 23–32 degrees north latitude and 127–134 degrees east longitude. The b-value situation in this region increases from 0.5 in the 10–30 km images to between 1.05–1.1 in the 30–100 km images and decreases from 1.1 to about 0.5–0.7 in the 30–100 km images for the Northern Ryukyu Islands.

When the images are at a depth of 100–1000 km, the areas where data can be acquired are mainly between 21–34 degrees north latitude and 121–132 degrees east longitude. There is no significant fluctuation in the overall data, mainly in the northern part of the Okinawa region of Japan where the b-value increases from 0.5 to about 0.8 overall, while the b-value in the Okinawa region decreases from 0.8 to about 0.5.

For the out rise of the Ryukyu subduction zone (to the right of the Ryukyu Island arc), the b-value is generally small (~0.5) at the depth range of 0–30 km. While below 10 km stronger along-strike variations are observed and between 10–30 km the out rise generally has a low-b-value feature. In 30–100 km, the b-value for the out rise of the Ryukyu subduction zone is generally getting large (0.8–1.1). For the Kyushu and Shikoku regions of Japan, the b-value is larger (0.8–1.0) at a shallower depth (0–10 km), and gradually gets smaller to 0.7–0.9 at 10–30 km and ~0.6 at 30–100 km.

4 DISCUSSION

The variation of b values is rich in geological significance. According to (Figure 2(a)) and (Figure 2(b)), we can see that b values are larger (between 1–1.1) in the area between 24–33°N and 127–134°E and the northern part of Kyushu, while they are smaller (between 0.5–0.8) in the whole of Taiwan. This implies that small earthquakes are more frequent and large earthquakes are less likely to occur within the same time scale. Possible reasons for this phenomenon include low-stress accumulation rate, low rock strength, and low friction coefficient. According to the GPS-derived velocity field along the Ryukyu arc (Nishimura *et al.*), we can see that the stress accumulation rate is greatest in Taiwan with small b values (between 50 mm/year and 70 mm/year), while large b values in the northern latitudes 24 to 33°N, 127 to 134°E, and Kitakyushu regions with large b values have small stress accumulation rates (between 5 mm/yr and 25 mm/yr). From the literature support and data found above, it is clear that the stress accumulation rate is slower in regions with large b-values, and the difference in stress accumulation rates leads to different b-values at different locations.

B-value also changes with depth, because of the variation of temperature and stress states. Our observation suggests that the Ryukyu subduction zone out rising region has a smaller b-value at a shallower depth while a larger b-value at a deeper depth. However, for the Kyushu region, the b-value variation with depth is the opposite of the Ryukyu arc. We speculate that for the Ryukyu subduction zone out rising, the oceanic lithosphere is thin, thus at shallow depth, the oceanic lithosphere has low temperature and brittle rock properties, which can accumulate more stress before failure and thus tend to support more large earthquakes, resulting in a smaller b-value. At the deeper depth, the subduction zone out of the rising region is entering the mantle which has a higher temperature and tends to be ductile, therefore only small earthquakes could happen. For the Kyushu region, the lithosphere beneath the island is thick (~60 km), which supports brittle failure (and large magnitude earthquakes). The b-value variation is more related to local stress states.

The b-value may also change with time especially before and after a major earthquake occurred (Hong *et al.* 2020). But the reason why the b-value changes after major earthquakes are still under debate.

5 CONCLUSION

We analyze the spatial distribution of the b-values in the earthquake occurrence frequency and magnitude relation of the Ryukyu subduction zone and the Philippine plate region. The clear along-strike variation of b-values correlates well with the subduction rate, and the along-depth variation of b-values differs in regions. Considering the shear stress state of the subducting plate, we mainly conclude that the different subduction rates of the plates lead to different stress magnitudes and thus different b-values.

REFERENCES

Fan, J., & Zhao, D. (2021). P-wave Tomography and Azimuthal Anisotropy of the Manila-Taiwan-southern Ryukyu region. *Tectonics*, 40(2), e2020TC006262.

Genschel, U., & Meeker, W. Q. (2010). A Comparison of Maximum Likelihood and Median-rank Regression for Weibull Estimation. *Quality Engineering*, 22(4), 236–255.

Hong, T. K., Park, S., Lee, J., & Kim, W. (2020). Spatiotemporal Seismicity Evolution and Seismic Hazard Potentials in the Western East Sea (Sea of Japan). *Pure and Applied Geophysics*, 177(8), 3761–3774

Nishimura, S., Hashimoto, M., & Ando, M. (2004). A Rigid Block Rotation Model for the GPS-derived Velocity Field Along the Ryukyu arc. *Physics of the Earth and Planetary Interiors*, 142(3–4), 185–203.

Wang S.P., Luo. G., Shi. Y. N., et al. (2020). Seismic b-value and Stress Field Characteristics in the Eastern Subduction Zone of Okhotsk Microplate. *Chinese J. Geophysics. (In Chinese)*, 63(4): 1444–1458.

Wiemer, S., & Wyss, M. (2002). Mapping Spatial Variability of the Frequency-magnitude Distribution of Earthquakes. *In Advances in Geophysics* (Vol. 45, pp. 259-V). Elsevier.

Wu G., Zhou Q. & Ran H., L. (2019). *The Maximum Likelihood Estimation of b-value in Magnitude-frequency Relation and Analysis of its Influence Factors*. Institute of Geology, China earthquake Administration, 41(1).

Xie M. Y. & Meng L. Y. (2021). Seismicity and Evolution Characteristics of b-values of Changning Area in South-eastern Region of Sichuan Basin. *Earthquake research in China*, 37(2), 494–507.

Water Conservancy and Civil Construction – Oke & Ahmad (Eds)

Research on construction technology and equipment of prefabricated structures in a subway station

Jianqiu Wu* & Wei Wang*
Engineering Research Institute, China Construction Eighth Engineering Division Corp., Ltd., Shanghai, China

Jing Guo*
Shield Technology Research Institute, China Construction Eighth Engineering Division Rail Transit Construction Corp., Ltd., Nanjing, China

Min Sun* & Lei Han*
Engineering Research Institute, China Construction Eighth Engineering Division Corp., Ltd., Shanghai, China

Xiaoli Sun*
Shield Technology Research Institute, China Construction Eighth Engineering Division Rail Transit Construction Corp., Ltd., Nanjing, China

ABSTRACT: The prefabricated components of subway stations are large in size and heavy in weight, which requires high-level construction technology and equipment, thus limiting the universality of this construction method. Based on the prefabricated station project of Huanghai College Station of Qingdao Metro Line 6, the structural system of the prefabricated station is analyzed from the aspects of structural selection, structural system, etc. Given the characteristics of high precision requirements and control difficulty of prefabricated components, the construction technology and auxiliary construction equipment, including finish leveling of foundation pit bottom, intelligent gantry crane, intelligent assembly trolley, and automatic tensioning control system, are studied. The engineering practice has proved that prefabricated structure construction is not only applicable to the construction of subway stations under various environmental conditions but also applicable to the construction of subway sections, pedestrian passages, road tunnels, underground pipe galleries, and other open cut underground structures, with a wide range of applicability.

1 INTRODUCTION

In recent years, to deal with environmental pollution, labor shortage, and other problems, the reform of prefabricated construction technology in the construction industry has been vigorously promoted. After years of development and policy promotion, the overground prefabricated buildings have made remarkable progress, the relevant technology and management system have been initially established, and the engineering application has increased year by year (Qian 2019). However, in the field of underground engineering in China, the research and application of prefabricated construction technology for large underground structures are still in their infancy, especially in the field of prefabricated construction technology of subway stations (Yang 2019).

*Corresponding Authors: jianqiuw1987@163.com, ww096014@foxmail.com, 214614626@qq.com, 7071186@qq.com, hanleiwell@163.com and ww119@vip.qq.com

DOI: 10.1201/9781003450818-82

The large size and heavy weight of prefabricated components put forward higher requirements for construction technology and construction equipment, thus limiting the universality of this construction method. In terms of design, the allowable range of component assembly accuracy error is within 2mm, which makes it difficult to control the assembly accuracy (Feng 2020; Yang 2020). This determines that the construction equipment must have the characteristics of high accuracy, micro difference adjustment, intelligent control, and adaptability. At present, there are two categories of construction equipment (Li 1995; Zhang 1997). One is that workers assemble the prefabricated components on the assembly frame after hoisting by existing lifting machinery. The combination of lifting machinery and assembly frame can improve the installation accuracy to a certain extent, which applies to the superposition and cast-in-situ structural forms with strong tolerance, but cannot meet the requirements of full prefabricated assembly with high accuracy. For the fully prefabricated assembly equipment, there is an integrated transport and assembly machine, which can meet the high-precision assembly work through the adjustable oil cylinder. However, the assembly accuracy and attitude adjustment are mainly carried out manually, lacking the means of intelligent measurement, positioning, and assembly to further improve construction efficiency and accuracy.

Based on the prefabricated station project of Huanghai College Station of Qingdao Metro Line 6, the construction technology and equipment of a fully prefabricated structure are studied.

2 STRUCTURAL SYSTEM OF THE PREFABRICATED STATION

2.1 *Structural selection of prefabricated station*

The research and application work are carried out based on the Huanghai College Station of Qingdao Metro Line 6. The total length of the Huanghai College Station is 218 m, which is an island station with two floors underground. The length of the small mileage of the cast-in-situ section is 42.1m, which is a double column and three-span cast-in-situ structure with a width of 18.5 m. The length of the large mileage is 27.9 m and that of the single column and double span cast-in-situ structure is 18.5 m wide. The length of the middle standard section is 148 m, designed as a single-column double-span prefabricated structure of 21.3 m wide and 74 rings in total, as shown in Figure 1. The irregular air duct structure at both ends of the station is integrated with the section shield shaft, and cast-in-situ concrete is used for construction. A complete standard structure section is formed in the middle of the station, and the prefabricated structure is adopted to effectively improve the overall assembly rate of the station.

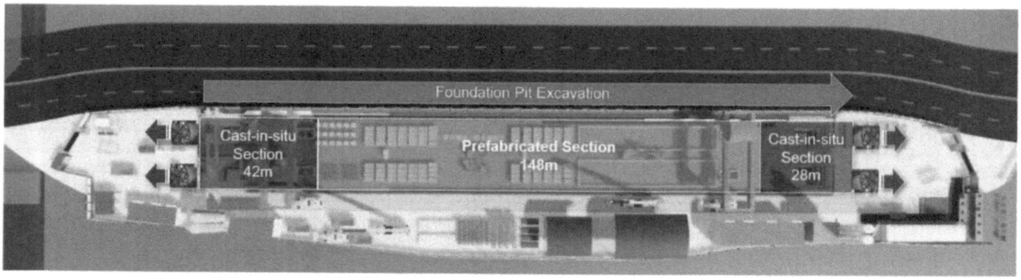

Figure 1. Plan of Huanghai college station.

The station structure adopts a single arch long-span tunnel structure. Compared with the traditional rectangular frame structure, the single arch long-span structure is simple in form and well-stressed (Zhang 2021), which is conducive to saving the amount of reinforcement and has a good architectural effect (Li 2019), which can effectively alleviate the sense of claustrophobia in the underground space (Figure 2). The disadvantage is that the arch camber of the structure leads to a large buried depth of the structure and increases the amount of foundation pit work. Due to the thrust of the arch foot, certain engineering measures should be taken to control the deformation of the arch structure and its impact on the surrounding environment of the station.

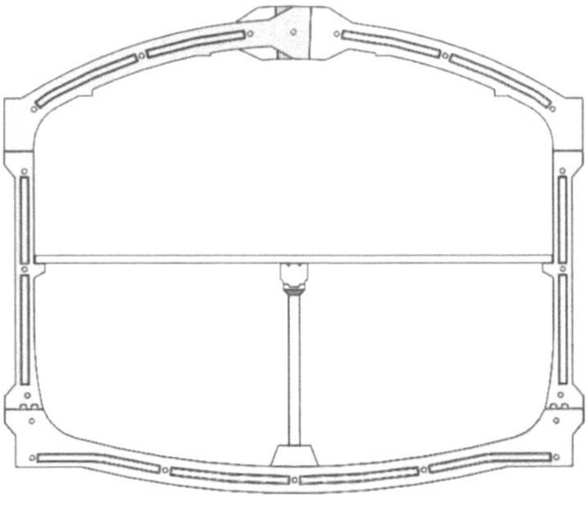

Figure 2. Single arch large-span tunnel structure.

2.2 *The structural system of the prefabricated station*

2.2.1 *Structure splitting method*

The structure volume of the subway station is huge (Yang 2021). In addition to the vertical structure, its structure ring also needs to be split into several standard components horizontally (Lin 2022; Wang 2022). For prefabricated underground structures connected by flexible joints, the structural disassembly method should follow the basic principles of structural stability as the basis, joint mechanical performance as the support, and component production, transportation, and hoisting as the premise.

The sectional dimension of Huanghai College Station is 20.5m (width) × 18.5m (height), 2m longitudinal ring width. The bottom plate structure is not disassembled and is transported by a 21m retractable flat trailer. The vault structure is divided into two components. The tenon and groove joints are set at the vault to form a stable three-hinged arch structure. The side wall, middle plate, and beam column are all prefabricated components. Therefore, the standard ring of Huanghai College Station is composed of six large prefabricated components with a ring width of 2m, as shown in Figure 3.

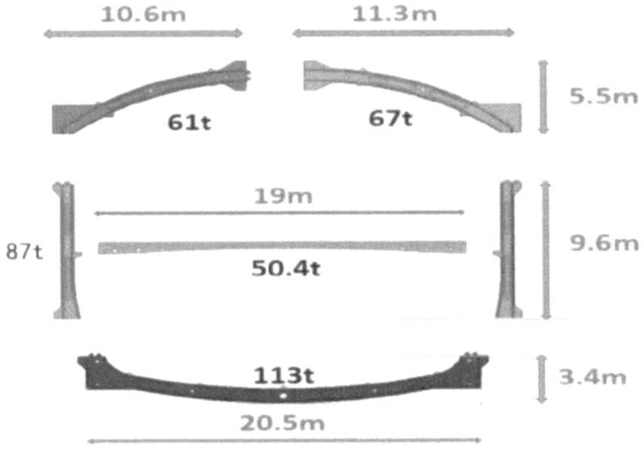

Figure 3. Structural diagram of the prefabricated station.

After the overall assembly of the lining structure is completed, a stable tunnel-bearing structure is formed under the constraint of the hard backfill material of the lateral fertilizer trough of the foundation pit.

2.2.2 *Structural assembly*

The structural self-waterproofing and joint waterproofing modes are adopted for the structural waterproofing of the prefabricated station, without the need to set an external waterproof layer. Two layers of composite expansion rubber gaskets are set at the joint seams, and modified epoxy resin is poured into the joint seams.

In each lining ring, the joints of the bottom plate components are compressed by prestressing, while the joints of the upper components are compressed mainly by their weight, and auxiliary bolts are set for connection; The joints between rings are tensioned and locked ring by a ring with relay prestressed reinforcement, as shown in Figure 4.

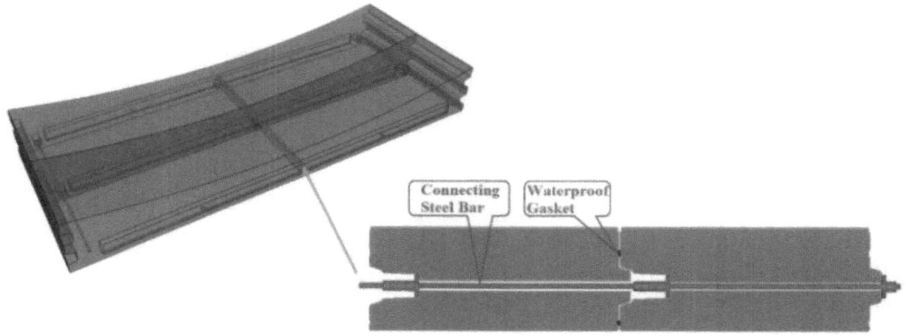

Figure 4. Schematic diagram of longitudinal prestressed tension connection.

3 CONSTRUCTION TECHNOLOGY OF PREFABRICATED STATION

3.1 *Fine leveling of the foundation pit bottom*

Two methods of fine leveling of foundation pit bottom, namely "fine leveling strip method" and "uniform leveling of foundation surface", are proposed, which create conditions for high-precision assembly of foundation pits under different geological conditions.

3.2 *Intelligent gantry crane equipment*

Aiming at the characteristics of stable operation, accurate positioning, and high precision requirements in the hoisting process of prefabricated components, an intelligent gantry crane is developed for the hoisting and handling of prefabricated buildings (Figure 5).

The gantry crane travels longitudinally, transversely, and vertically along the foundation pit, and the controlled displacement is accurate to 5 mm. The sensor positioning system based on photoelectric technology can accurately measure the relative position of the crane in the work site and the crane itself and can provide multiple groups of accurate three-dimensional coordinates for the crane. The positioning accuracy of the crane and trolley is 5 mm, the positioning accuracy of lifting height is 5 mm, and the attitude detection accuracy of special slings is 5 mm.

3.3 *Assembly operation trolley*

The prefabricated component assembly auxiliary trolley (Figure 6) has been developed, which realizes the comprehensive integration of multi-functional assembly operations, greatly facilitates assembly construction and improves the stability and safety of construction operations.

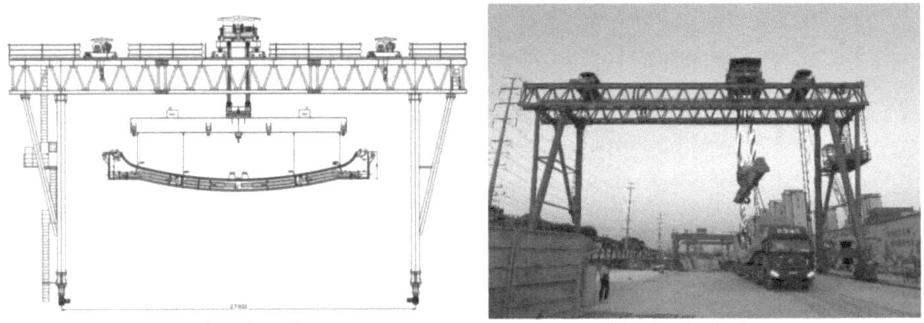

(a) Design drawing (b) Physical picture

Figure 5. Intelligent gantry.

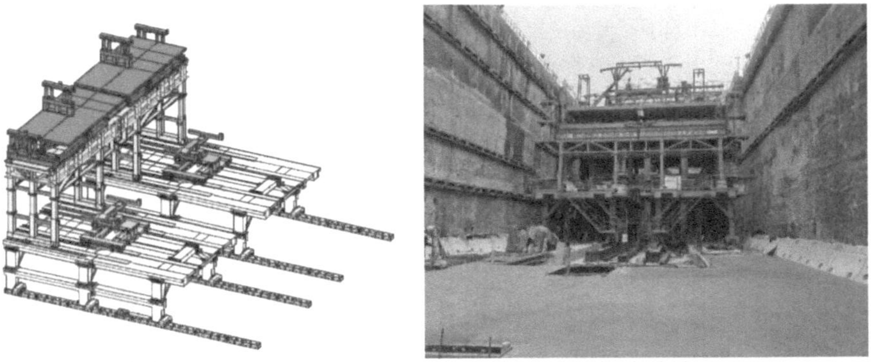

(a) Design drawing (b) Physical picture

Figure 6. Assembly operation trolley.

3.4 *The intelligent positioning system of the assembly trolley*

The intelligent positioning system of the assembly trolley in the subway station is developed (Figure 7), and the whole machine uses 22 sensors. The component positioning guidance system uses the first ring that has been installed as the benchmark, so the installation of the first ring requires manual detection of control accuracy. After the installation of the first ring is qualified, the perpendicularity, distance, altitude, and other parameters of each component are determined through sensors to achieve the precise installation of each component.

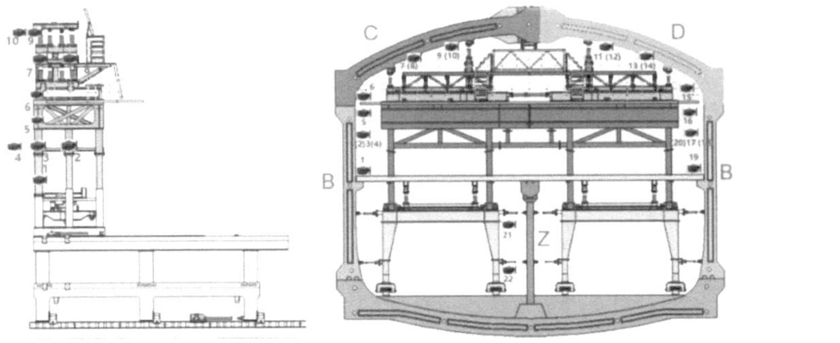

Figure 7. Intelligent positioning system.

3.5 *Automatic tensioning control system*

The automatic control system of tensioning is developed, which realizes high-precision automatic prestressed tensioning control and rapid assembly construction. The system solves the key technical problems of assembly, such as multi-point tensioning coordinated control of large prefabricated components, determination of dynamic tensioning load, and accurate control of joint width.

4 CONCLUSIONS

Based on the Huanghai College Station of Qingdao Metro Line 6, a series of core achievements in industrial construction including construction technology and equipment is achieved. The intelligent and accurate assembly equipment of prefabricated subway stations has been realized, and intelligent technology systems such as intelligent hoisting and mechanical automatic assembly process have been formed, providing a new construction mode for subway stations.

The prefabricated structure construction technology developed for subway stations is also applicable to other open-cut underground structures such as subway section structures, pedestrian passages, underground pipe galleries, etc. The structure can be single arch large-span tunnel structures or rectangular frame structures, with a wide range of applicability.

REFERENCES

Feng, Y. (2020). Research and Engineering Practice on the Prefabricated Composite Structure of Wuzhong Road Station of Shanghai Rail Transit Line 15. *Jiangsu Urban Rail Transit*, (4): 7–9.

Li, T. (1995). Design and Construction Experience of Minsk Metro Single Arch Station. *Metro and Light Rail*, (2): 44–48.

Li, X., Liu, C., Zhang, Q. (2019). Comparative Analysis of Mechanical Properties of Arched and Rectangular Subway Stations. *Construction Technology*, 48(16):1–4.

Lin, F. (2022). Mechanical Property Analysis of Top Arch Joint for Prefabricated Underground Metro Station Structure based on In-suit Monitoring. *Tunnel Construction*, 42(3): 430–436.

Qian, Q. (2019). Underground Space Utilization Helps Develop Green Buildings and Green Cities. *Tunnel Construction*, 39(11): 1737–1747.

Wang, X. (2022). Research on Key Technologies of Intelligent Design for a New Prefabricated Subway Station. *Railway Standard Design*, 66(02):117–123.

Yang, X., Lin, F., Huang, M. (2020). Research on Flexural Bearing Capability of Long Grouted Single Mortise Tenon Joints for Prefabricated Metro Station Structures. *China Civil Engineering Journal*, 53(4): 111–118.

Yang, X., Shi, Z., Lin, F. (2019). Research on Shear Capacity and Checking Method of mt·G-joint for Application in Prefabricated Underground Structures. *Advances in Materials Science and Engineering*, (10), 1–12.

Yang, Y., Lu, M., Liu, C. (2021). The Study on Concrete Crack Monitoring and Recognition in a Joint Experiment of a Prefabricated Metro Station. *Journal of Railway Science and Engineering*, 18(12):3303–3310.

Zhang, C. (1997). Construction Technology of Prefabricated Segment Assembly Method for a Long-span Tunnel in France. *Modern Tunnelling Technology*, (2): 10–18.

Zhang, J., Zhang, J. (2021). Research on the New Type of Subway Station Prefabricated Structure and Cost Index. *Journal of Railway Engineering Society*, 38(08):96–101.

Water Conservancy and Civil Construction – Oke & Ahmad (Eds)
© 2024 The Author(s), ISBN: 978-1-032-58614-4

Evaluation of Baihetan arch dam performance based on displacement separation method

Jinhua Guo
College of Water Conservancy and Hydropower Engineering, Hohai University, Nanjing, China

Jianrong Xu
Power China Huadong Engineering Corporation Limited, Hangzhou, China

Tongchun Li*
College of Water Conservancy and Hydropower Engineering, Hohai University, Nanjing, China

Yu Peng & Jianxin Wang
Power China Huadong Engineering Corporation Limited, Hangzhou, China

Lingang Gao
College of Water Conservancy and Hydropower Engineering, Hohai University, Nanjing, China

ABSTRACT: Based on the displacement monitoring data of limited measuring points, it is one of the effective means to deduce the variation law of the dam's spatial displacement field to evaluate the working performance of the dam. In this paper, based on the displacement separation method, the constrained deformation of the dam foundation and the elastic modulus of the dam body can be solved simultaneously. According to the research results of calculation examples, the spatial displacement field of the dam can be derived by using this method for a complex foundation. Based on the application research results of this method in the Baihetan Arch Dam Project, it can be seen that the relative error distribution of the spatial displacement field is reasonable and the deformation separation results are consistent with the actual situation, which is of great significance to the safety evaluation of arch dams.

1 INTRODUCTION

Displacement monitoring is an important means in the evaluation of dam working behavior (Shan 2022; Wei 2022). Based on the displacement monitoring data of the limited measuring points in the space (Zhao 2010), it is necessary to conduct in-depth analysis and research on the displacement of the arch dam body (Huang 2007; Liu 2020) to evaluate the safety of the whole cycle by summarizing its change rules, which can objectively evaluate the dam safety. Jin Xinxin *et al.* (Jin 2021; Xiao 2021) qualitatively and quantitatively analyzed the horizontal displacement evolution process and physical laws of Wudongde Arch Dam during the initial impoundment period, and then evaluated the working behavior of Wudongde Hydropower Station at the initial impoundment period. Lin Chaoning (Lin 2019) took the constraint force as an unknown quantity for deformation separation and made some improvements in the inversion method (Li 2014), and preliminarily discussed the applicability of the method in the case of complex foundation. However, the derivation of the formula is limited to two-

*Corresponding Author: ltchhu@163.com

DOI: 10.1201/9781003450818-83

dimensional and is only verified in gravity dams. This method reflects the influence of foundation constraint through the rigid body displacement at the dam centroid.

The displacement of any point of the dam body can be divided into two parts: the elastic deformation of the dam body caused by the load and the deformation caused by the constraint of the dam foundation. Guo Jinhua *et al.* (Guo 2022) proposed a new deformation separation method based on data driving. By using this method, not only the contribution degree of dam foundation restraint deformation and dam body elastic deformation in any measured point displacement can be analyzed, but also the change of elastic modulus of dam body in different stages can be analyzed, and the spatial-temporal variation characteristics of foundation restraint deformation can be found. This can be used to determine whether obvious damage occurs to the dam and foundation. These provide a scientific basis for rational evaluation of the working behavior of arch dams. In this paper, the rationality of this method is verified by constructing an arch dam example. In the safety evaluation study of Baihetan Arch Dam, the elastic modulus of the dam body can be inversely calculated by this method to determine whether the dam is in an elastic working state, and the changes of foundation mechanical parameters and other foundation problems can be judged by comparing the percentage changes of dam foundation constraint deformation separated in the previous and subsequent periods, to solve the problems that it is difficult to accurately locate the location of foundation weakening due to complex foundation conditions in practical projects.

2 DEFORMATION SEPARATION METHOD

2.1 *Principle of deformation separation method*

2.1.1 *Mechanical model of deformation separation method*

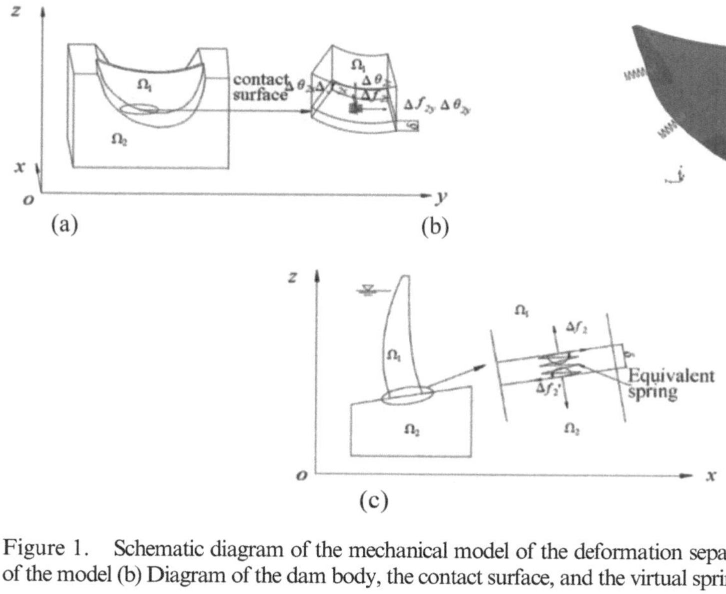

Figure 1. Schematic diagram of the mechanical model of the deformation separation method. (a) 3D sketch of the model (b) Diagram of the dam body, the contact surface, and the virtual spring (c) Model section diagram.

The displacement of any point of the dam body can be divided into two parts: the deformation caused by the constraint of the dam foundation on the dam body and the elastic deformation of the dam body itself. Therefore, this inversion method divides the entire model into two parts: the dam body and the dam foundation. The dam body consists of two parts: the internal part and the contact surface. The contact surface between the dam body and the dam foundation is equivalent to several springs, so the binding force of the dam foundation on the dam body can be equivalent to the binding force of the spring on the dam body, and then the displacement on the contact surface can be reflected through the deformation of the spring, to reflect the influence of the dam foundation on the dam displacement and to achieve the purpose of separating the deformation under the influence of the dam foundation constraint from the elastic deformation of the dam body itself. Further, the rationality and accuracy of inversion results can be improved. The specific model is shown in Figure 1.

2.1.2 *Solution of equation*

According to the deformation separation algorithm in reference [0], the dam body is divided into two areas: interior and contact surface. Finite element equations are established.

$$\begin{bmatrix} k_{11} & k_{12} \\ k_{21} & k_{22} \end{bmatrix} \begin{Bmatrix} u_1 \\ u_2 \end{Bmatrix} = \begin{Bmatrix} F_1 \\ F_2 \end{Bmatrix} + \begin{Bmatrix} 0 \\ f_2 \end{Bmatrix} \tag{1}$$

Where: u_1 is the node displacement of the internal point of the dam body; u_2 is the displacement on the contact surface; F is external load; f_2 is binding.

The spring deformation expression can be obtained by solving:

$$\{u_f\} = [c_f][k_f][R]^T([R]\{\bar{u}\} - \{u^*\}) \tag{2}$$

Where, $[c_f] = \left[[k_f]^T[R]^T[R][k_f]\right]^{-1}$ is the extraction matrix from all points inside the dam body to some measuring points; Let $[R]$ represent the deformation caused by the external load; $\{\bar{u}\} = [k_{11}]^{-1}\{F_1\}$ represent the transformation matrix between the constrained deformation of the spring and the nodes in the dam body; $\{u^*\}$ is the measured value array.

Then the predicted displacement $\{u_1\}$ of each point of the dam body can be solved.

In the process of elastic modulus inversion, the minimum objective function is as follows:

$$\min Q^2(E) = qerr/qabs \tag{3}$$

$qerr = \sum^{m} (u^* - u^c)^2$, $qabs = \sum_{i-1}^{m} (u^c)^2$, where u^* is the monitoring value and u^c is the finite element calculation value:

Artificial intelligence particle swarm optimization is applied to inversion analysis. When the initial value is given, the optimal solution of E is searched at a certain step in each gradient direction to obtain the minimum value of the objective function.

3 CALCULATION EXAMPLE

In this paper, an arch dam model with symmetrical left and right banks subjected to water load is constructed as an example for back analysis to verify the rationality of the deformation separation method.

3.1 *Calculation model and calculation parameters*

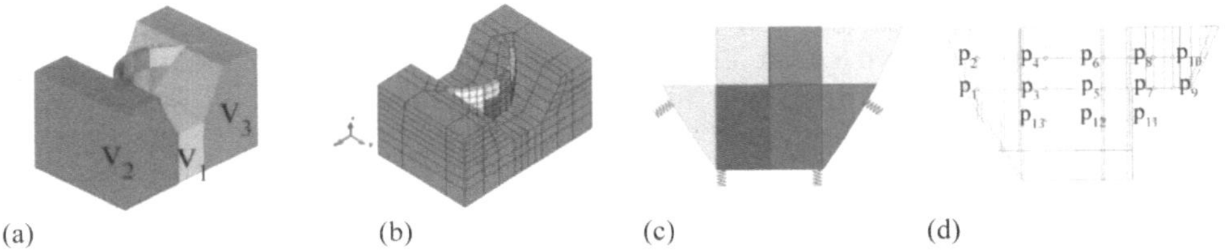

(a) (b) (c) (d)

Figure 2. Virtual spring model of arch dam and distribution of measuring points. (a) Foundation condition II (b) Foundation condition (c) Spring constraint II: 4 (d) Measuring point group a.

To accurately reflect the complex situation of the foundation, this paper establishes a foundation model to simulate the uneven elastic modulus of the foundation concerning the foundation conditions of the actual project. The material parameters of the foundation model are as follows: The foundation is divided into three groups (V1, V2, and V3) according to the quality difference of the dam foundation on the left and right banks. The elastic modulus from the left bank to the right bank is 19, 20, and 15 GPa respectively, and the Poisson's ratio is 0.167, as shown in Figure 2(a).

The arch dam model is 100 m high. The elastic modulus of the dam body is taken as 30 GPa and Poisson's ratio is 0.167. A hexahedral isoperimetric element finite element model is established, in which the range of foundation selection is 75 m from the upstream of the dam face, 150 m from the downstream, and 110 m below the foundation surface. During calculation and analysis, the foundation boundary is subject to normal constraints. The dam finite element calculation model is shown in Figure 2(b). In this model, the upstream direction (y-direction) is positive, while the upstream direction is negative; In the cross-river direction (x direction), it is positive to the left bank, otherwise, it is negative. A total of 13 measuring points at different elevations in the dam are selected for analysis. Each measuring point records the displacement in 3 directions. The distribution of measuring points in the dam and the position of the virtual spring is shown in Figure 2(d).

3.2 *Analysis process*

The calculation and analysis process is as follows:

1. The finite element method is used to calculate the deformation of the dam body under a load of upstream static water level (90 m).
2. Given the lack of actual monitoring data in the calculation example, it is assumed that the calculated value of the finite element displacement under the hydrostatic pressure at the above measuring points is the monitoring value. Two groups of virtual spring constraint groups and two groups of measuring point groups are respectively used to separate the dam foundation constraint deformation and inverse the elastic modulus of the dam body using the deformation separation method, and the displacement of all nodes of the dam body is inversely deduced from Formula (2).
3. Method rationality evaluation. The node displacement of the inverted dam body is compared with the calculated value of the overall finite element analysis, the relative error of each measuring point is analyzed, and the rationality of the application of this method to the arch dam analysis is verified. Because the result data is large, the data is imaged for analysis.

3.3 *Results and discussion*

Taking the objective function $Q = \sqrt{q_{\text{err}}/q_{\text{abs}}}$ in Section 2.1.2 as the evaluation index, according to the results of 39 groups of actual and predicted values of measurement point group a (3 directions) and 33 groups of actual and predicted values of measurement point group b (3 directions), the Q value is 0.57%. It can be seen that the calculated Q value is less than 3.00%. Therefore, it is feasible to calculate the overall displacement field of the dam and evaluate the dam safety through a limited number of measuring points. The calculated value of the dam body displacement finite element (the result of step 1) is compared with the predicted value of the dam body displacement of the deformation separation method (the result of step 2) and the relative error distribution of spatial displacement field is drawn, as shown in Figure 3. As shown in Figure 3, the error of dam body displacement is less than 3%, and the error is less than 1.5% when the left and right banks are about 20 m away from the dam foundation. Therefore, it is proved that the deformation separation method can predict non-measurement points of the dam, and these models are reliable and reasonable. Therefore, the overall displacement of the arch dam can be predicted and the displacement field of the arch dam can be constructed through a limited number of measuring points in the dam.

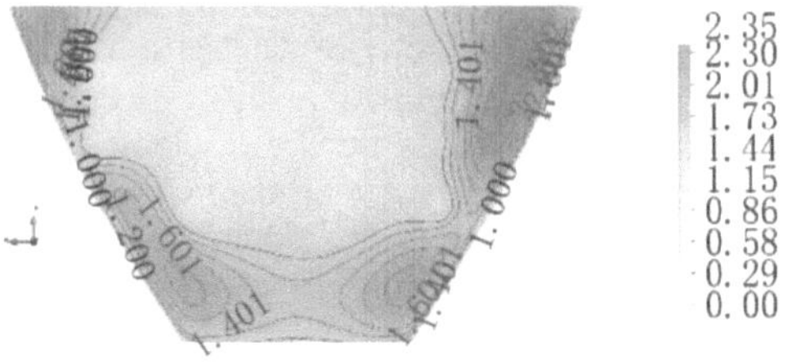

Figure 3. Error distribution of spatial displacement field along the river for three ground conditions.

The elastic modulus of the dam can be inversely calculated according to the deformation separation method and particle swarm optimization algorithm. The search range of elastic modulus is set to be 22~35 GPa. The results show that the inversion value of the dam modulus is close to the actual value and slightly higher than the actual value, which further explains that the deformation separation method applies to the uneven foundation.

4 ENGINEERING APPLICATION

4.1 *Project overview and calculation parameters*

Baihetan Hydropower Station is located in Ningnan County, Sichuan Province, and Qiaojia County, Yunnan Province, in the lower reaches of the Jinsha River. The barrage is a concrete double-curvature arch dam, with a crest elevation of 834.0 m and a maximum dam height of 289.0 m (Liang 2016). A relatively complete displacement monitoring system is arranged in the arch dam to monitor the radial and tangential displacement of the arch dam. A total of 27 normal vertical line measuring points and 5 inverted vertical line measuring points are installed at the elevations of 834 m, 795 m, 753 m, 703 m, 656 m, 600 m, and 579 m of the 7 #, 12 #, 13 #, 18 #, 22 #, 23 #, and 28 # dam sections of the arch dam. The displacement of the measuring points is obtained by superposition of the measured values of the inverted vertical line and the normal vertical line measuring points. Due to the asymmetric topographic and geological conditions on both banks of the super high arch dam site, the thrust distribution at the arch ended, dam displacement, and stress distribution are affected, reflecting the characteristics that the dam foundation on the left bank is slightly weaker than that on the right bank (Liang 2016), and the deformation amplitude of the dam near the left bank is more obvious than that on the right bank.

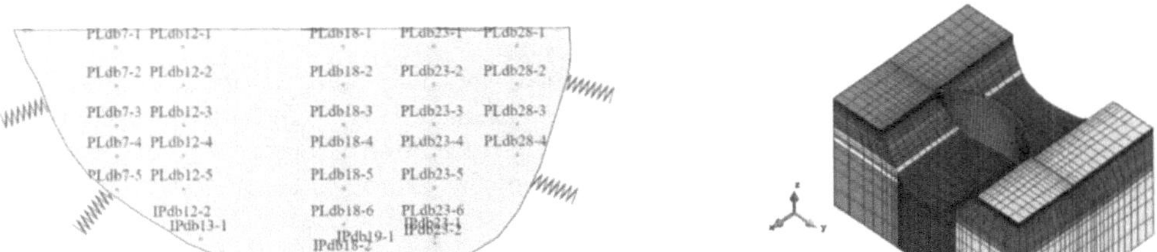

(a) (b)

Figure 4. Calculation model and measuring points of Baihetan. (a) Virtual spring position and measuring point distribution (b) Finite element calculation model.

In this paper, the detailed structure of the Baihetan Arch Dam is simplified, modeled, and analyzed. According to the results of the calculation example discussed in Section 2, the deformation separation method applies to an uneven foundation. There is a group of contact surfaces between the model dam body and the dam foundation with only one layer of thin layer elements, and the contact surface is simplified as seven springs, as shown in Figure 4(a). The four-node bilinear element is adopted as the finite element. The finite element grid includes 23728 elements and 28637 nodes. The overall finite element model of the arch dam is shown in Figure 4(b). It is specified that the downstream direction along the river (x direction) is positive, while the downstream direction is negative; In the cross-river direction (y direction), it is positive to the left bank, otherwise, it is negative. The material parameters, boundary conditions, and load conditions are as follows:

1. According to the mechanical parameters measured in the field test, the elastic modulus of the dam body is taken as 40 GPa, and the average value of Poisson's ratio is 0.163; The geological conditions of the dam foundation are complex, and the dam foundation is divided into 19 groups according to the left and right banks and elevation. The elastic modulus and Poisson's ratio are different.
2. Selection of the range of the foundation: it extends 325 m from the upstream of the dam face, 470 m from the downstream, and 400 m below the foundation surface. In the calculation and analysis, the normal constraint is applied to the foundation boundary;
3. Water load: we select the monitoring data after arch sealing and the corresponding water level for analysis, take the corresponding water level of 799.34 m on September 9, 2021, as the foundation water level, and use the displacement increment value of the corresponding water level relative to the foundation water level on other monitoring dates for analysis and calculation.
4. Temperature load: 199 temperature measuring points are arranged in dam monoliths 2 #~30 #. According to the temperature monitoring data on the corresponding monitoring date, the temperature field of the corresponding dam body is simulated and calculated and converted into temperature load to participate in the subsequent inversion calculation.

4.2 *Result analysis*

The monitoring data on September 21, 2021, is selected for the result analysis. According to the calculation process in Section 1.2, the elastic modulus of the dam body inversed by the deformation separation method is 41.43 GPa, which is consistent with the dam concrete test results. The relative error distribution of the spatial displacement field under the selected monitoring date is shown in Figure 5. It can be seen from this that the measuring points with large errors are mainly located close to the dam foundation, and the error of measuring points 50 m away from the dam foundation is about 10%. Considering the complexity of the actual stress in the simulation process, the error range is relatively reasonable. The above error is consistent with the simplification of the contact surface between the dam body and the dam foundation by the deformation separation method.

It can be seen from the proportion of the separated dam body deformation and dam foundation constraint deformation in the total displacement (Figure 6) that, along the river, the proportion of the dam foundation constraint deformation is higher at the measuring point close to the foundation, and the proportion of the dam foundation constraint deformation separated from the measuring point close to the dam foundation on the left and right banks is relatively higher, and the proportion of the left bank is higher. Considering that the deformation increment analysis is adopted in the calculation, the amplitude of water level change during the analysis period is not large, so the impact of elastic deformation caused by water level change is limited, which is roughly consistent with the results in the figure; the above phenomena show that the separation of restrained deformation and elastic deformation of the dam foundation of Baihetan Arch Dam is relatively good, and the influence of the dam foundation restraint is greater near the left and right banks, which is consistent with the actual situation that the elastic modulus of the bedrock of Baihetan Arch Dam is relatively low at the dam abutments on the left and right banks, and the dam foundation on the left bank is slightly weaker than that on the right bank.

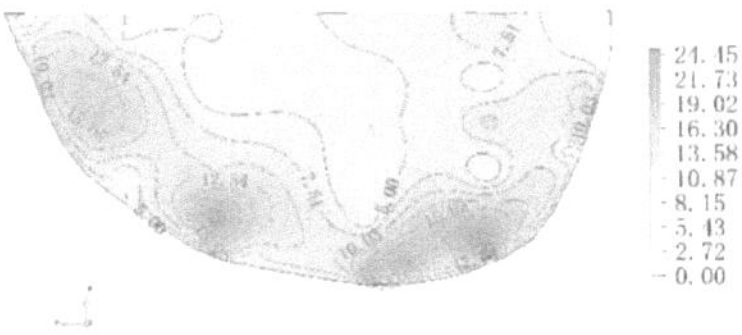

Figure 5. Error distribution of spatial displacement field along the river on September 18, 2021.

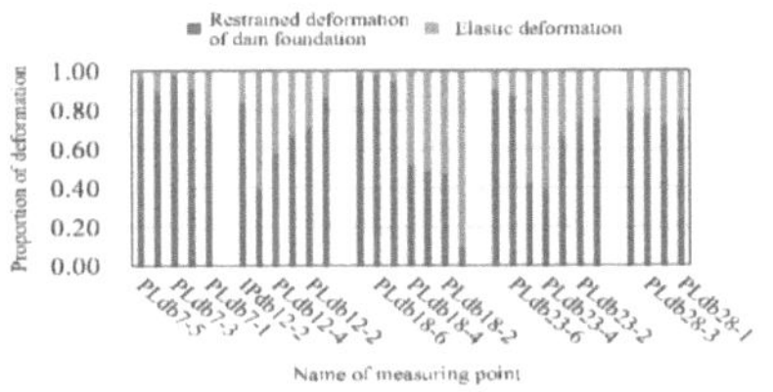

Figure 6. Proportion of two types of deformation on September 18, 2021.

5 CONCLUSION

The monitoring displacement of the arch dam body can be divided into two parts: the displacement generated by the load on the dam body and the constrained deformation of the dam foundation. The elastic modulus of the dam body can be inversed by the deformation separation method combined with the particle swarm optimization algorithm. By simulating different foundation conditions, the monitoring displacement of the measuring point and the resultant displacement error after separation are generally small and are suitable for complex foundation conditions. The error distribution in the application of the Baihetan project is reasonable, so the method is reasonable and feasible. The difference in the proportion of restrained deformation of the dam foundation separated from the measuring points near the left and right banks can also reflect the structural characteristics of the arch dam, the asymmetry of the foundation on the left and right banks, and the influence of the weak interlayer on the deformation. The research results are of great significance for dam safety evaluation.

REFERENCES

Guo Jinhua, Liu Xiaoqing, Li Tongchun, *et al.* Deformation Separation Method and Data-driven Arch Dam Performance Evaluation [J/OL]. *Journal of Hydropower*: 1–11[2022-11-20] http://kns.cnki.net/kcms/detail/11.2241.TV.20220915.1826.002.html.

Huang Yaoying. *Positive and Negative Analysis Method and Application of the Time-varying Effect of High Dam and Bedrock [D].* Jiangsu: Hohai University, 2007.

Jin Xinxin, Shang Yujie, Lu Zhengchao, *et al.* Analysis and Evaluation of Deformation Behavior of Wudongde Arch Dam at the Initial Stage of Impoundment [J]. *Water Resources and Power*, 2021, 39(12): 112–115.

Li Tongchun, Lin Chaoning, Zhao Lanhao, *et al.* An Interactive Method of Interface Boundary Elements and Partitioned Finite Elements for Local Continuous/discontinuous Deformation Problems [J]. *International Journal for Numerical Methods in Engineering*, 2014, 100(7): 534–554.

Liang Guohe, Hu Yu, Fan Qixiang, *et al.* Analysis of Valley Width Deformation Characteristics and Influencing Factors of Xiluodu High Arch Dam During Water Storage Period [J]. *Journal of Hydroelectric Engineering*, 2016, 35(9): 101–110.

Lin Chaoning, Li Tongchun, Liu Xiaoqing, *et al.* A Deformation Separation Method for Gravity Dam Body and Foundation Based on the Observed Displacements [J]. *Structural Control and Health Monitoring*, 2019, 26(2): e2304.1–e2304.18.

Liu Yi, Yang Bo, Zhang Jing, *et al.* One of the Research and Application of Ultra-high Arch Dam Design based on Behavior Simulation: The Development Status and Prospect of Structural Analysis Methods for Arch Dams in My Country [J]. *Water Resources and Hydropower Engineering*, 2020, 51(10): 41–54. Doi: 10.13928/j.cnki.wrahe.2020.10.006.

Shan Zhigang, Zhao Liuyuan, Ni Weida. Comparative Study on the Measurement Methods of Elastic Parameters of Rock Mass in Baihetan Dam Site Area [J]. *Journal of Hydroelectric Engineering*, 2022, 41(5): 103–114.

Wei Yilun, Hu Yu, Wang Yajun, *et al.* Mixed Model Method for Deformation Prediction During the First Storage Period of Baihetan [J]. *Journal of Hydroelectric Engineering*, 2022, 41(5): 84–92.

Xiao Maohao. *Research on Deformation Monitoring Model of 300m High Arch Dam in the Initial Stage of Impoundment [D].* Dalian: Dalian University of Technology, 2021.

Zhao Wenhao, Yan Weiwu. Research on Data-driven Fault Diagnosis [J]. *Microcomputer Information*, 2010, 26(28): 104–106.

Water Conservancy and Civil Construction – Oke & Ahmad (Eds)
© 2024 The Author(s), ISBN: 978-1-032-58614-4

Experimental study on flexural behavior of PVC formwork

Yeyi Zhu
Chang'an Dublin International College of Transportation, Chang'an University, Xi'an, China

Wenlong Song & Yuan Fang
Department of Civil Engineering and Architecture, Anhui University of Technology, Maanshan, China

Shuangshuang Bu
Chang'an Dublin International College of Transportation, Chang'an University, Xi'an, China
Department of Civil Engineering and Architecture, Anhui University of Technology, Maanshan, China

Changfeng Xie*
Department of Civil Engineering and Architecture, Anhui University of Technology, Maanshan, China

ABSTRACT: To study the flexural behavior of the PVC formwork, seven specimens are designed and experimentally studied. The effects of the formwork span, transverse rib spacing, and stiffening rib spacing on the failure mode, bearing capacities, strain development, and load-deflection are analyzed. The test results show that the formwork has a buckling failure. As the formwork span and stiffening rib spacing increase, the yield and ultimate bearing capacities of the formwork decrease. However, the increase in transverse rib spacing has little influence on the yield and ultimate bearing capacities. Meanwhile, increasing the formwork span, transverse rib spacing, and stiffening rib spacing can accelerate strain and deflection development and enhance the ultimate strains of the aluminum alloy frame and the stiffener bottom.

1 INTRODUCTION

Formwork is an important link in the construction project, which accounts for 20%-30% of the cost of concrete structure engineering, 30%-40% of the project consumption, and approximately 50% of the construction period. The use of building formwork directly affects the quality, cost, and benefit of the project.

In traditional formwork engineering, wood formwork (Wu *et al.* 2020; Xu *et al.* 2017; Zhou *et al.* 2019) is the most commonly used formwork, which has the advantages of lightweight, flexible assembly, and disassembly, while it has defects of large deformation, few turnover times, and high resource consumption. In contrast, steel formwork (Feng 2013; Xiao 2012; Wu 2014; Zhu *et al.* 2000) and aluminum formwork (Pan *et al.* 2017; Qin 2014; Wei 2019; Xin *et al.* 2017; Zhang 2016) show large strength and high utilization rate, while they exhibit problems, such as high cost and poor modification. Therefore, it is of great practical engineering significance to study and make new formwork.

A large number of scholars have carried out a series of investigations on the new formwork. Chen *et al.* (2014) studied the inner material mix ratio of PVC sandwich plastic formwork and came up with a raw material mix ratio that could meet the technical requirements of construction formwork specifications. Zhang *et al.* (2013) studied four plastic forms with different substrates,

*Corresponding Author: 514055036@qq.com

 DOI: 10.1201/9781003450818-84

such as PVC, PP, PE, and PP-R, and simulated the basic mechanical properties of the four types of forms under different hydration heat environments. The results showed that the average linear expansion coefficient of the PVC composite formwork was close to that of the steel bar and concrete, showing good coordination ability. Liang *et al.* (2016) designed a deformable PVC building formwork system, in which heating elements were added to the formwork and the temperature was controlled after electrifying, the formwork could produce corresponding deformation according to the characteristics of different structures. Zhou *et al.* (2017) systematically analyzed the construction technology of PVC composite plastic formwork and verified the feasibility of the application of PVC composite plastic formwork in practical engineering projects. Kuder *et al.* (2009) carried out a study on the mechanical properties of the non-removable PVC composite plastic formwork system and the results showed that the formwork could improve the mechanical properties of concrete, enhance the tensile deformation capacity of the section, and the flexural properties of the specimens.

Numerous experimental studies show that PVC formwork has excellent performance and feasibility of practical engineering application. Currently, as the whole formwork, PVC formwork mostly uses wood as the back, which has the defects of resource waste, low recycling rate, and difficulty of repair.

To save resources and improve the utilization rate of PVC formwork, aluminum alloy is added to the back of PVC formwork to form a new type of PVC formwork. Through the three-point loading tests, the influences of different factors (i.e., formwork span, transverse rib spacing, and stiffening rib spacing) on the mechanical properties of the composite formwork are analyzed, which provides a reference for the application of the new formwork in practical engineering and the compilation of relevant specifications.

2 EXPERIMENTAL PROGRAMS

2.1 *Specimen design*

Seven formwork specimens are designed and tested. As shown in Figure 1, the specimen is composed of a PVC formwork and an aluminum alloy back skeleton, in which the aluminum alloy back skeleton is composed of an aluminum alloy frame, transverse ribs, and stiffening ribs. In this study, the influence of three factors, including formwork span, transverse rib spacing, and stiffener rib

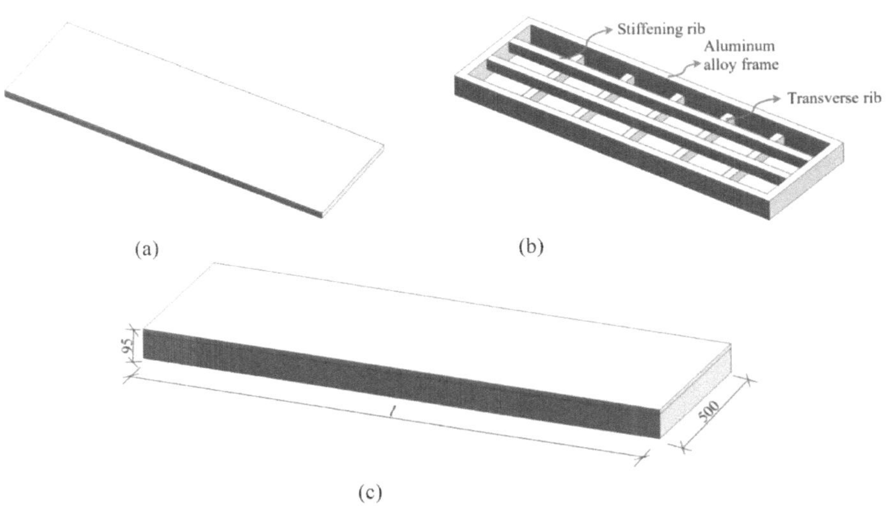

Figure 1. Design diagram of the PVC formwork. (a) PVC panel. (b) Aluminium alloy back skeleton. (c) PVC formwork.

spacing is considered. The width and thickness of the PVC panel used in the specimens are 500 mm and 15 mm, respectively. The specific parameters of formwork specimens are listed in Table 1.

Table 1. Design parameters and measured bearing capacities of the formwork specimens.

No.	l(mm)	S_h(mm)	S_z(mm)	P_y(kN)	M_y(kN·m)	P_u(kN)	M_u(kN·m)
PB-1	1200	200	150	23.0	4.6	32.0	6.4
PB-2	1600	200	150	19.0	3.8	21.5	4.3
PB-3	2000	200	150	17.0	3.4	19.0	3.8
PB-4	1200	150	150	27.0	5.4	31.8	6.36
PB-5	1200	300	150	21.0	4.2	31.0	6.2
PB-6	1200	200	100	29.0	5.8	34.0	6.8
PB-7	1200	200	250	21.5	4.3	24.0	4.8

Note: l is formwork span, S_h represents transverse rib spacing, and S_z denotes stiffening rib spacing, as shown in Figure 1(b). P_y and M_y represent the yield load and yield bending moment of formwork specimens, respectively. P_u and M_u stand for the ultimate load and ultimate bending moment of formwork specimens, respectively.

According to the test method suggested in "Thermoplastic Pipes–Determination of Tensile Properties–Part I: General Test Method" (GB/T8804.1-2003), the elastic modulus is 2.48×10^3 MPa, the ultimate strength is 4.6 kN, and the tensile strength is 30.7 MPa. Aluminum alloy is made of hardened aluminum alloy of 6061-T6 type. According to the production and test methods of metal material samples suggested in "Metallic Materials–Tensile testing–Part 1: Method of Test at Room Temperature" (GB/T228.1-2010), the yield strength and ultimate tensile strength are 156.5 MPa and 185.3 MPa, respectively. The elastic modulus is 6.9×10^5 MPa.

2.2 *Measurement arrangement*

To measure the deformation of the specimens, five displacement meters are adopted, two of which are used to monitor the deformation of the supports at both ends of the specimens, and the other three are arranged in the middle span of the PVC panels, transverse ribs, and stiffening ribs, as shown in Figure 2.

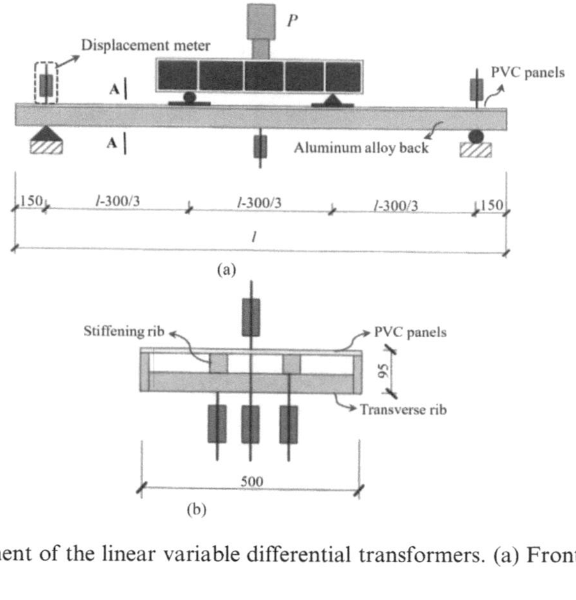

Figure 2. Arrangement of the linear variable differential transformers. (a) Front view. (b) A-A sectional view.

Meanwhile, several strain gauges are arranged at the intermediate of PVC panels, transverse ribs and stiffening ribs, and loading points. Five strain gauges are arranged in the mid-span section of the aluminum alloy longitudinal frame along the height direction, with an interval of 20 mm. Three strain gauges are arranged along the height of the aluminum alloy stiffening ribs and the distribution of strain measuring points is shown in Figure 3.

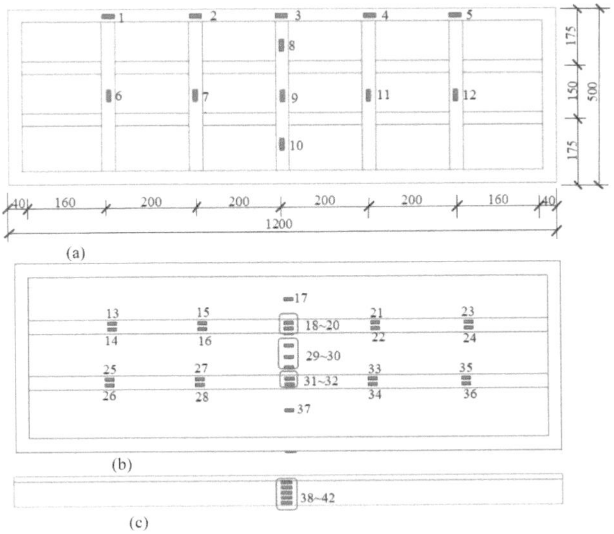

Figure 3. Arrangement of the strain gauges. (a) Layout drawing of the transverse rib and frame strain gauge. (b) Layout of stiffening ribs and sheet strain gauge. (c) Strain gauge arrangement along the height.

2.3 *Test setup*

A hydraulic jack is used to load in the tests and the test setup is shown in Figure 4. Before formal loading, a layer of medium-fine sand with a thickness of 100 mm should be laid on the surface of the formwork and the specimens should be preloaded to ensure good contact between the specimens. In the formal loading process, the hierarchical loading method is

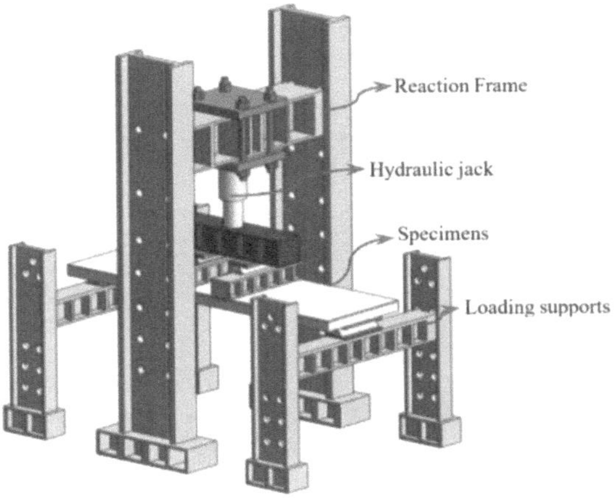

Figure 4. Diagram of the loading device.

adopted. Referring to the "Technical Specification for Combined Aluminum Alloy Formwork Engineering" (JGJ.386-2016), the uniformly distributed load is applied in the form of equivalent concentrated force. The uniform load value of each stage loading is q = 5 kN/m^2, which is equivalent to the value of concentrated force loading by using the formula ($Fl/6 = ql^2/8$). The load holding time is 15 seconds and the tests are terminated when the specimens are damaged.

3 EXPERIMENTAL RESULTS ANALYSIS

3.1 *Failure mode*

At the initial stage of loading, there is no significant change in the specimens and the PVC panel is under full section compression. The mid-span deflection and strain of PVC panels and aluminum ribs are small and linearly increased.

With the increase in load, the welding part of the aluminum alloy frame makes a slight sound, and the deformation of the bottom of the aluminum alloy frame increases. When the load increases to approximately 80% of the ultimate bearing capacity of the specimens, the mid-span deflection and strain of the formwork increase significantly, the frame yields, and the specimens enter the yield stage. A large bending deformation appeared at the bottom of the formwork, accompanied by a large "crack" sound, and the deformation at the bottom of the aluminum alloy frame further increases. As the load continues to increase, the strain of the formwork increases rapidly, the mid-span deflection increases significantly, one side of the aluminum alloy frame produces a curved deformation, one side shows a crack along the vertical direction, and the mid-span panel is slightly separated from the aluminum alloy frames. Eventually, the specimen is damaged by bending and buckling and the typical failure mode is shown in Figure 5.

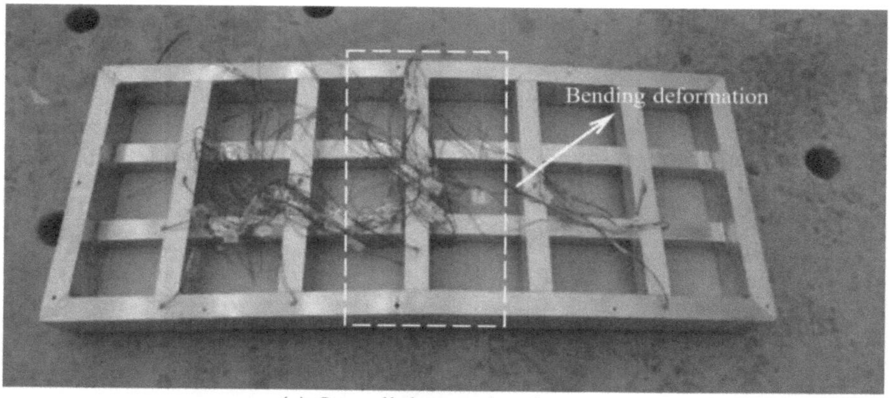

(a) Overall destruction form

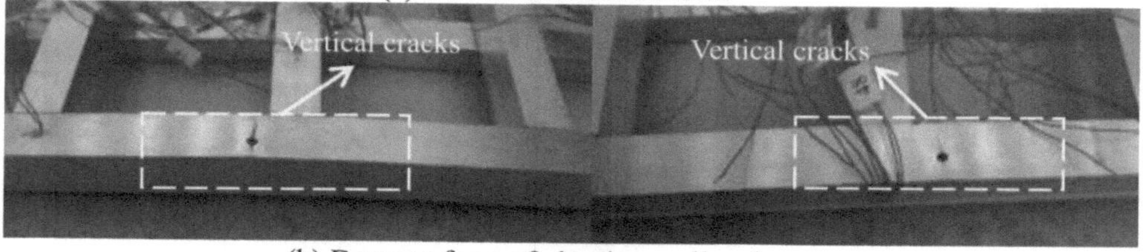

(b) Damage form of aluminum alloy frame

Figure 5. The failure mode of specimen PB-2.

3.2 *Bearing capacity analysis*

The bearing capacity test results of each specimen are shown in Table 1. The influence of formwork span, transverse rib spacing, and stiffening rib spacing on the bearing capacity of PVC formwork is shown in Figure 6. With the increase in the formwork span, the bearing capacity of the specimens decreases significantly, among which the most obvious one is PB-3. Compared with PB-1, its yield-bearing capacity decreases by 26%, and its ultimate bearing capacity decreases by 40.6%. This is because, with the increase in the formwork span, the bending moment of the specimen increases under the same load, the deformation development speed is accelerated, and the deflection in the span is increased.

With the increase in transverse rib spacing, the yield-bearing capacity decreases, while the ultimate bearing capacity increases first and then decreases. This is because, in the yield stage, the load transferred to the aluminum alloy stiffeners is shared by the aluminum alloy stiffening ribs and the transverse ribs. With the increase in the transverse rib spacing, the bearing capacity and flexural stiffness decrease, and the yield-bearing capacity decreases. With the further increase in load, the first failure of the aluminum alloy frame becomes the main failure factor of the specimens. At this time, the interface yield of aluminum alloy stiffening ribs and transverse ribs does not completely occur and the transverse ribs have little effect on the ultimate bearing capacity.

With the increase in stiffening rib spacing, the bearing capacity decreases. The PB-7 with a 250 mm span has the biggest decrease, and the yielding load capacity decreases by 25.9%, and the ultimate load capacity decreases by 29.4% respectively. The reason is that with the increase in stiffening rib spacing, the flexural stiffness decreases, the deformation resistance decreases, and the deformation development accelerates.

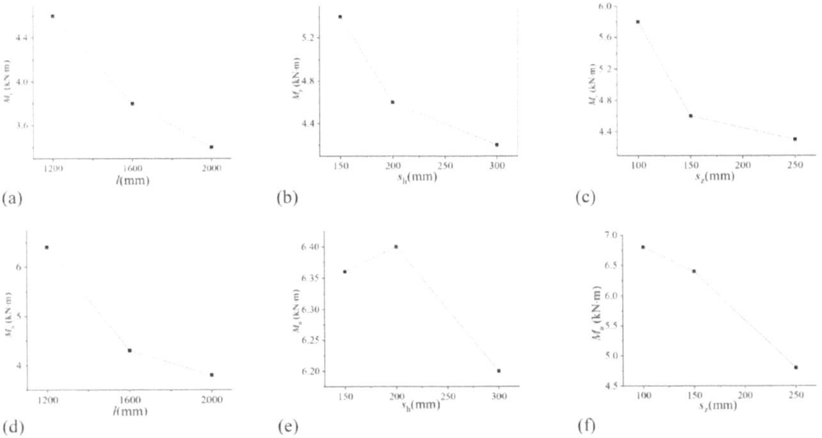

Figure 6. Effects of studied parameters on the bearing capacities. (a) Effect of formwork span on the yield bending moment. (b) Effect of transverse rib spacing on the yield bending moment. (c) Effect of stiffening rib spacing on the yield bending moment. (d) Effect of formwork span on the ultimate bending moment. (e) Effect of transverse rib spacing on the ultimate bending moment. (f) Effect of stiffening rib spacing on the ultimate bending moment.

3.3 *Strain analysis*

3.3.1 *Strain analysis of aluminum alloy frame and stiffener bottom*

The load-strain relationship of PVC formwork is approximately divided into the elastic stage, yield stage, and strengthening stage, as shown in Figure 7 and Figure 8.

At the initial stage of test loading, the specimen is in the elastic stage and the strain of the frame and stiffener bottom increases linearly. With the increase in the formwork span and the stiffening rib spacing, the slope of the load-strain curve of the aluminum alloy frame and the stiffener bottom decreases, indicating that the strain development rate increases. However, the variation of transverse rib spacing exerts little effect on the strain of the aluminum alloy frame and stiffener bottom.

When the load increases to approximately 80% of the ultimate bearing capacity of the specimens, the aluminum alloy frame is the first to enter the yielding stage and the tensile stress of the section is mainly borne by the aluminum alloy stiffening ribs and transverse ribs.

With the increase in load, the strain development of the aluminum alloy transverse ribs and stiffening ribs accelerates and the load-strain relationship of aluminum alloy stiffening ribs shows an inflection point. The loading increase rate slows down and the aluminum alloy stiffening ribs begin to yield. At this time, the panels have been slightly separated from the aluminum alloy frame. The compressive strain at the bottom of the panel decreases, while the tensile strain increases, but still maintains a linear growth.

The inflection point of the load-strain relation curve of the aluminum alloy frame appears earlier and the corresponding strain increases with the increase in formwork span, transverse rib spacing, and stiffener rib spacing. With the further increase in load, the strain of the aluminum alloy frame and stiffening rib increases continuously, the full section yield occurs, and the strain in the tension zone increases rapidly. The slope of the load-strain relationship of the aluminum alloy frame and stiffening ribs remains unchanged until the failure of the specimen. The strain of aluminum alloy transverse ribs increases with the increase in load, while it does not enter the yield stage.

The slope of load-strain curves of the aluminum alloy frame and stiffener bottom decreases with the increase in formwork span, transverse rib spacing, and stiffening rib spacing, indicating that the strain development rate increases.

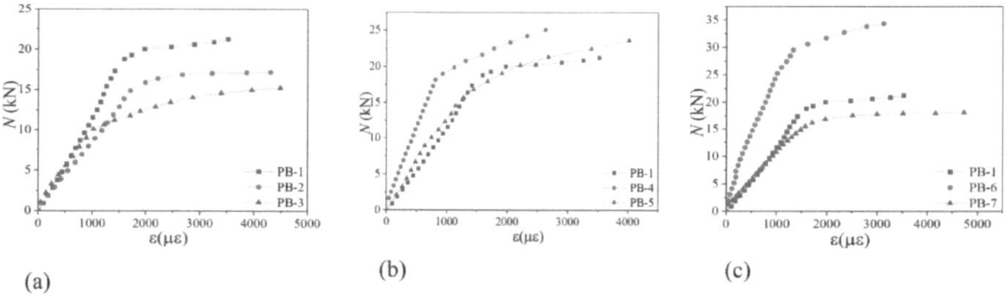

Figure 7. Effect of studied parameters on the strain of the aluminum alloy frame. (a) Effect of framework span. (b) Effect of transverse rib spacing. (c) Effect of stiffening rib spacing.

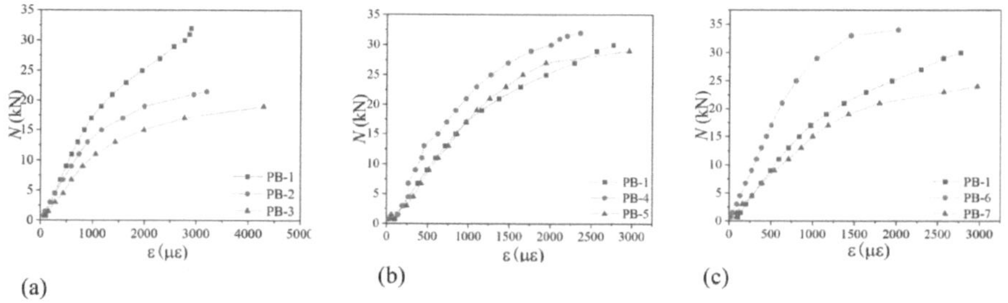

Figure 8. Effect of studied parameters on the strain of the stiffener bottom. (a) Effect of formwork span. (b) Effect of transverse rib spacing. (c) Effect of stiffening spacing rib.

3.3.2 *Strain analysis of stiffening ribs along the height*

Three strain gauges are arranged along the height of the aluminum alloy stiffening ribs, which are denoted as Point A (corresponding to the 0 mm scale line in the figure), Point B (corresponding to the 20 mm scale line in the figure), and Point C (corresponding to 40 mm scale line in the figure) from top to bottom. It can be observed from Figure 9 that the load-strain relationship of aluminum alloy stiffening ribs along the height can be divided into the elastic stage, yield stage, and strengthening stage.

In the elastic stage, the strain of the stiffening ribs along the height increases linearly. The load-strain curve at Point B has a large slope and the strain development is slow, while the load-strain curve at Point A and Point C has a small slope and the strain development is fast.

With the increase in formwork span, transverse rib spacing, and stiffener rib spacing, the slope of the load-strain relationship curve at Point A, Point B, and Point C along the height of aluminum alloy stiffener decreases, and the strain development rate increases.

As the load increases to about 80% of the ultimate bearing capacity, the neutralization axis moves upward, the height of the compression zone decreases and the strain growth rate along the height of the aluminum alloy stiffening ribs accelerates.

The load-strain curves at Point C, point A, and point B show an inflection point. The slope decreases and the load growth rate slows down. The yield occurs earlier at Point C and Point A and the corresponding yield strain is greater than that at Point B.

With the increase in the transverse rib spacing and stiffening rib spacing, the inflection point of the load-strain relationship curve of aluminum alloy stiffening rib along the height appears earlier, and the strain increases.

With the further increase in load, the strain at three points along the height of aluminum alloy stiffening ribs is always increasing. The slope of the load-strain curve at Point A and Point C is less than that at Point B and the strain of aluminum alloy stiffening rib increases rapidly. The slope of the load-strain relationship at all points basically remains unchanged until failure occurs, while the ultimate strain at Point A and Point C was greater than that at Point B.

With the increase in the formwork span, transverse rib spacing, and stiffening rib spacing, the slope of the load-strain relationship curve along the height of aluminum alloy stiffening ribs decreases, the strain development rate increases, and the ultimate strain increases.

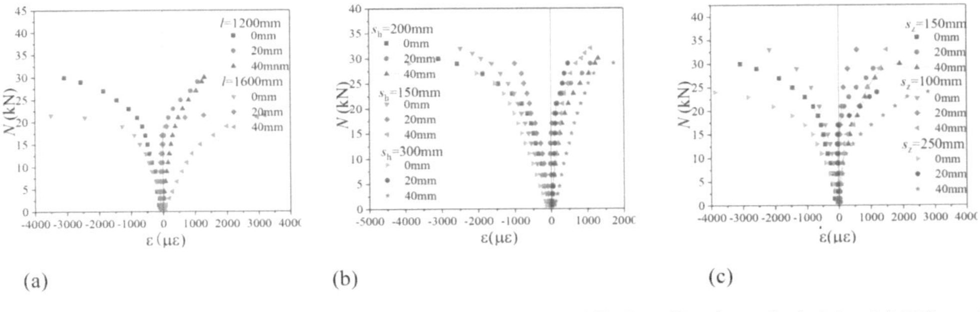

Figure 9. Effect of studies parameters on the strain of the stiffening ribs along the height. (a) Effect of formwork span. (b) Effect of transverse rib spacing. (c) Effect of stiffening rib spacing.

3.4 *Load-deflection relationship (N-Δ curves) analysis*

As shown in Figure 10, the load-deflection relationship is appropriately divided into the elastic stage, yield stage, and strengthening stage.

At the initial stage of test loading, the relationship curve $N - \Delta$ of the specimens increases linearly. With the increase in the formwork span, the deflection development rate increases, while the increase in the transverse rib spacing and stiffening rib spacing has little impact on the deflection.

When the load increases to approximately 80% of the ultimate bearing capacity, the aluminum alloy frame first enters the yield stage. Then, the aluminum alloy transverse ribs and stiffening ribs successively enter the yield stage and the $N-\Delta$ curve shows an inflection point and the slope decreases. With the increase in the formwork span and the stiffening rib spacing, the deflection increases rapidly.

With a further increase in the load, the flexural stiffness of the specimen decreases rapidly, resulting in a rapid increase in the deflection of the specimens. At this stage, the slope of the $N-\Delta$ curve remains unchanged until the specimen fails. With the increase in the formwork span, transverse ribs spacing and stiffening ribs spacing, the growth rate of the deflection increases, and the ultimate deflection of the corresponding panel increases when the specimen is damaged.

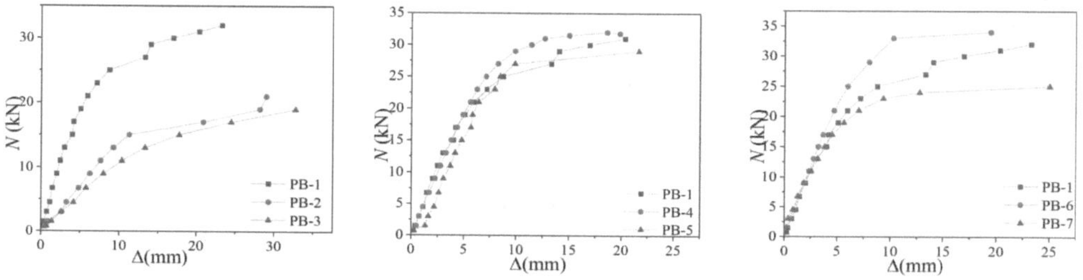

(a) (b) (c)

Figure 10. Effects of studies parameters on the load-deflection curves. (a) Effect of formwork span. (b) Effect of transverse rib spacing. (c) Effect of stiffening rib spacing.

4 CONCLUSIONS

The flexural tests of 7 PVC formwork are carried out and the influences of the formwork span, transverse ribs spacing, and the stiffening ribs spacing on the flexural performances of the specimens are analyzed. The following conclusions can be drawn:

1. The failure mode of the PVC formwork is mainly manifested as the bending and buckling of the aluminum alloy frame. The aluminum alloy frame shows a buckling phenomenon in the mid-span, with bending deformation on one side and vertical cracking on the other side. The mid-span of the panel and aluminum alloy frame behaves a slight separation.
2. The increase in the formwork span, the transverse rib spacing, and the stiffening rib spacing can reduce the bearing capacity of the PVC formwork. In comparison, the influence of the formwork span is the greatest, and the ultimate load-bearing capacity decreases by 40% when the formwork span increases from 1200 mm to 200 mm.
3. As the formwork span, the transverse rib spacing, and the stiffening rib spacing increase, the slope of the load-strain curves at each stage decreases, while the ultimate strains of the aluminum alloy frame and the stiffener bottom increase. More precisely, the top and bottom parts along the height direction of stiffening ribs yielded earlier, and both yield strain and ultimate strain are larger than those in the middle.
4. Initially, with the increase in the formwork span, the slope of the load-deflection relationship curve of the PVC formwork decreases. In the yield stage, the deflection of the specimen increases rapidly as the increase in formwork span and the stiffening rib spacing. In the strengthening stage, the growth rate of the deflection and the ultimate deflection increases as the formwork span, the transverse rib spacing, and the stiffening rib spacing increase.

REFERENCES

Chen, S. W. et al. (2014). Study on Proportioning and Production Technology of Inner Layer of PVC Plastic Sandwich-type Formwork. *Green Building*. 6 (5), 84–86.

Feng, K. (2013). *Engineering Application and Experimental Study of Novel Combined Steel Formwork in the Concrete Shear Wall*. Hunan university. Hunan Province.

GB/T 8804.1-2003. (2003). *Determination of Tensile Properties of Thermoplastic Pipes*. China Architecture & Building Press. China.

GB/T228.1-2010. (2002). *Tensile Test Room Temperature Test Method for Metal Materials*. China Architecture & Building Press. China.

G. J. 386-2016. *Combined Aluminum Alloy Formwork Engineering Technical Regulations*. Ministry of Housing and Urban-Rural Development of the People's Republic of China. China.

Kuder, K. G., *et al*. (2009). *Effect of PVC- Stay-In-Place Formwork on the Mechanical Performance of Concrete. J. Mater. Civ. Eng*. 21 (7), 309–15.

Liang, B. Q. (2016). *Research on Transformable Architectural Formwork of the PVC*. Jilin University. Jilin Province.

Pan, Q. F. et al. (2017). Experimental Research and Parameter Analysis of Flexural Behavior of Aluminum Alloy Formwork. *Indust. Constr*. 47 (11), 142–147.

Qin, L. (2014). *Study on Deformation Behavior and Carrying Capacity and Reliability of the Aluminum Formwork and Falsework System*. Harbin Institute of Technology. Heilongjiang Province.

Wei, S. L. (2019). *Optimal Design and Analyses of Mechanical Properties for Aluminum Alloy Formwork*. Shandong Jianzhu University. Shandong Province.

Wu, J. F. et al. (2020). Study on the Technology of Bamboo-wood Composite Building Concrete Formwork. *J. For. Environ*. 40 (3), 329–335.

Wu, T. Y. (2014). Design and Construction Application of Cantilever Truss Outer Frame Steel Formwork. *J. China & Foreign Highway*. 34 (5), 163–165.

Xiao, X. (2012). *Analysis and Application of Novel Channel Steel Formwork and Experimental Study of Column Template*. Hunan university. Hunan Province.

Xin, X. J. et al. (2017). Aluminum Alloy Floor Formwork Performance Test. *Journal of Henan University of Science and Technology(Natural Science)*. 38 (4), 54–59.

Xu, B.Q. et al. (2017). Study on the Construction Method for Space Polygonal Fair-faced Concrete Structure Using Wood Template. *Constr Technol*. 46 (14), 54–57.

Zhang, A. Q. (2016). *Research and Application of Aluminum Formwork Technology in Super Tall Building*. Changchun Institute Of Technology. Jilin Province.

Zhang, X. X. et al. (2013). Experimental Study on Basic Properties of PVC Plastic Building Templates. *Low Temp. Archit. Technol*. 35 (9), 38–40.

Zhou, J. (2017). Plastic Template (PVC) Applied Research in Construction Engineering. *Doors & Windows*. 11, 249.

Zhou, Y. W. et al. (2019). Research and Application of Cylindrical Wood Formwork Design and Force Analysis. *Value Eng*. 38 (27), 159–162.

Zhu, B. Y. et al. (2000). The Design and Manufacture of the Special-shaped Steel Formwork. *Constr Technol*. 29 (3), 31–32.

Water Conservancy and Civil Construction – Oke & Ahmad (Eds)
© 2024 The Author(s), ISBN: 978-1-032-58614-4

Numerical analysis on the influence of negative skin friction of pile group in collapsible loess sites

Bin Chen* & Qiong Xia
School of Civil Engineering, Lanzhou Jiaotong University, Lanzhou, China

ABSTRACT: The research about negative skin friction of pile foundations in collapsible loess is mostly focused on single piles, whereas the research on negative skin friction of pile groups is less. A 3 × 3 pile group model is built by using finite element software. The negative skin friction and neutral point position of the pile group are analyzed under the influence of different pile spacing, different collapsibility of soil around the pile, and different collapsible thicknesses. The results show that the negative skin friction of the pile group increases with the increase in pile spacing. The negative friction of pile groups increases along the pile shaft with the increase in the degree of collapsibility of the soil around the pile and the neutral point moves slowly down along the pile shaft. The negative skin friction of pile groups increases with the collapsible thicknesses and the neutral point position moves down sharply along the pile shaft. Under the same influencing condition, the distribution of negative skin friction and neutral point position of corner piles, side piles, and central piles also have obvious differences due to the influence of the pile group effect.

1 INTRODUCTION

With the extensive implementation of pile foundations in collapsible loess areas, it has great engineering significance to study the negative skin friction of pile group foundations in thick collapsible loess sites. The research methods on the negative skin friction of pile foundations in collapsible loess sites mainly focus on the field soaking test and model test and the relationship between the limited depth of collapsible loess and the location of the neutral point is analyzed by field immersion tests (Huang 2015). The change of negative skin friction along the pile is analyzed by the field pile foundation immersion test (Zhu 2018). The distribution law of negative skin friction and the law of neutral point location of pile foundation in collapsible loess are analyzed by centrifugal model test (Wang 2010). The bearing capacity is improved by slightly soaking the soil around the pile to eliminate loess collapse (Zhu 2013). Due to the large scale of the pile group project, it is difficult to implement it in the actual project. There are few studies on negative skin friction and the neutral point of the pile group in collapsible loess. With the wide application of finite element software in the theoretical research and practice of pile foundation engineering, numerical simulation has obvious advantages in analyzing pile swarms in collapsible loess considering the many factors affecting pile swarms and the difficulty and economy of field tests. The stress characteristics, pile-soil interaction, pile group effect, and negative skin friction are analyzed by numerical simulation (Du 2015; Fu 2015). The results are reasonable and reliable. However, many numerical simulation studies about negative skin friction of pile foundations focus on a single pile (Chen 2015; Yang 2010;) and there are few studies on negative skin friction of pile groups in collapsible loess sites under immersion. The negative skin friction of the pile group is expressed by the interaction of pile and soil, which is influenced not only by the changes in pile parameters but also by the changes in soil parameters.

*Corresponding Author: 1693766158@qq.com

 DOI: 10.1201/9781003450818-85

In this paper, based on the data of this collapsible loess site of Pucheng Power Plant, the collapsible behavior of the loess after immersion is simulated by the modulus reduction method, and the single pile model and 3x3 pile group model with different pile-soil parameters are established by using ABAQUS. The effect of pile spacing on the negative skin friction and neutral point location after immersion of self-weight collapsible loess is not only analyzed but also the effect of collapsible degree and collapsible thickness of loess on the negative friction and neutral point location after immersion of self-weight collapsible loess is analyzed.

2 THREE-DIMENSIONAL FINITE ELEMENT NUMERICAL MODEL

2.1 *Material properties used in the analysis*

The Pucheng Power Plant is the first soaking test of Lishi loess in China. The foundation soil of the site is a thick loess foundation with a loess and layered distribution of ancient soils. The physical and mechanical parameters of each soil layer are summarized in Table 1 (Li 1993). The parameters of piles and pile caps are summarized in Table 2.

Table 1. Physical and mechanical properties of the soil layer in Pucheng Power Plant.

Soil layers	Depth (m)	Cohesion (kPa)	Friction angle (°)	Natural unit weight (kN/m^3)	Saturated unit weight (kN/m^3)	Poisson ratio	Compressive modulus (MPa)
L_1	5.2	15	24	14.3	17.2	0.37	3
F_1	6.4	40	25	17.5	19.0	0.37	20
L_2	10.4	40	18	17.0	18.7	0.41	25
F_2	13.8	50	24	17.6	19.0	0.37	20
L_3	19.9	40	25	16.5	18.5	0.37	10
F_3	21.4	70	25	17.5	19.0	0.37	25
L_4	26.5	40	20	16.8	19.0	0.40	12
F_4	27.4	70	25	18.0	18.3	0.37	20
L_5	33.3	50	25	17.2	19.3	0.37	15
F_5	35.2	75	24	18.5	18.8	0.37	45
L_6	39.0	50	30	18.5	19.5	0.33	40
F_7	60.0	80	30	19.0	19.5	0.33	80

Table 2. Material parameters of pile and cap.

Type	Density (kg/m^3)	Elastic modulus (MPa)	Poisson Ratio
Pile	2500	2.5×10^4	0.18
Cap	2500	2.5×10^4	0.18

The soil size is 40 m × 40 m × 60 m the pile length is 32 m, the diameter is 1 m, and the cap size is 14 m × 14 m.

2.2 *Study cases*

(1) Different pile spacing
The effect of changing the pile distance on the negative friction of the pile group in loess is analyzed. The pile length and diameter remain unchanged, the collapsible thicknesses of soil are 32m, the collapsible degree is 100%, and the pile spacings are 3 m, 4 m, 5 m, and 6 m.

(2) Different degrees of collapsibility
The occurrence degree of self-weight collapsible loess is defined as = (self-weight collapsibility of a certain amount of water/saturated self-weight collapsibility) × 100%

(Zhao 2018). The pile length and diameter remain unchanged, the collapsible thicknesses of soil are 32 m, the pile spacing is 4 m, and the degrees of collapsibility are 25%, 50%, 75%, and 100%.

(3) Different self-weight collapsible loess layer thickness

The effect of varying the thickness of the collapsible loess layer with different self-weight on the negative friction of the pile group is analyzed while keeping other parameters unchanged. The pile length and diameter remain unchanged, the pile spacing is 4 m, the occurrence degree of collapsibility is 100%, and the self-weight collapsible loess layer thicknesses are 14 m, 20 m, 26 m, and 32 m.

2.3 *Modulus reduction method*

In this paper, the method of reducing the elastic modulus of soil (Wen 2008) is used to simulate the collapsible behavior of loess under soaking. The elastic modulus of soil after immersion is determined by the actual deformation of each soil layer.

Figure 1. Pile group model.

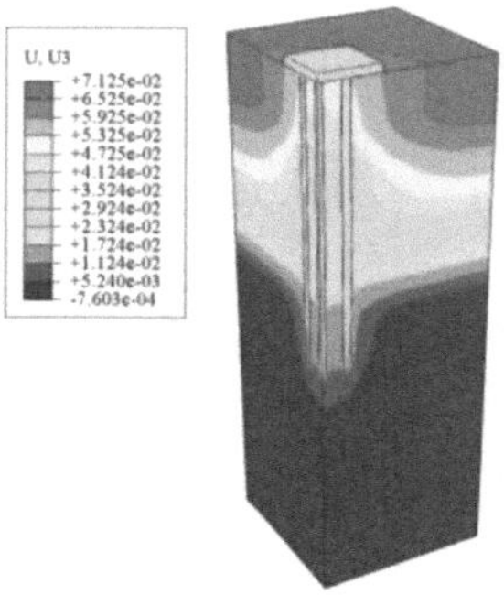

Figure 2. Cloud map of pile group after immersion.

3 ANALYSIS OF NUMERICAL SIMULATION RESULTS

3.1 *Influence of pile spacing on negative friction and neutral point of pile groups*

The curves of negative skin friction of pile groups with different pile spacing are shown in Figure 3.

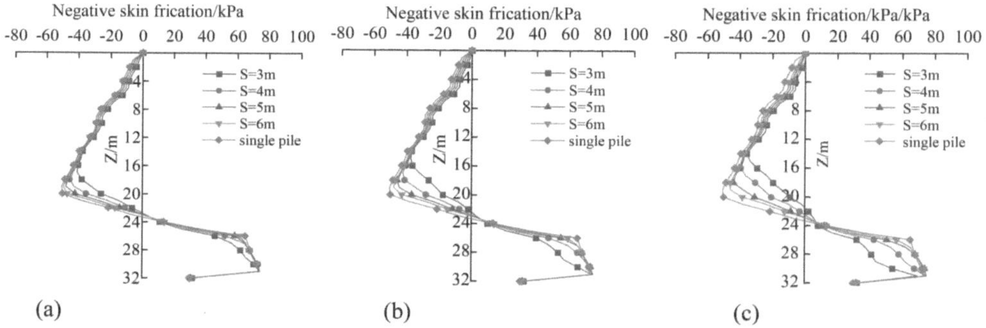

Figure 3. Variable curves of Nsf of pile groups with different pile spacings. (a) Nsf of corner pile. (b) Nsf of side pile. (c) Nsf of center pile.

It can be shown from Figure 3 that the negative skin friction of a single pile is greater than each pile of the pile group and the neutral point is also the deepest. At the same pile spacing, the corner pile has the highest negative skin friction, followed by the side pile, and the center pile has the lowest. The numerical values and distribution of negative friction of the corner pile are closest to the single pile. The pile spacing has a great influence on the center pile, followed by the side pile and the corner pile is the smallest. This is because as the pile distance increases, the interaction between the pile and the soil decreases. When the pile spacing increases to 6 times the pile diameter, the influence on the negative skin friction of each pile in the foundation is no longer obvious. The neutral point also moves down gradually with the increase in pile spacing and the neutral point of the corner pile is closest to the single pile.

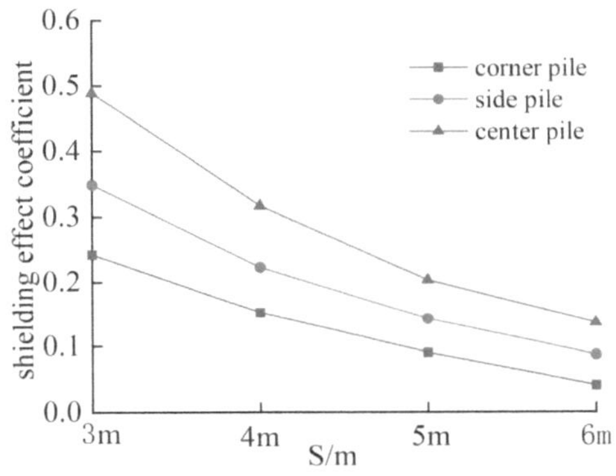

Figure 4. Shading effect coefficients under different pile spacing.

The influence degree of the pile group effect is expressed by the negative skin friction shielding effect coefficient (Lee 2006). Figure 4 shows the relationship between the pile spacing and the shielding effect coefficient. It can be seen from Figure 4 that the shielding effects of each pile in the pile group are different under different pile spacings. The pile spacings from large to small are 3 m, 4 m, 5 m, and 6 m, respectively. In the same pile group, the shielding effect of corner piles is the weakest, which shows negative skin friction is the largest, followed by side piles, and the center pile is the strongest. It can also be known that the shielding effect of the pile group gradually decreases with the increase in pile spacing. When the pile spacing increases to 6 times the pile diameter, the shielding effect of the pile group is very small.

3.2 *Influence of collapsible degree on negative friction and neutral point of pile groups*

Based on the collapsible displacement of saturated soil, the soil displacement for different collapsible degrees can be determined, and thus the soil elastic modulus for different collapsible degrees can be calculated. The effect of changing the collapsible degree on the negative friction of pile groups in loess is analyzed and the negative skin friction curves of pile groups with different collapsible degrees are shown in Figure 5.

It can be seen from Figure 5 that the negative skin frictions of the pile at each position in the pile group increase with the increase in the collapsibility and the peak position of negative skin friction also changes greatly. Taking the center pile for an example, the maximum negative skin frictions at the collapsibility of 25%, 50%, 75%, and 100% are 12.1 kPa, 19.4 kPa, 24.9 kPa, and 39.4 kPa, respectively. The peak positions of negative skin friction are at

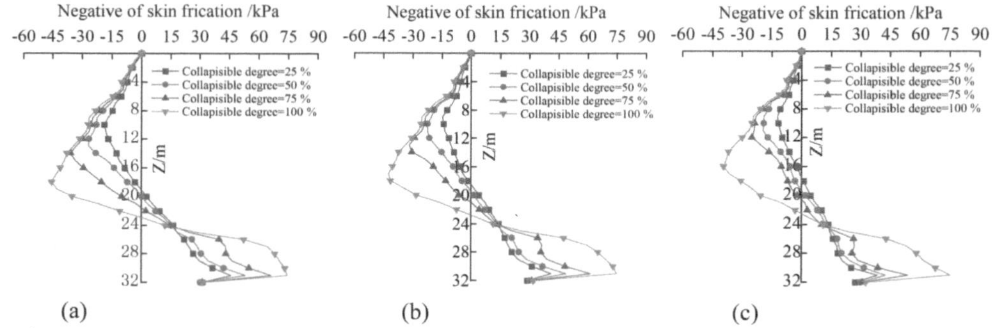

Figure 5. Variable curves of Nsf of pile group with different collapsible degrees. (a) Nsf of corner pile. (b) Nsf of side pile. (c) Nsf of center pile.

9 m, 11 m, 12 m, and 16 m, respectively. When the degree of collapsibility increases, the self-weight collapsibility of the soil around the pile increases, so the negative skin friction of the pile increases significantly and the peak position of the negative skin friction and the neutral point moves down obviously.

It can be seen from Figure 6 that the location of the neutral point also deepens with the increase in the degree of collapsibility. At the four degrees of collapsibility of 25%, 50%, 75%, and 100%, the distribution ranges of negative skin friction of the center pile are from 0 to 17.4 m, 0 to 18.8 m, 0 to 20.5 m, and 0 to 22.5 m, respectively. This is because the vertical displacement of the soil around the pile also increases with the increase in the degree of collapsibility of the soil around the pile. However, the displacement of the soil around the pile changes more than that of the pile, which leads to the location of the neutral point with the increase in the collapsible degree increases. It can also be seen from Figure 6 that the depth of the neutral point of the corner pile, side pile, and center pile gradually approaches with an increase in the degree of collapsibility.

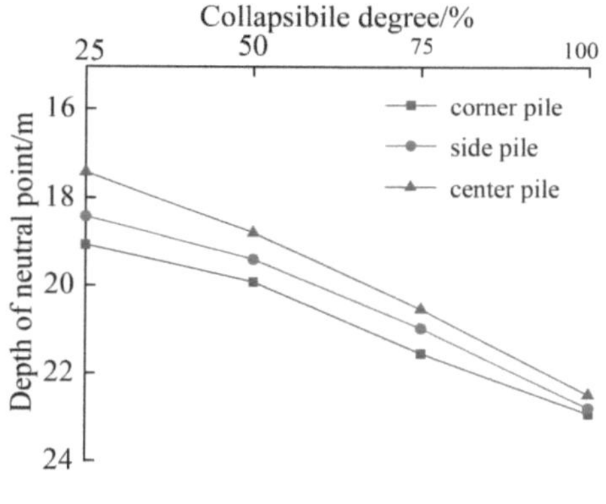

Figure 6. The position of the neutral point with different collapsible degrees.

3.3 *Influence of self-weight collapsible thicknesses on negative friction and neutral point of pile groups*

The curves of negative skin friction of the pile group with different self-weight collapsible loess layer thicknesses are shown in Figure 7.

It can be seen from Figure 7 that the negative skin friction and the peak position of the negative skin friction of each pile of pile group increase obviously with the increase in the

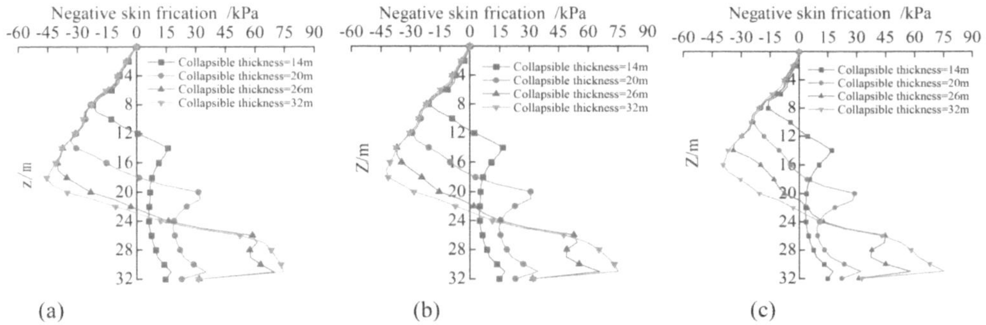

Figure 7. Variable curves of Nsf of group piles with different collapsible thicknesses. (a) Nsf of the corner. (b) Nsf of side pile. (c) Nsf of center pile.

collapsible thicknesses. Taking the corner pile for an example, when the collapsible thickness is 14 m, the maximum negative skin friction and the peak position are 22.9 kPa and 8 m. When the collapsible thickness is 20 m, the maximum negative skin friction and the peak position of the corner pile are 33.5 kPa and 13 m. When the collapsible thickness is 26 m, the maximum negative skin friction and the peak position of the corner pile are 40.1 kPa and 16 m. When the collapsible thickness is 32 m, the maximum negative skin friction and the peak position of the corner pile are 45.5 kPa and 18 m. Since the collapsibility of the soil mass is small when the collapsible thickness is small, the collapsibility of the soil mass increases with the increase in the collapsible thicknesses. As the immersion depth increases, the negative skin friction continues to develop along the pile body, and the negative skin friction and peak position change significantly.

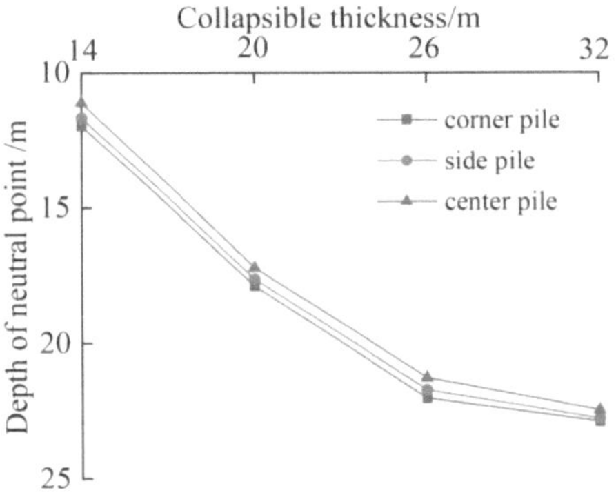

Figure 8. The position of the neutral point with different collapsible thicknesses.

It can be seen from Figure 8 that the collapsible thickness has a great impact on the neutral point position. Taking the center pile as an example, the neutral point depths are 10.9 m, 17.1 m, 20.7 m, and 22.5 m at the collapsible thicknesses of 14 m, 20 m, 26 m, and 32 m, and the displacement depths of the neutral point position are 6.2 m, 3.6 m, and 1.8 m. It can be seen that the neutral point position of each pile moves downward with the collapsible thicknesses. When the collapsible thicknesses increase, the soil collapsibility increases significantly and the neutral point position moves down significantly. However, when the collapsibility of deep soil is small, the collapsibility also decreases, and the displacement of the neutral point position decreases with the increase in collapsible thicknesses.

4 CONCLUSION

The following conclusions can be drawn from the numerical analysis of the pile group model in collapsible loess:

1. At the same pile spacing, the negative skin friction of the corner pile develops the fastest, followed by the side pile and the smallest of the center pile. The neutral point of the center pile is shallower than the side pile and corner pile. The center pile is most affected by the Pile group effect, the side pile is the second, and the corner pile is the least. With the increase in pile spacing, the negative skin frictions of each pile increase. When the pile spacing is 6 times the pile diameter, the pile group effect is already very small, and the value and distribution of the negative skin friction of the corner pile and the location of the neutral point are almost close to that of a single pile.
2. When the degree of collapsibility of soil around the pile increases, the skin friction of the pile increases, and the negative frictions of the corner pile and side pile increase rapidly. Since the center pile is subject to the pile group effect, the negative skin friction of the pile develops slowly. The neutral point position of the pile foundation gradually deepens with the increase in collapsibility.
3. The negative skin friction of the pile increases with the increase in the collapsible thicknesses, the peak position of the negative skin friction moves down, and the position of the neutral point gradually deepens obviously. At the same collapsible thicknesses, since the pile group effect is different, the locations of the peak negative friction of each pile in the pile group are different. The peak of negative skin friction of the corner pile is the largest, followed by the side pile and the center pile being the smallest.

REFERENCES

Chen Xinze, Hu Xin, Zhang Xihong. Experimental Study on Negative Friction of Pile Foundation in Collapsible Loess Area [J]. *Electric Power Survey and Design*, 2015 (01): 11-16-19.

Du Siyi, Shi Lei. Research on the Mechanical Characteristics of Pile Groups under Vertical Load [J]. *Journal of Zhengzhou University (Engineering Edition)*, 2015, 36 (04): 67–71.

Fu Guihai, Wei Limin, Deng Zongwei, Jiang Jianqing. Three-dimensional Numerical Analysis of the Influence of Pile Top Load on Negative Friction Behavior [J]. *Journal of Underground Space and Engineering*, 2015, 11 (02): 456–461.

Huang Xuefeng, Yang Xiaohui, Yin He, Liu Zilong, Zhou Junpeng. Study on the Relationship between the Lower Limit Depth of Collapsibility and the Neutral Point Position of Pile Foundation in Collapsible Loess Site [J]. *Geotechnical Mechanics*, 2015, 36 (S2): 296–302.

Lee, C. J., Lee, J. H., Jeong, S. The Influence of Soil Slip on Negative Skin Friction in Pile Groups Connected to a Cap [J]. *Géotechnique*, 2006, 56(1):53–56.

Li Dazhan, He Yihua, Sui Guoxiu. Research on Q_2 Loess Large Area Immersion Test [J]. *Journal of Geotechnical Engineering*, 1993 (02): 1–11.

Wang Changdan, Wang Xu, Zhou Shunhua, Wang Binglong. Centrifugal Model Test of Self-weight Collapsible Loess and Single Pile Negative Friction [J]. *Journal of Rock Mechanics and Engineering*, 2010, 29 (S1): 3101–3107.

Wen Hua. *Study on the Mechanism of Negative Friction of Rectangular Closed Diaphragm Wall Bridge Foundation in Collapsible Loess Foundation [D]*. Southwest Jiaotong University, 2008.

Yang Dezhi, Feng Shijin, Xiong Juhua, Hu Bin. Numerical Simulation of Pile Foundation in Collapsible Loess [J]. *Low-temperature Building Technology*, 2010, 32 (02): 98–100.

Zhao Ye. *Study on Negative Friction Characteristics of Piles in Thick Collapsible Loess Site [D]* Lanzhou University.

Zhu Yanpeng, Yang Kuibin, Wang Haiming, Yang Xiaohui. Preliminary Test on the Influence of Micro Immersion on Negative Friction of Pile Foundation [J]. *Journal of Geotechnical Engineering*, 2018, 40 (S1): 1–7.

Zhu Yanpeng, Zhao Tianshi, Chen Changliu. Experimental Study on the Change of Negative Friction of Pile Foundation Along Pile Length [J]. *Geotechnical Mechanics*, 2013, 34 (S1): 265–272.

Water Conservancy and Civil Construction – Oke & Ahmad (Eds)
© 2024 The Author(s), ISBN: 978-1-032-58614-4

Food security and deficit irrigation on potato production in arid regions of China

Zeyi Wang
College of Agronomy and Agricultural Engineering, Liaocheng University, Liaocheng, China
College of Water Conservancy and Hydropower Engineering, Gansu Agricultural University, Lanzhou, Gansu, China

Shouchao Yu
College of Agronomy and Agricultural Engineering, Liaocheng University, Liaocheng, China

Hengjia Zhang*
College of Agronomy and Agricultural Engineering, Liaocheng University, Liaocheng, China
College of Water Conservancy and Hydropower Engineering, Gansu Agricultural University, Lanzhou, Gansu, China

Chao Liang, Youshuai Bai & Xietian Chen
College of Water Conservancy and Hydropower Engineering, Gansu Agricultural University, Lanzhou, Gansu, China

ABSTRACT: As a crucial food crop, potato is of great strategic significance to ensure national food security. In the face of extreme global climate change and frequent drought disasters, the search for water-saving and high-yielding cultivation techniques for potatoes is a hot spot and focus of current research. This paper summarizes representative reports on the application of the Deficit Irrigation (DI) technique in different potato-growing regions in China. By analyzing the effects of water deficit on yield, water use efficiency (WUE), and quality at different stages of potato growth, it provides theoretical support for efficient potato cultivation and promotes the widespread application of DI technology in potato crop production.

1 INTRODUCTION

Food security (FS) is the access of people to sufficient, safe, and nutritious food to meet their food needs and preferences at all times through physical, social, and economic measures to enjoy a healthy life (FAO 2021). The issue of FS is a matter of national livelihoods and global development. Since 2015, the FS situation in the world has been reversed, food insecurity has become more acute, and the number of people at risk, especially the lives and livelihoods of vulnerable groups, has increased dramatically around the world. And since early 2020, the Newcastle pneumonia outbreak has continued to spread globally, further exacerbating the global food insecurity problem (Zhu *et al.* 2021). FS is not only about people's well-being but also about sustainable economic and social development, which is one of the core interests of the country.

*Corresponding Author: 596088683@qq.com

DOI: 10.1201/9781003450818-86

The potato industry is the backbone of agricultural production. Potato (*Solanum tuberous* L.) has the advantages of drought tolerance, cold tolerance, and adaptability, and has become the fourth largest food crop after rice, wheat, and maize (Liu *et al.* 2021), as well as a high-quality crop for both vegetables, feed, and energy. Therefore, it is widely grown in countries around the world. Currently, the total area under potato cultivation worldwide amounts to 18.33 million ha, with a total production of nearly 300 billion kg (Luo *et al.* 2022). Among them, China is the largest potato cultivation area and production, with a perennial cultivation area of 5.333 million ha and a total production of more than 90 million t as of 2018, accounting for 30% and 28% of the world, respectively (Wang *et al.* 2022). The potato is a shallow-rooted crop that is sensitive to changes in soil moisture and low water content can lead to a decrease in potato yield and quality (Men & Liu 1995). In addition, the scale of potato cultivation in China has been expanding over the years. The traditional large furrow irrigation model with a large volume of water in a single irrigation has resulted in a serious waste of irrigation water and low water use efficiency. Therefore, there is an urgent need to find efficient water management measures for potato cultivation to ensure FS, revitalize the rural regional economy, and enhance farmers' economic income in China (Huang 2021).

The world's available water resources are becoming increasingly scarce. Water for agriculture has been squeezed, especially for irrigation, which accounts for a large proportion of agricultural water consumption. Thus, there is an urgent need to find efficient irrigation techniques that can save water and improve quality with guaranteed potato yields. Eventually, the WUE will be enhanced to achieve drought resistance and increased effectiveness.

2 WATER-SAVING AGRICULTURE AND FS

2.1 *Deficit Irrigation (DI) profile*

As a predominantly agricultural country, China is faced with frequent climate extremes (especially drought problems) and an increasing shortage of available water resources. For this reason, vigorously promoting the construction of water-saving agriculture is a sure way to ensure healthy and sustainable agricultural development.

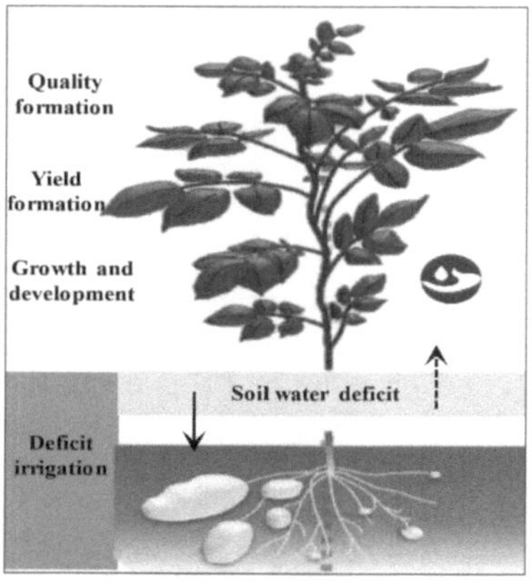

Figure 1. Diagram of deficit irrigation potato.

DI belongs to non-sufficient irrigation. Its theory was proposed in the 1970s as a mechanism for crop response under water stress. At the beginning of this century, it was concluded that DI could save water and regulate quality, Then it was combined with crop micro-irrigation technology to form the theory of collaborative water and fertilizer management (Mpelasoka, Behboudian & Green 2001). Today, the integration of big data and artificial intelligence has developed into the theory of data assimilation, which enables precise field moisture management and precise water and fertilizer management on a large regional scale. Therefore, it shows that rational irrigation control during potato production can achieve drought resistance, water conservation, efficiency, and quality regulation (Figure 1).

2.2 *Water-saving agriculture addresses the need for FS*

China is one of the world's water-poor countries, with a large proportion of water used in agriculture, low efficiency in the use of irrigation water, and a serious waste of water resources. Therefore, there is a large potential for water conservation in agriculture. Constructing high-efficiency water-saving agriculture cannot only effectively alleviate the conflicts between the supply and demand of agricultural water resources but also enhance the efficiency of agricultural irrigation water utilization and achieve stable or even increased production. As a proven irrigation technique, DI can both save water and improve efficiency and quality, which is very important to ensure FS.

3 EFFECTS OF DI ON POTATO PRODUCTION

3.1 *Water consumption*

The growth stages of potatoes are mainly divided into the seedling (SG), tubers formation (TF), tubers expansion (TE), and starch accumulation stages (SA). Tian *et al.* (2011) showed that the water requirement of potato seedlings in Liaoning was 10%–15% of the water requirement of the whole growth cycle, 23%–28% of the water requirement of TF, 45%–50% of the water requirement of TE, and 10% of the water requirement of SA. The study on soil water content during different growth periods showed that the optimum soil water capacity was 65% at SG, 75% at TF, 80% at TE, and 60%–65% at SA (Tian *et al.* 2011). Geng *et al.* (2019) showed that the water consumption modulus of potatoes in central Ningxia was about 20% at SG, 25% at TF, 35% at TE, and 20% at SA (Geng *et al.* 2019). Relevant studies are shown in Table 1.

Table 1. Statistics of reported studies on potato deficit irrigation (DI) in dry areas of China.

	FW		Y		WUE		IWUE		Q	
References/Province	CK	DI	CK	DI	CK	DI	CK	DI	CK	DI
Pan *et al.* / Qinghai	260.0	107.9	–	–	–	–	–	–	–	–
Hu *et al.* / Hebei	580	150.0	6.74	2.06	2.19	1.31	–	–	–	–
Cao *et al.* / Dingxi	710	680	3.70	3.52	15.26	21.59	–	–	–	–
Pan *et al.* / Zhangye	–	–	3.49	3.30	6.17	6.91	7.94	9.38	55.60	67.58
Liu *et al.* / Wuwei	–	–	3.50	3.25	6.95	7.46	9.24	10.46	61.52	68.73

Note: FW = fresh potato weight (g); Y = potato productivity ($t{\cdot}ha^{-1}$); WUE = water use efficiency ($kg{\cdot}m^{-3}$); IWUE = irrigation water use efficiency ($kg{\cdot}m^{-3}$); Q = quality (Starch, %); CK = full irrigation; DI = deficit irrigation; FC = Field capacity; and empty cells (dashes) = not reported.

3.2 *Water productivity and yield*

Water scarcity has become a major constraint on agricultural development in arid areas and irrigation is the most immediate solution. Pan *et al.* (2022) showed that soil water stress during potato tuber development led to a significant reduction in potato yield. The greater the degree of stress, the lower the yield (Pan *et al.* 2022). While Hu *et al.* (2021) showed that WUE decreased by 4.34%–40.15% and economic yield decreased by 25.59%–81.49% compared to CK in all treatments after water stress. However, there was a certain compensatory effect after rehydration (Hu *et al.* 2021). Cao *et al.* (2019) concluded that under mild DI, the yield per hectare of 'Green potato 9' had a compensatory effect, compared to normal irrigation, with increases of 22.79% and 11.71%, 41.48% increase in WUE, and 60.05% increase in irrigation efficiency (Cao *et al.* 2019).

3.3 *Quality*

With the improvement in living standards, consumers have gradually paid attention to the nutritional balance of food, and water conditioning has been considered the safest measure to improve crop quality. Pan *et al.* (2021) concluded that potato yield was not significantly different from the control when treated with a mild deficit at SG. DI at SG or TF improved potato quality with the most significant DI at SG and a detrimental effect on quality at the TE (Pan *et al.* 2021). Liu *et al.* (2018) showed that a moderate water deficit during the TF could significantly improve the potato fruit starch content, while water deficit at the TE and SA was not conducive to quality improvement (Liu *et al.* 2018).

4 RESULTS

In arid areas, water productivity can be maximized by DI, which is more economically beneficial to farmers than maximizing yield and is beneficial to production and ecology. For example, DI can maintain lower ambient moisture compared to full irrigation and effectively control the damage of mycorrhizal diseases (Cicogna *et al.* 2005). However, the active DI during the phenological period can reduce the leaching of nutrients in the surface layer, which protects the groundwater environment to a certain extent and reduces the input of agricultural fertilizer (Ünlü *et al.* 2006). If both water deficit and fertilizer application are considered, deficit irrigation is preferred (Fox & Rockström 2000, 2003). In addition, studies have shown that DI can affect the length of crop sowing date and growth cycle, thereby alleviating the water supply conflict in the region during the peak period of water demand (Corbeels *et al.* 1998; Farré & Faci 2006).

Of course, DI also has corresponding disadvantages. First of all, the application of timely and moderate water deficit is the premise of stable and even increased yield. However, it is difficult to determine the duration and degree of effective DI for different crops or the same crop, which needs years of basic irrigation experiments and research. Secondly, it is difficult to determine the specific critical period of water demand for crops in actual agricultural production (Geerts & Raes 2009). The irrigation technique is not easy to be mastered by the public. Meanwhile, for saline-alkali land, the application of DI will prevent the leaching of salts trapped in the cultivated land and may aggravate the saline stress of crops.

In the current study on potatoes, water deficit is mainly set by soil water content, which is generally considered to be 55%–65% of field capacity as a mild deficit, and the above work also shows that mild water deficit at potato SG has little effect on tubers yield and water productivity is improved. In addition, the water deficit at SG and TF can significantly improve potato quality.

5 CONCLUSION

Drought is a key factor restricting potato large-scale production. Potato is grown in China mainly in arid and semi-arid cooler regions, most of which lack irrigation, and even in irrigated growing areas, short-term droughts often occur due to water shortages. Therefore, there is an urgent need to study the response of potatoes to stresses, such as reduced water availability in different climatic zones, to identify similarities and differences in the response of the same crop to stress, as well as threshold indicators of response to stress, which have important theoretical and practical implications for improving potato water management and WUE.

ACKNOWLEDGMENTS

This work was financially supported by the Industrial Support Plan Project of the Gansu Provincial Department of Education (No. 2022CYZC-51), the National Natural Science Foundation of China (No. 52269008), and the Key Research and Planning Projects of Gansu Province (No. 18YF1NA073).

REFERENCES

Cao, Z. P., Liu, Y. H., Zhang, X. J., Shen, B. Y., Qin, S. H., Liu, Z., Wang, L., Li, C. Z., and Zhang, J. L. 2019. *Effect of deficit Irrigation on Growth Yield and Water Use of Potato Transactions of the Chinese Society of Agricultural Engineering*. 35. 114–123.

Cicogna, A., Dietrich, S., Gani, M., Giovanardi, R., and Sandra, M. 2005. *Use of Meteorological Radar to Estimate Leaf Wetness as Data Input for Application of Territorial Epidemiological Model (Downy Mildew-Plasmopara Viticola) Physics and Chemistry of the Earth*, Parts A/B/C. 30. 201–207.

Corbeels, M., Hofman, G., and Van Cleemput, O. 1998. Analysis of Water Use by Wheat Grown on Cracking Clay Soil in a Semi-arid Mediterranean Environment: Weather and Nitrogen Effects. *Agric. Water Manage.* 38. 147–167.

Farré, I. and Faci, J. M. 2006. Comparative Response of Maize (Zea mays L.) and Sorghum (Sorghum bicolor L. Moench) to Deficit Irrigation in a Mediterranean Environment. *Agric. Water Manage.* 83. 135–143.

FAO. 2021. The State of Food and Agriculture 2021: Making Agri-food Systems more Resilient to Shocks and Stresses https://www.fao.org/3/CB4476EN/online/CB4476EN.html last accessed 2022/9/3.

Fox, P. and Rockström, J. 2000. Water-harvesting for Supplementary Irrigation of Cereal Crops to Overcome Intra-seasonal Dry-spells in the Sahel. *Phys. Physics and Chemistry of the Earth, Part B: Hydrology, Oceans, and Atmosphere*. 25. 289–296.

Fox, P, and Rockström, J. 2003. Supplement Irrigation for Dry-spell Mitigation of Rainfed Agriculture in the Sahel *Agric. Water management*. 61. 29–50.

Geng, H. J., Yin, J., Wu, J., and Liu, Y. Z. 2019. *Effect of Different Irrigation Amounts on Growth, Water Consumption, and Yield Quality of Potato Water Conservation Irrigation*. (03). 43–47 + 58.

Geerts, S. and Raes, D. 2009. Deficit Irrigation as an On-farm Strategy to Maximize Crop Water Productivity in Dry Areas *Agric. Water Manage.* 96 1275–1284.

Geerts, S., Raes, D., Garcia, M. *Crop Water Use Indicators to Quantify the Flexible Phenology of Quinoa (Chenopodium Quinoa Willd.) in Response to Drought Stress Field Crops Res.* 108. 150–156.

Hu, M. M., Zhang, J. Z., Zhang, L. F., Liu, Y. H., and Huang, P. J. 2021. *Effects of Water Stress and Rehydration on Growth, Development, and Yield of Potato Agricultural Research in Arid Regions*. 39. 95–101 + 121.

Huang, A. Z. 2021. *Potato Industry Development Status and Countermeasures Agricultural Development and Equipment*. (08). 42–43.

Liu, P. L., Zhou, Y., and Zhang, W. J. 2021. Research on Production Efficiency and its Influencing Factors in the Main Potato-producing Areas of China in the Context of Staple Iodization. *Journal of Agronomy, Yanbian University*. 43. 93–100.

Luo, Q. Y., Lun, R. Q., Gao, M. J., and Liu, Y. 2022. *Strategic Paths for High-quality Development of China's Potato Industry From 2021 to 2025 China Agricultural Resources and Zoning*. 43. 37–45.

Liu, J. Y., Jia, S. H., and Liang, Z. E. 2018. *Effect of Oasis Sub-membrane Drip Irrigation to Adjust Deficit on Potato Growth and Quality People's Yellow River*. 40. 152–156.

Men, F. Y. and Liu, M. Y. 1995. *Potato Cultivation Physiology* (Beijing: China Agricultural Press).

Mpelasoka, B. S., Behboudian, M. H., and Green, S. R. 2001. *Water Use, Yield and Fruit Quality of Lysimeter-grown Apple Trees: Responses to Deficit Irrigation and to Crop Load Irrigation Science*. 20. 107–113.

Pan, N., Su, W., Zhou, Y., and Wang, J. 2022. *Effects of Soil Water Stress on Growth and Development, Photosynthetic Characteristics and Yield of Potato Hebei Agricultural Science*. 26. 70–75 + 94.

Pan, X. F., Zhang, H. J., Deng, H. L., and Li, F. Q. 2021. Effects of Deficit-regulated Irrigation on Potato Growth, Yield, and Quality at Different Fertility Stages in Hexi Oasis. *Agricultural Engineering*. 11. 130–136.

Tian, Y., Huang, Z. G., and Yu, X. Q. 2011. *Experimental Study on the Water Requirement of Potato Modern Agricultural Science and Technology*. (08). 91–94.

Ünlü, M., Kanber, R., Şenyigit, U., Onaran, H., and Diker, K. 2006. Trickle and Sprinkler Irrigation of Potato (Solanum tubersoum L.) in the Middle Anatolian Region in Turkey *Agric. Water manage.* 79. 43–71.

Wang, S. G., Lv, H. Z., Lu, T. Q., He, Y. K., Lv, C. X., and Yang, B. N. 2022. *Current Situation and Suggestions for the Development of the Potato Processing Industry in China Agricultural Engineering*. 12. 76–79.

Zhu, J., Zang, X. Y., and Li, T. X. 2021. China's food Security Risk and its Prevention Under the New Development Pattern. *China Rural Economy*. (09). 2–21.

Water Conservancy and Civil Construction – Oke & Ahmad (Eds)

Study on construction technology of cutoff wall in overflow weir section of sandy geological river

Xiaohua Feng*
China Railway 17th Bureau Group 3nd Engineering Co., Ltd., Shijiazhuang Hebei, China

ABSTRACT: Building a beautiful ecological water environment in urban rivers is essential. To ensure the performance of the cut-off wall, this paper studies the construction technology of the cut-off wall in the overflow weir section of the sandy geological river. In this paper, combined with the construction situation of the cut-off wall in river engineering, the raw materials of asphalt concrete under the river in the sandy section are allocated. According to different purposes, four raw material mix ratios are obtained by combining water-cement ratio, cement, water, natural sand, gravel, water-reducing agent, and sand ratio. In this paper, the construction technology of the cutoff wall is studied and the slot sections are divided. We prepare for river construction, build the cutoff wall, pour concrete of overflow weir, check the quality of the wall, and correct the deviation of hole position caused by the seepage flow of the cut-off wall. The quality of the cut-off wall is tested and the quality of six marked points is tested through the center temperature, wall width error, wall thickness error, and flatness of the cutoff wall center line. The test results of all test points are within the quality standard, which shows that the cut-off wall construction technology can be applied in practice. This method can ensure the engineering quality of the cut-off wall under a sandy geological river channel.

1 INTRODUCTION

There will be urban rivers in any city, which is also a primary component of the survival and development of urban residents. However, with the continuous development of the economy, the pace of urban modernization is gradually accelerated and the population density within the city is increasing. These reasons lead to increasing demand for urban water resources. Once a flood disaster occurs in the city, the threat to the local area becomes increasingly serious. Therefore, people began to set up water conservancy projects, such as flood control and drainage in city river sections. At the same time, people combined aesthetic ideas with the construction of safer and more comfortable ecological water environments in the rivers inside the city. The cut-off wall is the product of this concept. Cut-off walls, also known as continuous underground walls, are a kind of anti-seepage technology that integrates concrete pouring technology. In many buried river overburden layers, people often set up anti-seepage devices upstream and drainage decompression devices downstream to prevent river overflow. The weir is a kind of facility born from this phenomenon (Lv 2022). However, in sandy geology, the soil viscosity is poor and the construction of the overflow weir section becomes much more complicated than in other areas. Concrete pouring is the main form of cut-off wall construction. To ensure the quality of river construction and

*Corresponding Author: 5135107@qq.com

DOI: 10.1201/9781003450818-87

prevent flood disasters, this paper studies the construction technology of cut-off walls in the overflow weir section of a sandy geological river.

2 PROJECT OVERVIEW

The river where the project is located is around a reservoir in a county in China. The total capacity of the reservoir is about 315 million m^3, the length of the seepage control wall in the first section is about 1,356 m, the area of the wall is about 25,637 m^2, and the area of the wall is 26,748 m^2. All the facilities of the project should be constructed according to the target of the Class II project. The design flood standard is one in 100 years and can withstand an earthquake of intensity VIII (Wu 2021). In the project, the total length of the seepage control wall is about 6,354 m and the maximum depth of downward extension is about 41 m. The thickness of the seepage control wall is divided into two categories, 60 cm for the first category and 80 cm for the second category.

3 RAW MATERIAL ALLOCATION OF ASPHALT CONCRETE UNDER SAND SECTION RIVER CHANNEL

There are specific requirements for the quality of asphalt concrete when configuring the cut-off wall materials. The configured asphalt concrete needs to meet the anti-seepage requirements of the cut-off wall and have sufficient deformation capacity. The configured asphalt concrete needs to meet the anti-rheological standard of river overflow weir, keep enough stability for high-density water flow, be underwater for a long time, and have enough durability (Zhang, Zhang, He 2021). When configuring raw materials of asphalt concrete, asphalt materials with enough penetration index should be selected. In the asphalt mixture, the proportion of asphalt is about 10%–16%. The adhesive property of asphalt material needs to be excellent enough and its adhesive force needs to reach Level 4 (Wang *et al.* 2020). The bulk density ratio of asphalt concrete should be greater than 2.2 t/m^2, and the porosity of external inspection should be at most 4%. In contrast, the indoor sampling inspection should be at most 3%.

Asbestos cement should be selected as the cement in raw materials and its fineness should be less than or equal to 10.0%. The stability of the selected asbestos cement must be qualified, the density should be below 5g/m^3, the loss on ignition should be less than or equal to 5%, and the content of sulfur trioxide should be less than or equal to 4%. The initial setting time of the selected cement should be at most 50 minutes and the final setting time should be at least 10 hours. The compressive strength of the cement needs to be greater than or equal to 18.0 Mpa on Day 3 and 52.3 Mpa on Day 30 (Guo *et al.* 2020), and 4.5 Mpa and 6.5 Mpa on Day 3 and Day 30, respectively. The external additive needs to be selected as DH-9 air-entraining agent and the rate of taxation needs to be greater than or equal to 12%. The pH value needs to be determined between 7 and 9. The selected cement needs to have a compressive strength ratio that is greater than or equal to 140, 130, 120, and 115 on Days 1, 3, 7, and 30, respectively (Xie 2022). The compressive strength of each material within the concrete is calculated using Formula (1).

$$Q_n = \frac{1 - Q_w}{W_s} \tag{1}$$

Where Q_n represents the compressive strength, Q_w represents the actual test density, and W_s represents molding porosity. Aggregates need to have specific quality standards in the fineness modulus. The coarse aggregates are pebbles, small stones, and medium stones. Fly ash uses coal ash products with fineness less than or equal to 20% (Li 2021). In this project, four types of concrete need to be designed with different mix ratios, as shown in Table 1.

Table 1. Mix the proportion of concrete raw materials.

Material	Class I	Class II	Class III	Class IV
Water cement ratio	1.13	1.14	1.17	1.23
Cement	130	130	130	130
Water	270	250	270	270
Natural sand	476	445	498	456
Pebble	774	756	749	788
Water reducing agent /(kg/cm)	4.60	4.75	4.65	4.75
Sand ratio/%	50	53	54	56

Based on the raw material ratio in Table 1, we design the concrete for impermeable walls to meet the acceptance criteria.

4 STUDY ON CONSTRUCTION TECHNOLOGY OF CUTOFF WALL

In the construction preparation stage, CZF-1500 percussive reverse circulation drilling rig is used as the leading equipment for the construction of the cut-off wall, which is combined with a cleaner and sand pump to form a complete construction unit. The impact-broken filter residue is repeatedly circulated in the slot hole to form a mud purification facility with ample supporting space and high power consumption. In this way, the site space immediately becomes loose and the construction cost is reduced. The overall workflow can be roughly divided into five stages, dividing the channel section, preparing for river construction, starting the cutoff wall construction, pouring concrete of overflow weir, checking the quality of the wall, etc.

In the process of slot division, it is necessary to ensure the safety of hole-making, reduce the demand for wall anti-seepage, and ensure the rationality of construction. In the cut-off wall, the thickness is about 1–1.5 m. Since the sand geological channel is prone to collapse, the average number of holes in each slot is about 5–7 (Wang *et al.* 2021). The typical slot segments divided by slots are shown in Figure 1.

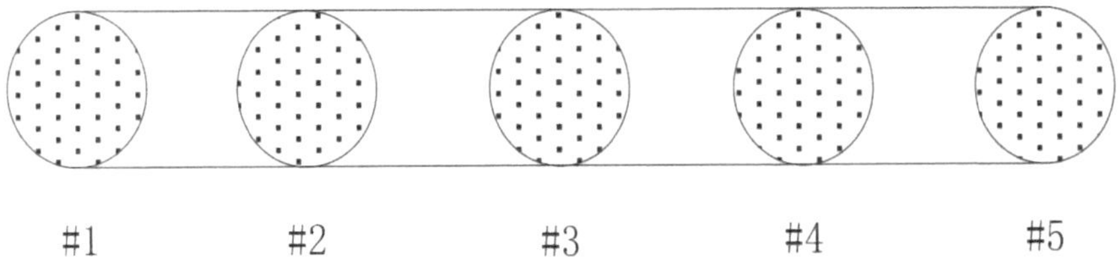

Figure 1. Typical slot division.

In Figure 1, there are five slots labeled #1, #2, #3, #4, and #5, respectively. In the case of neglecting the upstream dam, it will cause an inevitable head loss. At this time, the seepage flow of the cut-off wall can be expressed as follows.

$$Q_f = \frac{H_k}{2\sqrt{g_w}} \tag{2}$$

Where Q_f represents the seepage flow of the cutoff wall; H_k represents the permeability coefficient of the dam body; g_w represents the head loss of the downstream dam section.

The head loss borne by the cut-off wall at this time can be expressed as follows.

$$g_w = g_f - g_k \tag{3}$$

Where g_f represents the head loss of the clay core wall and g_k represents the head loss of the upstream dam foundation (Liu *et al.* 2020). In combination with the trenching construction, the center line of the guide wall must be located on the center line of the cut-off wall. During the construction of the central hole, it is necessary to strengthen the drilling efforts. People can correct the hole by repairing the angle in time when the hole position has deviated. The guide hole during construction can directly determine the quality of the single anti-arc line, so it can be combined with the construction process to control the quality of the anti-arc joint. When the deviated hole is corrected, if the angle of the guide hole cannot be corrected in time, the safety of the reaming step cannot be ensured. When reaming the hole, we should pay attention to the shaking amplitude of the wire rope. We cannot release the rope too fast and we check whether there is mud in the joint. When it is found that the entrainment leads to borehole deviation, it needs to be adjusted in time. If the deflection angle is too large, directional blasting is required to correct the deflection (Wang 2020).

The plastic concrete is poured after the manipulation and acceptance, while the joint pipe is cleared and accepted. During the pouring process, 5–7 slots are laid in the slot section and the setting degree of concrete in the slots needs to be measured at regular intervals. We measure the depth of the plastic concrete after pulling up the joint pipe. At this time, the construction integrity of the cut-off wall in the overflow weir section can be tested and the axis of the hole slot can be checked. After the construction, it is necessary to check the hole-sealing effect by water injection to ensure that the hole-sealing height and cut-off wall are in a horizontal state.

We suppose the slurry leakage occurs in the construction process due to the influence of sandy geological rivers. In that case, increasing the slurry level in the hole and groove can be ensured by increasing the reserve. We suppose the joint pipe cannot be lifted due to insufficient solidification time of plastic concrete. In that case, all the joint pipes can be pulled out in a cyclic way to realize the integrity and stability of the cut-off wall.

5 QUALITY CONTROL TESTING

For the construction technology of the impenetrable wall of the overflow weir section of the river with sandy geology designed above, the impenetrable wall of the overflow weir is established in the study area, as shown in Figure 2.

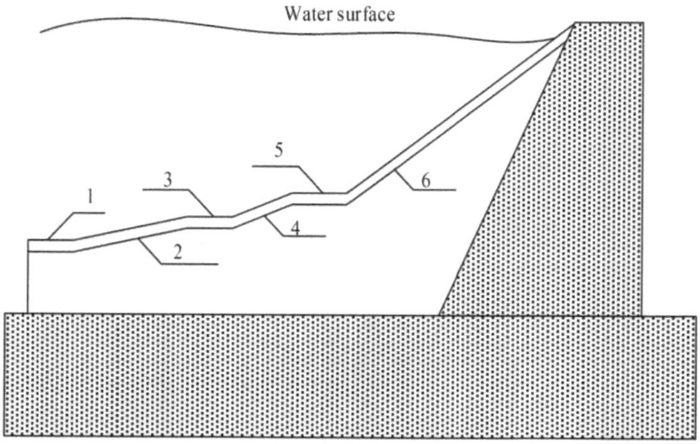

Figure 2. Schematic diagram of the impenetrable wall.

In this impenetrable wall, six marking points are places where the most concentrated forces are applied when constructing the impermeable wall. Five index indicators are set and their item names and quality control standards are shown in Table 2.

Table 2. Quality standards.

Serial number	Project name	Quality standard
1	Wall centerline	±6mm
2	Center temperature of cutoff wall	110°C–130°C
3	Wall width error	±2cm
4	Wall thickness error	±2cm
5	Planeness	\,le,1*cm*

Combined with the quality standards in Table 2, the quality control test of this project is carried out and the test results are shown in Table 3.

Table 3. Quality control results.

Marking point	Center line of wall/mm	Center temperature of cutoff wall/°C	Error of wall width /cm	Error of wall thickness /cm	Planeness/ cm
1	+3.455	127.14	+0.1144	−0.1568	0.4563
2	−2.146	124.56	+0.4544	−0.1749	0.4152
3	+3.744	125.85	−0.7637	−0.1525	0.6324
4	+2.151	121.74	+1.8255	+0.1356	0.5112
5	+3.215	112.25	+0.5412	−0.1152	0.4152
6	−3.745	124.34	+0.4554	+0.1152	0.3654

Combined with the five types of quality indicators in Table 3 and the six marking points, we can learn that the quality dismissals obtained in this example based on this impermeable wall construction technology are all within the indicator standards. The maximum center line of the wall is −3.745 mm, the temperature of the center of the impenetrable wall is between 110°C and 130°C, the maximum error of the wall width is 1.8255 cm, which is less than 2 cm, the maximum error of the wall thickness is 0.1749 cm, which is less than the standard of 2cm, and the maximum value of the flatness is 0.6324 cm, which is less than the standard of 1 cm. The technique designed in this paper works well and the impenetrable wall within the quality assurance can be obtained.

6 CONCLUSION

In summary, this paper designs a cut-off wall construction technology research method in the sandy geological river overflow weir section. In this paper, combined with the actual engineering situation, the cut-off wall construction technology under this condition is comprehensively designed. The research results can accumulate valuable theoretical knowledge for the construction of the cut-off wall of this type of river and create favorable conditions for the smooth implementation of the project.

ACKNOWLEDGMENTS

The study was supported by the Science and technology research and development plan project of China Railway Construction Corporation was supported by the "Study on the

Technology of Dredging and Reuse of Urban Rivers and Ecological Control of Water Conservancy Projects" (Grant No.19-C19).

REFERENCES

Guo, C., Yang, J., Shi, M. 2020. Improvement of Construction Equipment and Process of Polymer Ultra-thin Cutoff Wall. *Advances in Science and Technology of Water Resources*, 40 (03): 68–71 + 77.

Li, L. 2021. Analysis of Construction Technology of Concrete Cutoff Wall in Hydraulic Engineering. *Sichuan Cement*, (11): 29–30.

Liu, C., Jia, X., Wang, S. 2020. Design of Overflow Weir in Chahansulide Scenic Area of Hailiutu River. *Inner Mongolia Water Resources*, (07): 42–43.

Lv, X. 2022. Construction Technology for Building Permanent Anti seepage System on New Embankments. *Port & Waterway Engineering*, 2022(S2): 80–83 + 112.

Wang, X., Zhu, S., Zhou, T. 2020. Discussion on Construction Technology of High-pressure Jet Grouting Cut-off Wall of Qianping Reservoir. *Yellow River*, 42 (S1): 188–190.

Wang, K., Ji, Y., Wang, Y. 2021. Hydrodynamic Simulation and Erosion Analysis of Cascade Overflow Weirs at Xinquan Township Section of Yuanhe River. *Water Power*, 47 (04): 74–78.

Wang, L. 2020. Research on Key Technologies of Cutoff Wall Construction Under Strong Permeable Geological Conditions in Tidal Areas of the Estuary. *Pearl River Water Transport*, (03): 82–83.

Wu, C. 2021. Brief Discussion on the Construction of the Impervious Wall of the Cofferdam at the Inlet and Outlet of the Pumped Storage Power Station. *Yellow River*, 43 (S2): 222–223.

Xie, C. 2022. Discussion on Application and Management of Concrete Cut-off Wall Construction Technology in Water Conservancy and Hydropower Projects. *Development Guide to Building Materials*, 20 (16): 169–171.

Zhang, B., Zhang, L., He, P. 2021. Resistance to Hydraulic Fracturing of Basalt Fiber Reinforced TRD Cut-off Wall. *South-to-North Water Transfers and Water Science & Technology*, 19 (02): 400–408.

Water Conservancy and Civil Construction – Oke & Ahmad (Eds)
© 2024 The Author(s), ISBN: 978-1-032-58614-4

Research on design method of retaining structure of foundation pit near water

Kunpeng Wu, Fengshan Mao*, Junxing Luo, Mingxing Zhu & Youpeng Wen
CCCC Fourth Harbor Engineering Co., Ltd., Guangzhou, China
Key Laboratory of Environment and Safety Technology of Transportation Infrastructure Engineering, CCCC, Guangzhou, China
Southern Marin Science and Engineering Guangdong Laboratory (Zhuhai), Zhuhai, China

ABSTRACT: Different from the ordinary foundation pit, the direct water-facing foundation pit often faces more complex surrounding environment and construction conditions, such as wave action, unbalanced load on both sides, and limited construction space. In response to the above problems, relevant research has been carried out, and the deformation and internal force responses of the support structure under the change of external water level and wave load have been given. A method to determine the adjustment coefficient of the fixed point of the support under the unbalanced load has been proposed. The stiffness ratio of the truss plate support and the opposite support has been defined. The relationship between the stiffness ratio and the deformation of the support structure has been given. Based on this, the internal support can be preliminarily designed.

1 INTRODUCTION

As more and more major projects, such as cross-sea tunnels, cross-sea bridges, large wharves, and terminal buildings are put into construction, more and more new resorts, hotels, and other projects at the seaside and waterside have emerged due to the rapid development of the national economy and tourism. It is necessary to make full use of and develop underground space. A large number of waterfront foundation pits have also emerged accordingly. Due to the limitation of construction conditions and the influence of the surrounding environment, the foundation pit near water often has a riverside or seaside. During construction, it is subject to wave forces and produces asymmetric loads on both sides of the foundation pit. The unbalanced earth pressure can easily cause excessive deformation of the retaining structure towards the waterside, and even cause overall instability of the foundation pit, damage to the revetment, water gushing into the foundation pit, and other engineering accidents. The construction working surface on the other side is reduced due to the water conditions, so the beam slab support is often used as the excavation channel for waterfront pits. Under the combined action of horizontal force and vertical force, the bearing force of the support is more complex and the impact on the enclosure structure is more difficult to evaluate.

The stress conditions and design methods of foundation pit supporting structures affected by wave and other hydrodynamic forces are quite different from those of conventional foundation pits (Zhang & Gu 2020). Ying Hongwei (Ying et al. 2013) et al. analyzed the water and soil pressure distribution of the gravity retaining wall of the foundation pit in the sea area of the Gongbei Tunnel of Hong Kong Zhuhai Macao Bridge and the response of its stability to wave action. Ying Hongwei (Loehlin 1996) and others carried out foundation pit model tests using the wave flume system to study the response law of excess pore water pressure in clay silt

*Corresponding Author: wkunpeng@cccc4.com

DOI: 10.1201/9781003450818-88

foundations under wave action. Li Jian (Ying et al. 2016) et al. studied the seepage failure of foundation pit engineering in the sea environment and the seepage characteristics of the surrounding foundation. Although there have been some studies on the characteristics of supporting structure and pore water pressure of foundation pit under water level fluctuation, there are few studies on the internal force response of supporting structure under the direct action of waves. At present, most scholars use the method of direct or indirect adjustment of stiffness to calculate the bearing axial force under an unbalanced load. For example, Jiang Yan (Li & Fan 2019) calculated the asymmetric structure into a symmetric support structure by adjusting the support stiffness and iterating the stiffness to obtain the final stiffness. This method is not satisfied on the boundary where one side displacement is zero, such as the case where one side is an existing underground space structure. Jin Yabing (Loehlin 1997) proposed the concept and calculation formula of the earth pressure ratio and established the relationship between the earth pressure ratio and fixed point adjustment coefficient. This method establishes the relationship by introducing the intermediate variable earth pressure, which is tedious and does not throw away the boundary condition that does not satisfy the boundary displacement of zero. For beam plate support, the combination of side truss plate support and internal support plate support is often used to facilitate excavation and increase the excavation area of internal soil. It should be determined comprehensively according to the structural layout, service load, and other factors of the basement. However, due to the existence of unbalanced force in the foundation pit near water, its layout is closely related to the deformation and internal force of the support structure.

For the wave load, unbalanced force and the layout of beam plate internal support faced by the direct waterfront foundation pit, two water-facing foundation pits in Guangdong Province are taken as examples to analyze and propose corresponding solutions, which can provide a certain reference for similar water-facing projects.

2 INFLUENCE OF WAVE LOAD ON SUPPORTING STRUCTURE

Taking the foundation pit of a cruise ship home port as an example, the excavation depth of the foundation pit is 9.5 m, the excavation width is 100 m, and one side of the foundation pit faces the sea. The support structure of the seaside foundation pit is subjected to earth pressure, wave load, and hydrostatic pressure. The depth of the still water level at the water side of the foundation pit is 3.5-13.7 m. To obtain the response of the foundation pit under different wave loads, the static water level heights at the water side are taken as 0 m, 2 m, 5 m, 10 m, and 15 m, respectively, to calculate the response of the foundation pit support structure.

Figure 1. Site construction diagram of foundation pit near water.

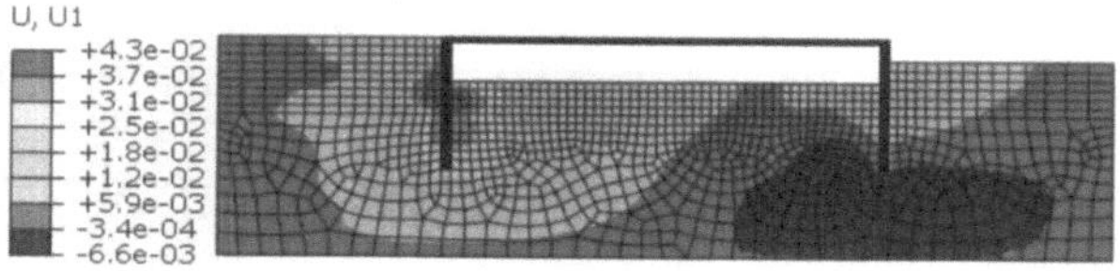

Figure 2. Schematic diagram of simplified analysis model of water adjacent foundation pit.

According to relevant literature, the average annual wave height of most sea areas near the Pearl River Estuary and offshore areas is 0.4-1.6 m and the maximum annual cycle is between 5-6 s. Therefore, different wave loads are taken for research, with wave heights of 1m and 2 m and wave periods of 5 s and 6 s. In the calculation process, the combination of water depth, wave height, and period in Table 1 is selected for analysis and the wave forces under different wave elements are obtained.

Table 1. Calculation of wave elements.

Wave height (H/m)	Wave period (T/s)	Water level depth (d/m)
1, 2	5, 6	0, 2, 5, 10, 15

To verify the accuracy of the model, an actual foundation pit model is established. The static water level depth is 3.5 m, as shown in Figure 6(a). The numerical analysis model is used to extract the deformation of the retaining wall and the results are compared with the field monitoring results of the inclinometer of the retaining wall. The overall trend of the numerical simulation and the measured curve is consistent and the relative difference between the two is within 10%.

2.1 *Analysis results*

When the depth of the static water level is less than 10 m, the bending moment of the landside retaining wall decreases significantly with the increase in the depth of the static water level. The reason is that the pressure acting on the right side of the supporting structure is small and the supporting structure as a whole deforms to the right. The maximum bending moment on the landside retaining wall is 2 m at the bottom of the pit. With the increase in the depth of the static water level, the position of the maximum bending moment gradually rises. For example, when the depth of the static water level is 0, the maximum bending moment is 2 m below the bottom of the pit. When the depth of the still water level is 10 m, the position of the maximum bending moment is 1 m above the pit bottom. When the static water level depth is 15 m, the bending moment of the landside retaining wall is equivalent to the bending moment when the static water level depth is 0. This is because when the static water level depth is within the interval 0.5L > d > 0.2L (where L is the wavelength), the deeper the static water level depth is, the greater the wave load is.

Since the sea-side retaining wall is directly subjected to wave load, the deformation mode is more complex, as shown in Figures 3–4. With the increase in the static water level, the bending moment of the retaining wall gradually transits from the "C" curve to the "S" curve.

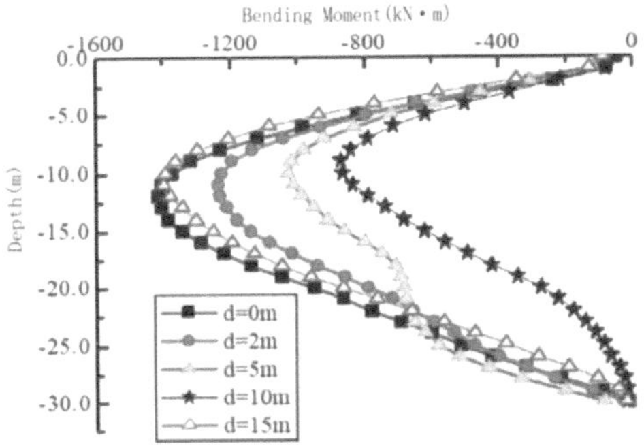

Figure 3. Bending moment of the landside retaining wall.

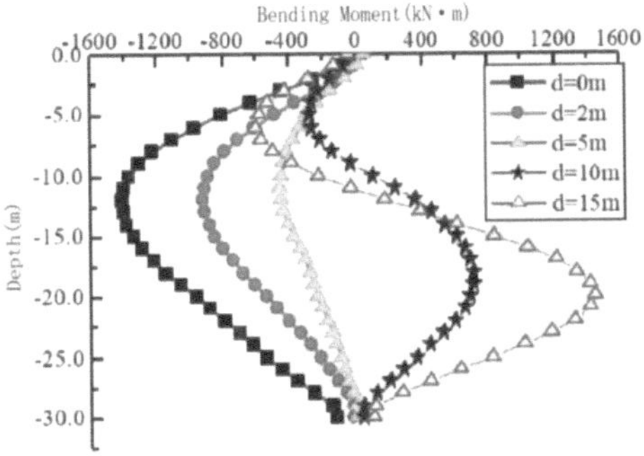

Figure 4. Bending moment of the seaside retaining wall.

The maximum bending moment decreases first and then increases. When the bending moment is the "C" curve, the retaining wall is under tension at the inner side of the pit. When the bending moment changes to the "S" curve, the bending moment varies with depth as a hyperbolic curve. The retaining wall is tensioned from top to bottom on the inner side of the pit, and then on the water side of the retaining wall.

When the depth of the still water level is between 0 m and 10 m, the horizontal displacement of the retaining wall toward the seaside gradually increases. When the depth of the still water level changes from 10 m to 15 m, the horizontal displacement increment of the retaining wall is very small. According to the bending moment of the retaining wall, when the depth of the static water level is 10 m, the horizontal displacement of the retaining wall at the seaside is the minimum and the bending moment of the retaining wall at the seaside is the minimum. When the depth of static water level is 15 m, the horizontal displacement of the retaining wall is deformed towards the outside of the foundation pit. The absolute value of the bending moment is relatively large due to the large change in the curvature of the retaining wall. This shows that the response of the foundation pit near water under wave load is more complex than that of the foundation pit not near water. The structural stiffness should be selected according to different water depths and wave conditions when designing the foundation pit.

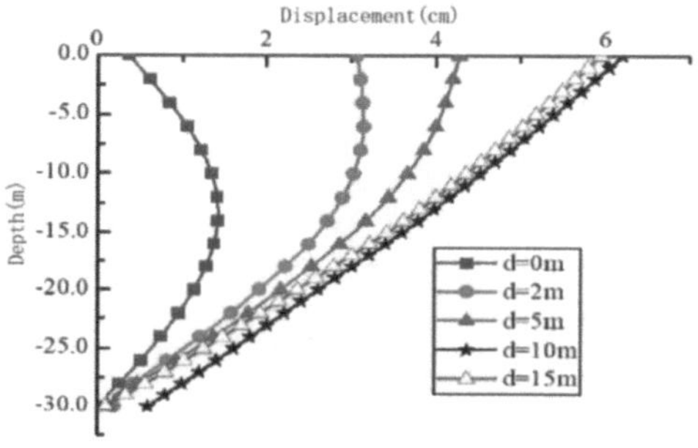

Figure 5. Displacement of landside retaining wall.

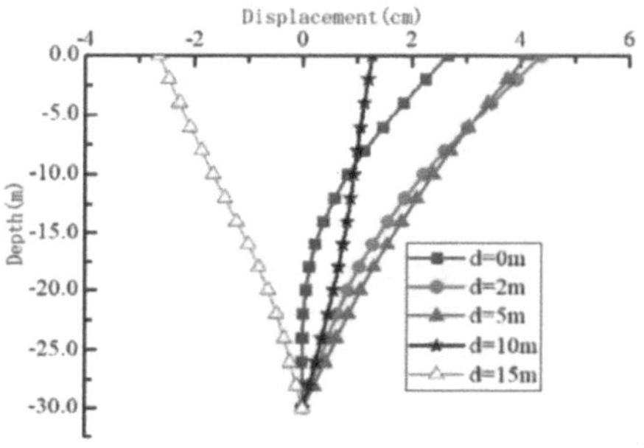

Figure 6. Displacement of sea side retaining wall.

In this project, the water level depth at the waterside is 2-10 m. The lattice diaphragm wall is used at local water level depths up to 13.7 m. The lateral stiffness is increased to reduce the effect of unbalanced force and ensure the stability and reliability of the foundation pit.

3 CALCULATION OF SUPPORT AXIAL FORCE UNDER AN UNBALANCED LOAD

3.1 *Problem description*

Calculating the internal force of the support structure requires the elastic support point stiffness coefficient, and the elastic support point stiffness system is closely related to the fixed point adjustment coefficient. According to Technical Specification for Building Foundation Pit Support (JGJ 120-2012), the soil property, depth, peripheral load, and other conditions of supporting two opposite side foundation pits are similar and the fixed point adjustment coefficient is taken as $\lambda = 0.5$. If the soil property, depth, peripheral load, and other conditions, or excavation time of the two opposite foundation pits are different, the value of the side with large soil pressure or the side excavated first is λ= 0.5-1.0, 1 is for the side with low earth pressure- λ. However, the calculation method of specific values is not clear and often relies on experience in engineering. Some scholars also define the earth pressure ratio according to the loads on both sides of different sizes and establish the relationship between the earth pressure ratio and the fixed point adjustment coefficient to determine the fixed point adjustment coefficient. It cannot be calculated under specific boundary conditions. For example, one side of the foundation pit is close to the existing underground structure and the earth pressure ratio cannot be calculated when there is no earth pressure on one side. This problem also exists when the definition of stiffness is directly used for calculation. According to the definition of fixed point, this method uses displacement as the adjustment coefficient of fixed point, which is more intuitive and accurate, avoids the error caused by introducing intermediate variables, and proves the applicability under specific boundary conditions.

3.2 *Fixed point adjustment coefficient calculation*

According to the definition of the fixed point (the position of the fixed point on the bearing when both ends of the bearing are moved by pressure at the same time)), the calculation method using displacement as the adjustment coefficient of a fixed point is more intuitive and accurate, which avoids the error caused by introducing intermediate variables and proves the applicability under specific boundary conditions.

(1) We assume l_b and l_s are the calculated lengths of the supports on both sides of the fixed point, respectively (subscript $b$refers to the side with larger load and subscript $s$refers to the side with smaller load), δ_b and δ_s are the total displacement on both sides of the inner support, respectively, and $N$is the support axial force. The following equation can be obtained.

$$\Delta l = \frac{F_N l}{EA}, \quad \delta_b = \frac{N l_b}{EA}, \quad \delta_s = \frac{N l_s}{EA} \tag{1}$$

The definition of adjustment coefficient according to support fixed point is as follows.

$$\lambda_b = \frac{l_b}{l_b + l_s}, \quad \lambda_s = \frac{l_s}{l_b + l_s} \tag{2}$$

We substitute the above two equations to get the following equation.

$$\lambda_b = \frac{\delta_b}{\delta_b + \delta_s}, \quad \lambda_s = \frac{\delta_s}{\delta_b + \delta_s} \tag{3}$$

It can be seen that the fixed point coefficient of the inner support is determined by the displacement at both ends of the inner support.

(2) The supporting structures on both sides of the foundation pit are calculated according to the symmetrical structure by using the standard method (the fixed point adjustment coefficient is taken as $\lambda_{b1} = \lambda_{s1} = 0.5$ at this time), and the deformations on both sides of the support are δ_{b1} and δ_{s1}. The axial force on both sides of the support is obtained and the first calculation result is obtained.

(3) As the foundation pit is in an asymmetric stress state, there will be $N_{b1} > N_{s1}$. The unbalanced axial force $\Delta N_1 = N_{b1} - N_{s1}$ can be applied to the side retaining structure with less load by pre-adding axial force and the side with less load can be recalculated by using the standard formula to obtain the new displacement $\delta_{s1\Delta}$ at the support, which is the second calculation. The first and second displacements of the side with a small load are $\delta_{s1} > \delta_{s1\Delta}$ and $\Delta\delta_1 = \delta_{s1} - \delta_{s1\Delta}$. The adjustment coefficient of the fixed point of the support is modified based on the following equation.

$$\lambda_{b2} = \frac{\delta_{b1} + \Delta\delta_1}{\delta_{b1} + \Delta\delta_{s1}}, \quad \lambda_{s2} = \frac{\delta_{s1} - \Delta\delta_1}{\delta_{b1} + \Delta\delta_{s1}} \tag{4}$$

The above correction formula is checked by using specific boundary conditions: when the structure is symmetrical, $\Delta N_1 = 0$. At the same time, t $\delta_{b1} = \delta_{s1}$, here is $\lambda_{b2} = \lambda_{s2} = 0.5$, which can be seen to conform to the actual situation. When the support at the side with a small load remains unchanged (i.e., it is considered as fixed support, for example, there is an existing underground space structure at one side of the foundation pit). At this time, $\Delta\delta_{s1\Delta} = 0$, $\Delta\delta_{s1} = \delta_{s1}$, then $\lambda_{s2} = 0$. It can be seen that it conforms to the actual and specification requirements. Therefore, the fixed point adjustment coefficient correction formula meets the actual boundary conditions.

The above steps (2)-(3) are the first round of calculation.

(4) Using the standard formula and the modified fixed point adjustment coefficient λ_{b2} and λ_{s2}, respectively, we calculate the support structures on both sides and obtain the corresponding deformation δ_{b2} and δ_{s2} on both sides of the support and the axial force on both sides of the support.

(5) We determine the axial force on both sides of the support: when $\frac{|N_{b2}-N_{s2}|}{(N_{b2}+N_{s2})/2} \le 5\%$, it is considered that the difference between the axial forces on both sides of the support is small, which is consistent with the actual situation and the calculation can be ended. We take $N = (N_{b2} + N_{s2})/2$ as the axial force value, when $\frac{|N_{b2}-N_{s2}|}{(N_{b2}+N_{s2})/2} > 5\%$, it is considered that the axial force error on both sides of the support is large. According to Step ②, the unbalanced force $\Delta N_2 = N_{b2} - N_{s2}$ is applied to the side with a small load, and recalculated. The adjustment coefficient of the fixed point of the support is corrected.

$$\lambda_{b3} = \frac{\delta_{b2} + \Delta\delta_2}{\delta_{b2} + \Delta\delta_{s2}}, \quad \lambda_{s3} = \frac{\delta_{s2} - \Delta\delta_2}{\delta_{b2} + \Delta\delta_{s2}} \tag{5}$$

Steps (4)-(5) above are the second round of calculation.

(6) We repeat Steps (4)-(5). When the nth round is calculated and $\frac{|N_{bn}-N_{sn}|}{(N_{bn}+N_{sn})/2} \le 5\%$ is met, the displacement of the support under the asymmetric load is δ_{bn} and δ_{sn}, respectively. At this time, the stress deformation of the retaining structure is the final stable state and the scheme design can be carried out according to this state.

3.3 *Calculation case*

Sections 2–2 and 10-10 of the pit are selected as the calculation sections, where section 2–2 is the slope of the water surface with a direct water depth of 3.5 m extending toward the sea (1:3 slope ratio).

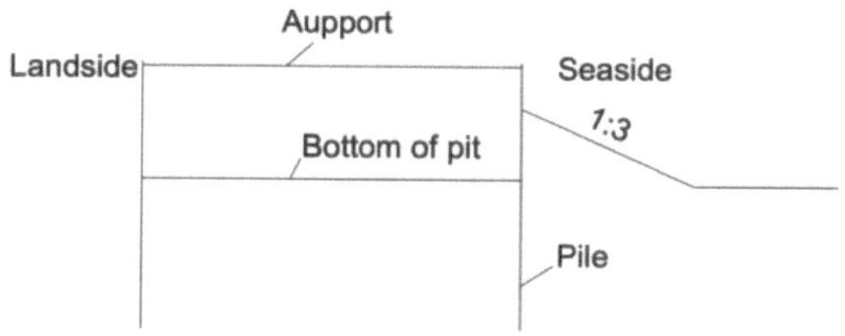

Figure 7. Section of calculation case.

Table 2. Calculation process.

Process	N_b (kN)	δ_b (mm)	Ns (kN)	δs (mm)	$\Delta\delta$ (mm)	ΔN (%)	λ_b	λ_s
STEP1	9253.37	21.72	6510.93	18.57	3.15	34.79%	0.62	0.38
STEP2	8830.93	23.2	7162.33	16.77	6.43	20.87%	0.74	0.26
STEP3	8640.31	24.63	7539.07	16.25	8.38	13.61%	0.81	0.19
STEP4	8155.26	25.45	7985.91	15.43	10.02	2.10%	0.87	0.13

The support axial force is calculated iteratively using the above method. After 4 iterations, the difference between the support axial forces is reduced to 2.1%. It can be considered that the support axial force is balanced at this time and the convergence speed is fast. At this time, the average value of 8070.59 kN of the calculated axial force on both sides is taken as the final calculated value, as shown in Table 2. According to the monitoring data of the third party of the project, the measured supporting axial forces at the measuring points ZC19 and ZC20 corresponding to the section are 8435 kN and 8439 kN, respectively, with the average value of 8437 kN, which is a close to the axial force value obtained by the method provided in this paper and the error is about 4.3%. It can be seen that this method can better calculate the supporting axial force under the unbalanced load.

4 INFLUENCE OF SUPPORT STIFFNESS RATIO ON SUPPORTING STRUCTURE

4.1 *Definition of stiffness ratio*

Since one side of the waterfront pit is adjacent to water, the construction space is reduced and the surrounding environment and geological conditions are often complex, the beam plate support is used as the excavation channel. Through the combination of internal support members, the stiffness of the support system can be improved, which is conducive to ensuring the overall safety of the pit. For the deep and large foundation pit with a large excavation area, when the bracing structure is used for support, the side truss is usually set between the braces. When the beam plate support (trestle) is used, the side truss structure can be set with plate members. At this time, the entire support system is composed of a beam plate structure.

The stiffness ratio of truss plate to support plate is introduced to describe their impact on the supporting structure and the stiffness ratio is calculated according to Equation (8).

$$K_N = \frac{EA_2}{L_2} = \frac{Eb_2h_2}{L_2} \tag{6}$$

$$K_B = \frac{384E}{8{L_3}^3} I_1 = \frac{384E}{8{L_3}^3} \frac{h_1 b_1^3}{12} \tag{7}$$

$$S = \frac{K_B}{K_N} = 4\frac{h_1}{h_2}\frac{L_2}{L_1^3}\frac{b_1^3}{b_2} \tag{8}$$

4.2 *Impact analysis of stiffness ratio*

We establish an analysis model to conduct structural analysis on the foundation pit under different stiffness ratios and set up bracing plates and truss plates, respectively. We take 16 m, 40 m, and 88 m as the net spacing of bracing plates. We adjust the width of truss plates to obtain different stiffness ratios. b2 is taken as 5 m, 5.1 m, 5.3 m, 5.6 m, 6 m, 6.2 m, 6.5 m, 6.6 m, 7 m, and 7.5 m, respectively. We set vertical loads on bracing plates and truss plates. The calculations were carried out by taking 0 kPa, 20 kPa, and 40 kPa, respectively, The established model is shown in Figure 8.

To analyze the influence of the maximum bending moment and horizontal displacement of the diaphragm wall with the change of stiffness ratio under different slab thicknesses, the bending moment and horizontal displacement corresponding to the same stiffness ratio and different vertical loads are averaged. With the increase in the truss plate brace stiffness ratio, the maximum bending moment and the maximum horizontal displacement of the diaphragm wall continue to decrease, showing a trend of rapid decrease first and then leveling off. It can

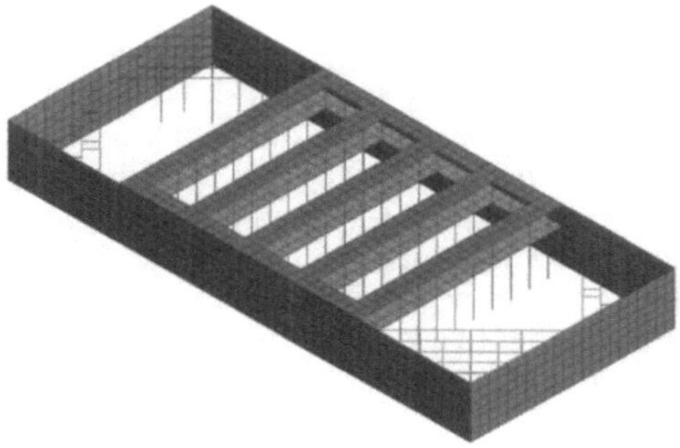

Figure 8. Analysis model.

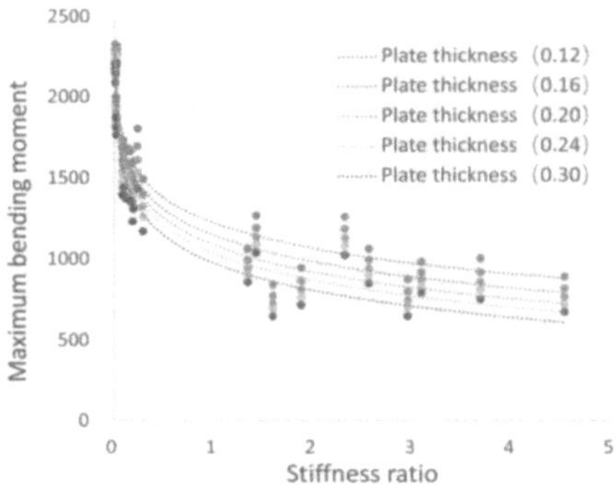

Figure 9. Relationship between pile bending moment and stiffness ratio.

be seen that when the lateral bending stiffness of the truss plate increases to a certain value, the reduction range of the bending moment and horizontal displacement of the diaphragm wall tends to be gentle. In addition, from the data fitting relationship, the maximum bending moment, the maximum horizontal displacement, and the stiffness ratio of the diaphragm wall all present a power function relationship with a high degree of fitting (the correlation coefficient exceeds 90%). The form is unified as the fitted equation of Equation (9). According to the nature of the power function equation of the fitting curve, as the stiffness ratio continues to increase, the change amplitude of the bending moment and horizontal displacement of the diaphragm wall continues to decrease. The curve of the fitting equation always passes through (1, | a |) and (1, | c |) points.

$$\begin{cases} M_{\max} = |a| \cdot s^{-|b|} \\ y_{\max} = |c| \cdot s^{-|d|} \end{cases} \tag{9}$$

Where Mmax is the maximum bending moment (kN · m/m); Ymax is the maximum horizontal displacement (mm).

According to the specification, the monitoring and early warning value of the horizontal displacement of the foundation pit support structure is 30-80 mm. The | c | and | d | in the fitting equation under different slab thicknesses are averaged to obtain the relationship between the stiffness ratio and the displacement, as shown in Equation (10).

$$s = \left(\frac{y_{\max}}{45.678}\right)^{\frac{1}{-0.448}} \tag{10}$$

According to the power function relationship between the maximum bending moment and horizontal displacement and the stiffness ratio in the calculation model, the selection range of bracing structure size can be initially determined according to the stiffness ratio 0.25-2.5 during the preliminary design of the beam-slab bracing system with truss (plate) members, and then the appropriate support structure size can be determined according to the actual engineering conditions, economic costs, and other requirements.

5 CONCLUSION

(1) With the increase in water level depth outside the foundation pit, the bending moment of the retaining wall at the seaside gradually transits from the "C" curve to the "S" curve.

The maximum bending moment decreases first and then increases. When the ratio between the depth of the water level outside the pit and the depth of the pit is less than 1, the bending moment of the side retaining wall near the water is smaller than that of the far water side. Therefore, no special treatment is required. When the ratio is greater than 1, the bending moment of the side retaining wall near the water increases rapidly and the maximum point moves downward. At this time, it is necessary to consider strengthening the side retaining structure near the water;

(2) A method based on displacement increment is proposed to calculate the adjustment coefficient of fixed point iteratively. Finally, the stiffness of both sides of the support under an unbalanced load is determined and the axial force of the support is obtained. The application of this method to practical projects matches well with the actual monitoring results;

(3) The ratio of the lateral bending stiffness of the truss plate to the axial compression stiffness of the support plate is introduced, which shows a power exponential relationship with the bending moment displacement of the support structure and further gives the expression of the relationship between the displacement and stiffness ratio of the support structure. At the design stage, the beam plate support can be initially arranged according to the displacement requirements and the specific size of the support structure can be determined according to the actual engineering conditions, economic costs, and other requirements.

REFERENCES

Jiang Yan, Yang Guanghua, Qiao Youliang, Zhang Yucheng. A Simplified Calculation Method and Application for Deep Foundation Pit with Asymmetric Supporting Structure [J]. *Guangdong Water Resources and Hydropower*, 2017(06):33–38.

Jin Yabing, Liu Dong, Sun Yong. Design and Calculation Method of Inner Support Structure in Deep Foundation Pit under Asymmetric Load [J]. *Chinese Journal of Underground Space and Engineering*, 2019 (6):1811–1818.

Jeng, D. S., Lin, Y. S. Response of the Inhomogeneous Seabed Around Buried Pipeline Under Ocean Waves [J]. *Journal of Engineering Mechanics*, 2000, 126(4): 321–332.

Li Jian, Fan Fangfang. Study on Seepage Characteristics of Foundation Pit in the Seaward environment [J.] *Water Resources and Hydropower Engineering*, 2019, 50(7): 202–208.

Loehlin, J. C. The Cholesky Approach: A Cautionary Note [J]. *Behavior Genetics*, 1996, 26(1): 65–69.

Loehlin, J. C. The Cholesky Spproach: A Cautionary Note [J]. *Behavior Genetics*, 1996, 26(1): 65–69.

Ou, C. Y., Chiou, D. C., Wu, T. S. Three-Dimensional Finite Element Analysis of Deep Excavations [J]. *Journal of Geotechnical Engineering*, 1996, 122(5):337–345.

Ying Hong-wei, Nie Wen-feng, Wang Qi-tong, Cheng Yong. Effect of Water Level Fluctuation on the Stability of Excavation with Gravity Retaining Wall by the Sea [J]. *The Ocean Engineering*, 31(04)48–54 (2013).

Ying Hong-we, Sun-wei, Zhu Cheng-wei. Experiment Research on the Response of Excess Pore Pressure to Wave Around Near-sea Excavation [J]. *Rock and Soil Mechanics*, 2016(s2):187–194.

Zhang Yi-fan, Gu Kuan-hai. Design on an Excavation in the Water Area of Water Intake of a Project in Macao [J] *Port & Waterway Engineering*, 2020(5):20–26.

Water Conservancy and Civil Construction – Oke & Ahmad (Eds)
© 2024 The Author(s), ISBN: 978-1-032-58614-4

Mechanism analysis of floor heave disease in operation period of phyllite highway tunnel

Chenke Sun*, Yilin Zheng, Huaizhong Qiu, Fei Li & Xiangbing Chu
The EXIBEI Highway Operation Management Co., Ltd. of Hubei Provincial Communications Investment Group, China

ABSTRACT: China's tunnel project has entered a phase where construction and maintenance are both important. Scientific management of tunnel operation period diseases has become a common research topic for many scholars. The floor heave during tunnel operation can lead to road heave and cracking and safety accidents are easy to occur when vehicles pass quickly. The floor heave disease occurred during the operation period of a phyllite highway tunnel. Field disease detection, drilling core analysis, and rock laboratory test are carried out to analyze the causes of the floor heave disease. In this paper, relying on an operating tunnel, the mechanism of floor heave disease is analyzed by combining field tests and indoor tests. The main factors causing the disease are the defects of the inverted arch structure, low strength of surrounding rock, obvious water absorption softening, and expansion.

1 INTRODUCTION

In recent years, China has increased its investment in the construction of modern transportation networks and the total mileage of highways has reached 5.3661 million kilometers (Hong & Feng 2020; Hong *et al.* 2022; Tian *et al.* 2020). According to the data (Rui 2019), more than 30% of tunnels in China have entered the disease development period. Therefore, the operation of tunnel maintenance workload is increasing.

To find out the mechanism of floor heave and formulate scientific and reasonable treatment measures, many scholars have carried out research and achieved certain results. Sui Yi (Sui *et al.* 2022) analyzed the reasons for the floor uplift of the 3# cross tunnel of a high-speed railway tunnel by numerical simulation and field tests. He proposed joint treatment measures to strengthen the bearing capacity of the surrounding rock of the basement and control the deformation of the surrounding rock. Song Jianjun (Song *et al.* 2022) summarized and analyzed the characteristics of floor heave disease of different types of operating tunnels and proposed a rapid treatment technology for floor heave disease of operating tunnels. Song Feng (Song 2021) studied the influence of swelling and shrinking strata on tunnel floor heave disease, screened out the influencing factors of floor heave disease, established the stress model of inverted arch surrounding rock, and obtained the optimal rise span ratio of tunnel inverted arch. Yang Jianmin (Yang *et al.* 2021) studied the relationship between the deformation of surrounding rock at the tunnel bottom and the deformation of the inverted arch structure through long-term monitoring and proposed treatment measures to increase the rise span ratio of the inverted arch and enhance the stiffness of the inverted arch. Wang Jinbo (Wang *et al.* 2017) studied the mechanism and treatment measures of the floor heave of the Mazui Tunnel. The research shows that the occurrence of the floor heave

*Corresponding Author: sck1227@126.com

 DOI: 10.1201/9781003450818-89

of the tunnel is mainly affected by the low strength of the surrounding rock and the unsealed inverted arch. He proposed a treatment scheme that combines the bottom bolt grouting with the inverted arch. Lu Junfu based on the stress characteristics of the flat bottom plate of the tunnel, supervised the mechanical model of the simply supported beam under vertical and horizontal two-way loads and revealed the relationship between the maximum bottom heave deformation of the tunnel invert and the elastic modulus, bottom plate thickness, vertical ground stress, horizontal stress, and tunnel span. Lu Junfu (Lu *et al.* 2021) and others carried out deformation monitoring of floor heave for a tunnel for more than 5 years. The method of combining the indoor rock shear creep test with depth learning is used to inverse the creep parameters of the rock mass at the tunnel bottom and the calculated value of the tunnel floor heave is further accurate.

2 ENGINEERING BACKGROUND

2.1 *Tunnel overview*

A four-lane expressway long tunnel with upper and lower sections is separated. During the operation period, the starting and ending stakes of the left section are ZK1478+341–ZK1482 +524, and the starting and ending stakes of the right section are YK1478+338–1480+508. The bedrock that the tunnel passes through is mainly phyllite. Affected by the fault zone, the rock stratum is twisted and folded and the joints are relatively developed. The groundwater is mainly bedrock fissure water and carbonate karst water, which are stored in rock fissures.

2.2 *Floor heave and its treatment during tunnel construction*

In December 2006, it was found that inversion heave disease occurs within the range of YK104+890–YK105+681 (791 m), YK105+721–YK105+926 (205 m), ZK105+280–ZK105 +470 (190 m), and ZK105+850–ZK105+948.5 (98.5 m). The diseased area should be treated by replacing the inverted arch with segmental spacing. The connection reinforcement should be embedded in the wall foundation on both sides of the inverted arch to strengthen the link strength between the inverted arch and the whole lining. The sinking area of side walls on both sides should be reinforced by grouting.

2.3 *Disease detection and treatment during operation*

The tunnel was put into use in 2007 and the pavement displacement occurred in 2012, resulting in deep cracks with a width of 10–15cm and long longitudinal cracks. The longitudinal cracks are caused by the reflection of cracks in cement concrete slabs. The pavement with deep cracks and longitudinal cracks has been milled, resurfaced, and repaired with asphalt potting.

3 MECHANISM ANALYSIS OF FLOOR HEAVE

3.1 *Current situation of tunnel diseases*

During the tunnel operation period, cracks and deformation appeared in the left tunnel pavement as cracks and leakage in the lining. Among them, pavement cracking deformation is relatively serious, which has a certain impact on driving safety.

1. Pavement disease

   The specific conditions of the road surface of the left tunnel are shown in Table 1. Figures 1 to 3 show the photos of pavement diseases.

Table 1. Statistics of important diseases of left tunnel pavement.

Station	Disease	Condition description
ZK1478+475	Crack	1 longitudinal crack, 20 m long
ZK1478+780~824	Crack	One longitudinal crack, 44 m long, 3 cm wide at the widest point, with slight settlement locally
ZK1478 +980~ZK1479 +100	Crack	1 longitudinal crack, 70 m long; The height difference between arch camber and pavement is 4.5 m, Maximum crack width 3 cm
ZK1479+162~260	Crack	1 longitudinal crack, 98 m in length, The maximum crack width is 3 cm
ZK1479+380~580	Arching	Road camber; Height difference with road surface up to 10 cm
ZK1479+766	Crack	Longitudinal crack, 3 m long
ZK1480+450~485	Crack	1 longitudinal crack, 35 m in length

Figure 1. Pavement cracks

Figure 2. Deformation of maintenance track.

(a) (b)

Figure 3. Site map of pavement uplift. (a) Road camber. (b) Pavement camber depth.

2. Lining disease

There are 31 cracks, 28 water seepage, and 6 concrete spalling in the left tunnel lining of the tunnel. The total length of the crack is 168.2 m and the maximum width is 5 cm. The site conditions are shown in Figures 4 to 6.

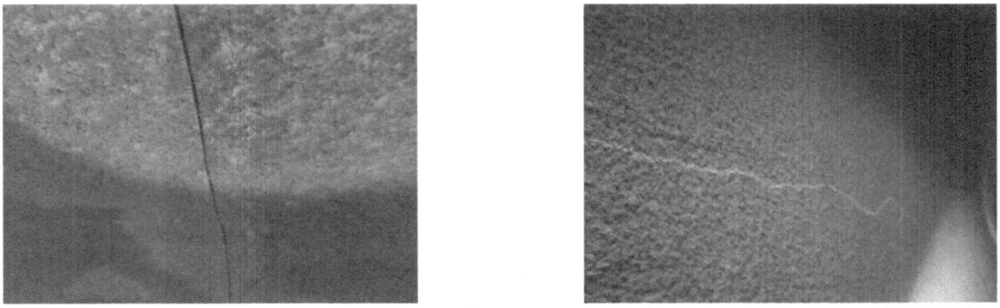

(a) Circumferential crack of the side wall (b) Longitudinal crack of the side wall

Figure 4. Tunnel lining cracks.

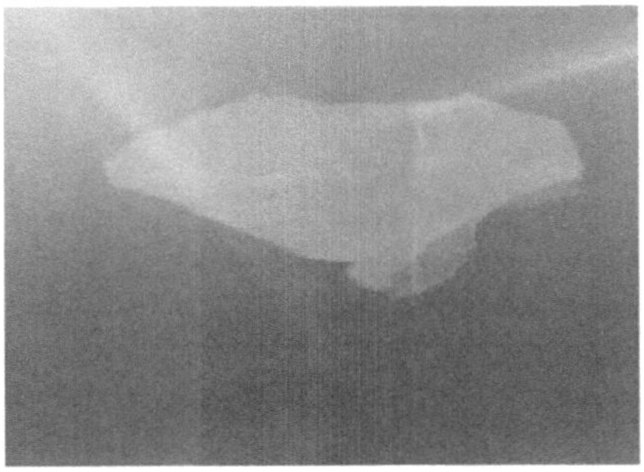

Figure 5. Flaking of fireproof layer.

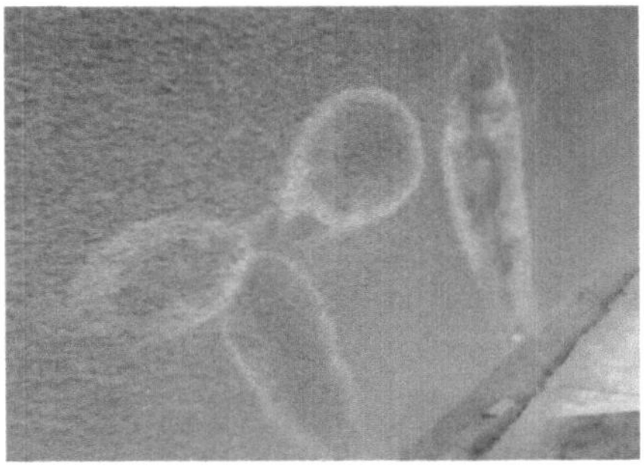

Figure 6. Sidewall seepage crystallization.

3.2 *On-site inspection*

1. Pavement drilling inspection
   A total of 22 sections are drilled and cored during the special inspection at the tunnel bottom and the core samples are shown in Figure 7. The test results show that the total length of coring in 6 sections meets the design requirements, while the total length of coring in 16 sections does not meet the design requirements. Among them, there is no inverted arch layer at the tunnel bottom of 6 sections and there is virtual slag in the concrete structure at the tunnel bottom of 8 sections.

(a) (b)

Figure 7. A sampling of tunnel bottom core drilling. (a) Sample drawing of tunnel bottom core. (b) Internal drawing of the coring hole.

2. Drill core analysis

To accurately grasp the lithology of the base course, seven boreholes are arranged in the area with relatively significant disease, with a single hole depth of about 15 m (more than 1.5 times the hole diameter). The drilling locations are ZK1478+796 (double hole), ZK1479+040 (single hole), ZK1479+210 (double hole), ZK1479+470 (single hole), and ZK1479+600 (single hole). The rock cores are shown in Figure 8.

(a) (b) (c)

Figure 8. Core sample. (a) 0-5 m. (b) 5-10 m. (c) 10-15 m.

It can be seen from Figure 8 that the lithology of the bedrock layer is phyllite, the rock mass is broken, the bedding fissures are developed, and the bedding combination is poor. According to the drilling histogram, drilling acoustic wave, and drilling television test results, the tunnel bedrock can be divided into five layers: asphalt pavement and concrete layer (0.6–1.8 m thick), backfill layer (0.7–1.9 m thick), broken phyllite layer (2.5 m thick), relatively broken phyllite layer (6 m thick), and relatively complete phyllite layer (more than 4 m thick).

3.3 *Indoor test*

According to the Standard for Engineering Rock Mass Test Methods (GB/T50266–2013), the rock sample is made into a cylinder test block with a bottom diameter of 50 mm and a height of 100 mm, as shown in Figure 9.

Figure 9. Phyllite sample.

1. Uniaxial compression test

   Phyllite is softened by water. The six rock samples are divided into two groups: one group conducts a uniaxial compression test under a saturated state and the other group conducts a uniaxial compression test under a natural state. The samples in the saturated state are numbered 1–1, 1–2, and 1–3. the samples in the natural state are numbered 2–1, 2–2, and 2–3. The basic mechanical parameters obtained from the test and calculation are summarized in Table 2 (the values are the average of the three samples).

Table 2. Basic rock mechanical parameters.

Sample status	Uniaxial compressive strength/MPa	Softening coefficient	Deformation modulus E_o/GPa	Elastic modulus E_e/GPa	Poisson's ratio
Saturated	1.53		0.80	0.75	0.30
Natural	3.20	0.48	0.96	1.72	0.29
Relative Difference (%)	52.00		16.67	56.40	3.45

It can be seen from the data in Table 2 that the uniaxial compressive strength of the phyllite sample in its natural state is 3.20 MPa, the uniaxial compressive strength after immersion saturation is 1.53 MPa, reduced by 52.00%, and the softening coefficient is 0.48. It can be seen that the bedrock of the tunnel is softened when encountering water. In the natural state, the deformation modulus of the phyllite sample is 0.96 GPa. After immersion and saturation, the deformation modulus is 0.80 GPa, which is reduced by 16.67%. The ability of phyllite to resist deformation decreases obviously after encountering water.

2. Analysis of rock mineral composition

The types and proportions of mineral compositions in rocks influence their characteristics of rocks. The mineral composition of the pyrochlore samples is analyzed by X'pert MPD Pro. The composition and proportion of various minerals are summarized in Table 3.

Table 3. Composition and proportion of various minerals.

Mineral composition	Illite	Montmorillonite	Chlorite	Quartz	Calcite	Dolomite	feldspar
Proportion	3%	55%	10%	15%	5%	2%	10%

It can be seen from Table 3 that the phyllite samples taken in the tunnel are mainly montmorillonite, accounting for 55%, and contain 3% illite. Montmorillonite and illite belong to hydrophilic mineral composition and have expansibility after absorbing water.

4 ANALYSIS OF THE MECHANISM OF FLOOR HEAVE

The surrounding rock in the area where the tunnel floor heave disease occurs is mainly phyllite and the groundwater depth is shallow. In combination with hydrogeological conditions, disease treatment plan in the construction period, maintenance measures in the operation period, and inspection and test results, the causes of tunnel floor heave are analyzed as follows:

1. The surrounding rock of the tunnel is phyllite. According to the indoor uniaxial compression test results, the pressure resistance of phyllite decreases after encountering water. According to the analysis of the mineral composition of rock samples, montmorillonite accounts for more than 50% and contains illite. Montmorillonite and illite show expansibility after absorbing water. The underground water in the tunnel floor heave area is shallow, the secondary lining cracks and seeps, and the waterproof and drainage system of the tunnel is partially blocked after long-term operation. The strength of phyllite decreases and expands after immersion, resulting in the uplift of the tunnel bottom.

2. The inverted arch structure has defects. Inverted arch structure defects can reduce the supporting force of tunnel closure and weaken the ability surround rock to resist elastic deformation, making the tunnel unable to achieve integrity under the design state. During the tunnel operation, the tunnel bottom structure is flat, which is not conducive to stress. Under the combined action of heavy vehicles and the upper load of the tunnel, the tunnel bottom can deform and bulge.
3. The surrounding rock is disturbed due to incomplete reinforcement during the construction and transportation period. In some areas where floor heave occurs, a single side locking anchor bolt is used for reinforcement, resulting in uneven stress of the tunnel. The weak side of the structure cannot meet the strength requirements and gradually deforms. The reinforcement during the construction and transportation period can disturb the surrounding rock and redistribute the ground stress. The surrounding rock at the tunnel bottom gradually deforms, leading to pavement uplift, cracking, and deformation.

5 CONCLUSION

This paper investigates the floor heave disease of an operating tunnel by on-site special inspection and indoor rock test and analyzes the causes of the tunnel floor heave disease. The main causes of tunnel floor heave are the absence of an inverted arch structure and poor surrounding rock conditions at the base. The joint of the surrounding rock at the tunnel base is developed and broken and the groundwater cannot be discharged in time, leading to softening of the surrounding rock and reducing the bearing capacity of the base. Under the load of the upper part of the tunnel, the bottom plate is damaged by buckling and the pavement is arched and cracked.

REFERENCES

CCCC Rui, Road Bridge Maintenance Technology Co., Ltd. *Example Set of Highway Tunnel Maintenance and Reinforcement [M]*. Beijing: People's Communications Press Co., Ltd., 2019: 1–4.

Hong Kairong, Du Yanliang, Chen Kui, et al. Full-face Tunnel Boring Machines (shields/TBMs) in China: History, Achievements, and Prospects [J]. *Tunnel Construction*, 2022, 42(5): 739.

Junfu Lu, Zheng Xiao, Yu Yu. et al. Research on the Back Analysis Method of Creep Parameters of Surrounding Rock at Floor Heave Section of Railway Tunnel [J]. *Journal of Railway Engineering Society*, 2021, 38(01): 66–71.

Kairong Hong, Huanhuan Feng. Development Trends and Views of Highway Tunnels in China Over the Past Decade [J]. *China Journal of Highway and Transport*, 2020, 33(12): 62–76.

Lu Junfu, Wang Mingsheng, Wang Kui, et al. Study on Mechanism and Analytical Method of Bottom Drum in a Horizontal Layered Mudstone Railway Tunnel. [J/OL]. *Journal of Railway Science and Engineering*: 1–15.

Siming Tian, Wei Wang, Jiangfeng Wang. Development and Prospect of Railway Tunnels in China (Including Statistics of Railway Tunnels in China by the End of 2020) [J]. *Tunnel Construction*, 2021, 41(2): 308.

Sui Yi, Wang Weijun, Yuan Chao, et al. On Mechanism and Control Technology of Drum Disaster in Soft Mudstone Tunnel [J]. *Mineral Engineering Research*, 2022, 37(01): 24–30.

Song Jianjun, Zhu Xiaoming, Zheng Wanpeng et al. Rapid Treatment Technology for Floor Heave of Highway Tunnels in Operation [J]. *China Building Materials Science & Technology*, 2022, 31(01): 109–113.

Song Feng. Research on Causes of Floor Heave and Invert Arch Structure of Weakly Swelling-shrinking Tunnels for Figh-speed Railway [J]. *Journal of Xi'an University of Science and Technology*, 2021, 41(03): 481–489.

Wang Jinbo, Huang Gun, Tang Jianxin, et al. Mechanism and Treatment Technology of Floor Heave in Horizontal Layered Shale Tunnel [J]. *Journal of Safety and Environment*, 2017, 17(06): 2246–2251.

Yang Jianmin, Xu Huairen, Shu Dongli, et al. Reflection on the Mechanism of Tunnel Floor Heave and Its Countermeasures [J]. *Journal of Railway Engineering Society*, 2021, 38(02): 74–79.

Water Conservancy and Civil Construction – Oke & Ahmad (Eds)
© 2024 The Author(s), ISBN: 978-1-032-58614-4

Study on dynamic assessment method of major risks in deep foundation pit engineering

Jiao Zhang*
School of Civil and Transportation Engineering, Shanghai Urban Construction Vocational College, Shanghai, China

Chun Meng
Academic Affairs Office of Shanghai Urban Construction Vocational College, Shanghai, China

Jiancheng Wang & Aibin Jiang
Digital Construction College, Shanghai Urban Construction Vocational College, Shanghai, China

ABSTRACT: Based on dynamically assessing the monitoring data of the deep foundation pit projects, the risk changes in different construction stages are analyzed according to the three construction stages of the foundation pit project, namely, the preparation stage before excavation (including the construction of the retaining structure), the excavation stage, and the construction of the underground structure. The possible risk factors and accident paths are disclosed. The major risk prevention, control measures, and emergency measures during the construction period are proposed. It can provide theoretical guidance for risk control measures and repair plans for similar deep foundation pits.

1 INTRODUCTION

With the vigorous development of China's construction industry, the demand for underground space is also increasing. Therefore, the number of deep foundation pit projects is increasing, which is facing more complex natural and environmental conditions. The scale and depth of deep foundation pit excavation are also growing. Due to the influence of construction conditions and the construction environment, the deep foundation pit project inevitably has the characteristics of complexity and uncertainty. Most of the foundation pit projects are temporary projects, with relatively small safety reserves, which are often ignored by the construction parties. Accidents of deep foundation pit construction happen frequently, especially in recent years. The occurrence of major foundation pit accidents in the subway has caused huge economic losses and negative social impacts. The safety of foundation pit projects has begun to be paid attention to by all aspects of construction. Therefore, the research on the dynamic monitoring and early warning methods of foundation pits based on the monitoring data is becoming increasingly urgent. And it is urgently needed to be applied in practical projects to solve the problem of the lack of an early warning control link in current projects.

In recent years, some early warning theories and methods in foundation pit engineering based on monitoring data have been investigated. For example, the deep foundation pit in soft soil areas according to the excavation monitoring data of a mining area was analyzed

*Corresponding Author: GGGLG@163.com

DOI: 10.1201/9781003450818-90

(Ding *et al.* 2018). A multivariate adaptive regression spline model for the maximum displacement in the monitoring of foundation pit support was established (Xiang *et al.* 2018). The mechanism of load transfer of pile groups by monitoring the settlement of pile groups near deep foundation pits in soft soil was studied (Tan *et al.* 2018). The automatic remote monitoring and early warning system of the foundation pit in terms of the layout of monitoring points, scheme selection, monitoring principle, and data analysis and processing were introduced (Yang *et al.* 2019). The centralized information management, real-time reading of monitoring data, automatic storage, query, calculation and analysis, and data analysis and early warning were studied to improve monitoring efficiency and security (Ye & Li 2019).

Based on the monitoring data, this paper developed a dynamic assessment method of major risks in deep foundation pit projects. The risk changes were analyzed in different construction stages, namely, the preparation stage before excavation (including the construction of the retaining structure), the excavation stage, and the construction of the underground structure. The monitoring items in each construction stage shall be specifically analyzed, and the possible risk accidents corresponding to the abnormal monitoring items shall also be analyzed. The possible risk sources that cause the change of monitoring data shall be found. Based on these, effective risk control measures and repair plans as well as the determination method of early warning for deep foundation pit works shall be proposed. It can provide a theoretical basis for early warning control of the risk in the construction of deep foundation pit works.

2 THE METHOD OF DYNAMICALLY ANALYZING MAJOR RISKS

2.1 *In this paper, the dynamic risk analysis is mainly based on three aspects:*

1. The change and development of risk factors

   The risk is constantly changing with the construction. When the risk factors begin to appear, they may not show any accident signs. There are a large part of unknown areas that are not recognized by people, which makes the risk analysis of the system biased or even makes people ignore the occurrence of major risk events, affecting the authenticity of the system's risk analysis results and the effectiveness of risk control and emergency measures. Different construction stages have different risk types and risk change characteristics. Some risk factors also have strong timeliness. During the construction process, the probability of major risk accidents is comprehensively calculated in combination with the occurrence of other unexpected engineering events.
2. Early warning of risks based on monitoring data

   Monitoring is an effective means of protecting the risk of foundation pit, which also reflects the risk change level of foundation pit engineering in real-time, and is also the basis for quantitative analysis of dynamic risks, on which the dynamic evaluation of major risks of foundation pit engineering is based. Engineering monitoring is the guidance of the design and construction of foundation pit engineering. For large and medium-sized foundation pits with large construction difficulties and complex projects, it is often difficult to predict the type of risk factors and risk changes, and there is no reasonable quantitative analysis method in theory, so it must rely on on-site monitoring data analysis.
3. Consideration of construction phases as the dynamic timeline

   Considering construction stages as the timeline, the dynamic risk analysis of foundation pit works is carried out. The construction process of the foundation pit is divided into the preparation stage before excavation, the excavation stage, and the underground structure construction stage. The middle excavation stage is the main stage of major risk accidents. The main works in the excavation stage are divided into earthwork, support construction, and dewatering works. Dynamic risk analysis and risk control are carried out from three aspects. At the same time, the monitoring data in the excavation stage shall be analyzed and fed back daily to calculate the significant risk value.

2.2 *The construction process of the deep foundation pit*

Based on the analysis of the monitoring data at each stage and different types of monitoring items, this paper summarizes the commonly used engineering early warning values, and fully considers the characteristics of monitoring items of environment and deep foundation pit support, which puts forward the corresponding calculation method of risk correction value, and obtains the determination method of risk early warning value for risk rating. The dynamic early warning method system of the foundation pit based on monitoring data is established. The specific construction flow chart is shown in Figure 1:

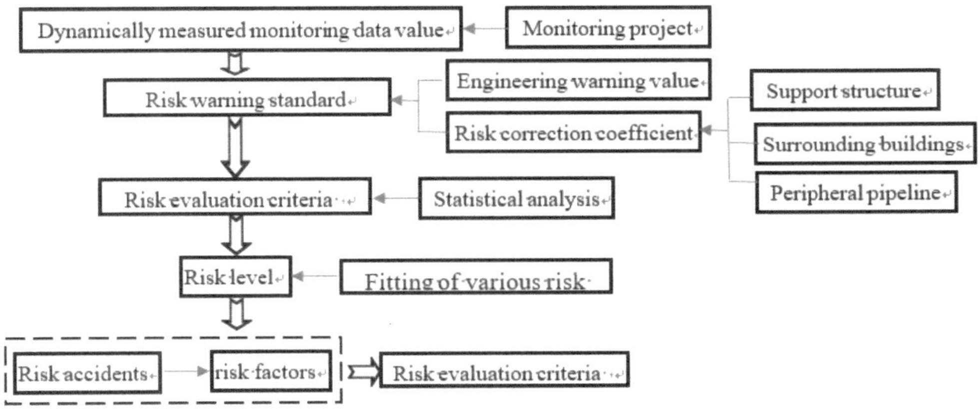

Figure 1. The construction flow chart of the deep foundation pit.

2.3 *Establishment of risk model based on foundation pit monitoring data*

The establishment of a risk model based on the monitoring data of the foundation pit, on the one hand, analyzes and calculates the cumulative value and change rate of the monitoring data of various monitoring items measured on the site, and obtains the risk indicators, evaluating them according to the aforementioned grading criteria.

The monitoring items in the foundation pit project are all carried out independently, however, each monitoring item also has a certain correlation. For example, the increase of the horizontal displacement of the wall top often leads to the increase of the horizontal displacement of the surrounding buildings at the same time; the increase of lateral deformation (inclinometer) of the enclosure wall often leads to the increase of ground settlement and vertical displacement of surrounding buildings and pipelines. Therefore, after obtaining the risk evaluation grade of each monitoring project, it is necessary to fit the risk indicators among the engineering monitoring projects to obtain a comprehensive indicator for evaluating the engineering risk.

Considering that the greater the risk level is, the greater the threat to the foundation pit is. Therefore, the monitoring indicators with higher risk levels should be given greater weight, and the resulting comprehensive indicators should also be correspondingly larger. Therefore, concerning the experience in risk assessment of shield tunnels, a risk model is established by using comprehensive risk indicators (a_t) of engineering monitoring:

$$a_t = \frac{\sum_{i=1}^{m} r_{i,r} a_i}{\sum_{i=1}^{m} r_{i,r}} \tag{1}$$

Table 1. Weighting coefficient of monitoring risk level.

Risk level r	1	2	3	4	5
$r_{i,r} = \frac{e^r}{e^5}$	0.018	0.050	0.135	0.368	1.000

3 PREVENTION AND CONTROL OF MAJOR RISKS DURING CONSTRUCTION

Studying the change law of risk factors is to better carry out risk prevention and control, the data collection method, and the literature method to master the methods of summarizing risk early warning values, the risk early warning standards of monitoring projects, the rating standards of dynamic risks, and other key issues. Based on risk assessment, the paper analyzes the possible risk accidents corresponding to the abnormal items of monitoring data and finds out the possible risk sources that cause the changes in monitoring data. Based on this, an effective dynamic early warning method for deep foundation pit is proposed.

The commonly used risk early-warning values of deep foundation pit engineering are summarized by using the inductive summary method (Xia 2021), and the monitoring items in each construction stage are specifically analyzed. Two major risk prevention and control measures for foundation pit engineering are proposed, among which one is to analyze the relevant risk factors of foundation pit collapse risk and to propose corresponding risk control measures, which are listed in Table 2.

Table 2. Risk factors and risk control measures for foundation pit collapse.

Risk factors	Control measures	Emergency measures for accidents
Continuous rainfall and rainstorms	Flood control measures are taken.	The earth at the collapse of the foundation pit shall be immediately backfilled, and the overload around the foundation pit shall be cleared. If the soil loss occurs at the back of the retaining structure, we should immediately fill it with sand or concrete. At the same time, the surrounding supports shall be re-checked to find out whether there is a support relaxation. If there is, we should immediately take precautions to prevent instability from spreading.
Insufficient insertion depth of the wall leads to instability of the support system.	The wall depth is detected.	
The dynamic load caused by large traffic flow is increased on surrounding roads.	The strict control of traffic flow	
Unknown soft soil layer	Reinforcement measures are taken after finding an unknown soft soil layer.	
In sudden excavation, an unstable soil slope breaks the support system when excavating to the bottom.	The excavation method is standardized.	
Burst underground water pipes or infiltrated other water sources	The surrounding water source is checked.	

On the other part, the surrounding environment is the focus of major risk prevention and control of foundation pit works. The relevant risk factors of collapse risk of nearby residential buildings are analyzed, and the control measures and accident emergency measures (Fu *et al.* 2020) are given in Table 3.

Table 3. Risk factors and risk control measures for the collapse of residential buildings.

Risk factors and accident occurrence path	Control measures	Emergency measures
We shall Ignore the building cracks. We shall chain the consequence of foundation pit collapse accidents.	The report of building crack measurement shall be strengthened. Control measures for the same foundation pit collapse are adopted.	People in the residential building shall be evacuated in an emergency, and the residential building shall take grouting reinforcement remedial measures.
We shall neglect the monitoring and alarm of uneven settlement of buildings.	The monitoring and alarm handling procedures shall be formulated, and the responsible person shall be specified.	

4 CONCLUSIONS

1. According to different types of major risk factors and the analyses of major risk factors and the study of treatment measures, it is necessary to take different risk response measures. It is necessary to timely grasp the impact of the surrounding buildings, underground soil layers, underground pipelines, and facilities in the construction. It is also key to timely finding dangerous situations and controlling the development of various accidents and dangerous situations. The practical basis for the control of major risk factors in foundation pit engineering is of the same importance.
2. Based on the monitoring data and the construction process, a dynamic evaluation method of major risks applicable to foundation pit engineering during the construction process is proposed. The dynamic analysis method of major risks takes the construction stage as the dynamic timeline and takes the monitoring data of different stages of foundation pit engineering as the main calculation basis to study the change and development of risk factors. The characteristics of various monitoring items such as the foundation pit self-support monitoring items and environmental monitoring items should be fully considered. It provides a theoretical basis for the risk grade assessment of the foundation pit excavation construction stage and the underground structure construction stage and combines deep foundation pit engineering cases with theoretical knowledge.
3. The potential directions for future deep foundation pit risk research are as follows. A sound safety risk management and control platform for the foundation pit is expected to be established. The combination of big data analysis and system safety engineering theory is anticipated to analyze the risk causes and propose risk control measures. With the utilization of information communication and Internet of Things technologies in the construction industry, the monitoring and detection of foundation pits will be strengthened with artificial intelligence and information management. A dynamic risk assessment method will be developed for risk analyses.

ACKNOWLEDGMENTS

This work is sponsored by achievements of the scientific research project at the school level of Shanghai Urban Construction Vocational College in 2023. Project name: Research on risk analysis method of immersed tunnel project based on Bayesian network.

Zhang Jiao: born in December 1981, female, professor, is mainly responsible for teaching and design of underground engineering structures.
Tel.: 13482236117
Email: ggglg@163.com
Mailing address: No. 60, Building 2, Lane 444, Miyun Road, Yangpu District, Shangha

REFERENCES

Fu Huimin, Wu Haiyang, and Zan Haizhu. (2020). Design of Deep Foundation Pit Support Structure of Qinghe Station of Beijing Zhangjiakou High-speed Railway. *Rail. Surv.* 46 (01): 87–94.

Tan Y., Lu Y., and Xu C. J., et al. (2018). Investigation on Performance of a Large Circular pit-in-pit Excavation in Clay-gravel-cobble Mixed Strata. *Tunn. Undergr. Sp. Tech.* 79, 356–374.

Xia Zeai. (2021). Risk Control and Technical Measures During the Implementation of Deep Foundation Pit Under Complex Environmental Conditions. *Buil Const.* 43 (06), 1166–1168+1179.

Xiang Y. Z., Goh A. T. C., and Zhang W. G., et al. (2018). A Multivariate Adaptive Regression Splines Model for Estimation of Maximum Wall Deflections Induced by Braced Excavation. *Geomech. Eng.* 14 (4), 315–324.

Yang Ao, Li Xiyin, and Xiang Shaopeng. (2019). Application of Automatic Remote Monitoring and Early Warning System in Deep Foundation Pit. *Hebei Transportation Education.* 16, (12), 33–36.

Ye Shuihua, Li Depeng. (2019). Monitoring and Numerical Simulation Analysis of Deep and Large Foundation Pit Excavation Under a Complex Environment. *Ksce. J. Civ. Eng.* 52 (2), 117–126.

Zhi Ding, Jieke Jin, Tong Chun Han. (2018). Analysis of the Zoning Excavation Monitoring Data of a Narrow and Deep Foundation Pit in a Soft Soil Area.: *J. Geophys. Eng.* 15 (4), 114–121.

Water Conservancy and Civil Construction – Oke & Ahmad (Eds)
 ISBN: 978-1-032-58614-4

Simulation of anchorage by bottom expansion filler material

Jinrui Wang
College of Traffic Engineering, Yellow River Institute of Transportation, Wuzhi County, Jiaozuo City, Henan Province, China

Yong Li*
National Energy Group Guoshen Huangyuchuan Coal Mine, Inner Mongolia, Ordos, China

Hua Nan & Longlong Guo
Department of Transportation, College of Energy, Henan Polytechnic University, Henan, Jiaozuo, China

ABSTRACT: For the present loose and soft surrounding rock roadway, the anchor system is not enough to provide the tensile force, and the anchor failure between the anchor and surrounding rock is easy to occur, which seriously affects the roadway stability and construction safety. In this paper, we simulate the home-made filler material by ANSYS to realize the homotopic modification of the filler material to the surrounding rock in the bottom expansion area, and improve the physical and mechanical properties of the surrounding rock in the anchorage section to increase the anchorage performance.

1 INTRODUCTION

In soft rock roadways, due to the weak nature of the anchoring envelope, the bond between the anchorage and the borehole envelope is weak and not sufficient to support the tensile strength, and slip effectiveness can occur. Shear damage usually does not occur at the anchor-filler interface. Different filler materials have different effects on the anchorage performance, and this study compares the anchoring effect on the surrounding rock by filling different ratios of filler materials when anchoring at the bottom of anchor holes.

2 CURRENT STATUS OF RESEARCH

In 1912, anchor rods were first used in coal mine roadway support in the United States, and became the main support method in rock engineering in the 1970s and 1980s. Anchor rod support is not only well used in coal mines, but also can be well applied and promoted in underground construction such as slope, surface support works of deep foundation pits in rock, and tunnels and quarries.

He, Du, Gong et al (2022) believed that when NPR anchor rod interacts with rock mass, it will change the mechanical behavior of rock mass, which can realize the dimensionality reduction and decoupling analysis of the mechanical behavior of rock mass. It can not only effectively control geological disasters such as large deformation and impact ground pressure of rock mass, but also can be controlled using small disturbances.

*Corresponding Author: 212002020060@home.hpu.edu.cn

DOI: 10.1201/9781003450818-91

Zhao and Li (2020) addressed the establishment of a prestressed anchor-perimeter rock interaction model and concluded that the preload force, nature of the surrounding rock, anchor and anchorage parameters are the main factors affecting the anchor shear stress distribution and anchor shaft force.

A large number of experts and scholars have researched the working principle, mechanism of action, and damage form and stability of anchor system of anchor rods from different aspects, and improved the anchor rod anchoring technology continuously, achieving fruitful research results in anchoring performance. Among them, borehole envelope reaming anchoring is one of the most common and mature techniques.

Xia, Ju, and Su (2020) used rotating jets to impact media, and hydraulic reaming hollowing formed artificial cavities in the borehole, which are larger than the diameter of the borehole through the cutting effect of high-pressure rotating water jets, greatly improving the effect of coal seam pressure relief and could significantly reduce the potential disaster caused by impact ground pressure in the roadway.

Nanhua, Nande, Su, and others (2020) developed a liquid-injected mining drill-type reamer, which adopts the way of adjusting water pressure for drilling and reaming operation and can realize the integrated operation of drilling, reaming, and filling. Its simple structure, easy operation, small size, lightweight, stable reaming, and retrievable tools greatly improve the anchor anchoring performance and ensure the safety of workers' underground operation.

Li (2021) adopted the optimized support technology of reaming anchor for 3# coal seam retrieval roadway in the Hongya coal industry and concluded that the root cause of insufficient anchorage force is the loose nature of the surrounding rock. The anchor force of the anchor rod can be effectively improved by means of the wedge-shaped reaming anchorage at the bottom of the anchor solid hole, which can better play its role of active support for surrounding rock.

Lu, Zhu, and Liu (2021) selected cement and epoxy resin for structural face grouting and carried out direct shear tests under different normal pressures, and concluded that grouting could increase the cohesive force between structural faces when changing the three-dimensional morphological characteristics of structural faces and increasing the shear strength. Fengxia Chi, Bo Han, and Yihan Sun et al (2022) investigated the effective law of mineral admixture type and content on the performance of cement-based-water glass and double-liquid grouting materials. It was concluded that fly ash could improve the fluidity and slag micronized powder could effectively enhance the late compressive strength of the agglomerate.

Liu, Li, Li, et al (2021) replaced the part of the ultrafine cement with micron-sized fly ash and used the external admixture nano-CaCO3 to improve the composite slurry properties. Nano-CaCO3 could improve the strength, but there was no significant change in viscosity, flow rate, and stone formation rate. The admixture of micron-grade fly ash reduced the viscosity and stone formation rate and strength and improved the flow rate appropriately. Fan Zhenwang, Zhang Lei, Chen Pengcheng, et al (2020) developed a low-viscosity ultrafine cement composite slurry with ultrafine cement, nano calcium carbonate, water-reducing agent, and ultrafine fly ash to meet the requirements of microfracture grouting and water plugging in the deep wellbore.

Li, Xiong, and Wang (2020) conducted performance improvement tests on high water filling materials and concluded that a water-cement ratio of 1.5:1, a foaming agent dose of 0.03%, a polypropylene fiber dose of 0.6%, and a compound early strength agent dose of 1%, which were the optimal ratios for high water filling materials with comprehensive properties such as good uniformity, swelling, and strength. Dai Yinshao, Tan Yuehu, Yang Qingheng, et al (2013) developed an anchoring material with fast hardening, micro-expansion, and high strength. The grouting material is modified by alumite and gypsum, and the cement mortar quickly generates a large amount of calcium alumina. The pore structure is gradually dense, which has the characteristics of rapid expansion in the early stage and stable expansion in the later stage.

Wang, Sun, and Wang, et al (2019) studied the water-rich material and added a quantitative amount of sodium silicate to the ternary system of silicate cement-aluminate and cement-gypsum dihydrate, which can make the crystalline transformation inhibit within the water-rich material, and ensure the strength decay at the later stage of hydration in the replacement session.

Ma (2016) developed a mining cement grouting material and concluded that if the cement slurry water-cement ratio is 0.5:1, the ACZ-1 modified additive ratio of 8% can meet the engineering needs, meeting the economic input requirements.

The main research of this thesis is the material of the filling body, which is filled in the part of reaming hole when the anchor rod is reamed bottom anchor. This research expects that the anchoring capacity of the anchoring system can be improved after using this material in the soft rock roadway.

Based on the above research by domestic and foreign scholars, it can be concluded that reaming anchorage is higher than normal anchorage, and the filling material has different effects on the anchorage effect. This research studies the effect of filling material on the anchorage effect.

3 NUMERICAL SIMULATION EXPERIMENTS

The numerical simulation used in this study is to compare the changes of stress and displacement between different materials for reaming, so as to determine the best filling material.

The soft armor used in this simulation is ANSYS, which is a finite element analysis software with analysis modules for meshing, solving, and post-processing. Each module is capable of studying the forces and displacements of the anchored objects.

3.1 *Modeling*

The model of numerical simulation was established separately for the anchor rod, anchoring agent, filler, and surrounding rock. The anchor rod was 22 mm diameter and 900 mm length left-handed rebar anchor rod; the anchoring agent was 20 mm inner diameter, 28 mm outer diameter, and the MSK2335 resin anchoring agent of 700 mm length; the filler was 28 mm inner diameter, 180 mm outer diameter, an inverted cone of 350 mm length; the surrounding rock was 800 mm long. The anchorage is required to closely cooperate with the anchor, filler, and surrounding rock within 700 mm anchorage length and not to produce mutual penetration. The grid is divided as follows:

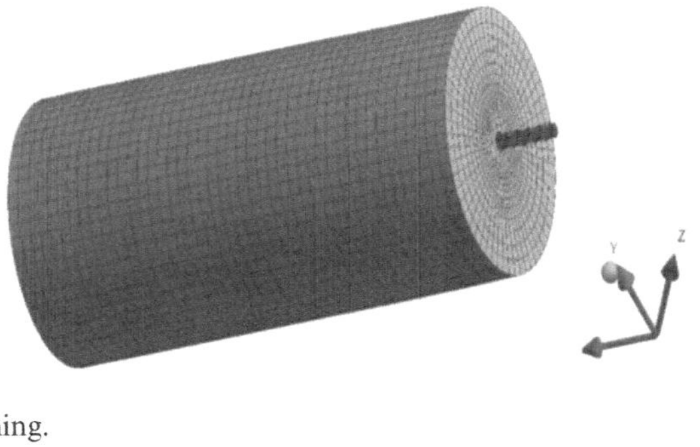

Figure 1. Model meshing.

3.2 *Model simulation process*

The experiment simulation is mainly for two groups. The first group is expanded bottom filler anchorage, and the filling body is a common material that is the comparison group. The second group is expanded bottom filler anchorage, and its filling material is self-made quick-setting material.

The experimental comparison should be quantitative, and this quantitation is to give 150 KN pulling force at the rod end of the anchor and analyze the change of anchor effect of different filling body materials under the same pulling force.

Table 1. Parameters of rock envelope and filler.

Name	Compressive strength/MPa	Cohesion/MPa	The angle of internal friction/°	Poisson's ratio	Shear modulus/MPa
Surrounding rock	3.2	2.56	34.3	0.24	4.71
General materials	21.3	18.04	25.4	0.34	47.59
Quick-setting materials	23.02	19.04	22.6	0.32	51.23

Table 2. The stress of each component for different materials.

Group	Anchorage/MPa	Anchor rods/MPa	Filling body/MPa
General material reaming anchorage	147.68	701.71	145.09
Self-made filling material reaming anchor	168.36	784.30	154.69

Table 3. Displacement of each component for different materials.

Group	Anchorage/mm	Anchor rods/mm	Filling body/mm
General material reaming anchorage	0.74	1.51	0.28
Self-made filling material reaming anchor	0.65	1.37	0.26

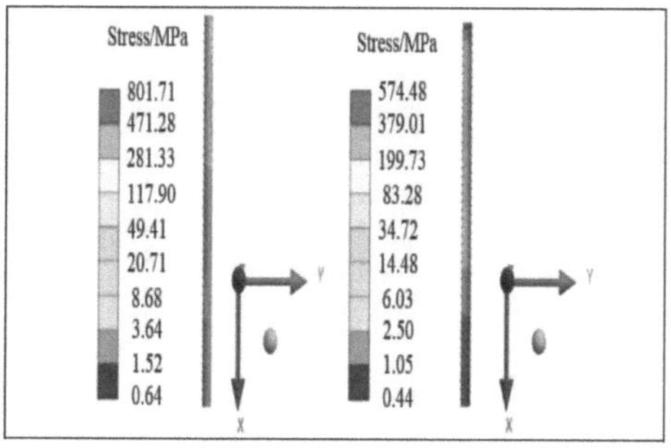

Figure 2. Anchor rod stress cloud.

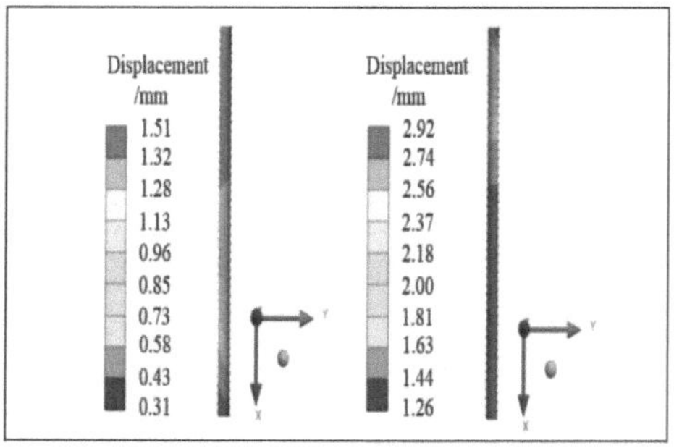

Figure 3. Anchor rod displacement cloud.

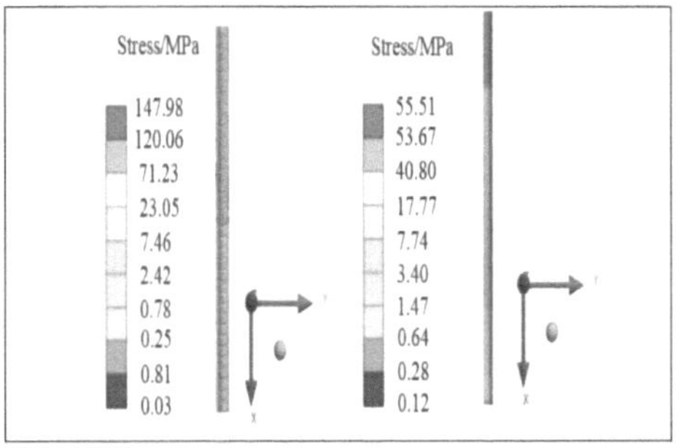

Figure 4. Anchor stress cloud.

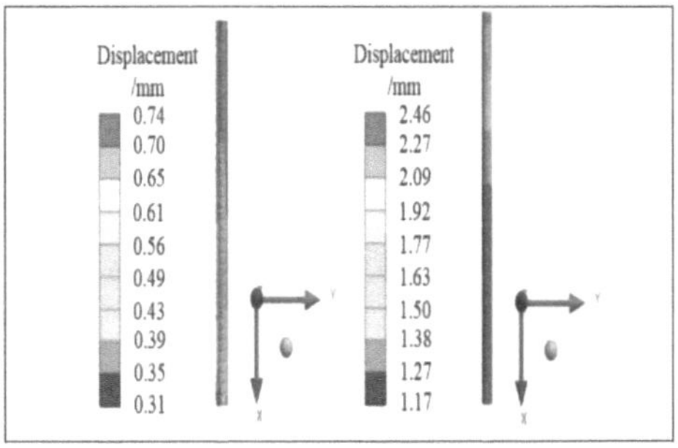

Figure 5. Anchor displacement cloud.

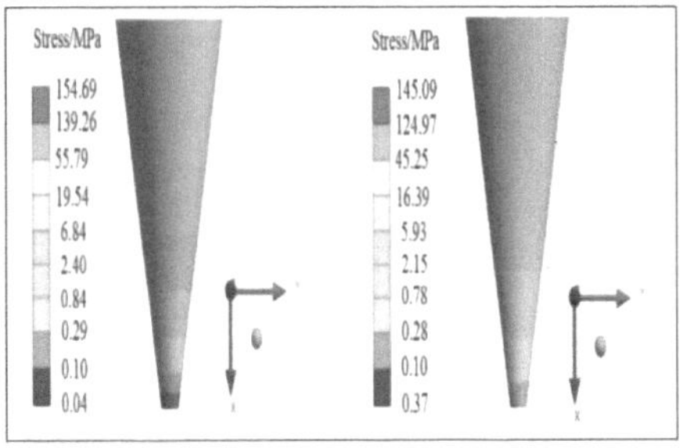

Figure 6. Stress cloud of the filler.

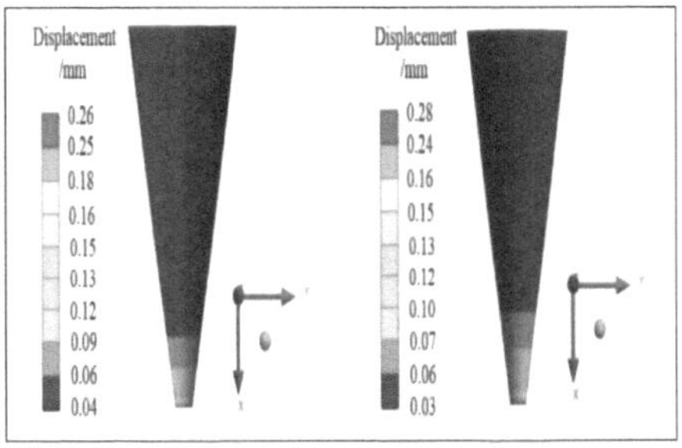

Figure 7. Displacement cloud of filler.

Under the load of 150 KN with the same pullout, the maximum stress value of the common filler material for the reamed filler anchor was 147.98 MPa; the maximum displacement value was 0.74 mm; the maximum stress value of the anchor rod was 701.71 MPa; the maximum displacement value was 1.51 mm; the maximum stress value of the filler was 145.09 MPa; the maximum displacement was 0.28 mm.

The maximum stress value of the anchor of the self-made filler material was 168.36 MPa with a maximum displacement value of 0.65 mm; the maximum stress value of the anchor rod was 784.30 MPa with a maximum displacement value of 1.37 mm; and the maximum stress value of the filler was 154.69 MPa with a maximum displacement value of 0.26 mm.

Compared with the common materials under the anchoring conditions of the expanded filler, the maximum stress value of the anchor is increased by 13.77%; the maximum displacement value of the anchor is decreased by 12.16%; the maximum stress value of the anchor rod is increased by 11.77%; the maximum displacement value of the anchor rod is decreased by 9.27%; the maximum stress value of the filler is increased by 6.62%; the maximum displacement value of the filler is decreased by 7.14%. This indicates that compared with the common materials, the quick-setting materials not only meet the engineering applications in time but also have a better anchorage effect than the common materials.

4 CONCLUSION

Under 150 KN pull-out load, the maximum stress value of the common material anchor increased by 166.58%; the maximum displacement value decreased by 69.92% compared with the conventional anchor; the maximum stress value of the anchor rod increased by

22.15%; the maximum displacement value decreased by 48.29% compared with conventional anchor; the maximum stress value of filler 145.09 MPa and the maximum displacement value 0.28 mm.

The maximum stress value of the quick-setting material anchor increased by 203.30%; the maximum displacement value decreased by 73.58% compared with the conventional anchor; the maximum stress value of the anchor rod increased by 36.52%; the maximum displacement value decreased by 53.08% compared with conventional anchor; the maximum stress value of filler was 154.69 MPa; the maximum displacement value was 0.26 mm.

REFERENCES

Chi F.X., Han B., Sun Y.H., Cheng Q.L., and Zhou W.J. "Effect of Mineral Admixtures on the Performance of Cement-water Glass Grouting Materials", [J]. *Science, Technology & Engineering*, 22 (02): 773–780 (2022).

Dai Y.S., Tan Y.H., Yang Q.H., Wang Z.H., Ding J.D., Xie J.N. "Development of Fast Hard Micro-expansion High Strength Anchor Grouting Material", [J]. *Civil Construction and Environmental Engineering*, 35 (04): 128–132 (2013).

Fan Z.W., Zhang L., Chen P.C., Zhang S., and Li Y.Z. "Research on Nano Calcium Carbonate Modified Ultrafine Cement Grouting Material", [J]. *Mining Research and Development*, 40 (05): 75–79 (2020).

He M.C., Du S., Gong W.L., and Nie W. "Mechanical Properties of Anchor Ropes with Negative Poisson's Ratio and Their Engineering Applications", [J]. *Mechanics and Practice*,44 (01): 75–87 (2022).

Li X.F., Xiong Z.Q., and Wang P. "Experimental Study on the Improvement of Mechanical Properties of Filling Materials Next to High Water Lanes", [J]. 30 (05): 95–100 (2020).

Li Z. "Research and Application of Anchor Reaming Anchoring Technology for Soft and Weak Surrounding Rock Roadway", [J]. *Journal of Shanxi Energy College*, 34 (01): 22–24+27 (2021).

Liu H.N., Li X.G., Li Y.Z., Zhang S., Chen P.C., Wu Y., Fan Z.W., and Tang C. "Research on the Performance of New Cement Slurry for Grouting and Plugging Water in Coal Mine Roadway Perimeter Rock". [J]. *Coal Mine Safety*, 52 (10): 57–63 (2021).

Lu H.F., Zhu C.D., Liu Q.S. "Study of Shear Mechanical Properties of the Structural Surface under the Action of Different Grouting Materials", [J]. *Journal of Rock Mechanics and Engineering*, 40 (09): 1803–1811 (2021).

Nan H., Nan D.Y., Su F.Q., and Wang B.J. *"Liquid-injected Mining Drill Type Reamer", [P]. CN108756757B*, 07–03 (2020).

Ma L.D. "Experimental Study on the Optimized Ratio of Cement Grouting Materials for Coal Mines", [J]. *Coal Mine Safety*, 25 (08): 40–42 (2016).

Wang Z.M., Sun Y.N., Wang Y.L., and Zhang S. "Strength Evolution Mechanism of Sodium Silicate Doped Water-rich Materials by XRD and FTIR", [J]. *Spectroscopy and Spectral Analysis*, 39 (10): 3199–3204 (2019).

Xi Y.X., Ju W.J., and Su S.J. "Experimental Study on Anti-scouring of Hydraulic Reaming and Hollowing of Impacted Ground Pressure Coal Seam", [J]. *Journal of Mining and Rock Control Engineering*, 2 (1): 013022 (2020).

Zhao C.X., Li Y.M. "Analysis of Force Characteristics and Anchorage Mechanism of Prestressed Anchors Based on Surrounding Rock Deformation", [J]. *Coal Mine Safety*, 51(07): 234–238 (2020).

Water Conservancy and Civil Construction – Oke & Ahmad (Eds)
© 2024 The Author(s), ISBN: 978-1-032-58614-4

Countermeasures and suggestions for accelerating the high-quality development of Jiaozuo public transport

Ziyan Zhao, Baohua Guo*, Mengjie Xu & Yan Wang
School of Energy Science and Engineering, Henan Polytechnic University, Jiaozuo, Henan, China

ABSTRACT: Urban public transport is an important urban infrastructure, which plays an important role in urban economic development and people's living standards. Promoting the high-quality development of urban public transport is an inevitable requirement to adapt to the new development and changes in China, and is also a necessary path to developing a strong transportation country. Based on the analysis of the current situation of public transport development in Jiaozuo City, this paper finds out the problems existing in the development of public transport in Jiaozuo City, and puts forward countermeasures and suggestions for the main problems, to accelerate the high-quality development of public transport in Jiaozuo City and create a better travel environment for the public.

1 INTRODUCTION

The 19th National Congress of the Communist Party of China and the Central Economic Work Conference made a major conclusion that "socialism with Chinese characteristics has entered a new era, and China's economic development has also entered a new era", pointing out that the basic feature of China's economic development in the new era is that China's economy has shifted from a stage of rapid growth to that of high-quality development. With the high-quality development of public transport in China and the continuous adjustment of relevant national policies, more and more domestic experts and scholars have studied and demonstrated the high-quality development of public transport, actively promoting the development of public transport in China. Jiang Hui (Jiang 2021) focused on the basic factors and weaknesses that restrict the high-quality development of public transport, deeply analyze the situation and requirements faced by the development of urban public transport, and put forward countermeasures and suggestions on governance system construction, legal system construction, service system construction, intelligent technology application, state-owned enterprise reform, personnel training, etc. under the development of public transport. Gao Wenting (Gao 2021), in combination with the situation of the development of Shanxi's transportation, puts forward countermeasures on how to build a transportation power and promote the high-quality development of Shanxi's transportation, from the aspects of deepening the supply side structural reform, accelerating green development, developing smart transportation, consolidating the safety foundation, and building legal transportation. Li Qingrui, Qian Junjun, et al (Li, Qian, Lu, Wu 2020) think it will provide strong support for the scientific formulation of high-quality "Fourteenth Five Year" transportation planning in China.

As socialism with Chinese characteristics has entered a new era, the main social contradiction in China has transformed into the contradiction between the people's growing need

*Corresponding Author: guobaohua@139.com

DOI: 10.1201/9781003450818-92

for a better life and unbalanced and inadequate development. The mission of urban public transport has also changed from meeting the people's travel needs to meeting the people's travel needs, which puts forward higher requirements for urban public transport. In recent years, Jiaozuo has made great progress in the development of public transport, but there are still many gaps compared with developed areas and high-quality development requirements. Therefore, based on the actual situation of Jiaozuo City, the research team carefully investigated the current situation of public transport development in Jiaozuo City, found the problems in the process, and analyzed the deep-seated causes. Combining with the actual situation of Jiaozuo City and the basic knowledge of traffic management, the research team finally put forward countermeasures and suggestions to accelerate the high-quality development of public transport in Jiaozuo City to create a better travel environment for citizens.

2 ANALYSIS OF THE CURRENT SITUATION OF BUS DEVELOPMENT IN JIAOZUO

2.1 *Line network planning*

2.1.1 *The status quo of the bus network in Jiaozuo city*

(1) Non-linear coefficient of bus lines
The non-linear coefficient represents the ratio of the actual operating distance to the spatial distance in the urban public transport operating line. The larger the non-linear coefficient is, the longer the detour distance will be. The excessive non-linear coefficient will result in a long travel time and uneven local passenger flow. The smaller the non-linear coefficient is, the straight the line will be. A too-small non-linear coefficient will cause inconvenience to transfer. China's "Urban Road Traffic Planning and Design Specifications (GB50220-95)" provides: the non-linear coefficient of public transport lines should not be greater than 1.4. So far, there are 44 lines in Jiaozuo City. Through calculation, the average non-linear coefficient of the current bus lines in Jiaozuo City is 1.67. The non-linear coefficient values of all bus lines exceed 1.4. Nine lines have non-linear coefficients between 1.0 and 1.4, 30 lines between 1.4 and 2.0, 4 lines between 2.0 and 2.5, and 1 line above 2.5.

(2) Areas not covered by the bus network
The survey found that in recent years, the communities around Wencheng Road, Shanyang District, and Jiaozuo City have been put into use in succession. Now there are three communities with a high occupancy rate. There are businesses and primary schools on the roadside. The number of permanent residents and the floating population is increasing. However, the whole section of Wencheng Road has no bus operation and no bus stops. The nearest station is in front of the office on Shanyang Road, and it is far from the main residential area of Wencheng Road. It takes 15-20 minutes to walk from the primary school on Wencheng Road to the bus stop on Shanyang Road. Donghuan Primary School is 1.1 km away from the nearest bus stop, making it inconvenient for citizens to travel.

2.1.2 *problems and cause analysis of bus network planning*

(1) Insufficient coverage of public transport network
From the regional analysis, the density of the bus line network in the core urban area is higher. Due to road restrictions, bus station layout, and other reasons, the density of the bus line network in the surrounding area of the central urban area is lower, the current bus line does not cover some of the city's peripheral areas, for example, Wencheng Road does not have bus line coverage, and the current bus travel demand mainly depends on walking to solve.

(2) The high non-linear coefficient of some bus lines
The average length of the line is large, some bus lines are too long, the coefficient of non-straightness is high, and the lines in the central urban area bypass seriously, resulting that the average travel time of residents' public transport is far higher than that of cars.

2.2 *Quality of service*

2.2.1 *The status of bus service quality in Jiaozuo city*

(1) Passenger satisfaction
This study investigates passenger satisfaction with Jiaozuo's public transport development. In this survey, about 100 residents were sampled and 93 valid samples were obtained. The scores of various indicators are shown in Table 1. After sorting out, the passenger satisfaction is 90.4%.

Table 1. Passenger satisfaction survey contents.

Serial number	Investigation content	Bus weight	The average score of valid questionnaires
1	Length of waiting time	1.5	8.5
2	Transfer convenience	1.5	8.5
3	Ride comfort	1.0	9.3
4	Waiting environment	1.0	8.6
5	In-car hygiene environment	1.0	9.5
6	Congestion degree in bus	1.5	9.5
7	Do drivers often have sudden stops, sudden rises, or sudden changes of lanes?	1.5	9.3
8	Bus driver‘s service attitude	1.0	9.3

According to the system table of evaluation index for public transport development in the literature (Qin 2019), the reference value of passenger satisfaction is shown in Table 2:

Table 2. Passenger satisfaction reference value.

Name of indicator	Indicator explanation	Reference value
Passenger satisfaction (%)	Comprehensive reflection of convenience, speed, reliability, and comfort	No less than 85

(2) Number of buses per million people
China's Code for Planning and Design of Urban Road Traffic GB5022095 stipulates that the planned ownership of urban buses and trams should be one standard car for every 800-1,000 people in large cities and every 1,200-1,500 people in small and medium-sized cities. According to the data of Jiaozuo City, there are 797 public transport vehicles, with a permanent resident population of 1,085,500 people (by the end of 2021). It is calculated that the number of public transport vehicles per 10,000 people is 7.34 standard vehicles per 10,000 people, with one bus per 1,363 people. According to Table 3, Jiaozuo is a big city.

Table 3. City classification table.

Grade	Small City	Medium-sized city	Metropolis	Megalopolis	Supercity
Division criteria (urban resident population)	Under 500,000	500,000 to 1 million below	More than 1 million and less than 5 million	5 million to 10 million below	Over 10 million

2.2.2 *Problems and cause analysis of bus service quality*

(1) Low operating efficiency and reliability
This paper investigates the satisfaction of Jiaozuo's residents with public transportation. Through analysis, it is found that the main reasons are long waiting times and poor transfer convenience. These residents do not choose public transportation because of the long walking time and inaccurate public transportation. Some people do not choose public transportation for other reasons, which shows that the level of public transportation operation service has an important impact on whether passengers choose to take public transportation. It can be seen that the slow speed and long waiting time of buses have become an important issue in the high-quality development of public transportation in Jiaozuo City. If the time of bus travel can be shortened, the operation efficiency can be improved, and the time gap with car travel can be narrowed, it will play a relatively large role in improving the competitiveness of bus travel and public satisfaction.

(2) No priority for vehicle supply
In terms of bus service hardware, there are few vehicles and insufficient effective supply. There are 797 bus vehicles in Jiaozuo. According to the survey, the ownership rate of bus vehicles per million people is only 7.34 units per million people, and the annual passenger volume is 100 million passengers. There is still a big gap with the advanced level.

2.3 *Infrastructure*

2.3.1 *Current situation of public transport infrastructure in Jiaozuo*

(1) Platform facilities
On the morning of June 2, 2022, investigators conducted a field survey on the Jiaozuo platform. Through sorting out and analyzing the questionnaire, it is found that Jiaozuo bus station facilities are complete, but there are also some deficiencies, such as ① The railway station platform: seats in the passenger waiting area are dirty. ② The information on the electronic station board is not updated in time, or even inconsistent with the actual situation.

(2) Operation status of the air conditioning vehicle
During the investigation, it was found that the air conditioners of Line 19 and Line 21 did not operate, which not only did not adapt to the hot climate of Jiaozuo City but also failed to meet the requirements of the masses.

(3) Ratio of new energy vehicles
The ratio of new energy vehicles refers to the ratio between the standard number of new energy buses and the total standard number of new energy buses. The new energy buses are low-carbon, environmentally friendly, efficient, and convenient. The higher the ratio is, the more adaptable they are to the needs of citizens. At present, the proportion of new energy buses in our city is only 53.4%, which is low in the field of public transport.

2.3.2 *Analysis of problems and causes in the development of public transport infrastructure*

(1) Inadequate maintenance of transport infrastructure
Although transportation infrastructure construction in Jiaozuo has made great progress, some areas of transportation infrastructure maintenance are still relatively insufficient. For example, the air conditioning is not running, the electronic bus stop information is not updated in time, the poor health of the passenger waiting in the area, and other issues, resulting in a poor passenger waiting experience.

(2) Low coverage of new energy vehicles
At present, to improve the attractiveness and comfort of public transport, public transport in many cities in China has vigorously developed new energy vehicles with high comfort, excellent performance, energy conservation, and environmental protection. New energy vehicles in Jiaozuo City have not yet reached full coverage, and it is necessary to continue to increase the investment in new energy vehicles. At present, the new energy buses that can improve the comfort of transportation only account for 53.4% of the total number of vehicles. In contrast, the proportion of new energy vehicles in Jiaozuo is far behind, which is not only incompatible with high-quality development but also fails to meet the requirements of the masses.

2.4 *Operation management*

2.4.1 *The status of bus operation management in Jiaozuo*

(1) Bus lane management
The purpose of building bus lanes is to reduce the interference of social vehicles on buses, improve bus speed and efficiency, and ensure the safety of bus operation (Song 2020). There are some problems in the management of bus lanes. Through the investigation, it is found that many social vehicles are driving on the bus lanes. The bus lanes are not dedicated to the bus, and the management of the bus lanes is not in place. The driving of social vehicles on the bus lanes interferes with the operation of the bus, which affects the efficiency of the bus operation and also causes hidden dangers to the safety of the bus operation.

(2) The status of bus driver wage in Jiaozuo
Jiaozuo Municipal Bureau of Statistics announced that the average annual wage of employees of the urban non-private unit in Jiaozuo in 2021 was 68,223 yuan, that is, the average monthly wage was 5,685 yuan. Jiaozuo Human Resources and Social Security Bureau issued a recruitment announcement in August 2021 to publicly recruit bus passenger drivers with a salary of about 3,500 yuan a month. According to the above situation investigation, the monthly average wage of employees of urban non-private units in Jiaozuo in 2021 is compared with that of bus passenger drivers in 2021, which is used as the standard to obtain the difference ratio (in percentage), to intuitively present the current situation of the wage of bus drivers in Jiaozuo (as shown in Table 4). It can be seen that the wage gap between bus drivers and employees of urban non-private units is relatively large, and the wage of bus drivers in Jiaozuo is relatively low.

Table 4. Comparison of wages of bus drivers and urban non-private sector employees

Index	The average monthly wage of urban non-private sector employees in 2021 (yuan)	Average monthly salary of bus drivers in 2021 (yuan)	Match number
Bus driver wages	5,685	3,500	-38.4%

2.4.2 *Problems and cause analysis of bus operation management*

(1) Those bus lanes that are not dedicated are still prominent.
Because no physical isolation or measures is relatively simple, the phenomenon of social vehicles occupying bus lanes is still high. On the road, motor vehicles and non-motor vehicles are mixed, and motorcycles grab lanes, especially battery cars. Due to the lack of effective control measures, the problem of road occupation is particularly prominent, which not only affects the running speed but also has a large safety hazard.
(2) Due to the long-term loss of business operations, work remuneration is difficult to increase, resulting in a serious loss of industry practitioners.

The reasons are as follows: ① Affected by the epidemic, buses with closed spaces are prone to infection; the implementation of shared bicycles; the proportion of citizens buying private cars has risen sharply. Multiple factors result in a decrease in passenger flow, a decrease in the income of the bus company, or even a loss, resulting in a lower salary paid by the bus company to the driver, or even unable to pay wages. ② the relevant departments did not pay subsidies on time, resulting in tight bus company funds and operational difficulties, and drivers can not be paid on time. ③ the income of drivers declines, or it is even unable to receive wages on time, then the driver resigned more.

3 COUNTERMEASURES AND SUGGESTIONS FOR HIGH-QUALITY DEVELOPMENT OF JIAOZUO PUBLIC TRANSPORT

3.1 *Bus network level*

1. We should improve the level of bus network coverage.
We can add new lines, increase the coverage of public transport, and solve the problem of no public transport coverage in some areas. It is recommended to add new bus stops on Wencheng Road.
2. We should optimize the adjustment of the non-linear coefficient.
We can optimize and adjust the lines with long mileages and large non-linear coefficients to further reduce the non-linear coefficient and bypass mileage. It is suggested to adjust the original bypass route, the No. 20 bus line. The optimization and adjustment of the No. 20 bus are shown in Figure 1. The blue line is the original No. 20 bus line, the red line is the adjusted No. 20 bus line, and the yellow line is the newly added bus line.

3.2 *Service quality level*

We should improve bus efficiency and travel reliability. Enterprises should always pay attention to the operation data, the line operation, and passenger flow changes, and find out the general rules of passenger travel and personalized needs. We can also adjust the urban roads, residential areas, factories, universities, and business districts, and make accurate route planning and design and operation schedules in time, to improve the operation efficiency of the bus network, fully meeting the travel needs of the people.

3.3 *Infrastructure level*

1. We should unify bus stop shelters and improve the waiting environment.
The waiting hall for buses is a public place frequently used by urban people, and the design must meet useful functions, such as shelter from rain and sun, temporary rest, understanding of bus routes, etc. Of course, we should also attach attention to the

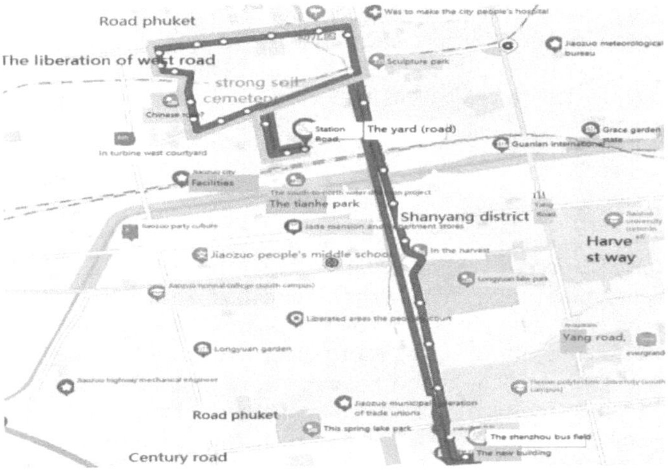

Figure 1. The diagram for optimization and adjustment of the No.20 bus route.

development and use of other ancillary functions, such as telephone equipment, wireless TV, and wireless network to realize information resource sharing, and the function of emergency power supply is outdoor emergency charging of electronic digital equipment (Zeng & Wang 2020).

2. We should increase the investment in new energy public transport vehicles and accelerate the construction of supporting facilities.

We should implement supportive policies, take measures to encourage the early retirement of old buses and trams and "yellow standard cars", and accelerate the application of new energy and clean energy vehicles. At the same time, we can further promote the pilot construction of Jiaozuo's public transport electrification, and accelerate the construction of supporting infrastructure for pure electric buses (Zeng & Wang 2020).

3.4 *Operational management level*

1. We should promote the construction and management of bus lanes, and accelerate the high-quality development of public transport in Jiaozuo.
   It is an indispensable and effective measure to promote the relevant competent departments to strengthen the enforcement of bus lanes. Otherwise, bus lanes are likely to become decorations. The method of "combining punishment with education" should be adopted. The occupation of bus lanes should not only be strictly punished but also the violators should be intensively studied for a certain time to ensure the effectiveness of law enforcement.
2. We should improve the price subsidy mechanism and the operating efficiency of public transport enterprises.

First, we should improve the price subsidy mechanism, comprehensively consider the social affordability, enterprise operating costs, and traffic supply and demand, and establish a multi-level and differentiated price system based on service quality, transportation distance, various public transport transfer modes, and other factors, reasonably define the scope of subsidies and compensation, and give appropriate subsidies and compensation to the policy losses caused by implementing low ticket prices, reducing or exempting tickets, and undertaking government mandated tasks. Second, Jiaozuo should adjust and optimize the bus operation routes according to the local conditions, and improve the bus load factor and the operating efficiency of public transport enterprises. Third, we should consider factors such as social affordability, operating costs of public transport enterprises, and local financial conditions, to reasonably determine the bus fare system to ensure the reasonable benefits of urban public transport.

4 CONCLUSION

Urban public transport plays a very important role in urban economic development and the improvement of people's living standards. Promoting the high-quality development of public transport in Jiaozuo is an inevitable requirement to adapt to the new development of China. Based on the current situation of public transport in Jiaozuo, the paper puts forward suggestions for speeding up the high-quality development of public transport in Jiaozuo from four aspects of public transport network planning, service quality, infrastructure, and operation management. I hope this study can provide some help for the high-quality development of public transport in Jiaozuo.

ACKNOWLEDGMENT

As the thesis is about to be completed, I would like to express my gratitude to all who have helped me. First of all, I would like to thank my tutor professor Guo Baohua, who gave me valuable guidance during my writing. I also want to thank my friends and classmates for their help and support. Secondly, I would like to thank the bus company for its help and data support for my paper. Finally, I would like to express my gratitude to the authors who quoted and used the relevant content for reference in this paper.

REFERENCES

Gao Wenting. Implementing the Outline of Building a Strong Transportation Country and Promoting the High-quality Development of Transportation in Shanxi [J]. *Transport Manager World*, 2021 (12): 119–122.

Jiang Hui. Discussion on High-Quality Development Strategy of Urban Public Transport from the Perspective of Traffic Power [J]. *Urban Public Transport*, 2021 (07): 31–35.

Li Qingrui, Qian Junjun, Lu Yi, and Wu Hui. "14th Five-Year" Ten Trends of High-Quality Development of Green Transportation [J]. *Journal of Management Cadre College of Ministry of Transport*, 2020, 30 (04): 14–17.

Qin Yuanyuan. *Research on the Problems and Countermeasures of Urban Public Transport Priority Development in Wanzhou District of Chongqing [D]*. Chongqing University, 2019.

Song Nachuan. Evaluation Method and Application of Urban Public Transport Development Level [D]. Hefei University of Technology, 2020.

Wang Zhiming. Suggestions on Public Transport Reform-Taking Linhe District of Bayannaoer City, Inner Mongolia Autonomous Region as an Example [J]. *Transport enterprise management*, 2021, 36 (03): 7–9.

Zeng Weidan and Wang Duoyang. "Chengdu Universiade" Longquanyi District Public Transport Countermeasure Research [J]. *China Management Informatization*, 2020, 23 (06): 185–189.

Water Conservancy and Civil Construction – Oke & Ahmad (Eds)
 ISBN: 978-1-032-58614-4

Analysis of quality problems and countermeasures in tunnel lining construction

Ge Kong
CTC Testing Holding Group Shanghai Co., Ltd, Shanghai, China

Jiao Zhang*
School of Municipal and Ecological Engineering, Shanghai Urban Construction Vocational College, Shanghai, China

ABSTRACT: In the past 30 years, tunnel engineering technology has developed rapidly in China. Since the geological conditions of tunnel engineering are complex, the requirements for engineering waterproof, impermeability and engineering life under different erosion conditions are relatively high (Tian Siming & Gong Jiangfeng 2020; Zhang Dejun 2020). The paper aims to improve the quality of tunnel lining construction, optimize the construction technology of tunnel lining (Li Dongyun *et al.* 2020; Zhu Xihao 2020), and reduce the accident rate of tunnel construction. Based on the collected various problems in tunnel lining construction at home and abroad, we summarize the problems mainly including the quality problems of tunnel lining construction joints and automatic pouring, vibration problems, and quality inspection problems of tunnel lining. The causes of tunnel lining construction problems are analyzed. Then we put forward the countermeasures to the quality problems of tunnel lining construction joints, the problems of automatic pouring and vibration of tunnel lining, and quality inspection problems of tunnel lining. The solutions to the quality problems of tunnel-lining construction during construction can provide a reliable theoretical basis for better tunnel-lining construction.

1 INTRODUCTION

With the rapid development of tunnel construction in the past 30 years, the average annual growth rate of highway and railway tunnels in China is up to 20%. The technologies for tunnel construction and operation have also been developed rapidly [1-3], especially the application of information technology in a lot of tunnel projects every year in China [4-5]. It brings new opportunities and challenges to tunnel projects. However, there will be a lot of problems in the construction process, like the construction quality of tunnel lining. Among them, the quality problems of tunnel lining construction joints, the problems of automatic pouring and vibration, and the quality inspection problems of tunnel lining are common at present. How to solve the problems has been puzzling for the construction of tunnel engineering.

Combining the information technology with the inspection standards issued by the Ministry of Housing and Urban-Rural Development of China, this paper puts forward the countermeasures to the above problems. It is of big significance for further improving the construction quality of tunnel lining and avoiding accidents in tunnel lining.

*Corresponding Author: GGGLG@163.com

DOI: 10.1201/9781003450818-93

2 PROBLEMS IN TUNNEL LINING CONSTRUCTION

2.1 *Quality problems of tunnel lining construction joints*

The overall control of tunnel lining structure construction is good, but there are many problems at the construction joints (Zhang Minqing *et al.* 2021), and water seepage is also common. Through collecting cases of quality problems of tunnel lining construction joints, and analyzing the causes of accidents, including construction control, design concept, and other aspects, the paper analyses the waterproof design drawings of the construction joints of a subway tunnel in a city and summarizes practical experiences of on-site construction for many years based on the actual situation of the site, thus finding prominent problems caused by the quality of tunnel linings construction joints, such as brick falling, holes, cracks, water seepage, and water leakage. The quality problems of tunnel lining construction joints can be divided into three categories: concrete "crescent" crushing cracks, exposure of ring-buried plastic waterproof belts, and loose concrete. How to solve these difficulties has always been the core and difficult problem of quality management of railway tunnel engineering. After randomly selecting 14 railway line tunnels, the paper found that the quality problems at the construction joints of tunnel lining works have existed for a long time, mainly shown as ring cracks, construction gaps, and particularly hollowing problems. Problems occurred in 10 tunnels, 3 of which were particularly serious. There were also quality problems in 7 tunnels, mainly invalid waterproofing caused by the circumferential plastic waterproof belt touching the concrete surface. In addition, there are many widespread circumferential cracks, symmetrically distributed on both sides, with a length of more than 3 m and a total width of about 2 mm. This quality problem not only endangers the waterproof quality of the tunnel but also endangers the operation safety of the tunnel. The quality problem of construction joints has a great impact on the normal operation of the tunnel, and the quality problem of tunnel lining construction joints has always been a concern for project professionals.

2.2 *The automatic pouring and vibration problem of tunnel lining*

After the quality of tunnel lining has been highly valued, people pay close attention to the normal operation of the railway, such as cavities, cracks, falling blocks, and more risky falling (Wei Jiazhi 2019). There are many inducements for quality problems, including improper management methods, insufficient subjective factors, low literacy of builders, outdated construction machinery, imperfect technology, unreasonable specifications, poor information content, and other factors. Therefore, to improve and enhance the quality of tunnel lining, it is necessary not only to deal with subjective factors but also to strengthen scientific research and eliminate the potential hazards of various factors. The pouring and vibration of concrete are important links to ensure the quality of tunnel lining. Traditional lining vehicles have problems such as formwork construction, fabric, difficult vibration, insufficient arch concrete, etc. at the construction site.

2.3 *The quality inspection problem of tunnel lining*

To ensure the efficiency and accuracy of tunnel lining detection, the paper analyzes the main potential lining quality hazards that may be caused due by its characteristics of high construction difficulty, long construction period, and high structural quality requirements (Gong Yanfeng *et al.* 2019). From the perspective of technology and management, the basic requirements are put forward according to the principles of advanced maturity, early intervention, phased development, and information management. Due to the limitation of construction technology and the lack of strict quality control, tunnel lining structures usually have quality problems and potential safety hazards. Sometimes the masonry problems cannot be directly revealed and must be confirmed according to the test indicators. At present, the inspection items of railway tunnel lining structures in China include the compressive strength of lining concrete, the thickness of lining concrete, the relative density of backfill after lining, the spacing of reinforcement, the thickness of reinforced concrete reinforcement protective layer, lining seepage and lining surface cracks, and the internal structure of tunnel lining.

3 SOLUTIONS TO PROBLEMS IN TUNNEL LINING CONSTRUCTION

3.1 *Countermeasures for quality problems of tunnel lining construction joints*

First of all, in order to better deal with the "crescent-shaped" crushing gap at the construction joint of the tunnel lining project, according to the selection of various plans and the vehicle positioning technology, a platform of the V-shaped groove zero lap rubber buffer lining was developed. The positioning technology of the V-shaped groove zero lap rubber buffer lining trolley is to set angle steel along the circumferential full arc at the front edge of the lining formwork trolley and to form a half V shape after the lining V-shaped chamfer. We should paste the triangular rubber strip on the half V-shaped chamfer, and the lining trolley is aligned with the sideline of the lining section and pressed on the triangular rubber strip, namely, the lining trolley is in place in the form of zero overlaps. Thus, the paper, combined with the tunnel construction process, studies a new technology of horizontal built-in stiffening buried rubber water stop.

This process is used in Project 18 of the Shanghai Metro Line, and the field experiment shows that the method has the advantages of good installation quality, convenience, high installation efficiency, and low comprehensive cost. The rubber water stop is put into the thick steel plate to improve the overall bending stiffness of the structure and the installation quality. After the lining construction, the external template shall be unloaded. The deformation caused by rubber water stops shall be strictly controlled, and the appearance quality is good.

At the same time, YDS high permeability epoxy chemical grouting material developed and produced by Guanghua Institute of Chinese Academy of Sciences is selected as the water cutoff and anti-seepage grouting material. Its physical and chemical performance indexes are shown in Table 1.

Table 1. Physical and chemical indexes of YDS high permeability epoxy chemical grouting materials.

Project		Index
Slurry performance	Initial viscosity (Component, 20°C) / (MPa/s)	2.2~12.0
	Surface tension (20°C) / (10^{-5} N/cm)	32~36
	Contact angle (SiO2, 20°C) (°C)	1~16
	The permeability coefficient of slurry (cm/s)	10
Mechanical properties of sand consolidation body	Compressive strength / MPa	70~90
	Tensile strength / MPa	8~14
	Shear strength / MPa	35~56

The construction process can be summarized as follows: installation of water stop → installation of grout stop hole → drilling of grouting hole → plugging of grouting section → grouting → restoration of finish and site cleaning.

Using the above construction process, some shield tunnel segment joints of Shanghai Metro Line 18 were treated for water interception and seepage prevention in the later stage. The seepage points in the regional section disappeared, meeting all the cancellation requirements and the subway tunnel operation acceptance standards.

We should observe the cast-in-situ concrete of the next layer of joint, drill holes on both sides of the construction gap, measure accurately with an electronic endoscope, and ensure that the average installation deviation is less than 1 cm. At the same time, we can use the geological environment radar detector to regularly check the construction gap, and no gap or hole is found. Compared with the traditional installation method, the installation time per circle is reduced from 7 hours to 3 hours, which is 50% higher than the existing installation method. Compared with the traditional installation method, the comprehensive cost is low.

3.2 *Solutions to problems of automatic pouring and vibration in tunnel lining*

First of all, after the concrete is poured into the formwork, the internal structure is loose and porous due to the cohesion between the stones and the sliding friction as well as the stones and the formwork. In the extremely unstable state, the content of voids and bubbles accounts for 5%~20% of the total volume of concrete, which fails to reach the specified relative density and will endanger the strength and durability of cement concrete. Therefore, the concrete must be vibrated and compacted to meet the required structure, size, and compressive strength of concrete. During the tunnel construction, the concrete of the support structure shall be poured symmetrically and vibrated hierarchically. At present, adhesive vibration is generally applied to outer wall vibration, and adhesive vibration is applied to the tunnel vault structure. Because the reasonable depth of concrete vibration is about 25 CM, the concrete construction of tunnel vault structure is a stage of weak tunnel construction support quality at present.

3.2.1 *Immersion vibration technology*

A. The main parts of vibration: the corners of the formwork, the places where the reinforcement is concentrated, under the main reinforcement, and between the reinforcement and the side formwork;
B. General insertion requirements: one-time insertion, multiple insertion, and tight insertion shall be adopted. It shall be inserted evenly and completely;
C. The regulations of vibration levels: when the concrete is vibrated in layers, the thickness of each layer shall not exceed 1.25 times the insertion depth of the vibrator, and shall not exceed 40~50 cm.
D. The regulations of left and right insertion: the flutter spacing of this method is 5~10 CM. It shall be inserted quickly to prevent the surface concrete from falling after vibrating, and to prevent the concrete cracks from separating.
E. Vibration duration: the vibration duration shall be fully controlled and not be too long or too short.

3.2.2 *Grouting process with formwork*

A. Number of grouting pipes: the grouting pipes are directly buried in the slurry of lining grouting vehicles. The buried length is generally 9~12 mm, and the total number of grouting pipe fittings should be 4. During the construction of reinforced concrete lining, the total number of pipe fittings can be appropriately increased according to the site conditions;
B. Grouting raw material: it shall own slight swelling, high flow, and high adhesion. Key performance parameters: bulk density of 2200~2300 kg/m^3, expansion coefficient of 0.3%~2%, and liquidity 32 cm for the time of 90 min;
C. Grouting time: after the initial setting of lining concrete (usually 2 h) and before the final setting (usually 12h);
D. Main grouting parameters: the final grouting working pressure is 0.2 MPa. The amount of formwork grouting is calculated according to the design requirements, namely, some cracks are not considered in the construction of the multi-tracked fast tunnel. And the gap size is 0.5~1 CM assuming that there is a gap in the arch.

At this stage, arc formwork grouting is still controversial, as most authoritative experts believe that this is also a good method for arch crown reinforcement, while others believe that the popularization of formwork grouting technology is likely to increase the difficulty of concrete pouring technology control at the construction site. Therefore, while using and promoting the technicality of formwork grouting, the grouting management scheme should be combined to further improve the lining level of tunnel construction.

3.3 *Countermeasures for quality inspection of tunnel lining*

General requirements are formulated based on the characteristics of tunnel detection:

(A) Inspection implementation suggestions are formulated from the aspects of cost control, efficiency, precision, and work reduction after fully considering the construction period and operation period;
(B) New projects are tested after considering geological environment standards, natural environment, design scheme characteristics, and raw material characteristics according to the specific conditions of the project;
(C) Problems in human inspection are dealt with to reduce the dependence of inspection on human resources after using automation technology, information management methods, and corresponding testing equipment;
(D) Machine and equipment shall be tested with comprehensive capacity and their total amount and efficiency shall be tested according to a single actual operation;
(E) Mobile phone software with intelligent identification (or auxiliary identification) shall be provided as far as possible to test the data processing method and identification efficiency (An Zheli *et al.* 2021).

Suggestions on technical standards such as test methods, instruments, equipment, preparation, actual operation, presentation process, analysis principle, and report according to the standards of comprehensive consistency, adaptability to the natural environment, stability, and accuracy is proposed. The hydrogeological case materials, construction drawings, engineering change materials, and engineering construction records of tunnel construction were collected, and the inspection scheme and performance parameter serial numbers were selected. The testing equipment shall meet the requirements of anti-pollution, and waterproofing, as well as the requirements of local air pressure, temperature, and natural environment. We should ensure to carry out regular maintenance, correction, and maintenance. The inspection shall take into account the performance of instruments and equipment, composition correlation, data processing methods, and the transmission of air pressure, temperature, and other elements in the measurement and evaluation, and create an adjusted solid model for quantitative inspection under the lining quality air pressure and temperature in sufficient test reports. The test results shall be tested by high-precision single-point detection methods, such as the borehole coring method, impact method, and acoustic method.

4 CONCLUSIONS

This paper analyzes the quality problems during tunnel lining construction and proposes countermeasures. The new technologies and equipment for the secondary lining construction quality inspection during tunnel construction in recent years are summarized. According to the requirements of mechanization, intelligence, and integration, it can promote a comprehensive, rapid, and accurate inspection of large tunnel groups in megacities. From the aspects of lining surface inspection, lining internal inspection, masonry quality inspection, and other new technologies, the lining requirements applicable to underground tunnel lining in megacities are also summed up. From the perspective of basic theories, work experiences, and characteristics analyses, some proposals for the test report, mode, and new technology application of metropolitan underground tunnels are put forward. However, it still needs to be further selected and improved in the combination with the actual working conditions. As for the detection equipment, to reduce the requirements for radar detection, it is necessary to develop and design unmanned detection equipment for the complex weather and natural environment in the plateau area.

In addition, some suggestions for future research are as follows. As the tunnel-lining structure will be affected by over-excavation and under-excavation, more reasonable technologies and construction management for tunnel-lining construction are important research directions. Informatization of monitoring projects is the second research direction. The collection, management, comments, and application of the test data must be carried out as early as possible. The integration of internet big data technology and artificial intelligence is another important research direction. Smart tunnel construction detection technologies and monitoring instruments are anticipated to become intelligent systems.

REFERENCES

An Zheli, Ye Yangsheng, Ma Weibin, et al. (2021). Discussion on Lining Quality Inspection Method and New Technology·During Construction of the Sichuan-Tibet Railway Tunnel. *Tunn. Constr*. 41 (7), 497–89.

Gong Yanfeng, Xiao Mingqing, Wang Shaofeng, et al. (2019). Review and Develop the Trend of Railway Tunnel Detection *Technology. Railw. Standa. Des*. 63 (5), 93.

Li Dongyun, Li Zhenfeng, Wang Yibo and Zhao Haitao. (2020). Research on the Real-time Monitoring System for the Transport Safety of Dangerous Chemicals in Expressway Tunnels. *Inn. Mongo. Highw. Transp*. (3), 51–55.

Tian Siming and Gong Jiangfeng. (2020). Statistics of Railway Tunnels in China by the End of 2019 (Chinese and English). *Tunn. Const*. 40 (2), 292–297.

Wei Jiazhi. (2019). Design and Application Study on Casting Concrete System with Layered and Window-by-window to Secondary Lining Side Wall in Tunnel. *Railw. Con. Tech*. (2), 14–17.

Zhang Dejun. (2020). Development Trend and Application Research on Tunnel Shield Construction Technology. *Eng. Cons. Des.*, (6), 185–186, 197.

Zhang Minqing, Jia Dapeng, Xin Weike, et al. (2021). New Technology for Quality Control of Construction Joints in Railway Tunnel Lining. *J. Railw. Eng*. 270 (3), 59–64.

Zhu Xihao. (2020). Tunnel Traffic and Electromechanical Integration Monitoring System Based on AI Video Detection. *China. Traff. Info*. (S1), 127–131+136.

Water Conservancy and Civil Construction – Oke & Ahmad (Eds)
© 2024 The Author(s), ISBN: 978-1-032-58614-4

Operational situation analysis on electromechanical facilities of Hong Kong-Zhuhai-Macao bridge based on safe and comfortable visual requirements

Zhong Wei
Nanping Administration Branch, Fujian Expressway Group Co., Ltd., Fujian, China

Ronghua Wang, Shangwen Qu & Jiangbi Hu*
Faculty of Architecture, Civil and Transportation Engineering, Beijing University of Technology, Beijing, China

ABSTRACT: Since Chinese and British standards (GB and BS) respectively are applied in different sections of the Hong Kong (HK) - Zhuhai (ZH) - Macao (MO) Bridge's electromechanical facilities, the effect presented to drivers' view varies. From the perspective of traffic operation safety, this paper conducts a comparative analysis of the safety visual recognition level of typical electromechanical devices - variable message signs (VMS) and tunnel lighting in different sections of the HK-ZH-MO Bridge where GB and BS adopted respectively. The research results show that the VMS in the HK section is conducive to drivers' safety and the comfort of visual recognition. The differences in the light environment characteristics between inside and outside the tunnel during the daytime at the exit of the Undersea Tunnel and Guanyinshan Tunnel are too obvious to cause the "white hole" effect. Some suggestions are provided to realize friendly coupling between the lighting in the tunnel and the natural light environment outside and energy-saving.

1 INTRODUCTION

The Hong Kong (HK) - Zhuhai (ZH) –Macao (MO) Bridge connects the three places of China, the two special administrative regions in the Guangdong (GD)-HK-MO Greater Bay Area, and nine cities of GD. It is a bridge-tunnel cluster project with the longest tunnel, the most complex technology, the most difficult construction, and the largest island in the world. The HK-ZH-MO Bridge starts from the HK Port Artificial Island near HK International Airport in the east, crosses the Lingdingyang in the west, connects ZH and the MO Artificial Island, and ends at the Hongwan Interchange in ZH, with a total length of 55 km. Among them, the ZH connecting section is 13.4 km long, the undersea bridge engineering section is 22.9 km long, and the island-tunnel section is 6.7 km long. The HK connecting section bounded by the border between GD, HK, and MO is 12 km long. The undersea bridge engineering is jointly constructed and managed by three places, and the ports and connecting lines of the three places are constructed and managed by the three separately.

Electromechanical (E&M) facilities are an important sub-project of expressway tunnel E&M engineering. The safety, rationality, and management-friendly characteristics of their service directly impact the operational safety and management of highways. The E&M facilities include monitoring, charging, communication, power supply & distribution, lighting, etc. Improving the service of E&M facilities helps to give full play to the road traffic capacity and to improve the economic and social benefits of the road. Monitoring facilities include video surveillance, dynamic information release & traffic guidance facilities, video detection, and event detection facilities, etc.

*Corresponding Author: hujiangbi@bjut.edu.cn

DOI: 10.1201/9781003450818-94

Among monitoring facilities, the variable message signs (VMS) as a facility directly serving road users, provide drivers with the road, traffic, and environmental conditions, etc. promptly, affect the drivers' cognition of road traffic environmental information, and assist them during the driving process, which causes stable and reasonable driving behaviors to improve traffic safety and efficiency [1]. Lu et al. [2] study the influence of factors such as light intensity, flicker frequency, and luminous form on the visibility of traffic signs by using a driving simulator. They found that the improvement from LED signs to retro-reflective signs in visual recognition performance is limited under good lighting conditions during the daytime, which is significant at night, and the visual recognition effect of semi-translucent luminous signs is better than that of dot matrix luminous signs. Liou et al. [3] used a highly reliable glare evaluation method to study drivers' visibility of LED traffic sign layouts and found that the lower the luminance ratio of the traffic sign is, the greater the glare experienced by the participants. When the luminance ratio is 6 200:2 066 cd/m^2, the sign layout has the best visibility to drivers.

The HK-ZH-MO Bridge is one of the world-class Chinese-made engineering projects, whose VMS of the mainland section are in accordance with Chinese standards [4–6] etc., and that of HK are designed under the British standard [7].

The characteristics of highway tunnels, especially long tunnels, cause high driving safety risks, accident severity, and great escape difficulties, affecting the traffic operational safety of tunnel sections. A good lighting environment enables drivers to drive safely and comfortably in the tunnel-affected area [8]. The lighting standards of tunnels all over the world mostly use a single luminance indicator, but the luminance values regulated by different countries as per different traffic flow and design speeds are different. Current research has proved that the analysis and evaluation of tunnel lighting quality using the mesopic theory is more effective, as it meets the requirements of human visual perception. But the related research mainly used the static test on different lighting segments of the tunnel, but few tests were adopted in a dynamic situation [9]. The lighting of the undersea tunnel in the mainland section of the HK-ZH-MO Bridge is according to Chinese standards [5].

The setting of VMS and tunnel lighting of the HK-ZH-MO Bridge represent the current technical level of construction in the Chinese and British systems. By comparing and analyzing the service effects of the operation of the E&M facilities, the VMS and tunnel lighting are set up in the different sections from the traffic operation safety and management, and this paper tries to provide a practical basis for management efficiency.

2 VARIABLE MESSAGE SIGNS (VMS)

2.1 *Visual performance parameter of VMS*

VMS are signs that display and convey information promptly on the changes in traffic, road, and climate conditions. They are generally used as a dynamic display of speed control, lane control, road conditions, driving behavior, etc. As the carrier of information dissemination, it has the characteristics of high efficiency and timeliness. VMS is pre-stored with information on various vehicle operating conditions in the system so that the management personnel can instantly control the system manually or automatically by the control system to display relevant information to inform drivers to take corresponding safe and reasonable driving measures when the roads, traffic, and climate environment conditions change, among which the luminous VMS, LED signs, and fiber optic signs are the mainstream technologies. LED signs are most widely used due to the advantage of a large amount of displayed information. The highway LED VMS are installed outdoors, so the huge impact of the all-weather working environment needs to be considered. To ensure safe and comfortable visibility in adverse weather conditions, avoid glare on rainy days and nights with low visibility, and reduce energy waste, LED VMS needs proper luminance and penetration performance with the function of automatically luminance adjusting according to the ambient illumination, including the factors affecting the driver's visual safety and comfort, such as sight distance, luminance uniformity, and beam angle [10]. Besides, color is also one of the factors causing traffic accidents, since it affects the drivers' psychophysiological conditions and driving efficiency. The physiological effect of color is mainly on visual ability and fatigue, while the psychological effect is changing people's emotions [11]. Red, yellow, and green are international traffic colors. Red is used for various prohibition signs, stop the danger, or information indicating firefighting facilities, yellow for conveying attention and warning information, and green for safety information. Accordingly, current Chinese standards have made corresponding provisions on the colors used for different information in VMS. Hence, besides the sight distance, luminance uniformity, and beam angle of the LED VMS, other factors also affect the driver's visual recognition, such as font colors, character height, character spacing, and information volume. The technical requirements for VMS *in* Chinese standards [5] are that the luminance of the layout in the tunnel is not less than 3 500 cd/m^2, the layout is free of glare, and the dynamic visual recognition distance is greater than 200 m with the function of fault self-checking. Chinese standards [4,6] specify that the LED VMS used outdoors should ensure the display quality and the driver's visual safety and comfort.

The VMS of the HK section is designed as per the EU standard [7], which regulates display color, beam angle, luminous intensity, and luminance ratio, and sets different levels for each performance parameter. The display colors are red, yellow, green, white, and blue, which are divided into 2 grades, C1 and C2, as per the purity, and the luminance is divided into 6 grades, which can be automatically adjusted as per the ambient illumination. The maximum (max.) luminance is 5 times the minimum (min.) luminance. When the external illumination reaches a higher level, the luminance value of the VMS should reach the corresponding level without external lighting, and the luminance ratio values are divided into 3 categories, R1, R2, and R3, as per visual clarity. The beam angle is divided into 7 grades from B1 to B7, and the size and spacing of text and symbols are divided into 5 grades from A to E as per the design speed and sight distance. The technical parameters of VMS in the HK section used the corresponding highest level. The optical lens used the concentrated light emitted by the LED and projects areas to be covered, while other areas are without light. Such concentrated light can achieve the required luminance with a small driving current. The overall design reduces the impact of stray useless light on road users and light pollution, saves energy, and achieves a long overall equipment life [6].

2.2 *Evaluation and analysis of the operational service effect of VMS*

The VMS on the whole line of the HK-ZH-MO Bridge are all LED VMS, of which the mainland section is designed as per relevant Chinese standards. The font colors are red, yellow, and green. The operation effect is shown in Figure 1.

During the daytime, besides the function of information transmission, the VMS in the mainland section can generate such long-wavelength and distinguishable colors as red, yellow, and green on the cone-shaped photoreceptor cells on the retina of the human eyes, making drivers' visual perception more sensitive [12]. However, from the perspective of the consistency of traffic color, the VMS should use the same color and the min. the luminance of the VMS on each reference axis is specified as 8 000 cd/m2, but the max. limits and luminance ratio requirements are not given. (Luminance ratio is defined in BS EN 12966-2014 [7] as a balance between light output and sign-reflection).

When the external illuminance is less than 1000lx, like on cloudy days, the min. the luminance of the VMS is too large, causing potential glare and energy waste. The luminance ratio is positively correlated with perceived sharpness. Under certain light environments, a proper luminance ratio can increase the perceived sharpness of VMS text and improve visual comfort without light pollution or energy waste. During nighttime operations, there is a face recognition camera installed above the gantry of some VMS in the mainland section. The light pollution caused by the high luminance of the supplemental light of the camera blurs

(a) Red information for temporary control
(b) Yellow information for traffic weather
(c) Green prompt information
(d) Red prohibition message in the tunnel
(e) Red message of minding the crosswind
(f) Red message of minding the weather conditions

Figure 1. VMS operation (a)-(d) for daytime and (e)-(f) for night. (a) Red information for temporary control (b) Yellow information for traffic weather (c) Green prompt information (d) Red prohibition message in the tunnel (e) Red message of minding the crosswind (f) Red message of minding the weather conditions

the VMS, causing difficulty for the driver to recognize accurately, as shown in Figure 2(b), which affects information transmission.

The VMS of the HK section is designed as per the relevant EU standards, its font color is white, and the operation effect is shown in Figures 2 and 3. The font color uses consistent explanatory white fonts with higher visibility compared with the black background, which

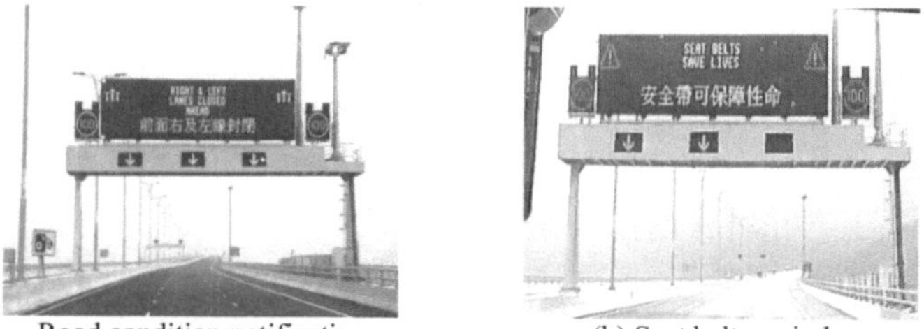

Road condition notification
(b) Seat belt reminder

Figure 2. VMS for daytime operation. (a) Road condition notification (b) Seat belt reminder

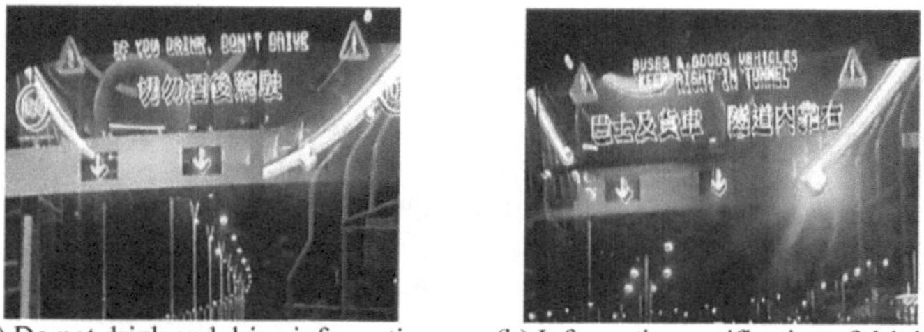

(a) Do not drink and drive information
(b) Information notification of driving lane requirements

Figure 3. VMS for night operation. (a) Do not drink and drive information (b) Information notification of driving lane requirements has a better effect on the driver's visual perception. For the luminance ratio, the setting is more refined. By setting a certain luminance ratio and the max. Luminance limit, can not only increase the perceived clarity of the text, and improve visual comfort, but also prevent light pollution such as glare or energy waste. For luminous uniformity, beam angle, and sight distance, the mainland section, and the HK section have similar regulations, while different regulations are stipulated in terms of character height, character interval, line spacing, etc., due to the differences in the characteristics of Chinese characters and letters. The visual recognition effect can meet the requirements of clarity and comfort.

3 TUNNEL LIGHTING

3.1 *Safety and comfort requirements for visual recognition of driving*

The purpose of tunnel lighting is to provide road users with a safe and comfortable visual environment so that drivers can see road traffic environment information at a safe distance to meet drivers' driving expectations and visual needs, ensuring that drivers can access, pass through, and exit the tunnel safely at a design speed. Presently, there are only luminance evaluation indicators in the tunnel lighting environment, but the actual light source characteristics should also include color temperature, color rendering index, etc., which may change with seasons, weather, and times of the day with a great impact on the dynamic visual recognition efficiency of the target [8]. Therefore, in different weather and periods, the tunnel lighting environment is quite different from the natural light environment outside the tunnel. When the difference is serious, it will lead to a "white hole" or "black hole" effect in the drivers' vision, forming a "blind" vision period [13]. The lighting in the tunnel can be in harmony with the natural light environment outside only if it changes according to the characteristics of the light source outside the tunnel, to ensure that drivers can safely and comfortably recognize the road traffic conditions ahead [1]. Therefore, the artificial light environment of the tunnel needs to fully consider the luminance, color temperature, color rendering index, and other characteristics of the light source. In addition, to provide drivers with good visibility and visual comfort, and avoid glare or flickering effects caused by alternating changes in luminance in the field of view, it is necessary to make corresponding requirements on the light distribution angle of the fixtures, the overall uniformity of road surface luminance, and the longitudinal uniformity of the midline luminance.

In terms of tunnel lighting control technology, the lighting control system should not only effectively monitor and manage the artificial lighting system, ensuring the characteristics of the natural light source in the approaching section of the tunnel and the artificial lighting source in the entrance section, the harmony of the characteristics of the artificial lighting source in the tunnel exit section, and the natural light source outside the exit. It shall also improve the safety level of tunnel operation to realize the energy-saving control and effectively protect the light source, prolong the service life of equipment, reduce the energy consumption of tunnel lighting, and achieve the unity of "safety" and "energy-saving". Chinese standards [5] and HK's "Public lighting design manual" both have relevant requirements for tunnel lighting and control systems.

3.2 *Evaluation and analysis of tunnel lighting operation service effect*

3.2.1 *Undersea tunnel of HK-ZH-MO Bridge*

The design speed of the undersea tunnel of the HK-ZH-MO Bridge is 100 km/h. The lighting system is designed in accordance with the regulations of Chinese standards [5], and two rows of LED lighting are installed in the entire section with uniform luminance as well as soft and comfortable light without glare. The tunnel entrance section is divided into two lighting sections of equal length, the luminance of the second section is half of that of the first section, and the lighting fixtures are symmetrically distributed outside the lane boundary, as shown in Figure 4(a). The transition section is divided into three transition lighting sections considering the CIE tunnel luminance adaptation curve. The luminance ratio changes in steps of about 3:1 and the length of each section is divided according to the curve, which is equivalent to the driving distance of 4 s, 4 s, and 6 s. The luminance of the middle section is based on the recommended values of the EU, CIE, and Japan's tunnel lighting regulations, and fully considers the actual operation of highway tunnels in China and the lighting effects test of human biological effect based on the detection distance of small objects, and transition and intermediate lighting fixtures are set above the lane dividing line, as shown in Figure 4(b). The exit section is divided into two lighting sections with a length of 30 m. According to the field dynamic illuminometer measurement, the luminance of the exit section 1 is 3 times that of the middle section, the luminance of the exit section 2 is 5 times that of the middle section, complying with the current standards, and lighting fixtures are symmetrically distributed on the outside of the lane line. The side walls of the entire tunnel are covered with white reflective tiles. When the tunnel light shines on the tunnel wall, the wall material reflects the light, which increases the road illuminance and reduces the investment and operation cost of tunnel lighting. At the same time, the reflective effect of the wall material also has an inductive effect, which can improve the driver's driving comfort.

Although the design of the lighting environment of the undersea tunnel considers the luminance and color temperature, the characteristics of the light environment inside and outside the tunnel at the exit during the day are too different to result in a "white hole" effect, see Figure 4(c). When tunnel lighting transitions from a yellow light source with a low color temperature in the entrance section to a white light source with a high color temperature in the middle section, a sudden change in light source affects the comfort of driving recognition, see Figure 4(d). To avoid glare in the tunnel and the "black hole" effect when driving out of the tunnel, which leads to potential safety risks, the enhanced lighting in the tunnel should be off at night. However, during the operation process, the tunnel lighting during daytime and night is not friendly coupled with the natural light environment outside the tunnel according to different control strategies, resulting in significantly higher lighting luminance levels in the entrance, exit sections, and transition

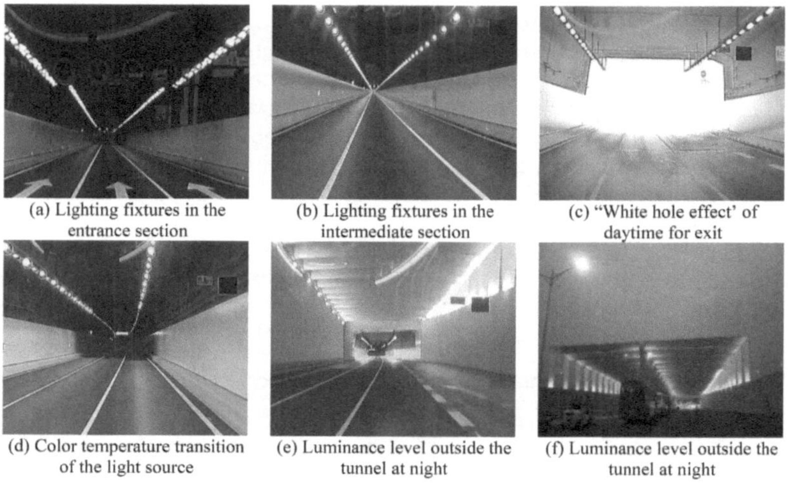

Figure 4. Undersea tunnel lighting operation status. (a) Lighting fixtures in the entrance section (b) Lighting fixtures in the intermediate section (c) "White hole effect' of daytime for exit (d) Color temperature transition of the light source (e) Luminance level outside the tunnel at night (f) Luminance level outside the tunnel at night

sections of the tunnel at night than the road outside the tunnel. When driving on this road section, the driver will feel uncomfortable with visual recognition caused by a sudden change in luminance, meantime waste of electric energy is caused. The lighting dimming cover outside the tunnel adopts the design concept of "seeing light but not lamp", which can provide the driver with light to adapt to the driving conditions, but its line shape is inconsistent with that of the lighting inside the tunnel with poor driving induction, as shown in Figure 4(e) and (f).

3.2.2 *The Guanyinshan tunnel at the HK section of the HK-ZH-MO Bridge*

The design speed of the Guanyinshan tunnel is 100 km/h, and the design principles of the lighting system followed the International Commission on Lighting (CIE-88: 2004) and the UK standard (BS 5489-2: 2003+1: 2008) and adjusted according to the "Public lighting design manual" of the Highways Department of HK Special Administrative Region to match local conditions. The whole line of the Guanyinshan tunnel adopts LED lighting sources according to the specifications, and the lamps are arranged above the edge line of the carriageway. The total length of the entrance section is equal to the stopping sight distance. In the first half of the length, the luminance is equal to the starting point value, while in the second half, it is reduced by a step ratio of 3:1, which is equal to 0.4 times the starting point luminance value at the end of the entrance section. During the daytime, the entrance section of the tunnel adopts the light source dense arrangement layout, and three rows of lighting fixtures with different intervals are arranged above the two-lane lines to approach the luminance level outside the tunnel, which effectively solves the "black hole" problem, as shown in Figure 5(a). The luminance of the transition section is changed according to the CIE tunnel luminance adaptation curve, the steps of the luminance ratio are not greater than 3, and the luminance of the last step section is not greater than 3 times that of the middle section. Four rows of lighting fixtures are installed in the transition section, and fixture spacing of the second transition section is large with a good luminance transition, as shown in Figures 5(b) and (c). According to the design speed in the middle section, it is required to maintain the luminance value of 8~10 cd/m2 during the daytime and 4~5 cd/m2 at night. Two rows of lighting fixtures are arranged in the middle section as shown in Figure 5(d). The lighting luminance in the exit section increases linearly from the middle section luminance to five times its luminance at a distance of 60 m from the exit. Six rows of lighting fixtures are

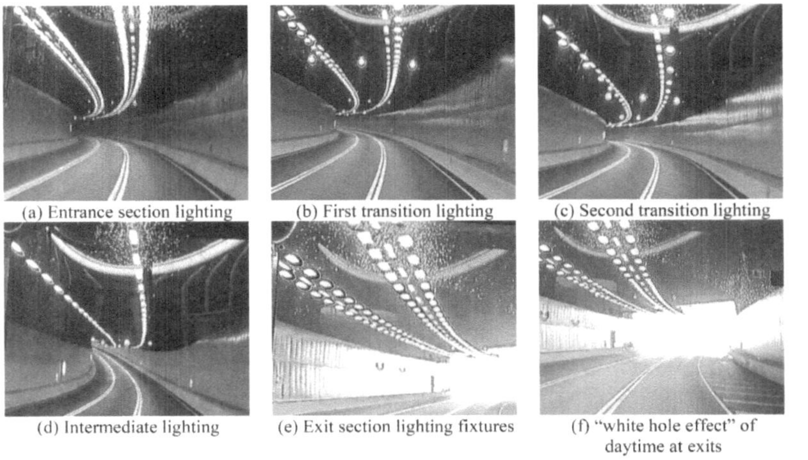

Figure 5. Guanyinshan tunnel lighting during daytime. (a) Entrance section lighting (b) First transition lighting (c) Second transition lighting (d) Intermediate lighting (e) Exit section lighting fixtures (f) "white hole effect" of daytime at exits

set in the exit section and entrance section, as shown in Figure 5(e). The side walls of the whole section of the Guanyinshan tunnel are covered with white reflective tiles, which reduces the investment and operating costs of tunnel lighting and improves drivers' driving comfort.

During the operation of the Guanyinshan Tunnel, the luminance level L20 (S) outside the tunnel is monitored in real-time by the luminance detector at the stopping sight distance outside the tunnel, and the luminance is controlled automatically or manually for hierarchical control. In the daytime, due to the large difference in the characteristics of the light environment inside and outside the tunnel at the exit, a "white hole" effect appears, as shown in Figure 5(f). At night, only three rows of lighting fixtures are turned on in the whole section of the tunnel through the control system and cover the structure of the dimming cover outside the tunnel. The luminance of the road in the tunnel is uniform and driving induction and the luminance level transition from inside to outside the tunnel are good, as shown in Figures 6(a) and (b).

(a) Lighting dimming cover at night

(b) Night lighting of Guanyinshan tunnel

Figure 6. Guanyinshan tunnel lighting at night. (a) Lighting dimming cover at night (b) Night lighting of Guanyinshan tunnel

4 ENLIGHTENMENT FROM THE CONSTRUCTION OF E&M FACILITIES

The E&M facilities of the mainland section and HK section of the HK-ZH-MO Bridge have adopted the Chinese and British standards respectively. The VMS and lighting facilities along the route can meet the needs of driving visual safety, but the VMS and lighting quality in tunnels of the mainland section can be further improved.

1. The VMS of the mainland section of the HK-ZH-MO Bridge are in red, yellow, and green, providing basic visual recognition and achieving corresponding functions as long as they meet the requirements of the regulations and are not interfered with by other facilities. However, from the perspective of the consistency of traffic color functions, a unified explanatory white font should be used. As to luminance and luminance ratio, the VMS in the mainland section only set a min. display luminance of 8000 cd/m2, while the max. luminance value and luminance ratio are not specified.
2. In the case of low illumination in the outdoor environment, the min. the luminance value of the VMS is large, which causes both the potential danger of glare and energy waste. The luminance ratio is positively correlated with perceived sharpness. Under certain ambient light conditions, setting the luminance ratio and the max. luminance limit reasonably can improve the visual comfort of variable information signs without causing light pollution or too much energy waste.
3. The lighting environment design for all sections only uses the luminance of the light source as the quantitative index. The quality of the light environment in the tunnel is quite different from the natural light environment outside, which is liable to cause many safety hazards in the tunnel section in terms of tunnel lighting design. By using the method of

visual efficacy, Zhang et al.[14] proved that drivers have different dynamic visual recognition efficiencies for small target objects in light environments with different combinations of three light source characteristic indicators of color temperature, luminance level, and color rendering index. Therefore, the lighting design index system, the lighting source meeting the lighting quality requirements, the monitoring equipment for the quality of the light environment inside and outside the tunnel, and the dimming strategy all need to be further systematically improved.

Based on the above comparative analysis, the next step for setting VMS is to carry out relevant research to clarify the selection of font colors conforming to traffic colors, and to study the max. luminance limit and luminance ratio threshold of VMS under different tunnel lighting environments. In terms of tunnel lighting settings and intelligent dimming, we aim to establish a complete lighting design index system and standard through real vehicle tests, and to study the dimming strategy under different working conditions in the tunnel based on the change of natural light outside the tunnel, so that the natural light outside is all-weather friendly coupling to the light environment in the tunnel to achieve safe operation, energy saving, and consumption reduction.

FUNDING

This work was supported by the scientific research project of Fujian Expressway Science & Technology Innovation Research Institute Co., Ltd.

REFERENCES

[1] BSI Standards Limited. (2014). *"Road Vertical Signs-variable Message Traffic Signs: BS EN 12966:2014"*. BSI Standards Limited, UK.

[2] Hu J., Jiang C., and Gao X. (2020). "Analysis of Tunnel Lighting Brightness in the High-altitude Area Based on Driving Workload." *Tunnel Construction*, 40 (S1), 17–24.

[3] Liou J. and Wu C. (2017). "Assessment of Drivers' Visual Perception of the Information Displayed in LED Traffic Signs at Road Construction Sites". *Journal of the Society for Information Display*, 25 (1), 53–60.

[4] Liu F. (2017). *"Research on Dimming Control Mechanism and Strategy of Expressway Tunnel"*. Beijing University Of Technology, Beijing.

[5] Liu W. (2012). "Primary Exploration in the Physical Form of Paresthesia Acquiring Color". *Value Engineering*, 26 (085): 319–321.

[6] Lu J., Xu T., Peng Y., *et al.* (2018). "Visibility of LED Active Luminous Traffic Signs". *China Journal of Highway and Transport*, 31 (4), 150–154.

[7] Ma W. (2015). *"Safety Research on the Tunnel Light Environment of Expressway Based on Bright to Dark Visual Recognition Demand."* Beijing University Of Technology, Beijing.

[8] Ministry of Transport of the People's Republic of China. (2014). *"Guidelines for Design of Lighting of Highway Tunnels: JTG/T D70/2-01-2014."* China Communications Press, Beijing.

[9] Ren F. (1993). *Traffic Engineering Psychology*. Beijing University of Technology Press, Beijing.

[10] Standardization Administration of the People's Republic of China. (2009). *"Light-emitting Diode Changeable Message Signs of Expressway: GB/T 23828-2009."* China Standard Press, Beijing.

[11] The Ministry of Public Security of the People's Republic of China. (2010). *"LED Variable Message Sign of Road Traffic Inducement: GA/T 484-2010."* China Standard Press, Beijing.

[12] Zhang G., Li D., Feng Z., *et al.* (2017). "Study on Night Visibility of LED Active Luminous Signs." *Journal of Highway and Transportation Research and Development*, 2, 254–257.

[13] Zhang X., Hu J., Wang R., *et al.* (2017). "The Comprehensive Efficiency Analysis of Tunnel Lighting Based on Visual Performance." *Advances in Mechanical Engineering*, 9 (4): 1–9.

[14] Zhu S. (2007). *"Study on Variable Message Signs Design"*. Beijing University of Technology, Beijing.

Water Conservancy and Civil Construction – Oke & Ahmad (Eds)
© 2024 The Author(s), ISBN: 978-1-032-58614-4

Development of seismic isolation structures in seismic applications

Chengxi Liu*
Architectural Engineering Institute, Xinjiang University, Urumchi, China

ABSTRACT: In construction engineering, the design of a building structure is closely related to stability, safety, and cost. In this regard, this paper briefly introduces the importance of seismic isolation structure design in building structure design, basic principles of seismic isolation design, seismic isolation of foundation, and frame shear wall system. Moreover, the characteristics and latest development results of foundation seismic isolation, lead core rubber seismic isolation support, and frame shear wall structure. At the same time, basic measures to ensure the seismic insulation performance of building structures are proposed. This paper summarizes various seismic structures, expresses the prospect, introduces the application of new materials in seismic isolation structures, and puts forward the development direction of seismic isolation structures and construction in the future.

1 INTRODUCTION

Accelerating the development of seismic isolation technology is an important basic work for the design of seismic-resistant structures of buildings [1]. First of all, the structural design of seismic-resistant buildings does not deeply analyze the development trend and daily use data of the seismic industry in a country [2]. If the seismic requirements of general high-rise buildings are directly applied to some high-rise buildings, it may directly cause many defects in the building projects during the design process [3]. Second, the technical concept of seismic system structure design does not keep up with the times [4]. Drawing on some advanced experience and successful application cases of international seismic system structure design, we can improve the design of seismic system structures of large buildings to eliminate major safety accidents [5,6]. Designers also need to fully grasp the application of advanced building anti-structure design technology concepts, improve the original seismic structure design technology guidelines for enterprises, and adopt more reliable advanced seismic design structure technology to effectively ensure the quality and safety of various construction projects [7,8].

2 FOUNDATION VIBRATION ISOLATION

2.1 *Principle of foundation seismic isolation*

The foundation seismic isolation system is a passive control system different from the traditional seismic structure system [9]. The laminated rubber support is set on the top of the foundation of the structure to extend the self-vibration period of the structure passes through a large horizontal displacement of the shock isolation layer, thus suppressing the

*Corresponding Author: 280042763@QQ.com

 DOI: 10.1201/9781003450818-95

transmission of horizontal seismic action to the superstructure [10]. The study illustrates that the basic shock isolation structure has a good shock absorption effect. Under the horizontal earthquake, the seismic isolation layer gives full play to the role of consuming seismic energy through large deformation, so that the deformation of the whole structure is concentrated in the isolation layer, inhibiting the transmission of the horizontal seismic action to the superstructure, so that the interlayer displacement reaction, the floor shear reaction, the floor acceleration reaction are significantly reduced, and the superstructure whether it is in the overall state of translation, reducing the horizontal seismic action of the structure [11–13].

2.2 *Lead core rubber vibration isolation bearing*

Lead core laminated rubber bearing (LRB) is set in the ordinary natural LRB lead core, under the action of ground vibration. The lead core provides greater damping and plastic deformation to absorb the energy into the earthquake and it can also increase the initial bearing stiffness to resist the structure in the wind load. Being a common metal, lead has certain advantages in the absorption of energy and plastic deformation during small earthquake action, making it suitable as an elastic-plastic body. Because of the perfect combination of lead core and laminated rubber, it can play both the role of lead core and laminated rubber. Thus, the lead core rubber bearing can be used directly in the engineering structure without needing to install other resistors, making it both economical and practical, as well as easy to design and construct [14].

Hakan Öztürk designed an irregular reference for the eight layers of the three dimensions architectural model, as shown in Figure 1 [15]. A seismic isolator model was developed to analyze the historical bidirectional nonlinear response of the two horizontal components of the seismic record simultaneously. Later, an architectural model with irregularities was created using a reference building model. To simulate their nonlinear behavior, they modeled columns and beams using fiber hinges for use in building models with seismic isolators and fixed brackets to simulate their nonlinear behavior. The structure with an isolator considers lead-rubber acted as support. Temperature analysis, taking into account strength deterioration due to cyclic movement of the lead core, and limit analysis were recommended by the code. At first, the strength disappearance of lead cores in conventional and irregular isolator structures was studied.

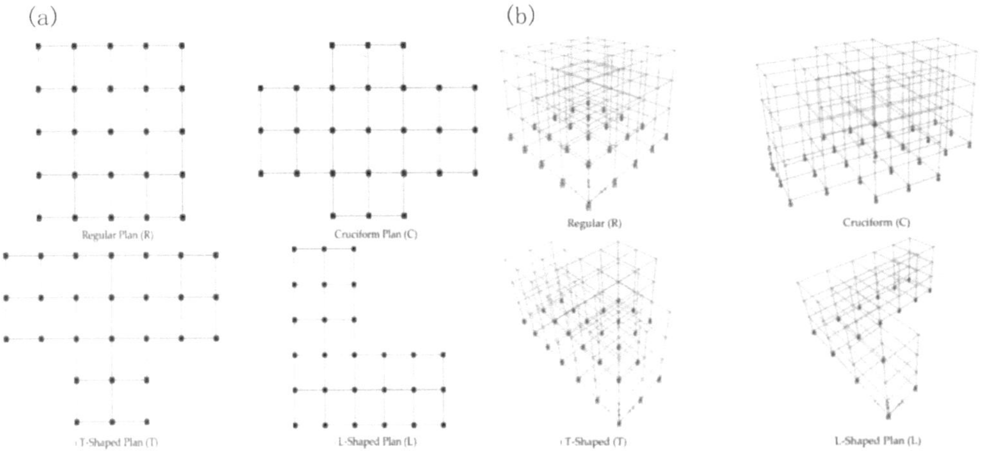

Figure 1. (a) The building plans and (b) Three-dimensional views of the structures [15].

2.3 *Laminated rubber bearing*

Laminated rubber support shock isolation technology is to set laminated rubber support between the upper layer and the foundation of the building to separate upper structure of the building and the foundation, thus changing the dynamic characteristics and dynamic action of the whole building structure. It is to properly extend the horizontal basic period of the upper structure of the rubber bearing. When there is an accidental load such as the occurrence of an earthquake, the structural system of the building is softened to weaken the transmission of the earthquake and strong shock waves [18], reducing the horizontal seismic action of the structure [16–18].

Zhang *et al* [19] Interpretation and analysis of the response of load-bearing bridges made of rubber properties under earthquakes with artificial neural network (ANN) were carried out, and the bearing principal structure model is concentrated on ANN technology was established by static cyclic tests on LRBs. In addition, the ANN-based bridge seismic demand model was combined with the ANN method and it is applied to the rapid interpretation and evaluation of bridge damage caused by earthquakes. Their results showed that the bearing size and vertical load were the major factors causing the bearing intrinsic structure model.

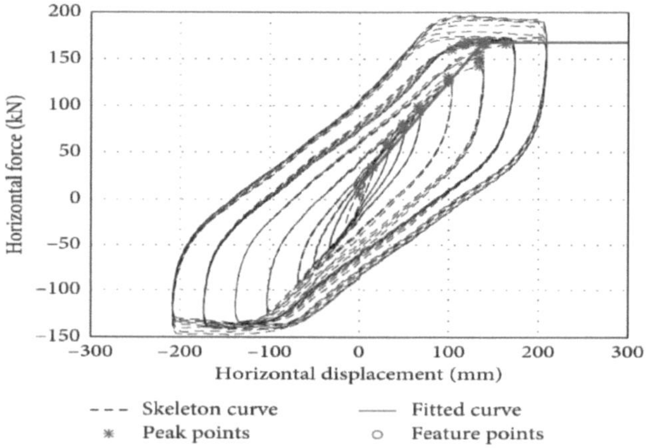

Figure 2. Specimen Y1-8-45 force-displacement curve [19].

Koo *et al.* [20] designed leaded small LRBs with a vertical design load of 10 kN. Beam structures with different natural frequencies (S1–S3) and four LRB-supported false masses were installed on shock absorbers to study their seismic isolation performance: (1) S1: structures close to isolated frequencies; (2) S2: the frequency of the input spectrum is close to the peak of the structure; (3) S3: high-frequency region structure. The test results were described according to the different seismic classes of OBE, SSE, and BDBE, and their analytical structures were compared to confirm that the lead-inserted small LRB had sufficient performance in isolating seismic energy and the dynamic characteristics expected in the LRB design. Figure 3 shows a comparison of the response spectra obtained from the testing and analysis of single girder construction and virtual masses on a shaker.

3 FRAME SHEAR WALL STRUCTURE SYSTEM FOR SEISMIC ISOLATION

3.1 *Principle*

The so-called frame shear wall technology refers to the combination of frame and shear wall, in which the beam and column use a rigid or hinged connection, and then constitute a

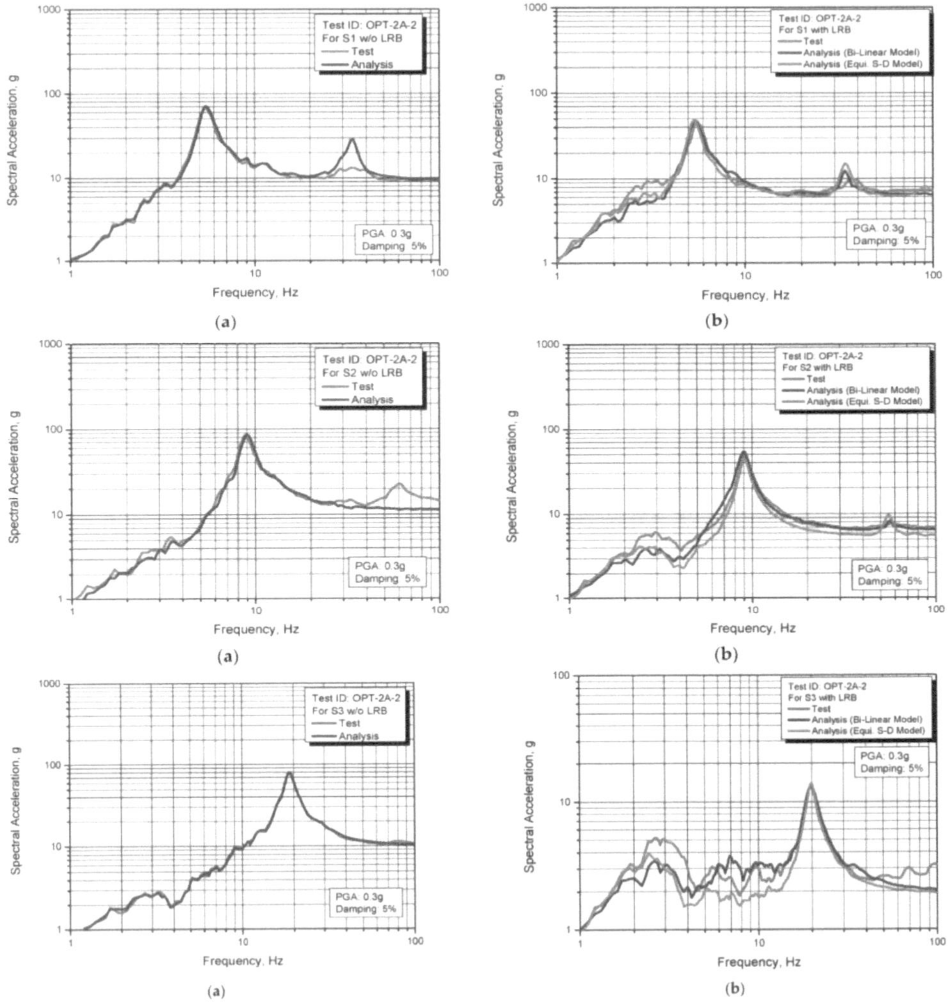

Figure 3. The comparison of response spectra of S1, S2 and S3 in structural experiments and analyses: (a) Non-isolated; (b) Seismically isolated [20].

load-bearing system of a structure [21]. In other words, by combining the beams and columns, the frame is formed to resist the horizontal and vertical load forces during the construction of the house [22]. At present, in the construction of housing buildings, the design of internal and external formwork, the fixing of internal and external formwork, and the lifting of wall formwork are mainly used.

3.2 *Applications*

Seon-Chee Park *et al.* [23] investigated the applicability and environmental protection of mixed-prefabricated composite structure systems in apartment buildings. The suitability of the hybrid precast composite structural system for application and environmental friendliness was verified by comparing it with the flat reinforced concrete Rahman frame. It is also proposed that the hybrid precast composite structural system can be used as a lateral intelligent frame and a gravity intelligent frame. According to the structural design results of the mixed prefabricated composite structure system, fewer materials are required per unit area compared with the slab and reinforced concrete Rahman frame structure. Therefore, this structural system can reduce CO_2 emissions and energy consumption and help to protect the environment.

Yun Chen Junzuo Li *et al* [24] hybrid coupled shear walls (HSW) with replaceable coupling beams (RCB) were investigated. The middle part of the coupling beam was replaced by a replaceable fuse. Through the cyclic loading test of four 1/2-scale coupled shear wall samples, the deformation process, component yield order, skeleton curve, and wall damage distribution were simulated, which were in good agreement with the test results. The main advantage of HSW is that the damage of the coupling beam is mainly concentrated on the replaceable fuses, while the other components remain unchanged. In addition, the damage to the wall piers is mitigated because the fuses can consume a large amount of energy. We have provided further suggestions for the conceptual design of HSW.

Aliaari *et al.* [25] developed the Seismic Filled Wall Isolator Subframe (SIWIS) system, as shown in Figure 4. Two vertical light steel nails and one horizontal light steel nails make up the SIWIS system, and there are "rigid and brittle" members in the vertical members. In order to reduce wind loads and buildings drifting under small to medium earthquakes, and to detach them during destructive events, SIWIS's design allows for infill wall frame interaction. In addition, the seismic performance of the SIWIS system was studied and analyzed in detail by using the nonlinear finite element model. The results show that this isolation system can improve the seismic isolation performance of walls filled with masonry because it protects the infill wall and frame from the damage of interaction.

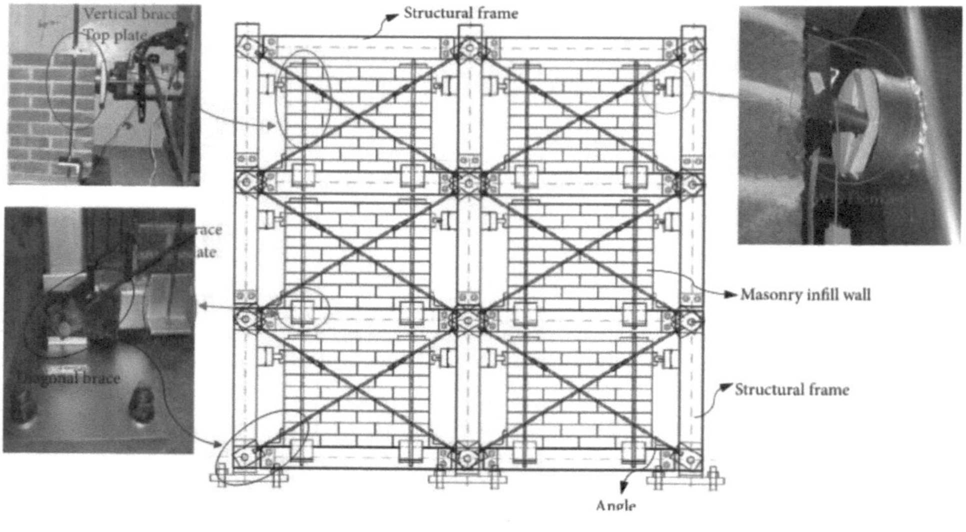

Figure 4. Schematic design of scaled two-bay three stories SIWIS frame [25].

3.3 *Replaceable structural originals*

Yuanqi *et al* [26] proposed a beam-column hinge joint with replaceable energy-consuming elements is proposed, wherein steel beams and columns are hinged with pins, increasing the replaceable energy-consuming elements at the corners of the pins. A section of high-strength and H-beam screw rigidly connect the heat dissipation element to the steel column and hinge it to the steel beam using high-strength bolts. They compared the effects of steel beam length-span ratio and linear stiffness ratio on joints with the seismic performance of welded steel frame beams and column joints and mainly analyzed the horizontal cross-section of energy-consuming components. The results show that in practical applications, the nodes can replace the energy-consuming parts, can be fully assembled, and can also control the nodes connecting the beams and columns. The deviation between the results of the numerical simulation and the results of the experiment is greater than 10%, so the agreement is better. In addition, the concept of "energy consumption is destroyed first and the ground is easy to replace" is particularly in line with the destruction mode of the node.

A novel fuse-coupled beam (FCB) was first suggested by Fortegni *et al.*[27], as shown in Figure 5. It was assumed that all the inelastic deformation will be centralized in the middle part of the beam (the fuse plane), to protect the two external steel beam members. Web plates with upper and lower flanges connect the fused segment to the outer steel beam segment, flange connections, and sliding key bolts. The comparative experimental results show that the FCB has an earlier loss of energy and lower stiffness than Classic beams connected by steel (SCB). However, the difficulty of repair/replacement and the cost after damage were minimized. A representative detail of the coupled steel beam is also shown in Figure 3, and the floor plan view of the prototype structure is drawn.

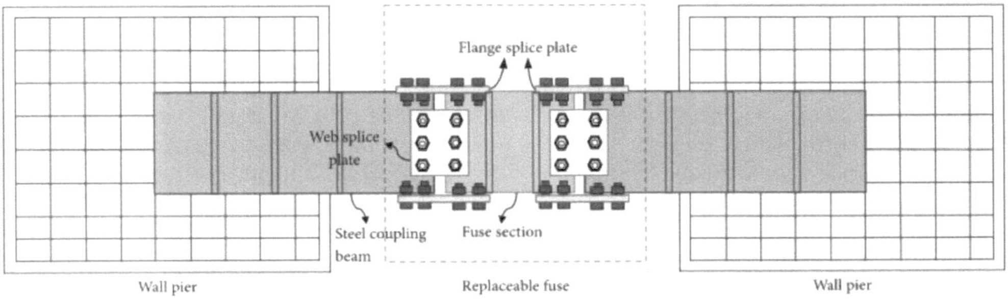

Figure 5. Details of replaceable Fuse steel connecting beam [27].

Mao *et al.* [25] suggested a shock absorber (SMA) manufactured using the properties of shape memory alloys, seismic energy was installed in the reinforced concrete frame shear wall structural system, and the coupling beam was consumed. An assembly schematic of the SMA is shown in Figure 6. The SMA shock absorber consists of four components made of steel (assembly parts I-IV) and two sets of SMA wires (groups A and B). When the relative vertical displacement device occurs between the ends of the cantilever shear wall, the seismic energy is dissipated by the SMA conductors located in parts II and III (or parts II and IV) around the SMA conductor, achieving tension dissipation and removal.

4 OUTLOOK

With the development of seismic isolation technology, seismic isolation structure is now widely used in the field of building civil engineering. The two most widely used seismic isolation bearings are LRBs and lead core rubber bearings, and the most widely used structure is the frame shear wall structure, where these two bearings are now widely used. However, both bearings have their own advantages and disadvantages. Ordinary LRB energy dissipation capacity can provide a large horizontal deformation capacity to improve the energy dissipation capacity, and must be combined with other dampers or energy dissipation components with it. On the other hand, lead core LRB, is in the middle of the

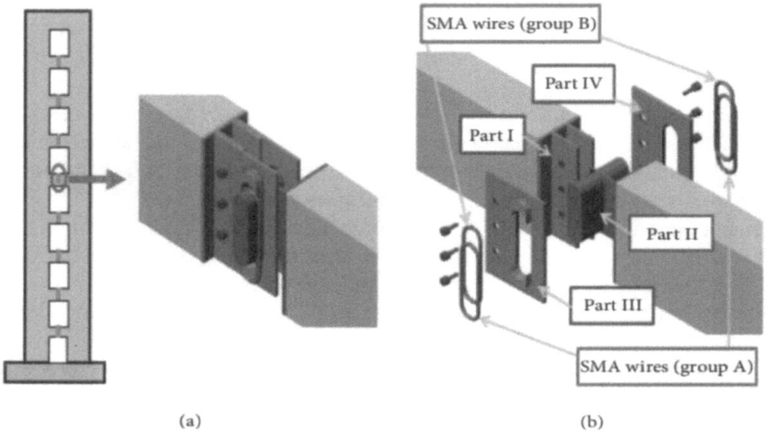

Figure 6. The SMA damper is installed in the position and assembly part in the coupling beam: position (a) of the SMA damper; (b) the damper assembly parts [25].

ordinary LRB open a certain space in the round hole into the lead core rod, but the lead core bearing energy dissipation capacity fully depends on the energy dissipation capacity of the lead core. Lead core deformation is difficult to restore the original state after the deformation of lead core, which directly affects the self-resetting ability of lead core bearing. Moreover, lead is a toxic metal element, if not disposed properly. It can cause environmental pollution or even poisoning incidents; meanwhile, the frame shear wall structure can still reduce the cost and improve the utility by adding new structural elements.

5 CONCLUSION

This paper summarizes two mainstream foundation seismic isolation and one seismic isolation structure and explains their basic principles and practical applications. The future development direction of the new seismic isolation technology is proposed; namely, the use of new materials and new structures, such as the use of replaceable coupled beam shear wall system, which has less post-earthquake damage compared with the traditional shear wall structure, improves the seismic capacity of the isolation bearings, and also enables the use of dampers of new materials for seismic isolation to be applied to the beams. This greatly improves the seismic isolation capacity of the building. At present, the research and application of the new structure are far from saturation, and they will be developed and applied in the future to make the seismic isolation technology play a greater role in protecting people's lives and properties.

REFERENCES

[1] Kubo, T.; Yamamoto, T.; Sato, K.; Jimbo, M.; Imaoka, T.; Umeki, Y., A Seismic Design of Nuclear Reactor Building Structures Applying Seismic Isolation System in a High Seismicity Region -a Feasibility Case Study in Japan. *Nuclear Engineering and Technology* 2014, 46 (5), 581–594.

[2] Loss, C.; Piazza, M.; Zandonini, R., Connections for Steel-timber Hybrid Prefabricated Buildings. Part II: Innovative Modular Structures. *Construction and Building Materials* 2016, 122, 796–808.

[3] Wang, D.; Zhou, J.; Jiang, W.; Wang, J.; Jiang, X., Research on Seismic Shear Gravity Ratio Limit for Super High-rise Buildings Higher Than 500m. *Building Structure* 2012, 42 (5), 24–27.

[4] Abou Alhaija, M.; Batali, L., Seismic Behavior of Micropiles and Micropiled Structures Used for Increasing Resilience: A Literature Review. *Applied Sciences-Basel* 2022, 12 (5).

[5] Rainieri, C.; Fabbrocino, G.; Cosenza, E., Some Remarks on Experimental Estimation of Damping for Seismic Design of Civil Constructions. *Shock and Vibration* 2010, 17 (4–5), 383–395.

[6] Hadad, Y.; Laslo, Z.; Ben-Yair, A., Safety Improvement by Eliminating Hazardous Combinations. *Technological and Economic Development of Economy* 2007, 13 (2), 114–119.

[7] Gerami, N. D.; Liaghat, G. H.; Rahimi, G. H.; Khazraiyan, N., The Effect of Concrete Damage on the Penetration Depth by the Tandem Projectiles. *Proceedings of the Institution of Mechanical Engineers Part C-Journal of Mechanical Engineering Science* 2018, 232 (6), 1020–1032.

[8] Nie, G.; Wang, W.; Du, K.; Wang, D.; Ma, J., A Review and Prospect of the Seismic Design Theory for Large Span Spatial Structures. *World Earthquake Engineering* 2020, 36 (2), 21–34.

[9] Guixiang, Z.; Hui, H., Vibration Isolation Technology and its Application to Concrete Structure Foundation. *Journal of Natural Disasters* 2008, 17 (2), 127–130.

[10] Du, D.; Wang, S.; Liu, W.; Sun, Z., Design Method and its Application in Hybrid Base-isolation of High-rise Buildings. *Journal of Southeast University. Natural Science Edition* 2010, 40 (5), 1039–1046.

[11] Wu, H.; Ding, J.; Chen, C., Research on Seismic Isolation Structure Design of Gymnasium Building in High Seismic Intensity Area. *Building Structure* 2020, 50 (3), 45.

[12] Wang, Y.; Chen, C.; Cui, M.; Xiao, C.; Feng, X.; Chen, F., Comparative Study of Calculation Methods for Effects Under Bidirectional Horizontal Seismic Action. *Building Structure* 2021, 51 (17), 10.

[13] Wang, P.; Wang, S.; Zhu, C., Seismic Dynamic Response Analysis of Urban Underground Utility Tunnel Structure. Journal of Northeastern University. *Natural Science* 2019, 40 (7), 1020–1027.

[14] Tan, K. C.; Hejazi, F.; Esfahani, H. M.; Chong, T., Development of Elastomeric Rubber Bearing Utilizing Core-and-Filler System. *Structures* 2022, 37, 125–139.

[15] Ozturk, H., Effects of Lead Core Heating on the Response of Isolated-Base and Fixed-Base Regular and Irregular Reinforced Concrete Structures. *Buildings* 2022, 12 (8).

[16] Park Sung-kui, Estimation of Aseismatic Performance of Laminated Rubber Bearing Through Shaking Table Tests. *Journal of the Korean Society for Railway* 2010, 13 (4), 440–446.

[17] Qiu, W.; Pu, Y.; Zhang, Q.; Zhang, X.; Wu, X.; Wang, K., Seismic Response Analysis of Arch Covered Bridge Considering Vertical Seismic Action. *World Earthquake Engineering* 2022, 38 (2), 169–180.

[18] Zhu, B.; Hu, W.; Li, C., Estimating Seismic Responses of Transmission Towers by Finite Element Method. *Earthquake Engineering and Engineering Vibration* 2006, 26 (5), 161–166.

[19] Zhang, B. Z.; Wang, K. H.; Lu, G. Y.; Guo, W. Z., Seismic Response Analysis and Evaluation of Laminated Rubber Bearing Supported Bridge Based on the Artificial Neural Network. *Shock and Vibration* 2021, 2021.

[20] Koo, G. H.; Shin, T. M.; Ma, S. J., Shaking Table Tests of Lead Inserted Small-Sized Laminated Rubber Bearing for Nuclear Component Seismic Isolation. *Applied Sciences-Basel* 2021, 11 (10).

[21] Liu, S.; Warn, G. P., Seismic Performance and Sensitivity of Floor Isolation Systems in Steel Plate Shear Wall Structures. *Engineering Structures* 2012, 42, 115–126.

[22] Wu, K. D.; Xing, Z., Stability of Imperfect Prestressed Stayed Beam-columns Under Combined Axial Load and Bending. *Engineering Structures* 2021, 245.

[23] Park, S.-C.; Hong, W.-K.; Kim, J. T., Application of Smart Frames to Tall Buildings with Dual Systems and with Building Frame Systems. *Indoor and Built Environment* 2014, 23 (1), 161–170.

[24] Chen, Y.; Li, J. Z.; Lu, Z., Experimental Study and Numerical Simulation on Hybrid Coupled Shear Wall with Replaceable Coupling Beams. *Sustainability* 2019, 11 (3).

[25] Wang, J.; Zhao, H., High Performance Damage-Resistant Seismic Resistant Structural Systems for Sustainable and Resilient City: A Review. *Shock and Vibration* 2018, 2018.

[26] Li, Y. Q.; Huang, B. H., Evaluation on Seismic Performance of Beam-Column Joints of Fabricated Steel Structure with Replaceable Energy-Dissipating Elements. *Sustainability* 2022, 14 (6).

[27] Fortney, P. J., Shahrooz, B. M., and Rassati, G. A., "The Next Generation of Coupling Beams," in *Proceedings of the 5th International Conference on Composite Construction in Steel and Concrete V*, pp. 619–630, zaf, July 2004.

Water Conservancy and Civil Construction – Oke & Ahmad (Eds)
 ISBN: 978-1-032-58614-4

Application analysis of digital survey technology in geotechnical engineering

Zhou Sha*
Sinochem Mingda Guizhou Geological Engineering Co., Ltd.
No. 82, North Section of Wanshou Street, Bozhou District, Zunyi City, Guizhou Province

ABSTRACT: In recent years, the development momentum of China's economic construction has become faster and faster, and technological innovation has promoted various industries to gradually move toward informatization and digitalization. The scope of application of digital technology in various industries is getting wider and wider, and the utilization rate is getting higher and higher. At the same time, it can effectively promote the sustainable development of various industries. As far as geotechnical engineering survey operations are concerned, the introduction of digital technology is very obvious, and geotechnical engineering is a major component of various projects in the construction industry, which has largely contributed to the recognition of digital technology in the application link. This paper conducts an in-depth analysis of the basic situation of geotechnical engineering digital technology, hoping to be of some help.

1 INTRODUCTION

In an environment where the development momentum of modern technology has been greatly improved, geotechnical engineering, as an important part of my country's construction industry, should receive full attention. Only by constantly updating geotechnical engineering survey technology, we can establish a more comprehensive understanding of the current geological conditions, establish a deeper understanding of the terrain and landforms in the area, and compare them in future engineering planning and design work to receive accurate information and data.

2 OVERVIEW OF GEOTECHNICAL ENGINEERING DIGITAL SURVEY TECHNOLOGY

Digital geotechnical engineering survey refers to the use of modern geographic surveying and mapping technology, data storage technology, computer technology, mobile communication technology, and CAD technology application, with the help of computer and system software, to collect all the information of a project (engineering survey, design concept, project progress, plans, changes, etc.) skillfully combined, so that the technical methods of engineering survey and design can be replaced by modern CAD technology,[1] to realize the digitalization of engineering survey data processing and intelligent graphics and text processing, digitized data collection, intelligent hardware architecture, and the establishment of a high-yield and intelligent engineering survey and design mechanism for multi-professional and multi-work processing.

*Corresponding Author: 434949998@qq.com

 DOI: 10.1201/9781003450818-96

3 APPLICATION ADVANTAGES OF DIGITAL SURVEY TECHNOLOGY IN GEOTECHNICAL ENGINEERING

3.1 *It is conducive to improving the coordination between the survey department and the design department*

Accurate and complete survey data is the prerequisite for the smooth development of the project when designing the project. However, due to the high complexity of geotechnical engineering in the survey operation, and the difficulty of understanding the survey results, many designers cannot effectively use survey data, which greatly increases the difficulty of design work. If such things continue, it will also cause uncoordinated work, poor cooperation, and low work efficiency between the design department and the survey agency. The use of digital geotechnical survey technology can effectively alleviate this situation [2]. The use of digital technology can build a 3D three-dimensional model for the survey project, and store all the data in the database. Designers only need to search and obtain the required data information intuitively and visually, and then effectively improve the design quality and efficiency to ensure the smooth implementation of engineering projects.

3.2 *Improve the coherence between digital maps and design systems*

The data and information marked in the geomorphic map are the premises of geotechnical engineering design. In actual application, the information on the geomorphic map must be communicated with CAD drawing software to complete the design task of the engineering system. However, the development of the survey technology in the past is not satisfactory on this point, resulting in the lack of connection between the geomorphic map information and CAD software access; thus, when the content of the geomorphic map is digitized, information distortion or data omission occurs,[3] affecting the overall construction project construction. With the help of digital survey technology, the data on the topographic map can be stored in the data system database using digital models and can be used in design tasks at any time to prevent information distortion.

4 SPECIFIC APPLICATION OF DIGITAL SURVEY TECHNOLOGY IN GEOTECHNICAL ENGINEERING

4.1 *Application of geographic information system*

The geospatial information system is mainly based on Internet information technology and has distributed deployment, universal access space, and operating system of the independent platform. The system mainly includes computer software technology, geology, and other professional knowledge, and with the strong support of computer hardware and software, it comprehensively analyzes and manages the geographical information data of the spatial physical data of geotechnical engineering investigation, and provides strategic decision-making and strategic planning for construction projects. It also manages issues related to credible relevant information,[4] creating great convenience for outdoor exploration operations. Although the integration of geospatial information systems and geotechnical engineering survey and integration are of different types, geospatial information is also covered in the mechanical structure of geotechnical engineering, most of which are related to spatial coordinates. Subsequent design tasks will all be implemented with the support of spatial data; that is to say, all-round geospatial information is the solid backing of geotechnical engineering investigation and design. Geospatial information system brings together the operating system for collecting data, data management, and data analysis, so the flexible application of geospatial information system to geotechnical engineering survey and design tasks can use the system in the collection, spatial analysis, and management modules to carry out surveys along the road. Operations also require scientific and reasonable analysis and management of information during the actual construction process. Compared with traditional survey technology, the geospatial information system has greatly improved. Specifically, the geospatial information system will be more efficient and rapid in the collection and processing of information data, and the collected data will have higher integrity and richer sources; the database system of the geospatial information system can accurately describe and express spatial entities and patterns. The integration with images is extremely precise, creating detailed information support for the subsequent survey and design data information, rational component design, analysis, and strategic decision-making [5]. Most importantly, the geospatial information system also has intelligent visual interaction functions, which makes the visual interaction of geotechnical engineering investigation and design possible.

4.2 *Application of geostatistics*

Geological applied statistics is derived from the premise of the basic theory of regional variables. Based on the use of various functions to interpret discretely distributed structured data in various spaces and abnormal changes in spatial layout, it is possible to conduct a professional evaluation of data and to study or simulate sample distributions of simulated data. Geological applied statistics include typically applied statistics and spatial summarization,[6] the core of which is to carry out rational analysis of geological landforms. In geotechnical engineering reconnaissance, geological application statistics have a very large correlation with geological history and stress conditions, and have a high autocorrelation, and such correlation can be obtained between any two points in the soil layer. It is shown that if the distance between two points is larger, the correlation between them will also decrease, and vice versa. Usually, at the level of describing the natural correlation of geotechnical space, it is based on the premise of any environment, and the variance reduction coefficient is used to infer the variability between the "point" and its environment to measure the related distance of geotechnical properties. It is a basic work in reliability analysis of geotechnical engineering investigation[7]. In the summary of geophysical parameters, the relevant distance is one of the most important technical parameters. If the geophysical properties of the soil layer are within the positive correlation distance, the geophysical properties of the two points will not be correlated with each other. Therefore, it is only necessary to calculate the distance between the geophysical parameters of the specific soil layer in the project to clearly know the geophysical properties of the rock layer. At the level of distance calculation, the calculation level includes the mean value method, correlation function method, and regressive simulation method. All kinds of calculation methods are indispensable for corresponding theoretical significance. There are also differences in the difficulty of use and different advantages[8].

4.3 *Build a geological model*

The geological modeling of digital geotechnical engineering investigation usually refers to the surface model method. The surface model method is a basic method generally used in the geotechnical engineering investigation industry. The specific modeling is to express the outer layer of the engineering geological body to express the homogeneous geological body. This type of modeling is widely used in geotechnical engineering investigations[9]. The relevant data in the surface model method all come from the initial observation point materials, which have geometric characteristics and attribute data characteristics. Through these observation point data, the interface of engineering geological bodies can be established, and then various types with the same characteristics. The scattered points are connected to form a grid-shaped arc surface, thus constructing the interface of the entire stratigraphic structure. There are two methods generally used in expressing the surface layer: the legend model method and the data model method. Taking the irregular network method in the analysis legend model method as an example to implement modeling, the irregular network method is based on many points in the survey range. An entire area is planned as a triangular network. Any point within the survey range can find the corresponding position on the triangular surface. If the point is not on the endpoint of the triangle, the data of any point can generally be obtained by means of interpolation attributes. The irregular network method [10] is a model based on segmented linear network three-dimensional space. Its network topology consists of various data storage methods. The relatively simple data storage method is to record the shape, edge, and network node of each triangle. The advantage of the class data storage method is given as a triangle. Its endpoint attribute data can be found, and the efficiency at the calculation level of the terrain profile line is relatively high. In addition, other changes can be added to the network topology according to the actual situation to improve the calculation' s efficiency.

5 CONCLUSION

To sum up, digital survey technology can effectively fill in the deficiencies of survey technology in the past, enhance the integration of survey work and architectural design work, and is extremely important to ensure the overall project quality. With the maturity and stability of digital technology and the increase in engineering project needs, its effectiveness will definitely increase day by day. However, in the specific application link, the establishment quality of the digital model and database must be guaranteed to further improve the effectiveness. At the same time, we should pay attention to the construction of a digital survey talent team to provide a guarantee for the digital establishment of subsequent geotechnical engineering survey tasks.

REFERENCES

[1] Fu Min. Discussion on the Technical Application of Geotechnical Engineering Survey Under Complex Geological Conditions [J]. *Western Exploration Engineering*, 2019, 31 (5): 26–27.

[2] Xu Yongliang. Discussion on Groundwater Issues in Geotechnical Engineering Investigation [J]. *Engineering Construction and Design*, 2019 (9): 39–40, 43.

[3] Jiang Guangwei, Liu Liang, Zheng Kai, *et al.* Research on Soil Classification and Engineering Economic Significance of Power Transmission and Transformation Line Engineering in Eastern Mongolia [J]. *Energy and Environmental Protection*, 2019, 41 (4): 96–103.

[4] Hu Zhaojiang, Cao Qizeng, Zou Yu. Research on Safety Management of Geotechnical Engineering Construction Under the New Situation [J]. *Housing and Real Estate*, 2022(13): 179–181.

[5] Wang Chunzhou. Analysis of Geotechnical Engineering Survey Technology based on Digitalization [J]. *Intelligent Building and Smart City*, 2022(01):90–92.

[6] Wang Zhen. Application of Digital Survey Technology in Geotechnical Engineering [J]. *Building Technology Development*, 2021,48(13):28–29.

[7] Chen Dan, Liu Zhe, Liu Jianyou, Fang Qian, Hailu. Status and Prospect of Intelligent Construction Technology of Railway Shield Tunnel [J]. *Tunnel Construction (Chinese and English)*, 2021,41(06):923–932.

[8] Gong Yalong. Application of Analytical Digital Technology in Improving the Efficiency of Geotechnical Investigation [J]. *Sichuan Cement*, 2021(04):196–198.

[9] Zhang Peng. Application of Digital Survey Technology in Geotechnical Engineering [J]. *China Metal Bulletin*, 2020(05): 279–280.

[10] Wang Jie. Application of Digital Technology in Geotechnical Engineering Investigation [J]. *Metallurgical Management*, 2020 (05): 143–144.

Water Conservancy and Civil Construction – Oke & Ahmad (Eds)
© 2024 The Author(s), ISBN: 978-1-032-58614-4

Research on influencing factors and countermeasures of construction quality of residential projects

Ye Yuan*
China Building Material Test&Certification Group (Shaanxi) Co., Ltd, Xi 'an, Shaanxi, China

Zhenquan Liao
AUPUP Electronic Commerce Co., Ltd, Shenzhen, Guangdong, China

Yue Zhao, Xiaodan Yu & Liang Zhang
China Building Material Test & Certification Group (Shaanxi) Co., Ltd, Xi 'an, Shaanxi, China

ABSTRACT: The construction quality of a residential project is the key to the final quality of a residential project. It determines the safety of the residential project's use process and the satisfaction of the owners. First, this research used literature analysis and meta-analysis to identify the factors affecting the construction quality of residential projects involved in the existing research on the basis of consulting domestic and foreign literature. Second, it adjust the identified influencing factors in the form of questionnaires combined with relevant residential project construction, supervision, and developer survey results. Third, the data collected by the questionnaire is used for exploratory factor analysis through SPSS 20.0 statistical software, the factors influencing the construction quality of engineering projects and their common factors are extracted and analyzed. The study found that construction management capabilities, quality improvement effectiveness factors, industry interaction factors, resource security factors, and macro-environmental factors will all affect the construction quality of residential projects. It is expected to provide a theoretical reference for the management of the construction process of residential projects.

1 INTRODUCTION

With the continuous acceleration of the pace of urbanization, residential projects have become the focus of current investment due to their large market demand and economic benefits. Under the catalysis of capital, a large number of large-scale residential houses have emerged in major cities in China, but some issues worthy of attention have also been exposed in the quality of construction, In January 2020, the large-scale collapse of Baden City Ecological Leisure Tourism Development Project in Jiangxia District, Wuhan City, caused a large number of casualties. In May 2020, a large accident occurred in the Far East Garden in Mabugang Town, Longchuan County, and the direct economic loss reached 10 million yuan. The relevant person in charge was responsible for criminal responsibility, which seriously affected the healthy and orderly development of the economy and society. In addition, because the construction process of a residential project is the main stage in which the design intent of the project is transformed into an engineering entity, and the construction process of a residential project has many related factors, quality concealment, and difficulty in final

*Corresponding Author: 15829277465@sohu.com

DOI: 10.1201/9781003450818-97

inspections, it results in the effectiveness of quality control in the construction process. So the factors affecting the quality of the residential project construction process had an important practical significance for the management of the residential project construction process and the improvement of the quality of the residential project.

Relevant scholars at home and abroad have conducted certain studies on the factors affecting the quality of the construction process of residential projects. Some scholars pointed out that the degree of guarantee of relevant resources is the decisive factor in the quality of construction projects[1]. Chan's research found that the cultural atmosphere of the construction unit of the residential project, the ability of the construction personnel, and the owner's degree of importance of the construction quality will all affect the construction quality of the residential project[2]. Some scholars pointed out that the feasibility of pre-construction plans for residential projects and the degree of quality control in the construction process are the keys to the construction quality of residential projects[3]. Ling's research found that the fund sufficiency of the construction unit, the degree of equipment resource satisfaction, and the strict degree of quality control of the construction process will affect the quality of the construction process of residential projects[4]. Domestic scholars have pointed out that the innovation in the management of the construction process of residential projects and the cooperation of various parties in the construction process will affect the quality of the construction process of residential projects[5]. Existing researches mostly refer to the factors affecting the quality of the construction process of residential projects from the theoretical level. The researches identify the factors that affect the quality of residential project construction. This article intends to use a combination of literature analysis and meta-analysis to explore the factors affecting the quality of the residential project construction process and provide a management reference to the quality management of the residential project construction process.

2 IDENTIFICATION OF FACTORS AFFECTING THE QUALITY OF THE CONSTRUCTION PROCESS OF RESIDENTIAL ENGINEERING PROJECTS BASED ON LITERATURE ANALYSIS

2.1 *Discovery of factors affecting the quality of the residential project construction process based on literature analysis*

This research uses residential projects, project quality, construction quality, etc. as keywords through document retrieval of well-known domestic and foreign journal databases such as CNKI China Academic Document Library, Wanfang Digital Platform, Elsevier, IEEE/IE, etc. Of 142 documents related to crisis factors, 40 documents that are inconsistent with the research focus of this article were removed, and 102 documents highly related to the research focus of this article were obtained. Through an intensive reading of the obtained documents, a total of 20 residential project constructions were unearthed. The details are shown in Table 1.

2.2 *Factor adjustment based on investigation*

Combined with the factors influencing the quality of the residential project construction process identified based on the literature analysis in Table 1, through the investigation of the management personnel of the construction enterprise, the on-site staff of the construction unit and the supervision unit and other staff who have business dealings with our unit, Amendments where there is ambiguity in the connotation, and supplements the factors that affect the construction quality of residential engineering projects identified in the literature analysis.

Table 1. Analysis table of factors affecting engineering quality.

Number	Influencing Factors	Definition	References
1	Perfection of laws and regulations	The more comprehensive the national laws and regulations on the construction quality requirements of residential projects, the construction party will strengthen the management of the residential project construction process and improve the quality of the construction process in order to avoid the risks caused by illegal activities.	[6][7][8]
2	Economic development	The economic development trend is related to housing demand, and the market demand for good economic development is also increasing. The market demands outlook conference prompts the construction unit to pay attention to the improvement in the quality of the construction process.	[1][9]
3	Technology advanced level	New technologies, new materials, and new processes are conducive to the improvement of the quality of residential projects	[9][10]
4	Industry Regulatory Intensity	Industry associations have established strict operating procedures for the construction process of residential projects, and the orderly market competition in the residential project industry is conducive to improving the quality of the construction process of residential projects.	[11][12][13]
5	Construction supervision	Relevant supervision units monitor the construction process	[14]
6	Coordination degree of each unit	Effectiveness of the cooperation between the construction unit and related parties such as construction unit, supplier, supervision unit, project schedule, and project funding	[15][16]
7	Perfection of the quality management system of the construction party	Whether the quality management system of the construction unit's quality manual, record preparation, and whether it has a comprehensive subcontracting inspection system	[8][9]
8	Technical clarification effectiveness	Effectiveness of the design clarification on the functions, features, design intent, and requirements of the building hosted by the construction unit and the effective degree of the internal construction design clarification of the construction unit	[17]
9	Rationality of construction organization design	The effectiveness of the construction plan, schedule, construction plan, and construction measures of the project	[18]
10	Technical ability of the construction party	Construction technology and technical methods used by the construction unit in the construction process	[19][20]
11	Effectiveness of material purchase and acceptance	Effectiveness of procurement, on-site inspection, and re-inspection	[21][13]
12	Managerial innovation	In daily management, the construction unit will set up a sound organizational structure and adopt a humane and flexible management model to stimulate the enthusiasm of em-	[22][20]

(continued)

Table 1. Continued

Number	Influencing Factors	Definition	References
		ployees, which will help improve the quality of the construction process of residential projects.	
13	Human Resource Management Ability	Participating in personnel qualification review, salary adjuster, and personnel training	[23][24]
14	Facilities and equipment management ability	Equipment, acceptance, installation and commissioning, maintenance of various machines	[25]
15	Financial resource management ability	The inadequate budget and actual guarantee of the funds required for the construction project led to the delayed arrival of the materials required during the construction process, improper process connection, and insufficient personnel security, which led to some construction quality problems in construction projects	[26][27]
16	Construction environment control ability	Suitability of environmental factors such as engineering technology environment, engineering management environment, labor environment, and humanistic environment	[28]
17	Monitoring and Measurement Normative	The accuracy and standardization of the finished product measurement of each node of the construction process by the construction unit will affect the satisfaction of the construction party and the owner, which will affect the quality of the residential project	[29]
18	Unqualified control degree of the construction process	The construction unit's management and control process and the division of powers and responsibilities for the unqualified construction process will affect the pro-activeness of the construction staff	[30]
19	Corrective and preventive measures to control the effectiveness	By analyzing the quality defects found in the construction process of the residential project, the industrial unit can formulate and implement effective corrective measures, which can improve the quality of the residential project construction process to a certain extent	[31]

3 QUESTIONNAIRE DESIGN AND DATA COLLECTION

3.1 *Questionnaire design*

According to Table 1 based on the factors influencing the construction quality of residential projects identified based on literature analysis, a questionnaire design method combining direct design and indirect design is adopted to design a questionnaire for the factors affecting the quality of the construction process of residential projects.

Direct design of the questionnaire: The direct survey questionnaire mainly takes the factors influencing the construction quality of residential engineering projects as shown in Table 1 and forms a direct survey questionnaire on the factors affecting the construction quality of residential engineering projects to reflect the research results of the theoretical circle.

Indirect design of the questionnaire: After conducting interviews with the company's technical personnel and technicians who inspected residential projects, the questionnaire items were adjusted according to their opinions, and an indirect questionnaire was formed.

The overall design of the questionnaire: After the above direct questionnaire and indirect questionnaire are summarized, a table of factors affecting the construction quality of residential engineering projects and the questionnaire used in the research process of this research are formed, The Likert five-point scale method is used to investigate the surveyed objects. The degree of agreement with the relevant items is used to obtain the data for the research.

Table 2. Summary table of factors influencing construction quality.

Number	Influencing Factors	Number	Influencing Factors	Number	Influencing Factors
1	Perfection of laws and regulations	2	Perfection of the quality management system of the construction party	3	Human Resource Management Ability
4	Economic development	5	Technical clarification effectiveness	6	Facilities and equipment management ability
7	Technology advanced level	8	Rationality of construction organization design	9	Financial resource management ability
10	Industry Regulatory Intensity	11	Technical ability of the construction party	12	Construction environment control ability
13	Construction supervision	14	Effectiveness of material purchase and acceptance	15	Monitoring and Measurement Normative
16	Coordination degree of each unit	17	Managerial innovation	18	Unqualified control degree of the construction process
19	Corrective and preventive measures to control the effectiveness				

3.2 *Selection of survey subjects*

In order to ensure the validity of the questionnaire survey data, this research selected some developers and construction units that have business dealings with this unit as the main survey samples of this research process. The subjects of the survey are project managers, technicians, and related front-line workers. The subjects of the survey are asked to rate the degree of recognition of the content described in the questionnaire.

3.3 *Collection of survey data*

The survey mainly relied on the on-site interviews, emails, and telephone interviews of the sampling staff for the unit to issue questionnaires. A total of 400 questionnaires were distributed, 371 questionnaires were recovered, 40 invalid questionnaires were eliminated, and 331 questionnaires were finally obtained.

4 DATA PROCESSING AND ANALYSIS OF FACTORS AFFECTING THE CONSTRUCTION QUALITY OF RESIDENTIAL ENGINEERING PROJECTS

4.1 *Project analysis*

SPSS 20.0 statistical software is used to calculate the critical ratio of each questionnaire item, and the score ranking is respectively regarded as the high group and the low group. I calculate the difference in the average number of scores of the high and low groups and delete the relevant items and their corresponding influencing factors that are not significant. The results show that the CR values of items 2 and 12 in the questionnaire did not reach a significant level, indicating that these 2 items are not discriminating. Based on expert opinions, this paper deletes 2 and 12 from the questionnaire items.

4.2 *Extraction and naming of common factors*

1) KMO and Bartlett test
   We used SPSS 20.0 software combined with valid survey questionnaire data and used KMO as the criteria to test the validity of the scale. KMO is suitable for applying factor analysis methods. The results are shown in Table 3 below.

Table 3. Results of validity analysis.

Kaiser-Meyer-Olkin Measure of Sampling Adequacy		0.767
Bartlett's Test of Sphericity	Approx. Chi-Square	3720.332
	df	153
	Sig.	.000

Table 4. Explanation of total variance.

	Initial Eigenvalues			Extraction Sums of Squared Loadings		
Component	Total	% of Variance	Cumulative %	Total	% of Variance	Cumulative %
1	4.721	27.772	27.772	4.721	27.772	27.772
2	2.673	15.723	43.495	2.673	15.723	43.495
3	2.490	14.647	58.141	2.490	14.647	58.141
4	1.455	8.558	66.700	1.455	8.558	66.700
5	1.022	6.012	72.712	1.022	6.012	72.712

Extraction Method: Principal Component Analysis.

2) Extraction and naming of common factors
   We used SPSS 20.0 to carry out an exploratory factor analysis on the influencing factors involved in the questionnaire and used the maximum variance method to carry out orthogonal rotation processing. It is found that there are 5 common factors of eigenvalues greater than 1, and the cumulative explanatory variable is 72.712%>70%, which indicates that the scale has good structural validity.
3) Analysis of reliability
   We used the Cronbach α coefficient as the criterion to test the reliability of the scale. The results of the Cronbach α coefficient of each common factor were all greater than 0.7, which indicated that the scale has good reliability.

Table 5. Results of reliability analysis.

Factor	Cronbach's Alpha if Item Deleted	Factor	Cronbach's Alpha if Item Deleted	Factor	Cronbach's Alpha if Item Deleted
1	.823	7	.860	13	.709
2	.841	8	.833	14	.764
3	.852	9	.842	15	.769
4	.811	10	.751	16	.869
5	.832	11	.865	17	.779
6	.825	12	.813	18	.759
19	.871		.882		
Cronbach's Alpha	0.822	Cronbach's Alpha Based on Standardized Items	.880	N of Items	19

4) The naming of common factors and discussion of results
   Through the aforementioned exploratory factor analysis, this study identified and extracted five factors affecting the construction quality of residential engineering projects, and standardized them with the connotation of each factor.

Definition of common factors:

Construction management ability: Technical clarification effectiveness, rationality of construction organization design, technical ability of the construction party, effectiveness of material purchase and acceptance, and managerial innovation.

Quality improvement effectiveness factors: Monitoring and measurement normative, unqualified control degree of the construction process, corrective and preventive measures to control the effectiveness

Industry interaction factors: Industry regulatory intensity, construction supervision

Coordination degree of each unit

Resource security factor: Human resource management ability, facilities and equipment management ability, financial resource management ability

Macro environment factors: Perfection of laws and regulations, economic development, technology advanced level

5 CONCLUSION

Based on the analysis of relevant theoretical research results of the factors affecting the construction quality of residential projects at home and abroad, the research has identified the relevant factors that affect the construction quality of residential projects mentioned in the theoretical circle, and combined with the investigation into relevant real estate developers and construction units. The identified influencing factors and their connotations have been improved and adjusted. At the same time, combined with the identified relevant influencing factors, the formal questionnaire used in this research was designed. Using SPSS 20.0 and the obtained valid survey questionnaire data, five key factors affecting the construction quality of residential projects were extracted. The research conclusions have the following enlightenment for the construction process management of residential projects:

First of all, the construction unit must fully understand the national laws and regulations and the requirements of the real estate industry during the construction process to prevent poor external macro-environment control from affecting the construction quality of residential projects. At the same time, it is necessary to pay attention to the monitoring and measurement of all aspects of the construction process and strengthen the control and correction of unqualified construction processes to avoid potential safety hazards to the quality of the project. In addition, the construction unit must actively integrate and configure all the resources required for the project, and establish a good communication channel with the responsible parties related to the construction to prevent construction problems caused by poor internal management.

When selecting a construction unit, the developer should focus on the qualification of the construction unit and the quality control ability of the construction process, and strictly control the construction contract, technical disclosure, organization, and other important links of the construction process. The construction unit should pay attention to the reasonable supply and deployment of key resources such as human resources, financial management, and facility resources required by the project during the construction phase, so as to effectively ensure the high quality of various activities throughout the construction project.

REFERENCES

[1] Ana Yang. Analysis of Influencing Factors in Construction Quality Control of Construction Projects [J]. Shanxi Architecture, 2013, 39(32): 209–210.

[2] Arditi D, Gunaydin HM. Factors that Affect Process Quality in the Life Cycle of Building Projects. J Construct Eng Manage 1998;124(3):194–203.

[3] Bing Jiang. Research on Key Technologies of Electrical Automation Control in Power Plants[J]. Technology and Enterprise. 2014, 29(02): 80–82.

[4] Bing Xu, Yuwen Liu, Haodong Peng. Discussion on the Control of Factors Affecting the Quality of Construction Projects [J]. Science and Technology Information, 2013, 20: 449.

[5] Chan APC, Wong FKW, Lam PTI. Assessing Quality Relationships Inpublic Housing: an Empirical Study. Int J Qual Reliab Manage2006; 23(8): 909–27.

[6] Daopeng Wang. Research on Management Methods of Construction Engineering Quality Problems *[D]*. Xi'an University of Architecture and Technology, 2009.

[7] Dingxuan Huang, You Jianxin, Du Bo. Empirical Analysis of Key Factors of Construction Projects from Different Role perspectives[J]. China Civil Engineering Journal, 2007: 40(4): 104–110.

[8] Dongjun Wan. Research on Project Quality Cooperative Supervision Mode Introducing Insurance Mechanism[J]. Journal of Engineering Management, 2011, 25(4): 373–377.

[9] Gaofeng Li. A Brief Talk on Construction Management and Quality Control Measures[J]. Science and Technology Innovation Herald, 2011, (18): 34–34.

[10] Guohua Zhou, Bo Peng. Analysis of Risk Factors of Construction Project Quality Management Based on Bayesian Networks—Taking the Beijing-Shanghai High-speed Railway Construction Project as an Example[J]. China Soft Science, 2009, 09: 99–106.

[11] Hongtao Xie, Wang Mengjun. An Empirical Study on the Status Quo of China's Major Project Quality Management and Quality Influencing Factors[J]. World Standardization and Quality Management, 2008, 08: 17–20.

[12] Jianhai Wang. Analysis of the Construction Management of Multi-storey Residential Buildings[J]. Building Materials Technology and Application, 2012, 12:40–42.

[13] Jimao Jiang. Research on Construction Quality Management of Construction Projects *[D]*. Nanjing University of Science and Technology, 2007.

[14] Korea Bo, Quanchen Gao. Several Noteworthy Issues in the Quality Control of Construction Projects [J]. Sichuan Building Science Research, 2007, 02:182–183.

[15] Ling FYY, Chan SL, Chong E, Ee LP. Predicting Performance of Design-build and Design-bid-build Projects. J Construct Eng Manage 2004; 130(1): 75–83.

[16] Ling FYY, Models for Predicting Quality of Building Projects[J]. Engineering, Construction and Architectural Management, 2005,12(1):6–20.

[17] Refaat H. Abdel-Razek. Factors Affecting Construction Quality in Egypt-identification and Relative Importance[J]. *Engineering*, Construction and Architectural Managent, 1998, 5(3): 220–227.

[18] Shixin Liu, Jianlin Liu. Customer Satisfaction Analysis of Real Estate Project Quality Based on ACSI [J]. New West *(Late. Theoretical Edition)*, 2012, Z1: 94+104.

[19] Wei Pan. ISM Modeling and Analysis of Influencing Factors of Construction Engineering Quality Accidents[J]. Journal of Engineering Management, 2012, 26(1): 79–83.

[20] Wei Xiong, Xiaobin Feng. An Empirical Study on the Relationship Between Quality Management Practice and Performance Based on Enterprise Characteristic Variables[J]. Journal of Zhejiang University (Humanities and Social Sciences Edition), 2012, 42(1): 188–20.

[21] Wenjie Wu, Changzheng Zhou. On the Quality Management Control of Construction Units[J]. Modern Business and Trade Industry, 2011, 23:315.

[22] Wuhong Nie, Bi Caixia. Analysis of the Causes of Construction Engineering Quality Accidents and Control Strategies[J]. Journal of Yangtze University (Self Science Edition) Science and Technology *Volume*, 2006, 03:10–101.

[23] Xijun Wu. In-depth Discussion on the Key Points and Factors of Construction Quality Management [J]. Science and Technology Information, 2009, 28: 132–133.

[24] Xin Yan. On the Quality Control of Residential Construction Management [J]. Building Materials and Decoration, 2013, (19): 240–241.

[25] Xueliang Hou, Hongliang Zhu, Gang Guan. Evidence-based Management Methods and Empirical Research on Quality Problems of Chinese housing projects—Analysis of Group Factors of Quality Problems[J]. China Civil Engineering Journal, 2008, 07: 92–97.

[26] Yiming Gong, Mingfang Ding, Jian Cui. Customer Demand Recognition and its Model[J]. Fudan Journal. 2003, 42(5): 718–720.

[27] Yueyi. Wu Construction Quality Management of Housing Construction Engineering [J]. Enterprise Guide, 2011, (5): 102–103.

[28] Yung P, Yip B. Construction Quality in China During Transition: A Review of Literature and Empirical Examination[J]. International Journal of Project Management, 2010, 28(1): 79–91.

[29] Yunna Wu, Yong Huang, Shuo Zhang, Yan Zhang. Quality Self-control and co-supervision Mechanism of Construction Agent in Public Investment Project in China[J]. Habitat International, 2012(36) :471–480.

[30] Zhangyu Liu. Construction Quality Management of Engineering Projects*[M]*. Wuhan: Huazhong University of Science and Technology Press, 2012.10.

[31] Zhangyu Liu. Construction Quality Management of Engineering Projects*[M]*. Wuhan: Huazhong University of Science and Technology Press, 2012.10.

Water Conservancy and Civil Construction – Oke & Ahmad (Eds)
© 2024 The Author(s), ISBN: 978-1-032-58614-4

Study on drilling speed increase technology of slim hole horizontal wells

Mengyi Liu*, Peng Wei & Hongwei Liu
No.4 Drilling Engineering Company, BHDC, Cangzhou, China

ABSTRACT: In recent years, domestic horizontal well drilling technology has also played an important role in oilfield production and made important progress in the research and application of horizontal well drilling technology, drilling speed increase, and other technologies. In view of the problems encountered in the process of slim hole horizontal wells in the study area, such as poor rock drillability, difficult selection of deviation points, complex horizontal section trajectory, and downhole safety caused by complex geological conditions, this paper makes a systematic study on the drilling of slim hole horizontal wells in the study area. Based on the theoretical analysis, combined with the field test application, this paper optimized the high-efficiency PDC bit suitable for rapid drilling in tight sandstone formation in the study area, optimized the trajectory of slim hole horizontal section, optimized the BHA and hydraulic parameters, and formed the drilling supporting technology suitable for slim hole horizontal well in the study area. The field application has achieved good results.

1 INTRODUCTION

A horizontal well is the extension and development of directional well drilling technology. Horizontal wells can use gas fields which cannot be developed by vertical wells. It has considerable economic benefits, high return on investment, a high gas rate well, and convenient management. Moreover, it can greatly shorten the drilling operation period and improve the drilling rate of the oil layer and oil and gas recovery. In the recent ten years, with the development of horizontal well drilling technology, the cost advantage of developing oil and gas resources in difficult-to-use blocks by slim hole horizontal well drilling will become more and more obvious. Slim hole drilling technology can change wellbore structure, reduce bit size and improve rock breaking efficiency of bit. Through slim hole drilling technology and horizontal well drilling technology, the cost is greatly reduced and horizontal well drilling technology of slim hole drilling has become a research hotspot of horizontal well drilling technology [1].

Horizontal well drilling technology also plays an important role in oilfield production in China. In recent years, horizontal well technology has developed rapidly, and the number of horizontal wells constructed by China National Petroleum Corporation exceeds 1000 every year. Important progress has been made in the research and application of horizontal well drilling technology and drilling speed increase technology. However, with the gradual deterioration of the quality of oil and gas resources, some new and difficult problems have gradually emerged in the application of horizontal well technology [2]. At the same time, the geological conditions of some oilfields are complicated, the poor drillability and strong abrasiveness of formation, coupled with the new requirements for the length of horizontal

*Corresponding Author: liumengyi@cnpc.com.cn

DOI: 10.1201/9781003450818-98

sections of horizontal wells, make the requirements of continuously improving drilling speed and reducing costs faced by horizontal well drilling technology constantly improve. How to apply horizontal well technology to realize the effective development of these low-grade resources has become a key problem that needs to be solved urgently [3].

Horizontal well drilling technology in China is in the stage of rapid development at present, and its drilling technology level has been in an advanced position in the world. However, compared with developed countries such as the United States and Canada, there is still a certain gap, especially in drilling equipment, tools and instruments, automatic drilling and integrated system, and comprehensive application of horizontal well technology [4]. Statistics show that, in recent years, the number of drilling wells in China has increased rapidly, with an annual drilling number exceeding 30,000, which has become the second-largest drilling country in the world. However, the number and proportion of deep wells, ultra-deep wells, and horizontal wells over 4500m, which reflect the technical level, are far less than those in the United States and Canada, and the average annual footage of drilling teams is only half of that in the United States. The overall drilling speed is slow and the technical level is still not high [5]. Moreover, the single-well production of horizontal wells for low permeability reservoir development is only twice that of vertical wells, so the overall effect is poor, and the technical level needs to be continuously improved [6].

2 BIT SELECTION

Bit selection plays a very important role in the whole drilling design and construction process. In the drilling process, the bit is a rock-breaking tool, and the drilling speed is directly related to the selected bit type and formation lithology. Choosing the proper bit can not only improve penetration rate but also reduce downhole accidents during drilling, to achieve the purpose of high speed, low cost, and safe drilling [7].

Acoustic time difference logging records the propagation characteristics of sound waves in underground rocks, which can be used to study the mechanical properties of formation rocks. The acoustic time difference recorded by acoustic logging curves is closely related to the strength, hardness, and elastoplasticity of rocks. Similarly, density logging also indirectly reflects the mechanical properties of formation rocks. The drillability of rock is the ability of a rock to resist the damage of drilling tools. It is also related to various mechanical properties of rocks. Therefore, it is usually necessary to do a large number of core mechanical tests in the laboratory and establish the relationship between rock mechanical parameters and logging parameters such as acoustic time difference and density, that is, the calculation models of rock strength, hardness, and drillability grade values, which can be used to calculate the drillability grade values of formation rocks in the same or similar gas fields in the future. Using logging data to determine formation lithology is an ideal method at present, which can replace some core tests.

In this paper, the formation drillability of the study area is analyzed, and the compressive strength, shear strength, indentation hardness, and drillability grade values of formation rocks are representatively calculated by using the acoustic time difference and density logging data of these three gas fields. The results are shown in Table 1.

The successful selection of the PDC bit is the key to determining the drilling speed of a well (Figure 1). Especially in directional wells, the stability of the tool surface of the selected PDC bit determines everything, otherwise the normal build-up rate will not be stable, resulting in the failure of directional construction. Because the PDC bit mainly depends on shearing rocks, the drilling footage has a great relationship with the matching of the selected bit type and the lithology of the drilled formation. Cretaceous strata in the study area are classified as second-class soft strata; Jurassic strata are classified as third-class and fourth-class soft strata; Triassic strata are classified as fourth-class soft strata and fifth-class middle strata. According to the formation characteristics and types, the domestic MD9535ZC bit

Table 1. Correspondence table of the extreme value of formation drillability in the study area.

Stratum		Extreme value of formation drillability	Corresponding stratum type
Cretaceous	/	2.5 ~ 3.3	Secondary soft stratum
Jurassic	Diazepam group	3.3 ~ 4.2	Third-grade soft stratum
	Zhiluo Formation	3.7 ~ 4.3	Third-grade soft stratum
	Yan'an Formation	4.1-4.9	Fourth-grade soft stratum
Triassic	Extension group	3.9 ~ 5.5	Fourth-grade soft stratum
	Paper Mill Group	4.8 ~ 5.5	Grade 5 middle stratum
	Heshanggou Formation	5.5 ~ 6.2	Grade 5 middle stratum
	Liujiagou Formation	5.1-6.1	Grade 5 middle stratum
Permian	Shiqianfeng Formation	4.5 ~ 5.9	Grade 5 middle stratum
	Upper Shihezi Formation	5.1 ~ 5.6	Grade 5 middle stratum
	Xiashihezi Formation	5.1 ~ 6.2	Grade 5 middle stratum

Figure 1. PDC bit.

and the foreign FX56R bit are selected for the vertical well formation. This type of bit is suitable for soft to medium hardness formation. The strata of the slope-making section and the horizontal section are all Permian, and the Permian strata are classified as five middle strata. According to the characteristics and types of this formation, domestic MD6633ZC and foreign FXD65D bits are optimized in the deflection section, which is suitable for medium soft to medium hardness formations. Domestic MD6543A and foreign FX55D bits are selected for the horizontal section, which is suitable for medium to medium hardness formation.

Horizontal sections are all in Shihezi Formation or Shanxi Formation, with gray-white sandstone mixed with gray mudstone and glutenite. Sandstone has high Shi Ying content, high hardness, and drillability grade 6, and this section has strong lithologic abrasiveness and poor drillability. The main consideration is to increase the abrasion resistance of the bit compound, work smoothly and attack strongly, increase the footage of a single bit to speed up the penetration rate, reduce the bit torque, and facilitate the control of wellbore trajectory. MD6543ZC is customized to reduce the cost, and the Halliburton FX55D bit is preferably used. The drill bit used has 5 spiral blades, a 13 + 16mm composite sheet with dense teeth in double rows, and the long parabolic crown has strong aggression and abrasion resistance.

FX55D bit adopts ultra-high quality 16mm ultra-high quality cutting teeth, five-blade double-row teeth design and double-row teeth design to improve diamond content in the bit shoulder. The design of vibration-damping teeth in the inner cone can reduce the vibration of the drill bit and improve its stability of the drill bit. DYD simulation software is used to adjust the bit stability in order to improve the penetration rate.

3 HYDRAULIC PARAMETER OPTIMIZATION

The optimization research of hydraulic parameters is classified and discussed by the formation and wellbore size. The depth of φ 374.6mm borehole is generally less than 500m. All drilled are loose strata in the upper part, which are not well cemented and have poor bearing capacity. In order to ensure sand carrying capacity, it is required to use double pumps for drilling, with displacement of 55l/s~60l/s and annulus return velocity above 0.6 m/s, which can not only clean the wellbore but also obtain a higher penetration rate. For a 241.3 mm borehole, the displacement must reach 35l/s~45l/s and annulus return velocity above 0.85 m/s in order to carry sand, which can clean the borehole and improve the penetration rate. For 215.9 mm borehole drilling, a single pump cylinder liner with a diameter of 170mm or 180mm is required, and the displacement is above 28l/s. Generally, the mud pump used in the field is 3NB1300C, and the displacement is between 28l/s and 32l/s under normal conditions, which can meet the construction requirements of various aspects. The φ 152.4mm borehole is a horizontal section. To meet the requirements of sand carrying, the displacement must reach 13l/s~15l/s, and the annulus return velocity must reach above 1.08 m/s, which can clean the borehole and improve the penetration rate

With the increase of annulus return velocity, the cuttings concentration in annulus will decrease obviously in both vertical and deviated wells. However, the annulus return velocity needs to be properly selected. The return velocity of the annulus is too small to reach the purpose of well washing, and the excessive return speed of the annulus is unfavorable to the protection of the shaft lining. On the other hand, it will cause excessive pressure loss in the annulus, which is unfavorable to the rational distribution of the pump power and will also reduce the penetration rate.

4 FIELD TEST AND TECHNICAL APPLICATION

4.1 *Trajectory control of a deviation-making section of slim hole horizontal well*

Roller bit: River drill φ 152.4 mm 517 and 537, MXL-DS55DX2 Hughes cone, recommended WOB 30–70 kN, rotational speed 40–80 rpm, reverse torque can be controlled at 60–100 kNm at 60–120 rpm rotational speed of φ 120 mm and φ 127 mm screw drilling tools, meeting the effective rock breaking efficiency under sliding and rotary drilling modes; most PDC bits adopt M1365RJBEST and Sichuan far-reaching DFJL1606BU special bits for directional wells or Hughes PDC. The recommended WOB is 20–40 kN, the rotating speed is 40–80 rpm, and the anti-torque can be controlled at 30–60 kNm at 60–120 rpm of φ 120 mm and φ 127 mm screw drilling tools, which is used for drilling in sand and mud intervals to improve rock breaking efficiency.

Screw drilling tool 5LZ120 × 7. 0-DWG 1.0 ~ 1.5°, structural parameters: outer diameter φ 120.7 mm (equivalent inner diameter φ 50.8 mm); the bending angle is about 1.0 ~ 1.5, and it is 0.9 m away from the lower end face; the centralizer has an outer diameter of φ 146 mm and is 0.52 m away from the lower end face; the length of standard drilling tools is 4.88 m, and the recommended working weight on bit is 55 kN (maximum 72 kN). The length of C drilling tools is 6.88 m, and the recommended working weight on bit is 55 kN (maximum 100 kN).

φ 152.4 mm bit + 7LZ120.7 × 7.0-DWG 1.0 ° (or 1.25 °) screw drilling tool + φ 101.6mm non-magnetic weighted drill pipe × 1 + LWD + φ 101. 6mm non-magnetic weighted drill pipe × 1 + φ 101. 6mm ramp drill pipe × several + φ 101. 6mm weighted drill pipe × several + φ 101. 6mm drill pipe.

In the drilling of each well, the upper stabilizer with a size below 146 mm is mostly selected by 1 angle screw, the build-up rate is kept at 3–6 °/30 m and the rotary footage is generally kept at 60–70% of the build-up section. If the geological horizon adjustment is too large (target point moves up or down), under the premise of considering smooth wellbore curvature and ensuring completion pipe string running, the adjustment of 1.25 screw bending angle is used as a reserve, and drilling parameters such as WOB and rotational speed are optimized to increase rotary drilling foot as much as possible to improve penetration rate.

4.2 *Trajectory control of a horizontal section of slim hole horizontal well*

Roller bits: Baxter M1365DPDC and Smith XR50Y cones, MXL-DS55DX2 Hughes cones, recommended WOB 30–70 kN, rotating speed 40–80rpm, and anti-torque can be controlled at 60–100 kNm at 60–120 rpm of φ 120 mm and φ 127 mm screw drilling tools, which meets the effective rock breaking efficiency under sliding and rotating drilling modes. Most PDC bits adopt M1365RJBEST and Sichuan far-reaching DFJL1606BU special bits for directional wells or Hughes PDC. The recommended WOB is 20–40 kN, the rotating speed is 40–80 rpm, and the anti-torque can be controlled at 30–60 kNm at 60–120 rpm of φ 120 mm and φ 127 mm screw drilling tools, which is used for drilling in sand and mud intervals to improve rock breaking efficiency (Figure 2).

Figure 2. Sichuan far-reaching DFJL1606BU special bits.

152.4 mm bit + 7LZ127 × 7. 0-DWG 1.0 ° (or 1.25 °) screw drilling tool + 148 mm undersized centralizer + 101.6 mm non-magnetic weighted drill pipe × 1 + LWD + 101.6 mm non-magnetic weighted drill pipe × 1 + 101.6 mm ramp drill pipe × several + 101.6 mm weighted drill pipe × several + 101.6 mm ramp drill pipe × several + 101.6 mm ramp drill pipe × several + 101.6 mm ramp drill pipe weighted drill pipe or drill collar is located in 0–30-degree interval to ensure the effective transmission of WOB to the bit, especially after the horizontal section reaches 800m, add drill collar 2–3 string properly in the vertical section to increase the effective weight of drill string to overcome the friction of horizontal section and keep the effective transmission of WOB. In the down-dip or down-dip horizontal well section, the outer diameter of the undersized centralizer φ 146mm and the distance from the bit can be adjusted to realize slightly increased rotation drilling.

After the trajectory control of the deep slim hole deflection section accurately hits the target, the optimized steering BHA can be fully used in stable drilling in the horizontal section. BHA 1 with a 148 mm undersized stabilizer is suitable for rotary drilling in horizontal sections with a strong ability to increase formation deviation. Although the well deviation increases slightly, the laboratory research results show that rotary drilling in a horizontal section can be realized by adjusting the structural characteristics of the stabilizer on guide drilling tool-placement position and diameter, drilling parameters such as WOB and rotational speed; BHA 2 is equipped with a φ 146 mm stabilizer according to the characteristics of weak formation deviation increase, which is suitable for the horizontal section with small relative slope increase of sand and mudstone, especially the first BHA in the horizontal section after entering the target. First, properly control the deviation increase trend, and second, rigidity transition of drilling tools to avoid sticking; BHA 3 adapts to the down-dipping section with deflection reduction force, and counteracts the downward trend of formation by the downward inclination trend of guided drilling tool rotation mode, to realize stable and flat rotary drilling in long horizontal section.

4.3 *Drilling speed increase results*

The average drilling cycle of 8 horizontal wells completed in 2020 is 175 days. Compared with conventional structure wells, the average length of the horizontal section is increased by 170 meters, and the drilling cycle is shortened by 9 days. The comprehensive speed increase effect of well W5 is particularly remarkable, with a breakthrough of 1266 m in the horizontal section and a drilling cycle of only 145 days (Figure 3).

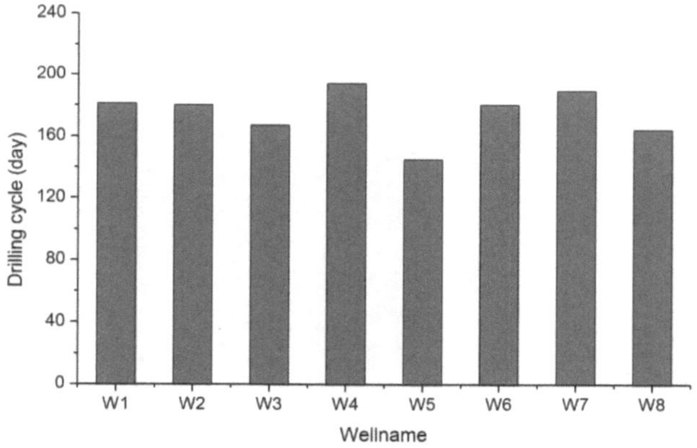

Figure 3. Drilling cycle of horizontal wells in the study area in 2020.

5 CONCLUSION

1. The roller bit has less footage and a low penetration rate, a long drilling cycle and high cost, and high construction risk. Based on the evaluation of formation rock drillability in the study area, the PDC bit sequence with imported Hughes PDC bit as the main bit and domestic PDC bit and roller bit as the auxiliary bit is optimized. This paper can meet the requirements of safe and rapid drilling of horizontal wells in the tight sandstone

formation of Denglouku Formation in Changling, and the penetration rate of the deflection-making section and horizontal section is significantly improved by field application.

2. Field test and application show that the matching technology of drilling speed increase for slim hole horizontal wells formed by the project research can meet the field construction requirements of horizontal wells in the long horizontal section of the research area, play an effective role in providing drilling speed increase, and provide effective technical support for efficient drilling of tight sandstone.

ACKNOWLEDGMENT

This work was not supported by any funds. The authors would like to show sincere thanks to those technicians who have contributed to this research.

REFERENCES

[1] Cui. H, Cao. S, Yang. C, Tang. H, Sun. L, Hole Trajectory Control Technology for Underbalanced Drilling of DP-19 Slim-hole Horizontal Well, 32nd ed., vol. 3. *Oil Drilling & Production Technology*, 2010, pp. 18–22.

[2] Guo. F, Sun. X, Luo. X, Completion Technique of Slim Hole Sidetracking for Horizontal Well Z3034, 35th ed., vol. 3. *Oil Drilling & Production Technology*, 2013, pp. 34–36.

[3] Rach. N.M, Slimhole Tools Offer Drilling, Completion Options, 103rd ed., vol. 44. *Oil and Gas Journal*, 2005, pp. 34–36, 38.

[4] Moelle. K, Young. J.D, (1970) On Geological and Technological Aspects of Oriented n-size Core Diamond Drilling, 4th ed., vol. 1. *Engineering Geology*, 1970, pp. 65–72.

[5] Wang. Z, Hao. T, Yong. L, Li. C, Dielectric Flushing Optimization of Fast Hole EDM Drilling Based on Debris Status Analysis, 97th ed., vol. 5–8. *International Journal of Advanced Manufacturing Technology*, 2018, pp. 1–9.

[6] Haba. B, Morishige. Y, Novel Drilling Technique in Polyimide Using Visible Laser, 66th ed., vol. 26. *Applied Physics Letters*, 1995, pp. 3591–3593.

[7] Yilmaz. O, Okka. M.A, Effect of Single and Multi-channel Electrodes Application on EDM Fast Hole Drilling Performance, 51st ed., vol. 1–4. *International Journal of Advanced Manufacturing Technology*, 2010, pp. 185–194.

Water Conservancy and Civil Construction – Oke & Ahmad (Eds)
© 2024 The Author(s), ISBN: 978-1-032-58614-4

Optimization analysis of children's schoolbag design based on Kansei engineering and KANO model—for children aged 7-12

Yuzhe Qi* & Yuchen Liu
Department of Industrial Design, Silla University, Sasang-gu, Busan, South Korea

ABSTRACT: With the continuous development of society, industrial manufacturing has been adhering to the people-oriented design process for product development. This study takes children's school bags as the product object. It uses the KANO model to analyze the target group to understand the various needs of users for the design and development of school bags. And the needs put forward by users are divided into basic, eager, excited, and indifferent needs. And determine the priority of product design and development according to the division level, which follows the basic needs > eager needs > exciting needs. Designers can create design points that better meet users' needs based on the demand levels obtained from the analysis. Secondly, through the research method of Kansei engineering, the user needs to be analyzed by the KANO model are transformed into tangible product design elements, which provide corresponding design inspiration and development assistance for subsequent designers. By associating the KANO model with Kansei Engineering, it is convenient to establish a relationship network between various design elements and user needs so that in the process of product design and development, the relationship between primary and secondary can be distinguished, and it is more conducive to grasping the consumption needs of consumer groups.

1 RESEARCH BACKGROUND

Our way of life has improved with the continuous development of science and technology. People's choices have become more diverse in terms of products. Under the premise of ensuring the product's basic functions, people are more inclined to choose products that are more suitable for them, whether in terms of product color, shape, etc., or product operating habits. In complex social groups, more product designs are carried out for young and middle-aged people, and relatively little attention is paid to the needs of children. Children are the future force of the developing country and society (el Thomas. 2014). Children's physical and mental health affects the country's and society's development. In terms of product development, it is worth looking forward to creating products that are more in line with children's needs, and it is also an important focus of the design. According to China's seventh population census results, children aged 0-14 accounted for 17.95% of about 253.38 million people. This is a huge consumer group for product design and development companies (Zeinab et al. 2012).

2 RESEARCH PURPOSE AND SIGNIFICANCE

This study takes children aged 7-14 as the research object, based on the special psychological activities and behavioral elements of children in this period as the benchmark for designing research experiments through the experimental analysis of Kansei Engineering and the

*Corresponding Author: qyz18762302734@gmail.com

 DOI: 10.1201/9781003450818-99

KANO model, to understand the real inner needs of children aged 7-14, so that in the design and development, we can create products that are more popular with users. As a concept and method of product development, emotional design has been applied in many fields. An appropriate design method is proposed for this user group to provide a reference for designers to develop such products.

3 MARKET RESEARCH AND USER ANALYSIS

3.1 *Market research on children's school bags*

With the gradual improvement of people's consumption level, people expect school bags to be strengthened in terms of practicality and better expanded in terms of decoration. The types of school bags are extremely diverse, including backpacks, single shoulders, safety bags, trolley bags, etc. According to changes in consumer tastes and the protection of children's physical and mental health, the materials of school bags are mainly leather, PU, polyester, cotton, and linen. Considering that children carry heavy items such as books on their backs, many manufacturers on the market also pay special attention to the carrying system of the schoolbag on the outside. On the one hand, it can provide extremely high toughness, preventing the schoolbag from breaking. On the other hand, the consideration of the physical development of children at this stage reduces the damage to the bones (Ademola James et al. 2014). In terms of price, the price of schoolbags sold by well-known brands on the market ranges from RMB 500-699.

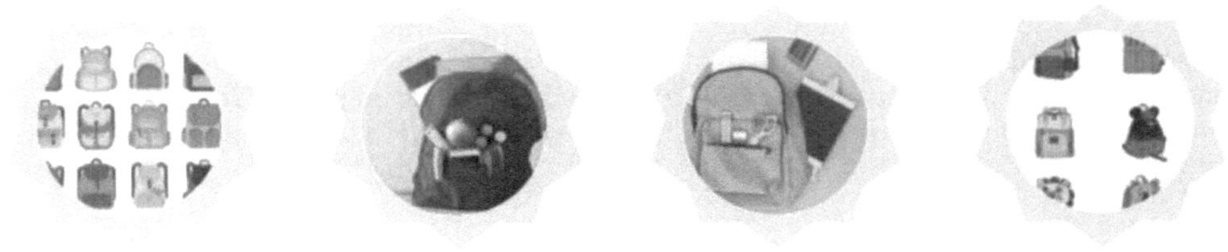

Figure 1. The caption of a typical figure.

3.2 *Analysis of psychological and physiological conditions of 7-14-year-old children*

Children in this age group are generally in elementary school, and their school adaptation includes five aspects: learning adaptation, behavioral adaptation, emotional adaptation, interpersonal adaptation, and attitude toward school. In terms of psychology, the self-evaluation of children in the first to the third grades showed a downward trend, and their self-confidence decreased. Children in the fourth to fifth grades can objectively evaluate themselves and realize their strengths, and their self-confidence has increased compared with those in lower grades. Children at this stage are eager to be praised and are particularly concerned about the evaluation of outsiders. In terms of intelligence, the development of children's attention, memory, observation, imagination, etc., is also gradually improved from the lower e upper grades (Thelma et al. 1987).

In terms of physiology, due to the characteristics of early puberty during this period, the height of girls grows faster than boys, but the overall growth trend is shown. During this period, the bones are in the development stage and are easy to bend, the movement coordination of the muscles is poor, and the control ability of self-behavior is relatively poor.

Table 1. The scope of children's intelligence changes with age.

Primary school children's intelligence changes			
Attention	7-10 years old Concentration for about 20 minutes (belongs to unintentional attention)	10-12 years old, about 25 minutes	12-14 years old About 30-40 minutes
Memory	Grade 1-2 mechanical memory	Grade 3 Conscious attention dominates	Grades 4-5 Will use finely crafted memory
Imagination	Intentional imagination increases and gradually occupies a dominant position.	The content of imagination tends to be complete, and the focus becomes more prominent.	
Observation	With the increased age and knowledge reserves, children's observation ability in middle and high grades has improved significantly.		

4 SCHOOLBAG DESIGN IDEAS AND METHODS

This research uses the KANO model to analyze user needs, understand the direction of product optimization, and then through the establishment of the Kansei Engineering model, transform the user's vague emotional semantics into clear design elements to design better products that meet the user's emotional needs.

4.1 *Analysis of users' needs based on the KANO model*

Noriaki Kano of the Tokyo Institute of Technology created the KANO model. It mainly explores the various needs of users for products and classifiesories. According to the relationship between different types of quality characteristics and user satisfaction, the characteristics of product service quality can be divided into five categories (Xu et al. 2009):

Must-be Quality—(M) One-dimensional Quality—(O) Attractive Quality—(A) Indifferent Quality—(I) Reverse Quality—(R) (Li. 2003).

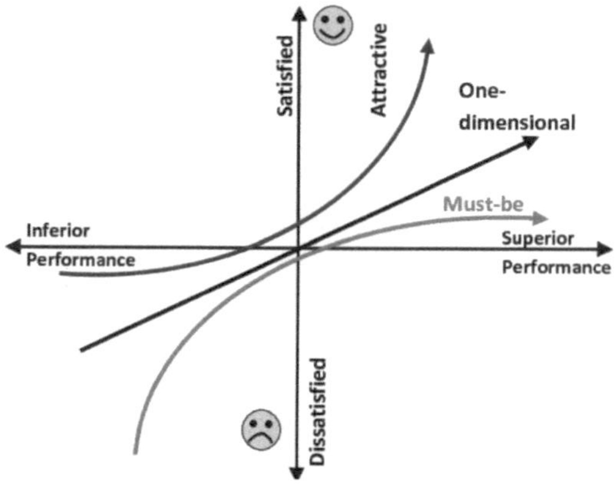

Figure 2. Schematic diagram of the KANO model.

Through consulting relevant information, online research, and other related channels, combined with the physical and psychological characteristics of children at this stage, we have sorted out six key needs for schoolbags for children aged 7-14. The six key requirements

Table 2. User requirements.

Scope	Demand	Serial number
Basic functions	Large capacity, clear and easy-to-find interlayer classification	S1
	The product is light in weight, more convenient to use, and cares for the development of the spine.	S2
	Waterproof, strong safety performance	S3
Exterior styling	Attractive styling, attractive styling and outstanding details	S4
	The material is softer and more sustainable, and environmentally friendly	S5
	Interactive elements are added to improve the appearance and decoration while inspiring children.	S6

are organized and classified into basic functional requirements and appearance requirements.

The user needs to be sorted out above were made into a KANO model experiment questionnaire. The satisfaction degree of the questionnaire was divided into five grades, 5 being very satisfied, 4 being reasonable, 3 being indifferent, 2 being reluctantly accepted, and 1 being very dissatisfied. And the division dimension of the questionnaire starts from both positive and negative aspects. According to the psychological analysis of children above, it can be known that children at this stage have relatively weak self-awareness and observation ability, so this survey mainly focuses on the guardians of this group, with a small number of target users mixed of 100 questionnaires distributed this time, and 96 were finally recovered after excluding questionable questionnaires. The questionnaire recovery rate was 96%, and the quantitative data of the questionnaire was summarized. The data in each area of the quantification table represent different user needs. The red area represents A excited type demand. The yellow area represents the O expectation type demand. Purple represents M basic type demand. Green represents I no difference type demand, light blue represents R reverse type demand, and the blank color represents the Q suspicious result (Tang & Long 2012).

Table 3. Each area type in the KANO model data.

Function		Provide this function				
		Very satisfied	As it should	Indifferent	Barely accept	Extremely dissatisfied
This function is not provided	Very satisfied	Q	M	M	M	A
	As it should	R	I	I	I	M
	Indifferent	R	I	I	I	M
	Barely accept	R	I	I	I	M
	Extremely dissatisfied	R	R	R	R	Q

According to the questionnaires, the quantification table data of each requirement was counted, and the attribute analysis of the KANO model was carried out.

During this period, it is calculated and analyzed with the help of the Better-Worse coefficient. Better represents the satisfaction coefficient after the demand is increased. The value is usually positive. The closer the positive value is to 1, the stronger the satisfaction improvement effect will be, and the faster the satisfaction will rise. Worse means user satisfaction will decrease if the product does not provide a certain function or service. The value

Table 4. Questionnaire data quantification table.

	KANO Model Property Classification						
Serial number	Q%	A%	O%	M%	I%	R%	Classify
S1	0	15.9	16.4	43.7	20.2	3.8	M
S2	0	27.1	39.3	23.5	8.7	1.4	O
S3	0	36.2	27.4	25.7	6.8	3.9	A
S4	0	13.5	13.7	52.8	16.2	3.8	M
S5	0	13.5	6.7	21.7	37.2	20.9	I
S6	0	22.3	37.6	28.5	6.5	5.6	O

is usually negative and closer to -1. It means that it has the greatest impact on user dissatisfaction and the stronger effect on reducing satisfaction. In drawing graphs for data analysis, the Worse value takes its absolute value. And summarize the data of the quantization table to get the average value of each function of S1-S6 (Simon et al. 2004).

Increased satisfaction factor: Better=(A+O)/(A+O+M+I)

Dissatisfaction coefficient after elimination: Worse=(O+M)/(A+O+M+I)*-1

According to the calculation of the Better-Worse coefficient, the user's needs are drawn into a matrix coordinate diagram for better analysis and observation. The origin of the quadrant is [|Worse|, Better], which is [0.59, 0.47] after calculation.

Through the above analysis, we can understand that S1 and S4 are the basic needs of users, S2 and S6 are the needs of desire, S3 is the need of excitement, and S5 is the need of no difference. In product development, designers should follow the order of basic needs > eager needs > exciting needs to develop products that better meet users' needs.

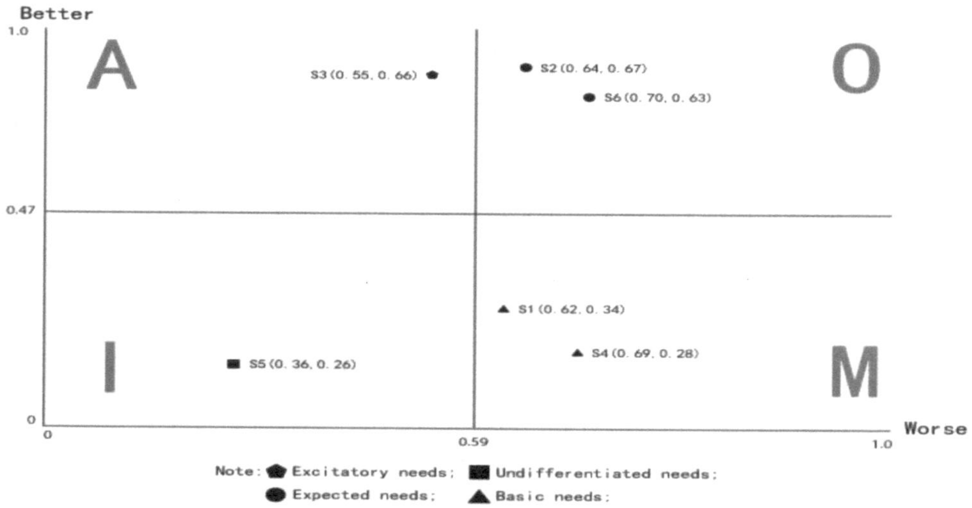

Figure 3. Quadrant diagram of user need type.

4.2 *Extraction of product elements based on Kansei engineering*

The experimental stage of Kansei engineering is divided into the stage of establishing sensibility semantics and the stage of collecting representative products. The established sensibility semantics vocabulary is connected with representative samples to obtain precise design elements (Josip et al. 2011). Through multiple channels such as the Internet, product books, and market research, 6 groups of emotional vocabulary and 10 representative products were finally screened out.

Table 5. Questionnaire data quantification table.

Safe - Dangerous	Interesting – Uninteresting	Beautiful - Ugly	Comfortable – Inconvenient	Large capacity - Small Capacity	Unique - General

Table 6. Sample library.

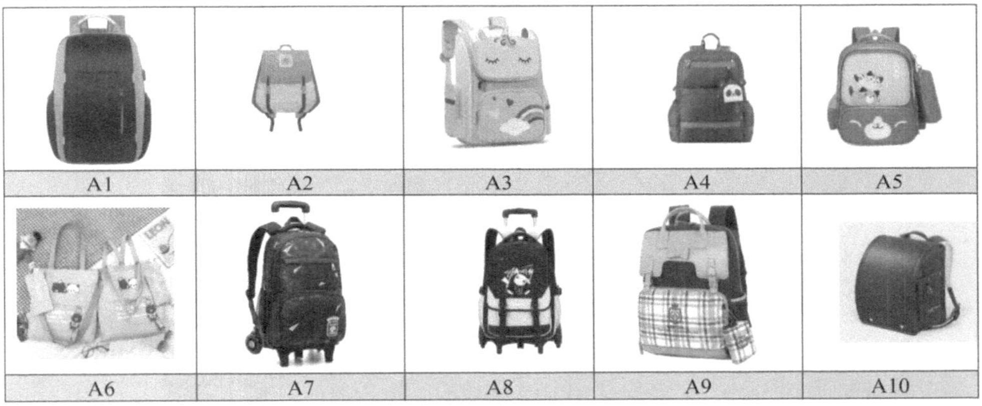

Through semantic differences, the final selected perceptual semantics and product samples are designed for survey questionnaires, questionnaires are distributed to users, and the survey results are collected for statistics. The relationship between perceptual semantics and product samples can be obtained to understand the relevant design elements (Mitsuo. 2004).

The above icons show that A1 and A10 are the safest in the statistical results, which is related to their sleek exterior shape and leather materials. Regarding fun, A3, A4, and A5 scored higher, and A4 and A5 scored the highest in aesthetics. Their common feature is cartoon images on the outside of the schoolbag, and the color matching is bright. A7 and A8 have the highest scores in comfort and large capacity, which is inseparable from the fact that they have pulleys at the bottom, which are easy to move and can load many items. A4, A5, and A8 scored the highest in terms of uniqueness. Their overall feature is that they are decorated with children's favorite cartoon images and are perfectly integrated with the school bags.

4.3 *Constructing the perceptual image model of the schoolbag*

Establishing a perceptual image model by combining perceptual semantics and representative samples with the user needs to be analyzed by the KANO model.

According to the classification of user needs in the KANO model above, we understand that in product development and design, the realization of user needs follows S1, S4>S2 and

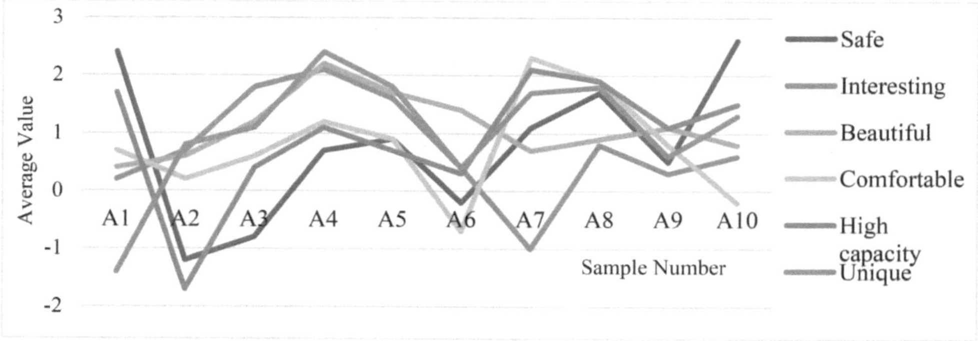

Figure 4. Perceptual semantics and sample relation table.

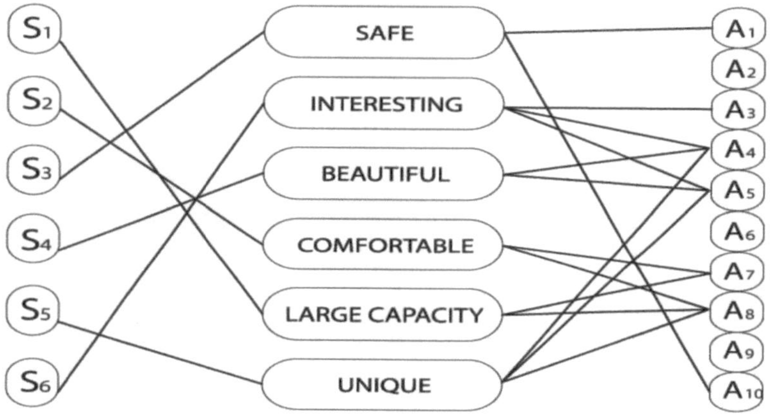

Figure 5. Perceptual image model.

S6>S3. The established perceptual image model connects perceptual semantics and representative samples according to the user demand level of the KANO model. After satisfying the basic needs of users in the design and development process, priority is given to comfort and fun as innovative development points to create more attractive products.

5 SUMMARIZE

Through the exploratory experiments of this research, it is helpful to help designers provide a favorable breakthrough in product development. The user demand level is obtained through the KANO model combined with the design elements extracted by Kansei Engineering. On the one hand, it is helpful to understand the impact of each product shape, material, structure, etc., on the inner emotional needs of users. On the other hand, it can clearly understand the user's desire for products and indirectly convert this virtual demand into tangible design elements, which provides inspiration for designers to develop products in the later stage. Ultimately, through the research method of this study, user needs are taken as the main goal of product development, which is conducive to the realization of the people-oriented core principle of product design.

REFERENCES

Adeyemi A J. & Rohani J M. & Rani M R A. (2014). 2014 Back Pain Arising from Schoolbag Usage Among Primary Schoolchildren[J]. *International Journal of Industrial Ergonomics*, 44(4): 590–600.

Cook D T. (2014). The Other "Child Study": Figuring Children as Consumers in Market Research, 1910s–1990s[J]. *Sociological Quarterly, 2000,* 41(3): 487–507.

Horn T S. & Hasbrook C A. (1987). Psychological Characteristics and the Criteria Children use for Self-evaluation[J]. *Journal of Sport and Exercise Psychology*, 9(3): 208–221.

Javadivala Z. & Allahverdipour H. & Dianat I, *et al.* (2012). Awareness of Parents about the Characteristics of a Healthy School Backpack[J]. *Health Promotion Perspectives*, 2(2): 166.

Mikulić J. & Prebežac D. (2011). A Critical Review of Techniques for Classifying Quality Attributes in the Kano Model[J]. *Managing Service Quality: An International Journal.*

Nagamachi M. (2004). Kansei Engineering[M] *Handbook of Human Factors and Ergonomics Methods.* CRC Press, 794–799.

Schütte S T W. & Eklund J. & Axelsson J R C, *et al.* (2004). Concepts, Methods and Tools in Kansei Engineering[J]. *Theoretical Issues in Ergonomics Science*, 5(3): 214–231.

Tang Zhongjun. & Long Yuling. (2012). Research on Personalized Demand Acquisition Method Based on Kano Model[J]. *Soft Science*, 26(2): 127–131.

Xu Q. & Jiao R J. & Yang X, *et al.* (2009). An Analytical Kano Model for Customer Needs Analysis[J]. *Design Studies*, 30(1): 87–110.

Yanzu Li. (2003). New Concept of Design: Kansei Engineering[J]. *New Art*, 24(4): 20–25.

Water Conservancy and Civil Construction – Oke & Ahmad (Eds)
 ISBN: 978-1-032-58614-4

Structural dynamics simulation of a vehicle-borne radar in transit

Zhan Hu* & Linfeng Hong*
Nanjing Research Institute of Electronics Technology, Jiangsu Nanjing, China

ABSTRACT: Vehicle-borne radars in transit face complicated vibrations and shocks, which poses great challenges for the stiffness and strength design of both bearing structures and installed equipment. This paper establishes the systematic finite element model for a vehicle-borne radar in transit, then analyzes the dynamic characteristics of the whole radar system, including model analysis, shock responses, vibration transmissibility, and random vibrations. By comparison of the rigid vehicle suspension, isolation performances for vibrations and shocks of the radar system with the flexible vehicle suspension are researched quantitatively. The structural strength of the whole radar system is verified. Besides, for installed equipment on the vehicle-borne radar, this paper can provide an important reference for environmental conditions of vibrations and shocks.

1 INTRODUCTION

Modern radar requires higher and higher maneuverability to improve survivability (Fang 2017; Hu 1999), and radar transportation is an important part of maneuverability (Hong 2022). As the widest transportation mode, road transport has become the basic requirement for ground mobile radars (Hong 2011), and vehicle-borne radar presents a variety of typical layouts (Hong 2021; Liu 2018). Due to the large differences in terrain and road conditions, the vehicle-borne radar will inevitably face complex dynamic load environments, such as different degrees of shocks and vibrations, which poses challenges to the structural design of the radar equipment. On the other hand, with a growing number of installed equipment on the radar, shocks and vibrations may cause loosening, cracking, debonding, and other failure problems (Peng 2018) and indirectly or directly affect the electrical performance. Therefore, analyzing the systematic structural dynamics of the whole radar vehicle in transit is of great significance.

The vehicle transport platform carries the vehicle-borne radar, so the structural dynamic analysis in transit is closely related to vehicle mechanical modeling. Many kinds of literature have simplified analytical modeling of vehicle dynamics systems, including the quarter vehicle model (Kulkarni 2018; Lu 2014a; Sharma 2020), the half vehicle model (Issa 2021; Li 2020), the seven-DOFs vehicle model (Kim 2014; Zhang 2011), the nine-DOFs vehicle model (Li 2010) and twenty-DOFs model (Li 1987), and researched dynamic loads on the road, impacts of speed bumps and random vibrations. Simulated analysis of a light truck (Lu 2014b) and a reflector antenna radar vehicle (Ren 2013) are studied based on establishing rigid-flexible coupling finite element models.

This paper establishes the systematic finite element model of a vehicle-borne radar in transport. It comprehensively analyzes its modal frequencies and shapes, shock responses, vibration transmissibility, and the PSD of random vibrations. This paper can verify the

*Corresponding Authors: huzhan@cetc.com.cn and hlf@sina.com

 DOI: 10.1201/9781003450818-100

radar's structural strength and quantitatively analyze the dynamic environment where the equipment is installed. Moreover, it can provide a significant reference for the design and dynamic environmental tests for the same type of vehicle-borne radars.

2 FINITE ELEMENT MODEL OF RADAR

The antenna array is folded and flattened when the vehicle-borne radar is in transit. The front end of the antenna structural frame is locked on the bearing platform through the transport support device. Its middle part and back end are connected with the antenna pedestal through a pair of hydraulic cylinders and pin shafts, respectively. All leveling support legs of the radar vehicle are contracted, and the anti-overturning legs are folded and locked at two sides of the platform, as shown in Figure 1.

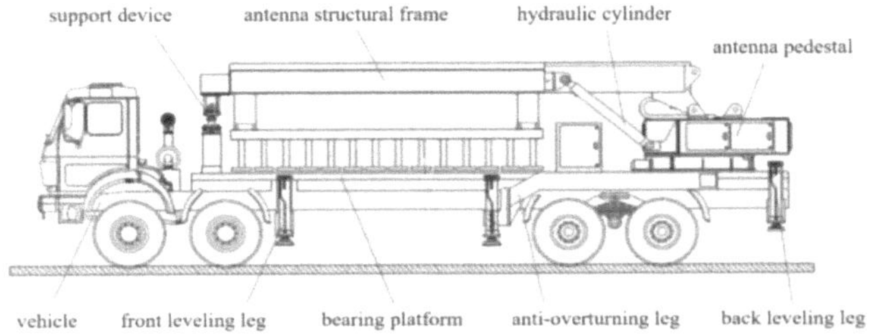

Figure 1. A diagram of the vehicle-borne radar is in transit.

The systematic finite element model of the radar vehicle aggregates 420,364 elements with a total weight of 40 t, as shown in Figure 2. The antenna array, bearing platform, antenna pedestal, and vehicle girder are modeled by shell elements. Solid elements model all support legs, and bar elements model hydraulic cylinders. The X-direction is longitudinal, the Y-direction is transverse, and the Z-direction is vertical. The transport support device is fixed on the bearing platform, and the fixed interface is flexible, allowing rotation around the Y-axis to eliminate machining and assembling errors. Therefore, the simulation model releases the degrees of freedom that the bottom of the transport support device rotates around the Y-axis.

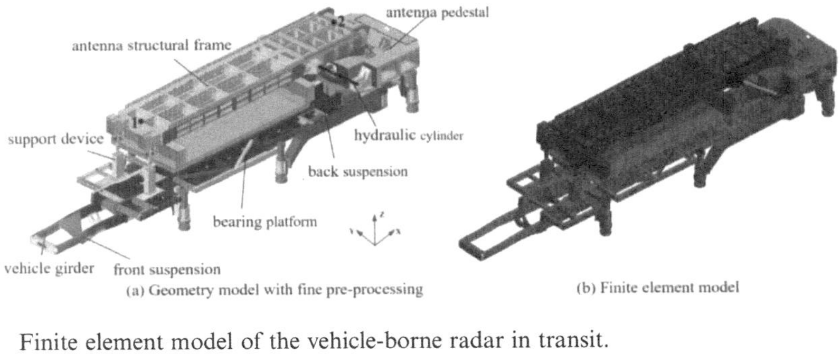

Figure 2. Finite element model of the vehicle-borne radar in transit.

The front and back suspensions of the vehicle are simplified as the equivalent spring-damping support boundary and are established by using BUSH elements. According to the actual characteristics of the vehicle suspension, we set the equivalent spring stiffness and damping parameters in different directions of the front and back suspension. Considering the absence of suspension, the rigid spring stiffness is set to compare and verify the vibration attenuation effect of the vehicle suspension on the radar vehicle.

The hydraulic cylinder is modeled by a two-force bar element with a circular section and an outer diameter of 130 mm. The hydraulic cylinders' linear bar element is used during the mode analysis, frequency response, and random vibration calculations. Impact loads may cause significant jumping on the antenna array. Considering the huge stiffness difference of the hydraulic cylinder between the tensile and constringent state, the hydraulic cylinder is established with the GAP element. Its stiffness expression is as follows:

$$K_{HC} = \begin{cases} EA/L_0 & (0 \leq l \leq L_0) \\ 0 & (l > L_0) \end{cases} \tag{1}$$

where E is the elastic modulus of the bar; A is the cross-sectional area; L_0 is the initial distance between the upper and lower end points of the hydraulic cylinder; l is the distance after deformation. When L_0 is greater than l, the hydraulic cylinder stretches, and the stiffness is set to zero. At this time, the GAP element is not stressed and does not transfer load.

3 MODAL ANALYSIS

Modal analysis is the basis of structural dynamics analysis. Modal analysis is conducted for the radar vehicle model in the transportation state. The first 30 modes are selected, and the cumulative effective modal mass ratio exceeds 95%. The first 6 natural frequencies (f_n) and modal shapes are shown in the following table and the comparison of the first modal shape is shown in Figure 3.

Table 1. Comparisons of the natural frequencies and modal shapes for the first six orders.

	Without suspension		With suspension	
Order	f_n /Hz	Modal Shape	f_n /Hz	Modal Shape
1	4.95	Translation of antenna frame in Y-axis	1.45	Overall translation in Y-axis
2	8.75	Translation of antenna frame in X-axis	1.69	Overall translation in X-axis
3	8.93	Torsion of antenna frame around the X axis	1.90	Overall translation in Z-axis
4	10.99	Bending of antenna frame around Y-axis	2.56	Overall rotation around X-axis
5	11.75	In-phase translation of anti-overturning legs in the Y-axis	3.17	Overall rotation around Z-axis
6	12.05	Out-phase translation of anti-overturning legs in the Y-axis	5.60	Overall rotation around Y-axis

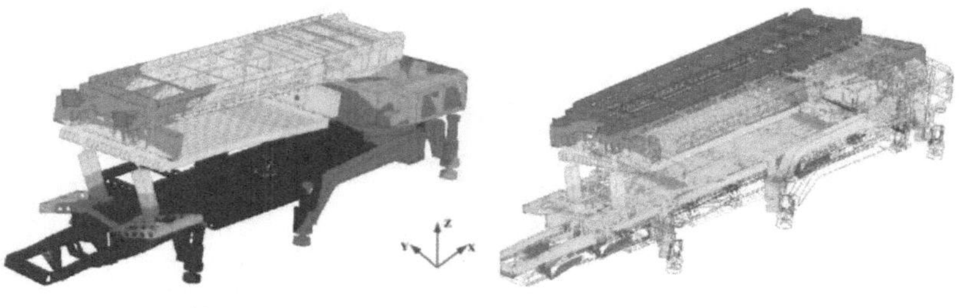

(a) without suspension (b) with suspension

Figure 3. Comparison of the first modal shape.

Through comparison, it can be seen that due to the existence of the flexible vehicle suspension, the first six-order modal shapes are third-order overall translations and third-order overall rotations of the whole vehicle, respectively. When the suspension is rigid, the antenna array structure in the large-span simple support state and the anti-overturning legs locked in the platform become the two main flexible structures of the whole system. Therefore, the natural frequencies introduced by the vehicle suspension are lower than those of the antenna array, which can provide good vibration isolation for the structures and installed equipment.

4 SHOCK RESPONSE ANALYSIS

Vertical, longitudinal, and transverse shock impacts are applied to the vehicle support boundary. According to the Laboratory environmental test methods (GJB.150A 2009), the back-peak sawtooth wave with a peak acceleration of 20 g and a duration of 11 ms is adopted. Because the stiffness of the hydraulic cylinder expressed by Formula (1) is non-linear, the shock response is calculated directly by the complete method. Since many important radar devices are installed in the antenna structural frame, the antenna structural frame is the most necessary structure to be checked in the shock simulation. Responses at the front end (node 1) and the back end (node 2) of the array structural frame (see Figure 2) are calculated. Comparisons to the rigid support conditions without suspension are presented in Figure 4 and Table 2.

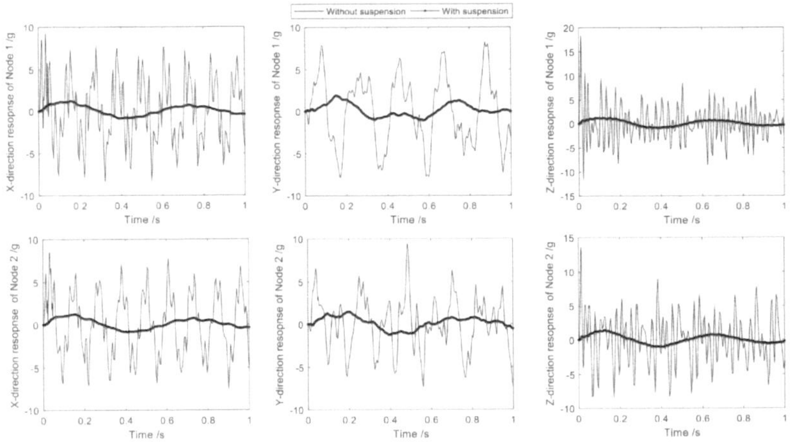

Figure 4. X-direction, Y-direction, and Z-direction shock responses of node 1 and node 2.

Table 2. Acceleration peak comparison.

			Without suspension		With suspension	
Direction	Shock peak /g	Node	Peak value /g	Attenuation ratio	Peak value /g	Attenuation ratio
X	20	1	9.25	53.75%	1.21	93.95%
Y	20		8.22	58.90%	1.92	90.40%
Z	20		18.4	8.00%	1.17	94.15%
X	20	2	8.55	57.25%	1.23	93.85%
Y	20		9.43	52.85%	1.44	92.80%
Z	20		13.6	32.00%	1.41	92.95%

It can be seen that when the radar vehicle has no suspension, acceleration responses of the antenna structural frame are very large. The maximum acceleration response in the Z direction reaches 18.4 g, close to the shock input's peak value. The maximum response occurs during the shock load. A simulation comparison of radar vehicles with suspension shows that the maximum value is less than 2 g, and the attenuation of the response peak relative to the shock input exceeds 90%. Acceleration responses of the antenna structural frame present gentle and low fluctuation and no obvious sharp shock. Therefore, the vehicle suspension plays a significant protective role for the radar structure and installed equipment.

The structural strength of the radar with the vehicle suspension under the shock load is analyzed. The maximum stress appears at the front arm of the antenna structural frame under the vertical shock, as shown in Figure 5. The maximum stress is 132 MPa, and its safety factor of yield strength is up to 2.61. Therefore, the structural strength of the radar vehicle has a sufficient safety margin under the shock in transport.

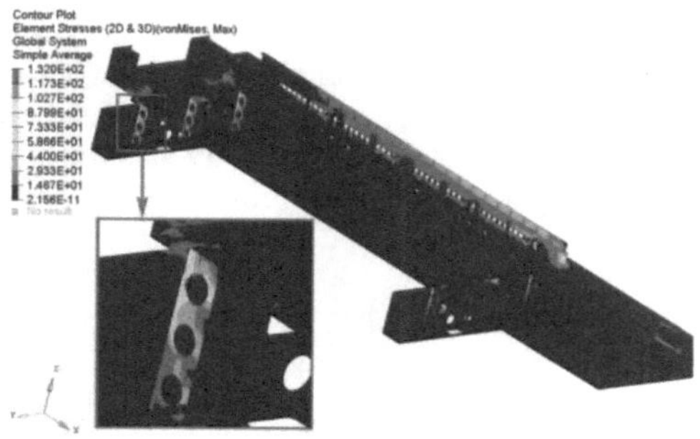

Figure 5. The stress of the antenna structural frame under vertical (Z) shock.

5 VIBRATION TRANSMISSIBILITY ANALYSIS

The vertical, longitudinal, and transverse sine-sweeping acceleration excitations are applied to support the boundaries of the vehicle suspension, and the frequency range is 0.1~500 Hz. Considering the first 30-order modes of the whole system, the mode superposition method is

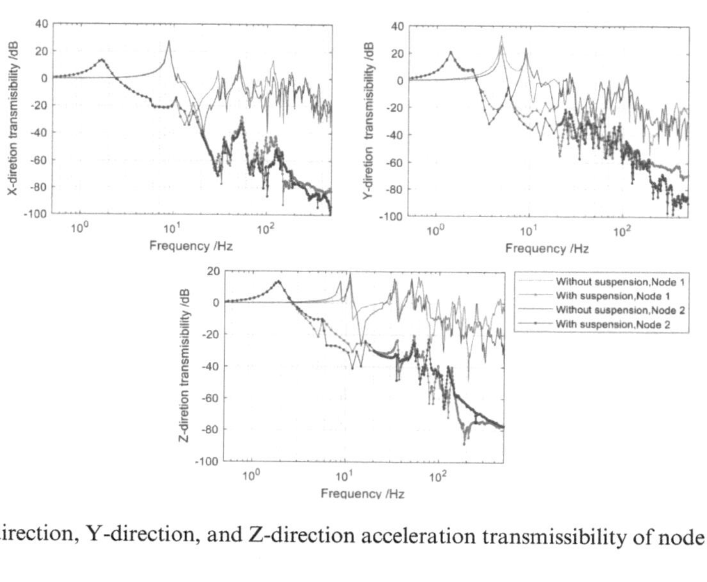

Figure 6. X-direction, Y-direction, and Z-direction acceleration transmissibility of node 1 and node 2.

used to calculate acceleration responses $A(f)$in the frequency domain. Finally, the acceleration transmissibility is calculated as follows:

$$T_a = 20\log_{10}(A(f)/A_0) \tag{2}$$

where A_0 is the amplitude of the sweeping excitation.

The acceleration transmissibility curves of node 1 and node 2 of the antenna structural frame are shown in Figure 6. The acceleration transmissibility curves without the suspension are relatively higher, and the attenuation performance in the high-frequency band is not significant. Several significant resonance peaks appear between 5 and 11 Hz. The maximum vibration transmissibility of X, Y, and Z reach 28.1 dB, 33.6 dB, and 19.2 dB, respectively, which correspond to the first four-order translation, torsion, and bending modes in transit (see Table 1). The acceleration transmissibility curves with suspension have three resonance peaks in the low-frequency band (1.5~2.0 Hz), corresponding to the first three-order vehicle translation mode caused by the flexibility of the vehicle suspension. When the excitation frequency exceeds 3 Hz, the vibration transmissibility quickly reduces to below 0 dB. That is, there is an obvious vibration isolation performance in the broadband. Compared with the no-suspension state, the attenuation of the vibration transmissibility curve with the suspension state reaches more than 20 dB in the frequency range of 5 to 500 Hz, and the maximum attenuation can reach 60 dB. Therefore, the vehicle suspension plays a very significant role in reducing the vibration transmissibility in the antenna structural frame of the radar vehicle.

6 RANDOM VIBRATION ANALYSIS

The road excitation of radar vehicles in transit can be regarded as an ergodic and stationary random process (Li 1987). This section focuses on radar vehicles' power spectral density (PSD) and calculates the stress using the three-interval method based on Gaussian distribution (Cao 2014). The acceleration spectrum of random vibration adopts the C.3 road spectrum in Annex of GJB150.16A-2009, with a frequency range of 5~500Hz. The RMS values for the X, Y, and Z-direction acceleration are 2.05 g, 1.62 g and 2.20 g, respectively.

Based on the frequency-domain response results in Chapter 4, the acceleration PSD curves of node 1 and node 2 are calculated, as shown in Figure 7. Trapezoidal numerical integration is used to calculate the envelope area of PSD curves, i.e., standard variance values for each above PSD curve can be obtained, see Table 3. In case of no suspension, there is a limited

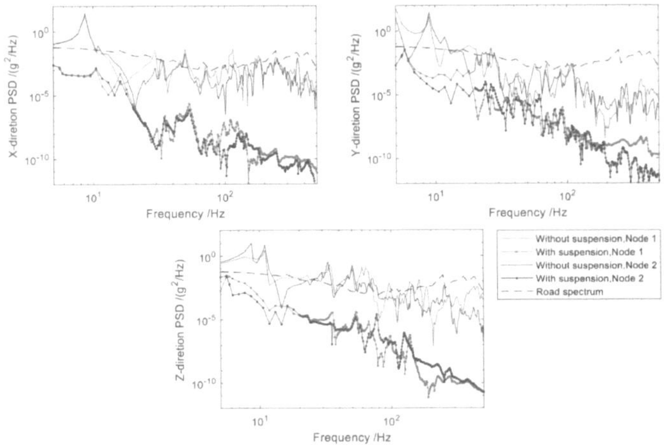

Figure 7. X-direction, Y-direction, and Z-direction acceleration PSD of node 1 and node 2.

Table 3. Comparison of the acceleration PSD standard variance.

			Without suspension		With suspension	
Direction	Road Spectrum /g	Node	RMS/ g	Attenuation ratio	RMS/ g	Attenuation ratio
X	2.05	1	3.906	−90.54%	0.075	96.34%
Y	1.62		0.428	73.58%	0.078	95.19%
Z	2.20		2.821	−28.23%	0.219	90.05%
X	2.05	2	4.626	−125.66%	0.067	96.73%
Y	1.62		0.572	64.69%	0.071	95.62%
Z	2.20		4.208	−91.27%	0.155	92.95%

Remarks: Negative sign indicates vibration amplification.

attenuation for the transverse (Y) PSD RMS, and the RMS of the longitudinal (X) and vertical (Z) acceleration PSD are significantly amplified. The vehicle suspension suppresses the random vibration of the radar system in the full frequency band. The acceleration PSD RMS attenuates by more than 90% relative to the road spectrum, so the vehicle suspension achieves an excellent vibration isolation performance.

In addition, the systematic simulation shows that acceleration PSD curves at node 1 and node 2 of the array structural frame are lower than the road spectrum in the full frequency band. These results are more accurate and proper as the random vibration inputs instead of the general road spectrum, which is helpful to improve the lightweight design of the installed equipment.

Besides, the RMS stress of the radar structure with the vehicle suspension is analyzed. The maximum stress appears at the back suspension boundary of the vehicle girder under the vertical vibration (see Figure 8). The maximum 3σ stress is 92 MPa, and its safety factor of the yield strength is up to 4.13. Therefore, the structural strength of the radar vehicle has a sufficient safety margin under random vibrations in transport.

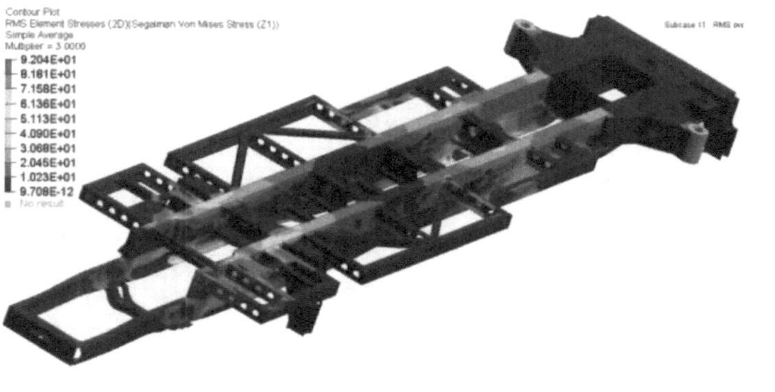

Figure 8. 3σ stress cloud map of the vehicle and the bearing platform.

7 CONCLUSION

This paper established a finite element model for the radar in transit, simplified the vehicle suspension with the equivalent stiffness and damping support boundary, considered the nonlinear characteristics of tension and compression stiffness of hydraulic cylinders under shocks, and completed the system dynamics analysis of the modal analysis, shock responses, vibration transmissibility and random vibrations. The whole radar vehicle structure meets the road transport requirements. The specific conclusions are as follows:

1) The suspension flexibility makes the first six natural frequencies of the whole system lower than the antenna array, which guarantees vibration isolation.
2) The vibration transmissibility shows that the vehicle suspension has a good vibration isolation effect for the vibration above 3 Hz, and the vibration suppression of the high-frequency band (> 100 Hz) exceeds 40 dB.
3) The shock response and random vibration simulation show that the response peak and random PSD RMS for the antenna structural frame attenuate by more than 90%. The safety factor of the yield strength is abundant for the whole radar structure.
4) The systematic dynamic simulation of the vehicle-bone radar can calculate installed equipment's fine environmental dynamic responses. It is more reliable than the direct reference of general standards, which greatly helps improve installed equipment's lightweight design.

REFERENCES

Cao B. & Wang Z.H. (2014). Research on Random Vibration Characteristics of the Bus Frame based on Displacement PSD, No.318 (06): 87–90.

GJB 150.16A. (2009). Laboratory environmental test methods for military materiel-Part 16: Vibration test.

Fang J. S., Zhang G. X. & Wang C. C. (2017). System structure design of a high mobility vehicle-borne radar. Electro-Mechanical Engineering, 33.04(2017): 6–9+17.

Hong L.F. (2011). Discussion on Transportability of Large or Medium Aperture Vehicle-borne Radar, *Electro-Mechanical Engineering*, 27(5), 1–3, 2011.

Hong L.F. (2021). The Layout Design of the Vehicle-borne Radar and Transportation Simulation, *Machine Design and Manufacturing Engineering*, 50(1), 1–4.

Hong L.F. (2022). Discussion on the Lift of Radar Motivation. *Machine Design and Manufacturing Engineering*, 51(07): 109–113.

Hu C.M. & Luo C.R. (1999). Discussion on Methods of Improving the Mobility of Ground Radars from the Viewpoint of Structure Design. *Electronics Machinery Engineering*. 81(5), 11–15.

Issa M., & Samn A. (2021). Passive Vehicle Suspension System Optimization using Harris Hawk Optimization Algorithm. *Mathematics and Computers in Simulation* (2).

Kim H. W., Min C. H. Lee, C. H., Hong S. & Oh J. W. (2014). Dynamic Analysis of Tracked Vehicle by Buoy Characteristics. *Ocean & Polar Research*, 36(4), 495–503.

Kulkarni, Ambarish, Ranjha, Sagheer A., & Kapoor *et al.* (2018). A Quarter-car Suspension Model for Dynamic Evaluations of an In-wheel Electric Vehicle. *Proceedings of the Institution of Mechanical Engineers, Part D. Journal of Automobile Engineering*, 232(9), 1139–1148.

Li C.D. (1987). The Multi-freedoms Model and Random Vibration Response Analysis of the Whole Vehicle System, *Automotive Engineering*, (02): 26–41+51.

Li J., Qin Y.Y. & Zhao Q. (2010). The Application of Multi-point Pseudo Excitation Method to Vehicle Random Vibration Analysis, *Automotive Engineering*, v.32; No.188 (03): 254–257+ 269.

Liu J. J., Hong L. F. & Lin Z. Y. (2018). Discussion on Integrated Design of Semi-trailer and Radar Rotating Platform. *Electro-Mechanical Engineering*, 34(03): 15–18.

Lu F., Chen S.Z., Liu C., Li M.H. & Zhao Y.Z. (2014a). Vehicle Vibration Velocity Estimation based on Kalman Filter, *Journal of Vibration and Shock*, 33(13), 111–116.

Lu S.S. (2014b). Simulation Analysis on Vibration Characteristics for a Light Truck based on Rigid-flexible Coupling Model. *Journal of Hubei University of Automotive Technology*. v.28; No.87(02): 17–20+32.

Peng H., Shan J.Y. & Yan W. (2018). Application of FMECA Technology in TR Module Design of Vehicle Radar, *Environmental Test Equipment*, 213(03): 116–120.

Ren K.L. & Li Z.G. (2013). Study on Dynamics of Running Vehicle Radar, *Equipment Environmental Engineering*, v.10; No.53(02): 62–66.

Sharma S., Sharma R. C., Sharma S. K., Sharma N. & Bhardawaj S. (2020). Vibration Isolation of the Quarter Car Model of Road Vehicle System using Dynamic Vibration Absorber. *International Journal of Vehicle Structures and Systems*, 12(5).

Zhang G.H., Kang H.G. & Zheng Y.X. (2011). Establishment of Vehicle Vibration Model and Calculation of Vehicle Dynamic Load, *Journal of Wuhan University of Technology (Transportation Science & Engineering)*, v.35 (04): 771–77.

Water Conservancy and Civil Construction – Oke & Ahmad (Eds)
© 2024 The Author(s), ISBN: 978-1-032-58614-4

Flame retardant and smoke suppression effect of lignin-Based flame retardant coatings

Zejian Jia*
XAUAT UniSA An De College, Xi'an University of Architecture and Technology, Xi'an, Shaanxi China

Boqiao Wang
China Jikan Research Institute of Engineering Investigation and Design Co., Ltd, Xi'an, Shaanxi, China

ABSTRACT: Wood faces severe fire risks when used as a building material. Currently, flame retardants are mostly used to reduce the risk of fire in wooden buildings. There are various types of current flame retardants, but there is an urgent need to develop a greener and more efficient one. In this paper, a nitrogen-phosphorus-modified lignin-based flame retardant was prepared, and the structure of the material was proven by elemental analysis. The flame retardant was tested by cone calorimetry, and the results showed that when the modified lignin content was 20%, the peak heat release rate, total heat release, peak smoke release rate, and total smoke release were reduced by 21.09%, 41.28%, 32.4%, and 31.91%, respectively, showing excellent flame retardant effects. The characterization results showed that the flame retardant resulted in faster and denser intumescent carbon layers, which was attributed to the phosphorus in the material that promoted the formation of carbon layers and the flame-retardant gases released from nitrogen during combustion that diluted the air content, thus achieving efficient green flame retardancy in cooperation with the dense carbon layers. This provides a precedent for the green and efficient development of bio-based flame retardants for a wide range of applications.

1 INTRODUCTION

Wood is an old building material and is one of the most widely used building materials today due to its unique material properties such as ease of processing, artistic grain, good physical and mechanical properties, aesthetics, and renewable nature. However, at the same time, due to its natural flammable properties, it increases the risk of fire in the context of widespread use. When a fire starts, it spreads quickly, and wood and other combustible materials that burn with heat and produce smoke gas are important factors that contribute to fire casualties, so applying appropriate flame retardant treatment to wood-based building decoration materials is critical. Flammable materials treated with fire retardants can significantly reduce the risk of fire and effectively slow down the spread of fire, which can win valuable time for rescuing people and fighting fires and effectively reduce the damage caused by fire. The intumescent flame retardant coating is a widely used and effective flame retardant treatment method in recent years. Its essence is to add intumescent flame retardants to the coating, which is made of a kind of paint with an expansion effect. Under the action of high temperature and flame, the intumescent flame retardant coating will undergo a series of chemical reactions, resulting in the formation of a dense layer of carbon on the surface of the

*Corresponding Author: jzj@xauat.edu.cn

 DOI: 10.1201/9781003450818-101

coating, which will isolate oxygen and temperature and play the role of a flame retardant. The effect of a fire retardant. Carbon-forming agents, dehydrating agents, and foaming agents are the three major components of the intumescent flame retardant system (Ge et al., 2022). The expansion flame retardant system is used in fireproof coatings, and the different ratios of different types of carbon-forming agents, dehydrating agents, and foaming agents will have a great impact on the flame retardant performance of fireproof coatings.

Ji et al. (2011) created intumescent fire-retardant coatings by combining a dehydrating agent, a carbon-forming agent, and a foaming agent with pentaerythritol and melamine diphosphate, among other ingredients. The performance characterization of the fire-retardant coatings revealed that the synthetic intumescent flame retardants had less effect on the mechanical properties of the coatings and that the flame retardant properties were further improved. Li et al. (2018) found that a blend of polyvinyl acetate resin and urea-formaldehyde resin could be used as a film-forming base for flame retardant coatings, so a class of intumescent water-based amino resin fireproof coatings was made with ammonium phosphate, urea polyphosphate, pentaerythritol, and melamine as intumescent flame retardant systems. It prolonged the ignited time, reduced the heat release rate and total heat release of the substrate, decreased the concentration of carbon monoxide and smoke release during combustion, and improved the mass fraction of the remaining material while having better flame retardant and smoke suppression effects. Yu et al. (2010) prepared intumescent fireproof coatings using epoxy resin emulsion as the base material and a synthetic pentaerythritol polyphosphate melamine salt intumescent flame retardant integrating a dehydrating agent, a carbon-forming agent, and a foaming agent as the expansion system. A characterization analysis of the fireproof coatings concluded that those with the addition of the synthetic intumescent flame retardant had a better fireproofing effect than those with the addition of the three agents alone. Wang et al. (2022) developed a water-based intumescent flame retardant coating with a highly intumescent carbon layer after combustion by adding kaolin and zirconium-containing ceramic fibers as reinforcing fillers to the intumescent flame retardant system. The best fire resistance of the coating was achieved when the content of kaolinite and ceramic fibers was 4% and 2%, respectively, which was attributed to the enhanced strength of the intumescent carbon layer due to zirconium-containing ceramic fibers.

At present, the development of intumescent flame retardant coatings still faces some problems; for example, the carbon formation effect of intumescent materials is not ideal, often requiring the addition of a large number of intumescent materials, resulting in poor dispersion of the whole system; the potential toxicity of the current traditional intumescent flame retardant coatings is also a major cause for concern (Rosace et al., 2018); and the high concentration of toxic gases generated after fire may cause injury to the relevant personnel. Among the many development directions of flame retardants, bio-based flame retardant materials are considered one of the most promising directions in the next generation of flame retardants because of their material sustainability, environmental friendliness, and unique flame retardant efficiency. In particular, a class of polymeric materials represented by lignin and its derivatives has great flame-retardant potential. As the second most abundant biomass material in nature, lignin itself is widely present in wood, and as a flame-retardant coating on wood, it greatly reduces the corrosion and contamination of the wood substrate compared with other chemical additives. It is also very rich in carbon content and can be used as a natural carbon source. Due to its richness in a variety of functional groups, it provides rich active sites for further chemical modification and chemical reactions (Sun et al., 2020; Anderson et al., 2018). However, there is no systematic study on the flame-retardant properties of lignin materials.

In this paper, we studied the effect of adding different contents of lignin on the improvement of the fire performance of intumescent flame retardant coatings and explored the flame retardant properties of lignin-based flame retardant coatings through performance characterization and experimental tests, which provided a certain database and theoretical basis for the development and application of related bio-based flame retardant coatings.

2 MATERIALS AND METHODS

The main materials used in the experiments were purified desalinized lignin, polyethyleneimine (99% PEI, Mw 600), ammonium polyphosphate (APP), and epoxy resin (EP, E-44), all of which are of analytical grade and can be used directly. According to the basic principle and synthetic route of the Mannich reaction (Ferry et al., 2015), the modified lignin materials with nitrogen-phosphorus elements were prepared by using the graft modification method using APP and PEI. The modified lignin powder was added to a small beaker containing 20 g of EP in the ratio of 0%, 10%, and 20%, and then the mixture was stirred for 2 hours. After adding the curing agent, the mixture was stirred for 2 hours to ensure uniform mixing. The mixture is placed in a vacuum oven to remove air bubbles from the material, and then the mixture is poured into a mold. The molds were placed in a blast drying oven and kept at 80 °C for 2 hours. Subsequently, the lignin-based intumescent flame retardant coatings were obtained by using modified lignin as the carbon source, APP as the dehydrating agent, PEI as the foaming agent, and EP as the coating curing agent.

Energy dispersive X-ray spectroscopy, scanning electron microscopy, and cone calorimetry tests were performed to investigate the flame retardant properties of the material and analyze the flame retardant mechanism. To better reproduce and simulate the real fire combustion scenario, the flame retardant performance of the material was tested by cone calorimetry according to the ISO 5660-1 standard (Shi et al., 2018). The sample was wrapped with aluminum foil on all sides except the heated side and placed horizontally on a sample holder.

3 RESULT AND DISCUSSION

3.1 *EDX elemental analysis and mechanical test*

Elemental analysis was used to determine the content of C, N, and P elements in the prepared dried lignin materials, and the results are shown in Table 1. After the modification, the content of N and P elements in the lignin material increased significantly, which proved the successful graft modification of N and P elements into the purified lignin and obtained good elemental ratios, which were conducive to the further flame retardant effect of N and P elements in the material.

The prepared flame retardant coatings were also subjected to relevant mechanical tests, as shown in Table 2. It can be observed that with the addition of lignin, the tensile properties of the material did not change significantly. This may be due to the unique nature of its macromolecular structure; although the addition of lignin can improve the mechanical properties of the material to a certain extent, when the lignin content reaches a certain degree, it will produce a spatial obstruction effect to reduce the cross-linking degree of the epoxy resin material; in addition, with the increase of lignin content, the co-mingled viscosity of the system increases, which may reduce the uniformity of the material, resulting in a certain defect at the material interface. This is also the reason why the mechanical properties cannot be enhanced further.

Table 1. Content of C, N, and P elements in modified lignin.

Element	Wt%	At%
CK	89.10	92.43
NK	06.52	05.80
PK	04.38	01.76

Table 2. Mechanical properties of flame retardant coatings.

Samples	Tensile Strength/MPa
0%-Lig.	32.2 ± 0.1
10%-Lig.	33.1 ± 0.2
20%-Lig.	32.9 ± 0.1

3.2 *Cone calorimeter test results*

The heat release rate (HRR), peak heat release rate (PHRR), total heat release rate (THR), smoke release rate (SPR), peak smoke release rate (PSPR), and total smoke release rate (TSR) obtained from the cone calorimetry test are important parameters for evaluating the flame retardant performance of the tested materials (Yu et al., 2019). It is important to evaluate the flame retardancy of lignin-based intumescent flame retardants and optimize the subsequent modification of the materials. In the experiments, cone calorimetric tests were conducted on test samples containing 0%, 10%, and 20% modified lignin, respectively.

3.2.1 *Heat release test*

HRR, PHRR, and THR are important parameters that reflect the heat release law of the material during combustion. The HRR and THR curves of the test samples are shown in Figure 1. It can be observed that the specimens without lignin incorporation reached PHRR at about 65 s and burned out at 220 s. The values of PHRR and THR during this process were 1814.74 kW/m^2 and 52.32 MJ/m^2, respectively. When 10% and 20% modified lignin were added, the overall HRR and THR curves became gentler, with the PHRR and THR reduced to 1632.21 kW/m^2, 37.43 MJ/m^2, and 1432.15 kW/m^2, 30.72 MJ/m^2, respectively, which were 11.12%, 28.4%, and 21.09%, 41.28% lower than the specimens without lignin incorporation. This proved that the flame retardant coating with the addition of modified lignin could effectively reduce HRR and THR during combustion, and the effect was more significant with the increase in modified lignin content, which effectively stopped the burning behavior of the material and reduced the heat release performance of the material, thus enhancing the flame retardant performance of the material.

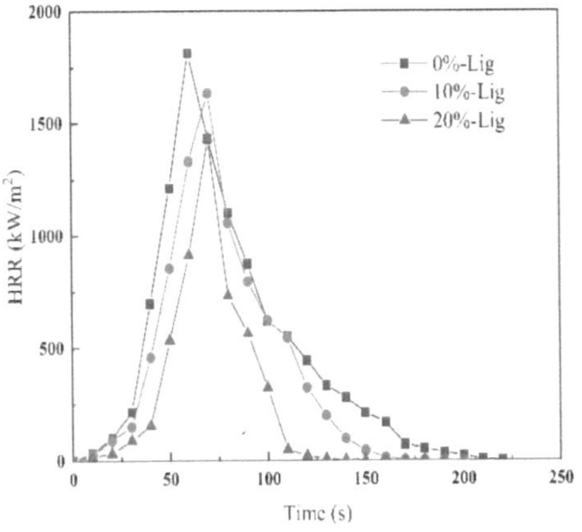

(a) HRR curves

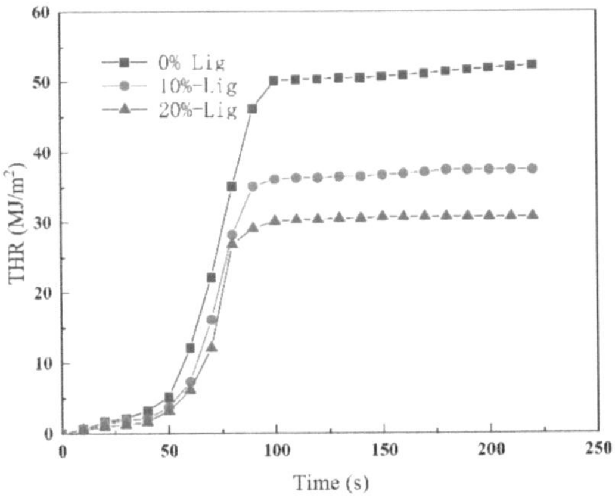

(b) THR curves

Figure 1. HRR and THR curves of specimens with different Lig. Contents.

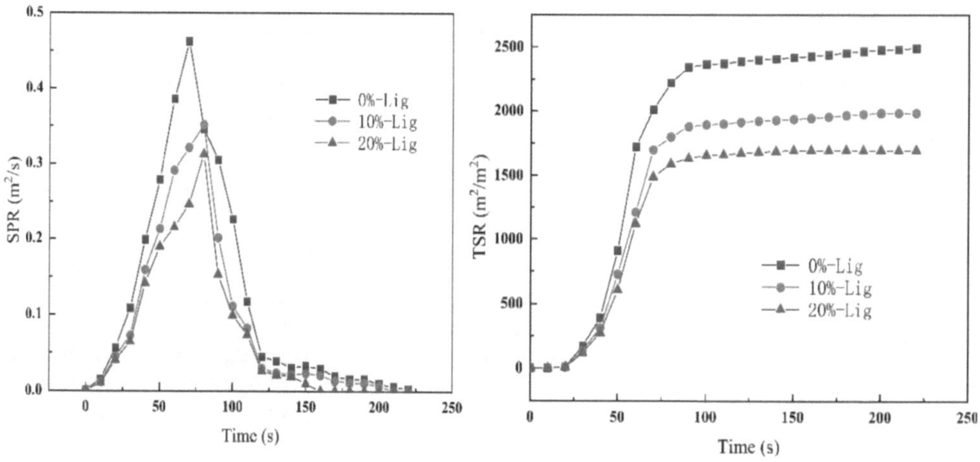

(a) SPR curves (b) TSR curves

Figure 2. SPR and TSR curves of specimens with different Lig. contents.

3.2.2 *Smoke release test*

Flame retardant coatings produce a mixture of solid small particles and gas smoke when used, which becomes one of the most serious factors endangering human lives in fires, so it is necessary to test and improve the smoke release of materials. Figure 2 shows the curves of SPR and TSR. As shown in the figure, the SPR curve becomes slower with the incorporation of modified lignin, and the PSPR decreases from 0.4624 m^2/s to 0.3512 m^2/s and 0.3126 m^2/s, respectively, which is a decrease of 24.04% and 32.4%. Meanwhile, TSR was reduced from 2492.12 m^2/m^2 to 1983.75 m^2/m^2 and 1696.71 m^2/m^2, a reduction of 20.4% and 31.91%, respectively. Such data indicate that the incorporation of modified lignin effectively improves the smoke emission during combustion and exerts a smoke suppression effect based on the flame retardant. In the case of fire, it can effectively reduce the danger of smoke to human life and health.

3.3 *Analysis of flame retardant mechanisms of materials*

After the conical calorimetric test, the morphology of the material was observed, as shown in Figure 3. The addition of modified lignin makes the surface of the material form a layer of stable and dense carbon layer structure after over-firing, which acts as a protective barrier for the material and prevents the transfer of heat to the substrate and the entry of air, thus slowing down the burning process of the material at the physical level. This is due to the excellent char formation performance of lignin, which is a powerful char source in the material and lays the foundation for the expansion and foaming of the material. A specific reaction process is shown in Equations (1) and (2). When materials containing phosphorus and nitrogen are burned, they combine with highly reactive radicals on the substrate to generate PO, HPO, HPO_3, H_20, NH_3, etc. The reaction of the substrate radicals is suppressed during combustion, where the phosphorus-containing substances act as dehydrating agents to promote the formation of a char layer on the surface of the material after heating, reflecting the flame-retardant utility of the condensed phase. The foam on the surface of the material is also caused by the addition of nitrogen to generate a flame-retardant gas. Not only does it fill the char layer more densely, but it also dilutes the content of flammable gases, and the water vapor generated cools the substrate, which exerts the flame retardant effect of the gas phase. Therefore, flame retardants play a flame retardant role in the combustion process by protecting and cooling the substrate, inhibiting thermal degradation reactions, and preventing heat transfer.

$$H_3PO_4 + Heat + O_2 \rightarrow HPO_2 + HPO + PO \rightarrow OH + PO \underset{\rightarrow}{\rightarrow} HPO + O \quad (1)$$

$$(NH_4)_2CO_3 + Heat + O_2 \rightarrow 2NH_3 + CO_2 + H_2O \quad (2)$$

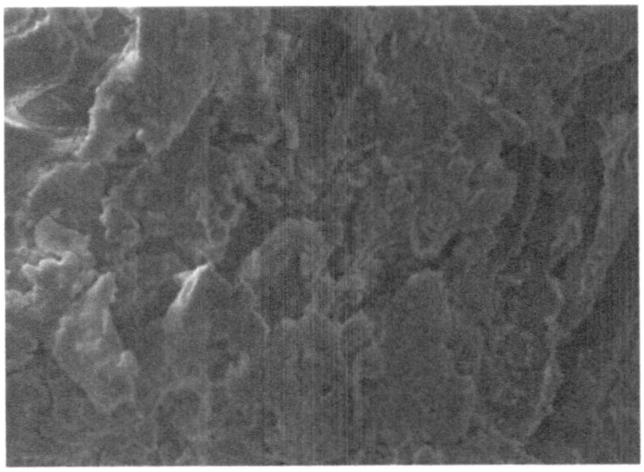

(a) 0%-Lig.

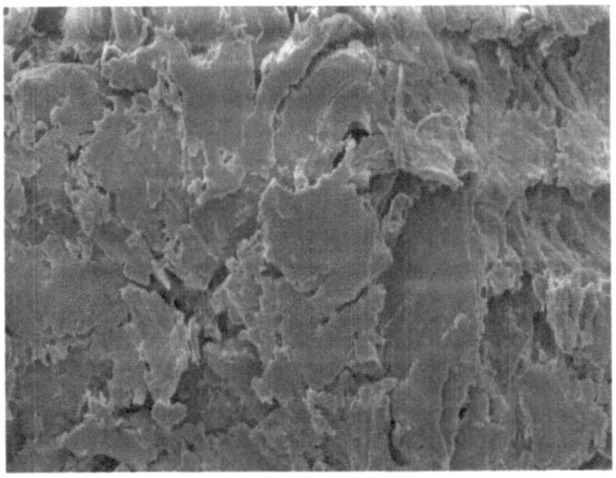

(b) 20%-Lig.

Figure 3. Apparent appearance of 0% Lig. and 20% Lig. specimens after conical calorimetric testing.

4 CONCLUSION

Based on the above test results and discussions, the following conclusions can be drawn:

(1) A nitrogen-phosphorus-modified intumescent flame retardant coating was prepared from lignin and epoxy resin. The addition of modified lignin could effectively reduce the HRR, PHRR, and THR of flame retardant materials. When the content of modified lignin was 20%, PHRR and THR were reduced to 1432.15 kW/m^2 and 30.72 MJ/m^2, respectively, which were reduced by 21.09% and 41.28%.

(2) The addition of modified lignin could effectively suppress the generation of toxic fumes at the same time. When 20% of modified lignin was added, the PSPR and TSR of the material were reduced to 0.3126 m^2/s and 1696.71 m^2/m^2, respectively, with a reduction of 32.4% and 31.91%, respectively. The moderate addition of lignin had little effect on the performance of flame-retardant coatings.

(3) After observing the apparent morphology of the material after overfire, it was concluded that the modified lignin material benefited from its excellent char formation performance and the synergistic effect of N and P elements. A continuous, dense carbon protective layer was formed on the surface of the material during the overfire, while the addition of N and P elements synergistically exerted the flame retardant effect in the gas phase and solid phase, respectively.

REFERENCES

Anderson, E. M., Stone, M. L., Hulsey, M. J., Beckham, G. T. & Roman-Leshkov, Y (2018). Kinetic Studies of Lignin Solvolysis and Reduction by Reductive Catalytic Fractionation Decoupled in Flow-Through Reactors. *ACS Sustainable Chemistry & Engineering*, 6, 7951–7959.

Ferry L., Dorez G., Taguet A., Otazaghine B. & Lopez-Cuesta J.M. (2015). Chemical Modification of Lignin by Phosphorus Molecules to Improve the Fire Behavior of Polybutylene Succinate. *Polymer Degradation and Stability*. 113 ,135–143.

Ge Y.Y., Qi Z.M., Sha D.S., Hu X.S. & Liu S.W. (2022). Durable Flame-retardant Cotton Fabric Modified by Water-soluble C-N-P Intumescent Flame Retardant. *Journal of Applied Polymer Science*. 139, (44).

Ji, Baohua & Wang, Mao-Yuan. (2011). Study on Synthesis and Properties of Pentaerythritol Phosphate Melamine Salts. *Fire Science and Technology* (10), 940–942.

Li, Jia-Peng, Liu, Ning, Wang, Feng-Qiang & Wang, Qing-Wen. (2018). Research Progress and Trends of Transparent Intumescent Amino Resin-based Flame Retardant Coatings. *Coatings Industry* (01), 83–87.

Rosace G., Castellano A., Trovato V., Iacono G. & Malucelli G. (2018). Thermal and Flame Retardant Behavior of Cotton Fabrics Treated With a Novel Nitrogen-containing Carboxyl-functionalized Organophosphorus System. *Carbohydrate Polymers*. 196, 348–358.

Shi Y.Q., Yu B., Zheng Y.Y., Yang J., Duan Z.P. & Hu Y. (2018). Design of Reduced Graphene Oxide Decorated with DOPO-phosphonamidite for Enhanced Fire Safety of Epoxy Resin. *Journal of Colloid and Interface Science*. 521, 160–171.

Sun Y.Q., Ma Z.W., Xu X.D., Liu X.H., Liu L.N., Huang G.B., *et al.* (2020). Grafting Lignin with Bioderived Polyacrylates for Low-Cost, Ductile, and Fully Biobased Poly (lactic acid) Composites. *Acs Sustainable Chemistry & Engineering*. 8(5), 2267–2276.

Wang, Q.H., Wang, X.J., Fang, J.J., Di, Z.G., Guan, Z.C. & Ma, S.J (2022). Preparation and Performance of Water-based Intumescent Fireproof Coatings for Steel Structures. *Coating Industry* (02), 42–48.

Yu B., Tawiah B., Wang L.Q., Yuen A.C.Y., Zhang Z.C., Shen L.L., *et al* (2019). Interface Decoration of Exfoliated MXene Ultra-thin Nanosheets for Fire and Smoke Suppressions of Thermoplastic Polyurethane Elastomer. *Journal of Hazardous Materials*. 374, 110–119.

Yu, L.-B., Huang, Z.-M., Kang, W.-B., Shi, Y.-Y. & Zhou, M (2010). Synthesis of Pentaerythritol Polyphosphate Melamine Salt as an Intumescent Flame Retardant and its Effect on Flame Retardancy of Epoxy resin. *Coating Industry* (03), 14–17.

Author index